Materials Engineering and Sciences Division 2016

Core Programming Area at the 2016 AIChE Annual Meeting

San Francisco, California, USA
13 - 18 November 2016

Volume 1 of 2

ISBN: 978-1-5108-3440-8

Printed from e-media with permission by:

Curran Associates, Inc.
57 Morehouse Lane
Red Hook, NY 12571

Some format issues inherent in the e-media version may also appear in this print version.

Printed by Curran Associates, Inc. (2017)

For permission requests, please contact AIChE
at the address below.

AIChE
120 Wall Street, FL 23
New York, NY 10005-4020

Phone: (800) 242-4363
Fax: (203) 775-5177

www.aiche.org

Additional copies of this publication are available from:

Curran Associates, Inc.
57 Morehouse Lane
Red Hook, NY 12571 USA
Phone: 845-758-0400
Fax: 845-758-2633
Email: curran@proceedings.com
Web: www.proceedings.com

TABLE OF CONTENTS

VOLUME 1

VOLUME 2

453440 Composite Materials: Mechanical and Tribological Property Improvement

Abstract:

Research Interests:

Solar cell panels exposed to dessert sands, harsh weather conditions, and contact cleaning method are subjected to scratching force. This factor can lead to loss of structural integrity, tremendously affect the performance of the panel system and greatly lower the energy conversion efficiency. Development of abrasion resistant coatings on solar cell panels can be achieved in composite systems containing both hard polymer and nano scale fillers. Halloysite nanotubes (HNT) encompass high aspect ratio, stiffness and strength while remain understudied about their coating usage. By appropriate treatment, HNT possess high dispersion quality and enable efficient interfacial adhesion to polymer after mixing. Using HNT in hardcoat polymers is supposed to increase strength, modulus and improve mechanical robustness (i.e., effective HNT modulus was estimated to reach ~300 GPa plus a half scratch depth of polymer without HNT). The reinforcement efficiency of HNT is quantitatively expressed using NanoIndentation, which showed up to 60% increase in modulus and hardness for composites with vertically aligned HNT. Anti-abrasion properties are characterized using abrasion test and followed by comparing morphological changes in a sample without undergoing the abrasive test. Consistently the composite coatings with out-of-plane particle orientations showed highest wear resistance. Scanning electron microscopy and atomic force microscopy have been used to verify the morphological and topographical merits after damage. This study aims at development of environmentally friendly and cost effective coating, which is highly potential to be commercialized.

Teaching Interests:

As a polymer material scientist and teacher, I believe it is my responsibility to challenge and support my students, in pursuit of three learning goals: (1) confront misconceptions about material science ideas; (2) master advanced polymer engineering skills; (3) cultivate an interest in science and understanding of how we do science and develop engineering. I developed my teaching practice through the MIT Kaufman Teaching Certificate Program (KTCP), undergraduate mentoring and junior researcher training service, and as a teaching assistant in three courses (1) Introduction to Material Science and Engineering, (2) Mechanics of Composite Materials, and (3) Structure, Property, and Processing of Polymeric Materials. Therefore I am capable and also enthusiastic in teaching materials science, polymer engineering and composite mechanics related courses.

463364 Nano-Engineered Functional Materials for Energy Storage and Biomimetic Applications

Abstract:

Nano-Engineered Functional Materials for Energy Storage and Biomimetic Applications

Samanvaya Srivastava

Institute for Molecular Engineering, The University of Chicago, Chicago, IL 60637, USA

samsri@uchicago.edu

A multitude of novel building blocks, including macromolecules and nanoparticles of myriad shapes and sizes, are now available to researchers owing to recent progress in material synthesis and fabrication techniques. However, as technology moves towards smarter materials and complex multifunctional devices, the need for understanding and controlling the interactions and assembly of these mesoscale units also becomes increasingly important. Driven by this motivation, my proposed research program will aspire to employ fundamental approaches from colloid, macromolecular and soft matter sciences to understand and tailor the physical processes that govern the structure-property relationships in these material building blocks and their assemblies. These studies will provide the groundwork for the overarching theme of my research, which is to design bottom-up approaches for assembly of mesoscopic building blocks into nano-engineered materials with *on-demand* properties.

My graduate research with Prof. Lynden Archer at Cornell University primarily focused on developing strategies for synthesis of inorganic-organic hybrid materials with tunable properties. Specifically, I developed strategies for controlling nanoparticle dispersion in polymer hosts, and consequentially synthesizing nanoparticle-polymer composites with desired mechanical and electrochemical properties. Further, in a collaborative effort with Lipson group at Cornell University, I worked on the development of functional inks for unique direct-write applications, including 3-D printed electronic and energy storage devices.

My current research as a postdoctoral researcher with Prof. Matthew Tirrell at The University of Chicago and Argonne National Laboratory involves designing and synthesizing block copolyelectrolytes that self-assemble via electrostatic interactions into polyelectrolyte complex (PEC) based micelles and hydrogels. At low polymer concentrations, oppositely charged triblock copolyelectrolytes form micellar networks that phase separate from the solution. These intriguing structures are understood to be driven via a combination of electrostatic and entropic interactions. At higher polymer concentrations, our investigations show that PEC based hydrogel morphologies span a wide range of features and are very sensitive to external stimuli, thus allowing for facile tuning of the hydrogel properties. I am also investigating the enhancements in the protein activity upon encapsulation in polyelectrolyte complexes. These studies, in collaboration with the Gardel group at The University of Chicago, are particularly important for intelligent design of PEC based micelles and hydrogels as drug carriers and tissue growth scaffolds.

My goal is to establish an independent research program that aims at designing novel bottom-up approaches for self-assembled material. Building on my previous work and harnessing my expertise in soft matter physics, I plan to pursue three areas of research that are set in a similar fundamental landscape, but will serve entirely different realms of applied research. The research themes that I intend to pursue are: i) developing new material platforms for redox flow batteries, which employ electroactive complex fluids that are flown in a parallel flow setup during the charging/discharging process and stored in reservoir tanks otherwise; ii) synthesizing surface-modified nanoparticle based hydrogels that will have superior tunability, stimuli responsiveness and tissue supporting functionalities, and iii) development of smart inks for additive fabrication (also known as 3-D printing) of responsive materials and integrated devices.

Selected Publications (17 total, 3 in preparation)

1. **S. Srivastava** and M. V. Tirrell, Polyelectrolyte Complexation, *Advances in Chemical Physics* **161** (2016), in print.
2. **S. Srivastava**, P. Agarwal, R. Mangal, S. Narayanan, D. L. Koch and L. A. Archer, Hyperdiffusive Dynamics in Newtonian Nanoparticle Fluids, *ACS Macro Letters* **4**, 1149 (2015).
3. R. Mangal, **S. Srivastava**, and L. A. Archer, Phase stability and dynamics of entangled polymer-nanoparticle composites, *Nature Communications* **6**, 7198 (2015).
4. **S. Srivastava**, J. L. Schaefer, Z. Yang, Z. Tu and L. A. Archer, 25th Anniversary Article: Polymerâ€"Particle Composites: Phase Stability and Applications in Electrochemical Energy Storage, *Advanced Materials* **26**, 201 (2014).
5. **S. Srivastava**, S. Narayanan and L. A. Archer, Structure and Transport Anomalies in Soft Colloids, *Physical Review Letters* **110**, 148302 (2013).
6. **S. Srivastava**, P. Agarwal and L. A. Archer, Tethered Nanoparticleâ€"Polymer Composites: Phase Stability and Curvature, *Langmuir* **28**, 6276 (2012).

Research Interests: Self-assembled hybrid materials, polymer and nanoparticle hybrid materials, X-ray based structure and dynamics characterization, polyelectrolyte physics

Teaching Interests: Polymer physics, fluid mechanics, advanced material characterization techniques

For more information, please visit: http://tirrell.ime.uchicago.edu/samsri

467733 Temporally Controlled Release of Platelet-Rich Plasma from Peg Microgels Having Tunable Biodegradation Rate and Size

Abstract:
Research Interests: Biomaterials, Drug Delivery

Teaching Interests: Â Biomaterials

Biomedical applications such as drug delivery and tissue engineering often require polymeric hydrogels which degrade predictably under physiological conditions. Biodegradable injectable drug delivery systems present attractive avenue for site specific and minimally invasive administration, coupled with precise temporal control over drug release, which can be employed for multiple applications. Specifically, poly (ethylene glycol) (PEG) has been thoroughly investigated for the design of biodegradable hydrogel networks because of its exceptional biocompatibility and inertness. In this study we designed biodegradable PEG microgels with a wide range of degradation kinetics using Michaelâ€™s addition chemistry. We further used these microgels for encapsulating and controlling delivery of proteins derived from platelet rich plasma (PRP). PRP is a concentrated assortment of growth factors and cytokines derived from the patientâ€™s own platelets. PRP has been proven useful in a number of regenerative medicine applications. However, an effective method to achieve sustained delivery of PRP is still required.

For the design of degradable PEG hydrogels we used multiarm PEG-acrylate (PEGAc) of varying molecular weight and number of arms (4-arm of 10 kDa, to 8-arm of 20 kDa) and three categories of dithiol crosslinkers, namely, ester containing, non-ester containing and dithiols with neighbouring functional groups of different electronegativities. The PEG hydrogel physical and mechanical characteristics could be modulated depending upon the reaction parameters and gelation time for multiarm PEGAc and dithiol crosslinker structure and molecular weight. By changing these parameters, hydrogels with controlled degradation ranging from 10 hours to 32 days and having mesh size in the range of 9 to 14 nm were obtained. We identified two sets of conditions which yielded fast gelling-fast degrading hydrogels and slow gelling-slow degrading hydrogels. Uniquely, the hydrogel storage moduli could be controlled by the dithiol crosslinker chemical identity independent of the degradation time or mesh size of the hydrogel.

Further, we used the optimized combinations of PEG hydrogels to make PEG microgels of sizes varying between 50 to 700 µm by manipulating several key electrospraying parameters. We also investigated various techniques and times suitable for storing protein-loaded PEG microgels. The microgels were stored at seven conditions: 25°C, 4°C, -80°C, Lyophilization (Lyo), -80°C/Lyo, -80°C/DMSO, and -80°C/DMSO/Lyo for 1, 4, 7, and 28 days. DMSO was introduced at 10% (v/v) in 1x PBS. Our results indicated that neither storage modulus nor the structure of incorporated model protein was affected post storage at different conditions. However, comparing the percent distribution of microgel size pre- and post-storage, we found an increase in the mean diameter indicating increased swelling post-storage. Furthermore, environmental scanning electron microscopy (ESEM) images showed that large pores were created by some of the conditions, mainly by lyophilization.

Finally, we tested the PEG hydrogels and microgels for encapsulation and delivery of PRP-derived growth factors. PRP at 2-10% w/v was encapsulated during gelation of the hydrogels or microgels. Release of growth factors was quantified by multiplexer analysis. Biological activity of PRP released from the hydrogels or microgels were tested *in vitro* using human dermal fibroblast (HDF) or by circular dichroism, respectively. Direct encapsulation of model proteins in the PEG hydrogels during hydrogel gelation preserved their secondary structure. Release kinetics for several model proteins was found to be dependent on the protein size, as well as mesh size and degradation rate of the hydrogel. The amount of total protein release increased over time, corresponding to increase in hydrogel mesh size. Multiplexer growth factor analysis of the releasates showed continuous release of growth factors (PDGF-BB, EGF, RANTES) from PRP-containing hydrogels until degradation. PRP releasates also demonstrated increased proliferation in HDF cells over a period of 72 hours, comparable to that of a bolus PRP injection. Furthermore, similar results could be translated to PEG microgels obtained by electrospraying.

Thus it can be concluded that we could obtain a range of biodegradable PEG hydrogels by a combination of number of PEG arms and dithiol crosslinkers and reaction pH. Further, these hydrogels could be made into injectable microgels which allowed localized and controlled delivery of PRP growth factors in the active form.

471788 Towards the Next Generation of Magnetic Resonance Spectroscopy: Harnessing Light and Spin

Abstract:

Research Interests:

With unmatched chemical specificity and the ability to probe materials across length- and timescales, nuclear magnetic resonance spectroscopy (NMR) and imaging (MRI) are two of the most powerful analytical techniques available to engineers, scientists, and medical professionals. However, they require costly infrastructure and are plagued by low sensitivity resulting from small nuclear spin magnetization. Typically, nuclear spin magnetization is generated by allowing the sample to equilibrate in a magnetic field. However, even with the largest available superconducting magnets, room-temperature equilibrium magnetization is only about 10^{-5} of the theoretical maximum. A general method to generate non-equilibrium nuclear spin magnetization would allow us to dispense with superconducting magnets, decrease sensitivity limits and experimental time by orders of magnitude, and enable a range of previously inaccessible applications.

With this motivation, my laboratory will be organized around three main thrusts:

1. **Enhancement of sensitivity for NMR.**
2. **Exploration of new magnetic resonance applications at low- and zero-fields.**
3. **Fundamental aspects of spectroscopy, chirality, and condensed-matter physics.**

Our primary goal is to harness the pure spin states of photons and electrons to achieve nuclear spin polarization. We will build upon my previous work with optical pumping in semiconductors [1] and nitrogen vacancy (NV^-) centers in diamond [2], where I have demonstrated a 170,000-fold increase in NMR signal from the ^{13}C nuclei in the diamond by coupling the electron and nuclear spin states with microwave irradiation. Future work involves the transfer of this nuclear magnetization to arbitrary samples of interest. This will require development of high-surface area diamond structures, fundamental studies of spin polarization transfer mechanisms, and modification of the diamond surface chemistry.

We also seek to extend NMR beyond the traditional model requiring a high-field superconducting magnet in a dedicated facility. Low-field NMR allows the use of cheap, portable magnets in benchtop and truly portable systems. Zero-field NMR allows the observation of spin-interactions that are invisible in the presence of a magnetic field, providing additional information about material structure, geometry, and dynamics [3]. Furthermore, zero-field experiments are free from the deleterious effects of inhomogeneous magnetic susceptibility, enabling high-resolution study of heterogeneous materials.

My laboratory will study fundamental aspects of spectroscopy including chiral effects of spin-spin couplings and spin dynamics in condensed matter. We anticipate ability to ability to directly resolve molecular chirality through antisymmetric spin-spin interactions without the addition of chiral derivatizing agents. Experimental and theoretical studies of spin dynamics in many-body solid-state systems are important to the understanding of solid-state NMR spectroscopy and also provide a prototypical test bed for concepts in condensed-matter physics.

1. **King, J. P. *et al*.** "Optically rewritable patterns of nuclear magnetization in gallium arsenide." ***Nature Communications*** 3:918 doi: 10.1038/ncomms1918 (2012).
2. **King, J. P. *et al*.** "Room-temperature ***in situ*** nuclear spin hyperpolarization from optically pumped nitrogen vacancy centres in diamond." ***Nature Communications*** 6:8965 doi: 10.1038/ncomms9965 (2015).
3. J. W. Blanchard, T. F. Sjolander, **J. P. King,** M. P. Ledbetter, E. H. Levine, V. S. Bajaj, D. Budker, and A. Pines. "Measurement of untruncated nuclear spin interactions via zero- to ultralow-field nuclear magnetic resonance." Phys. Rev. B **92**, 220202(R) (2015).

Teaching Interests:

My teaching philosophy can be stated concisely: effective teaching is the facilitation of effective learning. While this statement is simple, in practice it presents these challenges: How do we define effective learning? How do we assess if learning has occurred? What steps do we take based on this assessment? I will present my plan to address these challenges in the chemical engineering curriculum. I am interested in teaching all aspects of the core curriculum, but with my background would be especially interested in teaching thermodynamics, fluid dynamics, and transport phenomena at the undergraduate and graduate levels. I would also develop special classes in the theory and application of spectroscopic techniques.

474914 Towards a Greener and Scalable Synthesis of Sodium Titanate Nanorods and Its Application As Anode in Sodium Ion Batteries

Abstract:

Sodium ion batteries are accepted as LIB (lithium ion batteries) alternative to large scale energy storage, but its wide application is currently limited by both the synthesis and electrode materials. Thus, simple and scalable synthesis of electrode materials is desirable to lower the cost of synthesis. In this study, we report a surfactant-free, energy-efficient solid state synthesis to mass-produce uniform, single crystalline 1D metal oxide nanorods with tunable composition. To demonstrate the methodology, a series of single-crystalline $Na_2Ti_3O_7$/$Na_2Ti_6O_{13}$ nanorods with tunable composition are synthesized and applied in sodium ion batteries. By control of synthesis parameter, it provides an alternative low-temperature route to transform $Na_2Ti_3O_7$ into $Na_2Ti_6O_{13}$. The high theoretical capacity of $Na_2Ti_3O_7$ and the low volume expansion of $Na_2Ti_6O_{13}$ upon charge/discharge are synergistically exploited to achieve high electrochemical performance and stability. This synthesis method opens up a scalable, energy-saving route to mass-produce uniform 1D nanostructured electrode materials and fine-tune the composition and morphology according to performance demands.

The electrochemical performance of $Na_2Ti_3O_7$ and $Na_2Ti_3O_7$/$Na_2Ti_6O_{13}$ nanorods are compared to conventionally synthesized $Na_2Ti_3O_7$ irregular particles. $Na_2Ti_3O_7$ nanorods outperform $Na_2Ti_3O_7$irregular particles at all current rates in the range of 0.1 to 5C. Without surface modification, the nanorods can still manifest a reversible discharge capacity of 70 mAh/g at a high current rate of 5C. The enhanced electrochemical performance can be attributed to the high surface area of the nanorods as well as efficient ion transport induced in highly uniform one-dimensional nanostructure.

Research Interests: Nanomaterials; Li-ion, Li-air, Na-ion Batteries; Catalysis; Electrochemistry; In-situ Electrochemical Methods

Teaching Interests: Nanotechnology; Materials Science and Engineering

459307 Molecule Separation and Energy Storage Using Novel Porous Material Platform

Abstract:

Research Interests: Metal-organic framework membrane; gas adsorption and separation; energy storage; biomass conversion

Teaching Interests: Thermodynamics, transport, porous materials and applications

Metal-organic frameworks (MOFs) also known as coordination porous polymer are a series of novel porous materials which are composed of metal anodes and organic linkers through coordination bonds. MOFs usually have 3D porous structures, large surface areas and pore volume, which provide promising future in applications including molecule separation and energy conversion. Application example such as CO_2 capture from flue gas using MOFs will be shown in two scenarios, fixed bed and membrane. Unsaturated metal centers have been found to be the key for low partial pressure CO_2 adsorption. A robust hydrophobic MOF membrane showed high CO_2 permeance even under 80%RH, which is close to real flue gas condition. Proposed future work is focusing on oriented asymmetric MOF membrane for molecule separation and catalyst used in fuel cell and other electrochemical energy storage system. The unlimited pairs of metal and organic linker, chemical versatility, and tunable pore size and functionality open up great opportunity for MOFs in research frontiers of energy and environment related research.

473149 Designing Metal Oxide Materials for Reduction/Oxidation Reactions Based on a Fundamental Understanding of Their Behavior

Abstract:

Research Interests:

The release of anthropogenic pollutants such as greenhouse gasses, agricultural and industrial waste, and incomplete treatment of waste and sewage water are increasing and therefore having larger negative impacts on the environment. Dealing with these emissions, wither by avoiding them where possible, and remediating them when avoidance is not possible, is one of the major challenges of the 21st century. My research addresses both of these pathways through the understanding and design of metal oxide reduction/oxidation (Red/Ox) materials for solar thermochemical fuel generation and photocatalytic pollutant degradation. In solar thermochemical fuel generation concentrated solar heat is used to drive water and CO_2 splitting reactions using a metal oxide O-carrier to couple an O_2-generating half reaction to a H_2 or CO generating half reaction. The development of optimal materials is critical for this process. The metal oxide must reduce readily enough to operate at temperatures achievable with concentrated solar power(<1600°C), while not reducing so readily that it is incapable of driving the water or carbon dioxide splitting reaction. My research into solar fuels focuses on developing a fundamental understanding of what controls reducibility of metal oxides using computational chemistry (DFT) and, based on this understating, engineering of new materials with more favorable Red/Ox behavior. The computational design new materials are then verified experimentally. A seemingly different process, but fundamentally coupled research area is photocatalytic pollutant degradation. The photocatalytic remediation of pollutants is inherently a Red/Ox reaction, where a metal oxide, usually TiO_2 acts as catalysts for these reactions. My research in this area has focused on using quantum chemistry to develop a fundamental understanding the rate limiting step of this process, namely the adsorption of O_2 to the TiO_2 surface and simultaneous transfer of photo-excited electrons from the TiO_2 surface to molecular O_2. Based on this understand doping and other modification strategies for TiO_2 can be developed. In both of these cases the understanding developed can be ported to other similar processes where metal oxides facilitate RedOx processes. This includes very similar processes, such as chemical looping, or non-TiO_2photocatalysis or subjects further afield, such as solid oxide electrolysis and fuel cells, photolysis and development of catalytic converter materials. Through the research, my focus is first on a fundamental understanding of the material and its controlling processes followed by novel materials design and improvement.

Teaching Interests:

My approach to chemical engineering teaching is to convey the basics that every student needs to succeed in their careers, namely problem solving skills and a fundamental understanding of physical and chemical processes involved in chemical systems. As such I have interest in teaching at both the undergraduate and graduate levels and aim to instill a love and intuition of chemical engineering processes in the students. Of the major chemical engineering subjects, I am most interested in teaching chemical reaction engineering and transport processes at the undergraduate or graduate levels. However, I enjoy most of the chemical engineering curriculum and am happy to teach everything from basic chemistry through process design and graduate level thermodynamics. Having had the opportunity to design and teach half a semester of a masters level course on renewable fuels at ETH during my Postdoc, I am also interested in developing and teaching new graduate and undergraduate classes such as computational chemistry or renewable energy systems. I believe that the education and proper preparation of the students is the mission of a university and that this includes interactions outside of the classroom as well as within it. Therefore I plan to have an engaged research program where undergraduate as well as graduate students participate and can see the application of the subjects that they are learning about in class.

475158 Multi-Scale Modeling of Bulk Solutions and Solid/Liquid Interfaces

Abstract:

Research Interests:

Today, the close interplay between computational simulations and experiments has proven to be an invaluable tool in providing a fundamental understanding of key aspects of numerous engineering technologies and complex processes, as well as enabling the rational design of novel materials for diverse technological applications. In my proposed research program, I intend to use a multi-scale simulation approach to predict and understand the physicochemical behavior of bulk solutions and solid/liquid interfaces in order to gain a fundamental understanding of synthetic processes of complex materials, the knowledge of which can then be leveraged for cutting edge and novel applications in energy-storage devices. I plan on doing so a using multi-scale simulation approach which will provide indispensable information at the molecular- and mesoscale-levels to understand the physical/chemical behavior of bulk solutions and solid/liquid interfaces, which is crucial in the design and optimization of energy storage devices and the synthesis of complex materials. The development of such an understanding of solutions and interfacial properties is fraught with complications, but crucial to the design and optimization of energy storage devices and material syntheses methods for which experiments alone are insufficient due to difficulties associated with distinguishing between different phases, limitations for experimental resolution and the high cost associated with experimental work. Thus, the synergetic fusion of experimental input and multi-scale simulation will be a key factor in achieving a fundamental, comprehensive understanding of the delicate interplay between the macroscopic properties and events occurring at much smaller length scales.

Teaching Interests: My aim is to facilitate the development of students so that they can formulate and analyze the problems and use appropriate tools to obtain and interpret results. I believe that my students should be critical and

indendent thinkers with skills to work in teams. I have five years of teaching experience as teaching assiatant and I beleive that teaching must not be restrictied to mere transfer of facts and equations.

475595 "Click" Polymerizations: From Recycling Polymer to 3D Printing

Abstract:

Research Interests:

Smarter, greener and atomic efficient polymeric materials

Polymer science has seen an amazing century of development. Polymer materials have been mature and ubiquitous, and have completely changed the way we live today. Surprisingly I see vast research opportunities that could change our life for days to come, for good. My research is centered utilizing atomic efficient reactions in polymers that solve problems in cutting-edge areas from nano/micro structured materials, to recycling materials, to 3D printing products. Centered at implementing â€œclickâ€ reactions and many other dynamic covalent systems, I sought to answer tough questions with reliable and widely applicable mythologies. Along the way to pursue novel polymeric materials, I strive to mentor students in both undergraduate and graduate levels, as well as postdoctoral scholars for their professional standings as future researchers.

Teaching Interests:

Â My teaching experiences in graduate school as a teaching assistant and part-time instructor have gave me passion in pursuing life-long career in an academic position. My teaching philosophy is to embrace active classroom, encourage question/discussion among students, and follow each individual in the class. I would like to utilize online systems for sharing teaching materials, grading and assigning reading quiz. I would like to teach reaction engineering, (bio)kinetics, polymer materials, polymer chemistry for both undergraduate and graduate students.

475623 Designing Polymeric & Soft Material Systems Via Inverse Computational Methodologies

Abstract:

2nd Year Postdoctoral Research Fellow

Research Interests:

My research focuses on the design of polymeric and soft material systems where a new material can be designed for a specific application ***in silico*** with new properties not possible from conventional combinatorial experimentation. Computational resource capabilities have increased dramatically over the past few decades both in terms of processing speed and parallelizability, thus enabling simulation methodologies that combine multi-length scale virtual experiments with optimization algorithms to find new systems of interest faster than ever. There have been many successful examples of such design work in the areas of metallic and ceramic materials, but equivalent methodologies for polymeric and soft materials are still in their infancy. By combining and integrating the current state of the art computational tools in polymeric systems such as self-consistent field theory (SCFT), coarse grained Monte Carlo molecular dynamics, atomistic molecular dynamics, dissipative particle dynamics, and finite element methods, polymers can be designed for a given functional purpose considering effects at all length scales. By starting with the desired application such as advanced energy storage, ultrafast electronic devices, or biomedical interfacing materials, a chemical engineering system can be designed computationally to produce a given device with optimal properties to perform that application.

Postdoctoral Research:

– Awarded National Research Council Research Associateship Program Fellowship.

Proposal Name:

"***Enhancing the Measurement and Morphology Control of Thin Film Block Copolymers using Feedback Between Inverse SCFT Simulations and Resonant Soft X-Ray Scattering***"

Under supervision of R. Joseph Kline & Christopher Soles, Materials Science and Engineering Division, National Institute of Standards and Technology (NIST)

Doctoral Thesis:

"***Modeling and Theoretical Design Methods for Directed Self-Assembly of Thin Film Block Copolymer Systems***."

Under supervision of Caroline A. Ross and collaboration with Alfredo Alexander-Katz, Department of Materials Science and Engineering, Massachusetts Institute of Technology (MIT)

Research Experience:

As an undergraduate at the Georgia Institute of Technology (GT) I performed undergraduate research under a President's Undergraduate Research Award (PURA) examining the miscibility of fullerene based molecules with styrene based copolymers through both ultraviolent-visible spectroscopy and molecular dynamic simulations. In graduate school at MIT, I continued work in the self-assembly of copolymers. In particular, I examined thin film block copolymer systems and their incorporation into current manufacturing processes as a method to extend lithography methods for next generation microelectronic devices beyond the sub-10 nm node. I did this work by advancing SCFT simulations for examining directed self-assembly systems with topographical templates under solvent vapor annealing conditions. In this simulation framework, I developed an inverse design algorithm to find post template configurations necessary to produce given target complex morphologies. Currently at NIST I am investigating novel methods for characterizing nanoscale patterns fabricated by processes such as block copolymer directed self-assembly as a National Research Council postdoctoral research fellow. Using X-ray scattering techniques, I am optimizing the inverse computation methods needed to find the real space structure of underlying nanoscale patterns. I am also using physically integrated simulation methods (e.g. SCFT of block copolymers) that predict what structures are formed for different directed self-assembly boundary conditions.

Teaching Interests:

As an undergraduate at GT, I instructed introductory electromagnetism physics labs as well as served as a one-on-one tutor for the Learning Assistance Program. At MIT I taught organic chemistry labs, mechanical testing labs, traditional chalk lectures via recitations for a combined quantum mechanics and solid state physics course, and also helped develop course content for an in class demonstration driven thermodynamics course. I completed a teaching minor in Teaching Materials Science and Engineering for my teaching assistance accomplishments as well as a teaching certificate through the MIT Teaching Learning Laboratory. A copy of the letter for the teaching certification is available upon request. Based on these experiences as well as my research interest, I am well versed to teach any thermodynamic, processing, quantum mechanics, or organic chemistry related courses.

Future Direction:

There are many areas in designing polymeric and soft materials systems I would be excited to investigate in my future research. Here I highlight three particular areas. The inverse design paradigm I have used in my research is applicable to all of these areas. The areas examined in increasing order of difficulty and expected research timeline are (I) controlling the self-assembly of heterogeneous polymer, copolymer, and nanoparticle blends/composites, by finding the optimal processing conditions and chemical structures needed for a given application computationally, (II) characterizing designed polymer samples for computational model validation, and (III) determining the optimal chain sequence in proteins and other complex copolymers that can have precise control over the repeat unit sequence.

Area (I) is important as copolymer based systems naturally form ordered structures and complex morphologies essential to applications such as photovoltaics, light emitting diodes, memory storage, integrated circuits, and batteries. Area (II) is essential in both supporting area (I) by validating the systems designed for the given applications as well as enhancing the computational models needed to design such systems through direct comparison with the model results and experiment. Area (III) is the ultimate goal of any polymer scientist or engineer in being able to program desired properties via a carefully designed polymer sequence which can only be done with robust design algorithms. Each of these research areas naturally lead into the next and provide the pathway for the design, fabrication, and characterization of new polymeric and soft materials.

Selected Publications:

A.F. Hannon, D.F. Sunday, G. Khaira, J. Ren, P.F. Nealey, J.J. de Pablo, R.J. Kline. "Optimizing Self-Consistent Field Theory Block Copolymer Models with X-Ray Metrology," ***In Preparation***

A.F. Hannon, D.F. Sunday, D. Windover, R.J. Kline. "Advancing X-Ray Scattering Metrology Using Inverse Genetic Algorithms," ***Accepted by SPIE: Journal of Micro/Nanolithography, MEMS, and MOEMS***. (2016).

D.F. Sunday*, **A.F. Hannon***, S. Tein, R.J. Kline. "Thermodynamic and Morphological Behavior of Block Copolymer Blends with Thermal Polymer Additives," ***Macromolecules*** doi:10.1021/acs.macromol.6b00651 (2016). *Contributed equally

A.F. Hannon, W. Bai, A. Alexander-Katz, C.A. Ross. "Simulation Methods for Solvent Vapor Annealing of Block Copolymer Thin Films," ***Soft Matter*** **11(19)**, 3794-3805, (2015).

A.F. Hannon, Y. Ding, W. Bai, C.A. Ross, A. Alexander-Katz. "Optimizing Topographical Templates for Directed Self-Assembly of Block Copolymers via Inverse Design Simulations," ***Nano Letters*** **14(1)**, 318–325, (2014).

A.F. Hannon, K.W. Gotrik, C.A. Ross, A. Alexander-Katz. "Inverse Design of Topographical Templates for Directed Self-Assembly of Block Copolymers," ***ACS Macro Letters*** **2(3)**, 251-255 (2013).

A. Tavakkoli K.G., K.W. Gotrik, **A.F. Hannon**, A. Alexander-Katz, C.A. Ross, K.K. Berggren. "Templating Three-Dimensional Self-Assembled Structures in Bilayer Block Copolymer Films," ***Science*** **336**, 1294-1298 (2012).

475636 Thin Films and Two-Dimensional Materials for Energy Applications

Abstract:

2nd Year Postdoctoral Fellow

Research Interests:

New materials are constantly being designed to meet the growing and changing needs of society, such as energy storage, batteries, and electronic devices. Computers are now powerful enough that screening studies are practical, allowing us to do quick tests on a huge variety of materials. I have used density functional theory (DFT) to calculate important electrical and mechanical properties of these systems. Of special interest to me are two-dimensional materials, and interfaces between materials; interfaces allow fascinating phenomena such as two-dimensional conducting gases, and two-dimensional materials are so thin that their physical properties can be very different than that of bulk materials. In both cases, the properties of the system are vastly different for surfaces/interfaces/two-dimensional materials than their corresponding bulk cases.

Postdoctoral Projects: ***"Modeling of two-dimensional transitional metal carbonitrides (MXenes) for catalysis and energy storage"***

Under supervision of Jens Nørskov and Aleksandra Vojvodic, Department of Chemical Engineering, Stanford University and SUNCAT Center for Interface Science and Catalysis, SLAC National Accelerator Laboratory

PhD Dissertation: ***"First-Principles Studies of Perovskite Thin Films and Heterostructures"***

Under supervision of Alexander Demkov, Department of Physics, the University of Texas at Austin

Research Experience:

My academic career bridges physics, chemistry, materials science, and chemical engineering. I have always worked with experimentalists; I was fortunate in that my PhD group was dual theory/experimental. I have experience modeling mechanical, electric, magnetic and optical properties using DFT. I have had experience with multiple high-performance clusters, from massive nationwide level computer centers (TACC at the University of Texas at Austin, NERSC at Lawrence Berkeley National Laboratory) to midsize (Sherlock cluster at Stanford University) to a small cluster used exclusively by our research group (SUNCAT cluster at SLAC).

Teaching Interests:

I taught for four semesters at UT Austin; I taught a lab for premedical students taking physics, with multiple office hours per week to help students and encourage their learning, specially the application of physics towards medicine. UT Austin also gave opportunities to 'coach' students, i.e. a general area where students from any class, at any level, could come and ask questions about physics, which I also participated in for four semesters. I tutored ***pro bono***at my undergraduate and also taught on the side as a graduate student (from high school level to college physics).

Future Direction:

As faculty, I want to continue my research on both MXenes. These fascinating materials show vastly different mechanical and electrical properties that vary with functionalization. I would like to collaborate with experimentalists to show how their properties can be affected by different functionalizations, and demonstrated in the lab. I also want to understand better how to control the functionalization through experimental means (strain, temperature, pH, etc.) I also want to apply screening studies to interfaces (doping, interstitial atoms, vacancies), in order to model interfaces in practical devices such as field effect transistors, spin filters, and semiconducting heterojunctions. Because MXenes are so thin, they have interesting potential in device applications, and they are perfect for screening studies, because a small change in the composition can have drastic effects on conductivity, tensile strength, etc. The logical next step is to understand how their properties are affected by deposition/growth on real substrates.

Selected Publications:

"Two-Dimensional Transition Metal Carbides (MXene) as Efficient Electrocatalysts for Hydrogen Evolution", **K.D. Fredrickson**, Z.W. Seh, B. Anasori, J. Kibsgaard, A.L. Strickler, M.R. Lukatskaya, Y. Gogotsi, T.F. Jaramillo and A. Vojvjodic, submitted to ACS Energy Letters, **2016**

A.A. Demkov, K.J. Kormondy and **K.D. Fredrickson**, "Two-Dimensional Electron Gas at Oxide Interfaces," in ***Oxide Materials at the Two Dimensional Limit***, ed. F. Netzer and A. Fortunelli (Springer, New York) **2016**doi: 10.1007/978-3-319-28332-6

"Spin-polarized, orbital-selected hole gas at the EuO/Pt interface", **K.D. Fredrickson** and A.A. Demkov, J. Appl. Phys. 119, 095309 (**2016**). doi: 10.1063/1.4942837

"Carrier density modulation in a germanium heterostructure by ferroelectric switching", P. Ponath, **K.D. Fredrickson**, A.B. Posadas, Y. Ren, X. Wu, R.K. Vasudevan, M.B. Okatan, S. Jesse, T. Aoki, M.R. McCartney, D.J. Smith, S.V. Kalinin, K. Lai and A.A. Demkov, Nat. Commun. 6, 6067 (**2015**). doi: 10.1038/ncomms7067

"Atomic and electronic structure of the ferroelectric $BaTiO_3$/Ge(001) interface", **K.D. Fredrickson**, P. Ponath, A.B. Posadas, M.R. McCartney, T. Aoki, D.J. Smith and A.A. Demkov, Appl. Phys. Lett. 104, 242908 (**2014**). doi: 10.1063/1.4883883

475662 Layer-By-Layer Assembly for Water Desalination and Gas Separation

Abstract:

1st Year Postdoctoral Fellow

Research Interests:

Layer-by-layer (LbL) assembly is a bottom-up manufacturing technique that uses complementary interactions between components to deposit materials one layer at a time. Since a single layer is usually 1-100 nm thick, the structure of LbL assembly can be controlled with nanometer precision. This level of precision is unobtainable by traditional processing techniques, such as melt blending and solution mixing. Additional customization of the LbL assembly can be realized by changing its components and altering complementary interactions between the components. Besides being precise and versatile, LbL assembly is also very simple, allowing industrial-scale production through conventional dipping, spraying, or spin-coating procedures. These collective advantages make LbL assembly suitable for many applications, including anti-fogging, anti-microbial, drug delivery, and energy generation and storage. I am particularly interested in making advanced LbL assemblies for high-performance water filtration and gas separation applications.

Teaching Interests:

I will kindle students' interest in polymer science by relating classroom content with popular subjects among young people. Customized teaching methods will also be utilized to effectively reach every student as well as encourage the participation of underrepresented groups in science. Throughout my career, I plan to continually improve my teaching by seeking student feedback. Upon completing my course with a final group project, I expect my students to have a strong understanding of the fundamentals of polymers (or even fall in love with polymers), and have improved their critical thinking ability and communication skills. Based on this highly customizable teaching plan, my training in STEM teaching, and my background in polymer science and materials science, I am well prepared for teaching several polymer science and engineering courses, including polymer physics, membrane and separation technology, surface and interfaces, materials science and engineering, and multifunctional materials.

Completed Proposal:

I was a co-principal investigator for a proposal to the U.S. Department of the Interior, Bureau of Safety and Environmental Enforcement (E16PS00024); this proposal currently is under second-round review.

Postdoctoral Project:

"***Layer-by-Layer Assemblies with Advanced Gas Separation Properties***"

Under supervision of Dr. David Hopkinson, Department of Energy, National Energy Technology Laboratory

PhD Dissertation:

"***Improvements in Processing and Stretchability of Super Gas Barrier Multilayer Thin Films***"

Under supervision of Prof. Jaime Grunlan, Department of Mechanical Engineering, Texas A&M University

M.S. Thesis:

"***Modification of Immiscible HDPE/PA6 Blends using Carbon Nanotubes***"

Under supervision of Prof. Yong Wang, School of Materials Science and Engineering, Southwest Jiaotong University

Research Experience:

My research centers on structure-property relationships of polymeric materials. My first research project involved toughening immiscible polymer blends using carbon nanotubes (CNTs). In that project, I improved the toughness by selectively dispersing CNTs onto the interfaces between polymer domains. However, I was unable to control the dispersion of all of the CNTs, due to their strong Brownian motion in molten polymer blends. The desire to fully control the dispersion of nanofillers in polymer matrices propelled me to search for a new solution. I found the answer to this problem in LbL assembly, which can be used to disperse two-dimensional nanosheets in a polymer matrix with very high degrees of exfoliation and orientation. This unique layered-structure imparts exceptional gas barrier by creating a tortuous path for diffusion gas molecules, and improves the mechanical property of the thin film beyond predictions of the Halpin-Tsai model. By replacing rigid nanoparticles with soft polymers that combine with each other through H-bonding, I produced stretchy gas barrier coatings that remained functional under high-strain conditions. My current research on gas-separation membranes helps me better understand the transportation behaviors of different gases within polymeric thin films, and inspires me to develop new LbL assembly membranes for advancing separation technologies for gases and liquids.

Teaching Experience:

I actively sought opportunities to improve my teaching skills. I was a teaching assistant for an undergraduate-level Materials Science and Engineering course for one year, and enjoyed helping students solve problems. I also learned the 'dos and don'ts' of teaching, based on student feedback. I received training from a Science, Technology, Engineering and Math (STEM) teaching program, and guest-lectured an undergraduate-level class on ceramic materials. I have mentored several undergraduate students in my research groups, and almost all of my mentees continued their studies in graduate schools.

Future Direction:

I anticipate building a research group that uses state-of-the-art nanotechnology to counter pressing global challenges, such as water shortage, clean energy generation, and climate change.

Using LbL assembly for preparing water desalination membranes has attracted significant attention because LbL assembly features a multi-bipolar structure. This structure can be used to separate monovalent and divalent ions based on "Donnan exclusion". However, current LbL membranes are either too thick to permit high water flux, or too thin to provide sufficient salt rejection. Using rational post-treatment procedures, we can design a LbL assembly that features a nanoporous skin layer and a macroporous support layer, which impart high salt rejection and high water flux properties, respectively. This technology would improve our understanding on water desalination and should help reduce the global shortage of fresh water.

Besides separation in liquid media, I am also interested in gas separation. Specifically, I plan to produce multilayer membranes for hydrogen purification and carbon dioxide separation (the latter is also known as carbon capture). The hydrogen purification membrane exploits the difference between the kinetic diameters of different gas molecules. Specifically, graphene oxide sheets with tunable nanopores will be used as embedded molecular sieving nanofillers to generate high-purity hydrogen for fuel cells. For carbon capture membranes, I plan to use LbL assembly in conjunction with supported ionic liquid membranes to combine their advantages in mechanical and separation properties. Properly designed and developed films can be used to capture carbon dioxide to help counter global climate change.

The techniques and strategies summarized above can also be used to prepare other kinds of LbL assemblies with designed nanostructures and properties that are suitable for chemical sensing, self-healing, and self-cleaning purposes.

Select Publications:

F. M. Xiang, D. Parviz, T. M. Givens, P. Tzeng, E. M. Davis, C. M. Stafford, M. J. Green, J. C. Grunlan. Stiff and Transparent Multilayer Thin Films Prepared Through Hydrogen-Bonding Layer-by-Layer Assembly of Graphene and Polymer. ***Advanced Functional Materials***, 2016, 26, 2143.

C. Y. Cho, **F. M. Xiang (co-first author)**, K. L. Wallace, J. C. Grunlan. Combined ionic and hydrogen bonding in polymer multilayer thin film for high gas barrier and stretchiness. ***Macromolecules***, 2015, 48, 5723.

F. M. Xiang, T. M. Givens, S. M. Ward, J. C. Grunlan. Elastomeric polymer multilayer thin film with sustainable gas barrier at high strain. ***ACS Applied Materials & Interfaces***, 2015, 7, 16148.

F. M. Xiang, T. M. Givens, J. C. Grunlan. Fast spray deposition of super gas barrier polyelectrolyte multilayer thin films. ***I&EC Research***, 2015, 54, 5254.

F. M. Xiang, S. M. Ward, T. M. Givens, J. C. Grunlan. Structural tailoring of hydrogen-bonded poly(acrylic acid)/poly(ethylene oxide) multilayer thin films for reduced gas permeability. ***Soft Matter***, 2015, 11, 1001. Featured on Soft Matter Blog as a hot article for January 2015.

F. M. Xiang, S. M. Ward, T. M. Givens, J. C. Grunlan. Super stretchy polymer multilayer thin film with high gas barrier. ***ACS Macro Letters***, 2014, 3, 1055.

F. M. Xiang, P. Tzeng, J. S. Sawyer, O. Regev, J. C. Grunlan. Improving the gas barrier property of clay-polymer multilayer thin films using shorter deposition times. ***ACS Applied Materials & Interfaces***, 2014, 6, 6040.

For more information, please visit my personal website http://fangmingxiang.weebly.com

475702 Programmable Assembly and Deformation of Soft Matter

Abstract:

Research Interests:

While surface tension generally plays a negligible role on macroscopic scales, it is often the dominant force on nanometer to micrometer length-scales. The main goal of my research is to understand the role that surface tension plays on sub-millimeter scale objects, especially on soft material systems, and how to utilize this phenomenon to assemble and deform such objects. My research addresses several phenomena of nano-and micro-sized objects at fluid interfaces. On the nano-scale, amphiphilic block copolymers are used to explore interfacial behaviors due to their enhanced stability, mechanical properties, and tunability compared to other interfacially active materials such as small molecule surfactants or lipids. I have investigated the tailoring of amphiphilic block copolymer assemblies through deformation at the oil/water interface by inducing interfacial instabilities to incorporate inorganic nanoparticles into micelles, or by controlling osmotic stresses to prepare multi-compartment emulsions and capsules. Next, thin temperature responsive hydrogel sheets are used to probe elastic properties, buckling instabilities and capillary attraction at the micro-scale. I exploited the competition between surface energy and elastic bending energy to quantitatively determine key elastic properties of thin films that are otherwise challenging to measure. Additionally, I found significant edge imperfections due to the finite resolution of photolithography by observing interfacial deformations. The edge imperfections are then employed to drive the buckling of narrow photo-crosslinked hydrogel ribbons. Further, I introduced a new concept of capillary assembly of soft hydrogel sheets with programmed three-dimensional geometry. This offers opportunities to study correlations between elastic properties and surface tension, as well as capillary interactions between soft materials that can be extended to more complex multi-polar interface deformations.

I plan to form a research group to pursue,

- New lithographic techniques for programming internal heterogeneity in nanocomposite soft materials and characterization of their mechanical properties and photothermal effects.

- Deformation of liquids by external stimuli for the development of energy conversion devices.

Teaching Interests:

I am interested in teaching courses in Chemical Engineering at the undergraduate and graduate levels, as well as developing courses in polymer science and fluid mechanics.

475399 Biohybrid Materials for Applications in Human Healthcare and Sustainability -- Assistant Professor Candidate

Abstract:

R. Helen Zha – Faculty Candidate

Research Vision:

Synthesizing biohybrid materials for applications in human healthcare and sustainability, with focus on establishing relationships between molecular engineering, hierarchical material structure, and function.

Keywords: soft matter, self-assembly, nanomaterials, bio-inspired materials

Research Experience:

Adviser: Sam Stupp (Northwestern University, Chicago IL)

My Ph.D. research focused on the self-assembly of peptide-based amphiphiles into functional nanomaterials. I investigated the formation of ordered "hairy" membranes by peptide amphiphiles and biopolymers using an aqueous-aqueous interface. Through experimental study as well as collaboration with computational researchers, I identified the key factors in controlling membrane morphology and properties. Furthermore, I incorporated anti-cancer peptide motifs into the membrane self-assembly process to create versatile biomaterials for adjuvant cancer therapy. Additionally, I synthesized protein-mimetic nanostructures as potent angiogenesis inhibitors with potential applications in cancer therapy. (10/2007 – 10/2013)

Adviser: Bert Meijer (Eindhoven University of Technology, The Netherlands)

My postdoctoral research in the group of Bert Meijer investigated self-assembling supramolecular polymers and block molecules into ordered nanomaterials with feature sizes less than 5 nm. These materials comprised of a novel class of well-defined siloxanes oligomers with functional end groups. I studied the synergistic and antagonistic intermolecular forces exhibited by these oligomers in order to understand their fundamental self-assembly behavior as well as control material structure and properties. I further used this knowledge to create photoswitchable materials based on the isomerization of azobenzene. (11/2013 – 7/2016)

Adviser: Phil Messersmith (UC Berkeley, Berkeley CA)

My postdoctoral research in the group of Phil Messersmith will work towards developing materials inspired by spider silk and mussel adhesive proteins. (9/2016 - current)

Research Interests:

Molecular materials with defined nanostructure are vital in developing advanced technologies for a variety of emerging global needs. Self-assembly, the spontaneous organization of molecules into well-defined architectures, is key in controlling soft matter at the nano-scale. My research program will utilize self-assembly to create biohybrid materials for applications in human healthcare as well as energy and sustainability. These materials will incorporate versatile biological building blocks, such as peptides and nucleic acids, in conjunction with functional non-natural components. My research approach will incorporate molecular engineering with materials characterization on nano, meso, and macroscopic length scales. Thus, my research will be positioned at the interface between materials science, chemistry, biology, and engineering.

Teaching Interests:

I have gained significant teaching experience on topics in materials science & engineering, biomedical engineering, and chemistry. I look forward to teaching introductory as well as advanced courses in these fields. Furthermore, I have special interest in developing courses that will prepare students for working in nanotechnology, molecular engineering, and soft matter fields (e.g. Nanomaterials Engineering, Experimental Nano-Analysis).

Selected Awards, Publications, & Patents:

NSF Graduate Research Fellowship ($175,000) 9/2008 – 9/2011

Northwestern University Cabell Fellowship ($60,000) 9/2007 – 9/2008

Zha, R. H.; Sur, S.; Stupp, S. I. Self-assembly of cytotoxic peptide amphiphiles into supramolecular membranes for cancer therapy. ***Adv. Healthc. Mater*. 2013**, ***2***, 126-133. doi: 10.1002/adhm.201200118

Stupp, S. I.; **Zha, R. H.**; Palmer, L. C.; Cui, H.; Bitton, R. Self-assembly of biomolecular soft matter. ***Faraday Discuss*. 2013**, ***166***, 9-30. doi: 10.1039/C3FD00120B

Zha, R. H.; Sur, S.; Boekhoven, J.; Shi, H. Y.; Zhang, M.; Stupp, S. I. Supramolecular assembly of multifunctional maspin-mimetic nanostructures as a potent peptide-based angiogenesis inhibitor. ***Acta Biomater*. 2015**, ***12***, 1-10. doi: 10.1016/j.actbio.2014.11.001

Zha, R. H.; Velichko, Y. S.; Bitton, R.; Stupp, S. I. Molecular design for growth of supramolecular membranes with hierarchical structure. ***Soft Matter*. 2016**, ***12***, 1401-1410. doi: 10.1039/c5sm02381e

Zha, R. H.; de Waal, B. F. M.; Lutz, M.; Meijer, E.W. Endgroups of functionalized siloxane oligomers direct block copolymeric or liquid crystalline self-assembly behavior. ***J. Am. Chem. Soc.* 2016**, ***138***, 5693–5698. doi: 10.1021/jacs.6b02172 ***(ACS Editor's Choice Award)***

Zha, R. H.; Zhang, M.; Stupp, S. I. Anti-angiogenic molecules, nanostructures and uses thereof. US9,169,294, October 27, **2015.**

Research Interests

Understanding and Tailoring Novel Hybrid Nano Materials & Interfaces

Dr. Letian Dou *(ldou@berkeley.edu)*

Currently, one of the major topics in science and engineering is energy. There is a strong demand for new materials and technologies toward clean energy harvesting and energy-efficient devices (e.g. solar cells and light emitting diodes). Semiconducting materials play an essential role in such devices. Both inorganic and organic semiconductors hold great promise, but are currently limited by relatively more complicated fabrication and lower performance. By combining the best of two worlds, hybrid material provides a new solution. From a fundamental science point of view, the bonding between inorganic and organic materials and the charge transport across the interface at different length scales are important to understand. On the other hand, from a practical application point of view, hybridization may lead to superior properties and provide new challenges in materials design and manufacturing.

My research interests lie at the intersection of chemistry and materials science with the target of creating and understanding new hybrid materials. Particularly, I want to answer the questions of how to effectively interface inorganic and organic semiconductors **at different length scales**; how to manipulate the **charge, mass, and heat transport** for achieving better electronic and optoelectronic devices. The material systems of interests are inorganic halide perovskites and organic conjugated molecules/polymers. Using this model system, I will focus on ***materials chemistry*** to develop new hybrid materials with new functionality, investigate the assembly behavior and transport phenomenon at the ***interfaces***, and integrate them into ***electronic and photonic devices*** to improve the efficiency and reliability. The knowledge we obtain and the methodology we create will be generalizable to other material systems, therefore advancing our understanding in the basic chemistry and physics at the inorganic-organic interfaces. These research projects differ significantly from those of my Ph.D. and postdoctoral research advisors yet combine skills and knowledge that I learned from them. Chemical synthesis, advanced structural, optical, and electrical characterization, and electronic device fabrication techniques will be employed in the Dou Research Group.

Teaching Interests

My research background in chemistry, electronic materials and devices has given me experience in most chemical applications, and a vast background in chemical and physical methods and techniques. I am qualified to teach a broad range of topics in chemical engineering at *undergraduate level*; for example, basics in chemical engineering, thermodynamics, energy and mass transport, separation processes, kinetics, polymer science and engineering, etc. My interests in teaching at the *graduate level* are tied to my research work. Specific graduate-level teaching interests include conducting/semiconducting polymers, nano-materials for energy applications, chemical & physical properties of organic/hybrid semiconductors, and the fabrication and characterization of thin film electronic devices (light-emitting diodes/photovoltaics/field-effect transistors).

Letian Dou

315 Hildebrand Hall, Department of Chemistry
University of California, Berkeley, CA, 94720, USA
+1 310 409 5493 (phone); ldou@berkeley.edu
Website: https://letiandou.wordpress.com/

Educational Background

University of California, Berkeley (UC Berkeley)
Chemist Postdoc Fellow — 2014-present
University of California, Los Angeles (UCLA)
Ph.D. in Materials Science and Engineering — 2009-2014
Peking University, Beijing, P. R. China
B.S. in Chemistry — 2005-2009

Honors and Awards

- **Materials Research Society Graduate Student Awards** (2014);
- UCLA Materials Science Department Award (2014), one PhD student awarded annually;
- **The Link Foundation Energy Fellowship** (2013), three recipients in U.S. awarded annually;
- Chinese Government Award For Outstanding Students Abroad (2013);
- Materials Research Society Travel Awards (2013);
- Student Award by Society of Information Display (SID) – LA Chapter (2011);

Selected Publications ([†]equal contribution; full publication attached)

- **Letian Dou**[†], Andrew B. Wong[†], Yi Yu[†], Minliang Lai, Nikolay Kornienko, Samuel W. Eaton, Anthony Fu, Connor G. Bischak, Jie Ma, Tina Ding, Naomi S. Ginsberg, Lin-Wang Wang, A. Paul Alivisatos, Peidong Yang*
  Atomically Thin Two-dimensional Organic-inorganic Hybrid Perovskites
  ***Science 2015**, 349, 1518-1521.*

- **Letian Dou**, Yonghao Zheng, Xiaoqin Shen, Guang Wu, Kirk Fields, Wan-Ching Hsu, Huanping Zhou, Yang Yang*, Fred Wudl*
  Single-Crystal Linear Polymers Through Visible Light-Triggered Topochemical Quantitative Polymerization
  ***Science 2014**, 343, 272-277.*

- **Letian Dou**[†], Jingbi You[†], Jun Yang, Chun-Chao Chen, Youjun He, Seiichiro Murase, Tom Moriarty, Keith Emery, Gang Li and Yang Yang*
  Tandem polymer solar cells featuring a spectrally matched low-bandgap polymer
  Nature Photonics 2012**, 6, 180-185. **(Over 1000 citations)

- Jingbi You[†], **Letian Dou**[†], Ken Yoshimura, Takehito Kato, Kenichiro Ohya, Tom Moriarty, Keith Emery, Chun-Chao Chen, Jing Gao, Gang Li* and Yang Yang*
  A polymer tandem solar cell with 10.6% power conversion efficiency
  Nature Communications* 2013**, *4, 1446.* ***(Over 1500 citations)

- **Letian Dou**, Yongsheng Liu, Ziruo Hong, Gang Li*, Yang Yang*
  Solution-processed Low-bandgap Near-IR Conjugated Polymers/molecules for Organic Electronics
  Chemical Reviews* 2015**, *115, 12633-12665.* ***(Invited review article)

Research Experience

Chemist Postdoc Fellow, UC Berkeley/ Lawrence Berkeley National Lab/ California Research Alliance by BASF 2014-present

Advisor: Prof. Peidong Yang

- **Inorganic and hybrid perovskites nanomaterials.** Develop synthetic chemistry to achieve 0, 1, 2 dimensional nanostructures; study the fundamental optical and electronic properties; and apply the novel materials into photovoltaics, light emitting diodes, lasers, etc.
- **Ultra thin core-shell metal nanowires for transparent electrodes.** Design novel graphene-metal nanowire core-shell structures to achieve highly stable transparent conductors for displays and solar cells.
- **Grant proposal writing (to DoE, AFOSR, BASF)** and PhD student training.

Graduate Student Researcher, UCLA 2009-2014

Advisor: Prof. Yang Yang

- **High efficiency tandem polymer solar cells and visibly-transparent polymer solar cells.** Design and synthesis of novel low-bandgap polymers (near-IR absorbing) specifically for tandem and visibly-transparent solar cells. **Several research grants from NSF (2013-2015), AFOSR (2011-2014) and industry (EFL Tech., Sumitomo Chemical Inc.) were awarded** based on these works.
- Synthesis of fullerene derivatives as acceptor materials for polymer solar cells.
- Thin film morphology and charge transport characterizations in polymer solar cells.

Visiting Student, Univ. of California, Santa Barbara March-August, 2013

Advisor: Prof. Fred Wudl

- **Design, synthesis and characterization of novel polymer single crystals by photochemistry.** Discovery of the first visible-light-triggered topochemical polymerization and a new type of highly perfect polymer single crystal.
- Design and synthesis of new electron acceptors for organic electronic devices.

Teaching and Mentoring Experience

- **Teaching half of the courses** for MSE 248 (*Physics of Solar Cells*, preparing slides, giving lectures), 2014.

- **Teaching Assistant** for MSE 200 (*Principles of Materials Science*, leading discussions, answering questions, grading), 2013.
- **Teaching Assistant** of General Chemistry Lab (mentoring 20 students in the lab), 2009.
- **Mentoring** and training of many **junior graduate students**:
  i. Dr. Wei-Hsuan Chang [UCLA, Ph.D. 2016]: Selenium-containing low band gap polymers for organic photovoltaics.
  ii. Mr. Roger Zhao [UCLA, Ph.D. candidate]: Ladder type building blocks for high mobility conjugated polymers.
  iii. Ms. Ariella Machness [UCLA, undergraduate]: Synthesis of diketopyrrolopyrrole based low band gap polymers.
  iv. Mr. Minliang Lai [UC Berkeley, Ph.D. candidate]: Halide perovskite nanowires and nanosheets synthesis and optical properties.
  v. Mr. Teng Lei [UC Berkeley, Ph.D. candidate]: Controlled doping and alloying of halide perovskite nanowires.

Invited Talks and Conference Presentations

- *Atomically Thin Two-dimensional Organic-inorganic Hybrid Perovskites*
  Letian Dou, Andrew B. Wong, Yi Yu and Peidong Yang
  2016 ECS Meeting, May 2016, San Diego, CA. ***(Invited oral presentation)***

- *Low-Bandgap Polymers Designed for Tandem Polymer Solar Cells*
  Letian Dou, Jingbi You, Chun-Chao Chen, Jing Gao, Gang Li and Yang Yang
  2014 MRS Spring Meeting & Exhibit, April 2014, San Francisco, CA. ***(oral presentation)***

- *Single-Crystal Linear Polymers Through Visible Light-Triggered Topochemical Quantitative Polymerization*
  Letian Dou
  2014 Department of Materials Science Seminar, UCLA ***(oral presentation)***

- *Benzodithiophene- and Diketopyrrolopyrrole-based Low-bandgap Polymers Designed for Tandem Polymer Solar Cells*
  Letian Dou, Jingbi You, Chun-Chao Chen, Jing Gao, Wei-Hsuan Chang, Gang Li and Yang Yang
  2013 MRS Spring Meeting & Exhibit, April 2013, San Francisco, CA. ***(oral presentation)***

- *Recent progress in polymer solar cells and tandem polymer solar cells*
  Letian Dou, Jingbi You, Chun-Chao Chen, Jing Gao, Gang Li and Yang Yang
  ACS 244th National Meeting, August 2012, Philadelphia, PA. ***(Invited oral presentation)***

Patents

- *Active Materials for Electro-Optic Devices and Electro-Optic Devices*
  Yang Yang, **Letian Dou**, Jingbi You (WO2012135527, US20140020739 A1, licensed by

EFL Tech.)

- *Semi-Transparent, Transparent, Stacked and Top-illuminated Organic Photovoltaic Devices*
  Yang Yang, Rui Zhu, Chun-Chao Chen, **Letian Dou**, Gang Li (WO2014007867 A1, US20140084266 A1, licensed by EFL Tech.)

- *Conjugated Polymers for Electronic Devices*
  Yang Yang, **Letian Dou**, Wei-Husan Chang (US20150028265 A1)

- *Electro-Optic Device Having Nanowires Interconnected into A Network of Nanowires*
  Yang Yang, Rui Zhu, Chun-Chao Chen, **Letian Dou**, Gang Li (WO2014007867A1, US20140084266, licensed by EFL Tech.)

- *Ultra Thin Copper-Reduced Graphene Oxide Core-Shell Nanowires based Transparent Conductors*
  Peidong Yang, **Letian Dou**, Fan Cui (BK-2015-154, provisional filing)

- *Synthesis of Ultra-Thin Metal Nanowires Using Organic Free Radicals*
  Peidong Yang, Fan Cui, **Letian Dou** (UCALP073P/BK-2016- 136, provisional filing)

Service and Membership

- **Referee/reviewer** for many scientific journals including: *Journal of the American Chemical Society*; *Advanced Materials*; *Nano Letters*; *ACS Nano*; *Scientific Reports (by NPG)*; *Chemistry of Materials*; *Journal of Materials Chemistry A*; *Nano Energy*; *J. Phys. Chem. C*; *Macromolecules; Polymer Chemistry; Macromolecule rapid communications*; *Macromolecular Chem. & Phys.*; *Chemistry - A European Journal; Organic Electronics; Applied Surface Science; ACS Energy Letters; ACS Applied Materials & Interfaces.*
- **Committee** of 2016 Link Foundation Energy Fellowship.
- **Member** of American Chemical Society (ACS), Materials Research Society (MRS), Electrochemical Society (ECS), American Institute of Chemical Engineers (AIChE), SPIE, Society of Information Display (SID).
- Assistant of 2nd International Symposium on Materials Enabling Nanodevices (2010).
- Student chair of SID-LA chapter (2013).

Full Publication List (*over 6000 citations, H index = 25*)

Google scholar citation: *https://scholar.google.com/citations?user=nfkBN7UAAAAJ&hl=en*

2016

33. Yi Yu, Dandan Zhang, Christian Kisielowski, **Letian Dou**, Nikilay Kornienko, Yehonadav Bekenstein, A. Paul Alivisatos, Peidong Yang; "*Atomic Resolution Imaging of Halide Perovskites*", ***Nano Letters 2016,*** *submitted.*

32. Samuel W. Eaton, Minliang Lai, Natalie Gibson, Andrew B. Wong, **Letian Dou**, Jie Ma, Lin-Wang Wang, Steve Leone, Peidong Yang; "*Lasing in Robust Cesium Lead Halide Perovskite Nanowires*", ***PNAS 2016,*** *113, 1993-1998.*

31. **Letian Dou**, Fan Cui, Yi Yu, Samuel W. Eaton, Arky Yang, Joaquin Resasco, Christian Schildknecht, Kerstin Schierle-Arndt, Peidong Yang; "*Solution-processed ultra thin copper-reduced graphene oxide core-shell nanowire transparent conductors*", ***ACS Nano 2016***, *10, 2600-2606.*

2015

30. **Letian Dou**[†], Andrew B. Wong[†], Yi Yu[†], Minliang Lai, Nikolay Kornienko, Samuel W. Eaton, Anthony Fu, Connor G. Bischak, Jie Ma, Tina Ding, Naomi S. Ginsberg, Lin-Wang Wang, A. Paul Alivisatos, Peidong Yang; "*Atomically Thin Two-dimensional Organic-inorganic Hybrid Perovskites*", ***Science 2015***, *349, 1518-1521.*

29. **Letian Dou**, Yongsheng Liu, Ziruo Hong, Gang Li, Yang Yang; "*Solution-processed Low-bandgap Near-IR Conjugated Polymers/molecules for Organic Electronics*", ***Chemical Reviews 2015***, *115, 12633-12665.* ***(Invited review article)***

28. Fan Cui, Yi Yu, **Letian Dou**, Jianwei Sun, Christian Schildknecht, Kerstin Schierle-Arndt, Peidong Yang; "*Ultra-thin colloidal copper nanowires for transparent conductor with high performance*", ***Nano Letters 2015, 2015***, *15, 7610-7615.*

27. Dandan Zhang, Samuel W. Eaton, Yi Yu, **Letian Dou**, Peidong Yang; "*Solution-phase Synthesis of Cesium Lead Halide Perovskite Nanowires*", ***Journal of the American Chemical Society 2015***, *137, 9230-9233.*

26. Andrew B. Wong, Minliang Lai, Samuel W. Eaton, Yi Yu, Elbert Lin, Anthony Fu, **Letian Dou**, Peidong Yang; "*Growth and Anion Exchange Conversion of $CH_3NH_3PbX_3$ Nanorod Arrays for Light-Emitting Diodes*", ***Nano Letters 2015,*** *15, 5519-5524.*

25. Yang (Michael) Yang, Wei Chen, **Letian Dou**, Wei-Hsuan Chang, Hsin-Sheng Duan, Brion Bob, Gang Li, Yang Yang; "*Molecular Compatibility Enabling High-Performance Multiple Donor Bulk Heterojunction Solar Cells*", ***Nature Photonics 2015***, *9, 190-198.*

24. Wei-Hsuan Chang, Lei Meng, **Letian Dou**, Jingbi You, Chun-Chao Chen, Yang (Michael) Yang, Eric P. Young, Gang Li, Yang Yang; "*A Selenophene Containing Benzodithiophene-alt-thienothiophene Polymer for Additive-Free High Performance Solar Cell*", ***Macromolecules* 2015**, *48, 562-568.*

23. Yongsheng Liu, Qi Chen, Hsin-Sheng Duan, Huanping Zhou, Yang Michael Yang, Huajun Chen, Song Luo, Tze-Bin Song, **Letian Dou**, Ziruo Hong, Yang Yang; "A dopant-free organic hole transport material for efficient planar heterojunction perovskite solar cells", ***Journal of Materials Chemistry A* 2015**, *22, 11940-11947.*

2014

22. **Letian Dou**, Yonghao Zheng, Xiaoqin Shen, Guang Wu, Kirk Fields, Wan-Ching Hsu, Huanping Zhou, Yang Yang, Fred Wudl; "*Single-Crystal Linear Polymers Through Visible Light-Triggered Topochemical Quantitative Polymerization*", ***Science* 2014**, *343, 272-277.* (also see *Perspectives* by Nancy Goroff; "*A Clear Path for Polymer Crystallization*", ***Science* 2014**, *343, 258.*)

21. **Letian Dou**†, Yang (Michael) Yang†, Jingbi You†, Ziruo Hong, Wei-Hsuan Chang, Gang Li, Yang Yang; "*Solution-Processed Hybrid Perovskite Photodetectors with High Detectivity*", ***Nature Communications* 2014**, *5, 5404.*

20. Jing Gao, **Letian Dou**, Wei Chen, Chun-Chao Chen, Xuanrong Guo, Jingbi You, Brion Bob, Wei-Hsuan Chang, Joseph Strzalka, Cheng Wang, Gang Li, and Yang Yang; "*Improving Structural Order for a High-Performance Diketopyrrolopyrrole-Based Polymer Solar Cell with a Thick Active Layer*", ***Advanced Energy Materials* 2014**, *4, 1300739.*

19. Jing Gao, Wei Chen, **Letian Dou**, Chun-Chao Chen, Wei-Hsuan Chang, Yongsheng Liu, Gang Li, and Yang Yang; "*Elucidating double aggregation mechanisms in the morphology optimization of Diketopyrrolopyrrole-based narrow bandgap polymer solar cells*", ***Advanced Materials* 2014**, *26, 3142-3147.*

18. Wei-Hsuan Chang, Jing Gao, **Letian Dou**, Chun-Chao Chen, Yongsheng Liu, and Yang; "*Side-Chain Tunability via Triple Component Random Copolymerization for Better Photovoltaic Polymers*", ***Advanced Energy Materials* 2014**, *4, 1300864.*

17. Qi Chen, Huanping Zhou, Tze-Bin Song, Song Luo, Ziruo Hong, Hsin-Sheng Duan, **Letian Dou**, Yongsheng Liu, and Yang Yang; "Controllable self-induced passivation of hybrid lead iodide perovskites toward high performance solar cells", ***Nano Letters* 2014**, *14, 4158-4163.*

2013

16. **Letian Dou**, Jingbi You, Ziruo Hong, Zheng Xu, Gang Li, Robert A. Street, and Yang Yang; "*25th Anniversary Article: A Decade of Organic/Polymeric Photovoltaic Research*", ***Advanced Materials* 2013**, *25, 6642-6671.* ***(Invited review article)***

15. Jingbi You[†], **Letian Dou**[†], Ken Yoshimura, Takehito Kato, Kenichiro Ohya, Tom Moriarty, Keith Emery, Chun-Chao Chen, Jing Gao, Gang Li and Yang Yang; *"A polymer tandem solar cell with 10.6% power conversion efficiency"*, ***Nature Communications 2013**, 4, 1446.* ***(Over 1500 citations)***

14. **Letian Dou**, Wei-Hsuan Chang, Jing Gao, Chun-Chao Chen, Jingbi You and Yang Yang; *"A Selenium-Substituted Low-Bandgap Polymer with Versatile Photovoltaic Applications"*, ***Advanced Materials 2013**, 25, 825-831.* ***(Over 300 citations)***

13. **Letian Dou**, Chun-Chao Chen, Ken Yoshimura, Kenichiro Ohya, Wei-Hsuan Chang, Jing Gao, Yongsheng Liu, Eric Richard and Yang Yang; *"Synthesis of 5H-Dithieno[3,2-b:2',3'-d]pyran as an Electron-Rich Building Block for Donor-Acceptor Type Low-Bandgap Polymers"*, ***Macromolecules 2013**, 46, 3384-3390.* ***(Selected in SYNFACTS for its important insights in polymer chemistry)***

12. Chun-Chao Chen, **Letian Dou**, Jing Gao, Wei-Hsuan Chang, Gang Li and Yang Yang; *"High-performance semi-transparent polymer solar cells possessing tandem structures"*, ***Energy & Environmental Science 2013**, 6, 2714-2720.*

11. Jingbi You, **Letian Dou**, Ziruo Hong, Gang Li, Yang Yang; *"Recent trends in polymer tandem solar cell research"*, ***Progress in Polymer Science 2013**, 38, 1909-1928.* ***(Invited review article)***

10. Vladan Jankovic, Yang Yang, Jingbi You, **Letian Dou**, Yongsheng Liu, Puilam Cheung, Jane P. Chang and Yang Yang; *"Active Layer-Incorporated, Spectrally-Tuned Au/SiO2 Core/Shell Nanorod-Based Light Trapping for Organic Photovoltaics"*, ***ACS Nano 2013**, 7, 3815–3822.*

9. Yongsheng Liu, Yang Yang, Chun-Chao Chen, Qi Chen, **Letian Dou**, Ziruo Hong, Gang Li and Yang Yang; *"Solution-Processed Small Molecules Using Different Electron Linkers for High-Performance Solar Cells"*, ***Advanced Materials 2013**, 25, 4657-4662.*

8. Yongsheng Liu, Chun-Chao Chen, Ziruo Hong, Jing Gao, Yang (Michael) Yang, Huanping Zhou, **Letian Dou**, Gang Li, Yang Yang; *"Solution-processed small-molecule solar cells: breaking the 10% power conversion efficiency"*, ***Scientific Reports 2013**, 3, 3356.*

2012

7. **Letian Dou**[†], Jingbi You[†], Jun Yang, Chun-Chao Chen, Youjun He, Seiichiro Murase, Tom Moriarty, Keith Emery, Gang Li and Yang Yang; *"Tandem polymer solar cells featuring a spectrally matched low-bandgap polymer"*, ***Nature Photonics 2012**, 6, 180-185.* ***(Over 1000 citations; Top 5 worldwide scientific breakthrough of 2012 highlighted in USA Today; "HOT PAPER IN PHYSICS" highlighted by SCIENCEWATCH; and numerous media reports)***

6. **Letian Dou**, Jing Gao, Eric Richard, Jingbi You, Chun-Chao Chen, Kitty C. Cha, Youjun He, Gang Li and Yang Yang; *"Systematic Investigation of Benzodithiophene and Diketopyrrolopyrrole Based Low-bandgap Polymers Designed for Single Junction and Tandem Polymer Solar Cells"*, ***Journal of the American Chemical Society 2012**, 134, 10071-10079.* ***(Over 300 citations)***

5. Chun-Chao Chen[†], **Letian Dou**[†], Rui Zhu, Choong-Heui Chung, Tze-Bin Song, Yue Bing Zheng, Steve Hawks, Gang Li, Paul S. Weiss, and Yang Yang; "*Visibly Transparent Polymer Solar Cells Produced by Solution Processing*", ***ACS Nano 2012**, 6, 7185-7190.* ***(Over 200 media coverage including Time magazine, Los Angeles Times; most read article in ACS nano 2012)***

4. Jingbi You, Chun-Chao Chen, **Letian Dou**, Seiichiro Murase, Hsin-Sheng Duan, Steve Hawks, Tao Xu, Hae Jung Son, Luping Yu, Gang Li, Yang Yang; "*Metal Oxide Nanoparticles as an Electron-Transport Layer in High-Performance and Stable Inverted Polymer Solar Cells*", ***Advanced Materials 2012**, 24, 5267-5272.*

3. Youjun He, Jingbi You, **Letian Dou**, Chunchao Chen, Eric Richard, Kitty C Cha, Yue Wu, Gang Li, Yang Yang; "*High performance low band gap polymer solar cells with non-conventional acceptor*", ***Chemical Communications 2012**, 48, 7616-7618.*

2. Youjun He, Chun-Chao Chen, Eric Richard, **Letian Dou**, Yue Wu, Gang Li, Yang Yang; "*Novel fullerene acceptors: synthesis and application in low band gap polymer solar cells*", ***Journal of Materials Chemistry 2012**, 22, 13391-13394.*

2011 and before

1. Yijun Zheng, Jia Zheng, **Letian Dou**, Wenqiang Qiao, Xinhua Wan; "*Synthesis and characterization of a novel kind of near-infrared electrochromic polymers containing an anthraquinone imide group and ionic moieties*", ***Journal of Materials Chemistry 2009**, 19, 8470-8477.*

475884 Functional Polymers for Energy Generation and Storage: Donor-Acceptor Block Copolymers for Photovoltaics and Functional Polyimides for Dielectric Materials

Abstract:

Research Interests:

My research program will develop conjugated block copolymers for power generation and polymer dielectrics for energy storage. We will leverage expertise in polymer and organic chemistry as well as advanced characterization techniques, such as resonant soft X-ray scattering (RSOXS) and energy filtered TEM (EF-TEM). Our unique set of skills are crucial to develop new classes of materials that can serve as (1) model platforms for fundamental studies of optoelectronic and dielectric properties and (2) for demonstrations of breakthrough performance in device applications, furthermore, investigate fundamental structure-property relationship.

Wearable electronics utilized in health care, military, and communication applications require the development of a portable electrical power source. Flexible and light-weight organic photovoltaics are one of the most promising solutions, but challenges remain in performance and stability. My approach will develop a new class of active layer materials capable of exceeding the performance of state-of-the-art blend organic solar cells with improved stability. Donor-acceptor fully conjugated block copolymers (D-A BCPs) will be synthesized as single active layer materials, and their morphologies and solar cell performances will be investigated. Current synthetic approaches, including step-growth polymerization, leads to a mixture of products such as unreacted homopolymers, diblock and triblock copolymers. A focus of my group will be on developing polymerization strategies based on chain-growth mechanisms to prepare well-defined architectures and low dispersities. These well-defined D-A BCPs are expected to enable finer control of the microstructure. In organic electronics, donor/acceptor interfaces are crucial to overall performance, and block copolymers provide a unique approach to control such interfaces. D/A interfaces will be modified with a single bridge unit between donor and acceptor blocks. Furthermore, new chemical structures will be explored for the development of high performance D-A BCPs. Proposed chemical structures for new D-A BCPs are based on thieno-thiophene and cyclopentadithiophene donor blocks with naphthalene diimide acceptor blocks.

Polymeric dielectric materials suitable for electrical energy storage applications are promising because of the potential for flexible, lightweight, photo-patternable, scalable and robust properties compared to ceramic counterparts. But, polymeric dielectric materials have limited impact in various applications, such as hybrid and electric vehicles and aerospace power electronics, due to limited working temperatures. My approach will develop functional polymers to demonstrate improved dielectric properties for high temperature and high electric field applications. Two new diamine monomers of polyimide will be prepared associated with methyl sulfonyl and semi-fluoro alkyl groups. Methyl sulfonyl group is known as one of the strongest polar groups and is expected to enhance the dielectric constant. Semi-fluoro alkyl groups are expected to increase the break-down strength and provide stability in humid environments. A key strategy will be to prepare random copolymers consisting of diamine monomers with both methyl sulfonyl and semi-fluoro alkyl group and corresponding dianhydride monomers to optimize the performance in capacitors. Furthermore, crosslinking of functional polyimdes will enhance thermal stability.

My group will focus on the development of large-scale coating techniques for organic electronics and capacitors. My approach will fabricate solar cells and capacitors through blade or slot-die coating to capture the essential physics of roll-to-roll printing and we will investigate the microstructure within cast polymer films using in-situ X-ray scattering.

Teaching Interests:

I believe it is my responsibility to challenge and support my students to reveal their potential and enjoy their studies. Based on my experience as a student, I will pursue three learning goals in my teaching:

Conceptual learning: Establishing a strong foundation of basic concepts is most important, even more so than understanding other derivative content of a class, because students only understand and evaluate following concepts if they understand the fundamentals. Otherwise, students just memorize at best. I will devote more time, in particular at the beginning of the course, to ensure that students understand fundamental concepts.

Experiential learning: Learning needs momentum, and experience offers strong momentum for learning. When I was an undergraduate, I had a chance to observe a scattering experiment in a synchrotron laboratory. Before then, I had only superficial knowledge of X-ray scattering, but after the scattering experiment I was self-motivated to learn more. As a result, my experience at a synchrotron offered me a good chance to develop my passion for X-ray scattering. In experimental classes especially, I will provide opportunities for students to experience and see concrete examples as much as I can.

Stimulating learning: Students learn best when they are engaged in active inquiry. The level of motivation determines the final level of learning or understanding. It is not as exciting to carry out experiments by simply following instructions without sufficient information about the motivation or fundamentals of the experiments. When working with undergraduates for training or research programs during the summer, I tried to devote time to explain why we do undertake specific experiments and what we are expected to see as results. In case of organic reactions or polymerizations, I explained what we were going to do with the resultant materials in the future. In the same manner, I will help students accept the what and why of their studies, and encourage students to become engaged in classes.

As a researcher and a teacher in Chemical Engineering and Materials Science, I will continue to pursue the three learning goals as stated above, and hopefully contribute to the training of the next generation.

475966 Microfluidic Design of Multi-Phase Emulsion Drops for Functional Materials Production

Abstract:

Research Interests:

Emulsions are attractive for encapsulating, transporting, or controllably releasing a wide variety of chemical and biological active materials. Multi-phase emulsion drops are of particular interest due to the compartments of these emulsion drops that can be enlisted to impart functionality. By using the precise flow control of multi-phasic fluids in microfluidics, in conjunction with fluid-fluid interface design by proper selection of materials in each phase, I have produced novel functional materials that otherwise would have been inaccessible. For example, I developed a new strategy of combining bulk and microfluidic emulsification to achieve highly efficient encapsulation and enhanced retention of fragrance in polymer microcapsules. Additionally, I developed novel triple emulsion drops with an ultra-thin intermediate layer consisting of either hydrogel or fluorocarbon oil that separates the innermost cargo from the polymeric shell. These emulsion drops have interesting properties and great technological potential for encapsulation and controlled release of a wide variety of active materials. Moreover, I have worked on several collaborative projects through which we have designed ideal polymer network in emulsion templated microcapsules for encapsulation and triggered release of challenging actives, and semi-permeable polymeric shell in microcapsules for micro-sensor, and micro-reactor applications, just to name some.

Research Experience:

My academic career path has been a blend of many fields of science and engineering. My formal training at MIT (with Prof. Robert E. Cohen and Prof. Michael F. Rubner) was in Chemical Engineering, which I have studied the mechanisms that govern the structure and function of polymer thin films for developing a diverse range of functional films including antifogging coatings, switchable surfaces and stimuli-responsive hydrogels. Specifically, I have introduced a framework in which poly(vinyl alcohol) can be incorporated into multilayered thin films by investigating hydrogen-bonding interaction of polymers in aqueous media. By tailoring the interlayer diffusion of hydrogen-bonding polymers during the assembly, I also designed a complex hetero-structured architecture that sequentially dissolves with increase in local pH condition. Additionally, I have developed a new concept in controlling water condensation. By thorough characterization and analysis of a model system, a new physical theory, â€œzwitter-wettablilityâ€, was proposed whereby the film readily absorbs water vapor while simultaneously exhibiting hydrophobic character to liquid water. Moreover, this zwitter-wettable concept was further extended to real-life applications where the optimally designed film resulted in a significantly enhanced antifog and even frost-resistant behavior.

Teaching Interests:

Aside from my research career, I also have extensive teaching experience. I TAed an undergraduate course in Chemical Engineering at MIT in which I lectured undergrads, held weekly office hours, and developed course materials; I mentored undergraduate students as a part of their undergraduate thesis at MIT; I also mentored local community college students as a part of an educational outreach program from MIT CMSE; and, lastly, I am currently mentoring and supervising the research of graduate students on a large number of projects.

Future Direction:

As a new faculty, I would like to continue studying soft materials at interfaces, and engineer new hierarchical structures for industrial applications. In particular I would like to leverage the knowledge that I obtained from the two topics that I worked on during my Ph.D. and postdoc: polymer science/engineering and microfluidics, respectively. Many intriguing phenomena occur at fluid-fluid/solid-fluid interfaces and I believe my expertise will provide new perspective in designing functional materials.

My first thrust as a faculty is to take advantage of molecular rearrangement of polymers at surfaces to design novel polymer film structures. It has been demonstrated that configuration of a liquid drop on a given surface depends on the force balance across liquid-solid-air interfaces, which is governed by the surface topology as well as the surface chemistry. However, by separately tuning the surface and bulk properties of a film, unique surfaces can be prepared in which water drops exhibit lower equilibrium contact angle than typical hydrocarbon oils that have much lower surface tension than water. While these engineered surfaces have great potential for applications such as self-cleaning surfaces, and antifogging coatings, how to modulate the rate of rearrangement by materials design is still unclear. I believe that studying this in depth in conjunction with phase change heat and mass transfer will give rise to new applications such as contaminant separation from water vapor for reducing pollutant in air and water.

My second thrust is to utilize the middle phase that separates the innermost drop from the continuous phase in multi-phase emulsion drops to study polymers at interfaces and engineer new functional materials. It has been demonstrated that the middle phase in double-emulsion drops can serve as a template to form polymer microcapsules for encapsulation of actives. Yet, there has been little progress toward the production of microcapsules with membranes that fulfills the practical demands. Moreover, I believe that many well-known scientific findings observed in bulk such as micro-phase separation of block copolymers, and polymer network formation using various chemistries can be translated into this compartment provided by the multi-phase emulsion drops. This will span a wide range of new interesting scientific and engineering questions which will open the door in the field by creating a new route to design and fabricate advanced particles with unique functionalities.

Besides material design my interests include methods to scale up the fabrication of these emulsion drops to produce practical quantities. While microfluidic devices have been massively parallelized to produce highly monodisperse single emulsion drops in high-throughput, extending this framework to produce double-emulsion drops consisting of fluids that have high viscosities have been difficult to achieve. This problem will require extensive fundamental studies of hydrodynamics and interactions of fluids with surfaces. I will use simulation-guided 3D-microfluidic device geometry design, and develop a simple and robust method to selectively modify the surface wetting characteristics of a microfluidic device to solve this challenge. The results of this study will bridge the gap between lab-bench scale to industry-level scalable generation which is both fundamentally and practically important.

475967 Engineering the Surfaces of Tomorrow

Abstract:

Research Interests:

The interface between materials controls many processes of the physical world. Precise control of these interfaces will help solve many of the environmental and energy-related crises that humanity currently faces. Understanding the specific physical and chemical interactions occurring at interfaces allows for exciting, new properties to be realized. For example, controlling the stiffness of a coating between ice and, say, an aircraft wing, allows the interface to cavitate, opening up a new class of materials that cannot adhere to ice. Highly water- and oil-repellent textiles and surfaces can only be fabricated through precise control over the texture and surface energy of the components. Next generation materials cannot be designed by incremental improvements to the current status quo. Revolutionary paradigm shifts manifest when new fundamental principles are uncovered. Leveraging these fundamental scientific discoveries with robust engineering principles allows for exciting new materials to transition from lab-scale innovations to world-changing realities. Designer surfaces are a fascinating new field that, perhaps ironically, exist right at the interface between fundamental science and engineering design.

Research Experience:

I chose materials science and engineering because, even as the name suggests, the discipline combines science and engineering, discovery and implementation, curiosity and practicality. And because everything is made of something, my academic career path has been inherently broad and interdisciplinary. I am formally trained as a surface scientist, which has included engineering zeolite catalysts (ceramics), understanding high-strength fibers (composites), designing solid- and liquid-repellent surfaces (polymers), creating environmentally friendly textiles (natural) and fabricating drag-reducing coatings (metals and semiconductors). Use of these disparate materials and their corresponding interfaces has led to many interdisciplinary and collaborative projects, in the areas of fluid mechanics (Johns Hopkins, US Naval Academy and ONR), surface chemistry (AFRL), biology (University of Michigan and Yale) and rheology (MIT), as well as the many industrial collaborations in the areas of textiles, coatings and designer surfaces. For me, the most valuable research experiences have occurred when combining my core expertise with the expertise of others, leading to several high-impact discoveries (and publications).

Teaching Interests:

My main passion, whether in the laboratory or classroom, has always been teaching. I voluntarily TAed two laboratory classes at Cornell University during my undergraduate tenure. At the University of Michigan I taught several undergraduate classes as well as personally mentored incoming freshmen interested in pursuing a graduate degree. I was fortunate enough to teach under the tutelage of Steve Yalisove, who has been instrumental in pioneering a new type of formative-based learning. I wish to harmonize this new-age teaching style with the traditional methods as a future faculty member. As an NDSEG Fellowship recipient, I have also sat on several fellowship panels and mentored interested graduate students. Lastly, I have mentored many undergraduate and graduate students in my research group.

Future Direction:

As a faculty I will continue to combine fundamental scientific research with engineering principles in the area of surface science. In particular, I would like to further my thesis work on the design and application of liquid- and solid-repellent coatings. I am interested in taking an environmentally conscious approach to surface science, both on the design side as well as the implementation. Recently the EPA and the EU banned long-chain fluorocarbons, and the shorter-chain variants will soon follow. This is a ripe, untapped area of surface science: achieving extreme solid and liquid repellency without the use of fluorinated compounds. The design principles laid out in my previous publications will prove an excellent starting point for attacking this problem. I would also like to combine my graduate and undergraduate work by studying the interfacial phenomenon important to fiber composite body armor. As the failure of ballistic vests typically arises from poor interfacial stress transfer, taking a surface science approach to fiber dynamics should prove insightful. The reverse will also be fruitful: utilizing high-strength fibers in designer coatings will imbue mechanical durability and repellency concomitantly. As a third avenue, I am interested in applying many of the novel surfaces that I designed during my doctoral work as biologically active coatings. Almost every biological process is a surface phenomenon, making the interfacial dynamics critical but amenable to a surface science approach.

My philosophical approach to research is to attack unsolved puzzles in fun and interesting ways. Leveraging creativity, theory and experimentation, my research group will address any and all unanswered questions. These will expand to my broader research interests, which include polymer mechanics, fiber formation, drug delivery, solubility phenomenon, sustainable chemistry and biomimecry. Using my previous experience as a guide, I will forge as many collaborative ties as possible, taking an interdisciplinary approach to tackling problems in a wide variety of fields.

Selected Publications:

K. Golovin, S. P. R. Kobaku, E. T. DiLoreto, J. M. Mabry, and A. Tuteja, "Designing Durable Icephobic Surfaces". *Science Advances*, 2, 3, e1501496 (**2016**).

K. Golovin, D. H. Lee, J. M. Mabry and A. Tuteja, "Transparent, Flexible, Superomniphobic Surfaces with Ultra-Low Contact Angle Hysteresis". *Angewandte Chemie*, 52: 13007–13011 (**2013**).

K. Golovin and S. L. Phoenix, "Effects of extreme transverse deformation on the strength of UHMWPE single filaments for ballistic applications". *Journal of Materials Science:* 1-12 (**2016**).

K. Golovin, J. W. Gose, M. Perlin, S. L. Ceccio and A. Tuteja, "Bio-Inspired Surfaces for Turbulent Drag Reduction". *Philosophical Transactions of the Royal Society A*, 374: 20160189, (**2016**).

R. J. Stafford, K. Golovin, A. Dickinson, T. R. Watkins, A. Shyam and E. Lara-Curzio, "Comparison of Elastic Moduli of Porous Cordierite by Flexure and Dynamic Test Methods". *Ceramic Engineering and Science Proceedings*, 33, 6 197-203 (**2013**).

476006 Adventures in Liquid Crystals

Abstract:

Research Interests:

Liquid crystals are a unique class of synthetic materials with tunable long-range orientations. Cooperative reorientation of liquid crystals (LCs) in response to subtle molecular-level changes at the interface with external stimuli provides amplified responses, which makes them promising materials for numerous industrial applications.

In the first year of my postdocoral work with Prof. Juan de Pablo at the University of Chicago, my research was focused on designing responsive and active interfaces using liquid crystals. Developing a simple set up, we demonstrated that liquid crystal interface is capable of reporting aggregation of polypeptides at lipid-aqueous interfaces at nanomolar concentrations, and therefore offers the potential for early diagnoses of neurodegenerative diseases. The results of this work was published at *Advanced Functional Materials* and appeared on the journal cover page.

Furthermore, in collaboration with the Argonne National Laboratory, we perform synchrotron X-ray reflectivity measurements to obtain structural information of liquid crystal molecules at the interface. Our experimental measurements are accompanied by an advanced theoretical and computational analysis.

In the second year, I have been exploring the topological defects in nematic liquid crystals induced by colloidal particles. Specifically, we have examined experimentally and theoretically, the effect of edge curvature of colloidal particles with homeotropic surface anchoring on their defect configuration and self-assembly. Our studies indicate that the edges sharpness modifies the defect structure, which can significantly affect the interaction between particles and their eventual self-assembly in the nematic liquid crystal.

Blue phase (BP) Liquid crystals are another class of nematic liquid crystals that I am interested in. In fact, BPs are nematic liquid crystals with high chirality that represent unique ordered defect structures over narrow range of temperature. Specifically, I study polymer stabilized blue phases. Using geometrical confinement and organic reagents, I have been able to obtain the blue phases at room temperature and expand the blue phase temperature range.

Although my research in the future will focus mainly on studying fundamental aspects of liquid crystalline materials and their potential applications in electronics and biomedical devices, studying the physics of the collective migration of epithelial and endothelial cells is another field which I am interested to pursue. After accomplishing my Ph.D. degree in polymer physics at ETH Zurich and before joining the de Pablo lab, I have had an opportunity to work on different projects at Harvard University with Prof. Jeffery Fredberg and use my expertise in Polymer Physics and Particle Image Velocimetry to investigate the basic physical processes underlie collective migration of epithelial and endothelial cells in confined geometries and stagnation points. From these works, I have authored and co-authored several manuscripts in high-impact peer-reviewed journals.

Teaching Interests:

With the background in polymer engineering from the Amirkabir University and Polymer Physics from ETH Zurich, I have strong foundation on materials science in particular polymer science related courses including polymer physics, polymer chemistry, crystallization, colloidal systems, rheology and polymer characterization.

475720 Nanostructured Based Lab-on-Chips for Detection of Single Biomolecules

Abstract:

Research Interests:

Nanostructured materials and nanofabricated lab-on-chip devices can play pivotal roles in biomedical research applications including genetic diagnosis and sensitive pathogen detection. Identifying genetic disorders or disease states on-chip requires dynamic manipulation and efficient trapping of a small number of molecules of the target nucleic acid. Current technologies are limited either by sensor resolution or by trapping a low concentration of molecules at the detection sites. Inspired by the state of the art nanofabrication and nano-synthesis methods, my research is focused on designing and fabricating lab-on-chip devices and nanostructured electrodes to overcome these limitations.

In order to enhance the detection sites for biomolecules in low concentration analytes (e.g. Dopamine) I have developed metallic/metal oxide nanostructures with an extended surface area for electro-chemical and biochemical reactions. I have adopted a similar approach to develop nanostructured surfaces to efficiently immobilize DNA and proteins, as part of assays to indirectly detect a wide range of biomolecules. I use electrochemical and lithography techniques to fabricate nanotubular and nanowire structured electrodes as well as nano/microfabricated diagnostics platforms. The platforms have successfully been used for sensitive detection of biomolecules such as dopamine, glucose, uric acid and ascorbic acid as well as pathogens such as Escherichia coli and Staphylococcus aureus.
To further improve trapping efficiency of the target sample on the detection sites, I have developed a sample-delivery system based on reversible, tunable nanofluidic confinement of biomolecules based on the dielectrophoretic force, applied using Silicon Nitride and Indium Tin Oxide nano-patterned electrodes. The resulting device can concentrate and manipulate molecules at high throughput; and can be used to create open-access and simple designs that can be easily interfaced with microfluidic devices for buffer exchange and sample processing; and integrated with optical detectors or on-chip nanostructured electrodes to detect single biomolecules (such DNA fragments up to 100,000 base pairs in length). The ease of fabrication and uncomplicated instrumentation makes this technology a unique point of care instrument for optical detection of biomolecules.

To immediately initiate my research program as a faculty member, I envision three potential projects: (1) Graphene-TiO_2 electrodes for biological and chemical detection, (2) ITO/SiN_x device with programmable electrodes for Cell/DNA sorting and analysis, and (3) 3D Microelectrodes based lab-on-chip for study biochemical reactions.

Teaching Interests:

I would be interested in applying my teaching principles to a wide range of undergraduate and graduate courses in Thermodynamics, Kinetics Fluid Mechanics, Heat Transfer, Transport phenomena, Mass Balances, Biomedical image analysis, Materials properties and characterization, Mechanics of materials, Biomechanics, Fluid Mechanics, Biomedical Instrumentation, tissue engineering, Engineering design, Bio/nanoelectronics. I would like to pursue the development of more specialized courses in biomedical engineering such as Bio/nanomaterials, biosensors, design of medical devices and nanofabrication methods. The fields of materials science along with the current explosions in biomedical engineering and nanotechnology offer many opportunities for the development of interdisciplinary courses.

475739 Harnessing the Power of the Extracellular Matrix to Control Wound Healing and Tissue Regeneration

Abstract:

Research Interests:

Complications arising from current therapies and the increasing prevalence of disease necessitate the development of novel biomaterials for tissue engineering applications. The next generation of materials entering the clinic for cardiac applications includes biomaterials derived from decellularized cardiac extracellular matrix (ECM). While current *in vivo* investigations performed by our labs and others have shown the potential for these materials *in vivo*, the mechanisms by which these materials lead to improved patient prognosis are widely unknown. Classically, the concept of immune response to biomaterials has focused on the negative aspects of a foreign body response. My current postdoctoral work has led to the evaluation of composite silk-cardiac ECM patches for use in the treatment of heart disease. Specific investigations have focused on the role developmental tissue age plays in decellularized ECM composition, immunomodulation, and induced cell response. Using extensive data collected from RNA-SEQ and proteomics evaluation, the future work of the Stoppel laboratory surrounds the hypothesis that ECM proteins and peptides have the ability to modulate host response to biomaterials and wound healing cascades. The long-term goal of the laboratory will be to develop methods and strategies to better understand the biological mechanisms by which these responses occur. Our work will support the development of biomaterials that actively modulate immune response in a controllable and predictable fashion to rapidly advance the development and subsequent utilization of these naturally-derived materials in patients.

To accomplish these goals, we will aim to elucidate the mechanisms of host response to decellularized ECM-containing biomaterials under both healthy and diseased states, focusing on how the wound healing cascade varies in response to biomaterial formulations and ECM composition. A specific emphasis will be placed on therapeutic applications for muscle-based tissues to further advance the fields of cardiac tissue engineering, treatments for muscular dystrophies, and replacements for defects in soft muscle tissue. Students will have the opportunity for interdisciplinary training, as developing new materials requires bioreactor development and optimization via engineering strategies and computational modeling, material property evaluation, and assessment and optimization of designed platforms via traditional biological assays. The interdisciplinary nature of the group will necessitate strong collaborations, particularly with clinicians and immunologists, that will further enrich this training environment. In addition, the Stoppel laboratory will have a focus on development of a diverse group of engineers with strong communication and interpersonal skills to aid in career advancement and contribution to the scientific community.

My own training has included work with alginate hydrogels and cell encapsulation during my PhD in Prof. Susan Robertsâ€™ lab at the University of Massachusetts Amherst, along with postdoctoral training in the laboratories of Profs. David Kaplan and Lauren Black at Tufts University, developing new composite natural materials using silk fibroin and decellularized ECM for cardiac applications.

Teaching Interests:

In the Spring of 2016, I taught Honors 101: Artificial Organs and Prosthetics at UMass Boston as the professor on record as a part of my NIH IRACDA postdoctoral fellowship through the Training in Education and Critical Research Skills (TEACRS) Program at Tufts University. In addition, I have particiated in many workshops and short courses related to diversity and engagement in the classroom, and I hope to have a positive impact on the education of both graduate and undergraduate students. As a fundamentally trained chemical engineer with a focus on biomaterials and bioreactor design, I am excited to teach courses related to materials and energy balances, transport, reaction kinetics, reactor design, biomedical applications, or polymers/materials science.

475758 Polymer Science As a Tool for Materials Design and Biological Discoveries

Abstract:

Research Interests: I envision a research group focusing on emerging questions at the interface of polymer science and biology, with the vision to push the boundary of classical polymer science as well as to solve challenges in energy, environmental science and healthcare. Empowered by the knowledge of polymer science, I will focus on the development of new functional soft (bio) materials for energy harvesting and healthcare, elucidating the roles of biological gels in health and disease, and exploiting the use of synthetic biology in creating designer biomaterials.

Soft materials with extreme mechanical properties and their mechanics under large deformation. In the course of evolution, biological systems have optimized themselves with extraordinary properties. For instance, cells and tissues are extremely soft with typical Youngâ€™s modulus lower than 100kPa, muscles change their stiffness at will, and skins accommodate large deformation easily and self-heal upon damage. Seamlessly interfacing with them offers approaches to solve challenges in healthcare and improving the quality of life, but often requires materials with properties that are comparable, if not the same, to biological objects. Using relatively well-established polymer science as a tool, I will adopt a bottom-up approach to design materials on the molecular-level, enabling new materials with extraordinary mechanical properties and functions. Initially I will focus on (i) extremely soft, yet dry gels and (2) malleable, tough, self-healing polymers. In parallel, I will develop new, fundamental understanding of the mechanics of such compliant materials under large deformation and explore their applications in stretchable electronics and biomedical scaffolds from 3D printing.

Biological gels in health and disease. Biological gels, such as mucus and extracellular matrix, are the major components of mucosal surfaces that line respiratory, gastrointestinal and reproductive tracts. Small changes in biophysical and biochemical properties of these gels result in their abnormal functions that are associated with chronic obstructive pulmonary disease (COPD), cystic fibrosis, inflammatory bowel disease and reproductive tract infection. Restoring the functions of biological gels is critical to treatment of these diseases; this requires elucidating their complex biophysical and biochemical origin. To this end, I will harness relatively well-established knowledge and tools in physical and chemical science to answer emerging clinically and therapeutically important questions related to these biological gels. Using the concept of polymer brush established in polymer physics enables my recent discovery of a new paradigm for mucus clearance, the primary innate defense system for mammalian lungs. I will adapt this unique interdisciplinary approach to elucidate the biophysics and biochemistry of native rather than reconstituted mucus, to study the interaction between mucus and epithelial cells, and to explore the roles of mucus on functionalities of microbiome.

Synthetic biology as a tool for designer biomaterials. Classical polymer science is largely enabled by precise chemical synthesis the can create polymers with controlled structure and thus properties. Yet biopolymers such polypeptides not only have a prescribed structure, but also consist of motifs that possess different strength of interactions such hydrogen bonds, electrostatic and hydrophobic-hydrophilic interactions. Such interactions are relatively weak, yet enable biopolymers with robust and deterministic functions. Due to the diversity and complexity of these interactions, it is challenging to understand how they determine the properties of biopolymers; yet this offers opportunities to classical polymer science. In parallel, emerging synthetic biological techniques allow precise control over both the structure of biopolymers and the types of weak interactions. Harnessing the power of synthetic biology, I will design and create biopolymers with prescribed structure and weak interactions to explore the fundamental polymer science of such systems. Initial research will focus on amphiphilic biopolymers with ionic crosslinkable domains; the amphiphilic nature will allow the polymer to self-assemble to useful structures such as vesicles, whereas the crosslinkable sites can be used to fix the structures after self-assembly. In parallel, I will explore the application of such structures as synthetic vaccines.

Teaching Interests: With my diverse background, B.S. in Physics, Ph.D. in Materials Science, and Postdoctoral experience in Chemistry, Materials Science and Biology, I am well prepared to teach a broad range of topics in materials science. Examples of specific undergraduate courses include polymer chemistry, thermodynamics, mechanics, solid-state physics, and applied mathematics. Examples of advanced undergraduate or graduate courses include polymer physics and chemistry, advanced thermodynamics, materials characterization, biological physics and soft condensed matter physics.

I am also dedicated to develop interdisciplinary courses to contribute to the diversity of curriculum in classical Materials Science, Chemistry, or Chemical Engineering. As modern materials science and engineering is growing into interdisciplinary areas, the emphasis in materials science education should alter accordingly. In addition to fundamental courses, focuses should also be placed on interdisciplinary areas where materials science is coupled with physics, chemistry, and biology; an example of such area is soft matter and polymer science. Knowing the essential concepts and techniques in soft matter and polymer science greatly broadens the view and expertise of students and benefit to their research; according to my experience, however, there is almost no such course dedicated to provide students with an overview of essential techniques through hands-on experience, and only through research projects in laboratories students will have a chance to learn some of them. As a start, I would like to introduce a new course to cover a wide range of techniques and concepts in soft matter and polymer science in combination with hands-on experience. Specific course subjects will cover chemical synthesis, chromatography, light scattering, osmotic measurements, rheology, microscopy, wetting and de-wetting, cell culture, and computer simulation. Each topic will be covered by a two-week lecture; one week focuses on in-class discussion and the other focuses on laboratory training. Students will be required write a short scientific report about each topic and use the knowledge learned during the course to complete a collaborative final project selected from the frontiers of research.

476318 Towards an Understanding of Catalytic Synthesis and Application of Nanomaterials

Abstract:

Research Interests:

My research leverages the intersection between in-situ metrology (XPS, XRD, ESEM, ETEM), process engineering and material science to enable new developments in bottom up novel material design and synthesis. The key scientific questions in my field relate to understanding mechanisms that can be used to achieve atomic level control during bottom up synthesis of novel materials, while the engineering challenges center around developing scalable production processes and interfacing the new materials into functional systems for applications.

2D materials serve as ideal model systems that can be leveraged to develop a fundamental understanding of the mechanisms associated with atomic length scale control for material synthesis using scalable processes while simultaneously enabling high impact applications; e.g. atomically thin membranes (ATMs) offer transformative opportunities for desalination, dialysis, fuel cells, solvent separation, DNA sequencing and can enable in-situ metrology for solid-liquid systems. Such in-situ observations could offer unprecedented insights to help further fundamental understanding of heterogeneous catalytic processes, crystal nucleation/growth/dissolution, material performance in batteries, super-capacitors, bioprocesses, living cells and their interfaces.

My long term vision is to make significant contribution towards moving the design, synthesis and device integration of nanomaterials from their current state of trial and error based experimentation to real time, direct evidence based, fundamental understanding driven rational engineering approaches for atomic length scale control in large-scale manufacturing processes.

Teaching Interests:

I am comfortable teaching a wide range of courses, both introductory and advanced at the undergraduate and graduate levels in Process/Chemical Engineering, Mechanical Engineering, Materials Science and Nanotechnology but have a keen interest towards courses related to material synthesis, catalysis, reaction engineering, metrology and process design & development.

476072 Photovoltaic Processes

Abstract:

Research Interests: The price of photovoltaic (PV) power only depends on cost per area, power conversion efficiency, and rate of efficiency degradation. The microscopic processes and commercial production processes that give rise to those metrics are diverse, yet intimately connected. Research areas include substrates, transparent conductive contacts, absorbers, buffers, power degradation routes, encapsulation, edge sealant, power storage, and energy storage. My research will ultimately advance each of these 9 aspects through the simultaneous study of processes used to produce the PV devices, processes of their operation, and degradation processes. A deep understanding of each aspect will be necessary for long-term success in PV research, for academics and companies alike. In the near-term, several areas will be attacked: metal micro- and nano-grids for transparent conduction in solar cells, thin film absorber development, and new buffer compounds and deposition processes.

My postdoctoral research has been at the National Renewable Energy Laboratory, where I studied the selenization of Cu-Ga-In precursor films, the most popular industry deposition process for $Cu(In,Ga)Se_2$ (CIGS) thin film absorbers [1-3]. To investigate very recent world record CIGS power conversion efficiencies that used a potassium fluoride post-deposition treatment, I chose to do the first basic material science study of $Cu_{1-x}K_xInSe_2$ (CKIS) alloys [4]. I then used this foundational knowledge to fabricate an officially-measured 15.0%-efficient Cu-K-In-Se solar cell (matching the world record $CuInSe_2$efficiency, despite being grown at lower temperature, and without the 'gold standard' three-stage co-evaporation process).

My graduate research was at the University of Florida with Tim Anderson, where I studied thermodynamics of the Cu-Ga-In system. That work culminated in a paper that was recently nominated for the Cahn prize [5]. I also studied diverse aspects for halide chemical vapor deposition (CVD) of CIGS: complex chemical equilibrium calculations, finite element method fluid flow simulations of atmospheric pressure CVD, experimental Cu, Ga, and In volatilization by HCl reaction, with supporting empirical and analytical fluid flow modeling, and the first experimental CVD growth of Cu-Se, In-Se, and Cu-In-Se [6].

Teaching Interests: Along with research, I will teach transport, thermodynamics, electrical engineering, materials science classes . Where possible, I will add emphases on thin film growth, semiconductors, electron devices, and photovoltaics.

[1] **Muzzillo, C.P.**, Mansfield, L.M., DeHart, C., Bowers, K., Reedy, R.C., To, B., Noufi, R., Ramanathan, K., Anderson, T.J., (2014) â€œThe effect of Ga content on the selenization of co-evaporated CuGa/In films and their photovoltaic performance,â€ *40th IEEE PVSC Proc.*, Denver, CO, 1649-1654.

[2] **Muzzillo, C.P.**, Mansfield, L.M., DeHart, C., Bowers, K., Reedy, R.C., To, B., Ramanathan, K., Anderson, T.J., (2015) â€œDifferences between CuGa/In and Cu/Ga/In films for selenization,â€ *42nd IEEE PVSC Proc.*, New Orleans, LA, 1-6.

[3] **Muzzillo, C.P.**, Mansfield, L.M., Ramanathan, K., McGoffin, J.T., Sites, J.R., Anderson, T.J., (2016) â€œPhotovoltaic properties of selenized CuGa/In films with varied compositions,â€ 43rd IEEE PVSC Proc., Portland, OR, in press.

[4] **Muzzillo, C.P.**, Mansfield, L.M., Ramanathan, K., Anderson, T.J., (2016) â€œProperties of $Cu_{1-x}K_xInSe_2$ alloys,â€ *Journal of Materials Science*, 51(14), 6812-6823.

[5] **Muzzillo, C.P.**, Campbell, C.E., Anderson, T.J., (2016) â€œCu-Ga-In thermodynamics: Experimental study, modeling, and implications for photovoltaics,â€ *Journal of Materials Science*, 51(7), 3362-3379.

[6] **Muzzillo, C.P.**, (2015) â€œChalcopyrites for solar cells: chemical vapor deposition, selenization, and alloying,â€ Dissertation, Chemical Engineering, University of Florida, Gainesville, FL.

476078 Polymeric Materials for Biomedicine and Nanotechnology

Abstract:
Polymeric Materials for Biomedicine and Nanotechnology

Stephanie Christau,University of Michigan1[st]Year Postdoctoral Fellow

Postdoctoral Project:

Preparation of Polymeric Janus Particles for Stimulation of Immune Cells. Advisor:Joerg Lahann, Department of Chemical Engineering, Biointerfaces Institute, University of Michigan.

Dissertation:

Smart Hybrid Materials with Tailored Functions: Immobilization of Gold Nanoparticles into Stimuli-Responsive Polymer Brushes.

Advisor:Regine von Klitzing, Department of Chemistry, Berlin Institute of Technology (TU Berlin).Co-advisor: Jan Genzer, Department of Chemical Engineering, North Carolina State University.

Research Interests:

Although my degree is in chemistry, I have been exposed to different areas of research throughout my academic career. This includes general chemistry, polymer science and engineering, and applied physics. As my PhD thesis focused on the **development of responsive surface coatings** made of **polymer brush/gold particle hybrids,** I have a strong background in polymer and colloid science. In addition to receiving training in standard characterization techniques (ellipsometry, spectroscopy, electron microscopy), I have gained experience in x-ray and neutron scattering techniques. The latter includes writing proposals to get measurement time at the neutron facility, proper planning of the scattering experiments, and the data analysis which requires sophisticated fitting procedures. In my postdoctoral project, I am applying my skill set while also broadening my knowledge base in the area of biomedical science. Yet, I familiarized myself with the preparation of **polymeric particles and their characterization**, and **cell culture experiments**. I have invaluable experience in many research areas including **chemistry, material science, polymer and colloid science and engineering, biomedical engineering, and applied physics**.

Postdoctoral Research Fellow, January 2016 â€" present

Department of Chemical Engineering, University of Michigan, Ann Arbor, MI, USA.Â *Advisor:* Joerg Lahann

Development of polymeric nanoparticles for T cell activation

Electrohydrodynamic co-jetting of multicompartmental nano-/microparticles

Postdoctoral Researcher, October 2015 â€" January 2016

Department of Chemical and Biomolecular Engineering, North Carolina State University, Raleigh, NC, USA.Â *Advisor:* Jan Genzer

Preparation of polymer brush charge gradients on flat substrates

Ph.D. (chemistry). *DFG-IRTG Fellow,* July 2012 â€" October 2015

Department of Chemistry, Berlin Institute of Technology (TU Berlin), Germany.Â *Advisor:*Regine von Klitzing

Department of Chemical and Biomolecular Engineering, North Carolina State University, Raleigh, NC, USA *Advisor:* Jan Genzer.

Preparation of stimuli-responsive polymer brush surface coatings and immobilization of gold nanoparticles

Self-assembly and plasmonic properties of polymer brush/nanoparticle composite materials for sensor applications

Neutron reflectometry measurements and data fitting to resolve internal structure of polymer brush/nanoparticle hybrids

Preparation of polymer brush gradients (grafting density, charge)

Diploma (chemistry). *Researcher,* 2011 â€“ 2012

Department of Chemistry, Berlin Institute of Technology (TU Berlin), Germany.Â *Advisor:*Â Martin Schoen (TU Berlin)

Department of Chemistry, Eidgenoessische Technische Hochschule (ETH) Zurich, Switzerland.Â *Co-advisor:Â* Martin Quack (ETH Zurich)

High-resolution FTIR spectroscopy of the carbon dioxide hot bands and their contribution to the greenhouse effect (measurements and simulation)

Undergraduate Researcher, 2010 â€“ 2011

Department of Chemistry, Berlin Institute of Technology (TU Berlin), Germany.Â *Advisor:Â* Regine von Klitzing.

Structure properties of polymers in bulk and under confinement (viscometry and atomic force microscopy measurements)

Teaching Interests:

I mentored several undergraduate students at North Carolina State University, TU Berlin and the University of Michigan. I also mentored two bachelor theses; duties included developing a feasible project, organizing overall timeframe and experiments, and training in the laboratory (both, synthesis and characterization techniques) and in scientific writing. I also TAed and guest-lectured undergraduate courses in â€˜polymer and colloid scienceâ€™ at the TU Berlin. Duties included course material, lecturing, and office hours. Furthermore, I am participating in the course â€˜High Impact Principles and Practices for STEM Educationâ€™ at the University of Michigan. The purpose of this course is being prepared to teach effectively as future faculty member in science and engineering.

Successful proposals:

German Science Foundation (DFG, Deutsche Forschungsgemeinschaft) Postdoctoral Research Grant, 2016.

Future Directions:

As a faculty I would like to continue in the field of polymer and colloid science. The projects I am proposing include the **development of polymeric (bio)materials**for

Stimulation of T cells to elicit a specific immune response â€“ applications include cancer immunotherapy and cancer vaccines.

Controlled and sustained *delivery of cytokines/chemokines* â€“ applications include chronic wound healing and cancer immunotherapy.

Development *of composite materials with tunable plasmonic properties* through immobilization of metallic nanoparticles â€“ applications include the development of smart sensors (via polymeric particles or surface coatings) and photothermal therapy (*biodegradable materials*).

Publications:

Journal articles.

(1) Christau, S.; Möller, T.; Brose, F.; Genzer, J.; Soltwedel, O.; von Klitzing, R. Effect of gold nanoparticle hydrophobicity on thermally induced color change of PNIPAM brush/gold nanoparticle hybrids, *Polymer* **2016**, doi: 10.1016/j.polymer.2016.03.088

(2) Kesal, D.; **Christau, S**.; Krause, P.; Moeller, T.; von Klitzing, R. Uptake of pH-Sensitive Gold Nanoparticles in

Strong Polyelectrolyte Brushes, *Polymers* **2016,**8, 134; doi:10.3390/polym8040134

(3) Christau, S.; Möller, T., Yenice, Z.; Genzer, J.; von Klitzing, R. Brush/gold nanoparticle hybrids: Effect of grafting density on particle uptake and distribution within weak polyelectrolyte brushes. *Langmuir,* **2014,** *30,* 13033-13041.

(4) Christau, S.; Thurandt, S.; Yenice, Z.; von Klitzing, R. Stimuli-Responsive Polyelectrolyte Brushes as a Matrix for the Attachment of Gold Nanoparticles: The Effect of Brush Thickness on Particle Distribution. *Polymers (Basel)*, **2014**, *6*, 1877.

(5) Üzüm, C.; **Christau, S.**; von Klitzing, R. Structuring of Polyelectrolyte (NaPSS) Solutions in Bulk and under Confinement as a Function of Concentration and Molecular Weight. *Macromolecules*, **2011**, *44*(19), 7782-7791.

Reviews.

(1) Christau, S.; Genzer, J.; von Klitzing, R. Polymer brush/metal nanoparticle hybrids for optical sensor applications: From self-assembly to tailored functions and nanoengineering. *Z.Phys.Chem.,* **2015,** *229 (7-8),* 1089-1117.

476086 Design of Advanced Materials for Application in Clean Energy and Carbon Capture and Utilization

Abstract:

Design of Advanced Materials for Application in

Clean Energy and Carbon Capture and Utilization

Peter C. Psarras

Energy Resources Engineering, Stanford University **Introduction** This past half-century has brought much change to the landscape surrounding the chemical processes that sustain our global socio-economic framework. In the midst of this evolving climate, it has become increasingly apparent that new motivations, such as those for greater energy-efficiency and less carbon-intensive processes, are necessary to avert serious global consequences. Fortunately, this call to attention has re-energized the field of materials research, whereby promising new technologies can bridge the gap between out-of-date processes and those better-suited for consequence mitigation. Concurrent to this scientific revitalization has been a related emergence in its own right Ð the dawn of high performance computing environments; thus, theoretical methods represent a powerful approach to the design of materials aimed at resolving the shortcomings of essential industrial processes through revision and innovation. These methods are often favorable in situations where there exists a need to explore and screen a number of materials at low cost and risk, where conditions prove unsuitable or even unfeasible via conventional methods, or where an understanding is sought at molecular levels or timescales that are incompatible with experimental measurement. This presentation will introduce how different theoretical methods can be used to solve four issues related to climate change: nitrogen-selective membranes for indirect CO_2 capture, nitrogen-functionalized carbon sorbents for direct CO_2 capture, the assessment of carbon neutrality for irreplaceable industrial processes, and the effect of late-transition metal doping on methane selectivity in Fischer-Tršpsch (FT) catalysis. **Research Interests: Nitrogen-Functionalized Porous Carbons for Enhanced CO_2 Capture** Traditional amine-based solvent technology is notoriously energy intensive, owing to high regeneration temperatures. Additionally, this process is water-intensive, a factor that is becoming increasingly important in drought-ridden states like California. Solid sorbents represent an alternative method for CO_2 capture, with liquid solvents replaced by (typically) large surface area, carbon-based porous frameworks. These sorbents are generally low-cost, easily fabricated, and are far less energy intensive in terms of regeneration. Further, their design can be customized through the inclusion of surface functional groups. Unfortunately, current sorbents display CO_2 loadings that are deemed too low to be cost-competitive with solvent-based scrubbing. Nitrogen-functionalization of these porous carbons could result in more efficient CO_2 capture through the introduction of surface charge heterogeneity which can selectively interact with the CO_2 quadrupole moment. Similar studies have examined the effects of oxygenated functional groups on CO_2 uptake and selectivity over N_2. This study employs grand canonical Monte Carlo methods to explore the effect of quaternary, pyridinic, pyrrolic, oxidized-pyridinic, and pyridonic groups on CO_2 uptake in porous carbon sorbents. Doping amounts are varied to ascertain the optimum coverage for CO_2 capture. Additionally, gaseous mixtures of CO_2/N_2 and $CO_2/N_2/H_2O$ are examined to assess CO_2 selectivity. Theoretical performance is validated through comparison with experimentally obtained CO_2 and N_2 isotherms over fabricated hierarchical micro/mesoporous N-doped carbon sorbents. The combination of these methods will inform on sorbent design by illustrating which groups are most important for enhanced CO_2 uptake. Design to include more of a particular functional group can be achieved through, for example, a change in carbonization temperature. **Assessment of the CO_2 Capture Potential from Irreplaceable Industrial Sources**

In 2013, the US industrial sector emitted approximately 1.4 gigatonnes of carbon dioxide (Gt CO_2), or 21% of total US CO_2 emissions Ð the third highest figure for any economic sector behind transportation (27%) and electricity (31%). These emissions can be further categorized as direct (fuel combustion accountable on-site, e.g., stationary combustion), indirect (assigned to electricity purchased for power and off-site steam generation) and process (CO_2 liberated as a reaction by-product). Direct and indirect emissions generally constitute ca. 80% of total emissions, with process emissions making up the balance, though the relative contribution of process emissions to total emissions is known to vary by industry.

While these industrial process emissions constitute a smaller percentage of total US CO_2 emissions, they produce commodities like glass, cement, ammonia and steel Ð items that form the irreplaceable fabric of industrialized nations. Unlike the power sector, for example, where mitigation might be achieved through a transition to renewable forms of energy and adoption of best practice technologies, there are few CO_2-free alternative routes to product for most industrial commodities; thus, these irreplaceable industrial processes (IIPs) represent committed CO_2 emissions that remain largely unabatable. Further, these IIPs command attention for two important reasons: 1) there is a committed nature to these emissions by virtue of process chemistry, and perhaps more importantly 2) they can yield exhaust streams with higher CO_2 content when compared to the high volume flue exhausts of the power sector. As the cost of CO_2 separation scales inversely with initial dilution of a mixed feed stream carbon capture technology retrofits have the potential to efficiently and economically divert emissions from industrial exhaust streams to viable CO_2 utilization opportunities, such as enhanced oil recovery (EOR), food processing, refrigeration, and fertilizer production. To assess the capture potential of these irreplaceable industries, it is necessary to geo-reference these sources alongside all current and potential future CO_2 sinks (utilization or market opportunities), with the goal of making economically sound linkages between source and markets of comparable scale. This entails a cost analysis of on-site capture, compression, and transport costs. Geographic information systems (GIS) mapping can assist in identifying regions of high CO_2 demand, geographically-logical source-sink partnerships, and viable routes of CO_2 transport. The aim is to classify these IIPs based on carbon-capture ÒreadinessÓ, which is ultimately a combination of the industry-specific and site-specific factors listed above.

Vanadium-Ruthenium Alloys for Indirect CO_2 Capture Membranes represent a promising technology for carbon capture, yet several material challenges persist. Ideally, a membrane will need to demonstrate excellent CO_2 selectivity, strong mechanical and thermal stability, and high membrane flux. Here, investigate alloyed V/Ru membranes for selective nitrogen capture. In this technology, rather than separating out CO_2 directly, nitrogen is separated from the flue stream. Nitrogen is far more efficient to separate, owing to its greater flue concentration and corresponding higher pressure differential across the membrane. Consequently, after N_2 separation the remaining exhaust components become more concentrated, CO_2 included, making all downstream capture points more efficient by the same virtue. This study will prepare 14 V/Ru alloys in concentrations spanning those considered in real application. Vanadium is known to bind nitrogen strongly, making it a good candidate for activation of the N-N triple bond. However, this tight binding inherently slows membrane diffusion. Nitrogen permeability can be represented as a product of its solubility and diffusivity. The former is sought by considering the effect of Ru concentration on relative binding energies and vibrational frequencies of interstitially-bound nitrogen, while diffusivity can be examined by calculating all potential TS-pathways, tabulating their rates, and employing kinetic Monte Carlo methods. Together, these results will combine to predict nitrogen permeability as a function of Ru content. These findings can inform on membrane design, whereby a theoretically optimal Ru composition can be validated experimentally. **The Effect of Late Transition Metal Doping on Methane Selectivity in Fischer-Tršpsch Catalysis (Dissertation Work)** Fischer-Tršpsch synthesis has endured a long and industrious career as the premier chemical method for the synthesis of hydrocarbons from non-petroleum-based precursors. Historically, a collection of factors has dictated interest in the FT process, including foreign policy, availability of precursor(s), availability of alternate fuel sources and reserves, and environmental considerations. The effect of late transition metal substitution into Fe(100), Ni(111), and Co(0001) surface analogs is investigated using density functional theory (DFT) methods. The surface is modeled first using a 7-atom cluster, with perimeter atoms saturated with hydrogen atoms to approximate surface coordination and mitigate dangling bond artifacts, and next with plane-wave periodic models. *Ab initio* calculations are performed to predict adsorption energies and transition state geometries and energies for eight surface adsorbate species: *C + *H, *CH + *H, $*CH_2$ + *H, and $*CH_3$ + *H to represent the hydrogenating steps, while *C + *CH, *CH + *CH, *C + $*CH_3$, and $*CH_2$ + $*CH_2$ represent species involved in four competitive coupling pathways. A review of the effect of Cu, Ag, Au, and Pd on the reaction energies and barriers associated with these critical steps is discussed. Methane selectivity can be described by a single descriptor, the carbon binding strength; thus, a comparison of these energies over different surface/promoter analogs is an effective method to screen for promising candidates. In this representation, a lowering of carbon binding strength is equivalent to the lowering of methane selectivity. **Teaching Interests:** My background in Chemistry has prepared me in the fundamentals of thermodynamics, quantum mechanics, catalysis, molecular simulations and adsorption. However, my research direction has forced me to become as proficient in the understanding of transport phenomena, fluid mechanics, and separations. I feel comfortable to instruct in any of these areas with confidence. Additionally, I plan to design a course centered around the fundamentals of electronic structure methods, tailored for application to climate change mitigation Ð suitable as a standalone course or special topic offering. Additionally, I have experience teaching several undergraduate lecture courses, including Physical Chemistry I and Survey of Physical Chemistry, as well as several laboratory courses. I have also had the great privilege of mentoring several students, many who have since published in high impact journals such as *Journal of Physical Chemistry C* and *Advanced Energy Materials*. Through my experiences in teaching and mentoring, I have developed the valuable skill of balancing research and grant writing with class time and student development.

476124 Characterization of Polymer Particles in Biological Environments for Drug Delivery Applications

Abstract:

Research Interests:

Vehicles for drug delivery are an ideal treatment solution for many diseases, including cancer, where the dosage of drugs is limited by toxic side effects. Creating a successful drug delivery system is difficult, however, and despite much research effort in this field, there are very few examples of drug delivery systems being used in current therapies. This lack of success stems from a limited understanding of how drug delivery vehicles interact with the biological environment and leads to poor design of the drug delivery systems. I intend to establish a research group to investigate the interaction of drug delivery vehicles with the biological environment (such as cells, blood, and proteins) using physical chemistry techniques (such as light scattering and calorimetry) in novel ways. My background in drug delivery and polymer physics will allow my lab to address issues with drug delivery vehicles in unique ways. This work can lead to successful drug delivery vehicles and a better understanding of how materials interact with the biological environment.

Education & Training

2013 – Present Postdoctoral Researcher, Chemical Eng., Biointerfaces Institute, University of Michigan

Advisor: Joerg Lahann

Project: Electrohydrodynamic co-jetting of polymer particles for targeted drug delivery applications

2007 – 2013 **Ph.D.** in Polymer Science and Engineering, University of Massachusetts Amherst

Thesis Advisor: Thomas P. Russell

Thesis Title: Particle behavior at anisotropically curved liquid interfaces

2008 **M.S.** in Polymer Science and Engineering, UMass Amherst

2003 – 2007 **B.S.** in Chemistry, Massachusetts Institute of Technology

Research Advisor: Paula T. Hammond

Project: Layer-by-layer polymer electrolytes for lithium batteries

Fellowships and Awards

Best TA for 'Polymer Characterization' at UMass Amherst as voted by students, Dec 2010

International Center for Materials Research International Research Fellowship, Nov 2009

Graduate Assistance in Areas of National Need (GAANN) Fellowship, Sept 2007 – Aug 2008

Current Research

My current research focuses on electrohydrodynamic co-jetting of polymer nanoparticles to target breast cancer stem cells. By surface modifying the particles with antibodies and incubating the particles with a breast cancer cell line, I have observed an increase in the uptake of the nanoparticles by the cancer stem cells using flow cytometry. The breast cancer cells are a rare population (typically less than 1% of the total population) and very difficult to target, yet an important target for cancer treatment.

Another thrust of my current research is the development of a method for analyzing particle behavior in complex solutions such as blood plasma. By using nanoparticle tracking analysis I have quantified the aggregation of nanoparticles in blood plasma. This method will allow for more thorough testing of particles *in vitro*and allow for better design of drug delivery vehicles.

Teaching Interests:

I am interested in teaching core classes as well as developing more advanced courses about polymers, such as Advanced Polymer Physics and Characterization. I have guest lectured in some graduate level classes at the University of Michigan, and was a TA several times at UMass Amherst (voted best TA for the class in 2010). I have taken several classes on teaching at the college level and I have mentored 12 students (undergraduate, high school, and masters) on research projects in the lab. I have also been involved in many outreach activities involving teaching. For instance, I taught chemistry to blind high school students at a summer camp where I had to tailor my teaching to address non-visual learning styles. The blind student summer camp also gave me much experience in accommodating lessons for those with disabilities as I had to adapt the experiments for blind students. Some of these adapted lessons were published in the *Journal of Chemical Education* in 2009.

Publications

McEnnis, K.; LaPres, C.; Santos Moreira, A.; Burkland, D.; Lahann, J.; Characterization of Nanoparticle Aggregation in Biologically Relevant Fluids. In preparation.

McEnnis, K.; Dinsmore, A. D.; Russell, T. P. Solid Particles Adsorbed on Capillary-Bridge-Shaped Fluid

Polystyrene Surfaces. *Langmuir.* **2015**, *31*, 5299–5305.

Lutkenhaus, J. L.; **McEnnis, K.**; Serghei, A.; Russell, T.P. Confinement Effects on Crystallization and Curie Transitions of Poly(vinylidene fluoride-co-trifluoroethylene). *Macromolecules.* **2010**, *43*, (8), 3844-3850.

Serghei, A.; Lutkenhaus, J. L.; Miranda, D. F.; **McEnnis, K.**; Kremer, F.; Russell, T. P. Density Fluctuations and Phase Transitions of Ferroelectric Polymer Nanowires. *Small.* **2010**, *6*, (16), 1822-1826.

Supalo, C. A.; Mallouk, T. E.; Amorosi, C.; Lanouette, J.; Wohlers H. D.; **McEnnis, K.** Using Adaptive Tools and Techniques to Teach a Class of Students Who are Blind or Low-Vision. *Journal of Chemical Education.* **2009**, *86*, (5), 587-591.

Lutkenhaus, J. L.; **McEnnis, K.**; Hammond, P.T. Nano- and Microporous Layer-by-Layer Assemblies Containing Linear Poly(ethylenimine) and Poly(acrylic acid). *Macromolecules.* **2008**, *41*, (16), 6047-6054.

Lutkenhaus, J. L.; **McEnnis, K.**; Hammond, P. T. The Glass Transition and Ionic Conductivity of Poly(ethylene oxide)/Poly(acrylic acid) Layer-by-Layer Assemblies. *Macromolecules.* **2007**, *40*, (23), 8367-8373.

Lutkenhaus, J. L.; Hrabak, K. D.; **McEnnis, K.**; Hammond, P. T. Elastomeric Flexible Free-Standing Hydrogen-Bonded Nanoscale Assemblies. *Journal of the American Chemical Society.* **2005**, *127*, (49), 17228-17234.

Reviews:

McEnnis, K.; Lahann, J. Nanoparticles for Targeted Drug Delivery. Invited review article in preparation for *Journal of Materials Chemistry B.*

476167 Highly Energy-Dense Rechargeable Alkaline MnO2-Zn Batteries for Grid-Scale Applications

Abstract:

Research Interests:

Current Research

Manganese dioxide (MnO_2) and zinc (Zn) are key constituents of primary or single-use alkaline batteries such as Duracell and Energizer. Single-use batteries are referred to as such because they deliver their entire energy capacity just once, after which they must be discarded. Despite this limitation, primary alkaline batteries are ubiquitous in society on account of abundance, low cost and relative safety of MnO_2 and Zn, and their low cost of manufacturing. Nevertheless, as environmental regulations tighten, the manufacture of these convenient energy cells necessitates a rethink. Producing a primary MnO_2-Zn battery consumes more energy than what it can deliver, and the use of MnO_2-Zn batteries is now a point of contention owing to their large impact on the environment. To this end, electrochemical engineering of MnO_2-Zn batteries towards making these cells rechargeable would represent a quantum advance. Rechargeability greatly improves the overall energy efficiency of these cells, and preserves its cost advantages. However, this objective has proved to be a herculean challenge for a generation of battery researchers owing to fundamental limitations in the material and electrochemical properties of MnO_2 and Zn. Briefly, MnO_2 can theoretically deliver a capacity of approximately 617 mAh/g. It delivers this capacity through a 2-electron electrochemical reaction; wherein each electron provides roughly 308 mAh/g. MnO_2 has been found to be rechargeable when the capacity has been limited to around 5-10% of the 617mAh/g. Its crystal structure breaks as more of the capacity is accessed, and it inherently forms electrochemically irreversible phases. Similar problems are associated with the zinc electrode, where higher utilization of its capacity causes dendrite formation, morphology changes and formation of inactive zinc oxides that ultimately lead to electrode failure. If the entire 2-electron capacity of MnO_2 can be accessed and problems associated with the zinc electrode can be averted, MnO_2-Zn batteries could, in theory, reach energy densities that are comparable to lithium-ion batteries. My current research at the City College of New York has successfully addressed these requirements. Allied with its low cost of manufacturing and safety, a rechargeable MnO_2-Zn battery could be a disruptive technology for energy storage.

Our breakthrough that permits accessing the 2-electron capacity reversibly was achieved by precisely altering the crystal structure of MnO_2 with dopants using a wholly novel and vastly cheaper synthetic route, and we have cycled the MnO_2 electrode to well over 6000 cycles at rates that are of interest to the battery community. Besides innovative technology development, I also unearthed fundamental insights about the material and its complicated electrochemical mechanism, wherein the dopants participate in those reactions involving $MnO_2$to prevent the formation of the electrochemically irreversible phases. Significant performance improvements were also achieved for the Zn electrode through use of complexing agents that inhibit morphological changes in Zn and localize the active material near the current collector. Novel crystal structures doped with elements that play a dual role in reducing charge transfer characteristics and encapsulating Zn were also discovered. The application of these strategies have resulted in cycling the Zn anode at 15-20% of its theoretical capacity for >900 cycles, where Zn has been known to fail at 5-8% of its of theoretical capacity in less than 300 cycles.

The cumulative improvements in the material and electrochemical properties of MnO_2 and Zn have resulted in fabrication of the very first rechargeable alkaline cell that can deliver a record high energy density of ~160 Wh/L, where 100% of the 2-electron capacity of MnO_2 and 15% of theoretical capacity of Zn is utilized. Theoretical calculations have shown that at ~30% utilization of the Zn theoretical capacity, a highly energy dense cell of ~250Wh/L is possible. My current work seeks to achieve these figures, and my work has laid strong foundations for the market entry of rechargeable MnO_2-Zn batteries that are cheaper but equally efficient as lithium-ion batteries. Overall, my postdoctoral work has resulted in the filing of 8 patents and I have recently submitted two manuscripts to high impact factor journals. It is currently under peer-review.

Previous Research

Doctoral Thesis: Design and assembly of nanostructured complex metal oxide materials for the construction of batteries and thermoelectric devices My graduate research at Purdue University served as an excellent primer for my current work. My doctoral work focused on synthesis of one-dimensional, complex metal oxide nanowires for thermoelectric and lithium-ion battery applications. I bridged two distinct technical domains â€“ energy conversion and energy storage â€“ with nanotechnology. Some highlights of my graduate work include the conception of novel synthetic routes for production of ultrathin perovskite nanowires and formalization of template-based approached for fabrication of complex metal oxides such as calcium cobalt oxide ($Ca_9Co_{12}O_{28}$). Saliently, for the latter, my work was the first demonstration of its kind wherein nanowires were synthesized specifically use in thermoelectric devices. Moreover, the thermoelectric characteristics of the nanowires were a significant improvement over competing alternatives reported elsewhere in literature. In addition, I also extended my approach to the synthesis of nanowires and nanostructures of lithium cobalt oxide ($LiCoO_2$) and investigated the impact of nanostructuring on the electrochemical properties of the material. The theme of combining application-driven research with fundamental investigations that seek to expand our understanding about nanomaterials has been a staple of my work since my graduate tenure. My graduate work resulted in publications and 1 patent.

Research Interests

I am a chemical engineer specializing in materials science and electrochemistry. I have a proven track record of advancing current understanding and develop commercializable technologies for energy generation and energy storage. Moreover, I have unique experience in scaling up my research from the nano- to macro scales, as well as from the bench to market. My doctoral work focused on developing new approaches for synthesis of one-dimensional, complex metal oxide nanowires and investigating their suitability for thermoelectric and battery applications. My postdoctoral research, on the other hand, has targeted the enhancement of the electrochemical properties of battery systems through ingenious use of materials science and scaling up these batteries for use in grid-scale applications. My training, research interests and innovation record will be key building blocks of my research program on multiscale engineering for energy conversion and storage. My program will comprise three pillars: (1) materials nanoscience, (2) materials synthesis, (3) device integration and scale-up. Among other projects, I intend to develop new thermoelectric materials for high-temperature applications, and conduct fundamental investigations towards improvement of aqueous battery chemistries. Beyond energy research, I am also interested in the development of catalysts for water oxidation, mesoporous materials for CO_2capture, and water purification. These pursuits will draw upon similar experimental approaches as my research program on energy.

Teaching Interests:

I have considerable teaching experience and previously served as the lead teaching assistant for undergraduate and graduate courses on chemical reaction engineering and transport phenomena in the School of Chemical Engineering at Purdue University. Some of these courses were subscribed by as many as 200 students. I delivered several lectures and led all tutorial and review sessions for these courses. I was praised by the students for the clarity of my presentation and conveyance of learning objectives, and my empathy for student advancement. I was awarded the Magoon Award for Excellence in Teaching, which is the highest accolade for teaching by a student at Purdue University. I can teach any course in the core chemical engineering curricula and technical electives such as electrochemistry and materials science. I value the old virtue of using the blackboard for instruction but also appreciate the power of multimedia and internet-based teaching in broadening the scope and reach of a course.

476173 Harnessing Interfacial Phenomena to Design New Soft Materials

Abstract:

Research Interests:

The intersection between colloidal and interfacial sciences has led to the development of self-assembled colloidal arrays which have been profoundly impactful for lithography, responsive materials, energy migration, and bio-sensing. The field now faces the challenge to transition from synthesis and assembly of proof of concept samples to engineering scalable methods that integrate functional colloidal assemblies into materials over large surface areas. In my research program, I aim to pioneer the incorporation of colloidal assemblies into soft materials by directing interfacial polymerization via chemical vapor deposition (CVD) onto colloidal assemblies at fluid interfaces. This approach will strategically leverage trapped states of colloidal assemblies with dynamic polymer growth to independently control the organization and orientation of colloidal monolayers from the polymer matrix for customization of material properties. Here I present previous works on (i) controlling the morphology of polymers deposited onto liquid substrates and (ii) developing a modular synthesis of colloidal Janus particles with diverse chemical compositions which together provide the platform for realizing my vision to harness interfacial phenomena to build colloidal assemblies into soft materials.

Research Experience:

CVD methods have been exploited for many years to transform the surface chemistry of structured materials by depositing uniform and conformal polymer coatings that preserve the underlying geometry and surface roughness. The initiated CVD (iCVD) process is unique because it uses free-radical initiators which eliminate precursor fragmentation present in other CVD methods, and produces polymers analogous to solution phase free-radical polymerization. This enables the iCVD process to deposit functional coatings incorporating hydrophobic, hydrophilic, or stimuli-responsive polymers. To extend deposition via iCVD beyond conformal thin films, we explored deposition onto liquids with low vapor pressures (such as silicone oil and ionic liquids) to exploit the dynamic diffusion and wetting at liquid—vapor interfaces. We demonstrated that the deposition can result in the formation of either polymer particles, planar films, or microstructured films at liquid surfaces depending on the spreading coefficient of the polymer on the liquid. In addition, precursors that are soluble in the liquid can absorb into the bulk and polymerize resulting in the formation of polymer—liquid gels, layered polymer composite films, and encapsulating polymer shells. This work elucidates the growth mechanism of polymers on liquid substrates and provides an understanding of how process parameters can be controlled to incorporate chemical functionality into a variety of structures and morphologies for potential applications in optics, sensing, and separations.

Inspired by the diversity of chemical functionality that the iCVD technique provides, we sought to bring the same range of chemical compositions to the fabrication of amphiphilic Janus particles which are considered colloidal analogues to molecular surfactants. The chemistry of Janus particles is the fundamental parameter that controls their behavior in bulk solution and at interfaces. To enable their widespread utilization, scalable methods that allow for the synthesis of Janus particles with diverse chemical compositions and shapes are highly desirable. We developed clickable Janus particles that can be functionalized through thiol-yne click reactions with commercially available thiols which widen the palate of chemical compositions. The extent of functionalization can be used to control the particle morphology, and thus the type of emulsion stabilized, as well as to fabricate composite Janus particles through sequential click reactions. Functionalizing Janus particles through thiol-yne click chemistry provides a fast-reacting, scalable synthesis method for the fabrication of chemically diverse Janus particles. Currently, we are working on applying our ability to tune the chemical composition to study their behaviors at fluid—fluid and solid—solid interfaces.

Teaching Interests:

Chemical engineers provide unique perspectives because they are able to exploit fundamental chemical and physical phenomena within real-world systems to produce and transform materials. This was exemplified in my graduate research which showed that the morphology of polymers deposited onto liquid substrates via iCVD could be controlled by the interplay between intrinsic polymer—liquid interactions and deposition conditions. The iCVD technique is a perfect example of how core chemical engineering principles (transport phenomena, reaction kinetics, thermodynamics, and separations) work together to control the polymer composition, thickness, molecular weight, and morphology. My experience developing the design rules for understanding deposition onto liquids via iCVD has prepared me to teach the core curriculum in chemical engineering, and the utilization of iCVD in my research program will provide a hands-on experience for undergraduate and graduate students to apply concepts learned in the classroom.

As an educator of chemical engineering, I also aim to train students how to merge their backgrounds in the central sciences and transport phenomena to construct efficient chemical processes. I will do this by emphasizing ingenious chemical engineering solutions in our everyday lives (such as decaffeinated coffee, pressure cookers, and time-release pharmaceuticals) throughout lectures and course assignments. These themes will culminate by challenging students to work in teams to redesign a process or material for alternative applications. To further promote the integration between fundamental science and process design, I plan to develop a graduate course on CVD methods that is accessible to students across engineering and the sciences. The purpose of the course will be to demonstrate the capabilities and diverse applications of CVD techniques while promoting discussion to foster collaboration between students with different backgrounds. My goal as a faculty member will be to inspire students to look for engineering solutions at the intersections between fields to optimize process design and product performance.

Selected Publications (15 total, 7 first author, 211 citations):

L. C. Bradley, K. J. Stebe, D. Lee. Clickable Janus Particles. *Submitted to J. Am. Chem. Soc.* June 2016.

L. C. Bradley, M. Gupta. Microstructured Films Formed on Liquid Substrates via Initiated Chemical Vapor Deposition of Cross-Linked Polymers. *Langmuir*2015, 31, 7999-8005.

L. C. Bradley, M. Gupta. Copolymerization of 1-Ethyl-3-vinylimidazolium bis(trifluoromethyl-sulfonyl)imide via Initiated Chemical Vapor Deposition. *Macromolecules* 2014, 47, 6657-6663.

P. D. Haller, **L. C. Bradley**, M. Gupta. Effect of Surface Tension, Viscosity, and Process Conditions on Polymer Morphology Deposited at the Liquid—Vapor Interface. *Langmuir* 2013, 29, 11640-11645.

R. J. Frank-Finney,† **L. C. Bradley**,† M. Gupta. Formation of Polymer—Ionic Liquid Gels Using Vapor Phase Precursors. *Macromolecules* 2013, 46, 6852–6857. †Authors contributed equally to this work.

L. C. Bradley, M. Gupta. Encapsulation of Ionic Liquids within Polymer Shells via Vapor Phase Deposition. *Langmuir* 2012, 28, 10276-10280.

476185 Structure-Property of Polymer and Its Composites: Multiscale Experimental and Computational Studies

Abstract:

Research Interests:

Polymers play vital roles in nearly every application in our current era, which range from mundane food packaging and personal care to advanced energy devices and protective gears. Such versatility of polymer use makes it an attractive platform to incorporate inorganic fillers with novel functionalities, which either enhance the material property of an existing platform or process as an entirely new polymer composite material. Recent studies have found that a strategic placement of nanofillers within a polymer can dictate the composite material properties, whether electrical, mechanical, or thermal. Thus, predictions of the nanostructure and its structure-property relationship are vital to tailor the nanocomposites to advanced applications such as in batteries, molecular sensors, and protective gear.

As with other classes of materials, investigating soft materials by theory and simulation are crucial to contributing towards expedited discovery of new materials. However, computational studies of polymers present more challenges than other classes of materials. Unlike other crystalline materials with unit cells of 10 nm or less, polymers are often amorphous or semi-crystalline entities that can span over microns to millimeters[2]in length scale. These macromolecules also span over a much larger time scale to respond to an external force, which results in higher computational cost. Although predictive equilibrium assemblies of polymer and its composites, or â €œnanocompositesâ€, are still valuable studies, polymeric materials are almost always processed to become a final product. Thus, investigating the structural dynamics of the nanocomposites with nonequilibrium dynamics such as flow and crystallization are crucial to accurately predict its end-use material properties.

To this end, I have sought answers to the following two questions:

First, ***how does the polymer and its composite structure change under nonequilibrium dynamics***? Second, ***what is the structure-property relationship of the nanocomposites and what are the applicable advanced material applications***?

Using **coarse-grained molecular dynamics**, multiple combinations of polymer(homopolymer or block copolymers) and nanofillers(nanoparticles, nanorods, and nanotubes) composite structures under A) **shear and extensional flow** and B) **cylindrical nanoscale confinement** were investigated. Our study reveals that depending on the control variables of the aforementioned dynamics, we can induce novel functional structures such as highly oriented polymer and nanomaterials, evenly placed nanomaterials within polymer matrix, and interconnected nanorods.

As a comparative experiment, electrospinning was used to fabricate polymer/nanofiller nanofibers which undergoes both extensional deformation and nanoscale cylindrical confinement. The structural agreement between the simulation and the experiment result confirms the validity of the model. As such, some of the novel structures observed in simulation were also observed in experiment, such as well-dispersed and well-aligned nanorods within the nano-scale polymer fibers. The novel functional properties that arise from nanostructures of electrospun nanofiber are then readily utilized in various applications such as **molecular sensors** and **high performance battery electrodes**.

I utilized **structure-property relationship of polymer configuration and its molecular orientation** to significantly alter the **mechanical properties of polymeric fibers**. For instance, by improving the molecular orientation and degree of crystallinity from gel-electrospinning ultra high molecular weight polyethylene, I successfully fabricated ultrathin fibers with high stiffness that rivals the commercial protective vest such as Spectra. By defining a fundamental connection between the predicted nanostructure and its functionality, we can expand much further on the capabilities of electrospun nanofibers to significantly impact the polymer processing and its application as a whole.

Teaching Interests:

From my extensive teaching and research experiences, I believe I have the credentials to make significant contributions to teaching in the Chemical Engineering or Material Science Department. My life-long training as a chemical engineer prepared me to analyze open-ended ill-defined processes in a more well-defined and systematic manner. I believe my research and teaching expertise has prepared me to teach transport phenomena (heat, mass, momentum) as well as thermodynamics. Of course, I can also teach statistical mechanics, which is a basis for many computational studies. I believe effective teaching empowers the studentsâ€™ ability to critically think and create their own knowledge, and this ultimately benefits the research community.

476190 Integrating Catalysis and Separations for Energy-Efficient Conversion of Biomass-Derived Feedstocks

Abstract:

Research Interests:

My research career has spanned surface science and catalysis of complex multifunctional biomass-derived molecules to adsorption and separations with a variety of nanoporous materials. I believe my combination of skills will allow for design of unique materials and development of techniques that enable practical catalytic and separations processes, particularly focused around biomass valorization and utilization.

My Ph.D. work at the University of Colorado Boulder under Professor Will Medlin focused on elucidating the fundamental surface reaction mechanisms of biomass-derived molecules, such as furfural, on transition metal surfaces and utilizing those insights to design catalysts for atom-efficient selective transformations. Using surface science techniques, I proposed different surface intermediates that led to either the desired hydrodeoxygenation or the atom-inefficient decarbonylation reactions. This fundamental insight led to the design of surface-modified and bimetallic oxide-supported catalysts with superior hydrodeoxygenation selectivity.

During my tenure as a postdoc at the Georgia Institute of Technology under Professors Chris Jones and Ryan Lively, I have broadened my skill set in the areas of separations, adsorption, and materials synthesis, working with a variety of nanoporous materials from intrinsically porous polymers to metal-organic frameworks to structured porous oxides. The majority of my work has focused on designing sorbent systems for separation of CO_2 and other acid gases from dilute sources such as flue gas and ambient air and controlling the interactions between liquid adsorbent molecules and solid supports. These projects have required engineering at length scales from the atomic-level design of adsorbents and surfaces to macroscopic contactors such as honeycomb-shaped monoliths and hollow fibers.

Research in my independent career will combine these diverse skill sets to develop advanced materials platforms for valorization of biomass through practical catalysis and separations processes. Much of the challenge in directly using biomass-derived feedstocks lies in the diversity of functional groups, which presents multiple sites for substrate-material interaction, or the difficulty in using traditional thermal separation methods to isolate compounds. My research program will aim at simplification of these feedstocks through catalytic or non-thermal membrane separations means, to enable energy- and atom-efficient conversion to high-value products. Materials design will be guided by an understanding of the fundamental interactions between substrate and catalytic or membrane materials with the goal of producing industrially-relevant materials.

Selected Publications (15 total, 9 first author, 130 total citations (Scopus)):

1. S.H. Pang, C.A. Schoenbaum, D.K. Schwartz, J.W. Medlin "Directing Reaction Pathways by Catalyst Active-Site Selection using Self-Assembled Monolayers," ***Nature Communications* 2013**, 4:2448.
2. S.H. Pang, J.W. Medlin "Adsorption and Reaction of Furfural and Furfuryl Alcohol on Pd(111): Unique Reaction Pathways for Multifunctional Reagents," ***ACS Catalysis* 2011**, 1, 1272-1283.
3. S.H. Pang, N.E. Love, J.W. Medlin "Synergistic Effects of Alloying and Thiolate Modification in Furfural Hydrogenation over Cu-Based Catalysts," ***J. Physical Chemistry Letters* 2014**, 5, 4110-4114.
4. S.H. Pang, M.L. Jue, J. Leisen, C.W. Jones, R.P. Lively "PIM-1 as a "Solution-Processable" Molecular Basket for CO_2 Capture from Dilute Sources," ***ACS Macro Letters*, 2015**, 4, 1415-1419.
5. S.H. Pang, C. Han, D.S. Sholl, C.W. Jones, R.P. Lively "Facet-Specific Stability of ZIF-8 in the Presence of Acid Gases Dissolved in Aqueous Solutions," **2016, *submitted***.

Teaching Interests:

My teaching philosophy is centered on two major goals for student learning: 1) the students develop a fundamental understanding of the theories and concepts central to Chemical Engineering, and 2) they become critical thinkers with the problem solving and teamwork skills that will be necessary for them to become successful. The achievement of these two goals will be accomplished via a flipped classroom, where time traditionally spent lecturing will be replaced with activities designed to engage students.

Because of my emphasis on conceptual learning and problem solving, I am primarily interested in teaching chemical engineering courses that form the foundation onto which the other courses build, such as Material and Energy Balances and Thermodynamics, though would be comfortable teaching any core Chemical Engineering course. I am also interested in developing upper-level elective courses in Chemical Kinetics and Catalysis or Nanotechnology. Students will be given the opportunity to work in project teams to investigate a topic of their choosing. Over the course of the semester, they will teach their classmates, with guidance from the teacher, for one or two class periods, using the educational tools mentioned above, and give a presentation on their chosen topic. In addition to enhancing their understanding through peer instruction, they will gain valuable practice giving presentations.

476199 2D Materials Assembly for Stretchable Electronics and Smart Fabrics

Abstract:

Assembly and Patterning of 2D Material Building Blocks for Multifunctional Devices in Stretchable Electronics and Smart Fabrics

Po-Yen Chen, *Hibbitt Independent Postdoctoral Fellow*
Massachusetts Institute of Technology & Brown University

Postdoctoral Project:

"Design of Multigenerational Topographies of 2D Building Blocks via Sequential Deformation and Intercalation Templating"
- Under supervision of *Ian Y. Wong* & *Robert H. Hurt*, School of Engineering, Brown University

Ph.D. Dissertation:

"M13 Virus-Enabled Assembly of 3D Nanostructured Composites: Synthesis and Applications in Solar Energy Conversion and Electrochemical Energy Storage Devices"
- Under supervision of *Angela M. Belcher* & *Paula T. Hammond*, Department of Chemical Engineering, Massachusetts Institute of Technology (MIT)

Fellowships:

2015, 2016 Hibbitt Engineering Independent Postdoctoral Fellowship (Brown University)
2013, 2014 Eni Energy Fellowship (MIT)
2013 Taiwan National Study Abroad Fellowship
2010 Saudi Aramco Fellowship (MIT)

Research Experience:

Complex surface topographies are of fundamental importance to materials science, physics, biology, and chemistry, impacting applications from catalysis to energy storage and environmental mediation. However, sophisticated design on topographical structures across multiple length scales and over diverse material systems remains challenging. This poster will present our recent works about the creation of multigenerational wrinkled/crumpled textures of versatile materials via sequential deformation and intercalation templating. First, we developed a fabrication that enables extreme mechanical deformation of 2D nanomaterials (*e.g.*, graphene) through multiple cycles of substrate contraction. In each cycle, the compression can be biaxial or unidirectional, leading to a family of different hierarchical wrinkled/crumpled textures. This method achieves a degree of enhancement on structural complexity that far exceeds that of simple one-step deformation processes. The multiscale patterns of graphene show superhydrophobicity (>160°) and improve the electrochemical current densities over 20 times (in comparison with planar coatings). Next, these multigenerational graphene were further utilized as sacrificial templates, and the multiscale surface patterns can be transcribed into other material systems, even beyond 2D nanomaterials (*e.g.*, metal oxides). By harnessing the intersheet gallery spacing of these graphene templates, a wide variety of hydrated metal ions can be uniformly intercalated and immobilized with the functionalities of templates. Subsequently, annealing these composites results in metal oxide thin films that transcribe the original graphene topography with high fidelity. These new textured metal oxide films exhibit enhanced electrochemical and photocatalytic performance over planar films and serve as high-activity electrodes for energy storage and photocatalysis. Overall, these unconventional approaches are broadly applicable for large-area patterning of other 2D nanomaterials and providing a bridging technology that can transcribe textures into other material systems that cannot sustain large deformation.

Research Interests:

My future research interests will focus on large-area patterning of hierarchical surface structures by using various 2D building blocks and study their physiochemical and barrier properties under large mechanical deformation. 2D building blocks, ranging from carbon-based 2D materials to clays and transition metal oxides, display extraordinary chemical, physical and mechanical properties but remain challenging to manipulate them into higher dimensional architectures with large deformability. To address these challenges, I will first continue the work on organizing 2D materials structurally as soft materials while retaining exceptional physiochemical functionality as hard materials. I will emphasize interfacial and surface engineering to facilitate the self-assembly of multiscale, higher-dimensional architectures with highly convoluted features. In the meantime, a systematic approach will be conducted for achieving a control on the wrinkled/crumpled topographies, selective areas, wavelength/width, length scale, amplitude/height, and dynamic patterning. Through above fundamental understanding, I will develop multiple strategies for the fabrication of complex surface patterns of 2D building blocks: 1) wrinkling and crumpling by mechanical deformation, 2) "snapping" surface via confined swelling, 3) origami and kirigami assembly, and 4) 3D printing.

By understanding the factors that produce these higher dimensional structures, it become feasible to design synthetic materials with exactly the kinds of surfaces needed for specific applications, such as stretchable electronics and smart fabrics. By using the proposed fabrication methods, I will apply the convoluted structures of 2D building blocks with high deformability into electromechanical devices, and these surface patterns will retain exceptional physiochemical functionality as well as still provide high performance during reversible stretching. In addition, based on our recent finding, the intersheet gallery spaces of 2D nanomaterials remain unchanged under large deformation, enabling the design of stretchable environmental barriers for the advance of smart fabrics. The combination of mechanical principles (harnessing instabilities of thin films) and materials selection (from 2D nanomaterials to transition metal oxides) leads to a general technological route that can empower great potential to fabricate various flexible and stretchable devices capable of large deformations.

Teaching Interests:

Aside from my research career, I also have extensive teaching experience. As a part of a Hibbitt Postdoctoral Fellowship, I am currently an instructor at Brown University and teach "*Chemical and Biochemical Reactor Design*". The duties include providing semester-long lecture series, developing course/experiment materials, and holding office hours. Also, I TAed an undergraduate lecture and laboratory course "*Polymer Science Laboratory*" in Chemical Engineering in MIT (for which I was the highest-ranked TA (6.6/7.0) in the semester). In an effort to further improve my teaching skills, I have obtained the Graduate Student Teaching Certificate from MIT's Teaching and Learning Laboratory. Lastly, I have mentored over 10 grads/undergrads in both MIT and Brown University. I found each of these experiences to be very rewarding and believe they have helped prepare me for future mentorship and teaching roles.

I am comfortable with teaching any undergraduate core classes in Chemical Engineering; at the graduate level, I could teach *Thermodynamics* and *Kinetics of Chemical Reactions*. I would also like to develop a new subject on sustainable energy that focuses on the fundamentals of thermodynamics, chemistry, and transport in a context that is applied to energy systems. This class is ideal for bridging several engineering disciplines and demonstrating to the students the importance and need for multidisciplinary approaches to addressing energy related issues.

Selected Peer-Reviewed Publications:

- 21 peer-reviewed journal articles (11 first-authored papers), 2 patents, 1 book chapter
- *Served as corresponding author

1. **P.-Y. Chen***, M. Liu, I. Y. Wong*, R. H. Hurt*, "Multifunctionality of Pressure-Actuated Multiscale Graphene Surface Patterns with Reversible Stretchability." *Nano Letters*. (In Preparation)
2. **P.-Y. Chen***, M. Liu, T. M. Valentin, Z. Wang, R. S. Steinberg, J. Sodhi, I. Y. Wong*, R. H. Hurt*, " Hierarchical Metal Oxide Topographies Replicated from Highly Textured Graphene by Intercalation Templating." *ACS Nano*. (Under Review)
3. **P.-Y. Chen***, Z. Wang, R. H. Hurt*, I. Y. Wong*, "From Flatland to Spaceland: Higher Dimensional Patterning of 2D Nanomaterials." *Advanced Materials*. (Invited Article)
4. **P.-Y. Chen***, J. Sodhi, Y. Qiu, T. M. Valentin, R. S. Steinberg, Z. Wang, R. H. Hurt*, I. Y. Wong*, "Multiscale Graphene Topographies Programmed by Sequential Mechanical Deformation." *Advanced Materials* 28, 3564 (2016).
    1. *- Featured as Inside Back Cover*
    2. *- Highlighted in Brown News (1) & (2), IEEE spectrum, and ScienceDaily*
5. **P.-Y. Chen**, N.-M. Dorval Courchesne, M. N. Hyder, J. Qi, A. M. Belcher*, P. T. Hammond*, "Carbon Nanotube-Polyaniline Core-Shell Nanostructured Hydrogel for Electrochemical Energy Storage." *RSC Advances* 5, 37970 (2015). (Invited Article)
6. **P.-Y. Chen**, X. Dang, M. T. Klug, N.-M. D. Courchesne, J. Qi, M. N. Hyder, A. M. Belcher*, P. T. Hammond*, "M13 Virus-Enabled Synthesis of Titanium Dioxide Nanowires for Tunable Mesoporous Semiconducting Networks." *Chemistry of Materials* 27, 1531 (2015).
7. **P.-Y. Chen**, J. Qi, M. T. Klug, X. Dang, P. T. Hammond*, A. M. Belcher*, "Environmentally Responsible Fabrication of Efficient Perovskite Solar Cells from Recycled Car Batteries." *Energy & Environmental Science* 7, 3659 (2014).
    1. *- Highlighted in MIT News (1)&(2), IEEE spectrum, and Washington Post*
8. **P.-Y. Chen**, M. N. Hyder, D. Mackanic, N.-M. D. Courchesne, J. Qi, M. T. Klug, A. M. Belcher*, P. T. Hammond*, "Assembly of Viral Hydrogels for Three-Dimensional Conducting Nanocomposites." *Advanced Materials* 26, 5101 (2014).
    1. *- Featured as Front Cover*
9. **P.-Y. Chen**, X. Dang, M. T. Klug, J. Qi, N.-M. Dorval Courchesne, F. J. Burpo, N. Fang, P. T. Hammond*, A. M. Belcher*, "Versatile Three-Dimensional Virus-Based Template for Dye-Sensitized Solar Cells with Improved Electron Transport and Light Harvesting." *ACS Nano* 7, 6563 (2013).
10. **P.-Y. Chen**, R. Ladewski, R. Miller, X. Dang, J. Qi, F. Liau, A. M. Belcher*, P. T. Hammond*, "Layer-by-Layer Assembled Porous Photoanodes for Efficient Electron Collection in Dye-Sensitized Solar Cells." *Journal of Materials Chemistry A* 1, 2217 (2013).

11. **P.-Y. Chen**, C.-P. Lee, R. Vittal, K.-C. Ho*, "A Quasi Solid-State Dye-Sensitized Solar Cell Containing Binary Ionic Liquid and Polyaniline-Loaded Carbon Black." *Journal of Power Sources* 195, 3933 (2010).

476216 Structure and Transport in Polymer Membranes for Energy-Efficient Separations

Abstract:

Structure and Transport in Polymer Membranes for Energy-efficient Separations

Hee Jeung Oh

(1) Department of Chemical and Biomolecular Engineering, College of Chemistry, University of California, Berkeley, California 94720, USA

(2) Mcketta Department of Chemical Engineering, Texas Materials Institute, Center for Energy and Environmental Research, The University of Texas at Austin, Austin, Texas 78758, USA

Research Interests:

With the worldâ€™s population growing rapidly, the need for clean water is greater than ever. Among water purification methods, membrane-based separations draw attention because of their energy efficiency (they do not require an energy-intensive phase-change step as in distillation), small footprint, and ability to eliminate trace contaminants. However, only a few membrane materials and formation methods are used in industry today, and current membranes often do not have adequate efficiency (i.e., selectivity), productivity (i.e., permeability), and chemical stability to be used for all sources of water. To overcome these challenges, new polymer membrane materials and formation methods are needed to precisely tune pore size and pore size distributions and to control chemical stability for better separation performance.

Transport of small molecules and ions through polymer membranes is predominantly controlled by polymer nanostructure. Therefore, morphological control of the membrane structure by tuning chain architecture, synthetic routes, and post-synthesis treatment is a critical aspect of designing membrane systems. My graduate research focused on the water and salt transport properties of ion-containing polymers (sulfonated aromatic hydrocarbons) prepared by a new membrane formation method, i.e., solvent-free melt processing. My postdoctoral research focuses on extending my understanding of a polymerâ€™s physical structure by designing and synthesizing highly-structured polymeric systems and characterizing them (e.g., through X-ray scattering, tomography, scanning transmission electron microscopy). Additionally, these porous, crystalline, and conducting polymers can be used as polymer electrolytes in electrochemical systems and as blood filtering devices for biomedical applications.

My research interests are in understanding the fundamental structure-property relationships of water and salt transport in polymers and composite materials. These interests span disciplines from chemical engineering, chemistry, and materials science. The advancement of the technologies mentioned above is highly dependent on the development of polymer membranes, which selectively permeate only desired components while maintaining chemical stability. My goal is to develop correlations between design, synthesis, structure, and transport properties of polymers for targeting applications in energy-efficient separations, batteries and biomedical devices. Synthetic work will focus on highly-structured polymers such as ion-containing polymers, block copolymers, and random copolymers, and polymer characterization will focus on water purification, electrochemical systems, and biomedical devices.

Teaching Interests:

I believe teaching and mentoring the next generation of scientists is the most satisfying aspect of being in academia. Both in classroom instructions and in a research laboratory, my goal is to help students identify important problems and to arrive at intelligent solutions by teaching critical thinking skills.

My education and experiences have prepared me well to teach both undergraduate and graduate courses in the Chemical Engineering department. Also as a graduate student and a postdoctoral researcher, I have supervised research projects for three undergraduate students and several high school students. Seeing these students succeed has been amazingly rewarding to me, and I am looking forward to continue this mentorship with my future graduate students and post-docs.

476220 Engineering Soft Materials with Tunable Structure and Functionalities

Abstract:

Research Interests:

Soft materials encompass a wide range of materials, such as polymers, colloids, emulsions, gels and biological materials, whose properties arise from an interplay of processes across multiple scales. Conquering the major challenges, especially in food, pharmaceuticals and healthcare in the 21st century requires novel and smarter soft materials that can provide unprecedented properties and extraordinary functions, capable of exquisitely interfacing between engineering, chemistry, medicine and biology. The discovery and development of novel materials not only bring about new applications and improved performances, but also enable innovative approaches of material manufacturing in a sustainable way. My vision for this research direction is strongly motivated by the burgeoning need for better multi-scale understanding the structure - dynamic function relation of soft materials and also to develop novel processes for materials chemistry that enable production of such materials ranging in scale from nanocomposites to hierarchically structured polymer *particles* and *gels*.

As future faculty, my goal is to establish an independent research program that aims at designing and developing novel bottom-up approaches for soft materials with tunable architectures and functionalities. I will give emphasis on better understanding on how to precisely control the molecular transport in such materials and surface molecular interactions, besides the understanding their structure-function relation. I intend to leverage my diverse research expertise along with use of innovative approaches at the interface of chemical engineering, polymer and materials science, and emulsion and interface science. The research themes that I intend to pursue are, but not limited to: (i) Understanding and formation of novel structured emulsions with new and remarkable functionality; (ii) Development of nanostructured polymeric materials/ nanocomposites for the capture/separation of industrially, environmentally, and biologically relevant chemicals; (iii) Design and development of the next generation of intelligent hydrogel materials based- delivery devices for applications in (bio)pharmaceutics and healthcare. My poster will highlight the key aspects of some of my past and current research along with an outline of future projects in the above-mentioned platform.

Teaching Interests:

Aside from my research career, I have also substantial teaching experience. I was a Lecturer in the Department of Chemical Engineering at Bangladesh University of Engineering and Technology (BUET), where I had the opportunity to teach one chemical engineering core module and several laboratory courses. At National University of Singapore (NUS), I worked as a tutor in particle technology course. I mentored many undergraduate and grduate students during my graduate studies and postdoc at NUS. Organizing tutoring and review sessions for the students gave me valuable teaching experience, as it was often then that the students asked the most penetrating and insightful questions. At MIT, I participated in â€˜Kaufman Teaching Certificate Programâ€™ to further enhance my teaching skills. I believe that my academic experience at BUET, NUS and MIT has tremendously enriched my perspective, and I would like my students to benefit from it too.

I am well-equipped to teach the core-curriculum for chemical engineering, including separation science, particle technology, mass and heat transfer, process design, and reaction engineering.

476227 Polymer Based Hybrid Materials: From Molecular Design to Applications

Abstract:
Research Interests:

Organic-inorganic nanocomposites with controlled hierarchical structures offer a particularly attractive platform for development of soft functional materials for various applications ranging from health care to energy storage. Here I am going to present how complexation of organic macromolecules and nanomaterials at molecular length scales and under mild conditions, as used by nature, can form advanced materials with exceptional engineering properties. Working with Prof. Matthew Tirrell and Prof. Juan de Pablo at the University of Chicago, I have synthesized patent-pending stable amorphous hybrid nanoparticles of controllable size and surface charge under mild conditions. These hybrid nanoparticles are composed of a polyelectrolyte, either polycation or polyanion and amorphous calcium phosphate nanospheres which are a few nanometer in size. The ACaP nanospheres are formed during synthesis and are uniformly dispersed inside the polymer matrix. We have also shown that through mixing these hybrid nanoparticles with their oppositely charged counterpart, it is possible to prepare injectable pastes with tunable rheological properties. The presence of stable ACaP nanospheres as an inexpensive, bioactive material along with very good handling properties makes these materials beneficial for tissue regeneration and dental applications. Moreover, I have been working on graphene based electrically conductive aerogels with exceptional thermal stability and compatibility with living cells.

In collaboration with prof. Mezzenga from ETH Zurich, we have been able to design gelatinâ€“graphene conductive biopolymer nanocomposites with a record-low electrical percolation threshold of 3.3 Ã— 10â^'2Â vol%, which arises from the homogeneous dispersion of the graphene nanosheets within the gelatin matrix. The results of this study have been published in *Advanced Materials*.

I also had the opportunity to get involved in a pioneering work on the understanding of glass-like dynamic properties of living cells. I joined Professor Fredbergâ€™s lab at the Harvard University in 2012. As a member of a multidisciplinary team there, I was able to use my knowledge in physics of glass transition from my PhD work to study well-characterized endothelial and epithelial cell monolayer systems. We have shown that the mechanics of the cellular monolayer in the lung, and in other organ systems as well, may be dominated by glass transition and/or jamming as in polymeric and colloidal systems. Our findings which offer a unifying new lens into the biology of the cellular monolayers have been recently published in *Nature Materials* and *Differentiation*.

As a faculty member I would like to focus on understanding and fine-tuning the assembly of polymers and 2D nanomaterials into engineering systems and devices. Specifically, my main focus will be on developing compositionally and synthetically simplified organic-inorganic hybrid systems with lower production costs and competitive functionality in wide variety of applications including regenerative medicine and energy storage.

Teaching Interests:

I have several years of teaching experince of polymer courses including polymer physics, polymer chemistry and polymer characterization at both the undergraduate level and graduate level at the University of Tehran. I have also supervised more than 20 MSc and 2 PhD students in polymer science.

In addition to polymer courses, I would also welcome the opportunity to develop and teach classes related to colloidal science and hybrid materials.

476241 Design of Functional Polymeric Materials: From Ion Transport to Bio-Inspired Assembly

Abstract:

Research Interests:

Engineering of polymer-inorganic hybrid materials with structure control from the nanoscale to the mesoscale has remained elusive, especially in comparison to biological composites such as nacre and bone. However, the ability to synthetically design highly crystalline, hierarchically structured, and well-ordered inorganic and composite materials is desired in fields ranging from biomaterials to energy. My research platform will focus on the development of polymer systems with chemical and sequence control focusing on three areas: (1) understanding the fundamental interactions between polymers and inorganics during self-assembly process via *in-situ* techniques, (2) combining these fundamentals with polymer self-assembly to engineer high performance hierarchical inorganic materials for energy storage and catalysis, and (3) responsive assembly of polymeric nanoparticles for healthcare.

This research program builds upon my Ph.D. work, undertaken at UC Santa Barbara under the guidance of Profs. Glenn Fredrickson, Ed Kramer, and Craig Hawker, which elucidated the interactions between polymers and ions for lithium-ion batteries. I focused on the development of poly(allyl glycidyl ether) (PAGE) as an alternative platform for lithium battery polymer electrolytes. The versatility of allyl glycidyl ether as a building block was demonstrated by the preparation and evaluation of various AGE-EO macromolecular architectures that show superior performance to both PAGE and PEO by utilizing copolymerization and thiolene coupling chemistry. Furthermore, in order to develop design rules for polymer electrolytes, a library of structurally similar polyethers, poly(glycidyl ether)s (PGEs) was synthesized by anionic polymerization and investigated by impedance measurements, dielectric spectroscopy, and pulse field gradient NMR to understand the role of polymer chemistry on ion diffusion.

I have taken this interest in controlling polymer chemistry and its effect on composite components to my current postdoctoral position at Cornell University with Prof. Uli Wiesner and Prof. Lara Estroff, where we seek to combine the long-range periodic structure direction of block copolymers with the increased information content provided by sequence control in proteins to generate highly controlled composite materials. Knowledge of inorganic crystallization pathways has been developed through *in-situ* GIWAXS studies of early formation pathways in hybrid organic-inorganic pathways.

Teaching Interests:

My teaching interests are in polymer physics, polymer chemistry, thermodynamics, and introductory materials science.

476424 Engineering Discrete Functional Building Blocks at Molecular Scale for Human-scale Applications

Abstract:

Research Interests:

The preparation of synthetic polymers and nanocomposites with structures, functions and properties that are precise and accurate remains a grand challenge for polymer and material chemists. Synthetic polymers and other material building blocks prepared using bottom-up approach are generally polydisperse, and their use as a statistical ensemble of molecular weights/sizes can be highly problematic in biological and nanoscale applications due to size-dependent effects, aggregations and phase separations.

My research program will focus in the development of an efficient and scalable synthesis of discrete polymer-based materials for applications on a human scale. In specific, there are three areas of research. (1) Synthesis of discrete oligomers/polymers and the utilization of their pendent groups and/or chain-ends to produce discrete macromolecular architectures. Such structures are expected to exhibit accurate and precise properties leading to specific functions. (2) Engineering of discrete polymer building blocks in conjunction with molecules containing swellable, photoswitchable, or degradable moieties to construct designer stimuli-responsive structures (UV, pH, toxic chemicals). (3) Engineering of discrete polymer building blocks with inorganic complexes and nanoparticles to construct hierarchical nanostructures for applications in separation and sensing. Mid-long term goal includes the development of biocompatible nanocarriers containing discrete materials for drug-delivery applications. As such, my research program attempts to address fundamental and technological challenges at the interface of chemical engineering, chemistry, material science and biology.

Postdoctoral Projects:

Department of Materials, University of California Santa BarbaraSupervisor: Prof. Craig J. Hawker

- *"Synthetic Oligomers and Polymers with Discrete Structures and Hierarchical Architectures"*

(1 completed, several sub-projects ongoing)

- *"Polymer Brush Nanostructures Using Visible-Light-Mediated Metal-Free Atom Transfer Radical Polymerization "* (completed)

PhD Dissertations:

Department of Polymer Science and Engineering, University of Massachusetts AmherstSupervisor: Prof. Todd Emrick

- *"Functional Nanostructures from Nanoparticle Building Blocks"*

Research Experience:

Briefly, my postdoctoral work under the guidance of Prof. Craig J. Hawker at UCSB primarily involves the synthesis and development of novel polymer architectures via controlled polymerization techniques, and the preparation of discrete oligomers/polymers. In my PhD work under the guidance of Prof. Todd Emrick at UMass Polymer Science and Engineering, I focused on the synthesis and characterization of nanomaterials and multifunctional polymer-based ligands for making nanostructures using directed self-assembly and covalent assembly techniques. I also investigated the synthesis of zwitterionic polymer covered nanomaterials and their potential applications as nonfouling nanomaterials and photothermal therapeutic agents for *in vivo*studies. Before joining UMass, I was a Research Scientist at Schlumberger Research (Boston) developing optochemical and electrochemical sensing, membrane separation and microfluidic platforms for downhole technologies. I obtained my master and bachelor degree in Chemical Engineering from the University of Tokyo under the guidance of Prof. Takeo Yamaguchi, Prof. Steven Kraines and Prof. Hiroshi Komiyama.

Teaching Experience:

I served as a teaching assistant in the Polymer Synthesis Laboratory in the Department of Polymer Science and Engineering at Univ. of Massachusetts for 2 years, teaching various polymer synthesis to 1styear graduate students. Over my career I have been fortunate to directly mentor 7 undergraduates, 8 graduate students and am currently leading a team of several post-docs and PhD students.

Teaching Interests:

I am comfortable teaching Polymer Chemistry, Polymer Synthesis and Characterization, Polymer Physics, Transport Phenomena, Materials Science, Nanomaterials, and Computational Chemistry.

Awards:

- Dow Materials Institute Travel Award 2015

- Dow Materials Institute Travel Award 2016

- 6 US Patents

- Monbusho Full Scholarship, Japanese Ministry of Education (1998–2003, 2005–2007)

Publications:

1. Lawrence, J.; Lee, S.-H.; Abdilla, A.; Nothling, M. D.; Ren, J. M.; Knight, A. S.; Fleischmann, C.; Li, Y.;

Abrams, A. S.; Schmidt, B. V. K. J.; et al. A Versatile and Scalable Strategy to Discrete Oligomers. *JACS*, **2016**, *138*, 6306

2. Discekici, E.; Pester, C.; Treat, N.; Lawrence, J.; Mattson, K; Narupai, B.; Toumayan, E.; Luo, Y.; McGrath, A.; Clark, P.; Read d. A. J.; Hawker, C. A Simple Bench Top Approach to Polymer Brush Nanostructures Using Visible Light Mediated Metal-Free Atom Transfer Radical Polymerization. *ACS Macro Letters*, **2016**, *5*, 258.

3. Lawrence, J.; Emrick, T. Pentafluorophenyl Ester-Functionalized Nanoparticles as a Ver- satile Platform for Selective and Covalent Inter-Nanoparticle Coupling. *ACS Applied Materials and Interfaces,* **2016**, *8*, 2393.

4. Liu, Y.; Chen, Y-C.; Hutchens, S.; Lawrence, J.; Emrick, T.; Crosby, A. J. Directly Measuring the Complete Stress-Strain Response of Ultrathin Polymer Films. *Macromolecules*, **2015**, *48*, 6534.

5. Lawrence, J.; Yamashita, K.; Yamaguchi, T. Correlating electronic structure and chemical durability of sulfonated poly (arylene ether sulfone)s. *Journal of Power Sources,* **2015**, *279*, 48.

6. Puodziukynaite, E.; Wang, H.W.; Lawrence, J.; Wise, A. J.; Russel, T.P.; Barnes, M.; Emrick, T. Azulene Methacrylate Polymers: Synthesis, Electronic Properties, and Solar Cell Fabrication, *JACS*, **2014**, *136*, 11043.

7. Maiz, J.; Zhao, W.; Gu, Y.; Lawrence, J.; Arbe, A.; Alegria, A.; Emrick, T.; Mijangos, C.; Colmenero, J.; Russell, T.P. Structural and Dynamic Study Of Polystyrene-block-poly(4-vinylpyridine) Copolymer Confined In Self-Ordered Nanoporous Alumina, *Polymer*, **2014**, *55*, 4057.

8. Pham, J. T.; Lawrence, J.; Grason, G. M.; Emrick, T.; Crosby, A. J. Stretching of Assembled Nanoparticle Helical Springs, *Phys. Chem. Chem. Phys.*, **2014**, *16*, 10261.

9. Lawrence, J.; Pham, J.T.; Lee, D. Y.; Liu, Y.J.; Crosby, A. J.; Emrick, T. Highly Conductive Ribbons from Stick-Slip Assembly of Organosoluble Gold Nanoparticles. *ACS Nano*, **2014**, *8*, 1173.

10. Pham, J. T.; Lawrence, J.; Lee, D. Y.; Grason, G. M.; Emrick, T.; Crosby, A. J. Highly Stretchable Nanoparticle Helices Through Geometric Asymmetry and Surface Forces. *Advanced Materials*, **2013**, *25*, 6703.

11. Lee, D. Y.; Pham, J. T.; Lawrence, J.; Lee, C. H.; Parkos, C.; Emrick, T.; Crosby, A. J. Macroscopic Nanoparticle Ribbons and Fabrics. *Advanced Materials,* **2013**, *25*, 1248.

12. Laprade, E. J.; Long, R.; Pham, J. T.; Lawrence, J.; Emrick, T.; Crosby, A. J.; Hui, C.-Y.; Shull, K. R. Large Deformation and Adhesive Contact Studies of Axisymmetric Membranes. *Langmuir*, **2013**, *29*, 1407.

13. Chen, X.*; Lawrence, J.*; Parelkar, S.; Emrick, T. Novel Zwitterionic Copolymers with Dihydrolipoic Acid: Synthesis and Preparation of Nonfouling Nanorods. *Macromolecules*, **2012**, *46*, 119. (*: equal contribution)

14. Bae, J.; Lawrence, J.; Miesch, C.; Ribbe, A.; Li, W.; Emrick, T.; Zhu, J.; Hayward, R. C. Multifunctional Nanoparticle-Loaded Spherical and Wormlike Micelles Formed by Interfacial Instabilities. *Advanced Materials*, **2012**, *24*, 2735.

15. Lawrence, J.; Yamaguchi, T. The Degradation Mechanism of Sulfonated Poly(Arylene Ether Sulfone)s in an Oxidative Environment. *Journal of Membrane Science*, **2008**, *325*, 633.

16. Lawrence, J.; Robinson, K. L.; Lawrence, N. S. Electrochemical Determination of Sulfide at Various Carbon Substrates: a Comparative Study. *Analytical Sciences*, **2007**, *23*, 673.

476850 Flow and Jamming of Particulate Materials

Abstract:

Research Interests:

The majority of industrial processes involve particulate forms of material. The grain sizes range from nano and mico-scale (e.g. colloids) to macro-scale (e.g. granular materials, emulsions, and foams). Even with a comprehensive knowledge of the particle-particle interactions, the prediction of mesoscale and large scale behavior of a collection of these particles is not trivial. A significant effort has been dedicated to understanding the phenomenology, and more importantly, unifying the laws which govern the collective dynamics of particulate systems. These studies cross between several fields such as Physics, Chemical and Mechanical Engineering, and material science. I am in particular interested in the flow of particulate systems, and I would like to pursue this field with experimental studies on the rheology of granular materials and colloids.

Postdoctoral Research:

(i) Reversibility of a Colloidal System: I am currently studying a colloidal system adsorbed at an oil-water interface, which is subject to cyclic shear. The colloidal particles are sub-micron size, such that Brownian motion has significant contribution to the single particle dynamics. Through this study, we found that thermal noise significantly affects the reversibility of material in a nontrivial manner.

[1] S. Farhadi and P. Arratia, "***Shear-induced Reversibility for Thermal Colloids***", to be submitted (2016).

(ii) Dynamics of Active Systems: In active particulate systems, in contrast to passive systems, single particle dynamics is independently present even without external driving. We have designed two species of active granular systems: spinners, and translators. The intra-structure of particles (which are 3D printed) is

designed such that they either spin or translate in the presence of an upward air flow. In this ongoing study, we aim at describing the collective dynamics, and in particular, we search for an effective temperature which governs the dynamics. Subsequently, we will be able to extract an equation of state for these systems.

PhD Research:

My PhD research was an experimental study on the jamming and rheology of granular materials, with a specific focus on elliptical-shaped particles. I performed continuous Couette shear (rheology) and cyclic compression (jamming-unjamming) on both systems of circular and elliptical shaped particles. The identical experiment, on both circles and ellipses, enabled us to understand the role of particle anisotropy on jamming and the bulk rheology.

[2] S. Farhadi, A. Zhu, RP. Behringer, "***Stress and Structure Instability in Cyclic Compression of Ellipses and Disks***", Physical Review Letters, (2015).

[3] S. Farhadi, RP. Behringer, "***Dynamics of sheared ellipses and circular disks: effects of particle shape***", Physical Review Letters 112 , 148301 (2014).

[4] S. Farhadi, RP. Behringer, Alex Zhu, "***Slow dynamics for elliptical particles under continuous shear and cyclic compression***", AIP Conf. Proc., 1542, 879-882 (2013).

Future Directions:

I would like to pursue my future research in two directions:

1- I am interested in developing techniques to directly measure force distribution for a disordered matter, in its grain scale. There have been few successful experimental techniques such as photoelasticity and contact area size to measure grain scale stress. However, the filed of colloids lacks experimental tools for this purpose so far. Knowledge of the grain scale forces is useful in connecting the structural properties to the dynamics, and eventually obtaining insight on problems such as self assembly, where both structure and grain scale forces play role.

2- An emerging filed in physical sciences is 'machine learning with big data'. The idea is to extract information by obtaining a large collection of data from a specific system, and create learning models to predict the dynamical behavior. I would like to apply these techniques to understand the underlying mechanisms of flow and transport in particulate systems.

Teaching Interests:

As a graduate student of Physics, I have been involved with teaching several introductory courses, which included lab instruction. I also have experience of teaching nonlinear dynamics, which is a higher level undergraduate course. I should emphasize that I have an undergraduate major degree in Chemical Engineering, which helps me with teaching courses such as thermodynamics, fluid dynamics, and heat and mass transfer. Also, as an expert in granular materials, I can develop new courses such as granular matter, transfer in porous materials, and more mathematical based courses such as nonlinear dynamics, and chaos.

477008 Controlling the Structure of Systems Ordered via Block Copolymer Phase Separation: Simulations and Experiments

Abstract:

Research Interests:

Block polymers and other ordered soft materials have the potential to allow a vast array of structures to be ordered at lengths scales from a few nanometers to several micrometers. This property allows for a varied range of practical applications in, for example, drug delivery, microelectronic materials, advanced plastics, and catalysis. However, only a small subset of the extensive array of possible systems have been explored. This space can best be explored with a concerted effort using both experiment and simulation. My research will start in three main areas:

1. Ordering of nanoparticles via block polymer ligand directed assembly
2. Phase behavior of diblock blends
3. Development and application of protracted colored noise dynamics for polymer systems.

1. Ordering of nanoparticles via block polymer ligand directed assembly

Nanocomposites are widely used for technological and commercial purposes and they have been ordered via a variety of methods. However, precise control is still challenging and the controlling mechanisms are still not understood. Simulations of these systems would provide a theoretical underpinning by which experimental space would be reduced, resulting in greater understanding of the critical properties more quickly.

2. Phase behavior of diblock blends

Control of ordered block polymer structures is most commonly done via precise synthetic techniques, but blending of polymers is a significantly easier way to control structures if the process can be understood. By fundamentally understanding the effect that blends of diblocks have on morphology, domain size, and other properties, new applications and easier production methods could be discovered. **3. Protracted colored noise dynamics for polymer systems**

Molecular dynamics simulations of polymeric systems are significantly limited in terms of the time and length scales that can be studied. A new technique, protracted colored noise dynamics for polymer systems is designed to efficiently sample over energy barriers and can reduce simulation time to equilibrium by more than four orders of magnitude. This technique could be adapted to work with atomistic simulations and a plethora of polymer architectures allowing access to timescales which are currently inaccessible. Systems of particular interest are semi-crystalline polymers such as polyethylene or large polymer systems such as micelle forming systems or star polymers.

Teaching Interests:

As a faculty member in a chemical engineering department, I would be pleased to teach a range of courses that combine my training with the needs of the faculty and the students. I can bring strong preparation and an enthusiastic attitude to any core chemical engineering class. Leveraging my postdoctoral and graduate research experience in polymer physics and simulations, I am also excited to develop new coursework in areas that would fit into both the undergraduate and graduate curriculum related to soft matter engineering and simulation techniques. My experience as a teaching fellow where I co-taught a numerical methods and my experience as a teaching assistant for 3 semesters which required weekly recitation sections has given me the experience I need to be an effective instructor early on.

The reason I want to be faculty is because I love to learn. I love research because itâ€™s never-ending learning and satisfying of curiosity. In teaching I can give this love of learning to my students. By showing them my passion and enthusiasm and enjoyment of learning and of knowledge, I will promote enthusiasm and enjoyment in my students.

477819 Application of Zwitterionic Materials in Stem Cell Expansion and Immunosuppression

Abstract:

Research Interests:

Ultra-low fouling zwitterionic hydrogels offer unique advantages over conventional biomaterials in biomedical applications. By exclusively using zwitterionic monomer and crosslinker, pure zwitterionic hydrogels can be prepared with highly tunable mechanical properties and functionalities for diverse applications. Uniform zwitterionic hydrogels with controllable mechanical and/or chemical gradients can also be developed. My research focuses on the development and application of zwitterionic hydrogels. Examples include zwitterionic hydrogels allow differentiation-free expansion of stem cells, implanted zwitterionic hydrogels resisting foreign-body capsule formation in mice, zwitterionic hydrogel coatings significantly enhancing the longevity and sensitivity of glucose sensors, functionalized hydrogels encapsulating and preserving cells with high viability and long-circulating zwitterionic nanogels.

Research Experience:

1) Development and Applications of Zwitterionic Hydrogel

l Differentiation-free expansion of stem cells and T cells in zwitterionic hydrogels

l Directed differentiation of stem cells in functionalized zwitterionic hydrogels

l Zwitterionic hydrogels resistant to the immunogenicity and fibrous encapsulation in mice l Time-independent self-healing zwitterionic hydrogels

l Long-life zwitterionic material-coated biosensors in whole blood

2) Development of Long-circulating Zwitterionic Nanomaterials

l Long-circulating and therapeutic protein-encapsulated zwitterionic nanogels

l Immunogenicity-free zwitterionic nanoparticles

3) Development and Applications of High-strength Hydrogels

l Construction of ultra-high strength hydrogels

l High-strength hydrogel-based gene transfection platform

l Develop shape-memory effect for high-strength hydrogel

l Directed differentiation of MSCs on a shape-memory hydrogel

Teaching Interests:

In my five-year graduate student career, I have been served as teaching assistance for many different chemical engineering courses like transport and thermodynamic. I am interested in teaching any chemical engineering course as well as polymer science course.

AWARDS

2015 National Award for Outstanding Chinese Students Study Abroad ($6000)

SELECTED JOURNAL PUBLICATIONS (Total: 28, Citations: 800+)

1. **Tao Bai**, Fang Sun, Lei Zhang, Andrew Sinclair, Sijun Liu, Jean Rene Ella Menye, Ying Zheng, and Shaoyi Jiang, 'Restraint of the Differentiation of Mesenchymal Stem Cells by a Nonfouling Zwitterionic Hydrogel', *Angewandte Chemie International Edition,* 126 (2014), 12943-48.

2. Lei Zhang, Zhiqiang Cao, **Tao Bai**, Louisa Carr, Jean-Rene Ella-Menye, Colleen Irvin, Buddy D Ratner, and Shaoyi Jiang, 'Zwitterionic Hydrogels Implanted in Mice Resist the Foreign-Body Reaction', *Nature Biotechnology,* 31 (2013), 553-56.

3. Tao Bai, Andrew Sinclair, Fang Sun, Priyesh Jain, Hsiang-Chieh Hung, Peng Zhang, Jean-Rene Ella-Menye, Wenguang Liu, and Shaoyi Jiang, 'Harnessing Isomerization-Mediated Manipulation of Nonspecific Cell/Matrix Interactions to Reversibly Trigger and Suspend Stem Cell Differentiation', *Chemical Science*, 7 (2016), 333-38.

4. Tao Bai, Sijun Liu, Fang Sun, Andrew Sinclair, Lei Zhang, Qing Shao, and Shaoyi Jiang, 'Zwitterionic Fusion in Hydrogels and Spontaneous and Time-Independent Self-Healing under Physiological Conditions', *Biomaterials,* 35 (2014), 3926-33.

5. Tao Bai, Yanjiao Han, Peng Zhang, Wei Wang, and Wenguang Liu, 'Zinc Ion-Triggered Two-Way Macro-/Microscopic Shape Changing and Memory Effects in High Strength Hydrogels with Pre-Programmed Unilateral Patterned Surfaces', *Soft Matter,* 8 (2012), 6846-52.

6. Tao Bai, Peng Zhang, Yanjiao Han, Yuan Liu, Wenguang Liu, Xiaoli Zhao, and William Lu, 'Construction of an Ultrahigh Strength Hydrogel with Excellent Fatigue Resistance Based on Strong Dipole–Dipole Interaction', *Soft Matter,* 7 (2011), 2825-31.

7. Yanjiao Han, **Tao Bai (Co-first Author)**, and Wenguang Liu, 'Controlled Heterogeneous Stem Cell Differentiation on a Shape Memory Hydrogel Surface', *Scientific reports,* 4 (2014).

8. Yanjiao Han, **Tao Bai (Co-first Author)**, Yuan Liu, Xinyun Zhai, and Wenguang Liu, 'Zinc Ion Uniquely Induced Triple Shape Memory Effect of Dipole–Dipole Reinforced Ultra High Strength Hydrogels', *Macromolecular rapid communications,* 33 (2012), 225-31.

9. Peng Zhang, Fang Sun, Caroline Tsao, Sijun Liu, Priyesh Jain, Andrew Sinclair, Hsiang-Chieh Hung, **Tao Bai**, Kan Wu, and Shaoyi Jiang, 'Zwitterionic Gel Encapsulation Promotes Protein Stability, Enhances Pharmacokinetics, and Reduces Immunogenicity', *Proceedings of the National Academy of Sciences* (2015), 201512465.

10. Xiyang Dai, Yinyu Zhang, Lina Gao, **Tao Bai**, Wei Wang, Yuanlu Cui, and Wenguang Liu, 'A Mechanically Strong, Highly Stable, Thermoplastic, and Self Healable Supramolecular Polymer Hydrogel', *Advanced Materials* (2015).

478365 From Reactive Nano-Particles to Self-Healing Materials: Chemical Research with a Green Twist

Abstract:

Research Interests:

Current and past research interests

My research is at the crossroads between the chemical, oil&gas and environmental engineering sectors, and encompasses colloid, polymer, emulsion and surface science.

Current research interests at SABIC:

Emulsions. Fouling issues in steam crackers may arise due to a number of reasons, including emulsion formation in the quench water tower. Interfacially active species were extracted and fractionated into solubility classes. The study showed the differences in the mechanisms of emulsion stabilization with the different fractions, and highlighted the correlation between interfacial film composition and morphology, and emulsion stability.

Iron-catalyzed radical polymerization. Undesired polymerization may lead to fouling issues in crackers, and should thus be avoided. The research conducted highlighted the role of iron in catalyzing polymerization of styrene and showed the correlation between solvent type and kinetics of radical polymerization.

Corrosion inhibitors. Film forming amines. Film forming amines (FFAs) are commonly used to inhibit corrosion in steam condensate systems, since they form passivating films which can act as a hydrophobic barrier against corrosive species in water. In steam systems FFAs are typically used in conjunction with neutralizing amines, which increase their solubility in water. The research conducted compared the effectiveness of different combinations of FFA and neutralizing amines in inhibiting corrosion, and investigated the factors controlling their performance.

Previous research interests:

Reactive polymer coated nano-particle optimization for groundwater remediation (PhD research). Nano zero-valent iron (nZVI) remediation technology is used to treat aquifers contaminated by chlorinated compounds and heavy metals, which react with iron forming less harmful compounds. The success of nZVI technology relies on adequate particle delivery to the polluted areas, and this was the focus of my research. Particle transport is largely influenced by the surface properties of the particles. Therefore, I investigated the effect of polymer coatings used to stabilize iron particle suspensions on the transport of particles through aquifers having different mineral compositions and water chemistries. I probed the forces of interaction between the particles and the substrates, because large adhesive forces hinder particle transport. In my study I utilized wet-water chemistry to synthesize and electrodeposited iron nano-particles on AFM probes to measure the adhesion forces between single particles and model mineral substrates under no-flow conditions. Furthermore, I probed sorption phenomena under laminar flow conditions using QCM-D. The particles synthesized were characterized using SEM, EDX, XPS and XRD to assess their morphology and composition before and after prolonged exposure to aqueous media, simulating groundwater conditions. My research showed the importance of hydration forces in the interactions between nZVI and mineral surfaces, and demonstrated that polymer coatings with a cationic counter-ion could inhibit particle attachment onto silicate minerals in a wide range of water chemistries.

Emulsions in the upstream oil&gas sector. Problematic emulsions are common in the upstream sector, and specifically in the oil sands industry. The collaborative research I conducted at the University of Alberta demonstrated the role of a specific asphaltene sub-fraction in stabilizing emulsions, and highlighted the factors controlling emulsion stability when using model compounds of asphaltenes. My research further focused on the use of polymeric demulsifiers to destabilize problematic emulsions. My research correlated demulsification performance to the chemistry of the demulsifiers and to their effect on the interfacial films, namely on their compressional and shear rheological properties, and on their morphology (as probed with BAM at the liquid-liquid interface, and with AFM of Langmuir-Blodgett films). A similar approach was successfully adopted to study emulsions stabilization by pH and temperature responsive NIPAM (poly(N-isopropylacrylamide)) micro-gels.

High performance polymer-based non-Newtonian gels: effect of cross-linker and polymer type on the shear and thermal stability of the gels. I developed polymeric gels for enhanced oil recovery applications at high temperature and high shear regimes. I engineered self-healing gels with synthetic and natural polymers, cross-linked with a variety of cross-linkers, and determined the best combination of polymer, cross-linker and water chemistry yielding a highly viscous and stable gel at elevated temperature and high shear rates. Importantly, I was able to produce an additive capable of generating radicals that could break the gel on demand with controlled kinetics, thus reducing its viscosity to extremely low values after a desired time period.

Future research scopes

My objective is to draw knowledge from the oil&gas and chemical sectors to produce materials that can be used for the treatment and the containment of contaminants, investigating at the nano-scale level the factors controlling their performance. My interests further extend to fouling phenomena, which are a major cause for energy losses in industrial processes. I intend to expand on my current research and engineer antifouling surfaces for diverse applications, ranging from petrochemical processes to water treatment using membranes.

Teaching Interests:

As a doctoral researcher at the University of Toronto I had the opportunity to support teaching for a total period of four years in the following disciplines:

- Engineering Mathematics;
- Engineering Graphics;
- Groundwater Flow and Contamination;
- Hydrology and Hydraulics.

I thoroughly enjoyed the role of teaching assistant, and I would be thrilled to pursue my vocation for teaching in the years to follow, engaging in either advanced or undergraduate courses.

Publications

Kuldeep Wadhwa, Juul Hennissen, Sharankumar Shetty, **Erica Pensini**, Carlo Geijselaers. Influence of the substitution of functional groups on the inhibition efficiency of TEMPO analogues on vinyl polymerization. To be submitted.

Erica Pensini, Fabrice Cuoq, Roy van Lier, Wolfgang Hater, Patrick Kraft, Tobias Halthur. Enhanced corrosion resistance of metal surfaces by film forming amines: a comparative study between cyclohexanamine and 2-(diethylamino)ethanol-based formulations. Submitted to I&EC.

Erica Pensini, Leo Vleugels, Martijn Frissen, Kuldeep Wadhwa, Roy van Lier, Gerard Kwakkenbos. A novel perspective on emulsion stabilization in gasoline crackers. Submitted to Colloids and Surfaces A.

Fan Yang, Plamen Tchoukov, **Erica Pensini**, Tadeusz Dabros, Jan Czarnecki, Jacob Masliyah, and Zhenghe XuAsphaltene Subfractions Responsible for Stabilizing Water-in-Crude Oil Emulsions. Part 1: Interfacial Behaviors. Energy Fuels, 2014, 28 (11), pp 6897–6904.

Zifu Li, David Harbottle, **Erica Pensini**, To Ngai, Walter Richtering, and Zhenghe Xu. Fundamental Study of Emulsions Stabilized by Soft and Rigid Particles. Langmuir, 2015, 31 (23), pp 6282–6288.

Jiebin Bi, Fan Yang, David Harbottle, **Erica Pensini**, Plamen Tchoukov, Sébastien Simon, Johan Sjöblom, Tadek Dabros, Jan Czarnecki, Qingxia Liu, and Zhenghe Xu.Interfacial Layer Properties of a Polyaromatic Compound and its Role in Stabilizing Water-in-Oil Emulsions. Langmuir, 2015, 31 (38), pp 10382–10391.

E. Pensini, D. Harbottle, F. Yang, P. Tchoukov, Z. Li, I. Kailey, J. Behles, J. Masliyah, Z. Xu. Demulsification Mechanism of Asphaltene-Stabilized Water-in-Oil Emulsions by a Polymeric Ethylene Oxide –Propylene Oxide Demulsifier. Energy & Fuels, 2014, 28 (11), pp. 6760-6771.

E. Pensini, B.E. Sleep, C. Yip, D. O'Carrol. Effect of water chemistry and aging on iron - mica interaction forces: Implications for iron particle transport. Langmuir, 2012, 28 (28), pp. 10453–10463.

E. Pensini, B.E. Sleep, C. Yip, D. O'Carrol. Forces of interactions between bare and polymer coated iron and silica: effect of pH, ionic strength and humic acids. Environ. Sci. Technol., 2012, 46 (24), pp. 13401–13408.

E. Pensini, B.E. Sleep, C. Yip, D. O'Carrol. Forces of interaction between fresh iron particles and iron oxide (magnetite): Effect of water chemistry and polymer coatings. Colloids and Surfaces A: Physicochemical and Engineering Aspects, 2013, 433, pp. 104–110.

E. Pensini, B.E. Sleep, C. Yip, D. O'Carrol. Carboxy-methyl cellulose binding to mineral substrates: Characterization by atomic force microscopy based - force spectroscopy and quartz-crystal microbalance with dissipation monitoring. Journal of Colloid and Interface Science, 2013, 402, pp. 58-67.

E. Pensini, B.E. Sleep, C. Yip, D. O'Carrol. Forces of interaction between iron particles and aluminum-silicates: Effect of water chemistry, polymer coatings and aluminum-silicate composition. Journal of Colloid and Interface Science, 2013, 411, 8-15.

480140 Energy Solutions through Electrochemical Processing: Electronic Devices,Energy Storage Devices, and Extractive Metallurgy

Abstract:

Research Interests:

Grid scale electrochemical energy storage devices

Multi-valent batteries for automobile propulsion

Electrochemical fabrication for future electronics and optoelectronics

Metal extraction processes and production processes of semiconductor materials

Corrosion for prolonging the life of devices and processes

Research Experience:

In my graduate work at Waseda University, I developed an electrochemical fabrication process for bit patterned magnetic recording media for future hard disk drives. Bit patterned magnetic recording media consist of an ordered highly crystalline and uniform ferromagnetic nanodot array. To fabricate these arrays, I developed a combined process using nano-patterning (nano-imprinting lithography) and electrochemical depositions – electrodeposition, electroless deposition. The process enabled deposition of magnetic materials onto nano-patterned substrates at the single nano-meter scale. I designed deposition conditions by characterizing the crystal structure, magnetic properties, and morphology (nucleation and growth) of the deposits. To better understand the electrochemical deposition reactions, I developed an in-situ technique for characterizing catalytic reactions on substrates at the molecule scale (1-2 nm) using Surface Enhanced Raman Scattering effects. Through this work, I demonstrated the feasibility of electrochemical deposition processes to fabricate nano-structures with both well-controlled crystallinity and uniformity at the single nano-meter scale. I also applied these processes to fabricate Si nano-pillars from ionic liquids.

Currently at MIT, I am leading the research project of liquid metal batteries, which consist of liquid electrodes and a molten salt electrolyte, to address the need for a grid-scale energy storage device. Large-scale electrochemical energy storage devices are valuable for the future grid, enabling intermittent renewable energy technologies to be integrated into the base load and creating a variety of economic benefits via off-peak shifting, load leveling, and frequency regulation. The liquid metal battery is, in principle, a reversible electrorefining cell. In this work, I have investigated the thermodynamics and kinetics of the electrochemical charge/discharge processes at the interfaces of the molten salt electrolytes and liquid metal electrodes. As a result, the applicability of calcium metal and its molten salts for electrochemical energy storage devices was demonstrated (the first calcium metal battery). Furthermore, I identified why and how corrosion occurs in the batteries and developed a solution to prolong their lifetime. These efforts have been the basis for the practical use.

My work contributes new knowledge of the electrochemical behaviors at the interfaces of electrodes and electrolytes (nucleation and growth, alloying and dealloying) at a wide range of scales (atom ~ meter) and temperatures (room temperature ~ high temperature). Tailoring these processes to specific applications is an enduring challenge in electrochemical processes as well as in the operation of the electrochemical devices. The backgrounds behind these topics are multi-disciplinary fields, such as surface science, electrochemistry, electrometallurgy, and multi-scale fabrication processes.

Postdoctoral Project: "Liquid metal battery for grid scale energy storage"

Adviser: Donald R. Sadoway, Materials Science and Engineering, Massachusetts Institute of Technology

PhD Dissertation: "Electrochemical Fabrication of Ordered Ferromagnetic Nanostructures and Control of Initial Deposition Process"

Adviser: Takayuki Homma, Applied Chemistry, Waseda University

Teaching Interests:

As for teaching, I feel competent to handle most any course at the undergraduate and graduate levels based on my multidisciplinary research experiences and extensive teaching experience. I have taught in the Department of Applied Chemistry at Waseda University and in the Department of Materials Science and Engineering at MIT. At both Waseda and MIT, I applied hands-on and heads-on training. At Waseda, I supported several professors in the department as an instructor of undergraduate courses in physical chemistry, inorganic chemistry, and lab courses. I managed teaching assistants and was also responsible for grading exams and reports. At MIT, I worked as an instructor for an undergraduate lab course (3.014 Materials Laboratory) by designing a syllabus, creating modules, giving short lectures, and preparing exam. Furthermore, I completed teaching certificate program at MIT in 2014. In Professor Homma's research group at Waseda, I mentored several undergraduate and master's students in writing their theses while pursuing my own thesis work. In the research group of Professor Sadoway at MIT, I also mentored six undergraduates for their research projects, two visiting students for their short-term projects, and two graduate students for their PhD research.

Future Research Direction:

As a faculty I will aim to contribute to energy research program and materials research program with the success of developing energy storage devices, electronic devices, materials synthesis processes, coating technology, and metal extraction and recycling processes. To do this, I will use the following features of electrochemical processes: precisely control over the composition, crystal structure, and phases; formation of complicated 3D structures by area-selective deposition and etching; rapid production from atom to meter scale. I have the advantage of being able to set the goal of the project from the broader perspectives like in metal extraction and approach the individual issues from understanding and designing basic chemistries at the atomic and molecular levels.

Selected Publications:

1. **T. Ouchi**, H. Kim, B. L. Spatocco, D. R. Sadoway, "Calcium-based multi-element chemistry for grid-scale electrochemical energy storage," *Nature communication*, 7:10999, doi: 10.1038/ncomms10999 (2016).
2. K. Wang, K. Jiang, B. Chung, **T. Ouchi**, P. J. Burke, D. A. Boysen, D. J. Bradwell, H. Kim, U. Muecke, D. R. Sadoway, "Lithium–antimony–lead liquid metal battery for grid-level energy storage," *Nature*, 514, 348–350 (2014).
3. *3.* **T. Ouchi**, H. Kim, X. Ning, D. R. Sadoway, "Calcium-Antimony Alloys as Electrodes for Liquid Metal Batteries," *J. Electrochem. Soc.,* 161(12) A1898-A1904 (2014).
4. **T. Ouchi**, N. Shimano, T. Homma, "CoNiP Electroless Deposition Process for Fabricating Ferromagnetic Nanodot Arrays," *Electrochimica Acta*, 56, 9575-9580 (2011).
5. **T. Ouchi**, Y. Arikawa, T. Homma, "Fabrication of CoPt Magnetic Nanodot Arrays by Electrodeposition Process," *Journal of Magnetism and Magnetic Materials*, 320, (22), 3104-3107 (2008).

476252 Material Interactions and Synergies in Lithium-Air Batteries and Electrochemical Devices

Abstract:

Research Interests:

Energy storage technologies are essential to the security and sustainability of our powered infrastructure. Due to their exceptionally high energy densities, lithium-based batteries have been the beneficiary of intense research and development efforts that now underpin portable electronics, electric vehicles and (soon) the "smart" grid. Research to improve upon or disrupt state-of-the-art Li-ion battery technologies, however, has been painfully slow, especially when compared to advances in semiconductor manufacturing. The complexity of rechargeable batteries, often containing multiple active materials with different cell designs and operating under various load conditions, explains the difficulty of hewing to a straightforward design methodology. My efforts have sought to inject rationality into the development of next-generation battery technologies (Li-air/Li-oxygen systems) with various form factors (ultrathin, transparent Li-ion) by deconvoluting the influences of material interaction and applying those findings to better material selection. My work on Li-air/Li-oxygen batteries has encompassed many aspects of that technology from catalyst-electrolyte synergy, to electrolyte side reactions and anode stability, to *in situ* and *operando* studies of redox processes, and reactant transport during cell operation. My work on designing ultrathin Li-ion battery architectures has involved the development of automated, tunable self-assembly techniques with microgram precision. The experimental tools I use to understand material interactions in electrochemical systems are currently being paired with computer simulations (molecular dynamics, density functional calculations and finite element analysis) from my team at Sandia National Laboratories to build our predictive capabilities and anticipate full device behavior. This holistic approach to understanding and improving lithium batteries showcases the benefits of combining materials science, electrochemistry and chemical engineering disciplines to address design challenges in myriad electrochemical systems (i.e. capacitors, sensors, fuel cells, electrolysers).

Research Experience:

My research experience has involved a deep dive into electrochemistry and materials science with projects touching on many energy-related applications. My doctoral work with Prof. André Taylor at Yale University explored the cutting-edge chemistry of catalysts in Li-oxygen systems as well as nanomanufacturing of ultrathin batteries by layer-by-layer (LbL) self-assembly. I have collaborated on projects including the development of metallic glass electrocatalysts for polymer exchange membrane (PEM) fuel cells, electrodes for microbial fuel cells, as well as electrochemical sensors and conductive thin films. At Sandia National Laboratory, I serve as a co-principal investigator and lead experimentalist for our Li-air battery development team where I interface with computational team members to develop predictive tools for material selection. My experimental focus at Sandia encompasses electrolyte development, particularly involving ionic liquid design, and device prototyping.

Teaching Interests:

My experiences as a teacher and mentor have had a significant impact on my career goals and research program. For three semesters, I served as a teaching assistant in the Department of Chemical and Environmental Engineering at Yale University in the subjects of environmental engineering and professional ethics. Through these roles, which involved teaching weekly discussion sessions and critiquing student reasoning, I gained insight into methods for introducing and refreshing concepts throughout a term. I delivered guest lectures in graduate courses and an annual Climate and Energy symposium, where I sought to relate new concepts to my audience's non-technical knowledge. Most influential to my development has been the experience of mentoring numerous undergraduate students in a lab setting, four of whom I worked with for greater than one year and three of whom are cited as co-authors on peer-reviewed publications. With these students, I developed techniques for initiating newcomers into the discipline of electrochemistry and formulated a groundwork for skill development and research contributions. After my mentees transitioned to graduate programs, a battery startup company and the finance industry, I have continued my support. Mentoring fellow graduate students in my group and serving as a regular editor and internal reviewer for their manuscripts has further extended my impact with junior colleagues. I see mentorship and one-on-one interaction with students as critical to ensure difficult concepts are understood and retained. My future teaching will retain this passion for student growth and the formation of strong mentor/mentee relationships.

Successful Proposals:

Porous Liquid Electrolytes for Metal-Air Batteries, Laboratory Directed Research and Development (LDRD), Sandia National Laboratories, 2016 – Principal Investigator

Selected Publications (of 18):

- Ryu, W.-H.; **Gittleson, F.S.**; Li, J.; Tong, X.; Taylor, A.D. A New Design Strategy for Observing Lithium Oxide Growth-Evolution Interactions Using Geometric Catalyst Positioning. *Nano Lett.*, **2016**. (*in press*)
- **Gittleson, F.S.**; Ryu; W.-H.; Schwab, M.; Tong, X.; Taylor, A.D. Pt and Pd Catalyzed Oxidation of Li_2O_2 and DMSO during Li-O_2 Battery Charging. *Chem. Comm.*, **2016**, *52*, 6605-6608.
- **Gittleson, F.S.**; Hwang, D.; Ryu, W.-H.; Hashmi, S.M.; Hwang, J.; Goh T.; Taylor, A.D. Ultrathin Nanotube/Nanowire Electrodes by Spin-Spray Layer-by-Layer Assembly: A Concept for Transparent Energy Storage. *ACS Nano.*, **2015**, *9*, 10005-10017.
- Ryu, W.-H.; **Gittleson, F.S.**; Schwab, M.; Goh, T., Taylor, A.D. A Mesoporous Catalytic Membrane Architecture for Lithium-Oxygen Battery Systems. *Nano Lett.*, **2015**, *15*, 434-441.
- **Gittleson, F.S.**; Ryu, W.-H.; Taylor, A.D. *Operando* Observation of the Gold-Electrolyte Interface in Li-O_2 Batteries. *ACS Appl. Mater. Inter.*, **2014**, *6*, 19017-19025.
- **Gittleson, F.S.**; Sekol, R.C.; Doubek, G.; Linardi, M.; Taylor, A.D. Catalyst and Electrolyte Synergy in

$Li\text{-}O_2$ Batteries. *Phys. Chem. Chem. Phys.*, **2014**, *16*, 3230-3237.
- **Gittleson, F.S.**; Hwang, J.; Sekol, R.C.; Taylor, A.D. Polymer Coating of Vanadium Oxide Nanowires to Improve Cathodic Capacity in Lithium Batteries. *J. Mat. Chem. A*, **2013**, *1*, 7979-7984.
- **Gittleson, F.S.**; Kohn, D.; Li, X.; Taylor, A.D. Improving the Assembly Speed, Quality, and Tunability of Thin Conductive Multilayers. *ACS Nano*, **2012**, *6*, 3703-3711.

476276 Life at Interfaces: Understanding the Fluid Dynamics, Transport and Surface Translocation of Bacterial Biofilms

Abstract:

2ndYear Postdoctoral Fellow

Research Interests:

In many natural environments, bacteria collectively aggregate to exist in densely populated and structurally diverse communities called biofilms. Within these biofilms, cells exhibit a remarkable division of labor, with various subpopulations of cells involved in active motility by flagellar propulsion, production of extracellular polymeric substances and other functions that collectively results in the initiation, growth, propagation, adhesion and dispersion of biofilms to specific chemical and environmental signals.

By using a combination of experimental techniques such as time-lapse microscopy, constructing fluorescent transcriptional reporters for distinct cell types, confocal microscopy, flow visualization, surface texturing and chemical patterning, my research systematically investigates the collective response that governs the bacterial growth and surface interaction of biofilms under a number of different environmental conditions (such as temperature, pH, agar gel concentration, etc.) with ***Bacillus Subtilis*** as the model organism.

In addition, I also work on theoretically and computationally modeling the stability of thin viscous sheets under shear with a non-uniform temperature gradient. Such a model has practical applications ranging from the manufacture of ultra-shin glass sheets via the float glass technique, to understanding physical and biological phenomena such as the buckling of bacterial pellicles during colony growth, and geological processes involving wrinkling of lava sheets.

Postdoctoral Projects:

(Under the supervision of ***L. Mahadevan*** & ***S. Rubinstein***, School of Engineering & Applied Sciences, Harvard University)

- Buckling instabilities in thin viscous sheets during extensional flows (theory & simulation)
- Characterizing growth & surface translocation in developing ***Bacillus Subtilis*** biofilms (experimental)

PhD Dissertation: "*Exploring Plastron Stability and Fluid Friction Reduction on Robust Micro-Textured Non-Wetting Surfaces.*"

Under the supervision of Robert E. Cohen and Gareth H. McKinley, Department of Chemical & Mechanical Engineering, Massachusetts Institute of Technology.

Research Experience:

My PhD work involved both experimental and theoretical research at the interface of polymer physics, fluid mechanics, wetting and soft matter. In my thesis, I developed a precise and repeatable spray technique for rapid fabrication of drag- reducing liquid-repellent coatings that can be deployed on a large scale. I used these coatings to study the fundamental problem of how structured liquid-repellent surfaces, which support a composite liquid-air interface, can reduce skin friction in both viscous laminar and turbulent flows in a Taylor-Couette setup, and developed a theoretical model that accounts for the effect of wall slip in near-wall turbulent flows. In addition to this central experimental theme, I have also worked on a mathematical analysis of the gravity driven deformation of drops on non-wetting surfaces, correlating the wetting dynamics on the barbs and barbules of aquatic birds with diving and wing-spreading behavior, and on developing a hierarchical model to design oil-repellent woven fabrics.

Teaching Interests:

During my education at MIT, I was a Teaching Assistant for the Graduate Transport Phenomena Course. I organized weekly help sessions and presented extra lectures for a class of 50 graduate students on subjects including fluid mechanics, heat/mass transfer, partial differential equations and mathematical methods. I have also directly supervised undergraduate & graduate research students on specific projects during my postdoctoral fellowship. I am interested teaching modules in fluids mechanics, transport phenomena, solid mechanics, heat transfer, mass transfer and applied engineering mathematics, amongst other courses such as thermodynamics and statistical mechanics.

Future Direction:

As a research faculty, I would like to continue the philosophy of using experimental and theoretical tools to understand the physicochemical hydrodynamics in complex systems. In this regard, I believe the bacterial biofilm system I am currently studying provides a rich experimental toolbox, with a number of interesting and currently unexplored phenomena to investigate. Specific examples include studying the biofilm rheology and mechanical behavior (***e.g.***, the viscoelastic characterization of the extracellular polymer matrix), fluid dynamics (directional flows of surfactant laden films generated by the bacteria during swarming) to surface interactions (influence of structured coatings on bacterial growth)

The uniqueness of my approach is the ability to perform direct in-situ measurements and quantify the dynamic response within the biofilm (such as spatial localization & switching between various cell types, observing transient response to antibiotics and external stresses, measurement of velocities during swarming, etc.) both at a single cell level and at a colony scale, and directly test and develop biophysical mechanisms governing growth and spreading. The ultimate goal is in designing chemically patterned & textured surfaces that selectively inhibit the growth of living matter on surfaces, and has a number of potential applications such as in preventing biofouling in marine vehicles, or spreading in infectious tissue, etc.

Due to the interdisciplinary nature of my work, I envision a strong and active collaboration with fellow faculty in engineering and the basic sciences and/or work with them to setup specific techniques (such as constructing reporter strains, etc.) in my lab.

Selected Publications:

Srinivasan, S., Wei, Z., Mahadevan, L. (**2016**). Buckling instability in extensional flow of a thin non-uniformly viscous sheet. ***(under review)***

Wang, X., Koehler, S. A., Wilking, J. N., Sinha, N. N., Cabeen, M. T., **Srinivasan, S.**, ... others. (**2016**). Probing phenotypic growth in expanding Bacillus subtilis biofilms. ***Applied Microbiology and Biotechnology***, ***100***(10), 4607–4615.

Srinivasan, S., Kleingartner, J. A., Gilbert, J. B., Cohen, R. E., Milne, A. J. B., & McKinley, G. H. (**2015**). Sustainable drag reduction in turbulent Taylor-Couette flows by depositing sprayable superhydrophobic surfaces. ***Physical Review Letters***, ***114***(1), 14501.

Kleingartner, J. A*., **Srinivasan, S.***, Truong, Q. T., Sieber, M., Cohen, R. E., & McKinley, G. H. (**2015**). Designing Robust Hierarchically Textured Oleophobic Fabrics. ***Langmuir***, ***31***(48), 13201–13213.

Srinivasan, S., Chhatre, S. S., Guardado, J. O., Park, K.-C., Parker, A. R., Rubner, M. F., Mckinley, G.H, Cohen, R. E. (**2014**). Quantification of feather structure, wettability and resistance to liquid penetration. ***Journal of The Royal Society Interface***, ***11***(96), 20140287.

Srinivasan, S., Choi, W., Park, K.-C., Cohen, R. E., & McKinley, G. H. (**2013**). Drag reduction for viscous laminar flow on spray-coated non-wetting surfaces. ***Soft Matter***, ***9***(24), 5691–5702. (featured on cover)

Srinivasan, S., Chhatre, S. S., Mabry, J. M., Cohen, R. E., & McKinley, G. H. (**2011**). Solution spraying of poly (methyl methacrylate) blends to fabricate microtextured, superoleophobic surfaces. ***Polymer***, ***52***(14), 3209–3218.

Srinivasan, S., McKinley, G. H., & Cohen, R. E. (**2011**). Assessing the accuracy of contact angle measurements for sessile drops on liquid-repellent surfaces. ***Langmuir***, ***27***(22), 13582–13589.

476294 Triggerable Tough Hydrogels for Gastrointestinal Biomedical Applications

Abstract:

Research Interests:

My research interests mainly include polymer, biomaterials, nanomaterials, biomedical devices, drug delivery and nanomedicines. In this presentation I will focus on the gastrointestinal biomedical applications of hydrogels.

The introduction of gastric-resident dosage forms has demonstrated significant promise for enhanced drug therapy in humans. However, the lack of precise control over the gastric-retention time remains an important challenge for clinical adoption of these dosage forms. Here we report the development of novel triggerable tough hydrogels (TTHs) that not only show exceptional toughness to overcome gastric pressures achieving long-term residence in stomach but also exhibit remarkable triggerability to be dissolved rapidly with biocompatible agents when required. We demonstrate a proof of concept extended drug delivery from TTHs in a large animal model. Given the dramatic mechanical properties and capable of on-demand dissolution of TTHs we anticipate their implementation in a broad of gastrointestinal biomedical applications.

Teaching Interests:

I have teaching experiences in general material science and chemistry. I would like to design two new courses: a biomaterials course and a nanomaterials course. Both of them can be offered to graduate students or senior undergraduate students.

458653 Human Stem Cell Derived Neutrophils As a Primary Neutrophil Model

Abstract:

As the first responders and the most abundant innate immune cell, neutrophils are a crucial part of the immune system. During an immune response, neutrophils migrate to sites of infection and eliminate bacteria through a series of anti-microbial functions including phagocytosis, production of reactive oxygen species, and NETosis. Improper regulation of neutrophil behavior contributes to many human diseases. Improper neutrophil function leads to chronic infections when neutrophils do not arrive at a site of infection or chronic inflammation when neutrophils persist at a site of infection. Current models for evaluating neutrophil behavior *in vitro* include cell lines and primary human neutrophils. Unfortunately, cell lines do not always accurately model many important neutrophil functions such as NETosis and phagocytosis and primary human neutrophils are hard to genetically manipulate. Recently, stem cell derived neutrophils have emerged as a possible model for primary neutrophils that would allow for genetic manipulation and the modeling of human disease.

In order to properly model primary human neutrophils, hiPSC-derived neutrophils must display the same morphology, migration, and anti-microbial behavior as primary human neutrophils. We have found that the morphology of hiPSC-derived neutrophils is indistinguishable from primary human neutrophils. Furthermore, hiPSC-derived neutrophils have the characteristic tri-lobed nucleus common to primary neutrophils. In order to reach sites of inflammation neutrophils must be capable of migrating to known neutrophil chemoattractants. We studied neutrophil chemotaxis using a previously published microfluidic device that creates a stable chemokine gradient and found that hiPSC-derived neutrophils efficiently chemotax to a source of the bacterial peptide fMLP, a common neutrophil chemoattractant. Finally, hiPSC-derived neutrophils must be able to perform anti-microbial functions. We visualized hiPSC-derived neutrophils in co-culture with *pseudomonas aeruginosa* using time-lapse microscopy. In contrast to popular neutrophil cell lines that lack the ability to phagocytose bacteria, we found that hiPSC-derived neutrophils were capable of efficiently phagocytosing *pseudomonas* over an extended period of time. We have also found that, upon activation, hiPSC-derived neutrophils produce neutrophil extracellular traps, proving these cells are capable of performing several anti-microbial functions.

Overall, we have found that hiPSC-derived neutrophils accurately recapitulate many characteristics and can serve as an excellent model of primary human neutrophils. We have shown that hiPSC-derived neutrophils share the morphology, migration capabilities, and anti-microbial functions of primary human neutrophils. hiPSC-derived neutrophils have the characteristic tri-lobed nuclei, are capable of chemotaxis to fMLP, a physiologically relevant chemokine, and poses the ability to phagocytose bacteria. In the future, hiPSC-derived neutrophils as a model will allow for more in-depth study of primary human neutrophil function.

463229 Mechanistic Model of CD3ζ Immunoreceptor Tyrosine-Based Activation Motif (ITAM) Phosphorylation Sequence

Abstract:

Gaining a detailed understanding of T cell activation has become increasingly important as T cells are used in immunotherapeutic applications. For example, T cells engineered with modified T cell receptors (TCRs) or CD3ζ-bearing chimeric antigen receptors (CARs) have emerged as promising cancer therapies. The CD3ζ chain, a member of the TCR complex, contains three immunoreceptor tyrosine-based activation motifs (ITAMs), which are thought to have different signaling functions: the first N-terminal ITAM is largely stimulatory, while the second and third ITAMs each exhibit some inhibitory signaling characteristics [1]. Lymphocyte-specific protein tyrosine kinase (LCK) is able to phosphorylate two tyrosine residues on each ITAM, resulting in T cell activation. The "kinetic proofreading" hypothesis is the prevailing theory regarding regulation of ITAM phosphorylation. It presumes that the presence of multiple ITAMs on the CD3ζ chain allows T cells to differentiate between foreign and self-peptides; however, the specific mechanisms that control LCK catalytic activity and its phosphorylation of the CD3ζ chain are still not well defined. Therefore, we have developed a mechanistic computational model of CD3ζ phosphorylation by LCK. The model is paired with site-specific mass spectrometry phosphorylation data from both CD3ζ and LCK *in vitro* to better understand the key steps that initiate T cell activation.

We constructed the model using BioNetGen, a rule-based formalism that allows us to account for the many species that arise from the multiple phosphorylation sites on CD3ζ. The model is implemented as a set of ordinary differential equations in MATLAB and fit to experimental data generated using an *in vitro* reconstituted membrane system that mimics the two-dimensional interactions that occur in T cells [2]. We use mass spectrometry to measure the level of phosphorylation at each individual tyrosine residue in the system over time. To determine the order in which each of the six tyrosine residues on CD3ζ are phosphorylated, we first stimulate CD3ζ with a constitutively active form of LCK and fit our mechanistic model to this data. We then combine this model with a previously developed model of LCK autoregulation [3]. The combined model will be validated using experimental data of wild-type LCK autophosphorylation and phosphorylation of CD3ζ. As a result, we generate a model that matches experimental data and can predict data not used in the fitting.

The model predicts the order and rate at which different ITAMs on the CD3ζ chain are phosphorylated, helping to decifer the kinetic proofreading mechanism that is thought to precisely regulate T cell activation. Additionally, we have applied the model to investigate how changes to the CD3ζ chain, such as removing the third ITAM, can affect overall activation. The results predicted by the model can be implemented in CD3ζ-bearing CAR-engineered T cells. Thus, the model generates new hypotheses that can be tested experimentally, allowing us to quantitatively explore T cell signaling and guide the development of immunotherapies. The model is a quantitative framework that can be used to examine the dynamics of CD3ζ chain activation by LCK and its effects on T cell activation.

Citations

[1] Chae, W., et al. *Int. Immunol.*, 2004, 16(9), 1225-36.

[2] Hui, E. and R. Vale. *Nat. Struct. Mol. Biol.*, 2014, 21(2), 133-142.

[3] Rohrs, J. A., Wang, P., and S. D. Finley. *Cell Mol. Bioeng.*, 2016, In press.

463718 Adoptive Transfer of CAR-Engineered T Cells with Surface-Conjugated Synthetic Nanoparticles Containing Small Molecule Inhibitors forÂ Reversing Intratumoral T Cell Hypofunction

Abstract:

Adoptive T cell transfers of genetically modified cytotoxic T cells to express chimeric antigen receptors (CAR) have become a promising immunotherapy method. CARs specifically target tumor-associated antigens or cancer stromal antigens with high binding affinity. Activation of a CAR engineered T cell is independent of the major histocompatibility complex (MHC) and is achieved primarily through the single chain variable fragment (scFv) that is fused with intracellular signaling domains. Multiple researchers have shown that adoptive transfer of CAR T cells is successful in patients with B cell haematological malignancies, but is still in the earlier stages of development for treatment of solid tumors. One limiting factor to adoptive T cell therapy is the suppressive tumor microenvironment that inactivates tumor infiltrated T cell (TIL) function. Moon et al. (2014) showed that hypofunctional TILs express high levels of inhibitors of the intracellular pathways, such as the up-regulation of DGKa and DGKz that phosphorylates diacylglycerol. In addition to intracellular inhibitors, the tumor microenvironment also contains high concentration of TIL suppressor molecules such as adenosine that is up taken by the A2AR receptor expressed on the cell surface of CD4 and CD8 T cells. Adenosine is generated from extracellular ATP through CD39 and CD73 expressed on the surface of tumor cells and regulatory T cells. We hypothesized that *in vivo* delivery of pharmaceutical drugs, namely DGK and A2AR inhibitors, with crosslinked multilayer liposome vesicles (cMLV) would reverse TIL hypofunction. For delivery, cMLVs containing drugs were conjugated with adoptively transferred CAR T cells to target hypofunctional TILs residing in the tumor. We demonstrated that conjugation of cMLVs to CAR T cells does not alter key functions, such as T cell cytokine secretion and cytotoxicity. Furthermore, our *in vivo* study showed that combination therapy of small molecule inhibitors with CAR T cells were able to reverse TIL hypofunction and increase tumor infiltrating CD3+ T cells.

464271 Sequencing the Evolving Functional Antibody Repertoire in Rheumatoid Arthritis

Abstract:

Combining large-scale antibody repertoire sequencing data with a comprehensive analysis of antibody function can provide a more complete understanding of the role of an antibody response in disease pathology. For example, a hallmark of rheumatoid arthritis (RA) is the generation of an adaptive autoimmune B cell response against citrullinated antigens that result from a post-translational modification of arginine. However, the precise specificity and functional properties of the anti-citrullinated protein antibodies (ACPAs) in RA remain poorly understood. To address this, we applied a high-throughput single-cell barcoding methodology to sequence the entire variable region of paired heavy and light chain antibody genes expressed by individual plasmablasts (activated B cells). Bioinformatics analysis of these sequencing results precisely quantifies the degree of clonal expansion and identifies the key affinity matured, clonally expanded antibodies generated in an immune response. Importantly, with the paired heavy and light chain data, we can express these antibodiesÂ *in vitro*, thereby enabling the direct linkage of a patient-derived antibodyâ€™s sequence with its function (e.g. antigen specificity). Furthermore, we evaluated the antibody repertoire over time to characterize the evolving adaptive autoimmune response, elucidate the key autoantibodies, and provide insight into the pathogenic mechanisms underlying RA. We sequenced single plasmablasts from peripheral blood sampled at serial time points from 8 individuals with RA and used a tetrameric citrullinated peptide-based sort reagent to isolate B cells that produce ACPAs. We bioinformatically identified lineages that persisted across the serial time points and characterized the expansion and contraction of clonal lineages, some of which expressed multiple antibody isotypes, in relation to clinical metrics (e.g. swollen and tender joints). We selected antibodies representative of the identified B cell clonal families for recombinant expression and defined their antigen targets with planar and bead-based antigen arrays. By combining detailed sequence information with antibody function, this broadly applicable approach facilitates a comprehensive investigation of the evolution of the antibody response in disease, which can reveal novel diagnostic markers and therapeutic targets.

466794 Elucidating T-Cell Activation Threshold Using Precisely Defined Three-Dimensional Artificial Antigen Presentation

Abstract:

Elucidating T-Cell Activation Threshold Using Precisely Defined Three-Dimensional Artificial Antigen Presentation

Mason Smith, Fei Wen Adaptive immune response is shaped in large part by T cells recognizing and binding cognate antigen on the surface of antigen presenting cells (APCs). While much is known about the behavior of T cells during an immune response, the exact molecular mechanism that translates peptide-major histocompatibility complex (pMHC)- T-cell receptor (TCR) binding to T cell activation remains a controversial topic in immunology. Several studies using antigen presenting fluid lipid bilayers have attempted to shine a light on this mechanism by looking at the minimum number of pMHC required to initiate T cell activation. These studies have found that the early steps of T cell activation (TCR triggering) can be *initiated*by as few as three agonist pMHCs in a membrane, and that there exists a global pMHC density threshold below which T cell activation does not occur. While these studies have provided unique insights into TCR triggering, their predicted T-cell activation thresholds vary in planar intermolecular pMHC spacing from 34 nm to greater than 150 nm. Further, conflicting results regarding the importance of local pMHC density and total pMHC encountered have been reported. Planar fluid lipid bilayers provide robust and scalable experimental systems with incredibly fine control over molecular density; however, it is not immediately clear if their results can be extrapolated to describe T cell response to pMHC on a three-dimensional cell surface. Additionally, planar pMHC display systems assume sustained interfacial contact between a T cell and a flat antigen-presenting surface and are typically characterized by some uncertainty in the number of pMHC molecules immobilized in each defined nanoclusters.

We have engineered a novel artificial antigen presentation system that provides precise control over the local and global arrangement of pMHC molecules on the surface of a living cell using recombinant scaffold protein. Using this artificial antigen presentation system, we are able to display thousands of clusters of up to six individual pMHCs with single-molecule precision on a cell surface. Further, cells presenting unique local pMHC densities can be sorted using fluorescence activated cell sorting (FACS) to isolate cells presenting different global pMHC display levels. By using FACS to isolate cells with the same global pMHC display level but discretely different pMHC cluster size, we can study T cell response to a wide range of local and global pMHC densities in a three-dimensional cell-culture environment. Preliminary T-cell activation results using sorted artificial antigen presenting cells support the existence of a minimum global pMHC display threshold for T cell activation, which is different from that reported using planar lipid bilayers. More importantly, they suggest that local pMHC valency can compensate for a reduction in the global pMHC display level. These results suggest that while two-dimensional systems for *in vitro* T cell activation present an important, scalable research platform, quantitative thresholds for T cell activation derived from these systems may underestimate the activation threshold in physiologically relevant systems.

470621 Development of a Virus-like Particle Based HIV Vaccine Candidate

Abstract:

Worldwide, 35 million people are infected with HIV. Antiretroviral therapies have helped to decrease the mortality rate, but their expensive nature makes them inaccessible to many millions of clinically eligible individuals. Despite over 30 years of intense efforts, an HIV vaccine is still not in sight. A prophylactic HIV vaccine would prevent the infection and do so at a much lowered cost, thus increasing access, particularly in the developing world. Whole-killed vaccines, in which the virus is physically destroyed by chemicals, heat or radiation, are used for many other diseases, but do not lead to a relevant immune response against HIV. The mutation rate of HIV is very high, so variant strains that are resistant to vaccines prepared against individual strains emerge rapidly.

Our approach aims to create an HIV vaccine candidate targeting the process by which the virus infects cells. In order to infect a new cell, the HIV membrane must fuse with the cell membrane. This membrane-fusion event is mediated by a specific protein on the surface of the virus, called gp41, which is highly conserved at both the nucleotide and amino acid levels. Earlier work in the field has elucidated the overall features of this membrane-fusion event, including an intermediate state of gp41 called the pre-hairpin intermediate (PHI). An FDA-approved drug, Fuzeon, binds to the PHI, preventing membrane fusion and blocking HIV infection. Additionally, a monoclonal antibody targeting this PHI (called D5) has been shown to prevent infection of cells by various strains of HIV. Together, these results validate our approach to use the PHI as a vaccine target.

Our proposed vaccine candidate consists of an engineered virus-like particle scaffold displaying both a PHI mimetic and stimulators of the innate immune system. Using the engineered virus-like particle scaffold results in a vaccine candidate with the correct size for efficient trafficking to the lymph nodes. We have produced a PHI mimetic in a scalable cell-free protein synthesis system and have optimized the attachment of the PHI mimetic to the virus-like particle in an ordered, repetitive fashion similar to that of the natural virus. Co-attached innate immune stimulators will act as a non-denaturing adjuvant to further ensure a strong protective response. Consequently, our PHI mimetic will provide a broadly protective immune response to many subtypes of HIV. Promising vaccine candidates will be evaluated in guinea pigs for serum responses which provide viral neutralization in an *in vitro* assay.

449315 Correlations of Antibody Response Phenotype and Genotype Revealed By Immunoglobulin Repertoire Sequencing

Abstract:

It has long been possible to measure the phenotype of antibody responses (antigen-specific titers) through conventional serological assays (e.g., ELISA). In contrast, the ability to measure the genotype of antibody responses has only recently become possible through the advent of high-throughput antibody repertoire sequencing (Ig-seq), which provides quantitative molecular information on clonal expansion, diversity and somatic hypermutation. However, Ig-seq is compromised by the presence of bias and errors introduced during library preparation and sequencing and thus prevent reliable immunological conclusions from being made. By using synthetic antibody spike-in genes, we determined that Ig-seq data overestimated antibody diversity measurements by up to 5000-fold and was less than 60% accurate in clonal frequency measurements. To overcome this, we developed molecular amplification fingerprinting (MAF), which uses unique molecular identifier-tagging before and during multiplex-PCR amplification, enabling tagging of transcripts while accounting for PCR efficiency. Our method allowed us to develop a novel algorithm for bias correction, which improved clonal frequency measurements to an accuracy of up to 99%. Furthermore, we demonstrated nearly absolute error correction (98-100%) of full-length antibody diversity measurements. In order to probe the relationship between antibody genotype and phenotype, we have performed extensive *in vivo* experiments from immunized mice. Specifically, we modulated titer (phenotype) by varying the number of immunizations (boosts) and in parallel performed MAF Ig-seq to assess genotype through a series of quantitative metrics. These metrics describe clonal selection, expansion, and somatic hypermutation. Importantly, we also used different protein antigens, this allowed us to determine how epitope complexity impacted the genotype of antibody repertoires. Finally, by applying multivariate modeling using MAF-corrected data, we were able to predict the immune status of individual antibody clones. This extensive systems-based analysis demonstrates how accurate Ig-seq provides insight on the genotype and phenotype of humoral immunity.

470102 Regulation of Intracellular Delivery through Peptide-Based Nanocarrier Design (Invited Talk)

Abstract:

Nucleic acid cargoes offer unmatched diversity in gene regulatory potential and therapeutics, and understanding of nucleic acid functionality continues to expand rapidly and dramatically through seminal discoveries including RNA interference approaches and gene editing technologies. In nature, the basis for gene regulation is ultimately encoded by the exquisite specificity with which cells are able to control both the location and accessibility of nucleic acid constructs to govern their activation states. My research program seeks to understand and control gene activation using synthetic constructs through nature-inspired approaches to control and quantify cell binding interactions and stability in polymer and peptide nanocarriers. The basis of our approaches is the design of stimuli-responsive polymers and peptides whose interactions with nucleic acids and cells can be controlled dynamically by specific intracellular or external triggers. We exploit our ability to control nucleic acid binding/release and cellular processing to gain new mechanistic insights over nucleic acid delivery, leading to design advances including histone-inspired DNA targeting, light-responsive gene silencing, and collagen turnover-stimulated gene expression. This talk will highlight two ways we have used nature-inspired peptides to control gene transfer in regenerative medicine.

Our approaches are exemplified by our work in histone-targeted nanocarrier design. Histones have received great interest as potential gene carriers for several decades due to their seminal role in chromatin packaging and gene transfer, yet therapeutic efforts with histones have lacked both a well-controlled materials approach and a deeper knowledge of cellular processing mechanisms. Hence, histone-based carriers have failed to reach clinical efficacy. We have capitalized on newly recognized and highly pivotal roles for histone tails in native gene regulatory control to develop a gene transfer method that utilizes native, histone-based processing pathways *via* incorporation of post-translationally modified (PTM) histone tails within controllably-assembled DNA vehicles (polyplexes). Our efforts proved that polyplexes displaying PTM-modified histone tails promote nuclear accumulation, DNA release, transcription, and enhanced transfection. Moreover, our group has combined detailed nanostructure engineering with sophisticated cellular imaging to identify novel aspects in the cell biology framework regulating polyplex transport to the nucleus.

We have also focused on novel mechanisms to exploit natureâ€™s ability to harness extracellular matrix (ECM) proteins such as collagens to sequester and control delivery of bioactive nanostructures. Our specific approaches have capitalized on a class of peptides known as collagen-mimetic peptides, or CMPs, that have been recognized for their unique affinity for native collagen, which can be tailored through alterations in CMP amino acid sequence and molecular weight. CMPs incorporate themselves into the natural collagen triple helical structure via strand invasion, in a reversible process previously that has been used to modify extracted collagen *in vitro* and exclusively target remodeling collagen *in vivo*. In our studies, we employed a proline-rich CMP designed to act not only as an adjustable tether to regulate collagen-polyplex affinity, but also as an adhesive/endocytic ligand for polyplexes. The use of a collagen scaffold afforded our system structural support and innate bioactivity to encourage cellular ingrowth and proliferation, whereas altering the extent of the modification of our vector provided additional tunability to allow tailorable release for prolonged time periods. This CMP-based approach also consistently and fully maintained polyplex activity in the presence of serum for at least a week, whereas most bolus and substrate-mediated gene delivery approaches report rapid reductions within hours or a few days, and the level of transgene expression directly correlated with MMP-concentrations and the extent of collagen remodeling, demonstrating â€œon demandâ€ release. The ability to tailor release over extended periods via physical attachments, combined with the ability to provide cell-trigger release and collagen-mediated uptake, make this approach very attractive for many applications in regenerative medicine.

470495 Role of SNP Characteristics on the Endocytosis and Intracellular Trafficking of siRNA

Abstract:

Short interfering RNA (siRNA) therapeutics, have garnered considerable attention for their potential to treat a wide variety of disease-associated proteins through the use of RNA interference (RNAi). siRNA therapeutics have been developed for cancers, genetic disorders, and infectious diseases, but their progress through clinical pipelines has been limited by delivery vehicles that are toxic, immunogenic, or ineffective. Despite the variety of delivery vehicles that have been developed, the essential chemical and physical characteristics of effective delivery vehicles remain unknown. It is also unclear if RNAi is initiated more potently following endocytosis by particular mechanisms and if this variability is specific to a certain cell type or delivery vehicle chemistry.

Our goal is to define nanoparticle characteristics that are essential for potent siRNA delivery. Using silica nanoparticles (SNPs) for their tunable synthesis, we are investigating delivery criteria in four main categories: sequence specific mRNA degradation (silencing), membrane translocation, endosomal escape, and association with RNAi proteins. Using SNPs, vehicle size, structure, charge, and surface functionalization can be varied to determine optimal vehicle design criteria in a variety of cell types.

The presentation will discuss our novel live cell assay that allows independent intracellular trafficking of delivery vehicle, antisense, and sense siRNA. Using confocal microscopy, we track vehicle-siRNA complexes during endocytosis (whether by Clathrin, Caveolin, ARF6, GRAF1, Flotillin, and Macropinocytosis), endosomal trafficking (Early, Late, Recycling, and Lysosome), trafficking to the endoplasmic reticulum or golgi, and exocytosis. Our results to date demonstrate that multiple vehicle characteristics affect siRNA delivery and demonstrate the critical role of intracellular trafficking in maximizing siRNA potency.

462951 Engineering Periodic shRNA Delivery Systems with High Silencing Efficacy

Abstract:

Challenges in the delivery of small interfering RNAs (siRNAs) to disease sites have hampered clinical translation of RNA interference, particularly in cancer therapeutics. Due to its low valency and high rigidity, siRNA is not readily condensed into stable nanoparticles that can successfully protect it from nuclease degradation and deliver it into target cells. Periodic short hairpin RNAs (p-shRNAs), consisting of repeating RNAi sequences linked together, can potentially address these delivery barriers by improving complexation with cationic delivery materials. Ranging from under 200 to over 5000 nucleotides in length, p-shRNA can be generated in large quantities through rolling circle transcription, in which an RNA polymerase repeatedly traverses a small circular DNA template. We have engineered these p-shRNA molecules via template design and selective enzymatic digestion of their hairpin loops to generate an open-ended p-shRNA (op-shRNA) that induces greater gene silencing than p-shRNA and siRNA in multiple cancer cell lines up to nine days.

The high valency and flexibility of op-shRNA dramatically improves complexation with low molecular weight polycations compared to siRNA. To develop a delivery vehicle optimized for the biophysical properties of op-shRNA, we have synthesized a library of biodegradable poly(beta-amino ester)s (PBAEs) via factorial design. Our designed PBAEs condense op-shRNA into compact nanoparticles below 100 nm in diameter, and demonstrate high transfection efficiencies at polymer and RNA concentrations much lower than typically required for PBAE-mediated gene delivery. By varying the molecular weight and incorporation of alkyl chains in our PBAE library, we were able to develop correlations between these parameters and multiple parameters influencing op-shRNA delivery, including complex assembly, stability, intracellular disassembly, cellular uptake, and cytotoxicity. In particular, PBAEs of low to intermediate molecular weight and increasing alkyl chain content delivered op-shRNA *in vitro* most efficiently, generating levels of gene knockdown comparable to those by commercial transfection reagents. Through nucleic acid engineering and rational carrier design, we have successfully developed a stable, potent RNAi delivery platform that can trigger significant gene silencing at low doses, and potentially enable higher therapeutic efficacy *in vivo*.

460542 Polypeptide/Nucleic Acid Complexes As Delivery Vehicles

Abstract:

Electrostatically driven polymer self-assembly mechanisms, such as polyelectrolyte complexation, are vastly underexplored compared to amphiphilic-based assemblies, and yet offer unique opportunities for both the encapsulation of charged therapeutics and controlled delivery, via the tailoring of intermolecular interactions using pH and salt. Nanoscale polyelectrolyte complexes micelles can be created using block-copolymers that incorporate a neutral, yet hydrophilic block to stabilize the polyelectrolyte complex on the nanoscale. The modular nature of these assemblies enables the addition of a targeting ligand conjugated to the hydrophilic block to increase the efficacy of delivering nucleic acids, creating polyelectrolyte complex micelles with a targeting ligand outside the corona of the micelle. Ultimately, we demonstrate the use of targeted polyelectrolyte complex micelles formed using therapeutic nucleic acids that inhibit microRNAs involved in atherogenesis, with the goal of creating a treatment for atherosclerosis. In doing so, we create trifunctional polymers that contain a targeting peptide (that target fibrin or endothelial cells), a polyethylene glycol domain to facilitate microphase separation, and a polylysine domain used to complex the microRNA inhibitors. We characterize these nanoscale assemblies and proceed to explore the ability of these micelles to inhibit microRNA-92a, which regulates the expression of endothelial transcription factors involved in the progression of atherosclerosis. Also we explore the ability to inhibit microRNA-33a in macrophages, which is involved in the expression of ATP binding cassette transporters involved in the trafficking of cellular cholesterol to the liver. Ultimately, we show the effectiveness and biodistribution of our targeted micelle assemblies using a murine model.

462124 Highly Potent mRNA Delivery In Vivo with Intravenously-Administered Ionizable Lipid Nanoparticles

Abstract:

Introduction: Messenger RNA (mRNA) has tremendous therapeutic applications, including protein replacement therapy, cancer immunotherapy, and genomic engineering. However, successful intracellular delivery of mRNA remains challenging, as nucleic acids cannot readily cross cellular membranes alone due to their relatively large size, negative charge, and hydrophilicity. Over the past decade, researchers have developed lipid nanoparticles (LNPs) to encapsulate and deliver small interfering RNAs (siRNAs) to silence protein expression in the liver, but these LNPs have only recently been reported to encapsulate and deliver mRNA to produce protein *in vivo*. Because of the considerable size, stability, and charge differences between siRNA and mRNA, we hypothesized that LNP formulation parameters of historically-used siRNA-LNPs could be better optimized to more efficiently deliver larger mRNA payloads. To this end, we developed a generalized approach for optimizing LNP formulations for mRNA delivery *in vivo* using Design of Experiment (DOE) methodologies. After using this strategy to significantly improve mRNA-LNP potency *in vivo*, we then applied this mRNA-optimized formulation with new classes of ionizable lipid materials and discovered some of the most efficacious intravenously-administered mRNA delivery vehicles reported to date in the literature.

Materials / Methods: The most common siRNA-LNP formulations contain four components: an amine-containing ionizable lipid, phospholipid, cholesterol, and lipid-anchored polyethylene glycol, the relative ratios of which can have profound effects on the formulation potency. We first encapsulated mRNA encoding for Erythropoietin (EPO), a secreted serum protein, into the original siRNA-based LNP formulation (which employed the ionizable lipid C12-200) to establish a baseline from which to improve. LNPs were synthesized via microfluidic mixing of an ethanolic lipid phase with an aqueous RNA phase to form particles approximately 100 nm in diameter. Then, we employed DOE to modulate several formulation parameters simultaneously (e.g. lipid:mRNA weight ratio, phospholipid structure, molar composition of excipients). DOE allowed for the reduction of the total number of formulations required to deduce statistically significant trends in our large, multi-dimensional design space, which is important for the economical screening of LNP formulations due to poor *in vitro*/*in vivo* translatability and costs associated with screening large libraries *in vivo*. EPO mRNA-loaded C12-200 LNPs were injected intravenously at 15 μg total mRNA per mouse, and blood was collected 6 hours post-injection; an ELISA assay was then used to quantify serum EPO levels. Using Definitive Screening and Fractional Factorial DOE Designs, we screened a total of 38 distinct LNPs *in vivo*and characterized top-performing LNPs via their mRNA encapsulation efficiency, diameter, polydispersity, pKa, zeta potential, and morphology. Then, we applied the formulation parameters from the most efficacious C12-200 LNP formulation to a new class of ionizable lipid materials formed through a ring opening reaction between alkenyl epoxides and a polyamine core; unlike C12-200, which has fully saturated hydrocarbon tails, this new library of alkenyl amino alcohol (AAA) ionizable lipids contained degrees of unsaturation in its tails.

Results: After employing this DOE-based optimization process, we significantly ($p < 0.05$) increased the efficacy of intravenously-delivered liver-targeting C12-200 LNPs seven-fold for EPO mRNA delivery relative to the formulation previously used for siRNA delivery. At a 15 μg (0.75 mg/kg) mRNA dose, we achieved serum EPO protein concentrations in mice of approximately 7,000 ng/mL, which is 5-6 orders of magnitude greater than physiological levels. We found that increased lipid:RNA weight ratios and the use of the phospholipid 1,2-dioleoyl-sn-gylcero-3-phosphoethanolamine (DOPE) were the most crucial factors in determining LNP efficacy. Interestingly, when our mRNA-optimized C12-200 LNP formulation was loaded with siRNA, it silenced protein expression in hepatocytes nearly identically to the original siRNA formulation, suggesting that siRNA-loaded LNPs may be more tolerant than mRNA-loaded LNP of formulation design space differences. Lastly, when C12-200 was substituted with a library of novel alkenyl amino alcohol (AAA) lipids and formulated into LNPs with EPO mRNA, lead lipid OF-02 was discovered to afford EPO serum concentrations of approximately 14,000 ng/mL, the highest efficacy yet reported in the literature. Furthermore, mRNA delivered with OF-02 LNPs was expressed predominately in the liver and had a linear dose-response curve ranging from 0.75 to 2.25 mg/kg RNA dose.

Conclusions: In conclusion, we significantly increased the potency of liver-targeting C12-200 LNPs for mRNA delivery using Design of Experiment methodologies, while showing that the same mRNA-optimized LNPs did not confer additional potency when delivering siRNA. The optimized LNP formulation showed applicability to other types of ionizable lipids, such as the novel lipid OF-02, which made highly potent intravenously-delivered mRNA-LNPs. This work provides evidence that some technologies previously used for siRNA delivery may require further optimization for better mRNA delivery. Additionally, the general strategy presented here could be used to screen multi-dimensional design spaces in vivo for optimizing other biomaterials and payloads.

467122 Folate Receptor-Targeted Aminoglycoside-Derived Polymers for Transgene Expression in Cancer Cells

Abstract:

Targeted delivery of anticancer drugs can overcome the limitations of poor drug bioavailability, drug resistance, and unintended side effects associated with current chemotherapeutic regimens. Folate-receptors are overexpressed in several epithelial malignancies, including ovarian, renal, lung, triple-negative breast and bladder cancers, making these attractive targets for the targeted delivery of small molecule and/or nucleic acid therapeutics to cancer cells. Aminoglycosides are hydrophilic molecules that contain multiple sugars, amines and hydroxyls, and are attractive as monomers for generating diverse libraries of macromolecules. We first cross-linked aminoglycosides with diglycidyl ethers and subsequently derivatized these polymers with folic acid resulting in the formation of folate-targeted polymers. These folic acid-conjugated aminoglycoside-derived polymers were investigated for transgene delivery and expression in triple negative breast and bladder cancer cells at different polymer: plasmid DNA weight ratios. The targeted folic acid-conjugated polymers demonstrated significantly higher levels of luciferase expression in cancer cells compared to their parental polymers. Competitive inhibition of the folate receptor using free folic acid significantly inhibited transgene expression activity, indicating a role for the receptor in transgene expression. The targeted polymers were employed for the delivery of a newly constructed pEF-TRAIL plasmid, which expresses the TNF-alpha Related Apoptosis Inducing Ligand (TRAIL) protein, which is known to iduce death selectively in cancer cells. Cancer cell death following delivery of the pEF-TRAIL plasmid was investigated as a single-agent treatment as well as with other chemotherapeutics resulting in a combination treatment approach. Our results indicate that folate-targeted, aminoglycoside-derived polymers may be promising new materials for delivering nucleic acids to cancer cells in several malignancies.

460866 Sustained Transgene Expression Via Substrate-Mediated Gene Transfer Results from Multiple Transfection Events

Abstract:

Statement of purpose: Sustained delivery of therapeutic genes in vitro and in vivo has a wide range of applications in studying biology and in developing therapies to treat disease. Non-viral vectors such as cationic polymers still present promising approaches; however, bolus transfection methods with polyethyleneimine (PEI)-based DNA polyplexes suffer from considerable levels of cytotoxicity and short-lived transgene expression levels. Here, we designed and characterized a hyaluronic acid (HA)-based porous hydrogel system for non-viral gene delivery by loading with surface-associated DNA polyplexes (Figure 1A). With this, we observed enhanced and sustained transgene expression over 30 days of cell culture, with better cell viability, both marked improvements over comparable bolus transfection techniques. Finally, we investigated mechanisms thought to be responsible for the sustained expression profile.

Methods: HA hydrogels were formed via a Michael-addition reaction between acrylated HA and thiolated peptide crosslinkers with the addition of RGD peptide for cell adhesion. The sphere templating method was used to introduce 60-μm diameter pores. Plasmid DNA was complexed with polyethyleneimine to form DNA polyplexes, which were surface coated by electrostatic association to the pore surfaces of the hydrogel by incubating the formed hydrogel in the polyplex solution. DNA loading was assessed with ^{32}P-labelled DNA and scintillation counting. Transfection profiles were obtained by loading DNA encoding for the Gaussia luciferase gene, seeding D1 mouse mesenchymal stem cells in the hydrogels, and detecting the expressed protein with the associated assay.

Results: ^{32}P analysis of DNA-loaded porous hydrogels showed that up to 4 μg of DNA could be loaded into the scaffold. Although the DNA was electrostatically immobilized to the scaffold, minimal polyplex release was observed with only 5% of the initial loaded DNA released over 7 days. Cell culture within this scaffold resulted in sustained transgene expression over a period of more than 30 days at levels higher than those seen using bolus transfection techniques (Fig. 1B). In addition, adjusting the DNA loading concentration resulted in modulation of transgene expression (Fig. 1B). While repeated bolus transfections in 2D culture result in marked toxicity (Fig. 1C), culture in surface-coated hydrogels was significantly less severe (Fig. 1D). To determine if sustained expression was due to re-transfection events, it was observed that levels of internalized plasmid DNA in cells cultured in surface-coated hydrogels was sustained at higher levels over time than in cells cultured in non-coated hydrogels but administered a bolus polyplex transfection (Fig. 1E).

Conclusions: After DNA loading by surface coating, there was minimal basal DNA release over time, suggesting that this may serve as a robust system for long-term DNA availability and expression. Indeed, sustained expression was seen over thirty days of culture and could be enhanced by increasing DNA loading. Although repeated bolus transfections result in severe toxicity, repeated internalization events are possible in surface-coated scaffolds without the same toxic effects and are responsible for the sustained expression observed. This system is a promising means of sustained gene delivery for various biomedical applications without the need for genomic integration via viral methods, which have innate safety concerns.

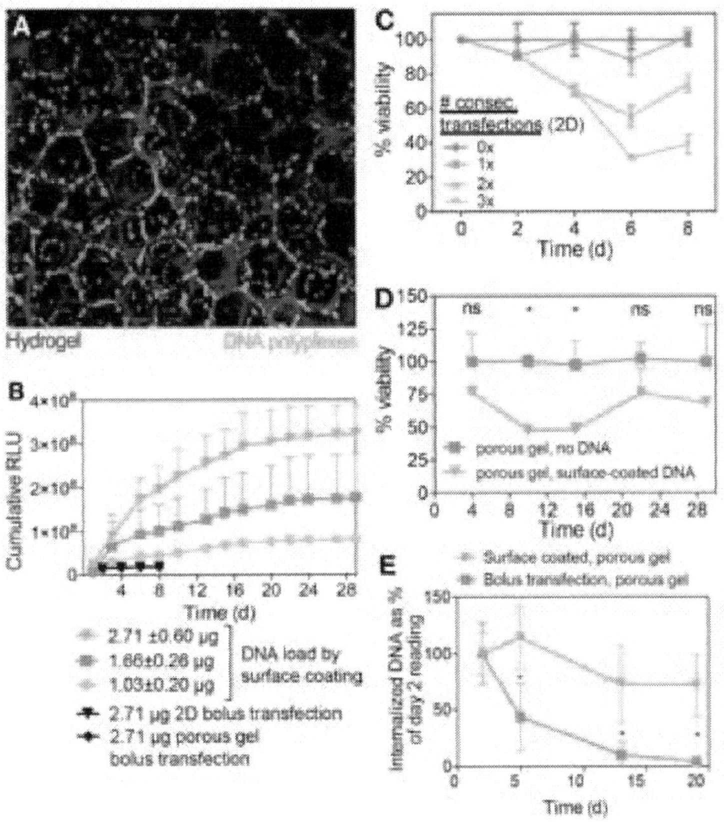

467956 Injectable Hydrogel Beads for Delivery of High Concentration Mab Formulations

Abstract:

Monoclonal antibody (mAb) based therapeutics are now the largest and fastest growing sector of the biopharmaceutical market. These products have shown excellent results in clinical settings as a result of the high binding specificity and minimal side effects of mAbs. However, our inability to predict protein-protein interactions poses a challenge to formulation efforts for these products. As an example, several recent studies have demonstrated that concentrated mAb formulations necessary for subcutaneous (SC) injection are plagued by high viscosities that are detrimental to this form of drug delivery. To mitigate these hindrances, we have developed a hydrogel delivery vehicle for the encapsulation of mAbs at very high concentrations while maintaining formulations with low viscosity suitable for SC injection. Monoclonal antibody encapsulated beads are produced using a thermally responsive nanoemulsion that can be cross-linked by UV curing. The temperature dependence can be utilized to tune the internal structure of the hydrogel bead and subsequently, control the local concentration and release of mAb. Additionally, the concentrated local environment within the hydrogel matrix is expected to help not only to stabilize the protein for long-term storage and but also to retain its native structure and function upon release.

471510 Development and Physicochemical Characterization of Tacrolimus-Loaded Nanocomposite Microparticles for the Treatment of Pulmonary Hypertension

Abstract:

Pulmonary arterial hypertension (PAH) is a progressive cardiovascular disease that leads to limited exercise capacity, right heart failure, and ultimately death. The primary treatment options currently available for patients with PAH are based on drugs with vasodilatory properties that improve cardiopulmonary function. Unfortunately, these drugs show no effect on prohibiting the progress of obliterative vascular pathology. Because of this, the only option for many patients to improve their health and quality of life is heart-lung transplantation. Therefore, new approaches to utilize compounds that focus on activating cellular mechanisms to reverse vascular remodeling are under investigation. Improving the function of the bone morphogenetic protein receptor-2 (BMPR2) signaling pathway is a potential direction in the development of improved PAH therapies, as the mutations of BMPR2 have been identified as the main cause of inherited PAH, accounting for 60 to 80% of familial cases. Since challenges exist in clinical application of BMPR2 gene therapy, tacrolimus (TAC) can be applied instead, as it increases signaling through the BMPR2 pathway. However, the application of TAC is hindered by its poor solubility, instability, and poor bioavailability. In addition to these limiting pharmacokinetic properties, systemic side effects such as neurotoxicity and nephrotoxicity also complicate the clinical use of TAC.

To overcome this problem, nanocomposite microparticles (nCmP) in the form of dry powder aerosols can be applied to deliver TAC more safely and effectively. This system is comprised of TAC-loaded acetalated dextran (Ac-Dex) nanoparticles (NP) entrapped in microparticle carriers using the excipient mannitol to allow for the delivery of TAC NP to the lungs. Upon pulmonary administration, the nCmP will deposit on the mucus in the respiratory tract, decompose into free NP and mannitol, and allow the nanoparticles to penetrate the mucus barrier and then release drug to the targeted site in a sustained rate. The goal of the described research was the initial development and physicochemical characterization of the nCmP systems with ability of targeted delivery, improved drug solubility, and potential of efficient penetration of mucus barrier and of controlled release via particle engineering.

The NP were approximately 200 nm in diameter with narrow size distribution both before loading into and after redispersion from nCmP. The NP exhibited smooth, spherical morphology and the nCmP were raisin-like spheres. High encapsulation efficacy was achieved both in the encapsulation of tacrolimus in NP and that of NP in nCmP. nCmP exhibited desirable aerosol dispersion properties allowing them to deposit into the deep lung regions for effective drug delivery. Overall, the designed nCmP system is a promising application in pulmonary TAC delivery for the treatment of PAH due to its novel features including targeted pulmonary delivery, improved solubility of tacrolimus, potential of permeation through mucus barrier, and of controlled drug release

468314 Nanoparticle-Mediated Inhibition of DNA Repair Sensitizes Brain Tumors to Radiotherapy

Abstract:

Radiotherapy (RT) is an integral component of the treatment for pediatric and adult brain tumors. However, survival is frequently accompanied by one or more radiation-induced adverse developmental and psychosocial sequelae, especially in children. Therefore, strategies that enhance RT in brain tumors while sparing adjacent normal brain are expected to improve life-long outcomes. The multifunctional DNA repair protein Ape1/Ref-1 has been implicated in conferring radiation resistance in brain tumors as Ape1 activity is inversely correlated with survival in brain cancer patients. Therefore, inhibiting Ape1 specifically in brain cancer cells should greatly enhance the effects of RT. We have developed a nanoparticle (NP) delivery vehicle that can stably bind, protect, and deliver nucleic acids to brain cancer cells and tumors. Here, we used our NPs to deliver siRNA against Ape1 (siApe1). We found that NP-mediated siApe1 delivery reduced Ape1 expression and activity, which was accompanied by a greater abundance of DNA damage. This diminished the resistance of brain cancer cells to RT. Therefore, NP-mediated inhibition of Ape1 may help enhance the therapeutic effect of RT in brain tumor patients and reduce treatment-induced morbidity.

461925 Theranostic Nanoparticles for Traumatic Brain Injury

Abstract:

Traumatic brain injury (TBI) is the leading cause of disability and death in people under 45. Currently, there are no effective treatment options available for the approximately 10 million new cases each year worldwide. Damage caused by TBI occurs in two phases: (1) the primary insult, and (2) the secondary phase where damage can spread beyond the primary injury leading to severe neurocognitive, physical, and psychosocial impairment. We have developed image-guided, Gd-conjugated, oxygen reactive polymer (ORP) nanoparticles to reduce the reactive oxygen species (ROS)-dependent secondary phase of the injury. ORP has a high ROS sponge capacity and is able to protect cell cultures from ROS-mediated damage. Magnetic resonance imaging (MRI) revealed high uptake and retention or ORP in the damaged region of brain in controlled cortical impact mouse models of TBI, which was attributed to a passive targeting effect similar to the enhanced permeability and retention effect observed in tumors. Importantly, ORP significantly reduced neurodegeneration and gliosis surrounding the primary injury in these mouse models. ORP shows initial promise as an effective therapy for TBI and will help provide a better understanding of nanomaterial interaction with damaged brain.

471209 Design, Synthesis, and Biological Evaluation of Novel Lipid Nanoparticle Materials for the In Vivo Delivery of Messenger RNA

Abstract:

The delivery of messenger RNA (mRNA) cargoes entrapped in lipid nanoparticles (LNPs) represents one potential strategy to replace deficient or non-existent proteins within target cell populations *in vivo*. Recent efforts have demonstrated that systemic administration of mRNA LNPs can induce functional protein expression in specific tissues in the body, but lead materials remain limited by their efficacy, biodistribution, and toxicity. Here, we employ rational design principles and synthesize novel LNP materials that can either *i.* induce higher levels of protein expression at a given dose or *ii.* modify the site of protein induction. Additionally, we establish critical structure-function parameters within our new mRNA LNP materials, demonstrating that *cis* alkenes play a pivotal role in improving the efficacy of both degradable and non-degradable mRNA LNP materials. *In vivo* experiments demonstrate batch-to-batch consistency, dose-response curves, and time course data for our lead mRNA LNPs. Toxicity studies also demonstrate that these materials are well tolerated *in vivo*. In sum, these studies not only establish highly potent materials for mRNA delivery, but also highlight the importance of employing rational synthetic design principles to better understand and improve our ability to deliver nucleic acids *in vivo*.

468226 Engineering Polymer Drug Conjugates to Synergistically Schedule Chemotherapeutics

Abstract:

Combination chemotherapy is commonly used to treat metastatic breast cancer; however, the success is limited due to systemic toxicity. To improve therapeutic efficacy, polymer drug conjugates carrying synergistic ratios of chemotherapy drugs can be used to selectively target cancer cells. While the importance of drug ratio on synergy has been previously investigated, the effect of temporal scheduling is not well understood. Here, we systematically evaluated the effect of temporal scheduling of doxorubicin (DOX) and gemcitabine (GEM) to determine both synergistic drug ratios and drug schedules. To engineer the temporal scheduling of DOX and GEM, hyaluronic acid drug conjugates with distinct linkers conjugating both DOX and GEM were optimized to control: (i) their individual release kinetics so as to match the optimized schedule and (ii) their molar ratios so as to optimize the synergistic dose. We show that polymer conjugates that deliver DOX and GEM in synergistic schedules and ratios are more effective at killing breast cancer cells compared to healthy epithelial cells. These results emphasize the importance in understanding the effect release rates have on the efficacy of synergistic drugs and provide the groundwork for the design of novel delivery systems capable of delivering exact molar ratios of synergistic drugs with temporal control.

462228 Hybrid Nanoparticles for Sequential and Controlled Delivery

Abstract:

Despite the great advance of nano-medicine in cancer treatment, it was reported that less than 10% of injected dosage could be delivered to the tumor sites. Co-delivery of drugs exploiting the blockage of multiple survival pathways in cancerous cells to achieve synergistic effect offers an alternative solution to this issue. In addition, recent growing knowledge of cellular networking points out that the release order of drugs plays a critical role in optimal therapeutic outcomes, which raises an another challenge in designing drug carrier. In this presentation, novel hybrid nanoparticles co-encapsulating hydrophilic and hydrophobic agents will be introduced. In this nanoparticle design, the release order of the agents can be programmed by loading them onto the surface or inside the nanoparticle. To validate the design of the nanoparticles, the size, structure, and drug release kinetics were assessed along with cytotoxicity tests.

455441 Coarse-Grained Modeling of Polymer Electrolyte Membranes

Abstract:

An attempt has been made to build a coarse grained (CG) model of polymer electrolyte membranes. We pursued a better predictive CG model for structure and morphology of the polymer electrolyte membrane, using the SDK CG water model. [1] The CG model using simple Lennard-Jones functions as in the SDK model is useful to optimize thermodynamic properties, like density and surface tension, at the cost of less accurate structure in the short distance. Alternatively, the structure-based mapping like inverse Boltzmann or reverse MC approach [2] is useful for reproducing pair-distribution functions with no guarantee for the thermodynamics properties. In the present study, we explored a possible combination of these two approaches, which can give a better model in predicting the morphology and detailed structure of polymer electrolyte membrane.

[1] W. Shinoda, R. DeVane, M.L. Klein, *Mol. Simul.* **33**, 27 (2007); *J. Phys. Chem. B*, **114**, 6836(2010).

[2] H. A. Karimi-Varzaneh, and F. Müller-Plathe, Multiscale Molecular Methods in Applied Chemistry, pp. 295-321, Springer Berlin Heidelberg, 2011.

455943 Precursor Effects on the Structure and Properties of Polymer Networks Synthesized Using Molecular Dynamics

Abstract:

The effects of precursor topology on the formation, structure and mechanical properties of polymer networks are studied using coarse-grained molecular models. Three different sets of precursors are designed in a way that in the ideal limit of complete conversion they would lead to an identical network structure. Molecular dynamics is then used to virtually synthesize crosslinked polymers from these precursors and the resulting material structures and properties are compared. Little difference is observed in commonly examined properties including the radial distribution function, macroscopic statistics of network connectivity, and glass transition behaviors. The elastic modulus of the final network however depends strongly on the precursor topology. The total number of elastic strands in the network is found to correlate strongly with the elastic modulus for most of the network formation process, except at the highly-crosslinked limit where substantial discrepancy is observed between networks from different precursors. Although these final networks contain a similar level of structural defects, the choice of precursor has substantial impact on the spatial distribution of the defects, which explains the precursor dependence of their elastic modulus. Our results indicate that the choice of precursor should be an important consideration in constructing molecular models for crosslinked polymers. It also provides insights into the experimental design of such materials.

457045 Multiscale CFD Simulation of Impregnation Die for Unidirectional Composites Production: Velocity Coupling Strategy Between Meso- and Macro- Scales

Abstract:

The permeability of a porous bundle of carbon fibers and impregnation rate are important factors of impregnation process of uni-directional carbon fiber composites (UD-CFC). Although the permeability of a porous medium is theoretically determined by Darcyâ€™s law , there is a lack of knowledge for multiscale interactions of a moving porous medium system in the UD-CFC impregnation process. In order to analyze the effects of the microscale structure and moving speed of carbon fiber in the presence of initial volatile material and polymer matrix on permeability and impregnation rate, a multiscale computational fluid dynamics (CFD) model including a microscopic 3D CFD model of a representative volume of tow segment with the real dimension of carbon fibers, and a macroscopic 3D CFD model for the whole dimension of the impregnation die was proposed. A moving frame reference approach for the tow domain was used in the macroscopic CFD model. The anisotropic and homogeneous permeability of the porous tow was calculated from the microscopic CFD model. The penetration velocity at the macroscopic CFD model was chosen as a convergence criterion for the multiscale iterations. Finally, a converged solution for permeability and impregnation rate of the porous CF bundle was obtained.

468338 The Dynamics of Tight Knots on Tensioned, Single Polymer Chains

Abstract:

In the limit of very long chains, coiled polymers almost always self-entangle and form knots. Here we perform Brownian dynamics simulations to study the motion of knotted polymers under high tension. In this regime, we find that knots exhibit glass-like dynamics. For example, instead of continuously moving along contour, the knot can occasionally get trapped in long-lived, metastable states. This caging behavior is highly dependent on the friction between self-entanglements, and we characterize how this arrested motion depends on the structure along the polymeric backbone. In short, we find that knot jamming arises from a 'rough' energy landscape along the polymer backbone due to short-range (excluded volume) interactions between polymer segments. When the potential wells in this landscape have a depth of several kT, the polymer reptates through the knot in slow, discrete hops. In the second part of the talk, we briefly discuss how this glass-like dynamics could lead to peculiar behaviors in knot relaxation. For example, we find that certain topologies swell normally when the chain's tension is released, but other topologies remain trapped in a tight conformation for long periods of time before relaxing. We discuss how the backbone 'roughness' alters this relaxation process and conclude by mentioning how knot jamming and relaxation could be important in various applications such as nanopore sequencing.

468937 Controlled Self-Assembly of Polymer-Grafted Nanoparticles

Abstract:

Two dimensional nanoparticle lattices can exhibit unique optical, electrical, and chemical properties giving rise to emerging applications for photovoltaic conversion, nanoelectronics and catalysis. In many applications, it is useful to be able to control the particle spacing, the crystal lattice formed, and the local composition of the lattice by co-locating nanoparticles of varying chemistry. However, control over all of these variables requires exquisite control over the interparticle interactions, and a large number of degrees of freedom affect the interactions. Achieving a particular structure by design requires solving the â€œinverse problemâ€, where one must optimize the chemistry to meet the structure or property that is desired. In recent years, a variety of examples have shown that one can finely control the interactions between nanoparticles through the use of polymers grafted onto the nanoparticle surface. In this talk, I will describe our efforts at solving the inverse design problem on polymer-grafted nanoparticles using an evolutionary design approach. By optimizing the polydispersity on a nanoparticle surface, we demonstrate that optimized polymer-grafting designs lead to square, honeycomb and hexagonal nanoparticle lattices. We show that surprisingly simple polymer chemistry can lead to these self-assembly structures. Finally, we present the investigation on the effective interactions of the nanoparticles and compare them against lattice forming pair-wise potentials from previous inverse design studies, showing the impact of many-body interactions.

469495 Systematic and Simulation-Free Coarse Graining of Polymer Blends and Block Copolymers

Abstract:

We propose a systematic and simulation-free strategy for coarse graining multi-component polymeric systems (e.g., polymer solutions, blends, nanocomposites, etc.), where we use the well-developed polymer reference interaction site model theory, instead of many-chain molecular simulations, for both the original and CG systems, and examine how the CG potentials vary with N (the number of CG segments on each chain) and how well the CG models can reproduce the structural and thermodynamic properties of the original system. Our strategy is quite general and versatile, is at least several orders of magnitude faster than those using many-chain simulations (thus effectively solving the transferability problem in coarse graining), and also avoids the problems caused by finite-size effects and statistical uncertainties in many-chain simulations commonly used in coarse graining. As two examples, we apply our strategy to the structure-based coarse graining of binary homopolymer blends and diblock copolymers. Our structure-based coarse graining preserves the critical point of symmetric binary blends and the mean-field order-disorder transition point of symmetric diblock copolymers at all coarse-graining levels, but cannot reproduce at any coarse-graining level the thermodynamic properties (i.e., the interchain internal energy per chain, the interchain virial pressure, and the change in the Helmoholtz free energy per chain from the athermal reference state) of the original system. The latter is simply due to the fact that the isotropic CG pair potentials (as commonly used) do not have the many-body nature of the potential of mean force caused by the coarse graining.

470226 In-Silico Tailoring Properties of Polylactide

Abstract:

Although significant progress has been recently made in the process-product development of bioplastics, they are still represented by only minute market share, mainly because of their higher cost and limited technical properties when compared to conventional petroleum-based polymers. Polylactic acid (PLA), or polylactide is a fully biodegradable thermoplastic polyester that is produced from renewable resources and offers significantly reduced carbon footprint when compared to its petroleum-based market competitors[1]. Nowadays, PLA is the most widely produced renewable and biodegradable polymer, having the third biggest market share among bioplastics[2], and steep increase in worldwide production capacity predicted for near future[3].

This contribution addresses development of multi-scale modeling platform as enabling technology for the simulation of PLA production and prediction of its rheological properties that affect the product processability and end-use-characteristics. The developed framework consists of three interconnected modules, starting with (i) macro-scale process simulator of PLA production by catalytic ring opening polymerization of lactide in a batch reactor, which provides dynamic evolution of important process characteristics (e.g., reaction mixture temperature, concentrations of reacting species) for (ii) micro-scale Monte Carlo (MC) simulation of polymer chains growth. MC simulation allows for prediction of important polymer characteristics such as full molecular weight distribution, and stores the information about the detailed microstructure of generated polymer chains for (iii) public domain software[4] enabling estimation of rheological properties of polymers with general branch-on-branch molecular structure. This approach allows for a direct mapping between the reaction conditions and application properties of PLA, and thereby opens the possibility for in-silico process optimization, virtual prototyping and tailoring of product properties.

During the development of the model-based enabling technology, attention was paid to the robustness and speed of the source codes (especially in the case of Monte Carlo simulation), so that they are suitable for potential use in the on-line model-based predictive control and optimization of PLA industrial production, enabling a big step towards economic and technological competitiveness of this promising biopolymer.

[1] A consortium led by E4tech (UK) Ltd (2015), *From the Sugar Platform to biofuels and biochemicals â€" Final report for the European Commision Directorate-General Energy*, Report No. ENER/C2/423-2012/SI2.673791.

[2] Sin L.T., Rahmat A.R., Rahman W.A. (2012), *Polylactic Acid: PLA Biopolymer Technology and Applications*, William Andrew.

[3] Carus M., Baltus W., Carrez D., Kaeb H., Ravenstijn J., Zepnik S. (2013), *Market Study on Bio-based Polymers in the World â€" Capacities, Production and Applications: Satus Quo and Trends Towards 2020*, nova-Institut GmbH

[4] Das C., Inkson N.J, Read D.J., Kelmanson M.A., McLeish T.C.B. (2006), *J. Rheol.*, 50, 207-235.

472292 Molecular Modeling and Simulation Studies of the Dehydration and Rehydration of Polymeric Porous Media

Abstract:

A biocompatible polymeric porous material was constructed and studied by coarse-grained molecular dynamics modeling and simulation that has a unique vectorial correspondence to the atomistic representation. The dehydration and rehydration phenomena of the polymeric porous material and those modified with a hydrophilic additive or cross-linkers were investigated using simulation approaches mimicking the actual processes. It was found that, in general, the strong interactions of water molecules with the polymer chains stabilize the porous structures of the polymeric materials and require the dehydration energy to be greater than that involved in pure water vaporization. The addition of the hydrophilic additive solutes and the incorporation of cross-linkers further enhance the water interaction energetics in the pore structures and increase the dehydration energy requirement. The water interaction energetics were found to be highly nonuniform in all three spatial dimensions, which has important implications in many fields. As dehydration proceeds, polymer conformational compaction and pore structural reduction occur due to downward stresses resulting from the surface tension of the descending water interface coupled with strong water-polymer and polymer-additive solute interactions. The stability of the porous structures was found to be enhanced by the presence of the hydrophilic additive because the additive molecules function as a filler in the pore space and interacts strongly with the polymer chains and water molecules so that they tend to be accumulated by the descending water level during dehydration at the vapor-liquid interface to form dense clusters of polymer chains at the interface. This phenomenon suggests that dehydration could be employed as a means to construct porous media with desirable spatial nonuniform density distributions of adsorption sites/ligands or catalyst sites/enzymes as the dissolved solutes. Cross-linkers provide additional steric support to the porous structures through the formation of three-dimensional networked structures that can sustain high structural stresses due to dehydration without minimal structural reduction. Extensively dehydrated porous structures were also found to have substantially improved stability so that the water molecules introduced during the rehydration process expand slightly the dehydrated porous structure of the systems examined

here, but could not recover the original pore structures that existed before the commencement of dehydration. The approaches presented here could form a basis for the systematic evaluation of the physicochemical effects of additive components and cross-linkers on the polymeric porous structures and their stability. It could also provide a molecular level design method of chemical engineering science for mechanistically controlling the structure and morphology of polymeric porous media as well as the distribution of additive solutes in such media during dehydration and rehydration.

449687 Entropic Control over Nanoscale Colloidal Crystals

Abstract:

Despite their broadly recognized utility in optical and catalytic devices, globally ordered colloidal crystal lattices remain challenging to assemble. Complex, hierarchical free energy landscapes, replete with small differences between local minima, often characterize ensembles of these structures. This makes it difficult to produce a single desired morphology without defects. Currently this problem is surmounted via either the "top-down", or "bottom-up" methodology whereby structures are engineered starting from the largest, or smallest relevant length scales, respectively. The ubiquitous design strategy in such approaches is energy minimization; many colloids have been computationally engineered with anisotropic pairwise interactions to achieve morphological control in theory. However, the complexity of these designs often makes experimental realization difficult. I will show how extensive computer simulations reveal that the introduction of polymeric co-solutes into crystallizing colloidal suspensions can be used to intelligently direct the resulting nano- and mesoscale colloidal structures by relying upon the polymer's entropic interactions alone. [1-4] These entropic interactions result entirely from the interplay between the polymer's internal degrees of freedom and the void structure of a material. This represents a novel third design paradigm that has the potential to significantly simplify control over colloidal polymorphism. I will elaborate on how to rationally design the co-solute structure to thermodynamically stabilize a single desired polymorph in a binary mixture, and the consequences that thermal perturbations have on this effect. [2] I will also offer insights into how to design temperature-dependent co-solute "switches" that allow the stability of a polymorph to be controlled via experimentally accessible parameters. [3] As a whole, this work represents a novel entropic route to polymorphic control that has not yet been explored.

References

[1] Mahynski, Panagiotopoulos, Meng, and Kumar, Nature Communications 5, 4472 (2014).

[2] Mahynski, Kumar, and Panagiotopoulos, Soft Matter 11, 280-289 (2015).
[3] Mahynski, Kumar, and Panagiotopoulos, Soft Matter 11, 5146-5153 (2015).
[4] Mahynski, submitted (2016).

460852 Engineering Macroscale Thermoelectric Transport By Chemical Modulation of Nanoscale Interfaces

Abstract:

Optimizing the performance of thermoelectric materials has historically proven difficult as their essential transport parameters are cross-coupled in bulk transport equations. However, new classes of hybrid inorganic/organic materials are emerging which present exciting opportunities as their transport behavior can strongly deviate from classical models. In this talk, I will present recent achievements in demonstrating new, interface-driven mechanisms for enhancing and modulating thermoelectric transport in nanoscale hybrid systems. Highlighted research will include p- to n-type conversion of transport type in Tellurium nanowires, subphase alloy scattering to boost power factor in Tellurium nanowires, phase transition anomalies in Cu_2Se nanocrystal arrays, and unexpectedly low thermal conductivity in hybrid perovskite nanowires.

472306 Doped Semiconducting Polymers As Solution-ProcessableÂ Thermoelectric Materials

Abstract:

Due to inherently low thermal conductivity and tunable electronic properties, semiconducting polymers are very promising solution-processable-materials for thermoelectric applications. Polymeric semiconductors are insulators in its pristine state, but are rendered conductive through chemical doping. Here, we studied the thermoelectric properties of poly(2,5-bis(3-tetradecylthiophen-2-yl)thieno[3,2-*b*]thiophene (PBTTT-C_{14}) doped either from solution or from the vapor phase. Overall, vapor doped films yield large Seebeck coefficients at high electrical conductivity (Power Factors ~ 100 Î¼$Wm^{-1}K^{-2}$). It important to note that the introduction of the chemical dopants can perturb the molecular packing and the morphology relative to the neat film. By leveraging various synchrotron X-ray scattering techniques that probe both the local and long range order, we discovered key structural features that can be correlated to the enhancement in the Seebeck coefficient and electrical conductivity for vapor doped films. Specifically, the introduction of the chemical dopant must limit the structural perturbation at small length scales while maintaining long-range correlation of polymer chains essential to efficient macroscopic charge transport. Lastly, to further elucidate the charge transport mechanism driving the thermoelectric properties, we report on the temperature-dependent measurements of both the Seebeck coefficient and electrical conductivity.

468413 Thermal and Thermoelectric Transport Coefficients in Graphene

Abstract:

Thermal and thermoelectric transport coefficients

in graphene

Enrique Muñoz, Physics Institute, Pontificia Universidad Católica de Chile, Vicuña Mackenna 4860, Santiago, Chile.

Abstract

Thermal, electronic and thermoelectric transport coefficients in graphene were studied. The effect of electron-phonon scattering processes, as well as the presence of C^{13} isotopes over the thermoelectric properties of extrinsic graphene were considered. Electron-phonon interaction is described within the deformation potential approximation. Electrical and thermal resistivity, as well as the thermopower, were calculated within the Bloch theory approximations. Analytical expressions for the different transport coefficients were obtained from a variational solution of the Boltzmann transport equation. The phonon-limited electrical resistivity $\rho e{-}ph$ shows a linear in temperature dependence at high temperatures, and follows a $\rho e{-}ph \quad T^4$ at low temperatures, in agreement with experiments. The phonon-limited thermal resistivity at low temperatures exhibits a $\quad T$ dependence and achieves a nearly constant value at high temperatures. The predicted Seebeck coefficient at very low temperature is $Q(T) \quad \pi^2 kB\ T/(3eEF)$, which shows a $n^{-1/2}$ dependence with the carrier density n, in agreement with experiments. On the other hand, the presence of a random distribution of C^{13} isotopes at fixed concentrations was studied as an strategy to reduce the lattice thermal conductivity, with the corresponding enhancement of the figure of merit for

thermoelectric applications.[3]

[1] E. Muñoz, Journal of Physics: Condensed Matter 24 (2012), 195302.

[2] E. Muñoz, J. Lu and B. I. Yakobson, Nano Letters 10 (2010), 1652.

[3] E. Muñoz, submitted (2016).

463547 Carrier Scattering at Alloy Nanointerfaces Enhances Power Factor in PEDOT:PSS Hybrid Thermoelectrics

Abstract:

This work demonstrates the first method for controlled growth of heterostructures within hybrid organic/inorganic nanocomposite thermoelectrics. Using a facile, aqueous technique, semimetal-alloy nanointerfaces are patterned within a hybrid thermoelectric system consisting of tellurium (Te) nanowires and the conducting polymer poly(3,4-ethylenedioxythiophene): poly(styrene-sulfonate) (PEDOT:PSS). Specifically, this method is used to grow nanoscale islands of $Cu_{1.75}Te$ alloy subphases within hybrid PEDOT:PSS-Te nanowires. This technique is shown to provide tunability of thermoelectric and electronic properties, providing up to 22% enhancement of the systemâ€™s power factor in the low-doping regime, consistent with preferential scattering of low energy carriers. This work provides an exciting platform for rational design of multiphase nanocomposites and highlights the potential for engineering of carrier filtering within hybrid thermoelectrics via introduction of interfaces with controlled structural and energetic properties.

465380 Combining Density Functional Theory Calculations, Supercomputing, and Data-Driven Methods to Design New Thermoelectric Materials

Abstract:

Density functional theory (DFT) simulations solve for the electronic structure of materials starting from the Schrödinger equation. Many case studies have now demonstrated that researchers can often use DFT to design new compounds in the computer (e.g., for batteries, catalysts, and hydrogen storage) before synthesis and characterization in the lab. In this talk, I will focus on how DFT calculations can be executed on large supercomputing resources in order to generate very large data sets on new materials for functional applications. First, I will briefly describe the Materials Project, an effort at LBNL that has virtually characterized over 60,000 materials using DFT and has shared the results with over 17,000 registered users. Next, I will talk about how such data can help discover new materials, describing how preliminary computational screening led to the identification and confirmation of a new family of bulk AMX2 thermoelectric compounds with measured zT reaching 0.8. I will outline future plans for how such data-driven methods can be used to better understand the factors that control thermoelectric behavior, e.g., for the rational design of electronic band structures, in ways that are different from conventional approaches.

450621 Phase Diagram, Microstructure and Thermoelectric Properties

Abstract:

Thermoelectric materials and devices have attracted intensive research and development interests due to their ability to convert waste heat directly into electricity. Energy usage efficiency could then be improved which is a very critical concern nowadays. Good thermoelectric properties require low thermal conductance but good electrical conductance. Phonon scattering caused by nanoprecipitates has been reported to be an efficient mechanism to enhance the thermal resistance but causing no significant hindrance on electrical conductance. Microstructures of thermoelectric materials are essential for their thermoelectric properties. Phase diagrams provide information of phase equilibria, and are fundamentally important for processing routes selection and understanding of materials microstructures.

Ternary thermoelectric materials systems, Bi-Te-In, Ag-Pb-Te and Co-Sb-Ga, are used as examples to illustrate the relationships between phase diagrams, microstructures and thermoelectric properties. Two kinds of ternary alloys were prepared. Ternary alloys of one kind were equilibrated at pre-determined temperatures, and the equilibrium phases were determined. Alloys of the other kind were quenched, and their primary solidification alloys were determined. Isothermal sections of ternary phase equilibria were proposed based on the results of determined ternary equilibrium phases together with the information obtained from their constituent binary phase diagrams, while the liquidus projections were determined based on results of the primary solidification phases.

According to the determined liquidus projections, ternary alloys with compositions at the ternary eutectics in the Bi-Te-In and Ag-Sb-Te systems, L (Bi- 13.0at%In-77.0at%Te)=Bi_2Te_3+Te+In_2Te_5, and L (Ag-40.at%Sb-36.at%Te)=$AgSbTe_2$ +Ag_2Te+δ-Sb_2Te, were prepared. Their microstructures were determined, and their thermoelectric properties were measured. Nanoprecipitates of Ag_2Te phase were observed. This ternary eutectic Ag-Pb-Te alloy has very low thermal conductivity (0.3w/mK at 415K), presumably due to the nano-sized lamellar structure and the nanoprecipitates enhancing phonon scattering. Very fine microstructures with nano- whisker-type precipitates were observed in the as-solidified ternary Bi-Te-In eutectic. These nanoprecipitates contribute to the alloys' thermoelectric properties.

Ternary alloys in the Bi-Te-In and Co-Sb-Ga systems with compositions close to the ternary solubility limits of Bi_2Te_3 and $CoSb_3$ phases were also prepared. The thermoelectric properties of the Ga-doped $CoSb_3$ alloys were improved with the concentration of Ga doping and the figure of merit increased to 0.7 with a 0.1at% Ga doping. Similar results were observed in the In-doped Bi_2Te_3 alloys. A small amount alloying would enhance the phonon scattering and reduce lattice thermal conductivity. However, secondary phases would form if the doping concentrations are higher than the solubility limits. Since the secondary phase might not have good thermoelectric properties, its existence would deteriorate the thermoelectric properties of the alloys. The phase diagrams are thus fundamentally important.

459105 Shaping the Spectrum of Thermal Radiation Using Nanostructured Materials for Efficient Thermophotovoltaic Power Generation

Abstract:

By manipulating the structure of materials at the nanoscale, we can tune the spectrum of thermal radiation. In doing so, we can selectively enhance or suppress certain modes linked to either desirable or undesirable properties. Such spectrally-selective transport holds promise for more efficient solar and thermal power generation.

Here, we enhance spectrally-selective radiative transport using nanostructured materials to improve the efficiency of thermophotovoltaics (TPVs). TPVs are solid-state heat engines that convert energy from a high temperature radiant source—the emitter—into excited charge carriers in a semiconductor diode—the cell. TPVs are a promising technology for applications in concentrated solar power and distributed combined heat and power because they are solid-state, scalable and compatible with a variety of high-temperature heat sources. To achieve high efficiencies, it is well known that spectral control over thermal radiation is critical. Nevertheless, several questions remain: 1) *What spectral features are most important in TPVs when accounting for parasitic losses and non-idealities?, and 2) How can nanostructured materials be designed to meet the stringent spectral and thermal requirements in TPVs?*

To address these questions, we developed radiative transport and energy conversion models to identify the most important spectral features in realistic TPVs. Based on these models, we show that suppressing radiative exchange between the emitter and the cell at energies below the bandgap of the cell (i.e., at long wavelengths) is critical for high efficiency. This is explained by the fact that, at the optimal operating temperature, the emissive power below bandgap is much higher than above the bandgap. Operation at higher temperatures is undesirable because of the increased high-temperature radiative losses and excited charge carrier thermalization in the cell.

Based on these insights, we developed thin-film and nanostructured materials for TPV applications where spectrally-selective transport is based on the interaction of thermal radiation with wavelength-scale structures. The structures are chosen to minimize radiative transport at energies below the bandgap and then optimized to achieve high efficiencies in the presence of realistic losses. By suppressing radiative exchange at low energies, we demonstrate a solar TPV device that is an order of magnitude more efficient than previous solar TPVs and has the potential to exceed the limit of single-junction solar cells. Further, we show how a scalable and thin-film cell design promotes spectral selectivity using a back-side surface reflector to return sub-bandgap radiation back to the hot emitter. Spectroscopy measurements confirm that the reflectance below the bandgap for these thin-film cells is significantly higher than for cells on bulk substrates. Based on these spectral properties we show that the heat engine efficiency can be significantly higher than the state-of-the-art, exceeding 30%.

This work provides a detailed understanding of the relative importance of spectral features in TPVs and offers strategies for how to best engineer thin-film and nanostructured surfaces to shape the spectrum of thermal radiation for TPVs. With integration of thermal storage, TPVs can become an important part of distributed energy systems and may lead to increased adoption of intermittent renewables.

467404 Vaporization of Nanoparticles in Dusty Plasmas

Abstract:

In the last 10 years, dusty plasmas have been increasingly explored for the synthesis of nanomaterials. Dusty plasmas are nonthermal discharges, which contain hot electrons and relatively cool neutral gas molecules in the midst of suspended nanoparticles. Dusty plasmas have been very successful for synthesizing high quality ceramic powders comprised of nanoparticles. Metal powders, however, have been more difficult to synthesize. It has been observed that when gaseous metal precursors are fed into the plasma, a metallic film is deposited on the reactor walls; but little, if any, particles are formed. Our hypothesis is that the intense ion bombardment experienced by particles in dusty plasma causes vaporization, despite the low gas temperature. Experimental and modeling results that support this hypothesis will be presented.

467470 Increasing Contact Area for Scaled up Processing of Solids Using Nonthermal Plasmas

Abstract:

Nonthermal plasma reactors have been very successful in the microelectronics industry for promoting various surface reactions such as thin film deposition and reactive ion etching. Nonthermal plasmas contain electrons with very high kinetic energy, as well as ions and neutrals that are near room temperature. Nonthermal plasmas are intriguing systems that offer many unique features. For example, surfaces in contact with plasmas can be subjected to radicals that would be difficult, if not impossible to produce by thermal processing. In addition, surfaces can be bombarded with ions that have controllable kinetic energy in the range from approximately 0.5 to 10 eV. These ions can be used to promote various surface rearrangements, desorption reactions, and in some cases even sputtering reactions. These reactions are fundamentally surface reactions, and therefore the reaction rate scales with surface area. Established techniques involve processing materials that have relatively small surface area, and therefore the reaction rates are relatively low. We will discuss our recent results on increasing the reaction rate of nonthermal plasma with solid precursors by increasing the plasma-solid contact area.

473010 Process Control for Atmospheric Pressure Plasma Systems

Abstract:

Atmospheric pressure plasma systems show significant promise in materials processing and medical applications, but challenges remain in device instrumentation and characterization, real-time process control, and dosage quantification. We discuss the design and implementation of a fully-automated atmospheric pressure plasma testbed for rapidly screening device designs and evaluating thermal, electrical, optical, and chemical instrumentation schemes and process control strategies.

465768 Topographically Selective Deposition of Dielectrics Using Ion Implantation

Abstract:

High quality thin films are important for diffusion barriers, adhesive buffer layers, etch stops and sacrificial layers in integrated circuits, MEMS and NEMS. Achieving maximum performance requires great precision in the fabrication technique. Advancements in technology and miniaturization of these devices require uniform, conformal coatings of complex, high aspect ratio features (steps, trenches, grooves and pores) on the nanoscale. Today's device fabrication is based on 'top-down' processes with multiple photolithography and etch steps which serve as the main bottleneck for device downscaling. Area-selective atomic layer deposition (AS-ALD) which combines surface modification and ALD in a 'bottom-up' approach to nanopatterning provides a promising alternative.

Previous studies have found that selectivity in AS-ALD relies on a change in hydrophobicity whereby more hydrophobic surfaces were found to cause an extended nucleation delay. Most reports on AS-ALD typically exploit this property, using tightly packed self-assembled monolayers (SAMs), which increase hydrophobicity, to modify the substrate surface. There are several challenges associated with the use of SAMs including degradation under ALD conditions, lengthy deposition times, and a limited number of suitable chemistries. To overcome these challenges, there is a need for new methods of surface modification to make AS-ALD a more viable technique.

In this presentation, a novel approach in AS-ALD will be described, whereby the substrate surface (Si) is modified through fluorocarbon implantation. It has been shown that through this process, a thin (~1.5 nm) fluorocarbon layer forms on the Si surface, making it hydrophobic (WCA ~104°). In a previous study by this group, this method was found to be successful in blocking Pt ALD up to 500 cycles. In this work, the implanted substrate's ability to block ALD of dielectrics (Al_2O_3, ZnO, HfO_2) is evaluated. Substrates with two and three dimensional features are also investigated. Nucleation on the implanted substrate (or regions) is investigated and compared to a non-implanted reference surface. Several techniques including, atomic force microscopy, scanning and transmission electron microscopy, X-ray photoelectron spectroscopy and Auger electron spectroscopy are used for analyses and characterization. The results show that there is reduced oxide growth on the more hydrophobic, implanted substrate. The potential and limitations of the ion implantation approach to AS-ALD will be discussed.

468289 Epitaxial Growth and Atomic Characterization of Fe-BTO(111) on SiC (0001) Using MgO Template Layer

Abstract:

Barium titanate is one of the most important ferroelectric materials for many applications in microelectronic, especially for ceramic capacitor production [1]. Recently, the thermoelectric properties of $BaTiO_3$ (BTO), previously overlooked, gained attention with the recent discovery of a high Seebeck coefficient (100µV/K) and low resistivity (200µΩ cm) in layered cobalt oxide, $NaCo_2O_4$ [2]. Because ferroelectric oxides are typically non-toxic, composed of commonly occurring elements, and available at relatively low-costs, much effort has been devoted to understanding the origin of the enhanced Seebeck coefficient in oxides, and to explore thermoelectric applications for perovskites oxides.

Many investigations have provided experimental evidence of spin and orbital degeneracy being the dominant contribution to the enhanced Seebeck coefficient in cobalt oxides [3]. $NaCo_2O_4$ is a structurally complex material in which edge-shared distorted octahedral (CoO_6) of oxygen ions form a 2D triangular lattice and sodium ions in a prism site between CoO_2 layers. It was found that the d-orbital of the cobalt ion located in the center of the octahedra is split in a way that allows the cobalt to always be in a low spin state regardless of its oxidation state. This results in orbital degeneracy and therefore high entropy due to spin state of the metal atom [2]. We are further exploring the microscopic origin of the enhanced Seebeck coefficient in the metal-oxides manipulating the spin entropy of a BTO through controlled doping with Fe. The goal is to enable tuning of the Seebeck coefficient while enhancing the conductivity and thus increasing the thermoelectric properties of the material.

The growth of high quality epitaxial Fe-BTO films with atomically sharp interfaces is first needed before investigating the spin state-thermoelectric relationship. Molecular beam epitaxy (MBE), which provides control of crystal growth at the atomic level, will be used to grow Fe-BTO on 6H-SiC. Through the use of

controlled parameters, the interface formation mechanisms will be investigated for effective heteroepitaxy of high-quality Fe-BTO on 6H-SiC by reflection high energy electron diffraction (RHEED) and x-ray photoelectron spectroscopy (XPS). Understanding the relationship between processing conditions and chemistry and structure will be used to link key electronic and thermoelectric properties. This understanding will enable the development of processing strategies to enhance the Seebeck coefficient and electrical conductivity of BTO.

In our previous work, RHEED studies supported the importance of a thin, 2.5 nm, crystalline MgO(111) layer for improved epitaxy and crystal structure of tetragonal BTO(111)[4]. Understanding from our previous results, will allow us to explain interface formation mechanisms for Fe-BTO/MgO/SiC heterostructures and influence the film crystal quality. In this study, the use of a magnesium oxide (MgO) template layer and the interface formation mechanisms will be investigated for effective heteroepitaxy of high-quality Fe-BTO films on 6H-SiC.

Author for correspondence: k.ziemer@neu.edu

[1]. E. Beaudrouet, A. Viviet, M. Lejeune, C. Santerne, F. Rossignol, C. Dossou-Yovo, M. Mougenot, R. Noguera, "Stability of aqueous barium titanate sus- pensions for MLCC inkjet printing", *J.Am. Ceram.Soc.*, 97 [4] (2014) 1248–1255.

[2]. I. Terasaki, Y. Sasago, K. Uchinokura *Large thermoelectric power in $NaCo_2O_4$ single crystals.* Physical Review B, 56, 1997.

[3]. G.D. Tang, X.N. Xu, C.P.Tang, L.Qiu, Y. Lv, Z. H. Wang *The spin entropy suppression induced by Fe^{3+} in $NaCo_2O_4$* Journal Of Applied Physics of Physics 2010, 107

[4]. T. L. Goodrich, Z. Cai, M. D. Losego, J.-P. Maria, L. F. Kourkoutis, D. A. Muller, and K. S. Ziemer, "*Improved epitaxy of barium titanate by molecular beam epitaxy*," Journal of Vacuum Science Technology B, vol. 26, p. 024803, 2008.

461976 SiC Substrate Cleaning for Epitaxy Using a Thermally Generated Atomic Hydrogen Beam

Abstract:

Wide bandgap semiconductors are becoming the new platform of choice for devices with greater power density and energy efficiency than silicon. In particular, Silicon carbide (SiC) is ideal for high power and high temperature (>150°C) electronic applications because of its high critical electric field voltage (2500 kV/cm) and good thermal conductivity (4.9 W/cmK) [1]. However, heteroepitaxial film growth of functional oxides on SiC, is a challenging and yet critical goal to achieve in order to fully utilize SiC in next-generation electronics.

Formation of an abrupt and effective interface is one of the basic requirements for integration of functional oxides on semiconductors. A reliable and effective cleaning procedure must be developed in order to produce consistent and well-characterized starting surfaces. Several SiC surface cleaning studies, including traditional wet cleaning, high-temperature hydrogen etching and hydrogen plasma treatment have been reported in order to remove contaminations and scratches made during polishing [2].

In our previous work, *6H*-SiC (0001) substrates were cleaned in an *ex-situ* hydrogen furnace. Tantalum foil strip is used as heating element during furnace operation where the *6H*-SiC sample is placed on it. The resulting *6H*-SiC surfaces are smooth with a uniformly stepped surface and a $\sqrt{3}\times\sqrt{3}R30^o$ surface reconstruction with a silicate adlayer, verified by reflection high energy electron diffraction (RHEED) and x-ray photoelectron spectroscopy (XPS) [3]. This is consistent with that reported by Starke and coworkers [4]. However, the high treatment temperature (~1600 °C) and sensitivity to small thermal gradients makes it impractical for a low-cost and high-throughput production.

In this study we demonstrate the effects of atomic hydrogen produced by a hydrogen atom beam source (HABS). The HABS is a thermal gas-cracking cell consisting of a tungsten filament set in high purity tungsten tube. It has been found that the atomic hydrogen irradiation is productive for removal of carbon and oxygen on Si surfaces at relatively low temperature without any ion damage to the substrates [5]. We compare the treatment of *6H*-SiC surfaces with hydrogen atomic beam source and the hydrogen flow furnace, in order to understand the fundamental H-atom interactions with the surface and then use that knowledge to develop an optimized surface cleaning process.

1. H. Lingqin, Zhu Qiaozhi, Gao Mingchao, Qin Fuwen, and W. Dejun, *Cleaning of SiC surfaces by low temperature ECR microwave hydrogen plasma.* Applied Surface Science, 2011. **257**(23): p. 10172-10176.

2. King, S.W., S. Tanaka, R.F. Davis, and R.J. Nemanich, *Hydrogen desorption from hydrogen fluoride and remote hydrogen plasma cleaned silicon carbide (0001) surfaces.* Journal of Vacuum Science & Technology A, 2015. **33**(5): p. 05E105.

3. Goodrich, T., J. Parisi, J. Leong, and K.S. Ziemer. *SiC surface preparation by hydrogen cleaning for high-temperature, high-power device integration.* in *Proceedings of the AiChE Annual Meeting.* 2005.

4. Starke, U., J. Schardt, J. Bernhardt, and K. Heinz, *Reconstructed oxide structures stable in air: Silicate monolayers on hexagonal SiC surfaces.* Journal of Vacuum Science & Technology A, 1999. **17**(4): p. 1688-1692.

5. Chun, Y.J., T. Sugaya, Y. Okada, and M. Kawabe, *Low temperature surface cleaning of InP by irradiation of atomic hydrogen.* Japanese Journal of Applied Physics, 1993. **32**(2B): p. L287.

456871 Chiral Templating of Self-Assembling Nanostructures By Circularly Polarized Light

Abstract:

Examples of chemical reactions affected by spin angular momenta of circularly polarized photons are rare and display low enantiomeric excess. High optical and chemical activity of nanoparticles (NPs) should facilitate the transfer of spin angular momenta of photons to nanoscale materials. However, such processes are currently unknown. Here we demonstrate that circularly polarized light (CPL) strongly affects self-assembly of racemic CdTe NPs. Illumination of NP dispersions with right- and left-handed CPL is shown to induce the formation of right- and left-handed twisted nanoribbons, respectively. Enatiomeric excess of such reactions was found to be greater than 30% which is ~10x higher than other CPL-induced reactions. In contrast, illumination with linearly polarized light and assembly in the dark led to straight nanoribbons. The mechanism of "templation" of NP assemblies by CPL is associated with selective photoactivation of chiral NPs and clusters followed by their photooxidation. Chiral anisotropy of interactions translates into chirality of the assembled ribbons. Their strong optical activity is confirmed by ensemble measurements, single ribbon CD spectroscopy, finite element modeling of the chiroptical properties based on Maxwell equations, and atomistic molecular dynamic simulations. The ability of NPs to retain polarization information, or the "imprint" of incident photons opens new pathways for the synthesis of chiral photonic materials and allows for better understanding of the origins of biomolecular homochirality.

468925 Anomalous ALD Growth per Cycle Under Precursor-Limited Conditions

Abstract:

Atomic layer deposition (ALD) is a thin film deposition process used to create highly conformal films with precise control of thickness and composition. ALD utilizes a cycle of sequential, self-limiting surface reactions to deposit the desired film one monolayer or sub-monolayer during each cycle. Typically, a binary sequence of gaseous precursors is used, with purge periods in between to prevent gas phase reactions. The self-limiting nature of the deposition half-reactions arises from surface saturation due to a finite density of surface reaction sites, or steric hindrance when precursor ligands remain after the chemisorption reactions.

Alumina ALD using trimethylaluminum and water as precursors is among the most widely studied ALD processes [1]. Under nominal deposition conditions, a growth-per-cycle (gpc) of 0.11 nm/cycle is typically found under self-saturating conditions. As one reduces either or both precursor dosages, one would expect to find that the gpc ultimately begins to drop from the value corresponding to the self-saturating plateau to a value of zero as the dosage drops to the same. Henn-Lecordier and coworkers [2], however, found anomalously high gpc values in the precursor-starved region of operation. In this paper, we will report similar findings in which gpc values that are approximately double that of the nominal value of 0.11 nm/cycle as the TMA does is reduced while holding the water dose constant. Furthermore, we find that film thickness remains spatially uniform in this region, up to the point where precursor depletion becomes important. Potential reaction mechanisms for the observed behavior will be presented.

[1] Puurunen, R. L., Surface chemistry of atomic layer deposition: A case study for the trimethylaluminum/water process, Appl. Phys. Rev. 97 121301:52 (2005).

[2] Henn-Lecordier, L., M. Anderle, E. Robinson, and G. W. Rubloff, J. Vac. Sci. Technol. A, 29 051509-1 (2011).

457744 Development of a Robust Computational Methodology for Value-Added Understanding of Selective Conducting Thin Film Deposition

Abstract:

There has been a growing interest in depositing conducting thin films by chemical vapor and atomic layer deposition (CVD and ALD, respectively) for various applications. Specifically, with rapidly decreasing feature sizes and more demand for integrated circuit speed, the need exists for depositing conducting thin films with low impurities, which will provide the required electrical properties. First-principles analysis of precursor chemistry and deposition mechanisms can significantly decrease experimental costs through initial virtual screening of potential organometallic precursors. This presentation will focus upon two key transition metals used in integrated circuits, Co and Cu, and relevant organometallic precursors used to deposit these metals. Particularly, in this collection of studies, Density Functional Theory (DFT) at the BLYP/DNP level of theory was employed to study ligand dissociation energies, decomposition mechanisms, adsorption/desorption, and surface reactions of various commercially available Co and Cu precursors on substrates found in common integrated circuits. Similarly, studies addressing challenges in transition metal surface wetting and potential agglomeration will also be discussed. Our studies have employed both gas-phase and periodic surface slab models, as well as cluster models, in efforts to balance computational time and accuracy. Modeling results were found to be in agreement with experiments and have provided insights on the parameters which should be considered in the design of new precursors for selective Co and Cu deposition. These results demonstrate how theoretical investigations are currently impacting precursor discovery and deposition understanding in industry.

471547 Simple Polymerization Reactions for Overcoming Long-Standing Challenges: From Fully Recyclable, Melt-Reprocessable Rubber Tires Containing Dynamic Covalent Bonds to Development of Broad-Temperature-Range Vibration and Acoustic Damping Materials (Invited Talk)

Abstract:

Innovative application of simple polymerization reactions can overcome long-standing problems as well as achieve solutions to critical needs associated with polymers. In the first example, I will demonstrate how innovative application of nitroxide-mediated polymerization, which was first developed in the 1990â€™s, can lead to recyclable, melt-reprocessable polymer networks or cross-linked polymers with 100% achievement (within error) of cross-link density after multiple reprocessing steps. This work relies on the reversible capping-uncapping reaction of the nitroxide radical that occurs at elevated temperature (e.g., ~120 degrees C in one incarnation) to yield dynamic covalent bonds. The approach described in this presentation can be applied to any polymer or monomer with at least one carbon-carbon double bond amenable to free radical polymerization, e.g., polymers such as polybutadiene and cis-polyisoprene used in the tire industry. I will show how a simple, one-step reaction can result in fully recyclable cross-linked polymers of the type used in the tire industry with complete property recovery after multiple recycling steps. I will also show how this simple chemistry can be used to design highly uniform polymer networks from monomer and describe how such materials may be useful in advanced engineering applications.

As a second example, I shall consider the challenge of designing polymeric materials via polymerization reactions that can yield excellent vibration and acoustic damping properties over a broad range of temperatures, from a minimum 60 degree C temperature (T) breadth to a maximum exceeding a 100 degree C T breadth. Several successful approaches will be described, all of which rely the production of polymeric materials that undergo nanophase separation with some phase mixing, thereby yielding materials with many compositions, and thus many glass transition temperatures, naturally present at equilibrium. Several cases will be detailed, including gradient copolymers yielding sinusoidal composition profiles at the nanoscale. Such gradient copolymers can be easily achieved by controlled radical polymerization methods, including nitroxide-mediated polymerization. Another case that will described will be polyhydroxyurethane (PHU) synthesized by aminolysis of cyclic carbonates. With PHU, the hydroxyl unit in the hard segment can undergo hydrogen bonding with oxygen atoms in the soft segment, thereby moderating the nanophase separation. Under optimal conditions, with enough but not too much such intersegment hydrogen bonding, excellent damping properties (as determined by tan delta > 0.3) can be tuned over exceedingly broad temperature ranges. The important role of molecular structure impacting nanoscale structure which in turn impact macroscopic properties is evident with PHUs: such properties cannot be achieved with polyurethanes lacking the hydroxyl unit present in PHUs.

474244 Ring-Opening Polymerization for 100% Renewables-Based Polyethylene Furanoate (PEF) for the "Green Bottle"

Abstract:

Along the shift of our societies from fossil-fuel based economies towards more sustainable civilizations, ring-opening polymerization (ROP) can be an alternative process to synthesize a novel class of renewable resource-based polymers. Furanic polyesters like polyethylene furanoate (PEF) have been ranked among the top potential chemicals by the US-DOE and may replace one of the world's dominant fossil-based plastics, polyethylene terephthalate (PET). Recent efforts and first commercial scale production by Dutch company Avantium and other research groups were invariably based on polycondensation, which is burdened with the necessary removal of condensation byproducts to reach high conversions and molecular weights, and thus typically requires reaction times in the order of days [1,2]. Cyclic monomers do not have endgroups and thus render byproduct removal unnecessary, which enables ROP as a living chain-growth mechanism to deliver high conversion to PEF within minutes. We present ROP as an alternative process to reach sufficiently high molecular weight PEF for commercial applications such as bottles, textiles, medical grafts, etc. [3,5].

Cyclic PEF monomers (cyOEF) can be derived from 100% renewables-based building blocks furandicarboxylic acid and ethylene glycol via different synthesis routes that have successfully been applied to PET, which all exploit the shift of the ring-chain-equilibrium to cyclics in dilution [4]. Synthesis pathways such as depolymerization of short PEF oligomers in 2-Methylnaphthalene have yielded >80% cyOEF with a ring-size distribution from dimer to heptamer within 4 hours, as part of other projects in our group. Maximization of material turnover as well as minimization of waste is possible by full recycling of unconverted linears and solvents [5]. Purification of cyOEF from residual linears via silica gel adsorption to yield >99% cycles is essential for ROP to deliver 1) sufficiently high molecular weights, 2) reproducible reaction control and 3) colorless products. The search for better and "greener" solvents is ongoing, aided by solubility predictions from COSMO-RS that show several aromatics and polyethers to possess promising solvent power and selectivities.

PEF applicable to typical commercial applications of PET can then be produced from high purity cyOEF. cyOEF were subjected to catalytic ROP to form PEF chains in yields of >95% and molecular weights equivalent to a PET bottle (~60'000 g/mol). Since absolute molecular weight determination is essential for such process development, all our samples were all analyzed with size exclusion chromatography (SEC) multi-angle light scattering (MALS) as well as diffusion (DOSY) NMR. The latter was developed especially for PEF, as it offers advantages over SEC-MALS such as reduced solvent use, analysis time <2 min and is less prone to hardware failure, while allowing for simultaneous absolute molecular weight and conversion analysis due to the different chemical shifts of cyOEF and PEF [6]. The reaction is usually complete after less than 20 min, which, even combined with cyOEF synthesis time, outperforms commonly applied solid-state polycondensation by about a factor of 20. Since the cyclic raw material exhibits a distribution of ring sizes, the individual species show melting points ranging from 270°C to 370°C and different reactivity. Successful conversion of each observed cyclic species from dimer to heptamer was achieved well below their melting point using 250-280°C with tin-catalysts such as FDA-approved tin octoate, the application of which could facilitate the entry to the FDA-regulated food packaging market. Lower temperatures are actually beneficial and allow for higher achievable molecular weights, due to the more limited impact of degrading side-reactions. This degradation became dominant usually after 20 min after reaching >95% conversion. Maximum achievable molecular weights also increased with monomer purity, as linear impurities reduce molecular weight through the introduction of additional end-groups. After having obtained PEF from ROP with favorable molecular properties, the material properties essential for bottle manufacture were analyzed. The higher glass transition temperature (85°C vs. 73°C) and lower melting point compared with PET (215°C vs. 260°C) indicate higher thermal stability and easier processing of the final bottles. An at least 5x higher oxygen diffusion barrier complements the advantageous properties of PEF, which can be explained with a higher molecular rotational energy barrier as derived from molecular dynamics simulations that we have recently performed.

While the scale-up to larger (kilogram) volumes and processing towards actual bottles is ongoing, the advantageous synthesis of PEF via ROP opens a new and promising pathway not only towards the highly anticipated "green bottle", but also enables advanced molecular architecture control of furan based polyesters through a "living" polymerization, e.g. for branching and block copolymers, which is infeasible with the current process based on PC-SSP.

References

[1] M Gomes, A Gandini, AJD Silvestre, B Reis, *J. Polym. Sci. A: Polym. Chem.* 2011, 49, 3759-3768
[2] L Sipos, E De Jong, MA Dam, JM Gruter, ACS Symposium Series 2012, 1105, P Smith(Ed.), 1-11
[3] D Pfister, G Storti, F Tancini, LI Costa, M Morbidelli, *Macromol Chem Phys* 2015, 216, 2141-2146

[4] R R Burch, SR Lustig, M Spinu, *Macromolecules* 2000, 33, 5053-5064
[5] JG Rosenboom, P Fleckenstein, G Storti, M Morbidelli, 2016, in preparation
[6] W Li, H Chung, C Daeffler, JA Johnson, RH Grubbs, *Macromolecules* 2012, 9595-9603

FIGURE 1

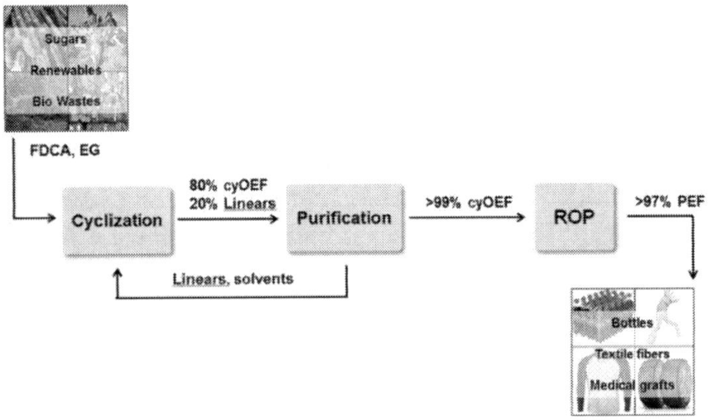

FIGURE 2

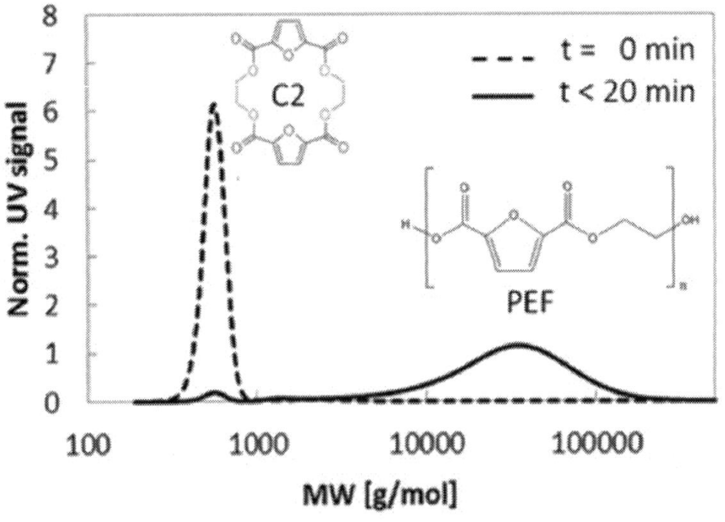

FIGURE 3

	Polycondensation (PC)	**Ring-opening polymerization (ROP)**
Raw material	Ethylene Glycol + Furandicarboxylic Acid (100% bio-based)	Ethylene Glycol + Furandicarboxylic Acid (100% bio-based)
Step 1	Polycondensation	Cyclization
Step 2	Solid-state post-polymerization	Ring-opening polymerization
Time step 1	5 hours	4 hours
Time step 2	3-7 days	< 1h
Molecular weight control	Time, vacuum	Recipe, initiators
Molecular weight achievable	> 60,000 g/mol (PET bottle grade)	> 60,000 g/mol (PET bottle grade)
Control of polymer structure	Low	High

Material Property	**PET**	**PEF**
Starting material	30% bio-based [10]	100% bio-based
Recyclability	100%	100%
Glass Transition T_G (°C)	73	85
Melting point $T_{M\,polymer}$ (°C)	260	215
O_2 gas diffusion (mL/m^2/d)	1950	392
Molecular stiffness (kJ/mol)	30	86

459745 General Route for the Preparation of Olefin Based Block-Copolymers

Abstract:

Block copolymers and their mesoscopic self-assembly remains a topic of intense research in polymer science. Potential applications of these materials includes separation, nanolithography, catalysis and plastic engineering. For the most part, block copolymers are produced by anionic or controlled radical polymerization, in which both polymerization techniques are not compatible with olefin monomers. Today, polyethylene and polypropylene account for more than 40% of the polymers produced worldwide. Therefore, developing an efficient synthesis route for the preparation of olefin-based block copolymers has high potential impact.

We will present a novel methodology that consists of converting any polyolefin made via catalytic polymerization into a macroinitiator compatible with anionic, ring opening polymerization and atom transfer radical polymerization. Our strategy yields olefin-based block copolymers with unprecedented structures. Finally, we will also present on the self-assembly and thermal properties of some of the materials.

468031 A New Synthetic Approach to Epoxide Polymerization

Abstract:

Polyethers are important in established and developmental technologies, fulfilling such diverse roles as biocompatible materials in drug delivery schemes, stabilizing polymers in nancomposites, and as solvent and heat resistant components in automotive applications. Historically, polyethers have been synthesized *via* an array of methods such as coordination-insertion or anionic ring opening polymerization. However, contemporary techniques have difficulty achieving precise control over one or many properties such as molecular weight, molecular weight distribution, or stereochemistry. Here we present a new, facile scheme to produce polyethers using a pseudo-ionic method that allows us to precisely tune a number of physical, chemical, and structural properties. We find that a chemically disparate set of monomers are compatible with our process making it remarkably generalizable. Furthermore, we show that average molecular weight can be easily tuned across a wide range while maintaining control over moleceular weight distribution. Finally, we demonstrate control over both the end groups and the stereo chemistry of the synthesized polyethers. Therefore, we have demonstrated a new polymerization platform for polyethers which provides a significant degree of control over the physical, chemical, and structural properties of the polymer.

472857 Organocatalyzed Atom Transfer Radical Polymerization Driven By Visible Light

Abstract:

The development of atom transfer radical polymerization (ATRP) is among the most significant technological advancements of polymer science. However, contamination of the polymer by the metal catalyst remains a major limitation. Organic ATRP photocatalysts have been sought to address this difficult challenge, but have not achieved the precision performance of metal catalysts. Here, we introduce diphenyl dihydrophenazines, identified through computationally directed discovery, as a new class of strongly reducing photocatalysts. These metal-free catalysts achieve high initiator efficiencies to synthesize polymers with tunable molecular weights and extremely low dispersities, using visible light, including sunlight.

459662 A Full Exploitation of the Pulsed Laser Polymerization Technique to Assess All Important Rate Coefficients in Acrylate Radical Polymerization

Abstract:

Free-radical polymerization (FRP) processes are extensively applied for the production of macromolecular materials, since they allow the synthesis of a large variety of vinyl polymers under mild reaction conditions and with a high tolerance toward impurities.[1] Despite the high industrial importance of FRP, the quantitative understanding of radical polymerization kinetics is still incomplete. Accurate knowledge of Arrhenius parameters and rate coefficients for key reactions such as chain initiation, propagation, and termination is a prerequisite for FRP design, particularly when conducted in large-scale reactors, in which temperature control is a key issue.

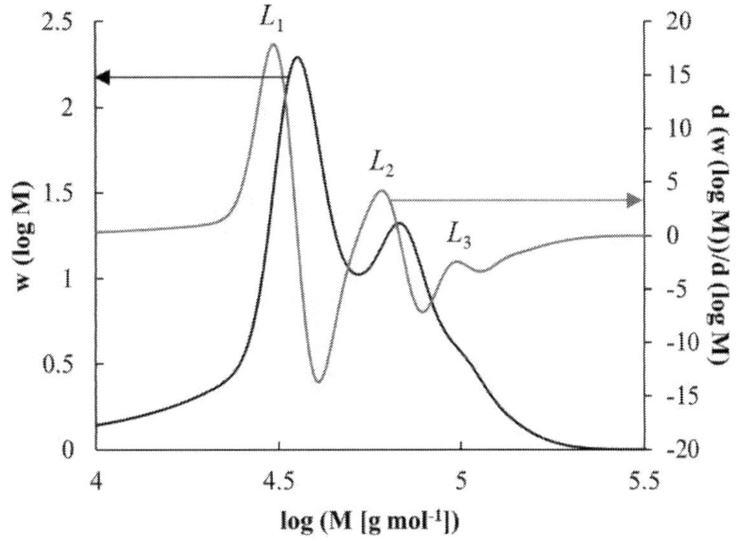

Figure 1. Determination of k_p via the characteristic inflection points of the PLP-SEC trace.

One of the most established experimental techniques to determine intrinsic rate coefficients in radical polymerization is pulsed laser polymerization (PLP; Figure 1). This technique, as originally introduced by Olaj and coworkers,[2] has been extensively used to obtain the propagation rate coefficient k_p. In PLP, photoinitiator radical fragments are consecutively formed via laser pulses with a frequency ν (or dark time $\Delta t = \nu^{-1}$), so that under well-defined conditions (*e.g.* limited monomer conversion) in the corresponding size exclusion chromatography (SEC) trace inflection points (L_j; $j = 1, 2, \ldots$) can be identified that are directly linked to k_p via (Figure 1):

$$L_j = k_p\,[M]_0\,(j\Delta t) \qquad (1)$$

Currently, the PLP technique has been used to accurately measure a wide range of k_p values.[3-7] For standard monomers such as styrene and methyl methacrylate, reliable values are already available. A special case is polymerizations with several radical types, for which the obtained k_p (Equation (1)) must be seen as an apparent averaged one ($k_{p,app}$). For example, in acrylate radical polymerization with applications *e.g.* in the coating and paint industry both secondary end-chain radicals (ECRs) and tertiary mid-chain radicals (MCRs) can be present (Figure 2), leading to the need for the determination of $k_{p,ecr}$ and $k_{p,mcr}$. The tendency of ECRs to switch to MCRs is expressed by the backbiting rate coefficient k_{bb} (Figure 2). Currently, a limited number of experimental methods exists for the determination of k_{bb}. One of the most promising methods was introduced by Nikitin and coworkers,[8] based on the observation that at high ν, Equation (1) relates to $k_{p,ecr}$, while at low ν, Equation (1) is influenced by both $k_{p,ecr}$ and $k_{p,mcr}$. From the onset of the sharp decrease in $k_{p,app}$ with decreasing ν, Nikitin and coworkers[8] assessed k_{bb}. Very recently Wenn and Junkers[9] suggested a more simplified procedure, based on the position of the sigmoidal fit to $k_{p,app}(\nu)$.

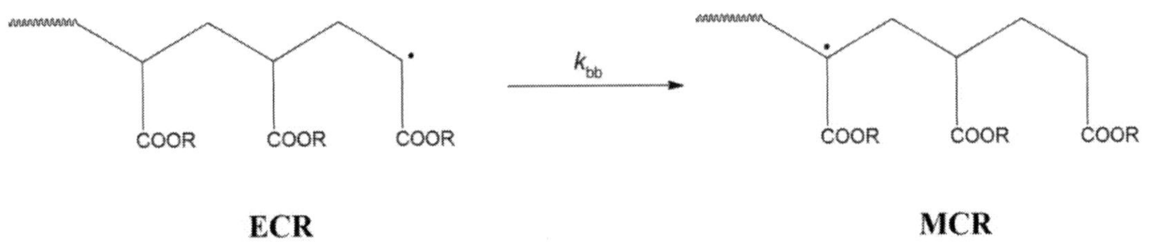

Figure 2. Backbiting reaction leading to a transformation of the radical nature from secondary to tertiary in acrylate radical polymerization; dominant path via a cyclic six-membered transition state.[10]

In the present work, a novel and highly accurate method to determine k_{bb} is presented, using a detailed kinetic Monte Carlo (*k*MC) model and considering 2,2-dimethoxy-2-phenylacetophenone (DMPA) as photoinitiator and *n*-butyl acrylate as monomer. For different solvent volume fractions (0-0.75), with this model regression analysis is applied to inflection point data in the low frequency domain (~100 s^{-1}), which can be easily scanned with less expensive PLP equipment. The novelty of the method lies in the variation of the solvent volume fraction, which allows to independently change the average MCR lifetime and to obtain a high sensitivity toward k_{bb}.

The developed *k*MC model is also capable of the accurate simulation of the complex SEC trace for acrylate PLP, hence, allowing a kinetic analysis transcending the correct simulation of only inflection points for k_p and k_{bb} determination. In particular, hidden information on chain initiation and short-long termination can be elegantly extracted. It is for instance confirmed that inhibition with one of the DMPA radical fragments as formed by photoinitiation is taking place by detailed analysis of the peak intensities. In addition, it is demonstrated that the extracted PLP data for short-long termination ($k_{t,app,ij}$; *i,j*: chain length) can be used to benchmark FRP diffusion models in the less studied regime of diluted conditions or low monomer conversions. Such information is crucial to design not only FRP processes but also novel radical polymerization processes, aiming at improved microstructural control.

[1] D. R. D'hooge, P. H. M. Van Steenberge, M.-F. Reyniers, G. B. Marin, *Prog. Polym. Sci.* **2016**, in press (DOI: 10.1016/j.progpolymsci.2016.04.002).

[2] O. F. Olaj, I. Bitai, F. Hinkelmann, *Macromol. Chem. Phys.* **1987**, *188*, 1689.

[3] K. B. Kockler, A. P. Haehnel, T. Junkers, C. Barner-Kowollik, *Macromol. Rapid Commun.* **2016**, *37*, 123.

[4] S. Beuermann, M. Buback, *Prog. Polym. Sci.* **2002**, *27*, 191.

[5] A. P. Haehnel, M. Schneider-Baumann, L. Arens, A. M. Misske, F. Fleischhaker, C. Barner-Kowollik, *Macromolecules* **2014**, *47*, 3483.

[6] T. Junkers, S. P. S. Koo, C. Barner-Kowollik, *Polym. Chem.* **2010**, *1*, 438.

[7] C. Barner-Kowollik, F. Gunzler, T. Junkers, *Macromolecules* **2008**, *41*, 8971.

[8] A. N. Nikitin, R. A. Hutchinson, M. Buback, P. Hesse, *Macromolecules* **2007**, *40*, 8631.

[9] B. Wenn, T. Junkers, *Macromol. Rapid Commun.* **2016**, *37*, 781.

[10] D. Konkolewicz, S. Sosnowski, D. R. D'hooge, R. Szymanski, M. F. Reyniers, G. B. Marin, K. Matyjaszewski, *Macromolecules* **2011**, *44*, 8361.

464664 High-Pressure Polymerization Process Technology: Modeling and Control of Polymeric Micro-Structure and Safety Considerations

Abstract:

The high-pressure process technology is extremely demanding with respect to invest, operation and safety considerations. Operation pressures up to 3000 bar and temperatures up to 300°C+ as well as production scales up to 450000 t per year and plant put special demands on material, operations and process safety. It is easily understood that at such operation conditions und scales explorations by trial and error are no longer an option. Even though one would accept the extraordinary costs, the safety aspect will tell to reject such ideas. Therefore, introducing new chemistry into processes such as the production of low-density polyethylene (LDPE) still the classical route is chosen with trials on mini-plants and on small and intermediate production scale. Examples for this will be presented at this meeting by Duchateau and Stimeier. However, understanding and improving the world-scale operation plants is also an issue. At this point modeling is an extremely helpful tool. It is this motivation why efforts have been taken in this field for quite some time and early on. Quite a remarkable level of quality in description has been reached up to now applying process modeling on the level of ODEs or Monte Carlo techniques to gather insight into the detail of polymeric micro-structure. In recent years a combination of modeling on the level of ODEs and Monte Carlo techniques together, so-called hybrid technology, has made it feasible to investigate complete production processes on a level of detail such as polymeric micro-structure, branching on the level of individual molecules together with their topology and incorporation of co-monomers on a molecular level. Reliable data for the kinetics and thermodynamics are indispensable for this as it will be discussed by Pflug at this meeting. All this helps building the bridge from control of micro-structure by process conditions towards application properties being characterized by rheology. Besides our group specifically the group of Kiparissides, Tobita and Reed generated remarkable progress in this field.

In the course of these developments elementary kinetics have been discussed with increasing detail. Subject of such discussions have been the propagation of terminal double bounds generating an excess of long-chain branches up to

networks, the effect of molecular topology onto the resulting structures after Î²-scission or degradation by shear forces inside the reactor. The quality of description was excellent in any of the demonstrated case studies. It is for the nature of such projects that the application is often restricted to specific cases of processes or selection of cases. A question that remains open for some instances is whether the found solution is unambiguous and no other mechanisms could generate similar results. Objects of inspection have been in such cases the high molecular weight shoulder of molecular weight distributions produced in autoclave reactors featuring attractive properties for coating material when having the right shape. A wider view onto the application of models such as tubular or autoclave processes, varying process licenses and/or production facilities should help to better understand the quality of a model with respect to universal application. It is this criterion that brings a model closer to its final target to be useful as design tool for processes and products. For such a purpose our group applied systematically simulation models onto various process applications. Results and conclusions about the relevance of the individual elementary kinetic processes for the different production strategies will be presented. It is interesting to see how these findings can be transferred to processes close to the reactor wall such as the laminar layer of flow.

Beyond product design by modeling, it would be favorable for the application of models to process optimization being able to describe the upper decomposition limit of ethene high-pressure mixtures within such models. The decomposition is an ultra-fast process imposing temperature trajectories of more than 1000 K/s and pressure trajectories of way beyond 1000 bar/s. Besides the loss of production running into such decompositions means significant stress for the operation equipment that should be prevented. Pioneers for the incorporation of decomposition reactions into models have been Zhang and Ray. For the sake of simplicity, they separated polymerization and decomposition kinetics to have a clear non-coupled action of parameters within their model. However, inspecting the two independent kinetic schemes one finds similar species and it is easily understood that the kinetics must be coupled also burning the produced polymer into ashes as it is observed for decompositions in practice. This contribution will inspect how far progress in coupling the kinetic mechanisms can be achieved although access to experimental sensors is limited. The experimental counterpart for this study are decomposition test under well-defined laboratory conditions. Decomposition temperatures and pressures are matched acceptable so that a potential can be seen that in near future simulation models may inherently incorporate limiting process conditions. At the moment the state of the art is that model based process suggestions have to be checked manually against experimental data.

The setup for experimental decomposition test on laboratory scale is equipped with pressure and temperature sensors for data acquisition on Î¼s base. This is sufficiently fast for pressure release experiments under well-defined laboratory conditions. Models may support analysis of pressure release experiments for a better understanding. First results will be shown for this.

455614 Simulation of Thermoset Polymerization

Abstract:

For decades the multi-billion dollar urethane industry has been founded on urethane formulations created primarily as an art by researchers with years of experience and laboratory trial and error. It is now possible to simulate the critical resin and foam forming processes, and to use that simulation to develop new formulations. The simulation has been used to estimate concentration, temperature, and density profiles.

The most recent advances have included the ability to simulate how changes in viscosity impact the mass transfer processes that limit cell growth (density) and slow down reactions as the gel hardens. These simulations are able to quantify underlying phenomena that till now were based primarily on speculation, such as: a) how some blowing agents are trapped in the resin because the resin cures before they can diffuse to the cell, b) the extent to which the pressure in foam cells decrease as temperature cools, c) the reasons why some foams shrink upon cooling, and d) the vapor pressure (and impact on emissions) of isocyanate monomers during the foaming process.

Simulation is now ready for use as a primary research and development tool for a) variations in foams to meet new criteria on density versus strength, b) adopting the use of new monomers, c) adopting the use of new catalysts, d) adopting the use of use blowing agents (or mixtures thereof), e) reducing emissions, and f) routine characterizing of monomer performance. This talk will summarize simulation capabilities and demonstrate the latest capabilities related to impact of viscosity on mass transfer processes.

461869 Development of a Coarse-Grained Model of Chitosan for Predicting Solution Behavior

Abstract:

Chitosan is a versatile biopolymer that has gained the attention of researchers in fields ranging from drug delivery and wound healing to water purification and oil spill remediation. It consists of a random sequence of glucosamine (GlcN) and N-acetylglucosamine (GlcNAc) monomers, which have drastically different properties in aqueous solution. GlcN monomers contain a primary amine that becomes protonated in solution, causing electrostatic repulsion between monomers, while GlcNAc monomers contain acetyl groups that introduce hydrophobic and hydrogen bonding interactions, leading to self-association between monomers. The amine group on GlcN can also be hydrophobically-modified to further influence its behavior in solution. We have developed a coarse-grained model of chitosan to use in designing self-assembled chitosan networks that can account for variation in the degree of acetylation (DA), charge density, and hydrophobic modification. Our results show that increasing the chitosan DA results in the formation of a percolated "gel-like" network in solution. Additionally, simulations of three different sequences of acetylated monomers (random, evenly spaced, and blocky) show that increasing the blockiness of the monomer sequence decreases the time required to form a network. The monomer sequence also affects the pore size distribution of the resulting chitosan networks, with blocky sequences leading to the largest pore sizes. We performed simulations of the diffusion of different sized particles within the chitosan networks, and showed that networks formed from evenly spaced monomer sequences (which result in the smallest pores) slow the diffusion of large particles but allow rapid diffusion of small particles. These results are being used to design chitosan hydrogels that could be used in the future to control the rate of drug release in the body. Our results also show that the addition of hydrophobic modification chains to chitosan leads to the formation of a "gel-like" network in the presence of hydrophobic particles. This indicates that hydrophobically-modified chitosan could be used as an effective additive for oil dispersants, causing oil to form a gel which could easily be removed from water following a spill.

452328 Bioinspired Materials Meet Microbiology: The Role of Ion-Containing Polymers and Structure-Property Relationships in Preventing Biofouling

Abstract:

Surface contamination by microorganisms is of great concern in a wide range of applications, including, wound healing scaffolds, medical devices, food packaging, and membranes used for water purification. As both microbial resistance and infectious diseases remain, new approaches to delay the onset of biofilm formation with less evolutionary pressure on pathogens are needed. Our research explores the effect that "green" bioactive agents, charge-containing polymers, and structure-property relationships have on the accumulation and inactivation of bacteria. Discussed here, I will highlight our use of antimicrobial compounds derived from edible plants that potentially, help prevent biofilms without encouraging the development of resistance genes. We have used cationic biopolymers and zwitterions, such as, chitosan and poly(methacryloyloxyethyl phosphorylcholine) (polyMPC), to render nanofiber mats and hydrogels antibacterial and antifouling. Additionally, we have demonstrated that even "greener" strategies including structure-property relationship can reduce microbial accumulation. For example, systematically, we investigated poly(ethylene glycol) dimethacrylate and agar hydrogels as a function of Young's moduli. Independent of hydrogel chemistry and incubation time, *Escherichia coli* and *Staphylococcus aureus* attachment decreased on softer hydrogels. Our recent investigations into synergistically pairing structure-property relationships with bioinspired charge-containing polymers will also be discussed. The overall goal of the talk is to illustrate some of our most recent findings and how these results can guide the green engineering of multifunctional antibacterial and antifouling materials.

469322 Conducting-Polymer Electrochromic Windows Powered by Transparent Single-Junction Organic Solar Cells

Abstract:

Lighting and cooling accounts for 20% of the energy consumption of residential and commercial buildings in the United States. Reducing energy needs associated with lighting and cooling will shave electricity peak demands and increase building efficiencies. In my talk, I will highlight the demonstration of a solar-powered electrochromic window that can be integrated into windows to increase building energy efficiency.

The solar-powered electrochromic window comprises a polyelectrochromic conducting polymer that is transparent in its reduced state and dark blue in its oxidized state. Integration with a semitransparent organic solar cell provides the necessary power to switch between its transparent and colored states. Unique to this approach is the use of a single-junction organic solar cell. By using materials that absorb exclusively in the ultra-violet and near-visible as photoactive layers of our solar cells, the resulting devices exhibit open-circuit voltages that are unprecedented for single-junction organic devices (> 1.4 V). As such, we are able to drive switching of the electrochromic window without the need to construct solar cells with complex tandem architectures. Importantly, the active layers of our devices are pinhole- and defect-free. Coupled with the inherently low resistive power losses in our devices, our device currents are scalable with areal coverage with minimal losses to device fill factor and maximum power. While device optimization of typical solar cells to increase photovoltage has almost always come at the expense of photocurrents, the scalability of photocurrents in our devices has allowed us to break this paradigm. We can tune the photovoltage per application needs through judicious selection of the donor/acceptor materials pair having prescribed energy levels and access the desired current levels by fabricating arbitrarily large devices.

451950 Ion Sorption, Diffusion and Transport in Charged Polymer Membranes

Abstract:

Charged polymer membranes are widely used for water purification applications involving control of water and ion transport, such as reverse osmosis and electrodialysis. Efforts are also underway worldwide to harness separation properties of such materials for energy generation in related applications such as reverse electrodialysis and pressure retarded osmosis. Additional applications, such as energy recovery ventilation and capacitive deionization, rely on polymer membranes to control transport rates of water, ions, or both. Improving membranes for such processes would benefit from more complete fundamental understanding of the relation between membrane structure and ion sorption, diffusion and transport properties in both cation and anion exchange membrane materials. Ion-exchange membranes often contain strongly acidic or basic functional groups that render the materials hydrophilic, but the presence of such charged groups also has a substantial impact on ion (and water) transport properties through the polymer.

We are exploring the influence of polymer backbone structure, charge density, and water content on ion transport properties. Results from some of these studies will be presented, focusing on transport of salt, primarily NaCl, through various neutral, positively charged and negatively charged membranes via concentration gradient driven transport (i.e., ion permeability) and electric field driven transport (i.e., ionic conductivity). One long-term goal is to develop and validate a common framework to interpret data from both electrically driven and concentration gradient driven mass transport in such polymers and to use it to establish structure/property relations leading to rational design of membranes with improved performance.

Ion sorption and permeability data were used to extract salt diffusion coefficients in charged membranes. Concentrations of both counter-ions and co-ions in the polymers were measured via desorption followed by ion chromatography or flame atomic absorption spectroscopy. Salt permeability, sorption and electrical conductivity data were combined to determine individual ion diffusion coefficients in neutral, cation exchange and anion exchange materials. Manning's counter-ion condensation models and the Mackie/Meares model were used to correlate and, in some cases, predict the experimental data.

461687 Instructing Cells with Programmable Peptide-DNA Hybrids

Abstract:

Instructing Cells with Programmable Peptide-DNA Hybrids

Ronit Freeman[1] *, *Nicholas Stephanopoulos*[1] *, *Samuel I. Stupp*[1]

1. Simpson Querrey Institute for BioNanotechnology, Feinberg School of Medicine, Northwestern University, Chicago, IL 60611, US, Email: s-stupp@northwestern.edu

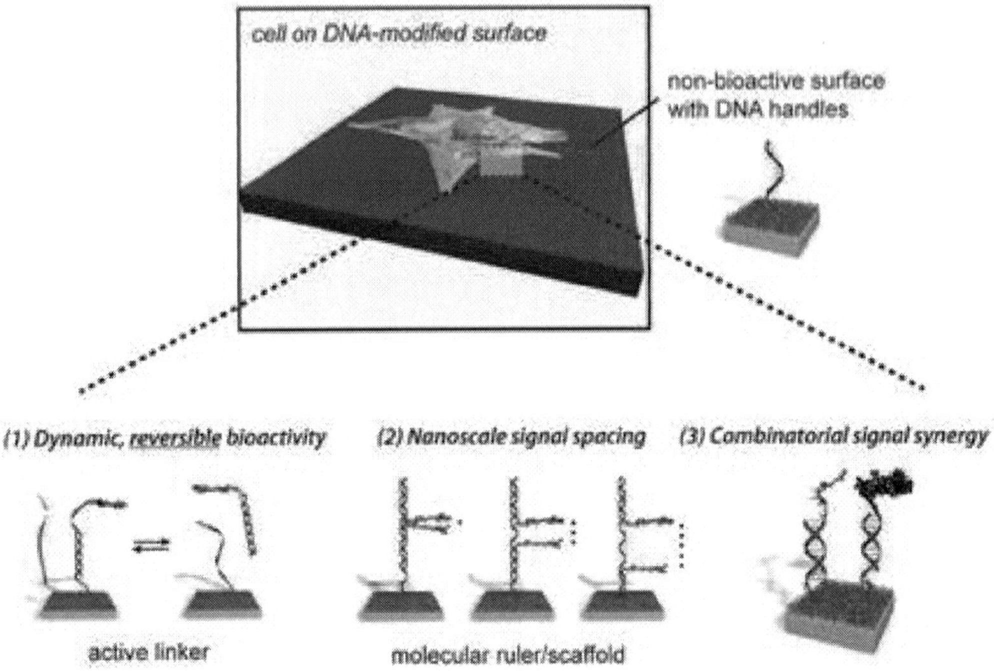

The native extracellular matrix is a space in which signals can be displayed dynamically and reversibly, positioned with nanoscale precision, and combined synergistically to control cell function. Artificial forms of this matrix for tissue regeneration need to recapitulate these three characteristics in a single molecular platform. Most efforts in this area have effectively addressed only one of these three key phenomena, and focused mainly on static cell adhesion or irreversible switching of bioactivity. Here we describe a molecular system that can be programmed to control the dynamics, spatial positioning, and combinatorial synergies of signals in extracellular matrices. In this approach, a biological signal is immobilized to a surface through DNA tethers, Figure 1. By engineering a series of tethers responsive to different stimuli, we show that cells adhered and spread on the surface reversibly. The DNA was also used as a molecular ruler to control the distance-dependent synergy between two adhesion peptides. Finally, orthogonal DNA tethers enabled the binding of both a peptide and a bioactive protein to the surface in order to promote cell differentiation. This peptide/protein-DNA (P-DNA) platform controls the way different signals are presented to cells and their relative spacing and confers superior reversibility by employing a set of bio-friendly stimuli. Multiple orthogonal DNA handles can be designed to allow for the selective presentation of different signals, with the ability to independently up- or down-regulate each over time. In principle, the anchoring strand can be attached to any type of material, including three-dimensional hydrogels, to better emulate the properties of the ECM.

Figure 1. Schematics of a DNA-modified surface used for the stimulus-responsive presentation of bio-active peptides and proteins in order to control cell adhesion, synergistic effects and differentiation.

472149 Delivery Materials to Induce RNAi in Bone Marrow to Control Hematopoietic Stem Cell Trafficking

Abstract:

Delivery Materials to Induce RNAi in Bone Marrow to Control Hematopoietic Stem Cell Trafficking

Michael J. Mitchell[1], Robert Langer[1]

[1]Department of Chemical Engineering, Koch Institute for Integrative Cancer Research, MIT, Cambridge, MA 02139

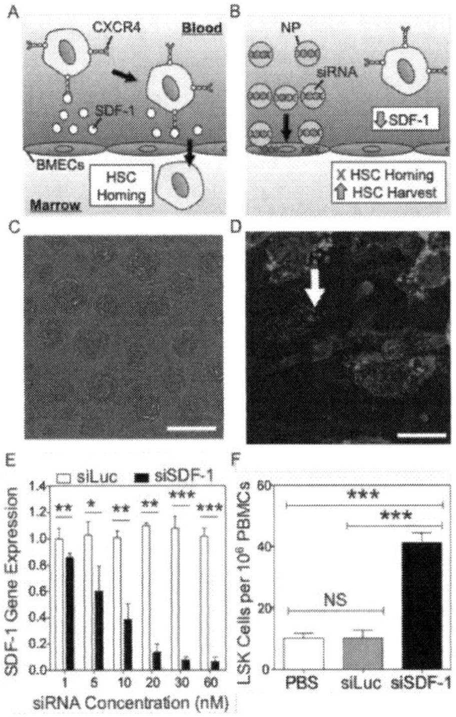

Introduction: Hematopoietic stem cell (HSC) transplantations (HSCT) are essential for restoring a patientÕs immune system destroyed by radiation or chemotherapy. However, HSCT remains a dangerous procedure for patients with life-threatening diseases, with a major complication being harvesting HSCs from blood to restore the recipientÕs immune system. Here, we developed a nanoparticle (NP) platform to induce RNAi in bone marrow (BM) *in vivo* to increase HSCs in blood for harvesting and transplantation (Fig 1A,B).

Methods: siRNA targeting stromal-derived factor-1 (SDF-1) and luciferase (Luc) were encapsulated in NPs consisting of lipids and modified polyethylenimine via rapid microfluidic mixing. NPs were characterized using DLS, SEM, and cryo-TEM. BM endothelial cells (BMECs) were treated with NPs at siRNA dosages of 1-60 nM *in vitro*. C57/BL6 mice were treated with NPs at a siRNA dosage of 1.0 mg/kg to determine *in vivo* SDF-1 knockdown. HSCs were harvested from blood 1-week after initial NP treatment and characterized via flow cytometry as Lin$^-$Sca-1$^+$c-Kit$^+$ (LSK) cells.

Figure 1: (A) HSC homing into marrow via SDF-1. (B) SDF-1 knockdown via *in vivo* NP-mediated RNAi reduces homing and increase HSCs within blood. (C) Cryo-TEM micrograph of lipid-polymer NPs encapsulated with siRNA. Scale bar: 100 nm. (D) NP-mediated siRNA (red) uptake in BMECs. Scale bar: 10 μm (E) NP-mediated SDF-1 knockdown in BMECs. (F) Lin$^-$Sca-1$^+$c-Kit$^+$ (LSK) HSCs isolated from blood per 10^6 peripheral blood mononuclear cells (PBMCs) 1-week post-SDF-1 knockdown. *P<0.05. **P<0.01. ***P<0.001. NS: not significant.

Results and Discussion: 60 nm polymer-lipid NPs containing siSDF-1 were formulated via microfluidic mixing (Fig 1C). NPs containing fluorescent siRNA were uptaken by BMECs as per confocal micrographs (Fig 1D). NPs containing siSDF-1 induced potent gene knockdown in BMECs, compared to siLuc controls (Fig 1E). *In vivo*, silencing of SDF-1 induced a nearly 300% increase in the number of HSCs harvested from blood, compared to controls (Fig 1F).

Conclusions: We have developed a NP platform that, potentially for the first time, induces potent gene knockdown in the bone marrow microenvironment *in vivo*. Clinically, this platform can be used to silence BM factors to increase the number of HSCs in blood available for harvesting and subsequent transplantation.

470144 Toll-like Receptor (TLR)-Functionalized Nanoparticle Adjuvant Carriers Toward Optimized Vaccine Formulations and Immune-Modulators

Abstract:

Adjuvants are molecules which enhance immune responses by stimulating various cellular pathways and are critical components of vaccine formulations. These molecules enhance the immunogenicity of synthetic antigens, which are otherwise unable to produce a robust response needed for immune protection. There are an astoundingly few clinically approved adjuvant molecules and there remains a huge unmet need for the development of novel adjuvant delivery approaches. Engineered particulate carriers offer a unique opportunity to deliver adjuvant molecules, as they can interface with key cells of the immune system for a variety of therapeutic applications, including vaccines and immunomodulation. While recent studies have demonstrated superior pre-clinical efficacy of nanoparticle vaccine formulations, surprisingly little work has been performed investigating the adjuvant component on the nanoparticle formulation. In this work, we engineer a series of particles to investigate the role of adjuvant density on the particle surface. We incorporate two toll-like-receptor (TLR) adjuvant molecules, CpG and LPS, on the surface of both cationic and anionic particle carriers and quantify the immune response in various biological environments as a function of particle characteristics. Our results indicate when dosed at equivalent particle concentrations with varied adjuvant surface densities, particles with the highest adjuvant density per particle resulted in the largest *in vitro* stimulation; however, when dosed at equivalent adjuvant mass per sample, particles of the lowest adjuvant density, corresponding to the highest number of particles dosed, resulted in superior stimulation. These results indicate a trade-off between adjuvant density and particle number. We believe this work provides a set of design rules which inform particulate vaccine formulations, to ensure both potency and safety, with potential future applications for novel treatments for cancer, inflammation, and allergy.

472993 Fabrication of Vaccines Based on Polymeric Microdisks to Enhance T Cell Immunity

Abstract:

Biomaterial vaccines offers unique opportunities to treat infectious disease and cancer. Interestingly, recent studies have found that the shape of synthetic vaccine carriers can impact their interactions with immune cells (e.g. macrophages, dendritic cells), leading to modulation of immune signaling and ultimately influencing the nature of immune response to a particular antigen.

In this work, we developed a disk-shaped polymeric vaccines through combining soft lithography with biodegradable polymers. The microdisks are generated by spin-coating poly(lactic-co-glycolic acid) (PLGA) solution mixed with toll-like receptor agonists 9 (i.e. CpG) and a model antigen (i.e. SIINFEKL) on a polydimethylsiloxane (PDMS) stamp with protruding micropillar array structures. Diameter of the microdisks, as well cargos (i.e. CpG and SIINFEKL) loadings, can be well controlled by the spin coating speed and the concentration of PLGA solution and cargos, allowing formation of microdisks with diameters of 1 ± 0.21 µm to 6.8 ± 0.35 µm. These values correspond to aspect ratios of 5 to 35. Importantly, these disk-shaped vaccines allow control over the relative loading of each vaccine component, provide a route to alter both the geometric properties and the immunological signals included in the disks. *In vitro* studies are testing the capacity of microdisks to alter the activation and presentation of SIINFEKL in primary dendritic cells (DCs) and T cell co-culture. The ability of these materials to drive antigen specific responses in mice and during tumor challenge will also be presented. This work allows us to investigate the role of each vaccine signals, as well as their relative ratios in T cell immunity.

467006 Temporally Controlled Release of Platelet-Rich Plasma from Peg Microgels with Tunable Biodegradation Rate and Size

Abstract:

Biomedical applications such as drug delivery and tissue engineering often require polymeric hydrogels which degrade predictably under physiological conditions. Biodegradable injectable drug delivery systems present attractive avenue for site specific and minimally invasive administration, coupled with precise temporal control over drug release, which can be employed for multiple applications. Specifically, poly (ethylene glycol) (PEG) has been thoroughly investigated for the design of biodegradable hydrogel networks because of its exceptional biocompatibility and inertness. In this study we designed biodegradable PEG microgels with a wide range of degradation kinetics using Michaelâ€™s addition chemistry. We further used these microgels for encapsulating and controlling delivery of proteins derived from platelet rich plasma (PRP). PRP is a concentrated assortment of growth factors and cytokines derived from the patientâ€™s own platelets. PRP has been proven useful in a number of regenerative medicine applications. However, an effective method to achieve sustained delivery of PRP is still required.

For the design of degradable PEG hydrogels we used multiarm PEG-acrylate (PEGAc) of varying molecular weight and number of arms (4-arm of 10 kDa, to 8-arm of 20 kDa) and three categories of dithiol crosslinkers, namely, ester containing, non-ester containing and dithiols with neighbouring functional groups of different electronegativities. The PEG hydrogel physical and mechanical characteristics could be modulated depending upon the reaction parameters and gelation time for multiarm PEGAc and dithiol crosslinker structure and molecular weight. By changing these parameters, hydrogels with controlled degradation ranging from 10 hours to 32 days and having mesh size in the range of 9 to 14 nm were obtained. We identified two sets of conditions which yielded fast gelling-fast degrading hydrogels and slow gelling-slow degrading hydrogels. Uniquely, the hydrogel storage moduli could be controlled by the dithiol crosslinker chemical identity independent of the degradation time or mesh size of the hydrogel.

Further, we used the optimized combinations of PEG hydrogels to make PEG microgels of sizes varying between 50 to 700 μm by manipulating several key electrospraying parameters. We also investigated various techniques and times suitable for storing protein-loaded PEG microgels. The microgels were stored at seven conditions: 25°C, 4°C, -80°C, Lyophilization (Lyo), -80°C/Lyo, -80°C/DMSO, and -80°C/DMSO/Lyo for 1, 4, 7, and 28 days. DMSO was introduced at 10% (v/v) in 1x PBS. Our results indicated that neither storage modulus nor the structure of incorporated model protein was affected post storage at different conditions. However, comparing the percent distribution of microgel size pre- and post-storage, we found an increase in the mean diameter indicating increased swelling post-storage. Furthermore, environmental scanning electron microscopy (ESEM) images showed that large pores were created by some of the conditions, mainly by lyophilization.

Finally, we tested the PEG hydrogels and microgels for encapsulation and delivery of PRP-derived growth factors. PRP at 2-10% w/v was encapsulated during gelation of the hydrogels or microgels. Release of growth factors was quantified by multiplexer analysis. Biological activity of PRP released from the hydrogels or microgels were tested *in vitro* using human dermal fibroblast (HDF) or by circular dichroism, respectively. Direct encapsulation of model proteins in the PEG hydrogels during hydrogel gelation preserved their secondary structure. Release kinetics for several model proteins was found to be dependent on the protein size, as well as mesh size and degradation rate of the hydrogel. The amount of total protein release increased over time, corresponding to increase in hydrogel mesh size. Multiplexer growth factor analysis of the releasates showed continuous release of growth factors (PDGF-BB, EGF, RANTES) from PRP-containing hydrogels until degradation. PRP releasates also demonstrated increased proliferation in HDF cells over a period of 72 hours, comparable to that of a bolus PRP injection. Furthermore, similar results could be translated to PEG microgels obtained by electrospraying.

Thus it can be concluded that we could obtain a range of biodegradable PEG hydrogels by a combination of number of PEG arms and dithiol crosslinkers and reaction pH. Further, these hydrogels could be made into injectable microgels which allowed localized and controlled delivery of PRP growth factors in the active form.

471818 Layer-By- Layer Encapsulation of Probiotics: Addressing the Challenges of Oral Delivery to Modulate the Microbiome

Abstract:

The gastrointestinal (GI) microbiome has been widely investigated for its role in many diseases, including diabetes, depression, and cancer; as such, efforts to modulate the microbiome are of particular interest. While many novel technologies for oral delivery have been developed, the challenges arising from the transport of live probiotics through the stomach to target areas in the GI tract (e.g. intestines and colon) have yet to be addressed. Here, we report a layer-by- layer (LbL) method for the encapsulation of probiotics to directly address these challenges by simultaneously: (i) protecting probiotics from GI tract insults to circumvent acid- or bile-mediated bacteria death, (ii) facilitating mucoadhesion and direct growth on intestinal surfaces, and (iii) controlling layer release via inclusion of pH sensitive polymers so as to facilitate targeted growth in pH-regulated areas of the GI tract. These effects translated in vivo for over 6-fold enhanced survival of LbL-probiotics as compared to non-coated probiotics. LbL-encapsulation of probiotics facilitates survival of probiotics through the GI tract to ensure adhesion and growth to intestinal tissues for potential applications in modulating and directing the microbiome.

472411 VE-Cadherin Signals and Substrate Stiffness Regulate Force Transduction through Endothelial Monolayers

Abstract:

The vascular endothelium forms a semi-permeable barrier, restricting the transfer of proinflammatory cytokines and endothelial hormones within body fluids to underlying tissues. Vascular endothelial cadherin (VE-cad) complexes are the main cohesion and force transducing proteins at endothelial adherens junctions, connecting cytoskeletons to form an integrated dynamic mechanical network, regulating endothelial function and permeability. Past studies showed that localized forces on VE-cad receptors trigger changes in global endothelial mechanics when cells were cultured on glass [Barry, A.K. *et al.*, *J. Cell Sci.* **2015**, *128*, e0705]. The signals further propagated through the monolayer to affect distant cells. In this study, we used traction force and monolayer stress microscopy to investigate the impact of matrix stiffness cadherin-activated cell responses and the long-range propagation of mechanically-activated signals.

Polyacrylamide gels (8.8 kPa and 40 kPa) with embedded fluorescent fiducial markers were prepared as described [Muhamed, I., *et al.*, *J. Cell Sci.* **2016**]. Cured gels were dried overnight and micro-spotted with fibronectin, to create circular 500 µm diameter patterns. Gels were then re-hydrated, UV-sterilized, and seeded with human pulmonary artery endothelial cells (HPAECs) for 24 hr, to yield a confined, confluent monolayer with 200-280 cells/array of 10 µm thickness. Next, carboxyl ferromagnetic particles functionalized with VE-cadherin were placed on the apical cell surface, and a twisting torque was applied to VE-cad receptors with a magnetic twisting cytometer (MTC) [Wang, N. *et al.*, *Science* **1993**, *260*, 1124–1127]. Time lapse images of the fiducial markers in the gels recorded during MTC experiments were post-processed to calculate cell-matrix (traction) and cell-cell (monolayer) stresses. Tractions were calculated using a custom-written software and a Fourier transform algorithm. Monolayer stresses were computed by finite element analysis [Tambe, D. T., *et al.*, *Nat. Mater.* **2011**, *10*, 469–475].

We first assessed monolayer tractions (Fig. 1a). On soft substrates (8.8 kPa), the population averaged tractions of unperturbed (-) and perturbed (+) monolayers were constant over the course of the experiment, as opposed to behavior on stiff substrates (40 kPa). The same trend was observed for monolayer stresses. Interestingly, computed stresses on perturbed monolayers on both substrate stiffness were initially higher than on unperturbed monolayers (Fig. 1b, c, and d). Additionally, on stiff substrates and perturbed cells, the monolayer stress distribution exhibited a rapid reduction in stress magnitudes with respect to unperturbed monolayers, due to the formation of endothelial gaps and loss of interendothelial connectivity. We consider how the substrate stiffness affects long-ranged monolayer disruption activated by VE-cadherin force transduction.

Our study indicates that basal (initial) traction and monolayer stresses anti-correlate and that external mechanical stimuli via VE-caderin shifts traction stresses from focal adhesions to adherens junctions. Collectively our data demonstrate how matrix stiffness and interendothelial stresses coordinate to regulate the collective integrity of endothelial monolayers.

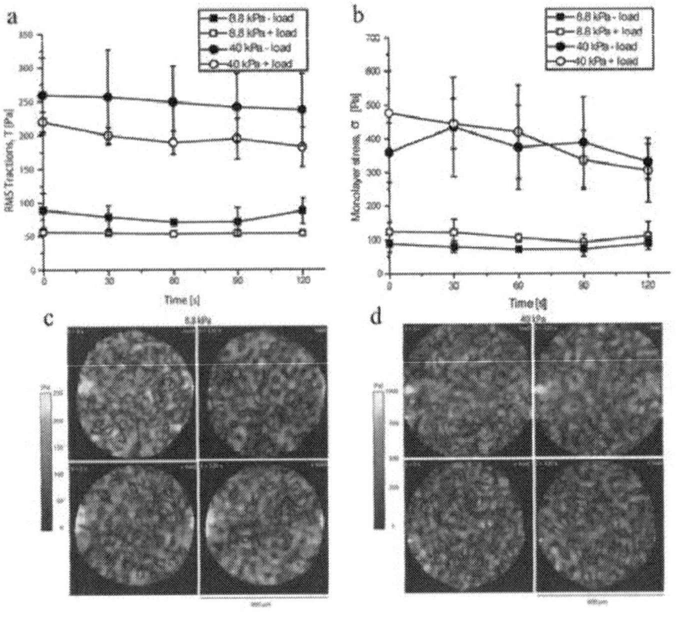

Fig. 1.a) time resolved (TR) RMS traction, b) TR magnitude stress distribution, and c) and d) initial and final heat maps displaying stress distribution at time points 0s and 120s for 8.8 kPa and 40 kPa -/+ load.

468550 Macroporous Polymer Scaffolds for the Transplantation of Embryonic Stem Cell Derived Beta-Cell Progenitors to a Clinically Translatable Site for the Treatment of Type I Diabetes

Abstract:

Type I diabetes mellitus, which affects an estimated 1.5 million Americans, is caused by autoimmune destruction of the pancreatic beta cells that results in the need for life-long insulin therapy. Although insulin therapy has been successful, hypoglycemic events and vascular complications persist. Allogeneic islet transplantation for the treatment of type I diabetes is a therapy in which donor islets are infused intrahepatically, which has led to the transient reversal of diabetes. However, allogenic transplantation has several therapeutic limitations, which include a shortage of donor islets, long-term immunosuppression, and high risk of tissue rejection. All of these limitations have led to the investigation of embryonic or induced pluripotent stem cells as an unlimited source of functional beta-cells. The ability to differentiate embryonic stem cells into pancreatic progenitors *in vitro*, with subsequent maturation to beta-cells *in vivo* has been demonstrated; however, these systems are not implemented in a manner that is clinically translatable. Therefore, our group has developed biomaterial scaffolds that support the *in vivo* maturation of pancreatic progenitors into mature beta-cells at an extrahepatic site that is clinically translatable (omentum). We have previously reported on the ability of microporous poly(lactide-co-glycolide) (PLG) scaffolds to create an environment that supports long-term function of islets at an extrahepatic site, and reversal of diabetes with a minimal islet mass. Herein, we propose to investigate the modification of both PLG scaffolds with extracellular matrix proteins, such as collagen, and the delivery of trophic factors, such as Exendin-4, to promote the *in vivo* development of human embryonic stem cell (hESC) derived pancreatic progenitors into mature insulin producing beta-cells. Our studies have demonstrated that PLG scaffolds support the differentiation to insulin producing cells, and that sustained exendin-4 delivery significantly increased C-peptide production and improved glycemic control compared to scaffolds without Exendin-4. In conclusion, the goal of this research is to develop a platform on which hESC-derived pancreatic progenitors can be

seeded and implanted into a clinically translatable site, such as the omentum, that will allow for efficient maturation of the cells into a population that can reverse diabetes.

467580 Anisotropic, Acellular, Silk-ECM Patches for Treatment of Myocardial Infarction

Abstract:

Heart failure is a major clinical issue that plagues both young patients born with congenital heart defects and older patients who have suffered from a myocardial infarction. These patients undergo reconstructive heart surgery or bypass surgery with positive short-term prognosis, but the disease often results in detrimental scar formation often limiting long-term patient outcomes due to the progression to heart failure. To ameliorate some of these complications natural, bioactive, degradable, and implantable biomaterial systems have been considered. The objective of this study was to improve current cell-free biomaterial designs for cardiac repair by utilizing silk fibroin scaffolds with uniaxially aligned pore architecture containing decellularized cardiac extracellular matrix (cECM) derived from adult or fetal porcine heart tissue (adult cECM and fetal cECM, respectively) with the goal of promoting a better healing response that limited or prevented the progression towards heart failure following MI.

To achieve these goals, silk sponges were formed as previously described,[1,2] with 5 study groups: sham (no repair), silk alone, silk + adult cECM, and silk + fetal cECM, and a commercially available collagen I (control). Briefly, *Bombyx mori* silkworm cocoons were used to obtain pure silk solutions, which was then mixed with ECM proteins derived from decellularized adult or fetal porcine heart tissue (adult or fetal cECM). Adult or fetal cECM was added to 3% wt/vol silk solution at 0.2 mg/mL prior to aligned sponge formation via directional freezing.[1] To generate the animal model, myocardial infarctions (MIs) were induced in 7-week old Sprague Dawley rats via permanent ligation of the coronary artery (Tufts IACUC approved). One week following MI, patches were implanted over the surface of the infarct, starting just below the infarct suture. Animals were monitored 1, 3, 5, 7, 9-, and 11-weeks post-implantation via echocardiography. Pressure-volume loops were acquired at 3- and 11-weeks post-implantation, in addition to analysis of tissue composition via immunohistochemistry, western blot, and qPCR.

Hemodynamics results from PV loops demonstrate that addition of the silk patch containing fetal cECM led to higher ejection fractions and larger stroke volumes at 11-weeks post-implant. Immunohistochemistry analysis showed that silk-fetal cECM patches led to a reduction in infarct expansion and ventricular wall thinning as compared to shams. Cell infiltration and degradation of the original material was improved in patches containing cECM at 11-weeks post-implant. Current work includes analyzing data from a second study of the same groups and also addressing mechanism by looking at cellular identity and gene/ protein expression to understand the role of cECM composition on the infiltrating cell population and the subsequent functional data.

Implantation of silk-cECM patches 1-week post-myocardial infarction appears to limit the progression to heart failure following MI as compared to sham controls. Ongoing work will increase the number of samples per group as well as compare to collagen sponge scaffolds in order to elucidate the role of cECM composition on patch integration, immune response, and maintenance of cardiac function. Future work will also assess whether silk-fetal cECM scaffolds seeded with human-derived induced pluripotent stem cells have greater potential for cardiac repair as compared to the cell-free approach via evaluation in a rodent model of heart disease.

References:

1. Stoppel, W. L., et al., *Biomed Mat* 2015, 10 (3).
2. Rnjak-Kovacina, J., et al., ACS Biomater Sci Eng 2015, 1 (4), 260-70.

467633 Characterization of Polymer Nanoparticle Aggregation in Biologically Relevant Fluids

Abstract:

Nanoparticles are often studied as potential drug delivery carriers, but further understanding of their behavior in the body is necessary prior to successful application. The surface chemistry of a nanoparticle is critical to its aggregation behavior. For targeted drug delivery applications, ligands on nanoparticles are often modified for specific targeting purposes with little known about how the ligands will affect the aggregation of the nanoparticles. In the case of targeted drug delivery carriers, nanoparticles that are injected into the blood stream must successfully circulate for a sufficient duration to find their intended targets. To simulate this particle circulation, researchers typically use animal models such as mice. In these experiments nanoparticles are often shunted to the liver and spleen by the immune system shortly after injection, and following this immune modulation, it is difficult to gain any meaningful knowledge. The use of animals for these experiments is costly and also raises ethical questions. As an alternative, a method of analyzing nanoparticles in biologically relevant fluids such as blood plasma has been developed using nanoparticle tracking analysis (NTA) with fluorescent filters. NTA tracks particles individually, and this particle-by-particle approach makes NTA an accurate method of determining particle concentration and size distribution, even in complex fluids such as blood plasma. Typically dynamic light scattering (DLS) is used to analyze particle size and thus aggregation behavior, but DLS cannot be used in this case because the components of blood scatter light as well. Using NTA with fluorescent filters, the aggregation behavior of the nanoparticles can be observed and quantified and specific ligand formulations can be tested for aggregation behavior in blood plasma before proceeding to in vivo studies.

In this work, NTA was used to analyze the aggregation behavior of fluorescently labeled polystyrene (PS) nanoparticles in blood plasma. The effect of surface chemistry on aggregation behavior was investigated by quantifying percentage of particles in aggregates at a given time point after adding to blood plasma. PS nanoparticles with various surface modifications were tested in blood plasma and the percentage of aggregation was quantified. The use of this characterization method will allow for better understanding of particle behavior once nanoparticles are injected into an animal, and potential problems, specifically aggregation, can be dealt with at this stage before investing heavily in *in vivo* studies.

469139 Application of Metamaterials and Rectenna for Capture of Blackbody Radiation

Abstract:

Metamaterials, a new synthetic set of materials with properties that do not exist on their own in nature, typically consist of various composite materials such as metals and dielectrics. Their repeating geometric structure and material set provide the metamaterial with a negative index of refraction. The negative index of refraction induces a primary function of metamaterials: their ability to manipulate electromagnetic waves. In this study, metamaterials are designed and fabricated to manipulate blackbody radiation into a well-organized, functional source of electromagnetic waves that are rectified to a DC voltage.

The metamaterial-rectenna unit is designed to convert heat into electricity using the optical properties of the blackbody radiation emitted from heat source. The blackbody radiation is optically manipulated by the metamaterial. In general, as the temperature of the object increases, the wavelength and intensity of the blackbody radiation increase. The metametarial-rectenna units fabricated in this study are designed to capture electromagnetic radiation with a frequency of 1 THz.

The constructed device uses a rectenna as the rectifying device to convert the alternating current (AC) induced from the electromagnetic waves emitted as blackbody radiation to a direct current (DC). The rectenna, containing a metal insulator metal (MIM) tunneling diode, utilizes the ordered electromagnetic radiation from the metamaterial. The metamaterial consists of copper and a photosensitive polymer in an alternating pattern that focuses the electromagnetic radiation over onto the rectenna. The metal of choice, copper, is selected due to its high thermal and electrical conductivity. SU-8 is selected as the photosensitive polymer due to its well-characterized optical properties as well as use in photolithography. The metamaterial is fabricated using various techniques of photolithography, including wet and dry etching of metals and oxides, thin-layer deposition of metals using evaporation, and electrodeposition.

In order to test the functionality of the metamaterial, a heat source applies heat at a set temperature to the metamaterial. Using test pads that have been built into the structure and connected to the rectenna, the diodes are characterized using four-point probe IV measurements. The thermal measurements utilize two of these probes to detect any voltage that is generated by the rectenna-metamaterial device when exposed to a controlled heat source. In this study, a proof of concept design of the metamaterial-rectenna units is achieved and a plan for future development is suggested.

469647 Testing Materials and Devices for Electromagnetic Energy Capture and Conversion to Electricity

Abstract:

The subject of this paper is terahertz (THz) recteenna. Rectenna are devices based on two fundamental components. There is an antenna that captures the photonic energy and a rectifier that transforms alternating current (AC) into a directional current (DC) for use in power systems. Research on our team is presently focused on developing efficient rectenna devices capable of capturing electromagnetic (EM) radiation is the form of infrared (IR) emitted waste heat. Materials-based solutions used in this regard include tailoring antenna geometry and composition and coupling metamaterials into the receiver. The antenna are coupled through a feedpoint to a metal-insulator-metal (MIM) type diode. MIM diodes possess the characteristic of being well-suited for high-speed aplications.

This bulk of this paper will center on testing strategies which are not trivial in the 30 - 300 THz region. Both free-space testing and thermal testing results will be presented. In order to truly quantify overall efficiency numbers, our reserach team uses computational finite element modeling (FEM). Thie paper will present results from FEM thermal transport models, optoelectronic models, and nanofabrication models. In summary, results will be provided about rectenna device performance at 1, 5, and 30 THz.

459954 Incorporation of Photo-Responsive Membrane Protein Species into Nanostructured Silica for Light-Driven Ion Transport

Abstract:

Membrane proteins are versatile biomolecules with diverse functionalities that impart sensing, signaling, transport, or catalytic properties that support the viabilities of biological cells. Such functionalities are highly selective to particular ions or molecules and often occur at high rates, which would be attractive for technological applications, such as chemical or biological sensing, separations, bioanalytics, and energy conversion. To effectively exploit membrane proteins for technological purposes often requires their incorporation into synthetic host materials that enable the proteins to function stably and be integrated into macroscopic devices. One interesting example is the membrane protein proteorhodopsin, which functions as a light-driven H^+-ion-pump that might be harnessed for photochemical energy conversion. Synthetic host membranes that contain macroscopically aligned proteorhodopsin species are expected to generate bulk ion gradients across host materials under illumination. Here, we present a solution-based synthetic protocol that allows high concentrations (up to 15 wt%) of active orientationally ordered proteorhodopsin species to be incorporated within nanostructured silica membrane hosts. Synthesis conditions and compositions were selected to stabilize proteorhodopsin molecules in the presence of the structure-directing surfactant and soluble network-forming silica species that co-assemble to form nanostructured silica host matrices, as established by small-angle X-ray diffraction analyses. Multidimensional solid-state NMR spectra show that proteorhodopsin molecules incorporated within nanostructured silica hosts retain native-like structures, though with some interesting differences. The optical absorbance behaviors of proteorhodopsin within the synthetic hosts are analogous to the photochemical reaction cycle of proteorhodopsin in native-like environments associated with the active transport of H^+-ions. The synthesis protocol is expected to be general and has been adapted to incorporate other functionally active membrane proteins within nanostructured silica host membranes. The versatile nanostructured silica-surfactant host materials open opportunities to integrate membrane proteins and their diverse functionalities into synthetic semi-permeable membranes.

461940 Suppression of Infrared Absorption in Nanostructured Metals By Controlling Faraday Inductance and Electron Path Length

Abstract:

While metals are widely used for electrodes in many optoelectronic devices, the device efficiency is limited by optical losses due to metals. As thinner devices are desired for reduced cost without sacrificing performance, the metal electrodes can be nanostructured to concentrate light in the photoactive material. A popular approach for the light concentration is to excite surface plasmon polaritons (SPPs). However, in such SPP-based devices, strong metal absorption is a major source of losses. The problem of this parasitic absorption becomes critical when a high device efficiency is desired. Moreover, if the device operates over a broad band as in solar cells, metal absorption needs to be controlled over the whole spectrum of interest and this poses a significant scientific and engineering challenge. Here, we investigate the loss suppression in metal nanocoil structures in the infrared range. Surprisingly, we find that the nanostructured metals can behave almost like a vacuum with a negligible optical loss in the infrared region. The loss is reduced in comparison to flat films by more than an order of magnitude over most of the very broad spectrum between short and long wavelength infrared. Our detailed analysis shows that this suppression of optical losses is due to both a large Faraday inductance and a long electron path. These two factors render the conduction electron in the metal effectively heavier than a hydrogen molecule. To investigate the usefulness of this effect in infrared detectors, we include a photoactive material in the coiled metal structures. In this case, the fraction of absorption in the active material increases by two orders of magnitude compared to non-coiled structures. Moreover, we demonstrate that the coiling of the infrared detector nanostrips increases the photoresponsivity by 15 times without additional electrical losses. These findings could benefit many metal-based applications that require low loss such as photovoltaics, photoconductive detectors, solar selective surfaces, infrared-transparent defrosting windows, and other metamaterials.

454155 Light Harvesting in Dye Sensitized Solar Cell Based on Consensitizer in Core-Shell Nanofiber Configuration Reducing Charge Recombination

Abstract:

Dye sensitized solar cell offers the potential of being a low-cost, high-efficiency photovoltaic device. However, the power conversion efficiency is limited as it cannot utilize all photons of the visible solar spectrum. A novel design of a core-shell photoanode is presented herein where a thin-shell of infrared dye is deposited on the core of sensitized TiO_2 nanofiber. Specifically, the ruthenium based dye (N719) sensitized TiO_2 nanofiber is wrapped by a thin-shell of copper phthalocyanine (CuPc). In addition to broadening the absorption spectrum, this core-shell configuration further suppresses the electron-hole recombination process. Instead of adopting the typical Förster resonance energy transfer, upon photons being absorbed by the infrared dye, electrons are transferred efficiently through a cascade process from the CuPc to the N719 dyes, the conduction band of TiO_2, the FTO electrode and finally the external circuit. This results in a high power conversion efficiency of 9.48% for the device. The proposed strategy provides an alternative method for enhancing the performance of DSSC for low-cost renewable energy in the future.

469311 Nanocrystal Doping Stabilizes the Perovskite Phase of Cesium Lead Iodide

Abstract:

All-inorganic, perovskite-phase cesium lead iodide is a promising material for photovoltaics. However, under ambient conditions it rapidly transforms to a non-functional, so-called yellow phase. Here we demonstrate more than two orders of magnitude improvement in the stability of nanostructured perovskite-phase cesium lead iodide thin films through a nanocrystal doping approach. The doped, functional perovskite phase is synthesized by co-assembling cesium lead iodide nanocrystals with cesium lead chloride dopant nanocrystals. The resulting doped nanocrystal solid is subsequently fused into a polycrystalline thin film by chemically-induced, room-temperature sintering. This process ensures nanometer-scale mixing even at compositions that exceed the bulk miscibility of the two constituents. Spectroscopy and X-ray diffraction indicate that some of this chloride is further dispersed during sintering and a polycrystalline mixed phase is formed. By comparison to DFT-calculated values, the relative change in band gap and the lattice contraction are shown to be consistent with a Cl:I ratio of a few percent in the mixed phase.

460501 Metal Oxide Electron-Selective Layers for Inverted Perovskite Solar Cells By Atomic Layer Deposition

Abstract:

Perovskites are promising candidates for low cost, high efficiency light absorbers with a large, tunable band gap. Recent advances in photostability and thermal stability from the introduction of mixed-cation lead mixed-halide perovskites in the form of cesium formamidinium (CsFA) perovskites has lead to significant interest in the development of perovskite-silicon tandem solar cells. Improvements in the deposition of transparent contacts, stability of organics within the device structure and blocking of moisture to the perovskite absorber must be made to realize highly efficient and stable perovskite-silicon devices. The improved thermal stability of the CsFA perovskite has opened up new processing options through atomic layer deposition (ALD) for the conformal coating of metal oxide materials on top of perovskite absorbers. In this work, we explore TiO_x and SnO_x ALD thin films deposited on top of CsFA perovskite devices in an inverted structure as dual-purpose layers for electron-selectivity and degradation protection. We will discuss the effects of ALD processing conditions on TiO_x and SnO_x electronic properties as well as on CsFA perovskite stability. We achieve stable >15% power conversion efficiency with 1.6 eV perovskite devices through maximum power point tracking and demonstrate the effectiveness of ALD metal oxide layers as buffer layers for the sputtering of indium tin oxide contacts for semi-transparent devices with >14% efficiency.

467509 Transparent Conductive Oxide Nanocrystals Coated with Insulators By Atomic Layer Deposition

Abstract:

Thin films comprised of transparent conducting oxide (TCO) nanocrystals are attractive for a number of optoelectronic applications. However, it is often observed that the conductivity of such films is very low when in contact with air. It has recently been demonstrated, somewhat surprisingly, that filling in films comprised of initially insulating TCO nanocrystals with another insulator by atomic layer deposition (ALD) dramatically increases the conductivity by many orders of magnitude. This work aims to elucidate the mechanism by which the ALD coating increases the conductivity. We used an *in situ* electrical probe to examine the effect on sheet resistance of removing two adsorbed oxygen species (physisorbed molecular water and chemisorbed hydroxide) and compared this result to thin films comprised of ZnO nanocrystals coated with Al_2O_3 and also HfO_2 by ALD. Although both insulating infills decrease the sheet resistance and increase the stability of the films, there is a stark discrepancy between the two. From the *in situ* measurements, it was found that coating with Al_2O_3 removes both physisorbed water and chemisorbed hydroxide resulting in a net reduction of the ZnO nanocrystals. Coating with HfO_2 only removes physisorbed water, which was confirmed by Fourier transform infrared spectroscopy. A similar phenomenon was observed when coating thin films comprised of Sn-doped In_2O_3 nanocrystals with Al_2O_3 by ALD; suggesting Al_2O_3 can be used to reduce and stabilize metal-oxide nanocrystals in general.

464634 Microstructure Development in $Cu_2znsn(S_xSe_{1-x})_4$ Thin Films during Annealing of Colloidal Nanocrystal Coatings

Abstract:

Copper zinc tin sulfoselenide is a promising material for thin film solar cells because it comprises inexpensive and abundant elements. Thin polycrystalline films of $Cu_2ZnSn(S_xSe_{1-x})_4$, or CZTSSe, may be formed by annealing nanocrystal coatings cast from colloidal dispersions in a selenium and/or sulfur atmosphere. This is a promising low-cost and high-throughput route for fabrication of CZTSSe-based thin film solar cells. During annealing, the nanocrystal films can transform into polycrystalline thin films with micrometer size grains. While benchmark efficiencies of CZTSSe solar cells have risen sharply in the last two decades, continued improvements in the solar cell performance and cost reduction require a fundamental understanding of the factors that influence the film microstructure. We have been investigating the microstructure evolution of polycrystalline CZTSSe thin films through systematic control of parameters such as chalcogen (sulfur or selenium) vapor pressure, annealing temperature, and heating rate in a closed system. In this talk, we will contrast microstructure development in CZTSSe films formed by annealing CZTS nanocrystal coatings in selenium vapor with microstructure development in CZTS films formed by annealing CZTS nanocrystal coatings in sulfur vapor. There are significant differences in the mechanisms of microstructure development despite the similar approaches. These differences are due to differences in thermodynamic phase equilibria of sulfur and selenium as well as due to differences in chemical reactions of sulfur and selenium with the carbon containing ligands that surround the nancrystals: these ligands are used to stabilize the nanocrystal dispersions and often end up in the coatings and films even after annealing. For example, annealing CZTS nanocrystal coatings at high selenium pressure (*e.g.*, 450 Torr) leads to the formation of a bilayer microstructure containing a microcrystalline CZTSSe layer on top of a nanocrystalline layer that is rich in carbon. In contrast, films annealed with sulfur do not show a distinct carbon-rich layer and most of the carbon volatilizes from the film during annealing. We suggest that carbon remains in the film due to the formation of a non-volatile carbon selenide polymer, the presence of which inhibits nanocrystal sintering and leads to the formation of a large-crystal top layer. The microstructure of films annealed in the presence of selenium liquid is influenced by liquid phase sintering. These films exhibit monomodal coarsening and densification into a compact film without the formation of a bilayer structure.

Multiscale study of the Self-Organized Vertical Concentration Profile of PEDOT:PSS for Work Function Optimization

Min Huang*
School of Chemical Science and Engineering, Tongji University
1239 Siping Road, Shanghai, China 200092

ABSTRACT

Conventional conducting polymers, poly(3,4-ethylenedioxythiophene):poly(styrenesulfonate) (PEDOT:PSS) has been widely used as the hole injection layer (HIL). It would be preferable to design a gradient profile of work function near the interface in vertical for individual applications. In the literature, tetrafluoroethylene-perfluoro-3,6-dioxa-4-methyl-7-octene-sulfonic acid copolyme (PFI) and nano Nickel Oxide were experimentally added into the PEDOT:PSS layer to create a self-organized vertical concentration profile. In this contribution, the mechanisms of the self-organized gradient effect were studied using multiscale analysis and dissipative particle dynamics (DPD) simulation. The Flory-Huggins parameters and intramolecular bonding parameters were obtained through reverse mapping of molecular dynamics simulations. DPD simulations with different additive molecular structure and concentration were performed, their influences on the self-organized concentration profile were discussed.

Keywords: multiscale, dissipative particle dynamics, PEDOT:PSS : PFI, NiOx nano particle, work function

1. INTRODUCTION

Transparent conductive polymer pss:pedot is very widely used in many applications such as organic solar cells[1,2], OLED[3], and recently perovskite solar cells[4]. it is often inserted in between indium tin oxide (ITO) and the active layer to serve as the hole injection layer (HIL). The work function of ITO is usually at about 4.8eV. Work function of the active layers according to their applications, maybe 5.8 eV. The work function of PSS:PEDOT can vary strongly according to its composition, different PSS:PEDOT ratio[5]. Therefore, in practice, optimized PSS:PEDOT ratio may be chosen to form a cascaded step-changed profile in work function. However, in some applications, such as OLED, this commonly used PEDOT:PSS has been found to have serious problems due to its inefficient hole-injection, inefficient electron-blocking, and substantial quenching of excitons close to the PEDOT:PSS. It was proposed that a smooth gradient profile can solve or at least lessen these problems. Many experimental studies aiming towards this goal, solvent surface treatment[6], NiOx nano particles[7], PFI diblock copolymer[8]

Modeling of the interface and additives interact with PEDOT:PSS is very important for mechanisms understanding, future device architecture designing and structure optimization. Classical molecular dynamics is a very powerful tool for it can provide atomistic detail of the interface, but it also has an equally great disadvantage, it can only apply to very small size dimension and short time duration, far from the length and time scales in real sense. This disadvantage maybe overcome through recently matured multiscale simulations. In a multiscale simulation mesoscale simulations were

* minontongji@tongji.edu.cn; phone 86 021 65982232; fax 86 021 65982287

performed to provide a bridge to relate microscopic view from molecular dynamics to the dimension and time in real sense. There are two popular mesoscale simulation tools, Coarse grain dynamics which uses parameters from direct mapping of the molecular dynamics, therefore, retained most features of molecular dynamics. The other method is the so called Dissipative particle dynamics (DPD). DPD was developed in the early 90's[9] and reformulated by Espanol[10], which combines coarse graining with a branch of thermodynamics, the Flory-Huggins theory for polymer[11,12,13]. Flory-Huggins theory for polymer mix states that the free energy of a polymer mix including three terms:

$$\left(\frac{\Delta G_{mix}}{RT}\right)\left(\frac{V_r}{V}\right)=\frac{\phi_A}{N_A}\ln\phi_A+\frac{\phi_B}{N_B}\ln\phi_B+\chi_{AB}\phi_A\phi_B \quad (1)$$

the first two terms is entropy of mix from ideal solution, the third term led by the interaction parameter χ_{AB} represents enthalpic interaction between monomer segments and/or solvents which can be calculated form cohesive energy density. The non-ideality of entropy of mix can be incorporated through chain flexibility using intramolecular bonding parameters. According to second law of thermodynamics, the energy will be at its minimum when the system reaches its equilibrium. Therefore, in this way, the DPD simulation has a thermodynamic constrain built in, the simulation process will go towards nothing but thermodynamic final state.

2. MODELS AND METHODS

Dissipative particle dynamics[14] (DPD) is a mesoscale dynamic simulation tool, therefore, atomistic details are not considered. This allowing it to give access to longer time and length scales than classical molecular dynamics (MD) simulations. In DPD simulation different beads with similar volume size and mass must be used to represent polymer monomers in a wormlike structure.

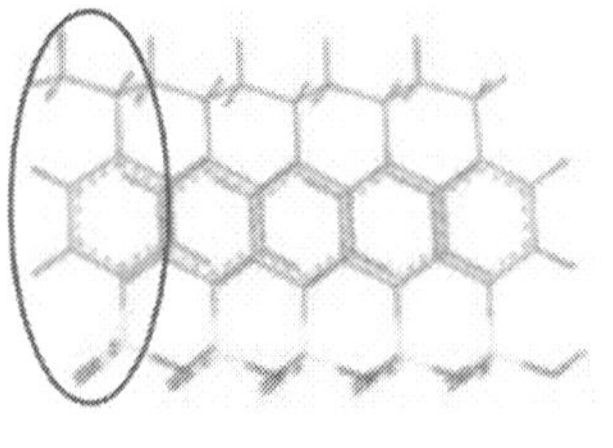

PSS

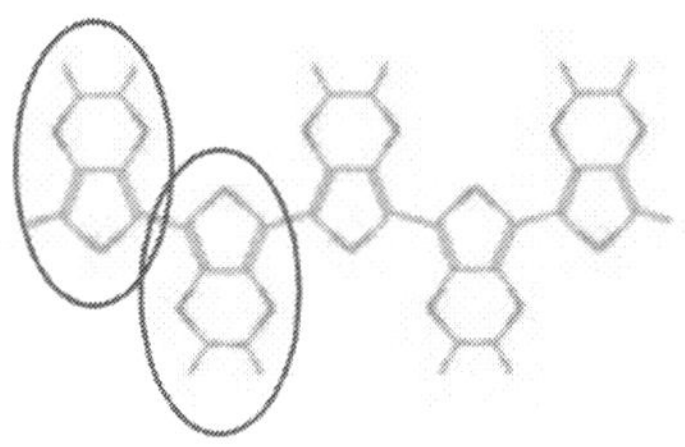

PEDOT

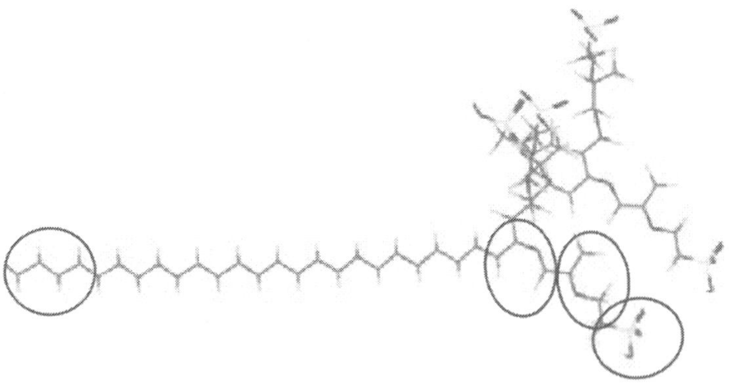

PFI

Figure 1. DPD beads representation of PEDOT, PSS and PFI

All monomers in this study were optimized using DFT, B3LYP[15,16,17] functional and 6-311G++2d,2p basis set[18]. The definition of PSS:PEDOT and PFI beads are shown in Figure 1. The polymer chain length is also very important parameter. In a condensed phase, a polymer chain may fold along its chain length, therefore forming many contact points with self and/or with other chains. As a result, much short chain model is needed. Previous studies, through solubility simulations, found that 15 or more monomers will be suffice. In this study, 25 were used. When block copolymer was involved, the molecular configuration also needs to be considered—there may be differences in degree of polymerization in each branch or alternating structure may exist.

2.1 Flory-Huggins Repulsion parameter

Enthalpic contribution χ_{ij} is obtained through a series of single and bi-component simulation boxes assembled according to bulk average concentration. Molecular Dynamics simulations were then performed. CFF force field was used for PSS:PEDOT:PFI system while UFF force field was used in PSS:PEDOT:NiOx system. Average cohesive energy density were extracted according to equation. Flory-Huggins interaction parameters χ_{ij} and DPD repulsion parameters were calculated accordingly.

$$\chi_{ij} = \left(\frac{\Delta E_{ij}}{RT}\right)V_{brad} = \left(\frac{\Delta E_{ij}}{8.314\times 298}6.022\times 10^{23}\times 10^{-30}\right)V_{brad} = \left(2.43\times 10^{-10}\right)\Delta E_{ij}V_{brad} \quad (2)$$

where V_{bead} is the volume of a bead.

$$\Delta E_{ij} = \phi_i\left(E_{coh}\right)_i + \phi_j\left(E_{coh}\right)_j - \left(E_{coh}\right)_{ij} \quad (3)$$

where E_{coh} represents cohesive energy.

$$\chi_{ij} = \left(\frac{\Delta E_{ij}}{RT}\right)V_{bead} \quad (4)$$

Once the Flory-Huggins parameters were decided, the DPD repulsion parameters of different beads can then be obtained according to Groot[19]

$$\alpha_{ij} = \frac{\chi_{ij}}{0.306} + \alpha_{ii} \quad (5)$$

2.2 Intramolecular bonding parameter

Simulation box contains all related components was assembled according to again the bulk average concentration. The generated trajectory was converted to obtain intra-molecular distances and angles distributions, the first peak was used to extract bond stretching and bending distribution function for each component, then convert to spring constants.

2.3 DPD simulation

The DPD simulation box were 10x10x20 DPD units for PEDOT:PSS:NiOx and 20x20x30 DPD units for PEDOT:PSS:PFI. Chain length of 25 repeated monomer beads were used, of which PEDOT(monomer):PSS equals to 1:2.5 weight ratio according to the literature[20]. Different PFI concentration were used in experiments in the same literature, low PFI concentrations have been published elsewhere[21], in this contribution, concentration of weight ratio from 5 and 8.5 of PFI will be presented. The concentration of NiOx nano particle is not as clearly defined in the literature, therefore, 0.1 and less were used.

In the experiment, PEDOT:PSS and PFI were in appearance, well mixed in the solution according to the standard device fabrication procedures. However, monomers confined to polymer chains do not have the same freedom as small molecules, total random mix condition at microscopic level is seldom met. The mixture will always be in some level of semi-ordered form. Therefore, DPD simulation boxes with stacking and random PEDOT:PSS and PFI initial conditions were generated side by side to mimic the processes change from ordered to semi-ordered and form random to semi-ordered form; and verify the results for each other. For PEDOT:PSS:NiOx, only random initial condition was used.

3. RESULTS AND DISCUSSES

3.1 DPD parameters

DPD parameters of PSS:PEDOT:PFI system were published in somewhere else[12]. Obtained PSS:PEDOT:NiOx DPD repulsion parameters are listed in Table 1, PEDOT and NiOx nano particle have strong attraction, while PSS and NiOx nano particles repulse mildly. Water molecule attracts to PEDOT and repulse to PSS mildly. Water attracts to NiOx nano particle very strongly. Their intramolecular bond stretching and bending results are listed in Table 2.

Table 1. PSS:PEDOT:NiOx DPD repulsion parameters α_{ij}

ij	EDOT	SS	NiOx	H_2O
EDOT	25			
SS	24.12	25		
NiOx	17.0	28.24	25	
H_2O	23.3	27.8	13.58	25

From Table 2, adding water to the PSS:PEDOT chain flexibility of both polymer have only little change. When NiOx is added, PEDOT chain become twice as rigid. This rigidness will contribute negatively to mixing and tend to form phase segregation.

Table 2. PSS:PEDOT Intramolecular bonding parameters

System	Interactions	AAMD Simu	
	Bond length	ks (kcal/mol/A^2)	r(A)
PEDOT:PSS	EDOT-EDOT	0.8158	4.4201
	SS-SS	0.0436	5.8071
PEDOT:PSS:H2O	EDOT-EDOT	0.6085	4. 3002
	SS-SS	0.0566	6.8432
PEDOT:PSS:NiOx	EDOT-EDOT	1. 6998	4.5438
	SS-SS	0.0179	9.1388
	Bend angle	k_θ (kcal/mol/rad^2)	θ_0 (degree)
PSS:PEDOT	EDOT-EDOT-EDOT	4.0538	128.7380
	SS-SS-SS	0.3159	31.1367
PSS:PEDOT:H2O	EDOT-EDOT-EDOT	4.4385	140.1322
	SS-SS-SS	0.1548	27.7342
PSS:PEDOT:NiOx	EDOT-EDOT-EDOT	2.1044	125.7708
	SS-SS-SS	0.3983	45.8840

3.2 DPD simulation results

DPD simulation from both initial conditions reached similar final state in all the PFI concentrations in study. The effect of concentration of different kinds on the vertical profiles from random initial condition are shown in Figure 2. PSS:PEDOT:PFI forms phase separation with PFI concentrated on the top surface, leading to self-organized gradient profiles for PSS and PEDOT.

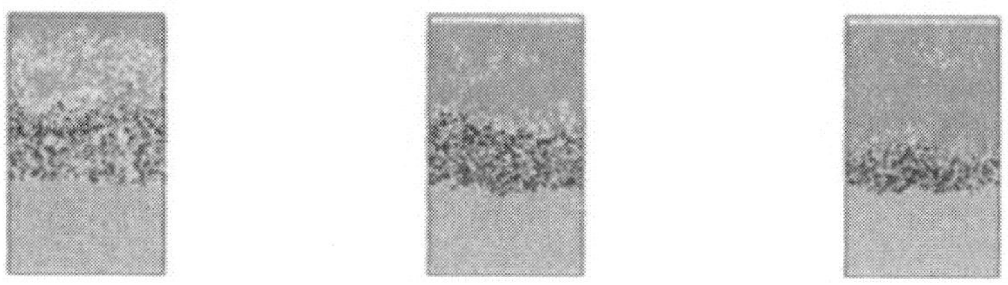

PEDOT:PSS:PFI 1:5:5 1:2.5:5 1:2.5:8.5

Figure 2. DPD simulation results of the self-organized gradient effect of ITO-PEDOT:PSS:PFI, random initio condition, where grey, blue, cyan and fulvous beads represent ITO, PEDOT, PSS and PFI respectively.

The same mechanism causing PEDOT to have steeper profile than PSS towards the surface, therefore, large eV value can be expected on PFI rich surface.

The effect of NiOx concentration on the vertical profiles are shown in Figure 3. From Table1 and 2, NiOx nano particles cling to PEDOT polymer chains, causing PEDOT chain to be more stretched and rigid, thus easier to form phase separation. At the same time, the strong attraction between NiOx nano particles and PEDOT pulls PEDOT together to form more condensed PEDOT phase, thus more condensed PSS phase on the opposite direction which lead to the self-organized PEDOT and PSS vertical concentration profile, which, through optimization, can be used to achieve gradient work function in vertical direction.

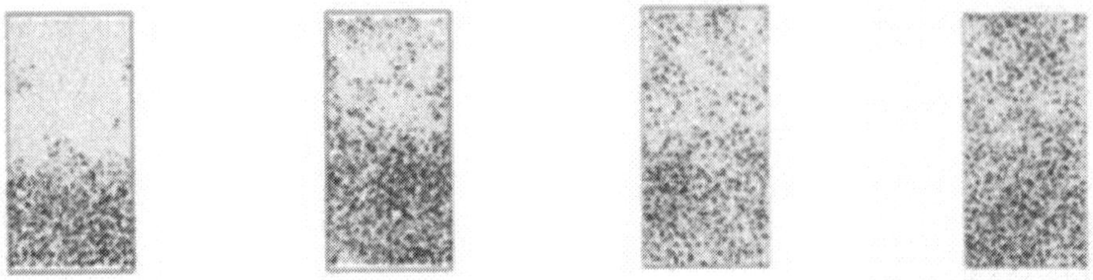

PEDOT:PSS:NiOx 1:5:0.1   1:2.5:0.05   1:5:0.04   1:2.5:0.025

Figure 3. DPD simulation result of the self-organized gradient effect of PEDOT:PSS:NiOx. Where blue, cyan and yellow beads represent PEDOT, PSS and NiOx respectively.

4. CONCLUSIONS

PSS:PEDOT:PFI forms phase separation in a wide PFI concentration region, with PFI concentrated on the top surface, leading to self-organized gradient profiles for PSS and PEDOT, The same mechanism causing PEDOT to have steeper profile than PSS towards the surface, therefore, large eV value can be expected on PFI rich surface. PSS:PEDOT:NiOx takes a different mechanisms. NiOx nano particles cling to PEDOT polymer chains, causing PEDOT chain to be more stretched and rigid, thus easier to form phase separation. At the same time, the strong attraction between NiOx nano particles and PEDOT pulls PEDOT together to form more condensed PEDOT phase, thus more condensed PSS phase on the opposite direction.

The model can be used for future numerical experimenting on effects of composition, structure configuration and additives for potentially tailored design and structure optimization.

REFERENCES

[1] Sun, K., Zhang, S., Li, P., Xia, Y., Zhang, X., Du, D., Isikgor, F. H. and Ouyang, J., "Review on application of PEDOTs and PEDOT:PSS in energy conversion and storage devices," J Mater. Sci: Mater Electron. 26, 4438–4462 (2015)

[2] Nie W., Tsai, H., Asadpour, R., Blancon, J.-C., Neukirch, A.J., Gupta, G., Crochet, J.J., Chhowalla, M., Tretiak, S., Alam, M.A., Wang, H.-L. and Mohite, A.D., "High-efficiency solution-processed perovskite solar cells with millimeter-scale grains," Sci. 367, 522-525 (2015)

[3] Wang, G.-F., Tao, X.-M. and Wang, R.-X., "Fabrication and characterization of OLEDs using PEDOT:PSS and

MWCNT nanocomposites," Compos. Sci. Technol. 68, 2837–2841 (2008)
[4] You J, Hong Z, Yang YM, Chen Q, Cai M, Song TB, Chen CC, Lu S, Liu Y, Zhou H, Yang Y, "Low-temperature solution-processed perovskite solar cells with high efficiency and flexibility", ACS Nano, 8(2):1674-80 (2014)
[5] Petrosino, M., Vacca, P., Miscioscia, R., Nenna, G., Minarini, C. and Rubino, A., "Effect of PEDOT:PSS ratio on the electrical and optical properties of OLEDs," Proc. SPIE 6593, 659310 (2007)
[6] I. Cruz-Cruza, M. Reyes-Reyesa, M.A. Aguilar-Frutisb, A.G. Rodrigueza, c, R. López-Sandoval, "Study of the effect of DMSO concentration on the thickness of the PSS insulating barrier in PEDOT:PSS thin films", Synthetic Metals, 160:1501–1506 (2010)
[7] Anil Kanwat, William Milne, Jin Jang, "Vertical phase separation of PSS in organic photovoltaics with a nickel oxide doped PEDOT: PSS interlayer", Solar Energy Materials and Solar Cells, 132:623–631 (2015)
[8] Chin, B.D., Choi, Y., Baek, H.-I. and Lee, C., "High efficiency phosphorescent OLEDs based on the heterostructured light emission and charge injection layers," Proc. SPIE 7415, 741510 (2009)
[9] P. J. Hoogerbrugge and J. M. V. A. Koelman. Simulating microscopic hydrodynamic phenomena with dissipative particle dynamics. Europhysics Letters, 19(3):155–160, JUN 1 1992
[10] P. Español and P. B. Warren. Statistical-mechanics of dissipative particle dynamics. Europhysics Letters, 30(4):191–196, MAY 1 1995
[11] Flory, P.J., "Thermodynamics of High Polymer Solutions," J. Chem. Phys. 10, 51-61 (1942)
[12] Huggins, M.L., "Some Properties of Solutions of Long-chain Compounds," J. Phys. Chem. 46, 151-158 (1942)
[13] Huggins, M.L., "Theory of Solutions of High Polymers," J. Am. Chem. Soc. 64, 1712-1719 (1942)
[14] Groot, R.D., and Warren, P.B., "dissipative particle dynamics: bridging the gap between atomistic and mesoscopic simulation," J. Chem Phys. 107, 4423-4435, (1997)
[15] Becke, A.D., "Density-functional exchange-energy approximation with correct asymptotic behavior," Phys. Rev. A 38, 3098-3100 (1988)
[16] Chengteh, L., Weitao, Y. and Robert, G.P., "Development of the Colle-Salvetti correlation-energy formula into a functional of the electron density," Phys. Rev. B 37 (2), 785-789 (1988)
[17] Vosko, S. H., Wilk, L. and Nusair, M., "Accurate spin-dependent electron liquid correlation energies for local spin density calculations: a critical analysis," Can. J. Phys. 58, 1200-1211 (1980)
[18] Ditchfield, R., Hehre, W.J., Pople, J.A., "Self-consistent molecular-orbital methods. IX. an Extended Gaussian-type basis for molecular-orbital studies of organic molecules," J. Chem Phys. 54, 724-728 (1971)
[19] Groot, R.D. and Madden, T.J., "Dynamic Simulation of Diblock Copolymer Microphase Separation," J. Chem. Phys. 108(20), 8713-8724, (1998)
[20] Lee, T.-W., Chung, Y., Kwon, O., and Park, J.-J., "Self-Organized Gradient Hole Injection to Improve the Performance of Polymer Electroluminescent Devices," Adv. Funct. Mater. 17, 390–396 (2007)
[21] Min Huang, "Multiscale study of the self-organized gradient effect of novel hole injection material PEDOT:PSS:PFI", Proc. of SPIE Vol.9941:994116, Organic Light Emitting Materials and Devices XX, 2016 SPIE Optics & Photonics, San Diego, California, USA, Aug. 29, 2016

472603 Study of Charge Transfer Dynamics in Spray Deposited Cu2ZnSnS4 (CZTS) Photoelectrodes for Performance Improvement

Abstract:

Thin film solar cell based on Cu_2ZnSnS_4 (CZTS) has the potential for large scale electricity production using solar radiation while all of the elements are earth abundant, inexpensive and environment friendly. The state of art CZTS solar cell has reached 12.6% efficiency while the commercial production will require at least 15% for economic feasibility. Here, we approach to understand the charge transfer dynamics in CZTS using ultrafast laser spectroscopy and focus on improving the efficiency of CZTS photovoltaics through the incorporation of a mesoscopic hole acceptor, nickel oxide. CZTS photoelectrode is prepared using spray deposition of precursor solution, and characterized using SEM, TEM, XRD, XPS and Raman Spectroscopy. The performance of the material is tested in three electrode photo electrochemical cell.

463080 Highly Efficient Solar Cells Made with $Cu_{1-x}K_xInSe_2$ Alloys: A Foundation for Engineering K in $Cu(In,Ga)Se_2$

Abstract:

For the first time, Cu-KF-In-Se co-deposition was used to form $Cu_{1-x}K_xInSe_2$ (CKIS) films over the full composition range ($0 < x < 1$). Compared to K-free $CuInSe_2$, the CKIS films exhibited increased majority carrier concentrations and wider band gaps, while $0 < x < 0.22$ compositions showed increased minority carrier lifetimes. Thin film solar cells were fabricated using soda-lime glass/Mo/CKIS/CdS/intrinsic-ZnO/Al:ZnO/Ni/Al stacks, where a Ni/Al finger grid with partial shading was used for the top contact. Photovoltaic (PV) performance was optimal at $x \sim 0.07$, which resulted in an officially-measured 15.0%-efficient deviceâ€"equal to the world record $CuInSe_2$ efficiency (grown at 575 Â°C), despite being grown at 500 Â°C, without the three stage process, and without device stack optimization. The devices were characterized with quantum efficiency, temperature- and light-intensity-dependent current-voltage, capacitance-voltage, and admittance spectroscopy. These data were then used to calculate charge carrier recombination rates in the bulk absorber, the depletion region, and at the interface. Bulk recombination was reduced for $x > 0$. Interface recombination was reduced at $x \sim 0.07$, while x â‰¥ 0.22 compositions had poorer performance due to interface recombination, relative to $x \sim 0$. These trends are consistent with both surface and bulk recombination affecting apparent lifetimes. The results showed that large K compositions deteriorated PV performance in CKIS alloys, while nevertheless, $Cu(In,Ga)Se_2$ solar cells with record efficiencies have exhibited large K compositions at the absorber/buffer interface. Processing techniques were employed to address this apparent discrepancy in material quality at high K content. Increasing the substrate's Na composition was found to increase the extent of CKIS alloy formation, relative to $CuInSe_2$ + $KInSe_2$ formation. Similarly, decreased substrate temperature was found to favor CKIS formation, relative to $CuInSe_2$ + $KInSe_2$. Substrate Na and temperature can therefore be used to engineer K bonding in $Cu(In,Ga)Se_2$ absorbers to enhance both initial and long-term PV power generation. Material growth

methods such as these may also be used to investigate the differences between CKIS alloys and Cu-K-In-Se films with very low $KInSe_2$ content.

468083 Study of Electron Transport Mechanism in Dye-Sensitized Solar Cell with the Effect of Morphology, Crystalline Structure and Electron Mobility

Abstract:

Dye-sensitized solar cells (DSSCs) are one of the most promising developments in photovoltaic devices. To obtain high efficiency DSSCs, extensive research has been carried out to consider the effect of dyes, electrolytes, and alternatives to titania.

The performance of photovoltaic devices could be improved by using nano-composites which can provide high electron mobility and efficiently collect photo-generated electrons through a better understanding of their fundamentals. One-dimensional nanomaterials including Single-walled carbon nanotubes (CNT) and metals, such as gold and silver, exhibit very high electron mobility.

However, the incorporation of such materials into DSSCs for efficient photovoltaic performance is still challenging. Also the fundamental of the above system is not well understood.

This study focuses on the synthesis of core-shell nano-composites made of TiO_2, various metals, and CNTs. Several fabrication techniques including chemical vapor deposition, electrochemical deposition, spin-coating, and electro spinning were applied to create various nanostructured composite materials such as one dimensional nanorods, nanowires, and nanotubes. Designing core-shell nanostructures of TiO_2in DSSC help us investigate and compare electron transfer properties of the different core materials.

The Mechanistic study of the DSSCs were carried out by using the following techniques: UV-vis scoptrocopy, time-resolved photoluminescence spectroscopy, impedance analyzer in combination. In addition, characterization methods such as TEM, SEM, XRD, and Raman spectroscopy were used. These studies will help to better understand the principle of DSSCs and better design efficient DSSC devices.

473712 Solventless Synthesis of Zeolitic-Imidazole Framework ZIF-8 Membranes Via Crystal-Specific Sintering Phenomenon

Abstract:

ZIFs, constructed by interconnecting divalent metal nodes (e.g., Zn, Co, Cd or their combinations) and imidazole linkers in periodic manners, have been one of the most intensively studied metal-organic frameworks (MOFs) primarily due to their exceptional thermal/chemical stabilities compared to other MOFs.[1] In combination with the ZIFsâ€™ stabilities, their pore apertures with molecular level dimension have led them to be considered as promising material candidates for energy-efficient membrane-based separations.[2]

When prepared, ZIF membranes have polycrystalline nature like zeolite membranes. Therefore their performances, considering the materials are properly chosen, depend on membrane microstructures, most importantly grain boundary structure. Since the membrane microstructures are generally governed by membrane processing technique, current research focus on ZIF membranes has been mainly about developing new processing techniques.[3] As results, newly designed methods, which are broadly classified into in-situ and secondary growth, enabled significant improvements in membrane performance, synthesis simplification, and potential scalability.[2]In spite of all these advancements, however, since the majority of the current techniques are still solution-based syntheses, there are still issues inevitably encountered such as producing wastes (e.g., environmentally harmful solvent, excessively-used expensive ligand, metal precursor) and putting a limitation on reducing membrane thickness which is very important for large quantity gas processing. Therefore, there are still needs of developing synthesis protocols which are sustainable and allow membranes with thin thickness.

Here, we report a new synthesis strategy which allows the formation of continuous ultrathin ZIF-8 films / membranes (~300 nm) in a solventless condition. This new approach is on the basis of our new discovery that intentionally-designed defective ZIF-8 crystals display sintering phenomenon without crystal degradation under specific vapor environment, leading to crystal merging and subsequently grain growth.

[1] K. S. Park, Z. Ni, A. P. CÃ´tÃ©, J. Y. Choi, R. Huang, F. J. Uribe-Romo, H. K. Chae, M. Oâ€™Keeffe, O. M. Yaghi, *Proceedings of the National Academy of Sciences* **2006**, *103*, 10186-10191.

[2] aJ. Yao, H. Wang, *Chemical Society Reviews* **2014**, *43*, 4470-4493; bB. Chen, Z. Yang, Y. Zhu, Y. Xia, *Journal of Materials Chemistry A* **2014**, *2*, 16811-16831.

[3] H. T. Kwon, H.-K. Jeong, A. S. Lee, H. S. An, J. S. Lee, *Journal of the American Chemical Society* **2015**, *137*, 12304-12311.

472934 Microplasmas for Substrate-Independent Deposition of Nanostructured Metals and Oxides

Abstract:

A general, substrate-independent method for plasma deposition of nanostructured, crystalline metals and oxides will be presented. The technique uses a flow-through, micro-hollow cathode plasma discharge (supersonic microplasma jet) with remote ring anode to deliver a highly-directed flux of growth species to the substrate. A diverse range of nanostructured materials, e.g., metals (Cu, Ni, Pd), oxides (CuO, SnO_2, NiO, Fe_2O_3, Co_xO_y), doped oxides ($Fe_xNi_{1-x}O$), and spinels ($NiFe_2O_4$) with different morphologies (nanoparticles, dense columnar films, or hierarchical nanostructures) can be deposited at room temperature on virtually any substrate (conductors, insulators, plastics, fibers, and patterns) in a conformal fashion. The effects of deposition conditions, substrate type, and patterning on film morphology, nanostructure and surface coverage will be highlighted. The microplasma synthesis approach presented herein provides a general and tunable method to deposit a variety of functional and hierarchical metal/oxide materials on many different surfaces. High surface area, conversion-type CuO electrodes for Li-ion batteries are demonstrated as a proof-of-concept example.

472002 Hierarchically-Structured Porous Carbon Films By Multiscale Templating and Interfacial Engineering

Abstract:

The promise of inorganic membranes such as carbon molecular sieves (CMS) to revolutionize continuous high-selectivity gas and liquid separations has been quelled by persistent struggles to break through commercial performance barriers associated with the well-known trade-off between permeability and selectivity. Respective synthetic challenges common to CMS and other inorganic membrane materials include the need to minimize defect-free membrane thickness while simultaneously tailoring membrane texture and function. In addition, whereas strategies exist for scaling polymeric membranes (e.g., hollow fiber spinning), the lack of corresponding strategies for facile areal scaling of inorganic membranes, commonly prepared as two-dimensional sheets, challenges their ability to meet stringent demands for industrial separations productivity. In this talk, we will describe our synthetic efforts to realize a class of ultra-thin, three-dimensionally corrugated, self-supporting carbon thin films with tunable meso- and micro-structure toward novel CMS membranes. We will describe a synergistic strategy that exploits nanoparticle (silica)-mediated continuous porous thin film formation and concomitant interstitial and interfacial confinement of polymerization and pyrolysis as a new strategy for simultaneously tuning thickness, geometry, and microstructure of selective carbon features.[1] Specifically, convective co-assembly of sacrificial silica nanoparticle templates with molecular carbon precursors results in tunable, 3D-interdigitated mesopore topologies with extensive specific surface areas.[2] Upon template sacrifice, the resulting films offer two types of independent but interdigitated pore structures contacted across ultra-thin (i.e., nanometer-scale) three-dimensionally distributed CMS pore walls. This 3D interdigitated pore structure represents a significant area-to-volume enhancement that holds promise as a platform for molecular separations. Beyond mesoscale structuring, we also find that template-mediated interfacial phenomena provide a new handle for simultaneously tuning molecular-scale structural characteristics (e.g., sp^2 hybridized carbon content) that ultimately impact film texture. This work aims to demonstrate how sacrificial, template-mediated hierarchical structuring may advance the viability of CMS membranes, and, more generally, may serve as a new framework for structuring other inorganic membrane materials.

References

[1] Z. Tian, M. A. Snyder*, "Hard-templating of symmetric and asymmetric carbon thin films with three-dimensionally ordered mesoporosity," *Langmuir,* 30 (2014) 9828-9837.

[2] Z. Tian, M.A. Snyder*, "Nanocasting of carbon films with interdigitated bimodal three-dimensionally ordered mesopores by template-replica coassembly," *Langmuir*, 30 (2014) 12411-1242

454878 Effect of the Chain Length of Alkylamine on Film Formation from Alcohol-Soluble Copper Complex Ink

Abstract:

Metal-organic decomposition (MOD) ink has a promising application for printed electronics technology. We synthesized self-reducible conductive ink with copper formate (Cuf) and primary alkylamines as ligand to form alcohol-soluble complex. Highly conductive copper films could be obtained on both rigid and flexible substrates at low sintering temperature under nitrogen atmosphere. The decomposition mechanism was also studied. It was found that amines could not only activate the dissociating of carboxyl from Cuf and thus decrease the decomposition temperature, but also retard the subsequent growth of copper nuclei. These two effects were dependent on the alkyl chain length of amines. With the increase of the alkyl chain length, the decomposition rate of copper-amine complex was slowed down, while the capping effect was enhanced resulting in narrow copper particle size distribution and less void space. The film from Cuf-Octylamine (Cuf-Octyl)showed the lowest resistivity, which was 10.12μΩ·cm sintered at 180 . Using blended amines with long-chain and short-chain could balance the dual effects, contributing to the formation of dense packed copper films with less organic residues and lower resistivity. The resistivity of copper film from Cuf -Octylamine-Butylamine (Cuf-Octyl-Butyl) ink was 5.25μΩ·cm sintered at 180 . Moreover, the solubility in alcohols and stability of Cuf-Octyl-Butyl were both improved compared to Cuf-Octyl. By adjusting the viscosity and surface tension, the prepared ink could meet the requirement of inkjet printing.

471516 Understanding of Diffusion Pathway of Cyclohexane through Nanoscale MFI Zeolite

Abstract:

Understanding of diffusion pathway of cyclohexane through nanoscale MFI zeolite

Xiaoduo Qi, Vivek Vattipalli, Wei Fan

Department of Chemical Engineering, University of Massachusetts Amherst, 686 N Pleasant Street, Amherst, 01002

Zeolites have been widely used in the fields of shape selective catalysis and separation due to their unique molecular sieving property. During catalytic reactions, reactants diffuse into pore network, react at active site, and diffuse out of pore network. To minimize the effects of internal mass transfer limitation on the overall reaction rate, nanoscale zeolite crystals have been designed and synthesized.[1] However, within small zeolite crystals, diffusion study has shown that the apparent diffusivity appears to be orders of magnitude smaller than that for large zeolite crystals[2]. This discrepancy of apparent diffusivity has led to the phenomena called "Surface Barrier", which is a general term to describe a mass transfer resistance near the surface of zeolites. For large zeolite particles, since the internal microspore diffusion is dominant, such a surface resistance is often negligible. While for small zeolite particles, the surface resistance can contribute significantly to the overall mass transfer rate.

To explain the origin of the surface barrier, pore blockage and pore narrowing mechanisms have been proposed[2]. However, with such hypotheses, Kinetic Monte Carlo simulation suggested 99.9% of the micropores on external surface have to be blocked[3]. In this presentation, we propose a new "Pore Re-entering" mechanism. According to this mechanism, molecules undergo a surface diffusion on the external surface during the desorption step. The surface diffusion is much slower than the Knudsen diffusion. Diffusion on the external surface might lead to re-entering of micropore openings, which significantly increase apparent diffusion length for the molecules. In order to demonstrate the existence of such effects, silcalite-1/silica nanoparticle mixtures were developed where silicalite-1 particles were isolated from each other by 35 nm solid silica nanoparticles. Mass transport study using cyclohexane suggested that the diffusion of cyclohexane is still controlled by the micropore diffusion with the same activation energy, but much faster in the mixtures, providing a strong evidence for the existence of the Pore Re-entering effect. The observation provides critical information for the rational synthesis of hierarchical porous materials for the applications in separation and catalysis. [1] Chang, C.-C.; Teixeira, A. R.; Li, C.; Dauenhauer, P. J.; Fan, W., Enhanced Molecular Transport in Hierarchical Silicalite-1. Langmuir 2013, 29, (45), 13943-13950.

[2] Teixeira, A. R.; Chang, C.-C.; Coogan, T.; Kendall, R.; Fan, W.; Dauenhauer, P. J., Dominance of Surface Barriers in Molecular Transport through Silicalite-1. The Journal of Physical Chemistry C 2013, 117, (48), 25545-25555.

[3] Teixeira, A. R.; Qi, X. D.; Conner, C. W.; Mountziaris, T. J.; Fan, W.; Dauenhauer, P. J., 2D Surface Structures in Small Zeolite MFI Crystals. Chemistry of Materials 2015 27 (13), 4650-4660

469390 Neutron Reflectometry Investigation of Hydrogen in Plasma Treated Hydrogen Doped Nanoporous TiO_2 Thin Films for Water Splitting Photocatalysis

Abstract:

Although hydrogen has long been accepted as the species responsible for photocatalytic activity of hydrogenated titania (also referred to as â€œblack titaniaâ€ because of its strong light absorbing properties), direct observation of this species has been elusive. In this work, we use neutron reflectometry (NR) to detect the hydrogen in plasma treated hydrogen doped nanoporous titania thin films. The cubic ordered nanoporous TiO_2 thin films were prepared by a surfactant-templated sol-gel method and were treated with hydrogen plasma, an approach hypothesized to capitalize on the high degree of disorder in the material and the high energy of the plasma species to achieve efficient hydrogen doping. UV-vis absorbance spectra indicated that the incorporated hydrogen atoms made TiO_2 films black with the enhancement of visible light absorption, and XPS analysis showed peak for Ti^{3+} state. The presence of hydrogen in hydrogen doped nanoporous titania (H-TiO_2) films is suggested by the scattering length density (SLD) profiles obtained from NR. The SLD of H-TiO_2 is much lower than that of TiO_2 film due to the negative SLD of hydrogen. Thus, for the first time, NR showed that the dramatic change in visible light absorbance of H-treated black TiO_2 is accompanied by significant hydrogen uptake and not just Ti^{3+} generation. The photocatalytic activity of hydrogen doped TiO_2 (H-TiO_2) films was evaluated by chronoamperometry. The effect of light sources on photoelectrochemical performance was explored using UV (365 nm) and blue (455 nm) LEDs. In photoelectrochemical water splitting, the H-TiO_2 showed about 27 times and 8 times higher photocurrent under UV and blue LEDs, respectively, compared to undoped TiO_2 films. Electrochemical impedance spectroscopy showed that better charge transfer is obtained in plasma treated hydrogen doped titania compared to undoped titania films. This study suggests that the dramatic change in visible light absorbance of hydrogenated black TiO_2 is due to significant hydrogen uptake, and plasma-induced doping is a promising strategy to enable the efficient incorporation of heteroatoms into surfactant-templated TiO_2 films, thereby leading to significant enhancement in photoelectrochemical hydrogen production.

462156 Synthesis of Tin(II) Monosulfide Nanoplates: A Potential 2D Material

Abstract:

Two dimensional (2D) layered materials such as graphene, metal dichalcogenides, and black phosphorus are of increasing interest because of their unique electronic properties. Tin(II) monosulfide (SnS) has the same crystal structure as black phosphorus and is a semiconductor with a 1.3 eV direct band gap, a 1.1 eV indirect band gap, and a high absorption coefficient (>10^4 cm^{-1}) in the visible range of the electromagnetic spectrum. SnS has potential applications in photovoltaics, photocatalysis, thermoelectrics, and lithium ion batteries. We present a facile chemical synthesis of SnS nanoplates via thermal decomposition of a single precursor in a coordinating solvent. Specifically, tin(IV) diethyldithiocarbamate ($Sn(dedtc)_4$) dissolved in oleic acid is injected into hot (300-340 °C) oleylamine to initiate rapid nucleation and subsequent growth of SnS after decomposition. Oleylamine complexes with $Sn(dedtc)_4$ and induces its decomposition to form SnS. Examination of the reaction product after 60 minutes at 340 °C with scanning electron microscopy shows thin plates that are 1 to 10 µm wide. Atomic force microscopy of these plates dispersed onto silicon show that the plates are 3-60 nm thick. Energy dispersive X-ray spectroscopy (EDS) indicates that the Sn-to-S ratio is one within the accuracy of EDS quantification. The crystal structure of the plates is confirmed to be orthorhombic (a=0.433 nm, b=1.119 nm, c=0.398 nm) Herzenbergite α-SnS phase with X-ray diffraction. When the plates are >1 µm wide, X-ray diffraction (XRD) from SnS plates cast onto substrates shows only the (040) diffraction as the plates settle on the substrate with their [010] axis aligned with the substrate normal. This textured XRD from plates cast on flat substrates and transmission electron microscopy indicate that SnS plates grow preferentially in the [100] and [001] directions with the [010] axis direction. Raman spectroscopy and confocal imaging show Raman scattering peaks at 96, 161, 186 and 217 cm^{-1}, all consistent with SnS. At low temperatures (300 °C) and short synthesis times (3 minutes) we also observe the presence of SnS_2 with both XRD and Raman spectroscopy. Raman spectroscopy shows a strong peak at 312 cm^{-1}. SnS_2 disappears at later times. This suggests that decomposition of $Sn(dedtc)_4$ first produces SnS_2 which is subsequently reduced, likely by oleylamine, to SnS. The SnS plate sizes depend on the oleylamine concentration, time, and temperature. Plate sizes increase with decreasing oleylamine concentration and increasing temperature. Plates also become larger and thicker with time. SnS nanoplate dispersions in toluene exhibit an optical absorption feature in the visible range of the electromagnetic spectrum, which we surmise to be of plasmonic origin. The nanoplate dispersions in toluene respond to electric fields.

461321 MoS_2-Passivated Bilayer Phosphorene Phototransistors

Abstract:

Despite its unique and promising properties such as high carrier mobility and conspicuous in-plane anisotropy that generated a tremendous amount of excitement for the past few years, BP has been suffering from its one significant drawback that may dim its potential as an active material in the next-generation nanoelectronics: severe crystal deterioration upon exposure to oxygen and water. Field-effect transistors (FETs) based on a MoS_2-passivated black phosphorus (BP) van der Waals (vdW) heterostructure could provide an effective solution to tackling this important challenge, while enhancing photoresponse of the device and retaining the conductivity of a BP channel almost intact. In this work, we fabricate 4 MoS_2-passivated BP FETs with varying BP thicknesses, 1.5 (bilayer), 5, 13, and 20 nm, to explore the effect of the MoS_2 passivation layer on the device stability and electrical characteristics both in the dark and under illumination (Î» = 600 nm). When in contact with a trilayer MoS_2, photoluminescence intensity of the bilayer BP crystal decreases by 25%, suggesting that a built-in electric field forms at the BP-MoS_2 p-n interface that could help dissociate photo-generated electron-hole pairs, thereby reducing the probability of the recombination events. The effectiveness of a few-layer MoS_2 as a vdW protection layer is tested by exposing BP-MoS_2 vdW vertical heterostructures to the air for up to 3 weeks as well as by annealing at a high temperature (350°C) in an inert argon environment. Then, the heterostructure FETs are fabricated so that direct comparisons between exposed and MoS_2-passivated BP regions on the same flake can be made. In general, the MoS_2 passivation layer gives rise to only an insignificant decrease in the transport characteristics in the dark, with the trend where its impact decreases as the thickness of the underlying BP crystal increases. Upon illumination, the MoS_2-passivated regions of the bilayer BP device shows enhanced photoresponse compared with the exposed regions by 75 % with no applied gate voltage. Also, a possible photogating effect is observed in the bilayer BP device while photoconductivity dominates the photoresponse process in the thicker devices. Thus, 2D MoS_2 2D thin films may aid in the realization of the future

transparent, flexible BP electronic and optoelectronic devices by acting not only as an atomically thin passivation layer but as a photoresponse enhancer.

474476 Regenerative Engineering: The Convergence Approach

Abstract:

Regenerative Engineering represents the Convergence of Advanced Materials Science, Stem Cell Science, Physical Forces, Developmental Biology and Clinical Translation.

The field works to capitalize on advances in these an other newly emerging fields and is especially suited to address grand challenges, and to challenge conventional norms and thinking on science.

The development of new technologies, new ways of thinking and new science through Convergence offers outstanding opportunities for creating new solutions to important problems.

474477 Regulation of Growth Factor Signaling and Regeneration by Modifications of Heparan Sulfate Proteoglycans in the Extracellular Matrix

Abstract:

Axolotls (and other Urodele amphibians) are unique among adult vertebrates in their ability to regenerate complex body structures after traumatic injury. We used an *in vivo* gain-of-function assay in the axolotl to demonstrate both qualitative and quantitate differences in extracellular matrix (ECM)-associated heparan sulfate proteoglycans (HSPG) regulate growth factor signaling that in turn regulates formation of limb skeletal structures. Most importantly, this activity is present in mammalian HSPG present in mouse skin in vivo as well as in commercially available, purified preparations of HSPG. This activity can both inhibit regeneration or induce *de novo* limb structure, thus regulation of HSPG is a mechanism for the regulation of regeneration, and establishes a crucial functional link between regeneration in salamanders and mammals. In light of recent findings that not all preparations of HSPG have activity in our regeneration assay, we have begun to screen for pro-regenerative HSPG activity between difference HSPG sources. Understanding the signals and responses of cells, whether they are pro- or anti-regenerative will be critical to achieving the goal of regenerative engineering to reform complex structure and restore function in humans.

474480 Cell, Peptide and Protein Delivery Problems in Regenerative Engineering

Abstract:

Convergence in the biomedical field rehires synergism of different factors, parameters, ideas and disciplines. This convergence allow for translation of bench-top research to bedside methods. We will discuss a range of recent efforts to provide intelligent cell, protein and peptide delivery systems with high output and exceptional precision for the treatment of diseases in the hospital or at home. We show how the basics of chemical engineering-based controlled release have led to a number of major improvements in the fields of cardiovascular diseases, diabetes, neurological diseases and autoimmune diseases.

475505 Nano- and Microfabricated Hydrogels for Regenerative Engineering

Abstract:

Engineered materials that integrate advances in polymer chemistry, nanotechnology, and biological sciences have the potential to create powerful medical therapies. Our group aims to engineer tissue regenerative therapies using water-containing polymer networks, called hydrogels, that can regulate cell behavior. Specifically, we have developed photocrosslinkable hybrid hydrogels that combine natural biomolecules with nanoparticles to regulate the chemical, biological, mechanical and electrical properties of gels. These functional scaffolds induce the differentiation of stem cells to desired cell types and direct the formation of vascularized heart or bone tissues. Since tissue function is highly dependent on architecture, we have also used microfabrication methods, such as microfluidics, photolithography, bioprinting, and molding, to regulate the architecture of these materials. We have employed these strategies to generate miniaturized tissues. To create tissue complexity, we have also developed directed assembly techniques to compile small tissue modules into larger constructs. It is anticipated that such approaches will lead to the development of next-generation regenerative therapeutics and biomedical devices.

453538 Development of Resorcinol Formaldehyde Aerogels with Enhanced Mechanical Properties Via Improved Particle Necking

Abstract:

A new highly porous resorcinol-formaldehyde (RF) aerogel with improved particle necking is presented in this work. This RF aerogel is developed under CO_2 supercritical drying conditions without any structural shrinkage. The water content and the catalyst percentage are varied to modify the solution's initial pH value and enhance particles connections. Firstly, the water content is fixed while the catalyst percentage increases in the solution to modify particles number and size. The number of particles increases with the catalyst content, and this generates a fine pearl necklace configuration with point-to-point connections that have brittle characteristics and a very low total thermal conductivity of 24.8 mW/m.K. The solid conductivity in this configuration is reduced at a high catalyst concentration due to the increased interface resistance and the reduced solid path. As the water content increases in the solution, the solution shifts towards the acidic region, and the chance to form a network decreases due to the increased void fraction. But as the catalyst percentage increases, the solution' pH value shifts towards the basic, and the particle number increases, resulting in strong particle-to-particle connections. This transforms the particle-to-particle point connections to neck connections (termed to be a semi-fibril-like structure in this paper) with a similar thermal conductivity (24.6 mW/m.K) as the pearl necklace configuration. The semi-fibril-like structural elasticity is improved substantially due to the neck formation between the particles. The study concluded that at high catalyst concentration the structure thermal conductivity is fixed while its elasticity is improved with the semi-fibril-like structural.

463063 Adsorptive Precipitation from Supercritical Solutions in Aerogels: A Way Towards Stable Amorphous Drugs

Abstract:

Properties of solid materials are fundamentally determined by their crystallinity. In pharmaceutical technology the transition from amorphous to crystalline state plays an essential role, since the bioavailability of a drug is influenced by its state. Amorphous drugs often have higher bioavailability compare to the corresponding crystalline form. Tendency towards recrystallization is a major factor precluding more wide use of the amorphous drugs in solid dosage systems.

To overcome this problem, we study adsorptive precipitation of drugs and drug-like organic solids in aerogels. Adsorptive precipitation consists in immersion of a porous carrier in a drug solution in supercritical carbon dioxide ($scCO_2$) followed by rapid expansion of the solution. The latter causes precipitation of the drug in the carrier.

To reveal the influence of the drug nature, adsorptive precipitation experiments with various drugs were performed. Biopolymer and silica aerogels were used as porous carriers. Crystallinity and loading were assessed by XRD and UV spectrophotometry, respectively. The results along with literature data are discussed in order to reveal a key parameter that governs the stabilization of the amorphous state. The conclusion is that the drug–carrier interactions are crucial for the stabilization of amorphous state.

To quantify the interaction in $scCO_2$/drug/aerogel system, supercritical fluid chromatography (SFC) experiments were employed. Retention time was interpreted in the frame of van't Hoff analysis. Due to intense interactions, elution by pure $scCO_2$ was often not possible. In this case, a modifier (alcohol) was used. Foe such quaternary systems ($scCO_2$/modifier/drug/aerogel) a novel thermodynamic model (mixed retention model) was developed and validated. SFC experiments were carried out in the CO_2 density range

of 0.40–0.85 g/cm^3. It was shown that CO_2 density plays an important role in drug–carrier interactions and changes the adsorption enthalpy by a factor of 2. Thus, by changing the density of the drug solution in sc-CO_2, it is possible to change the adsorption enthalpy of the drug on the carrier. Furthermore, it was found that the crystallinity of a model compound, naphthalene, measured by XRD is in an inverse relation to the adsorption enthalpy: the higher the enthalpy the lower the crystallinity.

To reveal the influence of the drug nature, SFC experiments were performed with various drugs. Biopolymer aerogels were used as carriers. The mixed retention model was applied to acquire enthalpy of adsorption from elution curves. Correlation between enthalpy of adsorption and crystallinity of the drugs is found and discussed along with literature data that considered to support our experimental findings.

463315 Protein Based Aerogels: Preparation, Applications and Potential for Food Engineering

Abstract:

In the last decades the interest toward porous materials based on natural polymers is continuously growing. A promising way to achieve mesoporous materials, having porosity up to 99% and correspondingly a huge internal surface area is the supercritical drying of gels resulting in aerogels. Natural polymer-based aerogels consisting of polysaccharides or proteins are promising for life science and food applications because of their biocompatibility and biodegradability. However, also the area of bio-based insulation could be targeted as well, since very low thermal conductivity is given.

In this work we focus on aerogels based on natural proteins, such as milk and egg white proteins, which are well-known as staple foods. Producing aerogels from natural food sources can open a new market for the aerogel in food, for instance as food additives. In this work mainly an application of protein-based aerogels as carrier material for sensitive and sensorial unpleasant compounds in food products is aimed. The carrier material is produced in different shapes, depending on the targeted application. Primarily, spherical particles are preferred due to simple handling and a later possible coating. The main requirements for such a carrier material are high specific surface areas coupled with high mechanical stability as well as a homogeneous particle size. Protein aerogels exhibit BET surface area up to 390 m^2/g depending on pH during the gelation. Aerogels formed at very low and very high pH values are found to have highest surface areas. Since the proteins have positive and negative charges the orientation and arrangement of the proteins vary with pH. Ordered proteins generate more ordered gel structures yielding aerogels with higher BET surface area compare to those gelated at the isoelectric point. The mechanical stability of the protein aerogels depends on the reactivity of S-H-groups. At higher pH more –S^- anions are formed and hence more S–S-bonds lead to stronger gels and aerogels.

The gel particle size distribution is depending on the droplet size distribution in the emulsion before gelation. It was optimized to obtain aerogel particle sizes between 50 to 100 µm.

Further, aerogels with the desired properties were loaded by two compounds, fish oil and ascorbic acid (vitamin C), *via* adsorption from supercritical CO_2. Remarkably high fish oil loadings were achieved without losing the free-flowing properties of the aerogel particles.

Further, storage and release behavior in simulated gastrointestinal fluids were investigated showing the potential of protein-based aerogels as carrier materials for food applications.

466138 Alginic Acid Foams with Hierarchical Porosity: Promising Materials for Dyes Adsorption

Abstract:

Alginic acid foams with hierarchical porosity: Promising materials for dyes adsorption

Nathalie Tanchoux[1*], Asja Pettignano[1,2], Luca Bernardi[2], Thierry Vincent[3], Eric Guibal[3], Françoise Quignard[1]

[1] Institut Charles Gerhardt, CNRS-UM2-ENSC-UM1, 8 Rue Ecole Normale, 34090, Montpellier, France.
nathalie.tanchoux@enscm.fr

[2] Department of Industrial Chemistry, UNIBO, V. Risorgimento, 4, 40136, Bologna, Italy.

[3] Ecole des Mines d'Alès, C2MA/MPA, 6 Avenue de Clavières, 30319, Alès, France.

Keywords: foams, hierarchical porosity, biopolymers, alginic acid, biocompatible adsorbent

Water pollution due to organic dyes is a serious environmental problem. Their release, even in low quantities, is highly undesirable because of their high visibility, low biodegradability and toxic effects on humans and marine organisms. Among the treatments proposed for water decontamination, adsorption is a popular method, because of its high efficiency and low operating cost. Therefore, the development of low-cost adsorbents is becoming extremely attractive and natural biopolymers are often presented as abundant, renewable and biodegradable resources for this application.[1]

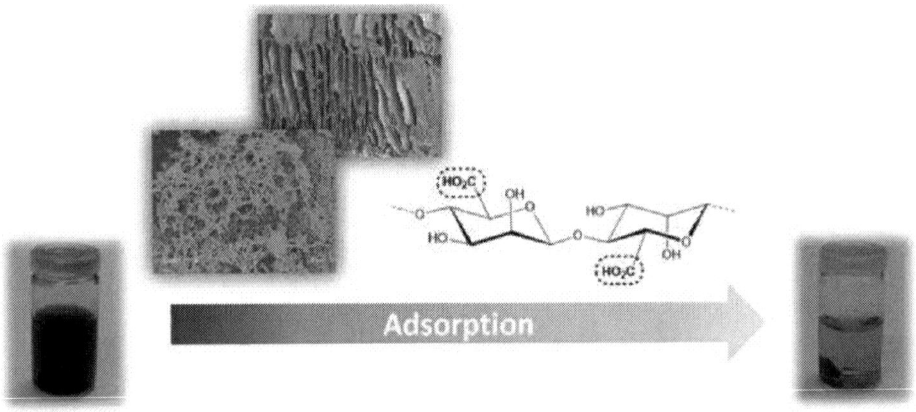

In our study, alginic acid, a biopolymer derived from brown algae, has been studied as a cost-effective replacement for conventional adsorbents. The presence of hydroxyl and carboxylic groups in its polymeric backbone confers: (a) the ability to associate, by physical and chemical interactions, with a wide variety of molecules [2] and (b) an excellent selectivity towards cationic dyes. Employing the tremendous manipulability of this polysaccharide, alginate foams with macro-meso pores have been developed with a simple procedure. The influence of preparation parameters (alginate composition, temperature, concentration, acidification procedure) on the final texture have been studied by SEM microscopy and N_2 adsorption, which confirmed the presence of a hierarchical porosity. The applicability of the newly developed material have been tested by examining its sorption behavior towards cationic dyes (such as methylene blue), by equilibrium and kinetic experiments, proving that alginic acid foams can be employed as biocompatible, green and low-cost adsorbents for wastewater remediation.

[1]. G. Crini, *Prog. Polym. Sci.* **2005**, 30, 38-70.

[2]. F. Quignard, F. Di Renzo, E. Guibal, *Top. Curr. Chem.* **2010**, 294, 165-197.

Thermoplastic Foaming Assisted by Microwave

Elham Rezvanpanah[1, 2], *S.Reza Ghaffarian Anbaran**[1], *Ernesto Di Maio*[2]
[1]*Department of Polymer Engineering & Color Technology, AmirKabir University of Technology, Tehran, Iran.*
[2]*Department of Chemical, Materials and Production Engineering, University of Napoli Federico II, Napoli, Italy.*

Abstract

Cell collapse can reduce the cell density and change the cell size of a foam structure. In particular, in conventional heating an optimal foaming temperature is selected. This temperature is a compromise between increasing the temperature to increase the polymer compliance and gas supersaturation; and decreasing the temperature to avoid weakening the polymer and limiting cell collapse. All of these processes must occur concurrently in the short time available for foaming. When attempting to foam large pieces, the described technique manifests its limits and, actually, the impossibility to achieve acceptable foam morphologies.

In this work, microwave heating is used, in comparison conventional heating, since it offers several advantages over conventional heating methods such as the use of a remote source, the relatively high speed of the process, and volume- and material-selectivity.

We used carbon nanotubes as the dispersed source of heating as well as their well-known ability as bubble nucleators. In this way, foaming will occur near the hot particles, where the polymer is sufficiently compliant, while remaining strong elsewhere. As a model system, we tested polystyrene/carbon nanotube foamed with carbon dioxide and achieved thick foams characterized by a uniform morphology and low density.

Keywords: Thermoplastic foaming, Microwave heating, Carbon nano tube

Introduction

Polymeric foams are important materials in a variety of applications due to their inherent low density, giving high specific strength, lightweight, materials savings, low thermal conductivity and good sound insulation and absorption [1]. Among the processing methods to produce foams, Physical Blowing Agent (PBA) foaming is one of the most utilized for its versatility and high throughput. In this method, the polymer is first saturated by a gas at a high pressure and then, after a pressure drop, it experiences a supersaturation state, which may result in the PBA escape to form of a new phase (gas bubble) [2]. To allow bubble formation, the polymer is softened above the melting point or the glass transition. In conventional heating systems (typically, by submerging the polymer/gas solution in an oil bath), heat is transferred from the hot oil to the material, through its surface, until the required temperature has been achieved throughout the sample (in particular, also at the very center of the sample). As plastics have low thermal conductivity, this process takes time and there often is a temperature gradient in the sample thickness. Inevitably, this induces strong differences in the foam thickness and restricts the maximum thickness, for foams processed with said method, to 1-2 mm. Above this limit, there is often a risk of having a collapsed structure at the hot, external sample parts or an unfoamed, bulk structure at the cold, internal part.

Preservation of cell density and prevention of the cell collapse or cell coalescence are key to maintain good foam properties and to keep the foam density low. To avoid this phenomenon, in the literature, different methods such as crystallinity, branching or nanoparticles are utilized to increase the melt index and elasticity [3-5]. Among these methods, polymer nanocomposite have received increasing attention in both scientific and industrial communities to control foam formation in terms of reduction of cell size and preventing cell collapse, and to include the functional properties brought about by the nanometric additive [6].

Microwave heating offers several advantages over conventional heating methods such as the use of a remote source, the relatively high speed of the process, and volumetric heating and material-selectivity [7]. Carbon Nano Tubes (CNTs) have already been investigated for microwave absorption applications due to their great thermal stability, reluctance toward chemical agents and excellent mechanical and electrical properties [7, 8].

Herein we report microwave heating as a new source of heating in thermoplastic foaming. As a model system, we tested polystyrene (PS) with CNT, foamed with carbon dioxide. We also proved effective MW foaming with thick samples (1cm), in comparison with oil bath heating.

Experimental

Materials:

PS (N2380) was supplied by Versalis S.p.A, Mantova, Italy. The average molecular weight, density and melt flow index, are 300 kDa, 1.05 g/cm3 and 2.0 g/10 min at 200°C and 10 kg, respectively, as reported by the producer.

Multi walled CNT (MWCNT) (average outer diameter: 10 nm, average length: 1.5 mm, surface area: 250–300m^2/g, carbon purity 90%, density = 1.44 g/cm^3) were supplied by Nanocyl^{TM}, Belgium, and were produced by Catalytic Carbon Vapor Deposition (CCVD).

CO_2 (99.95% pure) supplied by Sol Group S.p.A., Monza, Italy, was used as the physical blowing agent.

Preparation of PS/MWCNT nanocomposites:

PS/MWCNT 10 wt% master batch was supplied by Nanocyl^{TM}, Belgium. The nanocomposites were then prepared by mixing the master batch with pristine PS using a DSM twin-screw compounder (DSM Xplore MC15) at 200 °C and 100 rpm for 10 min. The final samples were prepared by compression molding at 200°C in 1mm and 10 mm thickness.

Foaming:

The samples were saturated with CO_2 at 10 MPa and 80 °C for 3h and 90h, for the 1mm and 10mm thick samples, respectively, in a high-pressure autoclave [9]. After the saturation, the pressure vessel was cooled to room temperature and pressure released to allow vessel opening. Saturated samples were then exposed to microwave radiation in a microwave oven for a determined time. All samples were placed exactly in middle of microwave cavity which base on literatures could receive the most uniform field [10]. To compare the microwave source system with conventional system, the water bath was considered. In this case, samples were kept in boiling water for a determined time and then air cooled.

Results and Discussion

As the MW radiation power was turned on, CNTs absorb microwave radiation and its temperature increases. Actually by using CNT as the sole microwave absorber particles in a clear polymer media, a localized heating system is introduced. In this case, the polymer close to the CNT gets warm and the supersaturated CO_2 can escape the polymer solution to nucleate the bubbles. The polymer/gas solution far from the particles is colder, conversely and no cell nucleation is possible. With time, the induced heat can transfer to the neighboring regions and cell growth can invade most of the polymer. In this way, foaming will occur only near the hot particles, with reduced risk of collapse. The Figure 1 is a schematic of cell nucleation and growth during MW assisted foaming.

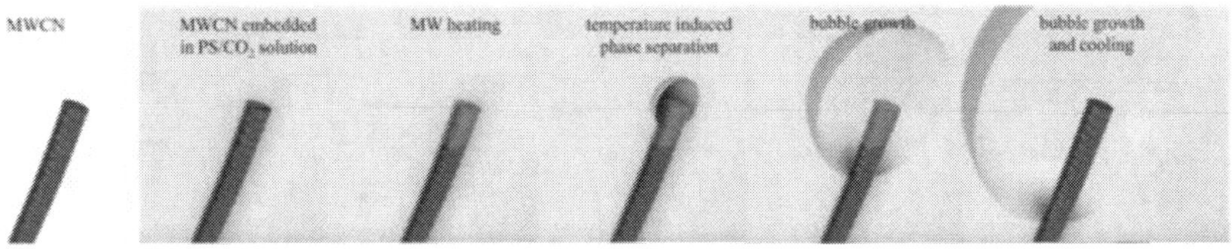

Figure 1. Schematic of cell nucleation and growth with MW radiation

Each unit (CNT and neighbor polymer matrix) has a very small size, is independent from others and is exposed to microwave simultaneously in all of the sample volume. This, of course, means more uniform cell structure, faster foaming and, most importantly, no dependency on sample thickness. As mentioned before the thickness is a limiting factor in other methods and usually thickness is around 1mm. In these experiment we successfully foamed, with uniform morphology, foams samples with 1cm thickness.

In the conventional heating system the heat must transfer from surface to the center of samples, which, by considering the low conductivity of polymers, takes time and induces a gradient in both local foam density and morphology. This phenomenon is obvious in water bath method (Figure 2- a1) foams. These samples experienced a notable gradient in the thickness, in contrast with foaming by microwave.

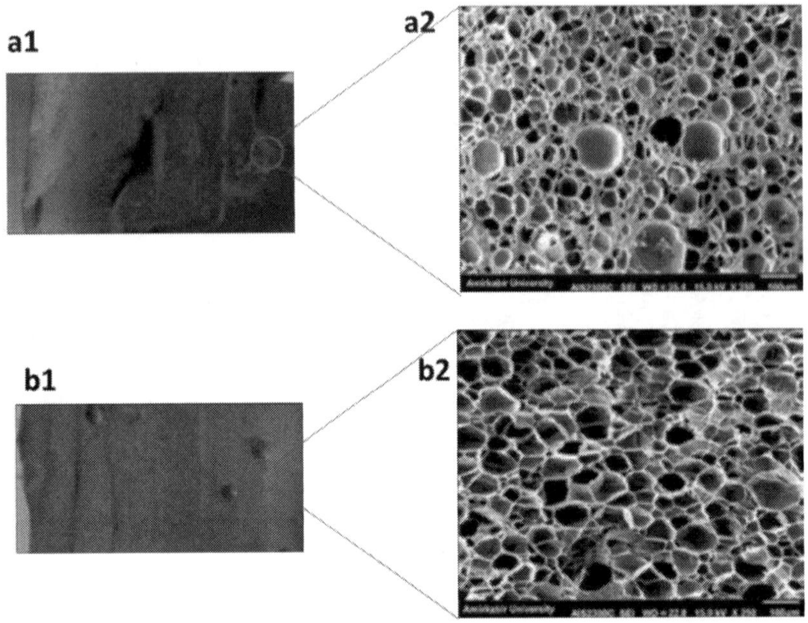

Figure 2 a: results from the water bath method, **1** optical image and **2** SEM image in the center of the foam. **b:** results from the MW method, **1** optical image and **2** SEM image in the center of the foam.

Table 1. Presents the density of foams of samples with different thickness. If we consider the density of foams as to verify foam quality and expansion, a nice improvement in expansion in MW method is evident. In both sample thicknesses, the foams with MW have lower density. We can conclude that with MW method the efficiency of the foaming process is almost independent of thickness.

Table 1.Foam densities for the two analyzed methods, sample thickness of 1 and 10 mm

	Density (g/cm^3)	
Thickness	1mm	10mm
Boling Water Bath@ 100 °C	0.14	0.34
Microwave Radiation @ 40 s	0.11	0.05

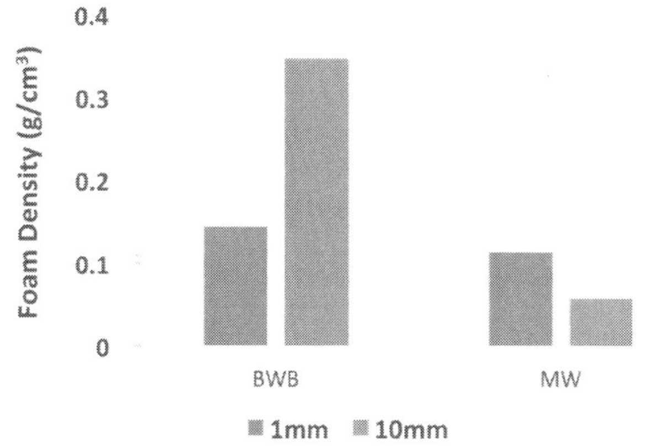

Figure 3. Density of samples at 1 and 10 mm thickness in water bath (BWB) and MW methods.

Conclusions

We used for the first time microwave heating as source of energy in PS/CNT nanocomposite foaming. CNTs react to microwave and transduce the radiation energy to heat. As this method used tiny dispersed heating source, more uniform morphologies are achieved with respect to neat PS and lighter foam are achieved. The MW foams have more uniform cell structure and can reach higher expansion ratio than the water bath foams, particularly at increased foam thickness.

References

1. Zeng, C., N. Hossieny, C. Zhang, and B. Wang (2010), "Synthesis and processing of PMMA carbon nanotube nanocomposite foams," *Polymer*, 51(3), pp. 655-664.
2. Tammaro, D., A. Astarita, E. Di Maio, and S. Iannace (2016), "Polystyrene foaming at high pressure drop rates," *Industrial & Engineering Chemistry Research*, 55(19), pp. 5696-5701.
3. Sharudin, R.W.B. and M. Ohshima (2011), "CO2-Induced Mechanical Reinforcement of Polyolefin-Based Nanocellular Foams" *Macromolecular Materials and Engineering*, 296 (11): p. 1046-1054.
4. Garancher, J.-P. and A. Fernyhough (2012), "Crystallinity effects in polylactic acid-based foams". *Journal of Cellular Plastics*,48 (5), pp. 387-397.
5. Moon, Y. and S.W. Cha (2007), "Study on viscosity changes with talc in microcellular foaming process" *Fibers and Polymers*, 8 (4), pp. 393-398.
6. Lee, L.J., C. Zeng, X. Cao, X. Han, J. Shen, and G. Xu (2005), "Polymer nanocomposite foams" *Composites science and technology*, 65 (15), pp. 2344-2363.
7. Wu, T., Y. Pan, E. Liu, and L. Li (2012), "Carbon nanotube/polypropylene composite particles for microwave welding" *Journal of Applied Polymer Science*, 126 (S2).
8. Savi, P., M. Miscuglio, M. Giorcelli, and A. Tagliaferro (2014), "Analysis of microwave absorbing properties of epoxy MWCNT composites" *Progress In Electromagnetics Research Letters*, 44, pp. 63-69.

9. Tammaro, D., Contaldi, V., Pastore Carbone, M.G., Di Maio, E., Iannace, S. (2016), "A novel lab-scale batch foaming equipment: the mini-batch, Journal of Cellular Plastics", *Journal of Cellular Plastics,* 52 pp. 533-543
10. Cheng, J., R. Roy, and D. Agrawal (2001), "Experimental proof of major role of magnetic field losses in microwave heating of metal and metallic composites" *Journal of materials science letters*, 20 (17), pp. 1561-1563.

471406 Polyurethane Foam Expansion, Polymerization and Bubble Pressurization

Abstract:

In this presentation, we discuss our new, experimentally driven, approach to foam expansion and polymerization modeling. We are working to develop high-fidelity computational models to be used to design and troubleshoot mold filling for polyurethane structural foam parts. We are studying PMDI polyurethane with a fast catalyst, such that filling and polymerization occur simultaneously. The foam is over-packed to twice or more of its free rise density, to reach the density of interest. Developing a relevant model to represent the expansion, filling, curing, and final foam properties is quite challenging. PMDI is chemically blown foam, where carbon dioxide is produced via the reaction of water and isocyanate. The isocyanate also reacts with polyol in a competing reaction, which produces the polymer. A new kinetic model is implemented, which follows a simplified mathematical formalism that decouples these two reactions. The model predicts the polymerization reaction via condensation chemistry, where vitrification and glass transition temperature evolution must be included to correctly predict this quantity. The foam expansion kinetics are determined by tracking the molar concentration of water and carbon dioxide in both the gas and polymer phases. A bubble-scale mathematical model has been developed that looks at the super saturation of carbon dioxide in the polymer and the concomitant bubble pressurization. The model is calibrated with bubble-expansion data from bubble growth experiments. It is used to predict the density of a foam bar and compared to to Xâ€"Ray CT data.

473436 The Role of Viscoelasticity in Bubble Breaking

Abstract:

The rupture dynamics of a bubble affects many phenomena in nature, e.g., magma fragmentation in volcanology, the opening of vesicles in medicine, and many processing technologies from the production of polymeric foams to the cooking of pancakes[1-2]. If the bubble is made of a viscoelastic fluid, it can store elastic energy during inflation that can be released during rupture, in turn affecting the breaking dynamics. In the present contribution, we investigate experimentally the way elastic energy (stored as a consequence of fast bubble inflation) drives rupture in a model viscoelastic fluid. A viscoelastic bubble is created via a pump and a homemade inflating system, a hot needle induces rupture and a fast camera is used to acquire images of the hole opening over time. The roles of the extensional rate and of the stored elastic energy on bubble rupture are investigated, and a new mathematical model to describe the rupture phenomenon is developed.

[1] de Gennes PG, Brochard-Wyart F and Quéré D, Capillarity and Wetting Phenomena, Springer (2004).

[2] Eggers J and Villermaux E, Physics of liquid jets, Rep. Prog. Phys., 71, 036601 (2008).

473667 The Use of the Viscoelasticity in Polymer Foaming to Obtain a Fully Opened Cell Structure

Abstract:

In this work a design tool to control cell-opening in gas foaming of thermoplastic semicrystalline polymers is developed. The sequence of events following bubble nucleation, namely, bubble growth and impingement, are modeled to gain a comprehensive, perspective view on the mechanisms of bubble wall rupture and on the conditions for achieving a fully opened cell morphology by flow induced crystallization. In particular, unlike the previously published literature, the polymer elastic recovery is recognized as an important factor for wall retraction, often dominant with respect to surface tension, which is typically considered as the only driving force determining the retraction of ruptured bubble walls. The new approach is experimentally validated on poly(e-caprolactone) (PCL), foamed with CO_2, as a model polymer/gas system.

474325 Bead-Milling and Post-Milling Recrystallization: An Organic Template-Free Methodology for the Production of Nano-Zeolite Catalyst

Abstract:

A new method for the production of nanosized zeolite powder by a top-down approach has been performed. [1,2] In this study, ZSM-5 (MFI type structure) was first milled to produce a nanopowder. This technique can destroy the outer portion of the zeolite framework, which lowers the micropore volume of ZSM-5 zeolite. To remedy this, the damaged part was recrystallized using a dilute aluminosilicate solution after bead milling. From the combined bead milling and post-milling recrystallization, nanosized ZSM-5 zeolite approximately 50 nm in size with high crystallinity was obtained successfully.

Commercial ZSM-5 was milled using a bead milling apparatus (Minicer, Ashizawa Finetech, Ltd., Tokyo, Japan) for 2-6hours. Recrystallization of the milled ZSM-5 was performed using dilute silicate solution with the composition of 0.0525 Na_2O : 0.117 SiO_2 : 10.0 H_2O. The importance of this particular ratio is that it provides a solution nearly in equilibrium with ZSM-5. This means that ZSM-5 neither undergoes macroscopic growth nor dissolution. Under these conditions, the remaining ZSM-5 crystallites act as seeds and the poorly crystalline parts of the milled ZSM-5 are more easily recrystallized back onto the ZSM-5, resulting in a more ordered product. Obtained samples were evaluated as acid-catalysts for cumene cracking. The test reactions were conducted so that diffusion through the zeolite pore structure was the rate-determining step of the catalytic reaction. As a result, ZSM-5 zeolite powder showed a higher catalytic activity in cumene cracking in comparison with the raw ZSM-5 zeolite. Furthermore, the decrease in crystal size suppresses catalyst deactivation through coke deposition during cumene cracking.

References

[1] T. Wakihara et al., Crystal Growth & Design, 11, 955-958 (2011).

[2] T. Wakihara et al., Crystal Growth & Design, 11, 5153-5158 (2011).

465376 Control of Al Sites in the CON-Type Aluminosilicate Zeolite

Abstract:

Al-containing CON-type zeolites has been synthesized by the post-synthesis method using the CON-type borosilicate zeolite ([B]-CON). Very recently, a method for directly synthesizing Al-containing CON-type borosilicate zeolites ([Al,B]-CON) has been developed. We have first found that the directly synthesized [Al,B]-CON exhibited a higher catalytic performance in the MTO reaction in terms of duration and the propene selectivity than the **MFI** and ***BEA**-type aluminosilicate zeolites. Based on the high-resolution 27Al MAS NMR, the Al distribution in the framework would be dependent on the preparation method, i.e., direct or post-synthesis ones. The purpose of this study is to synthesize Al-containing **CON**-type zeolites with different Al distributions and to investigate their MTO activity.

[Al,B]-CONs were directly synthesized from the mother gel containing N,N,N-trimeth-yl-(-)-cis-myrtanylammonium hydroxide (TMMAOH) as OSDA. Additionally, **CON**-type borosilicates ([B]-CON) and Zn-containing **CON**-type borosilicates ([Zn,B]-CON were also prepared. Al-containing CON-type zeolites were synthesized by post synthesis method using [B]-CON and [Zn,B]-CON zeolites.

The Al state was investigated by ^{27}Al MAS NMR technique in detail. The directly-synthesized [Al,B]-CON gave two peaks assigned to tetrahedrally-coordinated Al species at 58 and 54 ppm. the post-synthesized [Al]-CON using [B]-CON as parent (denoted as "[Al]-CON-P_[Zn]-CON") had only one peak at 58 ppm. Interestingly, the post-synthesized [Al]-CON using [Zn,B]-CON as parent (denoted as "[Al]-CON-P_[Zn,B]-CON") had peaks assigned to tetrahedrally-coordinated Al species at 60 and 54 ppm. These results imply that the Al distributions in the framework is strongly dependent on the preparation method. In general, Zn and B species are easily removed from the zeolite framework under acidic conditions. Considering that the ^{27}Al MAS NMR spectrum of [Al]-CON-P_[Zn,B]-CON was clearly different from that of [Al]-CON-P_[B]-CON,

Al atoms would be inserted into the Zn site, not B site.

Furthermore, we found that the Al distributions in the framework is varied with the B content of the [B]-CON as parent; [Al]-CON-P_[B]-CON was also prepared using [B]-CON (Si/B = 5). Its ^{27}Al MAS NMR spectrum was different from [Al]-CON-P_[B]-CON (Si/B = 25) and [Al]-CON-P_[Zn,B]-CON.

^{11}B MAS NMR spectrum of [B]-CON (Si/B = 25) showed one peak ascribed to tetrahedral B species, while [B]-CON (Si/B = 5) showed two peaks ascribed to trigonal B species in addition to tetrahedral B ones. Thus, we assumed that Al was inserted into the trigonal B site for [Al]-CON-P_[B]-CON (Si/B = 5) .

The directly prepared [Al,B]-CON catalyst showed a short life under W/F of 1.7 g h mol^{-1}; the conversion of methanol was kept at >99 % until only 3 h. [Al]-CON-P_[B]-CON (Si/B = 25) also showed a short life under W/F of 3.3 g h mol^{-1}, the conversion was drastically decreased at 1 h.

On the other hand, [Al]-CON-P_[Zn,B]-CON showed a much longer catalytic life but the trend of the products' selectivities was almost the same. These facts imply that the Al distribution of the **CON**-type aluminosilicate strongly affects catalytic life in the MTO reaction.

465471 Core-Shell Bulk BEA-Lamellar MFI Composite Prepared in One-Step: Integration of 3D and 2D Zeolites into Hierarchical Structures for Efficient Alkylation Reactions

Abstract:

The synthesis of hierarchical meso-/microporous zeolite materials with spatially controlled morphology, meso-/microporosity, and acidity is an expanding area of research interest for a wide range of applications. Here, we report a one-step dual template synthesis method for integration of 3-dimensional (3D) bulk BEA and 2D lamellar MFI into a new type of core-shell hierarchical meso-/microporous zeolite composite structures. Specifically, the 2D layered MFI nanosheets were laid over the surface of or interdigitated into the 3D microporous BEA particles in the core-shell bulk BEA-lamellar MFI (BBLM) zeolite structure, which generated a unique morphology of interconnected micropores and mesopores underneath the â€˜skinningâ€™ shell (~3-10 nm) of MFI nanosheets. The structure and morphology of the BBLM zeolites were elucidated by a combination of XRD, SEM, TEM, and Ar physisorption techniques. The composition and acidity were identified using ICP, XPS, solid state NMR, organic base titration, and NH_3-TPD measurements. The catalysis tests using conversion of benzyl alcohol in mesitylene showed that BBLM zeolites had an excellent activity and durability, which was resulted from the synergistic integration of the hierarchical structures in the BBLM composite. The core-shell BBLM zeolite provides a good model catalyst with integrated 2D-3D structures and meso-/microporosity for studying a series of important catalytic reactions in hierarchical zeolites. The one-step dual template synthesis method described herein is versatile and facile, which may prove to be a general platform for coreâ€“shell hierarchical zeolite design with potentially broader applicability to other porous materials.

467327 Effects of Zeolite Growth Modifiers on Different Stages of ZSM-5 Crystallization

Abstract:

Effects of Zeolite Growth Modifiers on Different Stages of ZSM-5 Crystallization

Wei Qin and Jeffrey D. Rimer

Chemical and Biomolecular Engineering, University of Houston, 4726 Calhoun Rd., Houston, Texas 77204 USA

Zeolite ZSM-5 is a common heterogeneous catalyst in refinery and petrochemical processes, owing to its excellent shape-selectivity, strong Brønsted acidity, and high thermal/hydrothermal stability. It has an MFI structure with 3-dimensional pores consisting of interconnecting straight and sinusoidal channels. Molecular diffusion is much faster in the straight channels oriented along the [010] direction. To this end, it is desirable to selectively control the [010] thickness of ZSM-5 crystals.

Here, we will present a novel approach to tailor ZSM-5 morphology that incorporates zeolite growth modifiers (ZGMs). The use of ZGMs has proven to be an efficient method to tailor the morphology of silicalite-1, which is the siliceous isostructural analogue of ZSM-5.[1] For our study of ZSM-5 synthesis in the presence of ZGMs, we assessed a library of modifiers previously screened for slicalite-1 to determine the influence of aluminum incorporation within the MFI framework. It was observed that subtle changes in the synthesis conditions can alter ZGM binding specificity to different crystallographic faces of ZSM-5. Using atomic force microscopy (AFM), we show that chemical force microscopy can be used to efficiently probe modifier-crystal interactions in order to elucidate ZGM molecular recognition for different zeolite surfaces.[2]

We also investigated the effects of ZGMs on the initial stages of amorphous precursor self-assembly and evolution (pre-nucleation) as well as crystal growth using a combination of two techniques: small angle X-ray scattering (SAXS) [3] and dynamic light scattering (DLS). We observed that polycations have a unique effect of promoting precursor aggregation during the induction period.

Through the judicious selection of ZGMs under different synthesis conditions, we were able to synthesize ZSM-5 crystals with various sizes and morphologies. We will present details of these syntheses as well as catalytic tests of ZSM-5 catalysts with a model reaction to gain a better understanding of the relationship between ZSM-5 morphology and catalyst performance. Collectively, this work provides quantitative guidelines for designing synthesis protocols to tailor ZSM-5 crystallization and improve catalyst properties for a wide range of applications.

(1) Lupulescu, A. I.; Rimer, J. D. *Angew. Chemie - Int. Ed.* **2012**, *51* (14), 3345–3349.

(2) Lupulescu, A. I.; Rimer, J. D. *Science.* **2014**, *729.*

(3) Fedekyo, J.; Rimer, J. D.; Lobo, R. et al. *J. Physical Chemistry C.* 2004, 108.

452031 The Generation of Novel Nanomaterials from the MCM-22-P Precursor Crystal

Abstract:

The creation of novel nanostructures from easy to produce precursor crystals has become a topic of significant interest in recent years. Methods including post synthetic functionalization, swelling, pillaring, and the introduction of metal nanoparticles to a precursor structure have all yielded new materials from these initial building blocks. One of the most promising methods to yield novel nanostructures from precursor crystals is the use of lamellar structures, such as the layered nanosheet MCM-22-P, to yield final inorganic crystals with tunable pore widths and metal oxide pillars between the layers. In our group the use of these nanostructures have currently yielded two materials of interest for adsorption and catalysis. Of particular interest is the use of clay chemistries to yield final structures composed of zeolite layers bridged by metallic oxide pillars, such as those containing titanium or silica oxide. Secondary functionalization by our group has also yielded materials prepared from the lamellar precursor structure that contain post synthetically impregnated amine groups into both the large and small pore spaces. These materials therefore can be used to create a wide range of interesting nanostructures, with both organic and inorganic functionalities, and utilizing relatively simple chemical methods.

460777 Mesoporous Metal Sulfides and Carbides

Abstract:

Mesoporous materials have attracted attention due to their great potential as catalysts, electrodes, and magnetic materials. Since the nanocasting method was introduced, a wide range of mesoporous transition metal oxides, such as Fe_2O_3, MnO_2, Co_3O_4, NiO, and TiO_2, with ordered pores and crystalline walls have been synthesized using mesoporous SBA-15 and KIT-6 silica as hard templates. Some unique behaviors of nanocast materials have been found for a number of applications. For example, we recently showed a photocatalytically active ordered mesoporous Co_3O_4 with extremely high surface area by nanocasting a mesoporous $MgCo_2O_4$ and then selectively leaching the magnesium. Through manipulation of the synthetic conditions and elemental compositions the resulting Co_3O_4 material exhibited a much higher activity in the photocatalytic water oxidation reaction compared to its bulk counterpart. This study demonstrated the importance of synthesis in preparing mesoporous materials with the desired composition and porosity. To continue addressing critical challenges with mesoporous materials new strategies to access materials that could not be synthesized previously are necessary.

Among all the transition-metal-containing materials, metal sulfides and carbides are of particular interest because of their potential applications in sensors, separations, catalysis, solar cells, fuel cells and batteries. A few attempts to synthesize mesoporous metal sulfides and carbides have been made. Pioneering work by Shi et al. showed the possibility to synthesize ordered mesoporous metal sulfides via the nanocasting method. In their report of mesoporous WS_2 and MoS_2, phosphotungstic acid ($H_3PW_{12}O_{40}·6H_2O$) and phosphomolybdic acid ($H_3PMo_{12}O_{40}·6H_2O$) were loaded into a mesoporous silica template, followed by a high temperature thermal treatment (600â„ƒ) under a flow of mixed H_2S and H_2. Although some important progress has been made, the range of ordered mesoporous metal sulfides and carbides that can be synthesized is still very limited.

In this presentation, we will discuss an oxide-to-sulfide transformation approach as a general synthetic method to prepare ordered mesoporous metal sulfides. Three first-row transition metal sulfides, FeS_2, CoS_2, and NiS_2, with highly ordered mesoporous structures and crystalline walls have been successfully synthesized for the first time. Preliminary investigation has been performed to explore the photocatalytic properties of the newly synthesized mesoporous metal sulfides. All the mesoporous metal sulfides exhibited higher activity than their bulk counterparts for the photocatalytic degradation of methylene blue dye under visible light.

In addition to metal sulfides, we will also show the syntheses of two ordered mesoporous metal carbides, Mo_2C and W_2C, using a nanocasting approach coupled with a simultaneous decomposition/carbonization process under a continuous methane flow. The as-synthesized mesoporous Mo_2C and W_2C have three-dimensionally ordered porous structures, large surface areas (70-90 $m^2\ g^{-1}$), and crystalline walls. In vapor-phase anisole hydrodeoxygenation (HDO) reactions, they exhibited excellent turnover frequencies under relatively low reaction temperatures (423-443 K) and ambient hydrogen pressures. Notably, the ordered mesoporous W_2C catalyst showed a greater than 96% benzene selectivity in anisole HDO, the highest benzene selectivity reported to date.

469972 Effect of Acid Gas Interactions on the Structure and Porosity of ZIF-8

Abstract:

Metal organic frameworks (MOFs) are a class of nanoporous materials that have received considerable attention for use in adsorptive and membrane separations. Zeolitic Imidazolate Frameworks (ZIFs) are well known amongst MOFs for their structural stability with respect to temperature and humidity. ZIFs can be synthesized utilizing different organic linker molecules and synthesis conditions, leading to a set of diverse materials with tunable surface areas, pore sizes, surface functionalities and possible applications. In many practical separation processes such as CO_2 capture from post combustion flue gas, landfill gas separation, natural gas sweetening and others, adsorbents are exposed to a variety of molecular species including acid gases that can have a detrimental effect on performance parameters such as selectivity or adsorption capacity, or can lead to irreversible structural changes in the adsorbent that may restrict its prolonged use. However, there is little knowledge of the effects of such acid gas exposure on the adsorption and separation properties of ZIFs.

In this study, the stability of a model ZIF material (ZIF-8) under SO_2 exposure in dry, humid, and aqueous environments was investigated in detail. To determine structural and textural changes, pXRD and N_2 physisorption were conducted before and after exposure. Furthermore, techniques such as EDX, XPS and FTIR were used to study the mechanistic aspects of these changes. ZIF-8 was found to be stable after prolonged exposure in both dry and aqueous SO_2 environments and also in the presence of humid air without SO_2. However interactions with ppm concentrations of SO_2 in the presence of high relative humidity led to its irreversible structural degradation over time as evinced by substantial losses in crystallinity, BET surface areas and pore volumes. EDX analysis showed a significant presence of S even after reactivation of the exposed ZIF-8. FTIR studies revealed the presence of S-O vibrations and the protonation of the nitrogen in the imidazole ring, suggesting cleavage of the metal-linker bond. A mechanism of ZIF-8 degradation combining the synergistic effects of SO_2 and humidity is proposed, which advances the present understanding of the stability of ZIF materials towards acid gas exposure.

470599 High Yield Stoichiometric Synthesis of ZIF-8 Nanoparticles Using Novel Reactor

Abstract:

Zeolitic imidazolate Frameworks (ZIFs) are chemical and thermally stable sub-class of metal organic frameworks. The desirable molecular sieving properties of ZIFs make these materials an interesting candidate for energy-efficient separation techniques, mainly membrane and adsorption techniques. The particle size becomes an important criterion for membrane applications as bigger particles can pose diffusion limitations. The main challenge for greater utilization of ZIFs is the difficulty to synthesize material with uniform particle size and high surface areas at large scales, while achieving high yields.

In this work, a scalable, solution phase synthesis is demonstrated for ZIF-8 nanoparticles with uniform diameters in the range of 40-80 nm. ZIF-8 is synthesized with 89% yield using stoichiometric precursor concentrations using a novel reactor. Turbulence created by the reactor provides homogeneous conditions in the reaction volume resulting in improved yield. The surface area and micropore volume of the synthesized materials confirm the high quality of products. The reactor was studied to determine the effect of different synthesis parameters including ligand to metal ratio, base concentration, and mixing intensity. The base concentration has the most effect on the product size, morphology, and yield. The versatility of the reactor is shown by synthesizing ZIF-67 with 79% yield. Further, a larger reactor was constructed and tested, demonstrating that the method can be scaled to increase productivity. Overall, this work presents a versatile and scalable route for nanoparticle synthesis.

462933 Modeling Framework for Predicting Phase Behavior and Transport in Oppositely Charged Polyelectrolyte Solutions

Abstract:

We develop a modeling framework for phase behavior and transport of oppositely charged polyelectrolytes (PE) in coacervates and Layer-by-Layer (LbL) assemblies that accounts for diffusion of both oppositely charged chains and their complexation. The core of the model is the development of a free energy model that includes free energies for ion pairing, counterion condensation, charge regulation, electrostatic free energy, as well as elastic energy of the network and Flory Huggins entropy and enthalpy. Combinatorial entropies for ion pairing, counterion condensation, and charge regulation are included along with reference state chemical potentials as adjustable parameters. A numerical approach is developed to carry out multi-component highly nonlinear flash calculations to determine phase behavior. We find a very strong influence of ion pairing and counterion condensation on phase behavior, and on the distribution of salt and polyelectrolyte species between the coacervate and supernatant phases. From this free energy model, and an extended Stefan-Maxwell flux law, diffusion of PE chains into a LbL film can be computed, providing a unified approach that connects phase behavior to transport in polyelectrolyte assemblies.

463554 Structure and Dynamics of Model Tapered Diblock Polymers

Abstract:

Tapered diblock polymers are akin to AB diblock polymers but contain a region with a gradient in composition inserted between the pure A and B blocks. We consider polymers in which the taper is a linear gradient from pure A to pure B, or B to A in the case of inverse tapers, and can be from 0% (diblock) to 100% (gradient copolymer) in length. We aim to show how adjusting the taper length and direction controls the interfacial and microphase structure, as well as the impacts on dynamics. Self-consistent field theory (SCFT) was used to study phase behavior while coarse-grained molecular dynamics (MD) simulations showed details of polymer conformations and diffusion, including diffusion of small molecule penetrants that have a favorable chemical interaction with only one of the polymer phases. In accordance with experimental results, adding a taper increases the interfacial region and decreases the order to disorder transition temperature of block copolymers. Interestingly, the region of the phase diagram where the bicontinuous double gyroid phase is preferred is widened significantly for short to medium length normal tapers, but not for inverse tapers. Inverse tapered polymers can stitch back and forth across the interface as would be expected of ABâ€™Aâ€™B type tetrablock copolymers, however, due to the mixed taper region, this behavior depends strongly on segregation strength. The interesting inverse tapered polymer conformations lead to nonintuitive trends in the domain spacing and dynamics, and in certain systems penetrans can diffuse faster through inverse tapered polymers than diblocks even while the polymer diffusion is slower.

470343 Innovation in Emerging Areas in Polymers

Abstract:

Markets trends of cost reduction, consumer convenience, sustainability & down-gauging drive the need for new and improved packaging solutions. Development of such solutions requires a look across the entire value chain â€" an Asset to Market look. This talk will focus on what is required for a successful packaging solutions to be launched in a competitive space. Key examples will be presented and will cover the trends related to recyclability and downgaging.

451186 Open Questions in Soft Matter: Is There an Ideal Glass Transition? Are Colloids Good Models for Molecular Glasses?

Abstract:

The glass transition event is frequently taken to be a manifestation of an underlying thermodynamic transition, perhaps 50 K below the laboratory measured glass transition temperature. The behavior of glasses manifests itself in both the thermodynamic and dynamic responses and these are frequently interpreted in terms of an 'ideal' glass transition.
Here we first examine the thermodynamics by using calorimetric measurements to determine the equilibrium heat capacity of a series of poly(α-methyl styrene) mixtures with results that show no evidence of a transition in the Ehrenfest sense as far as 150 K below the nominal glass transition temperature. The dynamic signature of glasses is found in the super-Arrhenius behavior of viscosity or relaxation time as a function of temperature. The so-called Vogel-Fulcher extrapolation of the measured dynamic property leads to an apparent finite-temperature divergence some 50 K below the nominal glass temperature. By performing experiments on glasses aged for very long times (Dominican amber of 20 million years) we have been able to explore the upper bounds to the equilibrium dynamics to some 43.6 K below the glass temperature and find strong evidence that the finite-temperature divergence does not exist, rather the glass exhibits an apparent Arrhenius behavior for the dynamics, suggesting that theories that are based on the finite temperature divergence of the relaxation times need to be re-evaluated. Finally, there is considerable activity in the Soft Matter community in which concentrated colloidal dispersions are used as models of glass-forming systems. Here we visit this problem using a novel series of concentration-jump experiments in PNIPAAM-particle colloids to mimic the classic temperature-jump experiments that were used by Kovacs to catalogue the kinetics of structural recovery in molecular glasses. Here we will use diffusing wave light scattering spectroscopy and classical rheological methods to examine specifically the "intrinsic" isotherm, asymmetry of approach, and the memory signatures and compare the colloidal system with the molecular glass. We find that the colloidal systems while exhibiting some of the features seen in molecular glasses, in detail show surprising differences in their structural recovery signatures.

459798 On Demand Release of Bacterial Biofilms Via Shape Memory Activation

Abstract:

Surface topography at nano- or micro-scale has been intensively studied as a promising antifouling strategy. However, fouling control by a static surface topography can be temporary, and biofilm can gradually develop over time. Surfaces that are able to change topography to remove established biofilms can help address this challenge. Here, we demonstrate that a shape memory polymer (SMP) with defined surface topography can both prevent bacterial adhesion and remove established biofilms upon triggered shape change, thereby offering more prolonged antifouling properties. For example, an SMP substrate with μm-scale hexagonal wells fixed with a temporary but stable uniaxial strain of >50% exhibited 48.1% reduction in biomass of 48 h *Pseudomonas aeruginosa* PAO1 biofilms before shape recovery compared with a static flat controls. Upon heating for 10 min at 40°C, the hexagonal wells recovered to the permanent shape, causing detachment of 99.9% of attached biofilm cells. Because rapid shape change can be obtained near body temperature (e.g., 10 min at 40°C in this study), biocompatible SMPs with appropriate surface topographies may be developed for future biomedical applications, and thus represent a new class of antifouling surfaces.

456013 Self-Assembly of "Hairy" Biomacromolecular Membranes Using an Aqueous-Aqueous Interface

Abstract:

Hierarchical self-assembly occurs ubiquitously in nature to form biomolecular materials with structure and function from nano to meso to macroscopic length scales. Here, we describe a dynamic self-assembly process which utilizes the supramolecular interaction between polysaccharides and peptide amphiphiles at an aqueous-aqueous interface to form membranes with perpendicularly aligned features. Through rational molecular design, we show that the chemical structure of the peptide amphiphile directs the self-assembly mechanism towards one of two regimes, resulting in vastly different membrane structure and properties. We investigate our system using molecular simulations, which identify the key role of interaction strength in directing the self-assembly process. Furthermore, we show the effect of membrane morphology on therapeutic function in anti-cancer applications.

465678 Biomimetic Surfaces for Drag Reduction in Turbulent Flow

Abstract:

A significant amount of fuel consumed by marine vehicles is expended in overcoming frictional drag. Any significant reduction in this frictional drag would result in noteworthy economic savings and reduced environmental impact. Superhydrophobic surfaces (SHSs) have shown promise in reducing drag in small-scale applications and/or in laminar flow conditions. The efficacy of these surfaces in reducing drag in turbulent flow, however, is not well characterized. Large, scalable SHSs in turbulent flow have shown drag increase, no change, or 10% to 30% drag reduction under certain conditions. In this work we discuss the design of mechanically durable, large-scale SHSs that cause significant drag reduction (> 50% reduction) in fully-developed turbulent flow. A large-scale flow facility was constructed specially to evaluate SHSs in fully-developed turbulence. We also discuss how the conventional methods of evaluating SHSs at ambient pressure cannot predict their drag reduction under turbulent flow conditions. Instead, we propose a new characterization parameter, contact angle hysteresis at higher pressure, which aids in the rational design of friction-reducing SHSs. Finally, we relate the drag reduction observed on our SHSs to the viscous length scales of the flow. Overall, we find that both the contact angle hysteresis at higher pressure, and the non-dimensionalized surface roughness, must be minimized to achieve meaningful turbulent drag reduction. In contrast, we show that even SHSs that are considered hydrodynamically smooth can cause significant drag increase if designed incorrectly.

477118 Self-Similar Dynamics of Extracellular Matrix Production during Bacterial Biofilm Colony Expansion

Abstract:

Introduction: Bacterial biofilms are multicellular microbial communities encased in a self-produced extracellular matrix. Understanding the growth and adhesion of biofilms and their interaction with surfaces is critically important in diverse industrial and medical scenarios, such as in preventing fouling of marine vehicles and halting bacterial growth in implants and medical devices. In this work, we use Bacillus Subtilis biofilms as a model organism to investigate the spatiotemporal dynamics of cell heterogeneity that gives rise to the collective growth and colonization on solid/air interfaces.

Materials and Methods: A triple transcriptional reporter (NCIB 3610 sacA::Phag-mkate2 amyE::PtapA-cfp ywrK::PsspB-citrus) was constructed to simultaneously visualize motility, matrix production and sporulation cell states in the developing biofilm. This strain was grown on a 1.5 wt% MSgg/Agar solid interface at 30 C. Simultaneous time-lapse brightfield and epifluorescence microscopy for each channel was performed every 10 minutes, for a duration of 72 hrs. The resulting movies were analyzed to determine the azimuthally averaged height profiles and transcriptional reporter activity levels for each of the cell types.

Results and Discussion: Two clear regimes were observed in biofilm growth. At the early stage (t<15 hrs), the exopolysachharide (EPS) matrix is globally produced. The radius of the biofilm exponentially increases with time. At t ~ 24 hrs, there is a transition to a second regime where the size of the colony scales as $R\sim t^{0.5}$. This regime is characterized by a well-defined propagating front where the active growth and production of extracellular matrix (EPS) in the biofilm is localized. The EPS production in the interior of the biofilm is down regulated, along with a simultaneous increase in sporulation activity that is indicative of nutrient starvation. The spatial profile of EPS reporter activity measured from the intensity profiles at the propagating front is self-similar and reaches a final characteristic asymptotic shape during late stage biofilm expansion. The width of this propagating front of the biofilm at the front can be increased by the introduction of fresh glutamate.

Conclusions: We have investigated the collective behavior and dynamics of B. Subtilis biofilms during colony expansion over a solid interfaces. Our results demonstrate the active growth in bacterial biofilms is confined to a locally propagating front. Manipulating the nutrient and environmental conditions at this local front can lead to potential techniques to control biofilm growth.

466090 Rational Design for Therapeutic Peptide-Amphiphile-Based Intracellular Delivery

Abstract:

Therapeutic peptides are highly promising tools to manipulate protein-protein interactions, because of their target selectivity and potential safety. Although, peptide amphiphile (PA) technology that combines a polar peptide with a hydrophobic tail efficiently delivers peptide payloads into cells and increases their stability in the circulation, obstacles remain for viable clinical translation. One major obstacle is that PAs frequently accumulate within endosomal compartments because of the greasy hydrophobic tail, which limits the ability of delivered peptides to target proteins within the cytosol. Thus, overcoming endosomal entrapment is a crucial step for the effective clinical transition of these peptide-based targeting agents. Additionally, intracellular trafficking and the ultimate fates of the hydrophobic tails and peptides are largely unknown.

In this study, we aim to overcome endosomal sequestration of peptide cargoes through a PA design that includes an endosomal cleavable linker between a diC16 hydrophobic tail and a previously validated biofunctional peptide sequence from the tumor suppressor p53 ($p53_{(14-29)}$). We use a cathepsin cleavable sequence, combined with FRET compatible double fluorescence dyes, to study the internalization, trafficking, and endosomal escape of intact PAs as well as individual lipid and p53 peptide moieties. Single molecule FRET measurements show efficient enzyme activity between the lipid tail and $p53_{(14-29)}$ while having no activity on the peptide itself. We find that cathepsin-cleavable PAs accumulate within the cell more rapidly than PAs lacking this sequence and very quickly loose intracellular FRET signaling indicating rapid cleavage of the $p53_{(14-29)}$ from diC_{16}. Using confocal microscopy, we show that cleaved $p53_{(14-29)}$ peptides and hydrophobic diC_{16} tails migrate to different parts of the cytosol. While a portion of diC_{16} tails appear to recycle to the cellular membrane, intracellular accumulation also exists. Conversely, PAs lacking a protease cleavage site retain FRET signaling over time, predominate within endosomes and exhibit far less intracellular accumulation over time.

We believe that cathepsin cleavable PAs will provide opportunities to functionally target numerous protein-protein interactions using a variety of therapeutic peptides while also providing a tool to better elucidate PA trafficking in real time within cells.

464169 Recapitulating Tissue Microenvironments with Biomaterials to Restore Immunity

Abstract:

Introduction: Deficiencies in the immune system can result in complications such as infections, autoimmune conditions and cancer. We investigated if the recapitulation of the bone marrow stroma with a biomaterial scaffold can drive immune cell development by increasing (i) the number of T- and B-lymphocytes and (ii) the immune repertoire. We used hematopoietic stem cell transplantation (HSCT) an an exemplary case of profound immunodeficiency, in which the pre-transplant conditioning regimen destroys immune cells and severely compromises the adaptive arm of the immune system. Immunological complications are common in patients receiving HSCT and are the primary cause of treaatment failure.

Materials and Methods: Injectable macroporous protein-polymer hydrogels, composed of alginate and bone-forming proteins were synthesized by low-temperature cryopolymerization. Differentiation of bone marrow isolated progenitor cells was used to assess the bioactivity of proteins. Sub-lethal and lethal irradiation in C57BL/6 mice was used to mimic immunodeficiency and myeloablative transplant conditioning respectively. Transplants were performed into the CD45.2$^+$ mouse strain and monitored for donor-derived cells from congenic CD45.1$^+$ B6.SJL mice. Hydrogels were injected subcutaneously to induce the development of a hematopoietic niche environment capable to generating lymphoid competent cells. The recovery of immune cells in peripheral blood was periodically monitored and sequencing of the T-cell receptor (TCR) was used to determine the diversity of the T-cell repertoire. An acute myeloid leukemia (AML) cancer model was used to test the efficacy of accelerated immune reconstitution. The generation of subset specific T-lymphocytes was investigated using the scaffold.

Results and Discussion: In vitro, the proteins within the hydrogel retained their bioactivity and differentiated Lin$^-$c-

kit$^+$Sca-1$^+$ (LKS) cells into naïve CD4$^+$ and CD8$^+$T-cells. In vivo, scaffolds induced the formation of a subcutaneous bony nodule in mice. A histological analysis of the injected hydrogel revealed the formation of a bony nodule with bone marrow, populated with lymphoid progenitors. In sub-lethally irradiated mice, the hydrogel accelerated the reconstitution of T- and B-lymphocytes by over two orders of magnitude within 3 weeks by recruiting and driving lymphocyte differentiation of the endogenous progenitor HSCs. In lethally irradiated mice that were transplanted, the hydrogel served as a preferential site for transplanted progenitor HSC engraftment and induced their differentiation into T- and B-cells. Recovery of the adaptive immune system followed a similar trend to that of the sub-lethally irradiated mice. The hydrogel-mediated expansion of the T-competent progenitor pool resulted in an increase in the thymic output of transplanted mice and corresponded to an increase in the naïve T-cells pool. Sequencing of the TCR revealed an increase in the frequency and diversity, as measured by the recombination of the variable (V) and joining (J) segments of the TCR gene, of the T-cell repertoire, mediated by the hydrogel. When combined with an anti-leukemia vaccine, accelerated immune reconstitution prevented AML relapse in mice. When combined with low dose IL-2, the scaffold rapidly expanded regulatory T-cells in vivo.

Conclusions: The results indicate that a programmable biomaterial with biological cues can recapitulate aspects of the bone marrow stroma. By increasing the available sites of donor cell engraftment and providing the cues for lymphopoiesis, the reconstitution of adaptive immunity can be accelerated after HSCT, which can potentially decrease HSCT-associated immunological complications and improve the treatment of immunological disorders.

Acknowledgements: The work was supported by the National Institutes of Health through grants U19HL129903 and R01EB014703

464390 Controlling Stem Cell Fate within Zwitterionic Hydrogels

Abstract:

We employed zwitterionic hydrogels as platforms to control the self-renewal and differentiation of human mesenchymal stem cells (hMSCs). Firstly, we demonstrate that nonfouling hydrogels composed of pure poly(carboxybetaine) (PCB) allow hMSCs to retain their stem cell phenotype and multipotency, independent of differentiation-promoting media, cytoskeletal manipulation agents, and the stiffness of the hydrogel matrix. In addition, we provided a photodynamic zwitterionic hydrogel capable of reversibly triggering and suspending the differentiation process via manipulating nonspecific interactions between cultured stem cells and the hydrogel. The state of differentiated stem cells can be altered when the hydrogel is exposed to a selected light program, while the differentiation can be immediately suspended when the hydrogel is converted into a pure zwitterionic form by light.

462234 Nanolayer Multi-Therapy Scaled Delivery from Implant Surface

Abstract:

One of the important problems in the field of orthopedic medicine is the ability to create a stable bone-materials interface with an implant. The most effective treatment requires the delivery of exacting amounts of therapeutics of different types over appropriate timeframes in and around the implant, while maintaining mechanical integrity of the implant materials and allowing for bone integration on their surfaces. This work presents **a novel next-generation implant coating for both eradication of an established biofilm within the bone cavity and accelerated bone repair** via the controlled delivery of antibiotic and growth factor in sequence from stable nanometer-scale coatings on the implant surface.

Infection is the most common reason for complications, which often lead to complete removal of implants (74%). Infection significantly increases morbidity, and places huge financial burdens on the patient and the healthcare systemâ €"projected to exceed $1.6 billion/year by 2020. Because infection is much more common in implant replacement surgeries, these issues greatly impact long-term patient care for a growing population. For revision arthroplasty of an infected prosthesis, a prolonged and expensive two-stage procedure requiring two surgical steps and a 6â€"8 week period of joint immobilization is todayâ€™s gold standard. A single-stage revision is preferred as an alternative; however, traditional bulk polymer systems such as bone cement cannot load sufficient amounts of therapeutic to eradicate existing infection, are insufficient or infeasible for the release of sensitive biologic drugs that considerably aid in bone regeneration, and lead to substandard mechanical properties and retarded bone repair.

These issues are addressed by conformal, programmable, and degradable dual therapy coatings (~500 nm thick) in a layer-by-layer fashion using the enabling nanofabrication tool of electrostatic multilayer assembly. The nanolayered construct allows large loadings of each drug, thus enabling ultrathin film coatings to carry sufficient treatment and precise independent control of release kinetics and loading for each therapeutic agent. The coating architecture was designed to enable both early burst and lower sustained release of antibiotics sufficient to eliminate infection, and a long-term sustained release of BMP-2 growth factor to induce more significant and mechanically competent bone formation than a short-term burst release. In rats, the successful growth factor-mediated osteointegration of the multilayered implants with the host tissue improved bone-implant interfacial strength by 15-fold when compared with the bare implant control, and yields a mechanical bond 17-fold higher than that created with the use of clinically available bioactive bone cement.

Here we focused on dual delivery of an antibiotic and a growth factor owing to the urgent need for enhanced infection-reducing *and* tissue-integrating strategies in orthopedic applications, but the excellent modularity of multilayers for incorporation and release of diverse therapeutics suggests this approach should be also applicable to different implant applications such as vascular graft and artificial heart implants for which the risks of infection are often ignored. Our findings demonstrate the potential of this layered release strategy to introduce a durable implant solution, ultimately an important step forward in the design of biomedical implant release coatings for multiple medical applications.

460480 The Role of ECM Biomechanics in Liver Progenitor Differentiation

Abstract:

Combinatorial ECM arrays reveal the role of biomechanics in liver progenitor differentiation

Andreas P. Kourouklis, Kerim B. Kaylan, Gregory H. Underhill

Department of Bioengineering, University of Illinois at Urbana-Champaign, Urbana, IL

Recent findings suggest that biomechanical signals within the liver microenvironment can regulate the differentiation of mature hepatocytes **[1]**. However, the role of ECM biomechanics in liver progenitor differentiation remains primarily unexplored, despites its potential importance in the processes of liver morphogenesis and regeneration. To examine these mechanisms, we created high-throughput cellular arrays with the capacity to reiterate combinatorial ECM cues and characterize the corresponding phenotypic expression. Moreover, we combined these arrays with substrates of modular stiffness and integrated them with traction force microscopy (TFM) to assess the associated traction stress. This strategy provides a novel avenue to examine cell differentiation and elucidate the role of combinatorial ECM cues in cellular fate.

With this objective, we fabricated high throughput BMEL cell arrays on soft (4kPa) and stiff (30kPa) substrates presenting all single and pairwise combinations of 5 ECM molecules (Fig. 1A). We found that the ECM composition along with the substrate stiffness coordinately control cholangiocytic differentiation (e.g. OPN+ cell fraction). Most notably, collagen IV (C4) and fibronectin (FN) exhibited distinct effects (Fig.1B). BMEL cells on FN expressed stiffness-mediated cholangiocytic differentiation, indicated by the higher OPN+ fraction on stiff substrates (Fig.1B), while the cells on C4 exhibited cholangiocytic differentiation indifferent of substrate stiffness. Moreover, BMEL cells on FN developed higher traction stress on stiff substrates (Fig.1B), while cells on C4 exhibited significant traction on both substrates. These data are suggestive of a mechanism by which liver progenitor differentiation is regulated by traction stress at the cell-ECM interface.

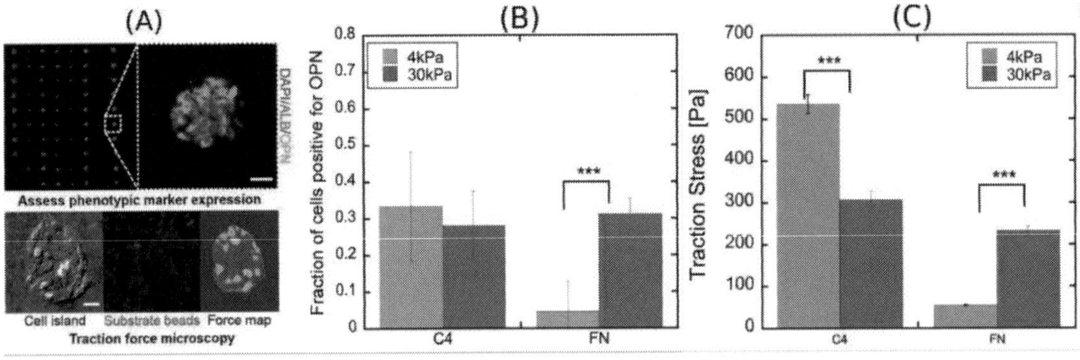

Figure 1: (A) Successful fabrication of cell arrays with BMEL domains of a diameter of ~150um. The BMEL arrays are immunostained for hepatocytic (albumin: ALB) and cholangiocytic (osteopontin: OPN) markers for high-throughput analysis. Lower raw demonstrates bead embedded gels for TFM with the corresponding phase contrast and heat maps. OPN expression (B) and traction stress (C) in response to stiffness and ECM composition.

In conclusion, we integrated microarray fabrication with TFM in order to reveal the role of cell biomechanics in the differentiation of liver progenitor cells. We showed that cells integrate combinatorial ECM cues using traction stress before they commit to their differentiated phenotype. Future experiments will assess the intracellular pathways that direct this differentiation response. Overall, this strategy provides new insight into liver progenitor fate decisions that could aid the investigation of liver disease mechanisms and the development of cell-based therapies.

References:

[1] Wells, R.G., Hepatology, 2008. **47**(4): p. 1394-400.

474206 Nanorod-like CH3NH3PbI3 for Planar Heterojunction Perovskite Solar Cell with Improved Performance

Abstract:

Hybrid organic-inorganic perovskite solar cell has recently attracted significant attention owing to the rapid improvement of the power conversion efficiency (PCE) as well as low cost manufacturing process. Great efforts involving high-quality perovskite film formation methodologies, compositional engineering of perovskite materials, and device architecture together with interface engineering have been made over the last six years, leading to a PCE leap from 3.8% in 2009 to the current world record of 22.1% in 2016 (NREL chart, http://www.nrel.gov/ncpv/images/efficiency_chart.jpg, Accessed 13.03.2016, 2016). The photo-to-electron conversion in perovskite solar cell involves several key steps, including light absorption by perovskite layer, charge transport in perovskite capping layer, and charge separation at n-type oxide/perovskite and perovskite/HTL interfaces. To achieve high photovoltaic performance, the morphology of perovskite is critically vital particularly for the plannar structured PSC, where the crystalline perovskite layer has to be self-standing without the support of a mesoporous scaffold.

This work reports a facile one-step HMT-assisted solution deposition method for reproducible fabrication of a high-quality nanorod-like $CH_3NH_3PbI_3$ thin films via controling over the $CH_3NH_3PbI_3$ crystallization in the presence of ammonia agent. SEM analysis review that a dense and grain-boundary-free perovskite capping layer can be prepared with orienting alignment of perovskite grains that fully cover the ammonia-modified ZnO substrate. XRD results demonstrate a high purity of $CH_3NH_3PbI_3$ thin film without no PbI_2 impurity. The perovskite film produces an appreciable enhancement in both photovoltaic performance and reproducibility as comparison with pristine perovskite, leading to an average PCE of 14.87±0.78%. The introduction of ammonia agent as nucleation-controlling and crosslinking agents and interface modifier provides a facile and effective approach for processing high purity and grain-boundary-free metal-halide perovskite thin film and tune the oxide/perovskite interface. The simplicity, ammonia agent-assisted and low temperature solution method can extend to construction of other perovskite-based hybrid optoelectronic devices.

473102 Lead-Free, Hybrid, Organic-Inorganic Halide for Light Harvesting

Abstract:

We will present the synthesis and optoelectronic properties of a lead-free, organic inorganic halide structures and show their integration as light harvesters in solar cells. Organic inorganic lead halide perovskite based solar cells have gained attention as a promising contender with nearly 20 % power conversion efficiencies. However, the environmental liabilities with the use and disposal of lead has led to the development of several non-lead based hybrid halides. The natural choice of many researchers has been tin (Group 14 , period 5) and to some extent bismuth (group 15, period 6). Although relatively new, the instability of tinâ€™s 2+ oxidation state has been a concern in spite of good semiconducting behavior. Here we will report the successful synthesis of a Ti based semiconductor with tunable bandgap based on the choice of the organic moiety. We will present and discuss the optical characteristics (ranging from 1.5 to 2.06) as well as the I-V characteristics of the of the various structures. While some of the synthesized products were clearly perovskites (based on XRD data), there are couple of structures whose crystalline structures have not yet been identified. The authors believe the XRD patterns to be a result of layered structures and expect to have it deciphered to present. The response of solar cell containing the hybrid halides coated onto a TiO_2 or ZnO will be presented.

463784 Improving Electron Transport in Nanostructured TiO_2 Electrode

Abstract:

Titanium dioxide (TiO_2) is one of the most widely used semiconductors in photovoltaics and photocatalysis because it is nontoxic, abundant, stable and photoactive. However, the wide bandgap, low electron mobility and short minority carrier diffusion length of TiO_2 limit its quantum efficiency in these applications. In this work, we present a solution chemical approach for making TiO_2 nanostructures for improving the electron transport in nanostructured TiO_2 electrodes.

461988 Symmetry-Breaking in Light-Trapping Nanostructures on Silicon for Solar Photovoltaics

Abstract:

While various materials have been investigated for photovoltaics, solar cells based on crystalline silicon (c-Si) dominate the current photovoltaics market. To reduce the cost of c-Si cells, wafer manufacturing companies have produced competitively priced thin c-Si films, ranging from a few microns to tens of microns, using a kerfless process. In such thin-film c-Si cells, light absorption becomes poorer than in thick films and light trapping is crucial to increase the photovoltaic efficiency. Han et al. have demonstrated that, among various light-trapping schemes, symmetry breaking in photonic nanostructures can approach the Lambertian light-trapping limit very closely. However, fabricating symmetry-breaking nanostructures in a scalable, cost-effective, manufacturable manner remains elusive. Here, we introduce a new approach to systematically break the symmetry in photonic nanostructures on c-Si surface. Using our approach, we fabricate low-symmetry inverted nanopyramid structures. Our method makes use of low-cost, manufacturable wet etching steps on c-Si(100) wafers without relying on expensive off-cut wafers. Our experiment and computational modeling demonstrate that the symmetry breaking can increase the Shockley-Queisser efficiency from 27.0 to 27.9% for a 10-micron-thick c-Si film. Further, our computation reveals that this improvement would increase from 28.1 to 30.0% with over-etching for a 20-micron-thick c-Si film.

464829 Computational and Kinetic Considerations for Morphology Prediction of Donor-Acceptor Oligomers for Organic Photovoltaics

Abstract:

Organic photovoltaic (OPV) solar cells are inexpensive, flexible, lightweight devices for sustainably generating electrical power. The efficiency and stability of OPV devices depends on their morphology, which we currently have poor synthetic control over. In this work, we perform molecular dynamics simulations of neat benzodithiophene-theinopyrrolodione (BDT-TPD) oligomers used in OPVs to evaluate the degree to which current simulation techniques are capable of predicting morphology in BDT-TPD films. The morphologies of simulated BDT-TPD volumes are characterized with simulated grazing-incidence X-ray scattering (GIXS) experiments and compared against experimental GIXS measurements.Â We find that the morphologies predicted from instantaneously quenched simulations better match experiments than structures that are allowed to equilibrate. This result highlights the competition between thermodynamics and kinetics in the structural evolution of polymer thin films.Â Further, we find that approximating the aromatic rings in BDT-TPD with rigid bodies rather than combinations of bond, angle, and dihedral constraints, results in better computational efficiency but poorer agreement with experimental structure. We determine the primary structural features in films of BDT-TPD are the stacking of oligomer backbones and the self-organization of backbone stacks into â€œribbonsâ€' with long-range spatial correlations. We conclude with a discussion of the shortcomings of current modeling tools and potential directions for improvement.

469339 Enhancing Dye Sensitized Solar Cell J-V Behavior By Integrating Nanoscale Polymer Films

Abstract:

Dye sensitized solar cells (DSSCs) are inexpensive, are easy to fabricate, have tunable optical properties such as color and transparency, and can be integrated into buildings and apparel easily. A DSSC is composed of three main components: (1) a photosensitive dye which converts sunlight into electrons, (2) a mesoporous TiO_2 photoanode which provides a large surface area for dye adsorption and a pathway for electron transport, and (3) a liquid electrolyte comprised of an iodide-triiodide redox couple which is used to regenerate the dye and acts as a hole transport medium. A major limitation of the current DSSC design is the liquid electrolyte, which is prone to leakage and evaporation along with being corrosive to metal contacts such as silver. Furthermore, electron recombination loss is significant at the TiO_2-electrolyte interface. Polymer electrolyte DSSCs can overcome these disadvantages and can potentially enhance the cell's photocurrent-voltage behavior [1]. However, due to the nanostructured, mesoporous nature of the TiO_2 photoanode and the effects of viscosity on fluid transport in such a structure, conventional polymer deposition techniques such as spin casting and dip coating lead to ineffective pore penetration. Often only the top 2 μm out of a >10 μm photoanode layer is in contact with the polymer with the remaining photoanode below completely void of polymer. This severely reduces device performance because a large portion of the adsorbed dye is not accessible and cannot be regenerated. To ensure optimal performing DSSCs, there needs to be intimate contact between the polymer electrolyte and the TiO_2 throughout the entire photoanode layer to ensure the redox couple can reach the adsorbed dye.

To overcome the limitations of liquid methods for incorporating polymer electrolytes into the mesoporous TiO_2, we employ a novel polymerization technique called initiated chemical vapor deposition (iCVD). iCVD is a solvent-free polymerization method that relies on gas-to-surface reactions at low pressures in which the reagents for polymerization, the monomer and initiator, are heated to vapors that can easily penetrate into the mesoscale voids of

the TiO
occurs conformally inside the pore surfaces. This makes it especially useful for cases that require penetration of sub-micron and nanometer-sized pores as issues of wettability and surface tension with liquid processing are absent. In addition, the lack of any solvent medium removes the possibility of liquids getting trapped in the photoanode, which may degrade performance, and are difficult to remove [2].

In this work, iCVD is used for the synthesis and integration of polymer electrolyte thin films within the mesoporous TiO_2 photoanode of DSSCs. Fourier transform infrared spectroscopy (FTIR), gel permeation chromatography (GPC), and X-ray photoelectron spectroscopy (XPS) analyses confirm that the polymers formed are stoichiometric in composition expected of linear homopolymers and chemically identical to ones formed with liquid solution methods. The surface reaction kinetics and mass transport have been found to depend on the fractional surface saturation of monomer, , which is an adsorption parameter that provides a measure of the surface availability of the monomer. By carefully controlling , conformal coatings throughout the TiO_2 photoanode have been achieved. To gain a better understanding on the effect of the polymer electrolyte on DSSC behavior, current-voltage, UV-VIS absorption, incident photon-to-current efficiency (IPCE) and electrochemical impedance spectroscopy (EIS) measurements have been employed. Experimental results indicate that DSSCs with ultrathin polymer electrolyte thin films integrated on the photoanode pore surface can enhance cell performance. For example, by applying poly(1-vinylimidazole) (PVIZ), power conversion efficiencies are found to be 27% higher than liquid-electrolyte DSSCs due to a simultaneous increase in short circuit current J_{sc} and open circuit voltage V_{oc} [3]. To our knowledge, this is the first reported demonstration of PVIZ as a polymer electrolyte in DSSCs.

1. Smolin, Y. Y.; Nejati, S.; Bavarian, M.; Lee, D.; Lau, K. K. S.; Soroush, M. *Journal of Power Sources* **2015,** 274, 156-164.

2. Nejati, S.; Lau, K. K. *Nano letters* **2010,** 11, (2), 419-423.

3. Kuba*, A. G.; Smolin*, Y. Y.; Soroush, M.; Lau, K. K. S. *Chemical Engineering Science. Special Issue: Energy Conversion and Storage* **2016,** (*in press.*) http://dx.doi.org/10.1016/j.ces.2016.05.007 (*equal contribution)

468024 Photocatalytic Photosystem I/PEDOT Composite Films Prepared By Vapor Phase Polymerization

Abstract:

Photosystem I (PSI) is an integral membrane protein complex contained within the chloroplast granum of plants and cyanobacteria. The product of over a billion years of systematic evolution, PSI is adapted to the earth's climate and sun's radiant energy profile. PSI transduces blue-violet and red light, sustaining 1.1eV of stored energy in <100ps by way of a cascade of irreversible internal reactions. Internal quantum efficiency for this process is nearly 100%, making PSI a model photodiode. As such, PSI has garnered interest as a macromolecule for use in photovoltaic and photocatalytic systems in a variety of roles. Recently, there have been efforts to include PSI within composite films, whereby the high potential and photocatalytic efficiency of PSI is coupled with an electrically conductive matrix.

We describe a versatile technique to fabricate photocatalytic composite films containing PSI and a chosen intrinsically conductive polymer (ICP). Iron (III)—a friedel crafts catalyst—is added to protein solutions by way of water soluble $FeCl_3$. Drop casting of these solutions on a substrate of choice is followed by contact with low temperature (40°C) monomer vapor of a well-studied ICP, poly(3,4-etylenedioxythiophene) PEDOT. In minutes, a ~200 nm thick polymer/protein composite is grown. Composite films retain the high conductivity of the used ICP and the characteristic absorbance and photocatalytic capabilities of PSI multilayers. Vapor phase processing allows versatility in choice of underlying substrate.

467643 Water-in-Water Emulsion Based Synthesis of Hydrogel Nanospheres with Tunable Release Properties

Abstract:

Poly(ethylene glycol) (PEG) micro/nanospheres have several unique advantages as polymer based drug delivery systems (DDS) such as tunable size, large surface area to volume ratio, and colloidal stability. Emulsification is one of the widely used methods for facile synthesis of micro/nanospheres. Two-phase aqueous system based on polymer-polymer immiscibility is a novel approach for preparation of water-in-water (w/w) emulsions. This method is promising for the synthesis of PEG micro/nanospheres for biological systems, since the emulsion is aqueous and do not require organic solvents or surfactants. Here, we report the synthesis of nano-scale PEG hydrogel particles using w/w emulsions using phase separation of dextran and PEG prepolymer. Dynamic light scattering (DLS) and scaning electron microscopy (SEM) results demonstrated that nano-scale hydrogel spheres could be obtained with this approach (**Figure 1**). We investigated the release kinetics of a model drug, pregabalin (PGB) from PEG nanospheres and demonstrated the influence of polymerization conditions on loading and release of the drug as well as the morphology and size distribution of PEG nanospheres. The experimental drug release data was fitted to a stretched exponential function which suggested high correlation with experimental results to predict half-time and drug release rates from the model equation. The biocompatibility of nanospheres on human dermal fibroblasts using cell-survival assay suggested that PEG nanospheres with altered concentrations are non-toxic, and can be considered for controlled drug/molecule delivery.

SHAPE * MERGEFORMAT

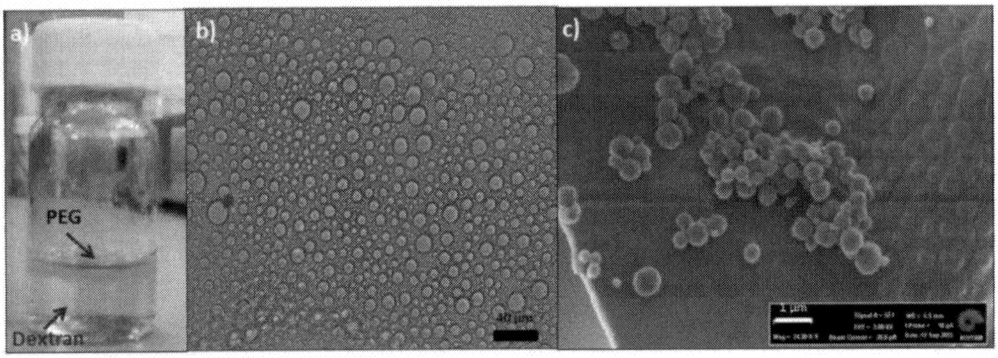

Fig. 1. a) Two-phase separation of dextran-PEG polymer b) Light microscope image of dextran-PEG w/w emulsion (scale bar: 40 µm) c) SEM image of dried PEG (scale bar: 1 µm)

471831 A Pharmacokinetic Model of a Tissue Implantable Insulin Sensor

Abstract:

While implantable sensors such as the continuous glucose monitoring system have been widely studied, both experimentally and mathematically, relatively little attention has been applied to the potential of insulin sensors. Such sensors can provide feedback control for insulin infusion systems and pumps and provide platforms for the monitoring of other biomarkers in vivo. In this work, the first pharmacokinetic model of an affinity nanosensor is developed for insulin operating subcutaneously in the limit of where mass transfer across biological membranes reaches a steady state. Using a physiological, compartmental model for glucose, insulin, and glucagon metabolism, the maximum sensor response and its delay time relative to plasma insulin concentration, are calculated based on sensor geometry, placement, and insulin binding parameters for a sensor localized within adipose tissue. A design relation is derived linking sensor dynamics to insulin time lag and placement within human tissue. The model should find utility in understanding dynamic insulin responses and forms the basis of model predictive control algorithms that incorporate sensor dynamics.

472055 Liposome-Encapsulated Synergistic Drug Combinations for Low Dose Chemotherapy

Abstract:

PEGylated liposomes revolutionized drug delivery for toxic chemotherapy drug doxorubicin, tokened as "red death", by reducing cardiotoxicity in patients. However, it remains unclear whether liposomal doxorubicin is therapeutically superior to free doxorubicin. We developed a novel PEGylated liposomal system, named DAFODIL (Doxorubicin And 5-Fluorouracil Optimally Delivered In a Liposome) which surpasses therapeutic effects of free drug administrations. DAFODIL delivers synergistic ratios of doxorubicin and 5-fluorouracil to tumors, and, with remarkably low doses, achieves greater than 90% reduction in tumor growth of a murine 4T1 mammary carcinoma model. Co-delivery of optimal, synergistic ratios ensured therapeutic effects at drug concentrations far below their maximum tolerable drug doses. Therefore, our approach improves both the therapeutic efficacy and toxicity of liposomal drug delivery systems, and provides a safe and effective means to deliver combination chemotherapies. This methodology is made feasible by extending the well-established ammonium-sulfate gradient encapsulation method to nucleobase analogues, a liposomal entrapment method initially only useful for anthracyclines. Therefore, our strategy can be utilized to efficiently evaluate various chemotherapy combinations in an effort to translate more effective combinations clinically.

469014 Photoexcited Quantum Dots Potentiate Antibiotic Activity in Multidrug-Resistant Bacteria

Abstract:

The rise of multidrug-resistant (MDR) bacteria is a growing concern to global health and is exacerbated by the lack of new antibiotics being developed. In order to treat MDR infections, new classes of antibiotics must be developed that are effective against already pervasive MDR infections. While current antibiotics are derived from natural products, investigating true synthetic antibiotics provides opportunity to design therapeutics for antimicrobial targets of interest. In this study we present a novel, nanoparticle based therapeutic that is able to inhibit MDR clinical isolates of *Escherichia coli*, *Salmonella typhimurium*, and *Klebsiella pneumoniae* both as a monotherapy as well as combination therapy with antibiotics that the isolates were previously resistant against. The nanoparticles used in this study generate specific, light activated reactive species (LARS) which alter the drug resistance phenotype and lower the level of resistance of the clinical strains to traditional antibiotics. We show that the redox homeostasis perturbing nanoparticles in combination with protein synthesis inhibiting antibiotics, namely chloramphenicol, streptomycin, and clindamycin, lower the IC_{50} of resistant MDR clinical isolates below the Clinical Laboratory Standards Institute (CLSI) breakpoint. We find that this phenomenon is not only limited to protein synthesis inhibiting antibiotics, but extends to other classes of antibiotics such as fluoroquinolones. Using a multiplicative model, we show that this combination therapeutic effect is additive and the effect is concentration dependent on both the antibiotic as well as the nanoparticle concentration. Interestingly the nanoparticle is able to have this drastic effect on antibiotic resistance at nanomolar concentrations highlighting the strong effect of the LARS produced. The bactericidal effect of the combination therapy is greater than that of each mono-therapy alone highlighting the potential utility of this combination therapy for treating MDR infections.

468253 Evaluation of the Cancer-Preventive Effect of Resveratrol-Loaded Nanoparticles on the Formation of Tumor Spheroids

Abstract:

Despite the recent advancements in the treatment of cancer, designing advanced therapeutics remains an enormous challenge for scientists. It is estimated in the U.S. that 1,685,210 new cases of cancer will be diagnosed in 2016 and 595,600 people will die from cancer, corresponding to approximately 1,600 deaths per day[1].

Chemotherapy has been extensively used to combat cancer, however its overall success is rather limited for lung cancer. An emerging alternative over the past few decades has been the use of chemopreventive phytochemicals, which are constituents of fruits and vegetables with potential pharmacological and/or nutritional functions. These agents may reverse or inhibit the carcinogenic process resulting in a decreased cancer risk [2]. The FDA has approved several agents for cancer prevention, including Tamoxifen for breast cancer and the Human Papilloma Virus (HPV) vaccine for cervical intraepithelial neoplasia. More than 200 other chemopreventive agents are being evaluated in clinical trials [3].

Resveratrol (3,5,4â€²-trihydroxystilbene, RSV) is a phytoalexin, a substance synthesized by plants to counteract pathogens [4]. It is abundantly found in red grapes and in lower amounts in peanuts and blueberries. In 1997, interest in RSV peaked when Jung *et al.* reported its chemopreventive effects on all stages of neoplastic activity (initiation, promotion and progression) [5]. *In vitro* studies show the promising properties of resveratrol; however, its limited bioavailability in the bloodstream makes it difficult to translate into clinical studies. A study by Dolfini *et al.* [6] showed that RSV induces drastic growth inhibition on a human breast cancer cell line in *in vivo* and *in vitro* assays. Also, RSV-loaded gel nanoparticles enhanced intracellular uptake and cytotoxicity in a NCI-H460 lung cancer cell line [7] and RSV-loaded nanostructure lipid carriers exhibited higher penetration into the skin in comparison to solid lipid nanoparticles [8], demonstrating improvement with RSV in nanoparticle systems.

The aim of this project was to evaluate the chemopreventive effect of RSV and RSV-loaded nanoparticles (RSV-NP) on the formation of tumor spheroids. We synthesized biodegradable nanoparticles comprised of the biodegradable polymer acetalated dextran loaded with RSV and we evaluated the efficacy of pure RSV and RSV-NP using A549 adenocarcinoma lung cancer cell line in 2D and 3D cell culture studies. 3D multicellular spheroids (MCS) reflect many of the properties of solid tumors and better mimic the barriers of drug diffusion, transport, and distribution within tumors [9, 10]. RSV and RSV-NP exerted an inhibitory effect on the growth of MCS, showing a promising alternative as a chemopreventive drug, which could palliate the incidence or mortality from lung cancer.

References

[1] Siegel, R.L., K.D. Miller, and A. Jemal, *Cancer statistics, 2016.* CA Cancer J Clin, 2016. 66(1): p. 7-30.

[2] Sporn, M.B., et al., *Prevention of chemical carcinogenesis by vitamin A and its synthetic analogs (retinoids).* Fed Proc, 1976. 35(6): p. 1332-8.

[3] Davis, J.S. and X. Wu, *Current state and future challenges of chemoprevention.* Discov Med, 2012. 13(72): p. 385-90.

[4] Singh, C.K., M.A. Ndiaye, and N. Ahmad, *Resveratrol and cancer: Challenges for clinical translation.* Biochim Biophys Acta, 2015. 1852(6): p. 1178-85.

[5] Jang, M., et al., *Cancer chemopreventive activity of resveratrol, a natural product derived from grapes.* Science (New York, N.Y.), 1997. 275(5297): p. 218-20.

[6] Dolfini, E., et al., *Resveratrol impairs the formation of MDA-MB-231 multicellular tumor spheroids concomitant with ceramide accumulation.* Cancer Lett, 2007. 249(2): p. 143-7.

[7] Karthikeyan, S., et al., *Anticancer activity of resveratrol-loaded gelatin nanoparticles on NCI-H460 non-small cell lung cancer cells.* Biomedicine & Preventive Nutrition, 2013. 3(1): p. 64-73.

[8] Gokce, E.H., et al., *Resveratrol-loaded solid lipid nanoparticles versus nanostructured lipid carriers: evaluation of antioxidant potential for dermal applications.* Int J Nanomedicine, 2012. 7: p. 1841-50.

[9] Sutherland, R.M., *Cell and environment interactions in tumor microregions: the multicell spheroid model.* Science, 1988. 240(4849): p. 177-84.

[10] Minchinton, A.I. and I.F. Tannock, *Drug penetration in solid tumours.* Nat Rev Cancer, 2006. 6(8): p. 583-92.

468082 Preparation, Characterization and *in Vitro* Validation of a Novel Paclitaxel Transport System to Target HER2-Positive Breast Cancer

Abstract:

Breast cancer is the second kind of cancer with the highest frequency worldwide, and the leading cause of oncological death among women in both, developed and developing countries [1]. Among the different clinical subtypes, the HER2-positive one, associated with a worse prognosis for patients, represents the 15-30% of all of the breast tumor cases [2].

One of the most efficient drugs used in the management of this subtype of cancer is paclitaxel, a natural terpenoid compound [3]. However, it has a poor aqueous solubility and an elevated toxicity, and these facts limit its clinical application.

With the aim of sorting out both disadvantages, a novel transport system for paclitaxel has been designed: polymeric nanoparticles of sodium alginate and piperazine, developed thanks to the electrostatic interactions that take place between both compounds at slightly acid pH values [4].

In such transport system, paclitaxel has been included in Î²-cyclodextrins in order to improve its aqueous solubility and in addition, trastuzumab, an antibody that binds specifically to the HER2 receptor, has been attached for avoiding the severe side effects that the terpenoid drug has. Both, the paclitaxel-Î²-cyclodextrins complexes and the antibody, have been attached to nanoparticlesâ€™ surface by a diimide-activate amidation process, and the conjugation efficiency of both of them to the nanoparticles has been analyzed. Later, the characterization of the transport system has been carried out by determining the average size and zeta potential of the nanoparticles, their stability over time under different conditions and the paclitaxel release kinetics. Finally, the effectiveness of this novel paclitaxel transport vehicle has been validated *in vitro*, by MTT assays performed with the BT474 cell line, with which positive results have been obtained. In this manner, a new door could be opened to improve the specific treatment of the HER2 positive breast cancer.

References

[1] Amaro, J., Severo, M., Vilela, S., Fonseca, S., Fontes, F., La Vecchia, C. et al. 2013. Patterns of breast cancer mortality trends in Europe. The Breast 22, 244-253.

[2] Figueroa-Magalhaes, M.C., Jelovac, D., Connolly, R.M., Wolff, A.C. 2014. Treatment of HER2-positive breast cancer. The Breast 235, 128-136.

[3] Singla, A.K., Garg, A., Agarwal, D. 2002. Paclitaxel and its formulations. Int. J. Pharm. 235, 179-192.

[4] Román, J.V., Rodríguez-Rodríguez, J.A., Martín del Valle, E.M., Galán, M.A. 2016. Synthesis of a new nanoparticle system based on electrostatic alginate-piperazine interactions. Polymer Adv. Tech. 27(5), 623-629.

472116 Adsorption, Stabilization and Recovery of Polyphenolic Flavonoids By TiO_2 Functionalized Mesoporous Silica Nanoparticles

Abstract:

Flavonoids such as quercetin and its derivatives, which are normally isolated from plants, are antioxidants with medicinal and therapeutic applications for the treatment of cancer, cardiovascular disease and other conditions because they act against tumors, inflammation, and reactive oxygen species. At the same time, because they are photoactive compounds, flavonoids are highly susceptible to degradation and activity loss after extraction from biological sources. Therefore, methods are needed to retain the activity of flavonoids during physicochemical processes and storage. A possible material of interest for binding, protection and release of a wide variety of compounds are engineered silica particles. Porous silica nanoparticles are of interest for the recovery, and more recently, the delivery of therapeutics. Recent progress in metal oxide chemistry allows the synthesis of silica nanoparticles of controlled size, pore properties, and surface functionalization. Titania (TiO_2) dispersed in mesoporous silica nanoparticles (MSNP) is hypothesized to have the potential to act as a high surface area adsorbent for quercetin due to its ability to bind dienols through chelation.

Here, the adsorption of quercetin and rutin (0.05-10 mg/ml in ethanol) is investigated on MSNPs (165 nm diameter) that are functionalized with submonolayer coverage (7-85%) of TiO_2. The amount of flavonoid adsorbed onto the particles was found to be a strong function of titania grafting density on the silica surface. As a function of quercetin concentration, strong binding occurs at low solution concentration followed by a gradual approach to saturation at higher concentration. Nanoparticles with approximately 60% of monolayer coverage of titania adsorb 230 mg quercetin/gram silica, which is more than 100 times the capacity of nonporous particles with similar titania coverage. The flavonoid recovery from nanoparticles was also investigated using biocompatible ethanolic citric acid solution (20% w/v) and a five step multistage desorption process were able to recover more than 90% of the bound flavonoids

cumulatively. The radical scavenging activity (RSA) of recovered and particle bound quercetin were measured and most of the RSA was retained after binding quercetin, and after recovery of bound quercetin into solution. The materials developed here have the potential to be used to strongly and directly bind flavonoids and related secondary metabolites from plant cells, and will serve as a synthetic platform to isolate, recover, and potentially deliver these compounds in biological systems.

473620 Degradation Kinetics of PLGA and PLGA Conjugated with Alendronate Nanoparticles

Abstract:

A drug carrier has to be able to target, transport and delivery drug to a specific place in the body. This carrier should be degraded by the body as quick as possible in order to avoid accumulation. Also the carrier have to be made from a biodegradable material which could be degraded by the body into nontoxic products. PLGA or poly(lactic-co-glycolic acid) is a biodegradable material used for drug delivery system. This material can be modified with different moieties like alendronate, in order to target to a specific place. Alendronate is a molecule with a higher affinity to bone tissue so, with this material could be possible to target an active compound like N-acetylcysteine to bone tissue. Although, a lot studies has been done about drug delivery systems, only few has been done about degradation of PLGA nanoparticles modified with a particular moiety, and we could not find reports about degradation kinetics of PLGA nanoparticles modified by alendronate (PLGA-ALE). The aim of this research was to compare the degradation kinetic of PLGA, PLGA-ALE and PLGA-ALE-NAC (nanoparticles made with PLGA-ALE and loaded with N-acetylcysteine) nanoparticles, the first one has been reported for many authors, the second and three one has not been reported.

Conjugation of PLGA with alendronate was made by carbodiimide chemistry, joining the acid terminated of PLGA with the amine of alendronate. The reaction was verified by NMR and FTIR. Nanoparticles were produced by nanoprecipitation method. Degradation kinetics was done suspended nanoparticles in PBS at 37 °C for 33 days, molecular weight of the polymer was measured by GPC.

Polymer degradation exhibited a first order kinetics, with values for the specific kinetics constant of 0.082, 0.0159 y 0.0163 days^{-1} for PLGA, PLGA-ALE and PLGA-ALE-NAC nanoparticles respectively. The first value was the highest of the three, which means that alendronate protect polymer chain from a fast degradation. Values of PLGA-ALE and PLGA-ALE-NAC are similar with a difference of 2.5% in favor to PLGA-ALE-NAC indicating a small effect of NAC in favor to polymer degradation. As a conclusion modification of PLGA with alendronate reduce the polymer degradation rate, but, NAC increase 2.5% the degradation rate.

463064 Confinement Facilitated Protein-Protein Stacking: As Investigated By Neutron Scattering

Abstract:

Confinement of biomolecules in structured mesoporous materials can offer many advantages such as chemical and thermal stability, and resistance to degradation in both biological and synthetic systems. This has led to the application of protein confinement in implementation of this platform to applications as diverse as pharmaceutical formulation, bio-catalysis and sensing. A fundamental understanding of protein behaviour under confinement is paramount to design the best mesoporous host for a particular application. While many studies have linked changes in confined secondary structure and enzymatic activity, the understanding of overall confined protein tertiary structure and orientation under confinement is only partially understood. Here, we discuss recent advances elucidated via small-angle neutron scattering (SANS), using protein-loaded mesoporous silica materials known to exhibit a link between a better preserved confined secondary structure and higher enzymatic activity in narrow pores in which the curved surface interacts electrostatically with the enzyme[1]. SANS enabled the general behaviour of two model proteins, representing distinct electrostatic conditions, to be observed over three confinement geometries in both the Henry and high concentration adsorption regimes. This analysis results in the observation of protein-protein stacking, which is driven by protein coordination that is facilitated by the geometry of the confining mesopores and the protein loading concentration.

References: [1] Sang L-C, Coppens M-O. Effects of surface curvature and surface chemistry on the structure and activity of proteins adsorbed in nanopores. Phys Chem Chem Phys. 2011;13(14):6689

455964 Fast, Efficient and Gentle Transfection of Human Adherent Cells in Suspension

Abstract:

The current strategy for delivering genetic materials to adherent cells using gene delivery vehicles (GDVs), involves incubating the GDV solution over a cultured cell monolayer, and its performance is limited by slow mass transport of GDVs to the cell surface. We demonstrate a novel suspension-based transfection method that provides efficient delivery of genetic materials to clinically relevant human cell types, such as induced pluripotent stem cells and fibroblasts, using lipid- and polymeric-based commercial transfection reagents (FugeneHD and Glycofect). Our method reduces the time required for transfection by one full day. The reason behind the success of our method is that suspension transfection leads to faster transport of GDVs to the cell surface. This resulted in maximum GDVs attachment to cell surface within 15 minute of transfection time, with atleast 2-fold higher binding efficiency of GDVs as compared to the plating method. Our method is also compatible with common techniques to improve gene delivery, such as cell cycle synchronization (6-fold increase in gene expression compared to control), facilitating easy adaptability in existing setups. Furthermore, to preserve cell physiology during transfection in suspension, we designed a microfluidic version of our method where the transfection takes place in a flow. In this device, the GDV and cell contact time is below one minute, and we achieved high binding of GDVs to cells (>90% GDV-bound live cells) and significant gene expression in several human cell types with minimal effect due to the device flow rates. The fast, efficient and generally applicable transfection methodologies that we will present can be used for rapid screening of different delivery systems and have significant potential for scale-up and high-throughput cell therapy applications.

Keywords: pluripotent stem cells, fibroblasts, suspension transfection, flow cytometry, microfluidics

462134 Poly(β-amino ester) Terpolymer Nanoparticles As Delivery Vehicles for mRNA and DNA

Abstract:

Introduction: A host of human diseases are caused by missing or aberrant proteins. Effective treatment of the root cause of these maladies therefore requires replacement of such proteins. The most common means of protein replacement therapy is simply to administer the protein itself, but protein stability concerns often limit the effectiveness of this strategy. Alternatively, one can deliver nucleic acids coding for the aberrant protein. Nucleic acids are more easily stabilized than proteins, and can be condensed and encapsulated within delivery vehicles which can facilitate their effective *in vivo* transport. To date, DNA has been the most-studied nucleic acid for protein replacement therapy due to its relatively high stability and low immunogenicity when compared to mRNA. However, mRNA does not pose a risk of insertional mutagenesis, and it need only reach the cytoplasm to be effective. Additionally, recent innovations in mRNA synthesis have yielded more stable and potentially less immunogenic nucleic acids, further increasing the interest in delivering mRNA for protein replacement therapy. Thus, there is now a significant interest in designing materials to deliver mRNA *in vivo*. Previously, our lab reported the synthesis of poly(Î²-amino ester) terpolymers (PBAE terpolymers), copolymeric materials capable of electrostatically condensing DNA into nanoparticles that can additionally be formulated with other hydrophobic excipients. In this work, we utilize PBAEs to deliver mRNA and DNA *in vitro* and *in vivo*.Â

Materials and Methods: PBAE terpolymers which had previously been synthesized for DNA delivery were synthesized via step polymerization, with synthesis conditions optimized for purification in diethyl ether to remove excess monomer. PBAE terpolymer nanoparticles were formed by nanoprecipitation via charge-based interactions with nucleic acids, as well as hydrophobic interactions between the PBAE terpolymer and a lipid-anchored PEG (PEG-lipid). Particle serum stability was assessed by doping nanoparticles with fetal bovine serum and measuring optical absorbance of the particle/serum solution over a 2-hour span at 37°C. *In vitro* nanoparticle efficacy was assessed by

formulation PBAE nanoparticles with luciferase-encoding mRNA or DNA and transfecting HeLa cells, with cellular luminescence measured as an indicator of luciferase protein production. *In vivo* experiments were performed by delivering nanoparticles intravenously to C57/BL6. In a typical example, a mouse was dosed with 0.5 mg/kg of luciferase-coding mRNA loaded into a PEGylated PBAE terpolymer, and luminescence was quantified after 24 hours using an IVIS imaging apparatus.

Results and Discussion: We demonstrate that PBAE terpolymers can also be used to encapsulate and deliver mRNA both *in vitro* and *in vivo*. Specifically, we show that mRNA-loaded terpolymer nanoparticles can be co-formulated with (PEG-lipid) to improve the serum stability of terpolymers, as well as to increase *in vivo* potency. Nanoparticles that were stabilized *in vitro* were also successful in delivering nucleic acid to the lungs following intravenous delivery to mice, although biodistribution experiments revealed that nucleic acid localization was spread throughout major organs. Additionally, we utilize PBAEs as a platform to compare the protein production of mRNA and DNA. Our data shows that certain aspects of PBAE structure, especially the diacrylate used to form the polymer backbone, influence whether the material will be more successful for mRNA or DNA delivery. We further compare the role of other factors, such as polymer molecular weight and formulation parameters, for PBAE-mediated mRNA and DNA delivery. Furthermore, we plan on taking advantage of the *in vivo* potency of PEGylated PBAE nanoparticles to compare the potency of DNA and mRNA-based protein production.

Conclusions: In total, our research has shown that PBAE terpolymers are a therapeutically-relevant material for nucleic acid delivery. PBAEs are also useful given that they can deliver both mRNA and DNA. Their ability to deliver both nucleic acids successfully, along with their facile synthesis, can be leveraged to investigate chemical and formulation parameters that favor delivery of one nucleic acid over the other, as well as protein production afforded by both nucleic acid types *in vivo*.

462851 Enhancement of Gene Therapy Via Inhibition of the Innate Immune Response

Abstract:

While the adaptive immune system detects extracellular pathogens, most cells also have an innate immune system that recognizes intracellular pathogens. For example, when viruses inject their DNA into cells, it is quickly detected by cytosolic â€œDNA sensorsâ€ that trigger a signaling cascade that leads to the expression of inflammatory cytokines, inhibition of translation, or even apoptosis to prevent the spread of the virus. Unfortunately, this defense mechanism can also significantly inhibit gene therapy, since most methods also introduce therapeutic DNA into the cytoplasm. In this work, we describe several different strategies to improve non-viral gene therapy by inhibiting the innate immune response. Specifically, we show that addition of small molecule inhibitors for several different enzymes (e.g. TLR9, TBK-1, Î²-catenin, and p38 MAPK) involved in this response can significantly improve transgene (luciferase) expression in cancer cells (e.g. prostate cancer, PC3 cell line). Alternatively, we have also enhanced transgene expression by constructing a plasmid that contains several repeats of a specific DNA sequence that is known to inhibit Toll-Like Receptor 9 (TLR9, a DNA sensor that binds DNA inside endosomes). ELISA assays show that the inhibitors also decrease the expression of inflammatory cytokines (e.g. IL-6), thereby confirming that the enhancement of transgene expression is due to inhibition of the innate immune response. Overall, these results clearly show that the innate immune response is a significant obstacle that must be considered in the development of gene therapy techniques and our work provides multiple alternative strategies to address this issue.

465582 Engineering Megakaryocyte-Derived Microparticles for Gene Delivery

Abstract:

Cell-derived microparticles (MPs) or microvesicles are small particles (100-1000 nm) shed from the plasma membrane from various types of cells, such as hematopoietic cells, endothelial cells, and embryonic stem cells (Cocucci et al., 2009). MPs play an important role in cell-to-cell communication by carrying and transferring various signaling molecules to target cells, such as RNAs, proteins or lipids. MPs have several key characteristics, and notably target specificity, biocompatibility and immunological inertness, that make them a vehicle for potential cargo delivery to desirable cell targets for therapeutic applications.

Megakaryocytic MPs (MkMPs) are the most abundant MPs in blood circulation, and were shown by our group to induce and promote the megakaryocytic (Mk) differentiation of human hematopoietic stem and progenitor cells (HSPC) (Jiang et at., 2014). We have recently shown that MkMPs have exquisite specificity in targeting only HSPCs, and preferably the more primitive HSCs. In this study, we aim to engineer MkMPs as vectors for delivering nucleic acids to HSPCs. We largely employed CHRF cell-derived MPs (CMPs) as a model MP in lieu of MkMPs, since CMPs can be produced readily and in large quantities, and are able to induce and promote Mk differentiation of HSPCs as well as MkMPs can. CHRF cells, a human megakaryoblastic cell line, have been validated as an excellent model of Mk differentiation of human cells. However, we also used primary human megakaryocytes and MkMPs when appropriate.

First, we desired to reduce, if not completely eliminate, the endogenous nucleic-acids (notably RNAs as most MPs contain no DNA) in CMPs so that we can load them with desirable cargo. We developed an RNase-treatment protocol so that the amount to RNA in CMPs decreased by 40 to 60 percent, and this was sufficient to abolish the function of CMPs in inducing and promoting Mk differentiation of HSPCs. This indicated that our RNase treatment protocol was effective in eliminating most of the endogenous RNA load. Then, we aimed to develop a reliable protocol for loading of desired cargo (and notably plasmid DNA (pDNA)) into CMPs. We examined two different methods for loading: endogenous (cargo loaded into parent cells before CMP generation) and exogenous (cargos loaded directly into CMPs). In endogenous loading, pDNA cargo was loaded into parent cells (both CHRF cells but also primary megakaryocytic cells were examined) by transfection, and MPs were collected after 48 hrs. In exogenous loading, pDNA cargo was directly loaded into MPs by electroporation at various conditions. The loading efficiency was quantitated by either measuring the fluorescent positive MPs if fluorescent-labeled pDNA was loaded, or quantitating pDNA amount from MPs and confirmed with DNA gel electrophoresis. We are detail our studies to optimize the loading efficacy and quantitate the efficiency of cargo delivery by CMPs or MkMPs to HSPCs.

Cocucci, E., G. Racchetti, and J. Meldolesi. 2009. Shedding microvesicles: artefacts no more. *Trends Cell Biol.* 19:43-51.

Jiang, J., D.S. Woulfe, and E.T. Papoutsakis. 2014. Shear enhances thrombopoiesis and production of megakaryocytic (Mk) microparticles that induce Mk differentiation of stem cells. *Blood*.

Acknowledgements: This work was supported by a State of Delaware Bioscience Center for Advanced Technology (CAT) grant (# 15A01570).

467718 Microfluidic Production of Ternary Polyplexes for Non-Viral Gene Delivery

Abstract:

Microfluidic Production of Ternary Polyplexes for Non-Viral Gene Delivery

Gene therapy has the potential to treat a variety of diseases by delivering DNA that codes for a therapeutic protein into human cells. Recombinant viruses are the primary vector choice in the majority of gene therapy clinical trials due to their high gene delivery efficiency. However, viral vectors suffer from serious disadvantages including poor transduction of many human cell types, immunogenicity and lack of targetability. Non-viral polymeric gene delivery vectors (polyplexes) provide an alternative solution but are limited by poor efficiency and cytotoxicity.

Improvements in polyplex-mediated gene delivery can be achieved through control of the polyplex structure and composition. Microfluidic nano-precipitation is an emerging field in which researchers seek to tune the physicochemical properties of nanoparticles by controlling the mixing of particle components during fabrication. Using this approach, several groups have demonstrated the successful production of enhanced polymeric gene delivery vectors. Controlled mixing of polymer and DNA in the laminar flow environment resulted in higher transfection efficiency and lower cytotoxicity. Other groups have demonstrated that incorporating a polyanion to form ternary polyplexes improves transfection efficiency and serum stability.

To advance non-viral gene delivery towards clinical relevance, we are developing a microfluidic platform that provides more controlled and reproducible layer-by-layer assembly of ternary complexes. Our approach provides diffusion-controlled mixing of plasmid DNA and polycations to form polyplexes, separation of excess free polycation, and diffusion-controlled introduction of polyanions to form ternary polyplexes in a sequential manner on a single device. We will describe fabrication and characterization of polyglutamate (PGA)/polyethylenimine (PEI)/DNA

ternary complexes. We will also compare these ternary polyplexes to conventional PEI/DNA and PGA/PEI/DNA polyplexes formed by conventional methods in terms of cytotoxicity and transfection in several human cell lines.

468252 Combinatorial Treatment Using Lipopolymer-Mediated TRAIL Gene Delivery and Kinase Inhibitors for Bladder Cancer

Abstract:

The estimated number of new cancer cases within the US for 2016 is 1,658,270 with 595,690 deaths projected during the same year. Urinary bladder cancer is the 6th most common type of cancer in men with 77,000 new cases and 16,400 deaths expected for both men and women in 2016. Treatments for bladder cancer depend on the type and stage of cancer and may involve surgical removal of the tumor, immunotherapy, chemotherapy (intravenous or intravesical), radiation, and/or surgery for bladder removal. Limitations of current chemotherapy regimens include poor drug bioavailability, drug resistance and toxicity, and additional side-effects on healthy cells. Therapies that can specifically target tumor cells can improve patient treatment and quality of life. Tumor Necrosis Factor-alpha Related Apoptosis-Inducing Ligand (TRAIL) induces programed cell death in several cancer types with minimal effect on normal cells, making it a desirable target for cancer treatment. However, resistance to TRAIL (either intrinsic or acquired) has been observed in cancers possibly due to overexpression of anti-apoptotic proteins or low expression of TRAIL receptors. Combination therapy using TRAIL may be a key strategy to sensitize cancer cells to TRAIL-induced apoptosis.

The current work investigates the treatment efficacy of using kinase inhibitors in combination with TRAIL gene delivery in bladder cancer cells. Protein kinases are involved in metabolism, transcription, cell cycle progression, and apoptosis and are linked to TRAIL resistance. The two kinase inhibitors chosen, TG101209 (a Janus kinase II inhibitor) and HMN-214 (a polo-like kinase inhibitor), were based on a previous kinase screen that showed enhanced transgene expression using these inhibitors. Novel aminoglycoside-derived lipopolymers were used for the delivery of a plasmid expressing TRAIL (pEF-TRAIL) in UMUC3 bladder cancer cells. These aminoglycoside-derived lipopolymers displayed enhanced transgene expression compared to parental polymers as well as PEI. A dose-response study suggested that the combined treatment of pEF-TRAIL and TG101209 or HMN-214 demonstrated significantly enhanced cell death compared to either treatment alone. In vivo studies confirmed a significant decrease in tumor volume in mice treated with the TRAIL plasmid alone. Experiments with the combination treatments in vivo are currently underway. This study demonstrates the effectiveness of aminoglycoside-derived lipopolymers for combinatorial treatments using pEF-TRAIL and kinase inhibitors, and suggests potential applications of this treatment in the clinic.

462095 Localized and Sustained Delivery of siRNA from Hydrogels Expedites Fracture Healing

Abstract:

Introduction: Impaired fracture healing, which stems from reduced mesenchymal stem cell (MSC) osteogenesis, is a major clinical challenge [1]. To augment MSC-mediated fracture healing, known inhibitors of bone formation can be downregulated using siRNA-mediated gene knockdown. For example, mouse knockouts of WW domain containing E3 ubiquitin protein ligase 1 (WWP1) exhibit robust fracture healing [2]. However, siRNA delivery suffers from inefficiencies that preclude fracture-targeted and sustained delivery of protected siRNA [3]. Thus, we developed a hybrid nanoparticle (NP)/hydrogel system where NPs protect and increase siRNA delivery efficiency and hydrogels provide localized and sustained delivery by controlling release of embedded siRNA/NPs.

Materials and Methods: Diblock copolymers composed of cationic dimethylaminoethyl methacrylate (DMAEMA) blocks (m~140) and endosomolytic blocks of DMAEMA, propylacrylic acid (PAA), and butyl methacrylate (BMA) (n~80) were synthesized by reversible addition-fragmentation chain-transfer (RAFT) polymerization [3]. Self-assembled siRNA/NPs embedded in hydrolytically degradable hydrogels prepared using 10 wt% poly(ethylene glycol)-b-poly(lactide)-b-dimethacrylate, 1 Î¼M siRNA/NPs, and 0.05 wt% lithium arylphosphonate (LAP) photoinitiator, following exposure to ~5 mW/cm^2 365 nm light for 10 min. Hydrogels with embedded siRNA/NPs were characterized via scanned electron microscope (SEM). Characterization of localization of both Cy5-siRNA and Cy7-PEG hydrogels at Mid-diaphyseal femur fractures (Balb/c mice) was accomplished using XENOGEN/IVIS imaging. Fractures were treated with WWP1 siRNA/NPs hydrogels, followed by RT-PCR, histology (Alcian blue & Orange G) and microcomputed tomography (ÂµCT) to assess fracture healing.

Results and Discussion: siRNA/NPs were loaded into degradable hydrogels for treatment of femur fractures. SEM imaging showed uniform distribution of NPs within hydrogels. According to IVIS imaging, siRNA and hydrogels were co-localized at fractures for ~21 days, whereas free siRNA/NP dispersed from injection sites rapidly. 14 days after WWP1 siRNA treatment, we observed significantly decreased WWP1 expression compared with untreated controls. Histological analysis showed enhanced bridging bone formation in siRNA/NPs hydrogel treated fractures, whereas extensive unmineralized cartilage and undifferentiated mesenchyme tissue were still present in non-treated and hydrogel only controls. µCT showed 1.3-fold increase in bone callus volume at week 3 in siRNA/NPs hydrogel treated fractures compared with untreated fractures, indicating accelerated bone formation and mineralization.

Conclusions: PEG hydrogels were designed for siRNA/NPs delivery to fracture sites in a sustained and localized manner. Target knockdown of Wwp1 using siRNA/NPs hydrogels showed significantly increased bone formation and accelerated healing. In sum, the development of the hydrogel system has outstanding therapeutic promise.

Acknowledgements: Funding: NSF-DMR1206219; NSF-CBET1450987 NYSTEM N11G-035; NIH-R01AR056696; NIH-R01 AR064200; NIHS10 RR027340.

References: [1] Hofmann, A., et al., Bone, 2008. 42(5): p. 894-906 [2] Shu, L., et al., J Bone Miner Res, 2013. 28(9): p. 1925-35 [3] Benoit, D.S., et al., Biomacromolecules, 2012. 13(11): p. 3841-9

470592 Solution-Processed, Lead-Free Metal Iodide Thin Films for Photovoltaic Applications

Abstract:

We have synthesized and deposited metal iodide films of materials such as BiI_3Â and SbI_3 thin films were using solution-based methods. We have found that Lewis bases such as THF and DMF readily form Lewis acid-base adducts with BiI_3, which gives them high solubility in these solvents at high concentrations (<400 mg/mL) suitable for spincoating thin films of BiI_3. We find that exposure to solvent during processing can have a dramatic effect on film morphology. For example, solvent annealing BiI_3 thin films under relatively mild temperatures (â‰¤ 100 Â°C) resulted in pronounced grain growth and recrystallization of the film. The thin films were chemically stable for at least several months and can withstand extended (several hours) annealing at temperatures below 150 Â°C. Annealing in air after spincoating was found to be beneficial to photovoltaic device performance due to the formation of a BiOI layer at the BiI_3 surface which facilitates hole transport. In non-optimized BiI_3 solar cells, we have achieved power conversion efficiencies of 1.0 % demonstrating the potential of BiI_3 as a nontoxic, all-inorganic metal halide photovoltaic absorber layer.

463241 In-Situ X-Ray Diffraction Studies on the Self-Assembly of Metal Halide Perovskite Thin Films

Abstract:

Metal halide perovskites (MHPs) are revolutionizing the solar cell research field - the record power conversion efficiency of MHPs based solar cells has reached 22%. This represents the highest efficiency among all solution processable materials and the fastest rate of efficiency improvement in the history of all photovoltaic materials. Based on this trend, MHPs have been called the â€œnext big thing in photovoltaicsâ€ and worldwide research efforts have grown explosively.

Understanding the nucleation and growth processes of the MHP thin films is a prerequisite for scale up of the device area, improvement in reliability and further increase in efficiency. Here we present our results on synchrotron based *in-situ* grazing incidence X-ray scattering studies on the MHP thin film formation processes. Our results reveal the sub-processes and mechanisms through which highly preferential crystallographic orientation of MHP films can be formed. We demonstrate methods to controllably tune the direction and degree of the preferential orientation. Impact of the degree of thin film orientation on solar cell performance will be discussed.

460508 Layer-Controlled Colloidal Dispersions of Two-Dimensional Organometal Halide Perovskites for Efficient Blue Light-Emitting Diodes

Abstract:

Layer-Controlled Colloidal Dispersions of Two-Dimensional Organometal Halide Perovskites for Efficient Blue Light-Emitting Diodes

Jakub Jagielski, Sudhir Kumar, and Chih-Jen Shih*

Institute for Chemical and Bioengineering, DCHAB, ETH Zurich, Zurich, Switzerland

*Author to whom all correspondence should be addressed. Email: chih-jen.shih@chem.ethz.ch

Abstract

Two-dimensional (2D) organometal halide perovskites exhibit the quantum confinement effect by reducing the thickness smaller than the Bohr radius, or ~10 unit cell thick (i.e., $n < 10$). Their colloidal dispersion is emerging as one of the most promising candidates for printable optoelectronics and nanocomposites. However, it is not yet possible to synthesize large quantities of high-photoluminescence-quantum-yield (PLQY > 20%) 2D perovskites with precisely controlled stacking, as is necessary for light-emitting diodes (LEDs). Here, we demonstrate a synthesis method for the large-scale production of monolayer ($n = 1$), trilayer ($n = 3$), and 5-layer ($n = 5$) 2D methylammonium lead bromide ($MAPbBr_3$) colloidal dispersions, which exhibit blue PL emissions at ~431 nm, ~456 nm, and ~485 nm, respectively, with high PLQYs (50% - 80%). The carefully-controlled ligand chemistry allows small lateral size and surface passivation, minimizing the effects of self-quenching and downhill energy transfer. The high PLQY has led to a record-high external quantum efficiency (EQE) of 0.2% in the blue LED devices, representing a >500-fold increase

compared to that of the best blue organometal halide perovskite LED reported so far. The solution-based process is expected to allow low-cost and high-throughput production, as well as the deposition of the new family of 2D materials onto arbitrary substrates.

466712 Dopant Diffusion and Implications for Device Stability

Abstract:

Organic light emitting diodes (OLEDs) and organic photovoltaic devices (OPVs) are typically fabricated from a series of different layers, each of which is designed to carry out a specific electronic or optical function. The P-I-N structure consists of a p-type doped or hole transport layer (HTL), an intrinsic emitting or absorbing layer, and n-type doped or electron transport layer (ETL). HTLs and ETLs are often a mixture of an intrinsic organic semiconductor and a molecular dopant that increases the charge density and adjusts the Fermi energy, just as n- and p-type dopants control charge transport in Silicon. However, in contrast so Silicon, the semiconductor has a glass transition temperature below 100°C and is almost never a perfect crystal. This means that organic dopants are subject to thermal diffusion and electric field induced drift at typical device operating conditions. Here we measure and model the diffusion of p-type dopants in typical HTL layers used for OLED and OPV devices. We demonstrate several different techniques (including fluorescence quenching, quasi-elastic neutron scattering (QENS), and near edge x-ray absorption fine structure (NEXAFS) spectroscopy) to determine that diffusion occurred between layers and for quantifying the diffusion within a material layers with thickness below 50 nm. Finally we address the negative implications for organic device longevity that can occur when the dopants diffuse from the HTL to the intrinsic layer.

469027 Linking Molecular Conformation to Charge Transport in Organic Materials

Abstract:

Solution-processed conjugated polymers are an important class of semiconductor, largely due to the ability to modify their optical and electronic properties through molecular design choices and their compatibility with low-cost, scalable manufacturing processes. An important property of these materials is the charge-carrier mobility, which often determines the performance of organic electronic devices, and depends sensitively on the active layer morphology. In order to manufacture the most efficient devices, it is therefore vital to optimize the morphology such that the charge-carrier mobility is maximized. However, understanding the relationship between morphology on the molecular level and the subsequent device-scale performance is a significant challenge spanning multiple length and time-scales.

We use coarse-grained molecular dynamics simulations to simulate a variety of pristine morphologies of thiophene-based conjugated oligomers with different side-chain architectures and under various annealing protocols. The range of morphologies considered allows the analysis of several ordered and disordered structures, including hexagonally packed cylinders, ribbons and Ï€-stacked lamellar structures, depending on the processing conditions. The charge transport characteristics are then determined for these morphologies using a combination of atomistic backmapping, efficient semi-empirical quantum chemical calculations that bypass the requirement for density functional theory, and kinetic Monte Carlo simulations to elucidate the hole mobility dependence on side-chain architecture and annealing temperature. Percolation analysis is also used to identify the structures that allow for the most efficient charge transport through the device. This demonstrates a highly-flexible, modular pipeline that can be used to connect sub nano-scale morphology to bulk thin-film behavior, informing future charge transport investigations and manufacturing processes alike.

Our data echoes experimentally obtained trends concerning the hole mobility, namely that increased annealing

temperature and regioregular side-chain placement improves charge transport. Using our methodology, we quantify this behavior and attribute these trends to regular side-chain architectures permitting the assembly into closely-packed Ï€-stacking crystals, amplifying the rate of inter-chain hopping, which is the limiting factor for efficient charge transport.

460419 Solution Shearing of Conjugated Polymer with Highly Aligned Nanofibrillar Structures for Organic Field-Effect Transistors

Abstract:

Control of polymer crystallinity and chain orientation typically rely on template-assisted techniques, mechanical rubbing and directional solidification, which require complex equipment setup and process optimization such as shearing speed or substrate temperature. Therefore, a simple solution coating method of organic semiconductors becomes attractive with the aim of industrial-scale roll-to-roll production. Here, we propose the use of solution shearing to induce long-range alignment in poly(3-hexylthiophene) (P3HT) thin films by simply tuning the aggregation of P3HT in solution. The synergistic combination of ultraviolet-irradiation and solution aging facilitates the self-assembly of P3HT into long nanofibrillar structures prior the coating process. The length of these nanofibers can be modulated via the sequential application of both solution treatments. The surface morphologies of deposited films reveal that the degree of polymer alignment was greatly improved with increased polymer assembly, which can be controlled by the solution aging time. The development of birefringence and increased dichroic ratio determined from polarized optical microscopy and UV-Vis, respectively, provide further support for the macroscopic ordering and polymer chain alignment. According to image analysis, which provides insight into the orientation distribution of polymer chains, the P3HT nanofibers are shown to align in the solution shearing direction, leading to significant charge carrier mobility anisotropy. The correlations between highly oriented nanofibrillar structures and their two dimensional charge transport properties were systematically investigated by shearing the pretreated solution parallel and perpendicular to the active channel. Spin coated thin films, which exhibit randomly oriented nanofibrillar structures will also be discussed. This facile solution coating method demonstrated an effective approach to investigate the charge transport behavior within highly oriented crystalline domains, which can be obtained by directly controlling their intrinsic solution properties without the need for extrinsic techniques such as grooved substrates or blade patterning.

461282 Automated Analysis of Orientational Order from Images of Fibrillar Thin Films

Abstract:

Nanofibers are a ubiquitous structural motif in a variety of functional materials. In the field of organic electronics, Ï€-Ï€-stacking of conjugated polymers leads to fibrillar morphologies with a wide array of fiber packing behavior. Fiber orientation and alignment are known to influence the charge transport properties of devices such as organic field effect transistors. The solution processing methods used to create these devices give rise to large variations in these structural parameters â€" however, they are only observable with imaging techniques such as atomic force microscopy (AFM). To bring more rigorous quantification of orientation and alignment to these materials, a comprehensive image analysis tool is introduced to quantify the two-dimensional orientation and alignment of nanofibers from atomic force microscopy (AFM) images. It has been developed in MATLAB and packaged as a standalone application so that researchers with no computational expertise can produce publication-ready figures directly from their images. AFM frequently yields images with low contrast and moderate noise, making quantitative feature extraction a significant challenge. In this protocol, each image is analyzed in the context of an Orientation Map, in which nanofibers are thinned to single-pixel width and an orientation is extracted for each of these pixels. The Orientation Map is obtained through a five-step process: fiber smoothing by anisotropic diffusion filtering, contrast enhancement by top hat filtering, binarization by adaptive thresholding, skeletonization, and recovery of orientations from the result of diffusion filtering. Each step involves parameters that can be set using physical heuristics, which are examined in detail. This Orientation Map yields an orientation distribution and a plot of S_{2D}, an orientational order parameter, as a function of frame size. The image analysis procedure is used to quantify differences in P3HT nanofiber morphology induced by various solution processing recipes, as well as the effect of spin-coating when used to deposit solutions of nanofibers. All examples presented in this protocol can be reproduced from beginning to end using the available software, with visualizations produced at each stage of processing.

468978 Computational Investigations of Perylene and Perylothiophene Packing

Abstract:

The self-assembly of small aromatic molecules into structures that exhibit long-range order is important to a variety of materials problems from fouling oil pipelines to the efficiencies of organic photovoltaics. We conduct coarse-grained molecular dynamics simulations of perylene (P1) and perylo[1,12-b,c,d]thiophene (P2) using flexible and rigid models to quantify how chemistry and model details influence self-assembly, phase behavior, and performance. We find melting transition temperatures in agreement with experiments for P1 and P2 and show that the rigid and flexible models produce identical morphologies at experimentally relevant state points. We find that ordered hexagonally-packed columnar phases are thermodynamically favorable over a wide range of densities and temperatures in both P1 and P2. We discuss similarities and differences between our simulated P1 morphologies and herringbone crystals observed in experiments. The sulfur in P2 leads to an increased melting temperature and tighter columnar packings compared to that of P1. We determine that the rigid model is more computationally efficient, with more timesteps accessible per second and generally faster relaxation and structural autocorrelation times. Yet, in a non-negligible number of cases the rigid model results in longer relaxation and autocorrelation times than their flexible counterparts.

454743 Harnessing Niche Concepts in Biomaterial Design

Abstract:

Advances in the field of tissue engineering are increasingly reliant on biomaterials that instruct, rather than simply permit, a desired cellular response. Instructive biomaterials hold significant promise for clinical applications as well as to enable mechanistic studies in the laboratory. As tissues can be dynamic, spatially-patterned, or inhomogeneous over multiple length and time scales, my lab is developing new approaches to engineer biomaterials at the structural and biomolecular level in order to replicate these heterogeneities. These efforts are providing new insight regarding the degree of biomaterial complexity required to instruct cell behavior in the context of development, disease, and regeneration. I will describe a collagen biomaterial under development to address current barriers preventing regeneration of orthopedic insertions such as the osteotendinous (tendon-bone) junction. Here we also use inspiration from nature (*e.g.*, porcupine quills, honeycombs, and plant stems) to better co-optimize bioactivity and mechanical competence. I will subsequently describe a microfluidic forming technique to create libraries of optically-translucent hydrogels containing overlapping patterns of cell, matrix, and biomolecule cues. We are using this platform to explore the coordinated impact of cell and matrix signals on (1) niche-mediated regulation of hematopoietic stem cell fate; and (2) malignancy and therapeutic response of human glioblastoma. I will show how these biomaterials can be used as rheostats to regulate processes such as self-renewal vs. differentiation; tissue regeneration and vascularization; and the etiology and malignancy of cancer.

Biography: Brendan Harley is an Associate Professor in Chemical and Biomolecular Engineering at the University of Illinois at Urbana-Champaign. He received a B.S. in Engineering Sciences from Harvard University (2000) and a Sc.D. in Mechanical Engineering from MIT (2006). He was a post-doctoral fellow in the Joint Program for Transfusion Medicine at Children's Hospital Boston and Harvard Medical School from 2006 – 2008. Dr. Harley co-authored the book 'Cellular materials in nature and medicine' (Cambridge University Press, 2010). He has received funding from the NSF, NIH, American Cancer Society, the U.S. Army, and the AO Foundation. He was a 2013 recipient of the NSF CAREER award, the 2014 Young Investigator Award from the Society for Biomaterials, and was elected a Fellow of the American Association for the Advancement of Science (2014). He also co-founded UK-based Orthomimetics, Ltd. (acquired by TiGenix, Ltd.), currently performing Phase I clinical trials on a material to repair osteochondral defects in the knee.

454746 Engineering Hemostatic Nanoparticles to Stop Internal Bleeding

Abstract:

Traumatic injury is the leading cause of death for both men and women between the ages of 5 and 44 worldwide, and blood loss is the primary cause of death at acute time points. Immediate intervention is essential to minimize mortality and yet there is a derth of technologies to address internal bleeding. We have developed functionalized nanoparticles based on poly(lactic-co-gycolic acid) (PLGA), poly(ethylene glycol) (PEG), and the arginine-glycine-aspartic acid (RGD) peptide which binds the glycoprotein IIb/IIIa receptor. The receptor is only exposed on activated platelets. Our particles leverage the specificity of biology and are designed to help form stable clots faster to halt bleeding after intravenous administration. We have tested these hemostatic nanoparticles in a range of injury models including the femoral artery model and liver injury models. We have also tested these particles in a blast injury model of polytrauma and a model of brain injury with hemorrhaging. In all cases, we saw a reduction in bleeding, and in the case of lethal injury models, we saw significant improvements in survival. In both the blast injury models we see that the administration of the hemostatic nanoparticles reduces bleeding and this, in turn, leads to better behavioral outcomes. This suggests that reducing bleeding may be a critical component of providing neuroprotection after injury. The functionalized nanoparticles or hemostatic nanoparticles reduce bleeding in a number of models of trauma. Following administration, the particles participate in clot formation and can be found at the injury site post injury. The chemistry of the formulation including the length of the PEG arms and peptide play significant roles in the efficacy and clearance of the particles and provide means to tailor the particle behavior. This treatment has the potential to greatly impact survival outcomes related to internal hemorrhage and may lead to better functional outcomes more broadly.

ACKNOWLEDGEMENTS

We would like to acknowledge the NIH Director's New Innovator Award DP2OD007338.

Biography: Erin Lavik is a professor of Chemical, Biochemical, and Environmental Engineering at the University of Maryland, Baltimore County (UMBC). Dr. Lavik's research focuses on engineering polymers to protect and repair the nervous system and treat trauma more broadly. The lab does this through a set of approaches including developing intravenously administered nanoparticles to stop internal bleeding, developing drug delivery systems for traumatic brain injury as well as diseases of the eye, and by developing novel materials that can not only direct stem cells but report on their health and function. Dr. Lavik received her bachelor's, master's, and doctoral degrees from MIT in Materials Science and Engineering. She has won a number of awards including the TR100 award in 2003 and the NIH Director's New Innovator Award in 2010. She became at Fellow of the American Institute of Medical and Biological Engineers in 2014. She is also an Associate Editor at Bioconjugate Chemistry. Beyond her research, Dr. Lavik has developed classes including Applied Tissue Engineering where the students make and test artificial arteries and learn about the issues involved in translating the technologies from the bench to patients.

454747 Nanolayered Staged Delivery Approaches to Wound Healing Â and Bone Regeneration

Abstract:

The biological proesses involved in wound healing and tissue regeneration, as well as the progression of a number of diseases, take place in stages that involve the release or presentation of distinct molecules and the chemical and physical signaling that ensues. Ideally one would be able to modulate the release of chemokines, growth factors, or siRNA and DNA that regulate genes that regulate such factors, to facilitate tissue regeneration in-situ. The use of alternating layer-by-layer (LbL) assembly enables a modular platform that can control release of therapeutics over a broad range of time scales, from several seconds to multiple months. We have developed systems that can address a range of different wound healing needs by taking advantage of this tunable platform and the ability to incorporate drugs programmed to release at a range of different rates and with staged release behavior. The nature of the layering process enables the incorporation of different drugs within different regions of the thin film architecture; the result is an ability to uniquely tailor both the independent release profiles of different therapeutics from the same film, and the order of release of molecules to targeted regions of the body. The staging of release of multiple growth factor proteins can achieve the regeneration of collagen and remodeling of tissue in wounds at appropriate time points to facilitate wound closure and tissue vascularization. Moreover, multilayered release coatings as thin as a half micron to several microns can be designed to deliver different growth factor proteins or siRNA and other biologic drugs directly to wounds to correct the dysregulation of wound healing processes that have gone awry, from burn and scar tissue to the closure of chronic wounds such as diabetic ulcers. These approaches are also applicable to the regeneration of bone in critical defects and for orthopedic applications.

461258 Understanding Relationships Between Molecular Structures and Thermomechanical Properties of Thermosetting Polymers with Novel Bio-Based Building Blocks

Abstract:

Thermosetting polymeric materials are widely used in protective coatings and structural applications. Many thermal and mechanical properties of thermosetting polymers are related to their specific monomer molecular structures and random network architectures. Although properties of thermosetting materials can be obtained experimentally, an ability to predict properties of new materials based on molecular structure will significantly facilitate new thermosetting material design. In this work, we use computer simulations to investigate the structure and the properties of thermosetting materials based on novel bio-based building blocks. Molecular dynamics simulations and experiments are used to study highly cross-linked epoxy networks comprised of a furanyl epoxy monomer, 2,5-bis[(2-oxiranylmethoxy)methyl]-furan (BOF), that is cross-linked by two furanyl amine hardeners, 5,5â€²-methylenedifurfurylamine (DFDA) and 5,5â€²-ethylidenedifurfirylamine (CH3-DFDA). A general constant-NPT MD procedure is used to identify the glass transition temperature (Tg) of the cross-linked polymers with two versions of general AMBER force field (GAFF) for parameters: one with original GAFF parameters and one with refitted dihedral parameters for specific parts. Important properties of these four systems including density and Tg are compared with experimental results. We also compare the simulated and experimental values for fully furan-based thermosetting materials to those using the conventional resin diglycidyl ether of bisphenol A (DGEBA) cured with the two furanly hardeners. Predicted Tg overpredicts experimental results but captures experimental trends between different epoxies and amines very well. Finally we analyze the effect of epoxy/amine monomer on Tg through microscopic details of the structures and flexibility of the polymer chains.

463556 Predicting the Binary Interaction Parameter Chi for Polymer Pairs from Oligomer Simulations

Abstract:

The binary interaction parameter chi plays a crucial role in predicting the phase behavior and thus materials properties of polymer blends and block copolymers. However, the determination of the chi parameter is not assumption-free. Current methods to estimate the chi parameter from theory, experiment, and simulation often involve the random-phase approximation. In this work, configurational-biased Monte Carlo simulations using the TraPPE-UA force field are performed to directly compute the free energy of mixing for a variety of oligomer blends. Extrapolation of these data to infinite-chain length allows to predict chi parameters, solubility parameters, and mixing irregularities. In addition, structural analysis from the simulation provides further insight into the mixing behavior of these blends.

472627 Uniaxial, Biaxial, and Shear Deformation of Simulated Amorphous Cis-, Trans-1, 4-Polybutadiene Chains

Abstract:

Rolling resistance in rubber tires leads to losses in vehicle fuel economy. We hypothesize the rolling resistance on the macroscale connects directly to deformation-induced changes in elastomer chain conformations on the microscale through changes in elastic free energy. To study this phenomenon, random chain conformations were generated using Floryâ€™s Rotational Isomeric State (RIS) Approach, which weights particular torsional states along the backbone of polymer chains. Polybutadiene chains were generated under unperturbed conditions where chain size and shape characteristics were studied. Ensembles of unperturbed chains were then uniaxially deformed along the x-direction and consequently compressed along the y and z directions. For comparative analysis, Gaussian chains were generated and compared with RIS chains. Variation in probability density distribution of chain vectors with deformation was observed. Changes in elastic free energy of the chain vectors under deformation were studied for chains of different number of repeat units at a single temperature and for chains at different temperatures for the same number of repeat units. Change in elastic free energy of a system increased with increasing deformation. At a particular deformation, RIS chains showed higher elastic free change than Gaussian chains. These changes in elastic free energy were then quantified to compute deformation force and stress acting on ensembles of chains. Good agreement was observed between deformation forces computed numerically and analytically for Gaussian chains. Under all conditions, RIS chains showed higher deformation force and stress as compared to Gaussian chains. Cis chains showed higher deformation force and stress than trans. For chains at a single temperature and varying number of repeat units, deformation force and stress decreased with increasing number of repeat units while for chains at a single repeat unit with varying temperature, the deformation force and stress decreased with increasing temperature. The study was then extended towards analyzing biaxial and shear deformations on ensembles of chains. Moduli of chains were computed under undeformed and deformed conditions.

470298 Choline and Phosphoryl Contributions to Hydration Structure and Dynamics of Poly(2-methacryloyloxyethyl phosphorylcholine)

Abstract:

We explore the hydration structure and dynamics of poly(2-methacryloyloxyethyl phosphorylcholine)50, or pMPC50, in solution using molecular dynamics simulations with an emphasis on isolating the contributions of the key chemical moieties. Surface bound films of this biocompatible polymer have been shown experimentally to produce tribological properties that surpass those of the human synovial joint [1] and are being developed for use in artificial joints [2]. The mechanism by which such materials are thought to provide ultra-low friction coefficients has been termed "hydration lubrication" [3]. Here, we further characterize the molecular level structure and dynamics of this hydration water around pMPC molecules that are experimentally grafted onto substrates for biomimetic lubrication. Previous results [4] showed that the charged choline groups of individual monomers tend to fold onto neighboring monomers yielding solvent facing, highly hydrated phosphoryl groups. Here, we further analyze the effect of the two outer chemical moieties, the choline and phosphoryl groups, by removing them and comparing the behavior of pMPC to poly(2-methacryloyloxyethyl phosphate) and poly(ethyl methacrylate). Preliminary results strengthen the previous hypothesis [4] that the choline moiety is responsible for stabilizing the polymer.

[1] M. Chen et al., "Lubrication at Physiological Pressures by Polyzwitterionic Brushes." Science 2009, 323, 1698–1701.

[2] M. Kyomoto et al., "Biomimetic hydration lubrication with various polyelectrolyte layers on cross-linked polyethylene orthopedic bearing materials." Biomaterials 2012, 33, 4451-9.

[3] J. Klein, "Hydration lubrication." Friction 2013, 1, 1–23.

[4] W.L. Roussell, C. Klein, C.R. Iacovella, P.T. Cummings, C. McCabe, "Molecular Origins of the Ultra-Low Friction Exhibited by Biocompatible Zwitterionic Polymer Brushes" accepted at Young Scientist

470146 Nonequilibrium Molecular Dynamics Simulations of Entangled Polymer Melts and Solutions Undergoing Planar Elongational Flows

Abstract:

Tube based models predict that concentrated polymer solutions behave roughly the same as entangled polymer melts, if they have the same number of entanglements and if stress, strain rate and time are made dimensionless appropriately. But there are a lot of experimental works that show entangled polymer melts and solutions exhibit different behaviors when they are subject to extensional flows at high extension rates. Specifically there is an upturn in extensional viscosity of polymer solutions at extension rates higher than the inverse of the Rouse time which has not been observed in polymer melts. Recent experiments of Wingstrand *et al.* [1], however, have demonstrated that using oligomers as solvent, polymer solutions could behave in a similar fashion as melts.

In this work we performed nonequilibrium molecular dynamics (NEMD) simulations of entangled $C_{700}H_{1402}$ and $C_{1000}H_{2002}$ linear polyethylene melts subject to planer elongational flows. We also simulated concentrated solutions of $C_{1000}H_{2002}$ in $C_{16}H_{32}$ as an oligomeric solvent with the same monomeric units in order to compare the behavior of polymer melts and solutions with roughly the same number of entanglements. Simulation results show that polymer solutions and melts exhibit roughly the same behavior at high extension rates. Specifically, similar to the melt, no upturn was observed in primary extensional viscosity of the solution, consistent with Wingstrand *et al.* experiments. Other system properties including chain orientation, chain stretch, and entanglement network properties and their potential connection to this finding will be discussed in this presentation. Preliminary results of extensional viscosity of concentrated solutions in small solvent molecules with various molecular architectures such as benzene will also be shown.

[1] Wingstrand, S.L., N.J. Alvarez, Q. Huang, O. Hassager, Phys. Rev. Lett. **115**, 078302 (2015).

466108 Using Molecular Simulation to Develop a Physically-Based Materials Genome for Semicrystalline Polymer Nucleating Agents

Abstract:

Molecular dynamics simulations were performed to assess the effect of various additives on the nucleation and growth of an oligoethylene. Heterogeneous nucleation was studied for a broad range of substrates by systematically adjusting the parameters that defined the additive material. A suite of techniques developed to study nucleation and growth in oligoethylene by molecular simulation was used to quantify the impact of different substrates on the resulting crystallization behavior. The time lag to steady-state crystal growth was identified as a key metric for evaluating nucleating efficiency. Increasing the surface attractive forces benefited nucleation, however with diminishing returns. A scan of the substrate lattice parameters revealed significant reduction in time lag when the substrate and polymer crystal unit cells were in crystallographic registry, suggesting an epitaxial mechanism. Substrate rigidity also influenced crystal phase nucleation, but its effect was tied to lattice matching between the substrate and polymer crystal. In this work we present a physically-based materials genome, to which molecular simulation is applied systematically to optimize the nucleating agent activity for a particular family of additives.

469171 Yield and Failure Behavior Investigated for Cross-Linked Phenolic Resins Using Molecular Dynamics

Abstract:

Molecular dynamics simulations were conducted to fundamentally evaluate the yield and failure behavior of cross-linked phenolic resins at temperatures below the glass transition. Yield stress was investigated at various temperatures, strain rates, and degrees of cross-linking. The onset of non-linear behavior in the cross-linked phenolic structures was caused by localized irreversible molecular rearrangements through the rotation of methylene linkers followed by the formation or annihilation of neighboring hydrogen bonds. The yield stress results, with respect to temperature and strain rate, could be fit by existing models used to describe yield behavior of amorphous glasses. The degree of cross-linking only indirectly influences the maximum yield stress through its influence on the glass transition temperature (Tg), however there is a strong relationship between the degree of cross-linking and the failure mechanism. Low cross-linked samples were able to separate through void formation, whereas the highly cross-linked structures exhibited bond scission.

467442 A Reactive Molecular Dynamics Simulation of the Thermal Decomposition in Graphene-Reinforced Polyethylene Oxide

Abstract:

In this work, we have used the Reactive Force Field (ReaxFF) to investigate the thermal decomposition of the virgin and graphene-reinforced polyethylene oxide (PEO). Our objective has been to elucidate the effects of the type and concentration of graphene platelets in the host polymer matrix, as well as the confinement effect of the PEO molecules in the galleries of the graphene platelets, on the kinetics and thermodynamics of the polymer thermal decomposition. We generated six different systems composed of 1) virgin PEO, 2) PEO/graphene with two graphene platelets, 3) PEO/graphene with four graphene platelets, 4) and 5) PEO/graphene oxide (GO) with two and four graphene platelets, respectively, and 6) PEO confined between two graphene platelets. Our first set of NVT simulations were comprised of cook-off simulations from 298 K up to 3,400 K for a total simulation time of 200 ps (a temperature increase rate of 16 K/ps). We used a time step of 0.05 fs for all simulations and controlled the temperature by the Nosé-Hoover thermostat with a damping parameter of 100 fs. We also performed ten independent 20-ps NVT simulations of the virgin PEO and high-concentration PEO/graphene and PEO/GO systems between 2,500-3,400 K with an interval of 100 K. We have quantified and experimentally validated the rate of polymer decomposition as a function of temperature. We have also determined the weight distribution and types of the species formed during the decomposition process. Our initial results show that the onset of PEO decomposition is shifted towards lower temperatures with increasing graphene and GO concentrations. Moreover, in the confined region between the two graphene platelets, the polymer chains decompose into larger species than those formed in the virgin PEO and low-concentration PEO/graphene and PEO/GO systems. For the first time, this study sheds new light on the effect of graphene nanoparticles on the kinetics and thermodynamics of polymer thermal decomposition using a reactive molecular dynamics simulation. Such molecular insight will enable the design of hybrid systems for use in extreme environments.

474230 Bone Target N Acetylcysteine Loaded in PLGA-ALE Nanoparticle to Osteoporosis Treatment. an *in Vitro* Test

Abstract:

N acetylcysteine (NAC) is a derivative of cysteine which is a precursor of well known antioxidant glutathione. This kind of components have shown the potential to attenuate osteoclasts activity *in-vitro* experiments. Osteoporosis is an illness characterized by an imbalance between osteoclast and osteoblast activity, so a component like NAC could be used as a therapeutic agent for its treatment. However, studies on the osteoporosis treatment with this agent have not yet shown beneficial effects. Despite there is no a conclusive explanation for this lack of activity, a possible cause is the low bioavailability of NAC (6% oral), preventing the required NAC doses to be delivered in the bone tissue to get a therapeutic effect.

In order to increase the bioavailability of NAC, and to selectively target the bone tissue, conjugated alendronate PLGA nanoparticles (PLGA-ALE) were used as carrier for NAC. This carrier should not react or destabilize the active compound, particularly as NAC denature by internally disulfide bond formation. Because of that, it is necessary to verify if the active compound retain its biological action after the loading process.

To confirm how suitable carrier are the PLGA-ALE nanoparticles to maintain NAC stability, this work developed an *in vitro* test to evaluate the NAC activity on osteoclast-cells like, previously generated by polyethylene glycol (PEG) induced macrophage fusion.

Synthesis of PLGA-ALE was done through carbodiimide chemistry. Nanoparticles were produced by the nanoprecipitation method and the *in vitro* test was done by fusion of U937 cells in the presence of PEG, following a fusion hybridoma protocol. Fused Cells were seeded on sterile bone slices for 4 to 6 hours, to attach the cells to the support. After that, a fetal bovine serum and free NAC supplemented culture medium was added and loaded within nanoparticles. Finally cells were incubated during 4 days, and subjected to reactive oxygen species (ROS), bone resorption pits and immunohistochemistry analysis. As osteoclastic activity surrogates, presence of different proteins were evaluated: matrix metalloproteinase (MMP), integrin aVb3 receptor (IB3), subunit of vacuolar ATPase (ATPase), DC-STAMP, Carbonic Anhydrase II (CAII), Cathepsin K, ADAM-12 and Calcitonin Receptor.

According with results both, free or polymer-loaded NAC, caused a decrease expression of the different osteoclast proteins (free NAC 74% and loaded NAC 71%). This experiments indicated that the polymer helps maintaining the stability of the active molecule up to 4 days. Taking into account that previous experiments demonstrated a high affinity of the polymer nanoparticles for the bone tissue, it is expected that a loaded NAC osteoporosis treatment to be more active than a free NAC one. Still *in-vivo* tests are necessary to confirm our findings.

As a conclusion, NAC loaded in PLGA-ALE nanoparticles retain the antioxidants properties of free NAC. The carrier did not have cytotoxic properties because cells grew in the present of it, and the entrapment process did not affect NAC stability.

472479 Inexpensive and Rapid Synthesis Unilammelar Liposomal Drugs for Targeted Delivery

Abstract:

Liposomes are rapidly emerging as one of the most advantageous nano-scale drug delivery systems. Their ability to enhance bioavailability through increased solubilities, lengthened clearance rates, and improved targeting capabilities has been well documented. However, current liposome synthesis approaches involve complex and tedious techniques limiting their investigation to specialized facilities. This represents a clear separation between basic liposomal pharmaceutical research, and their translation into therapeutic and theranostic platforms. Here we show a gentle method for reliably synthesizing, purifying, and concentrating unilamellar liposomes using common benchtop equipment. We use the ethanol injection method to encapsulate fluorescent molecules that vary in size (400-20k Da) within liposomes with a mean diameter of less than 150 nm. Using a standard benchtop centrifuge and Amicon Ultra filteration centrifuge tubes, we have characterized a simplified approach to gently and efficiently remove any unencapsulated drug while limiting aggregation or loss of sample, problems commonly encountered in ultracentrifugation and chromatographic approaches. We also demonstrate no loss in bioactivity following the purification process by incubating metastatic breast cancer cells with liposomes containing doxorubicin and monitoring cell viability over several days. It is our hope that with this efficient and accessible approach, the discovery of novel liposomal formulations may be better investigated and more widely implemented in clinical settings.

472629 Fabrication and Characterization of Fluorescently Labeled Polymeric Nanoparticles for Biodistribution Studies of Drug Delivery

Abstract:

The use of nanoparticles in drug delivery can address many of the difficulties faced by traditional methods such as systemic toxicity, transport limitations, and drug solubility. By loading drugs into biocompatible polymeric nanoparticles, drugs that were previously unusable due to their solubility or toxicity can now be used, and by fluorescently labeling the nanoparticles with a small molecule like 6,13-Bis(triisopropylsilylethynyl)pentacene (TIPS pentacene), the particles can be tracked for biodistribution studies in biological systems. TIPS pentacene is a fluorophore that has been primarily used in the optoelectronic device field. However, its high quantum yield and photostability make it a promising candidate as a fluorescent imaging agent for biomedical applications.

Using the Flash Nanoprecipitation process[1], different loadings of this fluorophore as well as a filler homopolymer of poly(L-lactic acid) (PLLA) have been incorporated into amphiphilic diblock copolymer nanoparticles of poly(ethylene oxide)-b-poly(D,L-lactic acid) (PEO-PDLLA). The effects of varying the TIPS pentacene loading as well as the homopolymer to diblock copolymer weight:weight ratio on the characteristics of the resulting nanoparticles such as size, zeta potential, and fluorescent intensity have been studied.

Nanoparticles were fabricated with hydrodynamic diameters in the range of 80-120 nm and polydispersity indices less than 0.3 as measured by dynamic light scattering. The zeta potentials of the particles were typically around -15 mV in phosphate buffered saline diluted 100X. Fluorescence studies show that the nanoparticles containing TIPS pentacene could be excited at wavelengths of 600 nm and below and had a principal emission peak at 650 nm. These nanoparticles were colloidally stable in cell-growth media supplemented with fetal bovine serum (FBS) for periods up to 12 hours. Cell culture studies with macrophage cells show that they were neither immunogenic nor cytotoxic. ELISA tests showed that the nanoparticles induced levels of the inflammatory cytokines interleukin-6 (IL-6) and tumor necrosis factor alpha (TNFa) in both M1 (pro-inflammatory) and M2 (anti-inflammatory) macrophage cells similar to those of mock negative control experiments. Further results will be presented on how these nanoparticles are internalized by macrophages.

1. Gindy, M. E.; Panagiotopoulos, A. Z.; Prud'homme, R. K., Composite block copolymer stabilized nanoparticles: Simultaneous encapsulation of organic actives and inorganic nanostructures. Langmuir 2008, 24 (1), 83-90.

470594 Polyelectrolyte Multilayer Films As Templates for Surface Modification to Design Liposomes Mediated Local and Sustained Therapeutic Delivery

Abstract:

The engineering of drug delivery platforms facilitating spatial and temporal release of a therapeutic is one of the key challenges in biomedical research that can ultimately lead to society-wide improvement in disease management. In recent years, substrate mediated delivery of cargo has shown great promise in applications including drug and gene eluding films/scaffolds, coatings for stents and other implantable devices, and controlling stem cell differentiation. Specifically, the drug delivery kinetics is particularly relevant when it is necessary to achieve effective dose and spatiotemporal release kinetics of the therapeutic agent at the intended site of injury. Delivery via immobilization of the therapeutic cargo by modifying the surfaces demonstrates higher translatable success compared to delivery using the free â€œbolusâ€ form by overcoming unfavorable burst kinetics, toxic offsite effects, and efficacy reduction due to systemic dilution. However, the lack in the development of a substrate mediated delivery platform that can exploit the advantages of nanocarriers to address bolus based therapeutic limitations has greatly hindered application. Here, we report the engineering of novel nanostructured substrate mediated therapeutic delivery system comprising of covalently functionalized liposomes (LNPs) with a surface decoration of hyaluronic Acid (HALNPs) embedded into an engineered Polyelectrolyte Multilayer (PEM) platform (HALNP-PEM) via ionic stabilization (Nature Scientific Reports, 2016). PEM provides an ideal surface modification tool that enables tuning the substrates to facilitate the immobilization of biomolecules including liposomes. The PEM base platform was constructed from poly-L-Lysine (PLL) and poly(sodium styrene sulfonate) (SPS) followed by adsorption of anionic HALNPs. An adsorption affinity assay and saturation curve defined the preferential HALNP deposition density for precise therapeutic dosing. A capping layer on top of the deposited HALNP monolayer was found to promote complete nanoparticle immobilization, support cell adhesion, and provide multi-dimensional nanoparticle confinement for predictable linear release profiles of both the nanocarrier and encapsulated cargo. Doxorubicin (DOX) was chosen as a model hydrophilic cargo and used to verify the structural integrity of the particles following deposition as well as to probe the therapeutic index of the HALNP-PEM platform. Furthermore, a direct comparison between bolus HALNP and the HALNP-PEM exemplified the effect of liposome confinement on intracellular uptake kinetics in patient derived metastatic breast cancer cells and confirmed sustained therapeutic release from the PEM platform. Lastly, we employed micropatterning to pattern both fluorescently conjugated HALNPs and DOX encapsulated in HALNPs into our PEM platform and found cell mediated uptake of HALNPs leading to micro-scale spatial control and sustained therapeutic release for over 60 hrs.

472948 Recognitive Methacrylated Alginate Nanoparticles for Protein Therapeutics

Abstract:

One of the main foci in biomaterial design is the development of novel approaches to specifically recognize biological cues and deliver therapeutics in a targeted mechanism. A variety of recognition molecules have been used to improve protein and drug delivery (e.g. antibodies, carbohydrates, peptides). However, these biomolecules are fragile, temperature and pH sensitive, and highly expensive, thus limiting their application worldwide. An alternative to these, are molecularly imprinted polymers (MIPs), which are biomaterials that recognize a predetermined template (e.g. proteins, small molecules). These systems are synthesized by crosslinking a polymeric matrix around the desired biomarker, therefore creating nanocavities that have complementary physicochemical characteristics to the template.[1] Among these systems are alginate particles, which have been previously used as drug and protein delivery vehicles.[2 Alginates are biocompatible anionic biopolymers, that can be ionically or chemically crosslinked, offering a wide variety of biochemical and physical characteristics.

The work presented herein was developed to provide insights on the protein/carrier interactions, and the importance that surface charge, chemical, and mechanical properties have on the recognition of proteins by molecularly imprinted natural polymers. Sodium alginate was functionalized with different methacrylate molecules (e.g. benzyl methacrylate, 2-hydroxyethyl methacrylate) using a carbodiimide reaction. Successful functionalization of the alginate was confirmed via Fourier transform infrared spectroscopy and NMR spectroscopy. Alginate nanoparticles were synthesized using water/oil microemulsions. Briefly, functionalized-alginate solution (up to 30% (v/v)) was emulsified in toluene with the biomolecular template, 1% (w/w) dioctyl sodium sulfosuccinate as a surfactant, and 1% (w/w)

Irgacure 184 as photoinitiator. Nanoparticle solution was crosslinked for 10 min using a 140 mW/cm^2 UV point-source. Resulting nanoparticles were purified and the template remains were removed by intensive washing steps. Physical characterization of the obtained particles was performed using light and electron microscopy, dynamic light scattering and zeta potential measurements confirming the morphology and charge of the functionalized alginate nanoparticles. Furthermore, the ability of these nanoformulations to recognize their template was evaluated using re-binding studies. Proteins similar to charge and size to the template used, were similarly recognized by functionalized alginate nanoparticles. However, by changing the functionalization molecule, it was shown that the physicochemical characteristics of alginate MIPs affected considerably their binding and recognition. These results exhibit the ability of molecular recognitive natural polymeric systems, such as alginate nanoparticles, to be tailored and enhance their specificity to be used as drug and protein delivery carriers.

[1] Peppas NA, Huang Y. Polymers and gels as molecular recognition agents. Pharmaceutical research. 2002;19:578-87.

[2] Lee KY, Mooney DJ. Alginate: properties and biomedical applications. Prog Polym Sci. 2012;37:106-26.

455389 Gold Nanoconjugates for Spinal Cord Injury Treatment: Recovery and Biodistribution

Abstract:

Due to their versatile surface chemistry, ease of imaging, and biocompatibility, gold nanoparticles (AuNPs) are employed as drug delivery devices in the treatment of spinal cord injury (SCI). In this talk, we will describe the synthesis of a tripartite nanoconjugate comprised of an AuNP drug carrier chemically conjugated to a transporter protein (wheat germ agglutinin coupled to horseradish peroxidase or WGA-HRP) and to theophylline, a SCI drug. The nanoconjugate structure was determined by UV-vis spectroscopy, transmission electron microscopy, dynamic light scattering, thermogravimetric analysis, and mass spectrometry. The therapeutic efficacy and biodistribution were studied in a validated animal model for hemi-diaphragmatic spinal cord paralysis. Our results show that a single administration of the nanoconjugate improved diaphragmatic activity at a much lower dosage than the effective systemic drug dosage and restored the respiratory drive. The effects lasted for 14 days. A biodistribution study using inductively coupled plasma mass spectrometry on the same animal model shows that the nanoconjugate was successful in targeting the respiratory neurons in the medulla. This research demonstrates the translational potential of our nanoconjugate design for SCI patients with respiratory dysfunction.

468582 Development of Drug Delivery Systems Based on a Fructose Polymer and 5-Fluorouracil

Abstract:

DEVELOPMENT OF DRUG DELIVERY SYSTEMS BASED ON A FRUCTOSE POLYMER AND 5-FLUOROURACIL

Alvaro G. Garcinuño, Antonio Tabernero, Miguel A. Galan, Eva M. Martin Del Valle

Department of Chemical Engineering. University of Salamanca.

There are new trends in biomedical engineering related with novel biomaterials which will be more compatible and more biodegradable than those that are being used at this time. For that reason, we use a polymer of fructose (levan) that has been successfully proved for forming nanocapsules by a self-assembled process.

Specifically, it is built a new drug delivery system (DDS) composed by levan and 5-fluouracil (a drug sometimes used in colorectal and hepatic cancer treatment). The work also presents results from a surface-modification of the initial levan (its carboxylate form).

The work shows how some parameters affect levan-nanoparticles synthesis, improving critical values such as size or Zpotential, in order to build a good DDS. The effect of pH and ionic strength are critical for nanoparticles assembly. Basic pHs are needed to obtain the best results (particle size between 300-400 nm with a Zpotential value around -20 mV). Those conditions avoid nanoparticle aggregation because the hydrocyl radicals of the medium take a hydrogen atom from the polymer that gets a negative charge. The ionic strength only affects the aggregation at high pHs (body blood ionic strength offers good results even at high pHs).

Moreover, carboxymethylation of the levan is performed to obtain carboxymethyllevan (CML). That surface-modification maintain good levels of size (between 300-400 nm), and improve Zpotential (-35 mV). That increment in Zpotential value prevents further particle aggregation, which is an interesting advantage for clinical or in vivo essays. Some studies of aggregation at different times were also performed. In both cases, at low ionic strength and basic pH, little aggregation was observed.

Results indicate that 5-fluorouracil is bound to either levan or CML surface by electrostatic interactions, obtaining the best results at basic pHs. According to drug release experiments, between 45-70% of the 5-fluoruracil is released from the levan in 23 hours depending on the pH, whereas only a low percentage of the drug is released at the initial times from the CML because of the stability of the amide bond that is produced on the surface. That fact emphasizes the good strategy of carboxymethylation for creating an effective drug delivery system.

469158 Characterization of Polyelectrolyte Membranes Loaded with *in-Situ* Grown Metal-Oxide Nanoparticles

Abstract:

Formation of nanocomposites by *in-situ* growth of metal-oxide nanoparticles (MONP) in polyelectrolyte membranes (PEM) has been quite appealing for several potential applications such as: fuel cells, water purification, and protective barriers against chemical warfare agents (CWA). PEM micro-structure can serve as capping agents for the nanoparticles and depending on PEM conditions (humidity, degree of swelling, solvents present) in-situ grown nanoparticles can adopt various shapes and sizes. Among several potential applications this paper is focused on PEM-MONP composites for self-detoxifying composite PEM barriers against CWA. These PEM have self-assembled interconnected hydrophilic and hydrophobic sub-domains that provide perm-selective properties and good transport of water vapor, while acting as a barrier against dangerous substances. PEM ion-exchange capabilities have been used for the *in-situ* formation of MONP within PEM at various conditions to observe the influence of PEM micro-structure on MONP growth. In addition, transport properties of PEM-MONP composites were analyzed by sorption and permeation experiments with water and chemical warfare surrogates (CWS). Sorption experiments results are compared with mathematical models and show the impact that MONP have on transport properties of PEM. Transmission and scanning transmission electron microscopy (TEM/STEM) were used to observe MONP growth and particle size distribution within PEM. X-ray diffraction (XRD) was used to analyze crystalline structure of MONP. Various spectroscopic techniques (NMR, XPS, EDS) were used to: assess chemical states of MONP within PEM, analyze permeation of CWS, uniformity of MONP distribution within PEM at various conditions.

Effect of nanofillers on properties of oilfield elastomers

Rostyslav Dolog, Valery Khabashesku

Baker Hughes 14990 Yorktown Plaza Dr., Houston, Texas, 77040, USA

ABSTRACT

This study describes how carbon nanotubes (CNTs) affect mechanical, thermal, and chemical properties of oilfield elastomers. Rubber compounds were investigated based on a custom peroxide-cured hydrogenated nitrile butadiene rubber (HNBR) recipe with various loading levels of carbon black (0–50 phr) and carbon nanotubes (0–15 phr).

Tensile properties of nanocomposites were assessed at room temperature and 177° C (350° F), and were highly dependent on the loading level of CNTs. Testing results indicated that the tensile strength and modulus of elastomers could be improved by combining CNTs and carbon black in one system, due to the synergy effect of two fillers with different particle sizes. Scanning electron microscopy (SEM) analysis was used to explore the quality of nanoparticles dispersion within a polymer matrix, the key issue in the preparation of elastomer nanocomposites.

Chemical properties and thermal properties of compounds attributed to the elastomer matrix, such as the glass transition temperature and decomposition temperatures, did not notably change due to incorporation of CNTs. However, nanocomposites demonstrated higher thermal conductivity and lower thermal expansion, compared to the original compounds. Furthermore, nanotubes affected the resistance of elastomers to explosive decompression and heat buildup under dynamic loads.

This research comprises a detailed examination of the effect of nanofillers on the properties of elastomers in light of oilfield applications.

KEYWORDS: Elastomers, carbon nanotubes, composites

INTRODUCTION

The oil and gas sector continuously raises requirements for elastomers used across the industry, imposing higher standards on the chemical resistance of elastomers and their mechanical properties at application temperatures. To meet the growing demand for high-performance rubbers in sealing elements, stators, and other critical applications, new elastomer formulations and chemistries are developed to pursue continuous improvement of properties. [1]The most common elastomers used in the oilfield tools are nitrile butadiene rubber (NBR), hydrogenated nitrile butadiene rubber (HNBR), ethylene propylene diene monomer (EPDM), fluoroelastomers (FKM), tetrafluoro-ethylene/propylene rubbers (FEPM), and perflouroelastomers (FFKM). Typical rubber compound recipes comprise raw elastomers, a curing system, reinforcing fillers and additives that facilitate processing or improve aging characteristics.

The performance of a cured rubber compound is usually adjusted by changing the recipe because each component is vital for achieving target properties. Mechanical properties of rubbers are greatly affected by the filler size and its aspect ratio, the quality of the filler dispersion, and the interface between the elastomer matrix and filler. The quality of the interface and the availability of chemical of physical bonds between the matrix and reinforcing filler define the efficiency of the load transfer to the filler.

[1] Elastomers for fluid containment in offshore oil and gas production: Guidelines and review. Health and Safety Executive 2005

Multiple studies have been conducted on the preparation, processing, and properties of elastomer compounds with nanofillers. However, most of those studies focused specifically on mechanical or electrical properties of such composites.[2] The goal of this paper is to present a complete picture of how CNTs affect properties that are relevant to oilfield applications. A custom peroxide-cured HNBR formulation, commonly used in oilfield applications, was selected as a base recipe for this research, due its relevance in the industry. This research is an important step towards the commercialization of elastomer products developed via nanotechnology. Research results help in understanding how rubber materials will perform in oilfield environmental conditions and provide an overview of advantages and disadvantages of nanofillers, compared to traditional fillers, such as carbon black and silica, which limit further improvement of elastomer technology.

Furthermore, our goal was to test the hypothesis[3] that combination of fillers of different size and structure, e.g., carbon black and CNTs, can create a synergistic effect, enabling significant improvements in matrix reinforcement and balancing the relatively high cost of CNTs compared to carbon black.

EXPERIMENTAL SECTION

Rubber compounds, coded as TX (X=1,2,...8), are elastomers based on a custom HNBR formulation with a peroxide curing system, featuring carbon black and/or CNTs as fillers. All these compounds were mixed in a C.W. Brabender internal mixer for 20 minutes (excluding the feeding time) at a 30-RPM rotor speed. A heating unit was not used. Because of the high shear stresses, however, the temperature has gradually been increased to 65–85° C, depending on the mixture viscosity. Each compound was molded into sheets and cured at 171° C (340° F) for 30 minutes in a hot press. Tensile properties of rubber samples were tested using dog bone specimens, according to ISO37 standard at room temperature and at 177° C (350° F), using an MTS Criterion Model 43 testing system. Five specimens per sample were pulled at a speed of 20 in./min. Strain was recorded using an extensometer, and a video extensometer system was used for high-temperature experiments.

Swelling tests in 3% KCl brine and LVT 200 oil were performed to assess the effect of fillers on the chemical resistance of elastomers. Standard one-inch rubber buttons were tested at 93° C (200° F) at ambient pressure for 10 days.

Thermal properties were studied using TGA Q50 and DSC Q2000 instruments. During thermo-gravimetrical analysis samples were heated at rate of 10° C in a nitrogen environment up to 600° C. Heating continued at the same rate to 800° C in an ambient atmosphere to burn off all organic material. The DSC experiment was conducted as follows: a sample was equilibrated at -75° C, and then heated during first heating scan to 150° C, followed by cooling to -50° C and a second heating scan to 150° C. The rate of temperature ramp was kept constant at 10° C/min.

The thermal expansion coefficient of rubber compounds was measured using a BAHR DIL806 dilatometer from TA Instruments between room temperature and 160° C. Samples were gradually heated at a rate of 2° C/min. and any change in sample length was recorded by the instrument. Thermal conductivity of elastomer composites was evaluated at room temperature with a C-Therm TCi Thermal Conductivity Analyzer by testing 0.08-in. thick elastomer samples using an in-built testing procedure for polymeric materials.

A Field Emission Scanning Electron Microscope JSM-7800F was used to analyze the quality of carbon nanotubes dispersion in elastomer matrix.

[2] Multiwall carbon nanotube elastomeric composites: A review, *Liliane Bokobza*, Volume 48, Issue 17, 10 August 2007, Pages 4907–4920.

[3] Synergy in hybrid polymer/nanocarbon composites. A review, *Urszula Szeluga et al.*, Composites: Part A 73 (2015) 204–231.

RESULTS AND DISCUSSION

HNBR formulations with peroxide curative system were prepared via melt mixing in a C.W. Brabender. A relatively long mixing time was selected based on the results of preliminary work, to ensure the dispersion of nanoparticles. All compounds were based on the same raw HNBR elastomer and featured identical loadings of peroxide, coagent, antioxidant, and plasticizer with respect to the base elastomer. The concentration of fillers, such as carbon black, CNTs and functionalized CNTs, varied from sample to sample.

Mechanical Properties

Temperature inside the C.W. Brabender mixing chamber and torque, which was applied to rotors to maintain a constant mixing speed of 30 RPM, were recorded and analyzed. An example of the mixing log from the Brabender is shown in the Figure 1. Components of the rubber formulation were carefully added to the mixing chamber. The total feeding time averaged at 5 minutes. Formulation was then mixed for the next 20 minutes. The sharp decrease in the torque at ~fourth minute was due to the addition of oil (plasticizer). The torque was gradually decreased from the seventh minute until the end of the experiment because the viscosity of the compound mixture decreased as temperature inside the chamber increased due to high shear. There was no reason to further increase mixing time. Possible improvements in dispersion of reinforcing nanofiller could not fully compensate for the negative effects caused by the reduction in polymer molecular weight due to chain scission, which can occur during processing step.

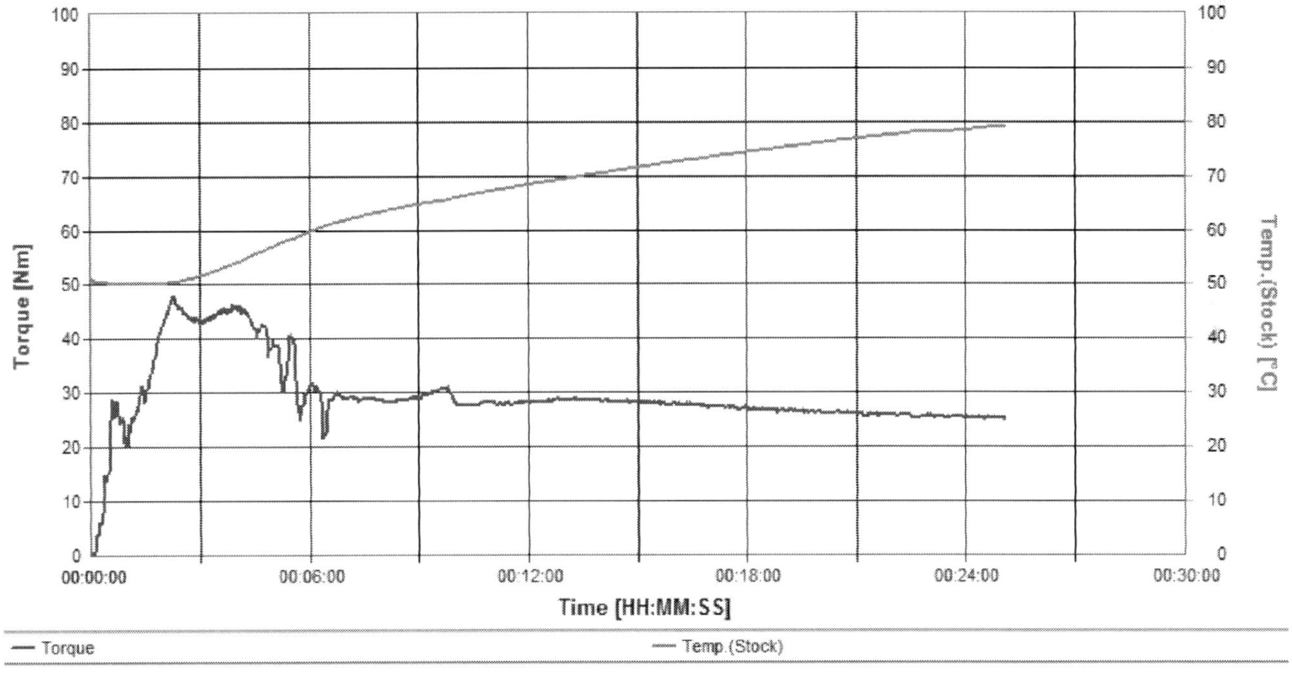

Figure 1. Mixing log of the T6 compound

Several observations can be made from obtained tensile data summarized in Table 1 and Figure 2. Incorporation of CNTs to the rubber compound significantly enhanced the modulus of the elastomer. The reinforcement effect of 5 phr of nanofiller was comparable to the reinforcement effect of 50 phr of carbon black until strain reached 50%. However, at high elongations carbon black proved to be a more effective filler. From comparison of samples T4, T5 and T6, incorporation of CNTs and carbon black to a rubber formulation enabled a synergistic effect in the modulus increase. Shore A hardness of the compound T6, which featured a combination of reinforcing fillers, was 8 points higher than that of compound T5, which contained only carbon black. However, adding CNTs decreased elongation at the break, which also negatively affected the ultimate tensile strength. It is worth mentioning that these compounds were developed to study the effects of the nanofillers on properties of oilfield elastomers, they have not been tuned for performance in high demanding applications.

Table 1. Filler composition, tensile properties and hardness of compounds at ambient conditions

Sample	CB (phr)	CNT (phr)	E_{25} (psi)	E_{50} (psi)	E_{100} (psi)	Tensile strength (psi)	Elongation (%)	Hardness Shore A
T2	0	0	70	96	120	856	472	48
T3	0	1	86	121	167	487	315	53
T4	0	5	168	256	403	717	188	64
T5	50	0	163	270	755	3531	293	68
T6	50	5	337	568	1352	2991	197	76

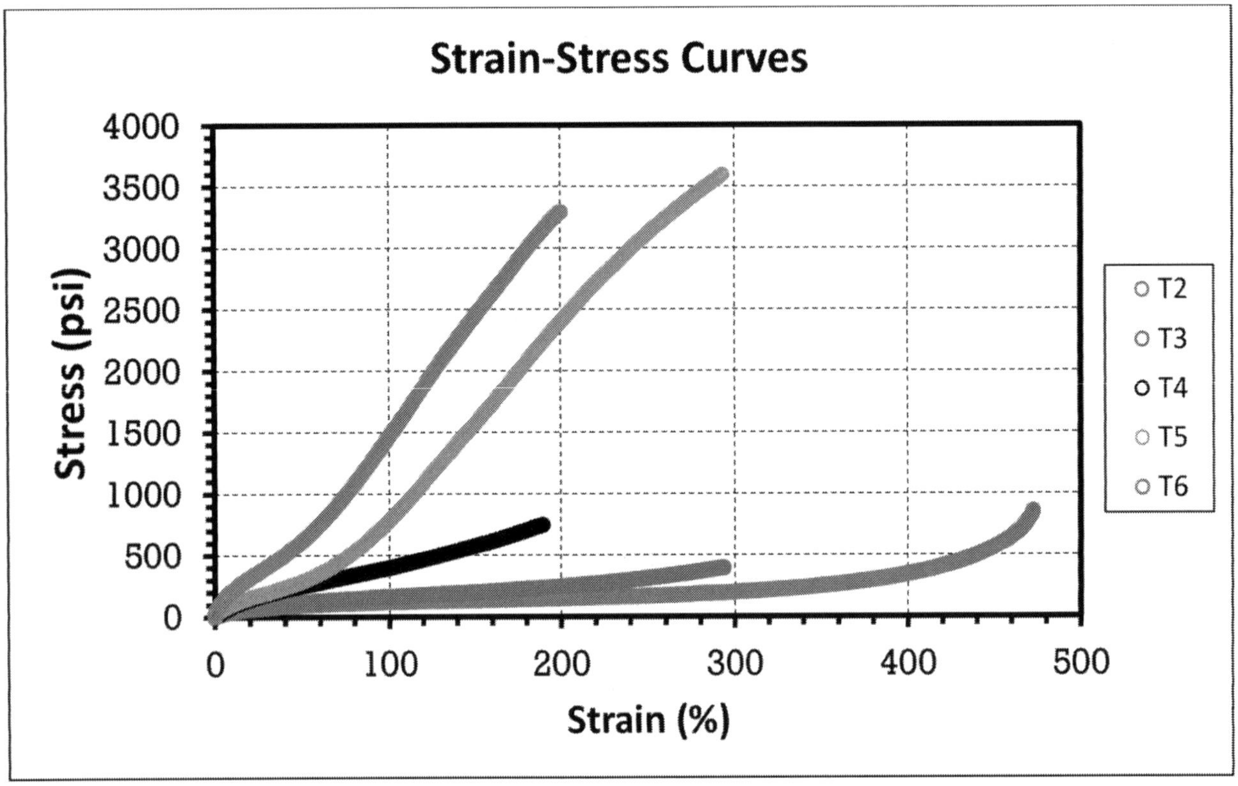

Figure 2. Examples of strain-stress curves at room temperature

Tensile properties at high temperatures are essential for elastomers in the oil and gas industry. The choice of a rubber compound for a specific application is defined by at least three factors: heat and pressure resistance, chemical resistance, and price. Tensile tests conducted at elevated temperatures enable evaluation of the performance of the rubber compound at conditions that resemble downhole environment in terms of temperature. High-temperature tests were conducted at 177° C (350° F). The selected temperature was at the upper limit for testing of HNBR compounds because elastomers are usually tested at the edge of their service temperature range. At least five specimens were tested per compound to obtain the presented data.

Table 2. Summary of tensile properties at 177° C (350° F)

Sample	E_5 (psi)	E_{10} (psi)	E_{15} (psi)	E_{20} (psi)	Tensile strength (psi)	Elongation (%)
T2	31	46			132	29
T3	41	58			108	15
T4	77	138			155	12
T5	61	130	194	261	832	44
T6	95	179	266	359	600	30

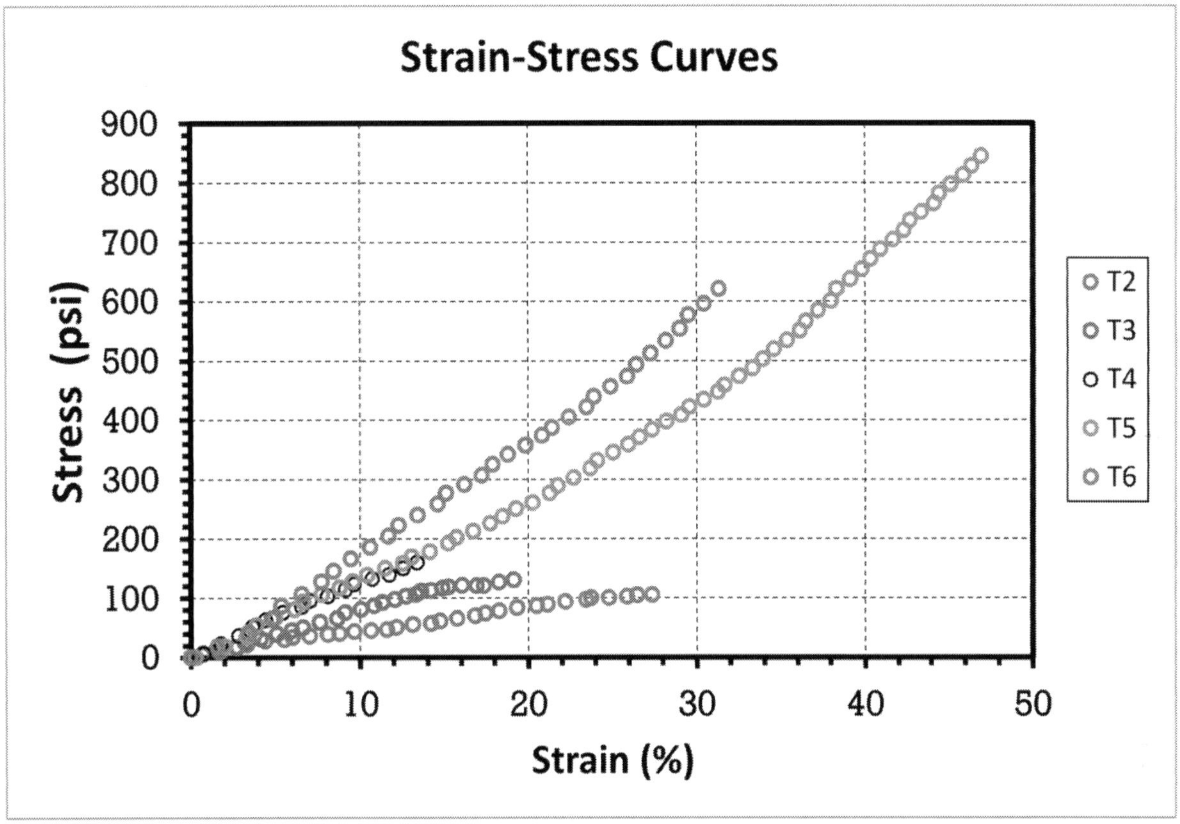

Figure 3. Examples of strain-stress curves at 177° C (350° F)

The summary of the test results, including modulus (e.g., stress at strain), tensile strength, and elongation at break, is provided in Table 2. In addition, one of the tests on each compound was plotted on a strain-stress curve. The first observation is that the general data trends from the high-temperature tensile test closely resemble the results of tensile test conducted at room temperature. However, at high temperature, rubber specimens break at much lower elongation, leading to a decrease in the measured tensile strength. This behavior is typical for elastomers and limits their application to a certain maximum service temperature. The modulus at 15% and 20% strain is not provided for sample T2-T4 because several specimen were broken before they reached such strains. However, it is possible get a general idea of their performance in comparison with other samples from Figure 3.

The main conclusions from the high-temperature (HT) experiment are the following:

1. At higher strains (over ~15% for HT tensile test) compounds with carbon black tend to demonstrate much higher modulus and overall tensile strength than samples reinforced with CNTs. The

difference in the behavior can be explained by the low affinity of not functionalized CNTs to the polymer. At higher strains, CNTs simply "disentangle" or "detach" from the matrix. Carbon black, however, has various functional groups and high structure surface that keep it connected to the matrix at higher strains.

2. Samples, which featured CNTs, tend to have lower elongation at break, even at small loadings, hence tensile strength also decreases.
3. A combination of CNTs and carbon black provides a certain synergy of properties. These compounds exhibit lower elongation at break and tensile strength, but they demonstrate a strong gain in modulus.

Formulations with high concentration of CNTs

Compounds P8 and P9 were prepared with a high loading of CNTs. These samples were mixed for 10 minutes (including feeding time) in a Brabender at 60 RPM and cured at 171° C (340 ° F) in a hot press for 30 minutes. The mixing procedure was modified based on the previous findings. These results are provided to demonstrate the potential of reinforcement by nanotubes, and should not be directly compared to the data obtained from testing of T series samples.

Table 3. Filler composition, tensile properties and hardness of compounds at ambient conditions

Sample	CNT (phr)	CB (phr)	E_{25} (psi)	E_{50} (psi)	E_{100} (psi)	Tensile strength (psi)	Elongation (%)	Hardness (Shore A)
P8	15	0	834	1594	2521	3029	161	77
P9	15	50	1002	1985	3586	3927	125	88

Results of tensile tests, at room temperature and at 163° C (325° F) are provided in Tables 3 and 4, respectively. Sample P8 and P9 demonstrated good overall performance, the presence of synergistic effect is visible from the tensile data, especially in high temperature experiment. Combination of 15 phr (~9 wt.%) of nanotubes with carbon black filler lead to all around properties improvement, when compared to the compound which only featured 15 phr nanotubes as a filler.

Table 4. Summary of tensile properties at 163° C (325° F)

Sample	CNT (phr)	CB (phr)	E_{25} (psi)	E_{50} (psi)	Tensile strength (psi)	Elongation (%)	Hardness (Shore A)
P8	15	0	460		847	44	77
P9	15	50	519	1169	1421	59	88

In summary, our results indicate that combination of high loadings of nanotubes (over 10 phr) and carbon black is more advantageous than using CNTs as a sole filler. Low loading levels of CNTs do not seem to be able to provide reinforcement, which would be sufficient for high demanding applications.

Thermal properties

Thermal properties of samples were assessed with DSC and TGA analysis. TGA enables evaluation of the thermal stability of the samples and re-evaluation of the actual composition of the samples, to account for possible losses of components during mixing. The results of the TGA analysis are summarized in the Table 5, and the example of the TGA curve is shown on Figure 4. Samples were heated at 10° C/min. to 600° C in N_2 environment and then subsequently heated at 10° C/min. to 600° C in air to burn off the carbon-based fillers.

Figure 4 illustrates an initial degradation of the polymer that started at 200° C and progressed to complete decomposition of aliphatic chains in the temperature range of 420-500° C with a corresponding DTA peak at ~460° C. The second rapid mass loss after 600° C was due to the burning of all remaining carbon material, including fillers, such as carbon black or nanotubes.

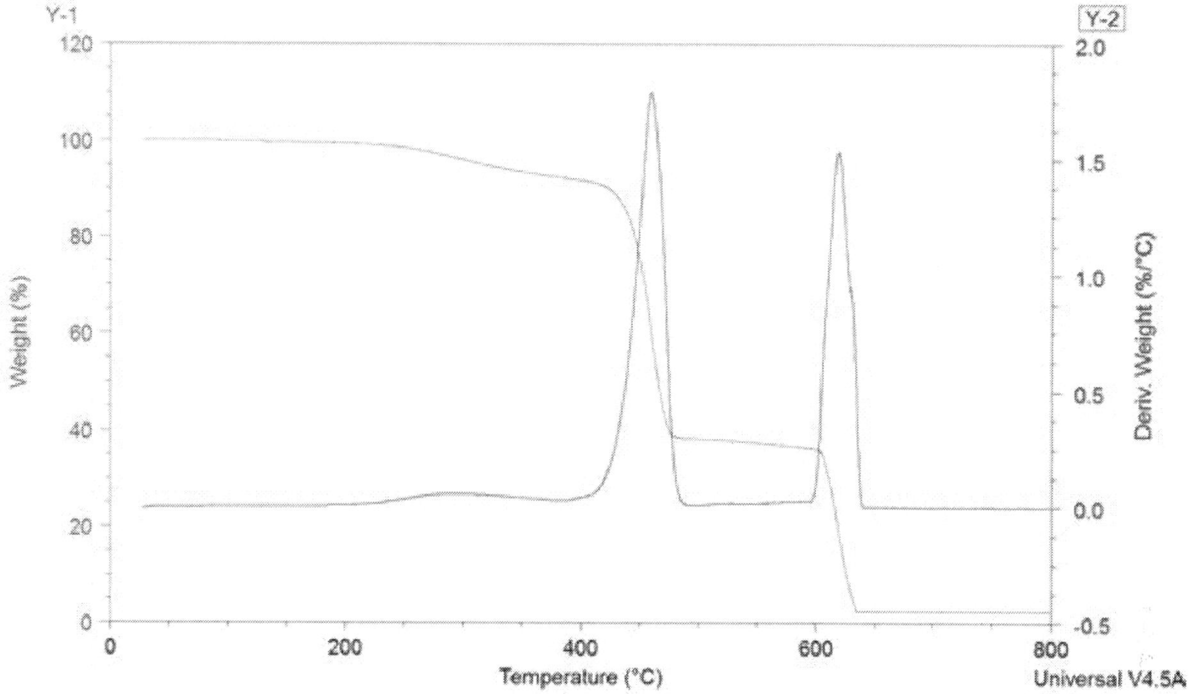

Figure 4. TGA/DTA curves for T6

There is an obvious increase of $T_{deg}1$ with the increasing concentration of CNTs for samples T2-T6. Due to the high thermal conductivity of CNTs they enable dispersion of the heat, local overheating, thus improving the thermal stability of the polymers. There is also an apparent higher ash content at 600° C in samples containing CNTs.

DSC experiments were conducted to determine whether there is a close interaction between the polymer matrix and nanofiller, in which case the glass transition (T_g) of the elastomer would be altered. A typical DSC curve (sample T6) is shown in Figure 5.

Table 5. TGA/DTA results

Sample	W_{600} (%)	W_{800} (%)	$T_{deg}1$	$T_{deg}2$
T2	2.4	2.4	452.8	-
T3	4.1	2.6	459.5	604.5
T4	8.1	3.3	467.6	606.6
T5	29.8	2.7	458.7	616.5
T6	36.3	2.6	459.5	618.3

The results of the DSC test were corrected using a melting point of an indium sample as a reference and are presented in Table 6. Apparently, at investigated level of filler loading and dispersion efficiency, the interaction between the filler and elastomer matrix was not strong enough to significantly affect the T_g of the elastomer.

The thermal expansion coefficient of rubber compounds was measured on 0.5-in. rubber cylinders that were heated to 160+° C at a rate of 2° C/min. and the change in sample length was recorded by the instrument. The values of thermal expansion coefficient were generated by the instrument's software for the temperature range ~30° C to 160° C and are presented in Table 7.

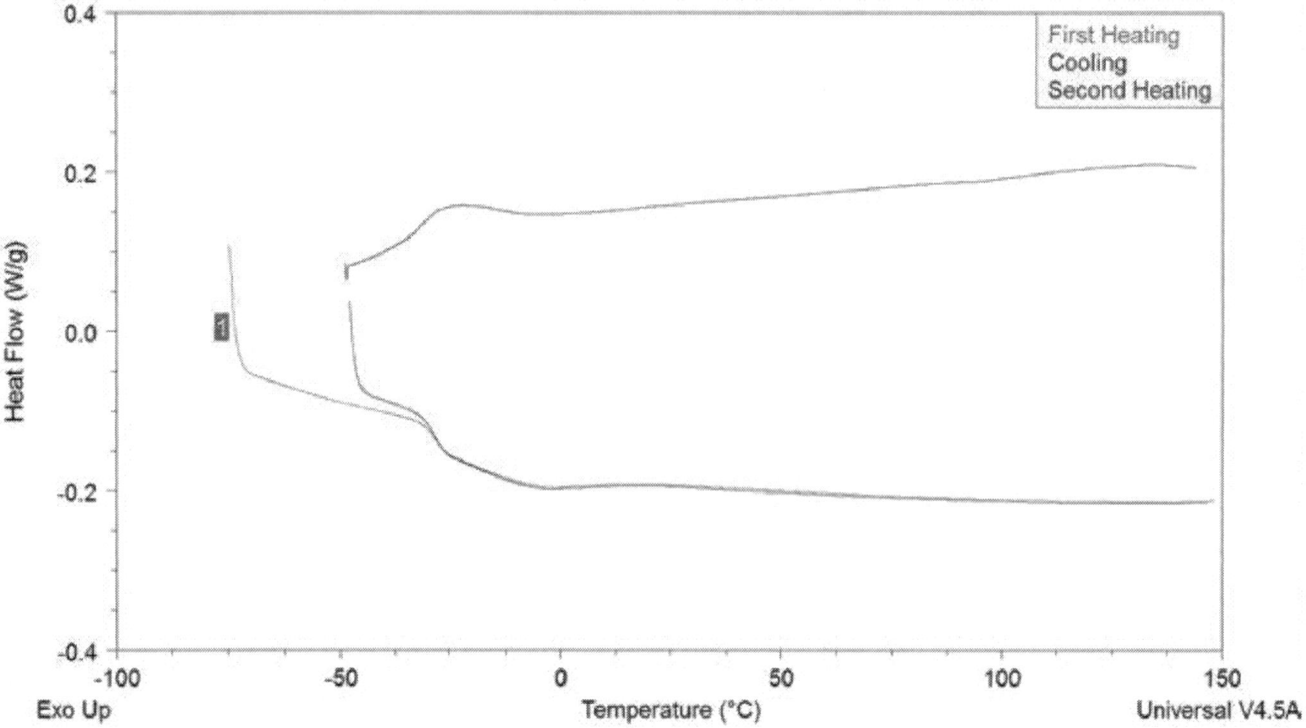

Figure 5. DSC curves of T6

Table 6. DSC results

Sample	T_g1	T_g2
T2	-29.4	-29.4
T3	-29.9	-29.2
T4	-29.2	-29.3
T5	-29.4	-29.8
T6	-29.2	-29.3

Table 7. Heat expansion coefficient of nanocomposites

Sample ID	α (10^{-6} K^{-1})
T2	205
T3	202
T4	225
T5	182
T6	135

The obtained thermal expansion coefficient is a fairly typical value observed for rubber compounds (~10^{-4} K^{-1}). A decrease in thermal expansion occurs with the increase of the filler loading. This effect is also expected due to the low heat expansion of carbon black and change in free volume with the addition of nanofiller. An increase in the modulus with rising filler concentration is another reason for the observed trend, because materials modulus is generally inversely proportional to the heat expansion coefficient.

Thermal conductivity of the elastomer composites was measured with a C-Therm TCi Thermal Conductivity Analyzer that employs the Modified Transient Plane Source (MTPS) technique in measuring the thermal conductivity and effusivity of material. The results are provided in Table 8. Thermal conductivity was significantly affected by the concentration of nanofiller. For example, incorporating only ~3wt% (5phr) of CNTs improved thermal conductivity by more than 50%, compared to the unfilled elastomer, and matched the thermal conductivity of a sample T5, which contained 50 phr of carbon black.

Table 8. Thermal conductivity measurements

Sample ID	Average λ, W/(m*K)
T2	0.292
T3	0.355
T4	0.444
T5	0.423
T6	0.528

Thermal conductivity of the elastomer is important because it is related to the thermal stability in real-life applications, especially in dynamic seals. High thermal conductivity can allow heat dissipation from local overheated spots in the elastomer that degrade first, thus, prolonging seal life in high-temperature dynamic applications.

A flexometer test is common for rubber industry and aims to analyze heat buildup inside elastomer compounds under dynamic loads, mainly due to the shear on the interface of blend components. Heat buildup is known to increase with increasing filler concentration and total surface area. Based on preliminary testing, we can conclude that heat buildup increases with the increasing filler content. However the materials' modulus grows faster than the heat buildup. In addition, CNTs also improve heat conductivity. Therefore, it seems that nanomaterials can be used to reduce damage in elastomers due to overheating.

Scanning electron microscopy

Sample T3 was examined with a SEM to evaluate the quality of the CNTs dispersion in an elastomer matrix in the compound containing 1 phr of CNTs as filler.

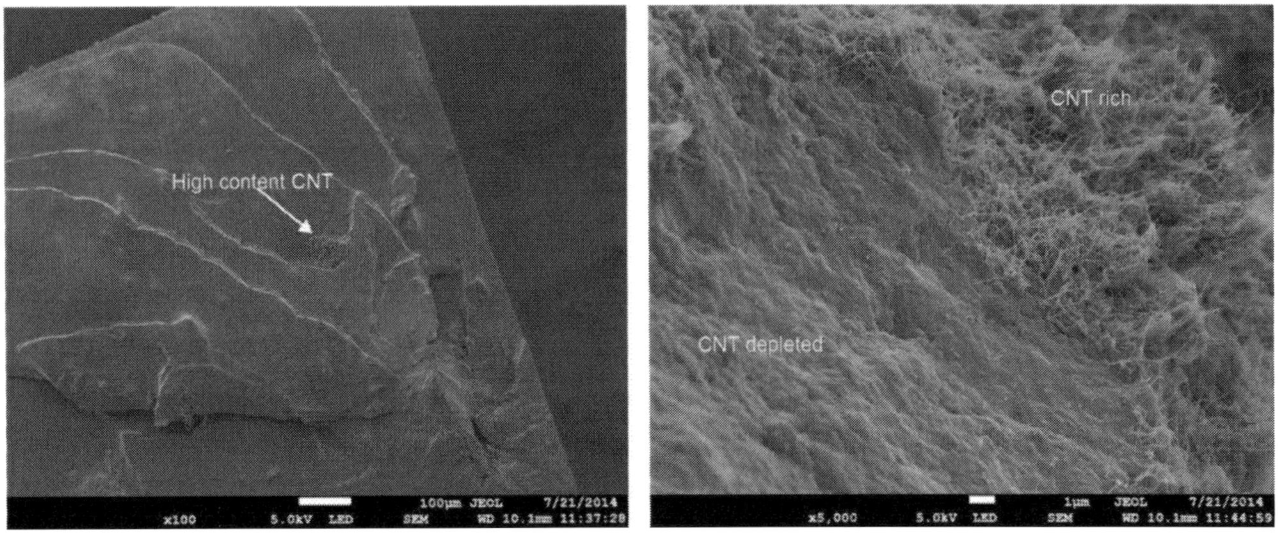

Figure 8. Large agglomerates of CNTs

The sample was prepared by fracturing a rubber band, then freezing it in liquid nitrogen, below its glass transition. The fracture surface was then coated with gold using a sputter coater and examined under the SEM. The conditions of the SEM analysis are provided on the respective images. Large agglomerates of CNTs have been detected, indicating insufficient mixing. Figure 8 shows large CNTs agglomerate at different magnifications. However, some nanotubes dispersed in the elastomer phase, providing the observed reinforcement effect, see Figure 9 (two different magnifications).

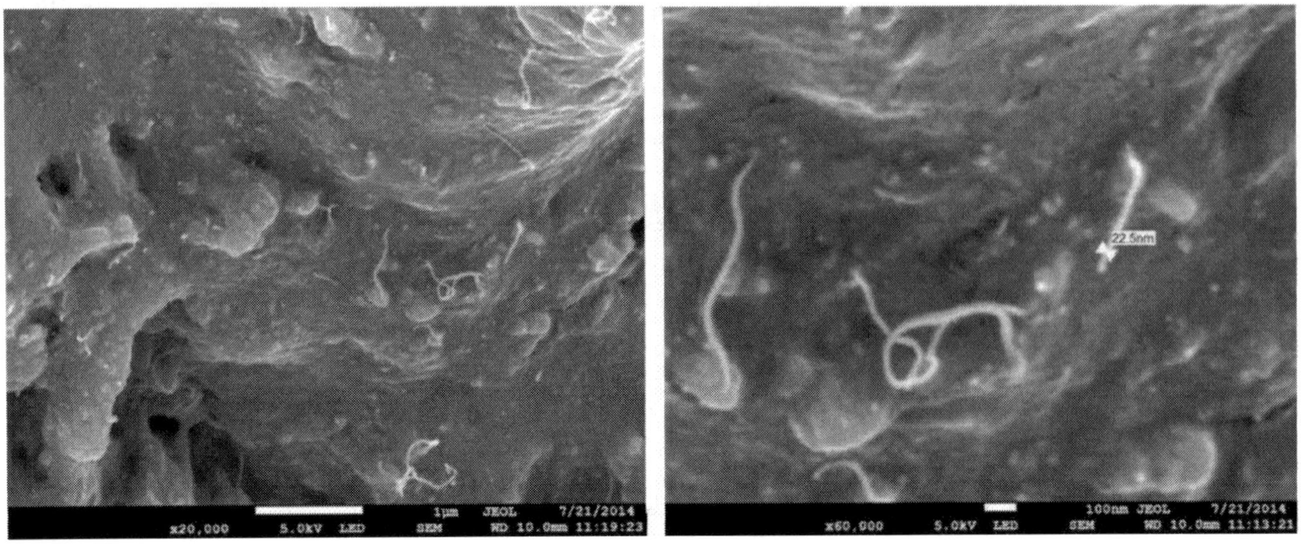

Figure 9. Nanotubes dispersed in the elastomer phase

Chemical resistance

Rubber compounds were swollen in 3% KCl Brine (a standard water-based fluid in the industry) and LVT200 (a standard oil-based fluid in the industry) in the oven at 93° C (200° F). The summary of the results is presented in Tables 9 and 10.

Table 9. Swelling test results (% volume increase) for samples tested in 3% KCl brine and initial density of the samples

Sample \ Time (h)	0	8	24	144	Specific Gravity
T2	0.0	0.7	0.7	0.6	0.96
T3	0.0	0.4	0.4	1.9	0.97
T4	0.0	0.5	0.4	0.8	0.99
T5	0.0	0.5	0.5	0.3	1.13
T6	0.0	0.5	0.3	0.7	1.14

Table 10. Swelling test results (% volume increase) for samples tested in LVT200

Sample \ Time (h	0	8	24	144
T2	0.0	6.6	10.3	13.0
T3	0.0	6.0	9.3	11.1
T4	0.0	5.3	8.4	10.7
T5	0.0	4.9	8.1	11.6
T6	0.0	3.7	5.9	6.1

There was no significant difference in the swelling from between samples T2-T5, which is expected, because chemical resistance is generally a function of the properties of elastomer matrix. Sample

T6 demonstrated much lower swelling in LVT200, compared to the rest of the sample group. This is believed to be due to higher strength and modulus of the sample. During swelling of elastomer, fluid continues to penetrate the matrix until it is thermodynamically favorable.

Explosive decompression

Explosive decompression, also known as a rapid gas decompression (RGD) test, simulates the field conditions that O-rings and other seals often encounter when oil tools are removed from the pressurized downhole environment to the ambient conditions at the surface. The test was conducted following the ISO guidelines (ISO/FDIS 23936-2:2011 (E), Annex B) for O-ring compression and pressure cycles. However, nitrogen gas was used instead of the ISO-recommended 10% CO_2 or H_2S gas mixture. Due to the substitution of recommended gas mixture with nitrogen testing conditions were significantly milder.

#204 O-rings made out of tested compounds were compressed to 15% strain and tested at 100° C in nitrogen atmosphere under ~ 1,000 psi pressure. A series of samples were soaked in a pressure vessel for 48 hours and then pressurized and depressurized once every day for 4 days following standard ISO procedures. When the RGD test concluded, all samples were depressurized, cooled down and removed from the vessel.

Figure 10. The most-damaged cross-section of O-rings from the second test

Table 7. Length of the longest crack in the specimen (in mm).

Sample	Cut 1	Cut 2	Cut 3	Cut 4	Assigned ISO Rating
T4	1.473	1.25	0.825	0.45	4
T5	1.179	0.875	0.272	0.11	2
T6	1.116	0.977	0.146	0.136	2
HNBR-70	1.375	1.359	0.175	0	3

Each O-ring was cut into 4 parts, and the cross-sectional surfaces were closely examined to choose the most-damaged specimen. That specimen was further studied under optical microscope to investigate for possible blisters, holes, and cracks. The length of the largest crack in the specimen was measured and the values are provided in Table 11. The pictures of the most-damaged cross-sections of each O-ring are shown on Figure 10.

All samples were evaluated according to the ISO damage rating system, which is based on the number, length and type of cracks present in the O-ring after testing (e.g., systems with assigned rating 0–3 were considered to pass the test, and assigned ratings 4 and 5 signifies that samples failed the test). Samples T2 (image and measurements are not shown here) and T4 failed due to the large number of cracks, with an overall length that exceeded the O-ring cross-section diameter by more than 3 times. Consequently, they were assigned rating 4. The RGD resistance of the compounds T5 and T6 was better than that of commercial HNBR-70. The HNBR-70 compound was chosen for comparison because its hardness is similar to the one of T5 and T6 and it is known that RGD resistance often scales with hardness of the rubbers.

Acknowledgements

I would like to thank Othon Monteiro (Baker Hughes), for conducting SEM experiment and assistance with its results interpretation.

Conclusions

Rubber compounds were prepared from a custom peroxide-cured hydrogenated nitrile butadiene rubber (HNBR) recipe, common to the oilfield industry, with various loading levels of carbon black (0–50 phr) and carbon nanotubes (0–15 phr). Mechanical, thermal, and chemical properties of elastomers nanocomposites were assessed. SEM revealed that melt mixing of investigated rubber compounds was not sufficient to fully disperse nanotubes in the polymer matrix, nonetheless, incorporation of nanotubes affected properties of the compounds.

Our results indicate that carbon nanotubes have significantly higher reinforcing potential, compared to the conventional carbon filler, however, the downside of the increased modulus is the decreased elongation and tensile strength. We suggest combining conventional fillers and carbon nanotubes in a single compound to achieve optimal results, due to demonstrated synergy of properties. This strategy enables improved properties at room and high-temperature environments, as well as improved RGD resistance of the compounds.

Chemical properties and thermal properties of compounds, their glass transition temperature and thermal stability (e.g., decomposition temperature) did not notably change due to incorporation of CNTs. However, nanotubes provided significant boost to thermal conductivity and decreased thermal expansion, compared to the original compounds. This can be advantageous in intensive dynamic applications in the oilfield.

468113 Characterization of Compressive Mechanical Properties of Hydrogenated Nitrile Butadiene Elastomers Reinforced with Three Different Types of Carbon Filler

Abstract:

A compression testing procedure, to understand the effect of filler morphology, concentration and surface chemistry, on the compressive strength, energy dissipation, and elastic recovery of elastomer nanocomposites, is developed.[1] Peroxide-cured hydrogenated nitrile butadiene rubber (HNBR) composites, consisting of carbon based fillers of different aspect ratios, were studied at temperatures in the range of 25 to 180 °C, and compressive stresses in the range of 500 to 1500 psi. Three different carbon based fillers, namely, polyacrylonitrile derived carbon nanofibers, graphite nanoplatelets, and carbon black, were investigated. A four-step force/compression profile allowed the determination of stiffness, stress relaxation, strain recovery, and permanent set in these elastomer composites. The three fillers differed greatly in terms of determining the modulus, fracture strength, and compression at break of the elastomer composite. A comparison with dynamic mechanical properties, and the tensile properties previously presented[2]will also be discussed.

1) Sankarasubramanian, M., "Filler shape and surface chemistry effects on compressive strength, energy dissipation, and elastic recovery of HNBR elastomer nanocomposites," Ph.D. Dissertation, Clarkson University, Potsdam, NY (May 2016).

2) Krishnan, S., Sankarasubramanian, M., Putnam, Z. A., and J. C. Moosbrugger, "Characterization of viscoelastic hysteresis and stress softening of filled-elastomers using cyclic loading experiments," presented at the 2015 AIChE Annual Meeting, Salt Lake City, UT (Nov. 8–13, 2015).

472807 Nafion Nanocomposite Membranes for Use in Prospective Large-Scale Energy Storage Devices

Abstract:

Due to their ability to suppress the crossover of vanadium ions, there is growing interest in the use of Nafion nanocomposite membranes as the ion conducting phase in vanadium redox flow batteries. However, little is known about the mechanism by which these nanoparticles reduce the crossover of undesired vanadium ions, an issue that reduces battery lifetime and efficiency. Furthermore, there is a limited understanding of how the surface chemistry of embedded nanoparticles, and in turn the interaction between the Nafion matrix and the nanoparticles, affects vanadium crossover in these membranes. This investigation seeks to elucidate how the presence of silica (SiO_2) nanoparticles in Nafion impacts both water/ion transport and polymer chain relaxation dynamics since these membrane properties are intimately coupled. Specifically, liquid water transport and chain relaxation dynamics in hydrated Nafion nanocomposites were characterized using *in situ* time-resolved Fourier transform infrared-attenuated total reflectance (FTIR-ATR) spectroscopy and neutron spin echo (NSE) spectroscopy, respectively. Non-Fickian diffusion was observed for the Nafion films, and the water uptake data were regressed to a diffusion-relaxation model to obtain both water diffusivity and water-induced polymer relaxation time constants. In addition to studies on neat (non-functionalized) SiO_2 nanoparticles, water and vanadium ion permeability, measured via ultra-violet visible spectroscopy, in Nafion-SiO_2 membranes were investigated for a variety of nanoparticle surface chemistries. Results from these experiments indicate that the surface chemistry of the SiO_2 nanoparticles impacts the vanadium ion permeability of the Nafion nanocomposite membranes.

Microstructural Characterization of GFRP Reinforcing Bars in Severe Environment

Hilal El-Hassan[a,*], *Abdelrahman Al-Sallamin*[a], *and Tamer El-Maaddawy*[a]

[a] *Department of Civil and Environmental Engineering, United Arab Emirates University, Al Ain, United Arab Emirates, P.O. Box 15551*

Abstract

Glass fiber-reinforced polymer (GFRP) bars are increasingly used as internal reinforcement in concrete structures because of their noncorrosive nature, lightweight, high strength, and ultraviolet resistance. The microstructure and composition changes of GFRP bars exposed to severe environmental conditions are investigated in this paper. Test specimens were placed in moist seawater-contaminated concrete for up to 15 months at temperatures of 20, 40, and 60°C. Microstructural analysis was performed using differential scanning calorimetry (DSC), Fourier transform infrared (FTIR) spectroscopy, scanning electron microscopy (SEM), and matrix digestion using nitric acid. A decrease in glass transition temperature (T_g) and matrix content of GFRP bars was recorded after conditioning. Conditioned GFRP bars exhibited matrix plasticization and chemical degradation. Findings from FTIR showed an increase in hydroxyl ions with conditioning at higher temperatures. Results of SEM indicated a separation at the fiber-matrix interface and matrix deterioration in conditioned specimens.

Keywords: microstructure, GFRP; concrete; severe environment

[*] Corresponding author. Tel: +971(56)350-4110; Fax: +971(3)713-5156

E-mail address: helhassan@uaeu.edu (Hilal El-Hassan)

1. Introduction

Of the many intrinsic characteristics of glass fiber reinforced polymer (GFRP) composites, corrosion resistance, high stiffness to weight ratio, and high strength to weight ratio are few properties that endorse GFRP as a replacement to steel reinforcement in reinforced concrete structures subject to harsh environments [1]. They are however susceptible to degradation in mechanical properties if subjected to alkaline, saline, or acidic solution, moisture/water, and elevated temperatures [2].

Several research studies were conducted to investigate the performance and microstructure of GFRP bars under different environmental conditions. Debaiky et al. studied the combined effect of immersing GFRP bars in water and alkaline solutions with sustained loading and elevated temperatures. A decrease in tensile strength was reported due to nonhomogeneous transfer of stress between adjacent fibers which was attributed to the delamination/debonding mechanism between fibers and matrix. While no chemical degradation of polymer matrix was detected, reversible plasticization was observed, leading to a lower glass transition temperature (T_g) [3].

In other cases, samples were preloaded up to 80% of the ultimate tensile strength (UTS) prior to conditioning. Results showed a decrease in T_g corresponding to plasticization. It seemed that the fiber, resin, and interface were unaffected by moisture and temperatures below 50°C [4]. Similarly, Dejke stated that glass fibers experienced more deterioration in the presence of water. Surrounding moisture acted as a plasticizer, significantly affecting the flexural strength and decreasing the T_g of resin [5]. At higher temperatures up to 350°C, decrease in tensile strength was associated with the thermal degradation of the polymer and matrix formation of microcracks. The conditioned GFRP bars experienced a 46°C drop in T_g [6].

Scanning electron microscope (SEM) imaging highlighted fiber debonding and deterioration of the fiber-matrix interface [7-10]. Upon conditioning, the polymer matrix and fibers were damaged progressively, resulting in voids in the microstructure and loss of tensile properties [11]. Fourier Transform Infrared (FTIR) spectroscopy had been also employed to study the degradation of polymer matrix through hydrolysis. While the carbon-hydrogen peak was not affected during conditioning, the OH band increased due to the formation of new hydroxyl groups. The ratio of OH to CH was used to quantify the changes in peak intensity [3, 4, 7, 12-14].

Numerous studies have addressed conditioning-induced changes in durability, mechanical and physical properties of GFRP bars immersed in alkaline or water solutions simulating concrete pore solutions. To better understand the behavior of GFRP bars in service and extend the lifespan of GFRP reinforced concrete structures, this paper investigates the microstructural changes in GFRP bars imbedded in seawater-contaminated concrete. Samples were exposed to elevated temperatures for durations up to 15 months. SEM, FTIR and DSC (differential scanning calorimetry) were used to characterize the development in GFRP microstructure. Matrix digestion was also employed to measure weight loss in the polymer matrix subsequent to conditioning.

2. Experimental Investigations

2.1. Test specimens and conditioning

Glass fiber reinforced polymer (GFRP) bars utilized in this study were made from high strength glass fibers with epoxy resin. The mass fraction of glass was 75.5% and was determined by matrix digestion using nitric acid in accordance with ASTM D3171 [15]. Thermogravimetry analysis (TGA) was also used following ASTM E1868 and resulted in a fiber content of 75% by mass [16]. The bars had inner and outer diameters of 8 and 9 mm, respectively and were 1200 mm long. To represent actual field conditions, test samples were surrounded by seawater-contaminated concrete at the middle third portion with a cross-section of 50 x 50 mm. The mixture proportions of the concrete are shown in Table 1. Ordinary Portland cement (Type I) was used alongside natural crushed stone coarse aggregates. Fine aggregates including natural crushed stone and dune sand were used (1:1 ratio by mass). The mixing water was seawater obtained from the Arabian Gulf. Prior to conditioning, polyvinyl chloride (PVC) pipes were installed along the exposed parts of the GFRP bars then blocked/sealed carefully at each end to protect these regions during conditioning.

The test specimens were immersed in temperature-controlled tap water baths for 5, 10, and 15 months. The conditioning temperature varied between 20, 40 and 60°C. The controlled reference was kept at ambient conditions for comparison with conditioned samples.

Table 1: Surrounding concrete mix proportions

Unit Weight (kg/m^3)				
Cement	Fine Aggregate	Coarse Aggregate	Seawater	w/c
400	580	1160	200	0.5

2.2. Microstructure evaluation

Microstructure and morphological changes in GFRP were examined using a JEOL-JSM 6390A scanning electron microscope (SEM). Specimens were cut, polished, and coated with a thin gold layer to ensure conductivity. The obtained micrographs were used to identify degradation in polymer and/or fibers.

Matrix digestion using nitric acid was utilized to evaluate fiber and matrix content of GFRP composite bars, according to ASTM D3171 [15]. The chemical reaction of GFRP bars with nitric acid results in digestion and dissolution of the epoxy resin. Powder samples of 0.5–1.5 g were collected and mixed with 50 mL of 70% aqueous nitric acid. The mixture was then placed in a controlled temperature bath for 6 hours at 80°C. The contents were filtered and washed with distilled water prior to drying in an oven for 1 hour at 100°C. Equation (1) expresses the matrix content in terms of mass measured:

$$\textit{Matrix content (mass, \%)} = \frac{\textit{Initial mass - Final mass}}{\textit{Initial mass}} \textit{ x 100\%} \qquad (1)$$

Fourier transform infrared (FTIR) spectroscopy was performed on GFRP samples to measure the degradation due to hydrolysis reaction. Specimens were interground with potassium bromide (KBr) and analyzed in transmittance mode from 400 to 4000 cm^{-1} at a resolution of 1 cm^{-1} using a Varian 3100 FT-IR spectrometer.

Powdered samples were examined by differential scanning calorimetry (DSC) using a DSC Q2000 calorimeter equipped with a refrigerated cooling system. Fifteen specimens, on the range of 10 mg, were sealed in aluminum pans and prepared for analysis. DSC curves were obtained by heating the samples from 20°C to 225°C at a heating rate of 5°C/min. The glass transition temperature (T_g) was determined according to ASTM E 1356 standard [17]. For each specimen, two scans were performed. The first scan was used to compare the T_g of conditioned samples to the unconditioned reference. A decrease in T_g was indicative of a plasticizing effect. The second scan was useful to identify the mechanism of degradation. If the T_g of conditioned sample, after the second scan, was in the same range as that of the reference, it indicated a reversible plasticizing effect due to the moisture absorption. However, a lower T_g compared to the reference signified an irreversible chemical degradation.

3. Experimental Results and Discussion

3.1. SEM Analysis

The longitudinal and cross-sectional micrographs of control and conditioned GFRP bar are shown in Figure 1 and 2, respectively. The effect of curing temperature on the microstructure was studied by altering the temperature while keeping the conditioning fixed at 15 months. Figure 1(a) of the unconditioned control sample identifies a rough surface of fibers with adequate bond to the resin. Specimens placed in water at 20°C for 15 days (Figure 1(b)) experienced crack formation and propagation along the length of the fiber-matrix interface. The increase in temperature to 40°C yielded a more visible opening in the GFRP microstructure as seen in Figure 1(c). Yet, it should be pointed out that the glass fiber remained unaffected by the water solution. The micrograph of samples immersed in water at 60°C is presented in Figure 1(d). Substantial disintegration of the matrix is noticed at the fiber-matrix interface with further crack development to a width of 6 μm. This degradation could result from crack propagation from the matrix to the interface as opposed to directly into the fiber [18]. Although the process did not significantly widen the crack, it produced a smoother fiber surface. It is an indication of matrix degradation due to hydrolysis reaction. In addition, deterioration of GFRP specimens was observed in cross-sectional micrographs of Figure 2. Although conditioning in water at 20°C initiated crack formation, the damage intensified at higher temperatures. Figure 2(d) shows that samples conditioned at 60°C experienced fiber debonding and matrix degradation. It is possible that placing GFRP bars above 40°C or 50°C could increase the thermal expansion coefficient in the transverse direction, which is influenced by resin expansion [19-21]. To study the effect on conditioning duration on GFRP morphology, samples subject to 40°C curing were compared at different ages as shown in Figure 3 and 4. It is evident from the micrographs that longer periods of conditioning could considerably affect the fiber-matrix interface and induce further crack propagation and widening.

3.2. Matrix Digestion Analysis

To evaluate the degradation mechanism and assess the conditioning-induced changes in fiber-matrix interface, matrix digestion by nitric acid was used. TGA was also employed, according to ASTM E1868, to validate the obtained results [16]. Table 2 shows the fiber and matrix content of GFRP bars after conditioning using matrix digestion. Longer conditioning and higher temperatures resulted in lower matrix content and a weaker fiber-matrix bond [8]. It is possible that the degraded matrix dissolved in the water solution or remained inside the concrete after conditioning. This is also evidenced by the micrographs of longitudinal GFRP sections (Figure 1) that showed a relatively clean surface after conditioning, indicating lesser matrix at the interface. A similar phenomenon was reported elsewhere with samples exhibiting a weight loss of 18% after a 2-hour exposure at 350°C [6]. Matrix digestion results are in good agreement with those obtained by TGA. The respective matrix content for the sample conditioned for 15 months at 60°C and the control were 16% and 25%.

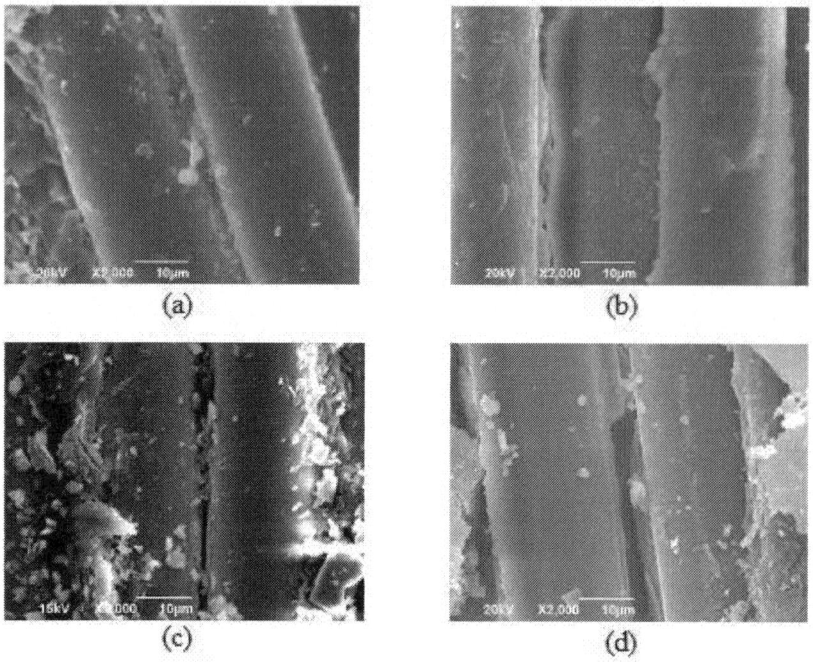

Figure 1: Longitudinal micrograph of GFRP bars: (a) control (b) immersed for 15 months at 20°C (c) immersed for 15 months at 40°C (d) immersed for 15 months at 60°C

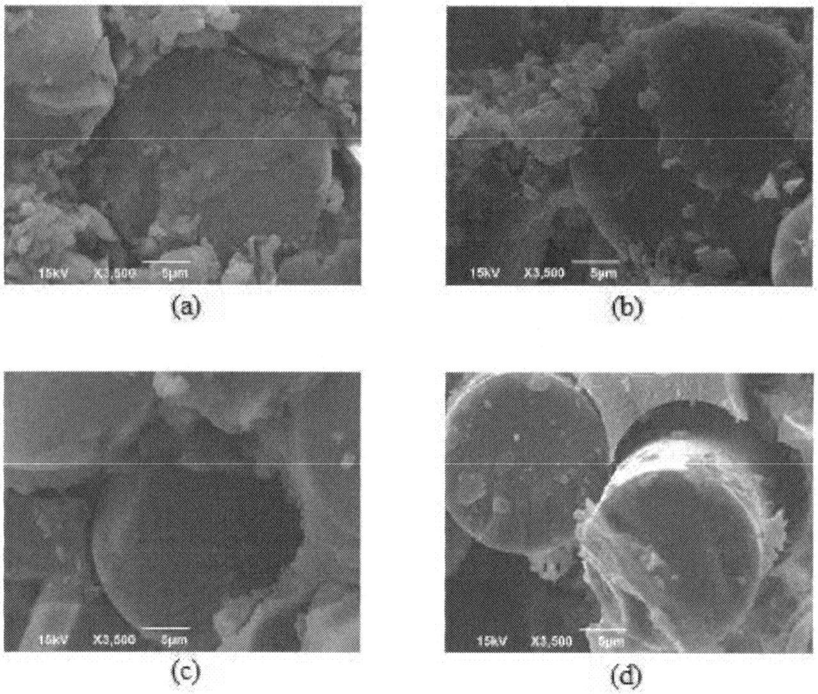

Figure 2: Cross-sectional micrograph of GFRP bars: (a) control (b) immersed for 15 months at 20°C (c) immersed for 15 months at 40°C (d) immersed for 15 months at 60°C

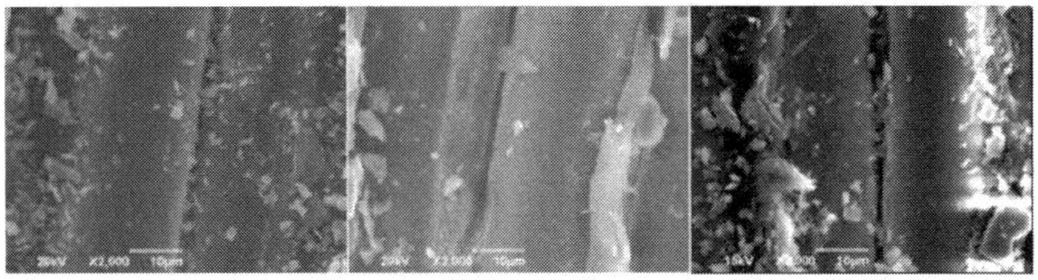

(a) (b) (c)

Figure 3: Longitudinal micrograph of GFRP bars: (a) immersed for 5 months at 40°C (b) immersed for 10 months at 40°C (c) immersed for 15 months at 40°C

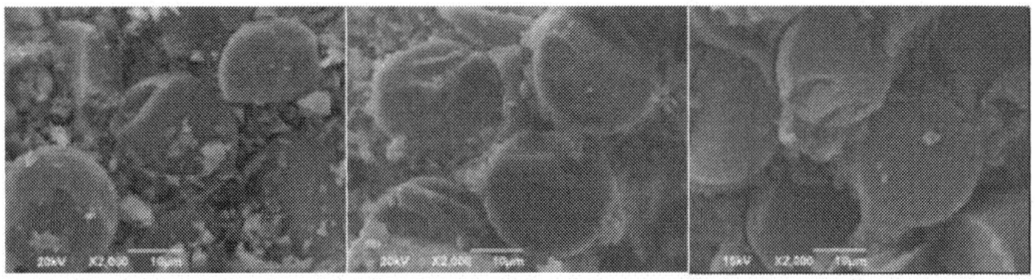

(a) (b) (c)

Figure 4: Cross-sectional micrograph of GFRP bars: (a) immersed for 5 months at 40°C (b) immersed for 10 months at 40°C (c) immersed for 15 months at 40°C

Table 2: Fiber and matrix content of GFRP bars subject to different conditioning

Conditioning		Content (%)		Matrix
Duration (months)	**Temp. (°C)**	**Fiber**	**Matrix**	**Retention (%)**
Control	Ambient	75.5	24.5	100.0
5	20	76.8	23.2	95.0
	40	79.4	20.6	84.3
	60	81.6	18.4	75.1
10	20	79.2	20.8	84.9
	40	82.3	17.7	72.2
	60	83.0	17.0	69.6
15	20	84.1	15.9	65.1
	40	84.1	15.9	64.9
	60	84.7	15.3	62.4

3.3. FTIR Analysis

FTIR spectra are used to evaluate the degree of hydrolysis reaction. In the process, two regions are studied: 2800–3000 cm^{-1}, representing carbon-hydrogen groups, and 3200–3600 cm^{-1}, representing hydroxyl groups. While hydrolysis leads to an increase in OH infrared band, it does not affect that of CH. The ratio of maximum peaks in each of OH and CH band characterizes the relative amount of hydroxyl groups in the specimen. Figure 5 is a plot of the FTIR spectra for samples conditioned for 10 months at different temperatures. The ratios for samples exposed to 20°C, 40°C and 60°C were 1.098, 1.121, and 1.146, respectively, compared to 0.586 for unconditioned counterparts. It is evident that the amount of hydroxyls had increased due to conditioning. Higher temperatures have slightly enhanced the reaction, associated with a 95% increase in hydroxyls compared to the reference. Ratios of other conditioned samples are presented in Table 3. A comparison between samples subject to 20°C for 5, 10, and 15 months showed no considerable changes in the OH/CH ratio. This implies that hydrolysis reaction did not significantly progress over time. When increasing the conditioning temperature for each duration, the OH/CH ratio did not greatly increase. It could be concluded that hydrolysis had initiated within the first 5 months and possibly plateaued thereafter.

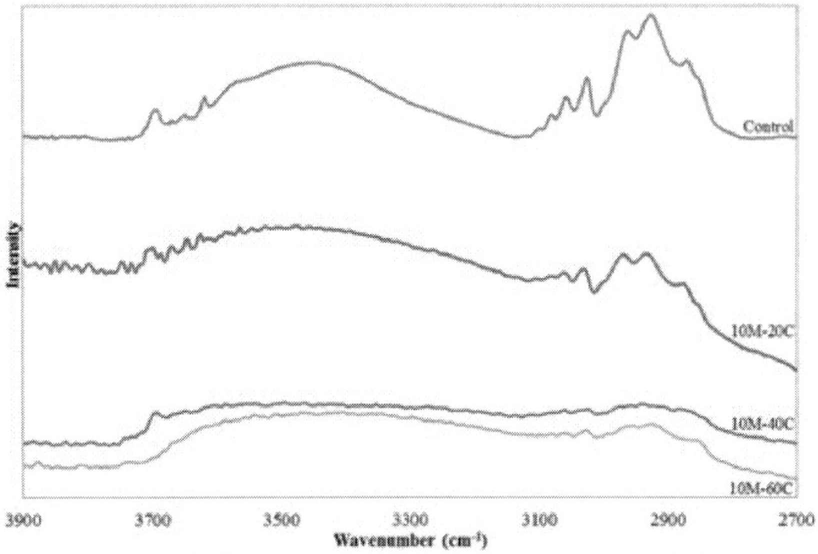

Figure 5: FTIR spectra of GFRP bars conditioned for 10 months at different temperatures

Table 3: Band ratios of conditioned and control samples

Conditioning		
Duration (months)	**Temp. (°C)**	**OH/CH**
Control	Ambient	0.586
5	20	1.096
5	40	1.119
5	60	1.139
10	20	1.098
10	40	1.121
10	60	1.146
15	20	1.113
15	40	1.147
15	60	1.153

3.4. DSC Analysis

The glass transition temperature (T_g) of control and conditioned GFRP samples are presented in Table 2. Two runs were conducted for each specimen. During the first run, a lower T_g was recorded for conditioned specimens compared to the reference. This was due to the plasticizing effect of water absorbed during conditioning which in turn reduced the T_g. As the sample was heated during the second run, the water evaporated and reversed the plasticizing effect. Thus, the T_g values were similar to those of the reference.

The influence of higher conditioning temperatures is studied. Samples subject to higher temperatures showed lower T_g after one scan. This may be explained by the degradation of the resin through plasticization. Evidenced by SEM imaging, higher temperatures amplified crack formation and propagation, leading to more voids in the microstructure. Such voids could increase the water absorption and thereby, reduce the T_g.

Another study involving duration of conditioning was furnished. After the first run, longer conditioning durations resulted in a lower T_g. However, after 10 months, no major change in thermal properties could be recorded by DSC. The plasticizing effect seemed to be more affected by exposure temperatures than durations subsequent to 10 months. Since the T_g of the second scan remained unchanged, the matrix did not undergo irreversible chemical degradation. It can be concluded that the deterioration mechanisms present in this study are degradation at the fiber-matrix interface, plasticization, hydrolysis and loss of the polymer matrix during conditioning.

Table 4: Glass transition temperature of GFRP bars using DSC analysis

Conditioning		**T_g (°C)**	
Duration (months)	**Temp. (°C)**	**1st Run**	**2nd Run**
Control	Ambient	125	125
5	20	115	126
5	40	105	126
5	60	98	126
10	20	100	125
10	40	94	126
10	60	90	126
15	20	99	126
15	40	94	128
15	60	90	126

4. Conclusions

The microstructure of GFRP reinforcing bars was characterized with respect to the combined effects of seawater contamination, water medium, and elevated temperatures. Multiple post-conditioning degradation mechanisms were reported. Although new hydroxyl groups formed (95% increase compared to the control GFRP), hydrolysis had apparently initiated within the first 5 months of exposure but did not progress over time nor with elevated temperatures. At the fiber-matrix interface, disintegration of the polymer matrix led to debonding/delamination of the glass fibers. Longer exposure at elevated temperatures intensified crack propagation and widening, resulting in more voids in the microstructure. Such voids allowed for more water absorption and consequently, caused plasticization of the matrix. This degradation phenomenon lowered the glass transition temperature (T_g) after one DSC scan. It could however be reversed upon heating the sample during the second run and evaporating the absorbed water. Elevated temperatures seemed to amplify the plasticizing effect, unlike longer exposure duration of 15 months. With a 10% mass loss of the polymer matrix, a relatively smoother fiber surface was identified using SEM imaging. Degraded matrix particles may have dissolved in the water medium or remained in the concrete after conditioning. Microstructural changes and tensile properties should be correlated to better understand the behavior and performance of concrete-wrapped GFRP reinforcing bars exposed to harsh environments. To simulate real-life applications, the collective effects of sustained loading and severe exposures should be also included in future work.

5. Acknowledgements

This project is supported by the United Arab Emirates University (UAEU) and Sultan Qaboos University (SQU) under grant no. 31N129. The contributions of the UAEU and SQU are greatly appreciated.

6. References

1. ACI Commitee 440, *Guide for the design and construction of concrete reinforced with FRP bars*, A. 440, Editor. 2006, ACI: Farmington Hills, Mich.

2. FIB, *FRP reinforcement in RC structures*, in *Bulletin 40*. 2007, The International Federation for Structural Concrete FIB: Lausanne, Switzerland.
3. Debaiky, A., et al., *Residual Tensile Properties of GFRP Reinforcing Bars after Loading in Severe Environments.* Journal of Composites for Construction, 2006. **10**(5): p. 370-380.
4. Robert, M. and B. Benmokrane, *Physical, Mechanical, and Durability Characterization of Preloaded GFRP Reinforcing Bars.* Journal of Composites for Construction, 2010. **14**(4): p. 368-375.
5. Dejke, V., *Durability of fibre reinforced polymers (FRP) as reinforcement concrete structures.* Chalmers Univ. of Technology, 1999. **Publication No. P-98**(18).
6. Robert, M. and B. Benmokrane, *Behavior of GFRP Reinforcing Bars Subjected to Extreme Temperatures.* Journal of Composites for Construction, 2010. **14**(4): p. 353-360.
7. Montaigu, M., et al., *Laboratory Characterization and Evaluation of Durability Performance of New Polyester and Vinylester E-glass GFRP Dowels for Jointed Concrete Pavement* Journal of Composites for Construction, 2013. **17**(2): p. 176-187.
8. Kim, H., et al., *Short-term durability test for GFRP rods under various environmental conditions.* Composite Structures, 2009. **83**(1): p. 37-47.
9. Robert, M., et al., *Temperature as an Accelerating Factor for Long-Term Durability Testing of FRPs: Should There Be Any Limitations?* . Journal of Composites for Construction, 2010. **14**(4): p. 361-367.
10. Al-Salloum, Y.A., et al., *Effect of harsh environmental conditions on the tensile properties of GFRP bars.* Composites Part B: Engineering, 2013. **45**(1): p. 835-844.
11. Mukherjee, A. and S.J. Arwikar, *Performance of Glass Fiber-Reinforced Polymer Reinforcing Bars in Tropical Environments—Part II: Microstructural Tests.* ACI Structural Journal, 2005. **102**(6): p. 816-822.
12. Robert, M., P. Cousin, and B. Benmokrane, *Durability of GFRP Reinforcing Bars Embedded in Moist Concrete.* Journal of Composites for Construction, 2009. **13**(2): p. 66-73.
13. Kamal , A.S.M. and M. Boulfiza, *Durability of GFRP Rebars in Simulated Concrete Solutions under Accelerated Aging Conditions.* Journal of Composites for Construction, 2011. **15**(4): p. 473-481.
14. Robert, M. and B. Benmokrane, *Effect of aging on bond of GFRP bars embedded in concrete.* Cement and Concrete Composites, 2010. **32**: p. 461-467.
15. ASTM, *Standard Test Methods for Constituent Content of Composite Materials*, in *D3171*. 2015, ASTM: USA.
16. ASTM, *Standard Test Methods for Loss-On-Drying by Thermogravimetry*. 2010, ASTM: USA.
17. ASTM, *Standard Test Method for Assignment of the Glass Transition Temperatures by Differential Scanning Calorimetry*, in *E1356*. 2008, ASTM: USA.
18. Mourad, A.H.I., et al., *Effect of Seawater and Warm Environment on Glass/Epoxy and Glass/Polyurethane Composites.* Applied Composite Materials, 2010. **17**(5): p. 557-573.
19. Gentry, T.R. and C.E. Hudak. *Thermal compatibility of plastic composite reinforcement and concrete.* in *2nd Int. Conf. on Advanced Composite Materials in Bridges and Structures*. 1996.
20. Gentry, T.R. and M. Husain, *Thermal compatibility of concrete and composite reinforcements.* Journal of Composites for Construction, 1999. **3**(2): p. 82-866.
21. Masmoudi, R., A. Zaidi, and P. Gérard, *Transverse thermal expansion of FRP bars embedded in concrete.* Journal of Composites for Construction, 2005. **9**(5): p. 377-387.

481350 Characterization of 3-D Printed Parts

Abstract:

Aerospace, automotive, medical, and many other industries are finding applications for 3-D printing of plastic parts, the primary advantages including fast turnaround when custom design is required (no need to fabricate a mold) and ability to manufacture nested parts. However, the Achilles heel for such processing is the low z-axis strength of the printed objects. The goal of this research is explore the characterization of compatibilized polymer blends of ABS and nylon on z-axis mechanical properties of 3-D printed parts. Experiments were carried out on two scales, the small scale being a MakerBot 2X Replicator, and the large scale being a â€œbig area additive manufacturingâ€ or BAAM printer located at Oak Ridge National Labs Manufacturing Demonstration Facility. In both cases a commercial ABS material was compounded with a compatibilizer, poly(styrene-maleic anyhydride) (SMA), and then compounded with the nylon. Tensile testing for the Makerbot scale samples showed that the Z axis modulus of the ABS decreased by over 12% on addition of up to 20% (w/w) SMA, but the X axis modulus was maintained. Several compatibilized nylon blends were made from nylon(60%)/ABS(40%)/SMA(5-20%) (w/w) and the modulus was tested in the x-direction print. The tensile testing for the compatibilized nylon showed an increase of the modulus for the 5% and 20% (w/w) mixtures with the 5% (w/w) mixture showing a modulus increase of 172%. Flexural testing (4 point bend) for the Blue Gantry scale samples showed that the X axis modulus of the polyamide (PA)/ABS/SMA blend increased on addition of ABS/SMA (200%) as compared to the base nylon X axis modulus. The fracture sites of the nylon materials were dull for all samples tested in the X axis direction. For those samples tested in the Z axis direction, some glossy features were noted in the fracture surface. Results for Charpy impact testing will be discussed as well. Characterization of the interfaces of printed parts will be discussed as well as novel failure modes.

459368 Effects of Pillaring Agents on Corrosion Resistance of Polybenzoxazine/Organoclay Nanocomposite Coatings with Intercalated Structure

Abstract:

Nowadays, a lot of efforts have been made to apply polybenzoxazine (PBz) as a novel coating material for protecting metals against corrosion, and outstanding performance were obtained based on these researches[1]. Further enhancement and some functionality favorable to anticorrosion can also be achieved when PBz combine with other constituents. In our previous studies, PBz/organoclay composite coatings with intercalated and exfoliated structure were successfully prepared by incorporating organoclay modified by different pillaring agents into PBz matrix. Results of SEM and AFM show that samples with intercalated structure possess a relatively uniform surface opposed to the ones with exfoliated structure, especially at higher clay content. Furthermore, it was found that intercalated structure of organoclays in PBz matrix is of great help to improve corrosion protection ability of composite coating. In particular, the sample with 3 wt% organoclay I.44P provides the best performance with corrosion rate reduced by 3 and 4 orders of magnitude compared to PBz and bare MS, respectively[2]. In this case, PBz/organoclay nanocomposite coatings with intercalated structure should be a promising coating material for corrosion protection.

Therefore, organoclays modified by two different pillaring agents were introduced in PBz coatings on mild steel. Composite coatings with intercalated structure were fabricated and characterized in terms of measurements to investigate their physicochemical properties and corrosion protection performance. The *d*-spacing values of organoclays were easily increased by blending with PBz monomer, which was evident from X-ray diffraction (XRD) patterns. Enhancement on corrosion resistance of PBz was also observed at all studied amounts. However, distinct degree of improvement for two systems were observed, which could be explained by the different interaction between PBz polymer chain and the pillaring agents involved.

1. "Changlu zhou, Zhong Xin." "Polybenzoxazine-based coatings for corrosion protection," in "Advanced and Emerging Polybenzoxazine Science and Technology," Ishida, H.; Froimowicz, P. Eds. Elsevier, Amsterdam (2016).
2. Changlu Zhou, Xin Lu, Zhong Xin, Yangfeng Zhang. Intercalated polybenzoxazine/organoclay composites with enhanced performance in corrosion resistance.Journal of Coatings Technology and Research. 2016, 13(1): 63-72

479664 Multiscale Characterization of Wear-Resistance in Epoxy/Nanoparticle Composites

Abstract:

Development of transparent and wear-resistant composite materials is of great significance in maintaining mechanical durability of optical surfaces (e.g., lenses, photographic film, windshields, goggles, energy cell panels, and packaging) [1]. Unlike the modulus and hardness values frequently measured from linear elastic regions in material systems, wear resistance parameters are more related to interactions between surfaces - specifically the removal and deformation of material on a surface as a result of mechanical action of the opposite surface [2]. The definition, measurement, and evaluation, as well as mechanism explanation of wear-resistance have been challenging [3-6]. For instance, wearing has been described variously in terms of abrasion, adhesion, corrosion, and erosion, and evaluated by gloss reduction, topography change, deformation, roughness, and weight loss, just to name a few [7]. In addition, it is difficult to obtain acceptable accuracy, variability, and reproducibility at multi scales due to limitations in different instrument types [8] and difficulties in micro- and nano-structural controls. The degree of wearing in nanoparticle-filled composites depends on several factors, namely, (i) physical and chemical compositions of the materials, such as dispersion quality, interfacial interactions, particle volumes, and particle orientations [9]. The measurement of wear resistance can become very unreliable for polymer composites [10]. At nano-scale, crystalline and amorphous regions in semicrystalline polymers and nanoparticles may generate divergent resistance to wearing [8]. At micro- and macro-scale, multiple combinations between polymer chain conformations and nanoparticle configurations may produce dissimilar or opposite trends [8]; and (ii) test environmental parameters, including scratching tip geometry, contact conditions determined by the applied normal and friction force, temperature, wearing rate, and so on [11, 12]. Therefore, it is of paramount importance to control both the material structure and test conditions for wear resistance consistency [13].

In this research, we studied the multi-scale wear-resistance in epoxy/nanoparticle composites prepared by spray coating method. Nano-scale scratching, micro-scale sand falling, and macro-scale Taber Abrasion characterizations were used to measure wear resistance. The nanoindentation, stylus profilometery, atomic force microscopy (AFM), and scanning electron microcopy (SEM), UV-Vis were used for characterizations. The tribological observations in this study have brought to light the potential of aligning nanoparticles in composites to increase the wear resistance and to preserve the transmittance, which is practically applicable in optical systems.

References:

[1] Aliofkhazraei, M. (2011) Characterization of nanostructured coatings. In Nanocoatings, Springer Berlin Heidelberg, Berlin, pp. 77-110.

[2] Ebert, D., Bhushan, B. (2012) Transparent, superhydrophobic, and wear-resistant coatings on glass and polymer substrates using SiO2, ZnO, and its nanoparticles. *Langmuir*, **28**, 11391-11399.

[3] Liu, J., Notbohm, J.K., Carpick, R.W., Turner, K.T. (2010) Method for characterizing nanoscale wear of atomic force microscope tips. *ACS Nano*, **4**, 3763-3772.

[4] Aoike, T., Uehara, H., Yamanobe, T., Komoto, T. (2001) Comparison of macro-and nanotribological behavior with surface plastic deformation of polystyrene. *Langmuir*, **17**, 2153-2159.

[5] Ren, S., Yang, S., Zhao, Y. (2003) Micro-and macro-tribological study on a self-assembled dual-layer film. *Langmuir*, **19**, 2763-2767.

[6] Huovinen, E., Hirvi, J., Suvanto, M., Pakkanen, T.A. (2012) Micro–micro hierarchy replacing micro–nano hierarchy: A precisely controlled way to produce wear-resistant superhydrophobic polymer surfaces. *Langmuir*, **28**, 14747-14755.

[7] Bhushan, B., Gupta, B.K. (1991) *Handbook of tribology: Materials, coatings, and surface treatments.* McGraw-Hill, New York, NY, Malabar, FL.

[8] Friedrich, K., Zhang, Z., Schlarb, A.K. (2005) Effects of various fillers on the sliding wear of polymer composites. *Composites Science and Technology*, **65**, 2329-2343.

[9] Song, K., Zhang, Y., Meng, J., Green, E.C., Tajaddod, N., Li, H., Minus, M.L. (2013) Structural polymer-based carbon nanotube composite fibers: Understanding the processing–structure–performance relationship. *Mater.*, **6**, 2543-2577.

[10] Myshkin, N., Petrokovets, M., Kovalev, A. (2006) Tribology of polymers: Adhesion, friction, wear, and mass-transfer. *Tribology International*, **38**, 910-921.

[11] Patnaik, A., Satapathy, A., Chand, N., Barkoula, N., Biswas, S. (2010) Solid particle erosion wear characteristics of fiber and particulate filled polymer composites: A review. *Wear*, **268**, 249-263.

[12] Biswas, S., Vijayan, K. (1992) Friction and wear of PTFE—a review. *Wear*, **158**, 193-211.

[13] Carpick, R.W., Salmeron, M. (1997) Scratching the surface: Fundamental investigations of tribology with atomic force microscopy. *Chemical Reviews*, **97**, 1163-1194.

469503 Electrophoretic Deposition of Nanocrystals Under Flow to Synthesize Semiconducting Thin Films with High Atom Economy

Abstract:

Electrophoretic deposition (EPD) of colloidal nanocrystals (NCs) under flow is explored as a general method for the fabrication of semiconducting thin films that is amenable to high throughput, low cost operation. EPD is readily employed for depositing ceramic or metallic coatings, which typically require hundreds of volts for deposition to occur. However, for photovoltaic applications, a low process voltage is critical to avoid damaging the accreting semiconductor. Here we report a continuous flow reactor design that can operate at reduced voltage compared to a traditional batch reactor. This design preserves the required high electrophoretic velocity of the NCs at low process voltages by utilizing narrow electrode spacing. In a batch reactor, the low volume-to-surface-area ratio dictated by such a narrow spacing of the electrodes would impose a limit on the mass of nanocrystals that are resident in the reactor and therefore the thickness of the films that can be deposited. This limitation is removed by continuously flowing the colloidal dispersion of NCs and thus thick films are deposited. Through modeling and experiment we elucidate the process parameters that control the uniformity of the resulting coating and the utilization of the NCs in the feed solution. The reactor design is compatible with large area substrates and is specifically designed to enable continuous, high-rate fabrication of the active layer of photovoltaic cells with high atom economy.

461291 Influence of Surface Reactions on the Infrared Localized Surface Plasmon Resonance of Indium Tin Oxide Nanocrystals

Abstract:

Heavily doped oxide nanocrystals exhibit a tunable plasmonic response in the infrared, a capability that is promising for future nanoscale photonic technologies. Nanocrystal carrier density, and thus spectral response, is adjustable via chemical reaction; however, the fundamental processes governing this behavior are poorly understood. Here, we study the oxidation and reduction of indium tin oxide (ITO) nanocrystals with O_2 and H_2, respectively, with in situ diffuse reflectance infrared Fourier transform spectroscopy (DRIFTS). We show that the main absorption feature redshifts upon oxidation more than 1000 cm^{-1} and blueshifts to its original position upon reduction. The same magnitude spectral shift is observed over many cycles of oxidation and reduction. A kinetic model, which includes surface reaction and bulk diffusion, allows us to quantitatively rationalize the process, revealing that surface reaction is rate limiting under the conditions studied. Our experiments provide a deeper understanding of the connection between surface reaction and carrier density in oxide nanocrystals, and open the door to *a priori* control of plasmonic response.

462988 Large-Scale Crack-Free Gold Nanoparticle Monolayers with Adjustable Electrical and Optical Properties

Abstract:

Gold nanoparticles are widely studied due to the ease of controlled synthesis, facile surface modification, and interesting physical properties. However, a technique for depositing large-area, crack-free monolayers on solid substrates is lacking. Herein we present a method for accomplishing this. Spherical gold nanoparticles were synthesized as an aqueous dispersion. Assembly into monolayers and ligand exchange occurred simultaneously at an organic/aqueous interface. Then the monolayer film was deposited onto arbitrary solid substrates by slowly pumping out the lower, aqueous phase. This allowed the monolayer film (and liquid-liquid interface) to descend without significant disturbance, eventually reaching substrates contained in the aqueous phase. The resulting macroscopic quality of the films was found to be superior to films transferred by Langmuir techniques. In addition, the interparticle spacing was adjustable from 1.4 nm to 3.1 nm using different length alkylamine ligands. Both the surface plasmon resonance (SPR) and the electrical conductivity of the films show a strong dependence on particle spacing. This approach allows the precise control of interparticle spacing over large-area NP films. The demonstrated self-assembly of the 2D NP superlattice could have an impact on the manufacture of high performance optical and electronic devices.

473585 *In Situ* Characterization of the Effect of Plasma Treatment on Transition Metal Dichalcogenides

Abstract:

Transition metal dichalcogenides (TMDCs) are layered materials with unique material properties that make them attractive choices for electronic and optoelectronic devices. Although plasma processing is commonly used in the fabrication of these devices, the inherent consequences of this treatment have not been fully explored. In this work, *in situ* X-ray photoelectron spectroscopy (XPS) is used to investigate the effect of mild plasma treatment on the chemical composition and material properties of MoS_2 and WeS_2. These results shed light on recent advances in the device performance of metal oxide semiconductor field effect transistors (MOSFETs), demonstrating how defect engineering can be exploited to improve electronic properties for next-generation devices.

461262 Observation of Switchable Photoresponse of a Monolayer WSe_2-MoS_2 Lateral Heterostructure *Via* Photocurrent Spectral Atomic Force Microscopic Imaging

Abstract:

In the pursuit of two-dimensional (2D) materials beyond graphene, enormous advances have been made in exploring the exciting and useful properties of transition metal dichalcogenides (TMDCs)—such as a permanent band gap in the visible range and the transition from indirect to direct band gap due to 2D quantum confinement—and their potential for a wide range of device applications. In particular, recent success in the synthesis of seamless monolayer lateral heterostructures of different TMDCs *via* chemical vapor deposition (CVD) methods has provided an effective solution to producing an in-plane p-n junction, which is a critical component in electronic and optoelectronic device applications. However, spatial variation of the electronic and optoelectonic properties of the synthesized heterojunction crystals—throughout the homogeneous as well as the lateral junction region—and the charge carrier transport behavior at their nanoscale junctions with metals remain unaddressed. In this work, we use photocurrent spectral atomic force microscopy (PCS-AFM) to image the current and photocurrent generated between a biased PtIr tip and a monolayer WSe_2-MoS_2 lateral heterostructure. Current measurements in the dark in both forward and reverse bias reveal an opposite characteristic diode behavior for WSe_2 and MoS_2, owing to the formation of a Schottky barrier of dissimilar properties. Notably, by changing the polarity and magnitude of the tip voltage applied, pixels that show the photoresponse of the heterostructure are observed to be selectively switched on and off, allowing for the realization of a hyper-resolution array of the switchable photodiode pixels. This experimental approach has significant implications toward the development of novel optoelectronic technologies for regioselective photodetection and imaging at nanoscale resolutions. Comparative 2D Fourier analysis of physical height and current images shows high spatial frequency variations in substrate/MoS_2 (or WSe_2) contact that exceed the frequencies imposed by the underlying substrates. These results should provide important insights in the design and understanding of electronic and optoelectronic devices based on quantum confined atomically thin 2D lateral heterostructures.

458340 Effect of Inorganic Nanoparticle Coating on the Optical Transmittance and the Surface Conductivity for Transparent Conductive Carbon Nanotube /Silver Nanowire Hybrid Film

Abstract:

For the application of the touch screen panel and transparent conductive electrode for the flexible display, highly transparent and electrically conductive film is required. To enhance the optical properties and the surface conductivity of carbon nanotube (CNT)/silver nanowire/polyethylene terephthalate (PET) hybrid film, inorganic overcoating has been applied by roll-to-roll coating processing. The maximum total transmittance of the hybrid film without washing and overcoating was less than 90%. The optimum thickness of the overcoating has been determine as 10 micro meter for the highest total transmittance, 92 %, while the sheet resistance of the hybrid film is 180 ohm/sq. For the further enhancement of the sheet resistance of the CNT/silver nanowire hybrid film to get less than 80 ohm/sq the thickness of the overcoating has been controlled as 30 micro meter, while the transmittance is more than 90%. The overcoating of nanoparticle can highly reduce the surface roughness of the CNT/silver nanowire hybrid film and enhance the optical clarity such as the total transmittance and the haze of the hybrid film. This work has been done using large scale roll-to-roll coating machine for the scalable product of the high performance flexible transparent conductive film.

464360 Electrostatic Double Layer Flash Memory Based on Two-Dimensional Crystals

Abstract:

A new type of flash memory will be presented based on the electrostatic doping of two-dimensional crystals using ions. The proposed device consists of two, 2D crystals separated by a 2D electrolyte through which the ions can pass. The top 2D crystal comprises the channel of a field-effect transistor (FET), while bottom 2D crystal is a backgate. When the ions are near the surface of the channel, they induce image charge in the channel and the device is the low resistance, or ON state. When the ions are pulled back to the backgate by an applied field, the device is in the high resistance or OFF state. The 2D electrolyte is cobalt crown ether phthalocyanine (CoCrPc) plus a salt, which can be deposited on the surface of 2D crystals simply by drop casting and annealing. The crown ethers solvate metal ions, and the ions can pass through the cavity of the crowns. Density functional theory (DFT) calculations show that the crown ethers present a small barrier to ion transport required for fast (nanosecond) switching, but the height of the barrier will be increased for long retention both by the image charge induced in the channel and by modulating the gate bias. The current-voltage characteristics of a simplified device architecture (monolayer CoCrPc:$LiClO_4$ on graphene) will be presented. The graphene FET can be reconfigureably programmed by the 2D electrolyte, achieving sheet carrier densities of 4 x 10^{12} cm^{-2} at a low lithium concentration of 20 crown ethers to 1 Li^+. Based on the geometric packing of the molecules, as determined by scanning tunneling microscopy, the doping density is predicted to increase to 5 x 10^{13} cm^{-2} at a crown ether to lithium ratio of 1:1. State retention measurements show that the two states can be retained for at least 30 minutes (maximum time measured to date) with a memory window of 10 uA. In future work, the graphene will be replaced with MoS_2 because both the doping density and memory window are predicted to increase.

This work was supported in part by the Center for Low Energy Systems Technology (LEAST), one of six SRC

STARnet Centers, sponsored by MARCO and DARPA, and NSF grant #ECCS-GOALI-1408425.

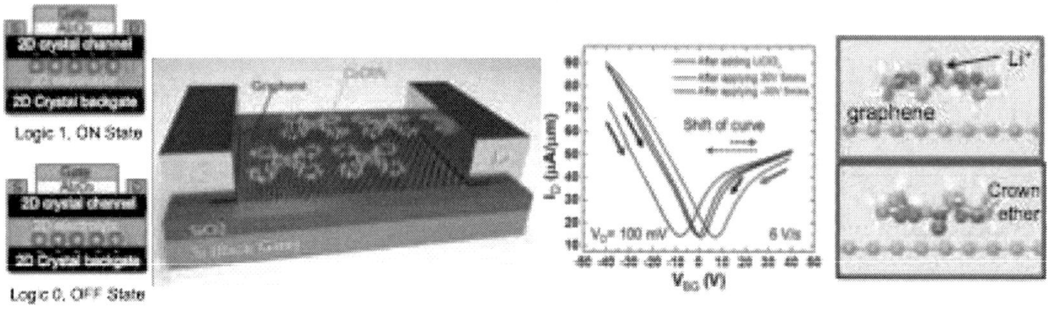

(a) (b) (c) (d)

Figure: Electrostatic double layer 2D crystal memory. (a) Cross section of proposed deice; ions induce image charge in the channel in the ON state, and are pulled back to the backgate in the OFF state. (b) Schematic of simplified device; 2D electrolyte and $LiClO_4$ on graphene FET (c) Transfer characteristics of simplified device at varying programming voltages; the shift in the Dirac point indicates reconfigureable doping by the 2D electrolyte corresponding to the two positions shown in (d).

459647 Multiscale Approaches for Modeling the Penetration of Field Effect in Two-Dimensional-Materials-Based Quantum Capacitors

Abstract:

Gate-tunable two-dimensional (2D) materials-based quantum capacitors (QCs) and van der Waals heterostructures involves tuning transport or optoelectronic characteristics by the field effect. Recent studies have attributed the observed gate-tunable characteristics to the change of Fermi energy in the first 2D layer adjacent to the dielectrics, while the penetration of the field effect through the one-molecule-thick material is often ignored or over-simplified. We present a multiscale analysis to model penetration of the field effect through graphene in a metal-oxide-graphene-semiconductor (MOGS) QC, including quantifying the degree of â€œtransparencyâ€ for graphene two-dimensional electron gas (2DEG) to an electric displacement field. We find that the space charge density in the semiconductor layer can be modulated by gating in a nonlinear manner, forming an accumulation or inversion layer at the semiconductor / graphene interface. The degree of transparency is determined by the combined effect of graphene quantum capacitance and the semiconductor capacitance, which allows us to predict the ranking for a variety of monolayer 2D materials according to their transparency to an electric displacement field as follows: graphene > silicene > germanene > WS_2 > WTe_2 > WSe_2 > MoS_2 > $MoSe_2$ > $MoTe_2$, when the majority carriers are electrons. Our findings reveal a general picture of operation modes and design rules for the 2D-materials-based QCs.

464902 Current-Driven Nanowire Formation and Nanopatterning on Crystalline Conducting Substrate Surfaces

Abstract:

The ability to form and precisely manipulate nanoscale features by controlling macroscopic forces is essential to advancing nanotechnology. For example, a macroscopic force in the form of an externally applied electric field can be used to drive atomic motion through the mass transport phenomenon of electromigration providing a paradigm shift in surface engineering and nanofabrication based on physical patterning.

Toward this end, in this presentation, we report a simulation study on formation of nanowires with precisely controlled widths, starting from single-layer conducting islands on face-centered cubic (fcc) crystalline conducting substrates under the controlled action of macroscopic forcing provided by an externally applied electric field that drives island edge electromigration. Our continuum-scale, fully nonlinear driven island evolution model with diffusional mass transport limited to the island edge accounts for edge diffusional anisotropy and island coalescence and breakup; the model is validated by comparison of its predictions with experimental results for Ag islands on Ag substrates reported in the literature. Numerical simulations based on this experimentally validated model and supported by linear stability theory show that large-size islands undergo a current-induced fingering instability, leading to nanowire formation after finger growth. Depending on the substrate surface crystallographic orientation, necking instabilities after fingering lead to the formation of multiple parallel nanowires per island. In all cases, the axis of the formed nanowires is aligned with the direction of the externally applied electric field. The nanowires have constant widths, on the order of 10 nm, which can be tuned by controlling the externally applied electric field strength. Achieving parallel arrays of ~10-nm-wide nanowires, dimensions barely accessible even by e-beam lithography, and demonstrating the ability to control their direction has important implications for developing future nanoelectronic patterning techniques applicable, e.g., to the interconnect structures of future electronic chips.

In addition, we demonstrate based on direct dynamical simulations that under specific conditions, and over a range of edge diffusional anisotropy parameters and nanowire edge orientation, the nanowire straight edge becomes unstable under the action of the externally applied electric field. This current-induced nanowire edge morphological instability leads to breakup of the nanowire into smaller-size daughter islands of practically uniform size and spacing to form regular patterns of nanoscale-size daughter islands. We have conducted a comprehensive parametric study to provide guidelines for generating such regular nanopatterns, which can be classified broadly into two categories depending on the geometrical orientation of the daughter islands in the regular array: aligned along a straight line or aligned along two intersecting straight lines to form a 'V'-shape. Moreover, using a linear stability theory we explain the origin of this nanowire straight edge instability; the theory is successfully validated by comparing its predictions for the number of daughter islands formed and the characteristic spacing between the islands with the results of the dynamical simulations. Our findings have direct implications for the development of novel external-field-driven physical nanopatterning techniques toward device nanofabrication.

466440 Computational Investigations of Drug Storage and Delivery in Bio-Compatible Nanoporous Materials

Abstract:

Encapsulation of drug molecules is required to improve the drug bio-distribution, the biological half-life of the active species and therapeutic effect of the drug.[1] Lipids, polymeric nanoparticles, metal clusters and carbon structures have been tested as drug storage materials but these materials have several limitations such as low drug loading capacities (<5 wt%), rapid drug release and toxicity.[2] Recently, metal organic frameworks (MOFs), a new group of nanoporous materials, have been studied in drug and cosmetic molecule storage applications due to their large surface areas, high pore volumes and tunable pore sizes and several MOFs are identified as promising materials for drug storage.[3] However, considering the very large number of available MOFs, it is not practical to assess the potential of each MOF in drug storage applications using purely experimental manners. Computational methods that can rapidly predict drug storage capacity of MOFs prior to experimental studies would be highly useful to identify the materials with the highest performances. However, there is almost no information about the drug storage performances of different types of MOFs. This lack of information also limits understanding the relation between drug storage performances of MOFs and their structural properties. In this work, we studied both adsorption and diffusion of representative drug and cosmetic molecules in various bio-compatible MOFs. Ibuprofen (an analgesic and anti-inflammatory drug) was chosen as a representative drug molecule, caffeine (lipo-reducer) and urea (hydrating agent) were considered as representatives of cosmetic molecules. Ibuprofen, caffeine and urea storage performances of 24 different bio-compatible MOFs were examined using configurational bias Monte Carlo (CBMC) simulations. Once a drug molecule is captured by a MOF, it has to be delivered to the target cells and released over a long period of time to increase its therapeutic effect. Short delivery time is not desired since in this case drug release occurs prior to reach to targeted cells. Therefore, release performance of materials must be also examined in order to identify the best performing materials as drug carriers. With this aim, we performed molecular dynamics simulations to examine ibuprofen, caffeine and urea diffusion in a MOF that showed high storage performance. Both rigid and flexible MOF structures were used in diffusion simulations to understand the effect of structure flexibility on the drug/cosmetic molecule diffusion. We also studied the effect of water on diffusion of drug and cosmetic molecules since these molecules are generally encapsulated in a liquid environment in practical applications. Finally, effects of MOFsâ€™ structural properties such as pore volume and pore size on the drug storage and diffusion were investigated to provide information on the materialsâ€™ structure-performance relationships.[4] This study showed that bio-compatible MOFs have strong potential in drug storage and delivery and they can be alternatives to traditional drug carriers. The computational methodology that we described here will be helpful to identify the most promising MOFs among many of them for drug/cosmetic storage prior to extensive experimental efforts. **Acknowledgment:** Financial supports provided by the TUBITAK 2211-C Scholarship Program. S.K. acknowledges TUBA-GEBIP Programme. *Corresponding author: skeskin@ku.edu.tr **References:** [1] C. Tamames-Tabar *et al.*, MOFs in Pharmaceutical Technology. In Bio- and Bioinspired Nanomaterials, Wiley-VCH Verlag GmbH & Co. KGaA: Germany, 2014. [2] P. Horcajada *et al.*, Chem. Rev. **112**, No. 2, 1232 (2011). [3] G. Férey *et al.*, Science, **309**, No. 5743, 2040 (2005). [4] I. Erucar and S. Keskin, Ind. Eng. Chem. Res., **55**, No. 7, 1929 (2016).

471914 Nanoharvesting of Polyphenolic Flavonoids from *Solidago Nemoralis* Hairy Root Cultures Using Functionalized Mesoporous Silica Nanoparticles

Abstract:

Plant secondary metabolites are produced in response to the environment of certain plants as a defense mechanism against predators and other stressors. Sometimes, these metabolites also bind to human receptors to induce favorable responses to diseases and pathogens, which can be enriched in a particular plant cell cultures by genetic and environmental manipulations. The conventional method of recovery of natural products from these genetically engineered plants is to grow the plants to a certain degree of maturity and then to harvest whole tissues, macerate the tissues to gain access to the compounds, and extract them. Nanoharvesting offers an alternative for continuous production of secondary metabolites from plant cell cultures, in which nanoparticles are designed to bind and carry molecules out of living cells. The carriers are to be designed so that they enter into plant cells and are released without inducing significant toxicity, and with specific binding of compounds of interest.

Here, nanoharvesting of polyphenolic flavonoids, enriched in genetically modified *S. nemoralis* hairy root cultures, is carried out using mesoporous silica nanoparticles (MSNPs) functionalized with both amines and titanium dioxide (TiO_2). Amine functionalization was performed to facilitate uptake of nanoparticles into plant cells, and TiO_2 functionalization to provide coordination binding sites for metabolites. Non-magnetic MSNPs (170 nm diameter) and magnetic MSNPs with Fe_3O_4 cores (70 nm diameter) with highly porous structures were synthesized and functionalized with amines and TiO_2. Both particles were shown to be taken up in *S. nemoralis* hairy roots, and recovered using external centrifugal and magnetic forces, respectively. Intracellular uptake and localization of the nanoparticles (0-1000 µg/ml in Murashige and Skoog media) in hairy roots were visualized by fluorescent imaging, after tagging the nanoparticles with rhodamine B isothiocyanate. Quenching of fluorescence in bulk solution using trypan blue was used to confirm intracellular localization of the tagged particles. Post-uptake viability of hairy roots was demonstrated by a concentration-dependent growth study. Proof of the flavonoid nanoharvesting concept was inferred from the color change (white to orange) of the recovered nanoparticles after exposure to hairy roots. Quantitative determination and analysis of nanoharvested flavonoids will be reported based on antiradical activity of metabolites bound to particles and HPLC.

451087 Modeling: A Tool for Experimentalists. Design, Synthesis and Evaluation of Self-Assembling Dendrons for Gene/Drug Delivery

Abstract:

Due to the relative easy synthesis and commercial availability, nanovectors based on self-assembling units are among the most utilized non-viral vectors for gene/drug transfer. Contextually, recent advances in molecular simulations and computer architectures allow for accurate predictions of many structural, energetic, and eventual self-assembly features of these nanocarriers *per se* and in complex with their genetic cargoes. In the present work, we aim at reviewing our own, most recent efforts in the field of computer-assisted design, synthesis, *in vitro*and in vivo activity of these (nano)biomaterials [1-9]. Specifically, we will discuss two cases:

1) Anticancer drug nanomicelles formed by self-assembling amphiphilic dendrimer able to overcome drug resistance in cancer cell [2]

Drug resistance and toxicity constitute challenging hurdles for cancer therapy. The application of nanotechnology for anticancer drug delivery is expected to address these issues and bring new hope for cancer treatment. In this context, we established an original nanomicellar drug delivery system based on an amphiphilic dendrimer (*AmDM*), which could generate supramolecular micelles to effectively encapsulate the anticancer drug doxorubicin (DOX) with high drug-loading capacity (>40%). The resulting *AmDM*/DOX nanomicelles were able to enhance drug potency and contrast doxorubicin resistance in breast cancer models by significantly enhancing cellular uptake while considerably decreasing efflux of the drug. In addition, the *AmDM*/DOX nanoparticles abolished significantly the toxicity related to the free drug. Collectively, our studies demonstrate that the drug delivery system based on nanomicelles formed with the self-assembling amphiphilic dendrimer constitutes a promising and effective drug carrier in cancer therapy.

2) Chameleon self-assembling nanovectors for efficient siRNA delivery *in vitro* and *in vivo* [4]

Small interfering RNA (siRNA) delivery remains a major challenge in RNAi-based therapy. Through the combined action of *in silico*/experimental efforts, we designed and produced for the first time an amphiphilic dendrimer *AD* able to self-assemble into adaptive supramolecular assemblies upon interaction with siRNA. The resulting nanovectors can effectively deliver siRNAs to various cell lines, including human primary and stem cells, thereby outperforming the currently available non-viral vectors. Most importantly, however, *AD*-mediated gene silencing was successfully achieved also *in vivo*.

References

[1] P. Posocco, E. Laurini, V. Dal Col, D. Marson, K. Karatasos, M. Fermeglia and S. Pricl, "Tell me something that I do not know. Multiscale molecular modeling of dendrimer/dendron organization and self-assembly for gene therapy," *Curr. Med. Chem*., vol. 19, pp. 5062-5087, 2012.

[2] T. Wei, C. Chen, J. Liu, C. Liu, P. Posocco, X. Liu Q. Cheng, S. Huo, Z. Liang, M. Fermeglia, S. Pricl, X. J.Liang, P. Rocchi, L. Peng, "Anticancer drug nanomicelles formed by self-assembling amphiphilic dendrimer to combat cancer drug resistance," *Proc. Natl. Acad. Sci. USA*, vol. 112, pp. 2978-2983. 2015.

[3] D. Marson, E. Laurini, P. Posocco, M. Fermeglia, S. Pricl, "Cationic carbosilane dendrimers and oligonucleotide binding: an energetic affair," *Nanoscale*, 7, pp. 3876-87, 2015.

[4] X. Liu, J. Zhou, T. Yu, C. Chen, Q. Cheng, K. Sengupta, Y. Huang, H. Li, C. Liu, Y. Wang, P. Posocco, M. Wang, Q. Cui, S. Giorgio, M. Fermeglia, F. Qu, S. Pricl, Y. Shi, Z. Liang, P. Rocchi, J. J. Rossi, L. Peng, "Adaptive amphiphilic dendrimer-based nanoassemblies as robust and versatile siRNA delivery systems," *Angew. Chem. Int. Ed. Engl.*, vol. 53, pp. 11822-11827, 2014.

[5] S. Kala, A. S. Mak, X. Liu, P. Posocco, S. Pricl, L. Peng, A. S. Wong, "Combination of dendrimer-nanovector-mediated small interfering RNA delivery to target Akt with the clinical anticancer drug paclitaxel for effective and potent anticancer activity in treating ovarian cancer," *J. Med. Chem*., vol. 57, pp. 2634-42, 2014.

[6] P. Posocco, X. Liu, E. Laurini, D. Marson, C. Chen, C. Liu, M. Fermeglia, P. Rocchi, S. Pricl, L. Peng, "Impact of siRNA overhangs for dendrimer-mediated siRNA delivery and gene silencing," *Mol. Pharm*., vol. 10, pp. 3262-3273, 2013.

[7] X. Liu, C. Liu, E. Laurini, P. Posocco, S. Pricl, F. Qu, P. Rocchi, L. Peng, "Efficient delivery of sticky siRNA and potent gene silencing in a prostate cancer model using a generation 5 triethanolamine-core PAMAM dendrimer," *Mol. Pharm*., vol. 9, 470-81, 2012.

[8] X. Liu, J. Wu, M. Yammine, J. Zhou, P. Posocco, S. Viel, C. Liu, F. Ziarelli, M. Fermeglia, S. Pricl, G. Victorero, C. Nguyen, P. Erbacher, J. P. Behr, L. Peng, "Structurally flexible triethanolamine core PAMAM dendrimers are effective nanovectors for DNA transfection in vitro and in vivo to the mouse thymus," *Bioconjug. Chem*., vol. 22, 2461-2473, 2011.

[9] A. Barnard, P. Posocco, S. Pricl, M. Calderon, R. Haag, M. E. Hwang, V. M. Shum, D. W. Pack, D. K. Smith, "Degradable self-assembling dendrons for gene delivery: experimental and theoretical insights into the barriers to cellular uptake," *J. Am. Chem. Soc*., vol. 133, pp. 20288-20300, 2011.

457771 Investigation of CNT-Induced Bacteria Lysis and Protein Release

Abstract:

Carbon nanotubes (CNTs) were investigated for their potential use as lysis agents for bacterial cells. Recombinant *Escherichia coli* BL21 (DE3) expressing b-lactamase and Green Fluorescent Protein (GFP) were exposed to various concentrations of CNTs under various levels of agitation and exposure times. Fluorescence assay for GFP, enzymatic activity for b -lactamase activity, and Transmission Electron Microscopy (TEM) were used to characterize the extent of lysis. A correlation was also developed that related the extent of protein release to the concentration of CNTs and agitation rate. The correlation indicated that the agitation rate and CNT concentration affect lysis in a similar fashion.

459693 Using Fluoroalkylated Polyethylene Glycol-Stabilized Perfluorocarbon Nanodroplet As Oxygen Carriers to Reduce Oxygen Inhibition Effect for Microalgal Growth

Abstract:

Introduction: Dissolved oxygen (DO) built up in the photobioreactor and/or its excessive accumulation in the medium have long been considered as one of the major limitations for closed and massive cultivation of microalgae. Although the DO may be removed via air sparging-based photobioreactor technology, disadvantages such as high cost, increased contamination, and severe cell damage caused by hydrodynamic stress remain obstacles for microalgal culture. Perfluorocarbon (PFC), a fluorine-substituted derivative of hydrocarbons which can dissolve large respiratory and other non-polar gases as compared to water, has been widely used as an oxygen/carbon dioxide transporter in various fields. To overcome the effect of oxygen inhibition on the microalgal growth, a strategy of using perfluorocarbon nanodroplets (PNDs) as oxygen scavengers was explored.

Materials and Methods: The PNDs were synthesized by perfluorooctyl bromide (PFOB) and fluoroalkylated methoxy polyethylene glycol (R_F-MPEG, Mw of PEG molecule = 2000 or 5000) with various formulations through single PFC-in-water emulsification in association with solvent evaporation approach and were characterized using Fourier Transform Infrared spectroscopy. The stability of each type of PND under dynamic environment was evaluated by analyzing their changes in size and concentration while circulating in the photobioreactor system for 7 days. The deoxygenability of each type of PND under 7-days oxygenated photobioreactor operation was determined based on the mass accumulation rate coefficient (K_La^{ac}) of DO in the medium by use of dynamic method. To verify the effectiveness of PNDs-aided photobioreactor on *Nannochloropsis Oculata (N. oculata)*growth under a hypercritical oxygen environment, the cellular growth rate in each group was examined through measurement of the cell concentrations by hemocytometry every 24 h for up to 7 days.

Results and Discussion: Our results showed that all of the PNDs expanded during circulation in the photobioreactor and the one prepared by R_F-MPEG-2000 (MW of PEG = 2000) and perfluorooctyl bromide (PFOB) with a volume ratio of 9:1 (PNDs-2000/PFOB (9:1)) exhibited the lowest size increase of 19% (from 200 ± 2.43 to 238 ± 1.99 nm) after 7 days. Moreover, the concentration of DO in the medium approximately decreased 25%, 15%, 9%, and 5% at day 7 where the value of K_La^{ac} was about 4.7, 5.92, 6.49, and 7.4 $\times 10^{-3}$ h^{-1} for the group with 1% (v/v) PNDs-2000/PFOB (9:1), PNDs-2000/PFOB (7:3), PNDs-5000/PFOB (9:1), and PNDs-5000/PFOB (7:3), respectively, as compared to the group without PND. These results clearly exhibited that the PNDs-2000/PFOB (9:1) can provide the highest stability and capacity of deoxygenation for the defined photobioreactor operation. Under oxygenated cultivation, the growth of *N. oculata* (μ) without PNDs was tremendously arrested (μ = 0.366 day^{-1}) within seven days, while the cells with aid of 1% (v/v) PNDs-2000/PFOB (9:1) successfully multiplied in the hyperoxygenated environment that the growth rate of 0.467 day^{-1} was comparable to the normal breeding with neither PNDs nor oxygen injection (μ = 0.498 day^{-1}), and the biomass production was five-fold higher than the O_2-treated group without PND after cultivation for seven days.

Conclusions: In summary, the developed PNDs-2000/PFOB (9:1) enabled to efficiently remove the DO in the photobioreactor that provided a feasible means for massive culture of microalgae in closed setting.

467494 Cyclic Polyethylene Furanoate As a Monomer from Renewable Resources for Ring Opening Polymerization

Abstract:

Introduction

The replacement of oil based chemicals by renewable resource based chemicals is one of the major issues chemical industry is facing right now. The alternative polymer poly (ethylene furanoate), PEF, is a possible substitute for poly (ethylene terephthalate), PET. Among different process alternatives, PEF can be produced at high polymerization rates and molecular weights via ring opening polymerization of cyclic oligo (ethylene furanoate), cyOEF. On the other hand, the effective production of these cyclic oligomers at high purity is still the major obstacle towards the process industrialization. In this work, different strategies aimed to produce these cyclic molecules from the two monomers Furan dicarboxylic acid (FDCA) and ethylene glycol (EG) are explored. And a model is presented to describe the equilibrium behavior of said cyclic oligomers.

Three pathways towards cyOEF

The three synthetic strategies shown in figure 1 have been investigated. In all cases, high dilutions in organic solvents are required to achieve high yields; the final product is a distribution of cyclic oligomers with different number of repeating units.

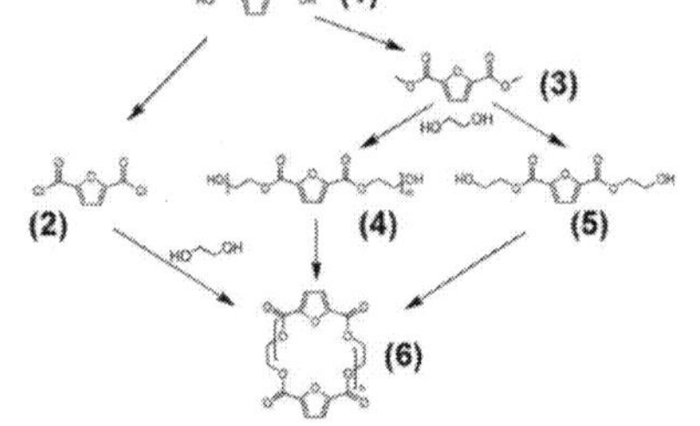

Figure 1 Reaction pathways for the production of cyOEF.

The first (scheme 1-2-6) is a fast cycle formation using the rapid esterification of acid chlorides with alcohols. Moderate selectivity towards cycles of around 50% are in contrast to a complex reaction setup and toxic chemicals involved for the chlorination. The product size distribution is kinetically controlled and consists mainly of C3, with C2 and C4 being the second most abundant species.

The depolymerization (scheme 1-3-4-6) and reactive distillation (scheme 1-3-5-6) are the two more promising routes towards cycles. For depolymerization FDCA and EG are converted to oligomeric chains of around 12 repeating units. This prepolymer is subsequently diluted in a high boiling organic solvent and converted to a mixture of cyclics and linears at 180-200 °C. In case of reactive distillation the FDCA and EG (in excess) are converted to very short chains (1-2 repeating units). The chains are then diluted in a high boiling organic solvent and finally converted to cyOEF by distilling a portion of the solvent together with the excess EG.

Both routes yield a similar distribution of cyclic species. Possible purification routes include precipitation and adsorption processes, which have both been applied to yield cycles with high purity starting from linear species (>99% by HPLC area). The yield of both reactions lies in the range of 60 to 80 % depending on the process conditions. One striking feature of this type of reaction is that it is possible to recycle the solvent and the linear side products in order to increase the yield of the process and furthermore make it more sustainable.

Modeling of the ring chain equilibrium

The equilibrium composition of cyclics and linears in 2-Methylnaphthalene and Dichlorobenzene (DCB) could be modeled using the Jacobson-Stockmayer model (Figure 2). Under high dilutions this equilibrium shifts towards the formation of cyclic oligomers. It is based on computing the probability of ring closure using the assumption of randomly coiled hydrocarbon chains for linear polymers. Two fitting parameters are involved: one is the equilibrium constant for the step growth polymerization, which is in equilibrium with the cyclisation reactions, and the second one is the effective length of the bonds in the polymer chains.

As shown in the figure below, our own results from depolymerization and reactive distillation experiments were fitted using the model. These results suggest that it is indeed possible to use this approach to describe the ring chain equilibrium. The model is currently under refinement in order to describe system-specific effects such as the complete absence of cycles with six repeating units.

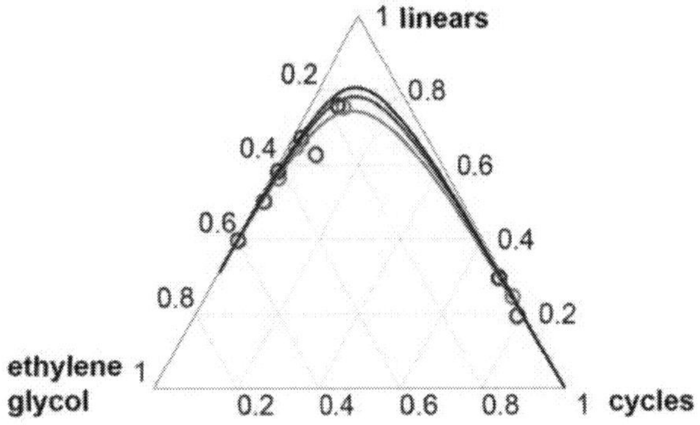

Figure 2 Equilibrium distribution of cyclic oligomers in DCB.

References

[1] Gomes M, Gandini A, *J. Polym. Sci., Part A: Polym. Chem.*, 49, 3759-3768 (2011).

[2] Burch, RR, Lustig, SR, Spinu, M *Macromolecules*, 33, 5053-5064 (2000).

[3] Stockmayer, WH, Jacobson H, *J. Chem. Phys.*, 18, 1600-1606 (1950).

453695 Polymeric Nano-Metal Composite Membranes for Water Remediation

Abstract:

Responsive polymers react to external stimuli like temperature, pH, radiation, ionic strength or chemical composition. These polymers in combination with nanomaterials provide novel hybrid properties such as bactericidal, electrical, hydrophilic, catalytic, selectivity, etc., and they can be self-supported or in a convenient matrix such as a membrane.

These membranes, functionalized with polymers having pH responsiveness (hydrogels), is a promising research field for numerous applications. The functional groups of the hydrogels gain or lose protons based on the pH, collapsing or extending their network, making them swell or un-swell solutions. These nanostructured materials can be exploited to synthesize metal nanoparticles and as ion exchangers. They also prevent the loss of nanomaterials to the environment and improve reactivity due to larger surface areas, expanding its range of applications to selective catalysis, remediation and mining.

The present work describes "green" techniques used to create nanocomposites based on polyvinilidene fluoride (PVDF) hollow fiber (HF) and flat sheet (FS) membranes, both sponge-like and thin. Hydrophobic HF membranes were synthesized with 57.2% porosity and 0.72 μm mean pore size. These membranes were then hydrophilized by a water-based process of cross-linking polyvinylpyrrolidone (PVP) by persulfate (PS). Though, hydrophilization of PVDF membranes using PVP is known in industry, as a post-process has only been implemented at lab-scale.

The PVP cross-linking is confirmed but also a ring-opening is present due to etherification. Hydrophilization was measured by contact angle changes (CA) (76.3° to 60.3°) and surface free energy calculations (40.3 to 49.8 $mJ \cdot m^{-2}$). HF and FS were subsequently functionalized with a poly(acrylic acid) (PAA) hydrogel through free radical polymerizations (thermal or redox) using PS as initiator. This work advanced FS membrane

functionalization from lab-scale to full-scale by modifications of the polymerization procedures.

Swelling tests were performed to hydrogels/functionalized membranes finding responsiveness up to 1676% swelling at pH 9.0. Also, mesh size values (ξ) were calculated showing an inverse relationship with membrane permeability, tuning the flux through the membrane. An increase of ξ from 5.89 to 8.33 nm (pH 3.0 to 9.0) represented a 93% permeability decrease (53.9 to 3.4 $L \cdot m^{-2} \cdot h^{-1}$) at pressures between 1.0 and 4.0 bar.

Throughout a double ion exchange of sodium/iron and a subsequent reduction, bimetallic Fe/Pd nanoparticles (60 to 300 nm) were synthesized. The Fe exchanged was almost 500 mg Fe/g PAA. Similarly, it was possible to use the exhausted accelerants of Fe ions of the redox polymerization to synthesize nanoparticles, but having only 9.2 mg Fe/g PAA of loading.

Finally, depending on the Pd concentration in the membrane, the catalytic activity in the reduction of trichloroethylene (TCE) was studied, having enhanced reactivities (up to ten times) in sponge-like membranes compared to previous studies. Using hydrogels alone, similar results to nanoparticles in solution phase were found without nanoparticle loss. 80% of reduction was achieved within 2 hours with chloride production near to stoichiometric values (3/1), demonstrating absence of intermediates.

This research is supported by the National Science Foundation EPSCoR, and by the National Institute of Environmental Health Sciences (P42ES007380). HF membranes were synthesized at the Singapore Membrane Technology Center, NTU (Singapore). Nanostone Water (Oceanside, CA) provided support and collaboration in the development of full-scale flat sheet membranes.

454787 Thermodynamic and Economic Assessment of the Production of Ethylene and Propylene from Bioethanol

Abstract:

Introduction

The increasing worldwide pollution and the environmental problems associated with the petroleum extraction and use, have led to the development of new energy resources and sustainable raw materials [1]. The production of bioethanol from biomass offers a sustainable biotechnology solution at competitive costs with high availability. In Colombia (South America), bioethanol production has increased because the local policies encourage its production, being the third producer in America [2]. However, the focus on bioethanol production has been related to its use as biofuel. Alternatively, there are some high-commercial-value compounds —mostly produced from petroleum— that can also be obtained from renewable sources [3]. Recently, bioethanol has been proposed as a source for producing commodities such as light olefins in the framework of the biorefinery concept [4]. Among these biomaterials, ethylene stands out as one of the most produced chemicals worldwide, having a projected production of 400 million ton in 2020 [1]. In addition, the demand for propylene is growing faster, having a projected production of 100 million ton in 2021[5]. Therefore, this study aims to propose a conceptual design for producing ethylene and propylene from bioethanol samples procured in Colombia, integrating the reaction and separation stages and assessing its technical and economic feasibility. Aspen Plus and Aspen Hysys were used as process simulators.

Methodology

Aspen Plus V7.3 and Aspen Hysys V7.3 (Bedford, USA) were used as process simulators throughout the study. In both, the Soave-Redlich-Kwong (SRK) equation of state was used as thermodynamic model. First, the bioethanol dehydration reaction was analyzed in the RGibbs reactor module in Aspen Plus. Water/ethanol (S/E) feed ratio and reaction temperature were varied in the module in order to assess their effect on products selectivity. Ethylene, propylene and *n*-butene were selected as possible products, and a selectivity response surface was obtained in Matlab R 2014 (Salt Lake City, USA). Aspen Hysys was used to perform the sizing of the process units, heat integration, and process yields. Products purification was modeled in the Distillation Column module and the energetic analysis was performed using Heat Exchanger and Heaters/Coolers modules in Aspen Hysys. Finally, a class-V-order economic evaluation was done to the final flowsheets by determining the capital expenses for all units in the processes [6].

Results and analysis

Figure 1a shows the surface response for ethylene selectivity during bioethanol dehydration simulation, when varying S/E ratio and temperature. The higher the temperature and S/E ratio, the higher the ethylene yield, with a selectivity exceeding 60% beyond S/E = 5:1 and 600 °C. On the other hand, Figure 1b shows a maximum peak of 45% selectivity for propylene at 400 °C, and a slightly positive effect of the S/E ratio. *n*-butene formation also presented a similar trend (Figure 1c), showing a maximum selectivity of 14% at 300 °C. Heavier olefins are formed by ethylene condensation, which is thermodynamically favored at lower temperatures [7]. However, the absence of ethylene at these lower temperatures reduces the amount of this compounds.

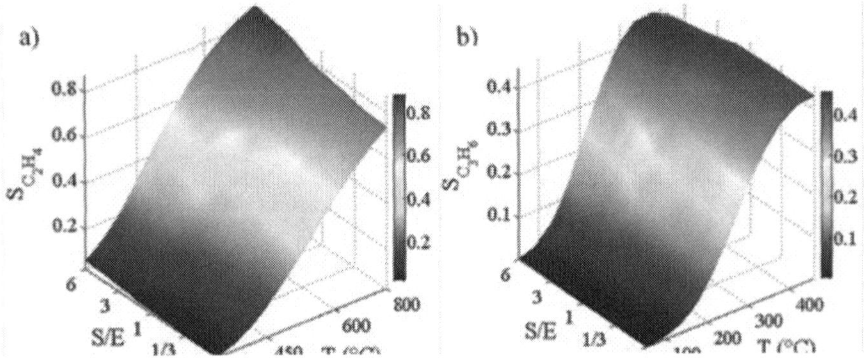

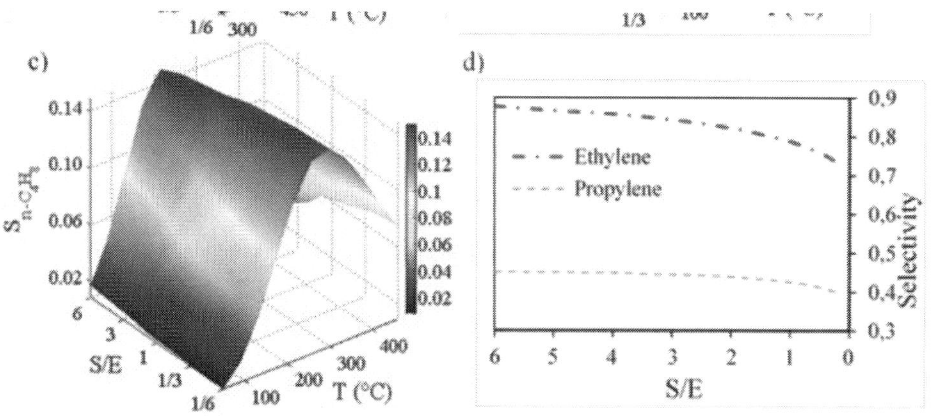

Figure 1. Matlab response surface from Aspen Plus simulated data for the production of (a) ethylene, (b) propylene, and (c) *n*-butene using different water:ethanol ratios (S/E) in the temperature range of 200–800 °C; and (d) the effect of the S/E ratio on the maximum ethylene and propylene yields.

Figure 1d shows the effect of the S/E on the maximum both ethylene and propylene selectivity. The excess of water is slightly positive for the ethylene formation because its acts as a quasi-inert diluent from a thermodynamic point of view [7]. In this way, a bioethanol dehydration reactor feed with a stream of S/E = 5 and operating at 615 °C was selected as the proper design to produce ethylene. In addition, a dehydration reactor operating at 400 °C was selected to produce propylene. These reactors were included in two different processes to analyze the conceptual design of two plants, one producing ethylene and another producing propylene from bioethanol.

Figure 2 shows the designed processes for producing (a) ethylene and (b) propylene from bioethanol. The processes begin with the feed stream of 300,000 ton/year of raw, fermented bioethanol (5 mol% ethanol, S1), which corresponds to the 10% of the national bioethanol production. This stream is heated to 40 °C before its further introduction to a flash separator V-100, which operates at 95 °C and 1 atm. These conditions were established according to the ethanol-water equilibrium to obtain an S/E = 5:1 (16.4 mol% ethanol). After V-100, the product stream rich in ethanol (S3) is heated to either 615 °C or 400 °C (reaction temperatures for ethylene or propylene production, respectively) by a recycle of the flash bottoms stream (S4). S4 is rich in water and is used to the process heat integration. The product streams (S7 in Figure 2a and S6 in Figure 2b) are pressurized to 19 atm in the compressor K-100. The pressurized streams are cooled to 18 °C by fresh water streams. These streams would deliver the energy duty for heating streams (S20 in Figure 2a and S19 in Figure 2b). Subsequently, the cooled streams are used as an inlet in a three-phases flash separator V-101 with the purpose of removing the residual water after the reaction. Finally, the product streams are introduced into distillation columns to separate the olefins.

The ethylene process delivered a stream of 99.7 mol% ethylene with an almost 100% recovery (S16 in Figure 2a), producing 0.60 mol ethylene/mol ethanol feed. In addition, propylene process delivers a stream with 92.0 mol% propylene in S15 in Figure 2b. This corresponds to the 70% of the propylene entering to the distillation tower, producing 0.43 mol propylene/mol ethanol feed.

A class-V economic assessment delivered investments of USD 2 million for producing ethylene and USD 2.23 million for propylene. These costs represent 159 USD/ton-year of ethylene and 327 USD/ton-year for propylene. True [8] reported investment costs of 1300 USD/ton-year for ethylene and 1450 USD/ton-year for propyylene produced from petroleum for a plant constructed in 2016. However, these costs include lumpsum, engineering, procurement, construction and commissioning turnkey contract; which not were included in the current analysis. Nevertheless, this class-V economic assessment shows that olefins production from bioethanol could be a profitable and sustainable alternative to petrochemical processes.

b)

a)

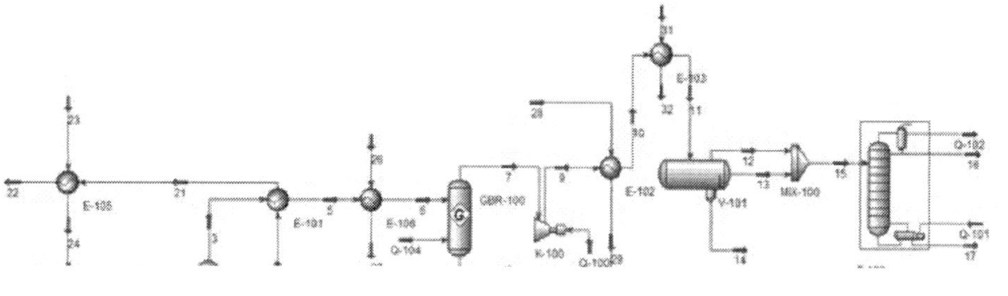

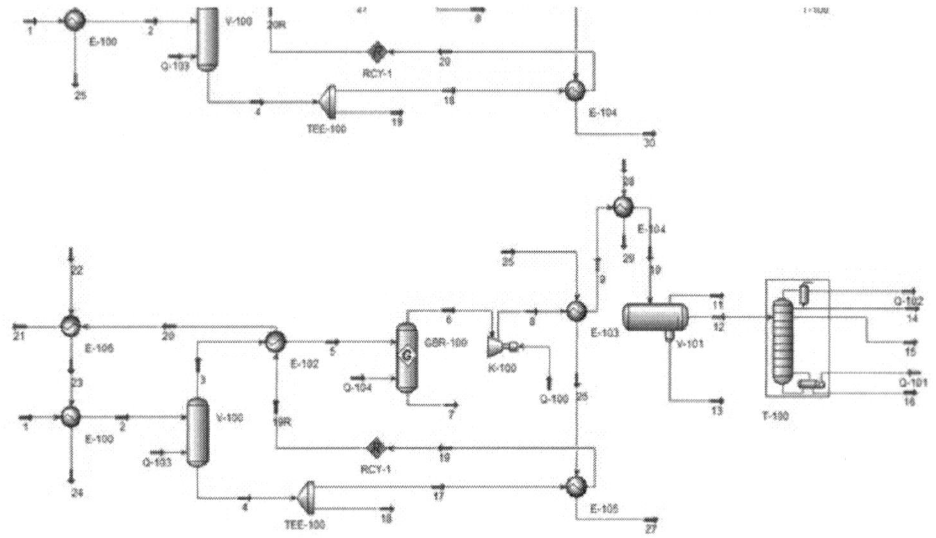

Figure 2. Process flowsheets designed in Aspen Hysys for producing (a) ethylene and (b) propylene from bioethanol. Blue lines represent mass streams and red lines represent energy streams. Equipment notation: E: Heat exchanger, V: Flash separator, K: Compressor, MIX: Mixer, T: Distillation tower, TEE: Splitter, and GBR: Gibbs reactor.

Conclusion

This study evaluated a conceptual design for producing light olefins from bioethanol base on thermodynamic predictions and using commercial software. To the best of our knowledge, it is the first time that reaction and separation stages are integrated and economically evaluated in two different processes, one to produce ethylene and another to produce propylene from bioethanol. Due to water showed a positive effect on the ethanol dehydration reaction, both processes start with a simple flash separator operating at 95 °C and 1 atm to reach the ethanol initial concentration needed in the reactors (S/E=5, 17.6 mol% ethanol). Two different reactors were designed, one operating at 615 °C for producing ethylene and another at 450 °C for producing propylene. The designed ethylene process delivered a stream of 99.7 mol% ethylene with a 60% yield and an investment cost of USD 2 million, equivalent to 159 USD/ton-year of ethylene. On the other hand, propylene production from bioethanol delivers a stream with 92 mol% propylene and 43% yield. Propylene process has an investment cost of USD 2.23 million (327 USD/ton-year propylene). These conceptual analysis show the technical, environmental and economic feasibility of the biorefinery concept development.

References

[1] H. Xin, X. Li, Y. Fang, X. Yi, W. Hu, Y. Chu, et al., J. Catal. 312 (2014) 204–215.

[2] G.P. Ortegón, F.M. Arboleda, L. Candela, K. Tamoh, J. Valdes-Abellan, Sci. Total Environ. 539 (2016) 410–9.

[3] O.J. Sánchez, C.A. Cardona, Bioresour. Technol. 99 (2008) 5270–95.

[4] R. Le Van Mao, T.M. Nguyen, G.P. McLaughlin, Appl. Catal. 48 (1989) 265–277.

[5] Ceresana, Market study: Propylene, (2014). http://www.ceresana.com/en (accessed May 1, 2016).

[6] American Association of Cost Engineers, (2015). http://www.aacei.org/ (accessed November 15, 2015).

[7] T. Lehmann, A. Seidel-Morgenstern, Chem. Eng. J. 242 (2014) 422–432.

[8] W. True, Oil Gas J. 110 (2012).

458322 Design and Economical Evaluation of Polygen Process to Co-Produce Synthetic Natural Gas (SNG), Methanol and Ethylene Glycol

Abstract:

Design and Economical Evaluation of POLYGEN process to co-produce Synthetic Natural Gas (SNG), Methanol and Ethylene Glycol

Bor-Yih Yu and I-Lung Chien*

Department of Chemical Engineering, National Taiwan University, Taipei 106, Taiwan

The steady state design and economical evaluation of poly-generation (POLYGEN) process to co-produce synthetic natural gas (SNG), methanol and ethylene glycol (EG) is studied in this work. POLYGEN has been a widely studied topic recently, in which several different chemicals can be produced parallel at the same time. In POLYGEN process, the relative production rates for different chemicals can be adjusted based on different market demands, daily usages, and also changing political strategies. This character enhances the flexibility of plant-wide operation and also the economic performance. Among the products, SNG holds very similar composition and heat value to typical natural gas, and can be used as a replacement in industrial and home usages. Methanol is one of the most important industrial feedstocks, which can be the basis of various kinds of chemicals. Ethylene glycol is also an important organic chemical, which can be used to synthesize polyesters, antifreezes, solvents, and so on.

The items covered by this work are illustrated in the following. In coal-based POLYGEN processes, syngas is first generated from coal gasification in a CO2-rich atmosphere, and the raw syngas is then passes the treating stage. Syngas treating stages contains two section, namely, sour water-gas-shift-reaction (SWGSR) and acid-gas-removal (AGR). In SWGSR, excess amount of steam is added to adjust the H2/CO ratio. After that it is cooled to low

temperature to enter AGR. In AGR, SELEXOL method is used to capture 99.7% of H2S and 92% CO2. Part of the captured CO2 is sent back to gasification section in order to reduce the carbon emission, and the other part is compressed and sequestrated. With proper arrangement in configuration, three streams with qualified composition to be the feedstocks for downstream chemical plants are produced.

For downstream chemical plants, the products produced will meet the industrial standard. The SNG has heat value higher than 965 (BTU/scf), the methanol purity is 99.5 (mol%), and EG purity is 99.8 (mol%). EG is traditionally synthesized through petroleum route, a novel route from coal-derived syngas has been widely considered a potential alternative as the oil shortages continuing. Thus in this work, the production of EG from syngas is adopted. Firstly, CO and H2 in syngas are separated, and are used in different reaction stages. In the first stage, CO is reacted to form dimethyl oxalate (DMO), and in second stage, DMO is further reacted with H2 to form ethylene glycol.

After the design, optimization works are studied based on minimized total annual cost (TAC), and the economical evaluation is performed to investigate the overall economic performance. Besides, some waste heat could be recovered by generating steam, and the steam is used to drive turbine to produce electricity for usage inside the plant-wide process. Thus through this work, a better understanding of POLYGEN process, and the design and operation of coal to ethylene glycol route is reached.

458992 Preparation and Properties of Soybean Oil-Based Composites Containing Natural Fillers

Abstract:

Biobased composites have been prepared, which consist of a vegetable-oil based matrix, with ground nutshell or eggshell as natural, rigid, filler materials.

Crosslinked polymers of acrylated epoxidized soybean oil (AESO) and styrene were first successfully produced across the entire composition range via bulk free radical polymerization. Mechanically, these copolymers varied from hard and rigid at high styrene contents, to soft and rubbery at high AESO contents.

The intermediate 50:50 copolymer was chosen as the matrix for composites containing varying amounts of pecan nutshell flour or powdered eggshells. Composites were produced by pre-blending the fillers into the comonomer mixture before reaction and solidification. Through the AESO and filler contributions, these composites contain substantial fractions of biobased starting materials.

Copolymers and their composites exhibited good structural integrity, and were characterized by flexural testing to fracture, dynamic mechanical analysis, SEM imaging, as well as hardness and impact testing.

Addition of the natural fillers to the matrix resulted in substantial increases in hardness and stiffness, with a nearly six-fold increase in modulus observed at 35 volume% eggshell filler. These increases were accompanied by modest decreases in breaking strength and impact energy, and no observed effect on the glass transition temperature of the matrix. Use of coupling agents [1] to increase adhesion and potentially further modify mechanical properties will be discussed.

General comparison of the filled polymer properties to a baseline of neat polystyrene shows improved or similar individual mechanical characteristics, indicating viability of such biobased composites as a replacement material.

[1] Nachtigall, S.M.B., Cerveira G.S., Rosa S.M.L., Polymer Testing, 26(5), 2007.

460812 Solvent Free Sucrose Esters Production in Reactive Systems Containing Emulsifiers

Abstract:

Surfactants are chemicals used in a wide variety of applications, mainly in the manufacture of consumer goods, industrial cleaners, food-feed, and cosmetic products. In the past, efforts have been made to develop biobased surfactants from renewable raw materials in order to reduce the environmental impact of traditional petroleum-based surfactants. Currently, most biobased surfactants available in the market are lipids-based, but still they contain polar moieties generated from fossil resources (e. g. sulfonic acid, ammonium, or poly-ethoxylated groups) (1). In contrast, in completely biobased surfactants (e. g. acylglycerols, alkyl-poly-glucosides, glycosides of fatty acids) both, the polar and non-polar structures are of renewable origin.

Among biobased surfactants, fatty acid sucrose esters stand out as important additives for the food industry (food ingredient number: E473). These are generally produced by transesterification of sucrose (from sugar cane) with fatty acid methyl esters (from vegetable oils). In addition to their renewable character, sucroesters are high value added, biodegradable and biocompatible surfactants, which can also be used in specialized applications such as cosmetics, clean and care products and pharmaceuticals. Recent estimations (2) indicate that the current market of sucroesters is around than 6000 t/year, with prices around 4 to 20 USD/kg depending on the nature of the product (i.e. lipid chain, degree of substitution, etc).

Sucrose is a polyalcohol with eight hydroxyl groups (See Figure 1), so theoretically, sucrose esters could be monoesters, diesters, triesters, and, so on (up to octaesters). As a result, tuning of reaction selectivity to a specific product is a major challenge. Besides, sucroesters properties and application are heavily determined by the degree of esterification. For example, sucrose monoesters are good water-in-oil emulsifiers, while sucrose octaesters are good oil-in-water emulsifiers. In general, the degree of substitution depends of different process conditions such as temperature, reactants molar ratio, catalyst nature and concentration, and the use of other additives (e. g. solvents, emulsifiers).

In addition to controlling the degree of substitution, a major challenge of the process is the incompatibility between reactants; only 0.1 g of sucrose can be dissolved in 100g of FAME at 140°C (3). This reduced solubility limits the mass transfer and reduces the reaction rate and conversion. A processing alternative to overcome this problem is the use of emulsifiers, which enhance contact and solubility of sucrose in FAMEs (4–7). Some reports indicate that fatty acid soaps (mono- and divalent), sucrose esters (the product itself), and mono- and diacylglycerols can be used as emulsifiers for the reaction. However, it is necessary to study the influence of such emulsifiers (nature and concentration) on the transesterification process (i.e. rate of reaction and esterification degree).

In this work, conversion and selectivity of transesterification reaction to produce sucroesters was experimentally evaluated using three different emulsifiers (potassium palmitate, sucrose palmitate and glyceryl monoester) at different concentrations (up to 20%w). Reactions were carried out in a 150 ml glass-jacketed reactor coupled to a methanol total condenser. The system was homogenized with an Ultraturrax® at 15000 rpm prior to reactions, and maintained under agitation at 600 rpm at constant temperature (110-140 °C) until no further evolution of methanol was observed. FAMEs consumption and sucroesters production were tracked by HPLC analysis of samples. Observed conversion were around 40% and mono substitution increased with the concentration of the emulsifying agents. Higher conversions were obtained when potassium palmitate was used as emulsifier, and reaction times were shorter when emulsifier's concentration was increased. Degree of substitution of the final product was higher when a mixture mono- and diacylglycerols was used as emulsifier and lower when sucrose palmitate was used. Results of this study can be used for preliminary process design in a solvent-free production of sucrose esters with different emulsifying properties.

Figure 1. Emulsifiers used to provide contact in sucrose and fatty methyl estermixtures (Molecular structures taken from ChemSpider)

References

1. HAYES, Douglas G. ***Biobased Surfactants : A Useful Biorefinery Product That Can Be Prepared Using Green Manufacturing***. Knoxville, TN, USA, 2012.

2. OTOMO, Naoya. Basic properties of sucrose fatty acid esters and their applications. In : HAYES, Douglas G., KITAMOTO, Dai, SOLAIMAN, Daniel K. Y. and ASHBY, Richard D. (eds.), ***Biobased Surfactants and Detergents***. Urbana, Illinois : AOCS Press, 2009.

3. ZHAO, Rongrong, CHANG, Zhidong, JIN, Qiong, LI, Wenjun, DONG, Bin and MIAO, Xiaowen. Heterogeneous base catalytic transesterification synthesis of sucrose ester and parallel reaction control. ***International Journal of Food Science & Technology***. 2014. Vol. 49, no. 3, p. 854–860. DOI 10.1111/ijfs.12376.

4. PARKER, Kenneth J., JAMES, K. and HURFORD, J. Sucrose Ester Surfactants — A Solventless Process and the Products Thereof. In : HICKSON, John L. (ed.), ***Sucrochemistry***. Washington, DC : American Chemical Society, 1977. p. 97–114.

5. FITREMANN, Juliette, QUENEAU, Yves, MAITRE, Jean-paul and BOUCHU, Alain. Co-melting of solid sucrose and multivalent cation soaps for solvent-free synthesis of sucrose esters. ***Tetrahedron Letters***. 2007. Vol. 48, p. 4111–4114. DOI 10.1016/j.tetlet.2007.04.015.

6. GALLEYMORE, Harry R, JAMES, Kenneth, JONES, Haydn F and BHARDWAJ, Chaman L. Process for the production of a surfactant containing sucrose esters. 4298730. 1981. United States.

7. RIZZI, George P and TAYLOR, Harry M. Synthesis of higher polyol fatty acid polyesters. 3963699. 1976. United States.

461686 Synthesis and Characterization of Biochar-Based Carbon Supported Metal Nanoparticles

Abstract:

Synthesis and characterization of biochar-based carbon supported metal nanoparticles

Sai Teja Neeli and Hema Ramsurn

Russell School of Chemical Engineering, University of Tulsa, Tulsa, OK 74104

Biochar is a carbon-rich solid formed after biomass has been carbonized. It is also a residue from a number of thermochemical conversion processes whereby biomass is converted to bio-oil and/or syngas. Therefore, this material is environmental-friendly and carbon-neutral. Traditionally, biochar has been used as soil amendments for the healthy growth of plants due to its retention capability of nutrients. This investigation aims to use biochar in a different manner: as a support for metal catalysts to promote green chemistry and engineering. The biochar used in this study is obtained from hydrothermal carbonization of model compounds of biomass namely lignin, cellulose and hemicellulose. This means that high pressure and temperature water is used as a benign solvent to depolymerize the model compounds to biochar. The biochar so produced is chemically activated with 0.1 M HNO_3 solution resulting in activated carbon with defined micropore size distribution, high micropore volumes and large specific surface areas. Metal precursors (like iron nitrate) are then added to the biochar and dried in a convection oven at 110 $^{\circ}$C for one week to obtain metal (iron) impregnated biochar. The dried mixture is thermally-treated in a quartz tubular reactor at 1000 $^{\circ}$C for one hour in a nitrogen flow of 500 ml/min. These synthesized biochar based metal catalysts will then be characterized using a number of techniques including scanning electron microscope (SEM), x-ray powder diffraction (XRD), inductively coupled plasma atomic emission spectroscopy (ICP-AES) and Brunauer-Emmett-Teller (BET)

surface area analyzer. Other metal catalysts such as platinum, gold and nickel can also be impregnated on the biochar in the same manner. It is expected to observe distinct XRD patterns corresponding to graphite for the thermally treated metal-impregnated biochar, due to the conversion of the carbonaceous materials to graphite. SEM images will help not only in sizing the metal nanospheres (about 20 -60 nm) but also in determining how the metal is supported: half embedded or completely enclosed in the porous char matrix. Based on the characterization results, these metal-impregnated biochar catalyst could be tested for a number of reactions. These supports can withstand high temperatures and can even be used for gasification reactions and upgrading bio-oil during hydrogenation reactions. There is also the possibility of using bifunctional metal/acid catalysts on the biochar nanoparticles for dehydration and hydrogenation reactions and for aqueous phase processing to produce selectively targeted alkanes from glucose. The ultimate goal is to use these green catalysts to upgrade the oxygenated biofuels by removing the oxygen through various mechanisms and hence increasing the quality and heating value of the desired product.

460527 Interactions and Complexation in Polyelectrolyte-Nanoparticle Systems (Invited Talk)

Abstract:

Polyelectrolyte (PE)-protein and PE-particle mixtures form an important class of systems in which their phase behavior and complexation characteristics play a crucial role in influencing their properties. Despite the practical importance of PE-particle mixtures, the parameters governing the interactions and their phase behavior characteristics have had much less theoretical work compared to their neutral counterparts. Recently, we have developed two pronged approach which allows us to study the effective interactions and the phase behavior in mixtures of charged nanoparticles (CNP) and polyelectrolytes (PE). On the one hand, we have adapted polymer self-consistent field theory approach to study the hierarchy of the two-, three- and multi-body interactions between CNP's. We have found that for the strong PE's and in the absence of polarization interactions, the CNP-CNP interactions are essentially pairwise and involves an interplay of depletion attraction and electrostatic repulsions. We have also analyzed of weak PE's and the presence of polarization interaction. In such systems, CNP's interactions remain qualitatively similar to the above systems, but the multi-body interactions appears to be significantly enhanced. To study the phase behavior of such systems, we have adapted the single chain in mean-field simulation approach to effect a multibody simulation of the PE-CNP systems. Using such an approach, we study the phase behavior and complexation characteristics in a variety of systems.

453054 Thermally-Driven Nanostructure Evolution of Phase-Change Elastomer Gels

Abstract:

Phase-change elastomer gels (PCEGs), which combine an elastomeric network with a phase-change (hydrocarbon, HC) additive, are promising candidates for shape-memory and self-supported passive temperature control materials. The use of a thermoplastic elastomer (TPE) further enhances the potential of these materials due to their ability to be repeatedly processed and easily recycled. However, the competing structural formation processes of TPE (i.e., block copolymer) self-assembly and HC crystallization on ultimate morphology are not well understood and may translate to a better understanding of macroscopic properties. In this presentation, PCEGs composed of poly[styrene-*b*-(ethylene-*co*-butylene)-*b*-styrene] (SEBS), a commonly utilized TPE, and crystalline HCs are considered. The shape-memory and thermal properties of SEBS / HC blends will be provided to motivate the current work, and then the focus will shift to characterization of the blends' nanostructural details discerned from a combination of small- and wide-angle X-ray scattering (SAXS and WAXS, respectively). Within the composition range considered, the molten-state blends ($T_{m,HC} < T < T_{g,S}$) exhibit a body-centered cubic (BCC) morphology in which S spheres reside in an EB / HC mixed matrix. Surprisingly, HC crystallization ($T < T_{m,HC}$) does not impact the shape of the S spheres, but completely distorts their lattice positions. This change, along with scattering from the HC crystals, can be tracked as a function of temperature to provide structural transitions in the blends. Further analysis of PCEGs containing two different HCs reveal that crystallization can be split into two separate, yet co-existing, crystal populations. These systems also indicate that the distortion of the BCC lattice is dependent on the overall crystallinity.

472968 Biopolymer–Solvent Phase Behavior for the Lignin–Acetic Acid–Water System

Abstract:

Lignin is unique among renewable biopolymers in having significant aromatic character, giving it potential applications for polymeric systems ranging from electrodeposited coatings to polyurethane foams to an inexpensive precursor for carbon fibers. Thies and co-workers (2015) have developed a technique for simultaneously solvating, fractionating, and purifying the lignin polymer recovered from by-product streams of the lignocellulosic biofuels and pulp-and-paper industries. Called Aqueous Lignin Purification with Hot Acids, or ALPHA for short, the process involves combining solid lignin with hot acetic acid–water mixtures to produce two liquid phases: a highly solvated, polymer (lignin)-rich phase and a solvent-rich phase. Surprisingly, the polymer can be easily partitioned by molecular weight into either phase simply by varying the acetic acid–water ratio (e.g., from 20/80 to 80/20). In addition to partitioning by molecular weight and generating low polydispersities, ALPHA allows one to both solvate and "ultra-clean" the high mol wt fraction of the polymer so that it is clean enough (ppm metals levels) for conversion to carbon fiber. However, as of yet the thermodynamics responsible for the creation of this fascinating polymer–solvent system have yet to be examined. To obtain a better fundamental understanding of how and why the acetic acid–water solvent system is so effective for processing lignin, the compositions of the polymer-rich and solvent-rich phases in our system were measured, with an emphasis on the partitioning of the acetic acid and water between the two liquid phases. In particular, liquid–liquid equilibrium compositions of each phase were measured for a softwood-derived lignin with acetic acid–water mixtures at temperatures ranging from 65 to 95 °C. The global pseudoternary phase-behavior characteristics of this polymer–dual solvent system were also elucidated and will be discussed, along with the implications of the observed polymer–solvent phase behavior for the development of a viable purification and fractionation process for the renewable biopolymer lignin.

471974 Large- Scale Structural Transitions in Supercoiled DNA Revealed By Coarse -Grained Simulation

Abstract:

Topological constraints, such as those associated with DNA supercoiling, play an integral role in genomic regulation and organization in living systems. However, physical understanding of the principles that underlie DNA structure and organization at biologically-relevant length-scales remains a formidable challenge. We develop a coarse-grained simulation approach for predicting equilibrium conformations of supercoiled DNA. With this approach, we study the conformational transitions that arise due to supercoiling across the full range of supercoiling densities that are commonly explored by living systems. Simulations of ring DNA molecules with lengths up to the scale of topological domains in the Escherichia coli chromosome (~ 10 kilobases) reveal large-scale structural transitions elicited by supercoiling. The structural transitions result in 3 supercoiling conformational regimes that are governed by a competition between chiral coils, extended plectonemes, and branched hyper-supercoils. We find that these conformational transitions exert a pronounced effect on the looping probability of genomically distal elements, and we consider the consequences of these effects on regulation of the genome by distal regulatory elements. The length-scales and supercoiling regimes investigated here coincide with those relevant to transcription-coupled remodeling of supercoiled topological domains, and we discuss possible implications of these findings in terms of the interplay between transcription and topology in bacterial chromosome organization.

466467 Thermodynamics of Sorption and Swelling in Polyethylene at Gas-Phase Polymerization Conditions

Abstract:

Thermodynamics of sorption and diffusion of monomer(s) and/or diluents in semi-crystalline polyolefins is one of the most important phenomena affecting catalytic polymerization processes, down-stream processing as well as application properties of these materials. In the atmosphere of suitable solvent the swelling also occurs. Swelling causes the prolongation of the diffusion path and thus it has to be investigated together with sorption and diffusion. Despite the importance, thermodynamically consistent and predictive tools for sorption equilibria are available only for a few specific cases and the attempts to predict the swelling of semi-crystalline polyolefins using equations of state have not been successful so far. Prediction of diffusion in polyolefins is another open problem. This situation is caused by several reasons: (i) scattered experimental data for the broad range of various PE grades, temperature, pressure and composition of penetrant mixtures, (ii) elastic constrains by tie chains connecting crystallites affecting the thermodynamics of amorphous phases, (iii) misunderstanding of the semi-crystalline morphology of lower crystallinity PE, (iv) neglecting the importance of relaxation dynamics in PE on the diffusion, and (v) temperature-dependence of crystallinity.

In our work, sorption and diffusion of low molecular weight penetrants is measured gravimetrically, where we measure the mass increase of polymer due to the gas sorption. The gravimetric data have to be corrected to the buoyancy force of the pendulous swollen polymer. Experimental measurement of swelling was carried out in video-microscopic apparatus. The method is based on the visual observation of a polymer particle by microscope with attached digital camera. The measured swelling isotherms were used in the evaluation of the gravimetric data.

This contribution presents results for sorption and swelling of ethylene, co-monomers (1-hexene, 1-butene) and inert diluents (propane) in various PE grades at conditions relevant to catalytic polymerization of olefins. This work stands out of similar studies in the open literature due to: (i) systematic measurement for a broad range of PE densities from 0.90 to 0.96 g/cm^3 covering practically all relevant grades, (ii) reporting a comprehensive set of swelling data for several penetrants, (iii) systematic measurements for several penetrants at a range of industrially relevant temperatures and pressures, and (iv) modelling the sorption data by the advanced PC-SAFT equation of state.

The experimental data and observed trends presented in this work contribute to the complex understanding of the mass transport and sorption equilibria of penetrants in polymers.

463934 Preparation of Micro and Nanocellular PBT/Ptmg Diblock Copolymer Structure By Using Supercritical CO_2 As the Blowing Agent â €“ Effect of Microphase Morphology

Abstract:

Preparation of micro and nanocellular PBT/PTMG diblock copolymer foam by supercritical CO_2 foaming ¨C effect of microphase morphology

Solid-state foaming process of segmented block copoly(ether esters) composed of polyoxytertramethylene (PTMG) ¡°soft¡± blocks and poly(butylene terephthalate) (PBT) ¡°hard¡± blocks using supercritical CO_2 ($scCO_2$) as the blowing agent was investigated. Plastic foams with nano/micro-scale cellular structures were prepared from PBT/PTMG copolymers. Two kinds of PBT/PTMG copolymers with different block ratio were used to investigate the influence of microphase morphology on foaming behavior. The CO_2 dissolved into soft segments (PTMG) and hard segments(PBT) were carefully characterized by using magnetic suspension balance (MSB), meanwhile the gas diffusion coefficient was fitted from the Fick¡¯s diffusion law. The results show that the solubility and diffusivity of CO_2 in soft segments (PTMG) is significantly higher than in hard segments(PBT). PBT/PTMG copolymers were foamed in a high-pressure chamber by using $scCO_2$ and the effect of temperature, block ratio, and saturation time on the cell morphology was investigated, respectively. The cell size and location were highly concentered in the dispersed soft segments by using the difference of CO_2 solubility and diffusivity between the hard segments and the soft segments. It was found that only soft segments could be foamed in presence of CO_2. The average cell diameter of the PBT/PTMG foaming samples were controlled to be 400¨C1800 nm on the preferential condition of block ratio, saturation pressure and foaming temperature. Furthermore, the cellular structure could be highly oriented in one direction by using the foamed soft segment in the microphase morphology as a template for foaming to obtain an alternate lamellar nanocellular structure.

Figure 1. The SEM image of PBT/PTMG foaming sample.

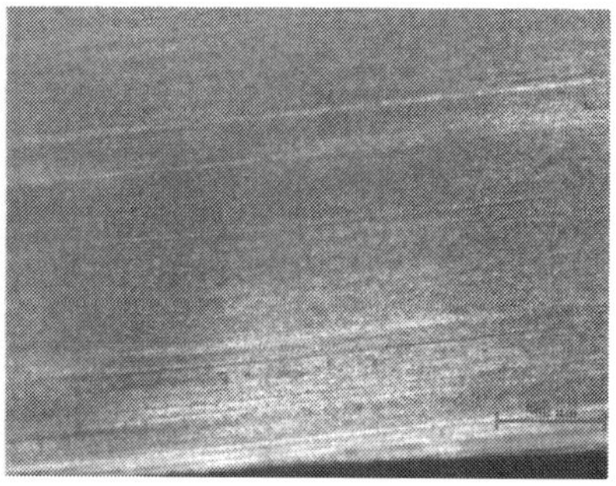

Figure 2. The POM image of highly oriented PBT/PTMG.

Figure 3. The SEM image of highly oriented PBT/PTMG foaming sample with alternate lamellar nanocellular structure.

466403 Relating Polymer Doping and Nanostructure Formation

Abstract:

It is well known that conjugated polymers like P3HT form nanocrystals in poor solvent solutions and that these same polymers â€œcrash outâ€ of solutions containing p-type dopants. We examine the relationship between polymer doping and nanostructure formation using a combination of small angle neutron scattering (SANS) and inelastic neutron scattering (INS). We use SANS to monitor the formation of P3HT nanocrystals in solution and examine how the presence of the p-type dopant (F4TCNQ) changes the crystal size, structure, and formation kinetics of P3HT nanocrystals. Next we use INS to determine P3HT configurations both inside and outside of crystalline domains for both doped and undoped polymers. These results demonstrate that only a small percentage of the volume is occupied by the ideal structure determined by x-ray crystallography and that the side chains are much more disorded than presented in literature. In addition, we fit the P3HT/F4TCNQ configuration using DFT modeling. We determine that the charging of the polymer backbone leads to increased side chain stiffness, which has significant implications for design of organic electronic devices based on thiophenes.

453786 Multiresponsive Poly(N-isopropylacrylamide) Copolymer Toward Metal Ion Recognition and Adsorption Via a Thermally Induced Sol-Gel Transition

Abstract:

Poly(N-isopropylacrylamide-co-maleic acid-co-1-vinylimidazole) (P(NIPAM-MA-VI)) has been prepared in water by free radical polymerization. The composition and molecular weight of copolymers were obtained by ^1H nuclear magnetic resonance (^1H NMR) spectroscopy and gel permeation chromatography (GPC). The aqueous solution of low molecular weight P(NIPAM-MA-VI) above a critical concentration experiences four distinct phases. Effects of monomer ratio, copolymer concentration and pH on phase transition temperatures were studied. The addition of divalent metal ions was found to promote gelation process, therefore the adsorption of P(NIPAM-MA-VI) to Cu^{2+} ions is further investigated. The adsorption behavior is pH-dependent and the adsorption isotherm is well fitted by the Freundlich model. In 3 mL of $Cu(Ac)_2$ solution containing 5.76 mg Cu^{2+} ions, the adsorption capacity reaches 19.1 mg/g dried polymer at 60 °C and pH = 5. The adsorbed P(NIPAM-MA-VI) can be regenerated by hydrochloric acid and reused in the following adsorption process. Synchrotron radiation small-angle X-ray scattering (SAXS) was used to investigate the effect of temperature and Cu^{2+} ions on the microstructure of P(NIPAM-MA-VI). This thermo-, pH-, and metal ion-responsive copolymer can be applied to ion recognition, water purification and enrichment of heavy metal ions.

454641 Modeling Controlled Release from Hollow Porous Nanospheres

Abstract:

The kinetics of diffusion-controlled release from hollow nanoporous spheres of varying thickness were investigated using classical mass transport theory. A new model was developed to describe the time-dependent mass transport through the shell. The model was compared with experimental kinetic profiles of transport of glyphosate in hollow silica nanosphere systems. The silica spheres were synthesized via a sacrificial method using Poly-(styrene-methyl acrylic acid) spheres as templates, thereby producing shell/core nanospheres with diameters ranging from roughly 367.6 to 406.8 nm and shell thicknesses spanning 22.0 to 32.8 nm. These nanosphere systems were loaded with glyphosate and then exposed to a reservoir of distilled water. Periodic sampling allowed determination of the glyphosate release kinetics as a function of shell thickness: the release rate decreased and the release period was prolonged by increasing shell thickness. The model was used to estimate the effective diffusivity and predict the concentration profile within the shell as well as the amounts of retained and released glyphosate. By rescaling the concentration profile in dimensionless form, it was observed that the amounts of released and retained glyphosate as functions of time could each be superimposed on a universal curve that was independent of shell thickness. Furthermore, the rate of the diffusive process with respect to shell thickness was determined by a single time constant that quantified the kinetics of the diffusive transport. It was also demonstrated that the associated diffusion coefficient of the mass transport must be calculated on a per-sphere rather than a per-mass basis. These results suggest that accurate predictions of time-released drugs and agrochemicals in diffusion-controlled nanosphere systems can be achieved via a single determination of an associated material rate constant. Predictions of the diffusion kinetics of hypothetical nanosphere systems were computed for a variety of experimental parameters, such as shell thickness and porosity.

455149 Atomistic Simulation Study on the Morphology of the Hydrated Perfluorosulfonic Acid Membrane

Abstract:

Molecular dynamics simulation was used to explore the morphology of hydrated Nafion membranes with different water contents. To quantitatively probe the segregated morphology of the hydrated Nafion membrane, Euler characteristic and surface-to-volume ratio were calculated to describe the morphology variation under the different water content. Simulation results showed that water and sulfonic groups of the membrane formed a porous connected water cluster at the low water content. With increasing water content, the connected water clusters gradually formed a water film to separate the hydrophobic region and then swelled in a single direction. The results well support a recent â €œfilm-like modelâ€ of the hydrated Nafion membrane [1].

Ref.

[1] Kreuer, K.-D.; Portale, G., A Critical Revision of the Nano-Morphology of Proton Conducting Ionomers and Polyelectrolytes for Fuel Cell Applications. *Adv. Funct. Mater.* **2013,** 23, 5390-5397.

455161 Parameterization of a Coarse Grained Model for Perfluorosulfonic Acid Polymer

Abstract:

A step-by-step approach was used to build coarse-grained (CG) models for hydrated perfluorosulfonic acid (PFSA) polymer membranes. The intramolecular bond potentials are described by using harmonic potentials and the parameters are determined by reproducing the bond and angle distributions obtained from all-atom molecular dynamics simulations. The nonbonded interaction is described by using Lennard-Jones function. LJ 12-4 function is used for the pairs involving water, while the LJ 9-6 and LJ 7-6 is used for any other pairs. Density, surface/interfacial tension and radial distribution function data are used for nonbonded parameter fitting. Comparison between the CG and AA simulations of PFSA membrane with different water contents has confirmed that the CG model can reasonably represent the density and structure properties of the membranes. The developed CG model will be useful in simulating PFSA membranes at a large-scale system

455559 Behavior of Dendritic Polymers in Solutions: DLS and NMR Study

Abstract:

Dendritic polymers possess unique properties. High degree of branching, globular architecture and a large number of surface functional group make them promising candidates for functional supramolecular materials. Dendritic polymers have a variety of prospective applications, e.g. functional coatings, adhesives, chemical sensors, photosensitive materials, nanocarriers in drug delivery, gene transfection, tumor therapy, biomimetic systems and the development of vaccines.

Our study were focused on experimental investigations of diffusion of a poly(propylene imine) dendrimer and a hyperbranched polyester in solutions.

A combination of field gradient NMR and DLS was used to study behavior of a poly(propylene imine) dendrimer of 5th generation (with amine end groups) in methanol-d4 solutions. The self-diffusion coefficient and the collective diffusion coefficient were determined as a function of concentration and temperature. The combination of both techniques permitted accurate estimations of the diffusion coefficient and the hydrodynamic radius at zero concentration, and from the concentration dependences the second virial coefficient.

Diffusion coefficients of dendritic polymers with end groups substituted with alkyl chains were also determined as a function of temperature and concentration. From the diffusion coefficient in zero concentration limit the hydrodynamic radii of the polymer molecules in solutions have been estimated. It was shown that at higher concentrations dendritic macromolecules with modified end groups form aggregates.

Analyzing the Chemical and the Physical Characteristics of Crosslinked Zein Gel Films Cast from Acetic Acid Solutions

Hazal Turasan, Emma Barber, Morgan Meiser and Jozef Kokini, Department of Food Science, Purdue University, West Lafayette, IN

Zein is the most abundant protein found in corn and large amounts are obtained as a byproduct of ethanol production [1]. Zein is amphiphilic and insoluble in pure water and is highly soluble in various ethanol/water solutions and acetic acid/water solutions. Zein has a wide range of utilization area. Its film forming capabilities makes zein one of the most commonly chosen plant-based materials for creating various analytical platforms. In this study biodegradable zein gel films were produced by solution casting and the mechanical, chemical and surface properties of the films were analyzed with different techniques. 70% acetic acid aqueous solution was used as the solvent due to its good dissolution properties for zein and due to its ability to produce very smooth surfaces with excellent surface topography [3], [4]. The concentration of zein was kept constant at 1:5 (g/ml) ratio. Oleic acid was used as a plasticizer at a 1:1 (g/g) zein to oleic acid ratio to increase flexibility and ductility of the film. For a better blending of the plasticizer into the zein matrix, monoglyceride at a 1:0.15 (g/g) oleic acid to monoglyceride ratio was used as the emulsifier. This study particularly focuses on the effect of crosslinking on the surface, mechanical and chemical properties of zein gel films. Glutaraldehyde was used as the crosslinker and was added to the zein solutions in different ratios and the solutions were allowed to crosslink for 20 hours after sonication. The solutions were cast on cured PDMS films and dried under the fume hood until complete evaporation of unbound solvent was obtained.

Chemical changes occurring due to crosslinking were studied with FTIR and FT-Raman spectroscopies. The surface properties of the contact side of the films with PDMS were analyzed with water contact angle measurements and Atomic Force Microscopy. The tensile mechanical properties were analyzed with a texture analyzer. FTIR spectra of zein gel films show typical characteristics of proteins at Amide I, II and III bands. In addition, some of the peaks were also seen as an indication of amino acid content of zein such as tyrosine or phenylalanine. The addition of glutaraldehyde at ratios of 4%, 8%, 10% and 12% with respect to the amount of zein were found to have a significant effect on the formation of peaks in the range 900-1100 cm^{-1}. There is a significant increase in the total intensity of this range. This increasing intensity, along with the increasing intensity in the 1230-1400 cm^{-1} range, suggests an increasing number of crosslinking bond formation between zein and glutaraldehyde. These changes suggest that the interaction between glutaraldehyde and zein happens through the OH groups of glutamine at the end of which C-O-C bond formation occurs [5]. Amide I bands of the FTIR spectra were further analyzed for secondary structure changes and the results showed that the addition of increasing amounts of glutaraldehyde resulted in conversion of some of the α-helix into β-turns consistent with the expectation that crosslinks would limit the formation of α-helix and in crosslink points these constraints would results in the rearrangement of zein in the form β-turns. Characteristics of zein protein were also studied with FT-Raman experiments. The formation of Amide I band at 1600-1700 cm^{-1} and Amide III band at 1200-1400 cm^{-1} are mainly due to the C=O stretching and N-H bending respectively [6]. The peak with the highest intensity was observed in 2800-3000 cm^{-1} range which represents C-H stretching [6]. The addition of crosslinker caused some changes in the FT-Raman spectra indicating disruption and formation of bonds during crosslinking.

AFM images showed that the surface of the crosslinked films were covered with needle-like structures while control films had fewer structures on their surfaces that were shorter and thicker. Control films had the smallest surface roughness values while the roughness increased as the ratio of glutaraldehyde increased. Films with 12% glutaraldehyde ratio had the highest roughness value. Water contact angle measurements did not show much of a trend between the degree of crosslinks and uncrosslinked films. Control films had an the highest average WCA value of 87.850 degrees while the lowest average WCA value was measured with 8% crosslinked films. Texture analysis measurements also showed that tensile strength of the films increased with increasing glutaraldehyde ratio. Also elongation of the films dropped with the addition of crosslinker.

In summary, this study focuses on the effects of crosslinking on the mechanical, chemical and surface properties of zein gel films cast from acetic acid solutions. The formation of new bonds due to the interaction between glutaraldehyde and zein molecules were clearly observed in FTIR results. FT-Raman results were consistent with FTIR results suggesting chemical changes due to increasing amount of glutaraldehyde. Increasing crosslinker amount increased the stiffness of the films as well as the surface roughness of the films. Lastly, surface wettability of the films were not consistently affected from the addition of crosslinker in different ratios.

[1] R. Shukla and M. Cheryan, "Zein: the industrial protein from corn," *Ind. Crops Prod.*, vol. 13, no. 3, pp. 171–192, May 2001.

[2] Y. Luo and Q. Wang, "Zein-based micro- and nano-particles for drug and nutrient delivery: A review," *J. Appl. Polym. Sci.*, vol. 131, no. 16, p. n/a–n/a, Aug. 2014.

[3] Y. Li, J. Li, Q. Xia, B. Zhang, Q. Wang, and Q. Huang, "Understanding the Dissolution of alpha-Zein in Aqueous Ethanol and Acetic Acid Solutions," *J. Phys. Chem. B*, vol. 116, no. 39, pp. 12057–12064, Oct. 2012.

[4] K. Shi, J. L. Kokini, and Q. Huang, "Engineering Zein Films with Controlled Surface Morphology and Hydrophilicity," *J. Agric. Food Chem.*, vol. 57, no. 6, pp. 2186–2192, Mar. 2009.

[5] D. J. Sessa, A. Mohamed, and J. A. Byars, "Chemistry and Physical Properties of Melt-Processed and Solution-Cross-Linked Corn Zein," *J. Agric. Food Chem.*, vol. 56, no. 16, pp. 7067–7075, Aug. 2008.

[6] E. C. Y. Li-Chan, "The applications of Raman spectroscopy in food science," *Trends Food Sci. Technol.*, vol. 7, no. 11, pp. 361–370, Nov. 1996.

455903 Polyolefin Toughened Polypropylene: Mixing Thermodynamcis

Abstract:

A series of poly(cyclohexyl ethylene-*ran*-ethylene) (CE) copolymers with similar statistical segment lengths were synthesized and then melt blended with isotactic polypropylene (*i*PP) at various loadings from 5 - 20 wt.%. Blend morphology was investigated using scanning electron microscopy (SEM). CE copolymers form well dispersed droplets with number average diameters of less than 0.4 Î¼m at a loading of 20 wt.%. Blend mechanical properties were examined with tensile tests. CE copolymers composed of 50 - 70 weight percent of cyclohexyl ethylene are effective in *i*PP toughening, increasing the break strain from 20% to 400% at a loading of 5 wt.%, which might be attributable to shear yielding of the matrix triggered by cavitation of the uniformly dispersed CE copolymer particles. Blend rheology results were fitted to the Palierne model and interfacial tensions were obtained from the fitting parameter. The small interfacial tension values between CE copolymers and *i*PP were found to be closely correlated to relatively small interaction parameter Ï‡ caused essentially by density mismatch. These results suggest that blends with controlled morphology and good mechanical properties can be prepared by carefully balancing the density and statistical segment length mismatch between the components.

457234 Development and Application of Newly Invented Polyimide Aerogel and Its Synthesizing Method.

Abstract:

The new polyimide aerogel synthesizing method was invented via using autoclave without super critical drying method, which largely reduced the economic cost and made the process much easier. The fabricated aerogel showed uniform porous structure and retained the great thermal, mechanical properties and chemical resistance. The polyimide aerogel is thought to be a material that can be widely applied in various industries for different objects, such as catalysts, semiconductor and electric devices. Recently, many kinds of approaches were taken place to improve and enlarge the usage of polyimide aerogel. The synthesizing method of polyimide aerogel was reformed and the porous thin film was also synthesized by the similar process. The porous film structure called sponge and it is now being tested for using in filtration and separation method. In addition, the ferrogel was synthesized by forming polyimide aerogel and super paramagnetic particles simultaneously which is formed inside of the aerogel structure. This structure shows great potential for appliance in drug delivery which largely increased the empty space of the carrier that the great amount of target particles could be contained. On the other hand, the polyimide aerogel also used in electric devices area by used in fabrication of low dielectric constant composite film, which greatly decreased dielectric constant and solved the problem of the water uptake on the film by its original hydrophobicity. The more approaches on polyimide aerogel are now being studied showing great potential impact on diverse areas.

457701 The Development of New Synthetic Approaches to Epoxide Polymerization

Abstract:

Progress in polymer science is built on a foundation of versatile synthetic methodologies, and quantitative conceptual tools for understanding macromolecular structure, and its relationship to physical properties that underlie a wealth of established and developmental technologies. The ring-opening polymerization of epoxide monomers provides a particularly versatile platform for synthesizing polymeric materials with controlled molecular weights, compositions, and chemical functionalities. This versatility is derived from its ring-strain driving force, which places few restrictions on the exact structure of terminal epoxide monomers used in anionic ring-opening polymerization. The desire to expand the palette of amenable monomer structures, accessible molecular weights, and bring about a commensurate expansion in the spectrum of available physical properties motivates the development of increasingly powerful synthetic methodologies. In this presentation, I will lay out a brief historical account of the development of synthetic approaches to polyether materials over the past 100 years, and how this precedent has led to our newest developments in the controlled synthesis of polyether materials.

458225 Fabrication of PBI Based Composite Membrane for Application in High Temperature Polymer Electrolyte Membrane Fuel Cells

Abstract:

Recently, the area of fuel cells have received increased attention from many researchers. One of those type is high temperature polymer electrolyte membrane fuel cells (HT-PEMFC), which utilizes phosphoric acid as a medium for proton conducting over conventional water. One of the main polymer series used for this application is polybenzimidazole (PBI). It has the necessary thermal and chemical stability, as well as affinity towards phosphoric acid, allowing it to be an excellent candidate for HT-PEMFC. Though it has the ability to surpass conventional low temperature PEMFC, one of its current limiting factor is its conductivity, which the basic PBI shows lower values (~0.1S/cm or less). To overcome this, PBI based composite membranes were developed for their application in HT-PEMFC by introducing Al-Si nanoparticles. The fillers were created via metal grafting the silica nanoparticle to improve its affinity towards phosphoric acid, one of the determining factors in performance of the membrane. A series of membrane containing different amount of filler was fabricated in order to study its effect on the membrane. Its basic characteristics were studied via thermogravimetric analysis (TGA), scanning electron microscope (SEM) and universal testing machine (UTM). Its performance as a PEM was studied via its phosphoric acid doping level, proton conductivity and a single cell test by fabrication of MEA.

458323 Influence of Sulfenamide Accelerators on Cure Kinetics and Properties of Natural Rubber Foam

Abstract:

The effect of molecular structure of sulfenamide accelerator, i.e., 2-morpholinothiobenzotiazole (MBS), N-t-butylbenzothiazole-2-sulfenamide (TBBS) and N-cyclohexyl benzothiazole-2-sulfenamide (CBS) on cure kinetics and properties of natural rubber foam was studied. It was found that the natural rubber compound with CBS accelerator shows the fastest cure rate and the lowest activation energy (E_a) because CBS accelerator produces higher level of basicity of amine species than other sulfenamide accelerators, further forming a complex structure with zinc ion as ligands in sulfur vulcanization. Because of the fastest cure rate of CBS accelerator, natural rubber foam with CBS accelerator shows the smallest and narrowest bubble size distribution. Natural rubber foam with CBS accelerator exhibits the lowest bulk density, thermal conductivity and thermal expansion coefficient as well as the highest compression set as a result of fast crosslink reaction.

458632 Chain-By-Chain Monte Carlo Method for Non-Linear Polymerization

Abstract:

Predicting chain microstructures is an important task for polymer scientists, that is made more interesting and challenging by the polydisperse nature of polymer molecules. In this work, a new method, called "Chain-by-Chain Monte Carlo Method" (CBC-MC), is presented for simulating chain microstructures one-by-one or chain-by-chain. CBC-MC is a new hybrid method that uses the mean-field background information as concentrations of polymer populations and small molecules from the *deterministic* solver to provide an environment in which we *stochastically*simulate chains one-by-one with kinetic Monte Carlo. The deterministic solver in this work uses method of moments. The main advantage of CBC-MC is that the use of the deterministic solver allows the elimination of the computational load associated with simulation of the whole ensemble. Method is suited for chemistries, or situations in which chain architecture develops slowly with respect to the background environment, such as controlled reversible-deactivation radical polymerizations. In this work, CBC-MC is applied to a non-linear copolymerization leading to gelation. Effect of a gradient distribution of pendant double bonds along the primary chains on the simulated portion of gel molecules is investigated. Primary chain results are compared with MOM and found to be in perfect agreement. Further investigations are done on primary chain microstructures to better understand multiple phenomena going on in these systems. It has been found that a gradient in PDB distribution along the primary chains can introduce heterogeneities in gel molecules in surface-bound type polymerizations where primary chains within gels are aligned in the same direction but these heterogeneities seem to be disappearing in bulk polymerizations where the chains alignments are random. Results confirm that if applicable, full information regarding the microstructure of chains can be obtained using this method with reduced simulation times and smaller sample sizes.

459761 Improving Risk Assessment of Color Additives in Medical Device Polymers

Abstract:

Many polymeric medical device materials contain color additives which could lead to adverse health effects. The potential health risk of color additives may be assessed by comparing the amount of color additive released over time to levels deemed to be safe based on available toxicity data. We propose a conservative model for exposure that requires only the diffusion coefficient of the additive in the polymer matrix, D, to be specified. The model is applied here using a model polymer (poly(ether-block-amide), PEBAX 2533) and color additive (quinizarin blue) system. Sorption experiments performed in an aqueous dispersion of quinizarin blue (QB) into neat PEBAX yielded a diffusivity D = 4.8 × 10-10 cm2s-1, and solubility S = 0.32 wt. %. Based on these measurements, we validated the model by comparing predictions to the leaching profile of QB from a PEBAX matrix into physiologically representative media. By considering the release of QB from PEBAX matrices over time and computationally estimating toxicity, we demonstrate that a significant enhancement in the size of device that would give rise to toxicological concern can be achieved. These findings suggest that an initial screening-level risk assessment of color additives and other potentially toxic compounds found in device polymers can be improved.

ESTIMATION OF KINETIC PARAMETERS OF ATRP POLYMERIZATION FROM MWD EXPERIMENTAL DATA USING THE PGF TECHNIQUE

C. Fortunatti,[a,b] V. Hanazumi,[a,b] C. Vitale,[c] A. Ciolino,[a,b] M. Asteasuain[a,b]
a) Planta Piloto de Ingeniería Química (PLAPIQUI), UNS-CONICET, Bahía Blanca, Argentina.
b) Departamento de Ingeniería Química, Universidad Nacional del Sur (UNS), Bahía Blanca, Argentina.
c) *Departamento de Química, Universidad Nacional del Sur (UNS), Bahía Blanca, Argentina.*

Introduction

The discovery of controlled radical polymerization (CRP) allowed producing polymers with unique molecular structures in mild operating conditions. CRP enables the synthesis of new materials with pre-specified properties without the stringent conditions required by anionic polymerization. This fact motivated a great interest in the scientific and industrial communities. Numerous scientific studies and patents, and several industrial products already developed show this point.[1]

Atom transfer radical polymerization (ATRP) is one of the most popular CRP techniques. Although it employs a transition metal complex as catalyst (that must often be removed from the product), it presents several advantages such as commercial availability of reagents, large range of polymerizable monomers and simple end-functionalization. These advantages have fostered an extensive study of its kinetic mechanism, which has ultimately led to the development of several ATRP variants that allow overcome most of the shortages of this technique.[2]

The production of differentiated materials requires a fine tuning of the polymer molecular structure. This can only be accomplished with a precise control of operating conditions. Mathematical models are useful tools for this task since they allow manufacturers to find the proper set of operating conditions necessary for synthesizing tailor-made polymers while minimizing the expenditure of resources on trial and error procedures.[3] For this purpose, accurate adjustment of the model kinetic parameters is needed.

Several authors have successfully dealt with the modeling of ATRP polymerization systems.[4-6] In all of these works, the estimation of model parameters was based on experimental data of conversion and/or average molecular weights. However, parameter estimation based on molecular weight distribution (MWD) data may be more accurate due to the higher amount of experimental information for the fitting process.

In this work we present a comprehensive mathematical model of the ATRP reaction able to predict average molecular properties as well as the full MWD. Average properties are predicted using the well-known method of moments. The MWD is modeled using the probability generating function (pgf) technique. This technique has great potential for parameter estimation from the MWD data using optimization-based procedures, since it allows a fast model execution and therefore a quick objective function evaluation.[7, 8] By means of this approach, activation, deactivation, propagation and termination kinetic constants of the ATRP of styrene (St) were estimated from experimental data of MWD. The experimental set up involved the synthesis of polystyrene at 105°C in solution of toluene using CuBr as the transition metal catalyst and hexamethyl-tris(2-aminoethyl)amine (Me_6TREN) as ligand and ethyl 2-bromo iso-butyrate (EBriB) as radical initiator. Our results show that the predicted MWD has very good agreement with the experimental data.

Methods

Materials

St (Aldrich, 99 %) was filtered in Aluminium Oxide (Al_2O_3). The monomer and solvent were purged with N_2 for at least 30 minutes. The initiator EBriB (Aldrich, 98 %), the catalyst (CuBr) and the ligand Me_6TREN and all other reagents were purchased from commercial sources and used without further purification.

Polymerization Procedure

The molar relation between monomer, initiator, catalyst, and ligand were [M]:[I]:[Cat]:[L] = [200]:[2]:[1]:[1]. Toluene (5 mL) and St (5 mL, 4.46×10^{-2} mol) were mixed in a Schlenk flask equipped with a stir bar. The solution became greenish-greyish when the catalyst CuBr (0.0312 g, 2.18×10^{-4} mol) was introduced. Subsequently, the ligand Me_6TREN (0.033 mL, 2.18×10^{-4} mol) was added, and the solution turned in a bright green color.

The mixture was purged with N_2 during an hour in constant agitation, in order to remove the oxygen from the system. After that time, the initiator EBriB (0.064 mL, 4.36 x 10^{-4} mol) was introduced with a syringe at ambient temperature. Then, the flask was placed in an oil bath and stirred at 105 °C.

Samples at 2, 4, 5 hours and 21.5 hours were taken. These samples were precipitated in cold methanol to stop the reaction. The resulting polymer was later dissolved in toluene to be analyzed.

Analyses of Samples

Molecular weights were measured on a gel permeation chromatography (GPC) system consisting of a Waters 515 HPLC pump, Rheodyne 7725i Injector, Polymer Laboratories 10 µm (PLgel) columns, and a Waters Refractive Index Detector 2412. Polystyrene standards were used.

Mathematical Model

The mathematical model considers the following kinetic mechanism:

Initiation:
$$I + CuBrL \xrightarrow{k_{act}^0} CuBr_2L + R^\circ \quad (1)$$

$$R^\circ + CuBr_2L \xrightarrow{k_{deact}^0} CuBrL + I \quad (2)$$

Thermal initiation of styrene:
$$3\,M \xrightarrow{k_{th}} R_1 + R_2 \quad (3)$$

Propagation:
$$R_n + M \xrightarrow{k_p} R_{n+1} \quad (4)$$

Activation:
$$CuBrL + D_n \xrightarrow{k_{act}} R_n + CuBr_2L \quad (5)$$

Deactivation:
$$R_n + CuBr_2L \xrightarrow{k_{deact}} CuBrL + D_n \quad (6)$$

Chain termination by coupling:
$$R_n + R_s \xrightarrow{k_t} P_{n+s} \quad (7)$$

The chemical species involved are: initiator EBriB (I), monomer St (M), the catalyst-ligand complex in reduced state (CuBrL), the catalyst-ligand complex in oxidized state ($CuBr_2L$), active radicals with n monomer units (R_n), dormant (inactive) radicals with n monomer units (D_n), and terminated (dead) polymer chains with n monomer units (P_n).

Only the kinetic parameter corresponding to the thermal initiation of styrene was taken from the literature.[9, 10] The other kinetic parameters were fitted in this work. It was assumed that the activation and deactivation reactions proceeded at the same rate both in the initiation and in the core-equilibrium, i.e. $k^0_{act} = k_{act}$ and $k^0_{deact} = k_{deact}$.

Average properties such as molecular weights, composition and conversion, were modeled using the well-known method of moments.

The MWD was computed using the 1D probability generating functions (pgf) technique. This technique consists in transforming the mass balances to the pgf domain. After solving the pgf balances equations, the pgf are inverted to recover the full MWD. The IFG inversion method was used in this work. Further details on this technique and its mathematical implementation may be found elsewhere.[11, 12]

The mass, moment and pgf balances included in the mathematical model of the process are shown below. For simplification, the catalyst-ligand complex in reduced state (CuBrL) is named "Red" and the catalyst-ligand complex in oxidized state ($CuBr_2L$) is named "Ox".

Mass balances:

Global:
$$\frac{d\left(\rho_{mix}V\right)}{dt}=0 \tag{8}$$

Generic species X:
$$\frac{d\left([\mathrm{X}]V\right)}{dt}=r_{\mathrm{X}}\ V \tag{9}$$

Density equations:[13]

$$\rho_{\text{mix}}=\left(\frac{x_{\text{St}}}{\rho_{\text{St}}}+\frac{x_{toluene}}{\rho_{\text{toluene}}}+\frac{x_{\text{I}}}{\rho_{\text{I}}}+\frac{x_{\text{Cat-L complex}}}{\rho_{\text{Cat-L complex}}}+\frac{x_{\text{polySt}}}{\rho_{\text{polySt}}}\right)^{-1} \tag{10}$$

$$\rho_{\text{toluene}}=889.9-1.064\ T(°\text{C})=\rho_{\text{I}}=\rho_{\text{Cat-L complex}} \tag{11}$$

$$\rho_{\text{St}}=919.3-0.665\ T(°\text{C}) \tag{12}$$

$$\rho_{\text{polySt}}=992.6-0.265\ T(°\text{C}) \tag{13}$$

Mass fraction of low molecular weight species Y: $x_Y=\dfrac{[Y]\,\mathrm{MW}_Y}{\rho_{mix}}$ (14)

Mass fraction of polymer:
$$x_{\text{polySt}}=\frac{\left(\lambda_1+\mu_1+\varepsilon_1\right)\mathrm{MW}_{\text{polySt}}}{\rho_{mix}} \tag{15}$$

where MW_Y is the molecular weight of species *Y*. Variables λ_1, μ_1, and ε_1 are defined below and correspond to the moments of order 1 of active radicals, dormant radicals and terminated polymer, respectively.

Reaction rates:

Initiator:
$$r_{\text{I}}=-k_{\text{act}}[\text{I}][\text{Red}]+k_{\text{deact}}\left[\text{R}^{\circ}\right][\text{Ox}] \tag{16}$$

Monomer: $$r_{\mathrm{M}} = -3k_{\mathrm{th}}[\mathrm{M}]^3 - k_{\mathrm{p}}[\mathrm{M}]\lambda_0 \tag{17}$$

Reduced catalyst-ligand complex:

$$r_{\mathrm{Red}} = -k_{\mathrm{act}}[\mathrm{Red}]([\mathrm{I}]+\lambda_0) + k_{\mathrm{deact}}[\mathrm{Ox}]([\mathrm{R}^\circ]+\lambda_0) \tag{18}$$

Oxidized catalyst-ligand complex:

$$r_{\mathrm{Ox}} = k_{\mathrm{act}}[\mathrm{Red}]([\mathrm{I}]+\lambda_0) - k_{\mathrm{deact}}[\mathrm{Ox}]([\mathrm{R}^\circ]+\lambda_0) \tag{19}$$

Initiation radical: $$r_{\mathrm{R}^\circ} = -k_{\mathrm{deact}}[\mathrm{Ox}][\mathrm{R}^\circ] + k_{\mathrm{act}}[\mathrm{Red}][\mathrm{I}] - k_{\mathrm{t}}[\mathrm{R}^\circ]([\mathrm{R}^\circ]+\lambda_0) - k_{\mathrm{p}}[\mathrm{M}][\mathrm{R}^\circ] \tag{20}$$

Moment of order a of active radicals:

$$\begin{aligned} r_{\lambda_a} &= k_{\mathrm{th}}[\mathrm{M}]^3(1^a+2^a) + k_{\mathrm{act}}[\mathrm{Red}]\mu_a + k_{\mathrm{p}}[\mathrm{M}]\left([\mathrm{R}^\circ] + \sum_{j=0}^{a}\binom{a}{j}\lambda_a\right) \\ &\quad -\left[k_{\mathrm{p}}[\mathrm{M}] + k_{\mathrm{deact}}[\mathrm{Ox}] + k_{\mathrm{t}}([\mathrm{R}^\circ]+\lambda_0)\right]\lambda_a \end{aligned} \tag{21}$$

Moment of order a of dormant radicals: $$r_{\mu_a} = k_{\mathrm{deact}}[\mathrm{Ox}]\lambda_a - k_{\mathrm{act}}[\mathrm{Red}]\mu_a \tag{22}$$

Moment of order a of terminated polymer chains: $$r_{\varepsilon_a} = k_{\mathrm{t}}[\mathrm{R}^\circ]\lambda_a + \tfrac{1}{2}k_{\mathrm{t}}\sum_{j=0}^{a}\binom{a}{j}\lambda_{a-j}\lambda_j \tag{23}$$

Average molecular weights can be computed from these moments as follows:

Number average molecular weight: $$\overline{M_n} = \frac{\lambda_1+\mu_1+\varepsilon_1}{\lambda_0+\mu_0+\varepsilon_0}\,\mathrm{MW_{St}} \tag{24}$$

Weight average molecular weight: $$\overline{M_w} = \frac{\lambda_2+\mu_2+\varepsilon_2}{\lambda_1+\mu_1+\varepsilon_1}\,\mathrm{MW_{St}} \tag{25}$$

The pgf needs to be evaluated at values of its dummy variable "z" determined by the inversion method. Once the MWD expressed in number fraction is computed by inversion of the overall pgf of order 0, the weight MWD is calculated from it. Therefore, it is not necessary to compute pgf of higher orders. The pgf balances included in the model are:

Pgf of order 0 of active radicals:

$$\begin{aligned} r_{(\lambda_0\mathrm{pgfR}_0(z))} &= (z+z^2)k_{\mathrm{th}}[\mathrm{M}]^3 + k_{\mathrm{act}}[\mathrm{Red}][\mu_0\mathrm{pgfD}_0(z)] \\ &\quad + k_{\mathrm{p}}\,z[\mathrm{M}]([\mathrm{R}^\circ] + [\lambda_0\mathrm{pgfR}_0(z)]) \\ &\quad - \left[k_{\mathrm{p}}[\mathrm{M}] + k_{\mathrm{deact}}[\mathrm{Ox}] + k_{\mathrm{t}}([\mathrm{R}^\circ]+\lambda_0)\right][\lambda_0\mathrm{pgfR}_0(z)] \end{aligned} \tag{26}$$

Pgf of order 0 of dormant radicals:

$$r_{\left[\mu_0 \text{pgfD}_0(z)\right]} = k_{\text{deact}}\left[\text{Ox}\right]\left[\lambda_0 \text{pgfR}_0(z)\right] - k_{\text{act}}\left[\text{Red}\right]\left[\mu_0 \text{pgfD}_0(z)\right] \tag{27}$$

Pgf of order 0 of terminated polymer chains:

$$r_{\left[\varepsilon_0 \text{pgfP}_0(z)\right]} = k_t\left[\left[\text{R}^\circ\right] + \tfrac{1}{2}\left[\lambda_0 \text{pgfR}_0(z)\right]\right]\left[\lambda_0 \text{pgfR}_0(z)\right] \tag{28}$$

Pgf of the overall polymer: $$\text{pgfT}_0(z) = \frac{\lambda_0 \text{pgfR}_0(z) + \mu_0 \text{pgfD}_0(z) + \varepsilon_0 \text{pgfP}_0(z)}{\lambda_0 + \mu_0 + \varepsilon_0} \tag{29}$$

Estimation of Kinetic Parameters

The simulations and estimation problems were formulated and solved in gPROMS v 3.6.0 (Process Systems Enterprise, Ltd.) in a standard PC. gPROMS contains a state-of-the-art Parameter Estimation capability. To perform the estimation this entity formulates a maximum likelyhood parameter estimation problem that maximize the probability that the mathematical model will predict the measurement values obtained from the experiments.[14]

Fourteen pre-defined points of each experimental MWD measured at 2, 4, 5 and 21.5 hours were inputted as data in the Parameter Estimation entity of gPROMS. Even though global optimality cannot be guaranteed, multiple starting points were tested in order to minimize the chances of falling in a local optimum.

Results and Discussion

The estimated kinetic parameters are presented in **Table 2**. Literature data is included in the table for the sake of comparison. It can be observed that the estimated parameters are in the range of previously reported literature values for similar polymerization systems.

Table 2. Estimated St ATRP kinetic constants compared with literature.

Reaction	Parameter	Our Estimation	Literature	Units
Activation ATRP equilibrium	k_{act}	0.81	20 – 0.04 [4, 15-18]	L mol^{-1} s^{-1}
Deactivation ATRP equilibrium	k_{deact}	2.1×10^5	$4\times10^5 - 1\times10^7$ [4, 15-18]	L mol^{-1} s^{-1}
Constant of ATRP Equilibrium	K_{ATRP}	3.9×10^{-6}	$4\times10^{-8} - 5\times10^{-6}$ [4, 15-18]	
Propagation	k_p	6.9×10^3	1.4×10^3 [10]	L mol^{-1} s^{-1}
Termination	k_t	2.7×10^7	3.3×10^8 [10]	L mol^{-1} s^{-1}

Figure 1 shows the MWD computed with the pgf model using the fitted parameters compared to the experimental MWD. It can be seen that the model predictions closely follow the experimental distributions. The difference between them is mostly within experimental error.

The computation time for each simulation was less than 10 seconds. Since the same model is solved several times to find the optimal parameters, this high computational speed helps to increase the estimation activity efficiency. However, the time taken to solve the parameter estimation is naturally affected by the location of the starting point. Depending on the starting point, the parameter estimation required between 70 to 600 seconds to be solved.

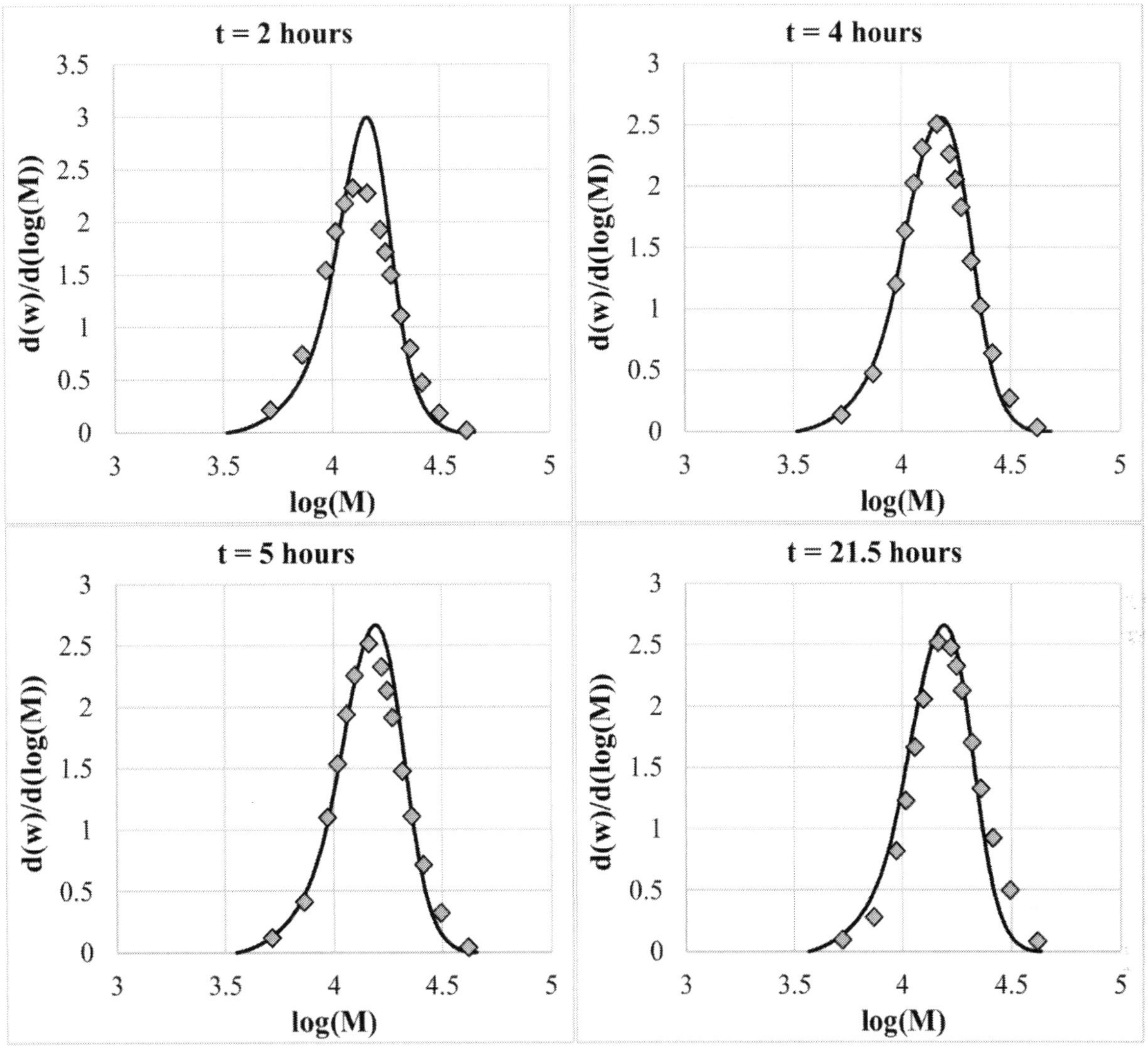

Figure 1. MWD of polymer synthetized by ATRP at different reaction times.
Full line: experimental MWD – Symbols: MWD estimated with pgf mathematical model.

Table 1 shows the average molecular weights computed with the moment equations (Eq.23 and 24) compared to the experimental data. A very good agreement is obtained. It should be noted that the average molecular weights were not explicitly included in the parameter estimation problem. Only data of the MWD function were used.

Table 1. Average molecular weights of polymer synthetized by ATRP at different reaction times.

	EXPERIMENTAL			MODEL				
Time	$\overline{M}_n$	$\overline{M}_w$	PDI	$\overline{M}_n$	$\overline{M}_w$	PDI	Error $\overline{M}_n$	Error $\overline{M}_w$
2 hs	12424	14003	1.13	11297	13800	1.22	**9.1%**	**1.5%**
4 hs	12767	14692	1.15	12435	15079	1.21	**2.6%**	**-2.6%**
5 hs	13261	15133	1.14	12709	15403	1.21	**4.2%**	**-1.8%**
21.5 hs	13126	14967	1.14	13631	16602	1.22	**-3.8%**	**-10.9%**

Conclusions

The kinetic parameters of the ATRP polymerization has been successfully estimated from MWD experimental data using the pgf technique. The obtained rate constants were in the order of the kinetic parameters reported in the literature and allowed to accurately reproduce the real shape of the distribution. What is more, the model requires little computational efforts and allowed to obtain accurate prediction of average properties even though these variables were not used to solve the parameter estimation problem.

In conclusion, the results show that the pgf technique has great potential for modeling, estimation and optimization of controlled radical polymerizations.

References

1. Matyjaszewski, Krzysztof and James Spanswick (2005), "Controlled/living radical polymerization," *Materials Today*, Vol. 8, pp. 26-33.
2. Braunecker, W. A. and K. Matyjaszewski (2007), "Controlled/living radical polymerization: Features, developments, and perspectives," *Prog. Polym. Sci.*, Vol. 32, pp. 93-146.
3. D'Hooge, D. R., D. Konkolewicz, M. F. Reyniers, G. B. Marin and K. Matyjaszewski (2012), "Kinetic modeling of ICAR ATRP," *Macromol. Theory Simul.*, Vol. 21, pp. 52-69.
4. Al-Harthi, M., L. S. Cheng, J. B. P. Soares and L. C. Simon (2007), "Atom-transfer radical polymerization of styrene with bifunctional and monofunctional initiators: Experimental and mathematical modeling results," *J. Polym. Sci., Part A: Polym. Chem.*, Vol. 45, pp. 2212-2224.
5. Kwark, Y. J. and B. M. Novak (2004), "Determination of the kinetic parameters of atom transfer radical polymerizations," *Macromolecules*, Vol. 37, pp. 9395-9401.
6. Pintauer, T., P. Zhou and K. Matyjaszewski (2002), "General method for determination of the activation, deactivation, and initiation rate constants in transition metal-catalyzed atom transfer radical processes," *J. Am. Chem. Soc.*, Vol. 124, pp. 8196-8197.
7. Asteasuain, M., C. Sarmoria and A. Brandolin (2002), "Recovery of molecular weight distributions from transformed domains. Part I. Application of pgf to mass balances describing reactions involving free radicals," *Polymer*, Vol. 43, pp. 2513-2527.
8. Fortunatti, C., C. Sarmoria, A. Brandolin and M. Asteasuain (2014), "Prediction of the full molecular weight distribution in RAFT polymerization using probability generating functions," *Comput. Chem. Eng.*, Vol. 66, pp. 214-220.
9. Asteasuain, M., S. Matheus, M. K. Lenzi, R. A. Hutchinson, M. Cunningham, A. Brandolin, J. C. Pinto and C. Sarmoria (2007), "‘‘Living’’ Free Radical Polymerization in Tubular Reactors. I. Modeling of the Complete Molecular Weight Distribution Using Probability Generating Functions,"

Macromol. React. Eng., Vol. 1, pp. 622-634.
10. Asteasuain, M., D. Covan, C. Sarmoria, A. Brandolin, C. Leite de Araujo and J. C. Pinto (2011Comprehensive Mathematical Modeling of Controlled Radical Copolymerization in Tubular Reactors), "Comprehensive Mathematical Modeling of Controlled Radical Copolymerization in Tubular Reactors," in *21st European Symposium on Computer Aided Process Engineering,* Vol. 29, pp. 51-55.
11. Brandolin, A. and M. Asteasuain (2013), "Mathematical modeling of bivariate distributions of polymer properties using 2D probability generating functions. part II: Transformation of population mass balances of polymer processes," *Macromol. Theory Simul.*, Vol. 22, pp. 273-308.
12. Brandolin, A., A. Assini Balbueno and M. Asteasuain (2016), "Improved numerical inversion methods for the recovery of bivariate distributions of polymer properties from 2D probability generating function domains," *Comput. Chem. Eng.*, Vol. 94, pp. 272-286.
13. Fortunatti, C. (2015), "*Simulación y Optimización de la Síntesis de Copolímeros de Estireno y Monómeros Acrílicos de Estructura Controlada*," Bahía Blanca, Buenos Aires, Argentina: Universidad Nacional del Sur.
14. gPROMS (2012), "*gPROMS v3.6 - Model Validation Guide*," Process Systems Enterprise Ltd.
15. Mueller, L., W. Jakubowski, W. Tang and K. Matyjaszewski (2007), "Successful chain extension of polyacrylate and polystyrene macroinitiators with methacrylates in an ARGET and ICAR ATRP," *Macromolecules*, Vol. 40, pp. 6464-6472.
16. Matyjaszewski, K., H. J. Paik, P. Zhou and S. J. Diamanti (2001), "Determination of activation and deactivation rate constants of model compounds in atom transfer radical polymerization1," *Macromolecules*, Vol. 34, pp. 5125-5131.
17. Vieira, R. P., A. Ossig, J. M. Perez, V. G. Grassi, C. L. Petzhold, A. Peres, J. M. Costa and L. M. F. Lona (2013), "Simulation of the equilibrium constant effect on the kinetics and average properties of polystyrene obtained by ATRP," *J. Braz. Chem. Soc.*, Vol. 24, pp. 2008-2014.
18. Fu, Y., M. F. Cunningham and R. A. Hutchinson (2007), "Semibatch atom transfer radical copolymerization of styrene and butyl acrylate," *Macromol. Symp.*, Vol. 259, pp. 151-163.

460063 Pil-Derived Carbon

Abstract:

The development of new carbon-based materials with exquisite control over surface chemistry, carbon structure and form, surface area, pore size, and conductivity is of significant interest for numerous applications, including energy storage (*e.g.*, capacitors). Recently, a few studies have demonstrated the ability to produce carbon from ionic liquids (ILs) and polymerized ionic liquids (PILs), both of which have unique physiochemical properties and result in carbon with various surface chemistries depending on the chemistry of the IL or PIL. However, to date, few studies have explored the relationship between the possible diverse PIL chemistries and the resulting PIL-derived carbon properties. In this study, numerous PILs with various backbone/cation pairings (backbones: ethyl methacrylate, styrene; covalently attached cations: butylimidazolium, trimethylammonium, piperidinium, butylpyrrolidinium, methylpyrrolidinium, pyrrolidinium) were synthesized as carbon precursors. The chemistry and molecular weight of the PILs were characterized using proton nuclear magnetic resonance (^{1}H NMR) spectroscopy, elemental analysis (EA), and gel permeation chromatography (GPC). PIL-derived carbons were produced from this set of PILs *via* pyrolysis at 10 °C/min. The surface chemistry, carbon structure and form, porosity, and conductivity of these PIL-derived carbons were characterized using EA, X-ray photoelectron spectroscopy (XPS), X-ray diffraction (XRD), scanning electron microscopy (SEM), transmission electron microscopy (TEM), electrochemical impedance spectroscopy (EIS), and gas sorption measurements. The relationship between PIL chemistry and PIL-derived carbon properties will be discussed, along with the subsequent implications on electrode materials for energy storage applications.

MODELING OF REVERSE ATOM TRANSFER RADICAL POLYMERIZATION IN MINIEMULSION INITIATED BY A WATER-SOLUBLE RADICAL INITIATOR

Ayelén Zurman[1,2], *Mariano Asteasuain*[1,2], *Claudia Sarmoria*[1,2] *and Adriana Brandolin*[1,2]
[1]*Planta Piloto de Ingeniería Química (PLAPIQUI), UNS-CONICET, Bahía Blanca, Argentina.*
[2]*Departamento de Ingeniería Química, Universidad Nacional del Sur (UNS), Bahía Blanca, Argentina*

Introduction

Controlled radical polymerization (CRP) has emerged in the last few years as one of the most promising polymerization techniques. It allows the synthesis of polymers with predetermined molecular weight, low polydispersity and controlled molecular architecture in mild reaction conditions. The latter has been one of the most important advantages of CRP compared to anionic polymerization, which requires stringent operating condition.

The main CRP variants include nitroxide-mediated polymerization (NMP), atom transfer radical polymerization (ATRP), and reversible addition–fragmentation chain transfer radical polymerization (RAFT). All of them have proven to be robust and capable of synthetizing polymers with defined molecular weight, topology, composition, and functionality.

ATRP has been widely studied both in academia and industry. It is based on a dynamic equilibrium between active and dormant chains. In this reaction an active chain ($P_n\cdot$) reacts with a deactivator (i.e $CuBr_2/L$), producing dormant species (P_nBr) and an activator ($CuBr_1/L$). These dormant species are incapable of propagating and predominate over radicals. When they are activated, they can propagate and grow or terminate. Therefore, polymer growth is controlled by this equilibrium, which is shown in Figure 1.

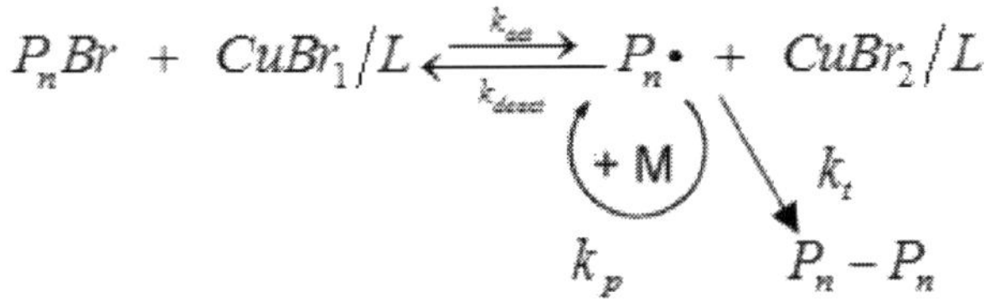

Figure 1. Scheme of ATRP equilibrium

ATRP offers distinct advantages such as the commercial availability of reactants, the simple synthesis of polymers with specific tailored functionalities and the wide range of monomers and temperatures that may be used.[1] Additionally, several initiating procedures have been developed for this technique, making ATRP suitable for different conditions and requirements. One of the procedures developed is the so called reverse ATRP. It has emerged as a response to the handling problem in direct ATRP caused by the high air sensitivity of Cu^{I}. Reverse ATRP employs more stable Cu^{II} complexes in the initiation step and a conventional free radical initiator, leading to a system easier to handle and more compatible with industrial scale processes.[2] Besides, due to these features reverse ATRP has been widely employed in dispersed systems.[3]

Dispersed systems respond to the need for making CRP more suitable for commercial use. This need has led to investigate the feasibility of CRP processes in aqueous dispersed systems, such as emulsion and miniemulsion.[2] Aqueous dispersions are a good alternative for large scale production.

They provide several benefits such as excellent heat transfer, process flexibility and ease of mixing and handling of the final product. However, aqueous dispersed systems present difficulties, such as partitioning of species between aqueous and organic phases, exit of radicals from particles and poor colloidal stability, among others. Although the first attempts were performed in emulsion polymerization, many of them failed leading to the use of miniemulsions. The miniemulsion system involves a dispersion of monomer droplets in water with much smaller droplets than in emulsion polymerization. Micelles are usually not present because surfactant concentrations are generally below the critical micellar concentration.

Miniemulsion systems have demonstrated to be robust for the different CRP techniques,[3] because droplet nucleation minimizes the requirement of mass transport of the control agent to the polymer particle.[2] They offer unique advantages over emulsion polymerization, since it allows the polymerization of highly water insoluble monomers and it is suitable for making particles containing additives (such as pigments or dyes). A disadvantage of miniemulsions is that they employ a low molecular weight hydrophobe (for avoiding diffusional degradation) that sometimes need to be removed from the final latex.[3]

ATRP in miniemulsion has been extensively studied.[4] Many experiments have been carried out using different monomers (St, BA, BMA), varying the initiating procedure and other experimental conditions (ligands, surfactant).[2,5–7] A significant progress has also been made in the modeling of ATRP system in miniemulsion, studying the compartmentalization effect and the partitioning of species in these systems.[5,8]

In the present work, a mathematical model was developed for a reverse ATRP in miniemulsion using a water-soluble initiator. This model is able to predict average molecular properties, such as number and weight molecular weights, as well as the full molecular weight distribution (MWD), and constitutes a useful tool for predicting and optimizing such systems.

Model Description

Description of the heterogeneous system

The system under study consists of a reverse ATRP in miniemulsion initiated by a water soluble radical initiator in a batch reactor at 70°C. The monomer is n- butyl methacrylate (n-BMA), the deactivator/ligand is $CuBr_2$/EH_6TREN and 2,2'-azobis[2-(2-imidazolin-2-yl)propane] dihydrochloride (VA-044) the initiator.

The initiator decomposes in the aqueous phase and generates radicals which propagate with the n-BMA available in this phase or terminate with other radicals. If the radicals do not terminate, they grow until a given length chain for which they become surface-active. Once they achieve this condition, they can enter into the organic phase where they may be deactivated by $CuBr_2$/EH_6TREN. This reaction produces dormant species and the activator $CuBr_1$/ EH_6TREN, establishing the ATRP equilibrium. These radicals also propagate, terminate or transfer to monomer inside the organic droplets.

In order to model the system described above, several assumptions were adopted:

i. The phase volumes are considered constant.

ii. Compartmentalization effects are insignificant since droplet size is large (particle diameter >200 nm).[8] Thus, the system can be treated as a bulk system.

iii. Partitioning of activator, deactivator and monomer take place. Additionally, it is considered that these species are at phase equilibrium throughout the polymerization.[5,9]

iv. The water is saturated with n-BMA.[10]

v. Only radical entry is accounted for; radical exit and reentry are neglected.[11]

Table 1. Kinetic mechanism of reverse ATRP in miniemulsion initiated by a water-soluble initiator.

Reaction	***Equation***	***Ref***
Initiator decomposition (aqueous phase)	$I-I \xrightarrow{k_d\ f} 2I\bullet$ (1)	2,10,12–14
Initiation (aqueous phase)	$I\bullet + M \xrightarrow{k_i} P_1\bullet$ (2)	
Propagation (aqueous phase)	$P_1\bullet + \mathrm{M} \xrightarrow{k_{P1}} P_2\bullet$ (3) $P_2\bullet + \mathrm{M} \xrightarrow{k_{P2}} P_3\bullet$ (4)	
Propagation (organic phase)	$P_n\bullet + \mathrm{M} \xrightarrow{<k_p>} P_{n+1}\bullet$ (5)	
Termination (aqueous and organic phase) By disproportionation By combination	$P_n\bullet + \mathrm{P}_m\bullet \xrightarrow{k_{td}} D_n + \mathrm{D}_m$ (6) $P_n\bullet + \mathrm{P}_m\bullet \xrightarrow{k_{tc}} D_{n+m}$ (7)	
ATRP equilibrium (organic phase)	$P_nBr + \mathrm{CuBr}_1/L \underset{k_{deact}}{\overset{k_{act}}{\rightleftarrows}} \mathrm{P}_n\bullet + \mathrm{CuBr}_2/L$ (8)	
Chain transfer to monomer (aqueous and organic phase)	$P_n\bullet + \mathrm{M} \xrightarrow{k_{trm}} \mathrm{D}_n + \mathrm{P}_1\bullet$ (9)	
Oligoradicals entry (aqueous-organic interface)	$P_3\bullet_{aq} \longrightarrow P_3\bullet_{org}$ (10)	9,11
Partitioning of activator, deactivator and monomer (aqueous-organic interface)	$CuBr_1/L_{aq} \underset{k_2}{\overset{k_1}{\rightleftarrows}} CuBr_1/L_{org}$ (11) $CuBr_2/L_{aq} \underset{k_4}{\overset{k_3}{\rightleftarrows}} CuBr_2/L_{org}$ (12) $M_{aq} \underset{k_6}{\overset{k_5}{\rightleftarrows}} \mathrm{M}_{org}$ (13)	9,10

Kinetic mechanism

The kinetic mechanism employed in this work is reported in Table 1. The participating species are: initiator (I-I), initiation radical (I·), monomer (M), radicals with n monomer units ($P_n\cdot$), dormant species with n monomer units (P_nBr), dead polymer chains with n monomer units (D_n), activator ($CuBr_1/L$) and deactivator ($CuBr_2/L$). The subscripts aq and org are employed for indicating whether the species is located in the aqueous or organic phase, respectively.

According to Tsarevsky and Matyjaszewski,[15] in protic solvents like water, the deactivator suffers a reversible dissociation and substitution of the halide ligand by water, leading to inefficient deactivation. Therefore, we consider that reactions of activation and deactivation do not take place in the aqueous phase.

We employ the usual assumption that the rate of initiation is much faster than the rate of initiator decomposition ($k_i >> k_d$).[16] For the organic phase, we use an average kinetic rate constant of propagation ($<k_P>$). However, for oligoradicals in the aqueous phase, two propagation rate constants are employed (k_{P1}, k_{P2}) with values larger than $<k_P>$. This is in agreement with studies that demonstrated that k_P has a chain length-dependency in the short chain-length region.[17,18] Additionally, we account for the chain-length dependency of k_t. Buback et al.[19] demonstrated that the dependence of k_t with monomer conversion consists of an initial plateau region (segmental diffusion) up to monomer conversion of 25-30%, followed by a decreasement region at higher conversions (translational diffusion and then reaction-diffusion control). This last region has been approximated with a linear dependence for its modeling. Pursuing a better numerical behavior of the optimization solver used for the parameter estimation, the dependence of k_t with conversion was implemented by the following smoothing function with continuous derivatives:

$$k_t(x) = \{1 - 0.5\cdot[\tanh(5\cdot(x-25))+1]\}\cdot 1.\,10^7 + 0.5\cdot[\tanh(5\cdot(x-25))+1]\cdot(1.3\,10^7 - 1.2\,10^5\cdot x) \quad (14)$$

Partitioning of Species

Partitioning of species is characteristic of heterogeneous systems. In this work the partitioning of the deactivator ($CuBr_2/L$), activator ($CuBr_1/L$) and monomer between aqueous and organic phases are included.

It is represented by using partition coefficients γ_i which are defined as[10]

$$\gamma_i = \frac{[i]_{org}}{[i]_{aq}} \quad (15)$$

where $[i]_j$ represents the concentration of the species *i* in phase *j* with *j*=org, aq.

As commented above we consider that phase equilibrium is maintained during the reaction. In order to achieve that, the rates of the forward and reverse mass transfer reactions between phases must be approximately equal, i.e.:

$$-k_{forward}[i]_{aq}V_{aq} + k_{reverse}[i]_{org}V_{org} \approx 0 \quad (16)$$

where $k_{forward}/k_{reverse}$ represents the ratio of the rate constants involved in Eqs (11), (12) or (13) (k_1/k_2, k_3/k_4 or k_5/k_6) and V_j is the phase volume with j= org, aq. From Eq. (16) a relationship between the transfer rate coefficients and the partitioning coefficient can be established:

$$\frac{k_{forward}}{k_{reverse}} = \frac{[i]_{org}}{[i]_{aq}}\frac{V_{org}}{V_{aq}} = \gamma_i\frac{V_{org}}{V_{aq}} \quad (17)$$

The pseudo-equilibrium is achieved in a numerical simulation by assigning large values ($>10^8$ s^{-1}) to the forward and reverse transfer rate coefficients.[9] Thus, we give a large value ($>10^8$ s^{-1}) to one of the transfer coefficient, and calculate the other from Eq. (17).

For the activator and deactivator, we use the partition coefficients reported in the literature.[10] In the case of monomer, it is determined from the saturation concentration of n-BMA in water ($[M]_{sat}$=3.4 10^{-3} M (T=60°C)[20], employing Eq.(15).

It should be mentioned that we obtained similar results when the partitioning of species was represented by equations derived from fundamental transport principals, as reported elsewhere.[21]

Radical entry

Once the oligoradicals from aqueous phase add a few monomer units, they become surface-active and can enter into the organic phase. Maxwell et al.[11] proposed an equation for calculating the degree of polymerization z in which oligoradical become hydrophobic and enter irreversibly the monomer droplets:

$$z = 1 + \text{int}\left(\frac{23\left[kJ\ \text{mol}^{-1}\right]}{RT\ \ln\left(\left[M_{aq,sat}\right]\right)} \right) \tag{18}$$

where the function int(.) rounds off the quantity in brackets to an integer value. The lower or the higher integer value has been used in the literature.[9,11] In this work we use the higher integer value which, for our data, yields a value of z = 3.

Modeling of average molecular properties and MWD

The model is based on the mass balance equations of the reacting species as reported below. A batch reactor is considered. Note that if a species is present in both aqueous and organic phases, it is treated as two different entities for mathematical modeling.

Initiator

$$\frac{d\left([I-I]\,V_{aq}\right)}{dt} = -k_d[I-I]V_{aq} \tag{19}$$

Monomer

$$\begin{aligned} \frac{d\left([M]_{aq}V_{aq}\right)}{dt} &= -2k_d f[I-I]V_{aq} - k_{P1}[M]_{aq}[P_1]_{aq}V_{aq} - k_{P2}[M]_{aq}[P_2]_{aq}V_{aq} \\ &-k_{trm}[M]_{aq}[P_1]_{aq}V_{aq} - k_{trm}[M]_{aq}[P_2]_{aq}V_{aq} - k_5[M]_{aq}V_{aq} + k_6[M]_{org}V_{org} \end{aligned} \tag{20}$$

$$\begin{aligned} \frac{d\left([M]_{org}V_{org}\right)}{dt} &= -<k_P>[M]_{org}\left(\sum_{n=1}^{\infty}[P_n]_{org}\right)V_{org} - k_{trm}[M]_{org}\left(\sum_{n=1}^{\infty}[P_n]_{org}\right)V_{org} \\ &+k_5[M]_{aq}V_{aq} - k_6[M]_{org}V_{org} \end{aligned} \tag{21}$$

Activator

$$\frac{d\left([CuBr_1/L]_{aq}V_{aq}\right)}{dt} = -k_1[CuBr_1/L]_{aq}V_{aq} + k_2[CuBr_1/L]_{org}V_{org} \tag{22}$$

$$\frac{d\left([CuBr_1/L]_{org}V_{org}\right)}{dt} = -k_{act}[CuBr_1/L]_{org}\left(\sum_{n=1}^{\infty}[P_nBr]_{org}\right)V_{org}$$
$$+k_{deact}[CuBr_2/L]_{org}\left(\sum_{n=1}^{\infty}[P_n]_{org}\right)V_{org} + k_1[CuBr_1/L]_{aq}V_{aq} - k_2[CuBr_1/L]_{org}V_{org} \quad (23)$$

Deactivator

$$\frac{d\left([CuBr_2/L]_{aq}V_{aq}\right)}{dt} = -k_3[CuBr_2/L]_{aq}V_{aq} + k_4[CuBr_2/L]_{org}V_{org} \quad (24)$$

$$\frac{d\left([CuBr_2/L]_{org}V_{org}\right)}{dt} = k_{act}[CuBr_1/L]_{org}\left(\sum_{n=1}^{\infty}[P_nBr]_{org}\right)V_{org}$$
$$-k_{deact}[CuBr_2/L]_{org}\left(\sum_{n=1}^{\infty}[P_n]_{org}\right)V_{org} + k_3[CuBr_2/L]_{aq}V_{aq} - k_4[CuBr_2/L]_{org}V_{org} \quad (25)$$

Radicals with n units of monomer-Aqueous phase

$$\frac{d\left([P_n]_{aq}V_{aq}\right)}{dt} = 2k_d f[I-I]V_{aq}\delta_{n,1} - k_{Pn}[M]_{aq}[P_n]_{aq}V_{aq}$$
$$+k_{Pn-1}[M]_{aq}[P_{n-1}]_{aq}V_{aq}\delta_{n,2} - \left[(k_{td}+k_{tc})[P_n]_{aq}\left([P_1]_{aq}+[P_2]_{aq}\right)V_{aq}\right] \quad n=1,2 \quad (26)$$
$$+k_{trm}[M]_{aq}\left([P_1]_{aq}+[P_2]_{aq}\right)V_{aq}\delta_{n,1} - k_{trm}[M]_{aq}[P_n]_{aq}V_{aq}$$

where $\delta_{n,i}$ is the Kronecker delta ($\delta_{n,i}$=1 if $n = i$, $\delta_{n,i}$=0 otherwise).

Radicals with n units of monomer-Organic phase

$$\frac{d\left([P_n]_{org}V_{org}\right)}{dt} = -<k_P>[M]_{org}[P_n]_{org}V_{org} + <k_P>[M]_{org}[P_{n-1}]_{org}V_{org}\left(1-\delta_{n,1}\right)$$
$$-(k_{td}+k_{tc})[P_n]_{org}\left(\sum_{m=1}^{\infty}[P_m]_{org}\right)V_{org} + k_{act}[CuBr_1/L]_{org}[P_nBr]_{org}V_{org}$$
$$-k_{deact}[CuBr_2/L]_{org}[P_n]_{org}V_{org} - k_{trm}[M]_{org}[P_n]_{org}V_{org} \quad (27)$$
$$+k_{trm}[M]_{org}\left(\sum_{m=1}^{\infty}[P_m]_{org}\right)V_{org}\delta_{n,1} + k_{P2}[M]_{aq}[P_2]_{aq}V_{aq}\delta_{n,3}$$

As mentioned before, we consider that all oligoradicals from aqueous phase with 3 monomer units enter the droplets, hence the last term corresponds to this phenomenon and equals the rate of generation of these radicals by propagation.

Dormant species with n units of monomer

$$\frac{d\left([P_nBr]_{org}V_{org}\right)}{dt} = -k_{act}[P_nBr]_{org}[CuBr_1/L]_{org}V_{org} + k_{deact}[P_n]_{org}[CuBr_2/L]_{org}V_{org} \quad (28)$$

Dead polymer chains with n units of monomer -Aqueous Phase (n=1, 2, 3 or 4)

$$\frac{d\left([D_n]_{aq}V_{aq}\right)}{dt}=k_{td}[P_n]_{aq}\left([P_1]_{aq}+[P_2]_{aq}\right)V_{aq}\left(\delta_{n,1}+\delta_{n,2}\right)$$
$$+\frac{1}{2}k_{tc}\left(\sum_{r=1}^{n-1}[P_r]_{aq}[P_{n-r}]_{aq}\right)V_{aq}\left(1-\delta_{n,1}\right)+k_{trm}[P_n]_{aq}[M]_{aq}V_{aq}\left(\delta_{n,1}+\delta_{n,2}\right) \quad (29)$$

Dead polymer chains with n units of monomer -Organic Phase

$$\frac{d\left([D_n]_{org}V_{org}\right)}{dt}=k_{td}[P_n]_{org}\left(\sum_{m=1}^{\infty}[P_m]_{org}\right)V_{org}+\frac{1}{2}k_{tc}\left(\sum_{r=1}^{n-1}[P_r]_{org}[P_{n-r}]_{org}\right)V_{org}+k_{trm}[P_n]_{org}[M]_{org}V_{org} \quad (30)$$

The well-known method of moments is applied to molar balances of species in organic phase (radicals, dormant species and dead polymer chains) in order to calculate the average molecular properties. The moment balances obtained are shown below:

Moment of order a (a =0, 1 or 2) of radicals $\lambda_a=\sum_{n=0}^{\infty}n^a[P_n]_{org}$

$$\frac{d\left(\lambda_a V_{org}\right)}{dt}=-<k_P>[M]_{org}\lambda_a V_{org}+<k_P>[M]_{org}\sum_{j=0}^{a}\binom{a}{j}\lambda_j V_{org}-\left(k_{td}+k_{tc}\right)\lambda_a\lambda_0 V_{org}$$
$$+k_{act}[CuBr_1/L]_{org}\mu_a V_{org}-k_{deact}[CuBr_2/L]_{org}\lambda_a V_{org}-k_{trm}[M]_{org}\lambda_a V_{org}+k_{trm}[M]_{org}\lambda_0 V_{org}$$
$$+k_{P2}[M]_{aq}[P_2]_{aq}V_{aq}3^a \quad (31)$$

Moment of order a (a =0, 1 or 2) of dormant species $\mu_a=\sum_{n=0}^{\infty}n^a[P_nBr]_{org}$

$$\frac{d\left(\mu_a V_{org}\right)}{dt}=-k_{act}[CuBr_1/L]_{org}\mu_a V_{org}+k_{deact}[CuBr_2/L]_{org}\lambda_a V_{org} \quad (32)$$

Moment of order a (a =0, 1 or 2) of dead polymer chains $\varepsilon_a=\sum_{n=0}^{\infty}n^a[D_n]_{org}$

$$\frac{d\left(\varepsilon_a V_{org}\right)}{dt}=k_{td}\lambda_0\lambda_a V_{org}+\frac{1}{2}k_{tc}\sum_{j=0}^{a}\binom{a}{j}\lambda_j\lambda_{a-j}V_{org}+k_{trm}[M]_{org}\lambda_a V_{org} \quad (33)$$

Please note that in Eqs. (21), (23), (25), (27) and (30), the expressions between brackets are the moments of order 0 of radicals and dormant species. The number $\overline{M_n}$ and weight $\overline{M_w}$ average molecular weights are calculated employing the following formulas:

$$\overline{M_n}=\frac{\left(\lambda_1+\mu_1+\varepsilon_1\right)}{\left(\lambda_0+\mu_0+\varepsilon_0\right)}M_{w,M} \quad (34)$$

$$\overline{M_w}=\frac{\left(\lambda_2+\mu_2+\varepsilon_2\right)}{\left(\lambda_1+\mu_1+\varepsilon_1\right)}M_{w,M} \quad (35)$$

where $M_{w,M}$ is the monomer molecular weight.

For computing the molecular weight distribution (MWD), the probability generating function (pgf) technique is employed. This technique is based on transforming the molar balances of species into the pgf domain. The pgf balances are solved and the MWD is recovered applying an appropriate inversion method.[22]

The pgf is defined as:

$$\phi_a(z) = \sum_{n=0}^{\infty} z^n n^a \frac{[X_n]}{\delta_a}$$

where ϕ_a is the pgf of order a of the distribution function $[X_n]$ corresponding to the polymer species X. In what follows we will call σ_a, φ_a, ε_a to the pgf of species P_n, P_nBr, D_n, respectively. Variable z is the dummy variable of the pgf. The pgf balances obtained:

Pgf of order zero of radicals ($\sigma_0(z)$):

$$\begin{aligned}\frac{d\left(\lambda_0\sigma_0(z)V_{org}\right)}{dt} &= -<k_P>[M]_{org}\left(\lambda_0\sigma_0(z)\right)V_{org} + <k_P>[M]_{org}z\left(\lambda_0\sigma_0(z)\right)V_{org} \\ &-(k_{td}+k_{tc})\lambda_0\left(\lambda_0\sigma_0(z)\right)V_{org} + k_{act}[CuBr_1/L]_{org}\left(\mu_0\varphi_0(z)\right)V_{org} \\ &-k_{deact}[CuBr_2/L]_{org}\left(\lambda_0\sigma_0(z)\right)V_{org} - k_{trm}[M]_{org}\left(\lambda_0\sigma_0(z)\right)V_{org} \\ &+k_{trm}[M]_{org}\lambda_0 z V_{org} + k_{P2}[M]_{aq}[P_2]_{aq}V_{aq}z^3\end{aligned} \tag{36}$$

Pgf of order zero of dormant species ($\varphi_0(z)$):

$$\frac{d\left(\mu_0\varphi_0(z)V_{org}\right)}{dt} = -k_{act}[CuBr_1/L]_{org}\left(\mu_0\varphi_0(z)\right)V_{org} + k_{deact}[CuBr_2/L]_{org}\left(\lambda_0\sigma_0(z)\right)V_{org} \tag{37}$$

Pgf of order zero of dead polymer chains ($\vartheta_0(z)$):

$$\begin{aligned}\frac{d\left(\varepsilon_0\vartheta_0(z)V_{org}\right)}{dt} &= k_{td}\lambda_0\left(\lambda_0\sigma_0(z)\right)V_{org} + \frac{1}{2}k_{tc}\left(\lambda_0\sigma_0(z)\right)^2 V_{org} \\ &+k_{trm}[M]_{org}\left(\lambda_0\sigma_0(z)\right)V_{org}\end{aligned} \tag{38}$$

Pgf of all chains forming the polymer (radicals, dormant species and dead polymer chains):

$$\Omega_0(z) = \frac{\lambda_0\sigma_0(z) + \mu_0\varphi_0(z) + \varepsilon_0\vartheta_0(z)}{\lambda_0 + \mu_0 + \varepsilon_0} \tag{39}$$

From the solution of these balances and by numerical inversion of the resulting pgf, the full MWD is recovered and compared against experimental data. The pgf technique has proven to be capable of predicting the MWD for several systems accurately and efficiently in terms of computational time, without making any simplifying assumptions or having any a priori information of the distribution shape.[23–25]

We only use pgf of order zero with the purpose of limiting the size of the system. From these pgf we obtain the MWD in number fraction. With the appropriate conversions, we express the MWD in others forms.

Simulations are performed in gPROMS (Process Systems Enterprise, Ltd.) using the built-in solver DASOLV. A standard desktop computer is employed with an Intel® Core™ i5-4460 3.20 GHz processor and 8 GB of RAM memory.

Kinetic constants

Most of the kinetic and diffusion parameters were taken from the literature, with the exception of the activation rate constant (k_{act}), deactivation rate constant (k_{deact}) and the propagation rate constants for short-length radicals (k_{P1}, k_{P2}) which were estimated from experimental data since no information was found in the literature for this system. The complete set of kinetic constants is reported in Table 2.

Table 2. Parameter used for reverse ATRP of n-BMA in miniemulsion at 70°C.

Symbol	*Value*	*Ref.*
k_d	$7.7\ 10^{-4}\ s^{-1}$	2
f	1	2
$<k_p>$	$1243\ M^{-1}\ s^{-1}$	26
k_{P1}	$6350\ M^{-1}\ s^{-1}$	This work
k_{P2}	$1625\ M^{-1}\ s^{-1}$	This work
k_{td}	0.9*kt	12
k_{tc}	0.1*kt	12
k_t	Eq. (14)	13,19
k_{act}	$0.22\ M^{-1}\ s^{-1}$	This work
k_{desact}	$3.24\ 10^5\ M^{-1}\ s^{-1}$	This work
k_{trm}	$0.0562\ M^{-1}\ s^{-1}$	17
γ_{CuBr1}	19	10
γ_{CuBr2}	0.4	10
γ_M	2030	Eq. (15)
k_1	$4.75\ 10^8\ s^{-1}$	Eq. (17)
k_2	$1\ 10^8\ s^{-1}$	9
k_3	$1\ 10^8\ s^{-1\ a}$	Eq. (17)
k_4	$1\ 10^9\ s^{-1}$	9
k_5	$5.07\ 10^{10}\ s^{-1}$	Eq. (17)
k_6	$1\ 10^8\ s^{-1}$	9

Results and Discussion

The model was employed to simulate the reverse ATRP of n-BMA in miniemulsion at 70°C. Simulations were carried out solving Eqs. (19)-(26), (29), (31)-(39), with the kinetic parameters of Table 2 and the initial conditions and other data reported in Table 3. The simulations were compared against experimental data from literature.[2]

Simulated and experimental kinetic plots are shown in Figure 2. As can be seen, the model reasonably predicts the experimental tendency. The linear increase with time implies that a controlled polymerization is obtained. However, in the initial part (time<20 min) the model predicts a lower conversion than the experiment does. The induction period is highly influenced by the decomposition rate of initiator, propagation rate of oligoradicals in the aqueous phase, entry of oligoradicals and the initial amount of deactivator $CuBr_2/L$ present in the organic phase. Because all of them have been modeled with parameters directly extracted from literature, it is expectable to obtain these differences.

Table 3. Initial conditions, monomer molecular weight and volumes of aqueous and organic phases used in the simulation of the reverse ATRP of n-BMA in miniemulsion at 70°C

Variable	*Value*	*Ref.*
$[I]_0$	$4.3\ 10^{-3}$ M	2
$[M]_{aq,0}$	$3.4\ 10^{-3}$ M	
$[M]_{org,0}$	6.9 M	
$[CuBr_1/L]_{aq,0}$	0	
$[CuBr_1/L]_{org,0}$	0	
$[CuBr_2/L]_{aq,0}$	$3.9\ 10^{-3}$ M	
$[CuBr_2/L]_{org,0}$	$1.6\ 10^{-3}$ M	
$M_{w,M}$	142.4 g/mol	
V_{aq}	20 mL	
V_{org}	5 mL	

Average molecular weight ($\overline{M_n}$) and polydispersity ($\overline{M_w}/\overline{M_n}$) evolution against conversion are shown in Figure 3. In the case of $\overline{M_n}$, a linear behavior is obtained. This again denotes that a controlled polymerization is achieved. Nevertheless, the simulated $\overline{M_n}$ is slightly higher than the experimental data, implying that the model predicts lower quantity of chains, probably because of higher termination of oligoradicals in aqueous phase and lower initial concentration of deactivator in droplets. In the other hand, the simulated polydispersity is in the range of the experiment ($\overline{M_w}/\overline{M_n} \sim 1.4-1.5$) since higher values of polydispersity are commonly obtained for CRP in aqueous dispersions compared to bulk systems.[3] However, the model predicts a slight decrease with conversion, not shown in the experimental measurements, which remains constant. As many factors can affect the polydispersity, we have not found a cause of the model behavior yet. Anyway, the prediction obtained for these average properties is quite remarkable, considering that most of the kinetic parameters were extracted from the literature for systems similar but not equal to the one studied in this work, and also, that the entry and partitioning processes were modeled in a simple manner in order to obtain a straightforward model.

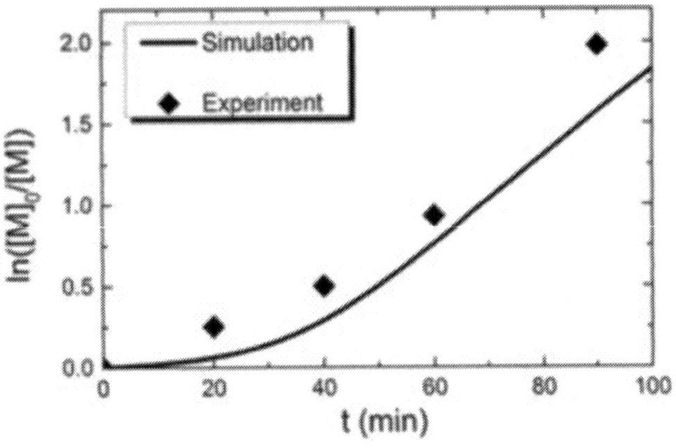

Figure 2. First-order kinetic plot vs conversion for the reverse ATRP of n-BMA in miniemulsion at 70°C. Initial conditions and other data reported in Table 3.

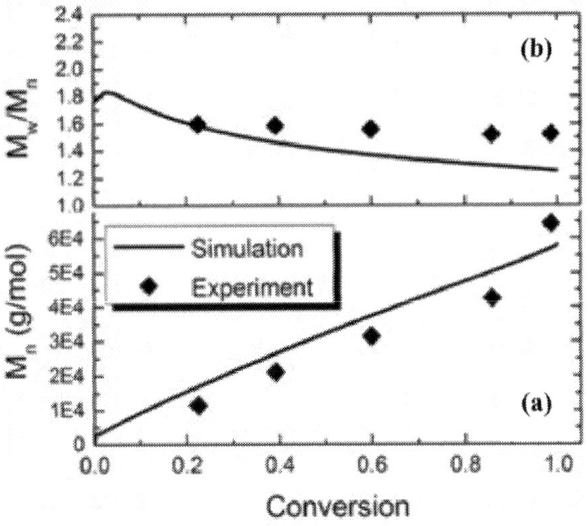

Figure 3. Number average molecular weight (a) and polydispersity (b) vs conversion for the reverse ATRP of n-BMA in miniemulsion at 70°C. Initial conditions and additional data reported in Table 3.

Finally, Figure 4 shows the predicted and experimental MWD at four different conversions. As can be observed, the agreement between simulation and experiment is quite noteworthy. As the conversion increases, the MWD shifts towards higher molecular weights. The differences observed corresponds to the deviations previously commented, especially in the polydispersity.

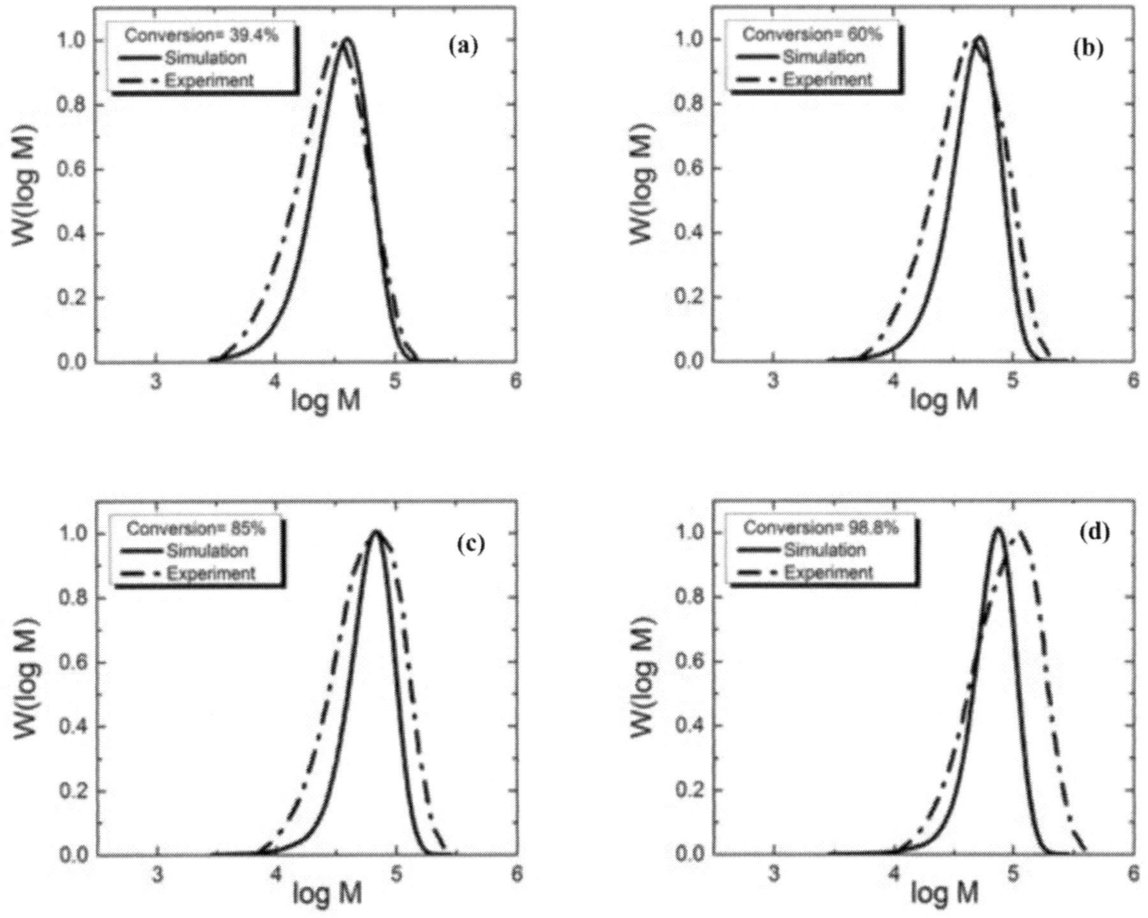

Figure 4. Molecular weight distribution at conversion of 39.4% (a), 60.0% (b), 85.0% (c) and 98.8% (d), for the reverse ATRP of n-BMA in miniemulsion at 70°C. Initial conditions and additional data reported in Table 3.

Conclusions

Reverse ATRP is an ATRP variant that allows synthesizing polymers with controlled structure, and can also be used in dispersed systems because it employs more stable Cu^{II} complexes and a conventional free radical initiator. Dispersed systems are of great interest since they constitute a good alternative for implementing controlled polymerizations on a large scale. As a consequence, several studies have been conducted in such systems, involving experimental and theoretical work. In this context, a tool that allowed experiment simulation and design of reverse ATRP in dispersed systems can be very useful for saving time and resources.

In the present work, a mathematical model was developed for a reverse ATRP in miniemulsion using a water-soluble initiator. This model is able to predict average molecular properties, such as number and weight molecular weights, as well as the full MWD. The model is derived from the deterministic balance equations of the reacting species. Average properties are predicted using the well-known method of moments and the MWD is modeled using the probability generating function (pgf) technique. Our results show that the predicted average molecular properties as well as the MWD agree well with the experimental information extracted from literature.

References

[1] Braunecker, W.A., and Matyjaszewski, K., ―Cotrolled/living radical polymerization: Features, developments, and perspectives," Progress in Polymer Science 32(1), 93–146 (2007).

[2] Li, M., and Matyjaszewski, K., ―Reverse Atom Transfer Radical Polymerization in Miniemulsion," Macromolecules 36(16), 6028–6035 (2003).

[3] Cunningham, M.F., ―Controlled/living radical polymerization in aqueous dispersed systems," Progress in Polymer Science 33(4), 365–398 (2008).

[4] Oh, J.K., ―Recent advances in controlled/living radical polymerization in emulsion and dispersion," Journal of Polymer Science Part A: Polymer Chemistry 46(21), 6983–7001 (2008).

[5] Kagawa, Y., Zetterlund, P.B., Minami, H., and Okubo, M., ―Atom Transfer Radical Polymerization in Miniemulsion: Partitioning Effects of Copper(I) and Copper(II) on Polymerization Rate, Livingness, and Molecular Weight Distribution †," Macromolecules 40(9), 3062–3069 (2007).

[6] Simms, R.W., and Cunningham, M.F., ―Reverse atom transfer radical polymerization of butyl methacrylate in a miniemulsion stabilized with a cationic surfactant," Journal of Polymer Science Part A: Polymer Chemistry 44(5), 1628–1634 (2006).

[7] Li, M., Min, K., and Matyjaszewski, K., ―ATRP in Waterborne Miniemulsion via a Simultaneous Reverse and Normal Initiation Process," Macromolecules 37(6), 2106–2112 (2004).

[8] Thomson, M.E., and Cunningham, M.F., ―Compartmentalization Effects on the Rate of Polymerization and the Degree of Control in ATRP Aqueous Dispersed Phase Polymerization," Macromolecules 43(6), 2772–2779 (2010).

[9] Ma, J.W., Cunningham, M.F., McAuley, K.B., Keoshkerian, B., and Georges, M., ―Nitroxide mediated living radical polymerization of styrene in miniemulsion—modelling persulfate-initiated systems," Chemical engineering science 58(7), 1177–1190 (2003).

[10] Qiu, J., Pintauer, T., Gaynor, S.G., Matyjaszewski, K., Charleux, B., and Vairon, J.-P., ―Mechanistic Aspect of Reverse Atom Transfer Radical Polymerization of *n* -Butyl Methacrylate in Aqueous Dispersed System," Macromolecules 33(20), 7310–7320 (2000).

[11] Maxwell, I.A., Morrison, B.R., Napper, D.H., and Gilbert, R.G., ―Entry of free radicals into latex particles in emulsion polymerization," Macromolecules 24(7), 1629–1640 (1991).

[12] Payne, K.A., D'hooge, D.R., Van Steenberge, P.H.M., Reyniers, M.-F., Cunningham, M.F., Hutchinson, R.A., and Marin, G.B., ―ARGET ATRP of Butyl Methacrylate: Utilizing Kinetic Modeling To Understand Experimental Trends," Macromolecules 46(10), 3828–3840 (2013).

[13] Wang, W., Hutchinson, R.A., and Grady, M.C., ―Study of Butyl Methacrylate Depropagation Behavior Using Batch Experiments in Combination with Modeling," Industrial & Engineering Chemistry Research 48(10), 4810–4816 (2009).

[14] Siddiqui, M.N., Redhwi, H.H., Verros, G.D., and Achilias, D.S., ―Evaluating the Role of Nanomontmorillonite in Bulk in Situ Radical Polymerization Kinetics of Butyl Methacrylate through a Simulation Model," Industrial & Engineering Chemistry Research 53(28), 11303–11311 (2014).

[15] Tsarevsky, N.V., Pintauer, T., and Matyjaszewski, K., ―Deactivation Efficiency and Degree of Control over Polymerization in ATRP in Protic Solvents," Macromolecules 37(26), 9768–9778 (2004).

[16] Rosen, S.L., and Brazel, C.S., [Fundamental Principles of Polymeric Materials] , Third, Wiley (2012).

[17] Sangster, D.F., Feldthusen, J., Strauch, J., and Fellows, C.M., ―Measurement of Transfer Coefficients to Monomer for *n* -Butyl Methacrylate by Molecular Weight Distributions from Emulsion Polymerization," Macromolecular Chemistry and Physics 209(15), 1612–1627 (2008).

[18] Beuermann, S., and Buback, M., ―Rate coefficients of free-radical polymerization deduced from pulsed laser experiments," Progress in Polymer Science 27(2), 191–254 (2002).

[19] Buback, M., and Junkers, T., ―Termination Kinetics oftert-Butyl Methacrylate and ofn-Butyl Methacrylate Free-Radical Bulk Homopolymerizations," Macromolecular Chemistry and Physics 207(18), 1640–1650 (2006).

[20] de Vries, R., Co, C.C., and Kaler, E.W., ―Microemulsion Polymerization. 2. Influence of Monomer Partitioning, Termination, and Diffusion Limitations on Polymerization Kinetics," Macromolecules 34(10), 3233–3244 (2001).

[21] Ma, J.W., Cunningham, M.F., McAuley, K.B., Keoshkerian, B., and Georges, M.K., ―Interfacial Mass Transfer in Nitroxide-Mediated Miniemulsion Polymerization," Macromolecular Theory and Simulations 11(9), 953–960 (2002).

[22] Brandolin, A., Assini Balbueno, A., and Asteasuain, M., ―Improved numerical inversion methods for the recovery of bivariate distributions of polymer properties from 2D probability generating function domains," Computers & Chemical Engineering 94, 272–286 (2016).

[23] Asteasuain, M., Sarmoria, C., and Brandolin, A., ―Recovery of molecular weight distributions from transformed domains. Part I. Application of pgf to mass balances describing reactions involving free radicals," Polymer 43(8), 2513–2527 (2002).

[24] Fortunatti, C., Sarmoria, C., Brandolin, A., and Asteasuain, M., ―Modeling of RAFT Polymerization using Probability Generating Functions. Detailed Prediction of Full Molecular Weight Distributions and Sensitivity Analysis: Modeling of RAFT Polymerization using Probability Generating Functions …," Macromolecular Reaction Engineering 8(12), 781–795 (2014).

[25] Fortunatti, C., Sarmoria, C., Brandolin, A., and Asteasuain, M., ―Prediction of the full molecular weight distribution in RAFT polymerization using probability generating functions," Computers & Chemical Engineering 66, 214–220 (2014).

[26] Beuermann, S., Buback, M., Davis, T.P., Gilbert, R.G., Hutchinson, R.A., Kajiwara, A., Klumperma, B., and Russell, G.T., ―Critically evaluated rate coefficients for free-radical polymerization, 3a. Propagation rate coefficients for alkyl methacrylates," Macromolcular Chemistry and Physics 201(12), 1355–1364 (2000).

460163 Robust Anion Exchange Membranes for Alkaline Fuel Cell Applications

Abstract:

Anion exchange membranes (AEM) can provide high ionic conductivity and facilitate enhanced transport phenomenon. However, one challenge is in the development of mechanically robust thin films. Chemical and mechanical degradation of thin films can lead to failure in fuel cells. In this study, an engineering polymer was used as the polymer matrix and blended with a polymer that could be functionalized into an AEM. Commercially available polymers were used as both components in these blends. Blend miscibility was examined using rheology and mechanical characterization. A prediction of the membrane integrity and its lifetime were measured under different hydration levels and humidity cycling. Dynamic mechanical testing of membranes was conducted using TA instruments ARES G2 instrument at a range of temperatures (30-90°C) and relative humidities ranging from 0-95% RH. Moduli changes with varying humidity saturation levels and temperatures are analyzed to understand the membrane mechanical characteristics. â€˜Water stressâ€™ tests are performed on the membranes to study tension and durability by quantifying the elongation at break during reversible water absorption-desorption cycles. In addition to the mechanical property assessments, the electrochemical conductivity was also studied. Various analytical and morphological techniques such as DSC, TGA, XRD, SAXS and FTIR were used to predict the performance and stability of the blend membranes. The microstructure analyses of the results were developed to understanding of the micro and nanoscale phase behavior of the membranes.

460401 Thermomechanical Properties of Polylactide/Oligomerized Castor Oil Blends

Abstract:

Polylactide, a commercially available bio-based, compostable polymer, is a renewable, environmentally friendly alternative to petroleum based polymers. A drawback of PLA is its brittleness, but this can be improved through compatibilized blending. PLA can be toughened by the incorporation of vegetable oils, but due to the large differences in viscosity and molecular weight, they are not very compatible. Oligomerized vegetable oils can potentially be more compatible. Blends of PLA with oligomerized castor oil (oligoCO) were produced via melt mixing using a Haake batch mixer. The oligoCO was synthesized via free radical polymerization in a supercritical carbon dioxide ($scCO_2$) medium to varying molecular weights. The effectiveness of the oligoCO as a toughening agent for PLA is investigated. The thermal properties of the blends were determined using thermogravimetric analysis (TGA) and differential scanning calorimetry (DSC). Scanning electron microscopy was used to analyze the morphology of the blends. The mechanical properties of blended samples are determined and assessed by tensile and impact testing as well as dynamic mechanical analysis.

460613 Kinetics of Prebiotic Depsipeptide Formation from the Ester-Amide Exchange Reaction

Abstract:

In this work, we introduce a kinetic model to study the effectiveness of the ester-mediated amide bond formation under prebiotic conditions. The results of the Miller-Urey experiments and meteorite investigations have provided evidence that amino acids, the building blocks of peptides, existed on the prebiotic earth.[1] However, a prebiotically-plausible process for the polymerization of amino acids is still unclear. In our previous study, we found a simple system composed of hydroxy acids and amino acids is capable of forming peptide bonds via esterification and the ester-amide exchange reaction.[2] To further understand the kinetic behavior of this complex copolymerization, we first designed a closed reactor to monitor the evaporation of water and lactic acid from the reaction mixture. The growth of initial species from a valine and lactic acid mixture was monitored by HPLC-UV/MS. A mathematical model was developed to simulate the reactions and to evaluate the rate constants at different temperatures. Our results reveal that copolymerization between valine and lactic acid is well-characterized by our model and a small number of physical parameters. We found that these reactions can be described by the empirical Arrhenius equation even when the reactions occurred at a dried state. Further calculations for the activation parameters show that the ester-mediated pathway facilitates amide bond formation primarily by having higher activation entropies. These results provide a theoretical framework that illustrates why the ester-mediated pathway for peptide bond formation would have been more favorable on the early Earth than peptide bond formation without the aid of hydroxy acids.

1. (a) Bernstein, M. P.; Dworkin, J. P.; Sandford, S. A.; Cooper, G. W.; Allamandola, L. J., Racemic amino acids from the ultraviolet photolysis of interstellar ice analogues. *Nature* **2002,** *416* (6879), 401-403; (b) Miller, S. L., A production of amino acids under possible primitive earth conditions. *Science* **1953,** *117*(3046), 528-529.

2. Forsythe, J. G.; Yu, S.-S.; Mamajanov, I.; Grover, M. A.; Krishnamurthy, R.; FernÃ¡ndez, F. M.; Hud, N. V., Ester-mediated amide bond formation driven by wetâ€"dry cycles: A possible path to polypeptides on the prebiotic earth. *Angew. Chem. Int. Ed.* **2015,** *54* (34), 9871-9875.

461267 Hybrid-Capacitors with Polyaniline/Carbon Electrodes Fabricated *Via* Simultaneous Electrospinning/Electrospraying

Abstract:

Hybrid-capacitors have the potential to synergistically combine the benefits of both electrochemical double layer capacitors (EDLCs) (longer cycle life) and pseudo-capacitors (higher capacitance). However, new processes that intimately combine the two primary materials from each capacitor (carbon and conductive polymer, respectively) within the electrodes in an ordered fashion on the nanoscale are needed to realize this potential. In this study, we report on a novel method (simultaneous electrospinning/electrospraying (E/E)) for fabricating hybrid-capacitors with high surface area electrodes consisting of polyaniline (PANI) and carbon particles (referred to as E/E electrodes). E/E produces a PANI/carbon nanofiber network with fiber diameters in the range of 500 to 2000 nm. The hybrid-capacitor with E/E electrodes exhibits an excellent specific capacitance of 235 F g^{-1} (*vs.* 138 F g^{-1} for capacitor with state-of-the art hybrid electrodes) at a current density of 1 A g^{-1}. Moreover, the capacitor with E/E electrodes retains approximately 84 % capacitance after 1000 charge-discharge cycles (*vs.* 67 % for capacitor with state-of-the art hybrid electrodes). A *post mortem* electron microscopy study of the E/E electrodes provides a rationale for the improved cycling stability. These results demonstrate the feasibility of producing E/E electrodes and their promise as future materials in hybrid-capacitors.

461412 Surface Coated Template Particles for Internally Functionalized Pores in Waterproof and Breathable Membranes

Abstract:

Surface coated template particles for internally functionalized pores in waterproof and breathable membranes

Authors: Mario Stucki, Christoph R. Kellenberger, Wendelin Stark

In recent years the public attention towards chemicals released to the environment has increased. In 2013, upon the discovery of multiple potentially dangerous and persistent chemicals in outdoor apparel[1], the global NGO Greenpeace initiated the campaign â€œDetoxâ€ to convince global outdoor brands to switch to alternative materials. Replacements were only partially given at the time, *i.e.* a completely porous, functional membrane was not amongst them. With the template removal method, shown in 2012[2], the optimal tool to find a new, environmentally benign way to produce fully porous, waterproof and breathable membranes, was available.

Commercial soft polyurethane was dissolved and mixed with template particles through ball milling. The resulting dispersion was spread on a flat surface and dried. Upon template removal a porous structure was obtained. With untreated template particles the requirements of the outdoor industry could not be met and additional practical problems like shrinking or collapsing of the porous network was observed. To overcome this, the template particles were coated with a single layer of stearic acid to increase hydrophobicity[3]. Using a high particles to polymer ratio (4:1 in weight) a sufficiently porous network was created to allow complete template removal, which is not obvious regarding the use of fillers in many polymers. The significant increase in water proofing (300%) showed the hydrophobic nature of the internally coated porous network. Surprisingly, the apparent diameter of the pores repelling water did not affect the diffusive pathway of water vapor, therewith breathability was kept at a constant high level.

Furthermore, less pore deformation was observed upon template removal. To prove the large scale feasibility, 5.8 m^2 of membrane were produced and laminated to a Nylon fabric. The final laminate was tailored into a functional rain jacket prototype.

[1] M. Santen, U. Kallee, Greenpeace, 2012.

[2] C. R. Kellenberger, N. A. Lüchinger, W. J. Stark, *WO2012097967 (A1)*, **2012**.

[3] M. Stucki, W. J. Stark, *EP15198440.8*, **2015**.

462556 Pore Size Tuning of Poly(styrene-co-vinylbenzyl chloride-co-divinylbenzene) Hypercrosslinked Polymers: Insights from Molecular Simulations

Abstract:

Hypercrosslinked polymers (HCPs) are highly porous materials that are attracting increasing interest due to their potential in energy applications. Furthermore, HCPs are considered low-cost and lightweight material with exceptionally high surface area, high thermal stability and relatively robust against chemical degradation. These attractive characteristics from HCPs make them an excellent candidate for important industrial applications such as gas capture, gas storage, catalyst applications, and separation processes.

In this study, the structure-property relationships of HCPs, poly(styrene-*co*-vinylbenzyl chloride-*co*-divinylbenzene), were investigated using atomistic simulations. A virtual polymerization algorithm, *Polymatic*, was used to synthesize the HCPs *in silico*. To ensure the conformity of our simulated samples, the structures were validated with experimental data including BET surface area, micropore volume, and gas loading (H_2 and CO_2). Furthermore, the influence of the degree of cross-linking on polymer structures was analyzed, and it was observed that, as expected, both the surface area and micropore volume increase as a function of the degree of cross-linking. We also investigated if divinylbenzene molar content in HCPs can be varied to tune the porosity of the polymer structure in order to improve H_2/CO_2 gas separation performance. Specifically, we varied the divinylbenzene content from 0 to 50 mol% and evaluated how structural and gas separation performance changes. Our results demonstrated that divinylbenzene can be used as a porosity tuning agent and it significantly affected the structure and properties of the polymer. However, no significant improvement in H_2/CO_2 gas separation in pressure swing adsorption process with respect to varying divinylbenzene molar content was observed, in contrast with experimental forecast.

462850 Preparation and Timed Release Properties of Self-Rupturing Gels

Abstract:

Polyelectrolytes attract widespread interest as building blocks for stimulus-responsive soft materials that can form, dissolve or change their properties on demand. Here, we exploit this stimulus sensitivity to prepare self-rupturing gels, which can serve as self-destructing soft devices. These gels are synthesized from an inexpensive commercial polyelectrolyte, poly(acrylic acid) (PAA), and have highly non-uniform crosslink densities. The disparate crosslink densities give rise to non-uniform osmotic swelling, which not only changes the gel shape, but also generates stresses that are high enough to induce gel rupture. The time required for this rupture depends on the variability in the crosslink densities, and the size and arrangement of the variably-crosslinked gel sections. Further, we show how these self-rupturing gels can be utilized in delayed/timed release applications by preparing self-rupturing hydrogels with capsule-like â€œcontainer and lidâ€ structures. The self-rupture of these containers removes the lid and releases the entire payload after a pre-programmed delay (which is on the order of days and can be tuned by varying the rupture time).

463053 Assessment of Wavelength Exponent Method for Monitoring Inverse Miniemulsion Polymerization of Acrylamide Using NIR Spectroscopy

Abstract:

Acrylamide is a solid, water-soluble monomer precursor of polyacrylamide, with different applications: flocculants, adhesives, viscosity control agent for oil, chemicals for the paper industry, mining processes. Polyacrylamide can be produced using differences polymerization techniques, such as solution polymerization, inverse emulsion, and inverse polymerization in microsuspension or micro emulsion. However few studies are reported in the literature related to inverse miniemulsion polymerization.

The compartmentalization of the radicals in submicron-sized particles provides several advantages for emulsion and miniemulsion polymerization techniques compared to bulk, solution and suspension polymerizations, such as higher reaction rate and higher molecular weights. In addition, the resulting product is a dispersion of submicrometric polymer particles and there are fewer problems due to heat transfer or mixing during the polymerization since the viscosity of the continuous phase is low.

It has been seen that, inverse miniemulsion polymerization of acrylamide shows a high reaction rate, due to high ratio between propagation and termination rate of acrylamide, which become the monitoring of proprieties such as, droplet/particle diameters, and conversion difficult. In addition, these properties are important to know the course of reaction, because it can be an indicative of loss stability of reaction. Thus, monitoring polymer properties during the reaction is important to eliminate or reduce variations in product quality or other failure during the process.

There is also a need for improving existing technology to reduce costs and increase process safety, maintaining or improving product quality, and reducing the amount of residual volatile organic compounds generated in conventional analytical methods. These conventional techniques usually are time-consuming and not suitable for *in-line* and *real-time* monitoring. Therefore, spectroscopic techniques, combined with optical fiber sensors, are in an outstanding position to monitor heterogeneous processes.

Among them, Near Infrared (NIR) Spectroscopy, which comprises the spectral region from 13000 cm-1 to 4000 cm-1, is a widely used technique to monitor polymerization processes. NIR spectroscopy techniques have the advantages of ease of handling and rapid measurements, the ability to monitor simultaneously different polymer properties, as well as the possibility of using a multiplexer system. However, it is an indirect method, so the fitting of multivariate calibration models is required to relate NIR spectra with properties of interest, such as monomer concentration, conversion, and droplet/particle diameters throughout the reactions. This calibration step is time-consuming and the fitted model is valid only for being used in the same conditions in which the model was built.

A method called Wavelength Exponent Method is shown in literature as a simple and fast technique for evaluate changes over time in properties of heterogeneous systems. The extinction of light by emulsions or dispersions is the result of light absorption by the continuous and dispersed phases plus scattering. For a nonabsorbing system, the measured spectral absorbance can be directly related to scattering by the suspended particles. The extinction efficiency of this scattering depends on the particle size parameter and the refractive index of both phases, evaluated at the correspondent wavelength. For dilute dispersions consisting of monodisperse spherical nonabsorbing particles significantly smaller than the wavelength of the incident light, scattering is described by the Rayleigh scattering regime.

Based on this theory, the so called Wavelength Exponent can be deduced and determined from spectroscopic measurements and its calculated value can be related to particle size. Therefore, his technique can be applied to evaluate changes in heterogeneous systems by observing changes in the exponent value. Usually this technique is applied to the visible range of the spectrum, but in this work it was observed that it can also be successfully used in the NIR range.

For using this method in the monitoring of a polymerization reaction, it is necessary to choose a wavelength range where there is no absorption of any components of the system, so the measured absorbance will be due to scattering only. This means that only physical information of the system is used, so changes in the chemical composition should not affect the results. This is an interesting advantage in the method, besides its simplicity, for the evaluation of particle size during the polymerization reaction. If no calibration step is made, the wavelength exponent method still can be used as a qualitative evaluation of changes in the properties of the reaction over time.

In the studied system, acrylamide polymerization by inverse miniemulsions, there are two types of particles present: monomer droplets and polymer particles. Therefore, if a droplet size analysis were made during the reaction, the measurements will not indicate the size of the monomer droplets or polymer particles separately, but a combination of both, making it difficult to monitor the reaction over time.

Although only a qualitative evaluation where made, the monitoring of the wavelength exponent over time, calculated from the measured NIR spectra during the acrylamide polymerization reaction, has shown interesting results. Besides the particle size dependency of the exponent, it was observed that, when the reaction conversion rate suddenly increases, there also is a sudden change in the behavior of the wavelength exponent. The founded results are supported by the behavior found in the droplet size measurements, indicating that this could be an interesting technique for monitoring such systems.

463618 Complexation Driven Assembly of Block Copolyelectrolytes into Spherical Micelles, Flower-like Micelles and Macro-Networks

Abstract:

Block copolymer amphiphiles, driven by solvophobic interactions, assemble into a range of micellar structures. Similar spherical micellar assemblies of block copolyelectrolytes can also be achieved by relying on electrostatic interactions between oppositely charged polyelectrolytes. These electrostatics driven assemblies, constituting of polyelectrolyte complex cores and neutral coronae, maintain a large fraction of water in their cores, thus making them particularly useful for encapsulation and delivery of charged therapeutics and nucleotides.

The generic micelle structures in complexation driven assemblies have been understood to be similar to those of amphiphilic micelles. However, recent investigations have unveiled salient differences between these solvophobicity and electrostatics driven assemblies. This presentation will highlight such differences by elaborating on both the typically acknowledged structure-defining contributions (via complexation) and additional structure-directing contributions (via intra-chain repulsion) of the charged blocks. Relying on extensive structural characterization from X-ray and neutron scattering experiments, specific examples of assembly of model oppositely charged diblock and triblock copolyelectrolytes that form spherical micelles and unique inter-connected networks of flower-like micelles, respectively, will be discussed. Further, the networks, at low polymer concentrations, phase separate from the solution, and our hypothesis on their structural features being primarily dictated by the conformations of the neutral block will be argued. In addition, conclusions from molecular dynamics simulations on the driving forces behind these unique assemblies and their direct comparisons with corresponding amphiphilic assemblies will be presented.

464146 Modeling of Microgel Synthesis By Precipitation Polymerization

Abstract:

Microgels are soft polymer particles with a diameter from 10 to 10,000 nm. Stimuli-sensitive microgels respond to environmental conditions such as changes in temperature or pH – This allows the use of microgels for a wide range of applications, such as drug delivery systems using temperature triggered release of guest molecules, or in regenerative medicine using the high degree of freedom in customization of the cross-linked polymer structure [1].

Good progress has been achieved in the study of correlations of the internal structure and the resulting properties [2]. However, tailoring microgels with desired properties demands an in-depth understanding of the synthesis process and the influence of its controlling factors on the internal microgel structure. This is best accomplished by mathematical modeling techniques to gain insight on the microgel synthesis and driving reaction kinetics. To obtain a validated process model, it is necessary to have suitable experimental data.

N-Vinylcaprolactam (VCL) based microgels can be synthesized by precipitation polymerization [3]. Initially, monomer and the cross linker *N*-*N'*-Methylenebisacrylamide (BIS) are completely soluble in water. Free radical polymerization of the monomer and cross linker is initiated by 2,2'-Azobis(2-methypropyonamidine)dihydrochloride (AMPA). Polymerization continues until radical oligomers reach a critical chain length, collapse due to their insolubility in water, and form precursor particles. The precipitation creates a second, polymer-rich phase where further polymerization reactions occur and contribute to particle growth. Size of the microgel particles is determined by several competing growth mechanisms, most importantly represented by precursor particle agglomeration and absorption of monomer, radicals as well as precursor particles by larger particles. The cross linker is central to establish a stable polymer structure and governs the microgel properties significantly.

In this work, a continuous kinetic model is developed for microgel synthesis by precipitation polymerization in an isothermal batch process. Mass balances for initiator, monomer and cross linker as well as oligomer and polymer are used to describe the liquid and the microgel phase. The incorporation of cross linker is described by the copolymerization mechanism. To account for the formation of cross-links by reactions of the pendant double bonds, additional balances for these species are introduced. The model covers a broad range of reaction steps which are assumed to occur in both phases. Assuming that the reactions are not affected by simultaneously occurring particle formation mechanisms as described above [4], the reaction kinetics can be determined.

To comprehensively determine the reaction kinetics, dedicated experiment series performed in a real-time calorimeter provide measurements of the heat flux from reactor content to the cooling jacket. This allows computing the heat of reaction by means of an energy balance for the entire reactor. Simultaneous Raman spectroscopy measurements provide mass fractions of monomer and polymer [5] of an average of liquid and gel phase.

Morphology of the microgel particles and their high water content makes a continuous distinction of the ratio of the phases and the growth of microgels to this time impossible. The ratio is, however, necessary to describe the transfer of polymer from the liquid to the gel phase. To account for the transfer of radicals by precipitation, the experimentally determined critical chain length is utilized. For the liquid phase, reaction rate coefficients for propagation were derived from quantum chemical calculations. This enables the prediction of the extent of the liquid phase as reaction locus which would have been difficult to determine experimentally. Partition coefficients from quantum chemical calculations provide additional information on concentrations of small molecules in both phases.

The proposed model enables a quantitative description of the synthesis for cross-linked VCL-based microgels. Reaction kinetics of homopolymerization as well as cross reaction coefficients were derived from a range of experiments with varying composition ratios of VCL and BIS. The determined propagation reaction rate coefficients of homopolymerization and cross reactions of monomer and cross linker differ significantly which leads to a faster consumption of cross linker. This aligns with the results of experimental characterization of VCL-based microgel particles to have an uneven distribution of cross-links in the microgel particle [6].

Furthermore, the model indicates that growth of the gel phase is also a product of absorption and reaction of monomers. Since the critical chain length for precipitation is small, concentration of oligomers in the liquid phase is low, which consequently limits the reactions in the liquid phase. With the partition coefficients from quantum chemical calculations indicating an accumulation of monomer in the gel phase, this implies that that the impact of polymerization reactions in the liquid phase is small.

For the complex microgel system, with limited experimental information of the complex polymerization behavior, integration of experimental measurements and quantum chemical calculation led to an accurate kinetic model. The model enables the reproduction of experimentally observed phenomena and allows predictability and insights in the kinetics of microgel synthesis. This builds the first step towards model-based process design and operation of synthesis of functional microgels.

[1] Saunders, Brian R.; Laajam, Nadiah; Daly, Emma; Teow, Stephanie; Hu, Xinhua; Stepto, Robert (2009): Microgels: From responsive polymer colloids to biomaterials. In: *Advances in Colloid and Interface Science* 147-148, S. 251–262. DOI: 10.1016/j.cis.2008.08.008.

[2] Boyko, Volodymyr; Richter, Sven; Grillo, Isabelle; Geissler, Erik (2005): Structure of Thermosensitive Poly(*N*-vinylcaprolactam-co-*N*-vinylpyrrolidone) Microgels. In: *Macromolecules* 38 (12), S. 5266–5270. DOI: 10.1021/ma050548w.

[3] Pich, Andrij; Richtering, Walter (2011): Microgels by Precipitation Polymerization: Synthesis, Characterization, and Functionalization. In: Andrij Pich und Walter Richtering (Hg.): Chemical Design of Responsive Microgels, Bd. 234. Berlin, Heidelberg: Springer Berlin Heidelberg (Advances in Polymer Science), S. 1–37.

[4] Arosio, Paolo; Mosconi, Matteo; Storti, Giuseppe; Banaszak, Brian; Hungenberg, Klaus-Dieter; Morbidelli, Massimo (2011): Precipitation Copolymerization of Vinyl-imidazole and Vinyl-pyrrolidone, 2 - Kinetic Model. In: *Macromolecular Reaction Engineering* 5 (9-10), S. 501–517. DOI: 10.1002/mren.201100020.

[5] Meyer-Kirschner, Julian; Kather, Michael; Pich, Andrij; Engel, Dirk; Marquardt, Wolfgang; Viell, Joern; Mitsos, Alexander (2016): In-line Monitoring of Monomer and Polymer Content During Microgel Synthesis Using Precipitation Polymerization via Raman Spectroscopy and Indirect Hard Modeling. In: *Applied spectroscopy* 70 (3), S. 416–426. DOI: 10.1177/0003702815626663.

[6] Schneider, Florian; Balaceanu, Andreea; Feoktystov, Artem; Pipich, Vitaliy; Wu, Yaodong; Allgaier, Jürgen et al. (2014): Monitoring the Internal Structure of Poly(*N*-vinylcaprolactam) Microgels with Variable Cross-Link Concentration. In: Langmuir 30 (50), S. 15317–15326. DOI: 10.1021/la503830w.

464342 Molecular Design of Bio-Based Plasticizers for PVC

Abstract:

There is growing concern for the toxicity of industrial polymer plasticizers to humans. Especially, phthalate derivatives are very common plasticizers and exhibit toxic effects by degradation. These plasticizers are particularly widely used in processing of polyvinyl chloride (PVC). In our study, various bio-based molecules with potential to replace phthalate derivative plasticizers for PVC have been modeled by employing atomistic scale simulations. Our approach depends on molecular dynamic simulations to determine polymer/plasticizer interactions and observe plasticization effects via the lowering of glass transition temperature (T_g). Structural stability was observed by solubility behavior of plasticizers in polymer and to describe T_g values and volumetric behavior of polymer/plasticizer mixtures, pressure-volume-temperature (PVT) relationship was used. Selected bio-based molecules including levulinic acid esters, succinic acid esters, oleate esters, dioctyl adipate and for comparison one of the phthalate derivative were constructed with TRAPPE-UA force-field, were optimized and their densities similar to literature were obtained Then, the atactic PVC homopolymer chains consisting of 150 monomers with equal fractions of meso- and racemic diads distributed randomly along the chain were constructed based on the Generalized Amber Force Field (GAFF), T_g and density of pure PVC was determined 344 K and 1.38±0.01 g/cm3. Polymer/plasticizer structures contained 10 to 30% plasticizer by weight to predict the performance of the plasticizers. Bio-based derivative esters, including levulinic acid esters and succinic acid, were also constructed with different chain length to observe chain length effect on plasticization. The variation of the polymer glass transition temperatures was investigated and compared as a function of plasticizer type and content by simulating PVT relationship and stability of the polymer/plasticizer mixtures were compared by calculated solubility values.

This work is part of the project supported by the Scientific and Technological Research Council of Turkey (TUBITAK) through the grant no. 114M178.

464353 Responsive Hydrogen-Bonded Polymer Nanocomposites Containing Discrete Layers of Gold Nanoparticles Prepared By Spray-Assisted Layer-By-Layer Assembly

Abstract:

The incorporation of nanoscale filler materials into layer-by-layer (LbL) assemblies can have dramatic effects on the properties of the resulting nanocomposites. The ability to control the location and relative concentration of the nanofiller within the LbL assembly throughout the fabrication process is desirable for producing products with specific, predictable, and reproducible properties. Here, we demonstrate for the first time the use of spray-assisted LbL assembly to produce hydrogen-bonded polymer nanocomposites of poly(ethylene oxide) (PEO) and poly(methacrylic acid) (PMAA) containing discrete regions of gold nanoparticles (AuNPs) vertically positioned throughout the film structure. Analysis of the internal structure using cross-sectional transmission electron microscopy (TEM) shows that the AuNP regions are clearly separated by "empty" regions of polymer with no drift or aggregation during or after fabrication. UV-Vis spectroscopy was utilized to confirm these observations. A brief study of the stimuli-responsive properties of the spray-assisted LbL assembled nanocomposites showed that the release of the AuNPs could be induced and modulated by elevating the environmental pH above the critical pH of the polymer system. UV-Vis spectroscopy was used to confirm/monitor the payload release. We propose that spray-assisted LbL assembly is a promising method for producing nanocomposites with tunable properties applicable to a wide range of specialty applications where strict properties requirements must be met.

464913 Charged Polysulfone Based Polymer Electrolytes for Lithium Ion Batteries

Abstract:

Polymer electrolytes have long been proposed as a means to solve many of the challenges inherent in Li-ion batteries. Numerous studies have highlighted promising polymer electrolytes that improve battery safety through mechanical suppression of dendrite growth, reduced flammability compared to typical liquid organic electrolytes, and high electrochemical stability. Within any binary salt system, including most traditional polymer electrolytes with dissolved salt, concentration gradients arise during battery operation due to a substantial fraction of current being carried by the anion, ultimately limiting the rate at which batteries can be cycled. These concentration gradients further limit the possible thickness of the porous electrode and thereby the possible cell-level energy density. To address this issue, covalently binding anions to the polymer backbone has been employed as a strategy to substantially suppress their mobility relative to Li^+. These so-called high transference number polymer electrolytes significantly reduce the effects of concentration gradients, potentially enabling higher discharge rates desirable for applications in automobiles, and increasing the possible amount of active electrode material relative to the amount of electrolyte and other passive cell components. This presentation will introduce a new class of high transference number polymer electrolytes based on the simple polysulfone condensation reaction between 4',4'-dichlorodiphenyl sulfone (DCDPS) and 4',4'-biphenol (BP). Polysulfone has long been known as one of the most thermally, electrochemically, and mechanically strong polymers, having an inherent T_g near 200 °C. A sulfonated, commercially available DCDPS monomer can readily be introduced to the polymer backbone and ion-exchanged to its Li^+-neutralized form. Short chain (M_n=1,000 Da), -OH terminated poly(ethylene glycol) can also be directly added to the initial condensation reaction and incorporated into the polymer backbone. The final copolymer is completely miscible, resulting in a single T_g that can be lowered well below ambient, and exhibits no crystallization, thereby providing a several order of magnitude increase in the ionic conductivity over the pure sulfonated polysulfone. A systematic study on ionic

conductivity will be presented as a function of PEG content and degree of sulfonation, demonstrating a promising new system for high transference number polymer electrolytes.

465410 Dynamic Bonds for Mechanically-Triggered Crosslinking and Healing

Abstract:

Covalently crosslinked polymers are generally classified as thermosets, which cannot be recycled, remended, or remolded. To develop responsive materials, dynamic bonds can be built into the network architecture, which can undergo bond rearrangements that impart desirable macroscopic properties such as the ability to relieve stress, reconfigure, reshape and heal [1, 2]. Often, these reactions are activated by an external stimulus such as light or heat. An alternative approach is to utilize mechanical deformation as a trigger via the incorporation of mechanophores, or mechanically-active groups, into the network. We have developed a new responsive material which heals after damage with enhanced local strength providing reinforcement where the material was once weakest [3]. Our approach incorporates a sacrificial bond into a polymer strand, which homolytically cleaves under a large, non-linear deformation. The cleaving of this polymer strand results in local polymerization and a concomitant increase in local material strength, thus mitigating macroscopic fracture. The energy from the damage event itself is therefore the trigger to heal and represents an important step towards developing responsive materials that heal without human intervention.

References:[1] Kloxin, C. J., and Bowman, C. N., "Covalent adaptable networks: smart, reconfigurable and responsive network systems," Chemical Society Reviews 42, 7161-7173 (2013).

[2] Kloxin, C. J., Scott, T. F., Adzima, B. J., and Bowman, C. N., "Covalent Adaptable Networks (CANS): A Unique Paradigm in Cross-Linked Polymers," Macromolecules 43, 2643-2653 (2010).

[3] **Gordon, M. B.**, French, J. M., Wagner, N. J., and Kloxin, C. J., "Dynamic Bonds in Covalently

Crosslinked Polymer Networks for Photoactivated Strengthening and Healing," Adv Mater 27, 8007-10 (2015).

466534 Thermodynamics of Sorption in Polyolefins in Liquid Media

Abstract:

Polyolefins, especially polyethylene (PE) and polypropylene (PP) are materials with very useful and consumer-attractive features, such as high chemical resistance, low weight, mechanical strength, flexibility, low cost, zero toxicity and ease of manufacture. Therefore, polyolefins represent major part of the polymer production. The polymerization can be operated either by gaseous or liquid-dispersion processes, whereas the thermodynamics of liquid penetrant sorption is relevant for the liquid-dispersion polymerization processes. Here, the polymerization reactor contains solid catalyst particle with growing polymer suspended in a gaseous or liquid diluent and gaseous monomer units, which pass through the polymer powder to the active site of the catalyst. The monomer concentration at the active catalytic sites is affected by the solubility of the monomer in polymer at the polymerization temperature and pressure, thus the monomer solubility and the associated monomer diffusion in the polymer powder influences the polymerization process. The polymerization process is also affected by swelling of the polymer particles. In general, the increasing volume of the polymer particle causes the increasing amount of the sorbed monomer, which can polymerize here and is resulting in an increasing polymerization rate. Also the increasing volume of the polymer particle causes the longer diffusion path of the monomer. The knowledge of sorption, swelling and transport properties in polyolefins is important not only for the polymerization rate, but also the subsequent processes are affected by these phenomena, thus it is important to determine the amount of sorbed diluent and unreacted monomer in the polymer particle after the polymerization in the reactor. This amount is affected by the solubility of the diluent and the monomer in the polyolefin particle.

In this work, the thermodynamic behaviour in liquid penetrant is studied using newly developed methods. The liquid-sorption measurement is based on recording the mass decrease during desorption, where the equilibrium sorption is extrapolated by the initial slope method. For the evaluation of swelling in liquid, modified video-microscopic method was used.

These measurements are not common in the literature, even though the diluents sorption and particle swelling in the liquid is more pronounced than in the gaseous diluents. Based on the new methodology of sorption measurements the solubility for a wide range of PE density and temperatures was measured. Also the swelling of polyethylene particles by liquid penetrant is presented in this work. The new experimental swelling data together with the solubilities are unique in the polyolefin research and could be usable as input data for the control and optimization of the polyolefin production process itself. In addition, the results are interpreted in the context of our recently proposed phase structure of PE [1]. We demonstrate that the phase structure of PE proposed in [1] offers a fundamentally new and self-consistent approach to the thermodynamics of penetrant sorption in PE.

References:

[1] ChmelaÅ™ J., PokornÃ½ R., Schneider P., SmolnÃ¡ K., BÄ›lskÃ½ P., Kosek J.: Free and constrained amorphous phases in polyethylene: Interpretation of 1H NMR and SAXS data over a broad range of crystallinity. *Polymer* 2015, 58, 189-198.

466541 High Pressure Co-Polymerization of Ethene by ATRP

Abstract:

The following contribution deals with the high-pressure copolymerization of ethene controlled by the atom transfer radical polymerization (ATRP) mechanism. Having a look at free-radical polymerizations, terminations occur frequently which leads to polymers with a broad molecular weight distribution and without an end functionality. Gaining control over the radical polymerization by establishing an equilibrium between the propagating radical chain and a dormant species one can decrease the radical concentration and hence, also reduce the radical termination processes. This leads typically to narrow molecular weight distributions. Furthermore, the resulting polymer possesses an end group functionality which can be used for post-polymerization reactions as well as the synthesis of block-copolymers.

Controlling radical polymerizations has already been undertaken to a wide range of monomers. However, it is still a challenge to gain control over the free-radical polymerization of less activated monomers like ethene due to their high radical reactivity.[1] Furthermore, especially ethene tends to show side reactions like inter- and intramolecular transfer while polymerizing. The reaction network of the ATRP of ethene can be seen in the figure below. The transfer reactions to polymer are not shown here not to overcrowd the scheme.

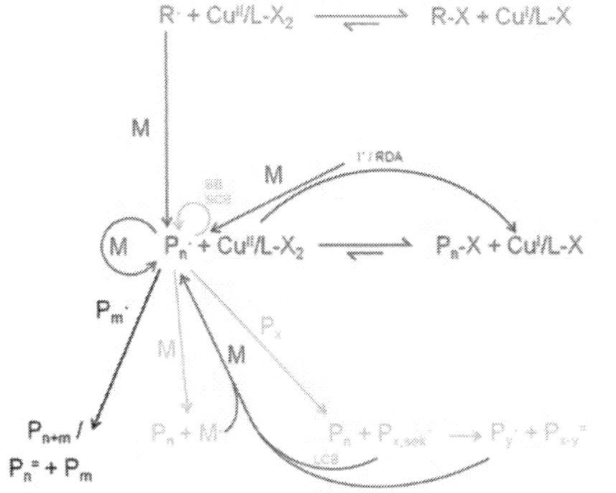

There have been several research activities on the field of controlled radical polymerization (CRPs) of ethene. Under mild conditions of 70 °C and 200 bar and in solution one could achieve slight control via the RAFT-mechanism, but the resulting polymers had low molecular weights.[2] In order to maintain high molecular weights pressure and temperature have to be increased significantly. At 2000 bar and 125-250 °C one could gain high molecular weighted polystyrene and polymethacrylates by introducing TEMPO derivatives, but no control could be achieved in the case of polyethylene.[3] Luft et al. found a decreased polydispersity index (PDI) for polyethylene when using other derivatives for nitroxyl mediated polymerizations (NMP), but they were still above 2.7. However, molecular weights were rather high (M_n>8.1*10^4 g/mol).[4] Slight control over the ethene polymerization could be achieved at 2000 bar and 175 °C by reversible addition-fragmentation chain-transfer polymerization (RAFT).[5] The advantages of high-pressure ATRP were only examined for (meth)acrylate homopolymerizations, yet.[6,7]

The present work deals with the control over the radical polymerization of ethene with comonomers by ATRP under 2000 bar and various temperatures in order to gain polymers with low polydispersities and functionalized end groups. In a first step the reagents are tested for their thermal stabilities under reaction conditions in a high pressure differential thermal analysis (DTA). Next, polymerizations are carried out in a stirred high pressure batch autoclave with injection system. They are conducted at different temperatures with various comonomers (butylacrylate, butylmethacrylate among others) and controlling agents (CuBr/Tris(2-pyridylmethyl)amine (TPMA) and CuBr/N,N,N´,N´´,N´´-Pentamethyldiethylene-triamine (PMDETA) complexes). Furthermore, reducing agents like 2,2´-Azobis(2-methyl-propionitrile) (AIBN), ascorbic acid and Tin(II) 2 ethylhexanoate ($Sn(EH)_2$) are introduced in order to regenerate the controlling agent.

The resulting polymers are analyzed regarding their molecular structure: the molecular weight distribution is obtained by high temperature GPC in trichlorobenzene; branching, chain end functionality and copolymer composition are examined by using inverse gated HT ^{13}C- and HT-^{1}H-NMR-spectroscopy.

In the experiments the PDI decreased compared to a free radical reference sample, synthesized under the same conditions. Furthermore, molecular weights are relatively high (M_n up to 5.2*10^4 g/mol). However, the detection of halogen chain ends by ^{13}C-NMR is challenging.

References

[1]Poli, R.; *Chem. Eur. J.*, **2015**, 21, 6988-7001.

[2]Dommanget, C.; D´Agosto, F.; Monteil, V.; *Angew. Chem. Int. Ed.* **2014**, *53*, 6683-6686.

[3]Minaux, E., Dissertation, Göttingen, **2011**.

[4] Luft, G., Patent WO 2005/090419 A1, **2005**.

[5]Roth, M., Dissertation, Darmstadt **2010**.

[6]Morick, J., Dissertation, Göttingen, **2012**.

[7]Wang, Y., Schroeder, H., Morick, J., Buback, M., Matyjaszweski, K., *Macromol. Rapid Comm.*, **2013**, 34, 604.

466628 Morphogenesis of Microstructured Polymer Foams Formed By Thermally Induced Phase Separation: Theoretical and Experimental Study

Abstract:

Polymer foam applications depend both on the polymer and the foam morphology. Generally, the tailored preparation of foams with various morphologies using a single method would open new application possibilities. A method enabling this tailoring is the thermally induced phase separation (TIPS). However, a deeper understanding of the TIPS preparation route is still missing. In this work, we present a validated model for the morphogenesis of micro-structured polystyrene (PS) foams formed by TIPS, in particular by spinodal decomposition [1]. The foam morphologies were predicted by the Cahn-Hilliard approach with the only input being the Flory-Huggins lattice model. To validate the model, we prepared PS foams via TIPS at various conditions using solvents with and without polar interactions (cyclohexane, cyclohexanol), characterized the morphologies and compared them with the predicted ones. Furthermore, using a thermo-optical method we determined the boundaries of the phase separation conditions, i.e., binodal curves. The morphology prediction is in a very good agreement with the experimental results. The model predicts correct length scales of close-cell morphologies for a variety of system compositions and temperatures in 2D and 3D. Moreover, it enables to predict different morphologies: (i) aggregated particles, (ii) open-cell structure, (iii) combination of open- and closed-cell structure, and (iv) closed-cell structure. Furthermore, we present our first experimental results of the dynamics of foam morphogenesis by TIPS. The addition of the real time scale into the model is the last step for the full description of the process dynamics and, thus, the preparation of foams of desired structures.

References

1. Vonka M., Nistor A., Rygl A., Toulec M., Kosek J.: Morphology model for polymer foams formed by thermally induced phase separation. *Chem. Eng. Journal* **2016**. 284: 357-371.

466717 The Task of Modeling Long Chain Branching in the LDPE Synthesis

Abstract:

Although the high-pressure LDPE process is very energy consuming and highly demanding concerning construction material, it is still widely applied in the plastic industry. In 2014 more than 90 million tons of polyethylene have been produced worldwide with a significant amount being LDPE.

One of the main reasons why LDPE stays competitive on the market despite the extreme process conditions of temperatures up to 300 °C and pressures up to 3,000 bars are the special properties of the product. The comparatively low density of the product can be ascribed to the short-chain branches (SCBs) present in the polymer, which are induced by the free-radical polymerization process. SCBs are created during an internal backbiting step of a growing polymer chain.

At the same time long-chain branches (LCBs) are formed when a radical functionality of a growing chain is transferred to another macromolecule and monomer is added subsequently. These LCBs account for the extraordinary rheological properties of LDPE. Thus describing this reaction step correctly when modeling the LDPE process is crucial to determine the LCB density as a significant product property. However, expressing it consistently over a wide range of reactors, process conditions and miniplants as well as industrial world scale plants has proved to be challenging. M. Busch demonstrated that the rate coefficient for the transfer of the radical functionality to a dead polymer molecule depends on the conversion in laboratory experiments.[1] In contrast T. Herrmann showed that this observation cannot be transferred to industrial processes directly.[2] Further investigations in this area are highly desirable, especially when the structure-properties relationship of LDPE is to be used for an enhanced and reliable simulation-based product design.

The aim of this work is therefore to gain a deeper insight into the transfer to polymer reaction and the underlying microscopic concept as well as the macroscopic effects/consequences. This ambitious target requires a combination of different approaches and disciplines. High-pressure polymerizations of ethene have to be conducted under well-defined conditions. The resulting product has to be analyzed analytically preferably including multi-angle light scattering, ^{13}C-NMR and rheometry in order to gain a deep insight into the LCB density of the polymer. Additionally, conducting matching simulations might prove useful to understand long-chain branching on a molecular level. While deterministic kinetic simulations yield average LCB densities, Monte Carlo simulations provide exact topologies of single macromolecules, which can be then used in rheological models.[3]

Combining these ideas/approaches several polymerization experiments will be examined with a special focus on the analytic evaluation of long-chain branching and the applicability of existing kinetic models.

References

[1] M. Busch, *Habilitation*, University of Göttingen **2003.**

[2] T. Herrmann, *PhD thesis*, Technical University Darmstadt **2011.**

[3] D. J. Read, *Journal of Polymer Science, Part B: Polymer Physics*, **2015**, *53*, 123-141.

466789 Anion Effects on the Structure and Ion Transfer Properties of Surface-Tethered Poly Ionic Liquids

Abstract:

Poly(ionic liquids) (PILs) refer to a special type of polyelectrolyte which carry an IL species in each of the polymer repeating units. PILs offer advantages over ILs in enhanced mechanical stability, improved processability, durability and spatial control over the IL moieties. The ability to grow surface-tethered polymer films with ionic liquid (IL) functionality as side chains can enable coatings with versatile and tailorable properties. Surface-tethered PILs were prepared via surface-initiated ring-opening metathesis polymerization (SI-ROMP) of norbornene-tethered ionic liquids to grow poly(ionic liquid) films on gold and silicon substrates. The PILs consist of the cations 1-norbornylmethylene-3-dimethylimidazolium (N1-dMIm$^+$) or 1-norbornylpropyl-3-dimethylimidazolium (N3-dMIm$^+$) with various anions that can be easily interchanged to tune the film properties. The choice of anion influences the structure and ion transfer properties of the films with the cation N1-dMIm$^+$ only able to accommodate certain anions. Copolymer PIL films were prepared via the anion exchange of multiple anions into the PIL films. Polarization modulation-infrared reflectance-absorption spectroscopy (PM-IRRAS) was used to quantify the extent of anion exchange in the films. Profilometry measurements were utilized to quantify the thicknesses of the films. The dependence of ion transfer through the film on the ionic liquid anion was characterized by electrochemical impedance spectroscopy and cyclic voltammetry.

466844 Study on Accelerated Aging of Biodegradable Poly(butylene adipate-co-terephthalate) Films for Mulch Applications

Abstract:

Poor weatherability is one of the bottlenecks for large-scale applications of biodegradable poly(butylene adipate-co-terephthalate) (PBAT) mulch films. Herein, PBAT-based films were subjected to accelerated weathering test under laboratory simulated conditions for investigation of their aging behaviors. During the aging, crosslink structures in the films were formed rapidly by recombination of free radicals formed via Norrish I chain scission. In addition, the content of butylene terephthalate (BT) units increased because water-soluble butylene adipate (BA)-rich segments were formed and washed away by water spay due to the susceptibility of aliphatic ester groups to Norrish II chain scission or photo-oxidation. Such structural changes resulted in significant performance reduction including a significant increase in T_g, decrease in T_m, and rapid deterioration of tensile properties. Incorporating a UV absorber 1164 into PBAT-based films effectively depressed both Norrish I and II chain scissions and the subsequent degradation reactions, thus greatly improving the weatherability of PBAT films. The gel formation, changes in composition and transition temperatures, and deterioration of mechanical properties were all effectively suppressed. Addition of either hindered phenolic antioxidant 1010 or hindered amine light stabilizer 944 improved the weatherability as well, but the effects were inferior to 1164 as they could only depress the degradation reactions following Norrish I.

466849 Synthesis of Comb-Branched Polyolefinic Elastomers Using a Tandem System with FI-Ti and CGC-Ti Catalysts

Abstract:

Using two combined catalysts in one reactor is an efficient way for synthesizing polyolefin copolymers with unique structures. Olefin block copolymers (OBCs) are one of examples which are produced by two distinct catalysts with different copolymerization ability in a single reactor. In our tandem system, a bis(phenoxyimine)-titanium (PTFI-Ti) was used as a macromonomer catalyst for the synthesis of linear polyethylene (PE) macromonomers with terminal vinyl groups, while a CGC-Ti catalyst (CGC-Ti) was served for copolymerization of ethylene, 1-octene, and PE macromonomer at a wide range of operation temperatures to form comb-branched polyolefin copolymers with long chain (LCB). The resulting comb-LCB copolymers with crystalline polyethylene as side chains and amorphous ethylene/1-octene random copolymers as backbone possessed good thermoplastic elastomer properties.

466877 Ferromagnetic Ink: Covalent Attachment of Charged Polymers on Magnetic Nanoparticles Enables Magnetic Deinking

Abstract:

Deinking of printed paper is within the most expensive steps in paper recycling. Application of a highly magnetic ink can simplify this step and might face current issues, as printed (magnetic) and non-printed (non-magnetic) paper parts can be distinguished. Screening and sorting of magnetic from non-magnetic wastes is already a routine procedure in large scale waste processing. Via previous screening step, the use of magnetic ink can minimize the paper amount which has to be fed into the expensive deinking process. Furthermore the deinking process itself can be facilitated using magnetic inks as magnetic pigments can be simply separated from paper fibers by application of a magnetic field[1].

An appropriate magnetic ink affords high saturation magnetization and dispersionstability in an environmental friendly solvent (e.g. water). Magnetic nanoparticle dispersions are usually made from superparamagnetic materials since in the absence of an external magnetic field, the lack of magnetic particle/particle attraction allows the creation of stable dispersions[2]. Common attempts to prepare stable dispersions of ferromagnetic particles (much higher saturation magnetization) failed since the strong magnetic particle/particle interaction can overcome repulsive effects from surfactants or steric stabilizers (typically polymers).

In the present work graphene protected metallic nanomagnets were modified with charged poly-(3-(methacryloylamino)propyl]trimethylammonium chloride (polyMAPTAC), in order to create highly stable dispersions of ferromagnetic nanoparticles. The covalently attached charged polymer brushes implement both steric and electrostatic repulsion forces which can counteract the magnetic attraction. It is demonstrated how the direct, covalent attachment of highly charged polymers on the graphene layers can circumvent stabilizer detachment[3].

More specifically, graphene-coated metal nanoparticles, prepared by flame spray synthesis, are covalently functionalized with initiator moieties for atom transfer radical polymerization (ATRP) *via* diazonium chemistry. Using surface initiated ATRP the ionic polymer brushes (polyMAPTAC) could be successfully grafted from the particle surface. As the charged polymer grow from the surface, the created brushes cannot collapse due to the mutual electrostatic repulsion of the charged polymer side chains. Thus, the polymer brushes are outstretched and evolve a significant steric repulsion force besides the intrinsic electrostatic repulsion of the charged polymers. Particle size distributions with an average diameter of 24 nm provide a ferromagnetic fluid with unprecedented stability in water over months. Due to the covalent character of the connection between the ion-containing polymer brushes and the particle surface, detachment of the hydrophilic polymer brushes is not observed. As prepared fluids can be printed by common ink-jet devices and act as ferromagnetic water based ink[1].

[1] M. Zeltner, L. M. Toedtli, N. Hild, R. Fuhrer, M. Rossier, L. C. Gerber, R. A. Raso, R. N. Grass and W. J. Stark, ***Langmuir***, **2013**, 29, 5093.

[2] E. Kita, T. Oda, T. Kayano, S. Sato, M. Minagawa, H. Yanagihara, M. Kishimoto, C. Mitsumata, S. Hashimoto, K. Yamada and N. Ohkohchi, *J. Phys. D. Appl. Phys.*, **2010**, *43*, 47401.

[3] M. Zeltner, R.N. Grass, A. Schaetz, S.B. Bubenhofer, N.A. Luechinger, W.J. Stark, ***J. Mater. Chem.***, **2012**, *22*, 12064.

467645 The Effect of Water on the Thermal Transition Observed in Poly(allylamine hydrochloride)-Poly(acrylic acid) Complexes

Abstract:

Polyelectrolyte complexes (PECs) are receiving increasing attention because of their stimuli-responsive behaviors with ionic strength, pH, and temperature. Of these, temperature is particularly intriguing in that PECs undergo a glass-melt transition of an unclear nature. Here, we present the thermal behavior of PECs containing weak polyelectrolytes poly(allylamine hydrochloride) (PAH) and poly(acrylic acid) (PAA) as influenced by water content and complexation pH. These are investigated using modulated differential scanning calorimetry (MDSC) and molecular simulation. MDSC reveals a glass-transition-like thermal transition (T_{tr}) that decreases in value with increasing hydration and with decreasing complexation pH. As complexation pH decreases, the PAA content in the PEC increases and the PAA ionization decreases. Added 1-proponal elevates the transition temperature, as does added urea. Atomistic detail molecular dynamics simulations show an increasing number of water-polymer hydrogen bonds within the hydrated complexes as water content increases. These results suggest that water has a plasticizing effect, and these results support the hypothesis that the transition is caused by a restructuring of the water hydrogen-bonding network within the PEC that then allows for subsequent chain relaxation.

467927 Online Monitor and Control of the Real-Time Composition of Ternary Gas Mixture in Batch Reactor By Three Cascade Flow Meters during Propene Polymerization

Abstract:

Abstract: The batch polymerization of propene is still widely applied in many industrial practice for production of polypropylene resins. However, since different monomers present different reactivities, composition drifting will take place in the copolymerization process, which in turn affects the properties of the copolymer products. Therefore, how to monitor and control the composition of monomers in batch reactor is of great significance.

In current work, we have developed a method for online monitoring and controlling the real-time composition of ternary gas mixture in batch reactor by three flow meters, a differential pressure volumetric flow meter, a thermal mass flow meter, and a Coriolis mass flowmeter in series.

We applied this technology to ternary copolymerization of ethylene/propene/1-butene and binary copolymerization of propene/1-butene with little H_2 as the chain transfer agent in a gas-phase batch reactor. We found that this technology could reliably monitor and control the real-time composition of ternary gas mixture in reactors as well as conveniently measure the consumption rates of different monomers, which enabled us to easily obtain some basic kinetic parameters such as chain propagation rate constants, reactivity ratios and so on.

Key words: propene batch copolymerization; composition monitoring and controlling; polymer kinetics

468117 Polymerized Ionic Liquid Triblock Terpolymers: Synthesis and Characterization

Abstract:

Polymerized ionic liquid (PIL) block copolymers are a distinct set of block copolymers that incorporate unique physiochemical properties of PILs (*e.g.*, high solid-state ionic conductivity, high chemical, thermal and electrochemical stability, and widely tunable physical properties) into block copolymer architecture, which allows for self-assembly into a range of nanostructures, where morphology type and domain size are tunable. To date, PIL block copolymers have been synthesized by a number of research groups and have shown distinct conductivity-morphology relationships as it relates to various applications. Several block chemistries have been explored, yet almost all reports are exclusive to (AB) diblock copolymers. Although AB diblock copolymers can provide orthogonal properties of high ion conduction and high mechanical strength in a solid-state material, there are limitations to AB diblock copolymers, such as limited set of morphologies, domain sizes, and a lack of means to achieve a combination of more than two properties simultaneously (*e.g.*, conductivity, strength, flexibility). Moreover, spatially connected 3D network morphologies, which often result in the highest ion conductivities, occur only over a small compositional range in AB diblock copolymers.

In this study, a PIL (ABC) triblock terpolymer was synthesized to explore a PIL-containing polymer with a richer compliment of properties, a wider diversity of cation/backbone chemistries, and a broader compositional range to obtain a variety of continuous 3D network morphologies. Specifically, poly(MMA-*b*-MEBIm-Br-*b*-HMA) was synthesized *via* sequential reversible addition-fragmentation chain-transfer (RAFT) polymerization and subsequently quaternized, where A block = methyl methacrylate (MMA), B block or PIL block = 1-[(2-methacryloyloxy)ethyl]-3-butylimidazolium bromide] (MEBIm-Br), and C block = hexyl methacrylate (HMA). Reaction kinetics of this PIL ABC triblock terpolymer were investigated and these results were utilized in scaling up the polymerization reaction to > 2 g quantities with narrow polydispersities and controllable block compositions. Chemical, thermal, morphological, and ion conductive properties were characterized using ^{1}H nuclear magnetic resonance (NMR) spectroscopy, elemental analysis (EA), gel permeation chromatography (GPC), differential scanning calorimetry (DSC), dynamic mechanical analysis (DMA), small angle X-ray scattering (SAXS), transmission electron microscopy (TEM), and electrical impedance spectroscopy (EIS).

468642 Donor-Acceptor Fully Conjugated Block Copolymers through Chain-Growth Polymerizations

Abstract:

Donor-acceptor fully conjugated block copolymers are composed of covalently bonded donor and acceptor conjugated polymers. Donor-acceptor block copolymers are a promising candidate in photovoltaics due to their ability to microphase separate at length scales commensurate with exciton diffusion lengths. Conjugated block copolymers can also serve as model systems to study the relationship between molecular structure, microstructure, and optoelectronic properties of conjugated polymers. The most widely used synthetic route for conjugated block copolymers is the polymerization of the donor block by Grignard metathesis (GRIM), followed by chain extension through a Stille or Suzuki-Miyaura polycondensation to grow the acceptor block. Due to the step-growth mechanism, this approach leads to a mixture of products including homopolymers, muliblock copolymers, as well as diblock copolymers. To minimize the formation of these impurities, different approaches are required such as synthesis of both blocks through chain-growth polymerizations. We prepared poly(3-hexylthiophene-2,5-diyl)-*block*-poly((2,6-bis(thiophene-2-yl)naphthalene-1,4,5,8-tetracarboxylic-*N*,*N'*-bis(2-octyldodecyl) diimide)- 5′,5″-diyl) (P3HT-*b*-PTNDI) through chain-growth polymerizations. After synthesis of P3HT by GRIM, the second block, PTNDI, was polymerized through a chain-growth mechanism in a condensation polymerization using active zinc. Chain extensions were tracked by NMR as a function of the reaction time, and formation of block copolymers was confirmed by GPC using a UV-vis detector. These well-defined donor-acceptor conjugated block copolymers are expected to serve as good model systems for studies of self-assembly in semiflexible copolymers and photophysical properties of donor-acceptor systems.

469100 Single Polymerization Determination of Reactivity Ratios Via in Situ Spectroscopic Techniques and a Simple Nonterminal Model for Chain Copolymerization

Abstract:

Here, we propose a new method for the determination of reactivity ratios based on a non-terminal model of copolymerization kinetics. We propose this non-terminal model of copolymerization as an initial method that should be utilized before more complex frameworks (e.g. terminal or penultimate model of chain copolymerization) are used to understand copolymerization kinetics. The ability to synthesize tailored polymer materials with specified functionalities and material properties is fundamental to understanding and ultimately controlling structure-property relationships. Through precise control over the distribution of monomers along the polymer chain, a wide variety of polymer architectures (block, random, gradient etc.) can be achieved with consequently varying and tunable physical properties. Importantly, during the simultaneous copolymerization of two (or more) monomers, the resulting polymer chains typically exhibit a composition gradient along the chain due to the monomer reactivity ratios straying from unity. However, as many ionic and coordination polymerizations exhibit ideal or close to ideal copolymerization, the product of their reactivity ratios is unity or close to unity. The implication of this behavior is that the enchainment of monomer is dictated by the monomers themselves and is independent or predominantly independent of the identity of the reacting chain end. Thus an analytical solution can be derived from the coupled ordinary differential equations and ultimately the determination of reactivity ratios can be achieved from kinetic polymerization data that span the full range of conversion. This nonterminal model removes the need to perform the more labor-intensive processes in order to perform techniques such as Mayo-Lewis or Fineman-Ross. We utilize monomer consumption data obtained with in situ ^{1}H NMR spectroscopy to obtain reactivity ratios for the aluminum-chelate catalyzed copolymerization of phenyl glycidyl ether and allyl glycidyl ether (r_{PGE} = 1.56 ± 0.01 and r_{AGE} = 066 ± 0.03). Furthermore, we apply this approach to other copolymerization whose experimental data is readily available in the literature. These copolymerization systems include a range of monomer types (styrenics, isoprene, lactones, lactide, and other cyclic ethers) and polymerization routes (anionic, coordination and zwitterionic). For systems where reactivity ratios have been determined via other methods we find good agreement using the nonterminal model proposed here.

469206 Exploring the Effects and Interplay of Elastin-like Polypeptide (ELP) Charge and Hydrophobicity on Mcherry-ELP Fusion Protein Self-Assembly

Abstract:

Enzymes are attractive catalysts for a variety of applications including pharmaceutical synthesis, commodity chemical production, and biosensors due to their substrate specificity, high catalytic rates, sustainable source, and their ability to operate at mild reaction conditions. However, fabrication of enzymes into biofunctional devices requires incorporation of the protein into a material. Block copolymers are well-recognized for their ability to self-assemble into a variety of different morphologies depending on temperature and block volume fraction. It has been established that similar phase behavior is observed when polymers are conjugated to globular proteins and, importantly, that the globular proteins maintain their original function. Recent work in the Olsen lab has shown that similar self-assembly occurs in elastin-like protein (ELP)-mCherry fusion protein systems, where ELP serves as a structural coil-like block and mCherry serves as a model globular protein. These fusion protein systems are of great interest due to the advantages they offer over polymer-protein conjugates of improved ease of synthesis and purification. However, there is little understanding of how protein design can be used to control self-assembly in these fusion protein systems. Furthermore, because the entire fusion protein is technically a single poly(amino acid), understanding of how amino acid identity changes propensity for microphase separation will inform the fundamental understanding of the physics of this self-assembly process.

In this work, we systematically explored the impact of the charge and hydrophobicity of the coil-like ELP on fusion protein self-assembly. This was achieved by varying the composition of amino acids in the ELP pentapeptide repeat sequenceâ€"Xaa-Pro-Gly-Yaa-Gly (XPGYG)â€"in which the first position, Xaa, can be either valine (V) or isoleucine (I), with isoleucine being more hydrophobic, and the fourth position, Yaa, can be any amino acid except proline. In this work, we used a five pentapeptide repeat sequence that varied hydrophobicity by tuning the I/V content in the first position and phenylalanine (F)/V content in the fourth position. Additionally, three different charge states (neutral, negatively charged, and zwitterionic) were studied by changing some of the amino acids in the fourth position to lysine (K) and/or glutamic acid (E). Combination of all possible permutations of charge and hydrophobicity generated 9 ELP sequences that were then fused to mCherry via genetic engineering. The self-assembly of these systems was studied using small-angle X-ray scattering (SAXS), atomic force microscopy (AFM), and light scattering.

IMPROVING COMPUTATIONAL COST OF MONTE CARLO MODELS IN RAFT POLYMERIZATION PROCESSES

Esteban Pintos,[a,b] *Claudia Sarmoria,*[a,b] *Adriana Brandolin,*[a,b] *Mariano Asteasuain*[a,b]

[a]*Planta Piloto de Ingeniería Química (CONICET-UNS), Camino La Carrindanga km 7, 8000 Bahía Blanca, Argentina.*

[b]*Departamento de Ingeniería Química, Universidad Nacional del Sur, Avda. Alem 1253, Bahía Blanca, Argentina*

epintos@plapiqui.edu.ar, csarmoria@plapiqui.edu.ar, abrandolin@plapiqui.edu.ar, masteasuain@plapiqui.edu.ar

Summary: Three implementations of the Monte Carlo method for the simulation of different mechanisms of the Reversible Addition-Fragmentation Transfer (RAFT) polymerization are presented. One of the most important steps from the point of view of computational performance of these algorithms is the reaction selection step. The different implementations presented in this work aim to increase the understanding of this reaction selection step and provide simulations with better computational performance and efficiency. In particular, our implementation proposes a modification of the Optimized Direct Method (Cao, Y.; Li, H.; Petzold, L. *J. Chem. Phys.* **2004**, *121* (9), 4059) in order to achieve a better performance for reaction systems where the most frequent reaction does not remain constant through the whole reaction time, like it happens in RAFT polymerization systems.

Keywords. RAFT polymerization, mathematical model, Monte Carlo, Julia

INTRODUCTION

Reversible Addition-Fragmentation Transfer (RAFT) polymerization has become one of the most important types of Controlled Radical Polymerization (CRP) due to its potential application for the synthesis of new materials using a wide variety of monomers and mild reaction conditions.

Even though a substantial number of scientific publications have been directed to understanding the mechanism of the RAFT polymerization, there are still some questions about underlying details of the basic steps of the RAFT mechanism that remain unanswered. For example, no decisive explanation has been reported for the slowdown of the propagation rate of some RAFT systems when the concentration of the Chain Transfer Agent (CTA) is increased. Three main theories attempt to explain this rate retardation effect: the Slow Fragmentation (SF)[1] theory, the Intermediate Radical Termination (IRT)[2] theory and the Intermediate Radical Termination with Oligomers (IRTO)[3] theory.

In this scenario, mathematical modeling becomes a valuable tool for helping to elucidate details of the reaction mechanisms. The time evolution of homogeneous chemical reactions is commonly calculated by solving a set of coupled differential equations. In this method, called the *Deterministic Method*, the species concentrations are represented by continuous functions of time and the predicted results are satisfactory in most cases. However, the formulation and solution of deterministic models usually involve some mathematical complexity. Moreover, these models lack of certain desired features that are important in polymerization processes. For instance, they do not take into consideration the stochastic nature of polymerization reactions. Besides, they are only able to provide limited information about the microstructure of the polymer chains, when compared to other simulation methods.

The other important approach for modeling polymer processes is the stochastic simulation. This approach is mainly represented by the Monte Carlo (MC) technique, and specifically by the Gillespie[4] algorithm. This algorithm, also called the Stochastic Simulation Algorithm (SSA), considers that the molecules are individually represented. It is relatively simple to implement and offers extremely detailed information about the polymer microstructure and chain topology. This information can provide a valuable insight about the mechanistic details of the RAFT polymerization.

Formulations of the SSA algorithm

There are several different formulations of the SSA. Two of the most important are the so called Direct Method (DM) and the τ-leaping[5] method. The DM is characterized by simulating a single reaction in each iteration step. For this reason it is considered an exact simulation algorithm. However, it has the disadvantage of being slower than the τ-leaping method and it also requires larger computational resources depending on the reaction system. On the other hand, the τ-leaping method divides the time history of the system into subintervals and calculates the number of times each reaction takes place in each subinterval. Hence, the method "leaps" through time from one subinterval to the next. The τ-leaping method is one of the most used strategies in order to decrease simulation time. The cost of this time reduction is a lower accuracy, compared to the DM. In this work we analyze different implementations of the DM aiming at reducing computational cost.

Regardless of the method used, the accuracy of these stochastic simulations depends significantly on the assumed size of the sample of the reaction medium. The size of this sample, which can be defined by the number of molecules it contains, directly affects computation times. The large values required for accurate and reproducible results can sometimes lead to prohibitive simulation times. Thus, the efficiency of the algorithm implementation becomes of critical importance, especially for the simulation of complex reaction systems.

In a previous work,[6] a Monte Carlo model of a RAFT polymerization system was developed. It was implemented in Julia,[7] a modern programming language designed to achieve high-performance in numerical and scientific computing. This model was used to obtain detailed information about the reaction system and the polymer molecular structure in RAFT polymerization processes. It provided an efficient method for predicting average properties and molecular weight distributions of all polymeric species present in the reaction system, including the bivariate Molecular Weight Distribution (MWD) of the intermediate two-arm adduct. The model produced accurate and reproducible results of the MWDs in approximately 5 minutes of computing time on a regular desktop computer equipped with an Intel® Core™ i5-3330 Processor (running at 3.20 GHz) and 8 Gb of RAM.

In the present work, different implementations of the DM algorithm are analyzed and compared for the simulation of RAFT reactions in order to improve the performance of our previous model.

METHODOLOGY

Julia Programming Language

Julia (http://julialang.org/) is a modern high-level dynamic programming language designed to achieve high-performance in technical, numerical and scientific computing. The language was designed and developed with the objective of providing a performance similar to that of C while using a high-level programming style. In order to achieve this, its developers combined several technologies, such as "multiple dispatch," "Just-In-Time compilation," and "Low-level Virtual Machine compiler infrastructure" among others.[7,8]

Julia can be regarded as an open source high-performance alternative to languages like MATLAB, R and Python, in addition to a high-productivity alternative to C, C++ and FORTRAN. Moreover, Julia is better suited for general purpose programming tasks than traditional scientific languages (C and FORTRAN), allowing it to be used not only for prototyping numerical algorithms but also for deploying them. On top of providing excellent performance, the resulting code is easier to work with, modify and maintain.

Julia was used in our previous work and compared to other programming languages, obtaining better performance results.[6]

Monte Carlo Simulation and Reaction Selection

The algorithm used in this work is based on the work by Gillespie[4] with several performance enhancing modifications and adapted to handle polymeric species as explained in our previous work.[6]

After profiling our previous code, we realized that one of the critical steps of the algorithm was the reaction selection step. In this step, all possible chemical reactions that can take place at a given instance during the simulation are listed and their reaction rates and occurrence probabilities are calculated. Based on these probabilities, one of the reactions is randomly chosen to be simulated next. This reaction selection step was found to be critical in the performance of the algorithm. Some strategies for dealing efficiently with this step have been proposed in the literature, such as the First Reaction Method (FRM)[9], the Next Reaction Method (NRM)[10], the Optimized Direct Method (ODM),[11] and the Sorting Direct Method (SDM).[12] Most of these models were proposed for the simulation of biological processes, but they can easily be extended for polymerization reactions or other chemically reacting systems.

In this work, we propose a modification of the ODM strategy to improve the efficiency of the reaction selection step of the DM algorithm. This strategy is originally based on the fact that in polymerization processes there are some specific reactions steps that occur considerably more frequent than others. The ODM algorithm takes advantage of this fact by placing the most frequent reactions at the beginning of the reaction search order. Since the MC simulation involves millions of iterations in which the reaction selection is repeated, a significant increase in the performance can be achieved by reducing the search depth. Unfortunately, in RAFT polymerization the sorting of chemical steps from most frequent to less frequent ones does not remain constant during the reaction time. This transient shift in reaction frequencies opens the possibility to further improve the performance of the ODM algorithm, by re-sorting the chemical reactions during the simulation.

The new proposed model updates the sorted reaction array at different intervals based on the extent of reaction. This easy to compute quantity provides a reliable means to identify the shifts in the reaction frequency order. It does not depend on the reaction system individually and can be generalized for different mechanisms. Thanks to this modification and other minor optimizations the new model considerably improves the performance of our previous one.

We tested 3 different search algorithms. The first called Linear Search is equivalent at the NRM. In our version of this algorithm reaction rates are listed from 1 to N (where N is the total number of reactions for the system) in the order that they appear in the kinetic mechanism. Then the algorithm proceeds to randomly select the reaction to be simulated by linearly searching from the first to the last in the order they were originally written.

The second algorithm called Ordered Search, equivalent to the ODM method, also performs a linear search but in this case the reactions are searched from the most to the less frequent reactions. In order to accomplish this ordering we need to pre-run the simulation and store the number of times each reaction is selected. This information is then stored on a vector with the indices of the array ordered according to the reactions frequency for the whole simulation time.

The third algorithm, called Improved Ordered Search, takes advantage of the pre-simulation run to obtain extra information about the reactions frequency. In this case the pre-simulation stores several reaction orders at different reaction times, the time intervals are dynamic and a new reaction order is stored only if one of the first 4 reactions in the search depth has changed with respect to the previous order. Theses search orders are then used on the real simulation at the appropriate intervals calculated by the pre-simulation.

RESULTS

Computer Hardware and Software

All the simulations described in this work were performed on a regular desktop computer equipped with an Intel® Core™ i5-3330 processor running at 3 GHz and 8 GB of RAM memory. Julia Language version 0.4.5 was used to run the simulations.

Test System

Given the theoretical nature of this work, the kinetic mechanism and the values of the kinetic constants used for the simulation are those typically found in the literature. The reactions involved for the three theories studied (SF, IRT and IRTO) are shown in Table 1.[13]

Table 1. Kinetic mechanism of RAFT polymerization.

Step	Reaction
Initiation	
(1)	$I \xrightarrow{f,k_d} 2I^{\bullet}$
(2)	$I^{\bullet} + M \xrightarrow{k_i} R_1^{\bullet}$
Propagation	
(3)	$R_n^{\bullet} + M \xrightarrow{k_p} R_{n+1}^{\bullet}$
Pre-equilibrium	
(4)	$R_n^{\bullet} + T \xrightarrow{k_a} {}^{\bullet}TR_n$
(5)	$I^{\bullet} + TR_n \xrightarrow{k_a} {}^{\bullet}TR_n$
(6)	${}^{\bullet}TR_n \xrightarrow{k_f} R_n^{\bullet} + T$
(7)	${}^{\bullet}TR_n \xrightarrow{k_f} I^{\bullet} + TR_n$
Core Equilibrium	
(8)	$R_n^{\bullet} + TR_m \xrightarrow{k_a} R_n\dot{T}R_m$
(8)	$TR_n + R_m^{\bullet} \xrightarrow{k_a} R_n\dot{T}R_m$
(9)	$R_n\dot{T}R_m \xrightarrow{k_f} R_n^{\bullet} + TR_m$
(10)	$R_n\dot{T}R_m \xrightarrow{k_f} TR_n + R_m^{\bullet}$
Termination	
(11)	$R_n^{\bullet} + R_m^{\bullet} \xrightarrow{k_{tc}} P_{n+m}$
(12)	$R_n^{\bullet} + R_m^{\bullet} \xrightarrow{k_{td}} P_n + P_m$
Cross Termination (IRT and IRTO only)	
(13)	$R_n\dot{T}R_m + I^{\bullet} \xrightarrow{k_{ct}} P_{(n+m)}$
(14)	$R_n\dot{T}R_m + R_p^{\bullet} \xrightarrow{k_{ct}} P_{(n+m+p)}$
IRT: $p = 1,\dots,\infty$	
IRTO: $p = 1,2$	

The chemical species involved are: initiator (I), monomer (M), active radicals with n units of M (R_n), chain transfer agent (T), inactive radicals with n units of M (TR_n), adduct radicals with two arms of different length (R_nTR_m) and dead polymer chains of length n (P_n). The same set of constants (k_a and k_f) was used for both the pre-equilibrium and the core equilibrium. If needed, using a different set for each reaction is straightforward. The initiation constant was assumed to be equal to the propagation constant ($k_i = k_p$), the initiator efficiency is denoted as f and was set to the value of 0.5. The initial concentrations of M, I and T were set to 5, 5×10^{-3} and 5×10^{-3} mol·L^{-1}, respectively. The final reaction time was 34 hours. The values of the kinetic constants were taken from the literature. [13]

Effect of the number of particles

As mentioned before, the selected number of particles affects the accuracy and performance of the simulation. This value has to be as small as possible for the simulation to run in a practical time while being large enough to provide reliable results. In our previous work,[6] we thoroughly analyzed the performance and results of the simulation with several number of particle values. We conclude that for the case of the SF theory it was found that a number of particles as low as 1×10^8 was able to predict accurate and reproducible results for average properties such as number average molecular weight (M_n), weight average molecular weight (M_w) and conversion while 5×10^8 molecules were needed to represent the MWD.

For the IRT and IRTO theories, simulations with 1×10^8 failed to produce accurate results. A number of at least 5×10^8 molecules was required to predict both average properties and MWDs with good precision.

Figure 1 shows the frequency of each reaction as function of the reaction time for the SF theory obtained by the pre-simulation. It can be seen that in the first 5 seconds of reaction time there are several reactions taking place that are not present in a significant proportion throughout the rest of the simulation. This reactions correspond to the initiation (1-2) and pre-equilibrium reactions (4-7). After the 5 second mark, the reactions taking place are mainly the propagation (3) and the core equilibrium (8-10) reactions.

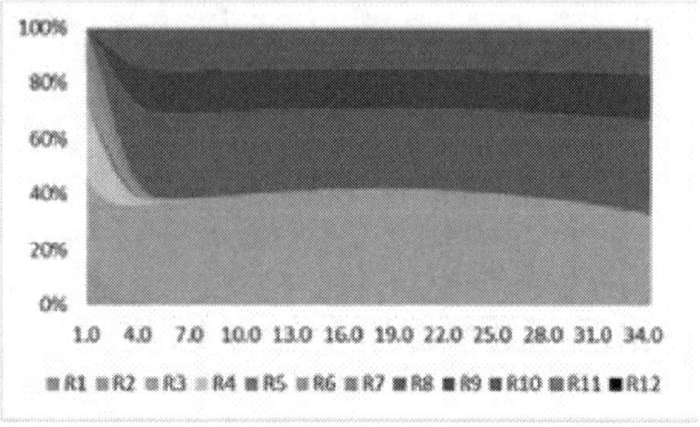

Figure 1. *Frequency of each reaction step as function of the reaction time for the SF theory.* $M_0 = 5$, $I_0 = 5\times10^{-3}$, $T_0 = 5\times10^{-3}$ $mol{\cdot}L^{-1}$, $tf = 34$ *h.*

Figure 2 increases the detail of the first 5 seconds of reaction time of Figure 1. It is clear that the propagation reaction (reaction 3) is one of the most frequent reactions throughout all the simulation time. In the first few seconds reactions number 2, 4, 6 and 7 are very frequent and we should gain additional computational performance by sorting the selection search order to take in consideration this behavior.

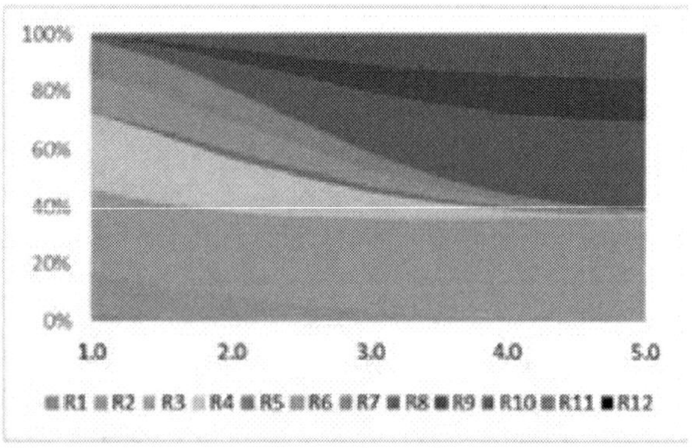

Figure 2. *Frequency of each reaction step as function of the reaction time during the first 5 seconds of reaction for the SF theory.* $M_0 = 5$, $I_0 = 5\times10^{-3}$, $T_0 = 5\times10^{-3}$ $mol{\cdot}L^{-1}$, $tf = 34$ *h.*

Similar results were obtained for the simulations of the IRT and IRTO theories.

The computations times for the different algorithms and the different values of the number of particles used are detailed in Table 2 and Table 3. These tables present the truncated mean result of 100 simulations. The truncated mean was used to discard 10% of the simulations times that were considerable higher than expected in order to obtain a more robust estimation.

Table 2. Computation times (in seconds) for the simulations performed with different algorithms using 5×10^8 particles.

Model	SF	IRTO
Linear Search	63	347
Ordered Search	62	325
Improved Ordered Search	56	287

We are presenting the data for the SF and IRTO theory only to emphasize the difference between the simulation times of these 2 theories The IRT theory is very similar to the IRTO and their simulation times are comparable.

It can be seen that for this particular reaction system the difference between the 3 algorithms is not very significant for the SF theory. This is mainly attributed to the fact that only the first 5 seconds of reaction time present variations on the selection order, while the other 29 seconds remain almost unchanged. Additionally the propagation reaction (3) is the most frequent across the whole reaction time, which is an early find in the search depth and this is the reason the algorithms for Linear Search and the Ordered Search present only a very small difference in their simulation times.

Table 3. Computation times (in seconds) for the simulations performed with different algorithms using 1×10^9 particles.

Model	SF	IRTO
Linear Search	139	745
Ordered Search	137	727
Improved Ordered Search	129	657

Even though the performance gain of using the proposed algorithm is small for the SF theory, the data suggest that for a slightly larger system (IRTO theory) the performance gain is more noticeable. This characteristic leads us to believe that the potential gain in performance for more complex systems like those with chain branching and copolymerization would favor in using this algorithm.

ACKNOWLEDGEMENTS

Authors wish to thank CONICET (National Research Council of Argentina) and UNS (Universidad Nacional del Sur) for financial support.

REFERENCES

(1) Barner-Kowollik, C.; Quinn, J. F.; Nguyen, T. L. U.; Heuts, J. P. A.; Davis, T. P. *Macromolecules* **2001**, *34* (22), 7849–7857.

(2) Monteiro, M. J.; de Brouwer, H. *Macromolecules* **2001**, *34* (3), 349–352.

(3) Konkolewicz, D.; Hawkett, B. S.; Gray-Weale, A.; Perrier, S. *J. Polym. Sci. Part Polym. Chem.* **2009**, *47* (14), 3455–3466.

(4) Gillespie, D. T. *J. Phys. Chem.* **1977**, *81* (25), 2340–2361.

(5) Gillespie, D. T. *J. Chem. Phys.* **2001**, *115* (4), 1716.

(6) Pintos, E.; Sarmoria, C.; Brandolin, A.; Asteasuain, M. *Ind. Eng. Chem. Res.* **2016**, *55* (31), 8534–8547.

(7) Bezanson, J.; Karpinski, S.; Shah, V. B.; Edelman, A. *ArXiv12095145 Cs* **2012**.

(8) Bezanson, J.; Edelman, A.; Karpinski, S.; Shah, V. B. *ArXiv14111607 Cs* **2014**.

(9) Gillespie, D. T. *J. Comput. Phys.* **1976**, *22* (4), 403–434.

(10) Gibson, M. A.; Bruck, J. *J. Phys. Chem. A* **2000**, *104* (9), 1876–1889.

(11) Cao, Y.; Li, H.; Petzold, L. *J. Chem. Phys.* **2004**, *121* (9), 4059.

(12) McCollum, J. M.; Peterson, G. D.; Cox, C. D.; Simpson, M. L.; Samatova, N. F. *Comput. Biol. Chem.* **2006**, *30* (1), 39–49.

(13) Zapata-González, I.; Saldívar-Guerra, E.; Ortiz-Cisneros, J. *Macromol. Theory Simul.* **2011**, *20* (6), 370–388.

469477 Synthesis and Single Molecule Studies of DNA-PNIPAM Copolymers

Abstract:

Polymer chain architecture encompasses a wide range of shapes and chain types, including linear polymers, branched polymers, ring polymers, and polymer networks. In addition to chain topology, the chemical composition of polymers also plays a key role in determining emergent properties. To this end, copolymers and block polymers with simple linear chain architectures can give rise to intricate structural ordering and assembly across multiple length scales. Polymer architectures play a critical role in supramolecular assembly of advanced materials with applications in encapsulation for drug delivery and catalysis, surfactants for emulsions, and materials for chemical separation. For example, PEG-PLA is an important drug delivery carrier that can encapsulate hydrophobic drugs such as camptothecin when mixed with drug solutions. Recently, the development of hybrid polymers that contain both synthetic and natural components have enabled new materials with higher compatibility, enhanced biodegradation, wide functionalities, and unprecented levels of sequence-control. A recent area of interest has focused on the self-assembly of hybrid polymers, especially DNA-based hybrid polymers, into various architectures such as micelles, vesicles, and tubes. Despite recent work, however, we still lack a complete understanding of this process.

In this work, we study the real-time transition of trigger-responsive hybrid polymers using single molecule techniques. In particular, we synthesize DNA-PNIPAM copolymers and study the supramolecular assembly of these materials using fluorescence microscopy. First, we use polymerase chain reaction (PCR) to produce 10 kb DNA with 1% substitution of chemically modified nucleotides containing DBCO, which is a strained alkyne for Cu-free click chemistry. In a separate reaction, we use reversible addition-fragmentation chain-transfer (RAFT) polymerization to synthesize fluorescent PNIPAM copolymerized with rhodamine monomers. In addition, the PNIPAM polymers contain a terminal azide group, which facilitates click reaction with the DBCO-modified DNA. In this way, we generate DNA-PNIPAM comb and bottle-brush polymers, and after purification, these materials show a sizable shift via agarose gel electrophoresis denoting the formation of the desired product. In this way, we effectively synthesize semi-sequence controlled DNA-PNIPAM polymers with different architectures, and we can directly study these materials using single molecule fluorescence microscopy, including the dynamic transition process for temperature responsive materials. In particular, upon increasing the temperature above the lower critical solution temperature (LCST) of PNIPAM around 32C, the material transitions from hydrophilic to hydrophobic character. In this way, we directly observe the spontaneous formation of higher order structures in solution by dynamically increasing the temperature above the LCST. As a control, we also used scanning electron microscopy (SEM) to characterize the self-assembled structures as a function of temperature, and we found micron-sized vesicle structures compared with the control group. Finally, we study the single chain dynamics of DNA-PNIPAM polymers using fluorescence microscopy, both in equilibrium conditions and under flow. Using this approach, we characterize single chain dynamics as a function of the temperature-sensitive dynamic transition of DNA-PNIPAM above and below the LCST of PNIPAM.

469949 Industrial Production of Styrene-Butadiene Rubber: Dynamic Modeling, Process Intensification, Sensitivity and Uncertainty Analysis

Abstract:

Styrene-butadiene rubber (SBR) is a synthetic elastomer with wide range of industrial applications, mainly as an essential component of car tires. Production of SBR latex by cold emulsion polymerization is usually carried out continuously in the cascade of six to fourteen tank reactors. Description of this process by means of mechanistic mathematical modeling is quite scarce in the literature, due to limited understanding of underlying phenomena, e.g. the influence of emulsifiers on the quality of produced copolymer, and the description of the thermodynamics of polymer particle swelling.

This contribution is focused on dynamic modeling of continuous SBR production by cold emulsion polymerization in the train of up to twelve tank reactors of industrial scale. The model is used by SBR manufacturer for deeper investigation of the production line behavior and mainly for off-line optimization of the production process[1]. The principal objectives of the developed simulation software are: (i) minimization of the off-spec product amount during transitions between different SBR grades, (ii) study of production line behavior with respect to varying quality of incoming monomers, (iii) investigation of possible technological improvements of the SBR production, and (iv) determination of reasons for undesirable quality of produced latex.

Reaction mechanism of free-radical copolymerization of styrene and butadiene[2] was processed by the moments of chain-length and chain-branching distributions (i.e., the polymer moments), allowing prediction of important characteristics of produced latex, such as conversion of monomers, copolymer composition, number-average and weight-average molar weight of copolymer. The last two mentioned quantities are empirically correlated to the most important characteristics of SBR latex, the Mooney viscosity[3]. The developed model is represented by a set of differential-algebraic equations (DAE) and validated by plant data both in the steady-state (profiles of important process/product characteristics along the reactor train) and in the dynamic regime (evolution of SBR characteristics during transition between different product grades). Strategy for optimum grade transition trajectory is suggested.

The study includes also sensitivity and uncertainty analysis of the optimization problem with respect to the feed composition and parameters of the copolymerization kinetics. A Monte Carlo technique with Latin Hypercube Sampling and correlation control (Iman Conover) is used for the uncertainty analysis, which is complemented by Morris screening and first order sensitivity index, Si (Sobolâ€™s method) for sensitivity analysis[4,5].Â The results provide importance ranking of feed versus kinetic model parameter uncertainty for model-based polymer process optimisation.

[1] A. Zubov, J. Pokorny, J. Kosek: Styrene-butadiene rubber (SBR) production by emulsion polymerization: Dynamic modeling and intensification of the process, *Chem. Eng. J.*, 207 (2012), 414-420.

[2] S. Saldivar, P. Dafniotis, W.H. Ray: Mathematical modeling of emulsion copolymerization reactors. I. Model formulation and application to reactors operating with micellar nucleation, *J. Macromol. Sci., Rev. Macromol. Chem. Phys.*, 38 (1998), 207-325.

[3] O. Kramer, W.R. Good: Correlating Mooney viscosity to average molecular weight, *J. Appl. Polym. Sci.*, 16 (1972), 2677-2684.

[4] J.C. Helton, F.J. Davis: Latin hypercube sampling and the propagation of uncertainty in analyses of complex systems, *Reliab. Eng. Syst. Safe*, 81 (2003), 23-69.

[5] G. Sin, K.V. Gernaey, M.B. Neumann, M.C. van Loosdrecht,W. Gujer: Global sensitivity analysis in wastewater treatment plant model applications: prioritizing sources of uncertainty, *Water Research*, 45 (2011), 639-651.

470279 Metal Nanocrystals Embedded in Polymer Thin Films for the Reporting of Material State

Abstract:

Stimuli-responsive materials able to detect thermal, mechanical, optical and chemical solicitation have been used to effectively develop sensors, probes and information displays by taking advantage of organic or inorganic chromophores such as metal derivatives, that is metal nanoparticles and nanocrystalline semiconductors, since they confer to their embedding material a variation in the opto-electronic properties as a response to the stimulus. Surface-initiated polymerization (SIP) is able to provide improved adhesion of the polymer film via chemisorption, superior control of the film thickness and composition, easy separation of the film from the monomer, tunability to multiple substrate compositions, and the ability to prepare uniform films on planar and nonplanar geometries. Among the various types of SIP, SI-ROMP has emerged over the past decade as a rapid and tunable SIP technique. Here we report the incorporation of CdSe nanocrystals and Au nanoparticles into a surface-tethered polymer film. The incorporation of the CdSe nanocrystals and Au nanoparticles into the polymer matrix enables the sensing of changes to the material state due to external mechanical and chemical stresses. In the case of the CdSe nanocrystals, external mechanical and chemical stresses elicit local damage to the polymer matrix, such as micro cracks or delamination, and causes the detachment of CdSe surface ligands and the aggregation of the CdSe nanocrystals leading to a change in the emission properties of the nanocrystals. The emission properties of the nanocrystals were determined by utilizing an integrating sphere. In the case of the Au nanoparticles, external mechanical and chemical stresses elicit local damage to the polymer matrix and causes the detachement of Au surface ligands and the aggregation of the Au nanoparticles leading to a change in the absorption properties of the nanoparticles. The absorption properties of the nanoparticles were determined by UV-VIS spectroscopy and polarization modulation-infrared reflectance-absorption spectroscopy (PM-IRRAS).

470436 Ultrathin Gel Electrolyte Layers for Interfacial Control of Lithium Insertion Electrodes

Abstract:

In lithium ion batteries, interfacial reactivity of the electrodes (both the anode and cathode) must be controlled to achieve high power capabilities and long cycle lives. The conventional lithium ion battery anode, graphite, forms a stable solid electrolyte interface (SEI) in initial cycles, which then kinetically hinders electrochemical reduction of the electrolyte in subsequent cycling. Lithium ion battery technology is moving towards higher capacity anodes (e.g. Si) and higher voltage cathodes (e.g. $LiNi_{0.5}Mn_{1.5}O_4$), and it appears that these materials do not naturally form stable SEIs. There is a need for engineered, synthetic solid electrolyte interfaces to accommodate the mechanical effects and electrochemical potentials associated with these new materials. We are developing polymer gel electrolyte coatings that provide mechanically and chemically stable interfaces for both anodes and cathodes. Ultrathin gel electrolyte layers were synthesized using a technique called initiated chemical vapor deposition (iCVD). iCVD provides exquisite control over composition and network structure in polymer thin films with thicknesses from 10 nm to several micrometers. Crosslinked poly(n-butyl acrylate) was synthesized directly on top of $LiCoO_2$ composite cathodes by iCVD. Adhesion promoters were used to covalently attach the polymer to the $LiCoO_2$ active material. The polymer layer imbibed standard liquid electrolyte (1M $LiPF_6$ in alkyl carbonate mixtures), and characterization by Fourier transform infrared spectroscopy, x-ray photoelectron spectroscopy and scanning electron microscopy showed that the gel remained attached to the electrode. The electrodes were assembled into half cells against Li metal and then cycled. The performance of the gel-coated electrodes compared to uncoated, state-of-the-art lithium ion battery electrodes will be presented.

470491 Exploration of Nanofiber-Nanoparticle Electrodes Fabricated *Via* Simultaneous Electrospinning/Electrospraying for Ultra-Low Platinum Fuel Cells

Abstract:

Recently, in our laboratory, a new simultaneous electrospinning and electrospraying (E/E) process was developed to produce unique nanofiber-nanoparticle electrodes for polymer electrolyte membrane (PEM) fuel cells and resulted in high power densities at ultra-low platinum (Pt) loadings. These results demonstrated the impact of electrode morphology on Pt loading. Although promising, these previous results did not explore the impact of all of the electrode morphology parameters (*e.g.*, fiber/catalyst content, fiber size, fiber/catalyst structure, porosity, *etc.*) on fuel cell performance (kW), Pt loading (mg), and subsequently Pt utilization (kW/mg). In this study, E/E electrodes were systematically fabricated to fundamentally investigate the impact of various parameters (nanoparticle aggregate size and distribution, nanofiber size and distribution, electrode porosity, nanofiber purity (ratio of conductive polymer Nafion to carrier polymer), nanoparticle purity (with and without Nafion)) on PEM fuel cell performance and Pt utilization. Results show that electrodes with low nanofiber purity using a conductive carrier polymer have similar Pt utilization to those with high nanofiber purity using a neutral carrier polymer. Electrodes with conductive carrier polymer nanofibers have higher Pt utilization with higher nanoparticle purity, while those with neutral carrier polymer nanofibers have higher Pt utilization with lower nanoparticle purity. At lower Pt loadings, electrodes with an average nanofiber size less than 100 nm have higher Pt utilization and those with an average nanofiber size greater than 100 nm have lower Pt utilization. Nanoparticle purity had no significant effect on nanoparticle aggregate size or electrode porosity, but changes in Pt utilization were observed due to the sole effect of mass transfer resistance from Nafion surrounding the nanoparticle. To our knowledge, previous studies have not been able to independently study mass transfer resistances, charge transfer resistances, and porosities of the catalyst layer on fuel cell performance, because of their coupled effects within the layer. Utilizing the E/E process, we can investigate the morphology in the catalyst layer and its effect on mass transfer and charge transfer resistances separately, which will highly impact the design of and elucidate various transport phenomena effects in the catalyst layer for ultra-low Pt loadings.

470577 Atomistic Simulation of Dynamics of Individual Molecules in Entangled Polymers Undergoing Homogenous Shear Flow

Abstract:

Nonequilibrium molecular dynamics (NEMD) simulations of entangled linear $C_{400}H_{802}$ and $C_{700}H_{1402}$ polyethylene melts were performed to investigate the chain dynamics over a wide range of shear rates $(\dot{\gamma})$. The rheological and dynamical responses of these liquids can be classified roughly according to three shear rate regimes; namely, $\dot{\gamma} < \tau_d^{-1}$, $\tau_d^{-1} < \dot{\gamma} < \tau_e^{-1}$, and $\dot{\gamma} > \tau_e^{-1}$, where τ_d and τ_e are the disengagement and entanglement time of the liquid. Under quiescent conditions, the liquid dynamics are described well by the reptation theory of Doi and Edwards. Over the intermediate shear rate range, shear-induced rotational motion of the individual chain molecules became a dominant relaxation mechanism as the number of chain entanglements decreased dramatically with increasing $\dot{\gamma}$, ultimately resulting in a plateau in the shear stress profile. A new timescale became evident that was associated with the period of the rotation/retraction cycles of the individual molecules. In the third region, the rotational motion of the chains became the sole relaxation mode of the system as the number of entanglements was reduced to a level too low to support conventional reptation theory. The decorrelation (τ_d) and rotational (τ_{rot}) relaxation times of the system were extracted by fitting a functional form of $a\exp(-t/\tau_d)\cos(2\pi t/\tau_{rot})$ to the end-to-end vector autocorrelation function data. Both the decorrelation and rotational characteristic times exhibited a shear-thinning behavior that scaled as $Wi^{-0.76}$ at high shear rates, regardless of the chain length. Wi is the dimensionless shear rate defined as $Wi \equiv \tau_d\dot{\gamma}$. The number of entanglements was also computed as a function Wi. It is shown that this characteristic timescale has a shear-thinning behavior which scaled as $Wi^{-0.35}$ at high shear rates for both liquids. The implications of these findings will be presented with regard to convective constraint release, Chain orientation, and chain stretch over a wide range of shear rates.

470878 Surface Functionalization of SiO_2 Nanoparticles in Nafion Nanocomposite Membranes for Use As Proton Exchange Membranes in Vanadium Redox Flow Batteries

Abstract:

Nafion nanocomposite membranes have received considerable attention for use as the proton conducting phase in vanadium redox flow batteries as a result of their ability to suppress vanadium ion crossover, an undesired event that occurs during battery operation, while maintaining a similar proton conductivity to that of the unmodified membrane. Although several studies have utilized nanoparticles with relatively simple surface chemistries, little has been done in the way of tailoring the surface functionalization of these nanoparticles to systematically control the interactions between Nafion and the nanoparticles. In this study, the effect of surface functionalization of silica nanoparticles on vanadium crossover in solution-cast Nafion nanocomposite membranes was investigated. Specifically, the permeability of vanadium ions across Nafion-SiO_2 membranes with sulfonic acid functionalized silica nanoparticles was measured using ultraviolet-visible spectroscopy. Different sulfonic acid supports were investigated, ranging from a relatively flexible, short saturated chain to a significantly less flexible chain containing aromatic groups. The nanoparticle loading and diameter were also varied in an attempt to better understand how nanoparticle-Nafion interactions act to reduce vanadium ion crossover in these nanocomposite films. In addition, the nanoparticle surface was grafted with polymers, e.g., polyvinylidene fluoride and polytetrafluoroethylene, to investigate the effect of surface chemistry on nanoparticle dispersion within the Nafion membrane. Results from this study indicate that the surface functionalization of the silica nanoparticles plays an important role in controlling the permeation of vanadium ions across these nanocomposite membranes.

470986 Zwitterion-Poly(ethylene glycol) Hydrogels Prevent Bacterial Adhesion

Abstract:

Developing strategies that delay and/or prevent the onset of biofilm formation without inducing evolutionary resistance of bacteria is a pressing challenge. While commercial antimicrobial agents and metallic nanoparticles have been proven effective at inactivating microorganisms, their overuse has led to the spread of antibiotic resistance. An alternative approach would be to fundamentally understand how intrinsic materials properties influence bacterial adhesion in order to improve the design of antifouling materials. We have demonstrated a statistically relevant correlation between hydrogel stiffness and bacterial adhesion. "Soft" and "stiff" model poly(ethylene glycol) (PEG) hydrogels were synthesized that had Young's moduli of 44 and 6,500 kPa, respectively. When challenged with the Gram-negative microbe ***Escherichia coli***, the "soft" PEG hydrogels exhibited a 15 fold reduction in biofouling than the "stiff" PEG hydrogels. Likely, this resulted because key cellular processes, including surface adhesion, are modulated in response to substrate stiffness, an intrinsic property of all materials. This phenomena, that fewer bacteria adhered to softer hydrogels, was also observed for a Gram-positive bacterial species, ***Staphylococcus aureus***. Further, this relationship was observed to occur independent of chemistry as mechanically similar agar hydrogels displayed the same correlation, suggesting that mechanical sensitivity should be factored into antifouling materials design. However, in order to withstand operational pressures, most medical devices are developed to be stiff, therefore, we developed a method to incorporate a zwitterionic polymer into the fully swollen hydrogels to improve antifouling performance. Through the distribution of positive and negative charges on individual polymer segments, zwitterionic polybetaines form a hydration shell that imparts strong fouling resistivity. Polymerization of a phoshobetaine zwitterionic polymer, poly(2-methacryloyloxyethyl phosphorylcholine) with polydopamine into the PEG-based hydrogels substantially enhanced the antifouling performance of all tested hydrogels. The stiff zwitterionic-PEG hydrogels exhibited the greatest improvement, after 24 hrs, ~90% less ***E. coli*** and ***S. aureus*** adhered to the hydrogels compared to the PEG only hydrogels. These results suggest that a fundamental structure-property relationship can be used

synergistically with zwitterionic polymers to further improve the fouling resistance of PEG-based hydrogels.

470997 Counteracting the Formation of β-Protein Fibrils By Natural Polysaccharides

Abstract:

Neurodegenerative diseases such as Alzheimer's disease (AD) and Parkinson's disease (PD) are characterized by aggregation and precipitation of normally soluble proteins: amyloid beta for Alzheimer's disease and α-synuclein for Parkinson's disease. Therefore, alteration of the aggregation pathways of these proteins is a promising therapeutic approach for neurodegenerative diseases. This research is focused on using the mucilage extracted from ***Opuntia ficus indica*** to inhibit aggregation of disease-related proteins.

Mucilage is a neutral pectin polysaccharide with a backbone of α-D-galacturonic acid and β-L-rhamnose and a branch of either arabinose or xylose extracted from cactus plants. Polysaccharides extracted from the mucilage are water-soluble, have a flexible backbone, and are amphiphilic in nature. Two different fractions of mucilage can be extracted from the cactus pads: pectin gelling extraction (GE) and non-gelling extraction (NE). We have previously demonstrated that both GE and NE fractions can be utilized to disperse heavy aliphatic and aromatic molecules. In this work, the effectiveness of mucilage in disturbing the formation of Aβ and α-synuclein fibrils was evaluated.

Aggregation kinetics was monitored by attenuated total reflectance Fourier transform infrared (ATR-FTIR) spectroscopy and time-resolved thioflavine-T (ThT) fluorescence spectroscopy. Transmission electron microscopy (TEM) was used to visually inspect the fibril morphology. We found that both extracts could significantly disturb the kinetics of amyloid proteins aggregation. Aggregation process was completely inhibited at approximately 1:20 mass ratios of GE extract/protein and 1:10 mass ratios of NE extract/protein. These represent much lower ratios than found with other natural inhibitors. Below these critical ratios, aggregation kinetics were dependent on the extract concentrations. TEM data confirmed that both extracts promoted formation of oligomeric protein aggregates instead of fibrils. FTIR data showed loss of β-structure for aggregates formed in the presence of extracts.

These experimental results show the effectiveness of cactus mucilage as an inhibitor of amyloid formation and suggest that it could be used for further drug development.

IN-PLANE STRESS MEASUREMENT IN POLYMER COATINGS AS INDUCED BY VAPOR SORPTION/DESORPTION PROCESSES

Elisa Pavesi, University of Bologna, Bologna, IT

Dr. Matteo Minelli, University of Bologna, Bologna IT

Prof. Ferruccio Doghieri, University of Bologna, IT

Abstract

The aim of this work is the design of an apparatus which allows to measure the in plane stress that arises in polymer coating of commercial polystyrene (N2380) below the glass transition temperature (Tg) as result of vapor sorption/desorption processes and its evolution in time. The in plane stress is measured using bending beam cantilever technique. The experiments were performed at different temperatures, using different fugacity of n-pentane inside the system. From the stress state, knowing the material proprieties, the change in the volume recorded during the sorption/desorption process was derived. Through the use ofthe non-equilibrium lattice fluid model (NELF) it is then possible to estimate vapor mass uptake/release in the coating.

Introduction

The knowledge of transport properties is a theme of considerable importance in a lot of productive sectors using polymeric materials. The maintenance of dimensional stability is an important key in many fields. The sorption of an external phase may induce swelling which causes material deformation, varying the characteristics with positive or negative effects depending on the application concerned.

Mass transport in polymeric thin films is influenced by numerous factors such as: penetrant/polymer characteristics, temperature and/or pressure, solvent fugacity in the system and thermal history of the polymeric film. Many elements characterize the analysis of volumetric properties and stress state originated from the mass transport in polymeric materials.

The aim of the present work is to design an apparatus which measure the in-plane stress that arise in polymeric coating below the glass transition temperature (Tg) as result of vapor sorption/desorption processes.

With the bending-beam cantilever technique is possible to measure the stress variation that mass transport induces in supported polymeric films [1-5].

One side of an unloaded beam is coated with a polymeric material. One end of this beam is then clamped whilethe other one is kept free. The magnitude of curvature that penetrant induces on the cantilever is finally measured by monitoring the displacement of the free end of the unloaded beam through a laser apparatus (Fig. 1).

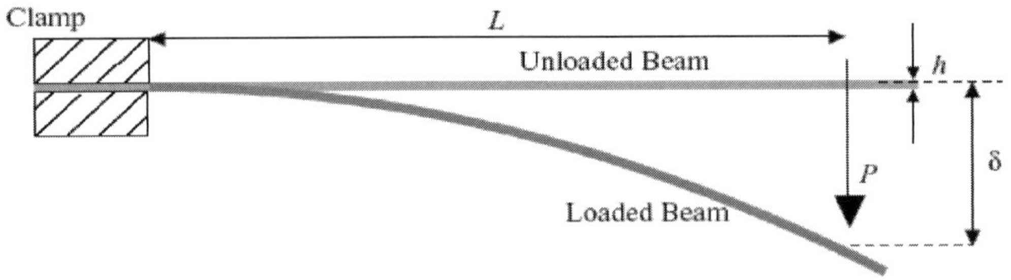

Figure. 1: Scheme of curvature in an unloaded/loaded beam

From the δ values (distance in μm between the end of the unloaded and loaded beam) is possible to obtain the variation in the stress state, as well as the change in volume inside the system.

By δ, using proper mathematical models, is possible to know the stress state of the polymer and some related information.

In particular this technique is useful for the investigation of:

- Stress state of polymer [6-9];
- Glass transition temperature of polymer [6,10];
- Beginning of breaking process [6];
- Elastic stiffness and thermal expansion coefficient for the polymer [11];
- Elastic modulus of the polymer [7];
- Fickian and non-Fickian diffusion of low molecular weight solvents in glassy polymers [12-14].

Materials and Methods

Experimental Apparatus

The experimental apparatus used is represented in Fig.2. The sample chamber is made in stainless steel and has two glassy windows that allow for the measurement through the external laser device. Inside the chamber the beam covered with polymer was located with a reference system. The bending beam was determined by the direct observation using a optical micrometer (Keyence LS-7030-M). The system is able to measure at the same time the distance between the free end of the beam and the reference system and its thickness to be sure about the stability of the system.

The entire system was hosted in a thermostatic chamber which ensures constant and fixed temperature during the test. The external pipe lines were wrapped with a resistance thermometer and protected with polyurethane foam to prevent penetrant condensation.

The system is able to work with static or dynamic flows of penetrant.

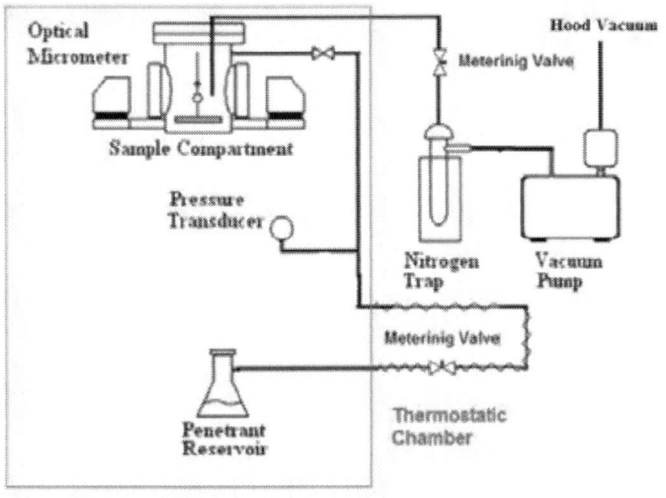

Figure. 2 Experimental Apparatus

Materials

The beam to measure the bending is usually a small and thin cantilever with a rectangular section. The beam geometry greatly influences the instrument sensibility: increases with its length and decreases with its thickness.

The characteristics such as surface finish, flatness of the substrates, elastic modulus, limit the possibility to use the same beam for many tests. For these reasons, the experimental tests were carried out on disposable cantilever made with readily available materials.

Tests to measure the sensibility of the bending signal were made on stainless steel and spring steel beams with size of 7,5X50 mm and thickness between 0,1-0,5 mm.

As a result of these tests, the spring steel beams with a thickness of 0,1 mm were finally selected. The signal detected with the optical micrometer was found to be linear with the thickness: doubling the polymer thickness, the deflection value doubles.

The polymeric films were made with commercial polystyrene N2380 product by Versalis, with characteristics reported in Tab.1. The polymer is deposited on the metallic beam through the solvent casting technique. This casting is following by 24h of evaporation under hood vacuum at ambient temperature. Subsequently a thermal treatment of annealing for 1h under vacuum and temperature higher than Tg is made. A treatment of controlled cooling under vacuum conditions follows the latter.

Annealing process is useful because it reduces the stress due to the solvent evaporation. During the substrate deposition, the polymer remains constrained due to the contact forces at the interface.

The optimal thickness of the polymeric film is found to be around 1-2 μm.

The solvent used in this work is n-pentane. All the experiments were carried out at different temperature between 30 and 45°C.

	Mw [g/mol]	ρ[g/cm^3]	Tg [°C]
Polystyrene N2380, Versalis	270000	1,05	103
n-pentane, Aldrich	72,15	0,626	36 (T_b)

Table 1: Chemical characteristics of polymer and solvent used

	Tickness [mm]	E[MPa] @ 25°C	Poisson Coefficient
Spring Steel	0,1	210598	0,305
Polysyrene N2380, Versalis	0,001	3350	0,35

Table 2: Mechanical characteristic of materials used in the bending tests

Stress state

Stresses in coating deforms the substrate on which the polymer is deposited, consequently there is a curvature of it.

This state of stress (σ) can be calculated from the knowledge of:

- Radius of curvature (r);
- Thickness of polymeric layer (t_c);
- Thickness of substrate layer (t_s);
- Elastic modulus of substrate (E_s);
- Poisson coefficient of substrate (v_s).

Assuming there is a goodadhesion between polymeric film and substrate, and accounting for the fact that the radius of curvature is much bigger than the substrate and film thicknesses and that the beam width is less than half of its length, the average stress state can be express such as [4-5]:

$$\sigma = \frac{E_S t_S^2}{(1-v_S)6rt_C} \quad (1)$$

The radius of curvature can be determined from the bending of the beam (□), applying the theorem of Pythagoras and considering that at the initial state the cantilever is horizontal ($r = \infty$).

$$r \approx \frac{L^2}{2\delta} \quad (2)$$

Where L is the length of the beam.

From the knowledge of the stress state of the beam, the relative variation of volume of the polymeric film between initial (V_0) and final (V_1)as induce by the sorption/desorption phenomena can be determined, based onthe elastic modulus of the polymer (E_c).

$$T_{xx} - T_{zz} = G(\alpha_x^2 - \alpha_z^2) = \frac{E_c}{3}\left[\left(\frac{V_0}{V_1}\right)^{\frac{2}{3}} - \left(\frac{V_1}{V_0}\right)^{\frac{4}{3}}\right] \quad (3)$$

Result and Discussion

The results of stress and volume change in time for the polymeric coating were obtained imposing sequences of increasing concentration steps of n-pentane in the system. In particular the fugacity of the system was incremented with steps of 5% every 24 h up to 20% as maximum fugacity. Examples of results that can be obtained from this kind of tests is offered in Fig. 3 and 4.Within the relatively small temperature range the thermostatic chamber allows to explore, the differences in terms of volume change are clear, although not so easy to express in quantitative terms.

Data for volume change estimated this way can be used to calculate polymer density as function of solute fugacity and from the latter information, through the use of NELF model [15-17], the mass of solute absorbed or released in the system can be ultimately evaluated at all steps examined in the procedure.This specific analysis is now in order and it will be performed in the follow up of this work, together with the comparison with mass uptake results which can be obtained by means of more specific techniques.

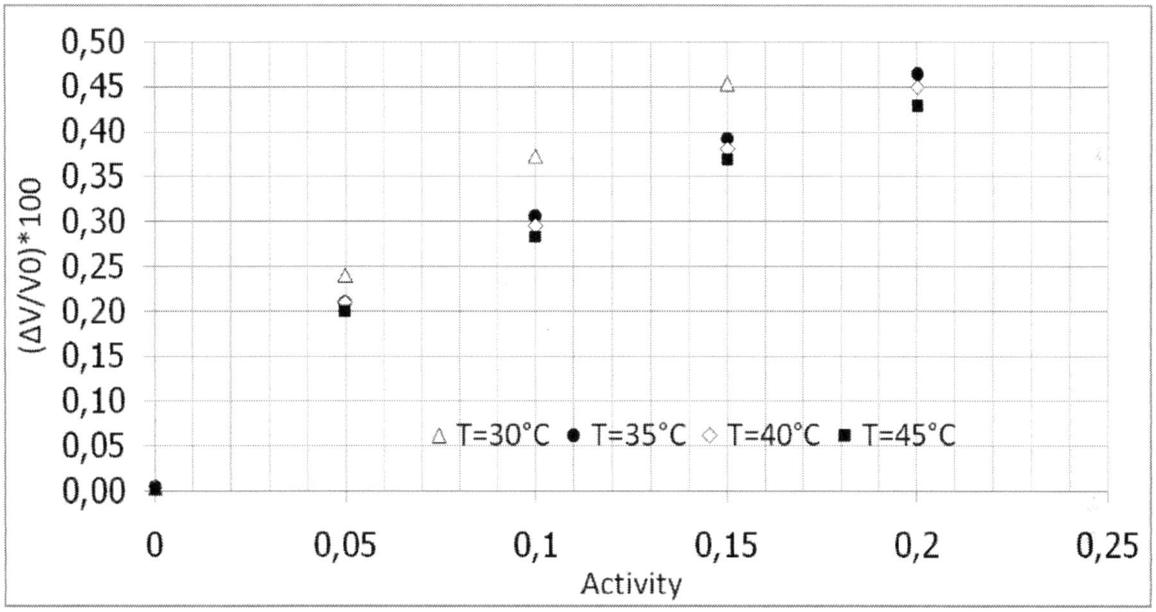

Figure 3: Changes in volume VS n-pentane activity at different temperature and with polymeric film thickness between 1.2-1.6 mm.

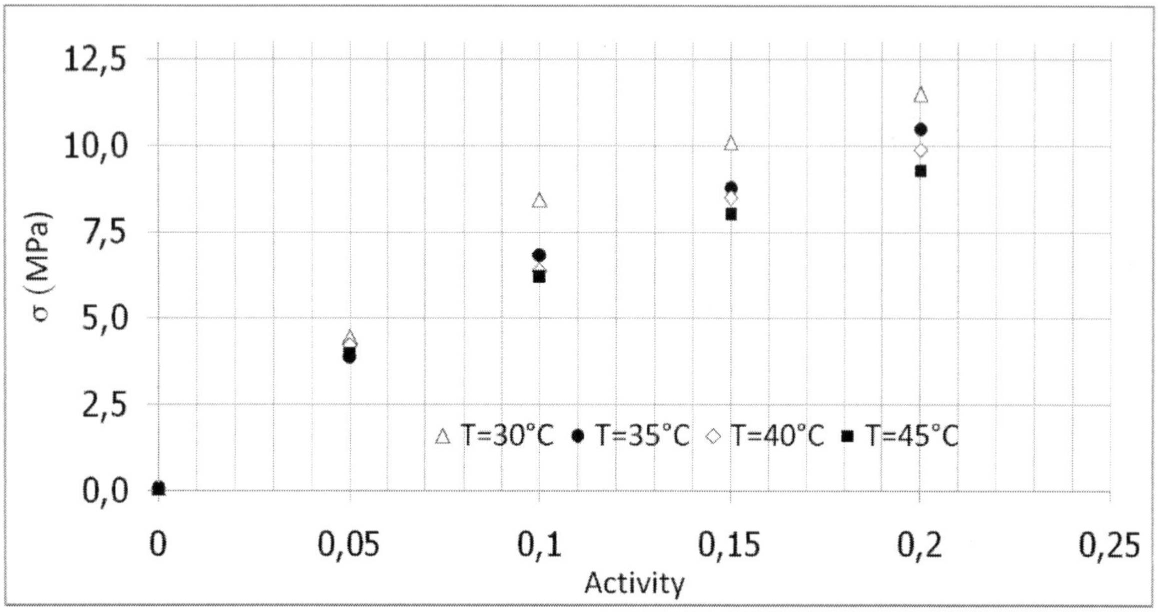

Figure 4: Stress VS n-pentane activity at different temperature and with polymeric film thickness between 1.2-1.6 mm.

References

1. Piccinini E. (2003), "Studio delle Proprietà Volumetriche di Membrane Polimeriche nel Trasporto di Materia", Tesi di Dottorato, Università di Bologna.
2. Campbell D. S. (1969) ,"Mechanical Properties of Thin Films", in *Maissel L.I., Glang R. Eds., Handbook of Thin Film Technology*, McGraw-Hill, New York, NY.
3. Tong H. M., Saenger K. L. (1990), "Bending-Beam Characterization of Thin Polymer Films", in *Tong H. M., Nguyen L. T., New Characterization Techniques for Thin Polymer Films*, Wiley Interscience, New York, NY.
4. Wilcox J. D., Campbell D.S. (1969), "A Sensitive Bending-Beam Apparatus for Measuring the Stress in Evaporated Thin Films", *Thin Solid Films*, **3**, 3.
5. Hoffman R.W. (1976), "Mechanical Properties of Non-metallic Thin Films", *in Dupuy C. H. S., Cachard A. Eds., Physics of Nonmetallic Thin Films*, Plenum, New York, NY.
6. Tong H. M., Hu C. K., Feger C., Ho P. (1986), "Stress Development in Supported Polymer Films During Thermal Cycling", *Polym. Eng. Sci.*, **26**, 1213.
7. Jou J. H., Hwang J., Hofer D. C. (1987), "In Situ Measurement of Temperature Dependance of Internal Residual Stress in Polymide Films Coated on Silicon Substrate", *IBM Research Report*, RJ 5984.
8. Han B., Tong H. M., Saenger K. L., Gryte C. C. (1987), "Mechanical Property Determination for Supported Polymer Films Using Double Bending-Beams", *Mat. Res. Soc. Symp. Proc.*, **76**, 123.
9. Berry B. S., Pritchet W. C. (1988), "Elastic and Viscoelastic Behavior of a Magnetic Recording Tape", *IBM J. Res. Develop.*, **32**, 682.
10. Chen S. T., Yang C. H., Faupel F., Ho P. (1988), "Stress Relaxation During Thermal Cycling in Metal/Polymide Layered Films", *J. Appl. Phys.*, **64**, 6690.
11. Astarita G., Sarti G. C. (1978), "Class of Mathematical Models for Sorption of Swelling Solvents in Glassy Polymers", *Polym. Eng. Sci.*, **18**, 388.

12. Berry B. S., Pritchet W. C. (1984), "Bending Cantilever Method for the Study of Moisture Swelling in Polymers", *IBM J. Res. Develop.*, **28**, 662.

13. Tong H. M., Saenger K. L. (1989), "Bending-Beam Study of Water Sorption by Thin Poly(methyl Methacrylate) Films", *J.Appl. Polym. Sci.*, **38**, 937.

14. Fu T. Z., Durning C. J., Tong H. M. (1991), "Simple Model for Swelling-Induced Stresses in a Supported Polymer Thin Film", *J. Appl. Polym. Sci.*, **43**, 709.

15. Doghieri F, Sarti GC (1996), "Nonequilibrium Lattice Fluids-A Predictive Model for the Solubility in Glassy-Polymers", *Macromolecules*, **29,** 2568-2580.

16. Sarti GC, Doghieri F (1998), "Predictions of the Solubility of Gases in Glassy-Polymers Based on the NELF Model", *Chem. Eng. Sci.*, **53,** 3435-3447.

17. Doghieri F, Sarti GC (1998), "Predicting the Low-Pressure Solubility of Gases and Vapors in Glassy-Polymers by the NELF model", *J. Membr Sci.*, **147**, 73-86.

471047 Relaxation Dynamics and Water Transport in Nafion-SiO_2 Nanocomposite Membranes

Abstract:

Perfluorinated ionomers, Nafion in particular, are the most widely utilized proton exchange membranes in vanadium redox flow battery applications due to its high proton conductivity and good chemical stability. However, Nafion suffers from high vanadium ion crossover, an undesired event that reduces energy efficiency and lifetime of the battery. While the introduction of SiO_2 nanoparticles into the Nafion membrane has proven to be effective at suppressing crossover of vanadium ions, little is understood about the mechanism by which this reduction in crossover occurs. In this study, the relaxation dynamics of Nafion-SiO_2 nanocomposite membranes, in the presence of water, was investigated using a number of spectroscopic techniques. Water transport experiments in these hybrid membranes were carried out using *in situ* time-resolved Fourier transform infrared-attenuated total reflectance (FTIR-ATR) spectroscopy. Unlike typical gravimetric based techniques, time-resolved FTIR-ATR spectroscopy provides molecular-level contrast between the diffusant and the polymer in real time during diffusion experiments, allowing for the simultaneous capture of both water transport and water-induced polymer relaxation (swelling). The diffusion of water through the nanocomposite membranes was found to be non-Fickian and the kinetic uptake data was modeled using a diffusion-relaxation model. In addition, the chain relaxation dynamics of the nanocomposite membranes was investigated using neutron spin-echo (NSE) spectroscopy, a technique that is used to study slow relaxational processes that occur in polymers. Scattering data for all membranes were attained at *q* â‰ˆ 1.2 Å…$^{-1}$, meaning that the measured dynamics of the nanocomposite membranes can be attributed to fluctuations on the length scale of the backbone of the Nafion. NSE spectroscopy data was collected atÂ two temperatures, 298 K and 348 K, and the results were compared to unmodified Nafion with varying thermal treatments.

471336 The Kinetics of Organic Molecule Diffusion in Water Swollen Keratin Fibres Using GC-MS

Abstract:

There has been considerable progress in understanding the structure and properties of hair. However, we still do not fully understand about the structure of hair and the penetration chemistry of actives in the hair, especially when hair is in a water-swollen state.

Hydration of hair, for example, is a fundamental process, which is of both a fundamental and practical importance. It governs swelling behaviour, structure, diffusion rates and mechanical properties of hair. Despite its importance, keratin water interactions (hydration) are poorly understood. This understanding is particularly true for hydration along single hair fibres and across hairs from different body sites, genders, ethnicities and ages. Addressing these key knowledge gaps will allow us to unlock critical insights in materials behaviour.

Hydration of hair, for example, is a fundamental process. It governs swelling behaviour, structure, diffusion rates and mechanical properties of hair. Despite its importance, keratin water interactions (hydration) are poorly understood. This is particularly true for hydration along single hair fibres and across hairs from different body sites, genders, ethnicities and ages. Addressing these key knowledge gaps will allow us to unlock critical insights to enable innovation.

In this study, the time dependent mass uptake of organic solutes in aqueous solutions by hair has been determined using Gas Chromatography- Mass Spectrometry (GC-MS). The effect of solute concentration and temperature on the kinetic uptake of monoethanolamine by hair fibres has been investigated. Diffusion coefficients have been determined using established first order mathematical analysis.

471533 Self-Assembly of Di-Fmoc-L-Lysine Based Supramolecular Hydrogels

Abstract:

Rheology and nanostructure of supramolecular hydrogels formed by self-assembly of a peptide-based amphiphile, di-fluorenylmethyloxycarbonyl-lysine (di-Fmoc-L-lysine) will be presented. The hydrogels were prepared by dissolving the gelator in various solvents such as DMSO, DMF, acetone, methylene, ethanol, and propanol, followed by addition of water. The mechanical properties of these gels were investigated using shear and cavitation rheology. The gels display strain-softening behavior at moderate strain values. Structural analysis were conducted using FTIR, electron microscopy, and atomic force microscopy. Self-assembled fibers with lateral dimensions ranging from 10 to 50 nm were captured in microscopy studies. FTIR results indicate Î²-sheet-like conformation of the peptides in the hydrogel.

4715383 Use of Chaotic Flows for Microfabrication of Complex Tissue-like Structures and Bioinspired Catalytic Surfaces in Hydrogels

Abstract:

Chaotic flows have the ability to create complex but predictable, controlled, and modelable 2D and 3D structures. These structures can vary from quite simple to extremely complex arrangements, depending on the "mixing recipe" and the time of mixing. In chaotic mixing, any flow filament grows exponentially over time, which makes this type of mixing a highly effective tool for quickly creating interfaces.

In tissue engineering, the creation of complex tissue-like structures remains challenging. We are presently using simple chaotic flows to create, in a reproducible and predictable manner, complex microstructures that resemble biological tissue. To that end, a drop of particles (microbeads or cells) is injected into a cross-linkable viscous liquid and dispersed under the action of the chaotic flow. We are currently using gelatin methacryloyl (GelMA) as a model fluid and cell suspensions or fluorescent microparticles as "ink." GelMA is a viscous fluid (as a pre-gel) that renders a hydrogel when crosslinked by exposure to UV light. This crosslinking therefore can preserve the high resolution microstructure originated by a chaotic flow. Chaotic printing can be applied to many relevant tissue engineering and organ-on-a-chip scenarios, ranging from the study of fundamental questions related to cell–cell interactions at material interfaces with different degrees of vicinity, to the development of vascularized functional tissues.

We also demonstrate the use of chaotic flows in the fabrication of convoluted bio-catalytic surfaces bioinspired by the endoplasmic reticulum (ER). A system of two sequential reactions, glucose oxidation and hydrogen peroxide conversion, is taken as our model. We independently immobilized biotinylated horseradish peroxidase and glucose oxidase in nanoparticles functionalized with streptavidin. We imprinted these two types of nanoparticles in a GelMA construct to fabricate a convoluted multi-sheet structure where both enzymes are immobilized in close proximity. The addition of glucose to the construct triggers an oxidation reaction, mediated by the glucose oxidase, to produce hydrogen peroxide, which then serves as the substrate for the peroxidase. The reaction is visualized by the development of red fluorescence due to the oxidation of Amplex™ red.

These two applications illustrate the potential of chaotic flow as a novel and powerful platform for the microfabrication of bioinspired systems.

458137 Polyvinyl Alcohol (PVA) As Protective Colloid for Polystyrene Nanoparticles Synthesis Initiated By Oil-Soluble Initiator

Abstract:

Compared with low molecular surfactants, polyvinyl alcohol (PVA) could afford latexes with particular tackiness as an important property for various applications. However, until now, except for some highly reactive monomers, most of monomers including styrene seldom use PVA as protective colloid in emulsion polymerization due to the lack of its stability. In this work, we used oil-soluble initiator, 2,2¡¯-azobisisobutyronitrile (AIBN), to accelerate the emulsion copolymerization of styrene and divinyl benzene in PVA colloids and synthesize polystyrene (PS) nanoparticles within 50 min at 70 °C successfully. Even at the monomer-to-water ratio as high as 4:1, the emulsion system could keep stable throughout the emulsion polymerization with only PVA as stabilizer, and the polymerization in PVA colloids was always dominant. The adjustment of the size of the PS nanoparticles in the range from 70 nm to 100 nm was achieved simply by changing the polymerization temperature, the addition of PVA, or the monomer-to-water ratio. The PVA adhering on the surface of the PS nanoparticles showed stable modification effect in the aqueous latexes, and also could be washed completely by dimethylsulfoxide (DMSO) to obtain PS nanoparticles without surfactant contaminant. This method was also exploited to synthesize poly(styrene-co-methyl methacrylate) and poly(styrene-co-4-chlorostyrene) nanoparticles, possessing good potentials in functional polymeric nanoparticles production with the advantages of high capacity and broad applicability.

472484 Molecular Simulations of Reverse Osmosis Aromatic Polyamide Membrane: Mechanical Properties, Desalination, Crosslinking Degree and Monomers' Isomer States

Abstract:

Reverse osmosis (R) is one of the leading desalination techniques used worldwide. Thin film composite membranes like polyamide (PA) membranes offer advantages in high flux, high rejection to salt and some low molecular weight organics, and stability at higher temperature and over a larger pH range. It is crucial to understand the mechanical properties of crosslinked polyamide membrane and its relationship with its performance in desalination. In this work, we use molecular dynamic simulation to estimate crosslinked polyamide membrane's mechanical properties (poison ratio and young's modulus), pore size distribution, water permeation and salt rejection capability as a function of crosslinking degree. We also examine the mechanical properties and pore size distribution by varying monomer's isomer states. The research will be important for future PA membrane experimental synthesis.

472022 Understanding the Influence of Different Interfacial Interactions on the Glass Transition Temperature and Self-Diffusion Coefficient in Unentangled Polymer Thin Films

Abstract:

In thin films, physical properties such as the glass transition temperature (T_g) and self-diffusion diffusion coefficient (D) are often different compared to those in thick films due to their higher interfacial area to volume ratio and other potential factors. These effects typically occur in submicron thick films and have been termed nanoconfinement effects. Understanding polymer dynamics under nanoconfinement is critical to advance nanotechnology; this knowledge could be exploited for tailoring material properties with applications ranging from nanocomposites to lithography, among others. However, there are several unsolved issues relating to nanoconfined mobility that are of primary interest such as the relationship between T_g and D. From a dynamic length scale point of view, T_g reflects the cooperative segmental mobility of 10-100 repeat units while D reflects the large-scale transport motion of the entire chain, and it is reasonable to expect they could be affected differently when polymer is nanoconfined. In this study, we employed fluorescence recovery after patterned photobleaching to evaluate the D of fluorescently labeled poly(isobutyl methacrylate) (PiBMA) as a model polymer. Films 16 - 300 nm in thickness were spin coated onto silica substrates possessing attractive polymer/substrate interactions and poly(cyclohexylethylene) (PCHE) substrates possessing weaker polymer/substrate interactions, then D and T_g were measured. In addition, the effects of free surfaces were investigated by capping the free surfaces atop PiBMA thin films with a PCHE layer. The effects of friction (e.g., between the substrate or capping layer and the confined polymer melt) on the T_g-D relationship will also be presented.

472744 Structural Evolution of a Polymeric Porous Medium Induced By Cross Flow or Tangential Flow of Solvent

Abstract:

Polymeric soft materials are ubiquitous in nature and in man-made matter. They comprise a variety of components and encompass a range of physical states with characteristic time and length scales over several orders of magnitude. Many of them are highly porous with solvents, solutes, and natural or artificial additives physically and/or chemically accommodated in the pore space. The resultant polymeric porous media and pore structures could be sensitive, by nature or by design, to physical and chemical changes that may arise externally or internally, and could exhibit significant structure-property relationships in the physicochemical, functional, and processing aspects of their properties. By filling the pore space and interacting with the skeleton polymers and other solute species, solvent molecules are involved during the synthesis and applications of polymeric porous media and capable of playing a very important role in determining the structures, properties, and functions. In fact, recent molecular modeling and simulation studies have shown that stable pore structures of polymeric porous media should be considered as being achieved by a balance between the polymer-polymer interactions, the solvent-polymer interactions, as well as the solvent-solvent and the solvent-solute interactions. This work is aimed at investigating the dynamical effects instigated by solvent flows on the pore structure and properties of a polymeric porous membrane. For this purpose, a biocompatible polymeric porous membrane was constructed by coarse-grained molecular dynamics (MD) modeling and simulation, and water was considered as the solvent to not only fill the pores but also form an exterior aqueous phase that submerge the model membrane. The whole model system was equilibrated for the membrane to develop a stable pore structure based on a balance between the polymer chains' flexibility, mutual steric support, and interactions with the solvent molecules under static condition. The pore structure was rigorously characterized by a molecular approach in order to establish a reference for evaluating the dynamical effects caused by different solvent flows. Tangential flows of solvent were imparted by moving the topmost layers of the water molecules in the exterior phase at different speeds in a direction parallel to the water-polymer interface. Cross flows of solvent were induced by applying different pressures on the topmost layers of water molecules in the exterior phase in a direction perpendicular to the water-polymer interface. Through the water-water and water-polymer interactions, the polymer molecules individually and the pore structure collectively experience the effects of the imposed momentum transfer and mass transfer and exhibit interesting and important changes including the membrane thickness, pore-size distribution, and the interaction energetics of water molecules in the pores. When the flows were stopped, the pore structures, however, were found not to be able to return back to the original structure. In general, the cross flows of solvent were found to cause greater pore structural reduction than the tangential flows and the resultant reduced pore structures were more dissimilar from the original one during and after the flow conditions. Other aspects of structural and energetic characterizations will also be compared and presented for a more complete elucidation.

473790 Organically Modified Clays As Rheology Modifiers and Dispersing Agents for Epoxy Packing of White LED

Abstract:

We develop a new class of polyetheramine-modified organoclays that enable the homogenous dispersion of inorganic phosphorescent phosphor powder, yttrium aluminum garnet $Y_3Al_5O_{12}$ (YAG), in epoxy, which can then be used for packing of light-emitting diode (LED) devices. The organoclays are prepared from the ionic exchange reaction of sodium montmorillonite (Na^+-MMT) with poly(oxypropylene) (POP) diamines of sufficiently high molecular weight, which are found to be effective for homogenizing YAG in the two-component epoxy system involving an anhydride curing agent and an aliphatic epoxy resin. By adding the POP-diamine-based organoclays, the viscosity of the anhydride/epoxy resins is greatly enhanced and the resin shows a non-Newtonian shear-thinning behavior. Because of the high viscosity that retards the molecular motion, the micrometer-scaled YAG particles can remain dispersed in the resins for a fairly long time without precipitation during the curing process. We investigate the rheological properties of the organoclays as the thickening agent and applied the organoclays to fabricate YAG particle-dispersed epoxy for packing of blue light LED that mixes the yellow light of the well-dispersed YAG particles to form uniform white illumination.

477006 Rapid End-Block Pullout in ABA Triblock Polymer Gels

Abstract:

The end-block pullout time of the ABA triblock polymer poly[styrene-*b*-(ethylene-*alt*-propylene)-*b*-styrene] (SEPS) has been studied via rheology. Gels of SEPS in squalane of varying concentrations were produced via a cosolvent method and then rheometry was performed at various temperatures. Master curves were produced by time-temperature superposition, and longest relaxation times were extracted. These relaxation times were then compared with the results of previous time-resolved small-angle neutron scattering (TR-SANS) experiments that focused on chain exchange kinetics in SEPS/squalane micelles. It was found that the relaxation time of the gels (associated with the end-block pullout time) was up to four orders of magnitude smaller than would be expected based on the TR-SANS experiments of equivalent diblock polymers. A model is proposed that accurately reproduces the triblock TR-SANS results by incorporating two additional factors: 1) a reduction in the energy barrier to end-block pullout due to the triblock architecture, and 2) a bias towards shorter chain lengths in the relaxation of the gels. The first factor was quantified by a recalculation of the energy barrier to end-block extraction using the gel relaxation time. The second was accounted for by the creation of an effective chain length distribution that assumes the shorter of two end-blocks will pull out before the longer block. These two effects are then incorporated into a model that accurately reproduces the triblock TR-SANS results and consistently explains the unexpectedly rapid pullout in ABA triblock rheology and TR-SANS experiments.

450609 Leveraging Computational Methods to Study the Properties of Star Di-Block Nanoparticles for Use in Targeted Drug Delivery

Abstract:

A star di-block copolymer consists of di-block polymer arms covalently bonded to a central core. The inner polymer block bonded to the core is hydrophobic, and the outer arm block is hydrophilic. The highly combinatorial nature of the star, e.g. variations in monomer chemistry combined with variations in arm length and arm loading, provide a vast landscape to characterize for drug-delivery applications, most of which remains unexplored. Polymeric materials are an attractive field for drug delivery due to the vast range of materials available and their interesting properties as a function of temperature and pH. We present a comparative simulation study of how the physical properties of model star polymers change by changing the chemistries of the hydrophobic and core material of the polymer. We have observed temperature sensitivity in some of the polymers studied and would like to leverage this property for further star polymer development. These new comparisons of chemical features and size are explored to understand what drives thermally induced structural changes, especially in regards to how cargo is released or the exposure of sites to hydrolysis reactions that promote biodegradation, all of which are important factors relevant to their suitability as a drug-delivery material.

455609 Dextran Sodium Sulfate Exposure Affects Intestinal Mucus Integrity

Abstract:

Dextran Sodium Sulfate Exposure Affects Intestinal Mucus Integrity

Jaclyn Y. Lock[1], Taylor Carlson[2], Albert Chen[2], Rebecca L. Carrier[1,2]

[1]Department of Bioengineering, [2]Department of Chemical Engineering, Northeastern University

Introduction: Dextran sodium sulfate (DSS) is frequently utilized to induce intestinal inflammation in animal models of colitis. This model is heavily used since it mimics the histological features of colitis, but the mechanism by which DSS induces inflammation remains unclear. Mucus, a natural biomaterial, covers the entire gastrointestinal tract and selectively controls the diffusion of molecules, particulate matter, and microorganisms to the underlying epithelial layer. Thus, we hypothesize that DSS may impact the mucus barrier. In this study, we investigated how DSS exposure affects mucus barrier properties through particle and microbe tracking, and micro-structural analysis. These studies may provide information on the potential role of an altered mucus barrier in progression of inflammation.

Materials and Methods: Native porcine intestinal mucus was harvested within two hours of sacrifice. Transport properties of green fluorescent protein (GFP) expressing *Escherichia coli* (*E. coli*), and 200 nm fluorescent particles with surface carboxyl, amine, and polyethylene glycol (PEG) functionalization were probed using multiple particle tracking technique. Briefly, microbes or particles were diluted in the presence of 1% DSS or maleate buffer (MB) control. 20 sec video particle trajectories were analyzed using a modification of a MATLAB script developed by Maria Kilfoil[3] to calculate mean mean-squared displacement (<MSD>) and effective diffusivity (D_{eff}), where MSD = $[x(t+\tau) - x(t)]^2 + [y(t+\tau) - y(t)]^2$ and D_{eff} = MSD * 4τ, x(t) and y(t) represent the particle coordinates at a given time and τ is the time scale. To image the changes in particle and microbe distribution within lectin-stained mucus exposed to DSS or MB control, z-stack images were obtained using a confocal microscope. Scanning electron microscopy was utilized to visualize micro-structural changes when 1% DSS was dosed to porcine intestinal mucus. StudentÕs T-test was used to determine significance between particle and microbe transport with $\alpha = 0.05$.

Fig 1: *Escherichia coli* mixed with MB control or 1% DSS and dosed to intestinal mucus.

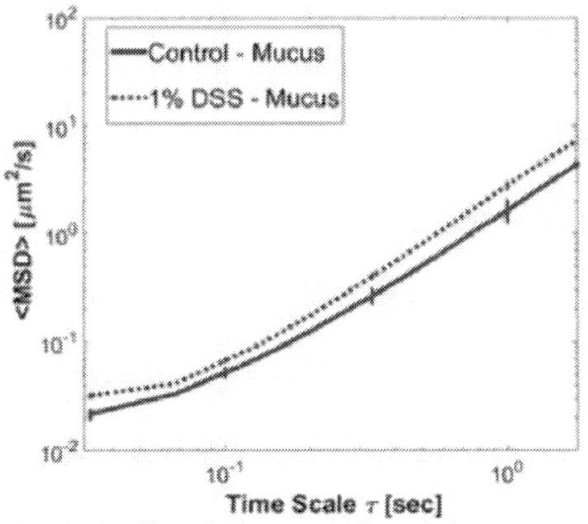

Fig 1: *Escherichia coli* mixed with MB control or 1% DSS and dosed to intestinal mucus.

Results and Discussion: Particle diffusion in mucus exposed to 1% DSS was significantly hindered. Specifically, there was a 8-, 3-, and 5- fold decrease in <MSD> for amine-, carboxyl-, and PEGylated particles when dosed with 1% DSS compared to MB control. Interestingly, for *E. coli* undergoing active transport, the microbes moved faster in the presence of 1% DSS in intestinal mucus and there was an increase in linear trajectories (Fig. 1).

Conclusion: These results indicate that a 1% DSS solution does alter mucus barrier properties, both with respect to particles and microbes. Further *in vivo* studies investigating impact of orally dosed DSS on a mucus layer on intact tissue will facilitate understanding the role of DSS in inflammation.

460080 Nanotopography Promotes Neuronal Differentiation of Human Induced Pluripotent Stem Cells

Abstract:

Nanotopography Promotes Neuronal Differentiation of Human Induced Pluripotent Stem Cells

Liqing Song[1], Kai Wang[2], Yong Yang[2], Yan Li[1]

1. Department of Chemical and Biomedical Engineering, FAMU-FSU College of Engineering, Florida State University, Tallahassee, FL 32310, United States

2. Department of Chemical and Biomedical Engineering, West Virginia University, Morgantown, WV 26506, United States

Abstract

Inefficient neural differentiation of human pluripotent stem cells (hPSCs), including human induced pluripotent stem cells (hiPSCs), motivates the investigations of the influence of biophysical microenvironment, in particular nanotopography, on hPSC fate decisions recently. However, the contributions of the shape and dimensions of nanotopography to neural lineage commitment of hPSCs have not been well understood. The objective of this study is to delineate the effects of the shape, feature size and height of nanotopography on neuronal differentiation of hiPSCs. Equally spaced, anisotropic nanogratings (500 and 1000 nm in linewidth) and hexagonally arranged isotropic nanopillars (500 nm in diameter, 450 nm in edge-to-edge distance), each having a height of 150 or 560 nm, were

seeded with human iPSK3 cells and induced for neuronal differentiation. The gratings of 560 nm height reduced cell proliferation, promoted cytoplasmic localization of Yes-associated protein (YAP), and enhanced neuronal differentiation (up to 60% Î²III-tubulin+ cells) compared with the flat control. Differential gene expression of TBR1, a cortical glutamatergic neuron marker, was observed for different substrates, while HOXB4, a marker for hindbrain/spinal cord, was less affected by different nanotopographies. The derived neuronal cells express MAP-2, Tau, Islet-1 (motor neuron progenitors), GABA and vGAT (GABAergic neurons), and glutamate (glutamatergic neurons), indicating the existence of multiple neuronal subtypes. The nanotopography also affected the gene expression of matrix metalloproteinases. This study provides the insights of how different nanotopography parameters (shape, feature size and height) modulate hPSC neural lineage commitment for applications in neurological disease modeling and drug discovery.

460338 Interfacial Engineering of Biodegradable Polyelectrolyte Multilayer Thin Films for Sequential and Sustained DNA Delivery

Abstract:

Polyelectrolyte multilayer (PEM) thin films containing anionic bioactive molecules such as DNA and synthetic polycations as carriers are promising biomaterials for localized nucleic acid delivery. Layer-by-layer (LbL) techniques of assembling PEMs have been widely applied for biomedical coatings due to their ability to encapsulate a wide range of biomolecules and their ability to deposit on various biomedical substrates such as Ti. Our research focuses on the synthesis and interfacial engineering of a novel bioreducible LbL film platform in order to achieve high efficiency and sustained DNA release in cellular microenvironments. Poly(amido amine) (PAA) containing the disulfide bond as gene delivery vector has been extensively investigated in our previous work. The disulfide bond in the polymer structure is predominantly reduced in target cells and cell nuclei due to the endogenous redox gradient in cellular microenvironments. Currently, we incorporated a highly transfecting monomer, 5-amino-1-pentanol (APOL), into the PAA molecular structure in order to further improve the transfection efficiency. In order to avoid bulk DNA release, we periodically inserted nonbioreducible polycations such as poly(ethyleneimine) (PEI) into the film as a barrier layer to slow down the film degradation rate. In order to understand the relationship between film assembly and film disassembly, we characterized the film structure using in situ atomic force microscopy, fluorescence spectroscopy, and dynamic light scattering. We tested DNA release from the film in vitro using HEK 293, MC 3T3, and NIH 3T3 cells. We found that film transfection can last more than one week. In addition, we incorporated two types of DNA plasmids into the LbL film at different layers.. This design is intended to demonstrate the dual stage release pattern of the LbL film, which is desirable for DNA vaccine delivery and tissue engineering applications. Our study contributes to the understanding of basic relationships among films assembly and disassembly, polycation molecular structure, and interfacial LbL film structure for improved use of LbL films as gene delivery vehicles.

460343 Gold Nanoconjugates: Targeted Drug Delivery, Drug Release, and Biodistribution

Abstract:

Gold nanoparticles (AuNPs) show great potentials in biomedical applications because of their versatile surface chemistry, ease of imaging, and tenability to take advantage of the blood-brain barrier transport mechanisms. AuNPs have also been found to be among the most benign nanomaterials partially attributed to the inertness of bulk gold. As a consequence, AuNPs are widely investigated for targeted delivery of drugs, genetic materials, antigens, and diagnostic agents. To act as a drug carrier, AuNP enables various conjugation chemistry with drugs and proteins. Also, the particle size, shape, surface chemistry of the AuNP can be precisely tuned. This poster describes a sample of a tripartite nanoconjugate comprised of an AuNP drug carrier chemically conjugated to a transporter protein (wheat germ agglutinin coupled to horseradish peroxidase or WGA-HRP) and to the spinal cord injury (SCI) drug, theophylline. Our results show that a single administration of the nanoconjugate improved diaphragmatic activity at a dosage lower than the effective systemic drug dose and restored majority of the respiratory drive. The effects lasted for up to 14 days. A biodistribution study using inductively coupled plasma mass spectrometry on the same animal model shows a successful targeting of the nanoconjugate in the respiratory neurons located in the medulla. By using of thermogravimetric analysis and laser desorption ionization mass spectrometry we determined the numbers of drug molecules conjugated to AuNP. All the AuNPs were characterized by UV-Vis spectroscopy, transmission electron microscopy, dynamic light scattering, and atomic force microscopy.

460578 A One-Step Method for Transferring Single Wall Carbon Nanotubes Onto a Hydrogel Substrate for Biomedical Applications

Abstract:

In recent years, significant progress has been made in elucidating the unique properties of carbon nanotube (CNTs)-hydrogel nanocomposites for biomedical applications. Due to the extremely hydrophobic nature of CNTs and the hydrophilic nature of the hydrogels, there are impediments to developing nanocomposites, including attaining a homogeneous dispersion of CNTs in the polymer solutions or alignment and patterning of CNTs in the hydrogels. Several approaches have been reported on patterning CNTs on various polymeric substrates. However, these methods include high temperature, high vacuum or the use of a sacrificial layer and hence, are incompatible with hydrogel substrates as they can lead to irreversible collapse in hydrogel structure. In this study, we designed a novel method to transfer CNTs onto the hydrogels with a single-step stamping technique. First, dense and aligned single wall carbon nanotubes (SWCNTs) were grown on single-crystal quartz substrate. Subsequently, polyethylene glycol (PEG), poly acrylamide (PA), and polyvinyl alcohol (PVA) hydrogel solutions were deposited on the CNTs grown on the quartz substrate. Upon hydrogel gelation, the CNTs were transferred from the quartz onto the hydrogel. The successful transfer of CNTs was confirmed by scanning electron microscopy (SEM). The transferring efficiency was found to be ~99%. Furthermore, results showed efficient cell attachment on the CNT-hydrogel composites that indirectly confirmed the successful transfer. CNT-hydrogel nanocomposites, such as the ones presented here, can be applied for various biomedical applications such as tissue engineering, drug delivery, and biomedical devices.

462695 A Proliferation Switch of Fibroblasts in Alginate Microcapsules By *in Situ* Conjugation of RGD Peptides

Abstract:

Extracellular matrix (ECM) greatly affects proliferation, differentiation, and functions of cells. Chemical composition and structure of ECM dynamically changes during many physiological and pathological processes such as development of embryos, wound healing, and tumor growth. In order to mimic these dynamics, we newly developed clickable alginate microcapsules, which can introduce thiol-containing peptides by *in situ* conjugation between maleimide and thiol groups. In fact, encapsulated fibroblast cells started proliferation by *in situ* conjugation of CRGDS, while free RGDS did not show any effect. In tissue engineering, accurate and economic methods for induction of differentiation of stem cells are required. We expect that these microcapsules can be utilized for future cell differentiation process.

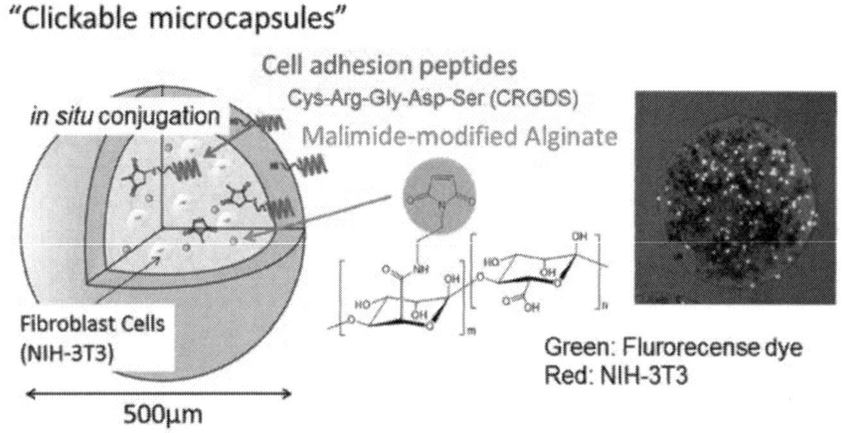

Fig 1. Schematic illustration of a clickable microcapsule. Malieimide-modified alginate microcapsules crosslinked with calcium ions can encapsulate NIH-3T3 fibroblast cells, which started proliferation by the stimuli of *in situ* conjugation of RGDS with maleimide residues of alginate matrix.

462736 New Nanostructured Oxygen Sensors for Biological Applications

Abstract:

Dissolved oxygen concentrations in live cells and cellular environments are critical for many physiological and pathological processes including metabolism, cell respiration/oxygen consumption, and tissue hypoxia. Among the many oxygen sensors, platinum(II)-5,10,15,20-tetrakis-(2,3,4,5,6-pentafluorophenyl)-porphyrin (PtTFPP) is known to be an excellent one due to its good responses to oxygen concentrations, high photostability, and high quantum efficiency as compared with others. Unfortunately, its extreme hydrophobicity limits its application in a biological environment. Herein, we will describe our results using various different amphiphilic block copolymers to incorporate typical oxygen probe of PtTFPP for enabling the application of the hydrophobic material in aqueous solution, achieving high brightness and water-soluble oxygen sensors. We will report the preparation of micelle-based oxygen sensors, their optical spectra, oxygen responses, and photo-stability. The sensors showed liner responses to dissolved oxygen with high reversibility, enabling their suitable application for measuring dissolved oxygen concentrations. The sensors did not show obvious cytotoxicity to cells. We have demonstrated that these sensors are applicable for cell respiration monitoring for microbes and eukaryotic cells with various cell densities. At a defined cell density, cell respiration velocity is relevant to cell viability, thus we have used the sensors for drug screening. For getting accurate analysis results in cell culture medium, we incorporated another fluorophore which does not respond to oxygen to prepare ratiometric fluorescent sensors.

464256 Self-Assembled Di-Block Polymersomes As Artificial Immune Cells

Abstract:

Introduction

Global healthcare in its current reactionary format is irrepressibly overwhelmed by a surplus of patients and tremendous medical expenses; associated treatment strategies are progressively insufficient. A paradigm shift centered on nanotechnology is anticipated to enable the gradual transition from reactionary to predictive medicine needed to enhance global wellness. In particular, it is hypothesized that the integration of synthetic immune cell mimics together with modern medical intervention will substantiate the concept of a smart vaccine capable of coordinating aggressive phagocyte and B-cell activities. Complex cellular systems, including relevant chemical accessories, are simulated through the design of smart polymeric materials called polymersomes (Ps). Ps are amphiphilic block copolymers that self-assemble into artificial vesicles via the hydrophobicity interactions of admixed aqueous and organic substances. The objective of the current study is to fabricate synthetic immune cells by exploiting tunable Ps membrane properties. The intelligent functionalization and digitization of Ps is postulated to mimic consummate immune cell behavior. Strategies are devised to supply an artificial vesicle affixed by several classes of antibodies and other species, including isolated mitochondria.

Successful functionalization requires the development of a bottom-up approach initiated through an assessment of the chemical bonds associating polymersome membrane properties with conjugated species and bacterial or viral shells. For the current application, chemical viability approximations suggest that a methoxypoly(ethylene glycol)-b-poly(D,L-lactide) (mPEG-PDLLA) diblock copolymer will be an ideal Ps prototype constituent[1]. The occurrence of polyethylene glycol within the hydrophilic block circumvents the probability of an Òautoimmune effectÓ in which artificial immune cells are dismantled by the bodyÕs natural defenses. The PDLLA block confers conditional stability to Ps stored outside the body at 4 ¡C and activity to those at physiological conditions.

Methods

A modified stirred-injection routine was applied for the self-assembly of Ps nanocarriers embedded with immunoglobulins, mitochondria, and the enzymatic species responsible for facilitating controlled Ps rupture. An organic solvent with a preferential capacity for generating nano-sized vesicles (i.e. tetrahydrofuran) was used to dissolve select block copolymers, before being infused with addendum species, including biotinylated homing devices. Hydration of the solution was achieved by the staggered addition of the organic solution to an aqueous mixture containing PBS, water, and other hydrophilic species, such as mitochondria. Theoretically, interfacial tensile forces between the hydrophobic regime and the hydration agent initiate spontaneous enclosure of the coupled Ps. Strategies were next devised to supply dynamic Ps that were conjugated with IgG primary antibodies along the peripheral hydrophilic regime and mitochondrial protein within the aqueous core. The direct introduction of IgG antibodies to a plain Ps solution yielded poor chemical attachment. Sufficient functionalization was achieved via a stable non-covalent conjugation associating the glycoprotein avidin with the water-soluble vitamin biotin. Tresylated polymer was subjected to an NMR chemical verification study prior to biotinylation in a biotin-lysine (biocytin) and methanol solution. The biotinylated Ps were washed in a secondary dialysis step that eliminated excess biocytin and organic solvent from the solution of interest.

The resulting vesicles were incubated overnight with an excess of avidin-rhodamine isothiocyanate before being dialyzed to remove free avidin from the Ps solution. Confocal microscopy images provide a validation of bilayer formation in addition to an optical quantification of the degree of avidin-biotin conjugation. Once avidin-biotin binding efficiencies were determined, unconjugated-avidin and biocytin functionalized Ps were incubated in a mixture of biotinylated IgG antibodies from human serum, before being dialyzed and introduced to a solution containing fluorescein isothiocyanate (FITC)-labeled mouse anti-human secondary antibodies. A final dialysis step ensured the exclusion of free immunoglobulins from solution. The relative degree of IgG surface attachment was measured by visual inspection of confocal microscopy images.

As part of a complementary Ps functionalization strategy, mitochondria were isolated from human dermal fibroblasts (HDF) via a rigorous centrifugation and homogenization routine. 120×10^6 HDFs were seeded 3 days prior to homogenization and fractionation. Structural integrity was confirmed using transmission electron microscopy (TEM) characterization. A subsequent Ps conjugation entailed the incorporation of mitochondria into the aqueous corona and catalase enzyme into the hydrophobic compartment prior to bilayer formation. Ps diameter, conventionally on the order of 100 nm, was tuned to accommodate for relatively large mitochondrial dimensions. Future bacterial and cytotoxicity assays will measure the viability of Ps that are functionalized with mitochondria and/or antibodies along the hydrophilic regime.

Results

Physiochemical size distributions of plain Ps were obtained using transmission electron microscopy (TEM). A droplet saturated with Ps was uniformly coated on a 300-mesh copper-coated carbon grid, allowed to settle for 1 minute, and dried in ambient temperature conditions. Negative staining with a 1.5% uranyl acetate solution enabled differentiation of Ps edges and features. TEM images at 2000× magnification revealed spherical Ps of about 80 nm – 150 nm in diameter (**Fig. 1**). The successful generation of enclosed Ps was superseded by a rigorous mitochondrial isolation procedure.

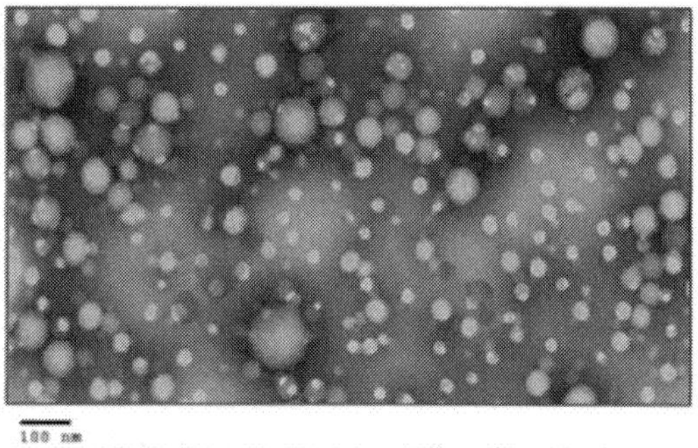

Figure 1: TEM of plain Ps. Scale Bar = 100 nm. Magnification = 20000×. HV = 80.0 kV.

A mitochondrial isolation routine was implemented on HDFs, and 500 μL of solution containing a pelleted sample were pre-fixed in 2.5% glutaraldehyde and maintained at 4 ¡C over a 48-hour time frame. The pellet was carefully removed from the fixing solution and sliced into six sections as a means of expanding surface area exposure and increasing the possibility of mitochondrial visibility. The samples were washed 2× in 0.1 M cacodylate buffer and post-fixed with 1% osmium tetroxide. As part of an ethanol gradient dehydration, samples were suffused in 30% × 1, 50% × 1, 70% × 1, 85% × 1, 95% × 1, and 100% × 2 ethanol, respectively. The samples were immersed in a custom resin before being embedded via a 48-hour heat treatment and thin sectioned. A 300-mesh copper-coated carbon grid was coated with the sample, which was negatively stained using 1.5% uranyl acetate. TEM of the samples revealed sufficient mitochondrial yield for incorporation into the Ps aqueous corona. Future work will focus on the integration of these two structural elements. **Fig. 2** shows a TEM image obtained at 15000× magnification of two neighboring mitochondria having diameters ranging from 400 nm – 450 nm. **Fig. 3** shows a TEM obtained at 20000× magnification of a longitudinally oriented 700 nm long mitochondrion.

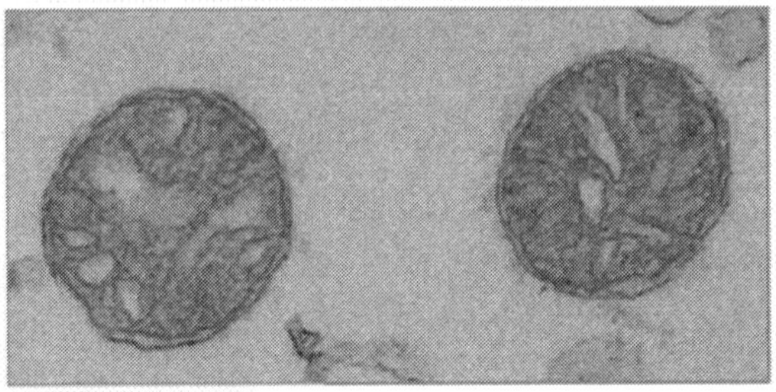

Figure 2: TEM of isolated mitochondria with diametric cross-section (into/out of the page). Scale Bar = 500 nm. Magnification = 15000×. HV = 80.0 kV.

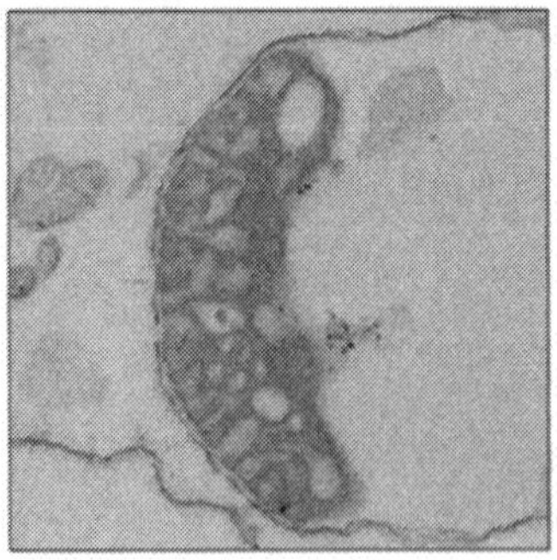

Figure 3: TEM of isolated mitochondrion with longitudinal cross-section. Scale Bar = 100 nm. Magnification = 20000×. HV = 80.0 kV.

Discussion

Experimental strategies were successfully implemented for the generation of Ps nanostructures and the isolation of mitochondrial protein. Extensive studies on the chemical nature of IgG surface attachment to an mPEG-PDLLA polymer will enable the design of Ps bounded externally by select antibodies. Specifically, surface functionalization is achieved exploiting high affinity avidin-biotin interactions. The terminal hydroxyl of the outer hydrophilic block is chemically modified through the application of a tresylation routine resolved by Nilsson and Mosbach[2] in 1984. Tresylation is accompanied by a substitution reaction in which the organic sulfonyl chloride 2, 2, 2 – trifluoroethanesulfonyl chloride (tresyl chloride) converts the hydroxyl group of the terminal Ps hydrophilic block into a good sulfonate leaving group. Nucleophilic biotin (and its derivatives) can subsequently react with the resulting Ps, forming stable linkages. Integration of polymersomes with isolated mitochondria and IgG antibodies is anticipated to contribute to delayed bacterial growth and minimal cytotoxicity. Size-distribution and morphology of resulting particles were assessed using TEM and DLS measurements.

In an alternate design scheme, the internal projection of variable-size immunoglobulins is conceived to diminish the bulkiness of the final construct while conferring plasma cell functionality. Biotinylated antibodies may be conjugated to the internal peripheral hydrophilic regime with the utility of the previously prescribed avidin-biotin interaction. Precise chemical triggers within the vasculature will prompt the inversion and self-assembly of relevant response architectures. The disintegration mechanism is followed by extrusion of the diseased composite from the organism through natural secretions. One mechanism for the controlled antibody release postulates the incorporation of mitochondria into the aqueous Ps core and catalase enzyme into the hydrophobic corona. Recent studies conducted by Jang et al.[3] have shown that catalase-embedded Ps may be engineered to release therapeutic agents in the presence of reactive oxygen species (ROS). The incorporation of isolated mitochondria into the aqueous Ps center will enable the localized production of hydrogen peroxide via the electron transport chain. Hydrogen peroxide will successively degrade into ROS and contribute to controlled Ps rupture.

Results of the prevalent analysis may be applied for the conception of a smart colloidal structure appended by several classes of homing devices. The emerging nano-structure is expected to drastically enhance the natural mechanisms associated with somatic hypermutation. Clinical injection of the composite Ps solution is hypothesized to activate an aggressive B-cell response in which thousands of antibodies disperse and mark antigens for destruction. Associated mitochondrial interactions are expected to drive controlled antibody release mechanisms. Proximate electrophysical studies will enable a self-powered device adept at sensing and tracking infected regions autonomously. Automated design is anticipated to permit the fabrication of a nano-device with response efficiencies and capabilities that exceed the healing properties shown by innate immune cells. The shift from reactionary to predictive intervention will prompt a new era in which early disease detection and treatment will contribute to enhanced longevity and productivity among patients.

Literature Citations

1. Geilich, B. *et al*. Nanoscale, 2015; 7:3511.

2. Nilsson K, Mosbach K. Methods Enzymol. 1984;104:56-69.

3. Jang, W-S. Soft Matter, 2016,12, 1014-1020

464272 Spatiotemporal Modeling and Rapid Sealing of Ruptured Tissue with Plasmonic Nanocomposites

Abstract:

Laser tissue welding is a tissue repair technique that utilizes a chromophore to convert photons into heat in order to seal apposed tissue edges by localized protein denaturation and interdigitation. Laser tissue welding has emerged as a promising alternative to sutures or staples conventionally used for sealing soft tissues, with the potential to limit surgical complications such as anastomotic leakage and wound dehiscence while also reducing healing times and scarring. However, insufficient closure strength, surgical inconvenience, and extensive thermal damage have hindered the clinical application of this technique, largely due to the unavailability of effective solders (surgical sealants). In this work, we describe the development of novel plasmonic nanocomposites for laser tissue welding that overcome these barriers in tissue welding. Several different gold nanorod-polypeptide were developed and investigated for welding ruptured tissues ex vivo. Leak pressure, burst pressure, and tensile strength measurements indicated that laser welding using plasmonic nanocomposites was able to restore significant mechanical integrity to the welded tissue. Additionally, we developed a spatiotemporal model for predicting local temperatures to aid in narrowing the parameter space and selecting optimal surgical parameters. Preliminary *in vivo* results in sealing colotomy incisions in mice are also presented.

464412 Functionalized Ferri-Liposomes for Hyperthermia Induced Glioma Targeting and Brain Drug Delivery

Abstract:

1. Introduction

Consisting of tight junctions between endothelial cells of the microvasculature of the central nervous system (CNS), the blood-brain barrier (BBB) keeps large microorganisms or hydrophilic molecules in the circulating blood from penetrating into the cerebrospinal fluid [1-2]. However, limited by current technologies, the delivery of therapeutic agents or imaging molecules in the brain is also blocked by these highly selective tight junctions. Currently, targeting of malignant brain tumors (such as glioblastoma multiforme (GBM)) and the delivery of therapeutic agents into the brain remains a big obstacle because of the existence of the blood-brain barrier and the difficulties in finding suitable candidates for specific tumor locating. In the current study, an optimized in vitro BBB model was established using mouse brain endothelial cells and astrocytes. This in vitro model was confirmed by an immunofluorescent (IF) stain and evaluating the trans-endothelial electrical resistance (TEER) before testing. In the meantime, superparamagnetic iron oxide nanoparticle (SPIONs) loaded thermosensitive liposomes were developed for better BBB penetration and selective targeting of the brain glioma under mild hyperthermia conditions introduced by an alternating magnetic field (AMF). To specifically target the glioma cells, a new cell-penetrating peptide (CPP) and an anti-glioma antibody (Ab) were also conjugated onto the liposome surface. With the encapsulation of SPIONs and doxorubicin (DOX) inside the functionalized liposome, a DOX drug release assay and cell uptake studies were carried out and the in vitro BBB model was used to indicate the permeability of these functionalized ferri-liposomes.

2. Materials and Methods

A. Material Synthesis and Characterization

CPP and Ab were first conjugated onto mal-DSPE-mPEG2000 by a well-established chemical reaction and the PEGylated ferri-liposomes were then prepared and encapsuled with iron oxide and DOX using the rehydration method [3]. Briefly, lipids were dissolved in chloroform and vacuum-dried overnight. Then, thin lipid films were rehydrated with an ammonium sulfate solution (PH=5.5) and extruded through 100nm filters to determine their diameters. The liposome solution was then gel-filtered with a HEPES buffer (PH=7.4) and DOX was loaded into the liposomes during an overnight incubation due to the pH gradient effect. The nanoparticles were then characterized by zeta potential to determine their charge and dynamic light scattering for hydrodynamic diameter. TEM was used to assess the iron oxide inner core diameter as well as the liposome diameter.

B. Drug Release Assay and Experimental Samples Tested Through the Blood-Brain Barrier Model

For the in vitro blood- brain barrier model, both mouse brain endothelial cells (b.End3, ATCC CRL-2299) and astrocytes (C8-D1A, ATCC CRL-2541) were first cultured in complete media (DMEM with 10% FBS and 1% P/S) in a flask at 37°C in a humidified incubator with 5% CO_2 to reach confluency before being moved to inserts. Then, astrocytes at a 10^5cells/cm^{2}density were seeded onto the bottom of the 24 well plates in complete DMEM media. After 24 hours of adhesion, endothelial cells were seeded onto the upper side of the inserts at the same density and the inserts were then placed in the 24-well plates containing astrocytes. The model was evaluated and confirmed using TEER and FITC-dextran transport. After such confirmation, the permeability and DOX drug release efficiency from several functionalized, traditional thermosensitive (TTSL), lyso-thermosensitive (LTSL) and non-thermosensitive (NTSL) liposomes were tested against the BBB model in an effort to increase BBB passage and deliver anti-tumor drugs for glioblastoma multiforme effectively while diminishing cytotoxicity. Each experiment was conducted in triplicate and differences between means were determined using ANOVA followed by student t-tests.

3. Results and Discussion

A. Material Characterization

Dynamic light scattering revealed that the hydrodynamic diameters of the samples ranged from 100nm to 150nm. TEM images show that the iron core was about 5-10nm for all of the samples.

B. Experimental Samples Tested Through the Blood-Brain Barrier Model

The permeability of FITC-Dextran across the model confirmed that the model was successfully established and the TEER value of this co-cultured model was around 150 Ohms/cm^2. Results showed that the functionalized ferri-liposomes had great permeability using the in vitro BBB model. In addition, compared to NTSL and TTSL, functionalized LTSL had significantly higher drug release efficiency and cell uptake ability towards tumor cells at its transition temperature (42 °C) induced by the AMF.

4. Conclusions

An *in vitro*model of blood-brain barrier was established using a co-culture method. The model was confirmed by comparing the permeability trend of FITC-dextran in serum-free medium and the TEER value with previous research. The results suggest a possibility to manipulate magnetic nanocarriers penetration across the blood-brain barrier by modifying the surface chemistry. Moreover, among all the thermosensitive nanocarriers, LSTL is suggested to be a good candidate for brain tumor drug delivery based on its good drug release results and BBB penetration results. Such data lay the foundation for the modification of liposomes to increase glioma targeting efficiency.

5. Acknowledgment

The authors would like to thank Northeastern University for funding this research.

6. References

[1] Christian P, et al. Magnetically enhanced nucleic acid delivery. Ten years of magnetofectionâ€"Progress and prospects, Advanced Drug Delivery Reviews. 2011; 63: 1300-1331.

[2] Richard G, et al. BOLD functional MRI in disease and pharmacological studies: room for improvement?, Magnetic Resonance Imaging. 2007; 25: 978-988.

[3] Li L, et al. Mild hyperthermia triggered doxorubicin release from optimized stealth thermosensitive liposomes improves intratumoral drug delivery and efficacy. J of Controlled Release. 2013;168(2):142-150.

466307 High Efficient 3D Cell Preservation and Retrieval with Biocompatible Magnetic Hydrogel Particles

Abstract:

Living cells are playing an increasingly important role in modern biomedical applications, such as diagnostics, tissue engineering, cell therapies, etc. However, facile and high efficient cell preservation methods are still high required in these applications. Cryopreservation have been routinely used, but it suffers from several problems: 1) cryopreservation can cause cell function loss and cannot work for many types of important therapeutic cells types, such as hepatocytes, cardiomyocytes, etc, 2) dimethyl sulfoxide (DMSO), the most frequently used cryoprotective agent, is an organic solvent and is intrinsically cytotoxic; 3) cryopreservation requires well-developed equipment and facilities as well as trained personnel, limiting its applications in many circumstances, such as remote locations.

In this work, we present a novel strategy for high efficient cell preservation and retrieval at physiological temperature or even room temperature. We proved that biocompatible magnetic hydrogel microparticles could assemble into a 3D environment mimicking ECM for cell accommodation. The cells stay in the interspaces among the hydrogel particles and are supported physically. In this circumstance, cells stop their proliferation while stay alive. Before use, the cells can be easily retrieved by applying an external magnetic field to remove the particles. This strategy can be highly useful for applications such as cell transportaion, in situ diagnostics, drug screening, etc.

467997 Self-Assembly of Semiconductor and Protein into Monodisperse Supraparticles

Abstract:

Terminal assemblies enable a simple means for fabricating complex bio-nano systems from a variety of components. They can also be called bionic assemblies because they combine the properties of inorganic and biological components. Self-assembly process that involves inorganic nanoparticles (NPs) and biological materials results in highly ordered terminal structures, which are potential analytical and drug delivery tools. The scientific value these systems are the possibility to integrate biological functions of proteins with properties of inorganic NPs in a way that leverages the attributes of both components. Such systems are fundamentally and technologically attractive due to their uniformity, versatility and simplicity of preparation. In this work we prepared various types of supraparticles (SPs), we combine ~5 nm protease protein with different NPs: ~3.4 nm FeS_2NPs, ~15 nm ZnO nanopyramid NPs, ~18 nm ZnO nanoplate NPs and ~5 nm ZnO nanosphere. We observed spontaneous formation of spherical supraparticles with a narrow size distribution containing both components. Assembly was originated from the competition between electrostatic repulsion and non-covalent attractive interactions. Non-covalent interactions between protease and like-charged NPs lead to drastically different self-assembly behavior previously unseen for each component individually, and which mimics the cooperative assemblies of proteins. We also demonstrate remarkable that the catalytic curve profile has changed, but the catalytic activity is retained in a determinate temperature and pH.

The FeS2/protein and ZnO/protein SPs were examinated by scanning electron microscopy (SEM) and transmission electron microscopy (TEM). These results reveal the formation of uniformly sized, spherical SPs with a TEM diameter about 150 ± 13nm which matches the diameter determined from Dynamic light scattering (DLS). Image of individual NP inside a SP can be distinguished by its crystal lattice. Besides the SPs were investigated by tomography 3D and the overall attractive potential between the similarly charged FeS_2 or ZnO NPs and Protease is investigated by theoretical calculations. Further were examined the effect of ionic strength and temperature of assembly on the diameter of the SPs. Various spectroscopy data were performed which shows the positions of all peaks in the UV-Visible / Circular Dichroism spectra remain unchanged, indicating that the electronic state and conformation of the protease molecules in the SPs are preserved.

468364 Morphogen Presentation within Micro-Fiber/Collagen Composites for Ligament Tissue Engineering

Abstract:

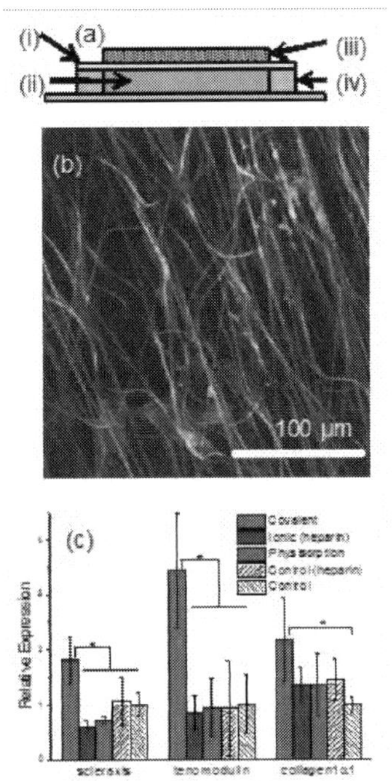

Figure 1: (a) a model composite (i) micro-fibers, (ii) acellular collagen, (iii) cellularized collagen, (iv) PDMS for support. (b) Confocal microscope image of MSCs in composites. Micro-fibers are green, cell nuclei are blue, and cell actin cytoskeletons are red. (c) Expression of scleraxis, tenomodulin, and collagen I after 7 days of culture in composites.

Tissue engineering holds promise in overcoming the limitations of existing autologous and allogeneic options. However, central to the realization of this goal is the achievement of instructive biomaterial scaffold materials that can guide the recruitment, proliferation, and differentiation of stem cells into organized functional tissues. In particular, we hypothesize that scaffolds that present anisotropic topographical and mechanical properties, as well as morphogenic signals are essential for guiding stem cell fate. To this end, we have developed model electrospun micro-fiber/collagen hydrogel composites – consisting of thin oriented micro-fiber meshes between collagen slabs (**Figure 1a**) – that allow us to selectively tune the mechanical, biochemical properties of both the fiber and hydrogel phases and assess their effects on stem cell proliferation and differentiation. For the goal of ligament tissue engineering, we have previously shown that mesenchymal stem cells (MSCs) attach to and align with the oriented micro-fibers (**Figure 1b**). Further, we found that systematic variation of the micro-fiber modulus affects mRNA expression of the tendon/ligament transcription factor scleraxis and the contractile protein α-smooth muscle actin (not shown). Currently, we are extending this work to determine the effect of tethering the morphogens (i.e., fibroblast growth factor (FGF)-2 and growth and differentiation factor (GDF)-5) to the micro-fibers.

To accomplish this, we are electrospinning thin (5-10 μm) aligned meshes of coaxial micro-fibers (~1 μm in diameter) consisting of a chitosan sheath phase and a polyurethane/polycaprolactone blend core phase on a rotating mandrel. The meshes are transferred to PDMS rings and functionalized with a mixture of FGF-2 and GDF-5 through covalent bioconjugation, ionic adsorption (to heparin-coated micro-fibers), or simple physisorption (to untreated micro-fibers). Resultant bioactive fiber meshes (1.2 cm diameter) are embedded within 0.5 wt% collagen gels containing 5×10^4 MSCs and cultured for up to 7 days and compared to samples without morphogens. Preliminary data shows that covalent bioconjugation results in enhanced expression of ligament markers scleraxis and tenomodulin (**Figure 1c**). In contrast, the incorporation of morphogens onto heparin-coated and uncoated micro-fibers does not enhance expression, perhaps due to denaturation of the proteins during adsorption or subsequent desorption and loss of the proteins from the micro-fiber surfaces. Concurrently, we are examining adsorption and desorption of morphogens from planar chitosan films. Initial data indicate covalent bioconjugation of FGF-2 from a 250 ng/mL solution results in a surface concentration of 58 ng/cm^2 with negligible protein loss over 7 days. In contrast adsorption of FGF-2 to chitosan and heparin-coated chitosan surfaces results in surface concentrations of 182 and 137 ng/cm^2, respectively.

Our results suggest that the covalent bioconjugation of morphogens onto the micro-fiber surfaces can have a strong positive effect on stem cell fate. We postulate that the enhanced expression of ligament markers (**Figure 1c**) is due the localization of morphogen/receptor complexes within or near focal adhesion complexes. This co-localization may mimic the native ECM, by supporting cell adhesion, migration, and guide behavior due to integrin association with both growth factors and adhesive proteins. Ultimately, our goal is to develop thick multilayered composites suitable for implantation, where the bioactive micro-fibers present a multitude of stimuli to recruit host cells and guide tissue formation in vivo.

470237 Variations in Chitosan/Tripolyphosphate Micro- and Nanogel Yield and Their Two Key Effects on Protein Uptake

Abstract:

Chitosan/tripolyphosphate (TPP) micro- and nanogels form spontaneously through ionotropic gelation in dilute aqueous solutions. Due to their biocompatible and mucoadhesive properties and mild formation conditions, they are widely explored as vehicles for ocular and intranasal delivery of protein drugs and vaccines. Aside from the knowledge that their protein uptake is enhanced by stronger protein/particle binding, however, factors that control protein association efficiency (*AE*) such as variations in the chitosan, TPP and protein concentrations used during protein loading, remain poorly understood. To this end, we hypothesized that many of the differences in the reported *AE*-values might reflect the largely-ignored variability in the particle yield (X_{Agg}), which also varies with the chitosan, TPP and protein concentrations, and is defined here as the fraction of the added chitosan that self-assembles into particles. To test this, we used bovine serum albumin (BSA) and Î±-lactalbumin (Î±-LA) as model proteins to analyze the correlation between *AE* and X_{Agg}. Consistent with our hypothesis, *AE* scaled linearly with the X_{Agg} (which increased with the TPP and drug-to-chitosan ratios) until all chitosan was aggregated into particles, and data collected at variable TPP and protein concentrations collapsed onto a single *AEÂ* vs. X_{Agg} line. These trends were also quantitatively independent of the protein/particle mixing procedure, suggesting that protein uptake was governed by its equilibrium partitioning. Further analysis of the protein/particle binding isotherms revealed the rise in *AE* with X_{Agg} to reflect two effects: (1) an increase in the number of protein binding sites within the particles (due to an increase in the particle mass); and (2) a decrease in the number of soluble (unaggregated) chitosan molecules, which formed soluble protein/chitosan complexes and (through this competitive binding) limited protein uptake into the particles. These findings highlight the need to carefully analyze the effects of formulation parameters on X_{Agg} when designing chitosan/TPP drug carriers, and can likely be extended to other biopolymeric nanocarriers that form through ionotropic gelation.

470877 Optimization of Cationic Nanogel Composition for Enhanced Co-Delivery of siRNA and Chemotherapeutics

Abstract:

Off-target toxicity and multi-drug resistance are two major limitations of current chemotherapeutic regimens as they lead to poor quality of life and low survivability. Targeted co-delivery of siRNA and chemotherapeutics offers a promising strategy to sensitize and kill cancer refractory to currently employed therapeutic regimens while minimizing side-effects. Here, we studied cationic nanogels as carriers of siRNA and chemotherapeutics for intracellular therapeutic delivery.

Biodegradable cationic nanogels were synthesized in an oil-in-water emulsion via activators regenerated by electron transfer atom transfer radical polymerization (ARGET ATRP) of methacrylate monomers. The ratio of hydrophobic/hydrophilic (tert-butyl methacrylate, 2-hydroxyl ethyl methacrylate, and 2-aminoethyl methacrylate) and aromatic (benzyl methacrylate) character versus overall cationic character of nanogels (2-(diethylamino)ethyl methacrylate) was systematically modulated to enhance therapeutic co-loading of siRNA and doxorubicin through balance of electrostatic, hydrogen bonding, hydrophobic, and aromatic interactions. Biodegradability was imparted through a disulfide crosslinker and ligand conjugation was conducted through carbodiimide chemistry.

Composition was verified using infrared spectroscopy and differences in tertiary amine content, volume swelling ratio, apparent surface charge, and dry size were analyzed using potentiometric titration, dynamic light scattering, zeta potential, and electron microscopy. Co-loading was optimized with respect to pH, ionic strength, and N/P ratio and compared with nanogel composition and volume swelling ratio. Material cytocompatibility in cancerous and healthy cell lines and a hemolysis assay were utilized to identify toxicity thresholds and optimal nanogel critical swelling pH. Transfection and cytotoxicity of loaded nanogels were investigated *in vitro* in drug resistant cell lines to demonstrate the co-delivery potential of this nanogel platform.

470894 Investigations on the Mechanical Forces Required for Mechanochemical Synthesis of Hydroxyapatite

Abstract:

Mechanochemistry is the field of science which studies chemical reactions induced by mechanical energy1. This paper focuses on the synthesis of hydroxyapatite, a bone substitute biomaterial, by mechanochemical synthesis, and in particular in determining the mechanical forces required for the solid state reaction to occur.

In recent years, mechanochemistry has become a popular alternative for executing chemical reactions, as opposed to traditional solvent processes, which make use of high temperatures and pressures in addition to high volumes of toxic solvents. Mechanochemical processing exhibits many advantages over traditional solvent heavy processes. The solid state reactions carried out in a solvent free environment are a greener alternative which improves the safety of the chemical reaction and limits the occurrence of undesired side reactions. Furthermore, the reaction may be carried out at ambient temperatures and pressures, with no further processing or solvent recovery required.

The mechanochemical process requires the input of mechanical energy to stimulate a chemical reaction. The mechanical energy initiates the process in which defects and dislocations develop in the material, which may cause a solid state change in the material. When mechanical activation has occurred, the mechanical energy supplied may potentially cause a change in solid state such as the formation of a metastable polymorphic form, the amorphous form, or alternatively, chemical reaction may occur.

For chemical reactions, the activation energy of the system, also known as mechanical activation, is the threshold energy that reactants must acquire for chemical reaction to occur. When the threshold energy level has been obtained, the reactants will convert to a higher energy transition state, prior to converting to a lower energy state with the formation of a product, with lower potential energy and higher stability. In the case of an exothermic reaction, when the activation energy of the reaction has been attained for a particular reaction and chemical reaction has initiated, energy is released, to form a product with a less energy, as described in figure 1. In contrast, an endothermic reaction involves the absorption of energy for chemical transformation to occur. Figure 2 demonstrates the absorption of energy for an endothermic reaction when the activation energy of the system has been achieved. The energy absorbed is utilised to form the final product with a higher energy state.

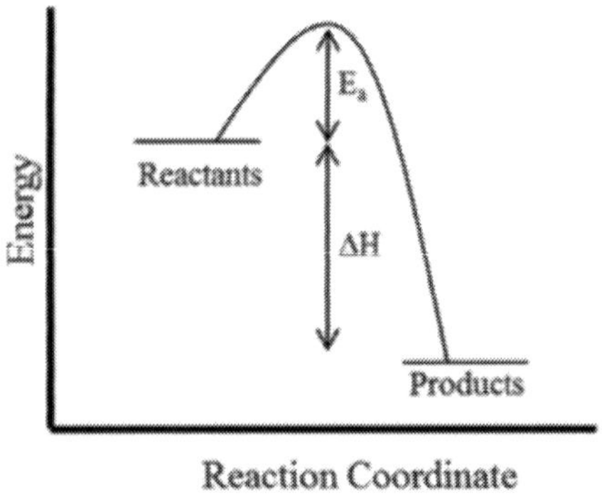

Figure 1

An exothermic reaction features a release in energy when mechanical activation has been achieved.

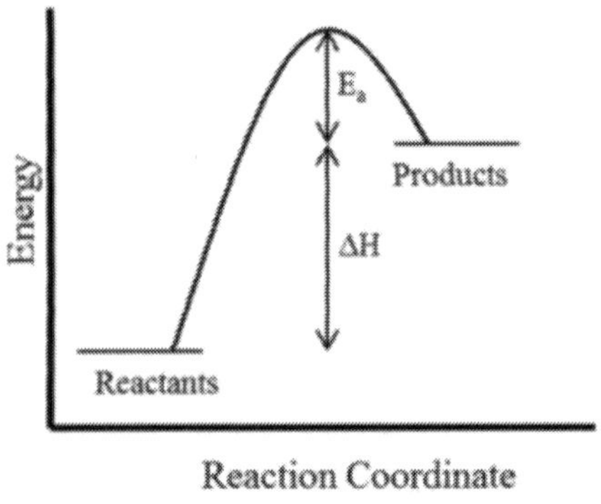

Figure 2

An endothermic reaction features absorption of energy when mechanical activation has been achieved.

Hydroxyapatite was synthesised from calcium hydroxide and diammonium phosphate in a Retsch Mixer Mill MM 400, varying both the reaction time and milling frequency. The solid state endothermic reaction was undertaken using molar ratios of calcium hydroxide and diammonium phosphate according to the following reaction equation:

$$10Ca(OH)_2 + 6(NH_4)_2HPO_4 \rightarrow Ca_{10}(PO_4)_6(OH)_2 + 12NH_3 + 18H_2O$$

$$\Delta H_r^{\,0} = 750.6\ kJ/mol$$

The reaction time for the mechanochemical synthesis of hydroxyapatite was measured at various milling frequencies between 5Hz and 25 Hz. The synthesis was carried out in a 25mL stainless steel milling jar, using a powder loading of 1g with one stainless steel milling ball.

Alternatively, an exothermic reaction investigating the use of calcium oxide as an alternative starting reactant was studied. Hydroxyapatite was synthesised using calcium oxide and diammonium phosphate under identical milling conditions defined above, according to the following reaction equation:

$$10CaO + 6(NH_4)_2HPO_4 \rightarrow Ca_{10}(PO_4)_6(OH)_2 + 12NH_3 + 8H_2O$$

$$\Delta H_r^{\,0} = -348.2\ kJ/mol$$

The formation of hydroxyapatite and reaction yield was analysed by powder x-ray diffraction (XRD) and Raman spectroscopy, which monitored the progression of hydroxyapatite with increasing reaction time and milling speed.

Scanning electron microscopy (SEM) was carried out to monitor a change in morphology and size of the milled powder. The change in morphology and size of the powder was analysed with respect to increasing reaction time and mill speed. Figure 1 below illustrates a change in particle size between the starting reactants calcium oxide and diammonium phosphate and the resulting milled powder containing calcium oxide and diammonium phosphate ball milled for 2 hours at a milling frequency of 20 Hz.

Figure 3

Physical Mixture of calcium oxide and diammonium phosphate prior to milling.

Figure 4

Ball milled powder mixture calcium oxide and diammonium phosphate milled for 2 hours at 20 Hz

Mechanochemical synthesis of hydroxyapatite involves the use of mechanical forces to induce a chemical change, as opposed to traditional methods which make use of solvents. The impaction forces present in the milling jar were estimated to determine the mechanical energy required for the mechanochemical reaction to occur. The impaction forces in the ball mill were estimated using Newton's Second Law of motion, which investigates the change in velocity or acceleration of an object, when subjected to a change in force. This may be applied to the ball milling process observed in the Retsch Mixer mill, in which the energy supplied is used to oscillate the milling jar in a lateral direction. This in turn transfers the milling ball in the mill jar, which promotes the mixing of powder. The impaction of the powder and milling ball occurs at the wall of the milling jar.

Newton's Second Law of Motion was applied to the milling process used in this study to estimate the impaction forces at the wall of the mill jar, which were calculated based on the mass and acceleration of the milling media in the mill jar, according to the following equation:

$$F = ma$$

$$m = mass\ of\ milling\ ball\ (kg);\ a = acceleration\ of\ milling\ ball\ (m/s^2)$$

The acceleration of the milling ball was calculated using the milling speed, the dimensions of the milling jar and the pattern of the milling media inside the mill jars. The mass of the mill ball remained constant for the purpose of this study. The acceleration of the mill ball varied depending on the mill speed utilised. The size of the milling jar used also remained constant throughout the study, and therefore had no impact on the acceleration of the milling ball.

Mechanical activation, or the activation energy of the reaction, denotes the mechanical energy essential to initiate a chemical reaction. In the case of the ball milling process, mechanical energy supplied from the ball mill is absorbed and mechanical activation will occur. Both the reaction rate constant and activation energy of the system are reaction specific.

The activation energy required in the formation of hydroxyapatite was estimated using the Arrhenius equation defined below:

$$K = A^{*}e^{\frac{-Ea}{RT}}$$

$K = Reaction\,Rate\,Constant \quad A = Pre - exponential\,factor$

$E_a = Activation\,energy\,(J/mol)\;R = Gas\,Consant\,(J/mol\,K)$

$T = Temperature\,(K)$

The reaction rates for the reactions specified previously were measured at a range of different temperatures in order to estimate the activation energy for the reaction.

The objectives of the study described aims to gain a better understanding for the mechanochemical process, in particular for mechanochemical reactions and mechanochemical synthesis of materials. The mechanical activation and mechanical energy necessary for initiating a mechanochemical reaction were estimated using the Arrhenius equation and Newton's Second Law of Motion. The reaction activation energy, enthalpy of reaction and impaction forces calculated characterise the mechanochemical process, and aid towards understanding the mechanochemical process.

References:

1. Baláž, P. *et al.* Hallmarks of mechanochemistry: from nanoparticles to technology. *Chem. Soc. Rev.* **42,** 7571–637 (2013).

470139 MIL-100(Fe) Synthesis and Characterization for Topical Treatment of Skin Diseases

Abstract:

Metal organic frameworks or MOFs are the latest class of porous materials witch are indeed very important solids because of their potential applications in gas storage, adsorption, separation, or catalysis. Among the tens of thousands of known MOFs, the MIL family built from trivalent metal centers and carboxylate bridging ligands has particularly attracted a great deal of attention due to their enhanced stability, enormous porosity, and very large pores. These porous inorganic-organic materials are potentially efficient to use as drug carriers in drug delivery systems. In the domain of health, there are important challenges to delivery of drugs in the body using non-toxic and biologically friendly nanocarriers. The synthesis of porous hybrid materials via hydrothermal or solvothermal methods requires several days for material crystallization, Since shorter reaction times are more attractive for industrial-scale production, syntheses using microwave irradiation can be a viable option. In this study non-toxic iron(III) carboxylate (MIL-100(Fe)) were synthesized by using microwave irradiation method via the reaction of ferric nitrate and trimesic acid (H3BTC) under HF-free conditions. The optimization of yield, crystallinity, and particle size was achieved through the control of synthetic parameters such as temperature, heating rate, and dwelling time. The â^¼400 nm particles were characterized using a variety of methods, including SEM, XRD, nitrogen adsorption measurements, TGA and particle size analyzer. Results show that the prepared samples have high crystallinity (>80%), high surface area (S_{BET}=1255 m^2/g), large pores, and the ability to incorporate functional groups into the framework.

471206 Development of a Microdevice-Based Human Mesenchymal Stem Cell-Mediated Drug-Delivery System

Abstract:

Development of a Microdevice-based Human Mesenchymal Stem Cell-mediated Drug-Delivery System

Junfei Xia[a], Ang-Chen Tsai[a], Teng Ma[a], Jingjiao Guan[a,b]

[a] Department of Chemical and Biomedical Engineering, FAMU-FSU College of Engineering, Florida State University

[b] Integrative NanoScience Institute, Florida State University

Abstract

Cell-mediated drug delivery allows to harness unique abilities of living cells to achieve controlled delivery of drugs, but traditional cell-mediated drug delivery techniques suffer drawbacks such as risk of intracellular degradation of drug payload or low drug loading capacity. These limitations can be potentially overcome by attaching a disk-shaped drug-laden microparticle termed microdevice to the external surface of a cell. On the other hand, human mesenchymal stem cell (hMSC) is a primary candidate for next-generation cell therapies. It is therefore an ideal cell type for cell-mediated drug delivery. In this presentation, we introduce production and characterization of a novel microdevice-based hMSC-mediated drug-delivery system for the first time. In particular, we present in vitro studies on how microdevice attachment affects viability, migration, stemness, capability of forming spheroid of the hMSCs. Successful development of this microdevice-based hMSC-mediated drug delivery system may lead to a new therapies for treating many human diseases.

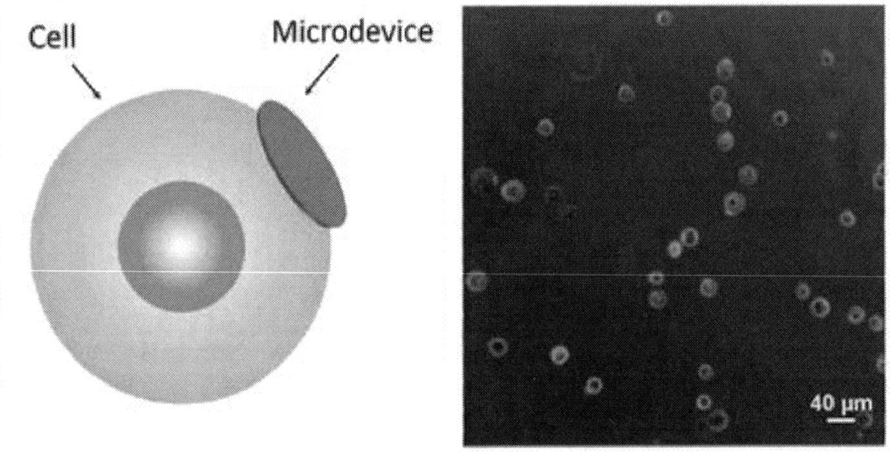

Fig.1 Left: Schematic illustration of cell-microdevice complex. Right: single human mesenchymal stem cell in suspension attached by microdevices.

Keywords: cell-mediated drug delivery, human mesenchymal stem cell, drug delivery

472088 Preparation, Characterization and Comparison of Nano-Composite Scaffolds Based in Chitosan, Poly(lactic acid), Poly(lactic acid-co-glycolicacid), and Hydroxyapatite for Tissue Engineering Applications

Abstract:

Scaffold design and fabrication are major areas of biomaterial research, and they are also important subjects for tissue engineering and regenerative medicine research. Scaffold plays a unique role in tissue regeneration and repair. Biocompatibility, biodegradability, mechanical properties and porosity are important parameters for selection of a desirable scaffold for tissue engineering applications. For instance, scaffolds are one of the necessary components for bone tissue engineering approach. They provide 3D growth space for cells as well as for transport of nutrients and metabolites for cell expansion and tissue organization. It is not easy to define an ideal scaffold since each tissue needs a particular matrix design with well-defined material properties. Therefore, it can be said that an ideal scaffold should at least meet several requirements depending on the application region such as biocompatibility, suitable mechanical strength, interconnected porous structure, and biodegradability. Nowadays, it has being recognized that some of the widely used scaffolds do not fairly recapitulate the cell microenvironment and that the extra cellular matrix (ECM) is a dynamic and hierarchically organized nano-composite that regulates essential cellular functions such as morphogenesis, differentiation, proliferation, adhesion and migration. As a consequence, researchers developed and used existing nanotechnological tools for tissue engineering to design advanced nanocomposite scaffolds that can better mimic the ECM and eventually assemble more complex and larger functional tissues. Thus, the advancement of nanotechnologies allows for the design of a large number of hybrid and multicomponent systems. In this study nano-composite scaffolds based in chitosan (CH), poly(lactic acid) (PLA), poly(lactic acid-co-glycolic acid) (PLGA) and hydroxyapatite (HAp) were prepared by freeze drying and gas foaming techniques in order to mimic the organic/inorganic nature of the bone. Chitosan (CH) is a bioderived polysaccharide obtained mainly from shells of crustaceans, corals or jellyfish. Due of its biocompatibility and a range of interesting properties, including

antimicrobial, antitumor, anti-inflammatory and immunity-enhancing ones, chitosan could serve a multifunctional purpose in biomedicine become it an attractive material for use in bone tissue engineering. On the other hand, poly(lactic acid) (PLA) and poly(lactic acid-co-glycolic acid) (PLGA) are FDA approved synthetic biodegradable polymers. Both polymers are biocompatible and their degradation products are low molecular weight compounds such as lactic acid and glycolic acid, which enter into normal metabolic pathways without causing any adverse effect. And Hydroxyapatite (HAp) $[Ca_{10}(PO_4)_6(OH)_2]$, the main inorganic component of hard tissues, has a variety of applications in bone fillers and replacements because its surface is highly reactive and leads to favorable attachment and bioactivity; it has osseo conductive and osseo inductive effects. Therefore, the enhancement of the properties of HAp proceeds along two main routes: (i) manipulation of microstructure; (ii) combination with other elements and materials. The composite is expected to have improved mechanical properties compared to the polymer and better structural integrity and flexibility than brittle ceramics. In fact the combination of ceramic and polymer could provide reinforced porous structures and controlled resorption rates. A composite material would also improve biocompatibility and hard tissue integration in a way that ceramic particles. Also, the choice of the processing technique is important. The scaffolds obtained were characterized in structure, morphology, porosity, degradation, water uptake, mechanical and thermal stabilities using diverse analytical techniques, such as: spectroscopy (FTIR), microscopy (SEM), thermoporometry (using DSC), thermal analysis (TGA, DSC) and physical and chemical analysis. All formulations were compared in order to assess the influence of each component that constitute the scaffold as well as the processing technique for the potential application of those nano-composite scaffolds in hard tissue treatments.

472758 Silica-Polysaccharides Nanocomposite Nanoparticles As pH-Responsive Controlled Drug Release Carriers

Abstract:

Biocompatible drug carriers with the ability of controlled therapeutic agent delivery is a promising way to overcome the limitations of conventional therapies. For example, the lack of target specificity and solubility of some drugs results in overdoses to achieve the sufficient therapeutic effects, which increases cost and side effects. Among all types of carriers, silica-based nanocomposites emerged as an encouraging candidate because of several properties of silica. First of all, silica shows acceptable biocompatibility and has high physical as well as chemical stability, which makes it an ideal support for nanomaterials. Secondly, silica has a low effect on light absorption and magnetic field interference, which allows the combined components to keep their original optical and magnetic properties. Finally, the hydroxyl groups on the silica surface give lots of potential for the modification of silica based materials for extended applications. Moreover, polysaccharides, such as alginate and chitosan, can be encapsulated into silica nanoparticles by sol-gel method to make silica based nanocomposite materials. These polymers can provide the pH controlled release ability due to their pH-responsive property. In this study, the pH-responsive silica polysaccharides nanocomposite nanoparticles were made by silica sol-gel method in water in oil microemulsion system, and were characterized by dynamic light scattering, transmission electron microscopy, and thermogravimetric analysis for their size, shape and polymer weight ratio. The release profiles at different pH buffer solutions for different drug molecules were also studied to test the pH-responsive property of these silica polysaccharides nanocomposites nanoparticles.

472169 Production of Monodisperse Polyacrylamide & Poly (n-isopropyl acryl amide) (PNIPAM) Nanoparticles Using Chemtor Fiber Reactor with High Throughput

Abstract:

Use of microfluidic flow regimes for large scale production of nanoparticles is important in the emerging fields of personalized and targeted drug delivery, fuel cell catalysis, LEDs, environmental remediation of spills, etc. Formulation of droplets with uniform diameters is a challenge which microfluidic devices have been reported to achieve. Thus microfluidics could be used to produce nanoparticles for drug delivery, but the throughput is inherently low. The overall goal of the present research is to scale capillary flows to a high throughput system using a patented Chemtor fiber technology. The fluid mechanics of such environments have been studied as a model system in a microfluidic â€œTâ€ junction apparatus for production of polyacrylamide and Poly (n-isopropyl acryl amide) (PNIPAM) nanoparticles. In first partby the laboratoryfiber reactor, monodisperse polyacrylamide nanoparticles with high through-put were produced- 662 gm per day throughput documented. Furthermore, poly acrylamide nanoparticles of 117nm and 107nm were produced at 20^0 Celsius and 15^0 Celsius respectively, so variation in size of formed Nano-particles with respect to change in experimental temperature condition recorded. Polyacrylamide nanoparticles produced in a batch process were comparable in terms of size and distribution. In next part PNIPAM nanoparticles with size 39 nm and polydispersity index 0.36 were successfully produced.PNIPAM is thermos-responsive polymer in aqueous solution it demonstrates LCST (lower critical solution temperature). This property of PNIPAM is very valuable in targeted drug delivery field. Particle size distributions were measured using dynamic light scattering analysis.

473211 The Interdigitated Electrodes Stimulation of the Proliferation and Differentiation of Living Cells

Abstract:

The electric and magnetic fields were reported to be useful on improving cell growth. Here we report a simple method of fabrication of self-doped sulfonated polyaniline (SPAN)-based interdigitating electrodes (SPAN-IDEs) which is applied on cell stimulation. The electrodes are fabricated through a layer-by-layer deposition of SPAN copolymers and in situ polymerization on PET film, which were used for murine bone marrow stromal cells (BMSCs) and pre-osteoblasts cells (MC3T3-E1). The samples were evaluated by alkaline phosphatase (ALP) activity and Von Kossa staining, which gave us promising results for SPAN-IDEs could be a potential biological scaffold on tissue engineering application.

455138 Nanoporous Metal Films Formed with Aqueous Organic Templates

Abstract:

Nanoporous metals have the highest practical metal surface area per unit volume. They are of potential value for surface-catalyzed chemical reactions; supercapacitors; and chemical separations based on electrical potential-dependent adsorption, such as capacitive deionization. Methods to grow nanoporous metals using soft templates have been published by many groups over the past two decades, but the growth mechanisms, optimal procedures, and properties of the materials are not yet fully clarified. We will present our efforts to make progress on these problems, with emphasis on palladium. The ability of palladium to reversibly store hydrogen is of value for small-scale metal hydride batteries, as well as hydrogen gas or pH sensors. These properties also lead to unique synthetic opportunities and challenges.

We have used liquid-cell transmission electron microscopy to study the mechanisms of metal growth by chemical reduction in the presence of surfactants. Our results suggest that Pd nanoparticles form and sinter around micelles present in the aqueous phase. Changing the size of the surfactant molecules affects the size of these micelles, and ultimately determines the pore dimensions. The stability of nanopores in metals is a strong function of temperature and the chemical state of the surface. Attention to these conditions is necessary when formulating a new material or synthetic method. Reactions that transport hydrogen between nanoporous palladium and the gas or aqueous phase are significantly faster than those involving lower-surface area materials.

Sandia National Laboratories is a multi-program laboratory managed and operated by Sandia Corporation, a wholly owned subsidiary of Lockheed Martin Corporation, for the U.S. Department of Energy's National Nuclear Security Administration under contract DE-AC04-94AL85000. The work was funded in part by Sandiaâ€™s Laboratory-directed Research and Development program. The Pacific Northwest National Laboratory is operated by Battelle for the U.S. Department of Energy under contract DE-AC05-76RL01830. SAND2016-4362 A

457359 Remotely Controllable Miniature Reactors: Magnetic Liquid Marbles

Abstract:

Multifunctional materials have gained a lot of interest in recent years, especially in the fields of biomedical engineering, catalyst, self-cleaning surfaces, gas sensing or drug delivery systems [1]. Liquid marbles, which are formed by rolling the water droplets over hydrophobic nano or micro particles, provide a unique manipulation method of water droplet with its non-wetting and non-stick properties. At the same time, stimuli-responsive liquid marbles response with the change of temperature, pH, magnetic field, UV radiation and available to use in various applications such as micro reactors or micro fluidics [2]. Numerous liquid marbles prepared by using magnetic particles, were studied in previous works. UV triggered magnetic nanoparticles have been synthesized providing a remotely triggered rupture of a liquid marble [3]. Fe_2O_3nanoparticles have also been dispersed in the core ionic liquids of PTFE particle based liquid marbles to provide magnetic manipulation [4].

Here we studied a new perfluorinated ligand to surface modification of magnetite nanoparticles and formed liquid marbles with water as core liquid. First, approximately 6 nm sized iron oxide nanoparticles were synthesized via co-precipitation of $FeCl_3$ and $FeSO_4$salts in water by adding NaOH solution. As-synthesized nanoparticles showed superparamagnetic properties and have 60 emu/g of saturation magnetization value. Then, water-soluble magnetic particles were washed 3 times with ultrapure water and redispersed in methanol. Perfluorinated ligand was added to the solution and interacted with particles under 5 minutes of sonication. To obtain liquid marbles, particles were dried, ground and various volumes of water droplets were rolled over this hydrophobic powder. In order to analyze the morphology, size and size distribution of the magnetite nanoparticles, High Resolution Transmission Electron Microscopy (HR-TEM) analysis was performed. Also, Dynamic Light Scattering (DLS), Vibrating Sample Magnetometer (VSM) and X-ray diffraction (XRD) were used to implement a detailed characterization of nanoparticles. Contact angle measurements were performed by the sessile drop method (Krüss DSA 100). Moreover, stabilization of liquid marbles with various volumes of core liquids were monitored by visualizing the marbles with a CCD (charge coupled device) camera for a certain time period.

Liquid marbles were able to keep their shape up to 75 minutes. The liquid marble that obtained with 5 microliters of water showed >150° of contact angle value on a glass slide. A linear relation was determined between the height and diameter values, that means formation of approximately spherical shaped small magnetic liquid marbles and gravitationally flattened puddles were obtained when the large volume of water droplets was used to form liquid marbles. Magnetic manipulation of marbles was performed by a permanent magnet and it was determined that the velocity of marbles under a magnetic field were considerably high. When a marble was on a glass surface and a magnet was approached towards under of the marble, magnetite nanoparticles got closer to the glass surface and top of the marble was opened. This opening/closing process of liquid marbles was reversible and can be repeated for more than 10 times. As a reactor, opening and closing process allows adding various reactants after the formation and transportation of marble. In this work, a gelation reaction was carried out as a model reaction. This gelation reaction provided a mechanical robustness for marbles. Thus, stability of marble was increased and it was possible to pick it up and press or stick a needle through it without rupture of the marble.

REFERENCES

[1] G. McHale,Â M. I. Newton, Soft Matter, 2011, 7, 5473.

[2] J. Sun, W. Wei, D. Zhao, Q. Hu, X. Liu, Soft Matter, 2015, 11, 1954.

[3] L. Zhang, D. Cha, P. Wang, Adv. Mater., 2012, 24, 4756â€“4760.

[4] S. Zhang, Y. Zhang, Y. Wang, S. Liu, Y. Deng, Phys. Chem. Chem. Phys., 2012, 14, 5132â€“5138.

458942 Fabrication of Patterned Metal Oxide Thin Film By Using a Combination of Microreactor and PDMS Microchannels

Abstract:

Microreactor-Assisted Nanomaterial Synthesis and Deposition (MANS&D) process combines the merits of microreaction technology and solution phase synthesis of nanomaterials. This technique uses continuous flow microreactors for the synthesis, assembly and deposition of nanomaterials. In synthesis, microreactor technology offers large surface-area-to-volume ratios within microchannel structures to accelerate heat and mass transport, which allows for synthesizing nanomaterials with various size and structures. To fabricating patterned thin films, MANS&D process was combined with polydimethylsiloxane (PDMS) microfluidic channels. In this study, ZnO and CuO nanostructure thin films were fabricated, and the process can be expanded to other patterned nanostructure thin films. The scaling up of these patterned nanostructured films was achieved by numbering-up the microchannel unit operations in parallel, allowing for the fabrication of these patterned films at various scales. This patterned structures can be potentially applied in solar cells, sensors, and thin film transistors.

462717 Synthesis Route to Sulfur Rich Porous Material for Thermal Insulation and Mercury Adsorption

Abstract:

The petroleum refining units produce excess amount of elemental sulfur annually via hydrodesulfurization; however the applications assosciated with elemental sulfur in bulk is limited except for production of sulfuric acid. Research activities contributing in finding useful applications for this excess sulfur is still in its initial stages. Here in we report synthesis route to chemically and structurally stable porous sulfur via sacrificial template and inverse vulcanization process. This methodology enabled processing of porous sulfur into various shapes with tunable thermo-mechanical properties. We also demonstrated that these porous materials exhbited low thermal conductivity comparable to that of commercially available thermal insulators (0.048 W/mK), along side a commercially comparable adsorption capacity of 150 ug/g.

463880 Microwave-Assisted Fast Rout to Achieve 1T to 2H Phase Reversion of MoS_2 in Solution: Processable Dispersions of 2H MoS_2 Nanosheets and 2H-MoS_2/Pt Composite

Abstract:

Exfoliated molybdenum disulfide (MoS_2) with 2H phase and semiconductor properties has unique properties and potential applications in a wide range of fields. However, the chemically exfoliated MoS_2 nanosheets are metallic 1T phase. Many methods have been developed to achieve 1T to 2H reversion. For example, Eda G. and co-workers demonstrated the successful phase reversion of MoS_2 nanosheets by annealing at temperature of 300 °C for 1 hour with a heating rate of 5 °C/min. We also revealed that this phase reversion could be achieved by NIR radiation with a commercial DVD optical drive in a previous study. Different from these studies limited to nanosheets deposited on solid substrates, Stanley S. Chou and co-workers have recently get processable MoS_2 and WS_2 dispersion with 2H phase by refluxing in inert high boiling point solvents like octadecene for 4 hours with a heating rate of ≈ 10 °C/min. However, searching for new strategies to achieve the phase reversion, especially a fast rout to get processable 2H-MoS_2 nanosheets and its nanocomposites remains an urgent task. This study reports a simple, fast and efficient microwave strategy to achieve 1T to 2H phase conversion of MoS_2. On the basis of the phase transformation, we also successfully realized the *in-situ* deposition of Pt NPs on the 2H-MoS_2 nanosheets. The chemically exfoliated MoS_2 nanosheets were transferred from water into the inert nonpolar solvent, o-dichlorobenzene (ODCB) according to the study by Stanley S. Chou and co-workers. Then the samples were treated by microwave irradiation. We also achieved the phase reversion and 2H-MoS_2/Pt nanocomposites with introducing the metal precursor. The microwave strategy described here should be easily adaptable to achieve the phase change of other exfoliated TMDs and to prepare other novel nanocomposite on a bulk scale.

462239 Metal-Organic Frameworks Coated Opto-Electronic Gas Sensor Incorporated with Copper Sulfide Nanostructured Thin Films

Abstract:

Near-infrared (NIR) gas sensing is an important technique that requires simple operation and low cost with high accuracy and quick response. However, the absorption band of gases at NIR region is typically low; therefore it is urgent to develop a new concept of sensing technology with high sensitivity. Low-cost materials consist of earth-abundant elements such as copper sulfide (Cu_xS) exhibit localized surface plasmon resonance (LSPR) throughout the NIR region and can be used as LSPR-based sensing material. In this work, Cu_xS thin films were fabricated by a modified successive ionic layer adsorption and reaction at room temperature. Dense and uniform Cu_xS thin films were obtained through low temperature annealing process in air. The annealed Cu_xS films showed nanoflake-shape and excellent opto-electrical properties that we could explore in the gas-sensing field. LSPR absorption is able to control by tuning the composition of Cu_xS. By introducing a thin layer of nanoporous material, metal-organic frameworks (MOFs), onto the surface of Cu_xS thin films, gas molecules can be concentrated within the LSPR field. The fabricated nano-composite films show an enhancement of CO_2 absorption in NIR region compared to that of samples with only one type of film. Interestingly, the I-V test results showed that the conductivity of Cu_xS films was decreased with the existence of CO_2 molecules. Based on our findings, MOF coated Cu_xS films can serve as an opto-electronic gas sensor in the NIR region.

463692 Crumpled P-Rich Metal Phosphide/Rgo Composite Powders for Advanced Li-Ion Batteries

Abstract:

Transition metal phosphides have been lately investigated as high capacity anode materials for lithium-ion batteries (LIBs) to replace commercial graphite electrodes that have low specific capacity. Depending on the amount of phosphorus, metal phosphides with various compositions can be synthesized. For example, nickel phosphides can exist in different stoichiometric ratios such as $Ni_{12}P_5$, Ni_4P_3, NiP, and NiP_2. Generally, it is known that transition metal phosphides are first transformed to metal nanograins (M) and amorphous-like Li_3P matrix, followed by an alloying reaction of the phosphorus nanograins with Li^+ during cycling. Therefore, NiP_2 material with high P content is attractive for anode materials because it can offer high theoretical capacity and low potential versus Li/Li^+. However, P-rich metal phosphide materials have been limited to a specific morphology so far, usually in the form of aggregated nanoparticles obtained from high energy mechanical milling. In order for P-rich metal phosphide materials to be applied as LIB anode materials, the development of a simple and cost-effective method to control the morphology of P-rich metal phosphides is crucial. In this study, we report a simple two-step method of systematically synthesizing reduced graphene oxide(rGO)/metal phosphide composite powders. The Ni/rGO composite powders prepared by scalable spray pyrolysis were then transformed into NiP_2/rGO composite powders by a simple phosphorization process in a vacuum-sealed ampoule. The synthesized NiP_2/rGO composite powders showed superior Li^+ storage properties.

465539 Hydrothermal Synthesis of TiO_2 Nanomaterials with Ethylenediamine

Abstract:

TiO_2 nanomaterials were synthesized by hydrothermal method under basic conditions using ethylenediamine (EDA) and the characterization of the resulting products was performed. In this study, the effect of EDA concentration, synthesis temperature, and difference in Ti source on crystallinity, crystalline structure, morphology and photocatalytic activity of the resulting TiO_2 was examined. Distilled water with low concentration of EDA was used as medium and titanium tetraisopropoxide (TTIP) or tetra-n-butyltitanate (TNBT) was used as Ti source. The results showed that distilled water with very low concentration of EDA (0.05 and 0.45 mol%) resulted in almost pure anatase TiO_2 with very high crystallinity at 180°C for both Ti sources of TTIP and TNBT. Less than or more than the above concentration of EDA resulted in partially amorphous pure anatase or mixed phase of anatase and brookite. With the increase of EDA concentration, ratio of brookite phase to anatase phase was increased and morphology was changed from sphere to rod like structure. Below 180°C, only amorphous phase TiO_2 was obtained. Difference in Ti source made little difference in most of the properties such as crystalline phase, crystallinity. Decomposition experiment of 2,4-Dinitrophenol (DNP) under UV irradiation demonstrated that photocatalytic activity of the resulting TiO_2 was much higher that P-25 which was known as high activity TiO_2.

466730 Development of a Sensing Device with an Integrated Plasmas Generation Unit

Abstract:

This work presents a sensing device with an integrated plasmas generation unit. This device is made of double-sided copper laminate with a 0.05-mm-thick dielectric layer in-between. The electrode patterns are defined by the toner-transfer method, which is an easy-to-fabricate process. The electrode pattern design contains three electrodes: one pair of electrodes patterned on the front side of the device and the third electrode patterned on the back side of the device. The sensing unit uses thick film ZnO, a commonly-used inorganic semiconductor, as the sensing element. The ZnO thick film is deposited at the gap between the paired electrodes on the front side by drop casting or screen printing. The plasma generation device generates atmospheric pressure microplasmas, plasmas with geometric dimension less than 1 mm. This microplasma is a dielectric-barrier-discharge (DBD)-type plasma operated at one atmosphere. The sensing unit is a conductometric-type sensor, which measures the DC resistance of the ZnO thick film across the paired electrodes on the front side. The microplasma is used to treat ZnO for annealing and/or regeneration purposes. When the resistance across the paired electrodes on the front side is measured, this device functions as the sensing mode. When the paired electrodes on the front side are ground and shorted and the back-side electrode is driven by high voltages, the DBD-type microplasma is generated at the gap between the paired electrodes on the front side, where ZnO thick film is deposited. In this mode, ZnO can be treated. Stable plasmas can be sustained in ambient air or argon with high voltage AC with the frequency of 21.5 kHz and the peak-to-peak voltage of up to 3 kV. Preliminary tests show that the resistance of the deposited ZnO thick film ranges from 10^6 to 10^9 Î©. After 15 minutes or less microplasma treatment, the resistance reduces by at least 1 order of magnitude. This proves the concept using the integrated plasma generation unit for in-situ and on-demand sensing element treatment.

470133 Gas Separation Performance Enhancement of Zeolitic Imidazolate Framework ZIF-8 Membranes Via Post Synthetic Ligand Exchange

Abstract:

Currently propylene/propane separation is performed with a highly energy-intensive cryodistillation process mainly due to their close boiling points. Although a variety of membranes with different materials have been studied for membrane-based propylene/propane separation, there are currently no membranes and membrane processes that are commercially available. This is mainly due to the fact that the majority of new membrane materials proposed failed to meet the separation performance requirements with respect to the propylene permeability and propylene/propane separation factor.[1,2,3]

Zeolitic imidazolate frameworks (ZIFs), a subclass of metal organic framework with zeolite topology, are composed of divalent metal nodes (typically Zn or Co) interconnected with imidazolate-based ligands, forming three-dimensional crystalline structures with pores and cavities in the scale of molecules.[4] ZIFs are considered very promising gas separation membrane materials because of their chemical and thermal stability in combination with their ultra-micropores (less than 0.5 nm). For example, membranes of ZIF-8, made of Zn and 2-methyl-imidazolate with a SOD structure, were reported having excellent propylene/propane separation due to the fact that its effective aperture size lies between propylene and propane.[5] Highly propylene-selective ZIF-8 membranes were synthesized using several methods including *in situ* counter diffusion and microwave-assisted secondary growth methods reported by Kwon et al.[6,7]

For industrial applications of ZIF-8 membranes, their long term stability is critically important. There are, however, only a few such studies have been reported.[8] In our long-term stability studies, we observed that time-dependent propylene/propane permeation and separation behaviors depend on synthesis methods.

Here, we would like to discuss the relationship between the membrane stability and inherent surface defects resulting from different synthesis conditions and to show a post-synthetic treatment as an effective means to stabilize membranes.

References and Notes

1 Koros, William J. and Fleming, G.K. (1993), "Membrane-based gas separation", *Journal of Membrane Science*, 83, pp 1-39

2 Baker RW., (2002) "Future Directions of Membrane Gas Separation Technology" *Industrial & Engineering Chemistry Research.*, 41 (6), pp 1393-1411

3 Tanaka, K., Taguchi, A., Hao, JQ, Kita, H., and Okamoto, K., (1996), "Permeation and separation properties of polyimide membranes to olefins and paraffins", *Journal of Membrane Science*, 121, pp 197-207

4 Park KS, Ni Z, Côté AP, et al., (2006), "Exceptional chemical and thermal stability of zeolitic imidazolate frameworks", *Proceedings of the National Academy of Sciences,* 103 (27), pp 10186-10191

5 Zhang C, Lively RP, Zhang K, Johnson JR, Karvan O, Koros WJ., (2012), "Unexpected Molecular Sieving Properties of Zeolitic Imidazolate Framework-8", *The Journal of Physical Chemistry Letters*., 3 (16), pp 2130-2134.

6 Kwon HT, Jeong H-K., (2013), "In Situ Synthesis of Thin Zeolitic–Imidazolate Framework ZIF-8 Membranes Exhibiting Exceptionally High Propylene/Propane Separation", *Journal of the American Chemical Society,* 135 (29), pp 10763-10768.

7 Kwon HT, Jeong H-K., (2013), "Highly propylene-selective supported zeolite-imidazolate framework (ZIF-8) membranes synthesized by rapid microwave-assisted seeding and secondary growth", *Chemical Communications,* 49, pp 3854-3856

8 Liu, D., Ma, X., Xi, H., Lin, YS., (2014), "Gas transport properties and propylene/propane separation characteristics of ZIF-8 membranes", *Journal of Membrane Science*, 451, pp 85–93

474052 Compressive Mechanical Properties of Metal-Organic Framework (MOF) Aerogels Prepared By Supercritical Drying

Abstract:

The compressive strength and properties of low density metal-organic framework aerogels are a result of their cellular architecture. A random cell construct generally gives rise to high specific surface area and interconnected pore channels for fluid transport. The stiffness of MOFs make the materials brittle, but by reengineering MOF into an aerogel creates a compressible material. Mechanical measurements shows that these low-density, light-weight MOF aerogels can recover its initial shape under 30 % compression. The Youngâ€™s modulus that follows E~Ï^2 behavior and it is anticipated that compressive MOF aerogels will make the material attractive in broad range of practical applications.

474070 Preparation of Graphene-MOFs Composite Aerogels

Abstract:

The self-assembly between graphene oxide nanosheets and metal organic frameworks was synthesized by using silk protein as cross-linker and investigated as a method to form three-dimensional porous graphene and MOFs objects with high surface areas and low densities. After freeze drying or supercritical drying of the assembled composite hydrogels, ultralight graphene-MOFs composite network aerogels with high surface areas, hierarchical and nacre-like pore structures, good electrical conductivity, and well-defined bulk shape were obtained. By adjusting the concentration of graphene oxide in the precursor mixture and using different kinds of drying method, aerogels with a density as low as 7.55 mg/cm^3 or a surface area as high as 523.5m^2/g could be prepared. The absorption performance was evaluated for the aerogel with the lowest density. Graphene aerogel can act as a matrix for immobilization and assembly of MOFs, which could efficiently minimize the potential release of MOFs into the environment. Due to the high porosity and hierarchical pore structure, the aerogel displayed fast absorption rates for malodor gasses as well as high absorption efficiencies. Active catalysts for oxidative treatment of VOCs, NH_3 and H_2S can be dispersed on the new aerogel materials. The graphene-MOFs aerogels are intrinsic bactericides and capable of reducing the viability of airborne bacteria.

468996 Raman Spectroscopic Characterization of the C-S-H and C-a-S-H Structures and Investigation of Their Behavior in Atmospheric CO_2

Abstract:

Raman spectroscopic characterization of the C-S-H and C-A-S-H structures and investigation of their behavior in atmospheric CO_2

Sinem Ortaboy[a,b], Jiaqi Li[c], Rupert J. Myers[c], Guoqing Geng[c], Paulo J. M. Monteiro[c], Roya[a] Maboudian, Carlo Carraro[a]

[a] *Department of Chemical and Biomolecular Engineering, University of California at Berkeley, USA*

[b] *Chemistry Department, Engineering Faculty, Istanbul University, Turkey*

[c] *Department of Civil and Environmental Engineering, University of California at Berkeley, USA*

Concrete is the most widely used construction material in the world. The primary constituent of this material is Portland cement (PC), which is mainly produced by mixing limestone (lime bearing) and clay (silica, alumina, and iron) and other ingredients such as alumina and iron oxide. The main reaction product of PC hydration is calcium silicate hydrate (referred to as C-S-H) gel, which has unique properties in the cement matrix as a solid binder phase[1]. The structure and composition of C-S-H gel greatly influences the strength, durability, and other physical and chemical properties of hydrated PC. The incorporation of aluminum into the C-S-H structure (referred to as C-A-S-H) is typical in modern PC-based cements, which are commonly synthesized using relatively more sustainable and aluminum-bearing supplementary cementitious materials e.g. fly ash[2]. Understanding the chemistry and structure of C-(A-)S-H is important in order to obtain concrete of desired quality and help in formulating processes to reduce the carbon footprint of the cement industry, which is responsible for 5-8% of global anthropogenic CO_2 emissions.

The uptake of CO_2 by C-S-H is generally regarded as a detrimental process because the carbonation process degrades the C-S-H gel structure[3] and potentially leads to corrosion of reinforcing steel and deleterious cracking of the reinforced concrete. However, the carbonation process is also a possible means to achieve CO_2 sequestration[4-6]. Regardless of the motivation, the ability to monitor CO_2 uptake in C-(A-)S-H can provide information about the atomic-scale and microstructural changes occurring in cement-based materials.

Raman spectroscopy is a powerful method to analyze functional groups through their molecular vibrations. This spectroscopic method provides detailed insight without significant damage to the sample and is sensitive to small changes in the chemical and structural composition of local atomic environments. In this paper, we report Raman spectroscopy results and use them to investigate the influence of temperature, Al content and atmospheric CO_2 uptake on the chemistries and structures of C-(A-)S-H gels prepared with different Ca/Si molar ratios.

REFERENCES

1. I. G. Richardson, Acta Crystallogr B Struct Sci Cryst Eng Mater, 2014, 70, 903-923.
2. R. J. Myers, E. L'Hôpital, J. L. Provis and B. Lothenbach, Cement and Concrete Research, 2015, 68, 83-93.
3. N. Ferralis, D. Jagannathan, J. C. Grossman and K. J. Van Vliet, Journal of Materials Research, 2015, 30, 2425-2433.
4. K. Vance, G. Falzone, I. Pignatelli, M. Bauchy, M. Balonis and G. Sant, Industrial & Engineering Chemistry Research, 2015, 54, 8908-8918.
5. G. Puerta-Falla, M. Balonis, G. Le Saout, A. Kumar, M. Rivera, G. Falzone, N. Neithalath and G. Sant, Journal of Materials Science, 2016, 51, 6062-6074.
6. A. M. Thiele, A. Jamet, G. Sant and L. Pilon, Energy Conversion and Management, 2015, 103, 374-386.

470043 Micro-scale cell patterning based on tunable CO_2 laser system

Abstract:

Micro-scale cell patterning based on tunable CO_2 laser system

Wenjun Zheng, Sichao Hou, Ming Su*

Department of Chemical Engineering, Northeastern University, Boston, MA 02115

Since cell patterning on 2D surfaces has emerged as an effective tool in cell biology, the development of easy patterning methods has attracted great attention. The lithography approaches, for instance photolithography and micro-contact printing, are often used for small scale pattering. Photolithography uses ultraviolet (UV) light to transfer a geometric pattern from a photomask to a light-sensitive chemical (photoresist) on the substrate. For micro-contact printing, an elastomeric stamp (Polydimethylsiloxane) forms patterns on the surface of substrate through conformal, transient contact. However, major issues of existing cell patterning methods are the strict operating conditions and the relatively complicated procedure without a matched accuracy. In this work, an easily operated, non-contact and high throughput method to produce micro-scale pattering with a tunable laser cutting machine is introduced.

Glass substrate was used to test the impact of parameters, including laser power (3~15W) and laser speed (7.7~20.0 mm/s), on the line width. The result indicated that the line width increased with the laser power and followed the Gaussian Equation. Specially, for small laser power (<5.0W), the increase of the line width with the laser power followed a linear relationship. A 2D theoretical modal was built to simulate the heat transfer in the laser cutting process. The simulation result indicated the laser penetration depth on glass substrate was independent of the laser speed, which was further used to predict the line width at the speed of 25.0 mm/s. The relative error between the experimental and simulated results is less than 2.0%. The prediction of line width on the polystyrene petri dish with a laser power ranged from 0.9~3.3W was conducted using the acquired model and the values of the parameters were deduced for a line width of 150 μm, based on which a pattern was machined for 2D cell patterning. This work demonstrates that the micro-scale patterning with a laser cutting machine is a promising method for refined cell coordinator and accurate cell positioning.

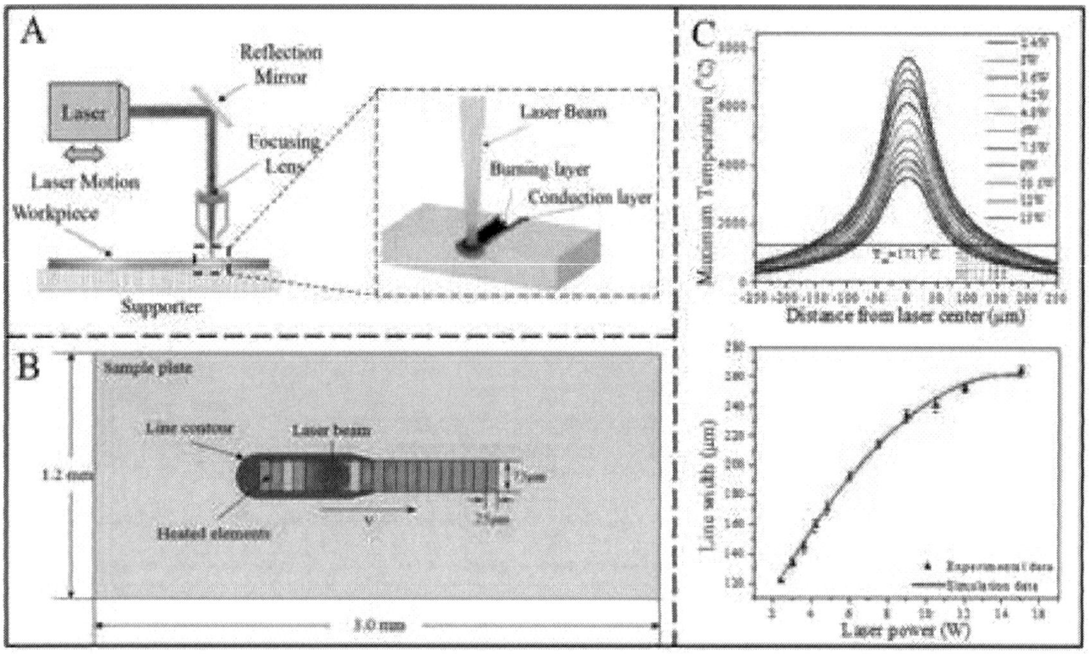

Figure 1. Schematic presentation of the laser cutting mechanism (A) and the 2D theoretical heat transfer model for laser cutting process (B). The line width was predicted using the theoretic modal (C).

Acknowledgement

<>This work was supported by the Department of Chemical Engineering, Northeastern University.

472845 RB5 Dye Removal Using a Mexican Natural Zeolite (MNZ) in Solution

Abstract:

Zeolite is a mineral belongs to the family of crystalline aluminosilicate, which is naturally found in large quantities. In Mexico, there are several sites throughout the country, being clinoptilolite the most abundant, due to its characteristics this mineral has been used as catalyst support, for drying gas, as absorbing gas and as a precursor for obtaining synthetic zeolites.

In this work, the analysis of the effect of the calcination temperature in a Mexican Natural Zeolite (MNZ), in its composition, textural, morphology and behaviour of the mineral in photocatalytic tests for degradation of the RB5 dye was performed. The composition of the zeolite does not change depending on the calcination temperature, the variations were due to the nature of the material. It was also noted that this zeolite is composed of a majority phase of clinoptilolite - heulandite, a secondary phase formed by erionita - mordenite and quartz and feldspar, which were considered as impurities in the material. With increasing calcination temperature the heulandite phase disappears after 500 °C, while the other phases remain to 700 °C. Furthermore, it is observed that the mineral particles are irregularly shaped of various sizes, because this was subjected to a milling process, the elemental analysis was performed on different areas, whereby the heterogeneity of the zeolite was tested. Finally, the photocatalytic tests for RB5 dye degradation were performed using the MNZ as the promoting agent, because the iron ore has in its structure in different arrangements, it was obtained the zeolite in suspension has a positive effect on the removal of dye, however, using high concentrations of mineral in suspension causes a screening effect, which inhibits the reaction.

475726 Developing highly sensitive chip nanocalorimeters based on the thermoelectric effect

Abstract:

Microcalorimeters have great potential as biosensors due to their ability to make measurements on very small samples and their short thermal equilibration times. They are not, however, sensitive enough to analyze nano/micro-scale samples or individual cells, and fluid handling is a critical issue. This work aims to break through the current limitations by measuring sub nano-Watt power levels quantitatively through use of micromachined sensors. Individual sensors consist of a combination of thermistors and thermopiles on a thin silicon nitride membrane that allow direct differential measurements between a sample and a reference. To maximize sensitivity, sensor geometry and materials were optimized based on the temperature field that develops around a typical sample and on calculations of the corresponding noise equivalent power and minimum detectable power (P_{min}). Specifically, thermopiles that consist of Constantan and Nichrome couples are capable of sensing below nW level power over a broad range of thermopile lengths, 10 - 10^3 μm. Arrays of calorimeter sensors were fabricated using silicon micromachining techniques and characterized as a function of thermopile dimensions. We use a similar thermopile design to fabricate a flexible sensor array based on polymer membranes to facilitate integration with various fluid handing techniques. This highly sensitive thermopile-based calorimeter offers particular promise for quantitative measurement of cellular bioenergetics and systems that are limited in analytic volume.

476689 Hull Cell Combinatorial Synthesis of Oxygen Evolution Catalysts

Abstract:

Discovery of novel materials for applications such as electrocatalysts, light absorbers, supercapacitors, and others involves the creation of large libraries of inorganic solid-state materials. Techniques to create these material libraries often involve the use of a series of masks and physical vapor depositions to create a library of material compositions. Electrodeposition methods offer an alternative, low-cost, and material-efficient way to create many novel inorganic solid-state materials. This research presents a high-throughput method for the combinatorial synthesis of novel electrocatalysts in a Hull cell, as well as methods for screening and characterization of these material libraries.

By creating electrodeposited libraries in the trapezoidally shaped Hull cell, the deposition conditions can be changed in a gradient across the working electrode. This allows for the creation of a library of testable materials in a single step. This research has focused on demonstrating this method to find a superior oxygen evolution catalyst by creating material libraries of mixed metal oxides, screening these materials for catalytic activity, and characterizing them with spectroscopic methods.

452010 Composite Metal Oxide/Nanocarbon Materials As High Performance Anodes for Next-Generation Automotive Li-Ion Batteries

Abstract:

Title:"Composite Metal Oxide/Nanocarbon Materials as High Performance Anodes for Next-Generation Automotive Li-Ion Batteries"

Alessandro Palmieri, Mengchen Liu, Ying Liu, Neil Spinner and William E. Mustain

Department of Chemical and Biomolecular Engineering, University of Connecticut, Storrs, CT 06269-3222

Lithium ion batteries play a significant role in mobilizing technology; however, to make the leap to automotive and grid storage materials with higher energy density are needed. Therefore, the Li battery anode material is a highly active area of research. Unfortunately, some of the most widely investigated material have either limited stability (silicon) or low capacity (graphite). Therefore new solutions have to be investigated to increase the energy density of lithium ion batteries (>300 Wh/Kg) while not sacrificing their stability and cycling performance compared to state-of-the-art systems.

In recent years, our group has focused on advancing metal oxide-based anode materials toward automotive targets. In this poster, we will show how metal oxide material chemistry and structure influences the reversible capacity and cycleability of Li-ion anodes through both physical and electrochemical characterization including identical-location TEM. We will also show how the electronic conductivity of the electrode plays a critical role in the achievable capacity and retention. This poster will focus on recent advances where several metal oxides have been enhanced by alloying, impregnation with metallic nanoparticles, or complexation with nanosized advanced carbons such as carbon nanotubes and graphene – leading to very high reversible capacity (> 700mAh/g) and long cycle life (several 100 cycles).

462206 Production of Monodisperse Lithium Carbonate NANO-Particles, to be USED As Sacrificial ANODE Template for Direct Acid FUEL CELLS

Abstract:

With increase in demand for energy the need for alternative power sources has become crucial. The advent of fuel cells has made it possible to power almost any portable device or machine that rely on batteries. Their main advantage is that they do not go dead eventually unlike a typical battery but will continue to produce energy as long as fuel and oxidant are supplied. By improving the porosity of the anode catalyst layer in these fuel cells, it is possible to improve the efficiency. Lithium carbonate is a good choice for sacrificial anode template (pore-former/porogen) especially for direct formic acid type fuel cells, due to its low solubility in the electrolytes being used. In such a case the size of the porogen particles plays the key role. The goal of this research is to be able to produce stable mono-disperse lithium carbonate nanoparticles. A patented CHEMTOR reactor, based on the principles of microfluidics, was used to achieve low poly-dispersity and better control over size of the particles. The product was characterized using dynamic light scattering technique (DLS). The fiber reactor resulted in average particle size of 109nm diameter while the batch process resulted in 839nm diameter. Future work involves in stabilizing the nanoparticles and size control by varying the operating temperature, flowrates and surfactant concentration.

462922 Rapidly Exfoliation of Layered Covalent Triazine-Based Frameworks into Quantum Dots for Selective Detection of Cu^{2+} Ions

Abstract:

Carbon quantum dots (CQDs) have attracted numerous attentions due to their unique physicochemical properties and a wide range of potential applications. The top-down strategy is widely used to prepare CQDs with low cost and simple operation. However, the starting materials are still limited to a few kinds of carbon structure, such as activated carbon, graphite, carbon nanotubes, graphene oxide, graphitic carbon nitride and so on. Very recently, fluorescent covalent organic framework (COF) have been synthesized by selectively choosing the fluorophore unit as the building blocks (such as pyrene, porphyrin and phthalocyanine). But most of them are insoluble bulk powers, which hinder their further applications in some promising field. Herein, we reports at the first time that the layered covalent triazine-based frameworks (CTF-1) could be developed as a potential candidate for the top-down fabrication of N-doped CQDs (CTFQDs). When exposed into the mixed solution of H_2SO_4 and H_2O_2, CTF-1 can be rapidly exfoliated into fluorescence nanosheets within several minutes. The obtained CTFQDs is approximately 50nm in lateral sizes and 1nm in thickness. The whole synthetic route only involve low-cost precursor and requires no hydrothermal treatment, which means that the production of CTFQDs can be easy scale-up. The as-prepared CTFQDs not only emitted intrinsic visible emission around 410 nm, but also showed a high fluorescent quantum yields (QY) of 21% (using Quinine sulfate as the reference) and large Stokes shift of 180 nm. Because of the unique PL properties and the good water solubility, the CTFQDs were further utilized as a photoluminescence probe for rapid, sensitive and selective detection of Cu^{2+} ion. The whole detection process can be completed in 1 min with a detection limit as low as 1 Î¼M and a dynamic range from 1 to 50 Î¼M. We expect that our effort will benefit future research on employing various kind of COFs for the CQDs preparation and applications.

463121 Utilization of MEMS Techniques to Deterministically Engineer High Power Li-Ion Battery Electrodes

Abstract:

Rechargeable, or secondary batteries, are required to power almost all of today's increasingly complex technology. Thus, new systems must be created to match the size and power needs of these devices. In most current commercial batteries, high charge and discharge rates lead to increased kinetic resistances and reduced capacity retention. Increasing electrode size and active material mass can increase energy storage but reduces power density. In order to optimize both energy storage and power performance, the battery electrodes must be designed to minimize the kinetic resistances of both electron and ion transport within the system while maintaining the infrastructure to hold a large amount of active material. Additionally, it is necessary for these electrodes to be deterministically engineered in order to have potential for high volume manufacturing. In this study, micromachining and electroplating techniques are utilized to design a fabrication process that deterministically engineers electrodes and achieves the goal of both high power and high energy density performance. The fabrication process maximizes the surface area of the electrodes, while the multilayer approach enables the manufacture to be repeatable and scalable, up to volumes on the order of a cm^3. The power performance of the electrodes is characterized through galvanostatic charge and discharge and analysis of the effect of C rate on capacity retention and cyclability.

463264 Harnessing Disorder in a Novel Nanomaterial for Light Harvesting Applications

Abstract:

Explorations of energy sources for the ever-increasing global population and its expected exponential increase in demand and consumption with the progression of this century have led to an array of creative designs. Of these, solar energy technology holds great promise given the now well-known calculation that within approximately one and a half hours, the energy incident upon the earthâ€™s surface has the capacity to power the world for one year. To harness this immense energy source, light harvesting technologies have expanded beyond the traditional silicon-based photovoltaic designs, mainly due to their thermodynamic restriction of an approximate 33% maximum efficiency, known as the Shockley-Queisser limit. Out of necessity, new designs have arisen including multi-junction, perovskite, plasmonic, and dye-sensitized solar cells. The last of these are promising because of their generally low cost materials and their flexibility, allowing them to be implemented in a variety of locations. Currently, however, they are experiencing an efficiency limit of ~10% due ultimately to charge transport issues, such as injected electron recombination with the electrolyte, instead of with the semiconductor material. Research with dye-sensitized solar cells is presently focused on improving the absorption capabilities of dyes and effective carrier transportation in order to circumvent this restriction.

A variety of light sensitive structures have been examined in our lab, including elongated ZnO rods in N719 dye, an array of nanoparticle systems, and TiO_2 inverse opals. Building on this foundation of fabrication experience, we now seek to explore a quantum dot sensitized solar cell. This novel material consists of â€˜ultrasmallâ€™ quantum dots, i.e. sub 2 nm diameters, which exhibit a phenomenon known as fluxionality, stemming from their high degree of disorder. Nearly 90% of the atoms in a QD of this size are on the surface and as a result these surface atoms switch bonds with their neighbors, altering their band gaps at femtosecond intervals. With the design proposed here, this fluxionality property will be used to absorb across the visible spectrum and into UV range as well by using the proximity of the UV-absorbing material to space the band gaps of the quantum dot farther apart. This in turn increases the carrier lifetime and allows for the capture of excited electrons.

Herein we report the behavior of this new quantum dot sensitized cell and compare its efficiency to that of N719 and other dyes commonly used in dye sensitized solar cells. We also note the key differences between the absorption capabilities of this novel structure and other quantum dot structures such as core-shell designs.

470553 Controlling the Absorption Spectra of Transition Metal Doped Nanostructures

Abstract:

Rare earth ions have played a prominent role in solid state lighting and optical displays due to their ability to emit at specific wavelengths. While this has traditionally been seen at a benefit, a new class of materials is needed in with adaptive luminescence is needed for long device lifetimes. Previously, organic materials have been used due to the ability of a single molecule to emit multiple wavelengths based on the phase of the material and the applied stress. However, for electronics applications it is difficult for these materials to repeatable convert between liquid and solid phase. Additionally, these materials are known to be carcinogenic and should be avoided in consumer applications. Alternatively, transition metals have been shown to have a wide range of emission and absorption spectra indicating the potential for a wide range of applications. Unfortunately, it is difficult to predict the appearance of the material based solely on the dopant as the energy level spacing is a function of the crystal field of the host lattice, as is the case for Cr^{2+} produce the red and green appearance in rubies and emeralds, respectively. In this work, transition metal doped thin nanostructures are used for adaptive luminescence, due to their wide absorption range, with the spectra tuned via external stimuli.

Specifically, 5-10 nm thick thin films of ZnO, TiO_2, and SiO_2 will be synthesized via common sol-gel techniques. Various first row transition metals, i.e., Cr, Mn, and Co, are readily incorporated into the film at various concentrations through the incorporation into the sol-gel solution film deposition. High temperature annealing was performed after film deposition to convert each material to the final oxide form, including anatase and rutile TiO_2. The three host materials where chosen to ensure different absorption spectra for the dopant based on the crystal field strength whereas a film below 10 nm prevents the ion shielding based on the distance from the interface. Nanoparticles approximately 20 nm in diameter and similar compositions were synthesized as an alternative material to the thin films. The dopant ions are seen to be readily incorporated into the polycrystalline thin films are lower concentrations based on X-ray diffraction and photoemission spectroscopy techniques. Absorption and photoluminescence measurements were conducted on all the various materials before and after annealing to quantify the shift in optical properties, indicative of the role of external forces on the luminescent particle. Based on these characterization methods it is possible to extract the local crystal field strength in order to predict the extent of absorption and emission shifts that are possible under stronger applied fields, such as electric and magnetic fields.

471173 Semiconducting Perovskite Oxides (ABO_3: A = La; B = Cr, Mn, Fe) for Photocatalytic Reduction of CO_2

Abstract:

Photocatalysis is one of the most researched protocols for harvesting solar energy. There have been several efforts towards solar photocatalytic water and CO_2 splitting. Though, titanium dioxide (TiO_2) has been one of the most popular photocatalysts, it has always suffered from the intrinsic inability to utilize a major range of solar spectrum due its large band gap, and hence, mostly suited for harvesting UV light. Thus, there is a significant interest in creating efficient photocatalysts that can capture the visible light.

Perovskite oxides of the form (ABO_3) is a unique material having a broad range of band gaps. They are mostly prevalent in cubic forms and suited for a vast range of applications from photocatalysis, electrocatalysis, photovoltaics, sensors etc. They are stable in varied conditions and exhibit high oxygen mobility. Usually they have a Group II element like calcium, strontium, barium or lanthanides as the "A" site component while they have transition elements as the "B" site component. These materials can accommodate different elements in their "A" and "B" sites and can form mixed metal oxide phases like $A1_{(1-x)}A2_xBO_3$, $AB1_{(1-y)}B2_yO_3$ and $A1_{(1\ x)}A2_xB1_{(1\ y)}B2_yO_3$. Thus, it presents a huge opportunity to control the material properties by simply compositional tuning.

We hereby, investigated the effect of varying the "B" site elements on the band gap and band edge positions. The different "B1" and "B2" elements in $AB1_{(1-y)}B2_yO_3$ were chromium, manganese and iron in different compositions while the "A" site had been fixed for lanthanum. We synthesized the materials via Pechini process and characterized through x-ray diffraction (XRD) and diffuse reflectance spectroscopy (DRS). These materials were henceforth, tested for photocatalytic CO_2 conversion using our lab-built reactor set-up. LEDs of different wavelengths were used to test the photo-reduction effect under visible and UV light. For better understanding of the reaction mechanism, we performed Diffuse Reflectance Infra-red Fourier Transform Spectroscopy (DRIFTS) experiments of CO_2 photo-reduction under different light sources. Theoretical calculations based on density functional theory (DFT) have been performed as well to probe the band-structure of these perovskite oxides. Results have been compared for various functionals like LDA, PBE-GGA, B3LYP, HSE06. The effect of composition on band gap tuning of $LaB1_{(1\ y)}B2_yO_3$ perovskite oxides and its implications towards photocatalytic reduction of CO_2 are thus being explored.

454848 Controllable Preparation of Ni-Co Nanosheets Covered Nanocages Via Acid Etching with Enhanced Electrochemical Properties

Abstract:

Controllable preparation of Ni/Co nanosheets covered nanocages via acid etching with enhanced electrochemical properties

Zijian Lv, Qin Zhong*, Yunfei Bu

School of Chemical Engineering, Nanjing University of Science and Technology, Nanjing, Jiangsu 210094, PR China

Email: lzj_0321@163.com; zq304@mail.njust.edu.com.cn; jpu441@yahoo.com.

Abstract

It has been well accepted that the performance of supercapacitor material highly depends on its morphology, thus the control on morphology is significant for acquiring advanced materials.[1] Herein, a facile method for controllable synthesis of nickel-cobalt layered double hydroxide (Ni/Co-LDH) with a unique hollow structure is presented. In which zeolitic imidazolate framework-67 (ZIF-67) nanocrystals are used as templates and etched in a nitrate solution. The morphology of Ni/Co-LDH shows an interesting variation with a different dosage of $Ni(NO_3)_2$ as depicted in Fig. 1(a-f). After reaction with $Ni(NO_3)_2$ in ethanol solution, the particles still retain a polyhedral shape under a low concentration of $Ni(NO_3)_2$ but lose the structural integrity with an excessive dosage of $Ni(NO_3)_2$. Specifically, Ni/Co-LDH-3 (Ni/Co-LDH-*x*, in which *x* indicates the mass ratio of $Ni(NO_3)_2$/ZIF-67) can maintain the polyhedral structure and obtain a hollow structure covered by interlaced nanosheets. Furthermore, it can be observed from the TEM images

(Fig. 1(g-h)) that Ni/Co-LDH-3 exhibits a hollow structure with nanosheets forming the shell.

Fig. 1 SEM images of ZIF-67 (a) and Ni/Co-LDH-*x* (b-f); TEM images of Ni/Co-LDH-3 (g, h).

Fig. 2(a-b) shows the electrochemical properties for as-prepared Ni/Co-LDH-*x* as supercapacitor electrodes. As depicted in Fig. 2(a), a pair of redox current peaks can be clearly observed in the cyclic voltammetry (CV) curves, indicating that the pseudocapacitive performance resulted from the surface faradaic redox reactions related to M-O/M-O-OH (M= Ni or Co).[2] It can be observed form Fig. 2(b) that Ni/Co-LDH-3 manifests an excellent electrochemical performance in terms of high specific capacitance (1580 F g^{-1} at 2 A g^{-1}). It might be attributed to the hollow structure which can provide interconnected pathways for both electrons and ions.[3] Furthermore, the interlaced nanosheets can serve as ion reservoirs, which might shorten the diffusion distance to the interior surfaces and accelerate the ion diffusion process in the electrode, so as to enhance the electrochemical property.[4] Moreover, the rate capability and cycling performance of Ni/Co-LDH-3 are investigated and the corresponding results are displayed in Fig. 2(c-d). Obviously, the Ni/Co-LDH-3 exhibits a favourable rate capability and displays high capacitance retention of 82.9% after 1000 charge/discharge cycles. It is attributed to the special hollow structure covered by interlaced nanosheets which can maintain the morphology and prevent agglomeration and deactivation during the iterative charge/discharge process.

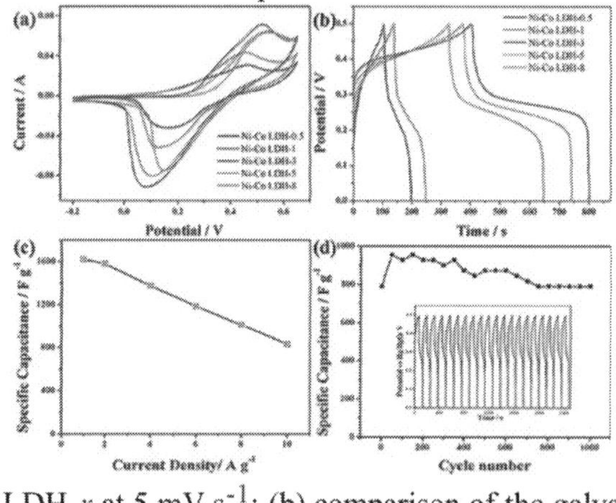

Fig. 2 (a) The CV curves of Ni/Co-LDH-*x* at 5 mV s^{-1}; (b) comparison of the galvanostatic charge-discharge results at 2 A g^{-1}; (c) specific capacitance of Ni/Co-LDH-3 electrode at different current densities; (d) Cycling performance of the Ni/Co-LDH-3 electrode at a current density of 8 A g^{-1}, the inset shows the charge/discharge curves of 20 cycles.

In summary, the variation of morphology indeed affects the electrochemical performance, and the Ni/Co-LDH-3 stand out from the as-prepared materials due to its unique hollow structure composed of interlaced nanosheets. This study provides an opportunity to explore the optimization of electrochemical properties by controlling the morphology in improving the electrochemical performances, which could provide further insights in the material design.

References

[1] H. Hu, B. Guan, B. Xia, X.W. Lou, *J. Am. Chem. Soc.* **137** (2015) 5590-5595.

[2] G.Q. Zhang, H.B. Wu, H.E. Hoster, M.B. Chan-Park, X.W. Lou, *Energ. Environ. Sci.* **5** (2012) 9453.

[3] C. Sun, J. Yang, X. Rui, W. Zhang, Q. Yan, P. Chen, F. Huo, W. Huang, X. Dong, *J. Mater. Chem. A.* **3** (2015) 8483-8848.

[4] Z. Jiang, Z. Li, Z. Qin, H. Sun, X. Jiao, D. Chen, *Nanoscale.* **5** (2013) 11770-11775.

473161 Graphene/Conducting Polymer Composite Fibers for Wearable Energy Storage

Abstract:

The advent of new electrode materials, particularly in the form of fibers, has enabled creative approaches to the fabrication of novel electrochemical cell configuration. Recently, fiber-shaped supercapacitors have attracted significant attention as wearable energy storage due to their ability to be woven or knitted into fabrics, or stitched into existing textiles. Long lengths of fiber electrodes are crucial to fabricate such fiber-shaped supercapacitors. Here, we report covalently bonded graphene fibers with high conductivity and mechanical strength through a wet-spinning method. Meters long fibers can be produced in a couple of hours. Conducting polymers, such as polypyrrole or polyaniline, are deposited on the graphene fibers to enhance the capacity using an electrochemical polymerization method. Two composite fibers are coated with polymer electrolyte and twisted together to form a solid-state supercapacitor. The fabricated supercapacitor shows good electrochemical performance and excellent flexibility, which make itself an ideal candidate as wearable energy storage device to be integrated into garments.

456687 Crosslinker Length Affects the Modulus of Poly(acrylamide)-Silica Nanoparticle Hydrogel Composites

Abstract:

Polymer hydrogels have shown to exhibit improved properties upon the addition of nanoparticles; however, the mechanical underpinnings behind these enhancements have not been fully elucidated. Moreover, fewer studies have focused on developing an understanding of how polymer parameters affect the nanoparticle-mediated enhancements. In this study, we investigated the elastic properties of silica nanoparticle-reinforced poly(acrylamide) hydrogels synthesized using crosslinkers of various lengths. Crosslinker length positively affected the mechanical properties of hydrogels that were synthesized with or without nanoparticles. However the degree of nanoparticle enhancement was negatively correlated to crosslinker length. Our findings enable the understanding of the respective roles of nanoparticle and polymer properties on nanoparticle-mediated enhancement of hydrogels and thereby the development of next-generation nanocomposite materials.

465560 Synthesis and Characterization of Poly(Acrylic Acid) Coated Magnetite

Abstract:

Poly(acrylic acid) (PAA) coated magnetite was prepared via simple one-step method and its properties were evaluated. The PAA coated magnetite particles were synthesized with co-precipitation of $FeSO_4$ and $FeCl_3$ in the presence of PAA of various molecular weight. The obtained samples were collected by permanent magnet, and characterized with XRD and SEM. The total coated PAA amount and free carboxy group in PAA were determined by TGA and neutralization titration, respectively. From the results of XRD and SEM, the magnetite nanoparticles were successfully synthesized regardless of the existence of PAA. The crystalline size of magnetite became large with increase of PAA molecular weight. In addition, both of the total coated PAA amount and free carboxy group increased with increase of PAA molecular weight.

465592 Nitrogen-Functionalized Graphene Oxide By Supercritical Ammonia for Carbon Dioxide Adsorption

Abstract:

Nitrogen-functionalized graphene oxide (N-FGO) was developed by reacting graphene oxide (GO) with supercritical ammonia ($scNH_3$). Functionalization was done by combining as-prepared GO with NH_3aq (28%) in an Inconel batch reactor for 1 hour by subjecting it to supercritical conditions with reaction temperature of 250^0C and operating pressure of 30mPa. At this condition, H_2O content of the NH_3aq, will still be in subcritical condition. To investigate the effect of H_2O content two methods were used. Assessment were based on the nitrogen content for each method after functionalization. In the first method, the system was in water-rich phase (denoted by N-FGO-1), GO was directly combined with NH_3aq, enabling H_2O to react with GO during functionalization. On the other hand, the second method was in ammonia-rich phase (denoted by N-FGO-2), where GO was placed in a holder to limit the reaction with H_2O. The successful functionalization was proven by characterizing GO properties before and after functionalization using Fourier transform infrared spectroscopy (FTIR), nuclear magnetic resonance spectroscopy (NMR), X-ray photoelectron spectroscopy (XPS), elemental analysis (EA), X-ray diffraction spectroscopy (XRD). Higher N content was achieved in N-FGO-2, however, both method were effective in the functionalization. Reaction mechanisms that describes the functionalization process were proposed based from characterization results. In this study, it was assumed that only replacement of terminal carbon occurred during the functionalization process which was due to the nucleophilic substitution reaction aided by the supercritical conditions. At supercritical condition, NH_3 forms radicals, such as $-NH_2$, -NH and â€"N. These radicals may react with C atoms forming new functionalities such as pyrrolic N, amino N (amine, imides and amides), and pyridinic N. Functionalization with carboxylic group involves subsequent reactions with $scNH_3$, dehydration and decarboxylation process. Reactions with epoxy group involves ring opening and hydroxyl group undergo dehydration forming aromatic amine.

Furthermore, carbon dioxide (CO_2) adsorptive properties were evaluated using thermal gravimetric (TG) apparatus. From results, N-FGO-2 resulted to the highest CO_2 adsorption capacity around 0.5 mmol CO_2/g adsorbent. While, N-FGO-1 can adsorbed around 0.45 mmol CO_2/g adsorbent and GO can adsorbed around 0.3 mmol CO_2/g adsorbent. Results indicate that the higher the N content of the adsorbent the higher the adsorptive capacity.

The study has presented promising results which can make notable contributions in the fields of carbon and graphene chemistry. The proposed novel process of graphene oxide functionalization with N groups can be applied to various types of materials and other carbon-based materials which in turn can be potentially useful for relevant applications such as electronics, energy conversion, storage devices, catalytic support and carbon capture.

466045 Synthesis of Porous Aggregates Made of Nitrogen-Containing Polymer Nanoparticles Incorporating Noble Metals for Catalysis

Abstract:

Polymers are known for their low cost of production and their ease of handling. Thanks to the broad choice of monomers, it is very facile to target a material with a specific composition or structure. Based on this principle, this contribution is focused on nitrogen-containing polymers, for which it is not necessary to add any post-synthesis modification since the nitrogen is present in the building block (monomer). The presence of this active species enables the synthesis of many different kinds of materials suitable for many applications, such as adsorption, catalysis, energy storage, etc.

Nanoparticles of polyacrylonitrile, a polymer with repeating ternary-coordinated nitrogen atoms, have been produced by emulsion polymerization (particle diameter: 120-160 nm). These functional groups can undergo cyclization when thermally treated under specific conditions, forming a microporous structure inside the polymer matrix [1]. Namely, a series of heat treatments with and without oxygen are required to induce cyclization reactions, needed to prevent polymer melting during the high temperature steps of the process. In addition, the removal of the non-carbonaceous part of the polymer by thermal degradation creates micropores and small mesopores inside the nanoparticles, with pore size ranging from 0.5 to 10 nm. Given the high sensitivity of the quality of the final material to the operating conditions of such treatments (temperature, gas flow and composition), they have to be selected very carefully. When copolymers are used (i.e., involving other monomers in addition to acrylonitrile), the extent of cyclization of neighboring cyanide groups is reduced, making the selection of the optimal conditions of the heat treatments even more difficult.

When considering these polymers as support for catalysts, wet impregnation is one of the most popular method to incorporate active species into the porous support. However, the loading might be incomplete as well as inhomogeneous and sintering of the catalytic species can be an important problem [2]. To mitigate such issues, in this work metal precursors were incorporated during the polymerization before any heat treatment, thus achieving a very high metal loading within the nanoparticles. As a matter of fact, the presence of nitrogen was somehow "binding" the noble metal precursors within the structure, thus embedding them into the polymeric structure: together with the formation of graphite-like carbon network during the pyrolysis, this embedding at the atomic scale is expected to (i) provide high catalytic activity and (ii) prevent or at least hinder the sintering of active species during the final application. Then, the resulting metal-containing colloidal dispersion was destabilized in a controlled manner by salt addition under stagnant conditions to form a gel-like structure. After drying, the aggregated polymer exhibited a limited macroporosity due to empty space in between the non-porous nanoparticles. Then, the produced monoliths have been gently crushed and sieved down to solid particles in the size range from 560 to 180 μm, with pores of about 100 nm. Finally, the collected powders have been thermally treated as mentioned to create the micropores needed to access the metal sites: this way, metal reduction could be applied directly in gas phase, or through carbothermal reduction, taking advantage of the pyrolysis's conditions, e.g. nitrogen atmosphere and high temperature [3].

In particular, Platinum-based precursors have been incorporated in polyacrylonitrile nanoparticles produced by miniemulsion polymerization. The experimental procedure involves the preparation of an oil-in-water emulsion by sonication of an aqueous phase containing surfactants and an oil-phase containing the monomer (acrylonitrile), an organic phase soluble initiator and the metal precursor, namely Platinum(II) acetylacetonate ($Pt(acac)_2$). The formed microdroplets entrap the metal precursor and the following polymerization reaction prevents metal diffusion to the continuous phase. This technique was successful, since almost 100% of Platinum incorporation (measured by ICP-OES) was reached [4, 5]. In addition, TEM-EDX experiments showed a very homogeneous distribution of the Pt(II) within the nanoparticle.

The final thermal treatment of the metal-containing polymer is currently under investigation. We expect to reach high values of surface area, in the order of 700 m^2/g, as was the case for polyacrylonitrile particles previously treated without incorporated metal. Similar to the approach used with single-site metal catalysts [6], the resulting materials will then be tested as atomic-sized metal catalyst for reduction reactions involving hydrogen.

[1] M. Rahaman, A.Ismail, A. Mustafa, *Polym. Deg. Stab.*, 92 (**2007**), 1421-1432

[2] D.P. Debecker, M. Stoyanova, U. Rodermerck, P. Eloy, A. Léonard, B. Su, E.M. Gaigneaux, *J. Phys. Chem. C*, 43 (**2010**), 18664-18673

[3] Yuzhen Yang, *Synthesis and Pyrolysis of Platinum Containing Polyacrylonitrile Latex Particles* (2011). *Theses and Dissertations*, Lehigh University USA. Paper 1337

[4] K. Landfester, M. Antonietti, *Macromol. Rapid Commun.*, 21 (**2000**), 820-824

[5] Y. Yang, E.S. Daniels, A. Klein, *J. Appl. Polym. Sci.*, 132 (**2015**), 41933 (1-7)

[6] G. Vilé, D. Albani, M. Nachtegaal, Z. Chen, D. Dontsova, M. Antonietti, N. López, J. Pérez-Ramírez, *Angew. Chem. Int. Ed.*, 54 (**2015**), 11265-11269

467164 Hydrothermal Synthesis of Composite Zeolite a and X Coatings

Abstract:

Different single- and multi-step hydrothermal synthesis methods have been developed to improve the zeolite coating properties [1,2]. The substrate heating method promotes the growth of relatively thick coatings in one synthesis step [1]. In this method, the metal is heated directly while the reaction mixture is kept at a lower temperature in a water/oil bath. Due to a temperature difference between the reaction mixture and the substrate, the crystallization in the bulk is suppressed, while that on the metal surface is promoted. Much thicker coatings of zeolite are thus obtained on the substrate, as compared to conventional synthesis.

Zeolites A and X are two of commonly investigated zeolites since they have favorable properties for different applications, ranging from separation and adsorption to ion-exchange. The use of the coatings of these zeolites in adsorption heat pumps has also received much attention. In this respect, the preparation of sufficiently thick and stable coatings has become a significant challenge. Pure zeolite A and X coatings with mass equivalent thicknesses of up to about 270 µm could be prepared on stainless steel plates by using the substrate heating method. The equivalent thickness mentioned here was calculated from (mass of zeolite deposited per coated area of substrate)/(density of zeolite). The void fraction in these coatings was quite high and actual average thicknesses up to 650 µm were observed by laser microscopy [3]. These coatings also exhibited an accessible sponge-like structure which seemed to result in promising adsorption kinetics.

The preparation of composite materials containing zeolite has already attracted interest for a few applications. Polymer-zeolite mixed matrix membranes have been obtained by dispersion of zeolite crystals in polymeric films for use in gas separation and pervaporation [4]. The quality of the interface between zeolite crystals and polymer are key points in related research studies. In a recent study, zeolite coatings prepared on stainless steel were covered by polyacrylic acid [5]. An interesting observation was that composite samples desorbed somewhat higher amounts of water compared to both pure zeolite and pure polyacrylic acid coatings. In another study, three-step syntheses were performed for forming rather low-silica-zeolite coatings on aluminum surfaces. Here, a high-silica-zeolite bottom layer, a bridging layer and a low-silica-zeolite top layer were formed consecutively.

In this study, composite coatings consisting of zeolite A and zeolite X were prepared. Different synthesis conditions were used for direct crystallization of these materials on stainless steel plates in two consecutive steps. Zeolite coatings were grown on stainless steel plates (AISI 316 grade). The plates had 1 cm diameter and 0.8 mm thickness. Coatings were prepared by substrate heating [1] and conventional synthesis methods for obtaining zeolite A and zeolite X coatings, respectively. The syntheses were carried out by using clear solution compositions represented by 50 Na_2O:Al_2O_3:5 SiO_2:1000 H_2O (zeolite NaA) [1] and 70 Na_2O:Al_2O_3:15 SiO_2:2100 H_2O (zeolite NaX) [6]. In order to produce zeolite A, a resistance temperature of 140 °C and a water bath temperature of 25 °C were used with the substrate heating method. The synthesis was carried out for 1 day after when the water bath temperature was increased to 60 °C and was kept there for 2.5 h. The crystallization of zeolite X was performed in an oven kept at 80 °C for 26 h of synthesis. Pure zeolite A, pure zeolite X and composite A+X and X+A coatings were prepared in this study. A+X indicates that firstly zeolite A was obtained on the substrate after which a second synthesis was performed to form an additonal layer of zeolite X on the metal plate. X+A was the opposite case where zeolite X was the first layer grown on the substrate followed by zeolite A formation. The coatings prepared were characterized by X-ray diffraction (XRD), scanning electron microscopy (SEM) and thermogravimetry (TG). TG analysis was performed to determine the water desorption between ambient temperature and 350°C under nitrogen flow. A heating rate of 20 °C/min was used.

It was possible to prepare composite crystalline zeolite coatings with different amounts of zeolites A and X in the samples, as verified by X-ray diffraction analyses. According to TG measurements, the pure zeolite X and pure zeolite A coatings exhibited the highest and lowest amounts of desorption, respectively. In general, a slightly higher amount of desorption occurred from the A+X composite zeolite coating when compared to the X+A composite. It should be noted that the ratio of zeolite X mass/zeolite A mass was higher in the former coating. It was determined that the samples exhibited almost the same amount of desorption in consecutive experiments, indicating crystallinities close to 100%.

References

[1]A. Erdem-Şenatalar, M. Tatlier and M. Ürgen, Micropor. Mesopor. Mat., 32 (1999) 331.

[2]V. Valtchev, J. Hedlund, B.J. Schoeman, J. Sterte and S. Mintova, Micropor. Mat., 8 (1997) 93.

[3] M. Tatlier, G. Munz, G. Fueldner and S. Henninger, Micropor. Mesopor. Mat., 193 (2014) 115.

[4] P.S. Goh, A.F. Ismail, S.M. Sanip, B.C. Ng and M. Aziz, Sep. Purif. Technol., 81 (2011) 243.

[5] A. Atakan, C. Atalay-Oral, M. Tatlier, T. Erciyes and A. Erdem-Şenatalar, A, Micropor. Mesopor. Mat., 156 (2012) 262.

[6] M. Tatlier, K.B. Cığızoğlu, H.K. Cigizoglu and A. Erdem-Şenatalar, J. Por. Mater., 15 (2008) 389.

467701 Design of Corrosion Inhibitors in Concrete Pore Solution: Modelling and Experimentation

Abstract:

Concrete, a complex composite material, is the single most widely-used material in the world. Its strength decreases when loaded in tension and, therefore, it is a common practice to reinforce concrete with steel to improve the tensile mechanical properties. Generally, all concrete structures contain reinforcing steel and degradation of this reinforced steel i.e., corrosion, is a major problem of worldwide significance. This steel degradation leads to plant shutdowns, resource waste, loss/contamination of product, reduction in efficiency and maintenance issues. Several corrosion inhibitors have been developed to address the problem of corrosion in different working environments. Usually, development of a corrosion inhibitor is an experimental approach where one uses prior knowledge to suggest potentially effective compounds. This development approach is normally time consuming and expensive. In order to accelerate the process of developing new, novel and (potentially) more effective corrosion inhibitors, we propose the use of a computer aided molecular design (CAMD) algorithm based on the Signature molecular descriptor (J. V. Faulon). In CAMD, we make use of computer-implemented algorithms to design molecule with optimal desired properties. A quantitative structure-property relationship (QSPR) model is created for the property of interest and it is employed to generate structures that have the desired properties (J. M. Faulon). We use this in developing corrosion inhibitors that can effectively inhibit corrosion of carbon steel (CS 1018) in a solution that mimics the environment inside a concrete pore.

Previously, CAMD has been used for designing chemical admixtures: shrinkage-reducing admixtures (SRAs) (Kayello HM). For SRAs the property of interest is the surface tension of compounds in aqueous solutions as this relates to the shrinkage in concrete. Five new compounds were identified, which performed similar to the commercial admixtures but had different chemical functionalities and better strength development.

Our property of interest in this work is corrosion inhibition efficiency (Soylev) and this tells us the percent that a compound inhibits the degradation of the carbon steel in a concrete pore solution. Our training set consists of ten commonly used and commercially available corrosion inhibitor compounds. We run electrochemical impendence spectroscopy (EIS) and poteniodynamic cyclic polarization for our concrete pore solution loaded with the different corrosion inhibitors to determine their respective efficiencies. Then we apply our CAMD algorithm to identify and, ultimately, confirm new corrosion inhibitors. References

Faulon, J.L.,Brown,W.M.,& Martin,S. "Reverse engineering chemical structres from molecular descriptors: How many solution?" *Journal of Computer-Aided Molecular Design* (2005): 637-650.

Faulon, J.L.,Churchwell,C.J.,& Visco,D.P.,Jr. "The signature molecular descriptor.2.Enumerating molecules from their extended valence sequences." *Journal of Chemical Information and Computer Sciences* (2003): 721-734.

Kayello HM, Tadisina, NKR, Shlonimskaya N, Biernacki JJ, Visco DP. "Application of Computer-Aided Molecular Design (CAMD) Using the Signature Molecular Descriptor-Part 1. Identification of Surface Tension Reducing Agents and the Search for Shrinkage Reducing Admixtures." *Jornal of the American Ceramic Society* (2013): 1-13.

Soylev, T.A., & Richardson, M.G. "Corroson Inhibitors for steel in concrete: State-of-the-art report." *Construction and Building materials* (2008): 609-622.

468152 Characterization and Synthesis of Polypropylene/Graphene Nanocomposite Materials

Abstract:

Polymer nanocomposites (PnCs) find a range of applications from generic to specific uses. Graphene being a 2-D material displays unique characteristics such as being the lightest, hardest and strongest. Additionally, it also exhibits great thermal and electrical characteristics which lead to its applications in variety of areas and it has hence also drawn immense interest in the scientific community. In this work the fabrication of PnCs with homogeneous dispersion of graphene to reinforce polypropylene (PP), was carried out. The preparation of the polymer nanocomposite was carried out using solution mixing method. Characterization studies were carried out to investigate the effects of graphene nanoparticles (GnP) on the PP matrix. Thermal studies included Differential Scanning Calorimetry (DSC) and Thermogravimetric analysis (TGA) which showed improved thermal stability of the PnC. Morphological characteristics were analysed using X ray diffraction (XRD) which showed that the crystalline structure of the PP matrix is only slightly affected by the GnP. Rheological behaviour showed little changes in the PnC while mechanical analysis using Dynamic mechanical analysis (DMA) gave out that the frictional losses incurred due to mobility of polymer chains were reduced with increase in the GnP loading from 0.5% to 5%. The aforementioned characterization studies led to the inference that PP/ GnP nanocomposite affords improved properties of the composite material. This specific PnC finds applications as a material for construction of storage tanks for fluids which are much stronger as well lighter than the conventional storage tanks.

473532 Surface Modifications of Superparamagnetic Iron Oxide Nanoparticles with Polylactic Acid-Polyethylene Glycol Diblock Copolymer and Graphene Oxide for a Protein Delivery Vehicle

Abstract:

Superparamagnetic iron oxide nanoparticles (SPION) have been used in the biomedical industry for such applications as magnetic resonance imaging (MRI) contrast agents or drug delivery for a number of years, however, none have included the use of graphene oxide as a protein/drug carrer or for reporting. In this study, after using a co-precipitation SPION synthesizing method, SPION is coated with polylactic acid-polyethylene glycol (PLA-PEG) diblock co-polymer through emulsion polymerization. The SPIO-PLA-PEG NPs are then coated with single layer of graphene oxide (GO). The biocompatable NPs can then be hydrogen-bonded surface coated with various proteins, without affecting the confirmation or active sites of the porteins.

ALIGNED HNT FOR MECHANICAL REINFORCEMENT, ABRASION RESISTANCE, AND TRANPARENCY IN SOLAR CELL COATINGS

Kenan Song [a], *Roberta Polak* [a], *Khalid A. Askar* [b,*], *Michael F. Rubner* [c,*], *Robert E. Cohen* [d,*]

[a] Department of Materials Science and Engineering and Department of Chemical Engineering, MIT, 77 Mass Ave, Cambridge, MA, USA, 02139. Email: songkn@mit.edu, rpolak@mit.edu

[b,*] Professor of MASDAR, Department of Materials Science and Engineering, MIT, 77 Mass Ave, Cambridge, MA, USA, 02139. Correspondence email: kaskar@mit.edu

[c,*] TDK Professor of Polymer Materials Science and Engineering, and Director of CMSE, MIT, 77 Mass Ave, Cambridge, MA, USA, 02139. Correspondence email: rubner@mit.edu

[d,*] St. Laurent Professor of Chemical Engineering Department, 77 Mass Ave, Cambridge, MA, USA, 02139. Correspondence email: recohen@mit.edu

ABSTRACT

Halloysite nanotubes (HNT), a kind of naturally formed ceramics, encompass high aspect ratio, stiffness and low cost while remain understudied about their coating usage. Theoretical and experimental results have shown a modulus as high as 400 GPa. In this study, HNT was included in polymer based thin films fabricated using coating method on glass substrate. Mechanical tests have shown a consistent increase in stiffness, hardness and scratch resistance as compared to polymer materials without HNT. Composite mechanics based calculations showed as high as 300 GPa in HNT modulus, which is close to theoretical predictions and top in literature. The high reinforcement as well as wear resistance was attributed to high tube orientations, as revealed from electron microscopy. By matching the refractive indices between segments in composites, the coating materials can display transparency up to 90% even at a high concentration of 20 wt% loadings of HNTs. This study aims at development of environmentally friendly and cost effective coating, which is highly potential to be commercialized in solar cell industries.

1. INTRODUCTION

Solar cell panel coatings require transparency. Among transparent polymers, thermoplastics including poly (methyl methacrylate) (PMMA), polystyrene (PS), polycarbonate (PC) and cyclic olefin copolymer (COC), and thermoset resin allyl diglycol carbonate (ADC) are the most frequently used materials [1, 2]. One of the major drawbacks limiting their uses is the poor scratch resistance[3]. To overcome such a problem, organic–inorganic hybrid composites have been studied for effective hard-coating materials. SiO_2, TiO_2, ZnO, ZrO_2, and AlN nano-particles mixed with organic compound based on sol-gel reactions have been reported[4, 5]. Nonetheless, the addition of non-organic particles very frequently affects the clarity and optical transmittance, therefore limits their usage in applications requiring anti-reflection properties (i.e., solar cells require 98% transmittance)[6-9]. In addition, lubricants have also been used either internally in the bulk or externally on the surface[10, 11]. However, introduction of lubricants frequently introduces swelling in polymers and makes the optical and mechanical properties worse than that without lubricants[11-16].

Here in this contribution hollow tubular ceramic particles, halloysite (HNT), were used as reinforcement fillers in transparent epoxy composites. A facile processing technique, spray coating method, has been used for fabrication of thin films on glass slides. Indentation results showed a consistent increase in modulus with higher loadings, given specific HNT orientation. The more highly oriented, the stiffer the composite coatings. Scratch resistance estimated from both scratch depth and wear rate showed lower values in composites with higher HNT concentrations and orientations.

2. EXPERIMENTATION

2.1 Materials

Dragonite HNT clay was obtained from Applied Materials (density 2.54±0.03 g·cm^{-3}, inner diameter 10-40 nm, out diameter 50-70 nm, aspect ratio ranges between 20 and 200. BHT pore volume 20%, surface area up to 100 m^{2}·g^{-1}, refractive index 1.534). Epoxy 142-112 (purchased from Epoxy Technology, Inc., density 1.18 g·cm^{-3}) and acetone (purchased from VWR, density 0.79 g·cm^{-3}) were used as obtained[17, 18].

2.2 Composites processing

The thin-film coatings were fabricated using spray coating method. HNT was mixed with epoxy and acetone solutions (acetone: epoxy ratios were 0, 1:20, 1:10, 1:5, and 1:1). For each acetone concentration, HNT loadings were 0, 1, 2, 5, 10, 15, 20 wt% (corresponding volume fractions are approximately 0, 0.5, 1.0, 2.4, 5.0, 7.7, 10.5 vol%). The mixtures were well dispersed and the bubbles were eliminated before spraying. The prepared batch solutions were then spray coated on glass slides, cured using 5000-EC flood lamp (working distance 2-6 ", manufactured by Dymax Incorporation) with a total UV energy output of 225 mW/cm^2 at 1.0" in direct emission (wavelength mainly from 350 to 450 nm). The samples were abbreviated as EH*y* (without acetone) and E/A*x*H*y*, with *x* as weight of epoxy per unit weight acetone (i.e., in samples mixed with acetone, *x* being 20, 10, 5, and 1) and *y* as HNT weight percentage (0, 1, 2, 5, 10, 15, and 20), as shown in Table 1.

Table 1 Nomenclature of HNT composites

Abbreviation	Examples
EH*y*	EH20, 20 wt% HNT in epoxy
E/A*x*H*y*	E/A5H20, 20 wt% HNT in epoxy. While in spray solution preparation the epoxy/acetone ratio is 5, for purpose of viscosity decrease as compared to pure epoxy.

To study the HNT alignment effects on mechanical properties, composites with random HNT orientations in film plane were prepared using spin coating method. 1 ml mixed epoxy/acetone (weight ratio between epoxy and acetone is 1:1) with various HNT concentrations from 0 to 20 wt% were spin-coated on glass slides at rate of 5000 rpm, and cured same as in spray coatings.

2.3 Characterization

2.3.1 Scanning Electron Microscopy (SEM)

A field-emission high-resolution scanning electron microscope (SEM) (Zeiss Supra 25, accelerating voltage 5 kV) was used to image the internal film structures. All samples were sputtered coated with a thin gold/ palladium layer (i.e., coating ranged from 15 to 20 nm) using a Gatan high-resolution ion beam coater for image analysis. To avoid the viscoelastic influence from epoxy, samples were fractured in liquid nitrogen (around -200 °C).

2.3.2 Indentation tests

Nanoindentation measurements were conducted using a TriboIndentator (Hysitron Inc.), equipped with a Berkovich diamond tip (tip diameter 1 μm). Indentation tests were operated in a displacement-control mode when the displacement applied is controlled according to a programmed loading function and the force continuously monitored with resolution of 1 nN. The loading function in this work consisted of a 5s linear loading, 5s unloading segments and 10s force dwelling at the peak load to reduce the influence of creeping effects. The maximum displacement was determined by the possible maximum indentation depth of the composite coating, being one tenth of the coating thickness (about 200nm) to exclude the influence from the substrate. A total of 36 indents with spacing of 2 um were taken to obtain average reduced modulus and hardness values on both control and composite samples coated on glass slides.

Nanoindentation hardness is defined as the indentation load divided by the projected contact area of the indentation. It is the mean pressure that a material will support under load. From the load–displacement curve, hardness can be obtained at the peak load as,

$$H = \frac{P_{max}}{A} \quad \text{(Equation 1)}$$

where A is the projected contact area. For an indenter with a known geometry such as the Berkovich tip used in this study, the projected contact area is a function of contact depth, which is measured by the nanoindenter *in situ* during indentation[19, 20]. Therefore, the projected area, A, can be measured and calculated directly from the indentation displacement.

The elastic modulus was calculated using the *Oliver–Pharr* data analysis procedure[21] beginning by fitting the unloading curve to a power-law relation.

$$E_r = \frac{\sqrt{\pi}}{2}\frac{S}{\sqrt{A_p}} \quad \text{(Equation 2)}$$

Where S is the slope of the tangent to the loading curve at maximal load and A_p is the contact area of the indenter. E_r is given by,

$$\frac{1}{E_r} = \frac{1-\upsilon^2}{E} + \frac{1-\upsilon_i^2}{E_i} \quad \text{(Equation 3)}$$

where E and ν are the elastic modulus and Poisson's ratio for the sample, and E_i and ν_i are the same quantities for the indenter. For diamond, E_i= 1141 GPa and ν_i= 0.07 [19, 20], and for epoxy used here, ν=0.3.

2.3.3 Scratch test

Here in this study, the tests were performed with a Rockwell-shaped diamond indenter tip (radius of 10 um) continuously loaded with a normal force of 200 uN. The transverse velocity of the sample to the diamond was 10 mm·min^{-1}, and the loading rate, dF_N/dx=10 N·mm^{-1}. The scratches were 10 um long. To ensure the homogeneity of HNT in composites, 5 scratches with spacing of 2 um were performed. An estimation of the specific scratch wear rate, w_r, can also be determined from the final parts of the scratches when considering the loss in volume of the material over an average short distance of the scratch lave, under an average normal load F_{Nave}, that acted over this distance,

$$w_r = \Delta V / (F_{Nave} l_{ave}) = C_r l_{ave} / (F_{Nave} l_{ave}) = C_r / F_{Nave} \quad \text{(Equation 4)}$$

where C_r refers to the cross-section of the scratch shortly before its end. The cross section can be assumed to have the shape of a segment of a circle, with a chord length equal to w (residual width) and height equal to the scratch depth h,

$$C_r = h(3h^2 + 4w^2)/(6w) \quad \text{(Equation 5)}$$

Assuming further that F_{Nave} is nearly equal to F_N at the end of the scratch, the specific scratch wear rate reduces to,

$$w_r = C_r / F_N \quad \text{(Equation 6)}$$

3. RESULTS

3.1 Halloysite Nanotubes

HNT is composed of alumina-silicate clay with micro hollow and nano tubular structure, and chemically similar to Kaolinite. The neighboring silica and alumina layers, and their waters hydration, curve and form multilayer tubes due to packing disorder. Monolayer HNT ($Al_2Si_2O_5(OH)_4 \cdot nH_2O$) refers to the so-called halloysite-10 Å (n=2). Heating HNT dehydrates it to halloysite-7 Å (n=0). A halloysite wall contains 15-20 bilayers (aluminum and silicon oxides). Chemically, the outer surface of the HNT has properties similar to SiO_2 while the inner cylinder core is related to Al_2O_3 (Figure 1). The charge (zeta potential) behavior of the halloysite particles can be roughly described by superposition of mostly negative (at pH 6-7) surface potential of SiO_2, with a small contribution from the positive Al_2O_3 inner surface (at pH 2-8). Three-point bending tests performed on individual nanotubes using an atomic force microscope tip was conducted [22]. The calculated elastic modulus goes up to 460 GPa, measured for a set of tubes with outer diameters ranging between 50 and 160 nm. Notice that the highest modulus for carbon nanotubes and graphene, the elastic modulus reaches 1000 GPa; however the price is by average 25000 higher. Therefore this kind of ceramic particle provides high potential in mechanical applications and this study provides one case about abrasion resistance.

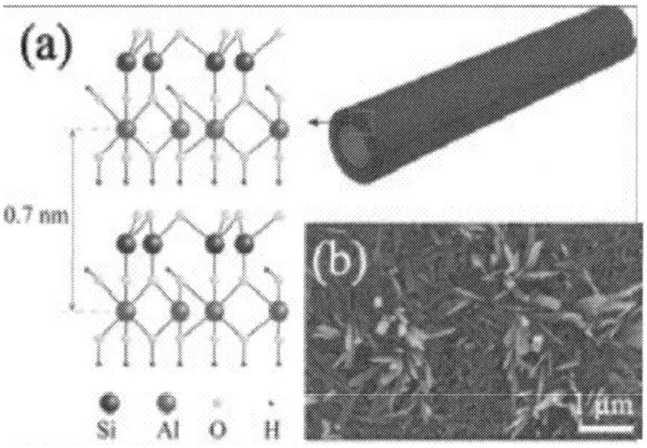

Figure 1 (a) Crystal structure of HNT and (b) SEM image of HNTs of various conformations

3.2 TriboIndenter: indentation and scratch

Figure 2 depicts the effects of the fiber volume fraction and alignment effects on hardness. Modulus showed the same trend but not listed here since hardness is generally believed to relate to abrasion resistance. From the figure it can be observed that disregarding the HNT orientation induced by various viscosities, the composites showed linear increases up to 1 vol% (i.e., 2 wt%). Beyond this concentration, the properties the hardness values displayed a plateau region, defining 1 vol% the percolation limit. Generally, for tubular particle filled polymer composites, particle length, volume fraction and orientation determine the final bulk mechanical properties[23]. For example, with an increase in HNT concentrations, the load carrying capability improved; however, at the same time dispersion quality and interactions between polymer and HNT deteriorated. A balance of these factors limited the consistent increases in both modulus and hardness parameters at high HNT loadings. For the HNTs used here, the dimensions and aspect ratios are fixed (length efficiency factor was calculated to be around 0.9) and thus have no effects in different samples. Therefore, orientation factor will be examined here regarding their influences on mechanical properties.

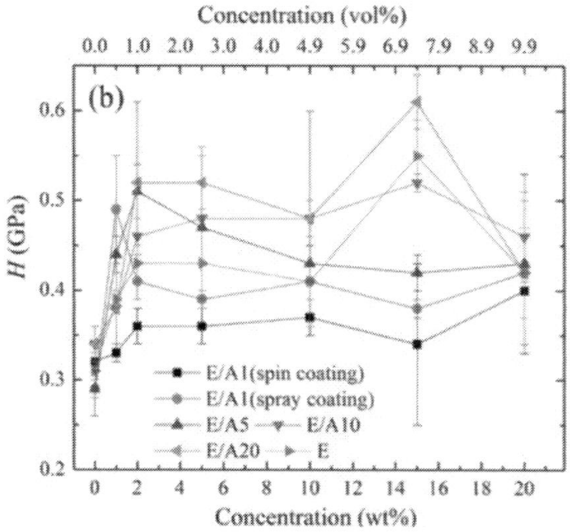

Figure 2 Hardness changes with HNT concentration; epoxy: acetone ratios ranges from 1 to 20 to pure epoxy. The increases of viscosity in sprayed solutions are helpful for the HNT alignment.

3.3 TriboIndenter scratching and Atomic force microscopy (AFM): topography and profilometry

The calculated wear rates were plotted in Figure 3. Very distinguishably, with the increase of HNT concentrations, the composites showed lower wear rates; in other words, the inclusion of HNT in all composites retards the abrasion procedure and improved the wearing resistance. On the other hand, with higher HNT alignment facilitated by higher viscosity, the abrasion was also improved. However, the degree of orientation needs to be further studied before the hypothesis of varied HNT alignment.

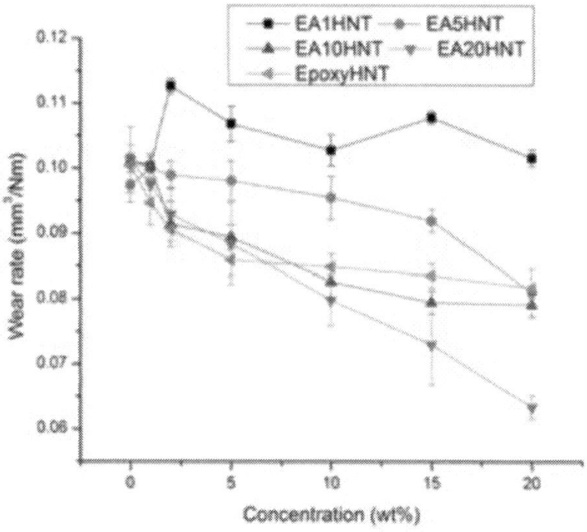

Figure 3 Wear rate vs. concentration relationship in all epoxy composites; epoxy/acetone ratios ranges to show the HNT alignment effects on wear resistance.

3.4 Scanning electron microscopy (SEM): dispersion and alignment

To examine the viscosity and concentration influences on HNT alignment in spray coating, SEM imaging in composites were conducted (Figure 4). It can be seen that with the increase of viscosity, the alignment of HNT was improved. The addition of acetone decreased the viscosity, and improved the ease of processing; unsurprisingly, the decreased viscosity reduced the constraining of HNTs and cannot retain the orientation induced by the aerodynamic flow. As shown in Figure 4, when samples change from epoxy/acetone ratio of 5 to pure epoxy, the alignment of HNT in most fracture surface alters from lying within plane to being parallel to the plane-normal direction.

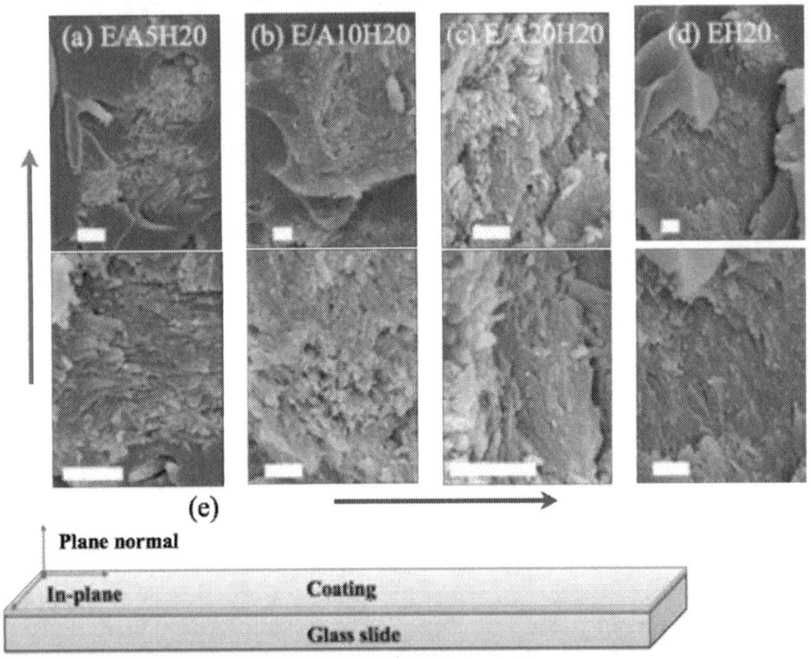

Figure 4 Viscosity influence on particle alignment in (a) E/A5H20, (b) E/A10H20, (c)E/A20H20 and (d) EH20. Viscosity increase from a to d signifies a higher constraining effect, which will retain the shear flow assisted HNT alignment. All concentrations of HNT are 20 wt% to exclude concentration influences.

4. CONCLUSIONS

The epoxy/HNT composites were prepared using spray coating method. This procedure takes advantage of aerodynamic flow to align the particles; while the presence of viscosity retains the HNT orientations. SEM images have shown the increase of alignment with higher viscosity and dampening property can also be used for proof of improved HNT orientations. The reinforcement in hardness has also been proven from indentation tests. Besides, scratching tests showed consistent trend. In summary, this study has focused on the HNT alignment effects on abrasion resistance and provides a new perspective for optimal coating design, simple processing and feasible characterizations.

5. REFERENCES

1. Althues, H., J. Henle, and S. Kaskel, *Functional inorganic nanofillers for transparent polymers.* Chemical Society Reviews, 2007. **36**(9): p. 1454-1465.
2. Zhang, Y., et al., *Tailoring polyacrylonitrile interfacial morphological structure by crystallization in the presence of single-wall carbon nanotubes.* ACS applied materials & interfaces, 2013. **5**(3): p. 807-814.
3. Schulz, U., et al., *Antireflection of transparent polymers by advanced plasma etching procedures.* Optics express, 2007. **15**(20): p. 13108-13113.
4. Malucelli, G. and F. Marino, *Abrasion resistance of polymer nanocomposites–a review.* Abrasion resistance of materials, Intech-Open Access Publisher, Rijeka, 2012: p. 1-18.
5. Tajaddod, N., et al., *Exfoliation of Boron Nitride Platelets by Enhanced Interfacial Interaction with Polyethylene.* Macromolecular Materials and Engineering, 2016. **30**(3): p. 315-327.
6. Raut, H.K., et al., *Anti-reflective coatings: A critical, in-depth review.* Energy & Environmental Science, 2011. **4**(10): p. 3779-3804.

7. Song, K., Y. Zhang, and M.L. Minus, *Using Low Concentrations of Nano-Carbons to Induce Polymer Self-Reinforcement of Composites for High-Performance Applications.* MRS Online Proceedings Library, 2015. **1752**: p. 137-144.
8. Song, K., et al., *Spectral Analysis of Lamellae Evolution and Constraining Effects Aided by Nano-carbons: A Coupled Experimental and Simulation Study.* Polymer, 2015. **75**: p. 187-198.
9. Meng, J., et al., *Forming Crystalline Polymer‐Nano Interphase Structures for High‐Modulus and High‐Tensile/Strength Composite Fibers.* Macromolecular Materials and Engineering, 2014. **299**(2): p. 144-153.
10. Wu, L., X. Guo, and J. Zhang, *Abrasive resistant coatings—a review.* Lubricants, 2014. **2**(2): p. 66-89.
11. Song, K., et al., *Lubrication of poly (vinyl alcohol) chain orientation by carbon nano‐chips in composite tapes.* Journal of Applied Polymer Science, 2013. **127**(4): p. 2977-2982.
12. Zhang, Y., et al., *Low temperature graphitization of interphase polyacrylonitrile (PAN).* Carbon, 2015. **91**: p. 479-493.
13. Song, K., et al., *Application of the Electron Density Correlation Function for Structural Analysis of X-ray Scattering/Diffraction Information from Polymer-based Nano-Composites.* MRS Online Proceedings Library, 2015. **1754**: p. 147-152.
14. Song, K., Y. Zhang, and M.L. Minus, *Polymer Interphase Self-Reinforcement and Strengthening Mechanisms in Low-Loaded Nanocomposite Fibers.* Macromolecular Chemistry and Physics, 2015. **216**(12): p. 1313-1320.
15. Luo, Y., et al., *Advances of drag-reducing surface technologies in turbulence based on boundary layer control.* Journal of Hydrodynamics, Ser. B, 2015. **27**(4): p. 473-487.
16. Song, K., et al., *Structural Polymer-based Carbon Nanotube Composite Fibers: Understanding the Processing–Structure–Performance Relationship.* Mater., 2013. **6**(6): p. 2543-2577.
17. Song, K., et al., *Spray-Coated Halloysite-Epoxy Composites: a Means to Create Mechanically Robust, Vertically Aligned Nanotube Composites.* ACS Applied Materials & Interfaces, 2016. **8**(31): p. 20396–20406.
18. Song, K., et al., *Improving Wear Resistance and Preserving Transparency in Composite Coatings By Tailoring Nanotube Orientations.* ACS Applied Materials & Interfaces, 2016: p. in Submission.
19. Hurang Hu, Landon Onyebueke, and A. Abatan, *Characterizing and Modeling Mechanical Properties of Nanocomposites-Review and Evaluation.* Journal of Minerals and Materials Characterization and Engineering, 2010. **9**(4): p. 275-319.
20. Bhushan, B. and X. Li, *Nanomechanical characterisation of solid surfaces and thin films.* International Materials Reviews, 2003. **48**(3): p. 125-164.
21. Oliver, W.C. and G.M. Pharr, *An improved technique for determining hardness and elastic modulus using load and displacement sensing indentation experiments.* Journal of Materials Research, 1992. **7**(06): p. 1564-1583.
22. Lecouvet, B., et al., *Elastic Modulus of Halloysite Nanotubes.* Nanotech., 2013. **24**(10): p. 105704-105711.
23. Halpin, J.C. and S.W. Tsai, *Environmental Factors in Composite Materials Design.* U.S. Air Force Tech. Rep. AFML TR, 1967. **1**: p. 1-62.

472025 Impedance Analysis of Ion Transport through Supported Lipid Bilayers on Accessible Mesoporous Silica Thin Films

Abstract:

As the barriers between cells and external environment, biological membranes exhibit highly selective permeability to molecules and ions, which is maintained by its main component, amphipathic lipid molecules and the embedded membrane proteins. Supported lipid bilayers allow the use of surface analysis techniques and are model systems for exploring the application of lipid bilayers in biosensing, selective separations and drug delivery. Mesoporous silica thin films are advantageous supports due their biocompatible properties and nanoporous architecture, which creates space for the incorporation of proteins and provides molecules or ions reservoirs underneath the lipid bilayer. Combining the high surface area silica thin film with the selective permeable lipid bilayers suggests potential applications for efficient selective aqueous solute separations and biosensing. While this biomimetic membranes hold great potential, an important step in these applications is to thoroughly examine the transport processes in composite synthetic membranes. In transport applications, the pore accessibility of silica thin film is a priority. In this work, silica thin films with orthogonally aligned hexagonal pore arrays, which offer continuous channels for molecule transport, are prepared. The thin film pore accessibility is characterized by electrochemical impedance spectroscopy (EIS) technique. Furthermore, the effect of lipid bilayer preparation methods on the silica thin film (in the form of pore enveloping, pore filling, tethering) on ion transport is explored, as a lipid bilayer with high electrically insulation is essential for detecting activity of proteins or biomimetic carriers in the bilayer. This study provides insights for making better barriers on mesoporous support for carrier-mediated membrane separation process.

469986 Electrochemical Gating of Charge Transport in Radical Polymers for Colorless, Transparent, and Ambipolar Organic Transistors

Abstract:

For more than 30 years, closed-shell macromolecules have dominated as the preferred charge transporting active layer materials for numerous organic electronic applications. While the progress that has been made in a number of organic electronic applications [e.g., organic photovoltaic (OPV) devices and organic field-effect transistors (OFETs)] has been impressive, there are still a number of challenges that are not readily overcome through the use of *π*-conjugated polymers. In order to address these critical needs, various non-conjugated oxidation-reduction (redox) active polymers are of considerable interest as a new class of charge transporting channels. Here, we demonstrate that many of these challenges can be addressed in a ready manner through the introduction of an emerging class of non-conjugated, but open-shelled, charge transporting macromolecules, radical polymers. Radical polymers are non-conjugated macromolecules that bear stable pendent radical groups. In the solid state, the populated unpaired electrons on the side chain of the macromolecule undergo rapid, but reversible, redox charge exchange without any breaking or formation of chemical bonds. Importantly, because these organic radicals of the valence shell electrons are thermodynamically stabilized, the electron transfer between radical sites is persistent even under ambient conditions, and our laboratory has established that the charge mobility and conductivity values of certain radical polymers are on the same order of magnitude as common conjugated semiconducting polymers. Moreover, because the radical polymers are not conjugated, they are colorless and transparent to the eye. Thus, there is a great promise in using these materials in myriad solid-state organic electronic applications; however, to date, the radical polymer community mainly has been focused on using these next-generation materials in wet-type devices for energy storage or conversion [e.g., organic radical batteries]. Therefore, it is of the utmost import to elucidate the fundamental charge transport mechanism associated with radical polymers such that new macromolecular designs and device applications emerge in a rapid manner.

We address this issue specifically in this work. Precisely, the charge transport mechanism of a model radical polymer, poly(2,2,6,6-tetramethylpiperidine-1-oxyl methacrylate) (PTMA), is definitely established for the first time through the introduce of an electrochemically-gated organic thin film transistor structure. In order to uncover the charge hopping mechanism of radial polymers in an unambiguous manner, an ion gel gate is applied to an organic transistor that has an active layer composed of PTMA. We fabricated the device structure by coating a PTMA channel layer on bottom gold electrodes, and "cut and stick" the ion gel (consisting of 1-ethyl-3-methylimidazolium tris(pentafluoroethyl)trifluorophosphate ([EMI][FAP]) organic ions and a poly(styrene-*b*-ethylene oxide-*b*-styrene) [PS-PEO-PS] triblock copolymer), which acts as a thin, transparent, and colorless gate. This gate electrolyte provides for electrochemical ion (cation or anion) penetration into the main channel layer through application of a gate bias. For instance, upon application of a positive gate bias, the nitroxide radicals of the PTMA undergo oxidation to oxoammonium cations, which act as dopant sites, and these sites facilitate hole charge (p-type) transport between the source and drain electrodes of the transistor. Furthermore, because of the ability of PTMA to undergo oxidation or reduction, application of a negative gate bias affords a transistor with n-type behavior as well. This first demonstration of using a radical polymer in an electrochemically-gated transistor reveals the potential of PTMA as a main charge carrier material, and stimulates the new paradigm of open-shelled, charge-transporting macromolecules. In addition, in either electrochemically doping instance, the conductivity of the PTMA layer is increased by a factor of 100 – 1,000 after application of relatively low gate biases (V_g = ± 2 V), which leads to ON/OFF current ratios of ~10^4. Therefore, this talk will both demonstrate the fundamental aspects of charge transport in a new class of electronically-active materials, radical polymers, and it will also prove the great potential that exists for these devices to be utilized as logic switches in completely colorless and transparent organic electronic applications.

465694 Direct, Single-Step Alignment of Solution-Sheared Donor-Acceptor Polymer Thin Films and Factors Influencing Their Deposition

Abstract:

Organic semiconductors (OSCs) have properties that allow them to fill an application space beyond silicon. Continuous fabrication processes, such as roll-to-roll printing, are highly desirable for lowering the industrial and economic barriers that impede a more widespread deployment of organic electronics in the marketplace. Solution shearing is a deposition technique developed by our group that not only can be easily adapted to such processes, but also has been demonstrated to produce highly aligned thin films of small molecule OSCs. Such alignment of OSCs in the active layers of electronic devices can impart desirable properties, such as polarized light emission or absorption, reduced device cross-talk due to charge transport, and enhanced charge transport due to better ordering.

While the alignment of small molecule OSCs has been widely demonstrated, we have also observed alignment in a donor-acceptor polymer based on a diketopyrrolopyrrole-terthiophene backbone deposited directly from solution without surface templating or complex pre- or postdeposition processing Ââ€“ a dichroic ratio up to ~7 was achieved with the solvent tetralin. The degree of alignment in the thin films were tuned by controlling the coating speed and was found to be a competition between the shear alignment of polymer chains in solution and the complex thin film drying process. Unlike previous reports, no charge transport anisotropy was observed because of the filmsâ€™ meshlike morphology, small crystallite size relative to the channel length, and the likelihood of more grain boundaries perpendicular to the coating direction. Coupled with the observed anisotropy in X-ray diffraction data, these results are consistent with polymer alignment in both the amorphous and crystalline regions of the films.

The shear speed at which maximum dichroism is achieved can be tuned with deposition parameters like temperature and substrate treatment. Longer polymer alkyl side chains were found to reduce the degree of alignment, and a change in molecular weight showed negligible effects on alignment. Solution shearing, our work shows, can be used to tune

polymer film alignment in a single-step process without relying on substrate patterning or postdeposition treatments.

467758 Influence of Crystalline Anisotropy on Localized Surface Plasmon Resonance of Semiconductor Nanocrystals

Abstract:

Doped semiconductor nanocrystals have been shown to host a tunable localized surface plasmon resonance (LSPR) over a wide optical range depending upon controlled dopant and carrier concentration. Metal nanoparticles such as gold and silver, have high free electron densities (~$10^{23}/cm^3$) creating resonances with Ï‰$_{lsp}$ in the visible region whereas, in doped semiconductor nanocrystals, a highly variable carrier density, (10^{18}~$10^{22}/cm^3$) enables Ï‰$_{lsp}$ over the entire infrared region.Â Employing anisotropic particle shapes in metal nanoparticles and nanostructures have given an additional control over Ï‰$_{lsp.}$ For instance, in metal nanostructures, it has allowed LSPR band spitting and realization of resonance in near-infrared for high aspect ratio nanorods. In semiconductor nanocrystals, studies so far have focused on tuning LSPR frequency (Ï‰$_{lsp}$) and the influence of anisotropic nanocrystal shape and intrinsic crystal structure remain poorly explored.

Unlike typical plasmonic metals like Ag or Au, doped semiconductors can have anisotropic crystal structures for example, $Cu_{2-x}S$ (layered) or Cs:WO_3 (hexagonal), which is the focus of our current study. Here in this study, we have shown that colloidally synthesized hexagonal phase Cs:WO_3 nanocrystals exhibit strong aspect ratio-dependent LSPR absorption peaks that can only be explained via a cooperative influence of crystalline and shape anisotropies. We demonstrate the dominant influence of crystalline anisotropy, which uniquely causes strong LSPR band-splitting into two distinct peaks with comparable intensities. This finding highlights the limitations of conventional treatments of LSPR that assume isotropic dielectric constants and attribute multimodal peaks uniquely to shape anisotropy effects. This understanding extends our ability to rationally tune LSPR lineshape and near-field enhancement via synthetic control of shape and crystalline anisotropies of semiconductor nanocrystals. In particular, the demonstrated multimodal

LSPR with near-equal intensities of h-Cs:WO_3 nanocrystals covers the near-infrared (NIR) region of great importance in photonic, solar, and clinical applications while maintaining high visible transparency due to its wide band gap.

456858 Inorganic Chiroptical Nanomaterials: Design Strategies and Origin of Homochirality

Abstract:

Recently, chirality of inorganic nanomaterials has been considered of much importance because they are of prime fundamental and practical interest due to the favorable power-law scaling of near-field enhancements[1]. The interest in the synthesis of chiral nanostructures has been fueled by the potential application of chiral nanostructures in biosensing, telecommunication, display technologies, diffraction-free patterning, and chiral catalysis[2]. Although chirality is often associated with biochemistry due to numerous chiral bio-molecules, today chiral inorganic nanostructures have attracted much attention as such their optical properties remain largely unexplored.

A further motivation to study chirality of inorganic nanostructures is to discover the origin of homochirality in natural compounds. The prevalence of only L-type amino acids and D-type sugars in nature is well-known example of homochirality. This dominant existence of only one of the two enantiomers among natural products has kept scientific attentions for decades. Several chemical routes are being debated, including chiral amplification and influence of circularly polarized light (CPL) from the cosmos. However, the circular polarization of cosmic light is dominantly observed in infrared range that is not absorbed by simple organic molecules.

In the talk, I will introduce a new strategy for synthesizing chiral semiconducting[2] and metallic nanostructures using CPL. This demonstration is showing the ability of nanoparticles that imprinting circular polarization information on their geometries, and also opens a door to understand the origin of homochirality.

References

1. Kuzyk, A. *et al.* DNA-based self-assembly of chiral plasmonic nanostructures with tailored optical response. *Nature* **483,** 311â€"4 (2012).
2. Yeom, J. *et al.* Chiral templating of self-assembling nanostructures by circularly polarized light. *Nat. Mater.* **14**, 66â€"72 (2015).

469343 Biomaterials for Human Pluripotent Stem Cell Derived Midbrain Dopaminergic Neuron Generation and Transplantation to Treat Parkinson's Disease

Abstract:

Cell replacement therapies have considerable biomedical potential; however, low survival of cells post-transplantation is a major hurdle hampering their efficacy. For example, cell replacement therapies have strong promise for the treatment of neurodegenerative diseases such as Parkinson's, which is a currently incurable, progressively debilitating disorder affecting ~10M people in the world. However, following striatal transplantation of midbrain dopaminergic (mDA) neurons for the treatment of Parkinson's disease (PD), less than 5% of the neurons typically survive pre-clinically. Such poor neuronal survival requires the resource-intensive generation and implantation of larger numbers of mDA neurons to compensate, and the widespread death of the majority of the graft could reduce the efficacy of the cells remaining. We took a two-pronged approach to resolve this issue by developing methods to first increase the yield of human pluripotent stem cell (hPSC) derived mDA neurons generated in vitro, and second, to increase the post-transplantation survival of neurons in vivo.

Human pluripotent stem cells (hPSCs) were cultured and differentiated into mDA neurons in 3D pNIPAAm-PEG gels. The authenticity of resulting neurons thoroughly characterized by immunocytochemistry, qPCR and voltage-sensitive dye based electrophysiology, and benchmarked against mDA neuronal development on 2D surfaces. Subsequently, a biodegradable, hyaluronic acid (HA) based hydrogel functionalized with ECM derived factors was developed for maturation and transplantation of mDA neurons generated on pNIPAAm-PEG. Post-transplantation survival of hydrogel-encapsulated neurons was then investigated 4.5 months after striatal implantation into rats.

mDA neuron production typically involves undefined components and difficult to scale-up 2D culture formats. Here, we used a fully-defined, 3D thermoresponsive biomaterial platform to derive mDA neurons from PSCs at substantially greater yield, quality, and maturity compared to 2D Matrigel coated surfaces. In particular, after 25 days of differentiation, ~40% of cells differentiated in 3D expressed tyrosine hydroxylase (TH), the rate limiting enzyme for dopamine production and a hallmark of mDA neurons, compared to 20% on 2D. Crucial for functional mDA neurons, ~ 5-fold higher proportion of neurons differentiated in 3D fired mDA-like action potentials. Interestingly, FOXA2 and EN1, markers known to improve survival of mDA neurons, were expressed at significantly higher levels in cells generated in the 3D platform compared to control 2D surfaces.

Low post-transplantation cell survival is often attributed to adverse biochemical, mechanical, and immunological stresses cells experience during and post-transplantation. To address these challenges, we developed a functionalized hyaluronic acid (HA) based hydrogel first to mature pluripotent stem cell (PSC) derived neural progenitors towards functional neurons in vitro in a biomimetic 3D environment, and second to effectively transplant hydrogel-encapsulated neurons into the central nervous system. We demonstrated that incorporating adhesive peptides to promote cell adhesion increased extended neurite density, and adding neurotrophic factors increased dopaminergic differentiation in our 3D HA gels. Furthermore, we observed higher survival post-harvest and injection in vitro for mDA neurons matured within hydrogels compared to those on 2D surfaces. By using an optimized 3D HA biomaterial matrix for neuronal maturation and transplantation, we significantly increased the post-transplantation survival of hPSC derived mDA neurons in the rat striatum by 5 fold compared to the current standard method of bolus injections of unencapsulated neurons.

The combination of a defined, scalable, and resource-efficient cell culture platform and an optimally engineered transplantation biomaterial significantly increased the yield of mDA neurons generated in vitro and improved their post-transplantation viability in vivo, thereby addressing crucial challenges in treatment of neurodegenerative disease.

469101 Engineering an Electroactive Hydrogel for Tissue Engineering Applications

Abstract:

Engineering an Electroactive Hydrogel for Tissue Engineering Applications

Andrew Spencer[1] Hicham Fenniri[1*], Nasim Annabi[1,2,3**]

[1] Department of Chemical Engineering, Northeastern University, Boston, MA, 02115-5000, USA

[2]Biomaterials Innovation Research Center, Brigham and WomenÕs Hospital, Harvard Medical School, Boston, MA, USA.

[3]Harvard-MIT Division of Health Sciences and Technology, Massachusetts Institute of Technology, Cambridge, MA, USA.

*h.fenniri@neu.edu;**n.annabi@neu.edu

INTRODUCTION

Hydrogels are hydrophilic three dimensional (3D) polymeric networks with physical, chemical and biological properties that can be designed to closely mimic the native extracellular matrix (ECM) of mammalian tissues. Because of their biomimetic properties, hydrogels are often used in various tissue engineering and regenerative medicine applications. For tissues whose function is highly dependent on electrochemical signaling between cells and that are responsive to an electric field, such as the heart, it is desired that the hydrogels are electroactive [1]. However, most hydrogels are inherently non-conductive. Numerous approaches have been taken to improve the conductivity of hydrogels using materials that have limited biocompatibility and biodegradability.

There exists a library of conducting organic polymers that offer an attractive alternative to other conductive materials for biomedical applications, due to their ease of processing, versatility, and enhanced biocompatibility both *in vitro* and *in vivo* [2]. Unlike many other conductive materials, such as carbon nanotubes, these conductive polymers can be easily dispersed in aqueous solutions Đ a requirement for biological applications. In addition to their practical advantages, they actuate upon electrical stimulation [3]. This enables simultaneous electrical and mechanical stimulation of cells, which may help to maintain their phenotype and facilitate their growth.

By dispersing a well-established conducting polymer complex, poly(3,4-ethylenedioxythiophene):polystyrene sulfonate (PEDOT:PSS), at various concentrations within a photo-crosslinkable naturally derived polymer, gelatin methacrylate (GelMA), we have synthesized an electroactive and flexible hydrogel that can be used for 3D cell encapsulation. The engineered 3D conductive hydrogel with tunable mechanical and electrical properties can be used for various tissue engineering applications, including muscle tissue regeneration and heart repair.

MATERIALS AND METHODS

Hydrogel formation and characterization: Gelatin methacrylate (GelMA) was synthesized according to a previously reported protocol [4]. PEDOT:PSS was obtained from Heraeus (Germany) in the form of an aqueous dispersion called CleviosTM PH 1000 (PEDOT:PSS ratio 1:2.5). The concentration of the dispersion in GelMA prepolymer after filtering through a 0.22 mm filter was determined to be 0.95 wt% after drying and weighing. Solutions of 7% GelMA at 0%, 0.1% and 0.3% PEDOT:PSS with 0.5% photoinitiator (Irgacure 2959) were prepared for all experiments. The hydrogels (thickness: 2 mm, diameter: 4 mm) were formed by exposing the GelMA/PEDOT:PSS solutions to UV light at 6.9 mW/cm^2 for 360 s. Then, the compressive properties of the engineered hydrogels were tested using an Instron 5944 mechanical tester. Compressive modulus was taken as the slope of the linear portion of the stress/strain curves at 0-5% strain.

3D Cell Encapsulation: C2C12 myoblast cells (ATCC) were cultured in complete media (DMEM + 10% FBS + 1% penicillin/streptomycin) in a humidified 5% CO_2 incubator prior to use. Cell were trypsinized and resuspended in solutions of GelMA/PEDOT:PSS precursor at concentration of 10^6 cells/mL of prepolymer solution. 7 mL of the mixture was then sandwiched between a glass slide and a petri dish with a 150 mm spacer and exposed to UV light for up to 30 s to crosslink to form a cell-laden hydrogel. Live/Dead, Actin/DAPI and PrestoBlue assays were performed on days 1, 3 and 5 to evaluate the cell viability, spreading, and proliferation in 3D hydrogels with varying mechanical and electrical properties.

RESULTS AND DISCUSSION

Mechanical testing demonstrated that the compressive modulus increases with increasing concentration of PEDOT:PSS. This might be due the interaction between the negatively charged PSS and the positively charged arginine or lysine groups in GelMA, adding physical crosslinking to the composite hydrogel in addition to chemical bonds generated during photopolymerization. The *in vitro* data also showed that incorporation of PEDOT:PSS into the hydrogels did not affect the metabolic activity of the cells, as confirmed by PrestoBlue cell viability assay. Live/Dead assay also confirmed that the incorporation of PEDOT:PSS into GelMA did not affect cell viability. In addition, the cells start spreading inside the engineered hydrogels at day 3 of culture. These results suggest that the incorporation of conductive polymer complex PEDOT:PSS created a biocompatible conductive hydrogel that will enable uniform electrical stimulation of electrically responsive tissues.

CONCLUSION

An electroactive hydrogel was developed via the incorporation of a conducting polymer complex into a naturally derived hydrogel. Compressive modulus was improved via the addition of PEDOT:PSS into the hydrogel. Cell viability, metabolic activity, and spreading were not affected by the incorporation of the conductive material into the hydrogel network as shown by PrestoBlue, Live/Dead and Actin/DAPI assays. These results suggest that the engineered composite hydrogel is suitable for various tissue engineering applications where the tissues are electrically responsive.

ACKNOWLEDGMENT

The authors would like to thank Northeastern University for funding this research.

REFERENCE

[1] N. Annabi, A. Tamayol, J. Uquillas, M. Akbari, L. Bertassoni, C. Cha, G. Camci_Unal, M. Dokmeci, N. Peppas, A. Khademhosseini, *Adv Mater* **2014.**

[2] R. Balint, N. Cassidy, S. Cartmell, *Acta Biomaterialia* **2014**

[3] M. R. Abidian, D. _H. Kim, D. C. Martin, *Adv Mater* **2006**

[4] W. Nichol, S. T. Koshy, H. Bae, C. M. Hwang, S. Yamanlar, A. Khademhosseini, *Biomaterials* **2010**

471892 Modulation of Inflammatory Response for Accelerated Tissue Vascularization and Bone Regeneration

Abstract:

There is a profound need for bone regeneration strategies due to trauma and various musculoskeletal diseases. Tissue engineering offer alternative approaches to regenerate tissues. The transfer of tissue engineering from the lab to the clinic is limited by the failure to adequately vascularize scaffolds and control immune response. The goal of this study is to harness the wound healing potential of macrophages and human mesenchymal stem cells (hMSCs) to enhance directed angiogenesis and bone formation. We explored the use of resveratrol, a natural compound commonly used in complementary medicine, to mediate M1 to M2 macrophage plasticity, impart natural release of angiogenic factors by macrophages and enhance osteogenic differentiation of hMSCs. To this end, we mapped the time-dependent response of macrophage gene expression as well as hMSC osteogenic differentiation to varying doses of resveratrol in 2D and 3D biomaterial environments. The results demonstrated that macrophages adopted a pro-healing phenotype in the presence of resveratrol, as confirmed by the expression of IL-6 and TNF-Î±. This effect was coupled with higher expression of anti-inflammatory markers VEGF and IL-10. In addition, hMSCs exposed to resveratrol demonstrated a higher level of calcium deposition and alkaline phosphatase activity. Our results altogether delineate the potential to synergistically accelerate angiogenic factor release and upregulate osteogenic signaling pathways by â€œdialingâ€ the appropriate degree of resveratrol release.

AUTHOR INDEX

AUTHOR INDEX

AUTHOR INDEX

AUTHOR INDEX

AUTHOR INDEX

AUTHOR INDEX

AUTHOR INDEX

AUTHOR INDEX

AUTHOR INDEX

AUTHOR INDEX

AUTHOR INDEX

AUTHOR INDEX

AUTHOR INDEX

AUTHOR INDEX

AUTHOR INDEX

AUTHOR INDEX

AUTHOR INDEX

AUTHOR INDEX

AUTHOR INDEX

AUTHOR INDEX

AUTHOR INDEX

AIChE
120 Wall Street, FL 23
New York, NY 10005-4020

ISBN 978-1-5108-3440-8

CURRAN ASSOCIATES INC.
proceedings
.com

Materials Engineering and Sciences Division 2016

Core Programming Area at the 2016 AIChE Annual Meeting

San Francisco, California, USA
13 - 18 November 2016

Volume 2 of 2

AIChE
The Global Home of Chemical Engineers

Materials Engineering and Sciences Division 2016

Core Programming Area at the 2016 AIChE Annual Meeting

San Francisco, California, USA
13 - 18 November 2016

Volume 2 of 2

ISBN: 978-1-5108-3440-8

Printed from e-media with permission by:

Curran Associates, Inc.
57 Morehouse Lane
Red Hook, NY 12571

Some format issues inherent in the e-media version may also appear in this print version.

Printed by Curran Associates, Inc. (2017)

For permission requests, please contact AIChE
at the address below.

AIChE
120 Wall Street, FL 23
New York, NY 10005-4020

Phone: (800) 242-4363
Fax: (203) 775-5177

www.aiche.org

Additional copies of this publication are available from:

Curran Associates, Inc.
57 Morehouse Lane
Red Hook, NY 12571 USA
Phone: 845-758-0400
Fax: 845-758-2633
Email: curran@proceedings.com
Web: www.proceedings.com

TABLE OF CONTENTS

VOLUME 1

VOLUME 2

463526 The Impact of Decellularization Agents on Renal Tissue Extracellular Matrix

Abstract:

The impact of decellularization agents on renal tissue extracellular matrix

Nafiseh Poornejad, Lara B. Schaumann, Travis Neuberger, Sarah Chambers, Beverly L. Roeder, Alonzo D. Cook

Brigham Young University, Provo, Utah

The combination of patient-specific cells with scaffolds obtained from natural sources could result in improved regeneration of human tissues. Decellularization of the native tissue is the first step in this technology. Effective decellularization uses agents that lyse cells and remove all cellular materials, leaving intact collagenous extracellular matrices (ECMs). Although complete removal of cellular remnants should be achieved to prevent the induction of an immune response, preserving the underlying structure and basement membrane components are of crucial importance to retain regenerative potential of a decellularized scaffold. Consequently, enough care should be taken in choosing cell-lysing agents and the procedure should be optimized for different tissues.

In this study, the impact of five decellularization agents (0.1 N NaOH, 1% peracetic acid, 3% Triton X-100, 1% sodium dodecyl sulfate (SDS), and 0.05% trypsin/EDTA) on renal tissue was examined to comprehensively compare a broad range of agents including acidic/basic solution, ionic/non-ionic detergents, and protease enzyme. The NaOH solution induced the most efficient cell removal (>90% DNA removal, p-value<0.05), and resulted in the highest amount of human renal cell viability and proliferation after recellularization, although it also produced the most significant damage to collagenous fiber networks, glycosaminoglycans (GAGs), and fibroblast growth factor (FGF). The SDS solution led to less severe damage to the ECM structure, but it resulted in less proliferation of reseeded cells.

Peracetic acid and Triton X-100 preserved ECM components intact. However, these agents alone were not efficient in removing cellular materials; especially peracetic acid, which left more than 80% of cellular remnants within the ECM. Trypsin enzyme didnâ€™t eliminate cellular materials and resulted in significant damage to the collagenous structure of ECM.

In summary, we suggest NaOH as the most efficient agent to remove cellular materials combined with non-ionic Triton X-100 as an optimum procedure for decellularization of renal tissue.

472692 Electrospun Silk with Selenium Nanoparticles for Antibacterial Skin Applications

Abstract:

Electrospun Silk with Selenium Nanoparticles for Antibacterial Skin Applications

Stanley Chung[1], Thomas J. Webster[1-3]
1. Department of Chemical Engineering, Northeastern University, Boston, MA
2. Department of Bioengineering, Northeastern University, Boston, MA
3. Center of Excellence for Advanced Materials Research, King Abdulaziz University, Jeddah, Saudi Arabia

Skin infections have accelerated in the past decade, often caused by *Staphylococcus aureus*, the leading cause of skin and soft tissue infections (SSTI). In addition, strains of *S. aureus* are developing resistance to most conventional antibiotic treatment options. Antibiotic resistance is a growing epidemic and is estimated by the Centers for Disease Control and Prevention to cost $20 billion annually in excess health costs and $35 billion in other societal costs. Thus, it is evident that new strategies are needed to combat this growing public health threat.

Selenium is a novel chemistry in biomedical applications and has shown beneficial properties for reducing bacterial infections. In particular, selenium nanoparticles have shown good antibacterial properties that selectively inhibit bacteria while maintaining mammalian cell growth.[1] In addition, the selenium nanoparticle coating process developed in this lab is facile and versatile, allowing researchers to deposit selenium nanoparticles as an antibacterial additive to a variety of biomaterials.[2,3]

Silk has been shown to possess many beneficial properties for skin regeneration by promoting collagen synthesis, re-epithelialization, wound healing, atopic dermatitis alleviation, and scar reduction. In particular, electrospun silk possesses morphology similar to that of native extracellular matrix in the body and is especially suitable for skin tissue engineering applications. Unfortunately, silk has also been shown to promote bacterial growth. Here, it is proposed that the combination of two biomaterials, selenium nanoparticles and silk, as a nanocomposite scaffold will promote healthy skin cell growth while reducing the bacterial load.

Silk fibroin was extracted from Bombyx mori silk worms and resuspended in formic acid at a concentration of 8% (w/v) for electrospinning. After electrospinning the silk scaffolds, selenium nanoparticles were deposited onto the silk scaffolds. *In vitro* skin cell viability and activity were then assessed by MTS assay (Promega) with passage 3-12 human dermal fibroblasts while *S. aureus* activity was assessed by an ATP assay (Promega). In all experiments, the control group comprised of cells seeded onto polystyrene tissue culture flasks. All experiments were conducted in at least triplicates and N=3.

Recent results on the cellular activity associated with electrospun nanocomposite scaffolds will be presented. Briefly, results show a statistically significant reduction in metabolic activity of *S. aureus*while promoting the metabolic activity of human dermal fibroblasts; for both the MTS and the ATP assay, addition of selenium nanoparticles further enhanced the response of the cells as compared to the control ($p<0.05$ for silk samples without nanoparticles and $p<0.01$ for silk samples with nanoparticles as compared to the control). Further studies involving microscopy and qPCR will elucidate the mechanism of action by which this bactericidal activity is achieved.

Acknowledgement:

We acknowledge the funding provided by Northeastern University.

References:

1. Tran, P. A. & Webster, T. J. Selenium nanoparticles inhibit Staphylococcus aureus growth. *Int. J. Nanomedicine* **6**,1553 (2011).

2. Wang, Q., Larese-Casanova, P. & Webster, T. J. Inhibition of various gram-positive and gram-negative bacteria growth on selenium nanoparticle coated paper towels. *Int. J. Nanomedicine* **10**,2885 (2015).

3. Wang, Q., Mejia Jaramillo, A., Pavon, J. J. & Webster, T. J. Red selenium nanoparticles and gray selenium nanorods as antibacterial coatings for PEEK medical devices. *J. Biomed. Mater. Res. Part B Appl. Biomater.* (2015).

470305 Engineered Cellulose-Based Cell Culture Platforms to Improve Human Health

Abstract:

Two-dimensional (2D) *in vitro* cell cultures do not mimic the microenvironment in the native tissues. Currently, there are several systems that are used to study cell cultures in 3D; however, these systems use complex instrumentation and expensive set-ups, require long and tedious optimization procedures, and do not yield in physiologically relevant structures. Other drawbacks of these constructs include heterogeneous distribution of cells, and potential mass transport limitations. Collectively, it is challenging to fabricate biocompatible constructs in cm-scale. We developed simple cellulose-based cell culture platforms that can be used for a wide range of applications for mammalian cells, bacteria, fungi, and plants cells. These new systems are simple, tunable, and low-cost.

We used cellulose-based materials as 3D scaffolds to support cells and hydrogels. We grew different types of cells including connective tissue cells, osteoblasts, neutrophils, gut cells, tumor cells, stem cells, plant cells, fungi, or bacteria in cellulose-based materials. Subsequently, we characterized the cell behavior by using viability assays, metabolic activity assays, immunocytochemistry, mechanical tests, X-rays scanning, and high-resolution imaging.

In this study, we successfully fabricated cellulose-based scaffolds to (i) investigate migration of primary human cells (tumor fibroblasts and lung cancer cells), (ii) induce template-guided biomineralization, (iii) generate 3D osteoblast cultures that are amenable to high-throughput sample preparation and analysis, (iv) form and control gradients of oxygen and biological molecules, and (v) develop new co-culture systems in multi-layered stacked cultures. In addition to primary cells and mammalian cell lines, we also cultured bacteria, fungi, and plant cells in cellulose-based materials. These cell cultures in cellulose resulted in high cell viability and high metabolic activity, and enabled differentiation of cells. In these experiments, we did not observe mass transport limitations to the cells. Our results have indicated that the cellulose-based materials allow for patterning and easy recovery of cells, can adapt modular configurations, and can provide physiologically relevant tissue models. The scaffolds we fabricated from thin layers of cellulose also yielded in free-standing structures in different scales (micron- to cm-).

To sum up, we developed new cellulose-based cell culture platforms to provide multicellular and compartmentalized tissue-mimetics for clinical applications. To overcome the limitations of the traditional tissue models, we developed a novel layer-by-layer approach to assemble tissue-like structures from cellulose-based low-cost materials. This strategy offers unique opportunities such as understanding fundamental biology, developing disease models for personalized medicine, and assembling different organs together.

471266 Modulating Cell-Matrix Interactions Is an Effective Method to Control Cell-Cell Junctions and Thus Improve the Predictive Capability of the Caco-2 in Vitro Model of Drug Permeability

Abstract:

Oral bioavailability is a key parameter in the development of any therapeutic compound, be it a biologic or a small molecule drug. The traditional preclinical method to screen for the oral bioavailability of a drug in vitro is the Caco-2 assay in which human colorectal adenocarcinoma cells (Caco-2) are cultured as a monolayer on top of collagen I-coated transwell inserts. While the Caco-2 assay has been a very successful in vitro predictor for many drug compounds, it has long been known to be ineffective at evaluating drugs that are transported via the paracellular route. Previous studies have demonstrated that paracellular transport is regulated by the presentation and organization of tight junction proteins at cell-cell contact points. The development and maintenance of tight junctions is influenced by focal adhesions at cell-matrix contact points, and tight junctions and focal adhesions are connected through the actin cytoskeleton. Therefore, we hypothesized that regulation of focal adhesions through matrix engineering could be leveraged to regulate tight junction formation and hence enable significant paracellular transport across Caco-2 monolayers. Protein engineering was used to synthesize a recombinant engineered extracellular matrix (eECM) based on an elastin-like amino acid sequence. Two independent matrix parameters, stiffness and density of integrin-engaging RGD ligands, were used to alter focal adhesion formation. As expected, the number of mature focal adhesions, as quantified by talin and vinculin staining, was significantly reduced on more compliant matrices and on matrices with fewer RGD ligands. Interestingly, the average Caco-2 cell area on the less adhesive eECM substrates was significantly less than that for cells grown on collagen I, and the eECM-cultured Caco-2 cells had similar cell area to primary native intestinal tissue epithelial cells. Furthermore, Caco-2 cells cultured on softer, lower integrin-engaging eECM exhibited a more small intestinal phenotype compared to collagen I-cultured Caco-2 cells, with increased expression of the tight junction marker claudin-2 and the small intestinal markers sucrase-isomaltase and aminopeptidase N. In drug permeability assays, the eECM cultured Caco-2 cells displayed a significant increase in paracellular transport of the model drugs inulin-FITC, ranitidine, famotidine, and nadolol. Notably, the use of eECM in this assay was found to not affect the permeability of dexamethasone and naproxen, drugs that are transported via transcellular routes. Our results demonstrate that the substitution of an eECM material for collagen I would be an easily implementable strategy for improving the physiological accuracy of the Caco-2 assay in preclinical screens for oral absorption of pharmaceutical drugs.

470786 Establishing Mechanically Active Synthetic Mucosal Interface in a Multi-Well Plate

Abstract:

Introduction: Human gastro-intestinal tract (GIT) accounts for more than 75% of the body surface that comes directly in contact with the external environment. This makes GIT the prime location for host-microbiome interactions, since 70-80% of immune cells are present along the tract. GIT epithelium consists of a monolayer of polarized cells arranged into microscopic features called villi and crypts, which increases the GIT surface area tremendously. Recent studies suggest that peristaltic motion is important for epithelial cell differentiation, polarization and structural arrangement into villi and crypts. Some progress has been made in recent times, there is still a growing need to develop model systems to integrate microbiome, epithelial cells, and immune cells in a single platform along with providing sufficient level of biological and mechanical complexity. To this end, we developed a mechanically actuated 3D tissue culture model system that can achieve the aforementioned integration of different cell types with multiple layer complexity in an easily adaptable well plate platform. Pneumatically powered mechanical actuator applies magnetic force that cause cyclic stretching of PDMS membrane supporting cell culture and can simulate peristaltic motion. Actuation frequency is easily controllable using Arduino software. In this study we demonstrate how this platform was used to mechanically stimulate colon carcinoma cells (HT-29, ATCC® HTB-38™) leading to polarization, vertical growth and 3D morphogenesis of these cells.

Materials and Methods: The in-vitro platform consists of an "insert" made of three parts, a PDMS membrane (100 µm thick, 100 µm pore size, and 400 µm pore separation), a medical grade stainless steel ball embedded in a PDMS peg, and a PDMS ring. The PDMS peg was bound to the PDMS membrane and the PDMS ring bound on the outer periphery of the membrane using plasma treatment (Harrick Plasma). The PDMS membrane was coated with collagen on the apical side and HT-29 cells were seeded. The 24 well plate was then placed on the mechanical actuator (at 37°C, and 5% CO_2) with mechanical actuation rate of 1 cycle/minute and cell culture medium was changed every day.

Results and Discussion: Cyclic stretching of HT-29 cells grown on top of a stretchable PDMS membrane cause vertical growth and formation of 3D morphological structures in HT-29 cells. Confocal fluorescence image confirms 3D growth of HT-29 cells. Commercial transwell inserts were used previously that allows for 2D culture of epithelial cells that were then exposed to pathogenic bacteria on the apical side and human peripheral blood mononuclear cells (PBMC) on the basal side. Preliminary results showed immune-modulatory effects of epithelial cells in terms of cytokine responses from PBMCs measured using ELISA (R&D systems). However, 2D cultures are not good representatives of the in-vivo physiology. Therefore, we replaced the 2D transwell system with aforementioned in-vitro platform. Currently, this system is being used to study bacterial interactions with functionally active epithelial barrier and immune cells.

Conclusions: A high throughput in-vitro platform was developed that allows for mechanical stimulation and polarization of epithelial cells via magnetic field. The system allows for compartmentalized co-culture of bacterial and mammalian cells to study the role of microbial invasion, epithelium, and cytokines in pathophysiology of human GIT.

471233 Recapitulating Liver Fibrosis in Vitro By Recreating the Fibrotic Milieu

Abstract:

Liver damage as a consequence of chronic stress can lead to fibrosis causing restructuring of the liver microenvironment (LME). The increase in the liver stiffness is a strong indication of the progression of liver diseases. However, the effect of increased stiffness on different liver cells has not been established. Clinical and animal studies have contributed significantly to our understanding of liver fibrosis biology but there exists a need for in vitro models to investigate the essential biochemical mechanistic pathways that associated with liver fibrosis microenvironment. Such models need to incorporate the dynamic changes in LME including the change in liver stiffness. In our study, we investigated the effect of substrate stiffness on the functions of cultured primary hepatocytes and liver sinusoidal endothelial cells (LSECs) utilizing a polydimethyl siloxane (PDMS) based substrate with tunable stiffness. We employed a soft substrate to represent the healthy liver tissue stiffness and a stiff substrate to represent diseased liver tissue and compared the cellular properties with the cells cultured on tissue culture polystyrene surfaces (TCPS). We observed that hepatocytes cultured on the soft substrate displayed a consistently more differentiated phenotype for a prolonged duration as compared to the hepatocytes that were cultured on the stiff substrate and TCPS. We further demonstrated that hepatocyte cultured on soft substrate showed better maintenance of hepatocyte-specific functions of urea and albumin synthesis. Similarly, cytochrome P450 (CYP) activity on soft substrate displayed a multi-fold increase in CYP activity, as compared to TCPS. In LSECs, we observed a stiffness mediated dedifferentiation in the cells, validated by defenestration of cells and overexpression of vascular endothelial cell markers. Together, these results indicate that the substrate stiffness plays a significant role in modulating hepatic cell behavior. Establishing mechanosensitivity of hepatic cells in liver fibrosis could result in the development of anti-fibrosis therapeutic approaches that target the mechanotransduction pathways of the cells to achieve functional restoration.

468907 Micropatterned Hydrogels to Promote Alignment in Co-Cultures Systems for an *in Vitro N*euromuscular Model

Abstract:

Introduction:

In vitro models are of immeasurable value to researchers and have become the gold standard in most tissue engineering and drug discovery applications. However, many of the currently used models do not fully capture the complex cell-cell interactions and 3D environment of multicellular organisms. The phenotype of cells can be vastly influenced by chemical and mechanical interactions within their microenvironment, therefore, it is critical to mimic native tissue environment when designing an *in vitro* models. **[1]**.

Primary cardiac muscle cells, cardiomyocytes (CMs), seeded on and inside of hydrogels, often lose their propensity to spontaneously contract, a visual cue of endogenous CM phenotyping, after only a few days **[2]**. However, the beating characteristics and phenotype of CM can be maintained *in vitro* for longer periods of time when CMs are subjected to external mechanical **[3]** or electrical stimuli **[4]**. Additionally, the presence of cardiac fibroblast in CM co-culture systems has been shown to prolong CM beating **[5]**. We hypothesize that in an *in vitro* model, the intrinsic cardiac neurons that help modulate heart rate **[6]** also may play a role in maintaining CM phenotype.

To test this hypothesis, we aim to construct a 3D co-culture system by encapsulating cardiac neurons and CMs in a methacrylated gelatin (GelMA) hydrogel. To align both cell types and mimic the architecture of native heart tissue, we have designed a two-step photolithography technique to encapsulate multiple cell types in a micropatterned GelMA hydrogel. This novel technique can be employed to study the interactions between the nervous system and the myocardium in a biomimetic environment.

Methods:

Encapsulated cells are aligned in the 3D hydrogel constructs through a multi-step photolithography technique modified from a single step technique developed by Nikkhah et al **[7]**. GelMA hydrogels are synthesized by photocrosslinking of pregel solutions using free radical chemistry. GelMA is produced by reacting methacrylic anhydride to the anime groups in gelatin as previous reported by Nichol et al **[8]**. Crosslinking is mediated via exposing the pregel solution: 10% (w/v) GelMA solutions containing 0.1% (w/v) free radical initiator (Irgacure 2959) to UV light (300-400nm, intensity 6.9mW/cm^2) for up to 75 seconds.

As a proof of concept, fluorescent beads were used as analogs for trypsinized cells. To pattern these beads, first a solution of a pregel-bead mixture is exposed to UV light though a photomask to form patterned lines of crosslinked GelMA (200 um widths with 200 um separation). Next, in order to create a contiguous hydrogel with distinctly aligned beads, the uncrosslinked suspension is washed away and a second pregel-bead solution is crosslinked in between partially patterned hydrogel by irradiating the entire sample. Hydrogels are imaged using fluorescent microscopy and the degree of gel segmentation in 3D is analyzed via intensity profiles of z-stacks taken at 10x on a Zeiss Axio Observer with NIH ImageJ software.

Results and Discussion:

We have successfully fabricated patterned hydrogels of two alternating type of fluorescent beads. Analysis of patterned hydrogels demonstrates two distinct patterns inside 3D GelMA hydrogels showed only a small percentage of leakage between line sections. However, this method is relatively efficient for spatially containing two individual cell suspensions. We have further plans to validate this process for a number of different line segment and spacing measurements.

Utilizing this technique, muscle cells will be patterned prior to neurons for our *in vitro*neuromuscular model. Future work will combine the neurons found in the heart and CMs into our patterned hydrogel to investigate if their interactions have an effect on neurite sprouting and growth as well as CM phenotype.

References:

1. Bhana, B., et al. (2010). Biotechnology and Bioengineering **105**(6): 1148-1160.
2. Bursac, N., et al. (1999). The American journal of physiology **277**(2 Pt 2): 44.
3. Radisic, M., et al. (2004). Proceedings of the National Academy of Sciences **101**(52): 18129-18134.
4. Zimmermann, W. H. H., et al. (2002). Circulation research **90**(2): 223-230.
5. LaFramboise, W. A., et al. (2007). American Journal of Physiology - Cell Physiology **292**(5).
6. Zarzoso, M., et al. (2013). Cardiovascular Research **99**(3): 566-575.
7. Nikkhah, M., et al. (2012). Biomaterials **33**(35): 9009-9018.
8. Nichol, J. W., et al. (2010). Biomaterials **31**(21): 5536-5544.

469318 Multi-Zonal Three-Dimensional in Vitro Culture Model of Growth Plate Cartilage Using Alginate Hydrogel Scaffolds

Abstract:

Organ regeneration using engineered tissues has the potential to revolutionize personalized patient care. However, precisely controlling the rate and direction of tissue growth remains a significant challenge for tissue engineers. Advanced tissue culture models that enable investigation of the mechanisms underlying anisotropic growth are necessary to overcome this challenge. During skeletal development, cartilaginous growth plates at epiphyses drive polarized bone growth, or endochondral ossification. This process is controlled by the precise regulation of chondrocyte differentiation via long- and short-range signaling mechanisms within the growth plateâ€™s architecture. As a result, the growth plate is organized in morphologically unique zones, each containing chondrocytes with unique phenotypes. While much has been described about the process of development of the growth plate, there is still a gap in the understanding of the interplay between specific signaling mechanisms and resulting architecture. To date, not enough is known about these intricacies to effectively modulate chondrocyte maturation in existing *in vitro* cartilage culture models. The design of an *in vitro* culture model of growth plate cartilage will be utilized to elucidate the structure and function of the regulatory network that coordinates chondrocyte maturation and leads to the formation of zonal architecture. This project focuses on the utilization of alginate hydrogels in the design of a novel, multi-zonal, three-dimensional (3-D) *in vitro* culture system aiming to recapitulate the supportive extracellular matrix and diffusional gradients of signaling factors found in growth plate cartilage. In order to test the hypothesis that the activation of the hedgehog (HH) signaling pathway is necessary for zonal arrangement, and in particular the formation of the columnar structures found in the proliferative zone of growth plate cartilage, radially layered alginate bead constructs are employed. We first show that an alginate bead culture method is supportive of cell viability and conducive to the presentation of factors known to modulate HH signaling, such as parathyroid hormone (PTH/PTHrP) and Indian Hedgehog (IHH), and the subsequent response of primary mouse growth plate chondrocytes to these factors. This response was observed with flow cytometry, ddPCR and western blot to monitor HH signaling. In order

to present these factors in a gradient, a layered bead construct was designed and optimized for the controlled response of chondrocytes to signaling factors. HEK293T cells transfected with a plasmid to produce PTH were encapsulated in an innermost alginate bead, while growth plate chondrocytes were encapsulated in an alginate layer around this inner bead, receiving PTH in a radially diffusing fashion. IHH was added to the media to set up a bidirectional diffusional gradient. The response of the chondrocytes to this gradient and the resulting zonal arrangement of the growth plate chondrocytes was monitored by the above methods along with the use of Fluorescent In Situ Hybridization (FISH) to visualize the characteristic, radially differential expression of genes specific to each of the defined zones. This scaffold design has proven to be a powerful tool in investigating the role of these specific factors on zonal development in the growth plate. With the knowledge gained from these experiments, we begin to form a better picture of the mechanisms behind growth plate cartilage maturation, which will lead to the development of better therapeutic techniques for patients suffering from growth plate cartilage disease or injury.

469666 High-Throughput Drug Screening Using 3D Micro Organotypic Liver Models

Abstract:

There are thousands of chemicals whose effects on human and animal populations are unknown or only partially understood. Specifically, the combinatorial effects of these chemicals to the liver are unknown. Traditional *in vitro* and *in vivo* studies, while informative, can be laborious and time-consuming. High-throughput screening (HTS) methodologies can rapidly test individual and cocktails of chemicals across a wide range of concentrations. This approach can identify chemical mixtures that are toxic to the liver. To date, HTS assays on liver cells have primarily been conducted on 2D monolayers or co-cultures. These analyses are useful, but do not capture the complex interplay that exists between hepatic cells and their *in vivo*microenvironment.

When the principal liver cell types, hepatocytes, are cultured in standard monolayers, they begin to lose their liver-specific functions within 24 hours. Moreover, these models fail to recapitulate the native stratified architecture of the liver. We have designed 3D liver models that contain all hepatic cell types and maintain the cellular compositions over extended time periods. The combination of HTS methodologies and these 3D liver models has the potential to transform current testing of pharmaceuticals and environmental toxicants.

We have assembled a novel *m*icro liver *o*rganotypic *c*ulture *m*odel assembled in a 96-well plate (denoted as mOCM) that can be used to rapidly screen individual and mixtures of chemicals. A novel aspect of the mOCMs is the use of automated procedures. An automatic cell dispenser was used to coat the substrates with collagen, seed the hepatic cell types, change medium, administer chemicals, and conduct assays within the 96-well plates. The mOCMs contained a biopolymeric membrane composed of collagen (type 1) and hyaluronic acid that separated the hepatocytes from the liver sinusoidal endothelial cells (LSECs) and Kupffer cells (KCs). This membrane mimicked the height and composition of the Space of Disse that separates the hepatocytes from the LSECs and KCs *in vivo*.Acetaminophen

(APAP), troglitazone and perfluorooctanoic acid (PFOA) were administered to the mOCMs across a range of concentrations (sub-lethal to highly toxic). Hepatocyte monolayer (HM) and collagen sandwich (CS) cultures served as 2D comparisons. The mOCMs exhibited up to 43 ± 6%, 33 ± 8% and 69 ± 3% decreases in viability when treated with 10 mM APAP, 50 mM troglitazone and 0.5 mM PFOA, respectively. When these concentrations were increased to 40 mM APAP, 200 mM troglitazone and 2 mM PFOA, cell death increased up to 79 ± 2%, 67 ± 12% and 84 ± 4%. In contrast, the HM and CS cultures only exhibited significant cell death at the highest concentrations of PFOA and APAP. The increased sensitivity of the mOCMs compared to the HM and CS cultures clearly demonstrate that these cultures maintain biotransformation capabilities of these chemicals.

Changes in viability, necrosis and apoptosis were simultaneously measured in the cultures using a high-throughput assay (ApoTox-Glo™ Triplex Assay). This HTS assay simultaneously provided information on cell death, apoptosis and necrosis by simply measuring the luminescence and fluorescence from each culture. In response to 40 mM APAP, there was a 6.9-fold increase in necrosis relative to controls in the mOCMs. The corresponding change was only 1.6-fold in HM and CS cultures. In response to 200 mM troglitazone, necrosis increased by 2.9, 6.4 and 16.1-fold in the HM, CS and mOCMs, respectively. Apoptosis increased up to 3.0-fold and 2.1-fold in the 2D cultures and mOCMs, respectively, relative to controls. *In vivo*, the expected mechanism of cell death for both APAP and troglitazone is necrosis. Therefore, these results indicate that not only are the mOCMs more sensitive than traditional 2D cultures, but the mechanism of cell death is more representative of the hepatic response *in vivo*.

Ongoing efforts are focused on the use of HTS assays to investigate changes in hepatic cellular functions, protein expression and functional markers in the presence of individual and cocktails of chemicals. By seamlessly combining HTS methodologies with a liver mOCM we seek to provide a platform for medium to high throughput testing of toxicants.

458646 Wnt-YAP Interactions during Neural Tissue Patterning of Human Induced Pluripotent Stem Cells

Abstract:

WNT-YAP INTERACTIONS DURING NEURAL TISSUE PATTERNING OF HUMAN INDUCED PLURIPOTENT STEM CELLS

Julie Bejoy, Liqing Song, Yan Li

Department of Chemical and Biomedical Engineering, FAMU-FSU College of Engineering, Florida Agricultural and Mechanical University, Tallahassee, Florida, USA

Abstract

Human induced pluripotent stem cells (hiPSCs) have special ability to self-assemble into mini-organ like structures (e.g., organoids, mini-brains). In this process, Wnt signaling impacts regional patterning and positional identity of hiPSC-derived neural progenitors. One important function of Wnt signaling is to regulate Yes-associated protein (YAP) expression (nuclear or cytoplasmic), the pivotal regulator of cell proliferation and differentiation during organ growth and tissue generation. However, the crosstalk between Wnt and YAP expression during neural patterning of hiPSCs has not been well investigated. The objective of this study is to reveal the capability of Wnt signaling in the regulation of YAP expression to modulate 3-D neural spheroid formation from hiPSCs. Human iPSK3 cells were induced toward neural lineages through embryoid body formation. Wnt signaling was activated using CHIR99021, which was found to induce nuclear localization of YAP. CHIR99021 treatment upregulated the expression of HOXB4, the marker for hindbrain/spinal cord, while in the absence of Wnt activation, the cells maintained rostral forebrain neural identity (expression of TBR1 and BRN2) (by RT-PCR analysis and immunocytochemistry). Modulation of YAP expression with cytochalasin D also influenced neural cell identity, indicating bi-directional interactions of Wnt signaling and YAP expression. The perturbation of neural patterning was also evaluated by incorporating the neural spheroids with microparticles fabricated with triblock copolymer polycaprolactone-poly(dimethylsiloxane)-polycaprolactone. This study should advance our understanding on the biological processes regulated by Wnt signaling and YAP activity during neural tissue patterning and brain organoid engineering. The results have the significance in neurological disease modeling, drug screening, neural tissue regeneration, as well as studying virus infection.

456522 3-D Spheroid Model of Human Adipose-Derived Stem Cells for Superior Adipogenic Differentiation

Abstract:

Development of *in vitro* models of adipogenesis is critical to find therapeutics for obesity and related comorbidities. Here we present the 3-D spheroid model using human adipose-derived stem cells (hASCs) and their subsequent adipogenic differentiation. Our work describes construction of a 3-D spheroid culture system atop positively charged elastin-like polypeptide-polyethyleneimine (ELP-PEI) coating to study the differentiation and maintenance process of hASC adipocyte spheroids when subjected to fatty acid treatment. Measurements taken from spheroid micrographs indicated spheroid size increased during the maintenance phase, though the spheroids treated with linoleic acid or oleic acid had smaller size compared to those cultured in the control maintenance medium. Increased spheroid size correlated with increased triglyceride accumulation and the linoleic acid or oleic acid treatments stimulated triglyceride accumulation in hASC cultures. The CD36 expression, representing cell competency for consuming extracellular fatty acids, was consistently found in hASC 3-D spheroids. Overall, we observed that, the hASC spheroids undergo slower differentiation along the adipogenic lineage and attainment of physiological functionality compared to the 3T3-L1 mouse spheroids. Taken together, we have shown the feasibility of adipogenic differentiation of patient-derived hASC spheroids, which with further development, may help elucidate key features in the adipogenesis process.

459049 Exploring the Dispersity-Enhanced Stimulus Response and Bacterial Release in Polyelectrolyte Brushes

Abstract:

Polyelectrolyte brushes attached to surfaces change their conformation when exposed to solutions of varying pH. As responsive materials, polyelectrolyte brushes are used in applications, including drug delivery, enhanced oil recovery, and antifouling coatings. These tunable surfaces are also widely employed to control wettability, adhesion, friction, and biofouling. Despite extensive use across a broad range of applications, the role of polyelectrolyte brush length and dispersity on the pH-responsive behavior is not well understood. Here, we synthesized poly(acrylic acid) (PAA) brushes with controlled brush length and dispersity using the grafting-from approach at a constant grafting density. We identified three key trends in the response of the PAA brushes to changes in pH. First, the thickness of dry brushes decreased as the pH was decreased. Second, the water contact angle measured at low pH increased with dispersity. Finally, disperse brushes exhibited hysteretic memory behavior (i.e. dependent on the direction of pH change) in the contact angle. The origin of hysteresis is consistent with dispersity-driven pH-dependent conformational changes in these polyelectrolyte brushes. Together, these results indicate that increasing brush dispersity significantly alters the pH-response. Next, we used this model system to promote cell detachment via an abrupt change in solution pH (from 4 to 9). In preliminary experiments, we found that bacterial adhesion (at pH 4) increased when brush dispersity decreased. Upon switching to pH 9 the bacteria began to detach from the brushes after a delay time of approximately 5 min, corresponding to equilibration of the brush properties at this pH. The delay time was unaffected by brush length or dispersity.

458523 Water Transport in PS-*B*-PEO Copolymer Membranes

Abstract:

Nanostructured polymer membranes are promising materials for CO_2 separation, water purification, and lithium air batteries. Poly(styrene)-*block*-poly(ethylene oxide) (PS-*b*-PEO) is an example of such nanostructured membrane with polar phase PEO component; mechanically robust PEO spherulites and hydrophobic PS component. Water transport is an essential parameter needed to properly model the performance of this membrane in various applications. We present the results of sorption, diffusion and desorption of water in PS-*b*-PEO at various water activities using gravimetric sorption measurements and Fourier transform infrared â€" attenuated total reflectance (FTIR-ATR) spectroscopy. A direct comparison with PEO homopolymer membrane shows that the presence of water affects both molecular structure and thermal properties of membranes at high water activities. Due to these structural changes, water sorption was found to increase exponentially at high water activity. The crystalline content of the PEO and the presence of glassy PS lamellae will be used to explain the transport results.

459920 Molecular Simulation of Flow-Enhanced Nucleation in Monodisperse and Bidisperse Alkane Melts

Abstract:

The high molecular weight tail has been widely observed to have a dramatic impact on the kinetics of polymer crystallization under flow. Experiments have shown that the crystallization rate drastically increases when the strain rate is large enough to stretch and orient the longest chains present in the melt. The mechanism by which these long chains accelerate crystallization at the nucleation stage, however, is not well-understood due the fact that experiments lack the spatiotemporal resolution to observe a nucleation event. Molecular simulation, on the other hand, has proven to be a useful tool for observing the nucleation stage in polymer melt systems. Using non-equilibrium molecular dynamics, we studied flow-enhanced nucleation in a bimodal mixture of 150 carbon (C150) and 20 carbon (C20) *n*-alkanes in order to observe and quantify the effect of presence of long chains. The mixtures studied ranged from 3-9 wt% C150. We found three regimes of nucleation associated with the relative magnitude of the applied strain rate, and longest relaxation rates of the chains in the melt. At strain rates below the inverse relaxation time of C150 and above the inverse relaxation time of C20, we did not observe a dramatic difference between the rates of nucleation in the bidisperse and monodisperse C20 melts. However, when the strain rate was larger than the inverse of the relaxation time for C150, but smaller than the inverse relaxation time of C20, we observed a templating effect in which the extended C150 chains served as templates for the nucleation of C20 crystals. This templating effect resulted in a dramatic increase in the nucleation rate as compared to nucleation under similar conditions for monodisperse C20 melts.

460507 All-Conjugated Block Copolymer Additives for Organic Solar Cells

Abstract:

Recent studies have demonstrated organic solar cells with power conversion efficiencies (PCEs) above 10%. However, the morphology of these devices is dictated by a kinetically trapped, metastable morphology, and strategies for achieving thermally-stable, co-continuous networks of donor and acceptor domains are desirable. Fully conjugated block copolymers (BCPs) with donor and acceptor blocks can be utilized as additives to both modulate the morphology and potentially improve the electronic properties of the active layer. The use of BCP additives represents an effective approach to control the donor-acceptor morphology, but prior studies mostly have focused primarily on block copolymers containing optically and electronically inactive polymer blocks or backbones. Here, we report a combined experimental and computational study of the impact of all-conjugated poly(thieno[3,4-*b*]-thiophene-*co*-benzodithiophene)-*b*-polynaphthalene diimide (PTB7-*b*-PNDI) BCP additives. We find that small amounts (2 wt%) of BCP additives lead to improved performance due to large increase in open circuit voltages (V_{oc}), and the V_{oc} is pinned to this higher value for higher BCP additive loadings. These results contrast prior results of ternary blend OPVs where V_{oc} is pinned to a lower value. We further demonstrate, by transmission electron microscopy and surface energy measurements, that voltage enhancement is likely due to the formation of a parallel bulk heterojunction which is formed after BCP addition to the PTB7/PCBM blends. In summary, this work demonstrates that all-conjugated block copolymers can be utilized as additives to both dictate morphology and enhance the electronic properties of organic photovoltaic solar cells.

460041 Chemical Stability and Ion Transport in Polymerized Ionic Liquid Anion Exchange Membranes

Abstract:

Non-platinum (low-cost), long-lasting, solid-state alkaline fuel cells (AFCs) require anion exchange membranes (AEMs) that have high alkaline chemical stability, high hydroxide ion conductivity, and adequate mechanical properties. In this study, numerous polymerized ionic liquid (PILs) were synthesized with various covalently attached cations (butylimidazolium, butylmethylimidazolium, trimethylammonium, butylpyrrolidinium, trimethylphosphonium) and various backbone/cation pairings (backbones: ethyl methacrylate, undecyl methacrylate, undecyl acrylate, styrene; covalently attached cations: butylimidazolium, trimethylammonium, butylpyrrolidinium). Their alkaline chemical stability was quantified using ^{1}H NMR spectroscopy, and the styrene/butylpyrrolidinium pairing proved to be highly chemically stable, with no degradation in 20 equivalents (0.5 M) KOH at 60 °C for 168 h.

Considering the high alkaline chemical stability of the styrene/pyrrolidinium-based PIL, the conductivity and chemical stability of a PIL block copolymer (ABCBA pentablock terpolymer) with a styrene/pyrrolidinium-based PIL block was investigated. A high hydroxide conductivity of 43.4 mS cm^{-1} at 60 °C in liquid water was achieved and no degradation or loss of conductivity was observed in the membrane after 168 h in 1 M KOH at 60 °C. PIL block copolymers utilizing a styrene/pyrrolidinium backbone/cation pairing in the PIL block represent a promising chemistry for producing highly conductive, chemically stable, robust AEMs for implementation in future solid-state AFCs.

ADVANCED MATERIALS
www.advmat.de

A Stretchable Graphitic Carbon/Si Anode Enabled by Conformal Coating of a Self-Healing Elastic Polymer

Yongming Sun, Jeffrey Lopez, Hyun-Wook Lee, Nian Liu, Guangyuan Zheng, Chun-Lan Wu, Jie Sun, Wei Liu, Jong Won Chung, Zhenan Bao, and Yi Cui**

Stretchable electronics have recently received intense attention due to their wide applications in various areas where devices must maintain intimate contact with curvilinear interfaces or undergo large deformation.[1] Lithium-ion batteries are the dominant power source for portable electronics and represent one of the most promising energy-storage systems for these stretchable electronic devices.[2] Two of the main challenges impeding the development of stretchable lithium-ion battery systems, however, are the limited stretchability and low specific energy. Recently, much effort has been devoted toward realizing the stretchability of battery systems. Several strategies have previously been successfully developed to achieve stretchable battery systems at the device level with stiff battery islands and stretchable interconnects (e.g., origami batteries[3] and serpentine interconnects[4]) and at the battery level through the use of buckled active materials on a prestrained substrate.[5] For the systems with device-level stretchability, the stability under stretching is still limited due to the rigidity of batteries. Worse still, the energy density of the overall battery systems is significantly reduced due to the large spacing between the small isolated battery islands. To achieve higher energy densities, the development of intrinsically stretchable electrodes is of key importance. Therefore, there is still an urgent need to explore electrodes with appreciable mechanical durability under strain for use in stretchable lithium-ion batteries.

Currently, $Li_4Ti_5O_{12}$ is the most widely used anode material for stretchable lithium-ion batteries due to its stable electrochemical properties.[5,6] However, it suffers from a low theoretical specific capacity (175 mAh g^{-1}) and high working potential (≈1.5 V), which greatly restrict the battery energy density. From this perspective, developing high-capacity anode materials with low working potentials is vital to achieve high specific energy in stretchable lithium-ion batteries. Little emphasis has been thus far placed on the development of high-capacity electrode materials for stretchable lithium-ion batteries, although success in this regard would allow much higher energy density for batteries. Among the anode materials, Si can deliver 24× theoretical capacity (4200 mAh g^{-1}) over $Li_4Ti_5O_{12}$ and has a much lower working potential (<0.5 V) compared with that of $Li_4Ti_5O_{12}$ (≈1.5 V).[7] Thus, it is highly desirable to explore Si anodes for stretchable lithium-ion batteries. Nevertheless, Si anodes have the serious issues of mechanical degradation and unstable solid-electrolyte interphase (SEI) due to the large volume expansion upon lithiation, which causes dramatic decay in battery performance.[8] Many attempts have been made to improve the electrochemical performance of Si anodes.[9] One of the most successful approaches has been the use of a self-healing polymer binder.[10] This soft and highly stretchable supramolecular polymer contains a large number of hydrogen bonding sites that allow it to mechanically heal at room temperature. The hydrogen bonding also helps the polymer adhere to the Si surface. This, combined with the self-healing ability, allows the formation of a stable SEI on Si during the large volumetric changes upon the lithiation and de-lithiation processes. This self-healing polymer is highly stretchable compared to traditional binder materials and has great potential to help realize stretchable Si electrodes via rational molecular design.

Normally, inorganic active electrode materials are stiff and rigid. The main approach to achieve stretchable electrodes is to create a stretchable structure out of normally non-stretchable active materials. Although a few of successful examples, such as buckled $Li_4Ti_5O_{12}$/carbon nanotube films on prestrained polydimethylsiloxane (PDMS) substrate[5a] and a spring-like carbon nanotube fibers,[5b] have been reported, there are still no reported high-capacity stretchable electrodes. Herein, for the first time, we demonstrate a high-capacity stretchable graphitic carbon/Si foam electrode enabled by conformal coating of a newly synthesized self-healing elastic polymer. Unlike the supramolecular self-healing polymer that was previously reported,[10] this material is much tougher due to a combination covalent crosslinks and sacrificial hydrogen bonds[11] and behaves elastically over a large range of strains. The amorphous Si layer on the graphitic carbon foam affords a high specific capacity, enabling a high overall capacity of 719 mAh g^{-1} for the graphitic carbon/Si composite electrode, four times that of the widely used $Li_4Ti_5O_{12}$ anode material for stretchable lithium-ion batteries, and maintained 81% of its

Dr. Y. M. Sun, Dr. H.-W. Lee, Dr. N. Liu, Dr. G. Y. Zheng, C.-L. Wu, Dr. J. Sun, Dr. W. Liu, Prof. Y. Cui
Department of Materials Science and Engineering
Stanford University
Stanford, CA 94305, USA
E-mail: yicui@stanford.edu
J. Lopez, Dr. J. W. Chung, Prof. Z. Bao
Department of Chemical Engineering
Stanford University
Stanford, CA 94305, USA
E-mail: zbao@stanford.edu
Dr. J. W. Chung
Samsung Advanced Institute of Technology
Yeongtong-gu, Suwon-si, Gyeonggi-do 443-803, South Korea
Prof. Y. Cui
Stanford Institute for Materials and Energy Sciences
SLAC National Accelerator Laboratory
Menlo Park, CA 94025, USA

DOI: 10.1002/adma.201504723

ADVANCED MATERIALS
www.advmat.de

Materials Views
www.MaterialsViews.com

capacity after 100 cycles. Our investigations indicate that the self-healing elastomer is uniformly coated on the 3D graphitic carbon/Si foam and endows the composite electrode with high stretchability (up to 88%) and endures 1000 stretching-releasing cycles at 25% strain without detrimental resistance increases.

The stretchability of the electrode is realized via the deformation of the foam structure with the help from the self-healing elastomeric coating. As illustrated in **Figure 1**a, the composite electrode deforms and recovers the foam structure along the direction of mechanical strain. Previously, graphene/graphite foams were used as the current collectors to realize (flexible) high-rate $LiFePO_4$ and $Li_4Ti_5O_{12}$ electrodes[12] or high-capacity Si-based electrodes[13] due to their high conductivity. In the as-designed electrodes, the 3D graphitic carbon foam not only functions effectively as a current collector and provides high conductivity for electron transport, but also works as part of the active material alongside the amorphous Si on the surface of the carbon, which enables fast reaction dynamics in the initial charge/discharge progresses in comparison with crystalline Si. The highly elastic self-healing polymer helps to build a robust interface that improves the electrochemical stability of the electrode and maintains the active carbon/Si foam electrode integrity during the stretching and releasing processes, rendering the whole structure stretchable.

Figure 1b illustrates the preparation and optical microscope images of the stretchable graphitic carbon/Si/polymer composite electrode. The nickel foams have a 3D skeleton structure, and a graphitic carbon layer was first grown on the nickel through a chemical vapor deposition (CVD) process. Different from the previous work on the growth of thin graphene,[14] we obtained a mechanically robust thick graphitic carbon foam with a 3D hollow skeleton structure using a different carbon source (vaporized hexane) at ambient pressure. After the removal of nickel, an amorphous Si layer was conformably grown on the surface of 3D carbon foams using silane as the precursor under a low-pressure condition. Due to the hollow carbon skeleton and porous 3D structure, the deposition of Si is uniform throughout the entire carbon framework. A graphitic carbon/Si/polymer foam electrode was achieved after drop-coating a layer of self-healing polymer on the foam. Figure 1c shows a schematic for the synthesis of the self-healing elastomer modified from previously reported structures.[10,15] A mixture of fatty acid dimers and trimers (30:70) was reacted with excess diethylene triamine (DETA) to yield amine-terminated oligomers. Instead of reacting with urea to give a

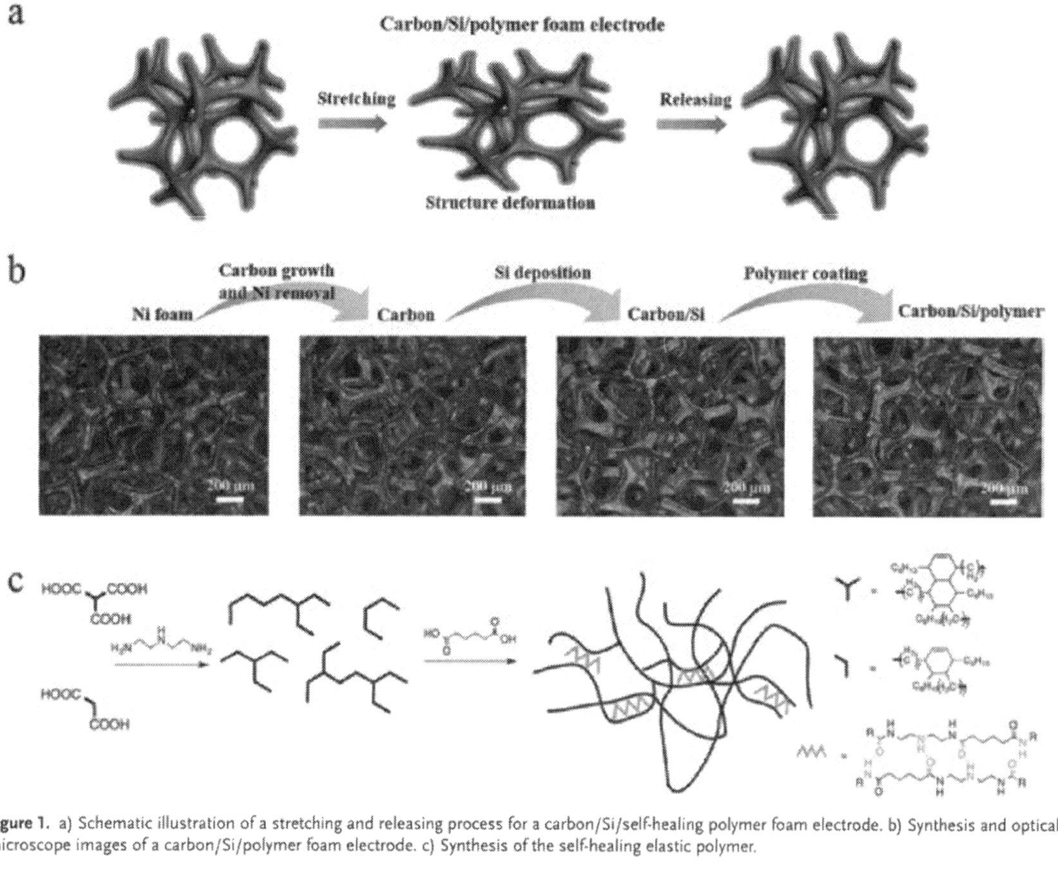

Figure 1. a) Schematic illustration of a stretching and releasing process for a carbon/Si/self-healing polymer foam electrode. b) Synthesis and optical microscope images of a carbon/Si/polymer foam electrode. c) Synthesis of the self-healing elastic polymer.

viscoelastic supramolecular polymer as in previous reports, oligomers were further polymerized and cross-linked with adipic acid to produce the final elastomeric network with strong interchain hydrogen bonding interactions through the amide groups formed. Before coating, the amine oligomers and adipic acid were dissolved in ethanol (1 g/10 mL). The thickness of self-healing polymer coating could be controlled by adjusting the amount of solution used. The successful preparation of such conformal graphitic carbon/Si/polymer structure provides a great opportunity to achieve a high-capacity anode for stretchable lithium-ion batteries.

The investigation on the structure and crystallinity of the carbon/Si foam is critical for understanding its electrochemical and mechanical properties. **Figure 2**a shows the X-ray diffraction (XRD) patterns of the as-obtained carbon and carbon/Si foams. A sharp diffraction peak around 26.4° is observed for the carbon foam, which is readily ascribed to graphite (JCPDS No. 41–1487). No new peaks are observed for the carbon/Si foam, indicating the amorphous nature of the CVD-grown Si. The crystalline nature of carbon and amorphous structure of Si in the carbon/Si foam are further confirmed using Raman spectroscopy (Figure 2b,c). The so-called G band (≈1582 cm^{-1}), D band (≈1335 cm^{-1}), D′ band (≈1619 cm^{-1}), and G′ band (≈2674 cm^{-1}) are observed in the Raman spectrum, which are the most prominent features of graphite (Figure 2b).[16] Usually, crystalized Si is located at ≈520 cm^{-1} in the Raman spectrum. The existence of a broad peak at ≈475 cm^{-1} indicates the amorphous Si structure (Figure 2c).[17] The digital images show that the grey carbon foam turns blue after the CVD deposition of Si layer (Inset of Figure 2d and Figure S4, Supporting Information). It is observed under scanning electronic microscope (SEM) that the foam structures of carbon and carbon/Si hybrid are well maintained (Figure 2d and Figure S5, Supporting Information), consistent with observation under an optical microscope (Figure 1b). Due to the conformal growth of carbon on the nickel foam, the carbon and carbon/Si skeletons possess a hollow structure after the removal of the nickel skeleton, which is confirmed by a fractured carbon/Si skeleton (Figure 2e). Clearly, the smooth surface of the foam becomes rough after the CVD deposition of an amorphous Si layer (Figure 2f). Furthermore, transmission electron microscopy (TEM) investigation reveals that the Si layer on the surface of carbon is uniform throughout the observed area (Figure 2g). The electron energy loss spectroscopy (EELS) spectra recorded for the area in Figure 2g are shown in Figure 2h,i. Figure 2h shows the Si L2,3-edges, confirming the amorphous Si structure.[18] The EELS spectrum with the carbon K-edge shows well-defined peaks due to transitions of 1s electrons to π^* and σ^* antibonding states, indicating the graphitic nature of the carbon in the composite (Figure 2i).[19]

The mechanical properties of the self-healing polymer alone were investigated to understand its effects on the carbon/Si foam. After crosslinking, the self-healing polymer is highly elastic and can be stretched for about 400% before breaking. The toughness and strength of the material can be changed by controlling the crosslinking time (**Figure 3**a). 105 min is found to be the optimal crosslinking time because it gives a material with the highest modulus. In Figure S6 of the Supporting

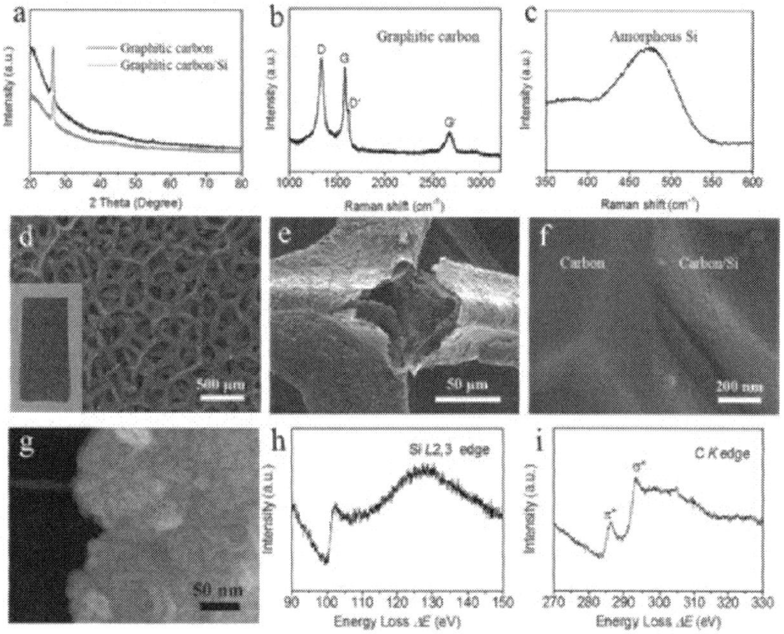

Figure 2. a) XRD patterns of the as-achieved carbon and carbon/Si foams. b,c) Raman spectra of the carbon/Si foam. d,e) SEM images of d) a carbon/Si foam and e) a fractured skeleton showing a hollow structure. f) High-magnification SEM images of the (left) carbon and (right) carbon/Si foams. g) TEM image of the carbon/Si foam. h,i) EELS spectra recorded for the area in g). The inset in d) shows a digital image of a carbon/Si foam.

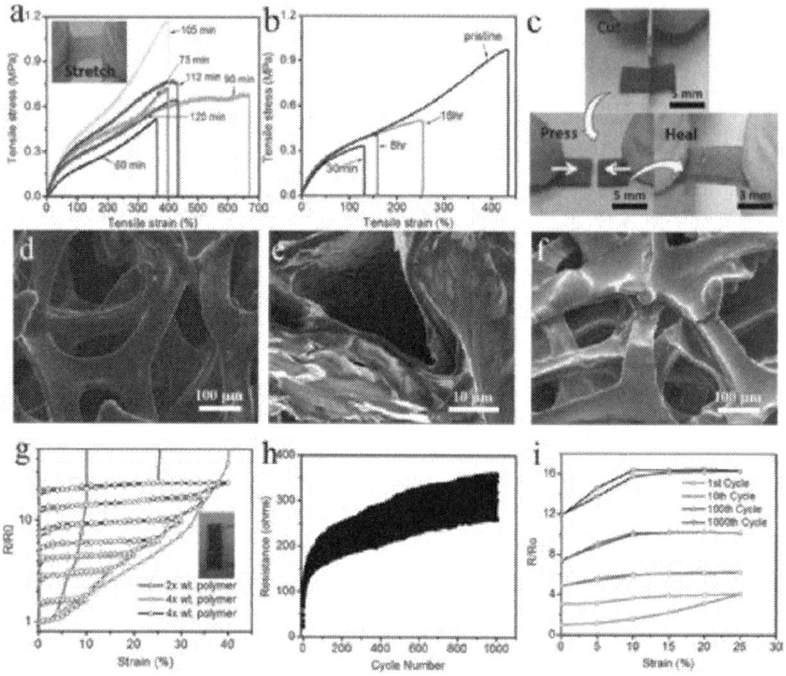

Figure 3. a) Stress–Strain curves for the bulk elastomer at different crosslinking reaction times. 105 min is found to be the optimal crosslinking time because it gives the material with the highest toughness. b) Self-healing of the bulk elastomer. Cut samples are allowed to rest for the prescribed time at room temperature and then compared to the performance of a pristine sample. c) Pictures of how the self-healing experiment is conducted. A sample is cut, pressed back together and then finally stretched until failure. d,e) SEM images of carbon/Si foams with 4× weight elastic polymer. f) SEM images of the carbon/Si foam electrodes with 4× weight elastic polymer after a ≈30% stretching-releasing cycle. g) Resistance vs strain curves of polymer coated carbon/Si foams supported on PDMS. Once stretched, resistance is stable at the final value until the foam is stretched further. h) Resistance measurements during strain cycling to 25%. Resistance increases over time. i) Highlighted cycles from h) show that resistance is relatively constant for any one cycle.

Information, the hysteresis of the bulk elastomer is shown. Strain cycles are applied consecutively, increasing by 100% each time. Very little hysteresis is observed up to 100% strain. Beyond that, more significant energy dissipation is seen and the material is not as tough upon re-stretching as in its pristine state. This energy dissipation comes from the sacrificial hydrogen bonds in the material, which break preferentially to the covalent crosslinks giving a much tougher elastomer.[11] The high density of hydrogen bonding sites also gives the elastomer a partial self-healing capability (Figure 3b). To test the self-healing, samples are cut into two pieces with a razor blade and then placed gently back together and allowed to rest at room temperature for the prescribed time (Figure 3c). Even after 30 min of healing at room temperature the modulus of the material is recovered and the healed sample can reach strains past 100%. Further healing was achieved up to 16 h, after which no further healing was observed. This incomplete healing is due to the irrecoverable fracture of the covalent crosslinks in the network. The hydrogen bonding sites are only capable of restoring a portion of the materials original mechanical properties. This bulk self-healing experiment is an extreme test of the healing ability of the material. In a device that is stretching to 50%, local strains large enough to fully fracture the elastomer will be rare and so the healing will instead act to restore the elasticity and strength of the material as it is strained and released. The self-healing elastic polymer possesses plenty of hydrogen-bonding sites to interact with the Si—OH on the Si surface (Figure 1c) and thus provides good adhesion to the Si layers.[10a] This is evidenced by the C=O stretching and the NH bending peaks in the Fourier transform infrared (FTIR) spectrum (Figure S7, Supporting Information), at ≈1650 cm^{-1} and ≈3250 cm^{-1}, respectively, that correspond to disordered hydrogen bonding of the amide groups, and a lack of a peak or shoulder near 1690 cm^{-1} that corresponds to "free" carbonyl groups.[20] The hydrogen bonding interaction with Si—OH on the Si surface has been observed in many other binders including CMC, alginate, and cyclodextrin.[21] The intimate contact between the elastic polymer layer and the surface of carbon/Si foam is confirmed by the SEM images (Figure 3d,e, and Figure S8, Supporting Information). Once the electrode undergoes strain, the self-healing elastic polymer can help to keep the foam structure intact.

Next, we quantified the stretchability dependence of the carbon/Si/polymer composite electrode on the amounts of the elastic polymer. Figure S9 of the Supporting Information shows stress–strain curves for the elastic polymer coated carbon/Si foams. As the amount of the elastic polymer increases, the foam gains increased stretchability because the connectivity of the polymeric network on the foam increases with thickness of the polymer layer. With 2× weight polymer the coating is thin and large cracks can be seen after 30% strain for the composite foam on a PDMS substrate (Figure S10a, Supporting Information). As the coating amount is increased to 4× weight and then 6× weight, the cracks get smaller and for the 6× weight polymer sample, polymer can be seen bridging different parts of the carbon/Si foam before and after cracking (Figure 3f and Figure S10b, Supporting Information). This increasingly connected polymer network allows for better stress dissipation through the sacrificial hydrogen bonds of the elastomer, and if polymer loading were to be further increased, the mechanical properties of the composite foam would eventually become similar to those of the bulk elastomer.

Over this stretching and releasing processes the electronic conductivity of the electrodes may change, which influences the battery performance. Elastic polymer films, such as PDMS, are widely used as the electrode substrates and packages for flexible and stretchable lithium-ion batteries.[4,5a,12a,22] In order to study the electronic conductivity of the composite foams and mimic the structure of a fully assembled stretchable battery under strain, the composite foams were loaded on an elastic PDMS substrate and strained. As the supported composite foams were stretched, small cracks formed across the length and width of the material to dissipate the stress and allowed the foams to maintain their electronic conductivity to strains far past values at which the unsupported foams would completely fail. For the 4× weight polymer composite foams, the average failure strain was 46.2% (±22.9%) and the best performing electrode reached a strain of 88% before it was no longer conductive (Figure 3g). This is also a significant improvement compared to our previously reported supramolecular self-healing polymer,[10] which only enabled an average strain of 8.5% (±3.5%) before electrode failure at the same wt% (Figure S11, Supporting Information). As the electrode was initially stretched, the resistance increased due to the new cracks that formed, but upon releasing the resistance was maintained at a constant value. The resistance stayed at this value until an even larger strain was applied, when the resistance again increased corresponding to the new strain that was applied. As further strain was applied, the hysteresis curve overlapped with that of a pristine electrode being strained for the first time. This again shows that an increase in resistance only occurs for strains past that already experienced by the electrode. Using these electrodes, it should be possible to achieve strains of 30%–50% in a full-cell stretchable device.

Since a stretchable lithium-ion battery has to undergo repeated stretching-releasing cycles over its lifespan, it is important to understand the stability of these electrodes during strain cycling. The electrodes were subjected to 1000 cycles of loading and unloading at a strain of 25% (Figure 3h). Confined by the elastic outer self-healing polymer layer, the overall electrode structure was kept intact. Over the many cycles, the resistance did drift upward to some extent but it did not reach values that are unacceptable for use as a battery electrode (<400 Ω after 1000 cycles). While drift was observed over many cycles, any one cycle from this data set did not show any large resistance increases (Figure 3i). If the cycling experiment is extended further, it is likely that the resistance would reach a plateau value as the system found an equilibrium state.

With the optimized amount of self-healing elastic polymer, the carbon/Si foam electrode affords remarkable battery performance. **Figure** 4a shows the reversible delithiation specific capacities of the carbon/Si electrodes coated with different amounts of self-healing polymer over 100 charge/discharge cycles. Here, all reported capacities are calculated based on the total weight of carbon and Si unless otherwise mentioned. The initial specific delithiation capacity of the pristine carbon/Si electrode reaches 1064 mAh g^{-1}, while the value for the bare carbon foam is 197 mAh g^{-1}. Considering that the Si amount in the carbon/Si electrode is ≈30%, the specific capacity with respect to Si is calculated to be as high as 3087 mAh g^{-1}. After coated with 2× and 4× weight self-healing elastic polymer, the carbon/Si electrodes deliver their initial delithiation capacities of 872 and 719 mAh g^{-1}, respectively. With a thicker self-healing polymer coating layer (6× weight polymer), the carbon/Si electrode only delivers 49% specific capacity (517 mAh g^{-1}) of the value that the pristine carbon/Si electrode achieves. Due to a relatively low ionic conductivity of 2.68×10^{-4} S/cm ($\pm 0.88 \times 10^{-4}$) in the dry state and 4.24×10^{-4} S cm^{-1} ($\pm 1.56 \times 10^{-4}$) in the swollen state (Figure S12, Supporting Information), a thick self-healing polymer coating layer may lead to increased lithium-ion diffusion paths, large electrode series resistance, and thus a relatively low capacity when compared with electrodes coated with a thinner polymer film. PDMS is the most widely used elastic matrix for stretchable devices. Here, we also tested PDMS as a potential elastic support for the carbon/Si foam, and while the PDMS formed a good coating on the foam (Figure S13, Supporting Information), the electrochemical performance was very poor due to the poor Li conductivity (Figure S14, Supporting Information). This shows that both the mechanical and chemical properties are important when selecting an elastomer support.

Besides the difference in the reversible specific capacities, as shown in Figure 4a, the electrodes coated with various amounts of self-healing polymer exhibit different cycle stability. The specific capacity of the pristine carbon/Si electrode after 100 charge/discharge cycles is 491 mAh g^{-1} with a low capacity retention of 46%. Coated with proper amounts of self-healing polymer, the electrodes possess much improved cycling stability. With 2× weight polymer, the carbon/Si electrode achieves a high specific capacity of 722 mAh g^{-1} after 100 charge/discharge cycles. The corresponding capacity retention is as high as 83% with 0.17% decay per cycle. When the amount of self-healing polymer increases to 4× weight polymer, the reverse specific capacity of the electrode is 584 mAh g^{-1} at the 100th cycle, which is still more than three times the theoretical capacity of $Li_4Ti_5O_{12}$, showing a similar capacity retention (81%) to the electrode with 2× weight polymer. The investigations on the morphology of the carbon/Si/polymer electrode and the pristine carbon/Si electrode after cycling show that the SEI layer on the surface of the pristine carbon/Si electrode is much thicker than that of the polymer coated electrode (Figure S15, Supporting Information), which explains the enhanced cycling performance of the self-healing polymer coated carbon/Si electrodes. Moreover, the graphitic carbon/Si/self-healing polymer electrodes after strain cycling still exhibit stable cycling only with a little faster decay in

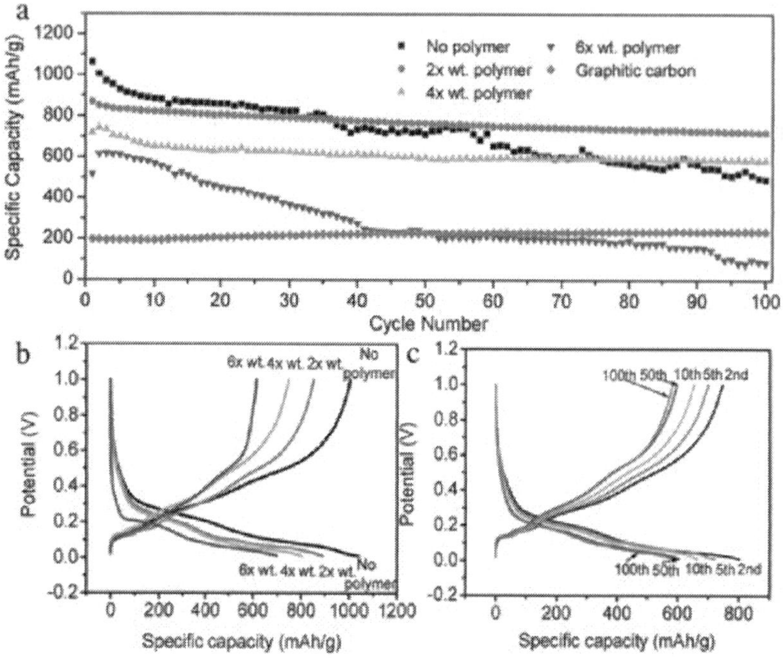

Figure 4. a) Specific delithiation capacities of the carbon/Si electrodes with different amounts of the self-healing polymer and the bare carbon electrode at the current density of 100 mA g^{-1} over cycling. b) Typical voltage profiles for the carbon/Si electrodes with different amounts of the self-healing polymer (second cycle). c) Voltage profiles of a carbon/Si electrode coated with 4× weight self-healing elastic polymer upon cycling.

capacity upon cycling, compared to the foam electrodes without undergoing strain cycling (Figure S16, Supporting Information). The much improved cycling performance is a strong indication of the structure enhancement and self-healing effect of the polymer helping keep the electrode intact. However, further increasing the amount of self-healing polymer (e.g., 6× weight polymer) leads to rapid capacity decay. The electrode almost loses all the capacity after 100 charge/discharge cycles (only 86 mAh g^{-1} for the 100th cycle). Regardless of their specific capacities, all the tested electrodes show stable Coulombic efficiency after the initial cycles (e.g., 99% from 12 to 100th cycle for the electrode with 4× weight polymer, Figure S17, Supporting Information). Therefore, to avoid low specific capacities and poor cyclability, a relative thin polymer coating layer is needed. Nevertheless, to obtain good stretchability, a relative thick polymer coating layer is preferred. In term of both electrochemical performances and stretchability, our results suggest that a 4× weight polymer amount is optimal and can meet the requirements of high capacities, stable cycling stability and good stretchability. Under the same electrochemical measurement condition, we also investigated the cyclability of the bare carbon foam electrode. The as-synthesized carbon electrode exhibits typical voltage profiles of graphite anodes with stable cycling (Figure 4a and Figure S18, Supporting Information).[23] It has been confirmed that the electric conductivity of graphite increases after the lithium insertion and can reach two orders of magnitude of the initial value.[24] Thus, the graphitic carbon skeleton of our electrodes can keep the entire electrode highly conductive during the charge/discharge cycles. The typical voltage profiles of the carbon/Si electrodes coated with different amount of self-healing polymer are shown in Figure 4b. With the increasing amount of the self-healing polymer, gradually decreased reversible specific capacities and slightly increased over potential are observed, arising from the low electronic conductivity and increased lithium-ion diffusion distance on the surface of the electrodes. Figure 4c shows the voltage profiles of a carbon/Si electrode coated with 4× weight polymer upon cycling. In spite of slight capacity decay, the charge/discharge curves remain steady and reversible even after 100 cycles.

In summary, stretchable lithium-ion batteries present unique opportunities for stretchable electronics but also limitations (e.g., low-energy density and limited stretchability). We have designed a conformal graphitic carbon/Si/self-healing elastic polymer foam structure to successfully achieve a high-capacity long-lifespan anode for stretchable lithium-ion batteries. Our demonstration here of a graphitic carbon/Si composite that has a high reversible specific capacity and a self-healing elastic polymer that enables long battery cycle life and good stretchability indicates a rational design for high-capacity stretchable electrodes. In comparison to the widely-used $Li_4Ti_5O_{12}$ anode material in stretchable lithium-ion batteries, the features of a high reversible specific capacity and low working potential are of key importance for the improvement of energy and power densities of the entire battery. This work presents an effective route in developing high-capacity Si-based anodes for stretchable lithium-ion batteries for the first time and paves the way for their promising applications in stretchable electronics. The successful stretchable structure design using a self-healing elastic polymer could also be extended to other stretchable material systems for various stretchable devices.

Supporting Information

Supporting Information is available from the Wiley Online Library or from the author.

Acknowledgements

Y.M.S. and J.L. contributed equally to this work. This work was supported by Samsung Electronics.

Received: September 25, 2015
Revised: November 21, 2015
Published online: January 27, 2016

[1] a) S. Xu, Y. H. Zhang, L. Jia, K. E. Mathewson, K.-I. Jang, J. Kim, H. R. Fu, X. Huang, P. Chava, R. H. Wang, S. Bhole, L. Z. Wang, Y. J. Na, Y. Guan, M. Flavin, Z. S. Han, Y. G. Huang, J. A. Rogers, *Science* **2014**, *344*, 70; b) D.-H. Kim, N. S. Lu, R. Ma, Y.-S. Kim, R.-H. Kim, S. D. Wang, J. Wu, S. M. Won, H. Tao, A. Islam, K. J. Yu, T.-I. Kim, R. Chowdhury, M. Ying, L. Z. Xu, M. Li, H.-J. Chung, H. Keum, M. McCormick, P. Liu, Y.-W. Zhang, F. G. Omenetto, Y. G. Huang, T. Coleman, J. A. Rogers, *Science* **2011**, *333*, 838; c) D. Son, J. Lee, S. T. Qiao, R. Ghaffari, J. Kim, J. E. Lee, C. Song, S. J. Kim, D. J. Lee, S. W. Jun, S. X. Yang, M. Park, J. Shin, K. Do, M. Lee, K. Kang, C. S. Hwang, N. S. Lu, T. Hyeon, D.-H. Kim, *Nat. Nanotechnol.* **2014**, *9*, 397.

[2] a) N.-S. Choi, Z. H. Chen, S. A. Freunberger, X. L. Ji, Y.-K. Sun, K. Amine, G. Yushin, L. F. Nazar, J. Cho, P. G. Bruce, *Angew. Chem. Int. Ed.* **2012**, *51*, 9994; b) M. Armand, J.-M. Tarascon, *Nature* **2008**, *451*, 652; c) B. Scrosati, J. Hassoun, Y.-K. Sun, *Energy Environ. Sci.* **2011**, *4*, 3287.

[3] Z. M. Song, T. Ma, R. Tang, Q. Cheng, X. Wang, D. Krishnaraju, R. Panat, C. K. Chan, H. Y. Yu, H. Q. Jiang, *Nat. Commun.* **2014**, *5*, 3140.

[4] S. Xu, Y. H. Zhang, J. Cho, J. Lee, X. Huang, L. Jia, J. A. Fan, Y. W. Su, J. Su, H. G. Zhang, H. Y. Cheng, B. W. Lu, C. J. Yu, C. Chuang, T.-i. Kim, T. Song, K. Shigeta, S. Kang, C. Dagdeviren, I. Petrov, U. Paik, J. A. Rogers, *Nat. Commun.* **2013**, *4*, 1543.

[5] a) W. Weng, Q. Sun, Y. Zhang, S. S. He, Q. Q. Wu, J. Deng, X. Fang, G. Z. Guan, J. Ren, H. S. Peng, *Adv. Mater.* **2015**, *27*, 1363; b) Y. Zhang, W. Y. Bai, X. L. Cheng, J. Ren, W. Weng, P. N. Chen, X. Fang, Z. T. Zhang, H. S. Peng, *Angew. Chem. Int. Ed.* **2014**, *53*, 14564.

[6] Y. Liu, S. Gorgutsa, C. Santato, M. Skorobogatiy, *J. Electrochem. Soc.* **2012**, *159*, A349.

[7] C. K. Chan, H. L. Peng, G. Liu, K. McIlwrath, X. F. Zhang, R. A. Huggins, Y. Cui, *Nat. Nanotechnol.* **2008**, *3*, 31.

[8] H. Wu, Y. Cui, *Nano Today* **2012**, *7*, 414.

[9] a) N. Liu, Z. D. Lu, J. Zhao, M. T. McDowell, H.-W. Lee, W. T. Zhao, Y. Cui, *Nat. Nano* **2014**, *9*, 187; b) N. Liu, H. Wu, M. T. McDowell, Y. Yao, C. M. Wang, Y. Cui, *Nano Lett.* **2012**, *12*, 3315; c) H. Wu, G. Chan, J. W. Choi, I. Ryu, Y. Yao, M. T. McDowell, S. W. Lee, A. Jackson, Y. Yang, L. B. Hu, Y. Cui, *Nat. Nanotechnol.* **2012**, *7*, 310; d) H. Wu, G. H. Yu, L. J. Pan, N. Liu, M. T. McDowell, Z. N. Bao, Y. Cui, *Nat. Commun.* **2013**, *4*, 1943; e) H. Kim, B. Han, J. Choo, J. Cho, *Angew. Chem.* **2008**, *120*, 10305; f) A. Magasinski, P. Dixon, B. Hertzberg, A. Kvit, J. Ayala, G. Yushin, *Nat. Mater.* **2010**, *9*, 353.

[10] a) C. Wang, H. Wu, Z. Chen, M. T. McDowell, Y. Cui, Z. N. Bao, *Nat. Chem.* **2013**, *5*, 1042; b) Z. Chen, C. Wang, J. Lopez, Z. D. Lu, Y. Cui, Z. N. Bao, *Adv. Energy Mater.* **2015**, *8*, 1401826.

[11] a) J.-Y. Sun, X. H. Zhao, W. R. K. Illeperuma, O. Chaudhuri, K. H. Oh, D. J. Mooney, J. J. Vlassak, Z. G. Suo, *Nature* **2012**, *489*, 133; b) E. Ducrot, Y. L. Chen, M. Bulters, R. P. Sijbesma, C. Creton, *Science* **2014**, *344*, 186.

[12] a) N. Li, Z. P. Chen, W. C. Ren, F. Li, H.-M. Cheng, *Proc. Natl. Acad. Sci. USA* **2012**, *109*, 17360. b) H. X. Ji, L. L. Zhang, M. T. Pettes, H. F. Li, S. S. Chen, L. Shi, R. Piner, R. S. Ruoff, *Nano Lett.* **2012**, *12*, 2446.

[13] J. Y. Ji, H. X. Ji, L. L. Zhang, X. Zhao, X. Bai, X. B. Fan, F. B. Zhang, R. S. Ruoff, *Adv. Mater.* **2013**, *25*, 4673.

[14] Z. P. Chen, W. C. Ren, L. B. Gao, B. L. Liu, S. F. Pei, H.-M. Cheng, *Nat. Mater.* **2011**, *10*, 424.

[15] P. Cordier, F. Tournilhac, C. Soulié-Ziakovic, L. Leibler, *Nature* **2008**, *451*, 977.

[16] M. A. Pimenta, G. Dresselhaus, M. S. Dresselhaus, L. G. Cancado, A. Jorio, R. Saito, *Phys. Chem. Chem. Phys.* **2007**, *9*, 1276.

[17] S. M. Kang, K. M. Ahn, B. T. Ahn, *J. Electrochem. Soc.* **2012**, *159*, H29.

[18] M. Schade, N. Geyer, B. Fuhrmann, F. Heyroth, H. S. Leipner, *Appl. Phys. A* **2009**, *95*, 325.

[19] A. A. El-Barbary, S. Trasobares, C. P. Ewels, O. Stephan, A. V. Okotrub, L. G. Bulusheva, C. J. Fall, M. I. Heggie, *J. Phys.: Conf. Ser.* **2006**, *26*, 149.

[20] M. M. Coleman, M. Sobkowiak, G. J. Pehlert, P. C. Painter, T. Iqbal, *Macromol. Chem. Phys.* **1997**, *198*, 117.

[21] a) J. S. Bridel, T. Azaïs, M. Morcrette, J.-M. Tarascon, D. Larcher, *Chem. Mater.* **2010**, *22*, 1229; b) I. Kovalenko, B. Zdyrko, A. Magasinski, B. Hertzberg, Z. Milicev, R. Burtovyy, I. Luzinov, G. Yushin, *Science* **2011**, *334*, 75; c) Y. K. Jeong, T.-W. Kwon, I. Lee, T.-S. Kim, A. Coskun, J. W. Choi, *Nano Lett.* **2014**, *14*, 864.

[22] Q. Sun, X. Fang, W. Weng, J. Deng, P. N. Chen, J. Ren, G. Z. Guan, M. Wang, H. S. Peng, *Angew. Chem. Int. Ed.* **2015**, *127*, 10685.

[23] N. A. Kaskhedikar, J. Maier, *Adv. Mater.* **2009**, *21*, 2664.

[24] W. Bao, J. Y. Wan, X. G. Han, X. H. Cai, H. L. Zhu, D. Kim, D. Ma, Y. L. Xu, J. N. Munday, H. D. Drew, M. S. Fuhrer, L. B. Hu, *Nat. Commun.* **2014**, *5*, 4224.

459844 Large Area Graphene Nanoribbons By Wetting Transparency-Assisted Block Copolymer Lithography

Abstract:

Graphene plasmons have gained considerable attention during the last decade due to their potential to overcome current challenges facing traditional plasmonic materials, such as goldâ€™s limited spatial resolution and extinction wavelengths. In particular, patterning graphene into nanoribbons (graphene nanoribbons, GNR) allows for tunability in the emerging fields of mid-infrared and terahertz spectroscopy based devices. However, the fabrication of GNR arrays for these plasmonic devices often includes a low-throughput electron beam lithography step that cannot be easily scaled to large areas. In this study, we developed a new and more scalable GNR fabrication method based on block copolymer (BCP) lithography. To pattern graphene into GNRs, lamellae-forming BCP domains must be oriented perpendicular to substrate and then selectively etched. However, perpenducular orientation can only be achieved when the surface energy of the underlying substrate is neutral or non-preferential to wetting either block of the BCP. Extensive contact angle and surface energy measurements indicated that the surface neutrality of the underlying substrate is mostly retained even with a single layer of graphene coating the top of the underlying substrate. This is due to the wetting transparency of graphene. Using wetting transparency-assisted BCP lithography, we successfully fabricated large-area (cm^2 scale) GNR arrays with 20 nm nanoribbon widths. The optical properties of the GNR arrays were studied by Fourier transform infrared spectroscopy, and a clear plasmonic extinction peak in the mid-infrared region (~1600 cm^{-1}) was observed. This GNR fabrication method could be useful for high-throughput production of a broad range of mid-infrared plasmonic devices, including modulators, biosensors, and photodetectors.

459331 Recyclable Cross-Linked Polymer Networks Via One-Step Controlled Radical Polymerization

Abstract:

Conventional polymer networks cannot be recycled into high-value products because of permanent, covalent cross-links. Rubber tires illustrate well the issues ranging from economic loss to environmental problems that arise with spent, cross-linked polymers. Research began over a decade ago to produce network polymers with dynamic or reversible bonds that can be self-healing or recyclable. Such networks can not only maintain good mechanical performance but also change their chemical structures or physical shapes in response to external stimuli (e.g., temperature) through reversible rearrangement reactions. Much of this research has focused on reversible reactions between functional groups, including thermoreversible Diels-Alder reactions,[1] transesterification reactions,[2] disulfide exchange reactions,[3] among others.[4,5] Such approaches often require complex synthesis of specific chemical units as precursors or suffer exchange inefficiency in the solid state. No previous study has achieved direct formation of recyclable networks from commercial monomer/polymer without pre-synthesis and pre-functionalization or demonstrated full property recovery (within error) for elastomers reflecting full recovery of cross-links after multiple melt-reprocessing steps.

Here, we developed a one-step strategy using nitroxide-mediated polymerization (NMP) to synthesize recyclable network polymers with dynamic alkoxyamine cross-links[6,7] that allows for reprocessing in the melt state through repeated bond formation and dissociation. With this method, we use stable nitroxide radicals containing a polymerizing moiety (carbon-carbon double bond) as both radical regulators during reaction and branch units on the polymers or cross-linker chains. We also use bifunctional initiators to generate chains with radical sites on both sides to connect the nitroxide branch units, resulting in cross-linked polymer. This NMP-based approach can be applied to any monomers or polymers that contain at least one carbon-carbon double bond that is amenable to radical polymerization. For

example, through this strategy, we designed a network polymer using polybutadiene and styrene monomer as a model for recyclable tires, showing full property recovery after multiple melt-reprocessing recycles; we also produced polymer networks from monomers with good reprocessability and relative uniformity. With nitroxide groups serving as thermally reversible cross-links and simple functionalization of monomers, this one-step strategy provides for both robust, sustainable recyclability of cross-linked polymers and design of networks for advanced technologies.

References

1. Chen, X.; Dam M. A.; Ono, K; Mal, A.; Shen, H.; Nutt, S. R.; Sheran, K.; Wudl, F. Science **2002**, 295, 1698.

2. Montarnal, D.; Capelot, M.; Tournilhac, F.; Leibler, L. Science **2011**, 334, 965.

3. Rekondo, A.; Martin, R.; Ruiz de Luzuriaga, A.; Cabañero, G.; Grande, H. J.; Odriozola, I. Mater. Horiz. **2014**, 1, 237.

4. Ghosh, B.; Urban, M. W. Science **2009**, 323, 1458.

5. Billiet, S.; De Bruycker, K.; Driessen, F.; Goossens, H.; Van Speybroeck, V.; Winne, J. M.; Du Prez, F. E. Nat. Chem. **2014**, 6, 815.

6. Otsuka, H. Polym. J. **2013**, 45, 879.

7. Wang, F.; Rong, M. Z.; Zhang, M. Q. J. Mater. Chem. **2012**, 22, 13076.

467975 Single Polymer Dynamics of Linear and Circular Chains in Semi-Dilute Solutions

Abstract:

Semi-dilute polymer solutions are commonly encountered in a wide array of applications, including inkjet printing and advanced 3D printing technologies. Semi-dilute solutions are characterized by large fluctuations in polymer concentration, wherein polymer coils interpenetrate but may not be topologically entangled at equilibrium. In non-equilibrium flows, it is generally thought that polymer chains can 'self-entangle' in semi-dilute solutions, thereby leading to topological entanglements in solutions that are nominally unentangled at equilibrium. Despite recent progress in the field, we still lack a complete molecular-level understanding of the dynamics of polymer chains in semi-dilute solutions. In this work, we use single molecule techniques to investigate the dynamics of dilute and semi-dilute solutions of lambda-phage DNA in planar extensional flow, including polymer relaxation from high stretch, transient stretching dynamics in step-strain experiments, and steady-state stretching in flow. Our results are consistent with a power-law scaling of the longest polymer relaxation time $(c/c^*)^{0.48}$ in semi-dilute solutions, where *c* is polymer concentration and c^* is the overlap concentration. Based on these results, we found an effective excluded volume exponent v = 0.56, which is in good agreement with recent bulk rheological experiments on semi-dilute DNA solutions.

We also studied the non-equilibrium stretching dynamics of semi-dilute polymer solutions, including transient and steady-state stretching dynamics in planar extensional flow using an automated microfluidic trap. Our results show that polymer stretching dynamics in semi-dilute solutions is a strong function of concentration. In particular, we observe a decrease in transient polymer stretch in semi-dilute solutions at moderate Weissenberg number *Wi* compared to dilute solutions, with the difference in polymer stretch between dilute and semi-dilute solutions decreasing for *Wi* > 1. Moreover, our experiments reveal a milder coil-to-stretch transition for semi-dilute polymer solutions compared to dilute solutions, which is interpreted in the context of a critical *Wi* at the coil-to-stretch transition. Interestingly, we observe a unique set of molecular conformations during the transient stretching process for single polymers in semi-dilute solutions,

which suggests that the transient stretching pathways for polymer chains in semi-dilute solutions is qualitatively different compared to dilute solutions due to intermolecular interactions. [1,2]

We further extended single polymer experiments to study the dynamics of ring polymers in ultra-dilute solutions[3] and in background solutions of semi-dilute linear polymers. This work represents the first single molecule measurement and direct observation of ring polymer dynamics in semi-dilute solutions under flow. Interestingly, we observe distinct "threading behavior" of single ring polymers in a sea of semi-dilute linear polymers, which results in qualitatively different dynamics compared to linear chains. Using fluorescently labeled circular DNA as ring polymers, we preformed a series of controlled strain-rate experiments on ring polymers in unlabeled semi-dilute linear background in planar extensional flow using precision microfluidics. We observe strong inhibition of polymer extension and *strikingly large fluctuations* in steady-state polymer extension for ring polymers in flow, demonstrating the strong interplay between polymer topology and polymer concentration. Taken together, our work aims to provide a molecular-level understanding of the role of polymer concentration and topology in non-dilute polymer solutions via direct observation of single chain dynamics in strong flow. [3,4]

[1] **K. Hsiao**, C. Sasmal, J. R. Prakash, and C. M. Schroeder, "Direct observation of DNA dy namics in semi-dilute solutions in extensional flow," *submitted* (2016).

[2] C. Sasmal, **K. Hsiao**, J. R. Prakash, and C. M. Schroeder, "Stretch relaxation of DNA molecules in semi-dilute solutions," *submitted* (2016).

[3] Y. Li, **K. Hsiao**, C. A. Brockman, D. Y. Yates, R. M. Robertson-Anderson, J. A. Kornfield, M. J. San Francisco, C. M. Schroeder, and G. B. McKenna, "When ends meet: Circular DNA stretches differently in elongational flows," *Macromolecules*, **48**, 5997-6001 (2015).

[4] **K. Hsiao,** C. M. Schroeder, and C. E. Sing, "Ring polymer dynamics are governed by a coupling between architecture and hydrodynamic interactions," *Macromolecules*, **49**, 1961-1971 (2016).

455889 Electromagnetic Wave Absorbing Polymer Nanocomposites

Abstract:

Multifunctional polymer matrix nanocomposites filled with nanofillers including iron nanoparticles, carbon nanotubes, graphene nanoplatelets, and the carbon nanotubes coated with magnetic iron oxide nanoparticles were prepared for microwave absorption at gigahertz frequency range. Surface treatment/functionalization with surfactant was used to improve the nanoparticle dispersion quality and then to enhance the microwave absorption performance. Relative complex permittivity and relative complex permeability are investigated and the corresponding microwave absorption efficiency were calculated within the tested frequency ranging from 2 - 20 GHz. Results indicate that silica coated iron core-shell nanoparticles were found to be able to reduce the eddy current effect of untreated iron nanoparticles in the polymer matrix. Meanwhile, carbon nanotubes coated with iron oxide nanoparticles could improve the matching between magnetic and dielectric loss for enhanced microwave absorption. Other fillers like graphene nanoplatelets with high microwave absorption efficiency were also studied and discussed in details.

References:

1) Electromagnetic Field Absorbing Polypropylene Nanocomposites with Tuned Permittivity and Permeability by Nanoiron and Carbon Nanotubes, ***Journal of Physical Chemistry C***, 118(42) 24784-24796 (2014). pdf

2) Electromagnetic Field Shielding Polyurethane Nanocomposites Reinforced with Core-Shell Fe-Silica Nanoparticles, ***Journal of Physical Chemistry C***, 115, 15304-15310 (2011). pdf

3) Magnetic and Electromagnetic Evaluation of the Magnetic Nanoparticle Filled Polyurethane Nanocomposites, ***Journal of Applied Physics***, 10, 09M511 (2007). pdf

456135 Bio-Inspired Sensitive and Reversible Mechanochromisms Via Surface Engineering

Abstract:

In nature, some marine organisms, such as *Vogtia* and *Cephalopods*, have evolved to possess camouflage traits by dynamically and reversibly altering their transparency, fluorescence, and coloration via muscle controlled surface structures and morphologies. To mimic this display tactics, we designed similar deformation controlled surface engineering via strain-dependent cracks and folds to realize four types of novel mechanochromic devices: (1) transparency change mechanochromism (TCM), (2) luminescent mechanochromism (LM), (3) color alteration mechanochromism (CAM), and (4) encryption mechanochromism (EM), based on a simple bilayer system containing a rigid thin film and a soft substrate. These devices exhibit a wide scope of mechanochromic response with excellent sensitivity and reversibility. The TCM device can reversibly and instantly switch between transparent and opaque state upon stretching and releasing. The LM can emit intensive fluorescence as stretched with an ultrahigh strain sensitivity in comparison to strain sensors based on electrical resistance change. The CAM can turn fluorescent color from green to yellow to orange as stretched within 20% strain. The EM device can reversibly reveal and conceal any desirable patterns. These novel devices are promising for applications in smart windows, dynamic optical switches, strain sensors, encryption, etc.

456436 Tuning the Dynamics of Moisture Responsive Wrinkles

Abstract:

Surface instability such as wrinkles commonly occurs in various materials with a wide scope of dimensions. Herein, we introduce three types of moisture responsive wrinkling surfaces with different dynamics as exposed to high humid environment based on a bilayer structure. All the three surfaces are initially flat and forming wrinkles as moisturized but with different responsive behaviors upon being further moisturized, dried and re-moisturized. In the first responsive dynamics, the highly opaque wrinkling surface can be rapidly generated as being moisturized and sustains for ca. 30 s followed by rewinding to transparent flattening surface upon further moisture exposure. No wrinkling surface is created again upon being re-exposed to moisture. This allows the design of an intriguing moisture responsive encryption device with the capability of â€œerase after readâ€. In the second dynamics, the wrinkling surface can be rapidly generated as being moisturized and keeps stable regardless of moisture exposure time. Upon drying, the winkles can release back to the original flattening surface. The wrinkling and flattening surface can be repeatedly and reversibly created upon multiple dry/moisture cycles. A corresponding novel application of breathing activated anti-counterfeit tab is demonstrated. In the third dynamics system, the wrinkling surface can be generated as being moisturized and keeps stable in moisturized and dried state. This surface can be applied as a water indicator for electronic circuit, anti-glare surface, and optical diffusor.

458009 Electrically Controllable Plasmonic Behavior of Gold Nanocube@Polyaniline Core/Shell Nanostructures

Abstract:

Noble metal nanoparticles, especially those made from gold and silver, have attracted significant attention because of their unique localized surface plasmon resonance (LSPR) properties that can be exploited in various applications such as photocatalysis, solar cells, and sensing. Even though specific plasmonic properties can be obtained through well-established synthesis techniques, it is difficult to modulate LSPR peak position and cross-section intensity in a real-time manner. Here, we demonstrate the electrically tunable plasmonic behavior of hybrid core/shell nanostructures composed of gold nanocube (AuNC) cores and electroactive polyaniline (PANI) shells. The oxidation state of the PANI shells was reversibly changed by applying an electrical potential, which resulted in a refractive index change of the PANI shells. Consequently, a significant LSPR shift was achieved with high reversibility and stability. Because of its unique core/shell structure, with AuNC cores and electroactive PANI shells in close contact, a high LSPR shift (24 nm) is achieved. Finite-difference time-domain (FDTD) computer simulations show that the electromagnetic field decay length of the core AuNC is an important parameter in the design core/shell nanostructures; the most efficient LSPR modulation was obtained for PANI shells thinner than the electromagnetic field decay length of the gold nanoparticles. Single particle studies showed that the PANI shell acts as a physical barrier between adjacent nanoparticles as well as an electroactive medium, which prevents plasmonic coupling while providing a variable refractive index medium.

470542 Wearable Sensor Comprised of Nafion/Single-Walled Carbon Nanotubes/Metal Oxide Nanoparticles for the Detection of the Chemical Warfare Agent Simulant DMMP

Abstract:

The imminent threat of odorless and colorless chemical warfare agents (CWA) such as organophosphorus containing nerve agents, has spurred intense efforts towards the development of gas sensors that can finely and selectively detect the presence of these fatal agents. Herein we present the development of wearable CWA sensing composite coatings comprised of defect-rich semiconducting MONP, SWNT and a semi-permeable polyelectroylyte membrane, i.e. Nafion. Nafion-SWNT dispersions are prepared and dropcasted onto wearable cotton fabric, and post-processing techniques are utilized to maximize the electrical conductance. Different candidates of semiconducting MONP (cerium, iron, nickel and zinc) are grown *in situ* at the hydrophilic ion exchange sites of the Nafion polymer. Composite materials are characterized by x-ray diffraction (XRD), Raman spectroscopy and scanning electron microscopy/electron diffraction spectroscopy (SEM/EDS). Finally, silver metal contacts are evaporated onto the sides of the wearable composite, and the resistance is measured as a function of dimethyl methylphosphonate (DMMP) vapor concentration.

470902 Study of Functional Polymeric Materials for Water Remediation

Abstract:

Protection of the environment from pollution is a challenging task of this century. Thereby treatment of industrial wastewater is the actual problem, since it is directly linked to the protection of water resources. In recent decades, flocculants are used as reagents for the purification of wastewater from the various contaminants. Their usage can improve the quality of the treated water and increase the productivity of treatment facilities. Despite the large amount of research in this area, the problem of natural and wastewater treatment cannot be considered to be solved.

This causes development of new and more effective as well as easily available flocculants and also modernization of existing water treatment technology.

Our research is to develop and study novel ampholytic functional polymers using 2-acrylamido-2-methylpropanesulfonic acid (H-AMPS), N,N-dimethyl-N,N-diallylammonium chloride (DMDAAC), N,N-dimethylacrylamide (DMAA), vinyl butyl ether (VBE) and vinyl ether of monoethanolamine (VEMEA) for water remediation. Synthesis of flocculants was performed by free radical copolymerization using initiators (AIBN, ammonium and potassium persulfates) in aqueous, organic solutions and in an inert gas (argon) atmosphere. Molar compositions of the synthesized copolymers were determined by elemental analysis, IR and NMR spectroscopy and potentiometric titration.

The kinetics of the reaction of copolymerization have been studied by dilatometry method. The synthesized copolymers were tested in the laboratory as flocculants for the treatment of wastewater of drilling fluids, suspensions of bentonite clay, hydrosol of iron hydroxide (III), solution of manure of animals and birds and etc. Experimental data showed that the synthesized copolymers exhibits excellent flocculating properties and could be used for purification of industrial wastewater from various highly impurities.

471623 Nanocomposite Kirigami As Strain-Tunable Optical Gratings

Abstract:

Nanocomposites with kirigami-inspired microstructures can create exciting opportunities for advanced technologies, ranging from stretchable electronics to adaptive optics. In kirigami techniques, patterned cuts are introduced in thin film composites for reducing strain concentration under macroscopic deformation and/or guiding formation of complex structures from simple 2D elements. Here we show that microscale kirigami of nanocomposites from stiff plastics, metals and carbon nanotubes can yield highly stretchable and tunable optical diffraction gratings. These reconfigurable optical gratings with over a 100% range of period tunability enable critical capabilities in beam steering and optical power modulation. The angular range of beam steering can be as large as 6.5 degrees for 635nm laser beam as compared to ~1 degree in previously reported surface-grooved elastomer gratings and ~0.02 degree in MEMS gratings. The versatility of the kirigami patterns, the diversity of the available nanocomposite materials, and their advantageous mechanical properties of the foundational materials open the path for engineering of advanced reconfigurable optical devices.

454107 Fast Growth of Single Crystal Graphene and Application of Graphene in Catalytic Reactions

Abstract:

Fast growth, with growth rate of 70 Î¼m/min, of single crystal graphene was demonstrated by preparing graphene in confined space. With this method, we increases the concentration of the reactive radicals that leads to boosted graphene growth rate while not affecting the nucleation density, which consequently allows the graphene growth into sub-centimeter size. Furthermore, we engineered well-distributed dicobalt phosphide (Co_2P) nanoparticles encapsulated in N,P-doped graphene (Co_2P@NPG) as electrochemical catalyst for hydrogen evolution reaction (HER). Our experimental results show significantly enhanced performance for such engineered core-shell nanostructure due to the synergistic effect from nanoparticles encapsulation and nitrogen and phosphors doping on graphene structures. The obtained Co_2P@NPG modified cathode exhibits small overpotentials of only -45 mV at 1mA cm^{-2}.

465526 The Influence of Surface Functionalization on Nanoparticle-Cellular Interactions

Abstract:

Owing to its exceptional antimicrobial activity silver nanoparticles (AgNPs) warrant the significant incorporation in the consumer and medical products to date. However, an important aspect when considering the use of AgNPs in for biomedical applications is the interaction of cells and proteins with nanoparticles. Proteins interact with NPs almost instantly, forming a protein-corona (PC) at the surface of the NPs. This PC remains associated with the NPs under in vivo conditions, thereby conferring the biological identity to the particles. Studies have been conducted on the interaction of NPs with human serum proteins, but there is a lack of information on the fate of these serum-coated NPs and how they interact with the proteins at the cellular level. The study aims to investigate the role of bioactive compounds (Curcumin and EGCG) in the AgNP-protein corona formation and its further interactions with cells. In my presentation I will specifically present the data on synthesis and characterisation of Curcumin and EGCG coated nanoparticles and discuss how these coated nanoparticles interact differently with cells and protein milieu. Further applicability of these bioactive coated nanoparticles for biomedical applications will be discussed.

468114 Synthesis and Characterization of Pegylated Self-Assembled Rosette Nanotubes As Drug Delivery Vehicles

Abstract:

Rosette nanotubes (RNTs), which are synthetic hybrid DNA nucleic-acid nanotubes, are self-assembled from fused guanine and cytosine motifs (GÎ›C motif). In RNTs, DNA Watson-Crick pairing takes place between the guanine side of the GÎ›C motif and its cytosine side. The spatial arrangement of these arrays leads them to form a six-membered supermacrocycle with high stability as a result of the extensive H-bond network and the inter-rosette Ï€-Ï€ stacking interactions. In addition, using synthetic approaches, varied functional groups have been grafted to GÃ™C such as amino acid, peptides, and crown ethers. Upon self-assembly, the functional groups will display on the surface of RNTs which imparted important physical and chemical properties.

RNTs have found numerous medical applications as a result of their high biocompatibility and low toxicity. In order to improve the solubility in aqueous and pharmacokinetics behavior of RNTs as drug delivery vehicles, we have developed polyethylene glycol (PEG) covalently grafted on the surface of RNTs through organic synthesis and self-assembly, in which PEG is non-toxic and widely used to improve chemical and physical properties of drug delivery vehicles. As a result, the surfaces of RNTs have been grafted PEG chains. Here we will present our synthetic strategy and self-assembly studies of this new material as well as *in vitro* cytocompativility studies.

461158 Photoexcited Quantum Dots for Killing Multidrug-Resistant Bacteria

Abstract:

Multidrug-resistant bacterial infections are an ever-growing threat because of the shrinking arsenal of efficacious antibiotics. Metal nanoparticles can induce cell death, yet the toxicity effect is typically nonspecific. Here, we show that photoexcited quantum dots (QDs) can kill a wide range of multidrug-resistant bacterial clinical isolates, including methicillin-resistant Staphylococcus aureus, carbapenem-resistant Escherichia coli, and extended-spectrum Î²-lactamase-producingÂ Klebsiella pneumoniaeÂ and Salmonella typhimurium. The killing effect is independent of material and controlled by the redox potentials of the photogenerated charge carriers, which selectively alter the cellular redox state. We also show that the QDs can be tailored to kill 92% of bacterial cells in a monoculture, and in a co-culture of E. coli and HEK 293T cells, while leaving the mammalian cells intact, or to increase bacterial proliferation. Photoexcited QDs could be used in the study of the effect of redox states on living systems, and lead to clinical phototherapy for the treatment of infections.

470547 Chitosan-Coated Selenium Nanoparticles and Their Affects on Bacterial Growth Kinetics

Abstract:

Introduction:

The prevalence of antibiotic-resistant bacteria has continued to grow in recent years, resulting from increased use (and misuse) of antibiotics in daily life. It is imperative, then, to develop novel bactericidal treatments and therapies to combat these pathogens. Unfortunately, the discovery of new antibiotics, in the traditional sense, has become increasingly difficult in recent years. Selenium is an essential trace material found in the body associated with antioxidant and metabolic mechanisms that play a role in preventing cell damage. Selenium nanoparticles (SeNP) have been previously found to be antibacterial without any apparent toxicity to healthy mammalian cells. Here, we have developed SeNP synthesized with a chitosan coating to improve colloidal stabilization and to introduce new surface chemistries for further targeting and delivery options. The chitosan-coated SeNP (Ch-SeNP) were able to conserve the size distribution and the antibacterial efficacy seen with previous iterations of SeNP. To compare these results, we employed the Gompertz kinetic growth models to analyze the changes in lag time (time in which it takes the bacteria to enter their exponential growth phase) and maximum growth rate, with respect to different concentrations of SeNP and Ch-SeNP. The results showed an increase in the lag time and a decrease in the growth of *S. aureus* bacteria with Ch-SeNP.

Materials and Methods:

Ch-SeNP were produced using a colloidal suspension method. Briefly, chitosan in a 1% acetic acid solution was added dropwise into a stirred sodium selenite solution. Ascorbic acid was then added dropwise to the solution and stirred until equilibrium was reached. The nanoparticles were centrifuged for 10 minutes to remove excess reactants, and

subsequently resuspended into DI water. DLS and TEM were used to characterize the Ch-SeNP to determine the size and surface charge. The growth kinetics were measured by incubating *S. aureus*with various Ch-SeNP concentrations in 3% TSB. The optical density (at 592 nm) was measured regularly for 24 hours throughout the growth assay. Gompertz and Richards growth models were applied to develop the growth kinetics to understand the role of the selenium nanoparticles on the bacteriaâ€™s proliferation. All experiments were replicated three times and statistical significance analyzed via ANOVA. The data obtained was compared with controls using a two-tailed, unpaired Studentâ€™s t-test, with $p<0.05$ considered to be statistically significant.

Results and Discussion:

The Ch-SeNP inhibited *S. aureus*growth over time. Most notably, the lag time of the growth curves increased while the growth rate decreased with increasing concentrations of Ch-SeNP treatments. Even for Ch-SeNP concentrations as low as 0.0023 mg/ml, both the lag time and growth rate changed significantly compared to untreated controls.

Conclusion:

Selenium nanoparticles coated with chitosan have easily reproducible batch-to-batch characteristics and are able to effectively inhibit the growth of *S. aureus*. Using the Gompertz growth model, the present study accurately calculated the growth kinetics of *S. aureus* with the chitosan coated selenium nanoparticles. Future studies can use the Gompertz and Richards model to accurately standardize the effectiveness of nanoparticles on the growth of *S. aureus*. Moreover, further studies regarding the mechanism of antibacterial action of selenium nanoparticles coated with chitosan are needed.

Acknowledgements:

The authors thank Northeastern University and the Department of Chemical Engineering for funding this research.

468698 Targeted Silver Nanoparticles for Selective Cytotoxicity of Tumorigenic Endothelial Cells

Abstract:

Silver nanoparticles (AgNPs) have widely documented cytotoxic properties that are influenced by size, shape, and surface chemistry. AgNPs introduced to a biological environment interact with proteins and cell membrane receptors, which influence cell uptake, gene expression and cytotoxicity. Thus, controlling the cell-NP interface has important implications in the application of AgNPs for therapeutic purposes. For example, AgNPs can be engineered to selectively interact with specific cell types by surface functionalization with biomimetic targeting moieties. The objective of this study was to investigate the effects of AgNP exposure on the cytotoxicity of hemangioendothelioma (EOMA) cells, which are a tumorigenic endothelial cell line. Cells were incubated with citrate stabilized AgNPs of variable diameter (10-100μm) at a 0.01 mg/mL. Absorbance based assays revealed that higher cell uptake and cytotoxic responses correlated with decreased particle size. Alternatively, AgNPs were functionalized with cyclic RGD peptides by either physisorption or covalent conjugation to target $\alpha_v\beta_3$-integrin receptors on the cell membrane of EOMA cells. The presence of $\alpha_v\beta_3$-integrin expression in EOMA cells was demonstrated via RT-PCR and confocal microscopy. EOMA cells exposed to RGD-AgNPs in serum-containing media did not show a significant enhancement in cell uptake or cytoxocity relative to non-binding RAD control peptides. In contrast, RGD-AgNP fabricated with NHS-ester reaction chemistry elicited a significant increase in cytotoxicity relative to non-targeted controls. DCFDA and ELISA assays suggest that enchanced cytotoxicity is correlated with increased ROS generation and an upregulation of inflammatory cytokine production. These results demonstrate the utility of functionalizing AgNP with RGD peptides for selective cytotoxicity of $\alpha_v\beta_3$-integrin expressing EOMA cells. Future work will be aimed at further elucidating the molecular mechanisms of RGD-AgNP induced cytotoxicity in EOMA cells.

473075 Cell Membrane-Camouflaged Nanomotors for Biodetoxification and Drug Delivery

Abstract:

Attempts to apply artificial micro-/nanomotors for diverse biomedical applications have inspired a variety of strategies for designing motors with diverse propulsion mechanisms and functions. However, existing artificial motors are made exclusively of synthetic materials, which are subject to serious immune attack and clearance upon entering the bloodstream. Herein we present a platform by using red blood cell as camouflage coating for micro-/nanomotors. As the first example, a red blood cell membrane-camouflaged nanowire that can serve as new generation of biomimetic motor sponge is described. The biomimetic motor sponge is constructed by the fusion of biocompatible gold nanowire motors and RBC nanovesicles. The motor sponge possesses a high coverage of RBC vesicles, which remain totally functional due to its exclusively oriented extracellular functional portion on the surfaces of motor sponge. These biomimetic motors display efficient acoustical propulsion, including controlled movement in undiluted whole blood. The RBC vesicles on the motor sponge remain highly stable during the propulsion process, conferring thus the ability to absorb membrane-damaging toxins and allowing the motor sponge to be used as efficient toxin decoys. The efficient propulsion of the motor sponges under an ultrasound field results in accelerated neutralization of the membrane-damaging toxins. Such motor sponges connect artificial nanomotors with biological entities and hold great promise for treating a variety of injuries and diseases caused by membrane-damaging toxins.
As another example, an elegant approach is used to directly turn natural red blood cells (RBCs) into functional micromotors with the aid of ultrasound propulsion and magnetic guidance. Since the RBC motors preserve the biological and structural features of regular RBCs, these motors possess a wide range of antigenic, transport, and mechanical properties that common synthetic motors cannot achieve and thus hold considerable promise for a number of practical biomedical uses. The ability of the RBC micromotors to transport imaging and therapeutic agents at high speed and spatial precision through a complex microchannel network is also demonstrated. Such ability to load and transport diagnostic imaging agents and therapeutic drugs within a single cell-based motor, in addition to a lower toxicity observed once the drug is encapsulated within the multicargo RBC motor, opens the door to the development of theranostic micromotors that may simultaneously treat and monitor diseases.

464400 Mesostructure and Pore Size Distribution of Mesoporous Silica / Anodic Alumina Hierarchical Membranes Tuned with Ethanol

Abstract:

Synthesis of a hybrid mesoporous silica / anodic alumina membrane is challenging, as many methods lead to overgrowth of silica on the external surface of the alumina membrane, incomplete alumina channel filling, non-uniform silica structures, or silica with a wide pore size distribution [1]. Silica nanorods with a controlled structure can be formed within anodic alumina channels, using the anodic alumina as a sacrificial template to facilitate the orientation of the silica pores. Mesopores of a controlled diameter can be created in the silica by a soft-templating method, using tri-block copolymer surfactant, introduced in the precursor solution. However, non-uniformity over macroscopic cross-sectional distances of the hierarchical silica/anodic alumina composite precludes use of these materials for reproducible membrane-based separations.

We show how ethanol content, in the silica precursor solution, has a significant effect on the type and order of the mesoporous silica phase. Varying the ethanol concentration controls the mesoporosity and the structure of the mesoporous silica over macroscopic distances, so that an improved mesoporous silica/anodic alumina hybrid membrane with controlled pore sizes is produced [2].

The hierarchically structured membranes were characterized by scanning electron microscopy (SEM), transmission electron microscopy (TEM), small-angle X-ray scattering (SAXS) and nitrogen adsorption/desorption measurements. These techniques provide evidence of the increasing degree of ordering and narrower pore size distribution obtained with decreasing ethanol concentration. Interpretation of the formed structures and the influence of ethanol are made possible by dynamic light scattering (DLS) experiments of the precursor solutions, which allow us to follow the evolution of the micelles that template the silica mesopores, for various ethanol concentrations.

The membranes are tested by using them to carry out protein separations, showing promise as a platform for membrane based applications. [1]. S. Meoto, M.-O. Coppens, *J. Mater. Chem. A*, **2014**, 2, 5640 [2]. S. Meoto, N. Kent, M. M. Nigra, M.-O. Coppens, submitted

470470 Creation of Highly Textured Metal Oxide Films Using Wrinkled and Crumpled Graphene As Intercalation Templates

Abstract:

Creation of Highly Textured Metal Oxide Films Using Wrinkled and Crumpled Graphene as Intercalation Templates

Po-Yen Chen[1,2]*, **Ian Y. Wong**[2], **Robert H. Hurt**[2]

[1]*Department of Chemical Engineering, Massachusetts Institute of Technology, Cambridge, MA, 02139*

[2]*School of Engineering, Brown University, Providence, RI 02912*

Email: pychen@mit.edu

Mismatched deformations in layered architectures of stiff coatings on soft substrates produce textured topographies. Recently this approach has been extended to create more complex, hierarchical wrinkled/crumpled textures of graphene coatings using multiple cycles of contraction, film removal, transfer, and further contraction. However, these complex multigenerational textures have only been realized in graphene-based films. We became interested in creating these complex multigenerational textures in other material systems, beyond graphene. One possible synthetic approach would be to start with the established multigenerational textured graphene, and translate the texture to other inorganic materials (*e.g.*, metal oxides) through sacrificial templating. In this study, a facile fabrication method is demonstrated for replicating complex, hierarchical graphene wrinkle/crumple textures in metal oxide films. The general synthesis approach is schematically illustrated in **Figure 1**. The fabrication involves: (1) spontaneous intercalation of metal ions into water-swollen GO multilayer films, (2) dehydration of cation-GO complexes and thermal decomposition of GO to dissociate the complexes, (3) calcination to remove the graphene template by O_2 oxidation and liberate the metal oxide. To explore the generality and versatility of this method a variety of mono- and multivalent metal ions were explored as intercalants (Ag^{+}, Zn^{2+}, Al^{3+}, Mn^{2+}, and Cu^{2+}), and a range of GO templates was applied with pre-fabricated wrinkle and crumple textures. We find that many but not all of the metal ions and synthesis conditions produce textured metal oxides, and the optimal precursors and concentration for successful replication are reported. Remarkably, even the finest structural features in the multiscale hierarchical GO templates can be replicated with high fidelity in ZnO, Al_2O_3, Mn_2O_3, and CuO films when using the optimal ion loading (**Figure 2**). Overall, this intercalation templating approach has broad applicability for the creation of complex, textured metal oxide films, and provides a bridging technology that can transcribes textures already realized in highly flexible graphene sheets into other material systems.

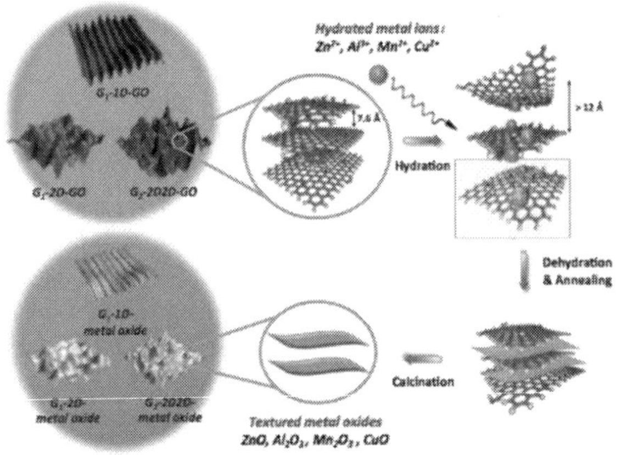

Figure 1. Schematic illustration of general synthetic process to generate textured metal oxides via using multiscale GO structures as versatile sacrificial templates.

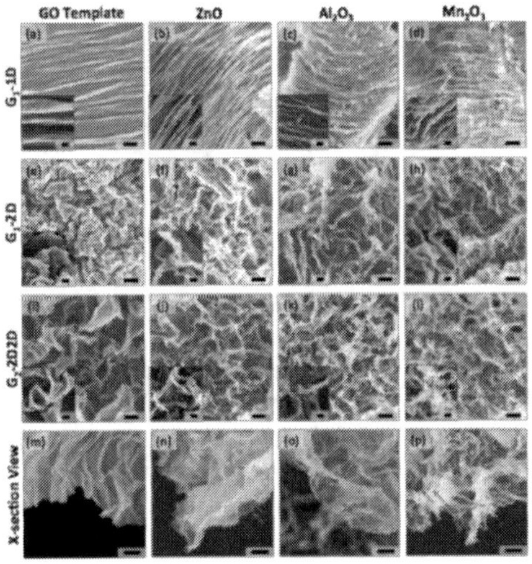

Figure 2. Metal oxide nanostructures templated by GO multiscale structures. The sequential deformation can generate multiscale GO films, which act as versatile templates to synthesize different metal oxide nanostructures. First row shows the SEM images of (a) G_1-1D-GO wrinkled template, (b) G_1-1D-ZnO, (b) G_1-1D-Al_2O_3, and (d) G_1-1D-Mn_3O_4 wrinkles. Second row shows the SEM images of (e) G_1-2D-GO crumpled template, (f) G_1-2D-ZnO, (g) G_1-2D-Al_2O_3, and (h) G_1-2D-Mn_3O_4 crumples. Third row shows the SEM images of (i) G_2-2D2D-GO hierarchical template, (j) G_2-2D2D-ZnO, (k) G_2-2D2D-Al_2O_3, and (l) G_2-2D2D-Mn_3O_4 structures. Fourth row shows the cross-sectional SEM images of (m) G_2-2D2D-GO hierarchical template, (n) G_2-2D2D-ZnO, (o) G_2-2D2D-Al_2O_3, and (p) G_2-2D2D-Mn_3O_4 structures. The scale bars in the first three rows are all 5 micrometers; the scale bars in the inset figures are 1 micrometer. The scale bars in the fourth row are 2 micrometers.

Continuous Flow Synthesis of Zeolite-A from Coal Fly Ash Utilizing Microwave Irradiation with Recycled Liquid Stream

S. Bukhari and S. Rohani
Department of Chemical and Biochemical Engineering
Western University, Canada
London, ON, N6A 5B9
Corresponding author, Email: srohani@uwo.ca, T: +1 (519) 661-2116

Abstract

Coal fly ash (CFA) was converted to zeolite using continuous microwave reactors while minimizing the wastewater produced during the zeolitization process. At bench scale, a continuous flow tubular microwave reactor was used to explore the effect of microwave irradiation on the crystallinity of the product. The waste water was reused in consecutive runs and ICP (inductively coupled plasma) analysis was conducted during and after each run to observe the concentration of the cations. An increase in the production of hydroxysodalite compared to zeolite-A was noted at higher levels of microwave energy. The XRD and SEM analyses were also conducted to corroborate the results. The CEC (cation exchange capacity) measurement showed the highest value of 0.405 meq/g when deionized (DI) water was used, the CEC dropped to 0.177 meq/g for thrice recycled waste stream. It was also found that higher microwave irradiation resulted in faster crystal growth, and the product crystallinity reached its maximum at 810 W of microwave irradiation in 60 min while 335 W of microwave irradiation resulted in the same crystallinity after 120 min of reaction. The design and specification of a pilot plant MW-based reactor system capable of converting CFA to zeolite, are presented at the end.

Keywords: Coal Fly Ash, Synthesis, Zeolite, Microwave, Continuous flow, Bench scale, Pilot plant

Introduction

Despite public and private investments and advances in renewable energies, coal remains the leading source of electricity generation in the world. Currently about 40% of the global electricity is produced by coal and it is predicted to account for 33% by 2035 according to International Energy Agency [1]. Even with this reduced share, coal will stay a major source of electricity production around the world. In 2010 alone, coal fired power plants produced about 780 million tons coal combustion products such as Coal Fly Ash (CFA) worldwide. Only about 50% of these products were recycled [2]. There has been much research conducted on producing value added products such as zeolites [3] from the accumulated CFA. Pioneering studies in zeolitization of CFA was reported using conventional heating [4–13].

The present work explores the utilizing of recycled waste water for the production of zeolite-A at larger scale unlike previous works focusing on microwave zeolitization of CFA at lab scale [7,8,14-19]. Moriyama et al. [20] used a 0.6 L reactor for the conversion of CFA to zeolite-P at an elevated pressure of 0.48 MPa and 153°C. The liquid to solid ratio was 1.1 L/kg in NaOH solution

(3 M). The zeolite-P produced, had a CEC ranging from 1.25 to 2.0 meq/g. Querol et al. [21] produced zeolite-P in a 10,000 L reactor. The reaction was conducted at 149.5°C and 0.35 MPa. The liquid to solid ratio in the slurry was 1.68 L/kg in 2.4 M NaOH solution. We report the work conducted in a bench MW reactor with a volume of 0.2 L and also a pilot scale reactor system with an 80 L MW reactor.

Materials and Method

Materials

Coal fly ash was obtained from a coal fired power plant (OPG, Nanticoke, ON, Canada) and was stored in a sealed container before use. Sodium hydroxide (Alphachem, Canada) and sodium aluminate anhydrous (Sigma-Aldrich, USA) were of analytical grade and used as received. Deionized (DI) water was used for the preparation of the solutions. Other chemicals used for characterization tests were of analytical grade.

Methods

Two sets of experiments were conducted; the first set of experiments were conducted to study the effect of MW power while the second set of experiments were to evaluate the effect of recycling the effluent stream.

Results and Discussion

X-ray analysis (XRD and XRD)

The main sources of Si and Al from CFA for hydrothermal zeolitization process are summarized in Table 1. The XRD data indicated that the main components of the CFA were amorphous aluminosilicate as well as quartz and mullite that existed as crystalline structures.

Table 1. XRF analysis of chemical composition of CFA

Parameter	Weight percent (%)
Major oxide	
SiO_2	41.78
Al_2O_3	19.61
CaO	13.64
Fe_2O_3	5.79
MgO	3.23
TiO_2	1.39
K_2O	1.1
Na_2O	0.94
P_2O_5	0.71
BaO	0.36
SrO	0.25
Cr_2O_3	0.01
MnO	0.02

Loss On Ignition	10.89
Total	99.72
Phases analysis	
Amorphous aluminosilicate	
Quartz (SiO_2)	
Mullite ($3Al_2O_3.2SiO_2$)	

The MW power experiments

X-ray Diffraction Analysis (XRD)

After the digestion in the mixing tank, the feed pump was turned on at a predetermined speed indicated in **Error! Reference source not found.** to continuously supply the CFA slurry through the microwave reactor. It took the system about 10 min to reach steady state. Following steady state, samples were collected every 30 min to conduct XRD analysis. The characteristic peak areas of the samples collected are indicated in Figure 1.

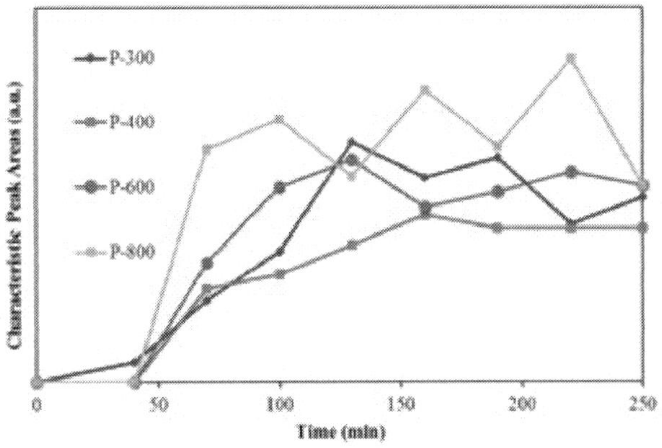

Figure 1. The progression of the XRD characteristic peak areas with time

The XRD analysis indicate that irrespective of MW power all the samples require at least 30 min before zeolitic crystals can be registered by XRD. Therefore, it indicates that the nucleation of zeolitic crystals is not a strong function of the MW power. However, once crystal growth starts, the samples irradiated under higher MW power grow faster compared to the samples irradiated with lower power. Once the area of the characteristic peak of a sample reached a maximum, it did not increase with the increases MW power input. All samples eventually had the similar crystallinity irrespective of the MW power input level.

Scanning Electron Microscope (SEM)

The SEM images of the solids were taken over the four h of microwave irradiation. The SEM images are illustrated in Figure 2. The SEM analysis of the microwave irradiated samples show that the surface of the CFA is the nucleation and crystal growth site for the zeolites. As the reaction progresses, small cubic LTA zeolitic crystals can be deciphered in the SEM images. These structures grow larger with increase in the time of microwave irradiation.

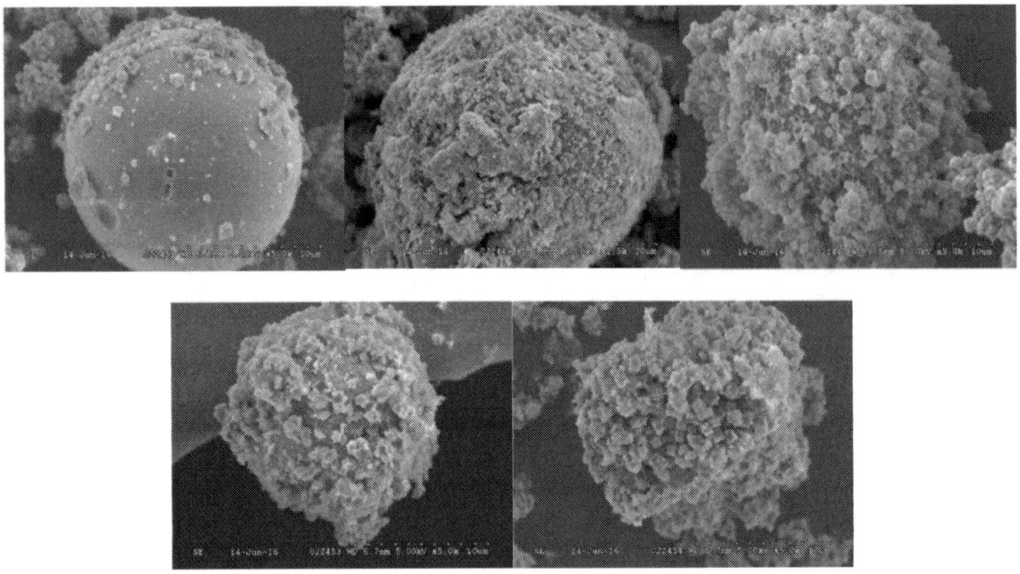

Figure 2. SEM images of zeolitized CFA at top left) 30 min, top middle) 60 min, top right) 90 min, bottom left) 120 min, bottom right) 150 min

Recycled stream experiments

X-ray Diffraction analysis (XRD)

The XRD analyses for the recycled stream experiments were conducted after the end of each run. The results are indicated in Figure 3.

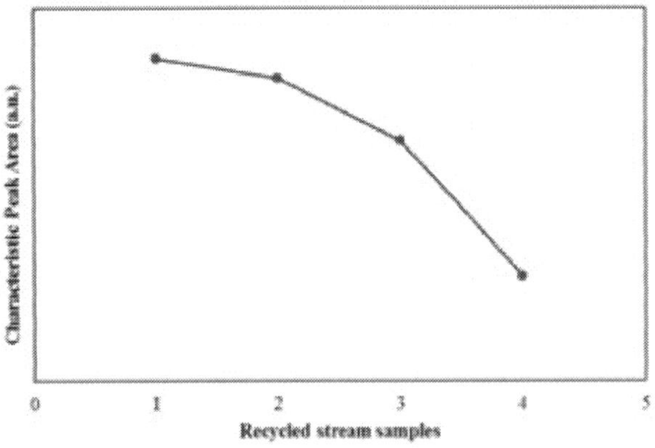

Figure 3. The XRD characteristic peak area of the recycled stream runs

It was observed that after each run the crystallinity of the final product was reduced, indicating that the recycling reduces the quality of zeolitic crystals compared to using fresh water. As the recycle stream is re-used the crystallinity of the product is further decreased. This can be explained due to the increased concentration of metallic ions with consecutive runs. This observation is later corroborated with the ICP results.

Pilot plant reactor system

Penetration depth of microwave irradiation is a measure of its efficiency to heat up a bulk material [25,26]. Penetration depth of MW is low in most materials which leads to difficulty in the scaling up of a microwave reactor. It is difficult to realize a continuous flow microwave system that can operate for volumes big enough to be compatible with the use of heterogeneous reaction condition and prevent the problem of line clogging. This has led to a stop-flow technique development, where reagent mixtures are pumped into a batch reactor under atmospheric pressure. The reaction takes place in autogenous pressure and a back regulating valve is used to keep the reactor mixture from escaping the reactor. After the completion of the reaction the product is pumped out [27]. However, this technic is not truly a continuous system.

The MW reactor overview

The MW reactor was designed as a continuous stirred tank reactor (CSTR) with the microwave introduced through slotted antenna. Utilizing MW antenna has the benefit of introducing the MW irradiation radially inside the reaction mixture unlike most MW systems where the electromagnetic waves travel from a single magnetron. The design of slotted antenna immersed in the reaction solution increases the surface and the liquid volume with which MW come in direct contact. Furthermore, the MW reactor vigorously mixes the reaction solution/suspension resulting in a homogenous MW heating. The design of the MW reactor with the slotted antenna and mixer is illustrated in **Figure 4**.

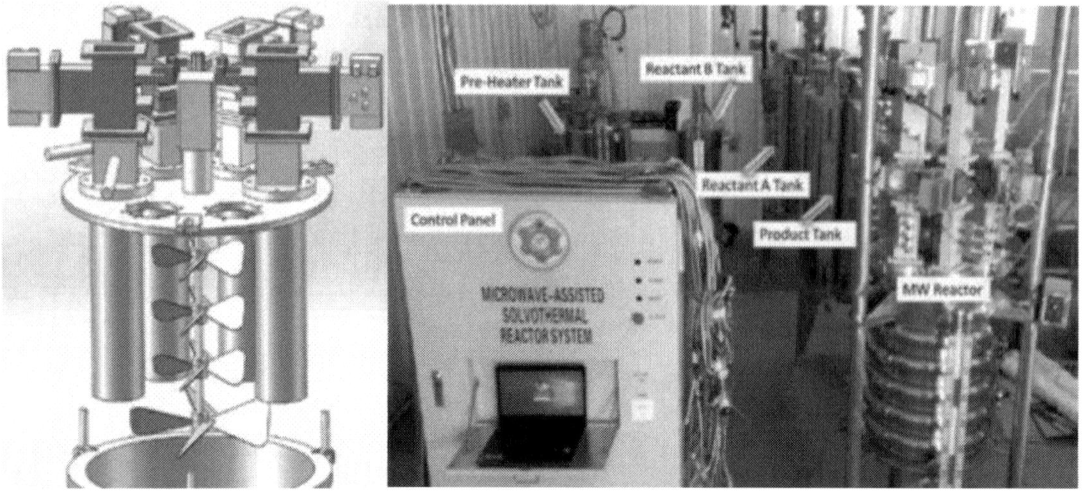

Figure 4. The 3-D illustration of the MW reactor with slotted antenna, and the pilot plant

The system was designed to utilize both single mode and multimode microwave irradiations. The reaction can also be conducted with a combination of both multimode and single mode regimes. In a single mode microwave, electromagnetic irradiation is directed through a precisely-designed wave guide that produces a standing wave, whereas in the multimode microwave there is no single standing wave but a mixture of many waves with different phase shifts. The microwave field density in a single mode microwave is much higher compared to a multimode microwave regime, therefore, similar energy intensity can be achieved in a single-mode microwave regime using lower power input [26]. Six specially designed slotted antenna produce 8 kW of microwaves. The reactor is jacketed and a heavy duty chiller is installed in order to control the temperature and prevent a temperature and pressure spike. The pilot plant consists of five tanks and one microwave reactor. There are also six peristaltic pumps. The pilot plant is illustrated in **Error! Reference source not found.**4.

Conclusions

Two sets of experiments converting CFA to zeolite were conducted utilizing the continuous flow microwave reactor. The first set of experiment was conducted to investigate the effects of microwave power on the zeolitization. It was observed that higher power of microwave irradiation resulted in a faster crystal growth while after 150 min of reaction it was observed that the crystallinity reached a maximum regardless of microwave power.

The second set of experiment was conducted with recycled waste liquid stream. The ICP analysis was conducted after digestion and after crystallization for each run. The concentration of silicon and aluminum significantly increased after each digestion step and decreased after crystallization. This signifies the dissolution of aluminosilicate constituents of CFA. The subsequent decrease of silicon and aluminum concentration in the slurry was due to the crystallizing of zeolitic crystals on the surface of CFA particles. However, each consecutive recycled stream showed lower amount of aluminosilicate. This indicates that the waste stream produced by the zeolitization process cannot be recycled indefinitely. This is also observed through XRD and SEM analysis. The XRD analysis indicated that the crystallinity of the zeolite produced with successive recycled streams was lower while the SEM showed a higher production of hydroxysodalite compared to zeolite-A. The CEC analysis further indicated that the zeolite produced with DI water had higher CEC of 0.405 meq/g compared to the runs with the recycled streams of 0.330, 0.305, and 0.177 meq/g.

References

[1] Agency IE, International Energy Agency, editors. World energy outlook 2013. Paris: International Energy Agency; 2013.

[2] Heidrich C, Feuerborn, H, Weir, A, Coal Combustion Products: a Global Perspective. World Coal Ash WOCA, Lexington, KY: 2013.

[3] Bukhari SS, Behin J, Kazemian H, Rohani S, Conversion of coal fly ash to zeolite utilizing microwave and ultrasound energies: A review, Fuel 2015;140:250–66.

[4] Holler H, Wirsching U, Zeolite formation from fly ash, Fortschr Mineral 1985;63:21–43.

[5] Shigemoto N, Hayashi H, Miyaura K, Selective formation of Na-X zeolite from coal fly ash by fusion with sodium hydroxide prior to hydrothermal reaction. J. Mater. Sci. 1993; 28:4781–4786.

[6] Hollman GG, Steenbruggen G, Janssen-Jurkovičová M, A two-step process for the synthesis of zeolites from coal fly ash, Fuel 1999;78:1225–1230.

[7] Behin J, Bukhari SS, Dehnavi V, Kazemian H, Rohani S, Using Coal Fly Ash and Wastewater for Microwave Synthesis of LTA Zeolite, Chem. Eng. Technol. 2014;37:1532–40. doi:10.1002/ceat.201400225.

[8] Bukhari SS, Behin J, Kazemian H, Rohani S, A comparative study using direct hydrothermal and indirect fusion methods to produce zeolites from coal fly ash utilizing single-mode microwave energy, J. Mater. Sci. 2014; 49:8261–71.

[9] Querol X, Alastuey A, López-Soler A, Plana F, Andrés JM, Juan R, et al., A fast method for recycling fly ash: microwave-assisted zeolite synthesis, Environ. Sci. Technol. 1997; 31:2527–2533.

[10] Bukhari SS, Rohani S, Kazemian H, Effect of ultrasound energy on the zeolitization of chemical extracts from fused coal fly ash, Ultrason. Sonochem. 2016;28:47–53. doi:10.1016/j.ultsonch.2015.06.031.

[11] Belviso C, Cavalcante F, Lettino A, Fiore S, Effects of ultrasonic treatment on zeolite synthesized from coal fly ash, Ultrason. Sonochem. 2011; 18:661–8.

[12] Wang S, Zhu ZH, Sonochemical treatment of fly ash for dye removal from wastewater, J. Hazard, Mater. 2005; 126:91–5.
[13] Musyoka NM, Petrik LF, Hums E, Ultrasonic assisted synthesis of zeolite A from coal fly ash using mine waters (acid mine drainage and circumneutral mine water) as a substitute for ultra pure water, Int. Mine Water Assoc., Aachen, Germany: 2011, p. 423–8.
[14] Belviso C, Cavalcante F, Lettino A, Fiore S, Zeolite synthesised from fused coal fly ash at low temperature using seawater for crystallization. Coal Combust Gasifica-Tion Prod 2009; 1:8–13.
[15] Hussar K, Teekasap S, Somsuk N, Synthesis of Zeolite A from By-Product of Aluminum Etching Process: Effects of Reaction Temperature and Reaction Time on Pore Volume, Am. J. Environ. Sci. 2011; 7:35.
[16] Musyoka NM, Petrik LF, Balfour G, Gitari WM, Hums E, Synthesis of hydroxy sodalite from coal fly ash using waste industrial brine solution, J. Environ. Sci. Health Part A 2011; 46:1699–1707.
[17] Musyoka NM, Petrik LF, Fatoba OO, Hums E, Synthesis of zeolites from coal fly ash using mine waters, Miner. Eng. 2013; 53:9–15.
[18] Dehnavi V, Luan BL, Shoesmith DW, Liu XY, Rohani S, Effect of duty cycle and applied current frequency on plasma electrolytic oxidation (PEO) coating growth behavior, Surf. Coat. Technol. 2013.
[19] Behin J, Bukhari SS, Kazemian H, Rohani S, Developing a zero liquid discharge process for zeolitization of coal fly ash to synthetic NaP zeolite, Fuel, 2016;171:195–202.
[20] Moriyama R, Takeda S, Onozaki M, Katayama Y, Shiota K, Fukuda T, et al., Large-scale synthesis of artificial zeolite from coal fly ash with a small charge of alkaline solution, Fuel, 2005;84:1455–61.
[21] Querol X, Umaña JC, Plana F, Alastuey A, Lopez-Soler A, Medinaceli A, et al., Synthesis of zeolites from fly ash at pilot plant scale. Examples of potential applications, Fuel, 2001;80:857–865.
[22] Treacy MMJ, Higgins JB, Collection of simulated XRD powder patterns for zeolites. Amsterdam, The Netherlands: Elsevier; 2007.
[23] Bain DC, Smith BFL, Chemical analysis. Handbook. Determinative Methods Clay Mineral., Glasgow: Blackie; 1987, p. 248–74.
[24] Aldahri T, Behin J, Kazemian H, Rohani S, Synthesis of zeolite Na-P from coal fly ash by thermo-sonochemical treatment, Fuel, 2016;182:494–501.
[25] Hasted JB, Liquid water: Dielectric properties. Phys. Phys. Chem. Water, Springer; 1972, p. 255–309.
[26] Buchner R, Barthel J, Stauber J, The dielectric relaxation of water between 0 °C and 35 °C, Chem. Phys. Lett. 1999; 306:57–63.
[27] Glasnov TN, Kappe CO, Microwave-Assisted Synthesis under Continuous-Flow Conditions, Macromol. Rapid Commun. 2007; 28:395–410.
[28] Morschhäuser R, Krull M, Kayser C, Boberski C, Bierbaum R, P. A. Püschner, T. Glasnov, K.C. Oliver, Microwave-assisted continuous flow synthesis on industrial scale, Green Process Synth. 2012; 1:281–90.

455873 Colloidal Solution Combustion Synthesis of Crystalline Mesoporous CeO_2 Catalyst with Tunable Porosity

Abstract:

Mesoporous metal oxides with uniform porosity are of considerable interest. In this work, we demonstrate for the first time a scalable, economic, energy and time efficient method for the synthesis of a crystalline mesoporous CeO_2 catalyst with adjustable porosity, by utilizing colloidal SiO_2 as a template. The size and amount of colloidal particles can tune the

porosity of the CeO_2 nanostructure and alter the heat transfer and energy balance of combustion. The CeO_2 prepared by colloidal solution combustion synthesis (CSCS) possesses uniform 22 nm pores and a 0.6 mL/g pore volume, which is among largest pore volume for CeO_2 reported in literature. The obtained mesoporous CeO_2 catalyst exhibited excellent activity for soot and carbon monoxide oxidation. In principle, the CSCS method can be applied to synthesize different high-porosity crystalline oxides. Results will also be shown for CSCS prepared mesoporous CuO.

[1] A. A. Voskanyan, K.-Y. Chan, and C.-Y. V. Li, *Chem. Mater.* 2016 DOI:0.1021/acs.chemmater.6b00505

472080 Template-Mediated Control of Hierarchical Structure in Crystalline and Amorphous Porous Materials

Abstract:

This talk will describe physicochemical insights into a templating strategy aimed at the synthesis of hierarchically structured, high-surface area, large-pore volume three-dimensionally ordered mesoporous (3DOm) materials. The accessible materials palette, spanning carbons to various metal oxides (e.g., TiO_2, ZrO_2), enables scalable synthesis of structured particles tailored for applications ranging among separations, photocatalytic decomposition of bulky molecules, and even enhanced supercapacitor and solar cell electrode technologies. The templating strategy employs synthetic size-tunable silica nanoparticles (ca. 5-50 nm) and their uni- and multi-modal colloidal crystalline assemblies for hard, sacrificial nanocasting of 3DOm replica structures. Ultimately, this approach enables multi-scale tailoring of replica properties. Specifically, at the mesoscale, topological diversity and 3D hierarchy of the templated pores can be controlled by exploiting multi-modal particulate porogen assembly or by â€˜one-potâ€™ co-assembly techniques (i.e., molecular replica precursors with particulate porogens). Emerging from the latter approach is a new class of materials with independent and interdigitated bimodal mesoporosity. Simultaneously, at the microscale, template-mediated interfacial control of replica polymorphism, allotropy, and even molecular scale function can also be achieved. This control across scales translates to enhanced molecular accessibility/transport and tailored function (e.g., catalytic, adsorptive, size selectivity) of the porous replica materials.

464087 Metal-Organic Framework Engineering: Directed Assembly from Molecules to Spherical Agglomerates

Abstract:

$[In(OH)(bdc)]_n$ (bdc = 1,4-benzenedicarboxylate) was used as the model metal-organic framework for showing the concept of directed assembly from molecules to spherical agglomerates in solution. The course of reaction crystallization at 60°, 80° and 100°C could be partially fulfilled by the Avrami equation. Noticeably, the portion which failed to be fitted by the equation was the secondary crystallization. Aging and editing effects during the course of secondary crystallization were responsible for the increase in crystallinity and structural rearrangements of $[In(OH)(bdc)]_n$ crystals. The spherical agglomerates of $[In(OH)(bdc)]_n$ crystals were assembled by the liquid bridges among the $[In(OH)(bdc)]_n$ crystals through the addition of a polyethylene glycol (PEG) binder. The growth of spherical agglomerates was traced and described by three curves: the depletion curve, the reaction progress curve, and the average agglomerate size versus mixing time curve.

472598 Self-Assembly of Designed Nanoscale Architectures from Assorted DNA-Framed Nanoparticles

Abstract:

Our work addresses an important question in nanoscale assembly: how to create arbitrarily designed nanoparticle-based structures using self-assembly? In this work we demonstrate a novel broadly applicable strategy for assembly of designed nanoparticle architectures by co-assembly of disimilar nanoparticles/DNA modules. Such modules integrate nanoparticles with DNA frames that provide diverse selective and anisotropic bonds. By adding various topological types of those nanoparticle modules and by controlling their stoichometry in the hetero-mixture we direct assembly process towards the desired architectures. We show that planar, both periodic and arbitrarily prescribed, nanoparticle assemblies can be generated on-demand.

473547 Self-Assembly and Chiroptical Response of Homochiral Semiconductor Nanohelices

Abstract:

Chiral semiconductor nanostructures provide a rich materials platform for polarization optics, photocatalysis, and biomimetics. However, unlike noble metals (such as Au or Ag), the relationship between the geometry of chiral semiconductors and their chiroptical properties remains enigmatic. Enantioselective preparation of semiconductor nanohelices from cadmium telluride nanoparticles (CdTe NPs) opens an opportunity to simulate and experimentally validate complex relationships between geometrical parameters and chiroptical properties of semiconductor nanostructures using circular dichroism (CD) spectroscopy. Simulated CD spectra not only accurately reproduced experimental results, but also provided spectral predictions for geometrical parameter modification. Comparisons between nanohelices made of CdTe and noble metals such as Au and Ag, revealed distinctly different simulated chiroptical responses. This is due to the occurrence of multipole plasmon resonance associated with relatively large size and aspect ratio nanostructures of noble metals.

465607 Transformable Liquid-Metal Nanomedicine

Abstract:

TRANSFORMABLE LIQUID-METAL NANOMEDICINE

Yue Lu[1,2], Quanyin Hu[1,2], Yiliang Lin[2], Frances S. Ligler[1,2], Michael D. Dickey[2] and Zhen Gu[1,2,]*

[1] *University of North Carolina at Chapel Hill, Chapel Hill, NC, 27599, USA;*

[2] *North Carolina State University, Raleigh, NC, 27695, USA*

Purpose

To date, numerous inorganic nanocarriers have been explored for drug delivery systems (DDSs). However, the clinical application of inorganic formulations has often been hindered by their toxicity and failure to biodegrade. We report here a novel liquid metal (eutectic alloy of gallium indium, designated EGaIn)-based nano-scale formulation for controlled drug delivery to achieve enhanced anticancer therapy.

Methods

To obtain liquid-metal based drug nanocarriers, an ¡°emulsion¡±-like ligand-mediated procedure is simply applied through ultrasonication at room temperature. During sonication, the thiolated ligands readily assemble onto the surface of the EGaIn, competing with the oxidation process and facilitating control of the particle size. The two ligands used here, thiolated (2-hydroxypropyl)-¦Â-cyclodextrin and thiolated hyaluronic acid, not only serve as capping agents during the formation of nano-scaled liquid-metal spheres, but also play roles of drug loading matrix and active targeting moiety, respectively. The final formulation (designated LM-NP/*L*) comprises three primary functional constituents: a cyclodextrin-based drug loading motif, the targeting ligand hyaluronic acid and an EGaIn core.

Results

The resulting nanoparticles loaded with Dox have an average diameter of 107 nm and demonstrate the capability to fuse and subsequently degrade under a mildly acidic condition, which facilitates release of Dox in acidic endosomes after cellular internalization. Equipped with hyaluronic acid, a tumour-targeting ligand, this formulation displays enhanced chemotherapeutic inhibition toward the xenograft tumour-bearing mice. This metal-based DDS with fusible and degradable behavior under physiological conditions provides a new strategy for engineering theranostic agents with low toxicity.

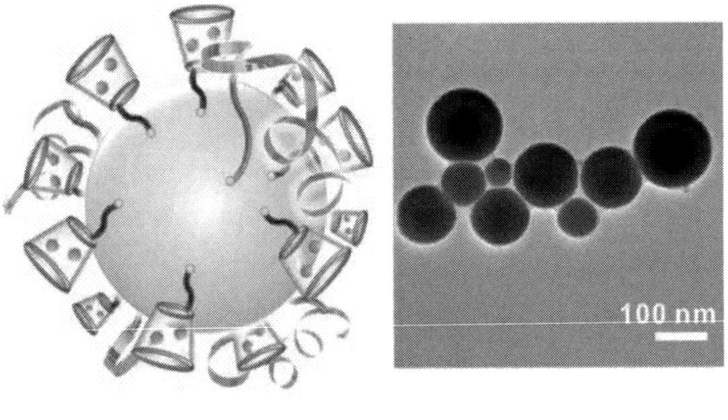

Figure 1. Schematic design and TEM image.

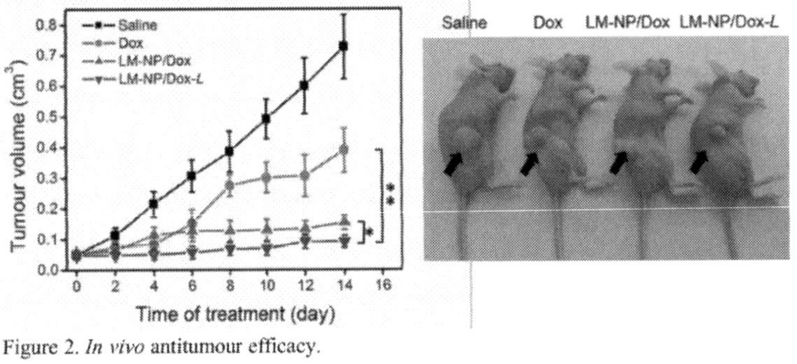

Figure 2. *In vivo* antitumour efficacy.

Conclusions

We have developed a new liquid metal-based drug delivery platform for anticancer therapy. The formulation can be easily formed and tailored *via* ligand-mediated self-assembly using facile ultrasonication. The resulting liquid-metal nanospheres are able to fuse for promoting drug release and eventually degrade under a mild acidic environment. After fusing, these nanospheres also display a contrast enhancing capability when imaged by X-ray, suggesting potential as a theranostic reagent. Systematic investigation of toxicology of LM-NP/*L* revealed no obvious toxicity at the treatment dose, favoring its biomedical applications.

Acknowledgements: NIH (CTSA, 1UL1TR001111) and NSF (CMMI-0954321, DMR-1121107).

References:

1. Yue Lu, Quanyin Hu, Yiliang Lin, Dennis B. Pacardo, Chao Wang, Wujin Sun, Frances S. Ligler, Michael D. Dickey and Zhen Gu (2015), ¡°Transformable Liquid-metal Nanomedicine,¡± Nature Communications, 6, 10066.

463628 Maintenance of Neural Progenitor Cell "Stemness" in 3D Hydrogels Requires Matrix Remodeling

Abstract:

While neural progenitor cells (NPCs) hold significant therapeutic promise, relatively little is known about the material properties required to maintain NPC stemness and regenerative potential. Previous studies using 2D substrates have suggested that matrix stiffness can bias NPC differentiation, but studies in 3D materials have conflicting results about the relative importance of stiffness. To determine the material properties required to maintain NPC stemness, we designed an elastin-like protein (ELP) hydrogel system in which matrix stiffness and degradability are co-varied independently of hydrogel swelling, microstructure, and nutrient diffusivity. NPCs cultured in gels with low stiffness and high degradability significantly improved their maintenance of stemness compared to cells in gels with high stiffness and low degradability. Quantifiable indicators of stemness included increased proliferation rate, enhanced metabolic activity, expression of the NPC markers Nestin and Sox2, and efficient differentiation into neurons and astrocytes. To assess if these changes in stemness were due to matrix stiffness, encapsulated NPCs were treated with small molecule inhibitors of cytoskeletal tension and classical mechanotransduction pathways. Interestingly, inhibition of mechanosensing resulted in no changes in NPC behavior relative to uninhibited controls. Furthermore, preventing integrin-mediated matrix engagement by using non-cell-adhesive ELPs also had no effect on NPC phenotype. These results indicate that the observed changes in NPC stemness were independent of matrix stiffness. To investigate the role of NPC-mediated matrix remodeling, antibody arrays and protease inhibitor screens were used to identify a disintegrin and metalloprotease 9 (ADAM9) as the protease primarily responsible for ELP degradation by NPCs. Reducing ADAM9 activity with small molecule inhibitors or with shRNA-mediated knockdown resulted in decreased hydrogel degradation and a significant reduction in NPC stemness. These results suggested that matrix remodeling, and not matrix stiffness, was the required material parameter for maintaining NPC stemness. To test the generalizability of this result, a second engineered biomaterial was designed using poly(ethylene glycol) crosslinked with ADAM9-cleavable peptides to create a family of hydrogels with independent tuning of stiffness and ADAM9-mediated degradability. When ADAM9-degradability was held constant, no differences in Nestin or Sox2 expression were observed regardless of the gel stiffness. In stark contrast, Nestin and Sox2 expression were significantly degreased in gels with 0% compared to 100% ADAM9-cleavable crosslinks, even in the softest gels. Taken together, these results have identified matrix degradation as a previously unknown requirement for NPC stemness maintenance in 3D hydrogels. Furthermore, these data demonstrate that this effect is independent of classical mechanosensing through cytoskeletal contractility, which was previously shown to play a critical role in NPC-matrix interactions in 2D. In the future, biomaterials that exploit ADAM9-mediated remodeling may increase the regenerative potential of NPC therapies.

469654 Combining Simulation and Spectroscopy to Determine the Structure and Orientation of a Carbohydrate Binding Module (CBM)-Inspired Model Peptide on Cellulose

Abstract:

Interfacial processes, such as the enzymatic conversion of an insoluble polysaccharide at the solid-liquid interface, are critically important to the design of new systems to convert renewable resources into green fuels and chemicals. Despite this fact, we often lack a molecularly detailed understanding of these interfaces that hinders the rational design of new and/or improved biocatalytic systems. Atomistic simulations of biomolecular adsorption using state-of-the-art multiscale molecular modeling tools have strong potential to combine with advanced, surface-specific spectroscopy to elucidate structural and mechanistic details at the interface. To this end, enhanced sampling methods based on the metadynamics family of methods have been combined with experimental data from sum frequency generation spectroscopy (SFG) to probe the equilibrium structure(s) and preferred orientation(s) of a 14-residue peptide fragment inspired by the carbohydrate binding module (CBM) of a cellulase. Experimental SFG data indicates the CBM fragment adsorbs to the cellulose surface and aligns along the fibers in a distinct orientation/conformation. We use a variety of molecular simulations to provide molecular level insight to this observation and add additional detail beyond the amide I backbone spectra, thus providing new mechanistic insights regarding biomass conversion. Additionally, simulation results are used to determine the energetic penalties for binding in alternate orientations, and to identify the specific reasons for the observed behaviors. In a broader sense, this integrated approach of combining computational methods with experiments aims to determine the role of the surface in modulating biomolecular structure, thereby permitting the rational design of new biocatalytic approaches.

472358 Award Session: Plasmonic Collagen Nanocomposites for Rapid Tissue Sealing and Repair

Abstract:

Laser tissue welding is a tissue repair technique that utilizes a chromophore to convert photons into heat in order to seal apposed tissue edges by localized protein denaturation and interdigitation. Laser tissue welding has emerged as a promising alternative to sutures or staples conventionally used for sealing soft tissues, with the potential to limit surgical complications such as anastomotic leakage and wound dehiscence while also reducing healing times and scarring. However, insufficient closure strength, surgical inconvenience, and extensive thermal damage have hindered the clinical application of this technique, largely due to the unavailability of effective solders (surgical sealants). In this work, we describe the development of novel plasmonic nanocomposites for laser tissue welding that overcome these barriers in tissue welding. Several different gold nanorod-polypeptide were developed and investigated for welding ruptured tissues ex vivo. Leak pressure, burst pressure, and tensile strength measurements indicated that laser welding using plasmonic nanocomposites was able to restore significant mechanical integrity to the welded tissue. Additionally, we developed a spatiotemporal model for predicting local temperatures to aid in narrowing the parameter space and selecting optimal surgical parameters. Preliminary *in vivo* results in sealing colotomy incisions in mice are also presented.

468629 Chitosan-Coated Selenium Nanoparticles for the Treatment of Skin Infections

Abstract:

Introduction: The skin is the bodyâ€™s strongest form of protection against infectious and pathogenic microbes, and when this barrier is perforated by wounds, lesions or other chronic skin conditions, the body is vulnerable to more serious diseases. Among these skin diseases, acne vulgaris is one of the most prevalent, affecting up to 95% of all adolescents and persisting in up to 10% of adults. *Proprionibacterium acnes* has been suggested as the main contributor to acne vulgaris. Our lab has synthesized selenium nanoparticles to combat bacterial infection, and have found them to be effective against several bacterial strains, and relatively non-toxic to healthy fibroblasts and other mammalian cells. Here, to make such particles suitable for skin applications, we employed a new synthesis of chitosan-coated selenium nanoparticles as an effective topical treatment for *P. acnes*, as well as common skin denizens, such as *S. aureus*, while preserving the healthy skin tissue surrounding the acne lesions.

Materials and Methods: Chitosan-coated selenium nanoparticles (Ch-SeNP) were synthesized in a precipitation reaction, in which chitosan (dissolved in 1% acetic acid) was added dropwise to sodium selenite solution, followed by an addition of ascorbic acid. The resulting Ch-SeNP are separated via centrifugation and resuspended in DI water. Characterization was performed via DLS and TEM. *P. acnes* (ATCC 11827) were cultured in modified reinforced clostridial broth medium or on tripticase soy agar plates with defibrinated sheep blood, while *S. aureus* was cultured in 3% tryptic soy broth or agar plates. For *S. aureus*, kinetic growth curves were tracked by measuring the optical densities of treated cultures via spectrophotometry over 24 hrs. *P. acnes*, an anaerobic bacterium, required anaerobic conditions to grow, and were thus tested via a colony-forming unit assay in generated anaerobic gas chambers over 4 days. All experiments were repeated three times, and statistical significance were analyzed via a two-talked Studentâ €™s T-test.

Results and Discussion: Ch-SeNP were found to be as effective, if not more, as antibacterial agents as the previously studied BSA-coated SeNP on *S. aureus*, with lag times lengthening as Ch-SeNP concentrations increased. At a mid-range concentration of Ch-SeNP (0.023 mg/ml), *P. acnes* growth was completely absent compared to untreated colonies, suggesting that Ch-SeNP were strongly effective at preventing *P. acnes* proliferation.

Conclusions: Chitosan-coated selenium nanoparticles were found here to be exceedingly more stable in solution than their BSA-coated predecessors without sacrificing their antibacterial efficacy. Additionally, we found the Ch-SeNP to be effective in preventing the proliferation of the anaerobe *P. acnes*, suggesting a novel SeNP application in topical skin infections.

Acknowledgements: The authors thank Northeastern University and the College of Engineering for funding.

472743 Ligand-Targeted Conjugate Systems for Delivery of siRNA to Tumors

Abstract:

Small interfering ribonucleic acids (siRNA) are a class of nucleic acid-based drugs that have the potential to knock down any gene with a high degree of specificity. They harness a natural gene regulation and antiviral mechanism for efficient destruction of any mRNA. This is useful for both the understanding of and treatment of diseases ranging from genetic disorders and viral infections to cancer. However, to be effective, the siRNA must travel from the injection/infusion site to the tissue of interest and then reach the cytoplasm of the desired cell type. Nucleic acids are highly charged and cannot cross the cell membrane unaided. Exposure to serum endonucleases also accelerates the destruction of the siRNA payload. Methods to cross these barriers include targeted delivery using vehicles ranging from lipid nanoparticles to polyplexes. Unfortunately, the delivery materials tend to be toxic, limiting the efficacy in patients, and also tend to localize to immune cells, the endothelium, or the liver. Currently, there are very few biomaterials that can successfully deliver siRNA to tissues outside of this group, and none are in Phase II clinical trials.

Recently, advances in siRNA stabilization technology have made unencapsulated siRNA conjugates viable, as chemical modifications to the RNA backbone have been used to reduce susceptibility of the siRNA to endonuclease degradation. These conjugates also have desirable treatment properties, including an in-human safety profile of up to 300 mg/kg and a clear regulatory path due to a well-defined chemical structure. Furthermore, due to their small size and thus ability to diffuse relatively quickly through extracellular matrix, the conjugates can be injected subcutaneously, a home procedure familiar to many diabetics on insulin, as opposed to nanoparticles, which must be infused intravenously at a clinic or hospital.

We have developed an azide-alkyne cycloaddition based system that allows for the delivery of ligand-targeted siRNA conjugates to models of glioblastoma (a form of brain tumor) *in vitro* and *in vivo*. The ligand used is based on chlorotoxin. *In vitro* studies using U87 (glioblastoma) cells show a similar degree of knockdown to commercially available transfection materials relative to a scrambled siRNA control. Confocal microscopy studies using derivatives of these ligands with fluorescently-labeled siRNA demonstrate a high degree of uptake relative to unconjugated siRNA in both cases. These conjugates are also able to traverse the bloodstream, deliver a nucleic acid payload into the tumor cells, and efficiently knock down expression of a housekeeping gene. Using fluorescent derivatives of our siRNA, we are able to observe localization of the siRNA to subcutaneous or intracranial tumor xenografts. Intravital imaging experiments demonstrate the ability of these conjugates to pass the blood-brain barrier and permeate the tumor. *In vivo* knockdown studies show a moderate degree of tumor-specific knockdown of a housekeeping gene in an intracranial model of glioblastoma using U87 cells.

These results demonstrate that siRNA knockdown of genes in tumors using these modular conjugate systems are feasible and present a way forward for using siRNA in the treatment of cancer. Future work involves using biologically-relevant siRNA to demonstrate anti-tumor effects in this and syngeneic tumor models.

Sustained Transgene Expression Via Substrate-Mediated Gene Transfer Results from Multiple Transfection Events

Truong, N., University of California Los Angeles

Segura, T., University of California Los Angeles

Originally presented on: 11/13/2016 17:36:00 - 17:54:00

Statement of purpose: Sustained delivery of therapeutic genes in vitro and in vivo has a wide range of applications in studying biology and in developing therapies to treat disease. Non-viral vectors such as cationic polymers still present promising approaches; however, bolus transfection methods with polyethyleneimine (PEI)-based DNA polyplexes suffer from considerable levels of cytotoxicity and short-lived transgene expression levels. Here, we designed and characterized a hyaluronic acid (HA)-based porous hydrogel system for non-viral gene delivery by loading with surface-associated DNA polyplexes (Figure 1A). With this, we observed enhanced and sustained transgene expression over 30 days of cell culture, with better cell viability, both marked improvements over comparable bolus transfection techniques. Finally, we investigated mechanisms thought to be responsible for the sustained expression profile.

Methods: HA hydrogels were formed via a Michael-addition reaction between acrylated HA and thiolated peptide crosslinkers with the addition of RGD peptide for cell adhesion. The sphere templating method was used to introduce 60-μm diameter pores. Plasmid DNA was complexed with polyethyleneimine to form DNA polyplexes, which were surface coated by electrostatic association to the pore surfaces of the hydrogel by incubating the formed hydrogel in the polyplex solution. DNA loading was assessed with ^{32}P-labelled DNA and scintillation counting. Transfection profiles were obtained by loading DNA encoding for the Gaussia luciferase gene, seeding D1 mouse mesenchymal stem cells in the hydrogels, and detecting the expressed protein with the associated assay.

Results: ^{32}P analysis of DNA-loaded porous hydrogels showed that up to 4 μg of DNA could be loaded into the scaffold. Although the DNA was electrostatically immobilized to the scaffold, minimal polyplex release was observed with only 5% of the initial loaded DNA released over 7 days. Cell culture within this scaffold resulted in sustained transgene expression over a period of more than 30 days at levels higher than those seen using bolus transfection techniques (Fig. 1B). In addition, adjusting the DNA loading concentration resulted in modulation of transgene expression (Fig. 1B). While repeated bolus transfections in 2D culture result in marked toxicity (Fig. 1C), culture in surface-coated hydrogels was significantly less severe (Fig. 1D). To determine if sustained expression was due to re-transfection events, it was observed that levels of internalized plasmid DNA in cells cultured in surface-coated hydrogels was sustained at higher levels over time than in cells cultured in non-coated hydrogels but administered a bolus polyplex transfection (Fig. 1E).

Conclusions: After DNA loading by surface coating, there was minimal basal DNA release over time, suggesting that this may serve as a robust system for long-term DNA availability and expression. Indeed, sustained expression was seen over thirty days of culture and could be enhanced by increasing DNA loading. Although repeated bolus transfections result in severe toxicity, repeated internalization events are possible in surface-coated scaffolds without the same toxic effects and are responsible for the sustained expression observed. This system is a promising means of sustained gene delivery for various biomedical applications without the need for genomic integration via viral methods, which have innate safety concerns.

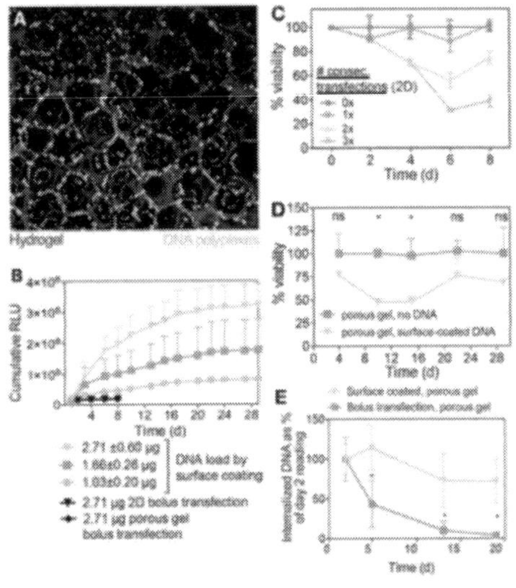

460626 Zwitterionic Gel Encapsulation Promotes Protein Stability, Enhances Pharmacokinetics and Reduces Immunogenicity

Abstract:

As an emerging class of therapeutics, protein drugs offer the advantages of higher specificity and potency than small molecules. However development of protein therapeutics is hindered by their poor stability, inadequate pharmacokinetic (PK) profiles, and immunogenicity. Polyethylene glycol conjugation (PEGylation) is the most successful strategy to date to overcome these shortcomings, and more than ten PEGylated proteins have been brought to market. However, anti-PEG antibodies induced by treatment raise serious concerns about the future of PEGylated therapeutics. Based on extensive studies of PEGylation, we believe that two issues are responsible for its immune response: the immunogenicity/antigenicity of PEG itself, and limited PEG coverage on the protein surfaces. By solving these two issues, we demonstrate a zwitterionic polymer network encapsulation (ZPNE) technology that effectively enhances protein stability and pharmacokinetics while reducing the immune response. Highly immunogenic fungal-derived uricase was used as a model therapy. In this study, uricase modified with a comprehensive zwitterionic polycarboxybetaine (PCB) gel network exhibited exceptional stability and a greatly prolonged circulation half-life. More importantly, the PK behavior was unchanged and neither anti-uricase nor anti-PCB antibodies were detected after three weekly injections in a rat model. This technology is applicable to a variety of proteins and unlocks the possibility to adopt highly immunogenic proteins for therapeutic or protective applications.

462847 Janus-like Meso-Porous Hybrid Frameworks for Super-Efficient and Cost-Competitive Water Desalination By Membrane Evaporation

Abstract:

Membrane evaporation (ME) is a novel desalination process, whereby a heat-conductive membrane is used to evaporate water in contact with the material. Similarly to membrne distillation, the difference of liquid vapour pressure across the pores of the membrane leads to vapour transport from the feed side to the permeate side, where vapours are thereafter condensed.

In contrast to other desalination techniques such as nanofiltration and reverse osmosis, ME offers a potentially low energy and high rejection route to the desalination of highly dirty or salty waters. However, to become competitive with other desalination technologies, specific membrane materials must be developed and tailored to allow for high flux, low fouling tendency and long term stability.

Specifically, in order to achieve efficient vapour transport from the hot to the cold side, a ME membrane must be highly porous, as thin as possible and not wettable by the process liquids. The pores need therefore to be large enough to facilitate vapour transport, while having sufficiently small dimensions to avoid liquid wetting and formation of a direct liquid bridge between the feed and the permeate sides. In addition, the membrane material must be heat conductive to allow for direct heating and evaporation of water in direct contact with the material surface. Most previous studies have investigated low porosity membrane materials formed from sintered stainless steel powders. While these materials are highly hydrophobic, they are also expensive and difficult to process. It is consequently important to investigate other alternatives as well as techniques for improving the process efficiency by modifying the membrane properties and structure.

Here, novel metal based scaffolds were decorated for the first time with ultra-thin layers of graphene materials by a direct electroplating. The micron-sized pores across the metal scaffolds were found by micro-Raman analysis to be entirely covred with controllable thicknesses of graphene, which latter was used for seeding specific functionalization and provide a strong interface as well as a corrosion protection barrier against chloride anions present in water. Both sides of the membranes were then specifically modified to alter the water evaporation and vapour diffusion and condensation characteristics. A Janus structure was prodcued by (i) grafting on one side of the membrane fluorosilane chains, used to generate a super-hydrophobic interface, while (ii) the second side of the membrane was coated with a nanoscale but yet thermally insulating and semi-hydrophilic alumina layer by Atomic Layer Deposition (ALD). The macro-properties of the materials, including their wettability, thermal conductivity, mechanical strength, as well as their surface charge and morphologies were systematically investigated and correlated to the electroplating conditions. The performance of the membranes for desalting model 3.5 wt% NaCl solutions were also linked to process parameters, including the liquid velocity within the module and the temperature difference between both sides of the membranes, as well as to the materials properties, fluorosilane functional groups densities and morphology of the alumina layer deposited.

In short, the membranes were found to exhibit Janus behaviours in nearly all aspects of their behaviours. The water contact angle values ranged from 160° to 60°, while the thermal conductivity of the alumina-coated side was found to be up to 25 times lower than that of the fluoro-sialne grafter graphene side. In addition, >99.99% desalination performance were achieved for flux as high as 145 LMH for a feed to permeate temperature difference of only 50°C. The hybrid Janus membranes were also found to be extremely stable and desalination tests for up to 5 consecutive days were performed without loss of performance. This work demonstrates the feasabiity of hybrid materials with controlled porosities and pore distributions for the competitive desalination of brines.

469836 Electrospun Nanocarbon Fibers for the Chromium Removal in Water

Abstract:

Uniform electrospun polyacrylonitrile / iron (III) nitrate nanofibers with small diameters were successfully prepared by systematically optimizing the electrospinning parameters including polymer concentration, salt loading, applied voltage, tip to collector distance, and flow rate. These parameters are shown to affect the fiber morphology and the addition of iron (III) nitrate salt was found to be able to reduce the fiber size and required electrospinning voltage. The improved thermal properties of the composite fibers were investigated and shown by thermal gravimetric analysis (TGA) and differential scanning calorimetry (DSC). The composite nanofibers were further carbonized to carbon nanofibers. Applications of the nanofibers for chromium removal have shown promise.

469583 Chemically Crosslinking Graphene Oxide and Chitosan for Scalable Water Treatment Membranes

Abstract:

Graphene Oxide (GO) has emerged as a next generation material for water treatment. Although numerous studies have focused on the use of GO as an adsorbent material, recent models proposed in the literature suggest that graphene oxide can be effectively applied toward membrane technologies. However fabrication of GO membranes for separation processes has been limited to small-scale vacuum filtration with an anodized aluminum oxide Anodisc®. This is due to the need for aluminum (Al) ions to crosslink the stacked GO layers preventing the membranes from breaking down from electrostatic repulsion in water. The acidic nature of GO causes Al ions to leach into the GO solution and crosslink oxygen functionalities between the GO lamellar layers keeping the membrane intact when submerged. The vacuum-assisted self-assembly (VASA) method is effective but limits the scale and scope in which these membranes can be studied and used.

To remedy this problem GO was crosslinked with chitosan (CSGO) to form a composite material that can be wet cast and fabricated at any size via convective deposition. The CSGO membraneâ€™s performance was tested in a crossflow reactor and challenged with a positively charged dye (methylene blue). The CSGO membranes were able to remove 100% of methylene blue from solution at concentrations ranging from 1-100 ppm. Mass balances obtained from the concentrate and permeate are indicative of size exclusion for the mechanism of removal. The flux rates for these solutions ranged from 2-4.5 LMH with a transmembrane pressure of 50psi. Future work involves increasing the flux across the membrane, while maintaining removal rates, and challenging CSGO composites with negatively charged dyes, radionuclides, divalent and monovalent salts.

453499 Metal-Organic Framework/Î±-Alumina Composite Withnovel Geometry for Enhanced Adsorptive Separation

Abstract:

UiO-66, as a prototypical zirconium-based metal-organic framework (MOF), provides a rapid uptake of arsenic from water when compared to other typical adsorbents with the same order-of-magnitude particle size. This fast kinetics allows an efficient adsorptive separation to be realized within a micro-space. Moreover, α-alumina can be specifically structured into a novel hollow fiber geometry: containing a plurality of open radial micro-channels on the shell side and a very thin barrier layer at the lumen. Through a facile vacuum filtration method, a MOF/α-alumina composite with novel geometry is developed and optimized in this study. The composite leads to a new concept for enhanced adsorptive separation: efficient adsorption occurs within numerous conical micro-channels with no loss of the active adsorbents during the process. As a proof of concept, this composite can effectively remediate arsenic contaminated water producing potable water recovery, whereas the conventional fixed bed requires eight times the amount of active adsorbents to achieve a similar performance. Looking forward, different functional composites can be prepared based on specific adsorptive applications, as a wide range of adsorbents can be loaded into the α-alumina hollow fiber, of which the micro-channel size and barrier layer pore size can be easily manipulated during fabrication.

462597 Multifunctional Epoxy Nanocomposites

Abstract:

In this work, the multifunctional epoxy nanocomposites filled with different nanofillers including magnetite, silica, and multi-walled carbon nanotubes were prepared. The conductive polymer polyaniline was introduced on the surface of nanofillers through surface initiated polymerization method to improve the interfacial interaction between nanofillers and epoxy matrix. The improved mechanical property of magnetite/epoxy nanocomposites was obtained due to the well-dispersed magnetite nanoparticles coverd by polyaniline. The tensile strength in polyaniline stabilized multi-walled carbon nanotubes/epoxy nanocomposites was enhanced 85% compared with that of pure epoxy. The introduction of phosphoric acid doped polyaniline and silica decreased the heat release rate peak of epoxy, providing the flame retardant property to epoxy.

468250 Graphene/Aramid Nanofibers Composite Electrodes for Structural Energy and Power

Abstract:

Flexible electronics, such as roll-up displays and wearable systems have been widely studied with the fast increasing demand for potable electronics. For these flexible, portable electronics, structural elements which can bear mechanical load and dissipate mechanical stress are required in the energy storage systems such as lithium ion batteries and supercapacitors. Flexible graphene or graphene oxide papers have attracted much attention as structural electrodes due to their extraordinary mechanical and electrical properties. The mechanical properties and the electrochemical performance of the graphene paper electrodes can be manipulated when combined with guest materials. In this presentation, we report on the composite paper electrodes containing graphene oxide and aramid nanofibers fabricated by flow-directed assembly. Recently developed aramid nanofibers, well-known for its high mechanical strength, have been used as the guest material for creating composites. The addition of the aramid nanofibers produced flexible, mechanically robust electrodes, where aramid nanofibers act as the guest material and interfacially interact with graphene sheets. In addition, the incorporation of the guest material increased porosity and alleviated the diffusion limitation for application as electrodes for supercapacitors. The composition, morphology, and the structure of the paper were characterized using scanning electron microscopy, Raman spectroscopy, and X-ray photoelectron spectroscopy. Mechanical and electrochemical performances of the composite electrodes were explored as structural electrodes.

453890 Composites Based on Biomimicry of the *Prunus* Spp. Seed Cyanide Defense System as Alternative to Pesticides

Abstract:

Many crops are ill-protected against insect pests during storage. Globally, there is still a significant loss of stored grains due to pest infestation especially in developing countries. To protect cereal grains from herbivores during storage, pesticides are often applied. While pesticides have an undoubtable functionality, increasing concerns are arising about their application.

Recently, we were able to mimic the principle of cyanogenic defense from plants such as bitter almonds (*Prunus dulcis*) and apply it onto wheat grains that naturally do not possess any cyanogenic defense.[1] Our motivation was to engineer a protective biomimetic coating material that defends the grain only upon triggering, is cost-effective, and free of conventional pesticides commonly used for grain protection during storage. To achieve an artificial cyanogenic defense mechanism similar to that found in seeds of *Prunus*species, we developed a novel onion-type grain coating consisting of separate layers containing either pure biodegradable polylactic acid (PLA), PLA mixed with the cyanogenic precursor mandelonitrile or the HCN releasing enzyme hydroxynitrile lyase. While the grains showed a cyanogenic potential high enough to theoretically have a effect on herbivores, their germination rate was reduced to approximately 25% compared to untreated grains, which was a clear drawback.

Here we would like to present the results from our recent study where we have significantly improved the cyanogenic grain coating design in terms of its effect on germination efficiency and have tested its actual performance against herbivore insects in laboratory and field experiments.[2] The optimized cyanogenic multi-layer coating consists of biodegradable polylactic acid with individual layers containing either the cyanogenic precursor amygdalin or Î²-glucosidase which is capable of degrading amygdalin to HCN. Cyanogenesis occurred only when the layers were ruptured *e.g.* by a herbivore attack. We could show that upon feeding coated cyanogenic wheat grains to the insect pest species *Tenebrio molitor* (mealworm beetle), *Rhizopertha dominica* (lesser grain borer) and *Plodia interpunctella* (Indianmeal moth), their reproduction as well as consumption rate were significantly reduced, whereas germination ability increased compared to non-coated grains. In field experiments, we observed an initial growth delay compared to uncoated grains which became negligible at later growth stages. Our proposed and experimentally tested strategy to artificially mimic and apply a naturally occurring defense mechanisms could be expanded to other crops than wheat and has the potential to replace certain pesticides with the benefit of complete biodegradability and increased safety during storage.

[1] Halter, J. G.; Chen, W. D.; Hild, N.; Mora, C. A.; Stoessel, P. R.; Koehler, F. M.; Grass, R. N.; Stark, W. J. Induced cyanogenesis from hydroxynitrile lyase and mandelonitrile on wheat with polylactic acid multilayer-coating produces self-defending seeds. *Journal of Materials Chemistry A* **2014**, 2, 853-858.

[2] Mora, C. A.; Halter, J. G.; Adler, C.; Hund, A.; Anders, H.; Yu, K.; Stark, W. J. Application of the *Prunus* spp. cyanide seed defense system onto wheat: Reduced insect feeding and field growth tests. *Journal of Agricultural and Food Chemistry* **2016** (in print).

451359 Preparation and Photocatalytic Properties of g-$C_3N_4/TiO_2/BiVO_4$

Abstract:

Preparation and Photocatalytic Properties

of g-$C_3N_4/TiO_2/BiVO_4$

Shen Xinlin, Wang Youliang, Li Kebin, Wang Fengyun

The Institute of Industrial Chemistry, Nanjing University of Science & Technology, Nanjing 210094, P.R.China

Keyword: g-$C_3N_4/TiO_2/BiVO_4$; preparation; MB degradation

Over the last few decades, the semiconductor-based photocatalysis has emerged as an environmentally benign technology for wastewater treatment due to its destructive ability towards a wide range of inorganic and organic pollutants[1-3] . TiO_2 is by far the most commonly being researched photocatalytic material owing to its many advantages for photocatalytic applications, e.g. high redox potentials, long-term thermodynamic stability, cost-effective and non-toxicity[1,4,5]. Whereas the bare TiO_2 suffers from several disadvantages including the limit to ultraviolet (UV) light absorption due to its wide band gap, as well as low quantum efficiency resulting from the rapid recombination of photogenerated electron¨Chole pairs. To overcome these drawbacks and enhance the photocatalytic activity of TiO_2, we constructed and prepared a ternary catalyst, i.e., mpg-$C_3N_4/TiO_2/BiVO_4$.

The mpg-C_3N_4 was obtained in a typical synthesis: dicyandiamide was used as raw material and SBA-15 as template agent and calcined under nitrogen atmospheres. The binary and ternary catalysts were prepared by hydrothermal. The XRD patterns of catalysts are shown in Figure 1. There are two peaks found in mpg-C_3N_4 sample at 12.28¡ãand 27.51¡ã, which can be indexed to (100) and (002) peak. The peaks at 25.5¡ã (101), 38.2¡ã (004), 48.2¡ã (200) and 54.9¡ã (105) indicate the existence of anatase TiO_2 phase (JCPDS No. 21-1272). The other diffraction peaks are agreement with monoclinic $BiVO_4$ phase (JCPDS No. 14-0688). No characteristic diffraction peaks of mpg-C_3N_4 can be observed for the ternary composite because of its low content and poor crystallization. But, its existence can be evidenced by FTIR spectra in figure 2. We have prepared the ternary mpg-C_3N_4/ $BiVO_4/TiO_2$ composite successfully.

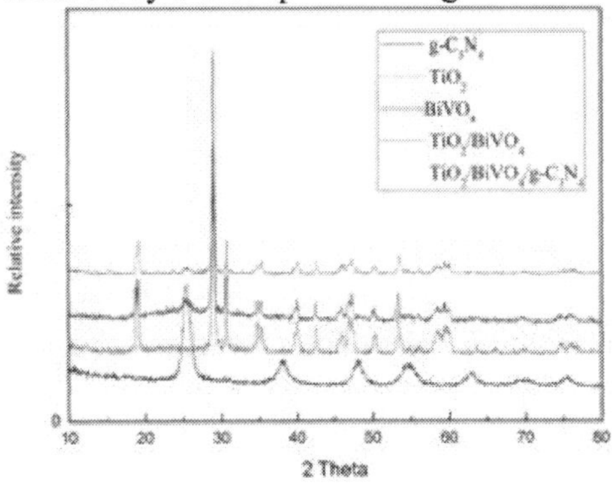

Figure 1 : XRD patterns of mpg-C_3N_4; $BiVO_4$; TiO_2; $BiVO_4/TiO_2$; mpg-C_3N_4/ $BiVO_4/TiO_2$

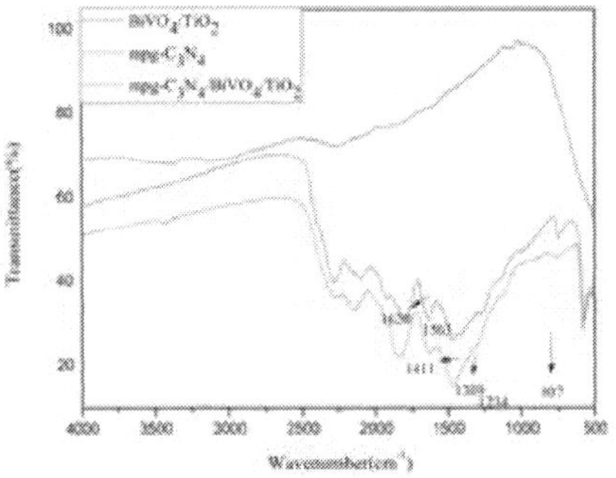

Figure 2 : FTIR spectra of mpg-C_3N_4; $BiVO_4/TiO_2$;mpg-C_3N_4/ $BiVO_4/TiO_2$

The TEM images of catalysts are shown in figure 3, we can see the binary and ternary catalysts are the formation of heterojunction. Although the size of the structure reach micron grade, the minimum structure of the hierarchical structure unit is still the nanoscale.

The effect of photocatalytic degradation for methylene blue (MB) under visible light radiation (¦Ë ¡Ý 420 nm) is shown in figure 4. It shows that the methylene blue is almost completely degraded by the ternary mpg-C_3N_4/ $BiVO_4/TiO_2$ composite after 2 h. It proves that the mpg-C_3N_4/ $BiVO_4/TiO_2$ composite is more efficiently than the binary and pure catalysts.

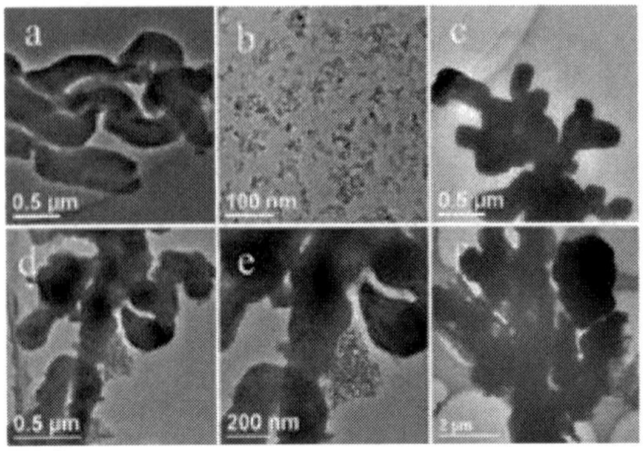

Figure 3 : The TEM image of catalysts : (a) mpg-C_3N_4 ; (b) TiO_2 ; (c) $BiVO_4$; (d-e)$BiVO_4/TiO_2$; (f)mpg-C_3N_4/ $BiVO_4/TiO_2$

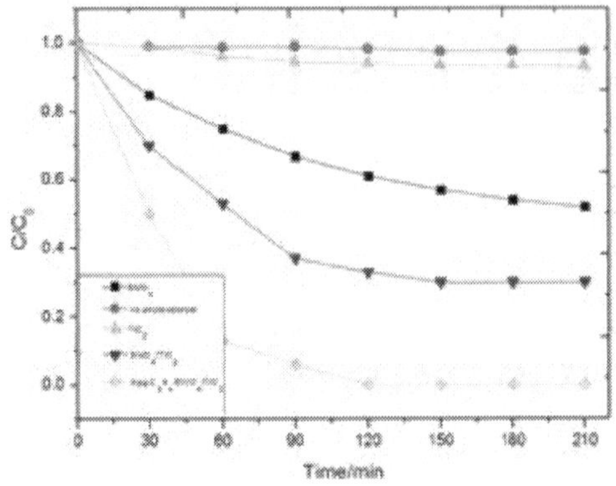

Figure 4 : The effect of photocatalytic degradation of methylene blue under visible radiation (¦Ë ¡Ý 420 nm)

The ternary mpg-$C_3N_4/BiVO_4/TiO_2$ catalysts have been prepared by calcination- hydrothermal successfully. It has been comfirmed of the formation of heterojunction by XRD, FTIR and TEM. The titled photocatalyst is efficiency for degradation of methylene blue under visible light radiation (¦Ë ¡Ý 420 nm).

References

[1] Bai X, Lv L, Zhang X, et al. Synthesis and photocatalytic properties of Palladium-loaded three dimensional flower-like anatase TiO_2 with dominant {001} facets[J]. J Colloid Interface Sci, 2016. 467: 1-9.

[2] Zhao W, Zhang Z, Zhang J, et al. Synthesis of Ag/TiO_2/graphene and its photocatalytic properties under visible light[J]. Materials Letters, 2016. 171: 182-186.

[3] Yu J, Liu Z, Zhai L, et al. Reduced graphene oxide supported TiO_2 as high performance catalysts for oxygen reduction reaction[J]. International Journal of Hydrogen Energy, 2016. 41(5): 3436-3445.

[4] Tan L-L, Ong W-J, Chai S-P, et al. Visible-light-activated oxygen-rich TiO_2 as next generation photocatalyst: Importance of annealing temperature on the photoactivity toward reduction of carbon dioxide[J]. Chemical Engineering Journal, 2016. 283: 1254-1263.

[5] Xu J, Wang G, Fan J, et al. g-C_3N_4 modified TiO_2 nanosheets with enhanced photoelectric conversion efficiency in dye-sensitized solar cells[J]. Journal of Power Sources, 2015. 274: 77-84.

472256 Diffusion of Ions in Charged and Uncharged Polymers (Invited Talk)

Abstract:

Charged polymer membranes are widely used for water purification applications involving control of water and ion transport, such as reverse osmosis and electrodialysis. Efforts are also underway worldwide to harness separation properties of such materials for energy generation in related applications such as reverse electrodialysis and pressure retarded osmosis. Additional applications, such as energy recovery ventilation and membrane-assisted capacitive deionization, rely on polymer membranes to control transport rates of water, ions, or both. Improving membranes for such processes would benefit from more complete fundamental understanding of the relation between membrane structure and ion sorption, diffusion and transport properties in both cation and anion exchange membrane materials. Ion-exchange membranes often contain strongly acidic or basic functional groups that render the materials hydrophilic, but the presence of such charged groups also has a substantial impact on ion (and water) transport properties through the polymer.

We are exploring the influence of polymer backbone structure, charge density, and water content on ion diffusion properties. Results from some of these studies will be presented, focusing on diffusion of salt, primarily NaCl, through various neutral, positively charged and negatively charged membranes via concentration gradient driven transport (i.e., ion permeability) and electric field driven transport (i.e., ionic conductivity). One long-term goal is to develop and validate a common framework to interpret data from both electrically driven and concentration gradient driven mass transport in such polymers and to use it to establish structure/property relations leading to rational design of membranes with improved performance.

461520 A Predictive Model for the Determination of Mixed Gas Transport and Solubility in Glassy Polymers

Abstract:

The description of the mixed gas transport and sorption in polymers is of fundamental importance for the development gas separation membranes, and for the evaluation of their performances under real conditions [1,2]. However, very few approaches are suitable to that aim, especially for glassy phases, for which the Dual Model Sorption model (DMS) is still largely used, in spite of its limited, if any, predictive ability [3].

The transport model recently proposed by Minelli and Sarti has already proved its ability in the description of the permeability and solubility of pure gases and vapors in glassy polymers, on the basis of a simple fundamental approach [4]. The model considers the diffusion coefficient as the product of the molecular mobility and the thermodynamic factor, related to the concentration dependence of the penetrant chemical potential. The model relies on the thermodynamic description provided by the nonequilibrium thermodynamics for glassy polymers, NET-GP (and the nonequilibrium lattice fluid, NELF, in particular), which provides the thermodynamic behavior of the polymer/penetrant system, in a predictive fashion [5]. In parallel, a simple exponential function of penetrant concentration is used to describe the mobility coefficient behavior, which contains the only two adjustable parameters of the model: the penetrant mobility at infinite dilution and the plasticization factor.

In this work, the transport model has been extended to describe the transport and the solubility of gaseous mixtures in glassy polymeric membranes. In particular, the experimental permeability data of CO_2/CH_4 mixtures in conventional as well as in innovative glassy membranes have been examined, and described by the present model, with no need of any additional parameters with respect to the pure gas case. The NET-GP approach can be properly employed to represent the multicomponent solubility in glassy polymers, as already proved for various systems [6], and the transport model accounts for the presence of a second component through its concentration dependence of the diffusion coefficient. The model is able to describe well the experimental data for all the systems investigated, and the permeability behaviors of both penetrant species are accurately represented in a wide feed pressure ranges.

Interestingly, the model is able to highlight completive sorption phenomena among the two penetrants, and the effects produced on both solubility and diffusivity coefficients. Furthermore, polymer dilation induced by CO_2 (or other swelling agents) can be simply identified, and its plasticization effect in case of real mixtures is quantitatively identified.

References:

[1] B. Freeman, Y. Yampolkii, Membrane Gas Separation John Wiley and Sons, 2010

[2] P. Bernardo, E. Drioli, Ind. Eng. Chem. Res. 48 (2009) 4638.

[3] W.J. Koros et al., J. Poly. Sci. B: Polym. Phys. 19 (1981) 1513.

[4] M. Minelli, G.C. Sarti, J. Membr. Sci. 435 (2013) 176.

[5] F. Doghieri, G.C. Sarti, Macromolecules 29 (1996) 7885.

[6] M. Minelli et al., Macromolecules 44 (2011) 4852.

462446 *In Situ* Pressure-Contact Time-Resolved Fourier Transform Infrared Attenuated Total Reflectance Spectroscopy: A New Method to Measure Liquid Diffusion in Free-Standing Polymer Films

Abstract:

A conventional method (*ex situ* gravimetry) for measuring liquid diffusion in free-standing polymer films typically consists immersing a film into a liquid, manually removing the film at a given time interval, removing the film and blotting the excess liquid from the surface of the film, weighing the film, re-immersing the film in the liquid, and then repeating this process many times to construct weight gain *versus* time data. However, this *ex situ* technique can introduce significant error when the sample size is small (typical of polymers synthesized at the academic laboratory scale) or take significant experimental time when the sample is thick, and this technique does not provide any molecular level detail. In this study, a new technique was developed to accurately measure liquid diffusion in free-standing polymer films with molecular level detail: *in situ* pressure-contact time-resolved Fourier transform infrared attenuated total reflectance (FTIR-ATR) spectroscopy. Previously, many studies have demonstrated the accurate measurement of liquid diffusion in polymer films using FTIR-ATR spectroscopy; however, polymers typically require solution casting on to an ATR element (or crystal) to ensure adequate adhesion or contact between the polymer film or crystal, where the infrared signal is measured inside the polymer close to the polymer/crystal interface. The new method developed in this study allows for an accurate measurement of liquid transport in polymer films for the case where the films cannot be easily solution cast or solution casting can alter the properties of polymer films that are processed using alternative methods (*e.g.*, extrusion). A custom-made cell was designed to introduce the liquid above the polymer film while also applying an adequate pressure between the polymer film and the ATR crystal to ensure adequate adhesion for accurate FTIR-ATR measurements. Liquid transport was investigated in free-standing polymer films with this new technique and compared with liquid transport measurements in free-standing polymer films using conventional gravimetry and liquid transport measurements in solution-cast polymer films using the conventional

FTIR-ATR technique. Experiments were performed on both rubbery and glassy polymer films and also as a function of temperature and film thickness. Non-Fickian diffusion was observed in glass polymer films, specifically diffusion-relaxation phenomena was observed, where the relaxation behavior differed for free-standing films compared to solution-cast films. The results from this new technique will be discussed in detail and in comparison to conventional gravimetry and conventional FTIR-ATR results.

457315 Penetrant and Ion Dynamics in Model Microphase Separated Copolymers

Abstract:

Nanostructured AB block copolymers have been of great interest for transport applications such as lithium ion battery electrolytes and CO_2 separation membranes. Various nanostructures (lamellae, cylinder, gyroid, sphere) form as functions of the fraction of A monomer, A-B interaction strength, and molecular weight, and one can choose the A monomer to allow transport of the target molecules while the B monomer provides mechanical robustness. Here, we study both diblock and tapered AB block copolymers (TBCs) where a "tapered" mid-block is inserted between two pure blocks of A and B monomers. The taper has a gradient in composition from pure A to pure B (or from pure B to pure A for an inverse taper) and taper length can be used an additional tuning parameter to control the nanostructure. A recent experiment shows a certain length of taper can improve both polymer and ion dynamics for battery electrolyte systems [1]. To provide physical insight into such results and predict which systems may improve transport, we perform molecular dynamics simulations using a simple coarse-grained bead-spring model. Monomer-sized penetrants with favorable interactions with one type of monomer are added to lamellar structures of normal and inverse TBCs at a range of taper length from 0% (diblocks) to 100% (gradient copolymer). We observe both polymers' and penetrants' dynamics increase as taper length increases for normal TBCs but non-monotonically change for inverse TBCs. For the inverse TBCs, nonintuitive trends in the dynamics arise from competing effects of local A-B mixing and chain conformations. We have also considered salt-doped systems (with full long-ranged Coulomb interactions) to explore how ion transport is different than that of uncharged small penetrant molecules.

[1] W.-F. Kuan, R. Remy, M. E. Mackay, and T. H. Epps, III, RSC Adv. **5**, 12597 (2015).

451322 The Kinetics of Diffusion in Keratin Fibres Using Mass Spectrometry

Abstract:

The Kinetics of Diffusion in Keratin Fibres Using Mass Spectrometry

N. Ali[1], D.R. Williams[1], J. Marsh [2] and S. Godfrey[3]

[1]Surface Particles Engineering Laboratory, Department of Chemical Engineering

Imperial College, Kensington, London SW7 2BY

[2] The Procter & Gamble Company, 11810 East Miami River Road, Cincinnati, Ohio 45252

[3] Researchand Development, Procter & Gamble, Darmstadt Germany.

Abstract

Over the years there has been considerable progress in understanding the structure and properties of hair. However, we still do not fully understand about the structure of hair and the penetration chemistry of actives in the hair.

Hydration of hair, for example, is a fundamental process. It governs swelling behaviour, structure, diffusion rates and mechanical properties of hair. Despite its importance, keratin water interactions (hydration) are poorly understood. This is particularly true for hydration along single hair fibres and across hairs from different body sites, genders, ethnicities and ages. Addressing these key knowledge gaps will allow us to unlock critical insights to enable innovation.

Hydration of hair, for example, is a fundamental process. It governs swelling behaviour, structure, diffusion rates and mechanical properties of hair. Despite its importance, keratin water interactions (hydration) are poorly understood. This is particularly true for hydration along single hair fibres and across hairs from different body sites, genders, ethnicities and ages. Addressing these key knowledge gaps will allow us to unlock critical insights to enable innovation.

In this study, the time dependent mass uptake of solutes has been demonstrated using Gas Chromatography Mass Spectrometry (GC-MS). The effect of solute concentration and temperature on the kinetic uptake of monoethanolamine by hair fibres has been investigated. Diffusion coefficient has been determined using Crankâ€™s equation for an infinite cylinder.

462941 A Combined Experimental and Modeling Approach to Study the Sorption and Diffusion Phenomena in Materials

Abstract:

A combined experimental and modeling approach to study the sorption and diffusion phenomena in materials

Hom N. Sharma, Stephen J. Harley, Yunwei Sun, and Elizabeth A. Glascoe

Lawrence Livermore National Laboratory

L288, 7000 East Ave., Livermore, CA 94550

925-423-5368

sharma@llnl.gov

Moisture sorption and diffusion in materials are of great interest in wide range of applications from fuel cells to microfluidics and pharmaceuticals.(1-3) The uptake and outgassing of moisture is also associated with aging and compatibility issues in a system or sub-system, which is important to establishing the lifetimes and viability of current assemblies and screening new materials for future designs. The process is dynamic and consists of different sorption modes and varies dramatically in different materials.(2, 4) Therefore, an in-depth understating of the moisture uptake and outgassing is essential.

In this study, we investigate the moisture sorption and diffusion using a combined experimental and modeling approach. Various materials including polymers are investigated over a wide range of temperature (30 – 70 °C) and relative humidities (RH) (0 – 90%) to quantify the moisture transport mechanism as shown Fig. 1. Further, a reactive transport model is developed, which includes (a) a triple-mode sorption model that includes absorption, adsorption, and pooling of species, (b) molecular diffusion, and (c) chemical reaction kinetics. Using a 1D or 3D model, we can simultaneously simulate the transport and chemical reactions of mobile species through materials. The equation of diffusion coupled with Langmuir adsorption and pooling sorption can be given as:

$$\frac{\partial C}{\partial t} = \frac{\partial (C_H + C_L + C_P)}{\partial t} = \frac{\partial}{\partial x}\left(Dx\frac{\partial C_H}{\partial x}\right) + \frac{\partial}{\partial y}\left(Dy\frac{\partial C_H}{\partial y}\right) + \frac{\partial}{\partial z}\left(Dz\frac{\partial C_H}{\partial z}\right)$$

where C, C_H, C_L, C_P, D are the total concentration of sample bulk mass, mass concentration in Henry mode, concentration in Langmuir mode, pooling sorption mode concentration, and effective diffusion coefficient, respectively.

Our results show that the particular mode could become dominant at certain RH regions indicating the material specific diffusion-sorption behavior. For example, Langmuir mode is dominant in Zircar RS1200 at the beginning while Henry mode gradually increases and reaches to its maximum. Above 50% RH, pooling (clustering) starts and becomes dominant sorption mode while Langmuir attains its equilibrium as shown in the figure below. In this presentation, we will discuss our combined experimental and modeling approach using diffusion-sorption mechanisms along with the parameter estimation and uncertainty quantification techniques in various polymeric and non-polymeric materials.

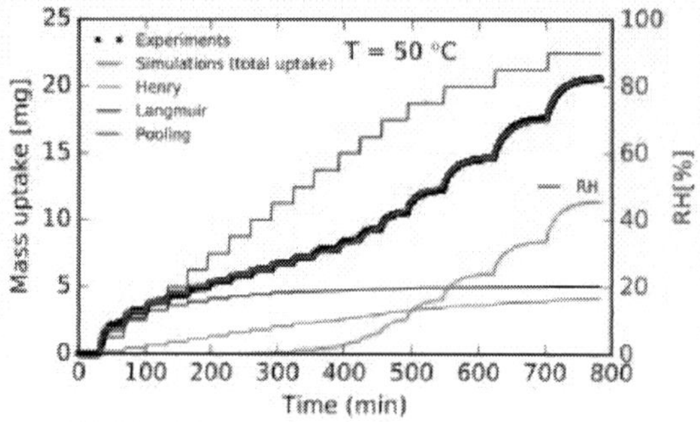

Fig. 1: Experiments and multi-mode diffusion-sorption modeling results of dynamic moisture uptake by Zircar RS1200 at a range of relative humidities at 50 °C.

This work performed under the auspices of the U.S. Department of Energy by Lawrence Livermore National Laboratory under Contract DE-AC52-07NA27344.

Reference:

1. Harley SJ, Glascoe EA, Maxwell RS. Thermodynamic Study on Dynamic Water Vapor Sorption in Sylgard-184. J Phys Chem B. 2012; 116(48):14183-90.

2. Harley SJ, Glascoe EA, Lewicki JP, Maxwell RS. Advances in Modeling Sorption and Diffusion of Moisture in Porous Reactive Materials. Chemphyschem. 2014; 15(9):1809-20.

3. Davis EM, Elabd YA. Water Clustering in Glassy Polymers. J Phys Chem B. 2013; 117(36):10629-40.

4. Sun YW, Harley SJ, Glascoe EA. Modeling and Uncertainty Quantification of Vapor Sorption and Diffusion in Heterogeneous Polymers. Chemphyschem. 2015; 16(14):3072-83.

458148 Triptycene-Containing Polybenzoxazole (PBO)-Based Membranes: New Synthesis and Gas Transport Properties

Abstract:

Triptycene-containing Polybenzoxazole (PBO)-based Membranes: New Synthesis and Gas Transport Properties

[1]Shuangjiang Luo, [1]Ashish Kushwaha, [2]Junyi Liu, [2]Haiqing Lin, [1]Ruilan Guo* (rguo@nd.edu)

[1]*Department of Chemical and Biomolecular Engineering, University of Notre Dame, Notre Dame, IN 46556-5637*

[2]*Department of Chemical and Biological Engineering, University at Buffalo, State University of New York, Buffalo, NY 14260-4200*

Polybenzoxazole (PBO)-based membranes, such as thermally rearranged (TR) polymers, hold great promise for gas separation that unite high gas permeability and high selectivity. It has been shown that gas separation performance of these thermally converted PBO membranes are largely determined by the precursor structures and the thermal treatment protocols. Herein we report the preparation and transport properties of two new series of PBO-based polymer membranes that were thermally derived from triptycene-containing *o*-hydroxy polyimide and polyamide precursors via thermal rearrangement (TR) process and thermal cyclodehydration (TC) process, respectively. Incorporation of triptycene units into poly(hydroxyimide) precursor structures led to significant increase of fractional free volume and created ultrafine microporosity in the converted PBO-based TR membranes, which enabled both high gas permeabilities and high selectivities. Although TC process of poly(hydroxyamide) precursor led to moderate improvement in the separation performance of the resulting triptycene-containing PBO membranes as compared to TR process, the PBO films converted via TC process exhibited excellent mechanical properties that are superior to many other TR membranes previously reported in the literature as well as the triptycene-containing TR membranes in this study. In particular, the PBO film thermally rearranged at 450 °C showed H_2 pure gas permeability of 807 barrer, CO_2 permeability of 269 barrer, and CO_2/CH_4 and H_2/CH_4 selectivities of 67 and 202, respectively, at 35 °C and 11 atm, which are far beyond the upper bound limits. In this talk, synthesis and characterization of these new PBO membranes will be presented. Fundamental transport properties including pure gas permeation data (with H_2, N_2, CO_2, and CH_4) will be discussed to illustrate the fundamental relationship between microscopic structures with macroscopic transport properties for these new triptycene-containing PBO membranes.

468297 Water Sorption and Diffusion in Nanofibrillated Cellulose Films and Composites

Abstract:

Nanocelluose is an innovative nano-sized material, which is attracting significant interest in view of its possible applications in various fields, such as, among the others, gas barrier materials or gas separation membranes. Indeed, nanocellulose films and nanocomposites are strongly barrier towards gases in dry conditions, while the permeabilities increase up to 2 or more orders of magnitude in humid environments, with significantly high selectivities toward CO_2. Water vapor can thus tune the nanocellulose properties from a barrier materials to a very promising membrane and the study of water sorption in these materials is crucial to completely understand the potentialities and current limitation of such new material for this kind of applications.

This work investigates the solubility and transport of water vapor in different types of Nanofibrillated Cellulose (NFC) as well as in polyvinylamine (PVAm)-NFC nanocomposites, by means of direct sorption and diffusion experiments in different experimental conditions.

In particular, NFC obtained by different cellulose pretreatments (carboxymethylation or enzymatic) prior to mechanical processing, were tested spanning over wide ranges of water activity and temperatures, between 16 to 65°C, while PVAm-NFC composite are investigated at various NFC loadings at 35°C.

Interestingly, the water diffusivity increased both with temperature and water activity, approaching a maximum value, independent on the operative condition considered for the tests. In addition, non-Fickian behaviors are observed especially at higher temperatures and R.H. inspected, clearly indicating the existence of different processes controlling the kinetics water sorption in the NFC film.

Similar behavior was also observed for in the case of nanocomposites, which also presented analogous water diffusivity, even if they showed significantly higher water uptake (larger than 50 % wt.) with respect to the pure NFC films that never exceed 10%wt.

The experimental data are then analyzed by appropriate modeling descriptions, seeking for the clear understanding of the process of moisture diffusion in nanocellulose based materails. Different approaches have been considered for solubility, accounting for both absorption and adsorption processes, and diffusivity, covering single or multiple Fickian diffusion, diffusion relaxation models, as well as the so-called parallel exponential kinetics model (PEK), often employed to represent the water sorption in cellulose fiber.

464834 Diffusion in Pebax Block Copolymers: Effects of Diffusant Mass and Polymer Composition

Abstract:

The rate at which small molecules diffuse through polyether-b-polyamide (PEBAX) copolymers is crucially important to their applications as controlled release media and gas separation membranes. PEBAX may also be blended with additives and used in medical devices, in which case accurate predictions of how fast those additives are released into the human body will serve the needs of public health. These uses have motivated us to experimentally measure a set of representative diffusion (D) and partition (K) coefficients over a wide range of diffusant masses (MW, ca. 50 â€“ 550 Da) and polyether contents (0 â€“ 84 mol. %) via a variety of gravimetric and spectroscopic sorption techniques.Â Here we will show that log(D) vs. log(MW) is linearly proportional with a slope of -4.7 within PEBAX 4033. The effect of polyether content on D and K can be well-accounted for by the effective medium theory of Sax and Ottino for transport in heterogeneous materials. These results broadly define upper limits on patient exposure to device additives, which will be useful in subsequently gauging the potential risks of being exposed to additives.

457102 Plasmon-Enhanced Photocatalytic CO_2 Reduction on Nanostructured Composite Electrodes

Abstract:

Using inputs of only sunlight, electricity, carbon dioxide, and water, photocatalytic CO_2 reduction could mitigate the effects of climate change caused by the burning of fossil fuels and produce valuable fuels or chemical feedstocks. However, current CO_2 reduction technologies suffer from high overpotentials and low selectivity, producing a mixture of carbon monoxide, methane, ethylene, formate, methanol, and other products.

Plasmon-assisted photocatalytic CO_2 reduction leads to greater selectivity and lower overpotentials by unlocking unique mechanistic pathways. It has been well-documented that the strongly localized near fields at the surface of plasmonic nanoparticles promote electron-hole pair generation in nearby semiconductors.[1–3] Plasmon decay generates an excited "hot" electron which can be transferred to a surface molecule for direct reduction or injected into an adjacent wide band gap catalyst, effectively limiting carrier recombination through charge separation and expanding the usable portion of the solar spectrum.[4–7] The electron dynamics in an irradiated plasmonic nanoparticle can alter the electronic coupling with surface adsorbed CO_2 and reaction intermediates, thereby changing the binding energy of these species and the catalytic properties of the plasmonic metals. This plasmon-enhanced effect was demonstrated when optically excited surface plasmons at the roughened surface of an Ag electrode in CO_2-containing solutions produced strong enhancement of the cathodic current.[8–9] The wide tunability of the plasmon resonance frequency with shape, size, and material confers fine control over these catalytic mechanisms, allowing for optimization of the photocatalytic performance.

To take advantage of plasmonic catalysis, nano-sized spheres, cubes, and wires of gold, silver, and copper

have been synthesized as these materials exhibit strong plasmonic and electrocatalytic behavior. For the semiconducting component of the photoelectrodes, nanostructured TiO_2 co-catalysts have also been prepared. Transmission electron microscopy (TEM) and scanning electron microscopy (SEM) were used to determine the shape and size distribution of the nanoparticles. X-ray diffraction (XRD) and optical spectroscopy were employed to characterize the electrodes. The CO_2 reduction ability of the plasmonic metal nanoparticles with and without a semiconductor co-catalyst has been tested in a custom glass photoelectrochemical cell with a quartz window that allows for front- and back-illumination of the electrode. The two electrode chambers are separated by a membrane and can achieve a gas-tight seal that allows for precise pressure measurements. Experiments were run both under isolation of the pressurized cell and with a constant CO
spectrometer (GC-MS) that allows detection and quantification of gaseous products. *Ex-situ* high performance liquid chromatography (HPLC) was used to quantify liquid products.

This material is based upon work performed by the Joint Center for Artificial Photosynthesis, a DOE Energy Innovation Hub, supported through the Office of Science of the U.S. Department of Energy under Award Number DE-SC0004993.

The work is supported by the National Science Foundation Graduate Research Fellowship under Grant No. 1106400.

Work at the Molecular Foundry was supported by the Office of Science, Office of Basic Energy Sciences, of the U.S. Department of Energy under Contract No. DE-AC02-05CH11231.

References:

(1) Thomann, I.; Pinaud, B. A.; Chen, Z.; Clemens, B. M.; Jaramillo, T. F.; Brongersma, M. L. *Nano Lett.* **2011**, *11*, 3440–3446.

(2) Ingram, D. B.; Linic, S. *J. Am. Chem. Soc.* **2011**, *133*, 5202–5205.

(3) Gao, H.; Liu, C.; Jeong, H. E.; Yang, P. *ACS Nano* **2012**, *6*, 234–240.

(4) Mukherjee, S.; Zhou, L.; Goodman, A. M.; Large, N.; Ayala-Orozco, C.; Zhang, Y.; Nordlander, P.; Halas, N. J. *J. Am. Chem. Soc.* **2014**, *136*, 64–67.

(5) Mubeen, S.; Lee, J.; Singh, N.; Krämer, S.; Stucky, G. D.; Moskovits, M. *Nat. Nanotechnol.* **2013**, *8*, 247–251.

(6) Naya, S.; Teranishi, M.; Isobe, T.; Tada, H. *Chem. Commun.* **2010**, *46*, 815–817.

(7) Lee, J.; Mubeen, S.; Ji, X.; Stucky, G. D.; Moskovits, M. *Nano Lett.* **2012**, *12*, 5014–5019.

(8) Kostecki, R.; Augustynski, J. *Chem. Phys. Lett.* **1992**, *194*, 386–390.

(9) Kostecki, R.; Augustynski, J. *J. Appl. Phys.* **1995**, *77*, 4701–4705.

459985 Photon, Electron, and Ion Management in Artificial Photosynthesis: Realizing Efficient Renewable Energy to Fuel Conversion

Abstract:

Direct production of fuels from sunlight, air (N_2 or CO_2), and water that can be stored, transported, and later converted into hydrogen or electricity to provide power for transportation and distributed energy generation, have received recent attentions worldwide. This technology could also provide chemicals as synthetic precursors or realize grid-level storage of intermittent solar energy. In artificial photosynthesis based on semiconductor photoelectrochemistry, production of chemical fuels generally requires the coupling of separated electrical charges with electrocatalysts for multi-electron chemical reactions. Meanwhile, the development of integrated, efficient and stable photoelectrochemical (PEC) systems requires the pairing of light-absorbing materials with an optimum bandgap combination. The development of such systems has been hindered in part by the lack of semiconducting materials that can simultaneously provide efficiency and stability in a corrosive aqueous electrolyte, typically strong acid or base. In this talk, I will present our recent progress in the development of efficient and stable PEC devices for H2 production from solar-driven water splitting. First, I will summarize some recent development of photon management in PEC devices and module designs from modeling to experiment. Then, I will discuss methods for heterogeneous interfacial energetic modification on covalence-bond semiconductors for efficient free carriers (electrons/holes) separation. Finally, I will present a novel approach to managing the ionic environment for creating a defect-tolerant condition, which leads to a record-setting solar water-splitting prototype.

462500 Tandem Core-Shell Si-Ta_3N_5 Photoanodes for Photoelectrochemical Water Oxidation

Abstract:

Tantalum nitride (Ta_3N_5
type doping, relatively small bandgap of 2.1 eV, and band positions that straddle the redox potentials of both the hydrogen and oxygen evolution reactions.[1] However, there are other challenges associated with Ta_3N_5 including poor charge transport properties, late photocurrent onset and photodegradation. To address all of these challenges, we fabricated efficient and stable core-shell heterostructured Si-Ta_3N_5 nanowire photoanodes coated with $CoTiO_x$ and NiO oxygen evolution co-catalysts.

Previous device architecture strategies for metal oxide semiconductors with charge transport limitations have included nanostructuring or scaffolding. In this work, a similar approach was applied to Ta_3N_5 to make a nanostructured core-shell Si-Ta_3N_5 device where the Si is a semiconducting, nanowire scaffold and the Ta_3N_5 is coated onto Si as a thin shell through which charge extraction should be efficient. With n-type Si as the scaffold, a heterojunction between Si and Ta_3N_5 is formed enabling a dual-absorber configuration resulting in a 200 mV cathodic shift for photocurrent onset due to the photovoltage gained from Si and therefore improves the overall performance of the Si-Ta_3N_5 device. The Si nanostructures were fabricated by a pseudo-Bosch deep reactive ion etching process, then tantalum oxide (Ta_2O_5) was coated onto Si by atomic layer deposition (ALD), and finally the Ta_2O_5 was nitrided in pure ammonia gas at 900 °C.

To gain a deeper understanding of charge transport within the thin Ta_3N_5 shells and to optimize the electrode

performance, we varied the thickness of the shell between 10 and 70 nm and tested for ferrocyanide oxidation as a facile photoelectrochemical reaction. We found that in a core-shell Si-Ta_3N_5 nanostructured device, photocurrent decreased with increasing Ta_3N_5 thickness, which indicates that charge extraction becomes problematic with thicker Ta_3N_5 films. To verify that decreasing photocurrent was due to poor charge transport in thicker films, we performed absorbed photon to current efficiency (APCE) measurements with Ta_3N_5 deposited on planar Si and quartz substrates. Again, we found that APCE decreased once the films were thicker than 30 nm, demonstrating that the minority carrier diffusion length is likely tens of nanometers.

Finally, to address the problem of Ta_3N_5 photodegradation and to demonstrate photoelectrochemical performance for water oxidation in basic electrolyte with the core-shell system, $CoTiO_x$ and NiO oxygen evolution catalysts were coated onto Si-Ta_3N_5 by dipcoating from a sol-gel and ALD, respectively. In terms of activity, we found that $CoTiO_x$ interfaces well with Ta_3N_5 and yields a high photocurrent density compared to NiO. With NiO we were able to achieve a photocurrent density close to that of $CoTiO_x$ after Fe incorporation into NiO (henceforth called Ni(Fe)O).[2] However, with $CoTiO_x$, the device is not stable; but with only ~1 nm of Ni(Fe)O we see a marked improvement in stability. The device loses 40-50% of its peak photocurrent over a 24 hour chronoamperometry test. Compared to the current Ta_3N_5 reports in the literature, the Ni(Fe)O catalyst affords noteworthy stability.[3-5] Therefore, in this work, we demonstrate a heterostructured Ta_3N_5 system with a tandem core-shell Si-Ta_3N_5 configuration that enabled earlier onset of photocurrent, efficient charge extraction from Ta_3N_5 and relatively stable long term water oxidation with a Ni(Fe)O co-catalyst.

1. Chun, W.-J.; Ishikawa, A.; Fujisawa, H.; Takata, T.; Kondo, J. N.; Hara, M.; Kawai, M.; Matsumoto, Y.; Domen, K. *The Journal of Physical Chemistry B* **2003,** 107, (8), 1798-1803.

2. Trotochaud, L.; Young, S. L.; Ranney, J. K.; Boettcher, S. W. *J Am Chem Soc* **2014,** 136, (18), 6744-6753.

3. Li, Y.; Zhang, L.; Torres-Pardo, A.; González-Calbet, J. M.; Ma, Y.; Oleynikov, P.; Terasaki, O.; Asahina, S.; Shima, M.; Cha, D.; Zhao, L.; Takanabe, K.; Kubota, J.; Domen, K. *Nat Commun* **2013,** 4.

4. Liao, M.; Feng, J.; Luo, W.; Wang, Z.; Zhang, J.; Li, Z.; Yu, T.; Zou, Z. *Advanced Functional Materials* **2012,** 22, (14), 3066-3074.

5. Liu, G.; Fu, P.; Zhou, L.; Yan, P.; Ding, C.; Shi, J.; Li, C. *Chemistry â€" A European Journal* **2015,** 21, (27), 9624-9628.

462505 Heterostructured c-Si/$BiVO_4$ Core-Shell Tandem Photoanode for Unassisted Photoelectrochemical Water Splitting

Abstract:

Photoelectrochemical (PEC) water splitting is a promising way of storing solar energy in hydrogen, which can be used as a carbon-free fuel or a chemical feed stock. In principle, a PEC electrode consists of a single active material capable of handling the key processes: absorption of solar photon energy, charge separation and transport, and electrocatalysis to make and break chemical bonds at the electrode surface. In practice, there is no existing material that can handle all of these functions simultaneously; thus, the multicomponent system that integrate different materials with more specialized functions is a more viable approach.

A tandem-junction PEC device offers a high theoretical solar-to-hydrogen (STH) efficiency due to the additive voltages across two photoabsorbers and the better utilization of the solar spectrum. Modeling STH efficiency as a function of band gap for a tandem device shows that over 20% STH can be achieved with band gaps of 1.2 and 1.8 eV.[1] One promising photoabsorber material combination is c-Si paired with a metal oxide, especially $BiVO_4$. In addition to the appropriate band structure alignment of both semiconductors, c-Si and $BiVO_4$ are low cost, earth abundant, and industrially scalable. However, the key challenges for these materials are the instability of c-Si in aqueous environment and the poor electronic properties of $BiVO_4$.

In this work, we present the fabrication of a heterostructured c-Si/$BiVO_4$ core-shell tandem based device capable of performing spontaneous water splitting without any precious metal. The wafer scaled device architecture was prepared by (1) nanostructuring of the c-Si substrate to serve as both a bottom absorber and a scaffold to improve the charge separation of $BiVO_4$; (2) tuning the c-Si p-n junction depth to maximize the device photovoltage; (3) engineering the SnO_2 interfacial layer to passivate the c-Si/$BiVO_4$ interface and protect the c-Si p-n junction; (4) conformally coating a $BiVO_4$ thin film on the nanostructured Si; and (5) decorating the c-Si and $BiVO_4$ surfaces with both oxygen and a hydrogen evolution catalysts.

(1) Seitz, L. C.; Chen, Z.; Forman, A. J.; Pinaud, B. A.; Benck, J. D.; Jaramillo, T. F. *ChemSusChem* **2014**, *7* (5), 1372-1385.

462617 The Fabrication of Cu_2o/g-C_3N_4/WS_2 Triple-Layered Photocathode for Photoelectrochemical Hydrogen Evolution

Abstract:

The Fabrication of Cu_2O/g-C_3N_4/WS_2 Triple-layered Photocathode for Photoelectrochemical Hydrogen Evolution

Xintian Xu, Yuanzhi Zhu, Xiaobin Fan, Guoliang Zhang, Wenchao Peng*

School of Chemical Engineering & Technology, Tianjin University, Tianjin 300072, China.

wenchao.peng@tju.edu.cn

Hydrogen is being vigorously pursued as a future energy carrier due to the fossil energy depletion. Photoelectrochemical (PEC) water splitting is an emerging hydrogen production technology by utilizing the solar light. This technology integrates in the same device both solar energy collection and water electrolysis. Both photo-active semiconductor and catalysts for hydrogen evolution reaction (HER) are necessary in this device. Cu_2O is a direct P-type semiconductor with a narrow band gap (~2.0 eV), which can be activated by the visible light. Moreover, Cu is a kind of earth abundant element, and the synthesis of Cu_2O is facile. Cu_2O is therefore one of the most promising low-cost photoactive semiconductor reported until now. However, the activity of bare Cu_2O is still poor, and it is not stable towards chemical oxidants and light corrosion. Combing with other semiconductors (e.g. TiO_2, ZnO) could improve the activity and stability of Cu_2O effectively.

As we know, Pt and other precious metals are the most efficient catalysts for hydrogen evolution reaction (HER).

However, these metals are rare and expensive to apply. Transition-metal dichalcogenides (TMDs), such as Molybdenum disulfide (MoS_2), tungsten disulphide (WS_2), are emerging HER catalysts that could be used as a substitute for noble metals in HER.

In this study, a novel Cu_2O/g-C_3N_4/WS_2 triple-layered photocathode was designed for the PEC hydrogen evolution. Cu_2O was electrodeposited on the FTO glass as the active layer for the light harvesting. A metal-free g-C_3N_4 was exfoliated and covered on the surface of Cu_2O as protection layer. G-C_3N_4 could also form p-n junction with Cu_2O to increase the activity. 1T-$WS_2$was obtained by lithium intercalation and used as HER catalyst for the further modification of the photocathode.

The obtained photocathode was then used for PEC HER test in the Na_2SO_4/H_2SO_4 solution (Î»â‰¥420nm, 100 mW/cm^2). With the irradiation, electron-hole pair will be generated in the p-type Cu_2O, and the excited electron could be then extracted through the g-C_3N_4 protective layer to produce hydrogen on the WS_2. The efficient extraction of excited electrons by this triple-layered photocathode leads to photocurrents of up to -6 mA/cm^{-2} at -0.4 V versus RHE (pH 6.0). The corresponding overpotential is -0.23 V (versus RHE) for the H_2 evolution. In sharp contrast, nearly no H_2 evolution happened under dark conditions. This photocathode is facile to synthesis and has great potential for developing low-cost PEC water-splitting device.

471805 High Chi Block Copolymers and Photopatternable Interfaces for Controlling Thin Film Structure (Invited Talk)

Abstract:

Block copolymer (BCP) self-assembly is showing great promise for enhancing the resolution of established, high-throughput photolithography techniques. This methodology has the potential to broadly impact device capabilities from magnetic data storage media to semiconductors. However, most device architectures possess the requirement that self-assembled BCP domains be confined only to prescribed areas. For example, templates for manufacturing magnetic storage media should contain regions with high density features for storing bits of data and, additionally, regions without any features. BCPs can self-assemble into features appropriate in size and density for data storage, but it is difficult to prescribe that features only appear in specific regions. We have recently developed a strategy to address this issue by demonstrating alignment layers (top and bottom adjoining layers to the BCP film) that have photoswitchable wetting characteristics for area selected control of BCP domain orientations. In this talk, several high-chi block copolymers capable of sub-10 nm patterning and their thin film self-assembly characteristics with these alignment layers will be highlighted.

464435 Role of Chain Semiflexibility and Density Fluctuations in Diblock Copolymer Phase Behavior

Abstract:

Block copolymer semiflexibility can arise from aromatic backbone and pi-bond conjugation in organic electronics applications. This poses challenges to apply theoretical predictions to experimental systems, since chain semiflexibility along with density fluctuation effects can dramatically influence the phase behavior of block copolymers. We investigate the phase behavior of a semiflexible diblock copolymer melt using random phase approximation and renormalization group theory. We present a chain rigidity-dependent free energy functional up to quartic order in density fluctuations. One-loop renormalization is used to account for fluctuation effects due to finite effective chain concentration. Using this approach, we examine the order-disorder transitions (ODT) and phase diagrams of semiflexible diblock copolymers beyond mean-field theory. We show that both the renormalized ODT Flory-Huggins parameters and phase diagrams are strongly dependent on diblock copolymer chain rigidity. Such theoretical formulation of semiflexible block copolymers broadens the applicability of theory to study nanostructured polymeric materials.

464361 Polymer Nanocomposite Films with Extremely High Filler Fractions Via Capillary Rise Infiltration (CaRI)

Abstract:

We present a novel approach to generate polymer nanocomposite films (PNCFs) with extremely high filler fractions (> 60 vol%) via capillary rise infiltration (CaRI) of polymer into a dense nanoparticle packing The CaRI process involves first generating a bilayer film of porous nanoparticle layer on a polymer layer, followed by annealing of the bilayer above the Tg of the polymer to induce polymer infiltration into the voids of the nanoparticle layer. By tuning the amount of polymer, we demonstrate that CaRI is also capable of generating fully filled or spatially homogeneous porous PNCFs. We utilize spectroscopic ellipsometry to characterize and monitor the polystyrene infiltration process into the titania nanoparticle packing in situ. The dynamics of CaRI is consistent with the Lucas Washburn model that has been used to describe capillary rise of liquid into porous media. However, we see a significant impact of confinement on the apparent viscosity of the polymer. In undersaturated CaRI (uCaRI), the infiltration process occurs in two stages. Upon annealing, we observe that the polymer layer is depleted rapidly via capillary-induced infiltration to form a dense composite at the base of the nanoparticle packing. Eventually, the front of this composite layer propagates throughout the nanoparticle packing, just as the refractive index of the composite decreases, indicating the redistribution of polymer throughout the nanoparticle matrix. We also demonstrate that the mechanical and optical properties of the CaRI composites can be readily tuned via varying the amount of polymer in the initial bilayers.

469646 Tuning Nanophase Separation Behavior in Segmented Polyhydroxyurethanes Via Judicious Choice of Soft Segment

Abstract:

Segmented polyurethanes (PUs) are composed of alternating sequences of soft and hard blocks. Depending on the composition of these blocks, they can exhibit properties ranging from soft elastomers to hard plastics while retaining the processing characteristics of thermoplastics. The soft blocks are typically long, flexible molecules with a glass transition temperature (T_g) below room temperature while the hard blocks have a T_g above room temperature and are composed of diisocyanate condensed with small molecule diol. Segmented PUs typically exhibit excellent nanophase separation due to the incompatibility between the soft and hard domains as well as the hydrogen bonding between polar urethane units in the hard domain. Varying degrees of nanophase separation are expected depending upon the structures of diisocyanates, soft segments, and chain extenders as well as overall material compositions.[1,2]

The increasing regulatory scrutiny on isocyanates in recent years has driven significant investigations for alternative pathways to PU or PU-like materials with cyclic carbonate aminolysis to produce polyhydroxyurethane (PHU) being one of the most promising candidates.[3] Polyhydroxyurethane is analogous to PU with the exception of additional primary and secondary hydroxyl groups adjacent to the urethane bonds. Unlike segmented PUs, the nanophase separation behavior in segmented PHUs is very different due to the presence of additional hydroxyl groups in the hard segment.[3,4] Torkelson and coworkers[5] recently reported the synthesis and characterization of segmented PHUs using several polyether-based soft segments. They showed that the hydroxyl groups caused significant phase mixing in PHUs with polyethylene oxide (PEO)-based soft segment due to a high degree of intersegmental hydrogen bonding to ether oxygen in the soft segment. This hydrogen bonding can be suppressed with polypropylene glycol (PPG)-based soft segments with sterically hindered oxygen atoms and polytetramethylene oxide (PTMO)-based soft segments with dilution of oxygen atom content. The suppression of hydrogen bonding to the soft segment produces nanophase-separated PHUs. In PTMO-based PHUs, the hydrogen bonding is suppressed but not completely eliminated leading to the formation of nanophase-separated systems with broad interphases having a wide range of local compositions with potential application as broad-temperature-range damping materials. To date, no study has yet demonstrated the ability to tune nanophase separation behavior in segmented, phase-separated PHUs to the level obtained in segmented PUs.

In this study, we demonstrate that nanophase separation behavior in segmented PHU copolymers can be significantly tuned via a judicious choice of soft segment. A series of PHUs were synthesized using two different soft segments namely PTMO and polybutadiene-co-acrylonitrile (PBN) and characterized with small-angle X-ray scattering (SAXS), dynamic mechanical analysis (DMA), differential scanning calorimetry (DSC), atomic force microscopy (AFM) and infrared spectroscopy (IR).

SAXS demonstrates that these PHUs possess nanophase-separated morphology with interdomain spacings of 10-16 nm independent of their soft segment compositions. DMA results also support the presence of nanophase-segregated structure in all PHUs with soft-segment glass transition temperature (T_g) well below room temperature and flow temperature above room temperature indicative of the hard segment undergoing glass transition. However, DMA reveals very significant difference between PTMO- and PBN-based PHUs. In PTMO-based PHUs, the storage modulus (E') show a very gradual temperature-dependent decrease and high tan δ (≥ 0.3) values over a broad temperature range (~70 °C) characteristic of nanophase-separated system with broad interphases having a wide range of local composition. In sharp contrast, DMA results of PBN-based PHUs display two distinct, step-wise transitions in E' with the corresponding sharp, well-defined peaks in tan δ indicative of nanophase-separated systems with a much sharper interphase similar to the case of isocyanate-based, segmented PUs and nanostructured block copolymer. DSC results also showed distinct soft segment and hard segment thermal transitions in the case of PBN-based PHUs while no discernible transition was observed for PTMO-based PHUs due to their broad interphases.

AFM phase imaging provides strong evidence of distinct soft and hard domains with sharp interphases for PBN-based PHUs whereas no distinct domain is observed in AFM images of PTMO-based PHUs due to the nature of their nanophase-separated structures despite evidences of nanophase separation from SAXS and DMA. IR characterization of the carbonyl region shows that PTMO-based PHUs has a comparable level of free, non-hydrogen bonded carbonyl (~1730 cm^{-1}) and hydrogen bonded carbonyl (~1700 cm^{-1}) while hydrogen-bonded carbonyl is the dominant absorbance in PBN-based PHUs.

The improved nanophase separation in PBN-based PHUs also leads to improved tensile properties. PBN-based PHUs exhibit tensile strength values from 0.8 to 3.4 MPa, which are higher than those obtained by PTMO-based PHUs from 0.4 to 0.8 MPa. Using soft segment with no possibility of intersegmental hydrogen bonding (as in the case of PBN-based soft segment) significantly improves nanophase separation in segmented PHUs relative to PTMO-based soft segment. This study demonstrates the importance of judicious soft segment choice to significantly tune nanophase separation behavior in segmented PHUs to yield systems with a much sharper interphase similar to isocyanate-based, segmented PUs.

References:

1) Yilgor, I.; Yilgor, E.; Wilkes, G. L. *Polymer* **2015**, 58, A1-A36.

2) Holden, G.; Kricheldorf, H. R.; Quirk, R. P., *Thermoplastic Elastomers.* (Chapter 2: Thermoplastic Polyurethane Elastomers) 3rd ed.; Hanser: 2004.

3) Guan, J.; Song, Y.; Lin, Y.; Yin, X.; Zuo, M.; Zhao, Y.; Tao, X.; Zheng, Q. *Ind. Eng. Chem. Res.* **2011**, 50, 6517-6527.

4) Blattmann, H.; Fleischer, M.; Bähr, M.; Mülthaupt, R. *Macromol. Rapid Commun.* 2014, 35, 1238-1254.

5) Leitsch, E. K.; Beniah, G.; Liu, K.; Lan, T.; Heath, W. H.; Scheidt, K. A.; Torkelson, J. M. *ACS Macro Letters* **2016**, 5, 424-429.

470653 The Effect of Nanoparticle Loading on Morphology and Function of Nanoparticle-Loaded Micelles

Abstract:

Nanoparticle-polymer composites combine the unique properties of nanoparticles and the mechanical strength of polymers to build tunable, multifunctional devices for healthcare, energy storage, and catalysis. Amphiphilic block co-polymers self-assemble in water, forming micelles with a variety of shapes and sizes that possess internal hydrophobic cores, which are stabilized by hydrophilic shells. The micelle cores serve as reservoirs for entraining hydrophobic nanoparticles, rendering them water-dispersible, whereas encapsulant properties are preserved and toxicity is minimized. High nanoparticle loading within micelles is desirable to improve functionality. However, loading nanoparticles into micelles can affect the shape, size and stability of the host micelles. Here, we studied the influence of nanoparticle packing within micelles to determine how these bulky cargoes effect final micelle morphology.

Quantum Dot (QD)-loaded micelles composed of poly(styrene-b-ethylene oxide) (PS-b-PEO) were synthesized using an emulsion-based approach developed by the Discher[1] and Hayward[2] groups. In this method, an organic solution of polymers and nanoparticles in a water-immiscible solvent is emulsified with water in the presence of a surfactant. Gradual extraction of the organic solvent by evaporation triggers co-assembly of the nanoparticles and polymers, yielding nanoparticle-loaded micelles. The QDs used in our study were non-spherical, with aspect ratios between 2.5 and 3 and were also larger than the polymer radius of gyration. We synthesized micelles with increasing QD loading by varying the proportion of QDs in the initial organic phase at constant polymer concentration. With increasing nanoparticle loading, there was a change in shape, size, deformation, and internal nanoparticle packing. The number of spherical micelles in the system decreased and more elongated structures, such as ellipses and cylinders, formed at higher QD loadings. Interestingly, the average sizes of the spherical micelles formed remained the same, whereas longer ellipses and cylinders formed with increasing QD loading. Beyond a critical QD loading, QDs appeared densely packed within nearly spherical composites as seen in transmission electron microscopy (TEM). Small angle x-ray scattering (SAXS) spectra of these densely packed composites revealed a Bragg shoulder, suggesting that the QDs adopt an ordered arrangement at high packing density.

As the use of QDs for labelling and tracking applications grows, non-spherical QDs such as nanorods, nanoplates, and tetrapods are being used for their unique shape-dependent properties. Here, we found that, despite their geometric dissimilarity, non-spherical QDs attempt to align within the spherical cores and deform overall micelle structures. Thus, non-spherical QDs can elongate spherical micelles, yielding ellipsoidal and cylindrical micelles, which could be detrimental in applications requiring uniform particle morphology. Such studies provide a general guideline for selecting optimum nanoparticle loadings for downstream applications and further the existing knowledge of nanoparticle packing within densely loaded micelles.

References

1. Geng, Y. and D.E. Discher, *Hydrolytic degradation of poly(ethylene oxide)-block-polycaprolactone worm micelles.* Journal of the American Chemical Society, 2005. **127**(37): p. 12780-12781.

2. Zhu, J. and R.C. Hayward, *Spontaneous generation of amphiphilic block copolymer micelles with multiple morphologies through interfacial instabilities.* Journal of the American Chemical Society, 2008. **130**(23): p. 7496-7502.

466728 Surface-Initiated Polymerization of Ionic Liquids

Abstract:

Ionic liquids (ILs) are organic salts that are liquid at or near room temperature (i.e. below 100°C). Ionic liquids possess unique materials and solvent properties that have led to their use in multiple applications, including as solvents, electrolytes, catalysts and polymer plasticizers. Poly(ionic liquids) (PILs) refer to a special type of polyelectrolyte which carry an IL species in each of the polymer repeating units. PILs offer advantages over ILs in enhanced mechanical stability, improved processability, durability and spatial control over the IL moieties. The ability to grow surface-tethered polymer films with IL functionality as side chains can enable coatings with versatile and tailorable properties. We report the surface-initiated ring-opening metathesis polymerization (SI-ROMP) of ionic liquids to grow poly(ionic liquid) films on gold and silicon substrates. The kinetics of film growth are rapid, with thicknesses approaching 200nm within 15 min of polymerization in a 0.1 M monomer solution and substantial film growth observed at monomer concentrations as low as 0.05 M.Â The ionic liquid monomer consists of the cations 1-norbornylmethylene-3-dimethylimidazolium (N1-dMIm^{+}) or 1-norbornylpropyl-3-dimethylimidazolium (N3-dMIm^{+}) with various anions that can be easily interchanged to tune the film properties. The dimethylimidazolium monomer was much more reactive toward SI-ROMP than the methylimidazolium analogue due to the acidic proton at CÂÂ methylimidazolium ring that can be removed to form a carbene complex with Grubbâ€™s catalyst rendering the catalyst inactive for SI-ROMP. Polarization modulation-infrared reflectance-absorption spectroscopy (PM-IRRAS) was used to quantify the extent of anion exchange in the films. Ellipsometry and profilometry measurements were utilized to quantify the thicknesses of the films. The dependence of ion transfer through the film on the ionic liquid anion was characterized by electrochemical impedance spectroscopy and cyclic voltammetry.

461018 Assembly of pH-Sensitive Gold Nanoparticles in Strong Polyelectrolyte Brushes

Abstract:

The modification of surfaces by means of polymer brushes has become an active area of research during the past few years due to numerous potential applications of such systems in nano- and biotechnology. The structure and conformation of a brush depends on external stimuli such as pH, temperature or ionic strength and can be manipulated by varying these attributes. Beside the responsiveness, polymer brushes have a positive influence on the steric stabilization of colloids, which make them suitable candidates as a matrix for the incorporation of inorganic nanoparticles like gold nanoparticles (AuNPs). AuNPs induce optical properties due to their surface plasmon resonance, which results in smart nanocomposite materials with tunable optical properties for application as sensors. In order to use the brush/AuNP hybrids for sensor applications, the arrangement of the AuNPs inside the brush has to be controlled. One important parameter is the charge of the AuNPs, which is assumed to drive the adsorption and penetration of AuNPs into the brush [1[2][3].

Here, we addressed this aspect by studying the effect of electrostatic interaction on the uptake and distribution of pH-sensitive, charge switchable AuNPs in poly-[2-(Methacryloyloxy) ethyl] trimethylammonium chloride (PMETAC) brush, a strong positively charged polyelectrolyte brush. The results showed that the AuNP uptake and spatial distribution depend strongly on the strength of the electrostatic interaction. Furthermore, changing the pH after preparation shows a strong effect on brush/AuNP hybrid thickness and optical properties [4].

References:

[1] S. Christau, S. Thurandt, Z. Yenice, R. v. Klitzing, *Polymers*, **2014**, 6, 1877-1896.

[2] S. Christau, T. Möller, Z. Yenice, J. Genzer, R. v. Klitzing, *Langmuir*, **2014**, 30, 13033-13041.

[3] S. Christau, J. Genzer, R. v. Klitzing, *Z. Phys. Chem.*, **2015**, 229, 1089-1117.

[4] D. Kesal, S. Christau, P. Krause, T. Möller, R. v. Klitzing, *Polymers*, **2016**, 8, 134.

459347 Interaction of PCL Based Nanopolymeric Micelles with Model Lipid Bilayers Via Explicit and Implicit Solvent Coarse-Grained Simulations

Abstract:

PCL based nanoparticles, typically formed from amphiphilic block copolymers consisting of water-soluble chains such as PEG, and functionalized with various ligands and other functional groups, are widely used in drug delivery applications. Nanopolymeric micelles formed from PEG-b-PCL copolymers have shown to evade the reticuloendothelial system, thereby providing improved circulation time to the nanocarriers, resulting in potential benefits to the betterment of human health. However, interaction of these nanocarriers with the cell membrane plays a crucial role in their endocytosis, potentially providing insights into their design and chemical make-up.

In this study, we perform coarse-grained molecular dynamics simulations to understand the interaction of micelles formed from MePEG-b-PCL copolymers, with a simple model DOPC lipid bilayer. Using existing models in the MARTINI coarse-grained force field for MePEG and PCL, we investigate the interaction of small micelles with a DOPC bilayer, in an explicit solvent setup. We found that the degree of hydrophobicity plays a crucial role in the interaction of the micelles with the bilayer, and also that the bilayer induced a change in morphology of the core-shell micelles to Janus particles during the internalization process. Key structural properties of the bilayer such as membrane thickness, lipid order parameter etc. were found to be largely unaffected during the internalization process, although dynamic and mechanical properties showed more changes when the bilayer was perturbed by the micelles. To explore larger spatio-temporal scales, the transferability of the developed models for MePEG and PCL to an implicit solvent setup using the 'Dry MARTINI' force field is also investigated. This paves the way for understanding the interactions of nanoparticles with large bilayers, at a much lower computational cost.

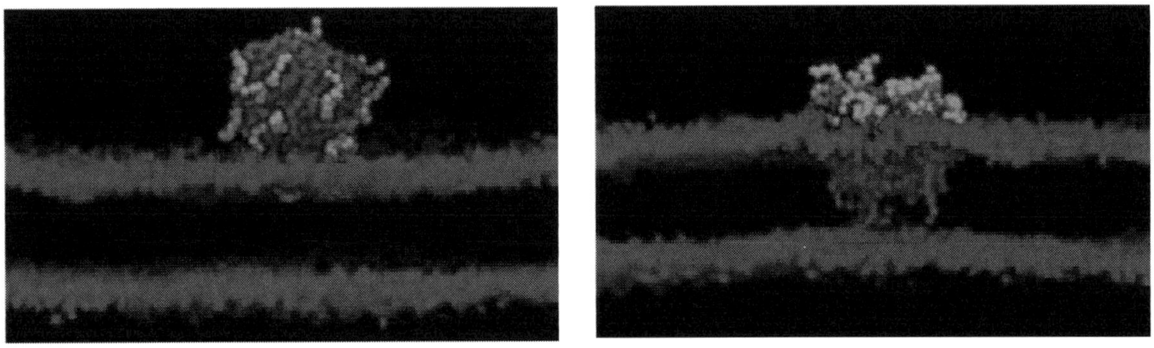

t = 0ns **t = 1000ns**

Fig 1. Bilayer induced change in morphology of core-shell micelles to Janus particles (color codes: green-MePEG, red-PCL, blue- DOPC; only the head groups are shown for better clarity)

465379 Novel Stimuli-Triggered Self-Healing and Strengthening Polymers

Abstract:

In nature, all living organisms can sense and respond to environment cues, such as light, temperature or pH. Examples of stimuli-triggered processes include contraction of eyes in response to bright or healing after a cut. These stimuli-triggered processes found in nature motivate the design of "smart" materials which can respond to environmental stimuli [1]. One such class of smart materials is self-healing polymers which have the capacity to recover their functionality after damage, thereby improving safety and increasing the service lifetime of the material. In particular, photo-active polymer networks are promising candidates for healing materials because the use of light offers spatiotemporal control and the response can be triggered remotely [2-4].

Here, we report a new class of photo-responsive polymer networks in which a dynamic bond is incorporated into the crosslinks of the network and undergoes a light-triggered, secondary polymerization which increases the modulus by two orders of magnitude while strengthening the network by over 100% [5]. Unlike traditional two-stage polymerization systems, in which the secondary polymerization is triggered by a leachable photoinitiator [6, 7], we incorporated a dynamic bond capable of initiating polymerization into the network backbone itself. By imparting functionality directly into the network architecture, our material is capable of initiating polymerization via the dissociation of its own crosslinks to become stronger in response to light. Specifically, a labile carbon-dithiocarbamate bond (or iniferter) was incorporated into the network backbone, which triggers a light-initiated, free-radical polymerization that increases the modulus by several orders of magnitude and approximately doubles the strength. The final modulus can be tuned *a priori*by modifying the concentration of the polymerizable group in the formulation. By adjusting the concentration, the material properties post-cure can be tailored to meet a wide range of specifications. The dynamic polymer network is readily transferrable to several applications. Specifically, we have demonstrated three properties: 1) photo-induced healing, reforming, and strengthening of a specimen after it has been severed,

2) spatial confinement of bulk property changes via photopatterning and 3) photo-curing the film into any particular 3D configuration.

References:[1] Stuart, M. A. C., Huck, W. T. S., Genzer, J., Muller, M., Ober, C., Stamm, M., Sukhorukov, G. B., Szleifer, I., Tsukruk, V. V., Urban, M., Winnik, F., Zauscher, S., Luzinov, I., and Minko, S., "Emerging applications of stimuli-responsive polymer materials," Nature Materials **9**, 101-113 (2010).

[2] Habault, D., Zhang, H. J., and Zhao, Y., "Light-triggered self-healing and shape-memory polymers," Chemical Society Reviews **42**, 7244-7256 (2013).

[3] Fiore, G. L., Rowan, S. J., and Weder, C., "Optically healable polymers," Chemical Society Reviews **42**, 7278-7288 (2013).

[4] Chatani, S., Kloxin, C. J., and Bowman, C. N., "The power of light in polymer science: photochemical processes to manipulate polymer formation, structure, and properties," Polym. Chem. **5**, 2187-2201 (2014).

[5] **Gordon, M. B.**, French, J. M., Wagner, N. J., and Kloxin, C. J., "Dynamic Bonds in Covalently Crosslinked Polymer Networks for Photoactivated Strengthening and Healing," Adv Mater **27**, 8007-10 (2015).

[6] Nair, D. P., Cramer, N. B., Gaipa, J. C., McBride, M. K., Matherly, E. M., McLeod, R. R., Shandas, R., and Bowman, C. N., "Two-Stage Reactive Polymer Network Forming Systems," Advanced Functional Materials **22**, 1502-1510 (2012).

[7] Ma, S. J., Mannino, S. J., Wagner, N. J., and Kloxin, C. J., "Photodirected Formation and Control of Wrinkles on a Thiol-ene Elastomer," ACS Macro Letters **2**, 474-477 (2013).

462411 Tungsten Oxide Decorated Zinc Telluride for the PhotoelectrochemicalÂ Water Splitting

Abstract:

Due to the increasing adverse effects on greener energy impacted by fossil fuel dependency and the resulting consequences on energy sustainability, a more succinct approach becomes inevitable for the continuous supply of clean and renewable energy. Tungsten Oxide (WO_3) has exhibited high photoactivities and electrochromic capabilities in the search for durable materials for hydrogen (H_2) production from photoelectrochemical water (H_2O) splitting. With a bandgap of 2.6 eV makes it an excellent candidate for harvesting the blue light of the solar spectrum with improved efficiency. Various experimental investigations have shown that electrochemically deposited WO_3 can be used to this effect. Sensitization with organic dyes that possess sunlight trapping capabilities have also been employed to enhance the performance of WO_3 as a photo catalyst for H_2O splitting. The successes in dye sensitization suggest that another metal compound, zinc telluride (ZnTe), which has an appropriate bandgap alignment for high utilization of the visible light of the solar spectrum can be used as a sensitizer as well. The fact that this material has not been studied in great detail in conjunction with WO_3 usage, makes it an interesting candidate of study in improving the performance of WO_3 as a photocathode for H_2 production. This material has a suitable bandgap of 2.26 eV for light harvesting and its conduction band edge positioned at -1.63 eV offers a large driving force for interfacial electron transfer from the semiconductor to the WO_3 acceptor. The objective of this work is to study the contribution of WO_3 decorated ZnTe in the splitting of H_2O. We will present the result of our investigation on cathodic electrodeposition of WO_3-ZnTe thin film on indium-tin oxide substrate, with its catalytic properties as an electrode for H_2 production.

468632 Molybdenum Silicide and Disulfide Protection Schemes for Silicon Photocathodes

Abstract:

Photoelectrochemical (PEC) water splitting offer a promising route to sustainable production of clean, renewable hydrogen.[1] In particular, silicon shows great potential as a small band gap photocathode in tandem photoelectrochemical (PEC) water splitting due to its appropriate band structure, energetic alignment for driving hydrogen evolution, and its band gap of 1.1 eV.[2] However, some challenges remain to ensure that silicon photoelectrodes possess the required catalytic activity and stability for use in highly effect PEC water splitting devices.

Molybdenum disulfide (MoS_2) is a low permeability material known for its excellent stability and HER activity in acid. We have previously demonstrated MoS_2 protected silicon photocathodes which operate for 100 hrs without significant loss of activity.[3] Here, we further this study to design strategies, such as the utilization of a molybdenum silicide interface layer, for improved activity and >800 hour stability with unchanged photoelectrochemical performance (onset potential, fill factor and light limited photocurrent density), which is among the best performance of any precious metal free photocathode.

Critical to developing future electrodes with enhanced stability, we probe the failure mechanism of the MoS_2 protected photocathodes. By XPS we observe that the MoS_2 is fully absent post stability testing. Detailed failure mechanisms of the photocathodes and electrolyte are uncovered by electrochemical methods, SEM, Raman microscopy and ICP. Furthermore, we implement increased light intensity, as well as light and dark cycling to probe accelerated durability testing protocols utilizing MoS_2 protected photocathodes. From such analyses we facilitate insightful design of future corrosion resistant protection schemes.

(1) Walter, M. G.; Warren, E. L.; McKone, J. R.; Boettcher, S. W.; Mi, Q. X.; Santori, E. A.; Lewis, N. S. Solar Water Splitting Cells. *Chem. Rev.* **2010**, *110*, 6446â€"6473.

(2) Seitz, L. C.; Chen, Z.; Forman, A. J.; Pinaud, B. A.; Benck, J. D.; Jaramillo, T. F. Modeling Practical Performance Limits of Photoelectrochemical Water Splitting Based on the Current State of Materials Research. *ChemSusChem* **2014**, *7* (5), 1372â€"1385.

(3) Benck, J. D.; Lee, S. C.; Fong, K. D.; Kibsgaard, J.; Sinclair, R.; Jaramillo, T. F. Designing Active and Stable Silicon Photocathodes for Solar Hydrogen Production Using Molybdenum Sulfide Nanomaterials. *Adv. Energy Mater.* **2014**, *4* (18), 1400739.

463462 Utility of Dual-Layer Photoanode for Photoelectrochemical Biomass Conversion

Abstract:

Photoelectrochemical (PEC) cells use solar energy to power desirable chemical reactions, such as splitting water into hydrogen, a fuel source, and oxygen. It has been shown that certain catalysts improve the efficiency of PEC water splitting, but less work has been done developing catalysts for chemical production in PECs. Here, we present the synthesis and characterization of electrocatalysts which enhance the efficiency of other oxidation reactions, where water oxidation is actually the competing reaction. Specifically, we have demonstrated improved performance of dual layer photoanodes on the oxidation of 5-hydroxymethylfurfural (HMF), a biomass derivative. Under 1.5 AM G illumination, we observe 21% conversion of HMF on a single layer photoanode while the addition of an electrodeposited catalyst increases HMF conversion to 33%. Both linear sweep voltammetry and chronoamperometry measurements indicate the catalyst positively affects the biomass oxidation reaction but not water oxidation. By adjusting reaction conditions and controlling the interfaces between the photoanode layers as well as the electrolyte, electrocatalysts may be fine-tuned to behave differently. This result opens new doors to the utility of PECs for applications beyond water splitting.

465069 Molybdenum Disulfide As a Protection Layer and Catalyst for Gallium Indium Phosphide Solar Water Splitting Photocathodes

Abstract:

Gallium indium phosphide ($GaInP_2$) is a semiconductor with a nearly ideal bandgap for solar water-splitting as the top absorber in a dual junction tandem absorber device. It has been used in conjunction with a gallium arsenide (GaAs) bottom absorber in an overall water splitting cell with 12.4% solar-to-hydrogen (STH) efficiency, one of the highest STH efficiencies for an integrated photoelectrochemical (PEC) water-splitting device reported to date.[1] However, $GaInP_2$ suffers from one of the biggest challenges facing the field: instability due to electrochemical corrosion in aqueous electrolytes. Molybdenum disulfide (MoS_2) nanomaterials can be used to both protect $GaInP_2$ and significantly improve its catalytic ability since it is resistant to corrosion and also possesses high activity for the hydrogen evolution reaction (HER). In this work, we demonstrate that $GaInP_2$ photocathodes coated with thin MoS_2 surface protecting layers exhibit excellent activity and stability for solar hydrogen production and we probe the details of failure mechanisms using a novel flow cell microscopy technique.

Our $GaInP_2$ photocathodes demonstrated no loss in performance (photocurrent onset potential, fill factor, and light limited current density) until 60 hours of operation which represents a five-hundred fold increase in stability compared to bare p-$GaInP_2$ samples tested in identical conditions.[2] We believe this to be one of the first successful attempts to stabilize $GaInP_2$ using a thin film protection layer scheme. Furthermore, as this protection scheme has previously been used successfully on silicon photocathodes[3], this work highlights the potential for MoS_2 to be used as a thin film protection layer for many different semiconductor water splitting devices that are unstable in acid. Using a custom-designed flow cell coupled with various microscopic techniques, we gained a greater understanding of the failure mechanisms of MoS_2 as a thin-film protection layer. We discovered that pinhole formation in the MoS_2 layer exposes the $GaInP_2$ substrate, which readily corrodes in the acidic conditions, ultimately leading to device degradation. The flow cell further allowed us to capture the time scale of this pinhole formation. These insights represent a deeper understanding of MoS_2 as a protection layer and can be leveraged to improve the stability of thin film protected semiconductor water splitting devices.

References

(1) Khaselev, O.; Turner, J. A., A Monolithic PhotovoltaicPhotoelectrochemical Device for Hydrogen Production Via Water Splitting. Science 1998, 280, 425427.

(2) Britto R.J., Benck J.D., Young J.L., Hahn C., Deutsch, T.G., Jaramillo T.F., Molybdenum Disulfide as a Protection Layer and Catalyst for Gallium Indium Phosphide Solar Water Splitting Photocathodes. Journal of Physical Chemistry Letters. *Accepted (2016)*

(3) Benck, J. D.; Lee, S. C.; Fong, K. D.; Kibsgaard, J.; Sinclair, R.; Jaramillo, T. F., Designing Active and Stable Silicon Photocathodes for Solar Hydrogen Production Using Molybdenum Sulfide Nanomaterials. Advanced Energy Materials 2014, 4, 18

468243 Bioelectricity Generation from a Carbon Soot Electrode Using a Paper Based Microbial Fuel Cell

Abstract:

Disposable energy harvesting devices on a paper based platform in Microbial Fuel Cell (MFC) can provide low cost and affordable power supply in resource limited regions[1]. Electrodes play crucial role in these devices. The tedious and multiple steps involved in the fabrication of electrodes currently limit the application of MFC. Present research work deals with a readily available eyeliner as a novel conductive ink for the electrode preparation in paper based MFC. The inherent capillary action of the paper matrix allows a quick absorption of the ink when coated[2]. This results in the formation of a highly porous integrated paper electrode. The application of these coated electrodes was shown in a paper based air-breathing MFC where both anode and cathode were prepared using such coated conductive papers. An MFC employs exoelectrogenic-microbes as catalysts in the anode chamber to harness power from wastewater. When an air-cathode is used in the device configuration, it yields water as the only byproduct[3]. Using such device on a paper platform, a maximum power of 12.5 W/m^3 was obtained from a single device with *Shewanella putrefaciens* as the biocatalyst. Such an electrode can further find various applications in flexible electronics, electrochemical sensors and analytical devices.

References

[1] F Sharifi, S Ghobadian , F R Cavalcantiand NastaranHashemi, *Renew. Sust. Energ. Rev.*, 2015, 52, 1453–1472.

[2] L. Hu, J. W. Choi, Y. Yang, S. Jeong, F. La Mantia, L.-F. Cui and Y. Cui, *Proc. Natl. Acad. Sci. U. S. A.*, 2009, **106**, 21490–21494.

[3] R. Veerubhotla, A. Bandopadhyay, D. Das and S. Chakraborty, *Lab Chip*, 2015, **15**, 2580–2583.

459762 Sustainable Power Sources Based on High Efficiency Thermopower Wave Devices

Abstract:

There is a pressing need to find alternatives to portable power sources such as Li-ion batteries, which contain toxic metals, present recycling difficulties due to harmful inorganic components, and rely on elements in finite global supply. Thermopower wave (TPW) devices, which convert chemical to electrical energy by means of self-propagating reaction waves guided along nanostructured thermal conduits, have the potential to address this demand. Herein, we demonstrate orders of magnitude higher chemical-to-electrical conversion efficiency of thermopower wave devices, in excess of 1%, with sustainable fuels such as sucrose and NaN_3 for the first time, that produce energy densities on par with Li-ion batteries operating at 80% efficiency (0.2 MJ/L versus 0.8 MJ/L). We show that efficiency can be increased significantly by selecting fuels such as sodium azide or sucrose with potassium nitrate to offset the inherent penalty in chemical potential imposed by strongly p-doping fuels, a validation of the predictions of Excess Thermopower theory. Such TPW devices can be scaled to lengths greater than 10 cm and durations longer than 10s, an over 5-fold improvement over the highest reported values, and they are capable of powering a commercial LED device. Lastly, a mathematical model of wave propagation, coupling thermal and electron transport with energy losses, is presented to describe the dynamics of power generation, explaining why both unipolar and bipolar waveforms can be observed. These results represent a significant advancement toward realizing TPW devices as new portable, high power density energy sources that are metal-free.

468865 Nanostructuring, Oxygen Anion Diffusion Study and Electrochemical Performance of Double Perovskite Electrode for SOFC

Abstract:

The state of the art requires the synthesis of nanostructured materials to be used as electrodes, which will show improved electro-catalytic activity and bulk transport, thus in-turn providing better electrochemical performance. Chemical synthesis methods show limitations in obtaining reduced particle sizes. For example on adopting a chemical synthesis method an average particle size of 40 nm was obtained of $GaBaCo_2O_{5+d}$ (GBCO). Herein an attempt is made to reduce the particle size further by using a relatively green and novel top-down bio-milling process. The GBCO powder was subjected to a fungus, *Sacchromyces cerevisiae*, which was cultured in 250 mL YPD medium for 96h at 301K. We report the reduction in particle size of phase pure (fig1a) GBCO to a size lesser than 20 nm as observed using HR-TEM. Thin-film of $La_{0.6}Sr_{0.4}Co_{0.2}Fe_{0.8}O_{3-d}$ (LSCF) electrodes (thickness ~500 nm), were fabricated using a spray deposition method. In order to test the electro-catalytic performance of the GBCO nanoparticles, symmetric cells of GBCO on gadolin doped ceria (GDC) electrolyte were fabricated and electrochemical performance measured in area specific resistance was compared to the chemically synthesized GBCO (fig1b).Molecular dynamics simulations were performed to study the transport of oxygen anion [1] in anisotropic GBCO. The diffusivity of the GBCO was observed to be thermally activated with an activation energy (E_a) of 50.8 kJ/mol (fig1d).The diffusion coefficient was calculated to be of the value of 5×10^{-8} at 873K in *a-b* plan which was observed to be higher by an order of magnitude in comparison to the diffusion *c* axis (3.33×10^{-9} cm^2 s^{-1}) (fig1c). Similarly, the effect of Fe dopant on diffusivity at B site was measured in case of $GdBa_{0.5}Sr_{0.5}Co_{2-x}Fe_xO_{5.5}$ structure for x = 0 to 1. Where, it was concluded that the oxygen anion diffusion was maximum in case of 50% Fe substituted composition as shown in Table1

Table 1: Oxygen anion diffusion coefficient of double perovskite materials

Material	Diffusion (cm^2 s^{-1}) (923K)	E_a (KJ/mol)
$GdBa_{0.5}Sr_{0.5}Co_2O_{5.5}$	4×10^{-8}	38.4
$GdBa_{0.5}Sr_{0.5}Co_{1.5}Fe_{0.5}O_{5.5}$	5.13×10^{-8}	40.6
$GdBa_{0.5}Sr_{0.5}Co_{1.5}Fe_{0.5}O_{5.5}$	7.5×10^{-8}	42.8

Figure 1: a)XRD and HR-TEM image of biomilled nanoparticle, b) Nyquist plot of GBCO nanoparticle doped electrode, c) oxygen density profile where, blue represent Ba and yellow and green represent Co and Gd atoms respectively and d) Arrhenius plot of oxygen diffusion calculated from MD.

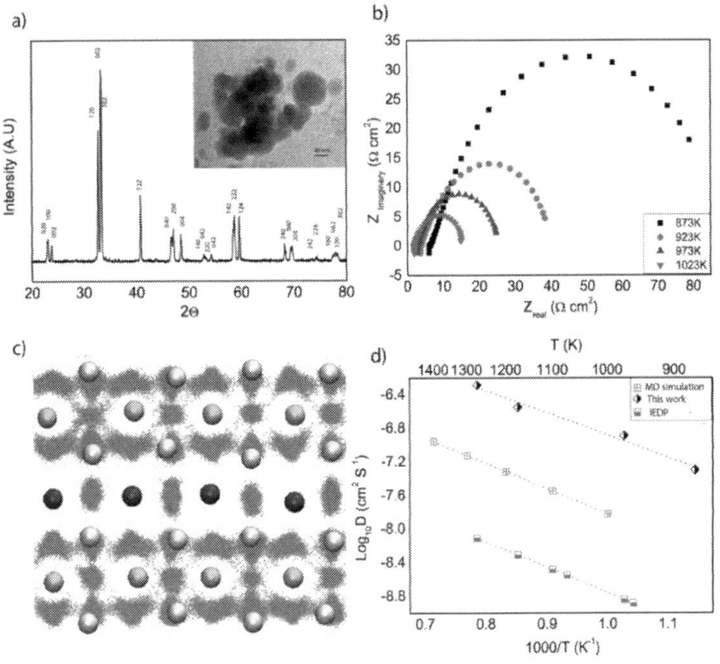

467700 Application of Photosystem I Multilayer Films for Photovoltage Enhancement in Aqueous Natural Dye Sensitized Solar Cells

Abstract:

Among the photosynthetic proteins, Photosystem I (PSI) is notable for achieving the most energetic reduction potential in nature and for its outstanding internal quantum efficiency. Stimulated by absorption of indigo, blue, and red photons, PSI facilitates electron excitation and sub-microsecond cross-membrane transport, all while retaining 1.1 eV of absorbed energy. As such, PSI is a model photodiode and the subject of broad scientific study and engineered emulation. In this study, we report photovoltage enhancement through application of PSI multilayer films atop TiO

glucoside—extracted from blackberries—and mediated by an aqueous $Fe(CN)_6^{4-/3-}$ redox shuttle. The enhancement phenomena is found to be dependent on both the amount of PSI deposited on the surface as well as the concentration of reduced redox group present in the electrolyte solution. We attribute the effect to PSI's photocatalytic reduction of photogenerated oxidized redox species in solution, which attenuates recombination rates at the dye/electrolyte interface. The blue and red absorption bands of PSI's light harvesting chlorophylls complement the strong green absorption of the sensitizing dye, allowing for better utilization of the visible spectrum.

468345 Compositions and Structures of High-N-Content Mesoporous Carbon Oxygen Reduction Electrocatalysts

Abstract:

Nitrogen- and transition-metal-containing carbon materials have been the subject of intense recent research interest, due to their desirable properties, including high oxidation stability, high conductivity, and high electrocatalytic activity, which enable their application as replacements for platinum-based oxygen reduction reaction (ORR) catalysts in fuel cells. Typically, these properties have been shown to be related to the N contents in the materials. However, N contents greater than ~4 wt% usually results in dramatically reduced electrical conductivity, and correspondingly reduced suitability for use as electrode materials. Nevertheless, a novel set of precursors have been identified which yield materials with order of magnitude greater N contents while retaining electrical conductivities up to 2.0 S/cm. When these materials are synthesized with high mesoporosities, surface areas in excess of 800 m^2/g, and the inclusion of iron-containing salts, they have been found to exhibit high ORR activities. Understanding the molecular compositions and structures of these functionalized mesoporous carbon materials aids the optimization of synthesis conditions, which yields improved performances of the electrochemical devices. Solid-state ^{15}N nuclear magnetic resonance (NMR) spectroscopy and X-ray photoelectron spectroscopy measurements demonstrate that the structures and compositions of these materials can be strongly influenced by choice of templating material, transition-metals included, and synthesis conditions, which result in physiochemical material differences that can be used to enhance electrocatalytic activities. Of particular interest are synthesis choices which can selectively alter surface chemistry without changing pore structure, and thus provide insight into the identity of the oxygen reduction active sites. Comparing and correlating the molecular compositions with the electrocatalytic activities crucially establishes that our materials with tunable compositions and cost-effective precursors exhibit ORR activities comparable to current Pt-based catalysts.

464466 High Throughput Analysis of Alloy Corrosion Across Composition Space: $Al_xFe_yNi_{1-x-y}$ ($x = 0 \rightarrow 1$, $y = 0 \rightarrow 1-x$)

Abstract:

Improving our fundamental understanding of the oxidation of multicomponent alumina-forming alloys is crucial to their ongoing development. In this work, high-throughput methods were developed to study oxidation of $Al_xFe_yNi_{1-x-y}$ alloys in dry air at 427 °C using composition spread alloy films (CSAFs) as combinatorial libraries ($x = 0 \rightarrow 1$, $y = 0 \rightarrow [1-x]$). CSAFs were deposited under ultra-high vacuum conditions using a rotating shadow mask, electron beam evaporation tool. The $Al_xFe_yNi_{1-x-y}$ CSAFs we first analysed using EDX to map their local compositions across the library, then exposed to oxidation conditions, before subsequent analysis using spatially resolved EDX, Raman spectroscopy, and XPS depth profiling. These allowing mapping of the oxygen uptake (Figure 1), the local phases formed and the distribution of the metal and oxide phases into the depth of the CSAF. This work used $Al_xFe_yNi_{1-x-y}$ CSAFs to determine the critical Al concentration, $N_{Al}^*(x, y)$, for establishment of a passivating Al_2O_3 scale continuously across $Al_xFe_yNi_{1-x-y}$ composition space ($x = 0 \rightarrow 1$, $y = 0 \rightarrow [1-x]$) in both dry air and a 10% H_2O/air mixture at 427 °C. The results divide the $Al_xFe_yNi_{1-x-y}$ composition space into four regions of phenomenologically distinct oxidation behaviour. At high an Al fraction, the alloy is passivated by either a surface Al_2O_3 scale or a subsurface Al_2O_3 scale. The boundary defining $N_{Al}^*(x, y)$ was determined across the entire continuous $Al_xFe_yNi_{1-x-y}$ composition space. The presence of H_2O vapour in an oxidizing environment increases the value of N_{Al}^* required to establish a passivating Al_2O_3 scale in multicomponent alumina-forming alloys. The $N_{Al}^*(x, y)$ in this environment was significantly higher across much of composition space than that measured in dry air. Physical insights from the observed differences between passivation in dry and humid air have been considered using a modified Wagner-Maak model.

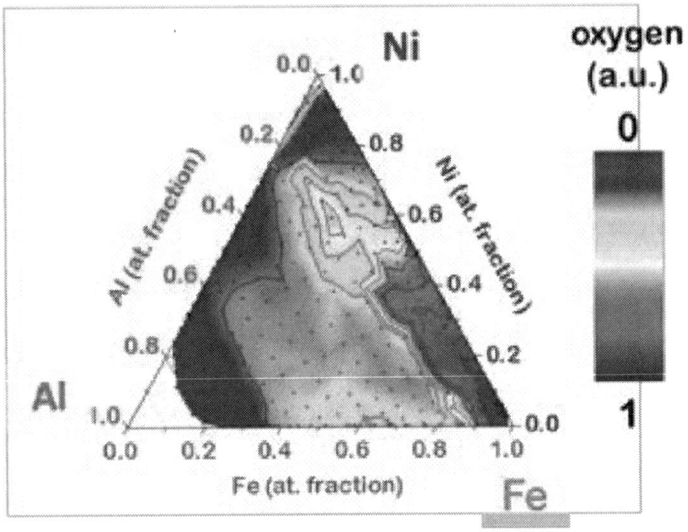

Figure 1. Map of oxygen uptake into a $Al_xFe_yNi_{1-x-y}$ CSAF following exposure to dry air at 427 °C for 4 hrs.

469063 A Novel Route to Phase Diagrams: Recovering 0 Kelvin Hamiltonian Parameters from High-Temperature Disordered Phases

Abstract:

Effective Hamiltonians, when used in tandem with statistical mechanics techniques, offer a rigorous connection between 0 Kelvin ab-initio simulations and finite temperature experimental observations. Specifically, cluster expansion Hamiltonians can extrapolate the complex, many-body electron problem of density functional theory into a series of simple site-wise basis functions (e.g., products of site coccupany variables) on an atomic scale. The resulting energy polynomial is computationally inexpensive, and hence suitable for the (tens of) thousands of calculations of thousand-site systems required by stochastic methods. We present a new method to run a cluster expansion "in reverse", using high-temperature observations and thermodynamic connections to predict the 0 Kelvin energy spectrum and associated ground states. By re-examining the cluster expansion formalism through the lens of entropy-maximization approaches, we develop an algorithm to select clusters and determine cluster interactions using only a few, high-temperature experiments on disordered phases. In this talk, we will develop our formalism and demonstrate multiple success using our approach. Specifically, we have explored three systems: two synthetic model Hamiltonians developed on a 2D triangular lattice, and an Au-Cu-like system in a 3D FCC lattice. In all cases, our new method can recover not only the stable ground states at 0 Kelvin, but also the full phase behavior through all of composition-temperature space.

471505 High-Throughput Prediction of Finite Temperature Free Energies of Solids

Abstract:

High-throughput, theory-driven computational materials design and discovery has become an essential tool in materials science and engineering. The data-driven and cost-effective framework for accelerated discovery introduced by the Materials Genome Initiative (MGI) has transformed the scale and rate of materials development by exploiting the predictive ability of quantum chemical computational methods. In turn, the drive to rapidly predict, screen, and optimize materials using first-principles calculations has led to large datasets and the construction of open-source databases that are populated primarily by low-cost density functional theory (DFT) total energy calculations (inherently performed at 0 K). Although these datasets are an invaluable resource, their predictive ability at finite temperatures is limited and current methods for evaluating the missing temperature dependencies are computationally prohibitive, limiting the success of MGI approaches for existing and emerging high temperature applications. In this report, we show that compound free energies of formation can be predicted with small errors at finite temperatures of up to at least 1800 K, and at the same computational expense as a DFT total energy calculation.

Our broadly applicable method for evaluating temperature-dependent compound free energies of formation employs readily computable material-specific descriptors and utilizes existing datasets of more than 50,000 inorganic compounds to enable rapid computational prototyping at elevated temperatures. We build upon prior work that has demonstrated the use of total energy calculations to approximate the compound enthalpy of formation at 298 K with mean absolute error relative to experiment (MAE) = 0.054 eV/atom with fitted elemental-phase reference energies (FERE). Our prediction requires only the DFT total energy, the FERE correction, and two minimal-computational-cost material-specific predictors to compute compound free energies of formation with MAE < 0.1 eV/atom up to at least 1800 K. At 1000 K, the MAE between our prediction and experiment is 0.039 eV/atom for oxides, 0.057 eV/atom for nitrides, 0.055 eV/atom for selenides, 0.049 eV/atom for arsenides, 0.044 eV/atom for phosphides, and 0.037 eV/atom for sulfides. These results bridge the temperature gap between existing MGI inspired computational approaches, which have historically been limited to low temperature, and high temperature materials applications, providing the equivalent of tabulated thermochemical data for more than 50,000 inorganic materials.

459508 First-Principles Based Design of Low Tolerance Factor Perovskites

Abstract:

We study the structural and vibrational properties of perovskite materials over a wide range of tolerance factors using first-principles calculations. We demonstrate how low tolerance factor materials that have been artificially constrained to the *Pnma* structure show ferroelectric instabilities. Our results are correlated with physical descriptors, such as ionic radii and electronegativity. The rules obtained from our first-principles analysis will provide useful guidance for design of the new ferroelectric materials.

472621 Descriptors and Approaches for Characterization and Screening of Inorganic Materials Databases

Abstract:

The use of databases in inorganic chemistry has become increasingly prevalent since 1990 [1]. Consequently, high-throughput (HT) database screenings are today routinely applied to find materials for emerging applications. In order to improve screening success rates accurate descriptors are required. However, accuracy and feasibility typically have to be balanced for a chosen HT approach [2]. Therefore, descriptors using different "levels of theory" and hierarchical approaches seem to be most promising. In this talk, we use simple structural and more complex energy as well as potential grid-based descriptors to characterize and screen different inorganic materials databases for various applications.

In the first case study, we scan the database of The Materials Project [3], which currently contains >60,000 inorganic crystalline structures obtained from relaxation with electronic density functional theory (DFT) [4,5]. We determine the number of basic structural motifs (e.g., tetrahedra, octahedra) occurring in each structure. A joint probability distribution (e.g., $P(N_{tet}, N_{oct})$) can then be used to construct a fingerprint of a given database. Because of the increasing number of available databases, such a quick visual access to similarities between databases is becoming more and more important.

Second, we analyze zeolite framework-type discoveries with a structural [6] and a thermodynamic [7] descriptor. We find that both descriptors are reliable indicators for the potential of a given zeolite to be deployable in a mature technology [8]. Our analysis also shows that synthesis trends around the new millennium yielded highly distorted, high-energy structures with low potential for technological use, and a similar trend can be currently seen [8].

Finally, we introduce a screening approach for intercalant diffusion in materials that are potentially useful as rechargeable batteries. To this end, we combine an extended version [2] of Voronoi decomposition [9,10]

with information from DFT to estimate diffusion barriers obtained from expensive nudged elastic band calculations. The resulting degree of correlation confirms this avenue as a promising path to screen hundreds to thousands of candidate structures.

[1] http://nisseshem.de/blogspot/databases-in-inorganic-chemistry-from-a-publication-statistics-perspective-2.html

[2] Pinheiro *et al.*, *CrystEngComm* 15, 7531–7538, 2013

[3] Jain *et al.*, *APL Mater.* 1, 011002, 2013

[4] Hohenberg and Kohn, *Phys. Rev.* 136, B864–B871, 1964

[5] Kohn and Sham, *Phys. Rev.* 140, A1133–A1138, 1965

[6] Zimmermann *et al.*, *J. Am. Chem. Soc.* 137, 13352–13361, 2015

[7] Kramer *et al.*, *J. Am. Chem. Soc.* 113, 6435–6441, 1991

[8] Zimmermann and Haranczyk, *Cryst. Growth Des.*, 10.1021/acs.cgd.6b00272, 2016

[9] Dirichlet, *J. Reine Angew. Math.*, 209–227, 1850

[10] Voronoi, *J. Reine Angew. Math.*, 97–178, 1908

465417 Towards More Complete Energy Landscapes of Zeolites: The Energetics of Zeolites with Comprehensive Ranges of Framework Topologies and Different Substituting Tetrahedral Atoms

Abstract:

When Si at the tetrahedral site (T site) of the zeolite frameworks is substituted by di- or trivalent atoms, negative charges at oxygen surrounding the substituting atoms are generated, thereby yielding anionic silicate-based frameworks. These negative tetrahedra require cationic charge-compensators that are typically active species for adsorption and catalysis. With respect to utilization of zeolites, as a result, the locations of substitutes are of great importance because they are directly related to active sites, for example, Brönsted acid sites when negative charges are counter-balanced by protons. Di- and trivalent substituting T atoms include Be, Zn, B, Al, and Ga.

The contents and locations of substitutes in the zeolite frameworks are one of the key factors determining the physicochemical properties of zeolites. Systematic evaluation of the characteristics of zeolites with a wide variety of framework topologies, a wide range of substituting contents, and various locations of substitutes is of great significance, but very challenging due to the limited ranges of the realizable chemical compositions in the existing zeolites as well as the limitation of the current analytical techniques.

Here, we present the systematic computational evaluation of the energetics of trivalent atom-substituted zeolites with 218 existing framework topologies at different substituting contents using the lattice-energy minimization technique with the aid of the Monte Carlo sampling. The results coincide well with the structural knowledge obtained experimentally. The relation between the relative framework energies versus the substituting contents varies in accordance with the framework topologies, suggesting that the relative thermodynamic stability of zeolites depends not only on the framework topologies, but also on the contents of substituting atoms [1]. The effects of types of the

substituting atoms (i.e., B, Al, and Ga) and the counterions (i.e., H^+ and Na^+) on the energetics of zeolites are also discussed in a context of their structure-directing functions

[1] K. Muraoka, W. Chaikittisilp, T. Okubo, *J. Am. Chem. Soc.*, doi: 10.1021/jacs.6b01341.

466306 Compilation and Analysis of the Computation-Ready, Experimental Metal-Organic Frameworks: Core MOF Version 2.0

Abstract:

We present an updated version of the Computation-Ready, Experimental Metal-Organic Framework (CoRE MOF) database, which includes additional structures (user-contributed, Cambridge Structural Database updates, and semi-automated re-construction of disordered structures using crystal generator) and new analyses (open metal site and duplicate detection) to provide an in-depth look at experimental MOF structures.

Open Metal Site. In preparing the database, we remove free and bound solvents to represent the fully activated structures. Depending on the coordination environment, the process can result in open metal sites (OMS), which can be used as active sites for catalysis and adsorption. In the recent work, we categorize the nature of OMS based on their molecular geometry and investigate trends in their elemental identity.

Duplicate MOFs. The CoRE MOF database contains experimental MOFs that are reported multiple times in the literature, such as IRMOF-1 and Cu-BTC, due to different groups reporting the same structure for their own purposes and due to differences in synthesis conditions. We used a â€œmolecular fingerprintâ€ approach to identify the duplicate structures to reduce redundancy in the database and to estimate the number of unique MOFs reported in the literature. We also used this approach to match a database of hypothetical MOFs with the corresponding experimental structures in the updated CoRE MOF database.

Additional MOF Structures. We have added approximately 1,000 new structures to the original CoRE MOF database [1], which includes structures that have recently been added to the Cambridge Structural Database, additional sources of atomic coordinates, and user-contributed corrections. We demonstrate the semi-automated use of a crystal generator to restore disordered structures and its implication on the future of *in silico*MOF discovery activity.

1. Y.G. Chung, J. Camp, M. Haranczyk, B.J. Sikora, W. Bury, V. Krungleviciute, T. Yildirim, O.K. Farha, D.S. Sholl, R.Q. Snurr, *Chem. Mater.*, **26**, 6185 â€“ 6192 (2014).

469122 Synthesis of MOF-74(Ni) Using Segmented-Flow and Microwave-Assisted Methods with Chemical Modulation

Abstract:

The control of nanoparticle uniformity is one of the main challenges in nanoparticle synthetic methods. Particle uniformity often dictates both the physical and chemical properties of the final products. Batch synthesis methods, such as hot injection and extended LaMer's method, lead to formation of products with very similar reaction history and consequent narrow particle size distributions.[1,2] Larger quantities of materials are required for industrial and precise control over size, and materials properties, becomes much more difficult. New scalable technologies are required for the precise control of nanomaterials during synthesis, and in this presentation we present a novel approach where we physically and temporally separate the nucleation and growth processes. For nucleation we use microwave radiation to minimize thermal gradients in the precursor solution while promoting high utilization of reagents with short reaction times.[3] A separate growth bath provides uniform heating of the nuclei during the growth stage, while segmented flow provides homogeneous mixing and constant reaction times throughout the process. A detailed understanding of each reaction step during colloidal nanoparticle synthesis is critical to obtain materials with the desired properties. There are several consecutive stages during nanoparticle synthesis, including: activation of precursors, formation of nuclei, growth of nanoparticles from the nuclei, and isolation of nanoparticles with the desired properties. The separation of the nucleation event from the growth stage is very important to the formation of nanoparticles with homogeneous size distributions. Using microwave radiation for fast and uniform heating for the nucleation zone, and using an oil bath at lower temperatures for the growth zone permits the distinct separation of nucleation processes from growth events. We will review the use of a segmented-flow, microwave-assisted reactor for the synthesis of a range of nanoparticles, including semiconductor nanoparticles for solar cells, quantum dots for lighting and display applications, as well as metal organic frameworks.[3–6] A common characteristic in all the nanoparticle systems studied, was narrower nanoparticle size distributions, substantially shorter reaction times, and high utilization of reagents.

The segmented-flow, microwave-assisted approach can be used to potentially support production of nanomaterials which still face commercialization difficulties due to the lack of scalable synthetic approaches, including metal-organic frameworks (MOFs).[7] MOFs are of considerable scientific and technological interest due to the flexibility in modifying their properties by changing either metal coordination species and/or organic linkers. MOFs can have three dimensional long-range order, high-surface area, large internal free volume space, and tunable adsorption properties. MOFs may be used in a wide range of applications including: gas storage, separation, catalysis, and more recently pharmaceuticals. However, the lack of scalable synthetic approaches still limits MOF production at industrial scales. In this study, we demonstrate the synthesis of MOF-74(Ni) using a segmented-flow, microwave-assisted reactor. The reaction chemistries were optimized so that the synthesis can be performed under relatively mild conditions. High yields (> 95 %) were obtained for short reaction times (~15 minutes), as opposed to days for typical batch reaction conditions. Optimization of reaction parameters for the microwave reactor has led to improvements in MOF crystallinity, conversion efficiencies, and production rates (> 4 g/h). We have found that chemical modulators significantly reduce the particle size distribution, increase the effective particle size, and enhance the adsorption properties of the MOFs. Characterization of the MOFs was performed using powder X-ray diffraction, scanning electron microscopy, transmission electron microscopy, BET isotherms, and UV-Vis spectroscopy to evaluate effects of microwave temperature and chemical modulator concentrations on MOF-74(Ni) properties.

References

1. Vreeland, E. C., *et al.*, "Enhanced Nanoparticle Size Control by Extending LaMer's Mechanism," *Chem. Mater.*, **27**,pp. 6059–6066 (2015).

2. Thanh, N. T. K., *et al.*, "Mechanisms of Nucleation and Growth of Nanoparticles in Solution," *Chem. Rev.*, **114**,pp. 7610–7630 (2014).

3. Kim, K.-J., *et al.*, "Continuous Microwave-Assisted Gas–Liquid Segmented Flow Reactor for Controlled Nucleation and Growth of Nanocrystals," *Cryst. Growth Des.*, **14**,pp. 5349–5355 (2014).

4. Hostetler, E. B., *et al.*, "Synthesis of colloidal PbSe nanoparticles using a microwave-assisted segmented flow reactor," *Mater. Lett.*, **128**,pp. 54–59 (2014).

5. Albuquerque, G. H., *et al.*, "Gas–liquid segmented flow microwave-assisted synthesis of MOF-74(Ni) under moderate pressures," *CrystEngComm*, **17**,pp. 5502–5510 (2015).

6. Fitzmorris, R. C., *et al.*, "Structural and optical characterization of $CuInS_2$ quantum dots synthesized by microwave-assisted continuous flow methods," *J. Nanoparticle Res.*, **17**,(2015).

7. Chung, Y. G., *et al.*, "Computation-Ready, Experimental Metal–Organic Frameworks: A Tool To Enable High-Throughput Screening of Nanoporous Crystals," *Chem. Mater.*, **26**, pp. 6185–6192 (2014).

471175 Injectable Hydrogels for Tandem Cell/Gene Transplantation

Abstract:

Cell transplantation by direct injection is a minimally invasive approach for the treatment of many ischemic diseases including stroke, peripheral arterial disease, and myocardial infarct. For example, adipose-derived stromal cells (ASCs) have been shown to promote neovascularization in preclinical models of ischemic tissues. Unfortunately, the direct injection of ASCs into ischemic tissue is limited by low cell viability and poor cell function. The transplanted cells are exposed to strong mechanical forces during injection (which results in cell damage) and low oxygen levels after injection (which results in further cell death and compromises cell function). To overcome these challenges, we have developed a family of injectable hydrogels that protects cells from mechanical forces during injection and provides mechanical scaffolding and biochemical cues to promote long-term cell survival and potency within ischemic tissues. At room temperature, all gel formulations are weak hydrogels (Gâ€™ <50 Pa) that form due to peptide-peptide heteroassembly to form a supramolecular network. These weak hydrogels undergo rapid shear-thinning to be easily hand injectable and rapid self-healing to localize transplanted cargo at the desired target site. Upon heating to 37Â°C, the hydrogels undergo a secondary crosslinking mechanism due to lower critical solution temperature (LCST) behavior. Simply tuning the weight fraction of the LCST polymer from 0-5 wt% results in hydrogels with tunable stiffness, Gâ€™ ~ 100 - 1000 Pa. Current research efforts focus on tuning the biochemical and mechanical cues provided by this delivery platform to optimize the hydrogel for tandem delivery of ASCs and therapeutic gene vectors to significantly enhance the angiogenic potential of transplanted cells.

469094 Tailoring the Mechanical Properties of Multi-Functional Polyampholyte Hydrogels for Tissue Engineering Applications

Abstract:

The development of novel polymeric scaffolds for tissue engineering depends on the successful demonstration of three critical features: resistance to nonspecific protein adsorption, the capability to deliver bioactive signaling molecules over both short and long term timelines, and the ability to easily tune the mechanical properties to match the tissue of interest. While there are numerous polymeric platforms that can address these three features individually, there are few, if any platforms that can adequately meet all three. Our work is focused on demonstrating the capability of polyampholyte polymers composed of equimolar concentrations of positively and negatively charged subunits to meet all three criteria. Polyampholyte hydrogels composed of varying ratios of [2-(acryloyloxy)ethyl] trimethylammonium chloride (TMA, positively charged), 2-carboxyethyl acrylate (CAA, negatively charged), and 3-sulfopropyl methacrylate (SA, negatively charged) monomers were synthesized. Regardless of the underlying composition, the hydrogels were demonstrated to be resistant to nonspecific protein adsorption from both negatively charged fibrinogen and positively charged lysozyme. All of the different hydrogel compositions were also demonstrated to be capable of covalently attached bioactive molecules using EDC/NHS conjugation chemistry. Finally, the mechanical properties of the resulting hydrogels were demonstrated to be dependent upon the composition, allowing for tunability in this property. When these results are coupled to the literature regarding polyampholyte polymers, it can be concluded that these systems represent a promising approach for tissue engineering applications.

465836 Cell Delivery Systems Via Complex Emulsion Templated Hydrogels

Abstract:

Multiphase emulsions with co-continuous and tunable internal microstructure are formed via colloidal silica self-assembly during spinodal decomposition in partially miscible binary solutions. Silica particles tuned to mutually wet the respective fluid phases strongly absorb at the interface at the onset of temperature-driven phase separation. Demixing comes to a kinetic halt once the interfacial area is sufficiently reduced to only accommodate particles, and the mixture undergoes a remarkable gelation transition. This now kinetically arrested, bicontinuous interfacially jammed emulsion gel (bijel) is comprised of co-continuous, fully percolating channels separated by a silica monolayer. Bijels provide a tunable template for synthesizing uniquely structured soft materials containing penetrating microchannels with consistent curvature. By selecting biocompatible polymers, the inherent properties of these templated materials can be utilized for probing applications in regenerative cell delivery and tissue engineering. We have synthesized a novel cell delivery system that contains fully penetrating and uniform void channels with tunable diameter for cell loading and unloading; exhibits robust mechanical properties brought about by the repetitive structure of the percolating polymer phase; and can be filled with common cell-friendly, soft matrix environments. These particular properties are exclusively afforded by this new class of soft materials because of their unique microstructure. In this talk, we will demonstrate how these bijel-templated hydrogels are utilized as cell delivery systems, with particular emphasis on cell seeding and unloading via chemoattractant gradients. Further, we will discuss the broader applications of these bijel-derived scaffolds for tissue engineering, particularly their application as biocompatible, densely vascularized membranes for therapeutic tissue implants.

463543 Non-Invasive Structural Investigation of Renal Scaffold By Magnetic Resonance Imaging (MRI)

Abstract:

Non-invasive structural investigation of renal scaffold by Magnetic Resonance Imaging (MRI)

Nafiseh Poornejad, Jonathan J. Wisco, Beverly L. Roeder, Alonzo D. Cook

Brigham Young University, Provo, Utah

Naturally obtained scaffolds combined with patient-specific cells could result in improved regeneration of human tissues. Scaffold fabrication is the first step in this technology, which is decellularization of the native tissue and sterilization of resulting scaffold. Decellularization, as the most effective procedure, could be done by exposing the native tissue to cell lysing agents that remove all cellular materials leading to the intact collagenous structure of extracellular matrices (ECMs). The majority of cellular remnants should be withdrawn to avoid immune response induction while the ECM should remain intact. In complex organs such as the kidney, the ultrastructure is of significant importance since it is essential for the filtration function of the kidney. Examining the fabricated ECM for structural integrity and intactness of the vascular network was pursued in this study via a novel method.

In this study, we decellularized whole porcine kidneys via our previously developed fabrication method (1-3) and investigated the ultrastructure of renal tissue by magnetic resonance imaging (MRI). Initially, we intended to perfuse a solution of nano iron particles, as contrast medium, through the vascular network to obtain high-resolution imaging with MRI. However, the nanoparticles would diffuse out of vessels in a decellularized ECM because of the lack of endothelial cells lining the vessels. So instead, human renal cells were loaded with nano iron particles and perfused through the renal artery (to image the vascular network) and renal ureter (to examine tubule structures) of native and decellularized porcine kidneys. The MRI scanning was performed on a 3 Tesla MR scanner (Trio, Siemens, Erlangen, Germany) with a 32-channel head matrix coil. The obtained results confirmed structural intactness of decellularized ECM compared with native tissue.

In summary, we developed a feasible method to investigate the vascular network and ultrastructural integrity of any decellularized organ including kidney, heart, liver, and lung by perfusing the organs with cells loaded with nano iron particles, and imaging the organs using MRI.

1. Poornejad, N., Frost, T.S., Scott, D.R., Elton, B.B., Reynolds, P.R., Roeder, B.L., andCook, A.D. Freezing/thawing without cryoprotectant damages native but not decellularized porcine renal tissue. Organogenesis **11**, 30, 2015.

2. Poornejad, N., Nielsen, J.J., Morris, R.J., Gassman, J.R., Reynolds, P.R., Roeder, B.L., andCook, A.D. Comparison of four decontamination treatments on porcine renal decellularized extracellular matrix structure, composition, and support of human renal cortical tubular epithelium cells. Journal of biomaterials applications, 0885328215615760, 2015.

3. Poornejad, N.M., N.; Salehi, A.S.M.; Scott, D.R.; Fronk, C.A.; Roeder, B.L.; Reynolds, P.R.; Bundy, B.C.; Cook, A.D. Efficient decellularization of whole porcine kidneys improves reseeded cell behavior. Biomedical Materials **11**, 025003, 2016.

467849 Bi-Modal Porous Poly(ε -caprolactone) Scaffolds Fabricated Via Two-Step Depressurization Supercritical CO_2 Foaming

Abstract:

Poly(Îµ-caprolactone) (PCL) is widely used as scaffold in bone tissue engineering due to its lasting degradation rate that matches the rate of natural bone growth. To mimic the extracellular matrix environment of natural bone tissue, emphasis has been laid on the pore morphology and structure of PCL scaffolds, and it demonstrates that a macroporosity of a few hundred microns is beneficial for cell adhesion, proliferation and differentiation; while a microporosity of tens of microns is necessary for nutrients and waste transportation. Therefore, formation of a bi-modal porosity is of particular interest in the fabrication of novel PCL scaffolds. Supercritical carbon dioxide foaming (sc-CO_2 foaming) is an alternative as solvent-free green process for scaffold fabrication. In this paper, a novel two-step depressurization sc-CO_2 foaming was designed, in which two different depressurization rates were utilized in one single foaming process to induce pores formation at different scales and attain bimodal porous scaffolds. Effects of important processing parameters, such as foaming temperature, saturation pressure, intermediate pressure, saturation time, holding time and depressurization rate etc., on pore morphology and structure of PCL scaffolds were investigated in details. Results indicated that intermediate pressure had dominant impact on the final pore morphology. Decreasing the intermediate pressure from 12 MPa to 5 MPa, the macropore size would increase from 130 Î¼m to 404 Î¼m. Interestingly, when intermediate pressure reached 9 MPa, only monomodal porosity was observed. Studies implied that nucleation size and macropore density was mainly determined in the first depressurization step by intermediate pressure and pore growth was more affected by the second depressurization. In summary, two-step depressurization sc-CO_2 foaming is verified available to produce bimodal porous PCL scaffolds with suitable pore structure and fine porosity. The plasticizing effect of sc-CO_2 during foaming causes the reduction of PCL polymer glass transition temperature (T_g) and melting point (T_m), which avails processing polymers at a relatively low operating temperature,

making it potential in protein growth factor incorporation for cells growth.

470724 Controlled Released Antibacterial Ag/Poly (L-lactic acid)/Poly(vinyl alcohol) (Ag/PLLA/PVA) Core Shell Nanofibers Prepared By Cold Atmospheric Plasma (CAP) Treatment and Electrospinning

Abstract:

Introduction: Electrospinning is a well-established process for the fabrication of nanofiber mats of high surface areas and high porosities. This process has been widely used for low-cost wound dressing production [1]. Silver is another widely-used and recognized effective anti-bacterial agent. Generally, the simplest and most commonly used method for incorporating silver nanoparticles into electrospun nanofibers is by the suspension of silver nanoparticles into polymer solutions. However, nanofibers produced by this method have reduced antibacterial properties due to silver nanoparticle aggregation. In order to avoid this problem, the objective of the present study was to utilize cold atmospheric plasma (CAP) [2], which is an ionized gas with ions, electrons, and excited atoms, as a reducing agent for photosensitive silver salts.

Materials and methods: Highly porous core-shell nanofibers were generated by rapid phase separation during the electrospinning process. A 10% PLLA solution was prepared using poly (L-lactic acid) (Sigma, U.S) with 1% silver nitrate (w/w) (Sigma, U.S). Then, the core-shell nanofiber was electrospun with a two-fluid coaxial spinneret (Inovenso Ltd. Co, Turkey), inner tube diameter of 1.2 mm, and outer diameter 1.4 mm, at a voltage ranging from 12-15 kV applied at the tip of the spinneret. The spun fibers were collected on a rotating collector at a speed of 300 rpm. Samples were then treated by cold atmospheric plasma at different times (30s, 60s, 90s, and 180s). Treated and untreated samples were observed by scanning electron microscopy (Hitachi, S-3700N, US). Anti-bacterial experiments and cell cytotoxicity studies will be conducted and reported later. All experiments were run in triplicate and repeated three times for each group with statistics analyzed by ANOVA followed by student t-tests.

Results and discussion: The topography of the fibers plays an important role in regulating initial cell behavior (such as cell adhesion and proliferation) due to changes in surface energy. SEM images of untreated and CAP treated nanofibers showed that the diameter of electrospun nanofibers ranged from 400 nm to 900 nm. Nanopores could be observed on the nanofiber surface, which indicated higher surface areas compared to nanofibers without nanopores. The diameter of nanopores on the untreated surface ranged from 5 nm to 20 nm. However, the nanopores became larger as CAP treated time increased. For 30s of CAP treatment, nanopores on the surface enlarged to about 30 nm in diameter, and 40 nm with 60s of CAP treatment. In addition, it is clear that the color of the electrospun polymer matrix gradually turned darker with increased durations of CAP exposure, indicating that silver nanoparticles were generated during CAP treatment.

Conclusions: In this study, a nanoporous core-shell nanofiber matrix was fabricated by an electrospinning method. Anti-bacterial silver nanoparticle reduction was successfully prepared by CAP treatment of the electrospun scaffold. In addition, results here showed for the fist time that the CAP treatment process is an effective tool to increase nanopore size on the surface of the nanofibers, resulting in increased surface areas, which will affect surface energy, initial protein adsorption events and cell adhesion as well as proliferation.

Acknowledgement: This study was supported by Northeastern University.

References: [1] D. Reneker, A. Yarin. Polymer, 2008, 49 (10), 2387â€"425.

[2] M. Wang, X. Cheng, W. Zhu, et al. Tissue engineering: part A, 2014, 5(20), 1060-1071.

461879 Peptide-DNA Hybrid Nanomaterials for Biology and Regenerative Medicine

Abstract:

Peptide-DNA Hybrid Nanomaterials for Biology and Regenerative Medicine

*Ronit Freeman[1], Nicholas Stephanopoulos[2], Samuel I. Stupp[1,3,4,5]**

(s-stupp@northwestern.edu)*

1. Simpson Querrey Institute for BioNanotechnology, Feinberg School of Medicine, Northwestern University, Chicago, IL 60611, United States

2. Current address: School of Molecular Sciences, Center for Molecular Design and Biomimetics, The Biodesign Institute Arizona State University, Tempe, AZ 85287, United States

3. Department of Materials Science and Engineering, Northwestern University, Evanston, IL 60208, United States

4. Department of Chemistry, Northwestern University, Evanston, IL 60208, United States

5. Department of Medicine, Northwestern University, Chicago, IL 60611, United States

Peptides and DNA represent two of the most attractive categories of molecules for the construction of nanomaterials for biology and medicine. Peptides provide a rich palette of biological functionality and self-assembly behavior, and DNA can be used to construct complex nanostructures with programmable, dynamic properties. We sought to merge the advantages of these two molecular platforms through the use of peptide-DNA (P-DNA) hybrid materials. Here we will outline recent progress in three distinct areas, each harnessing the unique potential of DNA as a novel biomaterial.

The first system used a DNA nanotube scaffold to present cell-adhesive peptides in a multivalent fashion (Figure 1A). Neural stem cells adhered to these DNA-peptide structures and differentiated selectively into neurons. Both the nanostructure and the biological signal could be independently controlled.

The second system involves the synthesis of peptide-DNA hybrid molecules that can be linked to non-bioactive substrates through Watson-Crick pairing to an immobilized complementary strand. The use of these molecular modifications on cell substrates has enabled us to manipulate different matrix functions, (Figure 1B): (1) induce reversible biological cell adhesion over "multiple cycles" using three orthogonal mechanisms; (2) optimize at the nanometer scale the distance between two signals in the matrix for synergistic signaling of the cell; and (3) probe the individual roles of two localized signals in the matrix. This system is the first to combine all three of these matrix functions using a single molecular platform. In particular, the platform described demonstrates the possibility of dynamic features on cell matrices not possible with previous approaches, which are either not reversible or use potentially harmful stimuli to the cell such as ultraviolet light.

This system provides a fundamentally new paradigm for DNA as a functional linker to mimic diverse ECM properties, and the orthogonality and programmability of DNA makes this platform highly attractive for a wide range of biological applications. In addition, this system can be readily interfaced with other DNA nanotechnology-based constructs, which have as of yet found only limited application in regenerative medicine and related fields.

A third project appended DNA or PNA(Peptide Nucleic Acids) to peptide amphiphiles (Figure 1C). These hybrid materials merge the dynamic properties of DNA with the bioactive and self-assembly properties of peptides. The resulting constructs were tunable and reversible in both their physical and biological properties due to the functionality of the DNA, allowing for dynamic signal presentation, reversible gelation and programmable hierarchical self-assembly of peptide-based filament structures.

The formation of 3D peptide amphiphile gels using the nucleic acids as cross-linkers involved the formation of long and dense bundled fibers that are enriched with nucleic acids while the non-bundled fibers are DNA-poor. The dimensions of the fiber bundles, and as a result, the mechanical properties of the gel network can be dynamically tuned by the number of DNA crosslinks. This dynamic 3D microenvironment provides new opportunities for dynamically altering different features and testing their effect on cell behavior.

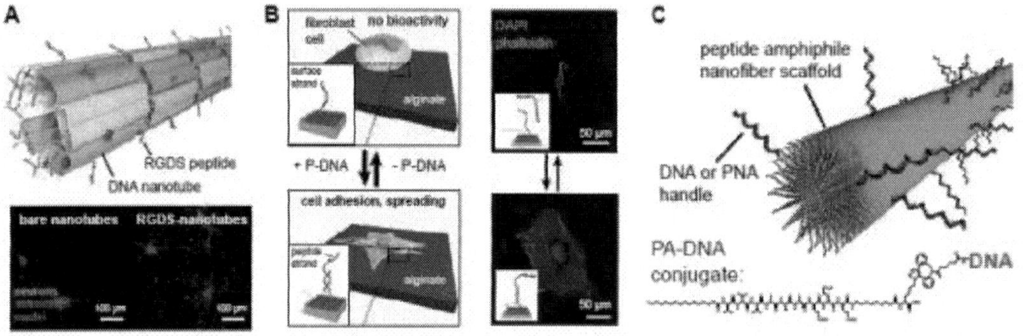

Figure 1. A) DNA nanotubes displaying the cell adhesion peptide RGDS promoted the attachment of neural stem cells and their selective differentiation into neurons, as compared with bare DNA nanotubes. B) DNA tethers were used to attach the RGDS peptide to non-bioactive alginate surfaces. The DNA allowed for reversible, dynamic presentation of the peptide by using toehold-mediated strand displacement. C) Peptide amphiphile (PA)-DNA molecules were co-assembled with unfunctionalized PAs to create hybrid nanofibers with handles for cross-linking into hydrogels or reversible presentation of biological signals.

464179 Programmable Immunotherapeutic Biomaterials to Potentiate Chemotherapy

Abstract:

Introduction: Acute Myeloid Leukemia (AML) has a very poor prognosis, with a 5-year survival of about 25%. While cytotoxic chemotherapy is successful at inducing remission of acute myeloid leukemia (AML), the disease has a high probability of relapse with chemo-resistant cells. One strategy to prevent relapse is to vaccinate mice against AML. Key to this strategy is to sustain a cytotoxic $CD8^+$T-cell response to eliminate residual cells after debulking by chemotherapy. In this study, we explored the development of a modular biomaterial-based vaccine with AML-specific antigens to prevent AML relapse.

Materials and Methods: Injectable macroporous alginate hydrogels were synthesized by low-temperature redox-induced free-radical polymerization. Granulocyte macrophage colony stimulating factor (GM-CSF), CpG ODN 1826, (5'-TCC ATG ACG TTC CTG ACG TT-3') and ovalbumin (OVA), cell lysate or an immunogenic peptide of Wilms' tumor protein-1 (WT-1) were added to the mixture prior to cryopolymerization. A GFP-luciferase expressing MLL-AF9 AML knock-in cell line from a C57BL/6J mouse was used to induce AML without irradiation. One version of the cell line was transduced to express ovalbumin. Cytosine arabinoside (Ara-C 100mg/kg) and Doxorubicin (Dox 4mg/kg) were used for induction chemotherapy. Hydrogels were injected subcutaneously and the recruitment of immune cells to the scaffold was monitored by periodically harvesting the scaffolds. Antigen specific $CD8^+$T-cells were monitored in blood. Secondary transplants were performed to test treatment efficacy.

Results and Discussion: In the MA-Alg/MA-PEG hydrogel, we observed an infiltration of neutrophils, dendritic cells and macrophages as early as 2 days after subcutaneous injection. There was an upregulation in activated dendritic cells at the site of the hydrogel, as well as in the draining lymph nodes. When used as an antigen prophylactically, the lysate of the bone marrow cells of leukemic mice prevented the engraftment of AML cells in mice. A western blot analysis of vaccinated mouse serum revealed the production of antibodies against a broad range of targets. When OVA was used as the antigen, paired with the OVA expressing AML cell line, the onset of leukemia was delayed significantly in prophylactically vaccinated mice, but not prevented altogether. SIINFEKL tetramer positive $CD8^+$ cells were detected in peripheral blood and were maintained over the duration of the study, until the mice were terminal. An analysis of the hematopoietic compartments revealed the gradual loss of OVA expression on AML cells, suggesting immune escape through loss of antigen. A bolus subcutaneous injection transiently increased SIINFEKL tetramer positive $CD8^+$, and delayed the onset of AML when challenged within that window – however it did not translate into increased survival. A known MHC class 1 restricted peptide fragment of a clinically defined AML antigen, WT-1, was incorporated in the vaccine as an antigen. When used prophylactically, WT-1 extended survival significantly, more than OVA, but not prevent AML associated mortality entirely. To mimic the clinical therapeutic management of AML, an induction chemotherapeutic regimen was used before the administration of the WT-1 vaccine. The chemotherapy significantly debulked the tumor, whereas the WT-1 vaccine prevented the relapse of AML in mice. WT-1 tetramer positive $CD8^+$cells were detected in peripheral blood over the duration of the study. Secondary transplants were performed after 150 days and confirmed the eradication of AML.

Conclusions: The results indicate the promise of a programmable biomaterial vaccine with clinically defined leukemia antigen targets. When used prophylactically, the vaccine alone can extend survival by sustaining the cytotoxic $CD8^+$T-cell response. When used in conjunction with cytotoxic chemotherapy, the biomaterial vaccine can potentially prevent the relapse of AML in a clinical setting.

Acknowledgements: The work was supported by the National Institutes of Health through grants U19HL129903 and R01EB014703 and R01EB015498.

464448 Surface Presentation of Adjuvants on Protein Nanoparticle Vaccines

Abstract:

Nanoparticle vaccines leverage the inherent immunogenicity of particulate matter to generate specific, adaptive immune responses to poorly immunogenic antigens. To examine the fundamental properties of effective nanoparticle vaccines, we have made nanoparticles directly from a crosslinked model antigen, ovalbumin (OVA), and examined the effects of nanoparticle size and coating on antigen uptake and processing, as well as dendritic cell (DC) inflammation and maturation. We observed that OVA nanoparticles enhance antigen uptake, experience attenuated endosomal acidification, and upregulate the DC maturation marker CD86 as compared to soluble protein. Dendritic cell inflammation was nanoparticle size- and coating-dependent. Small particles (270 nm) enhanced TNF-Î± production as compared to larger (350, 560 nm) nanoparticles, and 270 and 350 nm nanoparticles coated with an extra layer of OVA enhanced IL-1Î² production more than large coated nanoparticles, uncoated nanoparticles, or soluble OVA did. Based on these results, we have chosen coated, 270 nm nanoparticles as a platform for exploring other enhancements to nanoparticle vaccine adjuvancy.

In addition to coating with antigen, coating nanoparticles with molecular adjuvants can further enhance immunostimulation through several pathways. Molecular adjuvants can trigger immunostimulatory intracellular pathways, activate the extracellular complement system or target vaccines to specific cell types. Immunoglobulins can enhance immunity through all three of these mechanisms, and immunoglobulin G (IgG) and immunoglobulin M (IgM) have been identified as promising classes of adjuvants. Unlike traditional toll-like receptor ligands, such as flagellin, immunoglobulins are host-derived proteins, which greatly reduce the risk of an off-target immune response.

Building off of our prior work, we have coated OVA, flagellin, IgM, and IgG onto OVA nanoparticles. The activity of the coatings was confirmed by TLR-5/NFÎºB activation and complement activation studies *in vitro* for flagellin and immunoglobulin, respectively. Using an *in vivo* mouse immunization model, we examine the serum antibody responses and splenic T cell cytokine secretions in response to pathogen- and host-derived protein adjuvants. Our current work addresses three fundamental questions surrounding protein adjuvant use: (1) The benefit of antigen-adjuvant co-delivery, (2) The effectiveness of host-derived immunoglobulin in enhancing nanoparticle-based immune responses, and (3) Whether different immunoglobulin isotypes can trigger different types of adaptive immune responses. Our results contribute to current design strategies for molecular adjuvant incorporation onto nanoparticle vaccines.

467357 Acetalated Dextran Microparticulate Subunit Anthrax Vaccine Formulated Using Coaxial Electrospray Preserves Toxin Neutralization and Enhances Protection

Abstract:

Protein-based subunit vaccines are considered the safest type of vaccine, and their typically poor immunogenicity can be enhanced with well-characterized, potent Th1-skewing adjuvants. The use of polymer particles is an established approach to efficiently deliver both antigens and adjuvants to target antigen-presenting cells. However, particles composed of polyanhydrides and polyesters such as poly(lactic-co-glycolic acid) have detrimental acidic hydrolytic byproducts and relatively slow release kinetics. These substantial drawbacks emphasize the need for new polymer particles that exhibit pH-neutral hydrolytic byproducts and efficient intraphagosomal antigen/adjuvant delivery. Acetalated dextran (Ace-DEX) particles are a novel, acid-degradable formulation that fulfills these preferred design parameters. We have previously demonstrated Ace-DEX particlesâ€™ efficacy in several subunit vaccine formulations. These particles were formulated using the common batch solvent evaporation technique, and thus, some of the resultant in vivo immune responses were likely hampered due to the particle fabrication methodâ€™s denaturing characteristics.

An alternative polymer particle fabrication method such as coaxial electrohydrodynamic spraying (electrospray) is amenable to overcoming these formulation barriers. Coaxial electrospray processes do not create harsh denaturing conditions and have demonstrated gentle encapsulation of proteins in model studies. Despite these promising results, coaxial electrospray methods have yet to be applied to notable therapeutic or vaccine applications utilizing polymer particles. In our current work, we use coaxial electrospray to formulate an anthrax vaccine. We efficiently and gently encapsulate the labile recombinant protein antigen (antigen from *Bacillus anthracis*, causative agent of anthrax) and Th1-skewing resiquimod adjuvant (toll-like receptor 7/8 agonist) either in separate particle sets or co-encapsulated in the same particle. We compare our electrospray formulations in parallel with particles fabricated by solvent evaporation and the FDA-approved anthrax vaccine (BioThrax). Our electrospray formulations administered parenterally lead to increased Th1-skewing antibody production, preserved toxin neutralization, and enhanced protection against an inhaled challenge of the clinically virulent *B. anthracis* Ames strain. These promising results support electrosprayâ€™s potential to be a powerful platform technique for fabricating subunit vaccines and other protein-based drug delivery formulations.

469783 Engineering Lymph-Node Targeting Vaccines for Type 1 Diabetes

Abstract:

Vaccine approaches to restore antigen-specific immune tolerance to pancreatic beta-cell antigens without global immune suppression are potential therapeutic interventions for Type 1 diabetes (T1D). However, a method for realizing both efficacy and safety is yet to be developed. A major challenge in the development of effective vaccines is efficient delivery of vaccine components to antigen presenting cells (APCs) in lymphoid organs, where the orchestrations of immune cells are initiated.

We recently demonstrated that both the efficacy and safety of subunit vaccines can be dramatically enhanced by rational molecular design which target lymphoid organs via â€˜albumin-hitchhikingâ€™, a mechanism by which structurally optimized molecular vaccines are engineered with a lipophilic albumin-binding tail and follow subcutaneous injection, exhibit dramatic increases in LN accumulation via in situ complexation and transport with endogenous albumin, leading to 30-fold augmentation in CD8+ T-cell priming while greatly reducing systemic toxicity. Here we translate these findings to deliver immune suppressive vaccines to LNs, where they can be efficiently filtered by resident phagocytes and accumulate, promoting antigen-specific tolerance. Treatment of NOD mice with albumin-hitchhiking peptide vaccines significantly delayed and reduced the incidence of diabetes. This is associated with significant suppression of antoreactive CD4 T cells and increased frequency of regulatory T cells, as well as increased suppressive cytokine productions. Our findings suggest antigen delivery to lymph nodes via albumin-hitchhiking might be broadly applicable to treat autoimmune diseases.

466796 Silk Microneedle Skin Patches for Tunable HIV Subunit Vaccine Delivery

Abstract:

Development of a prophylactic HIV vaccine requires novel immunogen design and vaccine delivery strategies. Recent studies in our laboratories have shown that the kinetics of vaccine release modulate humoral immune responses. To control kinetics of prophylactic HIV vaccine, we developed a submicron-scale dissolving microneedle patch (MN) with release kinetics dependent on MN composition. We utilized silk protein to encapsulate an HIV envelope trimer (BG505 SOSIP) immunogen and adjuvants with poly(acrylic acid) (PAA) base. Upon dermal application, rapid dissolution of PAA is followed by silk beta sheet-dependent sustained release of encapsulated vaccine. In mice, whole animal imaging showed higher model antigen retention over one week in the skin with minimal local inflammation for silk MNs compared to equivalent intradermal injections. Antigenicity of SOSIP was maintained in silk as determined by binding to broadly neutralizing antibodies (bNAbs: PGT151, PGT145, PGT121, VRCO1) without binding to non-bNAb 14e. In mice, vaccine delivery in silk MNs promoted and sustained germinal center (GC) responses in lymph nodes with significantly higher GC B cells, plasmablasts and T follicular helper (Tfh) cells for upto 14 days following immunization compared to intradermal injections. Significant increases in MN induced anti-SOSIP IgG titers post boost correlated with Tfh induced increase in plasma CXCL13, a biomarker for strong antibody responses. Taken together, these results indicate the potential of silk microneedles for tunable HIV subunit vaccine delivery.

467119 Enhanced Cancer Immunotherapy By Microneedle Patch-Assisteddelivery of Anti-PD1 Antibody

Abstract:

Enhanced Cancer Immunotherapy by Microneedle Patch-Assisted Delivery of Anti-PD1 Antibody

Yanqi Ye[a, b], Chao Wang[a, b], Gabrielle M. Hochu[a], Hasan Sadeghifar[c], Zhen Gu[a, b, d *]

[a] Joint Department of Biomedical Engineering, University of North Carolina at Chapel Hill and North Carolina State University, Raleigh, NC 27695, USA;

[b] Division of Molecular Pharmaceutics and Center for Nanotechnology in Drug Delivery, Eshelman School of Pharmacy, University of North Carolina at Chapel Hill, Chapel Hill, NC 27599, USA;

[c] Department of Forest Biomaterials, North Carolina State University, Raleigh, NC 27695, USA;

[d] Department of Medicine, University of North Carolina School of Medicine, Chapel Hill, NC, 27599, USA

Introduction: For skin cancer treatment, immunotherapies have been intensively studied over the past several years).[1,2,3] The anti-PD-1 antibodies that target the inhibitory receptors have shown striking antitumor activity in phase II and III clinical trials of advanced melanoma). Despite the exciting clinical results of anti-PD-1 antibodies for the treatment of melanoma, the e cacy of the approach remains to be improved.

Materials and Methods: Here we present the physiologically self-degradable MN patch-assisted cancer immunotherapy for controlled delivery of aPD1 toward melanoma (Figure 1). Each MN is composed of biocompatible hyaluronic acid (HA) integrated with NPs that encapsulate aPD1 and glucose oxidase (GOx).[4, 5] With the GOx/CAT enzymatic system immobilized inside the NPs, the enzyme-mediated generation of gluconic acid promotes the gradual self-dissociation of NPs and results in the sustained release of aPD1.

Results and Discussion: We show that a single administration of the MN patch induces robust immune responses in B16F10 mouse melanoma model exceeding MN in the absence of the trigger element (GOx) or intratumor injection of free aPD1. Moreover, we demonstrated that the MN coloaded with aCTLA-4 and aPD1 resulted in synergistic treatment of melanoma. These results demonstrate that the MN patch-assisted system provides an innovative delivery strategy of aPD1 via a simple and safe technique that improves cancer immunogenicity and facilitates the clinical treatment of melanoma.

Conclusions: We have developed the MN patch-assisted immunotherapy that delivers aPD1 for the enhanced treatment of the skin cancer. Additionally, MN severs as a platform for combined therapy with other immunomodulators to enhance immunotherapy e ciency

Acknowledgements: This work was supported by NC TraCS (grant 550KR51307), NIH's Clinical and Translational Science Awards (CTSA, 1UL1TR001111) at UNC-CH, the NC State Faculty Research and Professional Development Award, and start-up package from the Joint BME Department of UNC and NCSU to Z.G.

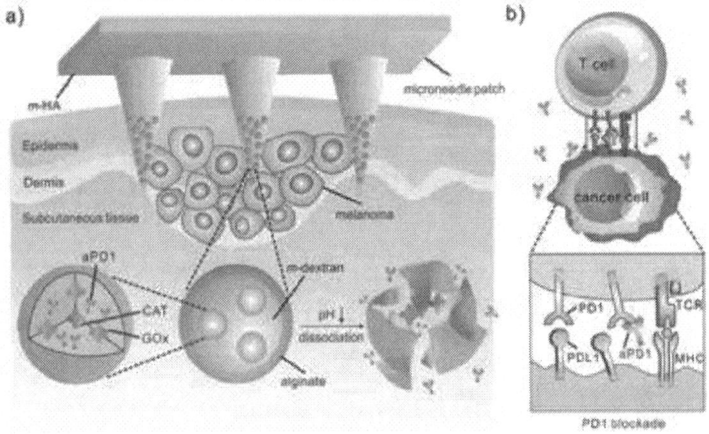

Figure 1. Schematic of the MN patch-assisted delivery of aPD1 for the skin cancer treatment. (a) MN patch loaded with self-dissociated NPs. (b) The blockade of PD-1.

References:

[1] Wang, C.; Ye, Y.; Gu, Z. *et al. Nano Letters* **2016**.

[2] Sullivan, S. P.; Prausnitz, M. R. *et al. Adv. Mater.* **2008**, 933–938.

[3] Gu, L.; Mooney, D. J. *Nat. Rev. Cancer* **2015**, 56–66.

[4] Yu, J.; Zhang, Y.; Ye, Y., Gu, Z. *et al. Proc. Natl. Acad. Sci.* **2015**, 8260–8265.

[5] Gu, Z.; Langer, R. S.; Anderson, D. G. *et al. ACS Nano* **2013**, 4194–4201.

464254 Targeting and Altering *In Vivo* macrophage Responses with Modified Polymer Properties

Abstract:

Targeting and altering *in vivo* macrophage responses with modified polymer properties

Kaitlin Bratlie

Departments of Materials Science & Engineering and Chemical & Biological Engineering, Iowa State University

Introduction: Two pathways for activating macrophages (MΦs) exist. One of these routes is termed the classically activated M1 pathway and is achieved through exposure to lipopolysaccharide (LPS). M1 MΦs are known as pro-inflammatory cells. The other pathway is reached through interleukin-4 (IL-4) and is known as the alternatively activated M2 pathway. M2 MΦ produce pro-angiogenic factors. Here, we successfully use *in vivo* imaging and histological analysis to identify the MΦ response and activation. We demonstrate the ability to induce various MΦ phenotypes with a change in material functionality as well as identify certain materials parameters that seem to correlate with each phenotype. This suggests the potential to develop materials for specific applications and predict the outcome of MΦ activation in response to new surface chemistries.

Ketone

Sulfone

Control

Acetal

Ketal

Oxime

Figure 1. *In vivo* fluorescence images of cathepsin activity in response to implanted polymer nanoparticles. (Left) Fluorescence image of cathepsin activity 7 days after implantation of 600nm pNIPAm particles. (Right) Implantation scheme.

Figure 2. *Ex vivo* arginase:iNOS levels for modified pNIPAm particles.

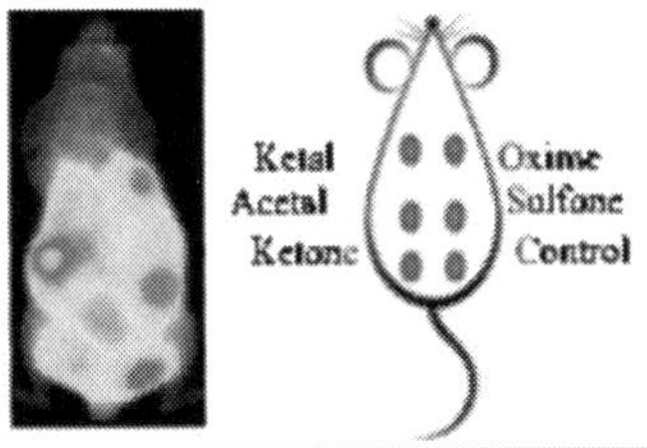

Figure 1. *In vivo* fluorescence images of cathepsin activity in response to implanted polymer nanoparticles. (Left) Fluorescence image of cathepsin activity 7 days after implantation of 600nm pNIPAm particles. (Right) Implantation scheme.

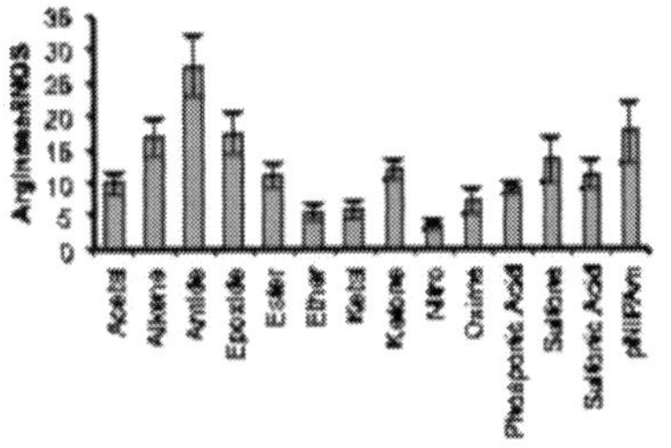

Figure 2. *Ex vivo* arginase:iNOS levels for modified pNIPAm particles.

Materials and Methods: The poly(N-isopropylacrylamide) (pNIPAm) (~600nm) particles were synthesized using established methods and modified through conjugation with surface modifiers through carbodiimide chemistry. MΦs were polarized to the M1 and M2 phenotypes and internalization of functionalized pNIPAm particles were measured. These materials were also subcutaneously injected in SKH1-E mice. Arginase:iNOS profiles have been consistently used as a measure of MΦ phenotype. Arginine activity was measured indirectly through a urea assay in which the lysate was activated, exposed to arginine, and conversion to urea was quantified. Nitrites were measured through a Griess reagent assay to indirectly determine the levels of inducible nitric oxide synthase (iNOS). Material properties of the modified particles including water contact angle, melting temperature, and zeta-potential were also measured.

Results and Discussion: SKH1-E mice were injected with the polymer particles and imaged for cathepsin activity (**Figure 1**). A range in cathepsin activity was observed for the different surface modifiers. The tissue sections from the mice in **Figure 1** were excised and homogenized. The arginase:iNOS displayed a spectrum between the M1 and M2 phenotypes for different surface modifiers as shown in **Figure 2.**

Distinct material parameters influence MΦ phenotypes. For example, increasing the hydrogen bonding of polymeric particles can polarize MΦs towards the M1 phenotype. The induction of multiple MΦ phenotypes confirms the ability to control MΦ activation *in vivo* and suggests the importance of chemical properties in the biocompatibility of materials. The results obtained suggest that changing properties can result in opposite MΦ phenotypes.

Conclusions: With this work, useful insight was gained about *in vivo* MΦ response to various material functionalities. Overall, this *in vivo* study was successful at drawing conclusions about the biomaterial parameters that will allow for tuning MΦ polarization responses to biomaterials and targeting MΦs.

463503 Chemokine Releasing Polymer Implants to Direct Immune Cell Migration in the Setting of Colon Cancer

Abstract:

Colorectal cancer (CRC) is the third leading cause of cancer related death in both men and women. The major cause of death is metastasis and frequently, the target organ is the liver. Successful metastasis depends on a microenvironment conducive to seeding and colonization by metastatic cells. Inflammatory immune cells are implicated in the formation of this microenvironment and interfering with current treatment options including chemotherapy. The objective of the current work is to control immune cell migration and accumulation *in vivo*, directing these cells away from target organs for metastatic disease. In an orthotopic model of colon cancer in which CT26 cells are injected into the cecum, liver metastasis occurs in 90% of mice. Histological analysis indicated that immune cells accumulate within the liver prior to the arrival of metastatic cells. Microarray analysis of pre-metastatic livers from mice injected with CT26 cells indicated that inflammatory chemokines CXCL1 and CCL6 were increased 5.6 and 6.3 fold, compared to mice receiving saline injection. This presentation will focus on the development of biocompatible polymer implants designed for sustained release of these chemokines from the subscapular space, a site where immune cells do not generally accumulate and colon cancer metastasis does not occur. We will discuss the ability of these implants to attract immune cells expressing CXCR2 and CCR1, prototypical receptors for CXCL1 and CCL6, respectively. We will identify the immune cell populations that reside within the implant site and report their numbers in the pre-metastatic liver and primary tumor in mice with and without polymer implant. Finally, the efficacy of these immunomodulatory implants to decrease liver metastatic load will be presented. Collectively, we expect the outcome of this work to impact our understanding of the underlying immunological mechanisms that promote metastatic cancer and aide in the efforts to treat it.

465305 Nanostructured Membranes and Functional Materials Based on Polymerized Ionic Liquids (Invited Talk)

Abstract:

Ionic liquids (ILâ€™s) have been suggested for applications as diverse as solubilizing cellulose, antimicrobial treatments, and electrolytes in batteries due to their molten salt properties. We have recently discovered that a polymerized cation (such as imidazolium) is an excellent host for *any* associated anion. As a result, polymerized ionic liquids are not just solid counterparts to ILâ€™s, but are excellent vectors for the inclusion of a massive variety of functionalities ranging from multi-valent ions for batteries to magnetic anions. Moreover, PIL block copolymers allow orthogonal control over mechanical and morphological properties, ultimately leading to a conceptual framework for processable, tunable, multifunctional materials. In this talk, I will discuss a class of protic polymerized ionic liquids (PILs) based on imidazolium cations which exhibit high ionic conductivities in the solid state. Further, the scaling of conductivity with IL loading, domain spacing, and temperature will be explored and polyelectrolyte block copolymers.

466781 *Dynamics in Weakly-Ordered Oppositely-Charged Polyelectrolyte Complex Solutions*

Abstract:

Recent experiments by Krogstad et al [1] determined the phase diagram of coacervate hydrogels formed by oppositely-charged triblock polyelectrolytes. Disordered and ordered phases were observed on the salt concentration-polymer concentration plane as determined by small-angle scattering methods and cryo-transmission electron microscopy. The ordered phases were in agreement with equilibrium structures calculated with an embedded fluctuation model by Audus et al [2]. Of current interest is in the dynamics of these highly-charged polyelectrolyte hydrogels. A close examination of the dilute disordered region by dynamic light scattering show strong clustering behavior. These large-scale clusters in the disordered region are a hierarchical assembly of micelles building blocks. The equilibrium between clusters and micelles was examined for well-defined 1:1 stoichiometric by charge mixtures of anionic and cationic functionalized poly(allyl glycidyl ether) end blocks with poly(ethylene glycol) middle block. Our new small-angle neutron scattering and dynamic light scattering results will be discussed to understand the solution structure on the temperature-concentration phase diagram plane.

[1] DV Krogstad et al. Macromolecules, 46, 1512 (2013). DV Krogstad et al. J. Phys. Chem. B, 118, 13011 (2014).

[2] DJ Audus et al. Soft Matter, 11, 1214 (2015).

460589 Ion-Containing Block Copolymers for Efficient Capture of a Chemotherapy Drug

Abstract:

Ion-containing polymer membranes are widely used in clean-water-related technologies (e.g., desalination, eletrodialysis), power generation (e.g., forward osmosis, fuel cells, batteries), and bioengineering (hemodialysis, drug delivery, controlled release). For biomedical applications, previous literature has focused on polymers used for drug delivery and controlled release. Sophisticated polymer design is required for biomedical applications since the drugs must remain in a carrier polymer until the target locates near the drug, and severe side-effects must not occur during this process, as a carrier polymer makes its way through the body.

In this work, we study ion-containing polymers for an emerging application that we call "drug capture". These polymers are incorporated in a new class of biomedical devices that we call "ChemoFilter", in order to increase the effectiveness of chemotherapy based cancer treatment. In this approach, the drug is injected at the target organ's artery. After the drug passes the target organ, it is captured by a polymer membrane placed at the exiting vein from the target organ using minimally invasive surgical methods (similar to approaches for introducing stents in patients having heart diseases). The goal is to capture all of the drug exiting the target organ using ion-containing polymer membranes, and the ChemoFilter is removed at the end of treatment.

Herein, we have prepared robust, ion-containing block copolymer membranes using a polystyrenesulfonate-*block*-polyethylene-*block*-polystyrenesulfonate (S-SES) triblock copolymer to remove a chemotherapy drug, Doxorubicin, from the blood stream. Doxorubicin is currently widely used in treating liver cancer, however, the current treatment is limited by adverse side-effects due to the interaction between the unused drug molecules and the human tissue. The goal is to remove most (>90%) of the drug in less than an hour to reduce toxic side-effects. The ion-containing polystyrenesulfonate microphase captures the drug, and the hydrophobic polyethylene microphase gives mechanical strength for the membrane.

In order to maximize the drug binding capabilities and fast binding kinetics of polymer membranes, we have systematically changed sulfonation level, block composition, and post-synthesis treatments in the polymer membranes, and this approach enables the tuning of polymer morphology. In this talk, the relationship between polymer morphology and the efficacy of the drug capture will be discussed.

464813 Oligonucleotide – Peptide Complexes: Phase Control By DNA Hybridization

Abstract:

Understanding the interactions of charged polymers is one of the key problems in polymer physics, and is intimately related to the classic biophysics problem of DNA condensation by basic proteins and small cations. This study presents results of complexation of single- and double-stranded oligonucleotides (10 – 100 nt) by cationic peptides and polyamines, bridging the gap between "short" and "long" polyelectrolytes. We observe several interesting behaviors, of which the most striking is that, while double-stranded oligonucleotides form solid precipitates when mixed with polycations, single-stranded oligonucleotides form liquid droplets (coacervates). Complexes are formed over a wide range of charge ratios, polymer lengths, and concentrations. When monovalent salt is added, precipitates formed by longer polymers 'melt' into coacervates at a concentration that appears independent of polymer length, but the complexes' (liquid and solid) stability vs dissolution is strongly dependent on length. Identical phase behavior is observed with RNA oligonucleotides, and when short polyamines are substituted for cationic peptides. Substitution of uncharged methyl phosphonate linkages indicates that polyelectrolyte charge density, rather than bending rigidity, determines the phase of the complexes. Single-stranded oligonucleotides remain competent for sequence-selective hybridization after coacervate formation, suggesting the possibility of environmentally-responsive complexes and nanoparticles for targeted therapeutic application or sensing.

472277 Magnetic and Biocompatible Polymerized Ionic Liquids

Abstract:

Polymerized Ionic Liquids (i.e., PILs) are an emerging class of materials with ionic liquid moieties covalently attached to a polymer backbone. As such, they synergistically combine the processability of polymers with the versatile chemical functionality of ionic liquids. PILs have been of interest for gas separation and electrochemical energy conversion due to the CO_2 solubility, ionic conductivity, and thermal and electrochemical stability intrinsic to ionic liquids. However, the ability to incorporate functionality through ionic liquid chemistry while retaining polymer properties turns PILs into a powerful platform for developing novel stimuli-responsive materials.

Herein, we present a magnetic and biocompatible PIL based on poly(ethylene glycol) (i.e., PEG) and the ionic liquid anion tetrachloroferrate(III) (i.e., $FeCl_4^-$). We illustrate a molecular design strategy based on a combination of epoxide ring opening anionic polymerization and thiol-ene click chemistry that allows for exquisite control over the molecular structure of the polymer and guarantees scalability appropriate for extensive physical characterization. PIL containing $FeCl_4^-$ is paramagnetic with a magnetic susceptibility at 30 °C of $30x10^{-6}$ emu/g.Oe, as characteristic of materials based on anions containing transition metals with partially filled d-orbitals. Additionally, the PIL significantly modifies the magnetic environment experienced by the ^1H of water, as revealed by a decrease in the spin-lattice relaxation time (i.e., T_1) of ^1H determined from Nuclear Magnetic Resonance (i.e., NMR) from 8 to 0.3 seconds. The structure-property relationships demonstrated in this work makes PILs based on PEG and $FeCl_4^-$ suitable as a T_1-weighted MRI contrast agent. The magnetic and biocompatible nature of PILs is promising for development of novel therapeutic agents (i.e., drugs, proteins, antibodies) based on polymer conjugates that exhibit controlled residence time within the human body; and magnetic functionality for MRI imaging, and facilitated targeting.

473017 Sequestering Small Moleculues in Polyelectrolyte Complex Coacervates

Abstract:

The strong association of oppositely charged polyelectrolytes in solution can lead to a liquid-liquid phase separation known as complex coacervation. This process leads to a dense, polymer rich phase with a different environment than the water rich phase in terms of hydrophobicity, dielectric constant, and so on. Bassed on these differences, small molecules present in the solution will partition selectively based on factors such as hydrophobicity, charge, ability to form hydrogen bonds or other types of secondary interactions. Complex coacervation is able to sequester high amounts of contaminants present in aqueous solution, including such molecules as pharmaceuticals or other compounds that exist more and more frequently in wastewater streams and are difficult to remove. In this work, it is shown that both electrostatically bonded complex coacervates as well as hydrogen bonded ones can be used to control the sequestration of small molecules based on relative interactions between the small molecules, water, and the polyelectrolytes. The small molecules investigated in this work include both dyes and compounds such as antibiotics. The ability to sequester these small molecules is shown over a range of pH values, small molecule concentration, and salt concentration as well. We show that secondary interactions are much more effective at sequerstering small molecules than are electrostatic interactions, which interfere with the formation of the coacervate. This work has implications for water treatment as well as for chemical microreactors, where one might imagine concentrating reagents in solution in order to enhance reactivities.

461585 Sequence Effects in Coacervate-Driven Self-Assembly

Abstract:

Oppositely-charged polyelectrolytes in aqueous solution can undergo associative phase separation in a process known as 'complex coacervation'. Originally discovered in biopolymers, and used in a variety of food and personal care products, the underlying physical process is now being harnessed for soft material self-assembly. We have used a combination of simulation and theory to demonstrate that structural features at the molecular level have a profound effect on coacervation. Motivated by this observation, we have developed a new hybrid simulation method that couples molecular Monte Carlo and Single Chain in Mean Field simulations to provide the multi-scale picture necessary to study coacervation-driven self-assembly. We demonstrate that this provides predictions that deviate significantly from prior theoretical results, and show that the sequence of charges along the backbone is a useful way to tune and design the self-assembly properties of charged block copolymers. Our investigaion has important implications for how monomer sequence may be used to design the next generation of self-assembled soft materials.

470898 Effect of Charge Patterning and Polymer Architecture on Polypeptide-Based Coacervates

Abstract:

Complex coacervation is an electrostatically-driven liquid-liquid phase separation that occurs when aqueous solutions of oppositely-charged polyelectrolytes are mixed. The resultant dense, polymer-rich coacervate phase has been utilized in a variety of different applications, including encapsulation such as gene and drug delivery. Furthermore, direct analogies have been drawn between biomolecule containing coacervate materials and aspects of the intracellular environment. Complex coacervation can be controlled by many factors such as the polycation/polyanion ratio, pH, temperature, architecture, and pattern of charges present. Among these factors, the effect of polymer architecture and charge patterning has been rarely studied due to limitations related to the synthesis of well-controlled polymers with defined branching and/or charge-patterned geometries. Here, we use polypeptides as a model polymer system to examine the effects of both polymer architecture and patterns of matched and mismatched sequences of charged residues on complex coacervation. Additionally, we also examine the effects of incorporating zwitterionic moieties, containing both positive and negative charges in close proximity, as copolymers into our polymers. These experimental efforts are supported by the parallel development of computational approaches for modeling and predicting the phase behavior of patterned polymeric materials. This systematic investigation of the effects of architecture and charge patterning will help to elucidate design rules to facilitate the tailored creation of complex coacervate-based materials with defined properties for a wide range of applications.

465412 Phase Behavior of Weak Polyelectrolyte Materials

Abstract:

Weak linear polyelectrolytes in dilute conditions undergo a first-order coil-to-globular transition whose character is intrinsically dependent on the mutual balance of solvation, charge strength and pH. Such transitions are of great interest in the creation of smart materials which can preferentially swell or deswell, evincing subtle changes in the environment. These applications necessarily utilize multiple chains together in a cross-linked network, or as part of a brush-functionalized surface. As such, it is imperative that the behavior of such systems at higher chain-densities be elucidated and understood. In this work, we utilize a coarse-grain hybrid grand-canonical Monte Carlo and Molecular Dynamics approach to explore the free energetic topography of a model hydrophobic polybase and explore the role of salt concentration and valency to influence the structure and phase behavior of weak polyelectrolyte chains, stars and brushes.

462862 1T-WS_2 on Graphite Foam As a Binder-Free Electrode for Enhanced Hydrogen Evolution

Abstract:

1T-WS_2 on Graphite Foam as a binder-free electrode for Enhanced Hydrogen Evolution

[1]Xiaomeng Guo, [1]Yuanzhi Zhu, [2]Junyi Ji*, [1]Xiaobin Fan, [1]Guoliang Zhang, [1]Fengbao Zhang, [1]Wenchao Peng*

[1]School of Chemical Engineering & Technology, Tianjin University, Tianjin 300072, China,

wenchao.peng@tju.edu.cn

[2]College of Chemical Engineering, Sichuan University, Chengdu, Sichuan 610065, China,

junyiji@scu.edu.cn

Hydrogen is regarded as a clean alternative energy carrier due to its high energy density and environmental friendliness. Electrochemical catalytic hydrogen evolution reaction (HER) is an efficient method for the production of hydrogen. Pt-group metals are the most effective HER catalysts due to their low overpotential and high efficiency. However, these metals are rare and expensive to apply on an industrial scale. To develop highly active and low-cost HER catalysts remains a big challenge. Molybdenum disulfide (MoS_2) and tungsten disulphide (WS_2) are emerging HER catalysts that could be used as a substitute for noble metals in HER.

In this study, a three dimensional graphite foam (GF) was synthesized on the Ni foam using chemical vapor deposition (CVD) method. After the removal of Ni foam with HCl aqueous solution, exfoliated 1T-WS_2 was dipped on the surface of GF to obtain the GF-WS_2 composite. Due to the 3D structure and strong mechanical strength of GF, the composite can be used as a binder-free electrode. The GF-WS_2 electrode was then tested for the HER performance in 0.5 M H_2SO_4 solution, and excellent HER activity was observed with a small overpotential of 0.1 V (vs RHE). Compared to the naked WS_2, the overpotential had been decreased ~0.1 V, and the cathodic current can increase ~10 times at a bias of -0.3V. This activity improvement should be due to the strong chemical and electronic coupling between the GF supports and WS_2 nanosheets. The electrical coupling could afford rapid electron transport from the less-conducting WS_2 nanosheets to the GF with more active sites. The GF-WS_2 is therefore a promising binder-free electrode with great potentials for hydrogen production.

462713 Flame Synthesis of Highly Transparent and Robust Nano-Layers for Enhanced Photo-Electrochemical Water Splitting

Abstract:

Direct chemical storage of solar energy is an attractive concept for off-the-grid renewable power generation. Currently the poor transparency of the thick low-cost catalyst layers required for water oxidation hinders the large-scale fabrication of efficient photo-electrochemical cells for artificial photosynthesis and hydrogen production. Flame synthesis of tailored nanoparticles and direct aerosol deposition is a scalable technology for the production of ultra-porous layers of photo-catalysts and optoelectronic devices[1]. Nevertheless the fragility[2] of this gas-phase self-assembly has limited their application in liquid environments with most studies requiring the in-situ high temperature annealing of the depositing particle to obtain sufficient mechanical stability [3]. This results in a drastic drop of the layer porosity and hinders the fabrication of optimal architecture for electrochemical systems.

Here, we report the engineering of robust high performance photo-electrodes with tunable porosity and composition. We discuss the critical parameters controlling the self-assembly of these nanoparticle layers and present a flexible approach for their mechanical and chemical stabilization. We apply this concept to the fabrication of photo- and electrochemical cells for water splitting demonstrating very high turn-over frequencies, controllable light absorption and efficient electron collection. We envision that this scalable synthesis approach can be readily implemented for the commercial production of low-cost devices for chemical energy storage and renewable fuel production.

References

1. Nasiri, N.; Bo, R.; Wang, F.; Fu, L.; Tricoli, A. *Adv. Mater.* **2015,**27, (29), 4336-4343.

2. Tricoli, A.; Graf, M.; Mayer, F.; Kühne, S.; Hierlemann, A.; Pratsinis, S. E. *Adv. Mater.* **2008,**20, (16), 3005-3010.

3. Tricoli, A.; Wallerand, A. S.; Righettoni, M. *J. Mater. Chem.* **2012,** 22, (28), 14254-14261.

468352 Investigation of the Use of Metal-Organic Frameworks for Combined Water Purification and Catalytic H_2 Production

Abstract:

Lately, a few studies have been reported on the use of metal-organic frameworks (MOFs) for water purification via adsorption of contaminants and/or via photocatalytic degradation. MOFs correspond to 3D crystalline nanomaterials networks consisting of metallic clusters connected via organic ligands. They are characterized by their high porosity and hence have been investigated for applications in water purification, especially for the removal of organics from water. It is interesting to notice that in parallel to those studies, recent work has also been devoted to using MOFs to photocatalytically produce H_2 from water and organic molecules. Up until now, organic contaminants adsorptive removal, photodegradation as well as H_2 evolution have been investigated independently of each other. Hence, an opportunity arises to combine water purification and energy generation (in the form of H_2).

In this work, we have focused on the investigation of two specific MOF families (UiO-66(Zr) and MIL-53) and investigated its adsorptive and photocatalytic properties. The UiO-66(Zr) material was synthesized along with its functionalized counterparts (X-UiO-66(Zr), with X = NH_2, NO_2, Br). MIL-53(M) was synthesized with different metals ions (M = Al, Fe and Cr). The materials were characterized by a number of analytical and spectroscopic techniques in order to determine the effects of functionalization on the band gap and conduction band. In addition, the materials were tested for the adsorption and photocatalytic degradation of organics in water and the simultaneous production of H_2. This work reports on the impact of the abovementioned structural and chemical modifications on the MOFâ€™s electronic properties and their performance as dual-purpose adsorbent/catalyst for water purification and H_2 production.

471808 $Ni(OH)_2$ As a Hole Mediator for Visible Light-Induced Urea Photo-Oxidation

Abstract:

Urea, a major component of human urine, has been proposed as a source of electrolytic hydrogen. We describe here CdS-sensitized TiO_2 assembled with urea electrocatalyst, $Ni(OH)_2$, on fluorine-doped tin oxide (FTO) coated glass (FTO/TiO_2/CdS/$Ni(OH)_2$) used as photoanode to realize simultaneous urea oxidation and water reduction using visible light irradiation. A photoelectrochemical reactor is employed to demonstrate the $Ni(OH)_2$ can serve as hole mediator to drive urea oxidation. The steady-state response of the semiconductor-catalyst electrode in the photoelectrochemical system is investigated, and the photoelectrode transient charge dynamics are studied by employing transient absorption spectroscopy. The obtained carrier dynamics can provide useful information on the competitive processes such as charge transfer and/or recombination within the photoanode.

470119 Study of Photocatalytic Activity of Nano Organic Hybrid Materials (NOHMs) for Photoelectrochemical Reduction of CO_2

Abstract:

A promising opportunity space for novel class of colloidal catalysts has emerged from recent advances in the synthesis of colloidal nanomaterials in which the composition of the inorganic core as well as the organic ligand shell can be precisely tuned. The prospect of tailoring the ligand shell, for optimized interaction with the surrounding fluid and the surface of the colloid, for catalytic conversion is particularly promising for photocatalytic reactions. We have investigated a new class of hybrid colloidal photocatalysts with combined functionality to capture CO_2 and convert it to liquid fuels. The photodriven CO_2-to-fuels reaction has profound technological implications as a prospect to harvest vast solar energy resources, capture CO_2 and store the energy in form of a liquid fuel. Scientifically, the photoreduction of CO_2 involves a complex interplay of coupled sub-processes including CO_2 transport to and adsorption on the catalyst surface, light absorption, proton-coupled multi-electron transfer, selectivity, and material stability. The photoelectrochemical reduction of CO_2 leads to various â€˜value-addedâ€™ products including methane, olefins, methanol, ethanol, formaldehyde, formic acid, formate, and carbon monoxide. Whereas access to novel colloidal nanoparticle catalysts has enabled advances in fundamental understanding of and control over their catalytic properties, progress towards deploying this know-how by integration with established CO_2 capture technologies (*i.e.*, amine scrubbing) has lagged behind. We embraced this challenge as an opportunity study basic processing-structure-performance relationships of nano organic hybrid materials (NOHMs) applied to the photoelectrochemical reduction of CO_2. We will describe the colloidal synthesis the inorganic nanoparticle core comprised of titania and metal co-catalyst and the CO2-philic ligand shell. We will discuss our recent advances in understanding and controlling the composition, morphology, ligand exchange, optical absorption, and CO_2 chemical and physisorption capacity. We will

relate structure and composition of the catalyst to the photocatalytic performance and will describe reaction products and conversion probed by gas and liquid chromatography.

472234 Nanostructured Silicon Photocathodes for Solar Water Splitting

Abstract:

Photoelectrochemical (PEC) water splitting is a promising route for synthesizing hydrogen using solar energy without direct CO_2 emissions.[1] PEC water splitting devices require highly active hydrogen evolution reaction (HER) catalysts to efficiently produce hydrogen at the photocathode.[2] Platinum and its alloys are the most active HER catalysts but the low availability of platinum may create problems for large scale implementation of solar water splitting.[3] There is a need to replace platinum with a non-precious metal earth-abundant alternative to enable cost effective, scalable devices.[4]

Recently, transition metal phosphides and sulfides have been shown to be among the most active nonprecious HER catalysts on a total electrode activity and turnover frequency (TOF) basis.[5] Though extremely active, the high surface area morphologies of these previously synthesized electrodes present additional challenges for integration onto PEC water splitting electrodes.[6] New methods are required to synthesize HER catalysts with a high density of active sites that also transmit a large portion of incident light.

We discuss our progress in synthesizing highly active transition metal phosphide and sulfide catalysts in nanostructured, low light-absorbing geometries. First, we compare nanostructured silicon semiconductors covered with optically thin catalysts to analogous flat silicon electrodes. These results provide insight into the general design considerations necessary to maximize the device efficiency for any given catalyst-absorber system. Next, we present a device structure which decouples the catalyst nanostructure from that of the silicon absorber. In so doing, we achieve low levels of recombination within the silicon and maximize the current-voltage characteristics of the device. Based on our findings, we discuss the application of our catalyst structure to other semiconductor systems and suggest general strategies for further improving the performance of solar water splitting electrodes.

[1] M. G. Walter, E. L. Warren, J. R. McKone, S. W. Boettcher, Q. X. Mi, E. A. Santori, N. S. Lewis, Chemical Reviews 2010, 110, 6446.

[2] L. C. Seitz, Z. Chen, A. J. Forman, B. A. Pinaud, J. D. Benck, T. F. Jaramillo, ChemSusChem 2014, 7, 1372.

[3] J. Greeley, I. Stephens, A. Bondarenko, T. P. Johansson, H. A. Hansen, T. Jaramillo, J. Rossmeisl, I. Chorkendorff, J. K. Nørskov, Nature Chemistry 2009, 1, 552.

[4] P. C. Vesborg, T. F. Jaramillo, RSC Advances 2012, 2, 7933.

[5] J. Kibsgaard, T. F. Jaramillo, Angew. Chem.-Int. Edit. 2014, 53, 14433.

[6] J. D. Benck, T. R. Hellstern, J. Kibsgaard, P. Chakthranont, T. F. Jaramillo, ACS Catalysis 2014, 4, 3957.

468999 Remarkable Enhancement of Photocatalytic Hydrogen Production in Sensitized and N_2/Ar Plasma Treated, Nanoporous TiO_2 Films

Abstract:

The optical and electronic properties of TiO_2 thin films provide tremendous opportunities in several applications including photocatalysis, photovoltaics and photoconductors for energy production. Despite many attractive features of TiO_2, critical challenges include the innate inability of TiO_2 to absorb visible light and the fast recombination of photoexcited charge carriers. In this study, we prepared ordered nanoporous TiO_2 films co-modified by graphene quantum dot sensitization and nitrogen doping (GQD-N-TiO_2) for hydrogen production from photoelectrochemical water splitting under visible light irradiation. First, cubic ordered nanoporous TiO_2 films were prepared by a surfactant templated sol-gel method. Then, TiO_2 films were treated with N_2/Ar plasma for the incorporation of substitutional N atoms into the lattice of TiO_2. GQDs were prepared by chemically oxidizing carbon nano-onions. The immobilization of GQDs was accomplished by reacting carboxyl groups of GQDs with amine groups of N-TiO_2 developed by the prior immobilization of (3-aminopropyl)triethoxysilane (APTES). Successful immobilization of GQDs onto N-TiO_2 was probed by UV-Vis, FT-IR, zeta potential and contact angle measurements. The nitrogen-doped TiO_2 (N-TiO_2) showed the significant reduction of band-gap (3.5 eV to 2.88 eV) and the unprecedented enhancement (240 times) of photocatalytic hydrogen production visible light illumination. The GQD-immobilized N-TiO_2 film showed about 2 times enhancement compared to N-TiO_2, showing the important role of GQDs. This outstanding enhancement is attributed to the high surface area of nanoporous films and synergistic effects of nitrogen doping and GQD sensitization resulting in enhanced visible light absorption, efficient charge separation and transport.

468774 Bifunctional Porous Materials for Combined CO_2 Capture and Catalytic Conversion

Abstract:

The rapidly-increasing concentration of Greenhouse Gases (GHGs) in the atmosphere has become a critical issue prompting the scientific and engineering community to find ways to efficiently mitigate emissions of GHGs, such as CO_2. Electricity generation using fossil fuels is one of the major contributors of CO_2 emissions and therefore, various end-of-the pipe technologies are currently proposed as carbon management media. A less energy intensive alternative than amine-based solvents are solid adsorbents which are capable of CO_2 capture at low temperatures as part of the Carbon Capture Utilisation and Storage process (CCUS). Another interesting development in CCUS research is the use of captured CO_2 as a chemical feedstock. Although this pathway will not eliminate the need for long term CO_2 storage, carbon utilisation could diversify the range of possibilities

Metal-organic frameworks (MOFs) represent a class of adsorbents with chemical and structural tunability, large porosity as well as high thermal and chemical stability. Owing to these features, MOFs are promising materials for CO_2 capture. In addition, recent research efforts point to the potential use of MOFs as photocatalysts, especially in the context of CO_2 reduction into useful products (e.g. CO and CH_4). Considering these two aspects, we are investigating the application of MOFs as bifunctional absorbent-photocatalyst materials for the combined capture and conversion of CO_2. This approach could potentially enable process intensification and improvement of the overall Carbon Capture Utilisation and Storage process sustainability. A series of MOFs were synthesised and fully characterised by a range of analytical and spectroscopic techniques (i.e. XRD, FT-IR, thermogravimetric analysis, N_2 sorption). These MOFs were tested as CO_2 adsorbents as well as heterogeneous CO_2 reduction photocatalysts in the gas phase using a purpose-built reactor. In the latter case, H_2 and H_2O as sacrificial agents under mild reaction conditions (ambient temperature and pressure). Products were analysed using a GC-MS and the MOFs were characterised after testing to identify any change in structure and/or chemistry. The results were used to identify mechanisms of reactions.

469270 Using Polymer-Grafted Nanoparticle Monolayers to Investigate Chain Conformations and Scaling (Invited Talk)

Abstract:

Predictions of polymer chain conformations and molecular weight scaling in solvents of varying quality and in the melt are well established.[1] They have been verified by measurements that have relied primarily on scattering techniques.[2] In this work, we take a different approach to directly measure polymer chain size and scaling using polymer-grafted nanoparticle monolayers imaged with electron microscopy. Our results are compared to predictions that include the effect of tethering to a surface.[3] Poly(ethylene oxide)-tethered gold nanoparticles (PEO-AuNPs) were prepared by grafting thiol-terminated PEO to Au NPs. The grafted NPs were then transferred into organic solvent and cast into a monolayer that self-assembled into a hexagonal close-packed (HCP) structure. The HCP lattice constant and the interparticle edge-to-edge distance were analyzed in detail by transmission electron microscopy (TEM) and small angle X-ray scattering (SAXS). The average edge-to-edge distance between adjacent Au NPs was found to be approximately equal to the PEO coil size. Using PEO of different molecular weights, the particle spacing was found to scale with the expected square root of the degree of polymerization. Solvent polarity was used to change the PEO chain conformations from random coil to collapsed globule, and these conformational changes were observed using the Au NP lattice spacing. The interparticle distances of the Au NP films formed under various experimental conditions have been found to agree well with theoretical predictions. This demonstrates a direct method to study the scaling behavior of polymer ligands grafted to nanoparticle surfaces.

References

1. Rubinstein, M.; Colby, R. H., *Polymer Physics*; Oxford University Press: USA, 2003, p 454.

2. Sperling, L. H., *Introduction to Physical Polymer Science*; Wiley Interscience: USA, 2006.

3. Zhulina, E. B.; Borisov, O. V.; Pryamitsyn, V. A.; Birshtein, T. M., Coil Globule Type Transitions in Polymers 1. Collapse of Layers of Grafted Polymer-Chains. *Macromolecules* 1991, *24*, 140-149, DOI: 10.1021/ma00001a023.

471773 Investigation of the Effects of Crosslinking on the Properties of Poly(ethylene oxide) Based Solid Polymer Electrolytes

Abstract:

Solid polymer electrolytes (SPEs) promise to provide many benefits for lithium ion batteries including enhancing safety, enabling the use of a solid lithium anode, and facilitating the development of flexible and stretchable batteries. Traditional SPEs use high molecular weight polymers cast into amorphous films. Two very important properties relating to the ionic conductivity of the SPEs are the degree of crystallinity of the polymer as well as the glass transition temperature. Unfortunately, these properties lead to tradeoffs in other important properties such as mechanical toughness and stability. Here, we demonstrate the use of UV crosslinked poly(ethylene oxide) and its derivatives to create a network polymer electrolyte structure. These networked polymers demonstrate interesting properties compared to traditional linear chain SPEs. Notably, the crosslinked polymer networks allow for tuning of the glass transition temperature, mechanical properties, and ionic conductivities of the polymers. This study will demonstrate a methodical and rigorous exploration of the effects of salt concentration, crosslink network density, and molecular additives on the properties of interest for SPEs. A general framework for designing network SPEs will be presented. These principles will be contrasted to those known for linear chain SPEs, elucidating further knowledge of the fundamental mechanisms of ion transport in SPEs. Preliminary results indicate that network SPEs allow the incorporation of conductivity enhancing plasticizers without experiencing substantial deterioration of mechanical properties, providing a substantial advantage to traditional SPEs.

473514 Nanoconfined Polymerization: Kinetics and Thermodyanics

Abstract:

The behavior of materials confined at the nanoscale has been of considerable interest over the past several decades. We have recently investigated the effects of nanoconfinement on polymerization kinetics and thermodynamics for various classes of polymerization, including trimerization of cyanate esters, free radical reaction of methacrylates, and ring-opening polymerization of dicyclopentadiene. Changes in reaction rates under confinement can generally be explained by changes in local packing, diffusivity, and surface effects, whereas changes in monomer/polymer equilibria reflect entropic effects. One goal of the work is to test reports in the literature for how diffusivity and chain entropy scale with nanoconfinement size and molecular size. For example, the early onset of autoacceleration in nanoconfined methyl methacrylate polymerization is well described by a model assuming that the diffusion coefficient scales with molecular size to the -3 power and with nanopore diameter to the 1.3 power. On the other hand, modeling the literature data from Yannopoulos and coworkers for the equilibrium polymerization of sulfur, we find that confinement entropy scales with molecular size to the second power and with nanopore diameter to either the -3.0 or -3.8 power, the former of which fits slightly better. The present paper will discuss new results with respect to scaling of chain entropy on nanoconfinement size and chain length for a series of methacrylates.

456151 Complexation of Zwitterionic Polyelectrolyte and Inorganic Nanocluster Macroion into Mechanically Strong Supramolecular Coacervate and Hydrogel

Abstract:

Complexation of polyanions and polycations has been investigated as the self-assembly of responsive polymeric materials ranging from layer-by-layer polymer thin films to polyelectrolyte complex coacervates and hydrogels. Instead of using two oppositely charged polyelectrolytes in salt solutions, we explore the coacervation of anionic tungsten oxide-based polyoxometalate (POM) macroions and zwitterionic polybetaine polyelectrolytes in LiCl salt solutions. The phase diagram of POM-polybetaine complexes is examined with varied POM-to-polybataine charge ratio and LiCl concentration. We are intrigued to observe a solution-coacervate-hydrogel transition as increasing POM-to-polybetaine charge ratios and broader coacervation region as increasing LiCl concentration. Importantly, the organic-inorganic macroion complexes in solution exhibit much enhanced viscoelastic properties with tunable temperature response in comparison to polyelectrolyte complexes. Bulk POM-polybetaine coacervates exhibit intriguing shear-thickening response, yet the polymer-rich liquid coacervates exhibit similar viscoelastic solid-like properties to those of the hydrogels formed at high POM-to-polybataine charge ratios. Zeta-potential and solution conductivity measurements suggest Li-cation medicated ion pairing between both net negatively charged polybetaine and POMs. The organic-inorganic macroion coacervation could open new material opportunities in controlling charge-driven phase behavior and polyelectrolyte assembly.

458631 Hybrid Stealth Liposomes: Addition of Pendant-Cholesterol Block Copolymers to Phospholipid Vesicles

Abstract:

The ability of block copolymers to self-assemble in aqueous solution, resulting in nanoscale structures, makes them viable platforms for drug delivery vehicles and therapeutic agents. In this thrust, a number of recent efforts have focused on chemically incorporating bio-inspired functionalities including cholesterol into the hydrophobic block of amphiphilic block copolymers. The molecular shape and intermolecular interactions (e.g. pi-pi stacking) in these materials can lead to intriguing structures such as disk and stacked-disk shaped micelles. Here, we extend the versatility of one such molecule, poly[(ethylene glycol)-*b*-(2-(5-methyl-2-oxo-1,3-dioxane-5-carboxyloyloxy)ethyl carbamate)] (PEG-P(MTC-Chol)), by combining a minority quantity with phospholipids capable of forming vesicles (also called liposomes). The P(MTC-Chol) blockâ€™s pendant cholesterol groups can collocate in the phospholipid bilayer, much like unbound cholesterol. The PEG block, on the other hand, prefers to extend itself into aqueous solution. A valid hypothesis might, therefore, be that the block copolymer chains will tether to the liposome bilayer through cholesterol insertion with the extended PEG sterically stabilizing the otherwise unstable liposome assemblies (as in stealth liposomes). In practice, this hypothesis remains true only if the number of cholesterol groups per chain is low. The block copolymer may prefer to form separate micelles or extract from liposomes as the P(MTC-Chol) block becomes too long, leaving the liposomes unmodified. In this work, the critical P(MTC-Chol) block length dividing the two cases was identified primarily through dynamic light scattering. Successfully formed hybrid stealth liposomes are further investigated in terms of their bilayer nanostructure and thermodynamics through a combination of cryo-electron microscopy, small angle scattering, and calorimetry.

450612 Structurally pH Responsive Nangel Star Polymers for Use in Drug Delivery Applications

Abstract:

Biomedical applications of diblock star polymers, globular nanoparticle architectures in which linear polymer "arms" are connected to a central polymer core, as a drug delivery system rely on the design of its polymer block chemistries and architectures to be thermosensitive and biocompatible, with the additional ability to be biodegradable over suitable time scales. Smart polymers that exhibit a pH induced structural response are important since different tissue environments and cell compartments exhibit a wide variation in pH, and thus offer the potential for a more targeted delivery of therapeutic drugs. The highly combinatorial nature of the star, e.g. variations in monomer chemistry combined with variations in arm length and arm loading, provide a vast landscape to characterize for drug-delivery applications, most of which remains unexplored. Here, we present a Small Angle X-ray scattering (SAXS) study of polyelectrolyte stars whose hydrophilic blocks consist of acidic, basic and neutral chemistries. Examining the structure of these stars over a range of pH has lead to interesting insights with respect to these polymer's linear analogues, and indicate that they undergo structural transitions over a narrow range of pHs, depending on the chemical composition of the arms.

472115 Tailoring the Thermo-Responsiveness of Elastin-like Polypeptides with Short Charged Sequences

Abstract:

Elastin-like polypeptides (ELPs) are widely used as functional domain in recombinant protein design for their entropy-driven thermo-responsiveness. When below the lower critical solution temperature (LCST), ELPs are soluble, and the solution stays in one homogeneous phase. However, when the temperature is above the LCST, ELPs phase-separate into a protein-rich phase, or coacervate. This coacervation occurs with a sharp transition (within 2 °C) and is reversible upon lowering the temperature below the LCST. Because of the LCST behavior, ELPs have been used as purification tags for recombinant proteins and to construct temperature responsive protein-based materials or drug delivery vehicles.

The LCST of an ELP is determined by several factors including protein concentration, salt identity and concentration in the solution, and pH of the solution, but is predominantly determined by the amino acid sequence of the ELP. Although the effect of amino acid composition of ELP sequences on the LCST is well studied, it has not been well established how other protein domains in an ELP-fusion protein can influence the LCST behavior. Here we report that the LCST behavior of an ELP-fusion protein can be significantly changed by short recombinant protein domains outside of the ELP sequence. We constructed two ELP-fusion proteins, c-ELP and v-ELP, with the same ELP sequence but with different N-terminal leading sequences consisting of 5-23 amino acids with different charge densities.

Despite the fact that the two N-terminal sequences comprise less than 10% of the total length of the ELP-fusion proteins, c-ELP and v-ELP showed a difference greater than 3 °C in their LCSTs in phosphate buffered saline (PBS) at the same protein concentration. We further quantified LCSTs of both ELPs at varying pH values using a tri-buffer system. c-ELP maintained a constant LCST of 45 °C from pH 5.7 to 9.0. In contrast, the LCST of v-ELP decreased from 65 °C at pH 5.7 to 35 °C at pH 9.0. To explain these large differences, we estimated the protein chain charge and found that both the ELP sequence and the leading sequence on c-ELP remained positively charged and had no significant change in charge over the tested pH range. On the other hand, the leading sequence on v-ELP changed from positively charged to negatively charged at pH 6.6. Based on the charge analysis, we hypothesize that the LCST behavior of ELP-fusion proteins can be modulated not only by the ELP sequence itself, but also by the charge sign and density on sequences outside of the ELP region. Moreover, when both ELP and the leading sequence have the same charge, the LCST will be higher to overcome the energy penalty from higher charge repulsion in the coacervate. When the ELP sequence is oppositely charged from the leading sequence, the charge repulsion in the coacervate will be lower, and the LCST will be lower.

Thus, our results provide insights into the design of ELP-fusion proteins with specific LCST behavior: the charge behavior of leading sequences should be taken into consideration and can also be harnessed as a new parameter in ELP-fusion protein design to fine tune the LCST. This allows us to have more flexibility in designing ELP sequences. A more precise control over the LCST behavior will allow us to build sophisticated ELP-based biomaterials for biological applications such as temperature-triggered hydrogel and pH-responsive drug delivery vehicles.

461363 Redox-Active Organometallic Polymers for Environmental and Energy Applications

Abstract:

Redox metallopolymers have been explored extensively for catalysis, energy storage and molecular recognition.[1] When properly designed, these charged polymers can become powerful vehicles to tune the redox-properties at an electrode interface, due to their fast Faradaic reaction and electron-transfer properties. Here, we present nanostructured electrodes functionalized with poly(vinyl)ferrocene/carbon nanotubes (PVF-CNT) as a attractive platform for the selective separation of organic micropollutants under competitive binding, with separation factors >150 based on specific functional group recognition for harmful carboxylates, sulfonates and phosphonates, and ion-capacity >200 mg/g.[2] The mechanism for this selectivity is explored through both spectroscopic, electrochemical and computational methods. In tandem, cobaltocenium-based polymer counter-electrodes (PMMAECoCp-CNT) were designed as efficient cathodes to increase pseudocapacitive charge and suppress side-reactions, especially the water reduction reaction, thus both enhancing energy storage capabilities (specific capacitance of 498 F/g) as well as increasing ion-selective behavior of the anode (98% current efficiency towards ion-selective process). The non-covalent functionalization of these organometallic polymers onto the electrodes grants cycling stability for over two days and >5000 cycles. Finally, further cathode design through ligand exchange has enabled us to develop cation-selective electrodes for both heavy-metal removal as well as aromatic organic cations – and in tandem, these asymmetric systems become highly energy-efficient systems for wastewater treatment, chemical purification and even bio-separations. This work highlights the potential of asymmetric polymer-functionalized redox-systems for electrochemical separations and energy storage, and the importance of organometallic design for selectively targeting molecular-level pollutants.

References

[1] G. R. Whittell and I. Manners, *Advanced Materials*, 2007, **19**, 3439-3468.

[2] **Su, X.**; Kulik, H; Jamison, T.F.; Hatton, T. A. 2016. Anion-selective redox electrodes: electrochemically-mediated separation with organometallic interfaces. *Advanced Functional Materials*. Advance Article Online. DOI: 10.1002/adfm.201600079

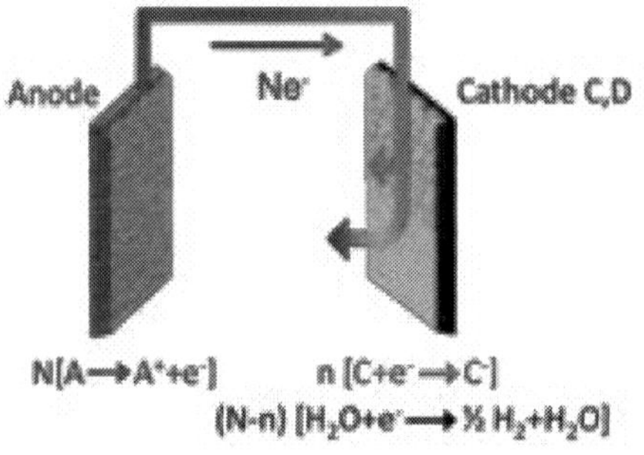

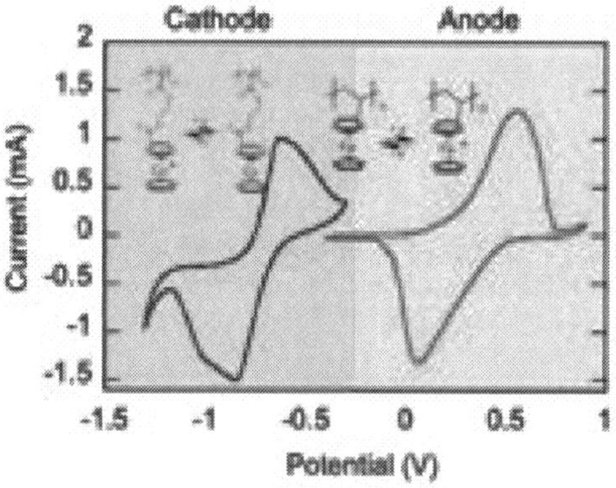

469414 Dynamic Asymmetry within Interfacial Polymer Layers on Nanoparticles and Their Consequences on Macroscopic Mechanical Properties of Polymer Nanocomposites

Abstract:

Miscible polymer blends with different glass transition temperatures (T_g) are known to form confined interphases between glassy and mobile chains. We recently showed that nanoparticles adsorbed with a high-T_g polymer, poly(methyl methacrylate) (PMMA), and dispersed in a low-T_g matrix polymer, poly(ethylene oxide) (PEO), exhibited a liquid-to-solid transition at temperatures above T_gâ€™s of both polymers. We reported that miscible polymers with asymmetric dynamics and composition reversibly stiffen upon heating with the inclusion of nanoparticles. In this talk, I will present the mechanical responses of other polymer combinations with varying T_g differences to reveal how the strength of particle-high T_g polymer attractions and crystallinity of the matrix polymer lead to their stiffening. In addition, I will discuss the rheological results on poly(vinyl acetate) (PVAc) and poly(methyl acrylate) (PMA) adsorbed nanoparticles to explore the strength of interphase interactions between particles and different polymers. Further, we studied the role of interfacial polymer on segmental and collective dynamics of the PEO matrix chains between sub-ns and 100 ns. Our spin-echo experiment results showed that the Rouse relaxation remains unchanged in weakly attractive composite systems compared to pure PEO. I will discuss tube dilation and disentanglements within weakly attractive composites (PMMA-adsorbed silica dispersed in PEO) and will present the decrease and recovery of the entanglements by vitrification and the mobility of the bound polymer, which is found to be in-line with the unusual softening-stiffening transition of our bulk polymer nanocomposites.

460667 Multiblock Ionomers for Membrane Applications

Abstract:

The water-energy nexus represents the confluence of two pressing engineering challenges facing todayâ€™s generation. The demand for clean and potable water continues to increase with the expansion of the global population and the rapid industrialization of developing nations. Correspondingly, new technologies to harvest, store, and discharge the energy required for the production of this valuable resource are critical. Membranes prepared from block polymers are useful as designer materials for both of these applications. For example, block polymers microphase separate to form task-specific phases: one that imparts mechanical strength and the other facilitates ion, electron, or water transport. Ionomers are increasing in popularity based on recent work that highlights their stable and unique morphologies, tailorable ionic conductivity, and high water flux with good salt selectivity. This presentation will highlight work from our lab on membranes fabricated from imidazolium-based triblock copolymers prepared using polymerizable ionic liquids or post-polymerization functionalization strategies for energy applications. Specifically, the influence of (i) charge placement and density in the central and outer blocks; (ii) thermomechanical properties; (iii) ionic liquid incorporation; and (iv) chemical composition of the cation and anion on the properties of imidazole-containing polymers will be discussed. Additionally, recent work has utilized these same materials for salt-selective water purification membranes. The structure-property-diffusion relationships that have been developed contribute toward a better understanding of the interplay between polymer morphology, water and salt sorption and transport characteristics, and thermomechanical properties.

469314 Realizing the Potential of Micro-Phase Separated Block Copolymer Electrolytes: Ion Domain Connectivity Plays a Prominent Role in Ion Conduction

Abstract:

Single ion conducting polymer electrolyte membranes (PEMs) are found at the heart of many electrochemical devices (e.g., fuel cells, flow batteries, electrolyzers, electrodialysis etc.).(*1*) The key requirements for such materials include high ionic conductivity, electron insulation, mechanical resilience, selectivity (i.e., high transference #), and chemical, thermal, and mechanical stability.(*2, 3*) Ionic conductivity is particularly important because it strongly influences the ohmic overpotential in the said electrochemical devices affecting overall device efficiency.(*1, 4*)

Block copolymer electrolytes represent an attractive set of PEM materials because their micro-phase architecture yields greater ion conduction over their random copolymer counterparts.(*5-7*) Furthermore, facile and controllable polymerization reactions allow researchers to access a variety of morphological architectures, while engineering of the block copolymerâ€™s self-assembly influences domain alignment.(*8-12*) However, there is lack of systematic studies correlating molecular level structural design to bulk material properties like ion transport.

In this work, a model lamellae-forming diblock copolymer electrolyte system was manipulated to examine the extent of ion domain connectivity on ion conduction.(*13, 14*) Careful control of the model systemâ€™s volume fraction gave diblock copolymer structures with slightly rich electrolyte domains or slightly rich hydrophobic domains. The slightly rich electrolyte domains had greater contiguous area fraction while simultaneously demonstrating substantially lower terminal defects. Having a contiguous area fraction from 0.95 to 1.0 resulted in a 2x improvement in ionic conductivity over a non-micro-phase separated block copolymer electrolyte or a micro-phase separated block copolymer electrolyte with very poor connectivity (i.e., a system with larger volume fraction of hydrophobic blocks). Incremental adjustment in the extent of connectivity revealed an exponential growth curve for ion conductivity as a function of contiguous area fraction. Furthermore, the benefits that a micro-phase separated block copolymer electrolyte affords in terms of ion conductivity were not realized when the block copolymer electrolyte had poor connectivity. The results of this work have far reaching implication into the rationale design of PEM materials based upon block copolymer designs. This talk will emphasize the importance of maximizing ion domain connectivity while taking great strides to minimize terminal defects to boost ion transport in PEM materials. This talk will close with future directions on how molecular level engineering of block copolymer electrolytes offer the potential to shed light on how other structural features (e.g., tortuosity(*15*), grain boundaries(*5*), and counterion condensation) alter ion transport in PEM materials(*16, 17*).

References

1. H. Strathmann, A. Grabowski, G. Eigenberger, Ion-Exchange Membranes in the Chemical Process Industry. *Industrial & Engineering Chemistry Research* **52**, 10364-10379 (2013).

2. H. Strathmann, *Ion-Exchange Membrane Separation Processes, Volume 9*. Membrane Science and Technology (Elsevier Science, Amsterdam, The Netherlands, 2004), vol. 9.

3. T. Sata, *Ion Exchange Membranes: Preparation, Characterization, Modification and Application*. (Royal Society of Chemistry, Cambridge, UK, 2004).

4. A. Z. Weber, J. Newman, Modeling Transport in Polymer-Electrolyte Fuel Cells. *Chemical Reviews* **104**, 4679-4726 (2004).

5. Y. A. Elabd, M. A. Hickner, Block Copolymers for Fuel Cells. *Macromolecules* **44**, 1-11 (2011).

6. N. Li, M. D. Guiver, Ion Transport by Nanochannels in Ion-Containing Aromatic Copolymers. *Macromolecules* **47**, 2175-2198 (2014).

7. Y. Schneider *et al.*, Ionic Conduction in Nanostructured Membranes Based on Polymerized Protic Ionic Liquids. *Macromolecules* **46**, 1543-1548 (2013).

8. H. Hu, M. Gopinadhan, C. O. Osuji, Directed self-assembly of block copolymers: a tutorial review of strategies for enabling nanotechnology with soft matter. *Soft Matter* **10**, 3867-3889 (2014).

9. M. Luo, T. H. Epps, III, Directed Block Copolymer Thin Film Self-Assembly: Emerging Trends in Nanopattern Fabrication. *Macromolecules* **46**, 7567-7579 (2013).

10. S. Ji, L. Wan, C.-C. Liu, P. F. Nealey, Directed self-assembly of block copolymers on chemical patterns: A platform for nanofabrication. *Progress in Polymer Science*, **54-55**, 76-127 (2016).

11. S.-J. Jeong, J. Y. Kim, B. H. Kim, H.-S. Moon, S. O. Kim, Directed self-assembly of block copolymers for next generation nanolithography. *Materials Today* **16**, 468-476 (2013).

12. M. P. Stoykovich, P. F. Nealey, Block copolymers and conventional lithography. *Materials Today* **9**, 20-29 (2006).

13. I. P. Campbell, G. J. Lau, J. L. Feaver, M. P. Stoykovich, Network Connectivity and Long-Range Continuity of Lamellar Morphologies in Block Copolymer Thin Films. *Macromolecules* **45**, 1587-1594 (2012).

14. C. G. Arges, Y. Kambe, H. S. Suh, L. E. Ocola, P. F. Nealey, Perpendicularly Aligned, Anion Conducting Nanochannels in Block Copolymer Electrolyte Films. *Chemistry of Materials* **28**, 1377-1389 (2016).

15. X. Feng *et al.*, Scalable Fabrication of Polymer Membranes with Vertically Aligned 1 nm Pores by Magnetic Field Directed Self-Assembly. *ACS Nano* **8**, 11977-11986 (2014).

16. K. M. Beers, D. T. Hallinan, X. Wang, J. A. Pople, N. P. Balsara, Counterion Condensation in Nafion. *Macromolecules* **44**, 8866-8870 (2011).

17. K. M. Beers, N. P. Balsara, Design of Cluster-free Polymer Electrolyte Membranes and Implications on Proton Conductivity. *ACS Macro Letters* **1**, 1155-1160 (2012).

466765 Structure/Property Relationships in Polymer Membranes for Water Purification and Energy Applications

Abstract:

Providing sustainable supplies of purified water and energy is a critical global challenge for the future, and polymer membranes will play a key role in addressing these clear and pressing global needs for water and energy. Polymer membrane-based processes dominate the desalination market, and polymer membranes are crucial components in several rapidly developing power generation and storage applications that rely on membranes to control rates of water and/or ion transport. Much remains unknown about the influence of polymer structure on intrinsic water and ion transport properties, and these relationships must be developed to design next generation polymer membrane materials. For desalination applications, polymers with simultaneously high water permeability and low salt permeability are desirable in order to prepare selective membranes that can efficiently desalinate water, and a tradeoff relationship between water/salt selectivity and water permeability suggests that attempts to prepare such materials should rely on approaches that do more than simply vary polymer free volume. One strategy is to functionalize hydrocarbon polymers with fixed charge groups that can ionize upon exposure to water, and the presence of charged groups in the polymer influences transport properties. Additionally, in many emerging energy applications, charged polymers are exposed to ions that are very different from sodium and chloride. Specific ion effects have been observed in charged polymers, and these effects must be understood to prepare charged polymers that will enable emerging energy technologies. This presentation discusses research aimed at further understanding fundamental structure/property relationships that govern water and ion transport in charged polymer films considered for desalination and electric potential field-driven applications that can help address global needs for clean water and energy.

460616 Fundamental Water and Salt Transport Properties in Zwitterionic Polymers and Their Use for Membrane Surface Modification to Enhance Antifouling Properties

Abstract:

Fouling by suspended solids and dissolved organic matters presents a great challenge for polymeric membranes for desalination and wastewater treatment. With superior hydrophilicity, zwitterionic polymers have been explored for membrane surface modification to enhance antifouling properties. However, there is a lack of understanding of relationship between polymer structure and water and salt transport properties. In this study, two series of zwitterionic polymers were prepared from sulfobetaine methacrylate (SBMA) and 2 methacryloyloxyethyl phosphorylcholine (MPC). These polymers were thoroughly characterized in terms of sol-gel fraction, density, glass transition temperature, contact angle, and water and salt transport properties. Interestingly, the zwitterionic polymers exhibit water sorption and permeability similar to non charged poly(ethylene glycol) (PEG) based materials. These zwitterionic polymers exhibit lower NaCl diffusivity and permeability and thus higher water/NaCl selectivity than the non charged PEG based materials. Moreover, we have modified ultrafiltration (UF) membranes using a solution containing dopamine and sulfobetaine methacrylate (SBMA). The dopamine and SBMA react and form a thin film on the surface, while dopamine provides adhesion to the membrane surface and SBMA provides superhydrophilic properties. The coating enhances the surface hydrophilicity, which is confirmed by contact angle measurement. The coated membranes demonstrate better stability and higher water flux than those of dopamine-coated and uncoated ones, when tested with water containing bovine serum albumin (BSA). For instance, the membrane coated with dopamine and SBMA exhibits water permeance 80% higher than the uncoated one, after 3 hour filtration using a 3g/L BSA solution in a crossflow system.

460548 Water and Salt Transport in Polymer Membranes Prepared By Solvent-Free Melt Processing

Abstract:

In industry today, polymer membranes are formed via solution processing (i.e., phase inversion techniques) that require large amounts of hazardous organic solvents. While effective from a processing standpoint, these membrane formation methods are highly material-specific and present challenges in controlling pore architecture (i.e., pore size, shape and distribution), which enormously affects the transport of small molecules and ions through the resultant membrane. Moreover, by avoiding the use of toxic and hazardous solvents for membrane formation, many benefits could be realized in terms of cost and environmental impact, and advancements in processing could also provide further insight on the next generation of membrane materials. Herein, work is presented on a new method of membrane formation using charged polymers on porous thin supports via a state-of-the-art multi-nanolayered technology, mainly aimed at producing membranes for desalination and water purification.

Robust, ion-exchange membranes based on sulfonated polymers were prepared by solvent-free melt processing, providing the first example of melt-processed membranes for applications in desalination. To investigate the polymer physics underlying the effects of processing history on these systems, transport of small molecules and ions has been investigated.

Interestingly, melt processed membranes (sulfonated polysulfone, BPS) prepared by different routes (i.e., plasticizer composition, processing temperature, and post-processing treatment) show different water and salt transport properties, compared to solution processed membranes. To understand the origin of this difference, ^{1}H NMR analysis and FT-IR analysis were conducted, but no chemical change was detected. Rather, differences in mass transport behavior result from processing methods having a significant effect on the polymer structures. These results imply that, in addition to varying the chemical structure of the polymer, varying processing methods also provides extra freedom to modify

polymer properties.

472275 Influence of Fixed Charge Group Concentration on Ion Sorption in Ion Exchange Membranes

Abstract:

Controlling ion sorption and transport in polymer membranes is critical for efficient operation of a number of membrane-based technologies such as reverse osmosis, electrodialysis, reverse electrodialysis, and fuel cells. Often, membranes for these processes are made from polymers having ionizable functional groups (fixed charge groups) bound to their backbone (i.e., ion exchange membranes, ionomers, etc.). Sorption of charged species (e.g., ions) in such materials is profoundly influenced by the fixed charge groups, due to Donnan exclusion. Consequently, the concentration of fixed charge groups can be used to tune, to an extent, sorption of ions in charged membranes. However, changes in the concentration of fixed charge groups are typically accompanied by changes in membrane water content (i.e., swelling), which also substantially influences ion sorption in membranes. Decoupling these two effects is critical for developing fundamental understanding of the influence of membrane fixed charge concentration on ion sorption in charged membranes, ultimately leading to strategies for rational design of high performance materials.

In this study, positively and negatively charged membranes having different fixed charge group concentrations but similar water content were synthesized. Ion concentrations in membranes equilibrated with NaCl solutions of varying concentration were measured using a desorption technique. Counter-ion (i.e., ion with opposite charge to that of the fixed charges) sorption increased with increasing fixed charge group concentration, since a larger number of counter-ions are required to electrically balance the fixed charge groups. Co-ion (i.e., ion with same charge as that of the fixed charges) sorption decreased with increasing fixed charge group concentration, presumably due to enhanced Donnan exclusion. However, despite similar changes in membrane fixed charge group concentrations, co-ion sorption in the negatively charged membranes was suppressed to a greater extent than in the positively charged membranes, presumably due to structural differences between the materials. The ion sorption data were interpreted using a model based on Donnan theory and Manningâ€™s counter-ion condensation theory. The model predicted the experimental ion sorption data in the positively charged membranes with no adjustable parameters. However, agreement between model and experimental data for the negatively charged membranes was poor, presumably due to membrane inhomogeneities introduced during polymerization. Use of the Donnan/Manning model for predicting ion sorption in inhomogeneous materials is discussed.

457973 Mixed Ion Sorption and Transport in a Cation Exchange Polymer Based on Sulfonated Polystyrene

Abstract:

Sorption and transport of sodium chloride, sodium fluoride, sodium nitrate, calcium chloride and their mixtures in a commercial cation exchange membrane based on cross-linked polystyrene with fixed sulfonate ions was investigated. For mixed salt sorption and permeation experiments, binary salt mixtures 50:50 wt were considered.

Pure and mixed ion transport parameters were correlated to the ion size and valence, as well to the membrane water content. The effects of the second component on salt transport parameters were investigated. When considering mixtures of salts having the same counter-ion (Na^+), mixed salt sorption and permeability coefficients do not change significantly relative to the values measured in the case of single salt, if they are reported as a function of salt activity coefficient in the contiguous, external solution. When considering mixtures of salts having different counter-ions (Na^+/Ca^{2+}), mixed salt transport parameters depart significantly from the respective values for single salt. The experimental findings were rationalized based on the Donnan theory [1].

[1] F.G. Donnan, Chem. Rev. 1924 1 73-90

471524 Using Inkjet Printing Devices to Fabricate Charge Mosaics from Chemically Tailored Copolymer Membranes

Abstract:

Charge mosaic membranes, which have distinct anionic and cationic domains that traverse the membrane thickness, allow dissolved salts to permeate rapidly. As such, they can be utilized in a variety of applications that require the selective permeation of ionic solutes (*e.g.*, piezodialysis). In this talk, we will introduce our efforts to prepare charge mosaics from copolymer membrane precursors using ink-jet printing devices. The copolymer precursors are chemically tailored so that well-defined counter-charged domains can be introduced on the surface of the membranes. Specifically, a polyacrylonitrile-based copolymer was synthesized *via* a free radical polymerization mechanism and its chemical composition and molar mass characterized using nuclear magnetic resonance (^{1}H NMR) and gel permeation chromatography (GPC), respectively. Copolymer substrates, which possessed nanopores with a diameter around 3 nm in size, were fabricated using a non-solvent induced phase separation (NIPS) method. The azide moieties incorporated into the copolymer precursor used to fabricate these membranes represent an attractive platform for the straightforward chemical functionalization of the membrane in the solid state. In particular, copper-catalyzed azide-alkyne click reactions enable rapid one-to-one covalent attachment of positively charged amine moieties or negatively charged carboxylic acid moieties to pore walls of the membrane. The properties of the charge-functionalized membranes were examined using hydraulic permeability, molecular weight cut-off, and streaming potential measurements. Subsequently, an ink-jet printer was used to generate well-defined domains of positive and negative charges within the pores of the copolymer substrate in a systematic manner. Membrane performance was optimized by coordinating the printing conditions with the membrane morphology. Results from Fourier transform infrared (FTIR) spectroscopy analysis demonstrate that up to 95% of the reactive sites can be converted into charged domains using this technique. The printed domains were distinguished and visualized using fluorescent microscopy. A series of piezodialysis experiments were conducted over a range of ionic strength using membranes functionalized with a single charge (i.e., cationic or anionic domains) as well as patterned mosaics of charge (i.e., cationic and anionic domains). Significantly, these experiments demonstrate that charge mosaic membranes are highly permeable to ionic solutes while single charged membranes reject up to 80% of the co-ions present in solution. The versatile and precise control of substrate chemistry at the nanoscale by printers reveal the potential application of resulted mixed mosaic membranes in targeted ion separation.

458775 Understanding and Engineering Heterogeneous Materials at the Molecular Level

Abstract:

Self-assembly or crystallization from solutions, melts, or solids involve complicated thermodynamic, kinetic, and mass transport effects that are challenging to elucidate and control, even more so in the presence of functional solute species. Such processes are central to the syntheses and resulting properties of diverse inorganic, organic, and inorganic-organic hybrid materials, the understandings of which are often elusive because of their heterogeneous, multicomponent, and/or non-equilibrium characters. The challenges are exacerbated by the complex order and disorder that additionally result from distributions of functional species and environments or the roles of surfaces, which can have large influences on macroscopic material properties and device performances. This is the case for many heterogeneous materials with important engineering applications, such as catalysts for hydrocarbon conversion or pollution abatement, advanced structural solids, membrane materials for electrochemical devices, and materials for harnessing solar energy. By using a combination of spectroscopic, scattering, and bulk property measurements, such materials can be probed over multiple length and time scales to correlate their compositions and structures with their macroscopic properties and functions. In particular, advancements in solid-state nuclear magnetic resonance spectroscopy, especially the availability of ultrahigh magnetic fields and dynamic-nuclear-polarization-enhancement techniques, enable new insights to be obtained on the local environments, interactions, and distributions of functional species in bulk solids or near surfaces. Recent results will be presented on the molecular-level properties of self-assembled or semi-crystalline inorganic, organic, and hybrid materials, which provide new understanding to guide their rational design and syntheses for chemical engineering applications.

463965 New Paradigms in Crystal Engineering: Tailoring the Physicochemical Properties of Materials for Chemical and Biomedical Applications

Abstract:

Crystal engineering is a broad area of research that focuses on methods of designing and/or optimizing materials for diverse applications in fields spanning energy to medicine. The ability to selectively control crystallization to achieve desired physicochemical properties requires detailed understandings of the thermodynamic and kinetic factors regulating crystal nucleation and growth. Combining this fundamental knowledge with innovative approaches to tailor crystal size, structure, and morphology has the capability to produce materials with superior properties. In this talk, I will discuss materials that grow by two disparate mechanisms: classical pathways involving 2-dimensional layer nucleation and advancement on crystal surfaces through monomer addition; and nonclassical pathways, termed *crystallization by particle attachment* (CPA), involving the formation of metastable precursors that play a direct role in crystallization. To examine these processes at near molecular resolution, my group has developed advanced atomic force microscopy (AFM) techniques to elucidate CPA processes *in situ* under solvothermal conditions. This unique AFM system is capable of capturing time-resolved dynamics of particle addition and post-attachment rearrangement, thus opening new pathways to probe crystal growth under realistic synthesis conditions. We also design â€œmodifiersâ€ to control crystal properties, such as size and morphology. Modifiers are molecules or macromolecules that interact with specific surfaces of crystals and regulate anisotropic growth rates. In this talk, I will describe how we use growth modifiers to control crystallization in two research areas: (1) the development of therapeutic drugs for crystals implicated in pathological diseases, such as kidney stones (calcium oxalate monohydrate) and malaria (hematin); and (2) the rational design of microporous zeolites with improved physicochemical properties for applications in catalysis, adsorption, and ion-exchange. Broad challenges associated with tailoring material properties will be discussed, along with the unique ways we investigate mechanisms of crystal growth and modification.

470448 Nanocrystal Electrochromic Smart Windows

Abstract:

Degenerately doped metal oxide semiconductors, like ITO, exhibit plasmonic resonance at near and mid-infrared wavelengths tunable by varying their composition. Nanocrystals of many such materials have now been synthesized and applications are emerging that leverage the responsiveness of their localized surface plasmon resonance (LSPR) to electronic charging and discharging. In this talk, Iâ€™ll focus on how we are applying this concept to develop electrochromic glass that can dynamically control heat loads and daylighting in buildings. We demonstrated that dual-band electrochromism (voltage control over near infrared and visible light transmittance independently) is achievable by embedding plasmonic ITO nanocrystals in a redox-active niobium oxide glass matrix. To develop a practical technology on the basis of this concept, the component materials and their mesoscale architecture can both be optimized so that we now can modulate a large fraction of incident solar radiation on demand. To enable low-cost manufacturing, weâ€™ve also developed low-temperature processing strategies and have now fabricated complete dynamic glass prototypes. The materials innovations needed to enable this progress will be emphasized.

465314 Electrochemical Effects in Thermoelectric Polymers

Abstract:

Solution processible, scalable polymers have the potential to have high electrical conductivity (*Ïf*), high Seebeck coefficient (also called thermopower, S), and low thermal conductivity (*Îº*), all necessary attributes for good thermoelectric performance. Organic materials also offer new opportunities to design materials with complex transport properties. In comparison to inorganic materials, ionic conductivities can be relatively high in organic materials near room temperature, e.g. as high as 1 S/cm. These high ionic conductivities can be obtained alongside electronic conduction leading to organic mixed ion-electron conductors. We suggest that this ability to simultaneously carry significant electronic and ionic charge can lead to unique thermoelectric properties, but the way that these transport properties influence each other within a single material is poorly understood. Conductive polymers such as PEDOT:PSS hold great promise as flexible thermoelectric devices, where solution processing techniques can lead to flexible devices in novel geometries. The thermoelectric power factor of PEDOT:PSS is small relative to inorganic materials because the Seebeck coefficient is small. Ion conducting materials have previously been demonstrated to have very large Seebeck coefficients, and a major advantage of polymers over inorganics is the high room temperature ionic conductivity. Notably, PEDOT:PSS demonstrates a significant but short-term increase in Seebeck coefficient which is attributed to a large ionic Seebeck contribution. In this talk, I will discuss how electrochemistry can be utilized to stabilize the Seebeck enhancement leading to stable improvements to power factor in mixed conductor thermoelectrics.

474713 Biomaterials in Regenerative Engineering: Immunoengineering and Translation

Abstract:

Regenerative medicine and tissue engineering strategies aim to provide tissue substitutes for repair of defects due to disease, trauma, or congenital abnormalities. Biomaterials play a key role in tissue engineering strategies where they are designed to provide a three dimensional environment for cells, present biological cues, and implant in a physiological milieu. Defining biomaterial properties that have direct therapeutic relevance has inspired the focus on the immune response and its impact on not only biocompatibility but also regeneration directly. Translation of key technologies to clinical application in musculoskeletal and soft tissue reconstruction will also be presented.

456821 Chiroptical Activity Enhanced Nanoparticles

Abstract:

Chirality can be considered as a ubiquitous relationship between structure and function in nature. Thus it is one of the cornerstones of todayâ€™s fundamental and applied research for the investigation of the intimate structure and property relationships of matter. Since matter structured on a length scale comparable to or smaller than the wavelength of light can exhibit unusual optical properties, chiral nanoparticles (NPs) have received a great deal of attention as promising candidates for exotic optical functions as in non-linear optics or negative refraction and so forth. The interdisciplinary character of NPs and the ease of their integration into composite materials also provide excellent opportunities to explore new dissymmetric materials. To achieve this goal, the high chiroptical activity of individual NPs is in demand as building blocks.

Here we demonstratedÂ a simple one-pot synthesis method which yields NPs having highly enhanced optical activity.Â The obtained NPs showed anisotropy factors of g=0.02 which is comparable to naturally chiral proteins, ~10x higher than conventionally synthesized NPs. Due to the high optical activity and simple synthesis procedure, this strategy offers wide range of applications as chiral sensors, catalysts and as metamaterials in advanced optical devices.

457533 Polymer Coated Nanoparticle As Additives for Interfacial Modification

Abstract:

Additives that can tailor oil-water interfacial properties are potentially useful for enhanced oil recovery processes. Here, we investigate polymer coated nanoparticles as additives for reducing oil-water interfacial tension and stabilizing oil/water emulsions. First, we investigate polymer-coated nanoparticles comprised of oxidized carbon black and covalently grafted amphiphilic polymeric chains. These polymer-coated nanoparticles can be dispersed in the aqueous phase, and when blended with surfactants the nanoparticles migrate to bicontinuous microemulsion phases. The domain size and characteristics of bicontinuous phases are characterized by interfacial tension measurements and small-angle X-ray scattering measurements. Second, we synthesize silica nanoparticles grafted with pH-responsive poly((dimethylamino)ethyl methacrylate) (PDMAEMA) chains. pH-responsive polymeric nanopoarticles can be used to stabilize oil-in-water emulsions under neutral conditions, and at low-pH triggers de-mulsificaiton of heavy oil emulsions. A combination of surface energy and absorbance measurements are used to analyze phase behavior of these emulsions. This work demonstrates the potential of stimuli-responsive polymer-coated nanoparticles for enhanced oil recovery processes.

458604 Multifunctional Magnetic Nanoparticles By Surface Initiated Atom Transfer Radical Polymerization

Abstract:

Superparamagnetic and ferromagnetic nanoparticles are of huge interest in various areas of science (*e.g.* chemistry, biology, bio-chemistry, medicine, sensor technology)[1]. The requirements for particles are as multifaceted as the areas, in which they are applied. Nanoparticles enjoy great popularity as particles at the nanoscale exhibit a favorable surface to volume ration in comparison to the same bulk material. However, this property is not that valuable, if other requirements do not fit.

For chemical applications, the particles most often have to be quickly separable, inert and stable against harsh environment like temperature, challenging pH-values and solvents.[2,3] In contrast to this, medical and biological applications usually afford dispersion stability in aqueous solutions and compatibility with biological relevant media (*e.g.* phosphate buffered saline (PBS) or Dulbeccoâ€™s modified eagleâ€™s medium (DMEM))[4]. For sensor application, the requirements to the magnetic nanoparticles contain huge dispersion stability, a high saturation magnetization and the possibility for further modification (*e.g.* attachment of a target substance *etc.*)[4,5,6]. Requirements to the magnetic nanoparticles which are highly demanded, often contain multiple combinations of different properties. This afford smart materials which contains amphiphilic character. Such smart materials can for example change the properties by an external trigger (temperature, pH, gas)[7].

Most of the different requirements can be achieved using carbon coated ferromagnetic nanoparticles as a platform. The graphene-like layers protect the magnetic metal core from oxidation and allows the covalent attachment of chemical functionalities *via* diazonium chemistry[8]. There are numerous functionalities which can be implemented by this approach. Amongst them, alcohol-, amine-, carboxy-, azide- or an initiator moiety for atom transfer radical polymerization (ATRP) can be listed, just to mention a few[7,9,10]. Adsorption of a polymer can usually not satisfy the requirements, as adsorbed polymers tend to form only single layers and desorption frequently occurs.

In contrast to this, surface initiated ATRP (SI-ATRP: polymer-growth from the surface) is suitable with a plethora of different monomers (desired functionality) and the polymerization can be well controlled (layer thickness and block-co-polymerization)[4,11]. Thus SI-ATRP gives the toolbox for various modification of the surface that allows achievement of several desired requirements.

Using charged monomers (*e.g.* 3-Sulfopropyl methacrylate potassium salt (SPM) or 3-(Methacryloylamino)propyl]trimethylammonium chloride (MAPTAC)) leads to highly stable dispersions in aqueous solutions. The stability can be tuned by the number of repeating units[4,6]. Using poly(ethylene glycol) methacrylate (PEGMA) as monomer the highly lipophilic graphene layers can be turned into hydrophilic and antifouling surfaces[12]. The use of glycidyl methacrylate (GMA) gives access to post-modification by epoxide ring opening.[4] Most challenging modifications require amphiphilic properties (dispersion stability vs. fast separation or lipophilicity vs. hydrophilicity). Such requirements need a coating with smart polymers, which can change their properties upon external trigger (temperature, pH *etc.*). By using N-isopropylacrylamine (NIPAM) as building block, the particles can reversibly change between hydrophilic and hydrophobic upon temperature change[7]. Here the promising potential of SI-ATRP on magnetic nanoparticles in terms of designing a desired nanomaterial for chemical, biological and medical applications is shown[13].

[1] A.H. Lu, E.L. Salabas, F. Schueth, *Angew. Chem. Int. Ed*, **2007**, *46*, 1222.

[2] A. Schaetz, O. Reiser, W.J. Stark, ***Chem. Eur. J.***, **2010,** *16*, 8950.

[3] Elia M. Schneider, Martin Zeltner, Niklaus Kranzlin, Robert N. Grass, and Wendelin J. Stark, *Chem. Commun.*, **2015**, *51*, 10695.

[4] C.J. Hofer, V. Zlateski, P.R. Stoessel, D. Paunescu, E. M. Schneider, R.N. Grass, M.Zeltner, W.J.Stark, *Chem. Comm.*, **2015**, *51*, 1825.

[5] Q. A. Pankhurst, J. Connolly, S. K. Jones and J. Dobson, *J. Phys. D. Appl. Phys.*, **2003**, *36*, R167.

[6] M. Zeltner, R.N. Grass, A. Schaetz, S.B. Bubenhofer, N.A. Luechinger, W.J. Stark, ***J. Mater. Chem.***, **2012**, *22*, 12064.

[7] M. Zeltner, A. Schaetz, M.L. Hefti, W.J. Stark, ***J. Mater. Chem.***, **2011**, *21*, 2991.

[8] Robert N. Grass, Evagelos K. Athanassiou, and Wendelin J. Stark, *Angew. Chem. Int. ed.*, **2007**, 4909.

[9] Alexander Schatz, Robert N. Grass, Wendelin J. Stark, and Oliver Reiser, ***Chem. Eur. J.*, 2008,27,8262.**

[10] Chun Ghee Tan, and Robert N. Grass, *Chem.Commun.*, **2008**, *36*, 4297.

[11] R. Barbey, L.Lavanant, D. Paripovic, N. Schuwer, C. Sugnaux, S. Tugulu, H.A. Klok, *Chem. Rev.*, **2009**, *109*, 5437.

[12] Elia M. Schneider, Martin Zeltner, Vladimir Zlateski, Robert N. Grass, and Wendelin J. Stark, *Chem. Commun.*, **2016**, *52*,938.

[13] Tong Bu, Tamotsu Zako, Martin Zeltner, Karin M. Sörgjerd, Christoph M. Schumacher, Corinne J. Hofer, Wendelin J. Stark, and Mizuo Maeda, *J. Mater. Chem.B*, **2015**, *16*, 3351.

465272 Gas-Phase Synthesis of Functional Nanoparticles for Energy Applicationsin

Abstract:

In this study we establish an aerosol-based spray pyrolysis system to generate high purity metal oxide nanoparticles (NPs) with tunable particle size and composition. Copper oxide-based nanoparticle (CuOx-NP) and copper-aluminum oxide nanoparticle (CuAlOx-NP) are chosen as the representative material, and the CO oxidation is used as the model reaction. Differential mobility analysis (DMA) was employed to in-situ monitor the secondary particle size of the synthesized NPs in the gas phase, and the scanning electron microscopy was used for providing the particle imagery of the electrostatically-deposited NPs and the primary size distributions of NPs. Thermo-gravimetric analysis was employed to determine the required temperature of the pyrolysis process, and x-ray diffractometry was used to correlate the crystallite size of NP with the pyrolysis temperature. Temperature-programmed reduction system is employed to measure the ability in the oxidation-reduction of CuOx-NPs and CuAlOx-NP, which can be correlated to the ignition temperature and the conversion rate of catalytic reaction. Result shows that we can successfully synthesize CuOx-NPs and CuAlOx-NP with tunable size and composition with ability of in-situ characterization of physical size in gas phase (i.e., by DMA). By tuning the concentration and the composition of Cu and Al precursor, we can control the particle size and the morphology of nanoparticle. The reduction temperature of CuOx and the corresponding light-off temperature of the CO oxidation reaction can be effectively improved. Addition of Al precursor was found to affect the ability of reduction and catalytic performance. Our work provides a prototype study to fabricate CuOx-NPs and CuAlOx-NP with an optimal performance in catalysis. The results can be used to develop a correlation of material properties versus the oxidation-reduction ability and the catalytic activity, which have shown to be a strong indicator to the performance in the energy release and the ability in catalysis.

465894 Monolayer Assembly of Monodisperse Nanoparticles for Magnetic and Catalytic Applications

Abstract:

With unique properties different from their bulk materials, nanoparticles (NPs) with controlled size, shape and composition have been extensively studied not only for scientific interests but also for technological developments. Organic solution-phase synthesis has been demonstrated to be a versatile and robust method for synthesizing these NPs. Over the years, assembling these well-defined NPs into functional devices through "bottom-up" approach has being actively pursued. In this talk, I will give two examples of synthesis and assembly of monodisperse NPs for magnetic and catalytic studies.

Specifically, I will discuss the controlled synthesis of monodisperse cobalt ferrite NPs with tunable magnetic properties and the large-area monolayer assembly of these magnetic NPs through the water-air interfacial self-assembly approach.[1] The monolayer arrays can be used as magnetic recording media, demonstrating great potentials in magnetic data storage application. This self-assembly approach can be extended to uniform cobalt NPs, which are of great interests for catalyzing OER in water splitting.[2]In this case, highly stable cobalt NPs are demonstrated to be more active and stable than the state-of-the-art iridium catalyst. More importantly, the well-defined monolayer catalyst of the cobalt NPs enables more detailed study of the intrinsic OER activity of cobalt NPs, showing 15 times higher turnover frequency and mass activity than the NPs deposited on conventional carbon black. These two examples demonstrate the importance of assembling well-defined NPs into ordered monolayer arrays for important technological applications and fundamental studies.

References

1. Wu, L.; Jubert, P. O.; Berman, D.; Imaino, W.; Nelson, A.; Zhu, H.; Zhang, S.; Sun, S., Monolayer Assembly of Ferrimagnetic $Co_xFe_{3-x}O_4$ Nanocubes for Magnetic Recording. *Nano Lett.* **2014,** *14*, 3395-3399.

2. Wu, L.; Li, Q.; Wu, C. H.; Zhu, H.; Mendoza-Garcia, A.; Shen, B.; Guo, J.; Sun, S., Stable Cobalt Nanoparticles and Their Monolayer Array as an Efficient Electrocatalyst for Oxygen Evolution Reaction. *J. Am. Chem. Soc.* **2015,** *137*, 7071-7074.

466107 Design, Synthesis, and Structure-Function Relationship of Amphiphilic Organic Ligands for Stable Nanoparticle Dispersions in Various Solvents

Abstract:

【Introduction】

Inorganic nanoparticles such as metals, semiconductors, and oxides have recently been recognized as promising functional materials which have physicochemical properties, including electrical, optical and mechanical properties. For example, polymer nanocomposites are one of the most attractive applications using nanoparticles. In order to take advantage of the unique property of nanoparticles, it is necessary to obtain stable dispersion of nanoparticles in several solvents, hopefully in high concentration. In most cases, the colloidal stability is achieved by modifying nanoparticles with a monolayer of organic ligands. However, the choice of suitable organic ligands for each solvent and particle generally requires trial-and-error approaches and can be problematic. In this context, we found that TiO_2 nanoparticles modified with a commercial phosphoric acid surfactant can be dispersed in both polar and less-polar organic solvents and applied this surfactant to prepare several stable colloidal dispersions. The surfactant has hydrophobic alkyl chain and hydrophilic ethylene glycol chain, which might be a key for the dispersion stabilities. However, the relationship between molecular structure and colloidal stability is unclear and the structure is not optimized for better dispersion. In this study, we designed and synthesized a series of novel amphiphilic organic ligands and used them to modify TiO_2 nanoparticles as a model. The resulting modified TiO_2nanoparticles were then dispersed in several organic solvents to study the relationship molecular structure and colloidal stability.

【Experimental method】

We designed the amphiphilic organic ligands with different lengths of alkyl and ethylene glycol chains and a phosphoric acid group as an anchoring moiety. A series of the ligands were synthesized in 3 steps from commercially available alkyl bromides. The ligands structures and purities were characterized by NMR analysis. TiO_2nanoparticle aqueous dispersion was purchased from Ishihara Sangyo Kaisha, Ltd. (STS-100, primary particle size is 5 nm) and were modified with the synthesized ligands by mechanical stirring. The modified nanoparticles were aggregated in the aqueous solution, which were then collected by centrifugation and washed with water to remove free ligands and dried in vacuum. The resulting modified nanoparticles were characterized by elemental analysis and were used for dispersion study in a wide variety of organic solvents using dynamic light scattering measurement.

【Result and Discussion】

We found that the loading amount of all synthesized ligands were approximately 5 μmol/m^2, suggesting that the stable monolayers were formed through the phosphoric acid-TiO_2 bonding. It should be noted that the excess addition (10 μmol/m^2) of the ligands caused clear aqueous solutions, while white aggregations were formed when lower amounts were used. This could be explained that the bilayers were constructed at the surface of TiO_2nanoparticles to show ionic phosphoric acid group, rendering the modified particles hydrophilic. The recovered modified particles were then dispersed into several organic solvents at 3.0 wt% using sonication. We confirmed that the dispersions were stable and the mean particle sizes were estimated to be 10-20 nm by DLS measurements in most organic solvents used. We found that the slight structural differences significantly affected the colloidal dispersion stabilities, which will be discussed in details in the presentation.

【Conclusion】

The novel amphiphilic organic ligands were designed and the synthesis routes were established. We found that the slight structural differences significantly affected the colloidal dispersion stabilities, leading to the optimization of the structure. The modified TiO_2 nanoparticles were nicely dispersed in various organic solvents at useful concentrations of 3.0 wt%.

469119 Hierarchical Nickel Carbide "Dandelion" Nanostructure: Controlled Synthesis and Potential Applications

Abstract:

Transition metal carbides (TMCs), *e.g.*, Ni_3C, Mo_2C, and WC, can exhibit high electrical and thermal conductivity combined with excellent thermal and mechanical stability due to their mixed covalent-ionic bonding. TMCs can be excellent catalysts with performance comparable to noble metals for some reactions, and can also serve as supports for noble metal catalysts. Rhombohedral nickel carbide, Ni_3C, is a common product of nickel carburization during catalysis using Ni. Research on the transformation of Ni to Ni_3C helps elucidate its formation mechanism. Recently we reported a hierarchical nickel carbide (Ni_3C) nanostructure that resembles a dandelion flower (Qiao *et al. Angew. Chemie*, 2016). A typical hierarchical nanocrystal (HNC) consists of one core and many radially-arranged branches, most of which are topped by a hexagonal cap. To understand the formation of the Ni_3C HNCs, we conducted time-resolved experiments. Quasi-spherical aggregates of primary particles, referred to as "cores", were observed after 75 min. The Ni content at this point exceeds that of stoichiometric Ni_3C. After 80 min, the radial arms have grown in one dimension, and the platelets have begun to grow in the perpendicular direction. From the 75th min to the 80th min, the cores blossomed into midsize HNCs, while also densifying to become non-porous. Further growth of both the arms and the platelets proceeds through the 90th min. The atomic composition of the HNCs matched the stoichiometry of Ni_3C, *i.e.*, 75% Ni and 25% C. During evolution from cores to dandelions, the specific surface area (SSA) increased from about 3 $m^2\ g^{-1}$ to over 60 $m^2\ g^{-1}$. We are presently testing this novel nanomaterial in electrocatalytic applications and will report these results along with the synthesis in this talk. In short, we developed a hierarchical Ni_3C "dandelion" nanostructures with specific surface area much larger than that of dense particles of comparable size. Insights into their growth mechanism can enlighten further research on the controlled synthesis of hierarchical nanostructures in

solution. In addition, we believe that our current efforts will demonstrate promising applications of this material.

457454 Graphene Nanoribbons As Conductive Pathways in Directly Deposited Silicon Nanofiber Anodes for High Performance Lithium-Ion Batteries

Abstract:

Silicon is the best candidate for the next generation of Li-ion battery anodes due to its high specific capacity of 4200 mAh/g .However, it expands to 300% of its original size upon insertion of lithium ions[1], which results in crack formation and pulverization of active material with cycling and thus capacity fading. Crack formation also results in constant solid electrolyte interface formation on the surface of the newly exposed silicon, which brings about unstable performance. Several strategies such as use of nanoparticles[2] and composites[3] have been employed by research groups to overcome aforementioned issues. However, all the modifications to silicon has to be done in an economical manner for the product to become an industrial realization.

In the present work we are engineering the nanostructure of a silicon composite with graphene nanoribbons (GNRs) and polyvinyl alcohol (PVA) to create a reliable anode for lithium ion batteries. Silicon nanoparticles and GNRs are dispersed along PVA nanofibers via a water-based electrospinning process to directly form functional fiber mats on top of the copper current collector. This method is compatible with a role to role production, and bypasses several long-winded steps in the conventional paste-based electrode making. Physical and chemical properties of the nanoparticles and their dispersion affect the fiber morphology and ultimately electrochemical performance of the composite material. Graphene nanoribbons, made by unzipping carbon nanotubes, offer unique electrochemical and structural properties when inside the composite fibers. It was observed that GNRs outperform their precursor CNTs in terms of overall capacity and retention. The Silicon-GNR-PVA fiber anode exhibited an outstanding 2000 mAh/g at a rate of C/15 and 1200 at C/5 over 200 cycles. Inclusion of GNRs improved the structural stability and conductivity of the fibers, while protection of nanoparticles from the electrolyte offered by the polymer resulted in stable cycling of

the directly deposited electrodes.

[1] C.-M. Wang, X. Li, Z. Wang, W. Xu, J. Liu, F. Gao, L. Kovarik, J.-G. Zhang, J. Howe, D. J. Burton, Z. Liu, X. Xiao, S. Thevuthasan, D. R. Baer, *Nano Lett.* **2012**, *12*, 1624.

[2] M. M., Ryu I, Wu H, Liu N, Hu L, Nix WD, Cui Y Yao Y, *Nano Lett.* **2011**, *11*, 2949.

[3] Y.-S. Hu, R. Demir-Cakan, M.-M. Titirici, J.-O. Müller, R. Schlögl, M. Antonietti, J. Maier, *Angew. Chem. Int. Ed.* **2008**, *47*, 1645.

467753 Si-Based Nanofiber Anodes for Li-Ion Batteries Prepared Using Particle/Polymer Electrospinning

Abstract:

The demand for rechargeable batteries with high energy and power densities has never been greater. The functionality of many portable electronic devices is limited by battery size and lifetime. Similarly, electric vehicle propulsion systems require significant improvements to satisfy consumer expectations for long drive distances and short recharge times. To meet the demands of energy-hungry consumers, new battery electrodes with high capacities and long cycle life at fast charging rates must be developed.[1-2] These requirements have motivated research efforts to develop next-generation Li-ion battery materials, such as nano-structured silicon anodes. There is also a need for new electrode architectures to replace conventional slurry cast materials. One approach for an electrode design is the use electrospun electrodes containing engineered void spaces to enhance Li^+ transport rates for high areal capacities at fast charge/discharge rates.

Silicon has received considerable attention as an active material for Li-ion battery anodes due to its high theoretical capacity (3,579 mAh/g vs. 372 mAh/g for graphite). However, silicon based anodes typically suffer from rapid capacity fade due to particle cracking and electronic isolation caused by large volumetric changes between the lithiated and delithiated states. While particle fracture can be eliminated by using nanosized Si, electronic isolation remains problematic. Another significant obstacle for commercialization of Si in Li-ion batteries is the irreversible consumption of Li during formation of the solid-electrolyte interphase (SEI) layer.

The present work is the latest installment by Pintauro and co-workers on the use of particle/polymer electrospun nanofiber mats for fuel cell and Li-ion battery electrodes.[1-5] Such electrodes are a viable and attractive alternative to electrospun mats prepared by carbonizing polymer fiber precursors at high temperature. In particle/polymer

electrospinning, there is no fiber pyrolysis; the electrospinning and post-processing are performed at room temperature which simplifies the process and preserves the beneficial characteristics of the polymer binder. Furthermore, the structure of the fibers can be easily altered for improved battery performance (e.g., by using core-shell or hollow-bore fibers or fibers with a high internal porosity). Due to these characteristics, high gravimetric, areal, and volumetric storage capacities can be achieved through the use of thick, densely packed particle/polymer nanofiber mats.

In this presentation, recent results will be presented on the fabrication and use of electrospun particle/polymer nanofiber anodes containing Si nanoparticles, carbon powder, and poly(acrylic acid) binder (Si/C/PAA). Charge/discharge properties of the nanofiber electrodes will be presented, and electrode capacities for different fiber mat thicknesses and mat porosities will be discussed. Notably Si/C/PAA nanofiber anodes have been prepared with high gravimetric, areal, and volumetric capacities of 1,100 mAh/g, 3.1 mAh/cm^2, and 560 mAh/cm^3, respectively, which were stable for 50 cycles at 0.1C. These results demonstrate that proper micron-scale design of electrodes can enhance Li-ion battery performance.

References

1. E. C. Self, R. Wycisk and P. N. Pintauro, *Journal of Power Sources*, **282**, 187 (2015).

2. E. C. Self, E. C. McRen, and P. N. Pintauro, *ChemSusChem*, **9**, 208 (2016).

3. W. Zhang and P. N. Pintauro, *ChemSusChem*, **4**, 1753 (2011).

4. M. Brodt, R. Wycisk and P. N. Pintauro, *Journal of The Electrochemical Society*, **160**, F744 (2013).

5. M. Brodt, T. Han, N. Dale, E. Niangar, R. Wycisk and P. Pintauro, *Journal of The Electrochemical Society*, **162**, F84 (2015).

463852 *In Situ* Activation of Nitrogen-Doped Graphene-Based Materials Anchored on Graphite Foam for High-Performance Energy Storage

Abstract:

Contact interface between the active materials and the current collector is essential for electron transfer rate and the mechanical properties of an electrode. With a traditional drop-casting electrode preparation method, the nano-sized active materials should mix with the nonactive and nonconductive binder, thus inevitably affect the total electrode resistance. Activation of graphene-based materials (e.g. graphene, carbon nanotube) by KOH can rebuild the crystalline structure to obtain continuous interface contact to facilitate electron transfer, while the molten KOH can act as the filler to enlarge the interlayer space to increase the electrolyte ion transfer rate.

Here, three-dimensional free-standing nitrogen-doped porous graphene-based materials/graphite foam composites (aNG/GF) were fabricated *via in situ* activation of nitrogen-doped graphene-based materials on highly conductive graphite foam (GF). During activation, the molten KOH reacts with the carbon to generate nano-holes as well as to rearrange the carbon, yielding a distribution of meso- and micro-pore channels and increases the specific surface area. After *in situ* activation, intimate 'sheet contact' was observed between the graphene sheets and the GF. The 'sheet contact' produced by *in situ* activation is found to be superior to the 'point contact' obtained by the traditional drop-casting method, and facilitates electron transfer. Due to the intimate contact as well as the abundant nano-holes and channels, the aNG/GF electrodes can exhibit high electron and electrolyte ion transfer rate, which is of great important for high-performance energy storage devices. The aNG/GF electrodes can deliver high gravimetric capacity and volumetric capacity with respect to the whole electrode mass and volume (including the active materials and the GF current collector). Furthermore, the tailored contact interface and the unique micro-structure could guarantee a high rate performance of the electrode even under ultrahigh current density, indicating its promising use as

a high-rate LIB anode.

470624 Solution Combustion Synthesis for High Performance $ZnCo_2O_4$ Anode in Lithium-Ion Batteries

Abstract:

Nanostructured, tailored $ZnCo_2O_4$ is synthesized by solution combustion synthesis using two fuels, glycine and citric acid, to explore how reaction conditions impact electrochemical performance as a Li-ion anode. Owing to the unique combustion reaction mechanism, the novel strategy of using mixed fuels and ammonium nitrate creates anode material with improved cycling stability and capacity. Characterization shows a mesoporous spinel product, composed of nanoparticles assembled into micron-sized secondary structures, with properties varied by the fuel ratio and type. A stable capacity of 1,200 mAh g^{-1} is achieved at C/10 and 1000 mAh g^{-1} at 1C, showing high energy and power capabilities as compared to the bulk material. Based on an understanding of the reaction mechanisms and influence of reaction parameters, solution combustion synthesis is a powerful and flexible technique to produce superior electrode materials.

473388 a General and Mild Approach to Controllable Preparation of Manganese-Based Micro/Nanostructured Bars for High Performance Lithium-Ion Batteries

Abstract:

To meet the requirements for large-scale applications in electric vehicles, hybrid electric vehicles and stationary energy storage, growing interest has been directed at the economical, sustainable and environmental friendly methods to produce electrode materials for rechargeable lithium ion batteries (LIBs) with high energy and power densities.

Recently, manganese-based electrode materials, such as layered LiNixCo1-x-yMnyO2, spinel LiNi0.5Mn1.5O4 and layered Li-rich xLi2MnO3â€¢(1-x)LiNi1/3Mn1/3Co1/3O2 (0 â‰¤ x â‰¤ 1) have received a flurry of research activities due to their advantages of high energy and power densities, low cost and high cycling performance.

In respect to thermal stability, larger particles with a higher tap-density would be less reactive in a highly oxidized state. On the other hand, an electrode consisting of uniform particles with an appropriate size could ensure a homogeneous intercalation reversibility from particle to particle over repeated cycles. Therefore, the control of particle size and morphology for tailoring lithium-ion battery electrode materials has been paid more and more attention in recent years. Till now, it has remained a challenge to develop a general strategy that allows the large-scale preparation of a wide array of 1D micro/nanostructured manganese-based electrode materials with a controllable phase, composition and shape, as well as a narrow size distribution.

Herein we have established a Chimie Douce approach --- by mixed solvent co-precipitation --- to selectively prepare an extensive series of 1D micro/nanostructured manganese-based electrode materials, both cathode and anode for LIBs with high energy and power densities, attesting to its generality and effectiveness. Based on the kinetic elaboration of crystallite nucleation and growth through mixed solvent mediation, the new route is easily controllable, repeatable, and generalizable to the fabrication of a large series of some highly promising manganese-based electrode materials with tunable 1D micro/nanostructures for enhanced performances in lithium ion batteries. As a significant difference from the conventional methods, the present method is conducted in ethanol/water mixed solutions at room temperature, without having recourse to long-chain surfactants or hydrothermal condition. In addition, it can be employed to produce different aspect ratios of microbars with mixed transition metal oxides by tuning the ethanol/water volume ratio in the mixed solvent. It is anticipated that the mixed solvent mediated co-precipitation strategy will open opportunities to heuristically and systematically design mixed transition metal oxides with controllable phase and composition, including not only manganese-based but also other metal-based materials, with desirable micro-/nano-structures for developing high performance lithium-ion batteries.

460764 Advanced Energy Materials Derived from Bijels

Abstract:

Pressing energy demands for national security, distributed power, and portable technologies require the development of next-generation batteries and supercapacitors that can simultaneously deliver high power and energy densities. In order to achieve this goal, the kinetics of ion and electron transport within the electrodes must be enhanced while maintaining a large capacity for energy storage. Therefore, the idealized electrode microstructure has been envisioned as a three-dimensional, co-continuous arrangement of percolating passages for efficient ion and electron transport that would also enable a large volumetric density of active materials. We report a novel method to produce this distinctive microstructure through the use of soft matter templates derived from bicontinuous interfacially jammed emulsion gels (bijels), which are a new class of soft materials with a co-continuous microstructure where two interpenetrating fluid domains are kinetically arrested through jamming of particles at the fluid-fluid interface during spinodal decomposition. Bijels can be converted to electrolytically active composites with co-continuous internal microstructure through simple chemical post processing steps. In this talk, we will present microstructured composite electrodes that have been synthesized by this technique and demonstrate how their electrochemical characteristics can be tuned over a wide range, allowing for concurrent delivery of high energy and power. The broader applications of this new class of soft materials in electrochemical energy conversion and storage, sensing, catalysis, and tissue engineering will also be discussed.

457746 Sulfur Self-Doped Micro/Mesoporous Carbon Derived from Lignin and Its Application onÂ Supercapacitor Â and Oxygen Reduction Reaction

Abstract:

The growing global concerns about energy needs, fossil fuels consumption and the related environmental issues have motived scientists to find new, green and sustainable energy resources and technologies. In this work, byproduct lignin was successfully converted into self-sulfur-doped graphitic carbons via in situ hydrothermal carbonization (HTC) by following the post-annealing treatment at elevated temperature. The surface, porosity, textural, structural and morphological properties of synthesized materials were investigated by SEM, XPS, Raman, FTIR and surface area analyzer. Electrochemical properties of the prepared materials were analyzed by cyclic voltammetry (CV), charge-discharge (CD), electrochemical impedance spectroscopy (EIS), linear sweep voltammetry (LSE) in a KOH electrolyte solution. Up to 3.5 wt. % sulfur intercalating on carbon molecules was successfully achieved. The resulting materials electrode displays high surface area (up to 550 $m^2 g^{-1}$) with mesoporous structure and partially graphitic/amorphous carbon structure. The obtained original electrode materials result in high electrochemical activity for supercapacitor applications. The SGC-850 electrode exhibited a specific capacitance value of 225 g^{-1} at a scan rate of 1 A g^{-1}. The SGC-850 electrode revealed a very consistent response over 10000 cycles at a harsh condition of 5 A g^{-1} current density. In addition to supercapacitor application, the obtained material was tested for oxygen reduction reaction (ORR) application. The ORR CV curve of the SGC-850 presents well-defined cathodic peak in the O_2 electrolyte. The SGC-850 goes under 4 electron pathway for the O_2 reduction. To sum up, metal-free SGC-850 is promising candidate electrode materials for supercapacitors and ORR applications.

460585 Towards Ambient Armor: Can New Materials Change Longstanding Concepts of Projectile Protection?

Abstract:

Materials that protect against the high strain rate loading of a projectile have traditionally been constrained by their use on the target of protection. Here we explore the possibility of new materials, particularly carbon nanocomposites, for projectile protection removed off of the target, and placed along the trajectory at an intermediate location. In the case of light projectiles with low kinetic energy, a mathematical model of such Ambient Armor (AA) separates the impact, capture and energy dissipation of an incident projectile into each of three stages respectively. A distinct scaling of the projectile to composite area ratio, and areal density ratio is derived for a given requisite stopping distance. We consider the case where the projectile becomes embedded into the composite, and also the gyroscopic destabilization that proceeds when penetration occurs. As an example, the model in the former case predicts that a 1 g and 800 m/s projectile could be decelerated to 3 m/s within 6-9 m using a 2.34×10^4 layer monolayer graphene nanocomposite at 0.02 volume fraction. For heavier projectiles with higher kinetic energy, a tandem system using a second composite body with the specified area density is explored with the idea of making the projectile gyroscopically unstable for rapid deceleration and subsequent interception. Off target reinforcement or AA may in general allow for the exploration of a broader range of material properties for projectile protection.

464168 Corrosion Protection of Aluminum Alloy Via Graphene-Polymer Nanocomposite Coatings

Abstract:

As hexavalent chromium (VI) is phased out of use as a corrosion inhibitor, the search for environmentally friendly non-toxic anticorrosion approaches continues. Here, we present graphene-polymer nanocomposite protective coatings made using a simple airbrushing technique. The nanocoating can be deposited over large areas onto a variety of surfaces from stock solutions of polymer and graphene oxide sheets, followed by chemical reduction. Sprayed onto aluminum alloy (AA) 2024-T3, the coatings bear a brick-and-mortar structure, in which graphene sheets lie parallel to the surfaces in a densely packed structure. The end result is pinhole-free multilayered composite coatings with improved barrier properties. The graphene sheets act as a physical barrier to oxygen diffusion, thus hindering the corrosion cycle. First, we present the spraying process, and what parameters contribute to making a pin-hole free nanocoating on a metal surface. Second, we present the final structure of the film and its physical properties. Lastly, we explore the suitability of this coating towards corrosion protection is evaluated using electrochemical impedance spectroscopy under varying salt conditions.

467837 Persistently Auxetic Materials (PAMs): Engineering the Poisson Ratio of 2D Self-Avoiding Membranes Under Conditions of Non-Zero Anisotropic Strain

Abstract:

Entropic surfaces represented by fluctuating 2D membranes are predicted to have desirable mechanical properties when unstressed, including a negative Poisson's ratio ("auxetic" behavior). Herein, we present calculations of the strain-dependent Poisson ratio of self-avoiding 2D membranes demonstrating desirable auxetic properties over a range of mechanical strain. Finite size membranes with unclamped boundary conditions have positive Poisson's ratio due to spontaneous non-zero mean curvature, which can be suppressed with an explicit bending rigidity in agreement with prior findings. Applying longitudinal strain along a singular axis to this system suppresses this mean curvature and the entropic out-of-plane fluctuations, resulting in a new, molecular scale mechanism for realizing a negative Poisson's ratio above a critical strain, with values significantly more negative than the previously observed zero-strain limit for infinite sheets. We find that auxetic behavior persists over surprisingly high strains of more than 20% for the smallest surfaces, with desirable finite size scaling producing surfaces with negative Poisson's ratio over a wide range of strains. These results promise the design of surfaces and composite materials with tunable Poisson's ratio by pre-stressing platelet inclusions or controlling the surface rigidity of a matrix of 2D materials.

457219 a Highly Active and Selective β-Nucleating Agent for Isotactic Polypropylene and Crystallization Behavior of β-Nucleated Isotactic Polypropylene Under Rapid Cooling

Abstract:

Over the past few decades, much attention has been focused on the preparation of the β-crystal form of isotactic polypropylene (β-iPP) under laboratory and industrial conditions. The structure, morphology, processing, properties and application of β-iPP have been extensively studied. In comparison to the traditional α-form of iPP (α-iPP), β-iPP exhibits improved elongation at break and better impact resistance. These unique performance characteristics of β-iPP, which are advantageous in many applications, justify the extensive basic and applied research efforts that have been focused on this material.

β-iPP can be produced in the presence of efficient, selective β-nucleating agents (β-NAs) under practical conditions. Many available β-NAs have been previously reported. The first efficient β-NA was γ-quinacridone (Dye Permanent Red E3B), which was reported by Leugering. A two-component β-NA (i.e., compounds of pimelic acid and calcium stearate) was introduced by Shi et al. Additional compounds with more defined structure that were similarly efficient included calcium salts of pimelic and suberic acid, as reported by Varga et al.. These two β-NAs have been categorized as a class containing certain group IIA metal salts or their mixtures with some specific dicarboxylic acids. *N,N′*-Dicyclohexyl-2,6-naphthalenedicarboxamide (NU 100) is also an efficient β-NA that was developed in the past decade. Although these substances can act as β-NAs for iPP, only a few aromatic amide compounds, such as NU 100, have been marketed. Unfortunately, NU 100 is not a highly selective β-NA, and α-iPP is always formed in its presence. Therefore, a highly active and selective β-NA for isotactic polypropylene is still required. In recent years, a novel group IIB salt of alicyclic dicarboxylic acid that acts as a β-NA was investigated in our group. In this study, we reported a novel IIB salt of aliphatic dicarboxylic acid (i.e., zinc adipate), which acts as a highly active and selective β-NA for iPP. Our results further enrich the family of IIB salts of dicarboxylic acid for use as β-NAs of iPP.

β-iPP is thermodynamically metastable, which has been reported by Varga in detail. The formation of β-iPP strongly depends on the crystallization process, especially the cooling rate. Both Varga and Wittmann attributed this "sensitivity" to the upper and lower critical temperature limit of the formation of β-iPP. The upper critical temperature limit is around 140 °C, and the lower limit is at 100-105 °C. Therefore, the growth rate of the β-phase is higher than that of α-phase between 100 and 140 °C. When the temperature is higher than 140 °C or lower than 100 °C, βα-growth transition will take place on the surface of a growing β-spherulite. Apparently, the content of β-iPP depends on the cooling rate. Therefore, the exploration of the dependence of the formation of β-iPP on the cooling rate under real processing conditions is technologically important and scientifically fascinating. Previous investigations have been extensively performed under laboratory conditions using standard differential scanning calorimetry (DSC). However, operational processes occur much faster compared to typical laboratory rates. Therefore, achieving controlled, constant heating and cooling rates that are much higher than the typical rates of standard DSC instruments, which range from 10 to 100 °C/min, is experimentally challenging. Recent advances in chip fast scanning calorimetry (FSC) by Mettler-Toledo (e.g., the Flash DSC 1 calorimeter, introduced October 2010) may provide the ability to investigate crystallization and melting behavior under rapid cooling rates. In this approach, the cooling rate and heating rate can reach 240,000 °C/min. Therefore, FSC is a powerful tool for investigating the crystallization and melting behaviors of polymers by mimicking realistic conditions. However, to the best of our knowledge, the literature contains few studies that address the dependence of β-crystals on the cooling rates under rapid cooling and supercooling.

In the current study, **Zinc adipate (Adi-Zn) was observed to be a highly active and selective β-nucleating agent for isotactic polypropylene (iPP). The effects of Adi-Zn on the mechanical properties and the β-crystals content of nucleated iPP were investigated. The impact strength of iPP nucleated with 0.2 wt% Adi-Zn was 1.8 times higher than that of neat iPP. In addition, wide-angle X-ray diffraction (WAXD) analysis indicated that the content of β-crystals in nucleated iPP ($k_β$ value) reached 0.973 with 0.1 wt% Adi-Zn, indicating that Adi-Zn is a highly active and selective β-nucleating agent for iPP. Furthermore, fast scanning chip calorimetry (FSC) studies using cooling rates from 60 to 13,800 °C/min revealed that the formation of β-crystals significantly depended on the cooling rates. At cooling rates below 3000 °C/min, only β-crystals existed. However, at cooling rates above 6000 °C/min, β-crystals failed to form. Moreover, a lower critical crystallization temperature that corresponded to the generation of β-crystals was investigated using cooling-induced crystallization, and the results are in good agreement with those of a previous study. Finally, an exponential decay of T_{cp} as a function of the cooling rate is proposed.**

460522 Layered and Scrolled Nanocomposites with Aligned Semi-Infinite Graphene Inclusions at the Platelet Limit

Abstract:

Graphene and other two-dimensional (2D) materials are distinct among nanoscale inclusions or fillers for composites in that they can potentially span the physical dimensions of the enclosing solid. However, alignment and assembly of continuous 2D components at high volume fraction and macroscopic dimensions remains an unsolved challenge in material science. Herein, we introduce a stacking and folding method that generates layer numbers that scale as 4^j where j is the number of successive quadrant segmentations of a 2D inclusion, to generate aligned graphene/polycarbonate composites with as many as 320 parallel layers spanning 0.032 to 0.11 mm thickness. An analogous transverse shear scrolling method generates Archimedean spiral nanocomposite fibers 0.10-0.16 mm in diameter and 2 cm in length. The process significantly increases the effective elastic modulus approximately 1.9-fold to nearly 1 GPa, approaching the limit of platelet filler theory, and increases the ultimate tensile strength to 40 MPa even at exceptionally low graphene volume fraction of only 0.00185. Graphene spiral fibers demonstrate exotic, telescoping elongation at break of 110%, or 30 times greater than Kevlar. Both composite types retain anisotropic electrical conduction along the graphene planar axis with a percolation threshold $V_G < 0.003$ vol%, and layer numbers less than 36 remain transparent with optical density < 42%. These results highlight new combinations of material properties available at this extreme platelet filler limit for nanocomposites.

468432 PMMA/TiO2 One-Dimensional Photonic Crystal Films with Structural Colors

Abstract:

Abstract: There are many examples of color that are not chemical in origin but rather physical in nature. Compared to pigment color, structural color has the characteristics of brightness, high saturation and fade-resistance. Artificial photonic crystals are the main source of structural color. Photonic crystals are a periodic array of media of different refractive micro-structures. Because one-dimensional photonic crystals have simple structure and facile preparation method, they have received widespread attention in recent years. In addition, as organic and inorganic complex photonic crystals may combine the advantages of both, they show promising prospect in application, such as coatings, naked eye inspection, displays, etc.

Here, nanoparticle-nanoparticle based organic-inorganic complex one-dimensional photonic crystal films were first designed with poly(methyl methacrylate) (PMMA) nanoparticles as low refractive index material and TiO_2 nanoparticles as a high refractive index one, and PMMA/TiO_2 stacks with structural colors were assembled through layer-by-layer spin coating. PMMA and TiO_2 nanoparticles with average diameter of about 40nm were first prepared. We used PMMA nanoparticles as low refractive index material and TiO_2 nanoparticles as the high refractive index one, and assembled PMMA/TiO_2 one-dimensional photonic crystals with structural colors through layer-by-layer spin coating. By adjusting the parameters of the nanoparticle dispersion and spin coating, such as the concentration of nanoparticles, rate and times of spin coating, and the stack number, we prepared angle-dependent structure colors of wide range from violet to nearly red. The morphology, reflective spectra and color properties of the one-dimensional photonic crystal films were investigated. This method can be effectively used for assembling other functional organic/inorganic complex one-dimensional photonic crystals for a wide usage.

462839 Facile Fabrication of Highly Omniphobic and Self-Cleaning Surface Based on Water Mediating Fluorinated Hybrid Nanocomposite

Abstract:

Liquid repellent surfaces are being promisingly applied in industry and our daily life. Herein we reported a simple and effective approach to fabricate omniphobic coatings with multi-scale composite structure by hydrolytic condensation of methyltriethoxysilane (MTES) and 3-[(perfluorohexyl sulfonyl) amino] propyltriethoxysilane (HFTES) and nanosilica sol. It is a one-step approach to precisely regulate the hierarchical structure and wettability of resultant surface via manipulate the aggregation of nanosilica by mediating water content. The fluorinated nano-sol can be easily coated on various hard and soft substrates by spraying or dipping methods, endowing the substrate with omniphobicity to different organic liquid. The optimized hybrid coatings exhibit outstanding superhydrophobic and oleophobic performances with static angle of 157.9°, 146.4°, 148.0°, 134.0° for water, diiodomethane, glycol and n-hexadecane respectively. The multi-scale roughness structure strengthens the water and oil repellency of the hybrid coatings significantly even under high temperature and keeps the similar hydrophobicity and oleophobicity after being treated at extreme acidic (pH=1) and basic (pH=14) conditions. Furthermore, the hybrid coating surfaces show outstanding self-cleaning performance and resistance to blood and egg white adhesion and have a great thermal stability below 350 â„ƒ. Owing to the facile method and its remarkable omniphobicity abilities in various conditions, the fluorinated nanosilica sol can be expected to have potential industry applications in many areas.

472854 Conjugated Polymer and Block Copolymer Additives in Bulk Heterojunction Opvs (Invited Talk)

Abstract:

Bulk heterojunction organic photovoltaics (OPVs) are based on an interpenetrating blend of donor and acceptor in the active layer. Additives can be used to tune the morphology and enhance electronic properties. Here, we present work in the development of conjugated polymer and block copolymer additives to both direct morphology and enhance photovoltaic performance. The morphological impact of the additives can be predicted based on estimates of the Flory-Huggins interaction parameter, and we find good agreement with self-consistent field theory calculations. Additives can also have a surprisingly large impact on electronic properties. Interfacially active all-conjugated block copolymers with donor and acceptor polymer blocks enhance the open-circuit voltage (V_{OC}) at low loadings (~ 2 wt %) but have little impact on the short-circuit current. Conversely, conjugated polymer additives compatible with both donor and acceptor phases enhance the short-circuit current and have only a small effect on the V_{OC}.Â We hypothesize that the additives can form either phase-segregated â€œparallel-likeâ€ channels or influence charge recombination by segregating to donor-acceptor interfaces. This work demonstrates that conjugated polymer and block copolymer additives, properly tailored, can significantly enhance the properties and performance of bulk heterojunction OPVs.

468195 Coarse-Grained Model of Conformational Disorder Effects on the Intra-Chain Electronic Properties of Polythiophenes

Abstract:

Polythiophenes are semiconductors with potential applications as active elements in organic electronic devices. Like other organic semiconductors, the widespread usage of these polymers is limited by a low performance relative to their inorganic counterparts. A key source of difficulties in developing high-performing polymeric semiconductors is the limited understanding of how conformational disorder affects their optoelectronic properties. A high level of conformational disorder is inherent to these soft materials. We develop a coarse-grained approach based on the tight binding approximation to model the electronic degrees of freedom of polythiophene chains, taking into account atomic coordinates changes. Particularly important are dihedral rotations and moiety distortions, which are known to disrupt extended electronic states. The initial stage of our model successfully captures the effects of dihedral rotations on the valence and conduction states of the chains. The model, which coarse-grains the chains at the level of a thiophene ring, is parameterized using density functional theory (DFT) calculations of the one-dimensional band structures for chains with imposed periodic variations in dihedral angles. In a first application, this model is used to compute the optical absorption spectrum of poly(3-hexylthiophene) (P3HT) chains in solution, and successfully reproduces the broadening of the absorption edge as a result of dihedral disorder. We extend the model to include the effects of thiophene ring vibrational distortions. The extended model combined with the variational method is used to predict the formation and properties of polarons and excitons in P3HT, in a cost effective way.

470286 Photo-Responsive Monomer for Light Mediated ROMP

Abstract:

Photo-responsive monomer for light mediated ROMP

Ishan Fursule, Qunfei Zhou, Matthew Beck and Bradley Berron.

Department of Chemical and Materials Engineering, University of Kentucky.

Stimuli-responsive polymers are changing the landscape of actuated materials in optoelectronics, molecular machines, and biomedical materials. Here, we combine the concept of Ring Opening Metathesis Polymerization (ROMP) with azobenzene photoactuation to achieve a new monomer with photo-mediated ROMP kinetics. This novel approach allows photoactuation of both a polymerization reaction and of the formed polymer. This will allow us to precisely control the thickness of thin photo-responsive polymer coatings and hence the penetration intensity of radiations in the film.

Internal ring strain of monomer is a prerequisite for ROMP to minimize the reversibility of polymerization and carbon-carbon double bond for catalyst to initiate the polymerization reaction and then also for propagation. To optically-control the ring strain to the molecule, we synthesized a family of cyclic molecules with a photo responsive azobenzene group. We fine-tuned the light-induced ring strain through computational molecular design, and we synthesized saturated closed ring monomers with photo-tunable ring strain. Photoresponse of these molecules is confirmed by UV-Vis and NMR. Molecules are stable in the cis and trans state, but can be photoswitched with appropriate light sources. We are now synthesizing the analogous family of unsaturated closed rings where mean ring strain is contingent upon both the ring size and the isomerization of the azobenzene group. The ROMP kinetics of the alkene containing monomers will be analyzed under UV or visible light.

454128 Highly Controlled Synthesis of Poly(3-hexylthiophene) Using Droplet-Flow Microreactors

Abstract:

Poly(3-hexylthiophene) (P3HT) is the mainstay semiconducting polymer for a broad range of organic electronic devices, and in particular thin-film organic photovoltaic (OPV) cells. Despite being over two decades since the first reported synthesis, there remains an unresolved problem with batch-to-batch consistency of P3HT in terms of electronic performance and solution rheology – both critical factors to the commercial success of printed OPV modules.

The primary cause of inconsistent performance is the inability to source materials with consistent molecular weight distributions and purities. To address this obstacle we have developed a droplet-flow reactor approach for the preparation of P3HT and other related conjugated polymers.[1–3] In contrast to conventional flask methods which suffer from scaling issues above the few-gram level, our flow approach may be scaled up to production levels in excess of 1 kg/day without loss of control over the chemistry, and is sufficient to meet current industrial demand.

In this talk we will discuss the background concepts behind the droplet-flow approach, and we will demonstrate how these reactors may be used to prepare well-defined P3HT-based materials with predictable properties. We will show how our new variant of the polymerization catalyst for P3HT provides a means of preparing polymers with a wide range of molecular weights while maintaining a high degree of control over the chain length polydispersity. We will further show how these droplet-flow reactors may be used to accurately tune the optoelectronic properties of P3HT by the controlled addition of an optically-distinct co-monomer into the polymer chain. To conclude we will discuss the use of bio-derived solvents in the P3HT synthesis, which enables full conversion to the product polymer to be achieved in just one minute - a significant reduction on flask methods that typically require several hours to reach completion.

(1) Bannock, J. H.; Krishnadasan, S. H.; Nightingale, A. M.; Yau, C. P.; Khaw, K.; Burkitt, D.; Halls, J. J. M.; Heeney, M.; de Mello, J. C. *Adv. Funct. Mater.* **2013**, *23*, 2123–2129.

(2) Bannock, J. H.; Al-Hashimi, M.; Krishnadasan, S. H.; Halls, J. J. M.; Heeney, M.; de Mello, J. C. *Mater. Horiz.* **2014**, *1*, 214–218.

(3) Bannock, J. H.; Xu, W.; Baïssas, T.; Heeney, M.; de Mello, J. C. *Eur. Polym. J.* **2016**.

452726 Approaches to Solution Processable n-Channel Ï€ -Conjugated Donor-Acceptor Co-Polymers and Device Applications

Abstract:

2,2'-bithiazole was synthesized in one step and copolymerized with dithienyldiketopyrrolopyrrole to afford poly(dithienyldiketopyrrolopyrrole-bithiazole), **PDBTz**. **PDBTz** exhibited electron mobility reaching 0.3 cm^{2}V^{-1}s^{-1} in organic field-effect transistor (OFET) configuration; this contrasts with a recently discussed isoelectronic conjugated polymer comprising an electron rich bithiophene and dithienyldiketopyrrolopyrrole, which displays merely hole transport characteristics. This inversion of charge carrier transport characteristics confirms the significant potential for bithiazole in the development of electron transport semiconducting materials. Branched 5-decylheptacyl side chains were incorporated into **PDBTz** to enhance polymer solubility, particularly in non-halogenated, more environmentally compatible solvents. **PDBTz** cast from a range of non-halogenated solvents exhibited film morphologies and field-effect electron mobility similar to those cast from halogenated solvents.

472879 Enhancing Charge Transport through Block Copolymer Architectures

Abstract:

Conjugated block copolymers provide an opportunity to control the active layer morphology of organic electronic devices. Previous work has demonstrated that photovoltaic devices with donor-acceptor block copolymers as the active layer outperform devices comprised of the constituent polymers. Self-assembly into lamellar morphologies leads to nanoscale domains on the scale of the exciton diffusion length while providing pathways for charge extraction. We have also recently explored linking a strongly crystalline block to a weakly crystalline polymer to intricately control local order. Promoting crystallization in one block can promote tight packing in the second block, potentially enhancing the electronic coupling between chains. Measurements of the charge transport characteristics of conjugated block copolymers reveals excellent charge mobilities for electron transport with aluminum source and drain contacts in a bottom gate, bottom contact transistor configuration. As a consequence of promoting close chain packing in the weakly crystalline block with the block copolymer architecture, we find that electron mobilities are higher for the block copolymer than for either of the homopolymers.

464881 Design of Nanostructured Polymer-Based Ordered Ferroelectric Diode Memory Elements

Abstract:

Memory elements that are compatible with flexible, mechanically-robust substrates will be crucial for the design of future organic electronic applications. While many different materials and devices exist for these applications, one of the most promising architectures is that of the polymer-based ferroelectric diode (FeD). In these systems a blend composed of a ferroelectric polymer and a conjugated semiconducting polymer are deposited as a thin film between two electrodes. While the nanoscale structure and phase behavior associated with the interpenetrating networks of the ferroelectric and conjugated polymers has been shown to be of utmost import with respect to final device operation and performance, few studies were performed with systematic tuning of these domains sizes. Here, we demonstrate the creation of an ordered ferroelectric diode (OFeD) using a combination of electron-beam (e-beam) lithography and solution-printing techniques. Importantly, the nanofabrication strategy does not alter the crystal structure of the ferroelectric polymer domains and does not negatively impact the charge transport ability of the conjguated polymer domains. Through the systematic design and fabrication of nanostructured active layers, we demonstrate that there is an optimal domain spacing (~400 nm) for the OFeD active layers that balances the ability for ferroelectric domains to impart a remembered (i.e., written) charge state into the conjugated polymer domains and the necessity for the conjugated polymer to transport charge out of the device (i.e., read-out the device). Moreover, we demonstrate that, due to the solution-processable nature of the ferroelectric-conjugated polymer blend, that the OFeD fabrication paradigm can be readily shifted to a more scalable platform (i.e., beyond e-beam lithography) and is compatible with common flexible substrates. In this way, we establish both the fundamental physics that dictates OFeD operation and present a ready means by which to scale these devices for next-generation organic electronic applications.

464899 Role of Thermal Fluctuations on Local Lattice Disorder and Charge Transport in Conjugated Polymers

Abstract:

Local order governs the electronic properties of conjugated molecules. Although small molecules can be strongly ordered, all conjugated polymers examined to date exhibit significant cumulative lattice disorder. Using atomistic molecular dynamics (MD) simulations, we investigate lattice disorder in single crystals of a commonly studied conjugated polymer, poly(3-hexylthiophene-2,5-diyl) (P3HT). We demonstrate that thermal fluctuations of thiophene rings lead to cumulative disorder of the lattice in the Ï€ stacking direction with an effective paracrystallinity g of about 5%. The effective coherence length of the crystal in the Ï€ stacking direction, bounded by the cumulative lattice disorder, is about 20nm, setting a maximum value to be expected for experimental measurements. By performing Kinetic Monte Carlo simulations on the disordered lattices, we can also demonstrate the effect of local disorder on charge transport within crystalline planes of P3HT.

472647 Directing Multi-Scale Assembly and Alignment of Printed Conjugated Polymers

Abstract:

Self-assembly and crystallization have played a central role in the development of modern electronics and energy materials. Recent years, conjugated polymers have emerged as a new class of electronic and photoelectronic materials that are light-weight, flexible and can be solution printed at low cost and high throughput. Conjugated polymers have demonstrated potential uses in a diverse range of applications from transistors, thermoelectrics, sensors, light-emitting diodes to solar cells etc.

Charge transport in conjugated polymers is influenced over orders of magnitude by morphology parameters across all length scales, in particular crystallinity and molecular order within the domains and degree of alignment between the crystalline domains. The crystallization and morphology characteristics of rigid, short conjugated polymers stand in contrast to the traditionally studied flexible long chain polymers. In addition, during solution printing, the crystallization process is strongly coupled with the mass transport, fluid dynamics, and interfacial properties. Despite the critical importance of crystallization to morphology and charge transport, conjugated polymer crystallization remains poorly understood and challenging to control, particularly in the context of solution printing.

In this talk, we present new insights and strategies we recently developed for controlling multi-scale assembly of conjugated polymers that are directly compatible with solution printing. With high degree of control over thin film alignment, molecular packing and mesoscale structures, we further correlate these morphological characteristics with anisotropic charge transport properties towards establishing structure-property relationships and for understanding the interplay between intrachain and interchain charge transport.

463627 Bio-Orthogonally Crosslinked, Engineered Protein Hydrogels with Tunable Mechanics and Biochemistry for Cell Encapsulation

Abstract:

Covalently crosslinked hydrogels are commonly used as 3D materials to study cell-matrix interactions and for therapeutic delivery of cells in regenerative medicine applications. While covalent crosslinking provides increased material stability and enhanced mechanical rigidity, most of the crosslinking chemistries also react with chemical moieties present on encapsulated cells and biomolecules. Recent progress in the development of bio-orthogonal reactions that eliminate these side reactions can be employed to produce hydrogels with highly specific crosslinking. However, previous hydrogel systems prepared using some of these chemistries are limited by a variety of factors, including copper cytotoxicity, non-ideal gelation kinetics, batch-to-batch variability, and the inability to decouple mechanical properties, cell adhesivity, and degradation. In response, we have developed an engineered protein hydrogel crosslinked via strain-promoted azide-alkyne cycloaddition (SPAAC) that gels in seconds, preserves cell viability and function, permits independent tuning of cell-adhesive ligand concentration and hydrogel stiffness, and allows for selective chemical functionalization of the gels in the presence of living cells and serum. Engineered elastin-like proteins (ELPs) containing either a cell-adhesive RGD domain derived from human fibronectin or a non-adhesive, scrambled RDG domain were functionalized with azides and bicyclononynes (BCN) to facilitate SPAAC crosslinking. To form a gel, ELP modified with azides is simply mixed with ELP modified with BCN. SPAAC crosslinking occurs within seconds, and gelation is complete within minutes. Rapid gelation kinetics facilitate uniform cell dispersion within the hydrogels and are desirable for applications such as cell transplantation through injection and bioprinting, in which fast gelation may increase material retention at the site of injection and improve the fidelity of 3D-printed structures. To emphasize the importance of choosing a chemistry with such rapid gelation kinetics, ELPs were also modified with triarylphosphines to permit crosslinking via the bio-orthogonal Staudinger ligation reaction. Staudinger-crosslinked ELPs gelled in tens of minutes, a timescale too slow for successful cell encapsulation. By varying the extent of BCN modification and the total protein content in the rapidly gelling SPAAC-crosslinked ELPs, the storage modulus (Gâ€™) of the material was varied from 180 Pa to 3300 Pa. Furthermore, at a fixed total polymer content, varying the extent of BCN modification and the ratio of ELP containing cell-adhesive RGD peptides vs. non-adhesive RDG peptides permitted independent tuning of Gâ€™ (from ~1200 to ~3300 Pa) and RGD concentration (from 0 to 5.3 mM). The cytocompatibility of the crosslinking chemistry was assessed by live/dead staining of encapsulated mesenchymal stem cells (MSCs), endothelial cells (ECs), and neural progenitor cells (NPCs) one hour post-gelation. All three cell types remained highly viable (97-99%). Furthermore, the SPAAC-crosslinked ELPs permitted phenotypic maintenance of the cells in culture, with ECs forming tubular networks with open lumen-like structures and NPCs retaining the capacity to differentiate into neurons and astrocytes. Additionally, the importance of decoupled mechanics and adhesive ligand concentration was demonstrated using MSCs. MSCs in 5.3 mM RGD gels exhibited well-spread morphologies, while cells in 0 mM RGD gels remained rounded, independent of hydrogel stiffness. However, in gels with an intermediate RGD density (0.53 mM), significant spreading was only observed in 1200 Pa gels, not in 3300 Pa gels. Finally, the bio-orthogonality of the SPAAC crosslinking was demonstrated by functionalizing excess azide residues in MSC-laden gels with a BCN-bearing fluorescent probe in serum-containing culture medium. Only gels treated with the BCN-containing probe, and not a control, amine-containing probe, displayed significant fluorescence. These results demonstrate that ELP hydrogels crosslinked via SPAAC facilitate independent tuning of matrix stiffness and ligand chemistry, and SPAAC-ELPs permit efficient cell encapsulation and phenotypic maintenance *in vitro*. The bio-orthogonal nature of the crosslinking chemistry makes SPAAC-crosslinked ELP hydrogels attractive materials for applications such as cell delivery and bioprinting, while also permitting highly selective functionalization of the material in the presence of living cells.

466174 Synthesis and Characterization of Hyaluronic Acid and Heparin Thiol-Ene Hydrogels for the Spatial Sequestering of Bioactive Signals

Abstract:

Polymer scaffolds serve a central role in the field of tissue engineering by directing cellular processes based on the mechanical as well as biochemical properties Hyaluronic acid (HA), a non-sulfated glycosaminoglycan (GAG) distributed throughout connective, epithelial, and neural tissue, plays a role in wound healing and angiogenesis. HA hydrogels are widely utilized to create scaffolds for tissue repair and to study cell behavior *in vitro*. Though these materials are commonly synthesized via Michael addition chemistry, we have found that these chemistries lead to hydrogels that have local areas of high and low crosslinking as a result of inappropriate mixing and fast gelation kinetics (Figure A). This heterogeneity leads to inconsistent cell behavior through the gel and between gels of the same condition. To circumvent this limitation, we report on light activated thiol-ene HA hydrogels, which can be thoroughly mixed prior to gelation. In addition, we report on the incorporation of heparin, through the same thiol-ene chemistry allowing for dosing of heparin within an HA gel. Heparin (HP), an important glycosaminoglycan (GAG) capable of sequestering growth factors and protecting them from proteolysis, mediates the spatial presentation of both bound and soluble growth factors *in vivo* and is utilized here *in vitro* to control the presentation and release of growth factors. We show synthesis of norbornene-functionalized HA (HA-Norb) and norbornene functionalized heparin (Hep-Norb) (Figure B) using completely aqueous chemistry avoiding the need to transfer HA and Hep to the organic phase. As expected, we find that the initiator type and degree of crosslinking influences the mechanical properties of the gels. We find that the ability to dose in Hep is critical to allow extensive cellular spreading and migration (Figure B) through out the scaffold indicating that through modulating growth factor presentation and retention, we can guide cell behaviour. This new hydrogel material more closely mimics the ECM of tissues with substantial GAG content such as the brain.

455445 Photo-Mediated Oxime Ligation As a Bioorthogonal Tool for Spatiotemporally-Controlled Hydrogel Formation and Modification

Abstract:

Photoinitiated free radical polymerization has served as a useful tool in hydrogel formation, providing spatiotemporal control over the location and extent of crosslinking, as well as post-gelation functionalization. However, free radicals employed in this polymerization method are prone to non-specific reaction with biomolecules, limiting their potential usage in biological systems and cell encapsulation. Here we present a unique photo-mediated oxime ligation, proceeding through a mechanism independent of propagating free radicals, as an improved tool for hydrogel photopolymerization. Spatial control over the traditional oxime bioorthogonal condensation between alkoxyamines and aldehydes is afforded through alkoxyamine photocaging. Upon cytocompatible UV light exposure, the photocage is cleaved to permit oxime ligation at user-defined regions in time and space. Solutions of four-arm poly(ethylene glycol) macromers separately functionalized with benzaldehyde and photocaged alkoxyamine moieties were found to successfully form hydrogels within minutes following short durations of UV light exposure. Modifying light intensity, exposure time, and aniline catalyst concentration enabled tunable gelation kinetics. Adjusting light intensity and exposure time additionally provided control over final hydrogel moduli. Futhermore, masked light exposure allowed precise control over patterned hydrogel geometry and biochemical functionalization. We expect hydrogel biomaterials formed and modified by photo-mediated oxime ligation will aid in establishing a more-complete understanding of cell-matrix interactions.

460175 Fabrication of Tailored Hydrogel Particles By Controlled Oxygen Inhibited Photopolymerization

Abstract:

Synthetic hydrogels of poly(ethylene glycol) diacrylate (PEGDA) have been widely used in tissue scaffolding, drug delivery, and single cell analysis applications. Microfluidic devices, in conjunction with in situ photopolymerization, are used to produce homogeneous and monodisperse micro- and nano- sized hydrogel particles. These devices are typically constructed of polydimethylsiloxane (PDMS), which has a high permeability to oxygen and other gases. Acrylated monomers undergo radical polymerization which is intrinsically vulnerable to inhibition by molecular oxygen, resulting in an incomplete cure and the formation of peroxyl radicals. In hydrogel droplets, this inhibition is manifested as an oligomer shell around a polymerized core as a consequence of a persistent steady state region of high oxygen concentration. While undesirable in some applications, we can take advantage of this phenomenon by washing away the oligomer shell and obtain a surfactant-free hydrophilic surface which can be chemically modified. Additionally, we can fabricate hydrogel particles that have a lower size boundary than the one determined by fluid dynamics within the device. Physical approaches such as purging with an inert gas and using higher light intensity are used to reduce oxygen inhibition and thus change the oligomer layer thickness. We have developed a time dependent model that compares photopolymerization kinetics to mass transport rates to predict the unpolymerized shell thickness under different operating conditions. We validated this model empirically, and have established the trends observed from varying inert gas pressure, purging time, and optical intensity. Using these results, we can easily modify the operating parameters to obtain polymerized particles with desired diameters from arbitrary-sized droplets.

469072 Injectable and Degradable Zwitterionic Polycarboxybetaine Hydrogels

Abstract:

Hydrogels constructed from zwitterionic polymers and crosslinkers based on polycarboxybetaine (pCB) offer unique advantages for drug delivery and tissue engineering applications. Because of their strong hydration and resistance to nonspecific interactions, they can act as scaffolds to maintain stem cell pluripotency, control stem cell self-renewal or precisely direct their fate [1], mitigate the foreign body reaction to implanted materials [2] and demonstrate self-healing abilities [3]. To realize injectable pCB hydrogels for these applications, we have developed chemical and physical crosslinking strategies that allow rapid *in situ* gel formation under physiological conditions and reversible viscoelastic changes. In one strategy, we exploited â€œzwitterionic fusionâ€ to form injectable hydrogels with dynamic mechanical properties that can reversibly change and self-heal in response to shear [3]. Several approaches to incorporate degradability into pCB hydrogels will be discussed and compared. Furthermore, cells, biologics or drug-loaded particles can be entrapped in these injectable zwitterionic hydrogels, which act as effective carriers and depots.

[1] T. Bai, F. Sun, L. Zhang, A. Sinclair, S. Liu, J.R. Ella-Menye, Y. Zheng and S. Jiang, "Restraint of the Differentiation of Mesenchymal Stem Cells by a Nonfouling Zwitterionic Hydrogel", Angewandte Chemie. Vol. 53, pp 12729-12734 (2014)

[2] L. Zhang, Z. Cao, T. Bai, L. Carr, J.R Ella-Menye, C. Irvin, B. Ratner, S. Jiang, "Zwitterionic hydrogels implanted in mice resist the foreign-body reaction", Nature Biotechnology. Vol. 31, pp 553-556 (2013)

[3] T. Bai, S. Liu, F. Sun, A. Sinclair, L. Zhang, Q. Shao and S. Jiang, "Zwitterionic fusion in hydrogels and spontaneous and time-independent self-healing under physiological conditions", Biomaterials. Vol. 35, pp 3926-3933 (2014)

467674 A Microfluidic-Based Cell Encapsulation Platform to Achieve High Long-Term Cell Viability in Photopolymerized Pegnb Hydrogel Microspheres

Abstract:

Cell encapsulation within photoinitiated polyethylene glycol (PEG) hydrogel scaffolds has been demonstrated as a robust strategy for drug delivery, tissue engineering, high throughput screening, and developing platforms to study cellular behavior and fate. The spatially and temporally specify PEG hydrogel properties, chemical functionalization, and cytocompatibility has advanced in recent years. Recent development in microfluidics has recently enabled the miniaturization of PEG hydrogels, thus making the fabrication of miniaturized cell-laden vehicles possible. However, the presence of oxygen dramatically inhibits the chain growth photopolymerization of polyethylene glycol diacrylate (PEGDA), thus limiting its application in microfluidics. Another promising PEG-based scaffold material, PEG norbornene (PEGNB), is formed by a step growth photopolymerization, which is not inhibited by oxygen. PEGNB is also more cytocompatible than PEGDA and allows for orthogonal addition reactions. The step-growth kinetics, however, are slow and therefore present a challenge to achieving fully polymerized microgel droplets within microfluidic devices. Here, we describe a microfluidic-based droplet fabrication platform that consistently generates monodisperse cell-laden water-in-oil emulsions. PEGNB droplets are collected and photopolymerized under UV exposure in bulk. In this work, we compare this microfluidic-based cell encapsulation platform with a vortex-based method in terms of controlling microgel size, uniformity, post-encapsulation cell viability and long-term cell viability. Several factors that potentially influence post-encapsulation cell viability are investigated. Finally, this platform is compared with a similar cell encapsulation platform using PEGDA in term of long-term cell viability. We show that this PEGNB microencapsulation platform is capable of generating cell-laden hydrogel microspheres at high rates with well-controlled size distribution and high long-term cell viability.

462160 *In Situ* Deposition of Cellular Hydrogels for Treatment of Inflammatory Bowel Disease

Abstract:

***In Situ* Deposition of Cellular Hydrogels for Treatment of Inflammatory Bowel Disease**

Meryem O. Pehlivaner[1], Theodore Lutkus[2], Adam K. Ekenseair[1]

[1]Department of Chemical Engineering, [2]Department of Bioengineering

Â Northeastern University, Boston, MA 02115, USA

Inflammatory bowel disease (IBD), most notably Crohnâ€™s disease and ulcerative colitis, currently afflicts millions of patients worldwide and leads to chronic inflammation of all or part of the digestive tract and damage to the intestinal lining. Current treatment plans focus on systemic or intestinal delivery of drugs; however these drugs frequently fail or are inadequate to prevent or reverse the damage. Over the long term, patients will require surgery to remove diseased sections of their bowels [1-5]. The purpose of this study is to develop a new class of colonoscope-based treatment options that specifically target the diseased area and locally deliver drugs and stem cells through sprayable application of a regenerative hydrogel during diagnostic procedures. Thermo-responsive hydrogel solutions can solidify immediately on warm body tissue after spraying and create a homogeneous therapeutic coating on the diseased tissue. The hydrogelâ€™s physical properties can be easily manipulated, and the use of spray deposition techniques will enable rapid and minimally-invasive application of homogenous, multilayer, and conformal coatings *in vivo*. Ultimately, the regeneration of diseased or damaged sections of the intestinal tract could prevent the need for bowel resection, lead to reduced symptoms, and provide a higher quality of life for patients.

In this study, poly(*N*-isopropylacrylamide)-based thermogelling and crosslinking polyamidoamine macromers were synthesized and combined to create a sprayable, *in situ* forming dual gelling hydrogel system for delivery of cellular hydrogel coatings onto warm tissues. The impact of spray technique, solution formulation, pressure, and viscosity on the viability of fibroblast cells encapsulated in hydrogels was investigated. Smooth, fast, and conformal hydrogel coatings were obtained when thermogelling macromers were sprayed with high polyamidoamine concentration at low pressure. Moreover, decreased early micronetwork formation, viscosity and pressure all improved cell viability in hydrogels. To further increase the survival rate of sprayed cells, a custom-designed airbrush was produced via 3D printing technology and viability of fibroblast cells in hydrogels was increased from 45% to 85%. *Ex vivo* tests showed that airbrush spraying is a versatile tool to create cellular hydrogel coatings onto intestinal tissue surfaces in less than a minute. The results demonstrated that sprayable, *in situ*forming hydrogels capable of delivering cell populations on diseased tissues offer promise as novel therapies for regenerative medicine.

References

[1] A. Sonnenberg, Occupational distribution of inflammatory bowel disease among German employees, Gut 31 (1990) 1037-1040.

[2] P.L. Lakatos, Recent trends in the epidemiology of inflammatory bowel diseases: up or down? World Journal of Gastroenterology 12 (2006) 6102.

[3] C.N. Bernstein, J.F. Blanchard, E. Kliewer, A. Wajda, Cancer risk in patients with inflammatory bowel disease, Cancer 91 (2001) 854-862.

[4] S.B. Hanauer, Inflammatory bowel disease: epidemiology, pathogenesis, and therapeutic opportunities, Inflamm. Bowel Dis. 12 (2006) S3-S9.

[5] C.N. Bernstein, A. Kraut, J.F. Blanchard, P. Rawsthorne, N. Yu, R. Walld, The relationship between inflammatory bowel disease and socioeconomic variables, Am. J. Gastroenterol. 96 (2001) 2117-2125.

467486 Microrheology As a Tool to Measure Cell-Material Interactions and Degradation of Covalently Adaptable Hydrogel Scaffolds

Abstract:

Microrheology measures material properties both spatially and temporally. This technique is an ideal measurement technique to understand the dynamic changes in a complex scaffold during degradation and how cells interact, remodel and degrade a 3D matrix. In this work we measure scaffold degradation of a covalently adaptable hydrogel, a poly(ethylene glycol) (PEG)-hydrazone scaffold, and cell-mediated remodeling of a PEG-peptide scaffold. We use multiple particle tracking microrheology (MPT) to characterize the spatial and temporally changes in both materials. In MPT, one micrometer fluorescently labeled probe particles are embedded in the matrix and the movement of these particles is tracked and related to rheological properties using the Generalized Stokes-Einstein Relation. Degradation of the covalently adaptable hydrogel is initiated by pushing the scaffold out of equilibrium by changing the pH of the incubating buffer. The scaffold is measured through time and oscillations in breakage and reformation of bonds is measured throughout the degradation reaction. The critical relaxation exponent and critical degradation time are identified with time-cure superposition. It is found that this scaffold undergoes breakage and reformation once during degradation at an acidic pH and repeatedly at physiological pH. Additionally, MPT is used to measure PEG-peptide scaffold degradation due to remodeling by human mesenchymal stem cells (hMSCs) encapsulated in 3D. In this work, MPT identifies that scaffold degradation is greatest furthest from the cell with the amount of cross-links inversely related to the distance from the cell. The sensitivity and ability to measure material properties both spatially and temporally with MPT has enabled the discovery that hMSCs will not degrade the scaffold directly under it to enable spreading and attachment prior to motility.

465633 Synthesis of Hollow Co_3O_4 Nanoparticles on Nitrogenâ€"Doped Porous Carbons for High-Performance Supercapacitors

Abstract:

Supercapacitors have been considered as next generation energy storage devices due to their high power density caused by rapid charge/discharge rates, long life cycle and eco-friendliness. In view of electrode materials, several strategies to improve performance of the supercapacitors were suggested and applied. We incorporated Co_3O_4 nanoparticles into nitrogen-doped porous carbons to investigate their potential as supercapacitors. The nano-sized Co_3O_4 particles grow and their morphology becomes hollow on the carbon support, which is attributed to the atomic diffusion (the Kirkendall effect) during the annealing. The cobalt-doped composites have high BET surface areas of about 3000 m^2 g^{-1} and possess mesopore-dominant structures caused by the hollow Co_3O_4 nanoparticles. As a result, the hollow Co_3O_4 in the composites simultaneously induces both enhanced pseudo-capacitance and efficient pore structure for energy storage. The specific capacitance of the composites is as high as 300 F g^{-1} in an aqueous solution of 6 M KOH at 1 A g^{-1}, and the material shows competitive specific capacitance of 210 F g^{-1} at 10 A g^{-1} for a high power density. The present study provides a unique insight into the synthesis of hollow nanoparticles on a porous substrate material for high-performance supercapacitance.

458006 Lecithin-Derived N-Doped Carbons for Supercapacitor Electrode Material

Abstract:

Soybean lecithin, plant derived surfactant, is a well-known natural surfactant in the animal tissue, food, and paints industry. Because of lecithin production is huge amount annually, it is urgent to find additional application of lecithin. In this work, an active supercapacitor electrode material is synthesized from lecithin-derived biomass material. Lecithin (as a carbon source) was first converted into bio-char via hydrothermal carbonization (HTC) then urea was introduced as a nitrogen source for biochar. Up to a 4 wt.% nitrogen intercalating into carbon was achieved. The morphology, thermal stability, and surface properties of lecithin, bio-char and N-doped carbon were examined. The resulting material show high surface area with micro/meso porosity. The capacitive characteristics of as-prepared materials were investigated by means of cyclic voltammetry (CV) (from 1.0 to 100 mV s^{-1}) and charge-discharge (from 1.0 to 20 A g^{-1} respectively. NGC-800 exhibited the highest capacitance value (411 F g^{-1} at 1 mV s^{-1} scan rate via cyclic voltammetry and 495 F/g at 1 A g^{-1} current density in 1 M KOH. Therefore, this study suggests that N-doped graphitic carbon materials synthesized from lecithin may be environmentally-sustainable candidate for energy storage applications.

452717 3D Porous Graphene Nanostructure Fabricated with a Simple, Fast, Scalable Process for Applications in High Performance Flexible Gel-Type Supercapacitors

Abstract:

A simple, fast, and scalable mix-and-heat process was developed for production of three-dimensional (3D) porous graphene nanostructure. The process involves only mixing and heating of starch and a graphene oxide (GO) suspension at 90 °C for 10 min to form 3D graphene monoliths, from which a three-dimensionally well-connected porous graphene nanostructure, starch/RGO, possessing a high specific surface area of 1519 m^2/g was obtained through common carbonization and activation treatments. No time-consuming hydrothermal or sol-gel chemistries and no energy-demanding supercritical fluid or freeze drying operations are needed to form the 3D graphene monoliths. The starch/RGO material was applied as the electrode material to fabricate flexible, gel-type symmetric supercapacitors of outstanding capacitive performances, delivering a high energy density of 19.8 Wh/kg at the power density of 0.5 kW/kg and exhibiting an excellent high rate capability of a high power density of 9.9 kW/kg at the energy density of 9.6 Wh/kg, among the highest for carbon material based, gel-type, symmetric supercapacitors. These excellent capacitive performances may be attributed to the advantageous structural features of the starch/RGO material, including the large surfaces exposed to the electrolyte ions for capacitance generation, highly porous structure for fast mass transfer of the electrolyte ions, and three-dimensionally well-connected conductive paths for the involved charge transport. The starch/RGO based supercapacitor exhibited outstanding mechanical stability with a retention rate of 90% in both energy and power densities at a large bending angle of 138°. It was also found that the starch/RGO based supercapacitor exhibited optimal performances in both energy and power densities at 80 °C and is thus applicable in a wide temperature range environment.

470638 Wet Spinning of Transition Metal Chalcogenide Fibers

Abstract:

Fibers comprising of molybdenum disulfide (MoS_2) were fabricated by a facile wet spinning procedure utilizing a coagulation bath of polyvinyl alcohol (PVA) for inclusion into flexible supercapacitors. Stable dispersions consisting of MoS_2, carbon nanotubes, and surfactant were prepared and characterized by Raman spectroscopy, transmission electron spectroscopy (TEM) and x-ray photospectroscopy (XPS). The latter of which indicates the presence of both exfoliated sheets and non-exfoliated nanoplatelets. Morphology of the fibers were characterized by scanning electron microscopy and electron diffraction spectroscopy (SEM/EDS) indicating that the exterior is principally comprised of MoS_2 nanoplatelets coated by carbon, the latter arising from the coagulation process. Meanwhile cross sections of the fiber reveal that the interior is predominantly MoS_2. The resultant fibers are durable, evident by the ability to be tied into knots, with mechanical properties similar to carbon nanotube fibers prepared in a similar fashion. Cyclic voltammetry measurements were performed in electrolyte solutions of H_2SO_4 and Na_2SO_4 in order to evaluate the ability of the wet spun fibers to serve as anodes in supercapacitor devices. Finally fibers prepared from tungsten disulfide (WS_2) were prepared to showcase the universality of this method towards other members in this class of transition metal chalcogenides (TMD).

468142 Dehydrogenation Mechanisms in Nanocrystalline Magnesium Hydride

Abstract:

Magnesium hydride is a promising material for on-board hydrogen storage due to its high gravimetric capacity (7.6 wt. %). However, its applicability is limited by slow dehydrogenation kinetics. Dehydrogenation mechanisms involve kinetics of several phenomena and the morphological changes of metal/hydride phase.

Dehydrogenation kinetics consists of nucleation and growth of magnesium phase by interfacial movement and/or by diffusion of H-atom through the hydride/Mg phase. Kinetics is studied using Johnson-Mehl-Avrami-Kolmogorov (JMAK) equation: $\alpha = 1\text{-}\exp(-kt^n)$, where α: converted fraction of Mg phase, k: kinetic factor involving nucleation density (N_0), growth velocity (U) and growth dimensionality (n) of Mg phase. n represents the morphological manifestation of Mg phase which can initially nucleate and grow as isolated colonies in 3D ($n \approx 3$). Eventual impingement of the neighbouring Mg colonies, can arrest growth in the lateral dimensions (i.e. $n < 3$). n and k along with other kinetic factors can explain the dehydrogenation mechanisms.

Dehydrogenation mechanism was investigated by performing isothermal experiments at 320-400 °C. The dehydrogenation curves show two distinct stages, namely, the incubation stage and post incubation stage. The incubation stage experiences slower kinetics and is observed predominantly at lower temperatures. Dehydrogenation mechanisms in the incubation and post incubation stages are studied by dividing the entire dehydrogenation curve into several small segments and applying JMAK analysis. This analysis yields discrete values of n and k which vary with time and their variation trends are different in the incubation and post incubation stages.

During the incubation stage, n increases from a negligible (n=0.01) to a significant (n=2.24) value. As n reflects the average growth dimensionality of the Mg phase, its increase can be attributed to the increase in number of growing Mg colonies with time. The activation energies for nucleation (10 kJ/mol) and growth (213 kJ/mol) are estimated from N_0 and U values, respectively. The activation energy of nucleation is one order lower than that for growth suggesting that Mg nucleation is fast while its growth is slow in the incubation stage.

In the post incubation stage, n values decrease with time from a high value ($n>1$) to a negligible value ($n<0.50$). The decreasing n can be attributed to the changing growth morphology of Mg phase during dehydrogenation. Microscopic observations (Dark Field TEM) and the estimation of volume fraction of Mg/MgH_2 phases shows that Mg phase nucleates and grows mainly near the particle surface and partly in the interior. During this stage, the interface velocity (U) changes by ~1 order. The estimated activation energies for nucleation and growth in post incubation stage are 55 kJ/mol and 152 kJ/mol respectively. This indicates that Mg phase growth limits dehydrogenation in the post induction stage.

The overall activation energy for dehydrogenation estimated using Raman spectroscopy is 158 kJ/mol. This is close to the activation energies for growth in both the incubation and post incubation stages, suggesting Mg phase growth as the controlling step during dehydrogenation of MgH_2.

473136 Effects of Surface Modified Ferrofluids on Energy Induction in Oscillating Heat Pipes

Abstract:

Co-precipitation and surface functionalization of ferrofluids containing nanoscale ferrous oxide and cobalt oxide was performed in order to gain stable ferrofluids without compromising magnetization capabilities. Both as synthesized and surface modified nanoparticles in an aqueous carrier solvent were used as the working fluids in oscillating heat pipes (OHP) for thermal energy transportation and electrical energy conversion. This novel scheme uses ferrofluidic induction to generate power via pulsating movement of ferrofluid, due to an axial thermal gradient and associated phase change, through an in-line solenoid between two magnets. To avoid nanoparticle aggregation in the ferrofluid during long usage periods and thermal cycling, a number of different ligands were examined as a surface modifiers to maintain the nanoparticles in solution, prevent agglomeration, and maximize the voltage signal from the solenoid. Characterization of the ferrofluid samples was performed using ATR-FTIR spectroscopy, dynamic light scattering (DLS), X-ray diffraction (XRD), transmission electron microscopy (TEM), atomic/magnetic force microscopy (AFM/MFM), and magnetometry. Cobalt ferrite nanoparticles surface-modified with a variety of different small molecule ligands (e.g., citric acid) demonstrated good magnetic strengths and also generated voltages close to those of the as synthesized (without surface modification) ferrofluids, while maintaining dispersion. A maximum effective thermal conductivity of 12.9 kW/m×K at 470 W power was measured for a non-harvesting ferrofluid test. The effectiveness of ferrofluids as OHP working fluids will be discussed along with the implementation of bias magnets in the harvesting configuration which enhanced OHP thermal conductivity by 11%.

471012 Heat Transfer in Metal Organic Frameworks during Gas Adsorption

Abstract:

Fast heat transfer during gas adsorption into metal-organic frameworks (MOFs) plays a key role in improving their efficiency for gas storage. Despite its importance, full understanding of heat transfer during gas adsorption into MOFs is still lacking. Using molecular modeling, we have investigated the transient phenomena related to heat and mass transfer within MOFs during gas adsorption. Our study reveals that heat transfer in MOF occurs much faster than gas diffusion. However, poor thermal transport within pure gas region prevents fast release of heat generated during adsorption. With respect to the effect of MOF size on its storage efficiency, we show that the density of adsorbed gas for bigger MOFs is lower due to higher temperature rises occurring during gas adsorption. Moreover, adsorption time is longer for bigger MOFs with respect to smaller MOFs.

465512 Capillary-Driven Clustering of Semiconductor Nanorods into End-to-End Network and Vortex Structures

Abstract:

Semiconductor nanorods in the strong quantum confinement regime exhibit both size-dependent energy gap and anisotropy-derived optical properties, such as linearly polarized emission. Self-assembly of these elongated nanomaterials into long range order has been pursued for long in the past, only to garner relatively marginal progress. In this presentation, the capillary force between nanorods at the assembly interface will be discussed in the context of long-range clustering of of CdSe nanorods into two-dimensioanl end-to-end network. In addition, nanorods that cluster into upright assembly often exhibit vortex structures, which will ultimately hamper perfect arrangement of nanorods into long range. The vortex formation has been discussed in previous literature; however, in most cases, the mechanism with which the vortices form has been left to speculation. In this regard, we investigated the effect of solvent polarity on the formation or prevention of vortex structures. In this presentation, we will discuss the implication of the vortexes from the viewpoint that the nanorod-to-nanorod distance will affect carrier dynamics via inter-nanorod energy transfer.

References

[1] Kim D, Kim W D, Kang M S, Kim S-H, Lee D C 2015 *Nano Lett.* **15** 714

[2] Kim W D, Chae W.-S., Bae W K, Lee D C 2015 *Chem. Mater.* **27** 2797

463010 Three-Phase Self-Assembly of Gold Nanoparticle Monolayers – Overcoming Ligand Size Limitations

Abstract:

A simple yet effective technique was developed to prepare centimeter-scale monolayer gold nanoparticle (Au NP) films of long-range order with hydrophobic ligands. Aqueous Au NPs were entrapped at an oil/water interface where the Au NP surface could be modified with various types of amine-terminated ligands, including alkanes and polystyrene. The Au NPs then spontaneously translocated to an air/water interface to form a monolayer film. The spontaneous formation of an Au NP film at an air/water interface was due to the minimization of the system Helmholtz free energy. The self-assembled Au NP films are of a hexagonal close packed structure. The interparticle center-to-center space was dictated by the amine ligand length. These Au NP monolayers exhibit tunable surface plasma resonance and excellent spacial homogeneity for surface-enhanced Raman-scattering. The "air/water/oil" self-assembly method developed here not only benefits the fundamental understanding of NP ligand conformations, but is also applicable to scale up the manufacture of plasmonic nanoparticle devices with precisely designed optical properties.

449692 Self-Assembly of Directionally Interacting Spheres and Rods

Abstract:

There has been considerable recent interest in understanding the self-assembly of mixtures of differently shaped objects. Non-directional entropic and energetic effects are known in the specific case of rods mixed with spheres to exclusively yield layered phases. [1,2] However, here we use computer simulations and theory to demonstrate that the introduction of directional energetic attractions, between pairs of rod ends and between rod ends and isotropically interacting spherical nanoparticles (NPs), can be used to intelligently stabilize specific crystal morphologies through the interplay of energetic and entropic effects. To illustrate, we rationally design rods and spheres to obtain either face-centered cubic (FCC) or hexagonal close-packed (HCP) NP crystals at will. By tuning the relative size of the rods and spheres we control the interaction between non-nearest neighbor NPs in these crystals allowing us to control their relative stability with respect to each other, and with respect to an amorphous phase. Experiments with mixtures of gold spheres isotropically decorated with single stranded DNA (ssDNA) and rods with complementary ssDNA ends, unequivocally verify this behavior. We therefore propose that directionally specific attractions, which reflect only some features of the building block anisotropies, offer an entrée into completely different classes of self-assembly behavior.

References

[1] Adams, Dogic, Keller, and Fraden, Nature 393, 349-352 (1998).
[2] Dogic, Frenkel, and Fraden, Phys. Rev. E. 62, 3925-3933 (2000).

473985 Recovering Dynamic Parameters of Nanoparticle Assembly from Disjointed Images of Nanoparticle-Polymer Composites

Abstract:

The incorporation of nanoparticles (NPs) into polymers constitutes a powerful strategy for enhancing their thermomechanical properties and for introducing new optical, electrical, and magnetic functionalities into the polymers. Here we will describe an approach for inferring dynamic parameters of NP assembly from spatially and temporally disjointed images of NP-polymer composites [1]. The approach involves adjustment of the parameters of a kinetic model of assembly until the size statistics of NP clusters computed from the model match those obtained from high-throughput analysis of the experimental images [2]. Application to shaped, metal NPs in polymer films reveals that NP structures grow via a cluster-cluster aggregation mechanism, where NPs and their clusters diffuse in a Stokes-Einstein manner and stick with a probability that depends strongly on the size and shape of the NPs and the molecular weight of the polymer.

[1] C. R. Murthy, B. Gao, A. R. Tao, and G. Arya, "Dynamics of nanoparticle assembly from disjointed images of nanoparticle-polymer composites," *Physical Review E*, 93, 022501 (2016).

[2] C. R. Murthy, B. Gao, A. R. Tao, and G. Arya, "Automated quantitative image analysis of nanoparticle assembly," Nanoscale, 7, 9793-805 (2015)

467512 Generalized Mechanistic Model for the Chemical Vapor Deposition of 2D Transition Metal Dichalcogenide Monolayers

Abstract:

Transition metal dichalcogenides (TMDs) like molybdenum disulfide (MoS_2) and tungsten disulfide (WS_2) are layered materials capable of growth to one monolayer thickness *via* chemical vapor deposition (CVD). Such CVD methods, while powerful, are notoriously difficult to extend across different reactor types and conditions, with subtle variations often confounding reproducibility, particularly for 2D TMD growth. In this work, we formulate the first generalized TMD synthetic theory by constructing a thermodynamic and kinetic growth mechanism linked to CVD reactor parameters that is predictive of specific geometric shape, size, and aspect ratio from triangular to hexagonal growth, depending on specific CVD reactor conditions. We validate our model using experimental data from Wang *et al.* (*Chem. Mater.*, 2014, 26 (22), 6371-6379) that demonstrate the systemic evolution of MoS_2 morphology down the length of a flow CVD reactor where variations in gas phase concentrations can be accurately estimated using a transport model (C_{Sulfur} = 9–965 μmol/m^3; C_{MoO3} = 15–16 mmol/m^3) under otherwise isothermal conditions (700 °C). A stochastic model which utilizes a site-dependent activation energy barrier based on the intrinsic TMD bond energies and a series of Evans-Polanyi relations leads to remarkable, quantitative agreement with both shape and size evolution along the reactor. The model is shown to extend to the growth of WS_2 at 800 °C and MoS_2 under varied process conditions. Finally, a simplified theory is developed to translate the model into a "kinetic phase diagram" of the growth process. The predictive capability of this model and its extension to other TMD systems promise to significantly increase the controlled synthesis of such materials.

467605 Structure-Dependent Stability of Magic-Number Thiolated Metal Nanoparticles

Abstract:

Magic numbers in the stability of $M_n(SR)_m$ nanoparticles (NPs) in Brust-Schiffrin-type syntheses have been recently identified with advancements in synthesis and characterization techniques.[1] The atomically-precise nature of these particles makes them attractive candidates for understanding the thermodynamic and colloidal stability of NPs in solution. Traditionally, this magic stability has been explained by electron shell closure methods such as the "superatom model".[2] However, this model has not held in rationalizing the stability of several recently-identified NP structures, such as the $Au_{20}SR_{16}$ and the $Au_{36}SR$ introduce a theory developed using density functional theory calculations relating the exceptional stability of thiolated-metal NPs to structure-dependent, thermodynamic descriptors. Using the identified descriptors we rationalize the stability of various monometallic and multimetallic atomically-precise nanoparticles both for charged and neutral systems. Our theory is able to capture a series of experimentally observed stability trends in nanoscale thiolate-protected metal NPs. The thermodynamic implications of our findings help rationalize the observed structures of NPs and help to inform thiolate ligand selection in NP synthesis. Our findings aid in accelerating the discovery of atomically precise, stable metal nanoparticles.

1. Jin, R. Atomically precise metal nanoclusters: stable sizes and optical properties. *Nanoscale* **15,** 1549–1565 (2015).

2. Walter, M. *et al.* A unified view of ligand-protected gold clusters as superatom complexes. *Proc. Natl. Acad. Sci. U. S. A.* **105,** 9157–9162 (2008).

3. Zeng, C., Liu, C., Chen, Y., Rosi, N. L. & Jin, R. Gold-Thiolate Ring as a Protecting Motif in the Au20(SR)16 Nanocluster and Implications. *J. Am. Chem. Soc.* **20,** 16–19 (2014).

466716 Rapid Microwave Assisted Synthesis of Zinc Oxide Nano-Forest for Solar Cell Applications

Abstract:

Nanomaterials based photo-electro-chemical solar cells require high surface area as well as high electron transport quality to achieve good efficiency. Nanowires or hierarchical nanostructured arrays fulfill these two requirements. Synthesis of such complicated structures are conducted either using vapor phase deposition technique or lengthy wet-chemical route. We have synthesized Zinc oxide nano-forest structures on FTO glass using microwave assisted wet-chemical route. FESEM images revealed formation of highly ordered nanoforests structure. The main stem of zinc oxide nanowires have diameter around 100-150 nm and length about 10 μm, whereas the branch nanorods have diameter less than 50 nm and length about 1 μm. Instead of days for preparation using conventional wet-chemical routes, such nano-forest structures were prepared within 4 h using microwave assisted synthesis. XRD study revealed such nano-structures are highly crystalline c-axis oriented materials. Photo-sensitizer dye molecules were loaded on such semiconductor nanostructures to fabricate photo-anode for dye-sensitized solar cells. ZnO nanoforests based such solar cells showed enhanced photovoltaic efficiency.

470005 Iron OXIDE Nanoparticle-Graphene Patterned Interfaces

Abstract:

IRON OXIDE NANOPARTICLE-GRAPHENE PATTERNED INTERFACES

Abhilasha Dehankar[1], Justin Young[2], Josh Goldberger[3], Ezekiel Johnston-Halperin[2], Jessica O. Winter[1,4]

[1]William G. Lowrie Department of Chemical and Biomolecular Engineering, [2]Department of Physics, [3]Department of Chemistry and Biochemistry, [4]Department of Biomedical Engineering

Nanoparticles, such as superparamagnetic iron oxide nanoparticles (SPIONs), exhibit unique properties because of their small size; and therefore, have significant potential to improve optical, electronic, and magnetic devices. Significant research has occurred in the synthesis of these particles; however, integration of these particles with higher order constructs, as required for device development, remains a challenge. Further, such integration could lead to the development of synergistic structures with novel properties. To harness the full potential of these composites, nanoparticles should be capable of being assembled in a specific, user-defined, and scalable manner on larger film-based materials. Therefore, the broad aim of this research is to develop facile and controlled techniques for generation of nanoparticle-film composites and to study the emergent properties of their interfaces.

As a model system, we are investigating graphene-SPION interfaces. Graphene is a diamagnetic single layer of carbon with excellent electronic properties. However, it lacks intrinsic magnetic ordering, which could be exploited for magneto electronic or spintronic functionality. Existing strategies for generating magnetism in graphene have disrupted native electronic properties. However, recent investigations have shown induced magnetism in graphene without impairing its electronic properties when magnetic materials are placed in close proximity with graphene films[1], [2]. Thus, we deposited SPIONs, which can locally modify the magnetic field felt by charge carriers in the graphene, with the goal of creating ordered assemblies with emergent magnetic properties. As a first approach, we deposited materials through direct deposition and determined that solvent properties significantly influence aggregation state. To create ordered structures, we first encapsulated SPIONs in micelles, that can in turn self-assemble on thin films following Langmuir-Blodgett[3] and spin coating deposition[4]. These materials were characterized using Atomic Force Microscopy (AFM) to evaluate composite surface topography. Magnetic and electronic properties of these composites would be evaluated. Knowledge gained from this research could enable novel 0D-2D interfaces. Similar techniques could be applied to a broad array of nanoparticles and surfaces to generate novel emergent behaviors.

1. Cavalcanti-Adam, E.A., et al., *Cell spreading and focal adhesion dynamics are regulated by spacing of integrin ligands.* Biophys J, 2007. **92**(8): p. 2964-74.

2. Wang, Z., et al., *Proximity-induced ferromagnetism in graphene revealed by the anomalous Hall effect.* Phys Rev Lett, 2015. **114**(1): p. 016603.

3. Spatz, J.P., et al., *Ordered Deposition of Inorganic Clusters from Micellar Block Copolymer Films.* Langmuir, 2000. **16**(2): p. 407-415.

4. Liou, J.-Y. and Y.-S. Sun, *Monolayers of Diblock Copolymer Micelles by Spin-Coating fromo-Xylene on SiOx/Si Studied in Real and Reciprocal Space.* Macromolecules, 2012. **45**(4): p. 1963-1971.

In bacterial assays, Gram-positive *S. aureus* (ATCC#25923) and MRSA (ATCC#43360) were incubated in 0.3% tryptic soy broth (TSB) media containing the self-assembled nanoscale supramolecules of peptide amphiphiles at concentrations from 20 to 100 μM at 37°C, and the growth curves were monitored by the optical density of a wavelength at 562 nm (O.D. at 562 nm) for 20 h. Also, to evaluate bactericidal effects, the self-assembled supramolecules of peptide amphiphiles at concentrations from 20 to 100 μM in 0.3% TSB were used to treat *S. aureus* and MRSA for 4 h in 37°C, and the viability of bacteria were quantified by colony forming units per ml (CFU/ml) after treatment. The viability of bacteria was evaluated by: [(CFU of treated sample)/ (CFU of untreated control sample)]*100%. All experiments were run in triplicate and repeated at least two times ($N \geq 2$) to demonstrate significance.

3. Results and Discussion

The TEM characterization showed successful self-assembly of the peptide amphiphiles discussed above. The self-assembled LEM nanofibers were about 10 nm in diameter and microns in length; the ACA nanorods had similar diameters but were shorter in length to about 100 nm, which probably resulted from the high cationic charge of the Cardin antimicrobial motifs. The hybrid nanofibers, which were combined by the LEM nanofibers and ACA nanorods, also exhibited diameters to about 10 nm and had a highly aggregated network structure, and the length of these nanofibers ranged from about 200 nm to 500 nm. The results of DLS measurements showed that although the hydrodynamic diameter of the LEM nanofibers (603 nm) was much larger than that of ACA nanorods (17 nm), the hybrid nanofibers had uniform hydrodynamic diameters (26 nm), indicating the successful co-assembly of the two peptide amphiphiles.

In bacterial assays, the ACA nanorods significantly inhibited the growth of both *S. aureus* and MRSA, as well as had high bactericidal effects towards these two bacterial strains. The 20 h bacterial growth curve of bacteria showed that the ACA nanorods at a concentration of 100 μM delayed the lag phase of *S. aureus* and MRSA to 18 h, whereas the untreated bacteria entered the exponential growth phase after the lag phase within 3 h. Also, the ACA nanorods at concentrations higher than 80 μM reduced the survival rate of both *S. aureus* and MRSA to as low as 10% compared with the untreated control sample. The results demonstrated for the first time that the self-assembled ACA nanorods had promising antibacterial properties against Gram-positive *S. aureus* as well as MRSA, suggesting that these self-assembling peptide amphiphiles could combat bacterial drug-resistance. In our ongoing studies, the antibacterial effects of the hybrid nanofibers in cooperation with a collagen-mimetic group, will be further elucidated.

459460 Design of 3-Helix Micelles with Tailorable Sizes and Shapes

Abstract:

Hierarchical self-assembled micelles from amphiphilic hybrid 3 helix peptide-polymer-lipid conjugates have shown great promise as nanocarriers in delivering therapeutic compounds in a targeted fashion. In order to realize the vast therapeutic potential of these micellar nanocarriers as well as the purpose of penetrating deep tissue or other biological barriers, there is a need to understand how PEG conjugation affects the micelle shape and size as well as self-assembly mechanisms. Atomistic molecular dynamics (MD) simulations of the self-assembly behavior in such length scales is challenging, which necessitates the development of a simpler coarse-grained (CG) model. First, we use a CGMD model based on the dissipative particle dynamics (DPD) technique to reveal the internal structure of 3-helix micelle and the behavior of conjugated PEG chains [1]. Next, we study the self-assembly patterns of the amphiphiles with different PEG molecular weights and conjugation positions on peptide bundles by considering the micelles aggregation number in the simulation box. We discover that the micelle size and stability is dictated by a competition between the entropy of confined PEG chains in the conjugation state, as well as intermolecular interactions among PEG chains that promote cohesion between neighboring amphiphiles. Furthermore, a computational phase diagram that can be used to design 3-helix micelles is conducted. This work opens pathways to control the morphology of self-assembled micelles with expectable and tailorable size, shape and stability as nanocarriers in drug delivery and other applications.

REFERENCES

[1] Ang, J.; Ma, D.; Lund, R.; Keten, S.; Xu, T. Internal structure of 3-helix micelle revealed by small-angle neutron scattering and coarse-grained MD simulation (In submission).

464817 Bioactive DNA-Peptide Nanotubes As Artificial Extracellular Matrices for Bone Tissue Engineering

Abstract:

Introduction: The extracellular matrix (ECM) is a complex network composed of collagen, fibronectin and other macromolecules that not only provides structural support but also relays crucial biochemical and biomechanical signals[1]. It dictates how cells attach, behave and interact with one another, thus, it plays a key role in tissue morphogenesis, differentiation and repair. In this context, the construction of artificial matrices that mimic the structure of the naturally occurring ECM and display biological signals might prove to be highly promising in controlling tissue regeneration[2]. Although electrospun fibers, polymeric hydrogels and self-assembled peptide amphiphiles have been widely investigated for cell adhesion and proliferation, thereâ€™s a need for scaffolds that are structurally programmable and easy to modify. DNA has emerged as one of the most promising building blocks for such nanoconstruction due to the high predictability of Watson-Crick base pairing, enabling the generation of one, two, three-dimensional assemblies for templated growth of nanowires, the organization of nanoparticles and as drug delivery carriers[3]. In the present study, DNA nanotubes were constructed by thermal annealing based on double-crossover motifs[4] and were further covalently functionalized with a BMP-7 derived peptide. The structure of the nanotubes was inspired by collagen with similar size and stiffness, and the BMP-7 derived peptide was selected because of its role in regulating the proliferation, differentiation and apoptosis of bone cells. The objective of the present study was to develop an extracellular microenvironment that can mimic the structural and regulatory characteristics of natural ECMs. Results showed that the DNA nanotube coated glass coverslips revealed a tangled network of fiber-like materials that resemble the morphology of natural ECM, suggesting DNA-peptide nanotubes could be a promising candidate as artificial extracellular matrices. Further experiments are still need to fully investigate their effects on osteoblast cell adhesion and proliferation.

Materials and Methods: Preparation of oligonucleotides stocks: All DNA strands were purchased with polyacrylamide gel electrophoresis (PAGE) purification from Integrated DNA Technology (IDT), dissolved at 500 ÂµM in IDTE buffer (IDT) and stored at -20 Â°C prior to use. For generating fluorescent tubes, the S3 sequence was ordered from IDT with a 3â€™-fluorescein (FAM) modification. Preparation of DNA-peptide hybrid conjugates: The BMP-7 derived peptide Ser-Asn-Val-Ile-Leu-Lys-Lys-Tyr-Arg-Asn (SNVILKKYRN) was conjugated to a DNA strand by Biosynthesis using a previously reported[2] Huisgen 1,3-dipolar cycloaddition reaction between a strained cyclooctyne and an azide. The resulting conjugate was purified by multiple high-performance liquid chromatography (HPLC) and characterized by mass spectrometry. DNA nanotube annealing: To form the DNA nanotubes, oligonucleotide stocks were further diluted to 50 µM in a TAE/Mg buffer (40 mM Tris, 20 mM acetic acid, 1 mM EDTA and 12.5 mM $Mg(OAc)_2$) and equal amounts of all constituent strands were mixed. The mixture was annealed from 95 °C to 25 °C over ~6 hours in a thermal cycler. After annealing, the solution was stored at 4 °C. Preparation of APTES-treated glass coverslips: In order to bind negatively charged DNA nanotubes, glass coverslips were treated with (3-Aminopropyl) triethoxysilane (APTES). The resulting coverslips had a positive charge at physiological pH. Briefly, to coverslips (12 mm in diameter) in a 12-well tissue culture plate, 100 µl of APTES were added into the spaces between wells (600 µl total). The plate was sealed with two layers of tape, wrapped in double layer of aluminum foil and then placed in oven at 90 °C for 18 hours. The coverslips were washed by ethanol and then DI water. Excess water was blown off by nitrogen gas and the coverslips were dried in a vacuum oven at 115 °C. Surface coating: To coat the surface of glass coverslips, 400 µl of the nanotube stock solutions were added on APTES-treated glass coverslips in a 24 well-plate and were incubated at 37 °C overnight to allow adhesion on the surface. Following incubation, the solution was removed and the coverslips were rinsed twice with sterile PBS before seeding cells. Transmission electron microscopy: To characterize size and morphology of DNA nanotubes, they were visualized using transmission electron microscopy operated at 80 kV. Briefly, 10 µl of the stock solution was applied on 300-mesh copper-coated carbon grids for 5 min. Following wicking and rinsing with deionized water, the grids were then stained with 10 µl of a 1.5% uranyl acetate solution twice for 10 s and dried for 15 min prior to imaging. Fluorescent imaging of the surface coating: Following annealing, the DNA nanotube solution was coated on APTES-treated coverslips according to the method described in the surface coating section. Images of the coatings were taken using a fluorescence microscope under 40X objective. A Z-stack was acquired to estimate the approximate thickness of the coating and a 3D reconstruction of the z stack was obtained using a 3D viewer plugin in ImageJ. Cell viability assay: To determine cell viability, human osteoblasts (HOB, Promocell) were seeded at a density of 20,000 cells/cm^2 in 96-well plate and subsequently co-cultured with the DNA nanotubes at various concentrations (0/1.25/2.5/3.75/5 µM) for 24 hours under standard cell culture conditions (37 °C, humidified, 5% CO_2/95% air) . MTS assays were used to determine cell density after incubation. Briefly, 20 µl of MTS dye solution was added per 100 µl of solution, and the absorbance readings were taken at 490 nm after 4 hours of incubation. Cell density was then determined with correlation to a standard curve. Statistical analysis: All experiments were conducted in triplicate and repeated at least three different times. Statistical differences were determined using analysis of variance (ANOVA) where $p< 0.01$ was considered statistically significant.

Results and Discussion: TEM images confirmed that the rationally designed DNA tiles can self-assemble and form long tubelike structures about 10-12 nm in diameter and often many micrometers in length with both walls and central pore clearly visible. Incorporating hairpin into one of the strands yields an almost indistinguishable structure, indicating that modifications at selected positions are not likely to affect tube formation. This offers the largest advantage of this system over other alternatives in that there is potential to vary the biological epitope without affecting the core structure. Stability of the nanotubes and nanotube-peptide hybrids against temperature, cation depletion and nuclease degradation were shown to be better than DNA double helices using TEM.The fibrillar morphology of the coating resembles that of extracellular matrices, indicating the potential of these DNA nanotubes to act as mimics of a native extracellular matrix. Finally, these nanotubes were shown to have no toxic effect on osteoblast cells with a concentration of up to 5 µM. Conversely, at low concentrations, specifically 1.25 and 2.5 µM, increases in cell density were observed.

Conclusions: Through above experiments, DNA nanotubes self-assembled from DNA tiles were successfully designed, constructed and characterized. The ECM-like fibrillar morphology and the minimal toxicity to osteoblast cells indicated their potential to be used as improved artificial extracellular matrices.

References:

1. Frantz C, Stewart KM, Weaver VM. The extracellular matrix at a glance. *J Cell Sci*. 2010;123(24):4195-4200.

2. Stephanopoulos N, Freeman R, North HA, et al. Bioactive DNA-peptide nanotubes enhance the differentiation of neural stem cells into neurons. *Nano Lett*. 2015;15(1):603-609.

3. Aldaye FA, Senapedis WT, Silver PA, Way JC. A Structurally Tunable DNA-Based Extracellular Matrix. *J Am Chem Soc*. 2010;132(42):14727-14729.

4. Rothemund PWK, Ekani-Nkodo A, Papadakis N, Kumar A, Fygenson DK, Winfree E. Design and Characterization of Programmable DNA Nanotubes. *J Am Chem Soc*. 2004;126(50):16344-16352.

464853 Colloidal Directed Assembly of Pi-Conjugated Oligopeptides for Supramolecular Electronics

Abstract:

The directed mesoscale engineering of nanoscale building blocks holds enormous promise to catalyze a revolution in new functional materials for advanced electronics. Bio-inspired systems can play a key role in this effort due to their inherent "programmable" function. In nature, biological polymers form chemically complex and functional heterostructures that underlie life processes. From this perspective, the ability to efficiently harness the assembly of synthetic peptides could offer effective pathways to produce functional materials. In this work, we study the self-assembly and directed assembly of non-natural peptides containing conjugated synthetic cores. In these materials, pi-conjugated units are flanked by symmetric oligopeptides with defined sequences, thereby yielding materials that can assemble into hierarchical structures via straightforward triggering with acidic solutions. We employ a general and facile strategy to induce self-assembly of synthetic peptides by simple evaporation of polar aprotic solvents. Notably, macroscopic peptide fiber arrays were crafted by confining the evaporation of peptide monomer solutions within preformed colloidal microchannels. Following assembly, we characterize the optical properties of peptide fiber arrays containing pi-conjugated cores using a variety of techniques. We further interrogate the electronic properties of these materials by integrating synthetic peptide assemblies as the active layer in a field-effect transistor (FET). In this way, the present study offers new insights into the design and fabrication of supramolecular electronics.

464997 A Self-Assembled pH-Responsive Multi-Component Platform for Oral Vaccination

Abstract:

Next-generation vaccine research must achieve safe, cost-effective and easily administered vaccines with 1) needle-free delivery mechanism, and 2) immunization with subunit vaccines. Oral vaccines improve ease of administration and confer protection at mucosal surfaces for more robust immune responses. Subunit vaccines (e.g. protein antigens) are desirable to overcome safety issues associated with current attenuated or inactivated oral vaccines, but are poorly immunogenic and absorbed, requiring the use of adjuvants. We have previously demonstrated success using anionic hydrogel carriers to deliver protein therapeutics, such as insulin and calcitonin, in a pH-dependent manner for targeted intestinal delivery. Successful vaccination strategies require direct antigen internalization and elicitation of immunogenic responses.

We have developed a system that combines two biomaterial platforms. In its core, polyanhydride nanoparticles (PNPs) based on 1,6-bis-(*p*-carboxyphenoxy)hexane (CPH) and sebacic acid (SA) serve simultaneously as adjuvant and delivery vehicle of subunit antigens. These are protected by pH-responsive polymers based on poly(ethylene glycol) (PEG) and poly(methacrylic acid) (PMAA), shielding a payload of PNPs through the harsh conditions of the gastrointestinal tract for selective release in the small intestine, proximal to antigen-sampling M cells. This platform is referred to as Polyanhdyride-Releasing MicroParticle Technology, or PROMPT.

PROMPT formulations were synthesized by pH-mediated anti-solvent precipitation to encapsulate PNPs. Briefly, the pH-responsive copolymer, denoted P(MAA-g-EG), was synthesized by UV initiated free radical polymerization of MAA and polyethylene glycol methacrylate and characterized by ^{13}C NMR, FT-IR, and GPC. PNP formulations were synthesized by anti-solvent nanoprecipitation to encapsulate model protein antigen, ovalbumin. P(MAA-g-EG) dispersions were flocculated in the presence of PNPs by desolvating the polymer through pH reduction. PROMPT

particle morphology and PNP encapsulation were evaluated by scanning electron microscopy and confocal microscopy, respectively. *In vitro* release kinetics were evaluated at physiological conditions and quantified using the micro bicinchoninic acid (microBCA) assay. Enzyme-linked immunosorbent assay (ELISA) was used to evaluate that antigenicity of ovalbumin released from PROMPT formulations. Cytocompatibility of self-assembled PROMPT formulations was evaluated in two cell lines: Caco-2 intestinal epithelial cells and RAW264.7 macrophages. RAW264.7 cells were also used to evaluate PNP internalization. Results indicated that PROMPT formulations can achieve targeted release of PNPs in intestinal conditions. Formulations preserve protein antigenicity, and nanoparticles released from PROMPT are readily internalized by macrophages, indicating PROMPT has potential as an adaptable platform for oral subunit vaccine administration.

Acknowledgements: This work was supported by a grant from the National Institutes of Health (5-R01-EB-000246-20) and a NSF GRFP. We thank the research group of Prof. Balaji Narasimhan from Iowa State University for their collaboration on this project.

468606 Filomicelles Self-Assembled from Degradable Di-Block Copolymers Deliver Retinoids and Chemotherapeutics in Durable Control of Carcinoma Cell Fate

Abstract:

Introduction: Conventional injection of semi-soluble drugs hit both healthy and tumor cells, causing side effects that limit dose. While this approach might be broadened with nanocarriers (such as flexible cylindrical micelles or filomicelles that have been previously shown to circulate longer and accumulate better *in vivo* [1]), the lack of aromatic groups in the core of the nanoparticles may hamper it from reaching peak loading of common aromatic chemotherapeutics.

Results 1: In order to improve the delivery of aromatic drugs in filomicelles, aromatic groups were integrated into the hydrophobic block of a degradable di-block copolymer. Aromatic filomicelles were formed by self-assembly of biodegradable amphiphilic di-block copolymer polyethylene glycol- poly-Î±-benzyl caprolactone (PEG-PBCL) with suitable block ratios. and loaded with a common chemotherapeutic, Paclitaxel. Aromatic filomicelles loaded more Paclitaxel than analogous aliphatic systems. Cell death and the aneuploidy in surviving cells (which indicates toxicity) were higher in both liver and lung carcinoma lines *in vitro* compared to aliphatic filomicelles. Importantly, the filomicelles themselves were bio-compatible and led to minimal cell death. Initial tests *in vivo* also suggested more potent tumor shrinkage, with minimal effects on thrombocrit and hematocrit.

Thus, aromatic filomicelles demonstrated a marked an improvement over the corresponding aliphatic nano-carriers [2]. However, despite inducing higher levels of aneuploidy in surviving cancer cells, the cells reverted back to the original proliferative state over several periods of the normal cell cycle. This is consistent with relapse commonly seen clinic. Retinoic acid (RA) and other retinoids regulate RA receptor transcription factors that induce differentiation and arrest

proliferation of many cell types, including cancer cells [3]. Combining RA with select chemotherapeutics has produced durable cures of select cancers, notably pro-myeloblastic leukemia (PML) where RA differentiates cells while chemotherapeutic kills the cancer stem cell [4].

Results 2: The above mentioned filomicelles were loaded with Paclitaxel, as well as RA, for testing effects on cancer cell lines *in vitro* and *in vivo*. Across multiple liver and lung carcinoma lines, we find dual treatment with RA plus Paclitaxel increases, aneuploidy, Lamin-A (a marker of differentiation) levels and cell death beyond those achieved by either drug single-handedly, with effects being durable. A month after treatment, relapse rates are low for RA-Paclitaxel treated cells (15%), compared to almost all (92%) for cells treated with Paclitaxel alone. Decreases in levels of key cell cycle factor Cyclin-D1 and particularly proliferation marker Ki-67 help clarify the basis for drug synergy. These effects are greatly enhanced by loading the drugs into filomicelles. Co-loading the drugs into filomicelles lead to a more potent system compared to separate loading, with no loss in the integration efficiency of drugs. Notably, relapse rates were ~2% three months after treatment, highlighting the improvement offered by nano-carriers. Ki-67 was the key affected protein, with levels half of those in free RA-Paclitaxel treated cells. Preliminary tests *in vivo* demonstrate sustained delivery for days as well as efficacy in shrinking tumors. Injections of RA-Paclitaxel combination are able to shrink tumors and arrest growth for over 15 days, a feat not achieved by Paclitaxel alone [2].

Conclusion: The above results highlight the irreversible synergy of killing cancerous cells while driving differentiation. Filomicelles accumulate better in tumors, thus increasing the efficacy of delivery and hence the treatment. The overall result is a system that efficiently delivers drugs to the tumor that leads to higher cell death, while arresting proliferation in surviving ones.

References:

1. Geng Y, Dalhaimer P, Cai S, Tsai R, Tewari M, Minko T, Discher DE. Shape effects of filaments versus spherical particles in flow and drug delivery. *Nature Nanotechnology. 2(4),* 249-255 (2007).
2. D. Christian, S. Cai, O. Garbuzenko, T. Harada, A. Zajac, T. Minko, and D. Discher Mol. Pharmaceutics, 2009 (6) 1343-1352
3. R. Connolly, N. Nguyen, and S. Sukumar. Clinical Cancer Research 2013 (19) 1651-1659.
4. Y. Jing, L. Wang, L. Xia, G. Chen, Z. Chen, and W. Miller. Blood 2001 (97) 264-269.

468777 Helping the Corneal Stroma Stay in Shape (Invited Talk)

Abstract:

The cornea of the eye is a remarkable collagen-based hydrogel that is 78%wt water, yet it has the strength to serve as a stable optical element despite continuous tension imposed by intraocular pressure. The corneaâ€™s balance of clarity and permeability with mechanical strength is maintained by the constant degradation and replacement of its constituents. When these are out of balance either due to excessive degradation or insufficient replacement, the cornea begins to bulge outward (e.g. keratoconus). Through a Caltech-UCSF collaboration we have developed a gentle method to reinforce the tissue using photoactivated crosslinking of the biomolecules already present in the stroma. The crosslinking is activated by visible light (using green light to activate the photosensitizer eosin Y), which shows low toxicity to the cells in the corneal stroma (keratocytes) and behind it (corneal endothelial cells). To progress to the clinic, we are exploring novel, minimally invasive ways to penetrate the corneaâ€™s epithelial barrier for drug delivery. This talk will present the visible-light photocrosslinking method, the reason that it motivates drug delivery to the stroma without disruption of the epithelium, and the approach we are exploring to achieve it.

457978 Self-Fluorescent Hyaluronic Acid-Based Gel for Dermal Applications

Abstract:

Combinations of polymer conjugates affording in situ gelation hold promise for treatment of pathological cavities (*e.g.*, arthritis) and sustained drug release. In particular, hyaluronic acid (HA) functionalized with reactive groups is regarded as an excellent biomaterial due to its tunable cross-linking kinetics and mechanical properties. HA-based reagents, however, can be irritating to surrounding tissues due to the reactivity of pendant groups, and their fast gelation kinetics can result in poor cavity filling. To overcome these issues, we have developed a biocompatible â €œclickâ€ reaction between cyanobenzothiazole (CBT) and D-cysteine (D-Cys) is employed to produce HA-based conjugates for in situ gelation. Rheological studies conducted on a gel obtained from the combination of HA-CBT and HA-d-Cys indicate optimal gelation time and mechanical properties. Further,Â *in vitro*Â studies on porcine skin demonstrate the ability of the gel to form in situ upon subcutaneous injection or topical application, and to act as a reservoir for sustained release of protein therapeutics. Finally, the safety of the HA-based conjugates is demonstrated on human keratinocytes. Our results demonstrate the applicability of the binary mixture for in situ gelation and the potential of the proposed system for a variety of biomedical applications.

470899 Dynamic Adhesion of *Staphylococcus Aureus* to Poly(ethylene glycol) Surfaces

Abstract:

Reversible adhesion of bacteria occurs on the time scale of seconds through physicochemical interactions of the bacteria with the surface. In flow, the translational velocity of biocolloids slows until either sufficient binding energy is achieved and the attachment occurs, or the biocolloid returns to bulk flow. This process of dynamic adhesion or "rolling" has been reported to enhance and spread bacterial colonization on antifouling surfaces. In this work, we investigated the dynamic adhesion of clinically relevant methicillin-resistance Staphylococcus aureus to five poly(ethylene glycol) (PEG) materials. We synthesized three PEG hydrogels with distinct mechanical properties and mesh sizes: stiff (6500 kPa, 10 Å), soft (310 kPa, 27 Å), and decoupled soft (220 kPa, 41 Å), as well as two planar brushes, 2k and 5k. In general, the subpopulation of dynamically adherent S. aureus sustained rolling for ~12 μm intervals on the brushes and hydrogels. We found that when S. aureus dynamically adhered to a surface, it was highly likely to experience subsequent adhesion events. This was especially true for S. aureus on stiff hydrogels and 5k brushes where >85% of the cells adhered at least twice. As the length of interaction remained constant over each encounter, the residence time of each S. aureus on the surface provided a universal metric to compare surfaces. For example, on average S. aureus interacted 5 s longer on the stiff than on the soft hydrogels (41 Å or 27 Å). This implies that stiff hydrogel surfaces are less effective at resisting bacterial attachment because there is an increased likelihood that S. aureus will permanently attach downstream. This fundamental study investigates how S. aureus interact and adhere with PEG materials holds potential to facilitate the design of next generation of antifouling surfaces.

457514 pH-Sensitive Mechanical Properties of Elastin-like Polypeptides

Abstract:

Recombinant elastin-like polypeptides (ELPs) are environmentally sensitive, elastomeric domains that have been used: as purification tags for proteins that are sensitive to denaturing purification methods, to encapsulate small molecules for drug delivery, and to coat scaffolds for improved cellular adhesion. ELPs are easily overexpressed, and their lower critical solution temperature (LCST) allows the protein to be purified in large volumes using thermal cycling. LCST behavior in ELPs is driven by charge interactions between the protein and the surrounding solvent and can be influenced by environmental factors such as ion concentration and pH. pH-sensitive protein systems can be useful for spatially selective drug release in acidic tumors, for pH-dependent valves in microfluidic systems, and as indicators for changes in pH beyond a desired region. In addition, pH-sensitive ELPs could be used for purification by altering the LCST through pH instead of temperature. Finally, pH-sensitive ELP coatings could easily be deposited or removed by altering the pH of the solution and thus the LCST and solubility of the ELP.

ELPs are composed of a repeating pentapeptide amino acid sequence containing one "guest residue" position that can be an amino acid of choice. This guest residue allows design flexibility. The inclusion of ionizable amino acids in the guest residue position confers pH-dependent surface charges. pH sensitivity can be achieved across a wide pH range if ionizable amino acids with a range of pKa values are included. Additionally, pH sensitivity may be induced by including ionizable amino acids before or after the ELP sequence in a protein tag. We expect that pH sensitivity will persist in hydrogels made from ELPs and that hydrogels containing pH-dependent ionizable groups will swell and soften when those groups become charged. However, to our knowledge, there are not yet any studies of pH sensitivity in ELP-based hydrogels. Thus, in this study, we seek to tune the mechanical properties of ELP-based hydrogels by varying the charged guest residues and tags.

We have created a family of recombinant ELPs with ionizable guest residues and ionizable protein tags. The ionizable guest residues include tyrosine (pKa ~10.4) and lysine (pKa ~10.5). The protein tags include histidine (pKa ~6.0) and aspartic acid (pKa ~3.7). The pKa of an amino acid indicates the pH at which it becomes ionized; the charge of the amino acid side chain changes from positively charged to neutral (lysine and histidine) or from neutral to negatively charged (tyrosine and aspartic acid). These shifting and competing charges contribute to the overall charge behavior of the protein. We hypothesize that lysine and tyrosine will contribute to the pH sensitivity at pH values above 7 and that chemically blocking these guest residues will reduce the pH sensitivity above 7. We also hypothesize that histidine and aspartic acid will contribute to the pH sensitivity at pH values below pH 7, and designs that do not contain the protein tag with histidine and aspartic acid will have reduced pH sensitivity below pH 7.

We have found that ELP hydrogels containing both ionizable guest residues and protein tags are sensitive to pH change both above and below pH 7. The Young's Modulus of these hydrogels decreased by 60% when submerged in a buffer at pH values below or above 7 compared to gels at pH 7. In buffers of pH 5 or below, the hydrogels increased in size, and gels at pH 3 were twice as swollen as gels at pH 7. When the lysine and tyrosine residues were selectively blocked through acetylation, these hydrogels lost their pH sensitivity and their Young's Modulus did not change at pH values above 7. Overall, our conclusions demonstrate that including ionizable guest residues or protein tags creates pH sensitivity in ELP hydrogels across a wide pH range and that this sensitivity can be altered through chemical treatments. Because the pH sensitivities of the ELP sequence and the protein tag exhibit distinct behaviors, it may be possible to use ionizable protein tags in recombinant protein designs using sequences besides ELPs in order to include pH sensitivity in the resulting hydrogel. Thus, ELP hydrogels with ionizable groups may be useful in designing materials for pH-sensitive drug release in areas such as acidic tumors or in environmentally responsive microfluidic valves.

465163 Rational Design of Charged Nanogels for Differential Protein Capture

Abstract:

Inspired by the natural phenomenon of molecular recognition, we have developed synthetic materials capable of binding specific proteins or protein classes. Polymers can be rationally designed to differentiate proteins based on their properties including isoelectric point, molecular weight, and glycosylation status. While these materials typically have lower specificity than antibodies, their significantly lower cost makes them attractive for use in biomedical and separation science applications. We have designed a set of crosslinked polymer nanoparticles containing varied ionic moieties and investigated how protein binding affinity and selectivity is affected by the amount and type of charge present in the nanogels. Nanogels were synthesized by the copolymerization of N-isopropylacrylamide (NIPAM) and methacrylic acid (MAA) with N,Nâ€™-methylene bisacrylamide as a crosslinker. Different charged groups were introduced into the nanogels via a carbodiimide mediated post-synthesis modification. Successful modification of poly(NIPAM-co-MAA) nanogels was confirmed by zeta potential measurements, titration, and FT-IR spectroscopy. The nanogels were incubated with a set of proteins, both individually and in mixtures, to assess their recognition capabilities. Our results suggest that beyond having complementary charge-charge interactions between the protein and polymer, the type of anionic and/or cationic group and charge density strongly influences polymer-protein binding preferences. Collectively these nanogels are able to distinguish among more than eight proteins, making them promising tools for use in separations or differential sensing applications.

464803 Engineering Click-Protein Hydrogel for Tissue Engineering

Abstract:

Protein-based hydrogels have many advantages over synthetic hydrogels for diverse applications owing to the many intrinsic properties of the proteins such as bio-compatibility, innocuousness, and biological activity. In addition, the ability of proteins to form self-organized, uniform and regulated structure with precise control facilitates the synthesis of hydrogel with the desired properties. However, strategies for protein hydrogel synthesis remain limited, leaving the field of protein hydrogel largely underdeveloped.

In this study, we present a robust, convenient and efficient click-hydrogel strategy that enables any protein to be transformed into bioactive protein hydrogels. This strategy is based on the copper-free click chemistry reaction that is highly efficient, stereo-specific and bio-orthogonal. The target proteins are first separately labelled with a desired functional group. Mixing of functionalized proteins under physiological condition leads to the spontaneous formation of a highly cross-linked protein network or hydrogel through click reaction.

Here, we demonstrate our click-hydrogel strategy using various target proteins of different sizes and quaternary structures. The mechanical properties of the hydrogel are affected by the protein concentration, quaternary structure and the density of the functional group. Protein hydrogels with compressive modulus from 2kPa to 200kPa, high stability in aqueous solution and high porosity were successfully synthesized. In addition, our click-hydrogels exhibit shape-memory properties and dehydrated hydrogels are able to completely and rapidly regain their original shape upon rehydration. Hydrogels made from proteins containing integrin-binding motif (RGD) are able to efficiently support mammalian cells attachment and proliferation with little to no toxicity. Our click-protein hydrogel should have broad application in biomaterial, tissue and biofuel cell engineering.

464712 Polyelectrolyte Complex Hydrogels: Self-Assembly, Structure and Rheology

Abstract:

Polyelectrolyte complexes (PEC) form when oppositely charged polyelectrolyte chains spontaneously associate and phase separate in aqueous media. Combining one or both of the polyelectrolytes with a neutral polymer restricts bulk phase separation of the PECs, and thus leads to self-assembled structures with PEC domains surrounded by neutral polymer coronae, forming micelles and hydrogels. The PEC domains in these assemblies can encapsulate therapeutics as well as genetic materials and thus have tremendous potential in drug delivery and tissue engineering applications.

We will present insights on the equilibrium structure of such PEC hydrogel assemblies comprising oppositely charged block copolyelectrolytes. Initial motivation will be provided by a brief discussion of structure and rheology of block copolypeptide hydrogels. It will be followed by detailed investigations of structure-property relationships of hydrogels comprising model block copolyelectrolytes with oppositely charged functionalized polyallyl glycidyl ethers as ionic and polyethylene glycol as neutral blocks. Large-scale ordering of the nanoscale PEC domains is observed at high polymer loadings, characterized by a disorder-order transition followed by morphological transitions with increasing polymer loading. Complementary X-ray and neutron scattering investigations providing a comprehensive structural description of these materials and strongly supporting our hypothesis on the charged block playing a structure-defining role and the neutral block undertaking a structure-directing role will be presented. In addition, the effect of key parameters such as polymer block lengths and salt concentration on the equilibrium materials properties, elucidated via detailed structure characterization and rheology studies, will also be discussed.

468173 Mussel-Inspired Injectable Gelatin-Methacryloyl Bioadhesive for Minimally Invasive Suregry

Abstract:

Wet tissue adhesion required in many surgeries such as those for internal organs. The current surgical adhesives are not ideal. Developing surgical adhesives with strong wet tissue adhesion, controlled mechanical properties, and excellent biocompatibility has been a significant challenge. Herein, we report for the first time visible light crosslinkable catechol conjugated Gelatin (GelMAC) adhesive for internal surgeries. Compare to commercially available adhesive, GelMAC showed superior wet tissue adhesion, controlled mechanical properties and high in vitro Cytocompatibility. Three different ex-vivo animal models were used to evaluate the feasibility and effectiveness of GelMAC as minimally invasive sealants in the absence of any additional conventional surgical methods, such as suturing or stapling. Our results show that GelMAC is highly translational and could have broad impact on surgeries where surgical tissue adhesives and sealants are used.

467991 Bio-Based Thiol-Ene Polymerized Gel-Polymers: Synthesis and Properties

Abstract:

Polymer gels usually consist of a polymer-solvent system prepared by chemical and/or physical crosslinking. Crosslinking is commonly used in the preparation of such polymers for cosmetic, energy, medical and food applications due to the resulting high degrees of mechanical stability, thermal resistance and molding ability. Growing environmental and economic concerns as well as the uncertainty that accompanies finite petrochemical resources has resulted in an increase in research and development of bio-based polymers. Lignin, a renewable feedstock, has captured much attention due to its abundance and ability to afford useful polymer building blocks when strategically depolymerized. In order to determine if lignin could serve as a basis for bio-based monomers for polymer gels, lignin-based molecules, vanillyl alcohol (4-(hydroxymethyl)-2-methoxyphenol) and gastrodigenin (4-(hydroxymethyl)phenol), were functionalized and subsequently polymerized with multi-functional thiol monomers. The polymer properties, including thermogravimetric, thermomechanical, and mechanical properties, of the resulting thin, flexible polymers will be presented. Moreover, the swelling behavior of the films, including dependence on polymer architecture, solvent type(s), and temperature, and comparison to non-lignin-based polymers will be presented. This study adds to the list of synthesis methods of lignin-based polymers and provides insight on how to design structurally and thermally desirable bio-based gel-polymers.

468408 High Modulus Thermoresponsive Elastin-like Polypeptide Gels As New Injectable Biomaterials

Abstract:

Injectable biomaterials have attracted a great deal of interest for facilitating minimally invasive medical procedures, enabling faster patient healing and lower medical costs. A major area of continued need is injectable materials that can achieve a very high modulus, greater than 1 MPa, in their hardened state, enabling their application in relatively stiff tissues such as cartilage. Here, we report a new high stiffness hydrogels generated from elastin-like proteins (ELPs) at 15-30 wt.% solids, in the range of the protein content in natural body tissue. Although ELP solutions are fluids below body temperature and typically form macrophase separated precipitates at high temperature, specific ELP sequences can instead form kinetically arrested macrophase separated structures that lead to the formation of extremely stiff gels upon heating to body temperature, enabling injection of a material that develops an ultimate modulus of more than 1 MPa with a rapid initial set in less than a minute. Small-angle neutron scattering and oscillatory shear rheology provide a detailed picture of the kinetically arrested microstructure and the time evolution of mechanical properties, and large amplitude oscillatory shear rheology is applied to provide insight into the way network structures break under larger deformation. Through simple oxidative chain extension to prepare very high molar mass proteins, the toughness of the gels may also be significantly increased, presumably due to the intertwining of single chains between many different kinetically aggregated domains within the gels. Finally, cell culture experiments performed using chondrocytes and mesenchymal stem cells demonstrate a high degree of biocompatibility and the potential of these materials as cell culture matrices for tissue engineering applications.

460233 Poly(ethylene glycol) and Elastin-like Protein Double-Network Hydrogels As Central Nervous System Extracellular Matrix Models

Abstract:

Central nervous system (CNS) tissue does not substantially regenerate from injury and disease because the extracellular environment in the damaged lesion is not conducive to neuronal regeneration. Most current in vitro models of the CNS extracellular matrix (ECM) utilize tissue harvested from rat or mouse brain and spinal cord making it difficult to isolate the effects that each individual biochemical cue has on cells. The challenges with using native tissue arise from the complexity of the CNS ECM with multiple biochemical cue expression changes occurring simultaneously during injury and disease progression, the presence of redundant effects to cells from separate ECM biochemical cues, and changes to the native tissue structural stiffness and topography from scar tissue formation.

We have developed a poly(ethylene glycol) (PEG) hydrogel with elastin-like protein (ELP) microdomains by mixing the two polymers and maintaining separate crosslinking mechanisms. PEG is crosslinked through the free radical reaction of methacrylate end-groups in the presence of the photoinitiator lithium phenyl-2,4,6-trimethylbenzoylphosphinate (LAP) and ultraviolet light. ELP is crosslinked by mixing with tetrakis(hydroxymethyl) phosphonium chloride (THPC); primary amines on lysine residues of ELP undergo Mannich-type condensation reactions with THPC. ELP displays a lower critical solution temperature (LCST) transition where it forms coacervates at physiological temperature which causes the two polymers to phase separate when mixed. Our crosslinking procedure fixes the polymers to form regions with high ELP concentrations and ones with high PEG concentrations. These PEG and ELP hydrogels (PEG-ELP) mimic the basic orientation of the neural interstitial matrix, which is the part of the CNS ECM found in deep tissue. The neural interstitial matrix contains a network of proteoglycans, hyaluronan, and tenascins with small amounts of fibrous proteins such as elastin and collagen. Our PEG-ELP hydrogels contain large areas of the bio-inert PEG with smaller regions of fibrous ELP, which contains RGD, a cell integrin binding motif, in the protein backbone sequence.

Preliminary data shows that PEG-ELP double-network hydrogels are more compliant than PEG hydrogels. The addition of 1 wt% ELP to a 6 wt% PEG hydrogel reduces the unswelled storage modulus from 253 Pa to 127 Pa. Oligodendrocyte precursor cells (OPC) encapsulated at 1e6 cells/ml survive well in 3 wt% PEG hydrogels with 3 wt% ELP (3/3 PEG-ELP) but do not survive in 6 wt% PEG hydrogels, as determined by live/dead imaging. Increasing the OPC concentration by an order of magnitude, 1e7 cells/ml, causes them to survive in a 6 wt% PEG hydrogel but they do not extend processes. OPCs extend processes in 3/3 PEG-ELP hydrogels over the course of 7 days. This work demonstrates that PEG-ELP hydrogels are the first step towards mimicking the neural interstitial matrix from a bottom-up approach, making them a suitable mimetic for studying the fundamental effect that ECM remodeling during CNS injury or disease has on cell recovery.

471537 Highly Elastic, Antimicrobial, and Sprayable Hydrogel for Wound Healing

Abstract:

Highly elastic, antimicrobial, and sprayable hydrogel for wound healing

Devyesh Rana[1], Ehsan Shirzaei Sani[2], Roberto Portillo Lara[2], Suzanne Mithieux[3], Anthony Weiss[3], Nasim Annabi[2,4,5]

[1] Department of Bioengineering, Northeastern University, Boston, MA, 02115-5000, USA

[2] Department of Chemical Engineering, Northeastern University, Boston, MA, 02115-5000, USA

[3]School of Life and Environmental Sciences, Charles Perkins Centre, University of Sydney, Sydney, 2006, Australia

[4]Biomaterials Innovation Research Center, Brigham and Women's Hospital, Harvard Medical School, Boston, MA, USA.

[5]Harvard-MIT Division of Health Sciences and Technology, Massachusetts Institute of Technology, Cambridge, MA, USA

Introduction

The skin is the largest functional organ in the human body. Its primary function is to serve as a protective barrier against foreign agents, both micro- (e.g. bacteria, viruses, etc.) and macroscopic (e.g. dirt, debris, etc.) (1). Hence, the risk of developing infections is amplified upon impaired skin integrity. Chronic wounds are estimated to affect 6.5 million patients. The annual health care cost associated with chronic wounds are estimated to exceed $25 billion (2). Burn wounds alone account for $11.3 billion (3). Patients suffering from severe burns and skin injuries still exhibit high mortality rates due to complications during the early stages of trauma. The first line of care in managing burn and trauma patients is to cover the wound properly to avoid excessive transdermal fluid loss, and to protect the wound area from environmental exposure to pathogens, in particular bacteria. The next step is to promote healing to restore the integrity of the skin and its barrier function (4, 5). However, large burns and chronic wounds often do not heal in a timely and orderly fashion, and bacterial infection remains a life threatening condition for these patients (6). In this project, we engineered a multifunctional hydrogel that can be sprayed on wounds to prevent bacterial growth and facilitate wound healing. This visible light-activated hybrid hydrogel is comprised of methacrylated tropoelastin (MeTro) and gelatin methacrylate (GelMA), which are conjugated with an antibacterial peptide (AMP). We hypothesize that the engineered hybrid gel will mimic the mechanical properties of the native skin and will adhere to the skin tissue upon photopolymerization to form an antibacterial and regenerative barrier on the wound area.

Materials and methods

All chemicals were purchased from Sigma Aldrich and used without further purification. GelMA and MeTro were prepared according to procedures described in our previous works [7, 8]. Different ratios of MeTro/GelMA prepolymer solutions (100/0, 70/30, 50/50, 30/70, 0/100) were interspersed in a solution of trimethanolamine (1.8% w/v) as a co-initiator, vinylcaprolactam (1.25% w/v) as a catalyst, and Eosin-Y (20% v/v) as a photoinitiator. An antimicrobial peptide (CPS Scientific) in the range of 0.01-0.03 % (w/v) was then incorporated into the MeTro/GelMA solution prior to polymerization via visible light (400-500 nm) for 120s. Mechanical properties such as tensile and compression moduli were tested using an Instron 5944 machine. Antimicrobial properties were also evaluated using colony forming and live/dead assays. ***In vitro*** biocompatibility of the engineered hydrogels was assessed by 2D and 3D culture of NIH 3T3 fibroblast cells.

Results and Discussion

The tensile modulus of the hybrid hydrogel was found to be tunable in the range of 5 - 35 kPa based on MeTro/GelMA ratios and final polymer concentrations. Similarly, the compressive modulus ranged from 4 - 65 kPa by varying polymer composition and concentration. The resulting hydrogels were sprayed onto porcine skin and were shown to adhere well to the tissue after photopolymerization based on wound closure tests. ***In vitro*** bacterial studies indicate that incorporation of 0.01% v/v antimicrobial peptide into a MeTro/GelMA hydrogel with the ratio of 70/30 was effective at killing both gram positive (Methicillin-resistant ***Staphylococcus aureus*** (MRSA)) and gram negative (***Escherichia coli***) bacteria. The number of colony forming units (CFUs) decreased from 38 to 15 for MRSA and from 13 to 2 for ***E.coli***. In addition, ***in vitro*** tests on the engineered hydrogels containing antimicrobial peptide showed more than 90% mammalian cell viability at days 1, 3, and 5 post-seeding. This result confirmed that the engineered hybrid hydrogels exhibit no apparent cytotoxic effect on 3T3 fibroblasts. The engineered sprayable and antimicrobial hydrogel has a remarkable potential to be used as a novel biomaterial for wound healing.

Conclusions

In this study, we developed a novel regenerative sprayable hydrogel with antimicrobial properties for wound treatment. In particular, the hydrogel formulation consisting of 70/30 MeTro/GelMA with a 15% final polymer concentration and 0.01% antimicrobial peptide, attained optimal mechanical, antimicrobial and biocompatible properties.

References

[1]. Epstein FH, Singer AJ, Clark RA. Cutaneous wound healing. New England Journal of Medicine. 1999;341(10):738-46.

[2]. Chandan K. Sen, Gayle M. Gordillo, Sashwati Roy, Robert Kirsner, Lynn Lambert, Thomas K. Hunt, Finn Gottrup, Geoffrey C Gurtner, and Michael T. Longaker. Human Skin Wounds: A Major and Snowballing Threat to Public Health and the Economy. Wound Repair Regen. 2009, 17(6), 763-771.

[3]. Latenser BA, Miller SF, Bessey PQ, Browning SM, Caruso DM, Gomez M, et al. National Burn Repository 2006: a ten-year review. J Burn Care Res. 2007;28(5):635-58. Epub 2007/10/31. PubMed PMID: 17969244.

[4]. Sen CK, Gordillo GM, Roy S, Kirsner R, Lambert L, Hunt TK, et al. Human skin wounds: a major and snowballing threat to public health and the economy. Wound Repair and Regeneration. 2009;17(6):763-71.

[5]. Atiyeh BS, Costagliola M, Hayek SN, Dibo SA. Effect of silver on burn wound infection control and healing: review of the literature. Burns : journal of the International Society for Burn [6]. Injuries. 2007;33(2):139-48. doi: 10.1016/j.burns.2006.06.010. PubMed PMID: 17137719.

[7]. Castilla DM, Liu Z-J, Velazquez OC. Oxygen: Implications for Wound Healing. Advances in wound care. 2012;1(6):225-30.[5]. Xin Zhao , Qi Lang , Lara Yildirimer , Zhi Yuan Lin , Wenguo Cui , Nasim Annabi , Kee Woei Ng , Mehmet R. Dokmeci , Amir M. Ghaemmaghami , Ali Khademhosseini, Photocrosslinkable Gelatin Hydrogel for Epidermal Tissue Engineering, Adv. Healthcare Mater. 2016, 5, 108–118.

[8] Nasim Annabi, Kelly Tsang, Suzanne M Mithieux, Mehdi Nikkhah, Afshin Ameri, Ali Khademhosseini, Anthony S Weiss, Highly elastic micropatterned hydrogel for engineering functional cardiac tissue, Adv. Funct. Mater. 2013, 23, 39, 4950-4959.

466838 Characterization of Collagen Type I and II Blended Hydrogels for Articular Cartilage Tissue Engineering

Abstract:

Abstract for AIChE:

Title: Characterization of Collagen Type I and II Blended Hydrogels for Articular Cartilage Tissue Engineering

Authors: Claire E. Kilmer [1], Nelda Vazquez-Portalatin [2], Alyssa Panitch [2], Julie C. Liu, Ph.D. [1,2]

Department/Affiliations: [1] School of Chemical Engineering and [2] Weldon School of Biomedical Engineering, Purdue University, West Lafayette, IN

Osteoarthritis (OA) is a debilitating condition that affects over 27 million people in the United States alone and is defined by degradation in articular cartilage extracellular matrix (ECM). Although there is no cure for OA, there are many treatment options including osteochondral grafting, autologous chondrocyte implantation, and marrow stimulation. However, these options usually promote the growth of fibrocartilage, which is inferior to the mechanical properties of native, hyaline cartilage. The overall goal of this project is to create a scaffold that will replace damaged cartilage and ease the pain of the condition while simultaneously preventing further cartilage deterioration. This study characterized and compared the properties of gels made of collagen type I and II blends, which have the potential to be implemented as a tissue-engineered scaffold.

Collagen type II makes up 90-95% of the collagen produced by chondrocytes in the ECM and is a promising scaffold material for use in articular cartilage. It has been shown that collagen type II hydrogels promote the differentiation of embedded mysenchymal stem cells to chondrocytes more efficiently than collagen type I gels. However, when compared to collagen type I, collagen type II forms fibrils of smaller diameter and exhibits poor mechanical properties when forming a hydrogel without crosslinking. It was hypothesized that hydrogels made with a blend of collagen type I and II will have superior mechanical properties compared to gels made with collagen type II alone. In addition, GAGs attach to proteoglycans within the extracellular matrix and allow articular cartilage to withstand compressive forces. Thus, we are interested in the interactions between a blend of collagen type I, collagen type II, and GAGs such as chondroitin sulfate (CS).

The hydrogels were prepared at different ratios of collagen type I to collagen type II (1:0, 3:1, 1:1, 1:3, and 0:1), and the amounts of collagen types I and II in the gel were measured. The collagen network structure was analyzed with cryo-scanning electron microscopy, and oscillatory tests on a rheometer were performed to analyze the mechanical properties of the scaffolds. From the experiments conducted, the 3:1 gels were able to incorporate both collagen type I and collagen type II within the gels. The 3:1 gels also retained a significantly higher amount of chondroitin sulfate than the other blends. Initial results show that the addition of collagen type II alters gel formation, network structure, and mechanical properties. Moreover, the 3:1 gels exhibited a void space percentage of 36.4% which was lower than the 1:1 gels (46.5%). The storage, loss, and complex moduli were larger for the 3:1 gels ($G' = 4.9 \pm 1.0$ Pa, $G'' = 1.2 \pm 0.1$ Pa, and $G^* = 5.0 \pm 0.9$ Pa) compared to the 1:1 gels ($G' = 1.2 \pm 0.2$ Pa, $G'' = 0.3 \pm 0.1$ Pa, and $G^* = 1.2 \pm 0.2$ Pa). From the different blends that were investigated, the 3:1 blend was able to form more consistent gels with superior mechanical properties as compared to the other blends. Thus, the 3:1 blend has the potential to be implemented as a scaffold for articular cartilage tissue engineering.

464123 Development of Icariin-Loaded Scaffold for Cartilage Tissue Engineering By 3D Bioprinting

Abstract:

Articular cartilage plays an important role in weight-bearing and movement. However, it is also more easily damaged in intra-articular fractures. The cartilage is hardly to repair itself once damaged seriously, since it has no vessel and nerve. Tissue engineering is a potential way to cure the cartilage damage, and the alginate gels, which have been proved that it can provide the environment of extracellular matrix, are commonly used as the base materials for 3D bio-printer. The printed channels and structures of the scaffolds are suitable for cells to adhesion and proliferation, which also ensures cells to handle nutrient uptake and waste disposal. Icariin, a Traditional Chinese Medicine, which has been demonstrated to be a promoting compound for extracellular matrix synthesis and gene expression of chondrocytes, was incorporated with gels and cells to construct Icariin-loaded bioscaffolds by 3D bioprinting. The scaffold morphologies were observed under the light and electron microscopes. Cell distribution, viability and proliferation were investigated and evaluated by Fluorescence Microscope, Laser Scanning Confocal Microscope and CCK-8 assay kit. The content of Glycosaminoglycan (GAG) released and Type II collagen secretion were determined by dimethylmethylene blue (DMMB) dye-binding assay and Rat Collagen Type II ELISA kit.

459876 Fiber-Reinforced Hydrogels: *In Situ* fabrication from Coextruded Polymeric Composites

Abstract:

Hydrogels are a highly desirable category of polymeric biomaterials for cell scaffolding, owing to their similarity to naturally occurring extracellular matrix. Hydrogels, which consist primarily of water are inherently weak, leading to exploration of new avenues of structural reinforcement in the form of embedded fibrous networks. One approach is the utilization of three-dimensionally (3D) printed scaffolds to provide a distributed support network; however, the feature sizes were restricted by the resolution of current 3D printing technology, limiting fiber flexibility and subsequent synergistic reinforcement. To achieve sub-micron scale fiber reinforcement, another strategy is the reliance on conventional electrospinning techniques with an additive manufacturing approach, impregnating gel around the fiber network. However, this additive technique often results in a tri-layer structure and poor fiber distribution throughout the gel matrix, also limiting synergistic reinforcement between the gel and fiber network. Utilizing novel matrix/fiber multilayer coextrusion technology, fiber-reinforced poly(ethylene oxide) (PEO) hydrogels with distributed, embedded, sub-micron scale fibers of poly(Îµ-caprolactone) (PCL) were fabricated *in situ* via a simple crosslinking scheme. By systematically varying melt pump rate, compressive stiffness was tailored ranging between 0.69 Â± 0.4 kPa and 1.94 Â ± 0.21 kPa, on par with values obtained from articular cartilage (1.61 Â± 0.17 kPa). Introducing a post-extrusion uniaxial drawing step increased the fiber modulus 9-fold, resulting in a 225% increase in PEO hydrogel stiffness with similar fiber loading. In a demonstration of material possibilities, rigid poly(L-lactic acid) (PLLA) was substituted for PCL during coextrusion, which increased gel stiffness 350% when compared to gels reinforced with PCL fibers. Hydrogels fabricated using the *in situ* fabrication technique displayed similar cell viability using the NIH 3T3 cell line when evaluated against hydrogels developed using electrospun PCL fibers and additive gel impregnation. Finally, PEO hydrogels reinforced with undrawn PCL, drawn PCL, and undrawn PLLA possessing significantly different moduli were evaluated for cellular response using the MC 3T3 pre-osteoblast cell line; it was observed that as gel stiffness increased, cells preferentially differentiated to osteoblasts. Advances in multilayer coextrusion technology have

provided a flexible platform for *in situ* fiber-reinforced hydrogel development with a wide range of material and architectural control for tailored mechanical properties and achieving targeted stem cell differentiation in a robust 3D hydrogel biomaterial constructs.

462954 Porous Structure Based Electrocatalysts for High Performance Fuel Cells

Abstract:

In this work, we systematically studied the effect of pore size and doping site position of metal-free catalysts in MEAs using well-defined ordered mesoporous carbon system to achieve highly active metal-free catalysts in MEA levels. At first, we synthesized mesosporous phosphorus doped ordered mesoporous carbon with large pore (~38 nm) by block copolymer assisted co-assembly. Then, addition nitrogen doping was conducted with 2 different conditions using different sized nitrogen precursors. Finally, we fabricated and tested MEAs that employ metal-free catalysts as a cathode catalyst. As a result, the MEA performance of synthesized catalyst with precisely controlled doping sites considerably exceeds that of previously reported metal-free catalysts.

465313 Design of Heterostructure Alloy Nanoparticles for Photocatalysis of CO_2 Reduction

Abstract:

Semiconductor nanocrystals of size in the strong quantum confinement regime, or quantum dots, exhibit size-dependent energy gap. The size-tunable shift of band-edge states can be a powerful tool in the control of exciton dynamics. For example, heterostructure nanocrystals with mere changes in geometric dimensions can lead to the band alignment from type I to type II, which could dramatically increase the exciton recombination lifetime.[1] In this presentation, I describe the design of heterostructure nanocrystals showing long exciton recombination lifetime. For the study, we synthesized various core/shell composite particles (e.g., PbSe/CdSe or CdSe/TiO2) and introduced CdS outer-most shell with varying thickness and morphology. [2] These heterostructure nanoparticles with prolonged exciton lifetime result in efficient photocatalysis of methylene blue reduction and hydrogen evolution from water.

In the way of addressing the heterostructures, we stumbled upon the observation that metal cocatalysts and surface can be engineered with relatively simple chemical approaches, rendering new photocatalytic systems suited in photocatalytic reduction of CO2 in gas phase. [3]

References

[1] Lee D C, Robel I, Pietryga J M and Klimov V I 2010 *J. Am. Chem. Soc.* **132** 9960

[2] Lee S, Lee K, Shin D J, Kim W D, Lee S and Lee D C 2014 *J. Phys. Chem. C* **118** 23612

[3] Jeong S, Lee S, Kim W D, Lee K, Lee S and Lee D C 2016 *ChemCatChem* in print

[4] Lee S, Jeong S, Kim W D, Lee K, Lee S and Lee D C 2016 *Nanoscale* in print

470971 Nanostructured Composite Intermediate-Temperature Solid Acid Fuel Cells Fabricated By Needleless Electrospinning

Abstract:

Fuel cells are the most efficient technology available today for the conversion of chemical energy into electricity. A new and promising fuel cell technology is the solid acid fuel cells (SAFCs), offering unique possibilities as a result of their intermediate-temperature region of fuel cells (100–300 °C). Their state of the art solid electrolyte, CsH_2PO_4 (CDP) is a solid acid compound with a superprotonic phase transition around 240 °C [1]. Under a dry condition the dehydration of CDP proceeds to form polymerization at temperatures above 200 °C, and CDP transforms into $CsPO_3$ [2]. Development of the catalytic performance can enable this type of fuel cell to be among the leading fuel cells on today's market.

Due to the pure ionic conducting nature of the electrolyte, the electrochemical reactions occur at the so-called triple phase boundaries, between the catalyst, electrolyte and the gas phase. Because of the triple phase, decreasing just the size of the catalyst does not increase the electrochemical reaction rates. However, in case the size of the electrolyte is also decreased, the electrochemical reaction rate is higher at the triple phase boundary resulting in increased power density of the fuel cell [1].

Needleless electrospinning technique was employed to fabricate nanostructured SAFC electrodes from CDP/$CsPO_3$-polymer solutions. Nanostructured CDP composite has the advantage of increased catalytic surface due to the large surface area. Needleless electrospinning is an inexpensive technique for producing nanostructured fuel cells at industrial scales. Instead of the commonly used nozzle, the needleless design employs a rotating drum immersed in a solution bath. Multiple Taylor cones form on the drum electrode surface resulting in rapid deposition of nanofibers over large areas. By applying $CsPO_3$ solutions, the phosphate groups form a chain, increasing the viscosity of the solution. This enables electrospinning with negligible amounts of polymers. The fibers produced from the solutions have diameters between 30 nm – 300 nm depending on the initial solution concentration, substrate temperature, and applied voltage. The deposition area is over 100 cm^2 and the deposition rate is up to 2 g/hr in the laboratory scale setup. Electrochemical characterization of the nano-composite electrodes, performed by A.C. impedance spectroscopy of symmetric cells under humidified hydrogen, confirmed the high activity of the electrospun nanocomposite electrodes. Fuel cell results using electrospun nanofibers from $CsPO_3$ solutions demonstrate that a 0.86 V cell voltage can be obtained at a current density of 50 mA/cm^2, which is ~10% higher than what can obtained using state-of-the-art techniques.

[1] C. R. Chisholm, D. A. Boysen, A. B. Papandrew, S. K. Zecevic, S. Cha, S., K. A. Sasaki, Á. Varga, K. P. Giapis, S. M. Haile, ***Interface***, 2009, ***18***, 53-59.

[2] Y. Taninouchi, T. Uda, Y. Awakura, A. Ikeda, S.M. Haile ***J. Mater. Chem.***, 2007, 17, 3182.

463690 Optimization of Pt/C Particles Electrocatalytic Activity By the Control of Carbon Nanostructures *Via* a Hybrid Aerosol-Colloidal Process

Abstract:

An optimum electrocatalytic performance of carbon-supported platinum (Pt/C) catalyst is determined by many factors, including the structure of the carbon support. In the study herein, we evaluated the effect of carbon nanostructures on the electrocatalytic performance of Pt/C catalyst and determined the most suitable nanostructure to deliver the best performance. Submicron carbon particles were prepared *via* spray pyrolysis of phenolic resin prior to being deposited with Pt nanoparticles using a microwave-irradiation process. Nanostructuration of the carbon particles was performed *in situ* owing to the colloidal interaction between phenolic resin and polystyrene latex (PSL) template, and was controlled by changing the properties of PSL. According to the results of the electrochemical analyses, Pt/C catalysts with porous carbon structure delivered the best performance compared to those with dense and hollow carbon structure. An evaluation on the pores size revealed the importance of a good balance between meso- and macropores for an effective electrocatalysis, which can be explained from the hydrodynamic profile in the catalyst layer.

469355 Highly Active Robust F Doped Transition Metal Oxide Based Solid Solution Electro-Catalyst for Acidic Medium Oxygen Evolution Reaction in PEM Based Water Electrolysis

Abstract:

The vast consumption of fossil fuels for meeting the energy demand has led to major unsustainable environmental concerns due to excessive greenhouse gas emissions warranting the identification and development of clean non-carbonaceous energy sources.[1-5] In this regard, hydrogen has been identified as a clean, non-carbonaceous and potential energy source with superior energy density than currently adopted carbonaceous energy sources for meeting the global energy demand. However, efficient and economic production along with cost-effective storage and distribution of hydrogen utilizing the clean non-carbonaceous approach is of paramount importance before universal adoption of hydrogen as a clean energy source in the impending non-carbonaceous fuel economy. Along these lines, electricity driven water splitting (water electrolysis) is a frontrunner among other clean pollution free approaches for hydrogen production (having low carbon footprint). The commercial development of water electrolysis is thwarted due to the need of expensive and precious noble metals based electro-catalysts (Pt, IrO_2) which exhibit excellent electro-catalytic activity and stability for oxygen evolution reaction (OER) in acid assisted proton exchange membrane (PEM) based water electrolysis. Hence, the identification and development of novel reduced noble metal containing electro-catalysts exhibiting excellent electro-catalytic activity and superior long term electrochemical stability similar/superior to state of the art OER electro-catalyst IrO_2 in harsh and highly acidic operating conditions of OER, will aid in the reduction of capital cost of water electrolysis cells and thus, its progression towards commercialization.[6]

With this aim in mind, exploiting theoretical first principles calculations of the total energies and electronic structures, fluorine doped transition metal (TM) oxide based solid solution electro-catalyst containing significantly lower noble metal content has been identified exhibiting surface electronic structure and electro-catalytic activity similar to that IrO_2.The thin films of 10 wt. % F doped $(TM_{1-x}Z_x)O_2$ (x=0.2, 0.3, 0.4; Z is a noble metal) coated on Ti foil were synthesized and studied as electro-catalyst for OER in acidic media. **Fig. 1** shows the SEM micrograph showing the nanoscale architecture of the representative composition $(TM_{0.8}Z_{0.2})O_2$:10F. Elemental x-ray maps show homogeneous distribution of elements throughout the material without any segregation at specific site.

Electrochemical characterization of the synthesized electro-catalysts has been carried out in a three-electrode configuration using 0.5 M sulfuric acid (H_2SO_4) solution as a proton source as well as the electrolyte, Pt wire as counter electrode and Hg/Hg_2SO_4 as the reference electrode (+0.65 V with respect to normal hydrogen electrode, NHE), with a scan rate of 10 mV/sec and at temperature of 40^0C. $(TM_{1-x}Z_x)O_2$:10F (for x = 0.2, 0.3, 0.4) electro-catalysts (total loading = 0.3 mg/cm^2) exhibit excellent electrochemical activity for OER with significantly reduced onset potential (~1.35 *vs* RHE) compared to that of IrO_2 (~1.43 *vs* RHE) (total loading=0.3 mg/cm^2). In addition, $(TM_{1-x}Z_x)O_2$:10F displayed outstanding ~8, ~14 and ~15 fold improved electro-catalytic activity at ~1.45 V (*vs* RHE) compared to IrO_2. Following chronoamperometry test conducted in 0.5 M H_2SO_4 solution at ~1.55 V (*vs* RHE) for 24 h, minimal loss in current density as well as excellent long term electrochemical stability was observed for $(TM_{0.8}Z_{0.2})O_2$:10F, similar to that of IrO_2.

These results show the promise of $(TM_{1-x}Z_x)O_2$:10F electro-catalyst portending ~80% reduction in noble metal content displaying significantly much higher electro-catalytic activity than IrO_2 and superior long term stability. In addition, these results show the promise of the formation F-doped solid solution in improving the electronic, physical, chemical and electro-catalytic properties of the doped TM oxide structures in acidic media. This will aid in significant reduction in the overall capital cost of PEM based water electrolysis for efficient and economic production of hydrogen. The results of these studies will be presented and discussed.

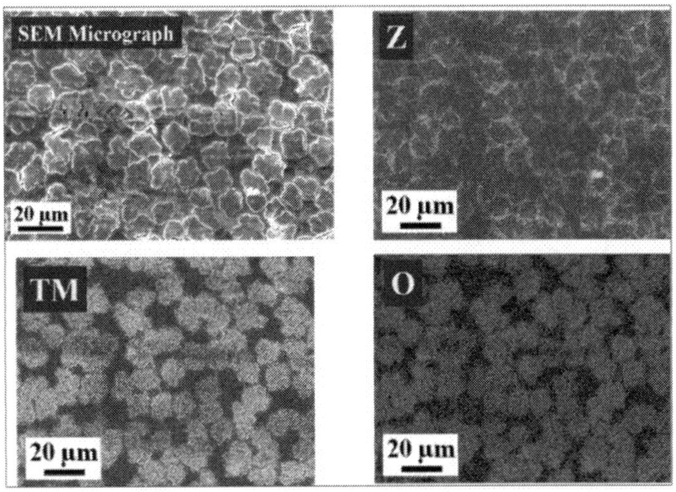

SEM micrograph with elemental x-ray maps of $(TM_{0.8}Z_{0.2})O_2$:10F electro-catalyst

References:

1 Patel, P. P. *et al.* Nanostructured robust cobalt metal alloy based anode electro-catalysts exhibiting remarkably high performance and durability for proton exchange membrane fuel cells. *Journal of Materials Chemistry A* **3**, 14015-14032 (2015).

2 Patel, P. P. *et al.* WO_3 based solid solution oxide–promising proton exchange membrane fuel cell anode electro-catalyst. *Journal of Materials Chemistry A* **3**, 18296-18309 (2015).

3 Patel, P. P. *et al.* High performance and durable nanostructured TiN supported Pt 50–Ru 50 anode catalyst

for direct methanol fuel cell (DMFC). *Journal of Power Sources* **293**, 437-446 (2015).

4 Patel, P. P. *et al.* Vertically aligned nitrogen doped $(Sn,Nb)O_2$ nanotubes – Robust photoanodes for hydrogen generation by photoelectrochemical water splitting. *Materials Science and Engineering: B* **208**, 1-14, (2016).

5 Patel, P. P. *et al.* Nitrogen and cobalt co-doped zinc oxide nanowires – Viable photoanodes for hydrogen generation via photoelectrochemical water splitting. *Journal of Power Sources* **299**, 11-24, (2015).

6 Datta, M. K. *et al.* High performance robust F-doped tin oxide based oxygen evolution electro-catalysts for PEM based water electrolysis. *Journal of Materials Chemistry A* **1**, 4026-4037, (2013).

Acknowledgements:

The authors gratefully acknowledge the financial support of NSF-CBET grant# 1511390. The authors also acknowledge the Edward R. Weidlein Chair Professorship funds and the Center for Complex Engineered Multifunctional Materials (CCEMM) for partial support of this research.

467668 Superwetting Nanoarray Electrodes for Gas-Involved Electrocatalysis

Abstract:

Superwetting Nanoarray Electrodes for Gas-Involved Electrocatalysis

Yingjie Li, Wenwen Xu, Zhiyi Lu, Xiaoming Sun*

Beijing University of Chemical Technology, Beijing 100029, China, Email: sunxm@mail.buct.edu.cn

Electrochemical gas-involved reactions can be divided into two types: gas-evolution reaction and gas-consumption reaction, both of which are crucial for a variety of energy conversion processes and industries (e.g. HER and ORR). For gas-evolution reaciton, if generated gas bubbles pin at the electrode surface and cannot escape from the surface in time, the accumulated bubbles will reduce the effective electrode surface area, increase diffusion resistance and enlarge polarization effect, resulting in more energy consumption. How to construct a novel electrode to promote gas bubble release is critical for improving the electrochemical efficiency besides activity improvement. Inspired from bio-inspired superwetting surfaces, we found that the interface behavior of electrode could be tuned by surface architecture construction, for example, transferring from aerophobic to superaerophobic by engineering a series of nanoarray electrode, eg. MoS_2 (Fig. 1), pine-shaped Pt, NiFe LDHs and Cu films [1-7]. This kind of superaerophobic electrodes could decrease the critical size of gas overflowing from the surface by cutting the three phase contact lines into discontinue dots, and thus reduce the diffused impedance and maintain the integrity of the solid−liquid interface that is necessary for electrocatalysis (e.g. water splitting and hydrazine fuel cells). On the other hand, for gas-consumption reaction, the construction of nanoarray ¡°superaerophilic¡± electrode could accelerate gas diffusion to reaction zone via gas-phase to solve the issue of low solubility and slow diffusion of gases in traditional electrocatalysiss reaction system with limited current density (eg., ORR). Therefore, the construction of superwetting nanoarray electrodes is imperative to improve gas transport at electrodes surface and to enhance activity and stability of electrodes.

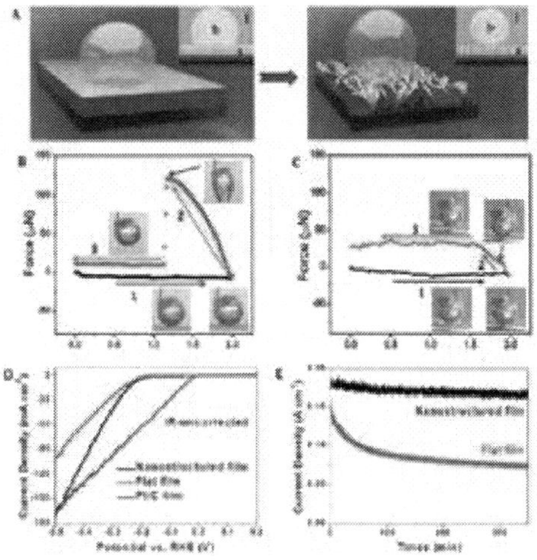

Fig. 1 (A) Schematic illustration of adhesion behaviors of gas bubbles on flat film (left) and nanostructured film (right), (B) and (C) Adhesive forces measurements of the gas bubbles on flat and nanostructured MoS_2 films, (D) Polarization curves (without IR-correction) of MoS_2 and Pt/C catalysts (E) Stability testing of flat and nanostructured MoS_2 electrodes

References

[1] Z. Lu, W. Zhu, X. Yu, H. Zhang, Y. Li, X. Sun, X. Wang, H. Wang, J. Wang, J. Luo, X. Lei, L. Jiang. Adv. Mater. 26 (2014), 2683-2687.

[2] Y. Li, H. Zhang, T. Xu, Z. Lu, X. Wu, P. Wan, X. Sun, L. Jiang. Adv. Funct. Mater. 25 (2015), 1737-1744.

[3] Z. Lu, M. Sun, T. Xu, Y. Li, W. Xu, Z. Chang, Y. Ding, X. Sun, L. Jiang. Adv. Mater. 27 (2015), 2361-2366.

[4] X. Liu, Z. Chang, L. Luo, T. Xu, X. Lei, J. Liu, X. Sun. Chem. Mater., 26 (2014), 1889-1895.

[5] M. Sun , Z. Lu , L. Luo, Z. Chang , X. Sun. Nanoscale, 8 (2016), 1479-1484

[6] W. Xu, Z. Lu, P. Wan, Y. Kuang, X. Sun. Small, 12 (2016), 2492¨C2498

[7] M. Jiang, Y. Li, Z. Lu, X. Sun, X. Duan. Inorg. Chem. Front. 2016, DOI: 10.1039/C5QI00232J.

Biography

Professor Xiaoming Sun gained his B.S. degree and Ph.D. in Department of Chemistry, Tsinghua University in 2000 and 2005, respectively. After postdoctoral work at Stanford University, he joined State Key Laboratory of Chemical Resource Engineering, Beijing University of Chemical Technology at 2008. His main research interests focus on separation and assembly of inorganic nanostructures, synthesis and separation of carbon nanomaterials and their composites, and structure control, opto-/electro-property investigations of nanoarrays and superwetting nanoarray electrodes for energy sciences. He has authored 63 journal articles (eg. ***J. Am. Chem. Soc., Angew. Chem. Int. Ed., Adv. Mater., ACS Nano.***), which have been cited >4600 times.

460470 Nanocatalysts for Water Splitting

Abstract:

Developing a clean and sustainable energy system is essential for the earth's sustainability. The hydrogen/water energy cycle, a clean and sustainable system, receives of great interest recently. Water splitting is one of the key reactions in the hydrogen/water energy cycle. Developing active and cost-effective nanocatalysts for water splitting is important to water splitting for the production of clean H_2 energy sources. This work will present various nanocatalysts for solar-driven and electric-driven water splitting reactions, especially for hydrogen evolution reaction (HER). The co-catalysts decorated Cu_2O/CuO photocathodes and MO/M heterostructured electrocatalysts were prepared by various interfacial engineering techniques. Combining advanced characterization and DFT computational techniques is capable of providing the insight of the heterostructured nanocatalysts for HER. It was found that the heterostructured interface plays a vital role and provides a highly active catalyst for HER. The design strategy for the HER catalysts can be developed via the understanding of the interfacial properties and reactions.

470723 Hydrogen Generation from Hydrous Hydrazine over Nickel-Doped Ceria Catalysts Prepared By Solution Combustion Synthesis

Abstract:

Hydrous hydrazine is considered to be a promising hydrogen storage material due to its high hydrogen content (8.0 wt.% for hydrazine monohydrate) and simple byproduct (only nitrogen) via complete decomposition. The key to exploit hydrous hydrazine in practical hydrogen generation applications is to develop efficient, selective, and economic catalysts for the decomposition. In this work, nickel-doped cerium oxides as new non-noble metal catalysts were prepared by solution combustion synthesis (SCS) of CeO_2, followed by impregnation of Ni or by one-step SCS of Ni/CeO_2. The effect of catalyst preparation conditions, including metal weight percentage (10%-80%), type of fuels (glycine, hydrazine, ethylene glycol, etc), ratio of fuel to oxidizer (0.5-3), and some additives on the properties of catalysts were investigated to optimize synthesis conditions and then their catalytic performance were tested. The experimental results demonstrate that the Ni/CeO_2 catalysts via SCS possess high catalytic activity and selectivity to generate hydrogen from hydrous hydrazine.

464096 Network Structure in Hybrid Solid Polymer Electrolytes (Invited Talk)

Abstract:

Batteries are of utter importance for energy storage and reducing carbon footprint to combat global warming and foster a sustainable future of our planet. To fabricate safer and more powerful energy devices including electric vehicles, one of the grand challenges is to fabricate mechanically robust and ionic conductive solid polymer electrolytes (SPEs) that form stable interface with electrodes. We herein report the design, synthesis and characterization of hybrid electrolytes with inorganic polyhedral oligomeric silsesquioxanes (POSS) as the crosslinker. The SPEs were prepared by a facile one-pot reactions between octakis[3-(glycidyloxy)propyldimethylsiloxy]silsesquioxane and bis(3-aminopropyl) terminated poly(ethylene glycol) (PEG) in the present of bis(trifluoromethane)sulfonimide lithium salt (LiTFSI). Uniform distribution of POSS and LiTFSI in the SPEs were confirmed by SEM-BSD and SEM-EDX. These hybrid network SPEs show promising ionic conductivity, mechanical properties, and lithium dendrite growth resistance.

455759 Design of High Transference Number Battery Electrolytes

Abstract:

High transference number electrolytes, those in which the electrochemically active lithium cation diffuses faster than its electrochemically inert counterion, have the potential to improve battery performance by limiting concentration polarization. High cation transference electrolytes are typically polymers with anions fixed to their backbone (ionomers), thereby only allowing cation diffusion over large length scales. In their dry state, such polymer electrolytes generally have substantially lower ion conductivity, particularly at room temperature, than liquid electrolytes, negating the advantages that polymers provide in terms of safety or Li^+ transference. Therefore, our research has focused on identifying routes to improve ion conductivity of ionomers through small molecule plasticization or improve Li^+ transference numbers of liquid electrolytes by using ionomers as the salt. Here, we have designed a series of poly(ethylene oxide) (PEO) based ionomers containing lithium sulfonate to explore these concepts. Our initial results on improving transference numbers of liquid electrolytes using these polymers as salts will be presented. By tuning the molecular weight of the polymer, solvent donor number, and concentration of polymer in solution, we begin to understand the relationship between electrolyte characteristics and performance. Preliminary results show that in the case of poorly dissociated ions, the ion and counterion move together, with a transference number of close to 0.5. However, in the case where ions are well dissociated, very high cation transference numbers (0.94) can be achieved.

461836 Effect of Transporting Enhancer in Fe_3O_4 Li-Ion Battery Anodes

Abstract:

Battery electrodes are complex mesoscale systems comprised of active material, conductive agents, current collector and binders [1]. For the improved design of a composite electrode, both electron and ion transport are critical factors to improve electrochemical performance. However, research rarely focuses on both factors.

The present work focuses on enhancing electron and ion transport in high capacity anode systems, by introducing a poly[3-(potassium-4-butanoate) thiophene] (PPBT) as a binder component and a polyethylene glycol (PEG) surface coating on Fe_3O_4 nanoparticles. PPBT is a water soluble carboxylate functionalized polythiophene, which seems favorable for forming porous structures. The PEG coating reduced the aggregate size and improved dispersion. Together the PPBT water system and PEG coating helped improve capacity retention by effectively decreasing charge-transfer resistance. PPBT and PPBT carbon composites also exhibited high electronic conductivity compared to PVDF, which was used as a reference due to its common use as a binder in Li-ion battery electrodes. SEM images illustrated that the PEG coating reduced the size of aggregates as compared to a Fe_3O_4/carbon/PPBT electrode from ~5mm to ~0.5-1mm. This improved dispersion, which positively contributed to battery performance. Furthermore, FT-IR and XPS provided evidence of chemical interaction between PPBT and Fe_3O_4 nanoparticles. Strong interactions between a binder and surface of active materials have been reported to improve stability [2]. The use of a PPBT binder and PEG surface coating improved both ion and electron transport, which provides a new approach for high-capacity energy materials [3].

An ongoing portion of this work takes a closer look at the active material, Fe_3O_4, and further studies the difference in particle sizes and PEG coating on battery performance. The PEG coating has already been shown to reduce aggregate size and improve dispersion [3]. Further testing is ongoing to study how crystallite size (10 nm vs. 20 nm) of nanoparticles affects battery properties. Preliminary SEM results indicate that at higher nanoparticle size (20nm), the aggregates from PEG are closer in size to the nanoparticles, which begs the question of how this impacts battery performance. Future testing is still needed to confirm this hypothesis and determine the "critical" nanoparticle size for greatest battery performance by enhancing the active material of the electrode.

[1] Y. H. Kwon, M. M. Huie, D. Choi, M. Chang, A. C. Marschilok, K. J. Takeuchi, E. S. Takeuchi, E. Reichmanis, *ACS Appl. Mater. Interfaces, 2016, 8* (5), 3452–3463.

[2] D. Mazouzi, B. Lestriez, L. Roué, D. Guyomard, *Electrochem. Solid-State Lett.* **2009**, *12*, A215.

[3] Y. H. Kwon, K. Minnici, M. M. Huie, A. C. Marschilok, K. J. Takeuchi, E. S. Takeuchi, E. Reichmanis. Electron/Ion Transport Enhancer in High Capacity Li-ion Battery Anodes, *submitted.*

466862 Integration of Ultrathin Polyaniline Films into Carbide Derived Carbon Supercapacitors Via Oxidative Chemical Vapor Deposition

Abstract:

The use of conducting polymers as the active materials for supercapacitors is a promising approach to improve the performance of supercapacitors[1, 2]. Conducting polymers enhance charge storage capacity through Faradaic redox reactions, and have the ability to be p- or n-doped. They are typically deposited via solvent-based techniques such as chemical bath deposition, electrodeposition, and casting from suspension. Nevertheless, challenges exist with making uniform conformal coatings via wet chemistry routes, especially with poor solubility of many conducting polymers and poor accessibility of the highly tortuous pore space within nanostructured electrodes. Oxidative chemical vapor deposition (oCVD) allows one to conveniently bypass these challenges as it is a solvent-free method where the reagents for oxidative polymerization, the monomer and oxidizer, are heated to vapor form that can easily penetrate into pores and polymerize (Figure 1). By understanding the oCVD polyaniline (PANI) synthesis parameters (reagent flow rates, reactor pressure, substrate temperature, saturation pressure), the reaction and diffusion rates can be carefully controlled to optimally coat the carbide derived carbon (CDC) electrode. By flowing vapors of aniline and antimony pentachloride (a strong oxidizer), a solid thin PANI film can be directly polymerized on a substrate. FTIR, UV-vis, and XPS confirm that oCVD produces an electrochemically active emeraldine state of PANI. The pore distribution of the CDC electrode, as determined by N_2 sorption, remains mostly unchanged before and after coating because the fine oCVD control over film thickness allows the pores to be coated without filling them, thereby maintaining the electrochemical double layer of the intrinsic carbon electrode, while imparting additional energy density through Faradaic redox reactions from PANI.

In this work, we are realizing a strategy proposed by Simon and Gogosti[3] to improve both the energy and power densities of electrochemical capacitors by integrating conducting polymers into nanostructures electrodes. In particular, we demonstrate that charge storage capacity can be improved significantly through the integration of ultrathin (<30 nm) PANI films into mesoporous Mo2C-derived carbide derived carbon (CDC, 800 °C synthesis temperature) electrodes using a single-step oCVD process. The ultrathin coating preserves to a large extent the native electrode surface area and pore distribution while simultaneously improving capacitance. The oCVD process allows PANI to be integrated into pores as small as 1.7 nm, and the oCVD PANI integrated supercapacitors have a specific capacitance more than 100% higher than that of bare CDC ones (136 F/g for 11 wt% of PANI in the CDC electrode vs. 60 F/g for bare CDC at 10 mV/s). This yields a PANI-only capacitance of ~690 F/g, which is close to the theoretical value of 750 F/g[4]. Even after high scan rates of over 100 mV/s, the added pseudocapacitance from PANI remains evident (Figure 1). The composite device exhibits excellent cyclability, which decreases only by 10% of the initially stabilized value (~100 F/g) after 10,000 cycles. To our knowledge, this work is the first reported synthesis of PANI via oCVD and demonstrates oCVD's advantages compared to other approaches in the integration of ultrathin PANI films into carbon supercapacitors.

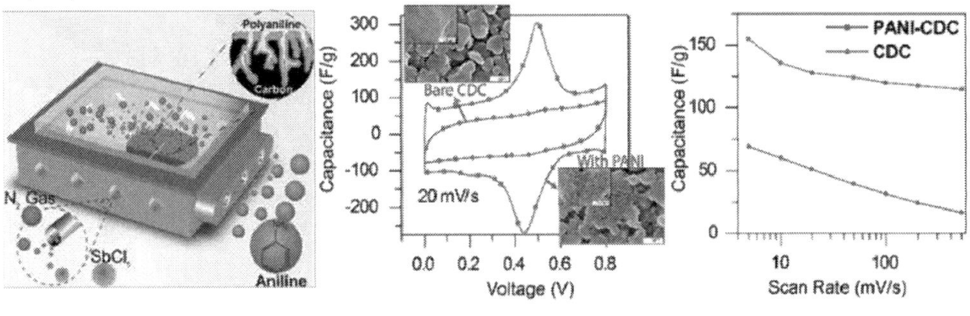

Figure 1. Left: oCVD process where the oxidant (SbCl5, orange) and monomer (aniline, green) vapors enter the reaction chamber and surface polymerize onto the electrode. Middle: A comparison of the cyclic voltammograms for both bare (red) and polyaniline (PANI) (green) electrodes at 20 mV/s. Inset shows SEMS of bare and PANI-coated CDC electrodes. Right: Rate performance of both devices based on cyclic voltammetry.

1. Laforgue, A.; Simon, P.; Sarrazin, C.; Fauvarque, J.-F. Journal of Power Sources 1999, 80, (1–2), 142-148.

2. Liu, D. Y.; Reynolds, J. R. ACS Applied Materials & Interfaces 2010, 2, (12), 3586-3593.

3. Simon, P.; Gogotsi, Y. Nature Materials 2008, 7, (11), 845-854.

4. Snook, G. A.; Kao, P.; Best, A. S. Journal of Power Sources 2011, 196, (1), 1-12.

472346 Controlling the Li-Air (O_2) Discharge Process with a Gel Polymer Electrolyte

Abstract:

The 2015 climate change conference in Paris espoused the need to stem the rise in global temperatures to below 2 °C. Therefore, there is a need to shift from fossil fuel use in electricity generation and transportation – industries responsible for the bulk of carbon dioxide (CO_2) emissions. Alternative energy conversion sources such as solar and wind are promising, but their intermittent nature affects long-term adoption. To complement these new conversion technologies, energy storage media such as batteries and fuel cells among others have been explored. The compact nature of batteries make them attractive over other energy storage media. Of commercially available batteries, lithium-ion batteries are the most energy-dense, and are used in electric vehicles and portable electronic devices. However, Li-ion batteries suffer from high cost limiting their use for stationary grid applications, and their energy densities are not high enough to allow for electric vehicles to compete with gasoline-powered cars in terms of mileage and cost. To address these limitations of Li-ion, other battery chemistries such as metal-air and metal-sulfur are being investigated. Lithium-air (O_2) batteries have the highest theoretical gravimetric energy density and have garnered tremendous research interest.

Li–O_2 chemistry is governed by the reduction of oxygen during discharge and the oxidation of the reduced oxygen discharge product during charge. In conventional Li–O_2 cells, non-aqueous liquid electrolytes are often used. The oxygen reduction chemistry is dominated by a 2 mol e^-/mol O_2 (peroxide) process that some have attributed to be partially responsible for the sluggish reduction and oxidation kinetics, limited current rate, and poor capacity retention of Li–O_2 batteries. Battery chemistries such as Na–O_2 and K–O_2 that utilize a 1 mol e^-/mol O_2 chemistry have been shown to support higher current rates and better energy efficiencies.

In this work, we incorporate ionic liquids in a polymeric matrix and show that controlling the lithium/ionic liquid molar ratio in the gel polymer electrolyte can allow for a 1 mol e^-/mol O_2 reduction process in a Li–O_2 battery. Ionic liquid cations has been shown to support a 1 mol e^-/mol O_2 process using cyclic voltammetry, but not in actual Li–O_2 cells, where a 2 mol e^-/mol O_2 process (and Li_2O_2) is observed. Furthermore, we use multiple spectroscopic tools to confirm for the first time the formation of a solid ionic liquid-superoxide discharge product. Knowledge gained from this work should spur development of newer and more stable ionic liquids and polymers that can allow for better long-term Li–O_2 cycling. In addition, the mechanism observed here could prove vital for other battery chemistries such as metal-air and metal-sulfur where controlling intermediate solubility is paramount.

460841 Water-Polymer Mobility and Distribution in Hydrated Aromatic Ionomer Thin Films

Abstract:

The confined state properties of ionomers are surprisingly different from bulk membranes and are still not well understood. Hydration of fuel cell ionomers in supported thin films leads to complex multimodal interactions among water, polymer chains and substrate and these interactions may lead to interesting mechanical and transport properties. In this work, thin films (~25-250 nm) of sulfonated aromatic ionomer, S-Radel were investigated to understand thickness and hydration effects on density, water-polymer distribution and mechanical properties. The density values obtained from quartz crystal microbalance and spectroscopic ellipsometry showed a similar trend with thickness to those obtained from neutron reflectometry. Thicker films were always more dense and less water-rich at the interface compared to thinner samples. A fluorescent rotor probe was incorporated into the polymer samples to predict the stiffness of the films using time resolved fluorescnece. The presence of voids rationalized greater mobility and lower stiffness in the thinner films in the dry state. When the same sample was hydrated, film density and interfacial water volume fraction significantly increased with the disappearance of voids. In addition, the water distribution near substrate interface in the aromatic s-Radel films was distinctly different from that in fluorocarbon based Nafion films seen in previous work. Antiplasticization, or stiffening of the films with the addition of water, was observed from fluorescence lifetime of rotor probes. The plasticization properties appeared to be controlled by film thickness, film density, and water mobility.

457801 Highly Proton-Conductive Polyelectrolyte Membranes with Supramolecularly Suppressed Water Swelling

Abstract:

Highly Proton-conductive Polyelectrolyte Membranes with Supramolecularly Suppressed Water Swelling

Joseph Aboki (jaboki@nd.edu), Shuangjiang Luo (sluo2@nd.edu), Ruilan Guo* (rguo@nd.edu)

Department of Chemical and Biomolecular Engineering, University of Notre Dame, Notre Dame, IN 46556-5637

Polyelectrolyte membranes based on sulfonated multiblock copolymers with alternating hydrophilic and hydrophobic blocks hold great potential as alternatives to benchmark Nafion®for fuel cell applications, due to their much better proton conductivity at low relative humidity (RH) levels and thermal stability. Previous studies suggest that long hydrophilic sequences (>10,000 g/mol) are necessary to effect sharp nanophase separation and form continuous proton conducting nanochannels enabling high proton conductivity at low RH, which, however, is always accompanied with excessive water swelling resulting in deterioration of dimensional stability and mechanical robustness of the membranes.

In this study, we report an innovative supramolecular strategy to suppress excessive water swelling by introducing triptycene structure units into the hydrophobic sequences of multiblock copolymers, which induce strong supramolecular interactions with the hydrophilic sequences via the so-called chain threading and interlocking effectively constraining the polyelectrolyte membranes from excessive swelling. Specifically, a series of triptycene-containing poly(arylene ether sulfone) multiblock copolymers were synthesized via a coupling reaction between phenoxide-terminated hydrophilic oligomers (BPS100) and fluorine terminated triptycene-containing hydrophobic oligomers (TRP0). The molecular weight (ranging from 5,000 to 20,000 g/mol) and end group functionality of oligomers were precisely controlled by adjusting the stoichiometric feeding ratio of monomers. Additionally, the ion exchange capacity (IEC) of the multiblock copolymers was controlled by coupling oligomers with even and uneven block lengths. The resulting multiblock copolymers yielded tough and ductile membranes via solution casting. In general, the multiblock copolymer membranes showed increasing water uptakes and proton conductivities with increasing block length. Anisotropic swelling behavior was observed for all the copolymer membranes suggesting nanophase separated lamellae-like membrane morphology. Most importantly, for copolymers with long hydrophilic sequences (> 10,000 g/mol), despite of their expectedly high water uptake, they all showed greatly reduced swelling ratios as compared to Nafion® and reported nontriptycene-containing multiblock copolymers. This observation indicates that the supramolecular chain threading and interlocking interactions induced by the triptycene units in the hydrophobic sequences are instrumental in suppressing membrane swelling and improving dimensional stability. For example, the multiblock copolymer with hydrophilic block length of 15,000 g/mol had a water uptake of 107%, an excellent proton conductivity of 0.155 S/cm in R.T. liquid water (almost 3 times that of a Nafion® 212) and a volume swelling ratio of just 27% (more than 65% reduction when compared to Nafion® 212 and previously reported multiblock copolymer systems). In addition, a super high liquid water proton conductivity (R.T.) of 0.224 S/cm was achieved by a multiblock copolymer with uneven block length (i.e., 10,000 g/mol of BPSH100 and 7,000 g/mol of TRP0). These multiblock copolymer membranes hold tremendous potential for applications in proton exchange membrane fuel cells and water desalination membranes.

468854 Role of Oligomeric Additives on P3HT/PCBM Domain Interfaces and Photovoltaic Performance

Abstract:

Controlling the nanophase separation of donor-type polymers and fullerene-based acceptors in bulk heterojunction (BHJ) thin films is crucial to optimizing the performance and enhancing the stability of organic photovoltaics (OPV) devices. A variety of strategies, including thermal or solvent annealing or incorporating small molecule processing aids or block copolymer compatibilizers, have been used as a way to mediate phase segregation and enhance OPV performance. In this presentation, we describe our efforts to characterize the influence of oligomeric donor-type materials based on phase segregation and performance of BHJ thin films comprising poly(3-hexylthiophene) and PCBM fullerenes. A combination of electron microscopy and X-ray and neutron scattering methods, augmented with insights from multiscale modeling, are used to develop a picture of how the size of low molecular weight P3HT additives and their loading level influences the distribution of the additives and their molecular orientation at donor-acceptor interfaces. These features of nanoscale structure are correlated with results from device performance testing, providing useful insights into the links between additive design, BHJ structure and device performance. While results are specific to P3HT/PCBM systems, we believe that the findings are generally applicable to systems where miscibility, crystallization, and packing at interfaces conspire to affect structure and performance of OPV devices based on the BHJ concept.

467773 Elucidating the Charge Transfer Mechanism in Conjugated Radical Polymers

Abstract:

Conjugated polymers bearing pendant nitroxide radicals have been an intriguing research topic in energy storage due to their dual electroactive nature with different redox potentials within the same polymer backbone. Recently, we demonstrated polythiophenes bearing pendant nitroxide radical groups such as the TEMPO radical (2, 2, 6, 6-tetramethyl-1-piperidinyloxy). Polythiophenes possess a high oxidation potential with tunable electronic conductivity, and the organic radicals possess a very high electron transfer rate. However these polymers exhibited worse battery performance than that of either PTMA (the polymeric form of TEMPO radical) or polythiophenes alone. One reason for this phenomenon is that conjugated polymers and nitroxide radicals are both redox active and that intra and inter-charge transfer is possible. As such, it is relevant to investigate the charge transfer pathways in such a system. In this study, we report the systematic study of polythiophenes carrying pendant TEMPO radicals. By engineering the alkyl chain spacer length between the thiophene ring and TEMPO radical, we tune the electronic and electrochemical properties of these electroactive polymers. In addition, the charge storage mechanism during the galvanostatic charge/discharge process is also interrogated and the respective contributions from each redox moiety are also discussed. Evidence of internal charge transfer is observed using a potentiostatic pulse technique. These results provide a path forward in the design of conjugated radical polymers for energy storage.

472062 Preparation and Enhanced Properties of Surface Treated Silicon Carbide Powders Reinforced Polypropylene Thin Films

Abstract:

Polypropylene thin films reinforced with various loading levels of surface treated silicon carbide (SiC) powders were prepared by extrusion process. The surface treatment of silicon carbide powders was successfully proved by Fourier Transform Infrared (FT-IR) spectroscopy. The dispersion quality of SiC powders with and without surface modification is monitored by an oscillatory rheological investigation. Differential scanning calorimeter (DSC) and X-ray diffraction (XRD) tests were applied to investigate the crystal structure and crystallinity of the thin films. And the thermal stability was studied by thermogravimetric analysis (TGA). The morphology of thin films with different SiC loadings were characterized with a field emission scanning electron microscope (SEM). In addition, the dielectric permittivity and optical property of the thin films were studied systematically assessed as well.

469435 Magneto-Responsive Bionanocomposite Hydrogels As Injectable Scaffolds for Osteochondral Tissue Regeneration

Abstract:

Osteoarthritis (OA) is a degenerative joint disease that occurs when articular cartilage begins to degrade and can result in severe disability in the lower extremities of the body. Every year, 3.1 million surgeries are done to treat the damage caused by OA[1], and the annual cost of non-steroidal anti-inflammatory drugs (NSAIDs) approaches $4 billion[2]. Current treatment options, such as osteochondral tissue grafting, microfracturing, and total knee replacements, are only effective in alleviating symptoms in the short-term and fail to promote the regeneration of fully functional articular cartilage. Thus, more effective therapies are needed to regenerate damaged tissue in a manner that can recreate the stratified, heterogeneous structure of articular cartilage.

Hydrogel scaffolds capable of delivering viable stem cell populations to the cartilage defect and responding to external stimuli, such as a magnetic field, could be used to influence stem cell behavior and guide the reconstruction of a heterogeneous tissue structure. In particular, polymer-based hydrogels made from thermogelling macromers, such as poly(*N*-isopropylacrylamide) (pNiPAAm), offer promise as *in situ* forming injectable scaffolds capable of delivering viable stem cell populations *in vivo*. Furthermore, recent studies have discovered that cytocompatible, hydrophilic polyamidoamine (PAMAM) crosslinkers can counteract pNiPAAmâ€™s natural tendency to undergo syneresis post thermogelation,[3, 4] thus enabling these materials to completely fill irregularly-shaped tissue defects and promote integration with surrounding tissue.

The aim of this study was to investigate the efficacy of using a magneto-responsive injectable hydrogel to deliver viable stem cell populations and then stimulate osteochondral tissue regeneration through magnetically-actuated mechanical force. Functional paramagnetic iron (III) oxide Fe_3O_4 nanoparticles were incorporated into a dual-gelling

pNiPAAm-based hydrogel with degradable PAMAM-based crosslinking macromers. The physical properties of the hydrogel were assessed by rheology and scanning electron microscopy (SEM), and magnetic properties of the hydrogel were determined by superconducting quantum interference device (SQUID) magnetometry and tangential force measurements. *In situ* delivery of viable cell populations in the bionanocomposite hydrogels was demonstrated by fluorescent microscopy and biochemical assays. Finally, the effects of magnetic stimulation on stem cell differentiation were investigated.

1. *Arthritis Brochure. American Academy of Orthopaedic Surgeons*. Available from: www.aaos.org.

2. Van Manen, M.D., J. Nace, and M.A. Mont, *Management of primary knee osteoarthritis and indications for total knee arthroplasty for general practitioners.* J Am Osteopath Assoc, 2012. **112**(11): p. 709-15.

3. Ekenseair, A.K., Boere, K.W.M., Tzouanas, S.N., Vo, T.N, Kasper, K., Mikos, A.G.,, *Synthesis and Characterization of Thermally and Chemically Gelling injectable Hydrogels for Tissue Engineering' Biomacromolecules.* BioMacromolecules, 2012. **13**: p. 1908-1915.

4. Ekenseair, A.K., et al., *Structure-property evaluation of thermally and chemically gelling injectable hydrogels for tissue engineering.* Biomacromolecules, 2012. **13**(9): p. 2821-30.

466963 New Aligned Coaxial Nanofibers for Neural Tissue Engineering

Abstract:

New Aligned Coaxial Nanofiber Structures for Neural Tissue Engineering

Rachel Martin[1], Michael E. Mullins[1], Feng Zhao[2], Zichen Qian[2]

1 Department of Chemical Engineering, Michigan Technological University

2 Department of Biomedical Engineering, Michigan Technological University

Much effort has been recently focused on applying electrospun polymers as scaffolds in many areas of tissue engineering, including skin, bone, cardiovascular and neural tissues. In the case of electrospun fibers for neural tissue applications, some success has been met *in vitro*. Our group has demonstrated directed axonal outgrowth of dorsal root ganglia (DRG) when cultured on highly aligned electrospun nanofibers made from poly-L-lactic acid (PLLA); yet, less success was achieved during *in vivo* studies. As an alternative approach for neural applications, we created novel, engineered nanofiber structures via a coaxial electrospinning method. Several coaxial spinning nozzle designs were built and tested with several core/sheath combinations. The objective of this study was to create structures with a conductive polymer core and an insulating PLLA sheath using these coaxial electrospinning. We have designed cell culture studies using these new structures to supply an electric micro-current during *in vitro* cell studies and to observe the effect on axonal outgrowth. It has been previously shown that electrical stimulation can improve axonal outgrowth *in vitro*, and that axonal reconnections may be influenced using electrical stimulation. Due to its stable conductive properties, low toxicity, and biocompatibility we have focused on electrospinning poly(3,4-ethylene dioxythiophene)

(PEDOT) with polystyrene sulfonate(PSS) as the counter ion to improve solubility. The creation of aligned, coaxial PLLA/PEDOT-PSS nanofibers has been explored, and the resulting fibers characterized via fluorescence microscopy and SEM. The conductive properties of the PLLA/PEDOT-PSS fibers can then be studied using a four-point conductivity test. Once the physical and electrical properties have been confirmed, the aligned fibers are tested during *in vitro* cell cultures using chick DRG cells. By developing a tissue scaffold stage that can be used with a regular six-well plate, electrical stimulation can be provided along the conductive fibers during *in vitro* cell cultures. The axonal outgrowth is then measured and compared to cells cultured without electrical stimulation to determine if the axonal outgrowth was improved using electrical stimulation.

464236 Silk-Gold Nanorod Nanocomposite Films for Rapid Tissue Repair

Abstract:

Laser tissue welding is a tissue repair technique that utilizes a chromophore to convert photons into heat in order to seal apposed tissue edges by localized protein denaturation and interdigitation. Laser tissue welding has emerged as a promising alternative to sutures or staples conventionally used for sealing soft tissues, with the potential to limit surgical complications such as anastomotic leakage and wound dehiscence while also reducing healing times and scarring. However, insufficient closure strength, surgical inconvenience, and extensive thermal damage have hindered the clinical application of this technique, largely due to the unavailability of effective solders (surgical sealants). In this work, we describe the development of novel plasmonic silk nanocomposites for laser tissue welding that overcome these barriers in tissue welding. Leak pressure, burst pressure, and tensile strength measurements indicated that laser welding using plasmonic nanocomposites was able to restore significant mechanical integrity to the welded tissue.

459949 Continous Extrusion of Microtextured Composite Films

Abstract:

Micro-textured polymeric composite films filled with graphite nanoplatelets (GNP) were produced by continuous melt extrusion with a novel micro-patterned die. GNP dispersed with a significant extent inside the linear low density polyethylene (LLDPE) matrix. GNP addition into the polymer resulted a 30-fold increase of the in-plane thermal conductivity of the nanocomposite (8.8 W/m.K *vs* 0.3 W/m.K for pure LLDPE), and 10-fold increase of through thickness thermal conductivity. The anisotropy of the thermal conductivity was explained by the parallel alignment of the particles to the film machine direction. The electrical percolation threshold of the micro-textured films was found lower than the non-textured films. Micro-textures aided to increase convective heat transfer rate as a consequence of increased surface area like as fins of a heat exchanger.

449553 Effective and Industrially Relevant Compounding of Natural Fiber-Reinforced Thermoplastic Composites Via Solid-State Shear Pulverization

Abstract:

Conventional composite materials are most often made with glass fibers due to their high stiffness/strength to cost ratio, while carbon-fibers are also used frequently in advanced applications for their unparalleled mechanical properties. Recently, fibers derived from natural plants have been also explored as a technical composite filler material of choice due to a combination of practical advantages such as low density, low cost, and low wear to processing equipment. In addition, natural fibers uniquely bring about their environmental benefits such as sustainable production, CO_2neutrality, and minimal energy embodiment [1]. Accordingly, natural fiber composite materials have gradually gained prominence in the manufacturing industry. Specifically, flax- and hemp-reinforced composites possess advantages beyond mechanical performance and environmental benefits, including acoustic, optical, and haptic properties [2].

While melt compounding process with twin-screw extrusion is a conventional way to mass-produce short fiber composites, a specialized solid-state compounding technique called solid-state shear pulverization (SSSP) offers additional capabilities, especially when natural fibers are used as reinforcement [3]. Due to its low-temperature processing, unwanted thermal degradation of the natural fiber component is avoided, allowing for a wider range of thermoplastic matrix options. The SSSP process relies on mechanochemistry, rather than molten viscous flow, to not only effectively disperse the fillers and prepare homogeneous material, but also favorably de-bundle natural fibers from their technical to elementary fiber formats. This fiber hierarchy transition is expected to yield a high aspect ratio filler state, leading to superior physical properties in the composites [4].

The paper presents the results of the first processing-structure-property study in which SSSP is applied to short flax fiber materials and polyamide 6, a model bio-based plastic composite system. Mechanical property measurements were conducted on injection-molded specimens. Thermal characterization involved differential scanning calorimetry and thermogravimetric analysis. Fiber structure was probed with optical and electron microscopy. Fiber morphology and its interaction with the matrix polymer play a key role in dictating the mechanical and other physical performance of the composite.

[1] A. K. Mohanty, M. Misra, and L. T. Drzal, eds, Natural Fibers, Biopolymers, and Biocomposites, Boca Raton, CRC Press (2005).

[2] L. Pil, F. Bensadoun, J. Pariset, and I. Verpoest, Composites Part A 83, 193 (2016).

[3] K. Khait, S.H. Carr, and M.H. Mack, Solid-State Shear Pulverization, Boca Raton, CRC Press (2001).

[4] H. L. Bos, M. J. A. Van Degen Oever, and O. C. J. J. Peters, Journal of Material Science 37, 1683 (2002).

467992 Assembling Freestanding Conductive Polymer Tube Arrays at Liquid/Liquid Interface

Abstract:

Patterned tube array structures are mainly synthesized by template-assisted and substrate-initiated methods so far, which meet significant challenges in scale-up manufacturing and non-destructive transfer to desired support. Herein, freestanding tubular array of conductive polymers were successfully synthesized without using template or substrate. A "water/oil/water" triphasic reaction system was employed to control diffusion and reaction at separate "water/oil" and "oil/water" interfaces. The former controls reactant feeding rate and the latter determines reaction rate. By synergistic integration of diffusion and reaction in one reactor, freestanding tubular-arrayed conductive polymer thin films were successfully fabricated. Molecular dynamic simulation was performed to understand the reactant diffusion across the two interfaces. This "water/oil/water" triphasic method opens up a new approach to synthesize other patterned three dimensional materials by careful designing the reaction system.

471582 The Development of Ceramic Metal Oxide Membranes By Means of Reactive Electrospinning

Abstract:

Experts agree that carbon dioxide is the leading cause of the increase in global warming. To combat global climate change, Carbon Capture and Storage technology has been developed and implemented to reduce the amount of carbon dioxide and other malignant gases in the atmosphere. In efforts to contribute to this avenue of sustainable progress, we aimed to produce ceramic metal oxide membranes via reactive electrospinning with the intention of applying these novel nanofibrous membranes to greenhouse gas capture. Magnesium oxide membranes are considered viable candidates for adsorption processes because they exhibit high surface area, low density, high porosity and resistance to high temperatures and corrosion. The electrospinning solution consisted of 4MDa polyethylene oxide, acetic acid, magnesium methoxide, methanol and dichloromethane. During electrospinning, a voltage of 5-20 kV was applied to the solution in a flow-controlled needle housed in a humidity-controlled chamber. Once critical charge density was reached at the tip of the needle the solution formed a Taylor cone which ejected a continuous jet of magnesium oxide nanofibers. Another solution pursued was composed a 4 MDa polyacrylic acid solution which substitutes for the dichloromethane and polyethylene oxide. Experimental data was collected for varying flow rates, supplied voltage, solution composition, substrate concentration and chamber humidity. Infrared spectroscopy followed for analysis of product chemical composition. Fiber diameter and porosity were measured by Scanning Electron Microscopy (SEM). Ideally, the nanofibers exhibit a ceramic quality denoted by a visible, consistent rigidity under SEM analysis.

451870 The Design of Micelles for Molecular Diagnostics

Abstract:

Imaging modalities such as magnetic resonance imaging (MRI) are routine tools to characterize disease progression in the clinic. However, MRI has limitations as it aims to assign physical traits to assess the molecular state of the disease. In an effort to image and quantify molecular markers in vivo to accurately determine disease progression, we report on gadolinium (Gd)-incorporated, fibrin-targeting micelles for cardiovascular diseases, specifically targeting atherosclerosis. Micelles were self-assembled with DSPE-PEG2000-CREKA, DSPE-PEG2000-Cy7, and one of three different Gd-chelators: DSPE-PEG2000-DTPA(Gd), BSA-DTPA(Gd), or DSPE-DTPA (Gd). While the latter two molecules are commercially available and commonly used in nanomedicine and MRI, DSPE-PEG2000-DTPA(Gd) was synthesized for the first time to develop a chelating molecule with the same hydrophobic tail as the other micelle components. The particles were characterized and their targeting efficacy assessed in vitro and in vivo. Overall, we sought to gain additional insights regarding how molecular and micelle design can affect targeting and diagnostic capabilities and the outcome is discussed.

Specifically, CREKA micelles consisting of DSPE-PEG2000-DTPA(Gd) were spherical and approximately 10 nm confirmed by transmission electron microscopy (TEM) and dynamic light scattering (DLS). CREKA micelles incorporating either DSPE-DTPA(Gd) or BSA-DTPA(Gd) formed a mixed populations of spherical and cylindrical nanoparticles. Upon addition to fibrin clots in vitro, CREKA micelles with DSPE-PEG2000-DTPA(Gd) showed statistically significant binding capabilities compared to its non-targeting counterpart (no peptide). In comparison, CREKA micelles with DSPE-DTPA(Gd) or BSA-DTPA(Gd) showed no difference compared to its non-targeting counterparts. From these results, CREKA micelles incorporating DSPE-PEG2000-DTPA(Gd), or CREKA-Gd micelles, were used for further studies. T1 values decreased with increasing Gd content (2278-897 ms for 100% DSPE-PEG2000-CREKA micelles-25/75% DSPE-PEG2000-CREKA/DSPE-PEG2000-DTPA(Gd)) confirming efficacy as contrast agents. Upon injecting into late-stage atherosclerotic mice, enhanced signal was found at the aortic arch via MRI by both types of micelles, further confirmed via ex vivo fluorescence imaging. Clearance of micelles was found to be mainly through the liver and spleen, or the reticuloendothelial system, and H and E results of organ sections showed no morphological abnormalities.

In sum, novel nanoparticles for MRI were designed using various gadolinium chelators. Our data supports the importance of molecular design of self-assembling particles as it relates to functional and biological efficacy for disease diagnostics. Future studies will incorporate therapeutic components and dual MRI/positron electron tomography capabilities to enhance signal sensitivity.

472495 Controlled, Self-Directed Assembly of Novel in-Situ Forming Biodegradable Nanostructures for the Delivery of Ocular Therapeutics

Abstract:

The physiological and anatomical complexity of the eye make it a highly protected organ, substantially limiting drug delivery despite the ability to treat topically. Designing effective drug delivery systems to target specific areas of the eye is a major research challenge. Currently, topically applied drugs, including eye drops, make up over 90% of the ophthalmic drug market. Eye drops are grossly inefficient delivery vehicles, since less than 5% of the administered dose reaches the intraocular tissues. Thus, there is a considerable unmet need for more efficacious and convenient delivery systems of ocular therapeutics. Our work concentrates on engineering novel physically crosslinked nanocarriers, that are easily formulated using FDA-approved biodegradable polymers. These tailorable carriers provide solubility and prolonged bioavailability and residence time of the ocular therapeutic.

We have developed, designed, and synthesized novel in-situ forming nanocarriers, consisting of amphiphilic PLGA-PEG-PLGA. These nanostructures self-assemble in aqueous environments via reverse thermal gelation. By varying the lactic acid (LA) to glycolid acid (GA) ratio of the hydrophobic PLGA block, the molecular weight of PEG, the PLGA to PEG ratio and the polymer solution concentration, we produce injectable, in-situ forming nanogels that are optically clear at physiological temperatures. Characterization studies include optical property assessments, critical gelation temperature (CGT), rheology, critical micelle concentration (CMC), in-vitro degradation, and controlled and sustained drug release kinetics.

We demonstrate that carriers with a LA/GA ratio between 1 and 15, at compositions between 10%(w/v) and 25%(w/v), allow for over 90% visible light transmittance, gel formation between 34 and 37°C, and sustained and controlled in-vitro polymer degradation and drug release kinetics for over one week. Below the CGT, the triblock copolymer of interest forms a homogeneous micelle therapeutic injectable solution, held together by hydrogen bonding interactions.

Above the CGT, the solution swells forming an ordered packing, with hydrophobic forces dominating the 3D nano-network of bridged micelles with the drug embedded within the PLGA units. The controlled self-directed assembly of the amphiphilic molecules allows visible spectrum wavelengths to pass through the core-corona structure, making the gel optically transparent. The self-assembled polyester/polyether matrix slowly degrades via the cleavage of ester bonds, allowing for continuous and controlled drug release.

Changes in the physical and morphological states of the hydrogel are rapid and controllable, making this novel thermosensitive gel attractive for the delivery of ocular pharmaceutics. Our technology offers a more effective and efficient method of ocular therapy by providing a controlled and extended release of medication and has wide and far reaching potential to improve a number of ocular therapies.

465886 Molecular Gel Formation As a First Order Phase Transition

Abstract:

Self assembling systems based on dipeptide molecules that form ordered fibrous structures are seen as promising new materials for applications in tissue engineering, drug delivery, biosensors, photoelectronics, etc. However, the gelation mechanism in these types of systems has been largely overlooked and remains poorly understood, limiting the development of comprehensive design guidelines for specific applications. We explore the gel transition of the aromatic dipeptide molecule fluorenylmethoxycarbonyl diphenylalanine (Fmoc FF). The addition of water to a solution of Fmoc FF in dimethyl sulfoxide (DMSO) results in the self assembly of Fmoc FF molecules into a space filling fibrous network. We provide evidence that gel formation is associated with a first order phase transition resulting in nucleation and growth of strongly anisotropic crystals with high aspect ratios. The strength of attraction between Fmoc-FF molecules as a function of water concentration is estimated from long time self-diffusion measurements using ^1H NMR diffusion ordered spectroscopy (DOSY). The resulting phase behavior follows that observed for a wide range of other crystallizing nanoparticles and small molecules, consistent with the short-range nature of the intermolecular attractions. Furthermore, a lag time in the formation of the new phase is observed, indicating the existence of a free energy barrier to the formation of a crystal nucleus of critical size. The application of classical nucleation theory for a cylindrical nucleus indicates that one-dimensional crystal growth is driven by an imbalance of the surface energies of the ends and sides of the fiber. We believe that the discussed concepts can be extended and applied to other similar molecular gel systems, thereby providing guidelines that would facilitate the design of novel molecular gelators.

468014 DNA-Programmable Assembly of Enzyme Superlattices

Abstract:

Enzymes are one of Natureâ€™s most diverse and powerful nanoparticles. Assembling them into multicomponent crystals has been of great interest and can be accomplished through a variety of means, typically based on protein surface chemistry. The programmability of DNA makes it an attractive structure-directing ligand for assembling nanoparticles into well-defined materials. Recently, DNA has emerged as an effective ligand by which catalytically active proteins can be assembled into extended crystalline superlattices of CsCl symmetry. In these systems, protein surfaces are covalently functionalized with nucleic acid ligands, which direct interparticle interactions through Watson-Crick base pairing. In this study, we predicted that under certain conditions the anisotropic shape and surface chemistry of the enzyme could induce directional bonding interactions resulting in the assembly of a different lattice symmetry. To probe this, the enzyme catalase was functionalized with DNA ligands of varying flexibility and co-assembled with gold nanoparticles possessing different numbers of DNA bonding ligands. A phase transformation was observed with decreasing DNA flexibility and increasing gold nanoparticle ligand number from CsCl to Th_3P_4 symmetry, with mixtures of the two lattice symmetries present at intermediate states. Through fitting of the SAXS diffraction patterns, the composition of the mixtures was quantified. Coarse grained molecular dynamics simulations were used to further understand the energetics of the two phases. This is the first instance in these systems that phase transformation has been observed by tuning ligand flexibility, expanding on the means by which well-defined functional materials can be assembled from proteins using DNA.

468197 Microreactors As a Tool for Producing Polymer Nanoparticles By a Self-Assembled Process

Abstract:

Microreactors as a tool for producing polymer nanoparticles by a self-assembled process.

A. Tabernero, A. González-Garcinuño, Miguel A. Galán, Eva M. Martín del Valle

Department of Chemical Engineering; University of Salamanca, Salamanca, Spain

Microreactors have been used for these last years for improving nanomaterials synthesis because several reaction parameters can be easily controlled. This type of reactors provides a continuous shear stress on nanoparticles dispersion, allowing a uniform outcome and controlling the nanoparticles formation kinetics. In this work, we propose the use of a microreactor to avoid nanoparticles aggregation of a fructose biopolymer (levan). This polymer is amphiphilic and is able to produce nanoparticles by a self-assembled process and shows a great potential for biomedical applications. Furthermore, the effect of several experimental parameters on biopolymer particles aggregation and Z-potential will be studied in order to determine the best conditions to avoid nanoparticles aggregation. Microreactors nanoparticle results will be in addition compared with the results obtained with a conventional methodology to produce levan nanoparticles.Results indicate that particle size decreases from 400 nm. to 100 nm by using microreactors, whereas the Z-potential remains more or less the same (between -25 and -30 mV). Moreover, we studied the use of microreactors with the carboxymethylated form of the polymer (carboxymethyllevan), obtaining the same tendency. Therefore, microreactors can be used as a tool for producing nanoparticles of materials that can self-assemble in nanoparticles due to its amphiphilic nature.

471763 Polymer-Induced Liposome Aggregation: Toward the Application of Naked-Eye Bio-Detection

Abstract:

Signal amplification has been one of the important issues for detecting proteins or pathogens at low concentration regime. Here, we investigate a liposome clustering process induced by an amphiphilic triblock copolymer through hydrophobic interactions. The clustering is visible and thus can be used as an amplification platform for naked-eye detection. In this study, the liposomes are composed of dipalmitoylphosphatidylcholine (DPPC, di-C_{16}) and dihexanoylphosphatidylcholine (DHPC, di-C_6) induced by triblock copolymer linkers, (propylene glycol)$_m$ – (ethylene glycol)$_n$ – (propylene glycol)$_m$ ($PO_m - EO_n - PO_m$, reverse pluronic copolymers). The aggregation occurs when the two PPO hydrophobic blocks insert in the bilayers of different liposome resulting in a liposomal clusters. The clustering is presumably facilitated by the mismatch of acyl chain between DPPC and DHPC, as evident in the rapid increase of UV absorption and hydrodynamic radius (R_H). In contrast, the clustering of DPPC liposomes is hardly observed in the presence of polymer linkers. The experimental results are further corroborated by the simulation work, indicating that less energy is required for the polymer insertion into the interfaces between DPPC and DHPC compared to the pure DPPC regime. Moreover, we observed distinct aggregation kinetics between polymer linkers with different hydrophobicity. Specifically, increase of the polymer concentration leads to more obvious clusters for the relatively hydrophilic polymer linker ($PO_{14} - EO_{24} - PO_{14}$), but inhibits the aggregation for hydrophobic polymer linkers (i.e. $PO_{25} - EO_7 - PO_{25}$ and $PO_{26} - EO_{20} - PO_{26}$). This phenomenon is explained by the mechanism that the hydrophobic polymers strongly anchor at and fast occupy the "active" (DHPC-rich) sites of bilayer compared to the hydrophilic ones, whose hydrophobic PPO may frequently associate with and dissociate from the "active" sites, enhancing the stringing process between liposomes.

472283 Information-Directed Assembly of Dynamic Covalent Molecular Ladders

Abstract:

Nucleic acids present the most versatile class of materials for producing nanostructures to date and, through careful consideration of their residue sequence, can be designed to self-assemble via the hybridization of complementary strands into arbitrary structures with nanometer precision. Unfortunately, the versatility of nucleic acid assemblies is tempered by their thermal and mechanical instability, attributable to the weakness of the hydrogen bonds that hold the strands together. The substitution of the canonical nucleic acid base pairs with covalent inter-strand interactions precludes this instability and yields a unique and powerful nanofabrication strategy wherein the complex, information-driven assembly of nucleic acids is combined with the strength of covalent bonds. Notably, the creation of exquisite nanostructures by self-assembly and the toughness imparted by covalent bonds are generally perceived as mutually exclusive owing to the prevalent irreversibility of covalent bond-generating reactions. Fortunately, several covalent interactions are known to be reversible under particular reaction conditions, enabling the error correction mechanism that is essential for the selective fabrication of supramolecular structures. As a result of the enormously greater strength and directionality offered by covalent bonds in comparison to the weaker interactions observed in biology, these 'dynamic' covalent chemistries offer an elegant approach to nanostructure assembly that combines complexity AND toughness. Here, we demonstrate the sequence-dependent self-assembly of peptoid-based oligomers to afford molecular ladders with up to sixteen rungs via scandium(III)-catalyzed imine metathesis. Moreover, we examine the ability of this approach to generate branched ladders that can assemble into larger, robust, multi-dimensional nanostructures.

472204 Halide Perovskite Nanoplatelets

Abstract:

Colloidal perovskite nanoplatelets are a promising new class of semiconductor nanomaterials, exhibiting bright luminescence, tunable and spectrally narrow absorption and emission features, strongly confined excitonic states, and facile colloidal synthesis. Here, we demonstrate the high degree of spectral tunability achievable through variation of the cation, metal, and halide composition as well as nanoplatelet thickness. We synthesize nanoplatelets of the form $L_2[ABX_3]_{n-1}BX_4$, where L is an organic ligand (octylammonium, butylammonium), A is a monovalent metal or organic molecular cation (cesium, methylammonium, formamidinium), B is a divalent metal cation (lead, tin), X is a halide anion (chloride, bromide, iodide), and n-1 is the number of unit cells in thickness. We show that variation of n, B, and X leads to large changes in the absorption and emission energy, while variation of the A cation leads to only subtle changes but can significantly impact the nanoplatelet stability and photoluminescence quantum yield. Furthermore, mixed halide nanoplatelets exhibit continuous spectral tunability over a 1.5 eV spectral range, from 2.2 eV to 3.7 eV. These results demonstrate the versatility of colloidal perovskite nanoplatelets as a material platform, with tunability extending from the deep UV, across the visible, into the near-IR. In particular, the tin-containing nanoplatelets represent a significant addition to the small but increasingly important family of lead- and cadmium-free colloidal semiconductors.

467326 Ultrafast Photoexcited Carrier Dynamics in Ligand-Exchanged PbSe Nanocrystal Films: Lifetime, Mobility, Diffusion, and Interfacial Charge Transfer

Abstract:

Colloidal semiconductor nanocrystals have been used as building blocks for electronic and optoelectronic devices ranging from field effect transistors to solar cells. Properties of the nanocrystal films depend sensitively on the choice of capping ligand to replace the insulating synthesis ligands. Thus far, ligands leading to the best performance in transistors result in poor solar cell performance, and vice versa. To gain insight into the nature of this dichotomy, we used time-resolved terahertz spectroscopy measurements to study the mobility and lifetime of films of monodisperse lead selenide (PbSe) nanocrystals prepared with five common ligand-exchange reagents. Non-contact terahertz spectroscopy measurements of conductivity were corroborated by contacted van der Pauw measurements of the same samples. The films treated with different displacing ligands show more than an order of magnitude difference in the peak conductivities and a bifurcation of time-dynamics. Inorganic chalcogenide ligand-exchanges with sodium sulfide (Na_2S) or ammonium thiocyanate (NH_4SCN) show high THz mobilities above 25 $cm^2V^{-1}s^{-1}$, which is desirable for transistors, but nearly complete decay of transient photocurrent within 1.4 ns. The high mobility with NH_4SCN and Na_2S exchanges is more than offset by their short lifetimes and results in diffusion lengths of only ~200 nm. In contrast, ligand exchanges with 1,2-ethylenediamine (EDA), 1,2-ethanedithiol (EDT), and tetrabutylammonium iodide (TBAI) show ~5x lower mobilities but much longer carrier lifetimes, with ~30% of photoexcited carriers remaining for >10 ns. The long lifetimes with EDA, EDT, and TBAI yield diffusion lengths of at least 500 nm, which is approaching the film thickness desired for strong light absorption in solar cells. This bifurcated behavior may explain the divergent performance of field-effect transistors and photovoltaics constructed from nanocrystal building blocks with different ligand exchanges. Moreover, we have investigated ligand-exchanged PbSe films comprised of two different nanocrystal sizes. We will report on the effects of band alignment, ligand chemistry, and structural order on interfacial charge transfer.

Reference: Guglietta, Diroll et al., *ACS Nano*, 2015.

470983 Band Edge Engineering of Hydroxide Nanoparticles for Semiconductor and Electrochemical Applications

Abstract:

Band Edge Engineering of Hydroxide Nanoparticles for Semiconductor and Electrochemical Applications

Matthias J. Young, Nicholas M. Bedford, Tatyana Kiryutina, Taylor J. Woehl

The band edge properties of a material largely determine its viability as an active material for optoelectronic, photochemical, and electrochemical applications, among others. Layered double hydroxides (LDHs) are a class of semiconducting materials which have garnered recent interest due to their (photo)catalytic, electrochemical, and corrosion-resistant properties. In this work, we examine the effect of cation composition on the band gap and band edge positions in a range of LDH structures. We employ atomistic materials simulations to model the band edge properties of LDH structures with structures comprised of various +2 cations (*e.g.* Co,Mg,Ni) and +3 cations (*e.g.* Al,Cr,V).This modelling is compared and contrasted with the experimental optical properties of LDH nanoparticles. We also explore the effect of substitutional dopants and point defects on the band edge properties of Mg-Al LDHs. LDH nanoparticles are synthesized using coprecipitation and their structural properties are evaluated using TEM and synchrotron characterization techniques. A subset of these LDH structures are also evaluated for their electrochemical and optoelectronic properties.

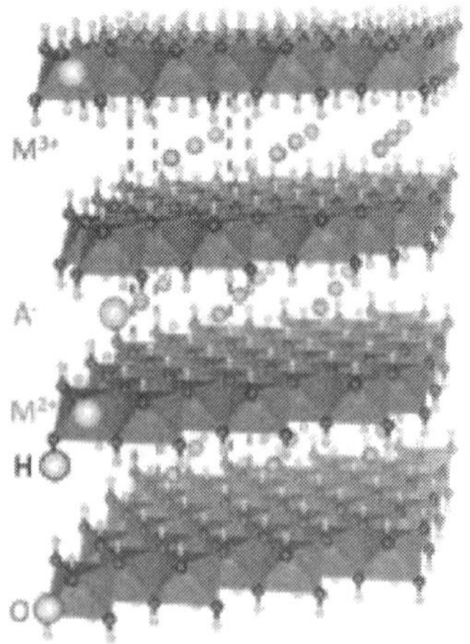

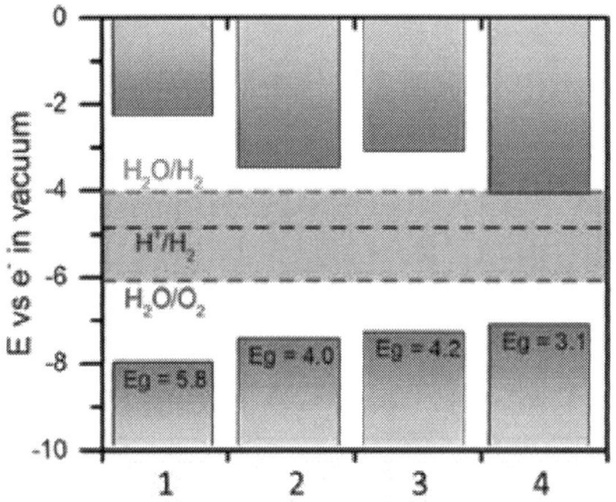

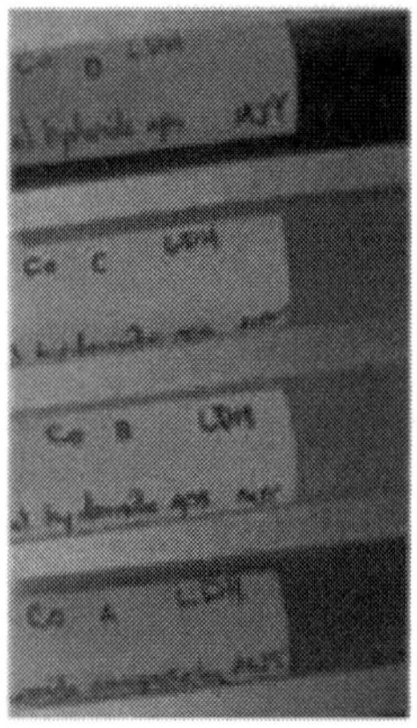

468498 Synthesis and Characterization of Cu_3SbS_4 Nanoparticles for Solution-Based Thin Film Solar Cells

Abstract:

The most efficient inorganic, thin film-based solar cells use CdTe or $CuInGaSe_2$ (CIGS) as absorber layers, both of which rely on toxic (Cd) and/or scarce elements (In, Te). The search for more sustainable processing of solar technologies has led to the identification of many Earth abundant and non-hazardous copper chalcogenide absorber materials. The Cu-Sb-S (CAS) system is promising in that it can be obtained in four major phases including $CuSbS_2$ (chalcostibite), $Cu_{12}Sb_4S_{13}$ (tetrahedrite), Cu_3SbS_3 (skinnerite), and Cu_3SbS_4 (famatinite), and the bandgaps for these compositions can vary from 0.5 to 2.0 eV.[1] The band-gap range and the high absorption coefficients (~ 10^5 cm^{-1} for Cu_3SbS_4 and $4x10^5$ cm^{-1} for bulk $CuSbS_2$) make these materials of interest as a p-type absorber for thin-film solar cells.[2]

There have been several attempts to prepare CAS thin films to fabricate single junction solar cells, where fairly low efficiencies between 0.46 % for sputtered thin film Cu_3SbS_4 solar cells and 3.22 % for $CuSbS_2$ have been obtained.[3,4] One of the challenges for the Cu-Sb-S system is the poor thermal stability of CAS particles. Decomposition of the most promising phases, such as Cu_3SbS_4 and $CuSbS_2$, during thermal treatments can significantly change the films electrical and optical properties, which then limits the solar cell efficiency.[5]There are still many strategies that need to be investigated for the CAS absorber layers to reduce film defects and film decomposition, which should increase efficiencies for CAS-based solar cell devices.

Typically higher efficiencies are obtained for photovoltaic devices prepared using vacuum deposition techniques. This solid state approach can be difficult since the fine control over the final film composition is challenging for CAS materials, since many phases can coexist due to their nearly identical thermodynamic stability.[1] However, for CAS and Cu_2ZnSnS_4(CZTS) the highest efficiencies have been obtained for solution processed devices. For these systems, the use of nanoparticle inks to prepare thin film based solar cells may be a cost-effective approach for producing higher performance solar cells.

There are many challenges related to the solution-based synthesis of uniform CAS nanoparticles with controlled properties. We have used the hot-injection method to produce CAS, since the method provides a high level of supersaturation of reagents for a very short instant of time causing rapid formation of nuclei.[6] This nucleation step and the use of proper capping agents lead to the formation of monodisperse and uniform nanoparticles. In this study, we have synthesized colloidal famatinite (Cu_3SbS_4) nanoparticles. X-ray diffraction results indicated that optimization of the synthesis conditions allows selective formation of the Cu_3SbS_4 phase without other impurities. Thin films were deposited by spin-coating nanoparticle inks, and their optical and electrical properties were studied. Optimization of post-deposition annealing conditions was performed to obtain large grain films of the desired phase. Characterization of the CAS nanoparticles and films was performed using scanning electron microscopy, transmission electron microscopy, Raman spectroscopy, and UV-Vis-NIR spectroscopy to evaluate effects of synthesis conditions on the nanoparticle properties and also to evaluate the optimization of films obtained by post-deposition annealing steps. The goal of these studies is to determine the most effective method to enhance CAS film optical and electrical properties to achieve high efficiency CAS thin film solar cells.

References

1. Ramasamy, K., *et al.*, "Selective Nanocrystal Synthesis and Calculated Electronic Structure of All Four Phases of Copper–Antimony–Sulfide," *Chemistry of Materials,* **26,** pp. 2891–2899 (2014).

2. Van Embden, J. and Tachibana, Y., "Synthesis and characterisation of famatinite copper antimony sulfide nanocrystals," *Journal of Materials Chemistry,* **22,** pp. 11466-11469 (2012).

3. Franzer, N. D., *et al.*, "Study of RF sputtered Cu_3SbS_4 thin-film solar cells," *Photovoltaic Specialist Conference (PVSC),* 2014 IEEE 40th, pp. 2326–2328 (2014).

4. Banu, S., *et al.*, "Fabrication and characterization of cost-efficient $CuSbS_2$ thin film solar cells using hybrid inks," *Solar Energy Materials and Solar Cells,* **151,** pp. 14–23 (2016).

5. Welch, A. W., *et al.*, "Self-regulated growth and tunable properties of $CuSbS_2$ solar absorbers," *Solar Energy Materials and Solar Cells,* **132,** pp. 499–506 (2015).

6. Kwon, S. G. and Hyeon, T., "Formation Mechanisms of Uniform Nanocrystals via Hot-Injection and Heat-Up Methods," *Small,* **7,** pp. 2685–2702 (2011).

473898 Mesoscale Modeling of Stress-Directed Compositional Patterning in Semiconductor Alloys

Abstract:

The ability to scalably fabricate periodic, large-area assemblies of Ge quantum dots (QDs) on Si or SiGe substrates with high degree of spatial and size uniformity would impact numerous technologies, including nano-/micro-electronics, optoelectronics, nanosensor arrays, and high-density patterned media for data storage. In this talk, we study computationally a process in which SiGe substrates are compositionally patterned over large areas using spatially-modulated elastic fields applied by a nano-indenter array [1]. In this process, an indenter array is pressed against a SiGe wafer and the entire assembly is annealed at high temperatures during which the larger Ge atoms are selectively driven away from areas of compressive stress. Compositional analysis of the substrates demonstrates that this approach leads to a transfer of the indenter array pattern into the near-surface compositional distribution.

We describe here a mesoscopic model of the â€˜stress transferâ€™ process. The model is based on a multiresolution lattice kinetic Monte Carlo (LKMC) simulation that is propagated using rates for atomic diffusion that depend explicitly on local values of stress, composition, and temperature [2]. The dependence of atomic diffusion on composition is regressed to experimental data while the stress dependence is described using the theory of activation volumes [3]. The stress field is computed using a continuum linear elastic description of the contact problem and is updated quasi-statically as the composition in the SiGe substrate evolves.

The model is used to investigate systematically the impact of several process parameters, including the shape of the indenters, the spacing between indenters, the annealing temperature, and the indenter strength. We find that certain process parameter configurations lead to compositional structures that may be useful for quantum confinement. Finally, the model performance is discussed, along with recent enhancements that lead to large increases in computational efficiency.

[1] S. Ghosh, D. Kaiser, J. Bonilla, T. Sinno and S. M. Han, Appl. Phys. Lett. 107 072106 (2015).

[2] D. Kaiser, S. Ghosh, S. M. Han and T. Sinno, Mol. Systems Des. & Eng. DOI:10.1039/C6ME00017G (2016).

[3] M. J. Aziz, Applied Physics Letters 70, 2810 (1997).

465803 Substitutional Doping in Nanocrystal Superlattice

Abstract:

Doping, a process in which atomic impurities are intentionally added to modify the electronic properties of a semiconductor, has revolutionized our world, making computers, transistors, detectors, solar cells and other devices possible. Artificial atoms, nanocrystals (NC) with electronic properties dictated by their size and shape, have also emerged as technologically important materials. In this contribution, we introduce the concept of nanocrystal (NC) doping by merging the two above mentioned concepts. We show that, by matching the size of monodisperse gold nanocrystals (Au NCs) with that of semiconducting nanocrystals of cadmium selenide (CdSe) or lead selenide (PbSe) quantum dots (QDs), it is possible to purposely dope the semiconductor superlattices and induce novel properties in these self-assembled materials. We demonstrate that, comparably to the case of atomic doping, the Au NCs take random positions in the superlattice and their concentration can be tuned over a wide range. We show that the electronic and optical properties of the superlattices are affected by the presence of the Au dopants, which increase the conductivity and the photoconductivity of bare CdSe films by several orders of magnitude. We anticipate that this approach can originate a wide variety of NC-doped structures with applications in several fields including electronic materials, solar cells, sensors and catalysis.

456759 Temperature-Dependent Modeling of Formation and Growth of II-VI Semiconductor Nanocrystals

Abstract:

Structural and spectroscopic characteristics of quantum dots (QD) have been experimentally studied for more than a decade owing to their unique physicochemical properties that enable a wide range of applications (e.g. photovoltaics, LED displays, and bio-imaging). Nevertheless, only few studies have attempted to model their formation.[1, 2] Typically, the average size of prepared II-VI QDs can be controlled,[3] but the field lacks the quantitative understanding resulting from comparison between experimental observations (e.g. size distributions and concentrations) and model predictions. Therefore, the aim of this work is to: i) develop a deterministic model describing the formation and growth of II-VI QDs; ii) validate the model against experimental results; and iii) provide insights into the key kinetic parameters and their temperature dependency.

The experimental model system selected is cadmium selenide (CdSe) because of the good control and reproducibility offered by this class of semiconductor nanocrystals. Performing microfluidic experiments *in-situ* absorption spectroscopy, we obtain average time-dependent properties of CdSe QDs (i.e. concentration, average size and distribution width) at temperatures between 160-220°C.[4] Based on population-balance equations, a kinetic deterministic model is developed and the formation of CdSe QDs is described in good agreement with the CdSe experimental data. The present model opens up the possibility to optimize the synthesis of II-VI QDs, while providing temperature dependent kinetic rates for different stages of the QD formation (e.g. nucleation, growth, and dissociation).

Literature:

[1] Abe S, Capek RK, De Geyter B, Hens Z. Reaction Chemistry/Nanocrystal Property Relations in the Hot Injection

Synthesis, the Role of the Solute Solubility. Acs Nano. 2013;7:943-9.

[2] Rempel JY, Bawendi MG, Jensen KF. Insights into the Kinetics of Semiconductor Nanocrystal Nucleation and Growth. J Am Chem Soc. 2009;131:4479-89.

[3] van Embden J, Chesman ASR, Jasieniak JJ. The Heat-Up Synthesis of Colloidal Nanocrystals. Chem Mater. 2015;27:2246-85.

[4] Abolhasani M, Coley CW, Xie LS, Chen O, Bawendi MG, Jensen KF. Oscillatory Microprocessor for Growth and in Situ Characterization of Semiconductor Nanocrystals. Chem Mater. 2015;27:6131-8.

469353 Single Enzyme Biomineralization of Size Controlled, Water Soluble Quantum Dots

Abstract:

Single Enzyme Biomineralization of Size Controlled, Water Soluble Quantum Dots

Robert Dunleavy, Leah Spangler, Zhou Yang, Li Lu, Christopher J. Kiely, Bryan W. Berger, Steven McIntosh.

Biological systems have evolved several unique mechanisms to mineralize, primarily structural, inorganic nanomaterials. These biomineralization processes are inherently "green", enabling low-cost and scalable production of nanomaterials under benign conditions in aqueous solutions. However, extending from these processes to achieve the regulated control necessary for reproducible, scalable biosynthesis of functional nanomaterials remains a central challenge. This is especially true of quantum dots (QDs), where precise size and crystal quality control is required to achieve the desired electronic properties. While several studies have described production of QDs from biological systems, the majority of these add a reactive chemical precursor, typically Na_2S, or do not provide the necessary tight control over particle size.

In this work, we describe a single enzyme capable of both the mineralization and templating of cadmium, lead, and zinc sulfide quantum dots within the quantum confined size range. The quantum dots are synthesized in the aqueous phase from normally unreactive metal acetate and L-cysteine precursors and are only formed upon the addition of the enzyme. The mineralization mechanism is proposed to consist of enzymatic turnover of L-cysteine to form reactive H_2S and is coupled to enzymatic templating of nanocrystal growth. We further demonstrate the flexibility of this approach to form core-shell and alloy nanoparticles. The biomineralized materials demonstrate aqueous phase band gap and quantum yield similar to chemically prepared materials.

470203 Self-Assembly of Antibody-Polymer Conjugates into Novel Sensing Materials

Abstract:

Affinity biosensors based on antibody binding represent one of the most successful and widely investigated categories of protein materials, having yielded technologies as ubiquitous in the US today as home pregnancy tests and often forming the basis for efforts at the development of low cost diagnostics. The sensitivity of these materials is fundamentally limited by the density of binding sites on the material surface, with the possibility of an order of magnitude boost in sensitivity of a transformative increase in surface binding sites could be achieved.

Here, we show that such an increase in binding sites can be achieved through antibody self-assembly into three-dimensional nanostructured materials. Using a site-selective chemistry, bioconjguates are prepared of IgG antibody with the synthetic polymer poly(N-isopropylacrylamide) (PNIPAM). Lyotropic self-assembly of the materials leads to the formation of large lamellar nanostructures that can be observed by both X-ray scattering and transmission electron microscopy. Studies of the self-assembly as a function of bioconjguate concentration and polymer length provide insight into the phase behavior of the system. This is the first such demonstration of antibody self-assembly into large structured arrays.

Using a simple flow coating and crosslinking process, the antibody bioconjguates can be immobilized in nanostructured thin films on different substrates. These films show a 3-4 times higher capacity for protein G binding than adsorbed antibody monolayers. As the size of the binding protein is increased, however, the difference is reduced, presumably because the large binder has trouble penetrating the polymer channels within the protein nanostructure, indicating a need for additional engineering of nanostructure size and transport rates. In addition to this, ongoing work is focusing on developing protein materials from smaller engineered protein binders such as nanobodies.

458100 Structural Studies to Determine the Mechanisms of Biobased Nanoparticle Synthesis

Abstract:

Protein and peptide directed synthesis of inorganic nanomaterials offers several advantages over traditional synthesis methods; however, despite being used to great effect, the fundamental mechanisms of how biomolecules control and drive nanomaterial synthesis is not well understood. We are using high-resolution structural biology techniques on a model gold binding and mineralizing peptide system to investigate how biomolecules are able to influence inorganic nanomaterial nucleation and growth. In this presentation we will present our current results, as well as discuss how this model gold mineralizing system is being used to develop an improved colorimetric ionizing radiation sensor for therapeutic use in cancer treatment.

460477 Biopolymers from a Thermophile: Production, Characterization, and Application

Abstract:

Exopolysaccharides (EPSs) are high molecular weight carbohydrate biopolymers that are composed of sugar residues linked together through glycosidic bonds, and secreted by microorganisms into the surrounding environment with certain properties and functions. The EPSs are one of the adaptations developed by various kinds of extremophiles in order to survive in the environments with extreme conditions, such as extreme temperature, salinity, or pH. Extremophiles are considered as potential producers of EPSs with novel and unusual characteristics which may hardly be found in more traditional polymers. Recently some extremophilic EPSs have already been characterized, and the properties and functions of EPSs from related extremophiles were also partly elucidated. However the accumulated knowledge concerning the structural and functional properties of EPSs secreted by extremophiles is still limited. In this research, a thermophilic strain, *Geobacillus* sp. (WSUCF-1) which produced unique type of EPSs has been studied. WSUCF-1 could utilize glucose as optimum carbon and energy source to produce significant amount of EPSs (382 mg/L). Two purified EPSs were obtained through ion-exchange and gel filtration columns, and both of them demonstrated higher degradation temperature (319.20 and 314.85 °C) than the thermophilic EPSs recently reported. Monosaccharide composition analysis showed that these two EPSs were composed by mannose and glucose with various relative molar ratios (1/0.2141 and 1/0.0270). Spectrum analysis also showed several unique features of the EPSs. EPSs from the relatively unexplored WSUCF-1 strain will likely provide a valuable resource for exploitation in novel biotechnological processes and opportunities in various industrial fields including bio-based polymeric materials.

471023 Genetically Encodable Acoustomagnetic Reporters for Background-Free Molecular and Cellular MRI

Abstract:

Technologies for non-invasive whole-organ imaging of molecular targets and cellular processes such as gene expression are essential to support biological studies, medical diagnostics, and the development of cellular therapeutics. While optical methods provide high-resolution images, they are fundamentally limited by the penetration depth of light in opaque tissues. In comparison, magnetic resonance imaging (MRI) and ultrasound are powerful and complementary non-invasive imaging modalities and have been widely used in clinics. Yet they have never been paired for molecular imaging. Such a pairing, especially in a way that one modality can control the image content from the other modality, could take advantage of the unique features of each modality to provide information not available via either technology alone. **Here, we describe the first contrast agents for acoustomagnetic imaging. These agents are based on gas vesicles (GVs), a unique class of genetically encoded gas-filled protein-only nanostructures formed by photosynthetic microbes as a means to regulate buoyancy.** These agents are genetically encodable, robustly detectable using susceptibility-dependent MRI sequences and "erasable" at specific ultrasound pressures. Thus they provide, for the first time, a direct method to eliminate the endogenous background, a major drawback for susceptibility imaging including functional MRI and superparamagnetic iron oxide (SPIO) nanoparticle imaging. In addition, clustering of these agents based on the presence of a specific analyte produces changes in MRI contrast that could enable the use of these agents as functional sensors.

We reasoned that the gaseous interior of GVs would have a different magnetic susceptibility from the surrounding aqueous media, and would thus induce a local magnetic field gradient that would alter the precession of nearby aqueous ^1H nuclear spins. Indeed, we show that GVs can be imaged at nanomolar concentration using gradient echo MRI and quantitative susceptibility mapping. Furthermore, GVs are "erasable" MRI contrast agents, since ultrasound pressure applied in a spatially selective manner can collapse these nanostructures, eliminating their gas compartment and thus the MRI contrast. This is very useful because, many endogenous tissue properties, such as the presence of blood vessels and the interface between different types of tissue, can produce hypointense contrast in gradient echo images, and the complication of endogenous contrast has been a major drawback to the design of susceptibility-dependent T2* MRI contrast agents. Due to their ability to be imaged before and after acoustic collapse, GVs enable, for the first time, the acquisition of background-free T2* contrast. We demonstrate the concept of acoustically erasable background-free imaging using GVs injected into mouse brains and collapsed using ultrasound.

Furthermore, acoustic collapse occurs at different specific pressures for GVs from different organisms, enabling multiplexed "acoustomagnetic" imaging using a pressure-scan paradigm. The paradigm could enable the tracking of multiple GV-labeled cell types using MRI. Finally, GVs can potentially be engineered as sensors of specific biological signals because the clustering of GVs produces a marked change in gradient echo images, as demonstrated ***in vitro*** by streptavidin-biotin cross-linking.

Overall, these results support the introduction of GVs as the first acoustomagentic reporters, which use the combination of ultrasound and MRI to enable background-free and multiplexed imaging. The ability of GVs to be genetically encoded and engineered opens the possibility of using this new form of contrast in a wide range of biological applications, especially in synthetic biology and cellular therapeutics.

474994 How Bolt Threads Is Using Chemical and Process Engineering to Grow Its Business of Engineering Silk (Featured Presentation)

Abstract:

Bolt Threads is a startup that believes that the combination of molecular biology, chemical engineering, and materials science can revolutionize how people think about and use natural materials. Bolt is developing the next generation of sustainable apparel by leveraging the latest advances in industrial biotechnology, automation, and materials science. The company is producing spider silk, a notoriously strong and lightweight material that scientists have tried to commercialize for decades. Boltâ€™s secret weapon is chemical and process engineering, which allows it to do what its predecessors couldnâ€™t: scale up its technology. From bench scale to industrial scale, process engineering is the foundation of building this high growth startup business. Leveraging the unique properties of its fibers, Bolt Threads will be delivering new products to consumers and brands worldwide.

470235 Efficacy of 5-Aminolevulinic Acid (5-ALA)-Mediated Photodynamic Therapy (PDT) Using Cold Atmospheric Plasma (CAP) As a Light Source for Anti-Tumor Applications

Abstract:

Introduction: Photodynamic therapy (PDT) is a treatment modality that takes advantage of the cytotoxic effects induced by a light source and a photosensitizer to generate singlet oxygen species [1]. It is well known that 5-ALA has been widely used in photodynamic therapy, which is a natural biochemical precursor for heme synthesis in living mammalian cells. In this study, cold atmospheric plasma (CAP) was used for the first time as a light source in photodynamic therapy. CAP is an ionized gas where the ion temperature is close to room temperature. It contains electrons, reactive oxygen species (ROS), reactive nitrogen species (RNS), radicals, and UV light [2]. The wavelengths for CAP are widely covered from 250 nm to 800 nm, which makes CAP a possible light source for photodynamic therapy.

Materials and Methods: 5-ALA and gold nanoparticles 2-4 nm in diameter were encapsulated inside polymersomes by self-assembly. After being encapsulated, TEM, DLS and SEM were used to measure particle size, zeta potential and particle morphologies. Both fibroblasts (ATCC, PCS-201-011) and melanoma A375 (ATCC, CRL-1619) cells were seeded at $2*10^4$ cells/ml on a 96-well plate maintained in DMEM (ATCC) supplemented with 10% FBS (ATCC) and 1% penicillin/streptomycin (Sigma). After 24 hours of incubation, polymersomes at different concentrations (from 100 to 500 Î¼g/ml) were added to the cells and were further co-cultured overnight. Then, both cells with nanoparticles were exposed under a CAP light source for 90s. Cell viability of the two cell lines was quantified by an MTS assay, and cell morphology was observed under confocal microscopy. All cells were incubated in a standard incubator at 5% CO_2at 37 Â°C. The cellular experiments were run in triplicate and repeated three times for each group.

Results and Discussion: Cell viability results with the 5-ALA mediated photodynamic therapy indicated that using CAP as light source can significantly decrease melanoma cell number when compared with non-CAP treatment. Specifically, at concentrations of 200 Î¼g/ml, the melanoma cell number reduced by more than 50% when compared to controls. More importantly, this treatment was non-cytotoxic to fibroblasts. In addition, confocal images indicated that the 5-ALA/PDT treated melanoma cells lost their cellular protrusions and showed a polygonal, nearly epithelial cell-like appearance in contrast to the elongated cells with multiple protrusions in untreated controls. Fibrillar actin formation in growth cones was reduced after ALA/PDT treatment, which showed a rearrangement of the actin cytoskeleton, resulting in a loss of cellular polarization. Also, focal adhesions decreased significantly observed from confocal images, which is related with cell integrin adhesion, signaling and actin cytoskeleton formation after ALA/PDT treatment.

Conclusions: 5-ALA/PDT with CAP as a light source treatment induced melanoma cancer cell apoptosis as well as suppression of cell growth and cell morphological changes. All of these results showed that 5-ALA/ PDT with CAP as a light source is a novel effective PDT method for anti-tumor applications deserving further study.

Acknowledgement: This study was supported by the Northeastern University.

References: [1] L. M. Rossi, P. R. Silva, L. R. Vono, et al. Langmuir, 2008, 24, 12534-12538.

[2] M. Wang, X. Cheng, W. Zhu, et al. Tissue engineering: part A, 2014, 5(20), 1060-1071.

455670 Repeatable and Adjustable on-Demand Local Anesthesia By Phototriggerable Liposomes

Abstract:

Current treatment of pain relies heavily on opioids and conventional local anesthetics. The former is associated with systemic side effects and addictionÂ (*1, 2*), while the latter provides short durations of anesthesia (normally 6 â€“ 12 h when typical post-operative pain lasts for a timescale of days) (*3*). Although sustained release systems can provide longer nerve block durations in animal models (up to days) (*4*), they have a major limitation: once administered, their drug release profile cannot be changed, regardless of the patientâ€™s changing needs. This is particularly troublesome when physical therapy or movement are desired, as motor block often accompanies the desired sensory block (*4*). An on-demand local anesthesia system where patients and physicians could noninvasively control the timing, intensity and duration of local anesthesia therefore highly desirable.

Here we demonstrate a phototriggerable local anesthesia system that can realize this goal. A near infrared (NIR) light-absorbing photosensitizer was encapsulated within liposomes composed of lipids labile to reactive oxygen species (ROS) (*6*). Upon irradiation with NIR light, the photosensitizer released singlet oxygen, inducing lipid peroxidation and enhancing the permeability of liposomal lipid bilayers, resulting in the release of encapsulated drug. A potent local anesthetic, tetrodotoxin (TTX), was loaded into these phototriggerable liposomes (Lipo-PS-TTX). *In vitro* experiments showed repeatable phototriggered TTX release with 730-nm laser irradiation (*7*). *In vivo* injections of Lipo-PS-TTX percutaneously onto the rat sciatic nerve induced an initial sciatic nerve block, and subsequent irradiations at 730 nm induced repeated sciatic nerve block. The timing, intensity and duration of nerve block were adjustable by the parameters of irradiation. Nerve block duration showed a clear relationship with irradiation energy density, demonstrating the controllability of the system. On-demand phototriggerable nerve block systems may allow pain treatment to be personalized and substantially improve pain management.

1. J. L. Apfelbaum, C. Chen, S. S. Mehta, T. J. Gan, *Anesth Analg* **97**, 534 (2003).

2. M. E. Lynch, C. P. N. Watson, *Pain Res Manag* **11**, 11 (Spring, 2006).

3. M. J. M. D. F. Fredrickson, A. F. Abeysekera, R. M. A. White, *Regional Anesthesia & Pain Medicine September/October* **37**, 495 (2012).

4. H. Epstein-Barash *et al.*, *Proc Natl Acad Sci USA* **106**, 7125 (2009).

5. A. V. Soldatova *et al.*, *Inorg Chem* **50**, 1135 (2011/02/07, 2010).

6. Y. Yorihiro, N. Etsuo, K. Yoshio, S. Hiroyuki, *Biochim Biophys Acta* **795**, 332 (1984).

7. A. Y. Rwei *et al.*, *Proceedings of the National Academy of Sciences* **112**, 15719 (December 22, 2015, 2015).

472751 Engineering Polymeric Nanostructures in Silicone Hydrogel Contact Lens Biomaterials for Controlled Release to Treat Glaucoma

Abstract:

Topical eye drops are the current standard of care for ocular therapeutics. Despite dominating the market, eye drops are only marginally effective with typically less than 1-7% of the applied drug being productively absorbed. Although the eye is easily accessible for treatment, there exist highly effective physiological barriers that make topical treatment extremely challenging. Low bioavailability and short residence time are combated with a rigorous eye drop regimen, making treatment highly susceptible to patient compliance issues. A number of ocular conditions rely heavily on prescribed topical treatment, sometimes leading to devastating consequences if treatment is inadequate. This is the case for glaucoma, the world's leading cause of irreversible blindness which is estimated to affect over 60 million worldwide and currently has no cure. If untreated or poorly treated, individuals experience progressive, permanent vision loss due to an increase in intraocular pressure that damages the optic nerve. Vision loss and ocular disease costs the United States approximately $139 billion per year, with an estimated 4.3 Americans estimated to develop glaucoma by 2030.

Our solution to replace eye drops for ocular therapeutics is the therapeutic contact lens. Macromolecular memory is one of the most promising concepts for therapeutic lens design, and our lab has shown (both *in vivo* and *in vitro*) this technology is capable of delivering drug at a controlled rate for a wide range of drug classes and molecules, lens types, lens modalities, and lens wear schedules. Engineering macromolecular memory involves molecular imprinting of flexible polymer chains. Functional monomers are selected to have high affinity to the drug via non-covalent interactions. Molecular complexes comprised of the functional monomers and drug molecules are crosslinked into the hydrogel matrix during polymerization. These nanostructures remain after removal of the drug and provide interaction points throughout the polymer that delay release. Lenses are specifically designed for the selected drug because the chemistry and structure of the drug molecule dictate which types of non-covalent interactions are utilized. This study highlights a selected combination of functional monomers interacting with two forms of latanoprost, a prominent glaucoma medication prescribed today.

Silicone hydrogel and hydroxyethyl methacrylate (HEMA) contact lens formulations were developed to release the prostaglandin analogue, latanoprost, in both its active form (LPA) and prodrug form (LP). The major components of our silicone hydrogel lenses were methacryloxypropyl-tris-(trimethylsiloxy) silane (TRIS), dimethyl acrylamide (DMA), and methacryloxypropyl terminated polydimethylsiloxane (DMS-R11). HEMA was the main constituent of the HEMA lenses. Poly(ethylene glycol) 200 (PEG200) and ethylene glycol dimethacrylate (EGDMA) served as crosslinkers. Functional monomers were selected to specifically cater to the chemistry of both forms of latanoprost. For LP, hydrogen bonding was the focus with acrylic acid (AA) and methacrylic acid (MAA) to form variations of a poly(AA-*co*-MAA-*co*-TRIS-*co*-DMS-R11-*co*-DMA-*co*-EGDMA-*co*-PEG200DMA) network and poly(AA-*co*-MAA-*co*-HEMA-*co*- PEG200DMA) network. For LPA, ionic bonding potential was taken advantage of with 2-(diethylamino)ethyl methacrylate (DEAEM) and diallyldimethylammonium chloride (DADMAC) for the development of poly(DEAEM-*co*-DADMAC-*co*-TRIS-*co*-DMS-R11-*co*-DMA-*co*-EGDMA-*co*-PEG200DMA) network and poly(DEAEM-*co*-DADMAC-*co*-HEMA-*co*-PEG200DMA) network. UV free-radical photopolymerization was used to form the contact lenses from solution. Drug binding studies were performed to show successful incorporation of macromolecular memory. Dynamic release studies were done in two ways: a large volume release system (200 mL at 34°C) and microfluidic devices designed to mimic physiological volumes and flow rates (3μL/min).

One of the most significant variables to consider in these formulations is the molar ratio of the functional monomers (M) to the template drug molecule (T)—the M/T ratio. In this study, various combinations of the M/T ratio were tested for both drug types (20/1, 50/1, 100/1) in the silicone hydrogel and HEMA lenses. Ratios between the monomers varied between 1 and 3 to show how the monomer content and proportions directly impact the nature of the memory sites formed and the resultant release properties from the lenses. Previous studies have shown that the release rate and overall release profile can be controlled through the adjustment of the M/T ratio and the diversity of functional monomers used. For both HEMA and silicone hydrogel lenses produced, higher M/T ratios resulted in a slower release of drug over time. Both LP and LPA were successfully released for over a week under *in vitro*physiological conditions in both lens types. Optical clarity and mechanical properties were also tested to show that the polymers had appropriate physical properties for use as contact lenses.

There exists a great need for an improved, more efficient and effective method of topical drug delivery to the eye. Macromolecular memory has the potential to revolutionize ocular therapeutics for a wide range of ocular conditions and diseases through the incorporation of nanostructured memory sites specifically catering to the chosen drug molecule. This technology has been proven to provide tailorable release rates with versatility in application, while also being capable of fitting seamlessly into current manufacturing schemes. Our study shows how this technology is applied for the first time to the global issue of glaucoma as a possible remedy for current drug delivery methods. Improved treatment via macromolecular memory could ultimately reduce incidence of glaucomatous blindness and greatly improve patients' quality of life.

465938 Core Shell Chitosan Microcapsules for Programmed Sequentialdrug Release

Abstract:

A novel type of core-shell chitosan microcapsules with programmed functions of sequential drug release are designed and developed by microfluidic technology for acute gastrosis therapy. The microcapsule is composed of a cross-linked chitosan hydrogel shell and an oil core containing both free drug molecules and drug-loaded poly(lactic-co-glycolic acid) (PLGA) nanoparticles. The resultant microcapsules can keep structural integrity in neutral medium without leakage of encapsulated substances. Based on the acid-triggered decomposition of chitosan shell and the sustained release of drug from PLGA nanoparticles, our proposed microcapsules successfully achieve the programmed sequential drug release performance, i.e., burst release first and then sustained release. In acid medium, the encapsulated free drug and drug-loaded PLGA nanoparticles can be rapidly released within 60 s to achieve the first burst-release. Later, the drugs loaded in PLGA nanoparticles are slowly released for a few days to achieve the second sustained-release due to the synergistic effect of drug diffusion and PLGA degradation. Such core-shell chitosan microcapsules with programmed functions of sequential drug release are promising to achieve a more rational drug delivery and controlled-release for the treatment of acute stomach illness. In addition, the designed strategy of programmed sequential release in this work is of versatility for various biomedical applications.

471991 Engineering of Virus-like Particles Functionalized with CD47 Extracellular Domain to Improve Targeted Delivery By Avoiding Macrophages

Abstract:

Targeted drug delivery is important to both improve the drug efficacy and reduce side effects. Virus-like particles (VLPs) have been engineered to be used as targeted drug delivery vehicles. VLPs are self-assembled, homogeneous nanoparticles derived from the viral capsid and contain no genetic material. The Swartz Lab has engineered Hepatitis B core VLPs to be stable during manufacture and storage, but to lose stability under cytoplasmic conditions, allowing us to release the drug cargo once the VLP is inside the cells. Additionally, the Swartz Lab has developed a process to functionalize the VLP surface by incorporating non-natural amino acids (nnAAs) using cell-free protein synthesis technology. By incorporating a nnAA with an azide group on the VLP and a nnAA with an alkyne group on the ligands, we can decorate the VLP surface with multiple ligands using the copper(I)-catalyzed azide-alkyne cycloaddition ("click") reaction. With this approach, each VLP can accommodate up to 120 ligands including targeting molecules such as antibody fragments on the surface.

One major challenge in using VLPs as a drug delivery vehicle is avoiding phagocytic clearance before reaching the targeted cells. For this purpose, the CD47 extracellular domain will be displayed on the VLP surface. CD47 is known as the "marker of self", because its extracellular domain interacts with SIRPα on phagocytes to inhibit phagocytosis. Several CD47 extracellular domain variants have been produced using cell-free protein synthesis and displayed on the VLP surface using the "click" reaction. These VLPs will be further functionalized with targeting molecules for specific delivery to prostate cancer cells. The CD47-functionalization will enhance the delivery efficiency of the VLPs by avoiding macrophage engulfment.

460473 Zwitterionic Gel Encapsulation Promotes Protein Stability, Enhances Pharmacokinetics Andreduces Immunogenicity

Abstract:

As an emerging class of therapeutics, protein drugs offer the advantages of higher specificity and potency than small molecules. However development of protein therapeutics is hindered by their poor stability, inadequate pharmacokinetic (PK) profiles, and immunogenicity. Polyethylene glycol conjugation (PEGylation) is the most successful strategy to date to overcome these shortcomings, and more than ten PEGylated proteins have been brought to market. However, anti-PEG antibodies induced by treatment raise serious concerns about the future of PEGylated therapeutics. Based on extensive studies of PEGylation, we believe that two issues are responsible for its immune response: the immunogenicity/antigenicity of PEG itself, and limited PEG coverage on the protein surfaces. By solving these two issues, we demonstrate a zwitterionic polymer network encapsulation (ZPNE) technology that effectively enhances protein stability and pharmacokinetics while reducing the immune response. Highly immunogenic fungal-derived uricase was used as a model therapy. In this study, uricase modified with a comprehensive zwitterionic polycarboxybetaine (PCB) gel network exhibited exceptional stability and a greatly prolonged circulation half-life. More importantly, the PK behavior was unchanged and neither anti-uricase nor anti-PCB antibodies were detected after three weekly injections in a rat model. This technology is applicable to a variety of proteins and unlocks the possibility to adopt highly immunogenic proteins for therapeutic or protective applications.

472682 Drug-Bearing Supramolecular Filament Hydrogels for Local Treatment of Diseases

Abstract:

We report here the direct use of therapeutic agents to create self-assembling supramolecular filaments that can form hydrogels in physiologically relevant conditions. Our strategy is to chemically conjugate a short peptide onto a therapeutic agent to form a drug amphiphile for self-assembly into one-dimensional (1D) nanostructures. Through the choices of the peptide sequences and the manipulation of the assembly kinetics, we are able to tune the gelation conditions, the drug release rates, and other properties important for local delivery. In one specific example, we found that rationally designed camptothecin (CPT) amphiphiles could spontaneously form a viscous solution in water containing self-assembled CPT filaments. This solution gels immediately upon contact with body fluids or cell medium. We carefully studied their drug release kinetics and in vitro efficacy against brain primary cells, and evaluated its in vivo efficacy using a resection brain tumor mouse model. Our results suggest that these drug-bearing supramolecular hydrogels have enormous potential to improve the current standards for local treatment of many diseases including brain tumor.

455388 Drug Release Study of a Tripartite Gold Nanoconjugate for Spinal Cord Injury Treatment

Abstract:

Gold nanoparticles (AuNPs) show great potentials for biomedical applications with regard to both diagnostic and therapeutic utilities. As a drug delivery carrier, AuNPs enable versatile conjugation chemistry with drugs and proteins with the ease of tuning particle size, shape, and surface chemistry. Our work focuses on the engineering of AuNPs/drug conjugates for treating respiratory dysfunction resulted from high-level of spinal cord injury (SCI). Respiratory problems are common and often lead to fatal consequences caused by SCI. In SCI patients, systemic administrations of drugs that improve respiratory function also cause harmful side effects. By chemically conjugating the AuNP drug carrier to a transporter protein (wheat germ agglutinin coupled to horseradish peroxidase or WGA-HRP) and to theophylline (a SCI drug or THP), a single administration of the nanoconjugate improved diaphragmatic activity at a much lower dosage than the free drug dosage. This talk will describe the synthesis and characterization of the nanoconjugate, particularly, the use of thermogravimetric analysis (TGA) and laser desorption ionization mass spectrometry (LDI-MS) to determine the numbers of drug molecules conjugated to AuNP. This talk will further describe the drug release profiles of the gold nanoconjugate in buffer and *in vivo*. For the *in vivo* study, recovery was found 72 hours after injection, which was 24 hours after the detection of the nanoconjugate in the targeting site. The *in vivo* results are correlated with the results from the drug release study in buffer and artificial spinal cord fluid using mass spectrometry. This study demonstrates a novel nanoconjugate capable of targeted drug delivery for the treatment of respiratory dysfunction associated with SCI.

463271 Controlling Structure within Hydrogel-Based Synthetic Extracellular Matrices through Self-Assembly and Light-Mediated Reactions (Invited Talk)

Abstract:

Hydrogel-based biomaterials increasingly are used as extracellular matrix (ECM) mimics for cell culture and regenerative medicine applications. Synthetic hydrogels have been designed to afford precise control of mechanical and biochemical properties for directing cell function and fate. However, these materials often are homogeneous on the nanoscale unlike the hierarchically structured native ECM. To address this, we have synthesized self-assembling peptides decorated with reactive groups for incorporation within hydrogels formed by light-mediated thiolâ€"ene click chemistry. Specifically, collagen mimetic peptides (CMPs) based on variants of Proline-Hydroxyproline-Glycine repeats [$(POG)_n$] have been functionalized with alkenes. These multifunctional CMPs assembly to form fibrils (~10-250 nm wide, TEM and AFM) and physical gels (G~100 Pa, rheometry). Pendant groups on these multifunctional CMPs have been reacted with various thiols (e.g., thiol-functionalized peptides, PEG) by photoinitiated, radically-mediated thiolâ€"ene reaction. This approach is promising for controlling the nanostructure within well-defined hydrogels for various biological applications, including controlled cell culture and regenerative medicine.

471521 An Antibacterial and Photocurable Hyaluronic Acid/Elastin like Polypeptide Hybrid Hydrogel for Cartilage Repair

Abstract:

An antibacterial and photocurable hyaluronic acid/elastin like polypeptide hybrid hydrogel for cartilage repair

Ehsan Shirzaei Sani[1], Iman Noshadi[1,2,3], Roberto Portillo Lara[1],Wendy Yu[1], Benjamin Geilich[1], Thomas J. Webster[1,4,5], and Nasim Annabi[1,2,3]

[1]Department of Chemical Engineering, Northeastern University, Boston, MA, 02115-5000, USA.

[2]Biomaterials Innovation Research Center, Brigham and Women's Hospital, Harvard Medical School, Boston, MA, USA.

[3]Harvard-MIT Division of Health Sciences and Technology, Massachusetts Institute of Technology, Cambridge, MA, USA.

[4]Center of Excellence for Advanced Materials Research, King Abdulaziz University, Jeddah, Saudi Arabia.

[5]Wenzhou Institute of Biomaterials and Engineering, Wenzhou Medical University, Wenzhou, China.

Introduction

Cartilaginous tissues play an important role in supporting mechanical loads and energy dissipation in joints of the musculoskeletal system. They are composed of specialized cells called chondrocytes, and a dense extracellular matrix (ECM) comprised primarily of type II collagen and proteoglycans. These tissues are characteristically avascular and alymphatic, and exhibit low cell densities, which limits their ability for self-repair after injury. Hyaluronic acid and elastin are also key components of the ECM in connective tissues, and play important mechanical and biological roles in cartilage repair. Elastin fibers are major ECM macromolecules that are critical in maintaining the integrity, elasticity, and the mechanical properties of articular cartilage. Hyaluronic acid on the other hand is an important component of cartilage and synovial fluid in the joints, and is involved in cell migration, differentiation, and proliferation, as well as regulation of ECM organization and metabolism [1, 2].

In this work, hybrid hydrogels containing methacrylated hyaluronic acid (MeHA) and an elastin-like polypeptide (ELP) were engineered for cartilage repair, by photocrosslinking different ratios of MeHA and ELP. ELPs are thermoresponsive, elastic artificial proteins, whose macromolecular structure can be tailored for different biomedical applications, by modifying their aminoacidic sequence through DNA recombinant techniques. The ELP sequence used in this study contained cysteine residues, in which the thiol groups were able to form disulfide bonds upon exposure to visible or UV light [3]. Antimicrobial ZnO nanoparticles were incorporated into the engineered hybrid hydrogels to impart antibacterial properties. The antimicrobial and mechanical properties of the engineered hybrid hydrogel, as well as pore sizes and swelling ratios, could be fine-tuned based on the ratio of MeHA/ELP, final polymer concentration, and crosslinking conditions. Furthermore, the biocompatibility of the engineered MeHA/ELP hydrogels was investigated *in vitro*.

Materials and Methods

All chemicals were purchased from Sigma-Aldrich and used without further purification. MeHA and ELP were prepared according to procedures described in our previous works [3, 4]. Briefly, MeHA/ELP prepolymers at different compositions were added to a 0.5% (w/v) photoinitiator solution. The ELP concentration varied from 0 to 20 % (w/v) and the concentration of MeHA varied from 0% to 2% (w/v). The solutions were then mixed with various concentrations of antibacterial ZnO, ranging from 0.1-0.3% (w/v). The mixtures were then sonicated for 60 min and photopolymerized for 120 sec under UV light (intensity: 6.9 mW/cm^2). The compression modulus, swelling ratio, and pore size characterization of the hydrogels as well as cell studies were performed based on procedures described previously [3]. Cell viability tests were performed using 3T3 fibroblast cells (ATCC® CRL-1658™) and cells were cultured in DMEM 1X medium with 10% FBS and 1% penicillin/streptomycin and incubated at 37 °C with 5% CO_2 (details described in [3]). Bacterial growth on hydrogel samples was evaluated using a colony forming unit (CFU) assay with *methicillin-resistant Staphyloccocus aureus* (MRSA, ATCC® 43300™). The plate colony-counting method was employed as described previously [5]. In order to perform statistical analysis, GraphPad Prism 6 software package was used. A two-way ANOVA test was performed to characterize statistical differences between mean ± standard deviation from every measurement.

Results and Discussion

The results of compression tests showed that by increasing the ELP concentrations from 0% to 15% in a 2% MeHA prepolymer, the compressive modulus of the resulting hydrogel was increased 3-fold from 13.9 kPa to 39.9 kPa, respectively ($p < 0.0001$) which shows a good improvement compared to previous studies. Although, the compressive modulus of cartilage is 0.45–0.80 MPa, the engineered hydrogels showed sufficient mechanical strength to support cell encapsulation and growth [6]. In addition, the swelling ratios of MeHA/ELP hydrogels decreased 9-fold from 5900% to 630% by increasing the ELP concentration from 0% to 15% (w/v), respectively ($p < 0.001$). This result suggests that by changing the composition of hybrid hydrogels, the swelling properties of the engineering hydrogels are tunable. Therefore, it is possible to control the pore sizes and swelling ratio of the hydrogel which are key properties for cell spreading and differentiation [7]. Based on the *in vitro* studies, MeHA/ELP hydrogels with 2% w/v MeHA and 10% w/v ELP showed more than 90% cell viability at days 1, 3, and 5 post-seeding ($p < 0.05$). The antibacterial results showed that after addition of ZnO nanoparticles, the number of MRSA CFUs decreased consistently after 24 hours of culture. In particular, the optimal formulation corresponded to a ZnO concentration of 0.2%, where the number of CFUs on the hybrid hydrogels decreased significantly from 40 to 25 per cm^2 compared to the sample without ZnO as a control ($p < 0.05$).

Conclusions

In this study, we developed an antimicrobial photocrosslinkable hybrid hydrogel based on MeHA/ELP/ZnO with tunable physical properties. The engineered hydrogels were biocompatible *in vitro* and also prevented bacteria growth at optimal ZnO concentrations. In summary, this study showed that the presently engineered hydrogels have the potential to be used in biomedical applications especially as an antibacterial hydrogel for cartilage tissue repair.

References

[1] G. C. Yeo et al., Adv Healthc Mater. 2015, 4, 16, 2530-56.

[2] J. Yu et al, J Anat. 2010, 216, 4, 533–541.

[3] Y. Zhang et al., Adv. Funct. Mater. 2015, 25, 4814–4826.

[4] G. Camci-Unal et al., Biomacromolecules, 2013, 14, 4, 1085-1092.

[5] N. C. Verissimo et al. J Biomed Mater Res Part A 2015:103A:3757–3763.

[6] I. L. Moss et al., Spine, 2011, 36, 13, 1022-1029.

[7] H. Park et al., Biomacromolecules. 2009, 9; 10, 3, 541–546.

471703 Biomaterial Scaffolds for Destruction of Disseminated Cancer Cells through Non-Invasive Local Hyperthermia

Abstract:

Breast cancer is the most common type of cancer affecting women in the U.S., with the main cause of death resulting not from the primary tumor but from metastasis to vital organs. Currently, there are very few therapeutic options for treatment of metastatic disease, as it often remains undetected until the burden of disease is too high. Focal therapies, such as localized application of heat, have shown promise in the treatment of various types of tumors. While these strategies are effective at treating solid tumors, they have not been successfully applied to disseminated tumor cells. We have previously developed microporous polymeric biomaterials that attract metastasizing breast cancer cells *in vivo* early in tumor progression. Recruitment of disseminating cells to the scaffold results in reduced tumor burden in the lung and liver, typical metastatic sites. In order to enhance the therapeutic potential of these scaffolds, we sought to modify them such that non-invasive local hyperthermia could be applied to disseminating cells by collecting them at a specific site in the body. To accomplish this goal, we incorporated metal disks into polycaprolactone scaffolds to generate heat through electromagnetic induction by an oscillating magnetic field within a radiofrequency coil. The amount of heat generation was modulated by varying the size of the metal disk or the strength and frequency of the magnetic field. For a magnetic field strength of 10 kA/m and a frequency of 190 kHz, we were able to vary the resulting temperature elevation from 4 to 30 °C by changing the diameter of the metal disk. When implanted subcutaneously in mice, the scaffolds were biocompatible and became properly integrated with the host tissue. In order to identify optimal conditions for *in vivo* application of scaffold-based focal therapy, fully integrated scaffolds were removed from mice three weeks post-implantation. *Ex vivo* characterization of heat transfer dynamics within the tissue-laden biomaterial was performed through a combination of mathematical modeling and experimental measurements of the temperature distribution throughout the scaffold upon inductive heating. To verify that inductive heating could kill cells within the scaffold, we demonstrated that increasing the temperature of the tissue from 37 to 52 °C for 10 seconds was sufficient to cause an 85% reduction in metabolic activity of the tissue, as determined using a WST-1 viability assay. This electromagnetic induction strategy can be readily applied non-invasively and repeatedly *in vivo*. The combination of capturing metastatic cells and destroying them very early in disease progression has the potential to dramatically improve the treatment of metastatic disease.

470917 Reversible Crosslinking of Engineered Hyaluronic Acid Hydrogels

Abstract:

Extracellular matrix (ECM) mechanics are well known to influence cellular phenotype; however, the effects of dynamic changes in ECM mechanics on cellular behavior are less understood. Recent work in biomaterials has led to many in vitro cell culture substrates with dynamic properties to isolate the effects of changes in mechanics. While these substrates span a range of polymeric chemistries and viscoelastic properties, there are few with the ability to revert to the initial condition. To address this issue, we designed hyaluronic acid-based hydrogels with photoresponsive crosslinkers. Specifically, hyaluronic acid polymers were modified with 4-(4-(1-(acryloyloxy)ethyl)-2-methoxy-5-nitrophenoxy)butanoic acid, which provides acrylate moieties for crosslinking via Michael-type addition reactions with thiols. Upon exposure to 365 nm light (5 mW/cm^2), degradation of the hydrogel network occurs via lysis of the nitrobenzyl ether group. Additional crosslinks can be generated using unconsumed acrylate groups by introduction of a photoinitiator such as eosin-Y and visible light. Using this method, the modulus can be switched by at least 2-fold in a biologically relevant range (initial modulus of approximately 1-4 kPa). Importantly, the high density of functional groups along the backbone of linear hyaluronic acid polymer chains allow for this tunability in mechanics, even when using irreversible cleavage reactions. In addition, the plethora of functional groups allows for the incorporation of cell adhesive moieties (such as RGD peptides) independent of the crosslinking sites. Ongoing work will address cellular morphology and markers of mechanotransduction in response to the changes in ECM mechanics.

463430 Covalently Adaptable Elastin-like Protein – Hyaluronic Acid (ELP – HA) Hydrogels with Secondary Thermo-Responsive Crosslinking for Mesenchymal Stem Cell Delivery

Abstract:

Shear-thinning, self-healing hydrogels are promising vehicles for therapeutic cargo delivery due to their ability to be injected using minimally invasive surgical procedures. We present an injectable hydrogel using a novel combination of dynamic covalent crosslinking with thermo-responsive engineered proteins. *Ex situ* at room temperature, rapid gelation occurs through the formation of dynamic covalent hydrazone bonds by simply mixing two components: hydrazine-modified elastin-like protein (ELP-HYD) and aldehyde-modified hyaluronic acid (HA-ALD). This shear-thinning network provides significant mechanical protection to encapsulated human mesenchymal stem cells (MSCs) at multiple injection rates during syringe needle injection and rapidly recovers after ejection to retain the cells homogeneously within a 3D environment. *In situ* at physiological temperature, ELP undergoes a thermal phase transition within the first network, confirmed by the observation of ELP-rich thermal aggregates using Coherent anti-Stokes Raman scattering (CARS) microscopy. The formation of the secondary network further reinforces and stiffens the network. During *in vitro* culture post ejection, MSCs are observed to maintain their ability to differentiate into multiple lineages, including chondrogenic, adipogenic, and osteogenic cell types. Together, these data demonstrate the promising potential of these ELP-HA hydrogels for injectable stem cell transplantation and tissue regeneration applications.

464252 Plasmonic Collagen Nanocomposites for Rapid Tissue Sealing and Repair

Abstract:

Laser tissue welding is a tissue repair technique that utilizes a chromophore to convert photons into heat in order to seal apposed tissue edges by localized protein denaturation and interdigitation. Laser tissue welding has emerged as a promising alternative to sutures or staples conventionally used for sealing soft tissues, with the potential to limit surgical complications such as anastomotic leakage and wound dehiscence while also reducing healing times and scarring. However, insufficient closure strength, surgical inconvenience, and extensive thermal damage have hindered the clinical application of this technique, largely due to the unavailability of effective solders (surgical sealants). In this work, we describe the development of novel plasmonic nanocomposites for laser tissue welding that overcome these barriers in tissue welding. Several different gold nanorod-polypeptide were developed and investigated for welding ruptured tissues ex vivo. Leak pressure, burst pressure, and tensile strength measurements indicated that laser welding using plasmonic nanocomposites was able to restore significant mechanical integrity to the welded tissue. Additionally, we developed a spatiotemporal model for predicting local temperatures to aid in narrowing the parameter space and selecting optimal surgical parameters. Preliminary *in vivo* results in sealing colotomy incisions in mice are also presented.

465474 Novel Bio-Ionic Liquid Functionalized Conductive Hydrogel for Cardiac Tissue Regeneration

Abstract:

Conducting polymers (CPs) constitute a new generation of organic materials, which exhibit electrical and optical properties that are similar to those of metals and inorganic semiconductors. They also possess the advantageous properties associated with regular polymers, such as ease of synthesis and processability, when compared to metals [1]. CPs are widely used in microelectronics for the manufacture of batteries, photovoltaic devices, light-emitting diodes and electro-chromic displays, as well as in biomedical applications including neural probes and prostheses, and controlled release systems. CPs have also been shown to modulate cellular processes such as cell growth and migration through electrical stimulation (conductivities from 10-4 to 9 S/cm). These properties make CPs and their derivatives remarkably suitable for tissue engineering applications using excitable cell types, including nerve, muscle, and cardiac tissue. However, there are still several technical challenges associated with the use of CPs for tissue engineering applications. The main drawbacks associated with the existing, CP-based biomaterials are: poor polymer-cell interactions due to the absence of cell binding sites, high hydrophobicity, poor solubility and processability, non-biodegradability, as well as the inability to optimize their mechanical properties. Therefore, the synthesis of biomaterials that are both electroactive and biodegradable is highly desirable. Electroconductive properties have been incorporated into biomaterials through the addition of nanomaterials (e.g. silver nanowires, gold nanoparticles, carbon nanotubes (CNTs), and graphene oxide) or conductive polymers (e.g. polyaniline, polypyrole, polythiophene) to the polymeric matrices. However, these approaches suffer from difficult and prolonged processing procedures, harsh reaction conditions, difficulties in the incorporation of nanomaterials into the polymer networks, limited solubility in aqueous media, and cytotoxicity. Therefore, there is an unmet need for the development of biocompatible, biodegradable, and conductive biomaterials, with tunable conductive and physical properties [1].

Here, we introduce a new class of conductive polymer-based biomaterials through conductive bio-ionic liquid (BIL) functionalization of different biopolymers. BIL are low melting organic salts that exhibit several technical advantages, including low volatility, high ionic conductivity and electrochemical stability, and excellent dissolution capabilities. BILs have been previously used as biocompatible and biodegradable materials for various applications such as cancer therapy, multi-responsive drug delivery systems, sensors, batteries, and biomedical implants [2-5]. Functionalization of biopolymers with BIL can provide conductive properties to the polymer networks, while preserving the biological and physical characteristics of the biopolymers. This property makes them unique candidates for various tissue-engineering applications. We performed chemical conjugation of BIL to various synthetic and natural polymers including gelatin, tropoelastin, elastin like polypeptides (ELP), collagen, poly(ethylene glycol) (PEG), hyaluronic acid (HA), poly(glycerol sebacate) polymer (PGS) and combinations of them, in order to generate conductive biomaterials with high biocompatibility and tunable physical and electrical properties. For example, we engineered a novel 3D gelatin-based hydrogel by using a visible light-activated gelatin prepolymer, functionalized with a BIL for heart tissue regeneration. FTIR and NMR analysis confirmed the successful functionalization of BIL to the biopolymer structure. The mechanical and electrical properties of the engineered hydrogel were then optimized to mimic the properties of the native heart tissue. Our *in vitro* tests confirmed that the engineered conductive scaffolds provide a suitable conductive and mechanical substratum for the culture of cardiac cells. This led to the formation of 3D functional cardiac patches, which can potentially be used for heart tissue regeneration.

References:

[1]- Kaur G , Adhikari R, Cass P , Bown M, Gunatillake P. RSC Adv., 2015, 5, 37553-37567

[2] Brock M, Nickel A-C, Madziar B, Blusztajn JK, & Berse B (2007) Brain Research 1145(0):1-10.

[3] Kumar SSD, Surianarayanan M, Vijayaraghavan R, Mandal AB, & MacFarlane DR (2014) European Journal of Pharmaceutical Sciences 51(0):34-44.

[4] Curto VF, et al. (2014). Physical Chemistry Chemical Physics 16(5):1841-1849.

[5] Dias AMA, et al. (2013) ACS Sustainable Chemistry & Engineering 1(11):1480-1492.

458427 Covalent Surface Modification of Carbon Coated Nanomagnets Allow Stable Dispersions in Aqueous Solutions and Specific Post-Modification

Abstract:

For many applications of magnetic nanoparticles such as for medical diagnostics, biotechnology, biomedical applications and magnetic biosensors, stable dispersions of magnetic nanoparticles in aqueous solutions are crucial[1]. Ferromagnetic pure metallic nanoparticles possess a very high saturation magnetization but also show a considerable tendency to agglomerate. Thus, it is highly desirable to create a material which combines the high saturation magnetization but still forms stable dispersions in aqueous solutions.

Here, we describe how specific covalent surface coating can lead to the formation of a metal cobalt nanoparticle based ferromagnetic fluid with unprecedented dispersion stability and chemical functionality [2]. More detailed, graphene-like carbon protected metallic nanomagnets, prepared by flame spray synthesis[3], which are covalently functionalized on the particle surface *via* diazonium chemistry are shown. A block-copolymer is grown directly from the particle surface by SI-ATRP. This block-copolymer consists of a negatively charged polymer block that is responsible for the colloidal stability of the particles and a second polymer block, which allows post-modification of the particles. The length of both polymer blocks is crucial for the colloidal stability of the particles and is studied in this work. Post-modification of this material by introduction of azide-functionality followed by the well-known "click"-reaction, allows the attachment of various desired target molecules. These nanoparticles form stable dispersions in various aqueous solutions, including biological relevant buffer systems.

1 A. H. Lu, E. L. Salabas and F. Schuth, *Angew. Chem.-Int. Edit.*, 2007, **46**, 1222.

2 C. J. Hofer, V. Zlateski, P. R. Stoessel, D. Paunescu, E. M. Schneider, R. N. Grass, M. Zeltner and W. J. Stark, *Chem. Commun.*, 2015, **51**, 1826.

3 R. N. Grass, E. K. Athanassiou and W. J. Stark, *Angew. Chem.-Int. Edit.*, 2007, **46**, 4909.

462710 Scalable Manufacturing of Nanostructured Particles By Gas-Phase Deposition Techniques

Abstract:

Core-shell nanoparticles and other nanostructured particles have high potential in applications such as catalysis, energy storage and pharma. However, a hurdle in their utilisation is that typically large amounts of such materials are required. Current liquid-phase and gas-phase synthesis methods often lack the high precision required or do not lend themselves to large-scale production. Gas-phase coating can be used to provide the surface of a particle with either a thin continuous coating or a decoration of nanoclusters. Coating techniques that can be used are chemical vapour deposition (CVD) [1], atomic layer deposition (ALD) [2], and molecular layer deposition (MLD) [3].

When carried out in a so-called fluidized bed, gas-phase coating is an attractive way of producing nanostructured particles with excellent scale-up potential. In such a system, the particles are suspended in an upward gas flow. We can do this both for nanoparticles and micron-sized particles. Nanoparticles â€" contrary to what is typically observed for larger particles â€" are fluidized as very dilute agglomerates with distinctive fluidization characteristics. I will discuss the challenges related to coating of nanoparticles using CVD, ALD, and MLD with such a system. Moreover, I will give a number of examples of the applications of nanostructured particles produced in this way.

[1] Valdesueiro, D., Meesters, G., Kreutzer, M., and van Ommen, J.R., 'Gas-phase deposition of ultrathin aluminium oxide films on nanoparticles at ambient conditions', Materials 8: 1249-1263 (2015).

[2] Goulas, A. and van Ommen, J.R., 'Atomic layer deposition of platinum clusters on titania nanoparticles at atmospheric pressure', Journal of Materials Chemistry A 1(15): p. 4647-4650 (2013).

[3] Vasudevan, S.A., Xu, Y., Karwal, S., van Ostaay, H.G.M.E., Meesters, G.M.H., Talebi, M., SudhÃ¶lter, E.J.R.,

and van Ommen, J.R., 'Controlled release from protein particles encapsulated by molecular layer deposition', Chemical Communications 51: 12540-12543 (2015).

468768 Manipulating Ligand-Nanoparticle Interactions and Catalytic Activity through Organic-Aqueous Tunable Solvents Recovery

Abstract:

Dispersed gold nanoparticles demonstrate much higher catalytic activities compared to their supported counterparts. Current methods of nanoparticle recovery are difficult to scale up, can alter the morphology of the nanoparticles, require large amounts of energy or solvents, and/or need highly specialized syntheses. Herein, we propose a facile system for the recovery of nanoparticles from reaction mixtures using Organic-Aqueous Tunable Solvents (OATS). OATS are homogeneous mixtures of an organic solvent and water which have the inherent property to phase separate into a heterogeneous mixture under moderate CO_2 pressure. A 60 vol% acetonitrile and 40 vol% water mixture was chosen to disperse gold nanoparticles for these experiments. Poly(vinylpyrrolidone) (PVP) was selected because it is water soluble and is believed to allow access to the entire metal surface for reactive compounds. We synthesized gold nanoparticles of average size 9.4 ± 1.4 nm and performed four different thermal treatments (no thermal treatment, 40, 50, 60ĒšC). The phase separation of the OATS mixture was determined to occur between 9.6-11.3 bar of absolute CO_2 pressure. Complete recovery of gold nanoparticles was achieved for all thermal treatments based on UV-vis absorbance of the localized surface plasmon collected from each phase. Catalytic activity of the nanoparticles was benchmarked using the hydrogenation of 4-nitrophenol. Reduction of catalytic activity occurred after the nanoparticles underwent thermal treatments or were subjected to pressure separation in OATS, but still remained highly active. The decrease in catalytic activity can be attributed to additional PVP functional groups having passivated the surface of the nanoparticle and blocked active sites as the system minimized free energy indicating that PVP does not allow access to the entire nanoparticle surface. The additional surface-bound ligands also prevented precipitation and growth in OATS, but slowed dynamic surface restructuring, ultimately decreasing the apparent catalytic activity.

472077 Protein Amyloid Fibrils As Template for the Synthesis of Silica Nanofibers

Abstract:

The preparation of silica fiber-like materials has captured the attention of scientists for the last decades. Their preparation requires the use of suitable templates. In this work, we show how protein fibrils can be used as templates for the preparation of long silica fibers, with a variety of aspect ratios. Protein amyloid fibrils are self-assembled structures resulting from partial denaturation of certain proteins, and their resulting aggregation mediated by β-sheets interactions. These structures are characterized by an extremely large aspect ratio, with a diameter that changes depending on the age of the fibrils, and on the degree of association of the different strands. Amyloid fibrils have an abundance of amino acid side chains at their surface that permit their dispersion in water and offer the chance to perform a wide range of simple derivatizations. We take advantage of these favorable features of amyloid fibrils and present a very simple procedure to coat them with silica. Starting with b-lactoglobulin fibrils with typical diameters of about 20nm , and a length of several micrometers, and by following two different strategies to grow silica (either in water at acidic pH values, or in ethanol-water mixtures under Stöber conditions) we have been able to control both the thickness and the roughness of the silica layer. Silica nanofibers with a thickness ranging from a few nanometers to hundreds of nanometers have been prepared. As an example of the flexibility offered by our approach, some of the rough silica-coated nano-fibers have been further functionalized to render them hydrophobic.

Such fibers have been further used for the fabrication of lotus-like surfaces. The surfaces have been prepared with a process consisting of four simple steps. First, ten microns colloidal particles have been arranged in either a single colloidal crystal, or in a binary colloidal crystal by combining them with smaller sub-micrometer particles. Large scale substrates have been coated with colloidal crystals. The hydrophobicity of the substrate has been measured by water contact angle, and found to be superior to that of the papillar structure of the lotus leaf deprived of the wax tubular structure. Then, silica nanotubes have been prepared by coating b -Lactoglobulin amyloid fibrils with silica by means of a Stober process. These silica nanotubes have been deposited on the binary colloidal crystal surfaces through a layer-by-layer deposition procedure, using positively-charged Ludox silica particles to reverse the sign of the charge on the surface of the substrate. Finally, the substrates have been hydrophobized by binding an alkyl chain on the surface of the fibrils. The final surfaces are superhydrophobic, with a water contact angle of 165.5 degrees, which is higher than that of the original Lotus leaf. The presented approach has two main advantages. It is easily scalable, and is based on a multi-step approach, which allows one a rational design of the nanostructure and microstructure impact on the final hydrophobicity of the surface.

463672 Conformal Layer-By-Layer Coatings on Spiky Hedgehog Particles

Abstract:

In order to disperse inorganic catalytic nanoparticles in apolar media, chemical camouflage with surfactants is required which greatly hinders catalytic activity. Due to this barrier, there has been little investigation of inorganic catalysts in apolar solvents. This study seeks to create functional dispersible catalysts by utilizing surface corrugation to circumvent this requirement. Hedgehog Particles (HPs), recently developed in the Kotov lab, consist of polystyrene microparticles with nanoscale zinc oxide spikes that result in marked reduction of Van der Waals interactions and enable dispersion in both polar and nonpolar solvents. Layer-by-layer (LbL) assembly allows for the production of functional thin films on microscale colloids incorporating a diverse set of catalytic nanoparticles and polyelectrolytes. By expanding LbL films to more complex colloidal surface geometries featuring high corrugation such as HPs, one can engineer a new generation of colloids with high dispersion stability, a unique balance of Van der Waals, electrostatic, and other interactions as well as a diversity of chemical/biological functionalities. Conformal coating of nanocatalysts and other functional materials was achieved by LbL films, as demonstrated by electron and confocal microscopy. Furthermore, zeta potential measurements confirmed the deposition of alternately charged LbL layers. Dispersion stability in nonpolar solvents is maintained with these surface modifications indicating promise in forming a new class of omnidispersible materials. These hybrid HPs are being investigated as a novel catalytic system which can greatly expand on catalysis in hydrophobic solvents.

470252 Non-Covalent Functionalization of Single Wall Carbon Nanotubes with Engineered Proteins for Targeted Subcellular Delivery

Abstract:

Single wall carbon nanotubes (SWCNTs) can be non-covalently dispersed with surfactants, polymers and proteins. We have observed that protein coated SWCNTs enter numerous cell types by the millions, more than an order of magnitude more than polymer coated SWCNTs. In addition to increasing uptake, protein coatings can impart function to nanomaterials. While many contemporary studies have focused on using proteins to target SWCNTs and nanomaterials to specific cells, we utilize proteins to target SWCNTs to regions within cells. Here, we will show the utility of modified albumin proteins and engineered native cellular proteins as dispersing agents for SWCNTs. To localize SWCNTs to the nucleus we dispersed SWCNTs with the tail domain of a nucleoskeletal protein lamin B1 (LB1) engineered from the full-length *LMNB1* cDNA. The low molecular weight globular protein has a central hydrophobic core for nanotube association and stabilization and an exposed nuclear localization sequence to promote active nuclear import. SWCNTs-LB1 enter HeLa cells and localize to the nucleus of cells; we visualize localization with Raman spectroscopy and NIR fluorescence imaging of SWCNTs and interaction with DNA by fluorescence lifetime imaging microscopy.

We have used similar techniques to show cellular uptake of bovine serum albumin (BSA)-coated SWCNTs within cells. We see localization of SWCNTs-BSA within the endosomal and metabolic processing compartments. Similarly to LB1, BSA is a small globular protein, and the hydrophobic cleft of the BSA binds to SWCNTs. BSA has smaller hydrophobic regions, independent of the hydrophobic cleft, that can be loaded with drugs or fluorophores. In vitro, we show that denaturation or enzymatic processing of the BSA releases the small molecules. SWCNTs-BSA in which the BSA are pre-loaded with rhodamine drastically increases small molecule delivery in culture. We have determined spatial and concentration distribution of rhodamine within the cells, and signal is both coincident and distinct from SWCNTs as measured by NIR fluorescence and Raman. We demonstrate efficacy of this approach by delivering a fluorescent chemotherapeutic drug daunomycin that reduces proliferation in cancer cells. Together, our results demonstrates a pathway to increase the delivery of a wide variety of drugs to cells through SWCNTs coated with albumin pre-loaded with drug molecules.

The complexity of protein structures allows for multimodal modification and manipulation of cellular processes in addition to SWCNT dispersion. The modification of native cellular proteins as non-covalent dispersing agents to provide specific transport opens new possibilities to utilize both SWCNT and protein properties for multifunctional subcellular targeting applications. Specifically, nuclear targeting will allow delivery of anticancer therapies, genetic treatments, or DNA, which in turn will promote development of novel cellular therapies. Using the transport properties of albumins and intracellular processing by enzymes allows delivery of native molecules and drugs.

460880 Effect of PMMA Particle Size on Coating Behavior in Combined Gliding Arc Discharge and Spouted Bed with Solid Precursor

Abstract:

A novel particle coating method using zinc particle as a solid precursor and gliding arc discharge was used to coat poly methyl methacrylate (PMMA) particles in spouted bed. An AC voltage was applied to two diverging electrodes to generate the arc discharge. A flow rate of 1 $L{\cdot}min^{-1}$ of argon gas was used to elongate the arc discharge as well as to establish a spouted bed condition inside the reactor. By using a combination of gliding arc discharge and spouted bed condition, a particle coating performance was achieved. Coating process was performed in a small scale reactor with a dimension of 75 mm × 30 mm × 3 mm. Both zinc solid precursor and PMMA particle were pre-loaded inside the reactor before the coating process. The coated samples were collected after a certain time intervals and were analyzed by scanning electron microscope (SEM) to observe the coating morphology and thickness. The energy dispersive X-ray spectroscopy (EDS) was also used to calculate the coverage of the coating layer, while an inductively coupled plasma optical emission spectrometry (ICP-OES) was used to measure the amount of zinc coating and indirectly used to calculate the thickness of the coating layer.

In previous study, two coating mechanisms were proposed for this novel particle coating process, which are the coating by physical contact with the partially melted zinc particle and coating by vapor deposition. For this research, we will focus on confirming the proposed coating mechanisms by studying the effect of PMMA particle size on the coating behavior. Two different sizes of PMMA particles were used, 0.3 and 0.6 mm in diameter. The increase in the PMMA particle size decreased the terminal settling velocity of zinc over PMMA (TSV) ratio from 1.91 to 1.35. The frequency of collision between zinc and PMMA particle will be affected by the change in the TSV ratio. Two applied voltages, 1.5 and 4.5 kV, was also studied to observe the changed in the coating mechanisms.

A fine layer of zinc coating was observed by SEM after 5 min. The results showed that all conditions tested exhibited similar trend for the amount of zinc coating as a function of coating time. The amount of zinc coating increased in the early stage of the coating process and became relatively steady afterward due to the peeling off of the coating layer. The amount of zinc coating dropped by approximately 50% when the diameter of the PMMA was doubled, regardless of the applied voltage. This is due to the decrease in the surface area to volume ratio of the PMMA particle.

The coating behavior depended significantly on the applied voltage. At 1.5 kV, the amount of zinc coating was higher than that of the coated sample obtained at 4.5 kV. Moreover, the coating seemed to be thicker when 1.5 kV of applied voltage was used to generate the gliding arc discharge. This is because, the energy provided can partially melt and partially vaporized the solid precursor allowing a mixed of coating mechanisms, coating by physical contact with the partially melted zinc particle and coating by vapor deposition. Due to the significant amount of coating obtained by physical contact with the partially melted zinc particle, it resulted in thick and rough coating layer. In addition, at low applied voltage, the amount of zinc coating was observed to increase when increasing the TSV ratio as the frequency of collision between zinc and PMMA was increased when TSV ratio was increased. This result in an increase in the rate of coating by physical contact with the partially melted zinc particle. On the other hand, at 4.5 kV, the energy provided by the gliding arc discharge was high enough to totally vaporize the solid precursor allowing mainly coating by vapor deposition. Hence, at this applied voltage, the amount of coating as well as coating thickness was independent of the TSV ratio.

The coated samples had an average thickness in the range of 100 to 200 nm. In the case of zinc percentage coverage, it is indicated that both applied voltage and TSV ratio did not have significant effect on the zinc percentage coverage. This means that the major contributor for the peeling off of the coating layer was the collision between PMMA particles and attack from gliding arc discharge rather than the collision between zinc and PMMA particles.

474016 Long Life Cycle Lithium–Oxygen Battery Using Molybdenum Disulfide Nanoflakes

Abstract:

Long Life Cycle Lithium–Oxygen Battery Using Molybdenum Disulfide Nanoflakes

Mohammad Asadi, Baharak Sayahpour, Amin Salehi-Khojin

Department of Mechanical and Industrial Engineering, University of Illinois at Chicago, Chicago, IL, 60607, USA.

Abstract:

Today, the world transportation system largely relies on liquid hydrocarbons (e.g., gasoline) which are limited in source and cause severe environmental issues. Although electrical motors are very clean and efficient compared with internal combustion engines, their utilization in transportation units is still pending on the development of electricity storage technologies with comparable power densities to those of liquid hydrocarbons[1–4]. Recently, limited numbers of electric vehicles (EVs) have been manufactured using existing rechargeable battery systems i.e., Li-ion battery. However, this battery technology suffers from low energy density (250 Wh/kg) that needs to be improved by a factor of five to make long distance (at least 500 miles) drive of EVs feasible on a single charge[4].

Here, we report a rechargeable Lithium-Oxygen (Li-O_2) battery system with an energy density of 1800-

2100 Watt-hour per kilogram of catalyst resulting in a driving range of more than 500 miles based on the efficiency of commercial EVs. The Li-O_2 battery cell integrated with molybdenum disulfide (MoS_2) nanoflakes as a catalyst and ionic liquid (IL) as an electrolyte preforms remarkably well evidenced by its high round trip efficiency (85%), small discharge/charge polarization gap (~0.8) as well as excellent stability and cyclability tested up to more than 700 cycles. Various in-situ (e.g., Differential electrochemical mass spectroscopy) and ex-situ (e.g., X-ray diffraction and Raman spectroscopy) characterization methods were used to elucidate the cell performance. We believe that this system with excellent reversibility without clogging issue can provide a unique platform to extend the Li-O_2 concept to the real air system (with impurities) and find a solution for EVs' energy storage systems that can operate at much higher efficiency and cyclability than available battery systems.

References:

(1) Lee, J. S.; Kim, S. T.; Cao, R.; Choi, N. S.; Liu, M.; Lee, K. T.; Cho, J. Metal-Air Batteries with High Energy Density: Li-Air versus Zn-Air. *Adv. Energy Mater.* **2011**, *1*, 34–50.

(2) Armand, M.; Tarascon, J.-M. Building Better Batteries. *Nature* **2008**, *451*, 652–657.

(3) Noorden, R. Van. The Rechargeable Revolution: A Better Battery. *Nature* **2014**, *507*, 26–28.

(4) Bruce, P. G.; Freunberger, S. A.; Hardwick, L. J.; Tarascon, J.-M. Li–O2 and Li–S Batteries with High Energy Storage. *Nat. Mater.* **2011**, *11*, 19–29.

457405 The Next Generation High Power, $LiFePO_4$ Cathode Material

Abstract:

<span" roman"="Roman"" new="New">The use of lithium-ion batteries for transportation applications can be broken up into two segments, low and high voltage. Low voltage systems encompass applications such as 12 V starter batteries and start-stop platforms, while high voltage systems are generally associated with PHEVs and EVs. Thus, when considering the use of rechargeable lithium-ion batteries in the transportation industry, and the development of active materials for this industry, it is important to consider the application to be served. For example, low voltage systems require high power densities with the ability to sustain high current densities at low temperatures. Conversely, high voltage applications main technology driver is energy density while enhancing cell safety and reducing cost. The focus of this presentation is on the development of active materials well suited for use in lithium-ion cells for low voltage applications. This is achieved by engineering electrode materials with irregular surfaces which result in high interfacial surface areas and short characteristic diffusion lengths. This can be accomplished through a porous secondary particle structure and/or nanoscale primary particle size. A study using X-ray Diffraction (XRD) has been utilized to measure the primary particle size and investigate the plane orientation in the crystalline structure and its correlation to performance. Knowing that Li diffusion is more favorable through specific crystalline planes may assist in material engineering development for better and faster Li-ion transport. These attributes are expected to provide lithium-ion batteries with high power densities. Additionally, this presentation will explain how the above key points were approached and resulted in high powder cathode materials with a Direct Current Resistance (DCR) < 9 ohm at -20°C; a key target set by USABC that enables Li-ion starter battery applications.

<span" roman"="Roman"" new="New">

<span" roman"="Roman"" new="New">

471084 Solid State Thermal Reaction of NaOH and Mn3O4 Drives the Formation of Sodium-Manganese Oxide Birnessite for Aqueous Electrochemical Energy Storage

Abstract:

Sodium-ion aqueous electrochemical energy storage devices have been widely studied recently due to their low-cost and environmental-friendliness compared with commonly used non-aqueous Li-ion batteries. Here, sodium-manganese oxide materials were synthesized *via* a solid state reaction by thermal treatment of Mn_3O_4 and NaOH at various molar ratios. Energy-dispersive x-ray spectroscopy (EDS) measurements confirmed that the atomic ratio of sodium to manganese varied from 0.02 to 0.29. The atomic structures of sodium-manganese oxide were studied with synchrotron X-ray and neutron scattering using pair distribution function analysis and Rietveld refinement. The results showed the structural evolution from monoclinic Mn_5O_8 nanoparticles to layered MnO_2 birnessite with the increasing sodium content. Cyclic Voltammetry (CV) measurements showed $Na_{0.29}MnO_2$ had a specific capacitance of 211 F g^{-1} at a scan rate of 5 mV s^{-1} in a 0.1 M Na_2SO_4 electrolyte in a three-electrode half-cell, 60% higher than that of pure manganese oxide (132 F g^{-1}). *In-situ* XRD measurements showed crystalline structure change and the *d*-spacing of (001) diffraction peak of sodium-manganese oxide varied 4% during charge and discharge processes, demonstrating its ability to achieve high capacity. Electro-kinetics analysis showed that more capacitive contribution was observed in sodium-manganese oxide compared with that of pure manganese oxide, indicating its high-rate power performance. Our synthesis methods as well as structure and electrochemical studies of sodium-manganese oxide materials provided an insight into the development of innovative electrode materials with high capacity and high power performance for aqueous electrochemical energy storage.

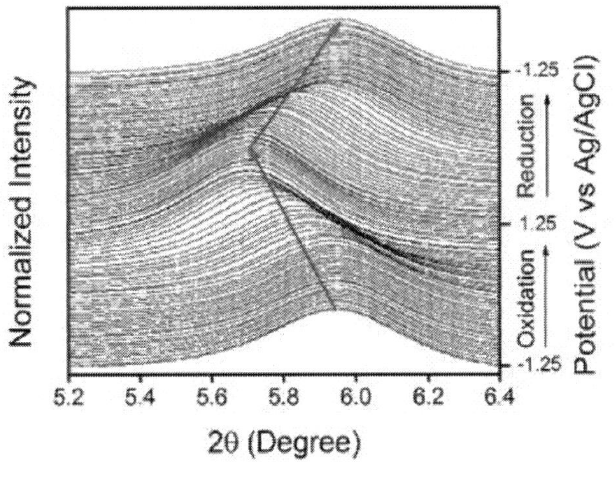

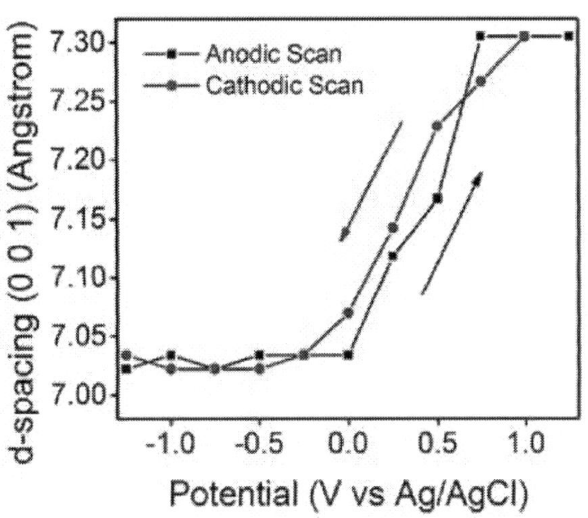

470774 Capacitive Behavior of Natural Biomaterials in High-Performance Renewable Supercapacitor

Abstract:

Supercapatior, as a promising high performance energy-storage device, has drawn great attention of researchers due to its outstanding advances, like high power, rapid charging/discharging rate and reliable runnability. Theoretically, the capacitive performance of a supercapacitor is dominated by two factors, electrochemical double-layer and redox reaction. Recently, an increasing critical requirement from environmental viewpoints has motivated the development of sustainable energy-storage materials. Many biomaterials and their derivatives own unique insights to be applied as supercapacitor. In this work, we successfully prepared biomass-based supercapacitor with good performance and investigated the effects of biomass in detail. Our preliminary results demonstrated the importance of biomass in renewable supercapacitor and indicated that it contributed to the capacitive behavior with more than one mechanism.

459708 Bipolar Plates for Redox Flow Batteries: Relating Conductivity to Morphology of Carbon-Based Polymer Composites

Abstract:

The increasing expansion of renewable sources of electricity stimulates the need for a large-scale storage of energy. A very promising candidate in this respect is the family of redox flow batteries (RFB) due to their good scalability and the possibility to decouple the capacity and power. For the broader commercialization, the price per kilowatt-hour of the energy stored by these devices need to be competitive with other technologies, such as pumped storage hydroelectricity, conventional rechargeable batteries or hydrogen-based storage. The relatively high price of the state-of-art RFB technologies is caused by their expensive components. Therefore, there is a large motivation for the development of cheaper materials with high performance and durability for the application in RFB. This work is focused on the composite bipolar plates used in RFB and their development and optimization using the combination of experiments and mathematical modelling.

Carbon-based polymer composites represent highly conductive materials with corrosion stability (even in acidic environment) and good mechanical properties at a reasonable price. Due to these properties they are widely used as bipolar plates in various electrochemical energy convertors, such as RFB and polymer electrolyte fuel cells. These composites consist of a polymer matrix and carbon fillers. Carbon fillers are responsible for the transport of electrons through the material. The resulting conductivity of the composite is strongly influenced by the nature of the carbon-based fillers and their content in the material. In order to obtain good mechanical properties and processability of the material it is crucial to minimize the total content of the fillers, while preserving high conductivity of the composite.

The composites of polypropylene with various fillers (graphite, black, fibers) were prepared using laboratory compounder Plasticorder Brabender PL2000. The effect of the filler content on the electrical conductivity of the composite plate was observed by the measurements of in-plane and through-plane conductivity. The morphology of the composites was observed using AFM, SEM and X-ray micro tomography. The mathematical model is based on the Poissonâ€™s equation and computes the effective conductivity of the composite plate from different conductivities of the matrix and fillers and from the morphology of the plate. This morphology can be either obtained from 3D-reconstructed images of the experimental samples or generated artificially using algorithms for the sphere or fiber deposition.

Our results indicate that a significant increase of the plate conductivity can be achieved by the careful tuning of its composition. For example, the addition of the carbon black into polymer-graphite composites resulted in a dramatic decrease of the percolation threshold due to synergetic effect of the fillers. Moreover, even with the same total filler content, different morphologies of the composite can further increase the conductivity, an effect observed in both our experimental and simulation results. Namely, the shape of the filler particles largely determines the effective conductivity. Anisotropic particles were found to be more efficient in achieving high effective conductivities, but at the same time their orientation becomes an important issue leading to the anisotropy of the whole composite.

The basic idea of this work is to improve the properties of a complex composite material through the understanding and subsequent tuning of its morphology. The developed combination of the experimental and theoretical approach offers a promising tool for the further optimization of carbon-based composite materials.

458505 Fabricating of High-Performance Functional Graphene Fibers for Micro-Capacitive Energy Storage

Abstract:

Although graphene is a typical two dimentional material, it was converted to multi-dimentional stuff to enhance special properties. The one dimensional graphene fiber is an example. Here we fabricated graphene fibers by utilizing ionic liquid as coagulation and functional diamines as cross-linkers to connect graphene oxide layers. The fibers showed excellent mechanical properties. The tensile strength of the resulant fibers reaches ~729 MPa after annealing at 2800 °C. All-solid-state flexible micro-capacitors are fabricated as well with promising energy storage performance. The fibers show a specific volumetric capacity as high as ~225 F/ cm^3 (measured at 103.5 mA cm^{-3} in a three-electrode cell), as well as a long cycle life of 2000 times. The initial results have confirmed that these fibers can be complement or even replacement of batteries in miniaturized portable electronics systems.

450678 Solid Dispersion Flow Battery Material Synthesis and Battery Characterization

Abstract:

Flow batteries are energy storage systems that store reversible electrochemical energy within active material species contained within the liquid electrolyte. Conventional flow battery systems are limited in their energy density by the solubility of the active materials within the electrolyte because above the solubility limit electrochemically inactive solid particles will precipitate. In addition, most conventional flow battery systems involve aqueous electrolytes, which limit the operating voltage due to the electrochemical stability window of water. Recently, researchers have explored using solid particles as the active material in flow batteries because starting with solid electroactive materials bypasses solubility limitations and results in substantial increases in the energy density of the flow battery system. In addition, often these semisolid flow batteries are comprised of lithium-ion battery materials dispersed in organic electrolytes. Using organic electrolytes and lithium-ion battery electrode materials increases the operating voltage of the battery system and further improves the energy density of the flow battery. One of the major challenges within this system, however, is that carbon additives are used to form interconnected particle structures within the fluid. This structured solvent results in high electrical connectivity and thus high power density in the semisolid flow batteries, however, the tradeoff is extremely high viscosities in the system making the electroactive fluids difficult and energy-intensive to pump.

Within our research group, we have developed a flow battery that consists of a solid dispersion of particles in the electrolyte. This solid dispersion flow battery consists of solid particles, and thus has the improved energy density that comes with bypassing the solubility limitations of conventional flow batteries. However, in contrast to many other semisolid flow batteries reported in the literature, by avoiding the use of carbon additives the viscosity is kept much lower in these fluids. The tradeoff of having a lower viscosity is that the power density is lower because electrochemical reactions proceed via collisions with the current collector, rather than throughout and interconnected carbon structure.

In this talk, recent efforts in our group to synthesize and characterize materials for the solid dispersion flow battery will be described. In particular, we have developed a scalable templated synthesis of active material nanoparticles of lithium-ion battery cathode materials that are electrochemically active and small enough to remain stably dispersed within the flow battery electrolytes. Recent efforts to characterize and understand the electrorheological coupling within this system, and the limits of the power capability within the flow battery architecture, will also be discussed.

465564 Highly Cyclable and Energy Dense Manganese Dioxide Cathodes for Advanced Alkaline Batteries

Abstract:

Manganese dioxide (MnO_2) and zinc (Zn) are one of the most abundant, safest and cheapest materials available. Together, they are found in common household batteries like Duracell, Energizer, etc. as small cylindrical alkaline cells. These cells or batteries are used as primary batteries, i.e., as single use batteries, where the entire capacity of the battery is delivered once and then discarded. The disadvantage of primary batteries is that it takes a lot of energy to produce the battery than the energy that can be actually obtained from it, and also, it creates environmental waste. However, the manufacturing of primary cells has still been rampant due to the cost of manufacturing MnO_2-Zn cells being very cheap. In terms of improving the overall energy efficiency, reducing waste and maintaining its cost advantage, it makes good sense, economically and environmentally, to make MnO_2-Zn cells rechargeable. However, the main deterrent to this direction has been the fundamental material and chemical problems of the main raw components, i.e., MnO_2 and Zn.

Manganese dioxide can theoretically deliver a capacity of approximately 617mAh/g. It delivers this capacity through a 2 electron electrochemical reaction (each electron providing around 308mAh/g). MnO_2 has been found to be rechargeable when the capacity has been limited to around 5-10% of the 617mAh/g. It suffers a crystal structure breakdown as more of the capacity is accessed, and it inherently forms electrochemical irreversible phases. If the entire 2 electron capacity can be accessed then theoretically it can reach energy density numbers near lithium-ion batteries. Similar problems are associated with the zinc electrode, where higher utilization of its capacity causes dendrite formation, shape change and formation of inactive zinc oxides that ultimately lead to electrode failure. These are the main deterrent to a cheap and safe battery that could be a disruptive technology in the energy storage field.

At the City College of New York, we have made a breakthrough in accessing the second electron capacity by altering the crystal structure of manganese dioxide through dopants and novel and cheap synthesis routes. We have cycled the MnO2 electrode to well over 6000 cycles at rates that are of interest in the battery community. Also, in this talk I will go through the various breakthroughs that have been made at the City College of New York with regards to solving other fundamental problems and the scope of these batteries in the current market.

472297 Field Alignment and Localized Field Screening in Block Copolymer Films Using Sub-1T Magnetic Fields (Invited Talk)

Abstract:

Magnetic fields offer significant promise for directed self-assembly of block copolymers. Their space-pervasive nature enables facile processing in arbitrary geometries over large areas, and without the dielectric breakdown concerns which attend the use of electric fields. Magnetic field alignment is predicated on the presence of a sufficiently large magnetic susceptibility anisotropy in the system. Such anisotropy is commonly achieved by integrating liquid crystallinity into the block copolymer by covalent or supramolecular attachment of mesogens to the polymer backbone. Although alignment can be readily achieved, to date, the required field strengths are typically on the order of several tesla (> 3 T), mandating the use of superconducting magnets or highly complex resistive systems.

Here we report on the use of a simple mesogen blending approach which dramatically reduces the critical field strength required for producing well aligned block copolymer nanostructures during a given processing timeframe. Appropriately chosen mesogens effectively co-assemble with the existing mesophases formed by covalently attached moieties in two side-chain LC block copolymers. The resulting systems display rapid alignment upon cooling from disordered states in magnetic fields as low as 0.5 T. We show that the improved alignment response is principally due to enhanced mobility in the system due to the presence of non-bonded mesogens, rather than due to the increased grain sizes observed in the blended system. Sub-1 T field alignment permits two important developments. First, block copolymer alignment can be performed using low cost permanent magnets, and second, the applied fields can be effectively screened using nanoparticles of simple magnetic materials such as cobalt. We demonstrate field screening by nanoparticle chains as well as by arrays of magnetic nanostructures at interfaces with block copolymer films. The localized field screening by these magnetic materials results in localized variation of the orientational order of the block copolymer structure. These developments offer a new route to local control of block copolymer morphology in thin films.

467974 The Effect of Trivalent Counterions to the Structure of Highly Dense Polystyrene Sulfonate Brushes

Abstract:

Surface tethered polyelectrolyte brushes are scientifically interesting and technologically relevant to many applications, ranging from colloidal stabilization to responsive and tunable materials to lubrication. Many applications operate in environments containing multi-valent ions, media in which our scientific understanding is not yet well-developed. We synthesized high-density polystyrene sulfonate (PSS) brushes via surface initiated atom-transfer radical polymerization, and performed neutron reflectivity (NR) and surface forces apparatus (SFA) measurements to investigate and compare the effects of mono-valent Na^+, and tri-valent Y^{3+} counterions to the structure of the densely tethered PSS brushes. NR and SFA results demonstrate that in monovalent salt solution, the behavior PSS brush agrees with scaling theory well, exhibiting two distinct regimes, the *osmotic* and *salted* brush regimes. Introducing trivalent Y^{3+} cations causes an abrupt shrinkage of the PSS brush due to the uptake of Y^{3+} counterions. The uptake of Y^{3+} counterions and shrinkage of the brush are reversal up on increasing the concentration of monovalent salt. Our findings demonstrate that the presence of multi-valent counterions can significantly alter the structure of polyelectrolyte brushes, in a manner different from mono-valent ions, which has implications for the functionality of the brushes.

466634 Dopant Induced Solubility Control Patterning of Conjugated Polymers

Abstract:

The solubility of semiconducting polymers can be "switched off" using addition of a high electron affinity molecular dopant. Spontaneous charge transfer with the dopant generates an organic salt that is completely insoluble in non-polar solutions. Here we demonstrate both chemical and optical mechanisms by which the doping can be reversed and the solubility of the polymer is "switched back on." Using these techniques, we are able to *vertically stack* and *laterally pattern* mutually soluble polymer layers, which are vital processing steps needed to expand the use of organic semiconductors in device applications. Optimization of these techniques has yielded diffraction limited film patterning with regular features of 200-300 nm with only solution processing steps and direct write laser patterning. Comparison of patterned and initial samples shows no change in the optical, electrical or chemical properties of the polymer. This means that the film is quantitatively dedoped after the patterning process is complete yielding intrinsic semiconducting polymer that is nano-patterned. Dopant induced solubility control (DISC) patterning offers a new avenue to process semiconducting polymers with applications in all areas of organic electronics.

463218 Controlling Ordering and Orientation in Nanostructured Thin Films through Combined Thermal and Solvent Annealing

Abstract:

Block polymer (BP) thin films potentially can be used to template large arrays of nanopatterns for advanced nanotechnologies. However, the practical utilization of directed BP self-assembly typically requires guide patterns of relatively small size scales. Two areas of recent progress in our research group include (1) the use of solvent vapor annealing to manipulate nanoscale morphologies and achieve well-oriented nanostructures, and (2) the use of engineered blade coating approaches to manipulate nanostructure orientation. In the first case, we are exploring the macroscopic alignment of BP nanostructures on a template-free substrate through raster solvent vapor annealing combined with soft shear (RSVA-SS). Our spatial control over nanoscale structures allows one to uniquely 'write' macroscopic patterns with microscopically aligned BP nano-features. This method significantly improves on previous approaches by eliminating the correlation between the BP alignment and shearing-pad shape, allowing continuous fabrication of highly ordered BP patterns in a two-dimensional manner. We demonstrate that complex patterns such as dashes, crossed lines, and curves can be imparted easily to thin films on featureless and untreated substrates. Furthermore, the ability to expand or shrink the "writing" size of the oriented regions from millimeters to centimeters provides a unique handle for manipulating pattern formation in an on-demand fashion. In the second case, we are exploring methods for the simultaneous deposition and nanoscale orientation control of block copolymer thin films. These techniques facilitate the generation of well-defined nanoscale features in a manner that is adaptable to continuous processing.

459881 Polyelectrolyte Brush Conformations in Multivalent Ion-Driven Brush Collapse

Abstract:

Charged polymers grafted at one end to a surface and packed densely will extend away from the surface, forming a brush-like structure. These polyelectrolyte brushes have found significant applications in stabilizing dispersed particles and modifying surface properties. Of great interest has been the response of the polyelectrolyte brush structure to changes in the environment, namely the solvent quality and salt concentration. The distance the polymer chains extend from the surface, the brush height, will depend on the environmental conditions and is instrumental to the performance of the surfaces modified with these brushes. In the presence of salt ions, the brush has been shown to contract due to a decrease in the brush osmotic pressure, and in the presence of a poor solvent, the brush undergoes a rapid and strong shrinkage to a collapsed brush state. Less well-understood is the response of the brush conformation to the presence and concentration of multivalent ions. Previous studies have shown that the brush height undergoes a rapid and strong shrinkage at a critical concentration of multivalent ions, behavior that is suggestive of brush collapse in a poor solvent, but expected to be caused by attractions between polyelectrolyte chains due to bridging by the multivalent ions. We have used an energy balance approach to provide a theoretical understanding of bridging-driven brush collapse, considering the sum of the electrostatic, polymeric and entropic mean field terms and including additional parameterized phenomenological terms for counterion condensation and bridging. The theory examines the dependence of the brush collapse on the multivalent ion concentration as well as the potential structures of the brush in the collapsed state, considering homogeneous, pinned surface micelles and bundled brush conformations. Understanding the role multivalent ions play in inducing brush collapse and the structure of the collapsed state is essential to application of these materials in the complex environments commonly found in industrial formulations and biological systems and may lead to the use of multivalent ions as a tuning parameter for new stimuli-responsive materials.

462981 Release Kinetics of Nisin from Chitosan-Alginate Complex Films

Abstract:

Understanding the release kinetics of antimicrobials from polymer films is important in the design of effective active packaging films that can increase food safety and shelf-life. The release kinetics of nisin from chitosan-alginate complex films prepared using various fractions of alginate (33, 50 and 66%) were investigated into an aqueous release medium. Films containing higher alginate fractions showed significantly lower degree of swelling in water. The total amount of nisin released from films decreased significantly with an increase in alginate concentration. The mechanism of diffusion of nisin from all films was found to be Fickian, and diffusion coefficients varied from 0.872×10^{-9} to 8.034×10^{-9} cm^2/s. Strong complexation was confirmed between chitosan and alginate polymers within the films using isothermal titration calorimetry and viscosity studies, which affects film swelling and subsequent nisin release. Complexation was also confirmed between nisin and alginate, which limited the amount of free nisin available for diffusion from films. These low-swelling biopolymer complexes have potential to be used as antimicrobial packaging films with sustained nisin release characteristics.

454977 Analysis of Deformation Wear in Polymer Coatings

Abstract:

Most common types of wear of polymers are abrasion, adhesion, and fatigue. Addition of inorganic hard material especially nano-filler (silica, alumina, graphite) in a polymer coating for electronic devices can result in improving its abrasion resistance, and in wear rate of glassy polymer coatings(1,2). Parameters found to be significant in the wear is a ratio of hardness to elastic modulus (related to the elastic strain to failure) and H3/E*2(related to the resistance to plastic deformation) are more important than hardness alone in determining the wear resistance of a material. Effective Young's modulus E* is measured by instrumental nano-indentation and is defined as E*=E/(1−v2), where v is the Poisson's ratio of the material. In digital printers, photoreceptor is a central device which is subjected to abrasion during blade and brush cleaning of the photoreceptor after image transfer. The abrasion of the top layer is found to be caused by embrittlement of the top layer. Bulk and surface properties were measured by preparing materials by spin coating process. Hysitron was used for measuring surface properties such as indentation Young's modulus and hardness.. Addition of nano-filler silica in the polymer coatings either in polycarbonate or polyester increased strain to break increases hardness to elastic modulus ratio decreasing coating wear. Compared to polycarbonate, polyester showed higher strain to break when internal plasticizer (a diamine) was added. Addition of nano-filler silica resulted in higher hardness to modulus ratio consistent with previous studies on other materials.

472389 Optimized and Tested Zein Film for Utilization As an Effective SERS Sensor

Abstract:

Surface Enhanced Raman spectroscopy (SERS) is a technique of growing interest to the research community. Through the implementation of nanophotonic structures and nanoparticles, SERS sensors have the ability to increase the Raman signal by 10^8 times [1]. Zein is a prolamine protein from corn. Its film forming capabilities make it a prime candidate for a number of different applications. One of these applications is as a biodegradable, SERS sensor. Gizem et al. found that a zein based sensor could increase the raman signal of analytes 10^4 times. However, the limit of detection for the sensor was above that which could be used as an appropriate sensor for food allergens and toxins [2]. This study worked to optimize and test a zein substrate for this particular application.

The first phase of the work was to optimize the formulation for the zein film. Zein films solubilized in solvent alone create a very brittle film. Therefore, a number of different additives are incorporated into the film matrix to optimize the various properties of the films for particular applications. In this case, we wanted a flexible yet sturdy film that would promote a high fidelity for imprinted nanostructures. To achieve this optimization, both oleic acid (OA) plasticizer and glutaraldehyde crosslinking agent (CL) were included into the film formulation at different ratios. All films were formed with a 70% ethanol solvent with a 1:5 w of zein/volume ratio. Mono and diglyceride emulsifier was also incorporated at 1:0.05 w OA/w. Besides looking at the fidelity of nanostructures with scanning electron microscopy (SEM) the films were also analyzed with various characterization techniques that looked at the physical properties (indentation and rheology), chemical properties (Fourier Transform Infrared Spectroscopy (FTIR) and FT-Raman), and surface properties (water contact angle and atomic force microscopy) of the films.

The physical properties showed that both plasticizer and crosslinking content have a strong impact on the zein films. Indentation shows that an increase in both plasticizer and crosslinking content decrease the Vicker's hardness of the material while increasing the indentation area. The rheology showed a deviation between the formulations for both gelation time and reaction rate. The 0.8:1 OA films stayed together, forming a "maramalde" type texture in 60 hours, while the 1.2:1 OA films took over 100 hours to create the same texture. The 1:1 OA films split, having the 4% CL films coincide with the high OA films and the 8% agree with those of the 0.8:1 OA. The rate at which the film solutions reached the "marmalade" state followed a first order reaction for lnG' vs. time. However, at the high oleic acid concentration, the Pearson correlation coefficient shows a deviation from the first order model.

The chemical properties described a mechanism for crosslinking that became increasingly ineffective with high plasticizer content. The main peak difference within the FT-Raman spectra came from a peak located at 881 cm^{-1}. This particular peak corresponds to a rise in C-N-C stretching vibration [3]. In order for this sort of bonding to increase within crosslinked films, the glutaraldehyde must react with an amine group. This sort of reaction is supported by the FTIR spectra and a decrease in both the Amide II band at 1540 cm^{-1} and the peak at 1080 cm^{-1}. The former corresponds to the N-H bending of amide groups [3], while the ladder is responsive to N-H stretching of primary amines [4]. Both of these peaks decrease in intensity with an increase in crosslinking concentration. A peak at 970 cm^{-1} also arises within the FTIR spectra with crosslinking. This peak is strengthened by out of phase C-O-C ring vibrations. Concluding that while the glutaraldehyde is creating C-N-C bridges between molecules, it is also oligomerizing and forming hemiacetal rings. The degree of oligomerization increases with plasticizer content.

The surface properties only showed a significant change in topography and SEM. As the films became more crosslinked the topography consisted of much larger features than the films with only plasticizer. The SEM images demonstrated that highest oleic acid ratio was too malleable to produce high fidelity nanostructures, while the low OA films had too high of an affinity to the lithographic medium to be effectively removed from the mold. It was at the midrange plasticizer content and the 4% and 8% crosslinking ratios that created the best films for this application. However, even with this change in topography and affinity to the lithographic mold, the water contact angle measurements showed no difference between the formulations.

The final phase of the research utilized the optimized formulations to produce and test a SERS sensor. Each of the zein films were molded with inverted nanopyramidal structures and coated with gold to create the SERS enhancement. The surfaces were further altered with the incorporation of gold nanoparticles a top of the nanophotonic structures. The sensor was tested both with rhodamine 6g and with gliadin, the protein responsible for the allergic response to gluten.

In conclusion this study focused on the application of zein corn protein films as a substrate for a SERS sensor. The zein film formulation was optimized with the inclusion of both oleic acid plasticizer and glutaraldehyde crosslinking, and then characterized with a wide variety of techniques. The 2 formulations that gave the highest fidelity of nanostructures were then chosen to be tested as a SERS sensor with the utilization of both Rhodamine 6G and gliadin protein. The study found this to be a successful application for zein films.

1. Cunningham, Brian. "[Illinois] ECE 416 Surfance Enhanced Raman Spectroscopy I," 2013.
2. Gezer, P.G., G.L. Liu, and J.L. Kokini. 2016. *Development of a biodegradable sensor platform from gold coated zein nanophotonic films to detect peanut allergen, Ara h1, using surface enhanced raman spectroscopy*. Talanta: 150, 224-232.
3. *The Handbook of Infrared and Raman Characteristic Frequencies of Organic Molecules*. Accessed April 6, 2016. https://books.google.com/books/about/The_Handbook_of_Infrared_and_ Raman_Chara.html?id=bYWNSi6abvwC.
4. Larkin, Peter.*Infrared and Raman Spectroscopy*, 213–15. Oxford: Elsevier, 2011. http://www.sciencedirect.com/science/article/pii/B978012386984510014X.

471848 Printability, Pinch-Off Dynamics and Extensional Rheology of Complex Fluids (Invited Talk)

Abstract:

Liquid transfer and drop formation/deposition processes underlying printing, jetting, spraying and coating involve complex free-surface flows, and the formation of columnar necks that undergo spontaneous surface tension-driven instability, thinning and pinch-off. The progressive self-thinning of the neck is often characterized by self-similar profiles and scaling laws that depend on the relative magnitude of capillary, inertial and viscous stresses for simple (Newtonian and inelastic) fluids. Stream-wise velocity gradients that arise within the thinning columnar neck create an extensional flow field that can orient and stretch macromolecules, contributing extra elastic stresses and extensional viscosity that change thinning and pinch-off dynamics. It is well established that polymeric complex fluids exhibit a much larger resistance to flow in an elongational flow field than Newtonian fluids with same shear viscosity and even O(~100) ppm amount of added polymers can lead to delayed pinch-off affecting printability of multicomponent fluids. However, extensional rheology measurements are far from routine and require bespoke instrumentation not available, or easily replicated, in most laboratories. We show that visualization and analysis of capillary-driven thinning and pinch-off dynamics of the columnar neck in an asymmetric liquid bridge created by dripping-onto-substrate can be used for characterizing the extensional rheology of complex fluids. Using a particular example of dilute, aqueous PEO solutions, we show the measurement of both the extensional relaxation time and extensional viscosity of weakly elastic, polymeric complex fluids with low shear viscosity Î· < 20 mPa.s and relatively short relaxation time, Î» < 1 ms. Characterization of elastic effects and extensional relaxation times in these dilute solutions is beyond the range measurable in the standard geometries used in commercially available shear and extensional rheometers (including CaBER, capillary breakup extensional rheometer). As the radius of the neck that connects a sessile drop to a nozzle is detected optically, and the extensional response for viscoelastic fluids is characterized by analyzing their elastocapillary self-thinning, we refer to this technique as dripping-onto-substrate (DoS) extensional rheometry. We show that dripping-onto-substrate methodology can be implemented using an off-the-shelf digital camera to

distinguish between pinch-off dynamics and printability of a wide variety of complex fluids. In this talk, we will elucidate how polymer composition, flexibility and molecular weight determine the kinetics of capillary-driven thinning and pinch-off in our experiments. Both effective relaxation time and transient extensional viscosity are found to be strongly concentration dependent even for dilute solutions. Lastly, we show how finite extensibility of polymers dramatically changes the kinematics from elastocapillary to viscocapillary under strong extensional flow fields that can lead to coil-stretch transition.

459256 In-Line Near-Infrared Spectroscopy for Optimization of Radical Modification Reactions of Polyolefins

Abstract:

Many analytical challenges are associated with in-line reaction monitoring to support continuous process optimization. To that end, techniques such as Raman, rheology, and IR[1] spectroscopy have been recently applied for the monitoring of flowing streams in "real time". In the present work, NIR was selected because it is a non-destructive technique, compatible with the used chemicals, substantially cheaper/more readily available, and it does not require the use of solvents.

Several monomers were successfully grafted onto polyolefins in an intermeshing co-rotating twin-screw extruder via radical modifications. This study demonstrates the use of in-line NIR in monitoring radical reactions in order to optimize process conditions for obtaining higher grafting efficiencies (GE).

Besides grafting, side reactions are observed resulting in low grafting efficiencies. Reducing the monomer concentration as well as increasing the mixing level by improving the screw design was shown to suppress side reactions and results into higher grafting efficiencies. It was also observed that initiator concentration controls the overall conversion, degree of grafting (DG) as well as side reactions.

References

1 J.A. Covas, O.S. Carneiro, J.M. Maia, S.A. Filipe, A.V. Machado, Evolution for chemistry, morphology and rheology of various polymer systems along a twin-screw extruder, The Canadian Journal of Chemical Engineering 80 (2002) 1065â€"1074.

463562 Rheological Properties of Hdpe/CO_2 and Ldpe/CO_2 Solution Under High Pressure

Abstract:

Rheological properties of HDPE/CO_2 and LDPE/CO_2 solution under high pressure

A high-pressure rheometer contained a rotor was used to determine the effect of temperature, static pressure and CO_2 concentration on the viscosity of linear high-density polyethylene (HDPE) and branched-chain low-density polyethylene (LDPE). First, the molecular parameters of the two polymers were characterized by high-temperature Size-Exclusion-Chromatography (SEC). The linear HDPE possessed narrow molecular distribution of the branched chain while branched-chain LDPE present broad distribution. The CO_2 solubility and diffusivity in the two polymers was investigated by using magnetic suspension balance (MSB). The gas diffusion coefficient was used to estimate the time required to saturate the sample in the gap according to the one-dimensional diffusion equation of Fick¡¯s second law.

Considered the potential existence of signal delay in the magnetic driven process and the frictional resistance, the test reliability was inspected by comparing the results of the plate geometry with the rotor using dynamic frequency sweep test. The results indicated that at low frequency range where is critical to the zero-shear viscosity, the rotor generate a little higher viscosity than the plate, however, the mechanical delay may overlaps the signal from the sample results to large error at high frequency range. Another problem that hampered the test was the stability of the sample after about 38 hours saturation at high temperature. The molecular parameters were selected to compare with the virgin pellets, indicated that the degradation was minor and acceptable.

The effect of temperature on the viscosity was studied using the rotor in the high pressure chamber with diluent-free polymer though the test may inaccurate compared with the plate geometry. The activation energy of LDPE is slightly higher than HDPE due to the relatively high free volume fraction. An inert gas, helium, was used to characterize the effect of static pressure on the rheological properties. The pressure shift factor was fitted with Barus equation, shows that there¡¯s turning-point for HDPE which is also put forward firstly to the authors¡¯ knowledge. At low pressure (<10 MPa), the pressure coefficients is slightly higher than LDPE, while at high pressure the rheological properties is insensitive to the pressure increase. This phenomenon can be attributed to the fractional free volume. Moreover, the pressure coefficients of HDPE decrease as temperature increase which is the exact opposite of LDPE. In addition, The combined effect of CO_2 pressure and concentration on the rheology was investigated after more than 10 hours¡¯ saturation. Compared the thickening effect of static pressure, the plasticization of dissolved gas is predominant of the polymer rheology. Lastly, the effect of gas concentration and pressure was assumed to be separated, and the F-K model was used to describe the effect of gas concentration on the viscosity alone.

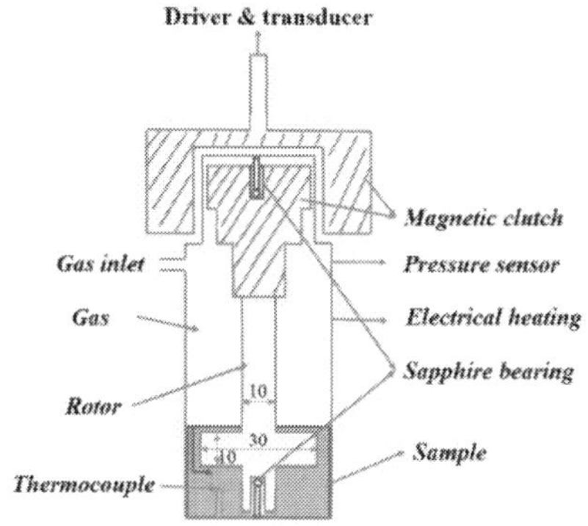

Figure 1. Schematic of high pressure attachment for the rheometer, which consist of a magnetic clutch, a pressure cell, a rotor and stator (dimensions in mm)

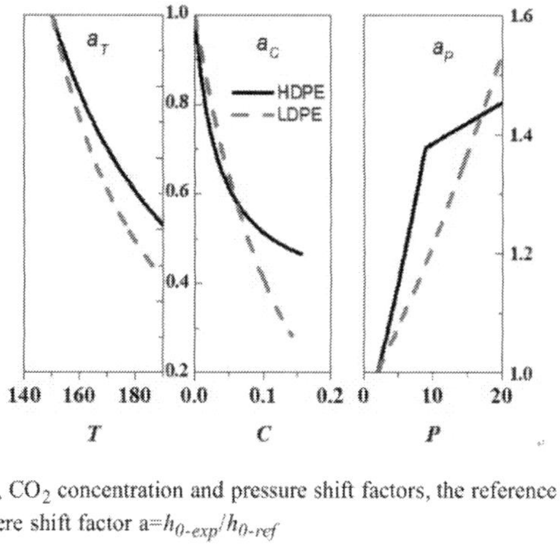

Figure 2. The temperature, CO_2 concentration and pressure shift factors, the reference temperature T_r = 150 °C, pressure P_r =1.0 MPa. where shift factor a=$h_{0\text{-}exp}/h_{0\text{-}ref}$

465818 Rheological Investigation of High Performance Polymer-Surfactant Systems for EOR Applications in Carbonate Reservoirs

Abstract:

The demand of oil has increased in last few decades and it will remain a major source of energy in next few decades. To meet the increasing demand of oil, it is important to increase the current production level by either exploring new fields or maximizing the production from existing fields. To maximize the production from a field, various enhanced oil recovery (EOR) techniques are employed which includes thermal EOR, gas EOR and chemical EOR.

In chemical EOR, different chemicals such as surfactants, polymers and/or alkalis are used to recover the residual oil from the reservoir. Surfactants are added to lower the interfacial tension between water and oil. Alkalis can reduce the adsorption of surfactants and adjust the pH. Water-soluble polymers are used to increase the viscosity of displacing fluid (water). Partially hydrolyzed polyacrylamide is most widely used polymer for EOR application due to low cost, high viscosity and availability. However, it is not suitable for high-temperature and high-salinity reservoir due to thermal degradation that leads to huge reduction in the viscosity In this work, novel acrylamide-based terpolymers and amidosulfobetaine surfactant systems were evaluated using rheology for application in carbonate reservoir. Rheology is an important tool for screening different polymers. One of the polymer was terpolymer of acrylamide, acrylamido tertiary butyl sulfonate, and acrylic acid while other polymer was terpolymer of acrylamide, N-vinylpyrrolidone and acrylamido tertiary butyl sulfonate. The acrylic acid-based terpolymer showed higher viscosity compared to N-vinylpyrrolidone-based terpolymer. In-house synthesized surfactants were compatible with all three polymers. The added surfactant has negligible effect on the rheological properties of the polymers. Detailed results will be presented and discussed.

The investigated polymer-surfactant systems showed good potential due to good thermal stability, rheological properties and compatibility at reservoir conditions.

471329 Structure and Nonlinear Rheology in Large Amplitude Oscillatory Shear of Polypropylene-Layered Silicate Nanocomposites

Abstract:

The nonlinear FT-rheology of polypropylene-layered silicate nanocomposites with a low nanoparticle loading has been investigated with large amplitude oscillatory shear (LAOS) tests over strain amplitudes ranging from 0.5% to 500% and at several frequencies from 0.5 rad/s to 5 rad/s. A sampling rate of 100 kS/s was used on each signal – the strain and the torque. An oversampling number of 3000 was used to collect data over sixteen cycles at each strain amplitude – cf. Merger and Wilhelm[1]. The matrix polymer was a copolymer of propylene with 3 mole% ethylene and functionalized with maleic anhydride (PO1015 from ExxonMobil). The volume fraction of onium ion exchanged montmorillonite (I.44P from Nanocor) was 0.02. This work follows up on recent results reported[2] from our laboratory which related the linear viscoelastic behavior of such nanocomposites to the extent of coupling between nanolayers and the polymer chains.

The relative third harmonic content $I_{3/1}$ plotted against strain amplitude displayed quadratic dependence up to a strain amplitude of 100% at several frequencies. This would indicate the absence of a filler network in the nanocomposite melts investigated here. Hence the viscoelastic behavior can be interpreted in terms of the breakup of two different entanglement networks – one involving particle attached polymer chains and another involving only the free polymer chains. Comparison of the storage modulus plots against strain amplitude for the nanocomposites and the matrix showed two distinct transitions in the case of the nanocomposite. The plots of relative harmonic intensity also exhibited a plateau followed by another power law region of increasing intensity; the transition between these regions was found to vary with frequency. The relative harmonic intensity variation for filled elastomers was related by Leblanc[3] to a Weibull probability distribution for breakage of network links. A similar analysis was carried out in this study to extract parameters for the network formed by chains attached to particles entangling with the free polymer chains. The strength of attachment between the chains and the particles can be evaluated and compared for nanocomposites with different coupling agents in this way.

1. Wilhelm, M., Reinheimer, P., Ortseifer, M., Rheol. Acta, 38, 349-356 (1999)
2. Ren, W., Chaudhary, A.K., Jayaraman, K., Ind. Eng. Chem. Res., 54 (16), 4264-4273 (2015)
3. Leblanc, J.L., J. Appl. Polym. Sci., 109, 1271-1293 (2008)

473221 Novel Viscoelastic Material Consisted of Polymeric Lignin

Abstract:

Lignin, a renewable resource available from lignocellulosic biomass, has the potential to replace petrochemical feedstock in the production of aromatic polymers. Applications of lignin as a polymer material, however, is hindered by brittleness and thermal instability of the material. Unless blended with other polymers (e.g. PLA, PP, PVA, PET) or added as a filler in composites, technical lignins are not suited for material applications. The complexity in lignin structure and depolymerization chemistry also complicates the use of lignin as a source of polymer building blocks.

We present a new process that converts enzymatic lignin to a viscoelastic material that contains over 80% lignin by weight. In this process, lignin is oxidatively functionalized and repolymerized using renewable chemicals (hydrogen peroxide, methanol, and DMSO) under mild conditions (ambient pressure, under 100 degrees C), yielding a homogeneous polymeric material. The synthesized material contains native beta-aryl ether linkages in lignin, while also exhibiting new viscoelastic properties distinct from the starting material. Rheological analyses of this new material has demonstrated stress response characteristics that have never been reported on lignin-rich materials. In an effort to enhance the viscoelasticity of the lignin-rich material, our studies on lignin functionalization chemistry revealed the association between polymerization parameters and mechanical properties.

449925 Enhanced Polymer Transport in a Crowdier Nanoslit Micropost Array

Abstract:

We found enhanced transport of DNA molecules in more crowded nanoslit micropost arrays, contrary to the expectation of slower diffusion in more crowded environments. The nanofabricated post arrays are weakly attractive and emulate crowded interactive environments, resulting in weak DNA adsorption. The coupling between DNA diffusion with adsorption to the microposts results in increased mobility with increasing post density, in contrast to studies of decreased DNA/protein mobility in dense gels. Hidden Markov analysis was employed to identify the apparent free energy barriers, which is in agreement with theoretical estimates of based on polymer conformation fluctuations between posts. We also performed Langevin dynamics simulations of macromolecular transport in micropost arrays to verify the enhanced diffusion in denser arrays. Faster DNA diffusion and hopping in crowded systems have important implications for biological systems and for technological applications such as macromolecule purification, and targeted drug delivery.

472822 Novel Method for Protein Stability and Delivery through the Formation of Complex Coacervates

Abstract:

Polyelectrolyte complexes and complex coacervation offer a novel means for the delivery of proteins and other bioactive materials. Polypeptides are biocompatible materials whose tunable properties lend themselves as viable vehicles for the delivery of drugs and, in particular, proteins. We also investigated the encapsulation of green fluorescent protein (GFP), bovine serum albumin (BSA), and lysozyme as model protein systems into polypeptide-based complex coacervates. These polypeptides are capable of mimicking the milieu found in cells, particularly cellsâ €™ natural characteristics of compartmentalization, crowding and soft interactions. The phase separation that occurs with coacervate formation gives way to compartmentalization and the high concentration of molecules simulates the biomolecules content, which also favors soft interactions between involved species. Additionally, these coacervates are aqueous, just like cells, and have tunable chemistry for furthering their cell-like characteristics. As such, the ability for peptide based coacervate systems to partially mimic this biological environment presents a viable means for potentially allowing proteins to exist in a wide array of environmental conditions without concern for their degradation. To further investigate protein encapsulation, we propose the creation of polypeptide-based block copolymers built from poly-lysine and polystyrene. We will focus on the characterization of polyelectrolyte complex micelle formation between various combinations of polypeptide and polystyrene. Previous work has shown that different ratio of hydrophobic to hydrophilic parts has a large impact on the geometry of micelle formation. By controlling the molecular weight of the block copolymer, the behavior of micelle stability will be investigated, which is very useful in protein encapsulation. It has been shown that the ionic strength and pH of the environment heavily influences coacervate formation and affects the encapsulation efficiency of proteins, thus, various mixtures of polycation/polyanion/protein mixtures were investigated. Trends from these environmental factors will then used to design delivery systems in the future. Key features of these mixtures that were varied were total monomer concentration, the geometry of micelle formation when using complex coacervate micelles, ratio of polycation to polyanion and initial concentration of protein. The stability

and geometry of micelle will be determined by light scattering and transmission electron microscopy. Encapsulation efficiencies were determined using a Bradford assay and UV/Vis spectroscopy and stability data were acquired using circular dichroism. Polypeptide-based coacervates are a powerful platform technology to understand the role of the chemical environment on the stability of a protein formulation. Ultimately, we look to harvest these materials to enable the development of highly stable therapeutics and biologics such as refrigeration-free vaccines. This novel system is a major stepping-stone for the effective treatment of therapeutic regiments, providing not only a milieu that mimics the biological environment, but a means for stabilization of bioactive materials.

472681 Silica Based pH-Responsive Nanocomposite Nanoparticles As Controlled Drug Release Carriers

Abstract:

Biocompatible drug carriers with the ability of controlled therapeutic agent delivery is a promising way to overcome the limitations of conventional therapies. Among all types of carriers, silica-based nanocomposites emerged as an encouraging candidate because of the biocompatibility, high thermal stability of silica materials. Moreover, silica based pH-responsive nanocomposite materials can be produced by encapsulation of natural or synthetic pH-responsive polymers, such as alginate, chitosan, poly (methacrylic acid) and poly (acrylic acid). In this study, the pH-responsive silica based nanocomposite nanoparticles were made by silica sol-gel method in water in oil microemulsion system with different polymer (alginate, chitosan, etc.) solutions as water droplets. Nano sized particles between 20-80 nm were detected by transmission electron microscopy. Different surface zeta potentials were showed by dynamic light scattering with different silica polymer nanocomposites. Silica-alginate nanocomposites showed negative zeta potential in acidic solution, however, silica-chitosan nanocomposites had positive zeta potential in the same solution. The polymer weight ratio was detected by thermogravimetric analysis. The results showed increased polymer weight ratio with higher concentration polymer solution used in the water in oil microemulsion. Different drug molecules, such as timolol maleate and nadolol, were used to test the release profile of the silica-polymer nanocomposite nanoparticles at different pH buffer solutions. For silica-alginate nanocomposites, more drugs were released at physiological pH solution compared to acidic solution. However, the release profile of silica-chitosan nanocomposites was opposite compared to the silica-alginate one. Consequently, the aim of this study was to investigate the different silica-polymer nanocomposite nanoparticles and tested their pH-responsive controlled release property.

454276 Porous Organic Cages for Drug Delivery

Abstract:

Porous materials have shown significant promise in drug delivery applications due to their high surface areas for adsorption, tunable pore size and chemistry, and regular, well-defined porosity. While several inorganic or hybrid organic-inorganic porous materials have been studied as drug carriers, concerns exist regarding the potential toxic effects of metals present in these systems. Porous organic cages are individual organic molecules with intrinsic porosity that, under specific conditions, self-assemble window-to-window with close packing to yield all-organic crystals with continuous, interconnected channels. In this work, we present the first-ever evaluation of these porous organic cages, specifically CC3 and CC5, for their drug delivery properties and biocompatibility. Crystalline particles of CC3 and CC5 have been loaded with cancer therapeutics (dichloroacetate and Î±-cyano-4-hydroxycinnamic acid), caffeine (useful for medicinal and cosmetic applications), and calcein (a model fluorescent molecule), with loadings of up to 25% by mass achieved, as determined by thermogravimetric analysis. Drug release from loaded CC3 and CC5 over several days was tested under simulated biological conditions in phosphate buffered saline (PBS), and drug release curves were obtained via UV-vis spectrometry to determine the kinetics and efficiency of drug delivery. The materials have also been shown to be stable in both water and PBS for at least one week, with no loss in crystallinity observed in powder x-ray diffraction patterns. Finally, two cytotoxicity assays have been conducted using HeLa cells to determine the effect of these materials both on cell membrane permeability and on metabolic activity. Neither CC3 nor CC5 compromised cell membrane integrity up to a concentration of 3 mg/mL, and while no reduction in metabolic activity was observed after incubation with CC3, the effects of particle size and potential breakdown products of CC5 on cytotoxicity are being investigated.

461136 Contact Lenses Prepared By Using the Layer-By-Layer Method for Sustained Drug Delivery

Abstract:

It is known that a lot of the people are suffering from eye diseases. Approximately 90% of the drug formulation is implemented as eye drops in topical implementation[1]. However, bioavailability of the eye drop is too low and only 1-5% of the drug can reach the target tissue. The remaining drug is thrown away from the nasal cavity and discarded drugs can cause serious side effects[2]. Researchers has developed various approaches to eliminate these disadvantages of the use of eye drops in recent years. Increasing the viscosity and permeability of eye drops or loading drugs into micro and nano particles or into contact lenses are some of the approaches that have been studied. Among the others, contact lense based drug delivery systems have two basic advantages such as minimizing the side effects of drugs used excessively and elimination of drug loss/waste[3].

A Synthetic polymer p-HEMA is used frequently for controlled release applications as well as being used as a soft material. There are few studies that focus on usage of p-HEMA contact lenses to achieve prolonged drug release. Even though, they are more successful than eye drops and ointment, sustained drug release cannot be provided as desired. Drug loaded p-HEMA contact lenses release the drug in few hours. There are some methods used to slowdown the drug release[4]. One of these methods is utilizing the layer by layer method in order to modify the surface of the p-HEMA hydrogels. This method is based on the affinity between complementary functional groups[5].

In this study it is aimed to use the drug loaded p-HEMA contact lenses for the treatment of the conjunctivitis, which is one of the common eye diseases. Conjunctivitis is inflammation of conjunctiva and the membrane layer of eyelid on inner surface. Dexamethasone is a hydrophobic drug used for conjunctivitis treatment which was added into hydrogels. The surface of hydrogels was modified via "Layer by Layer" (LbL) technique. By using this technique, it is possible to obtain up to ten layers. Moreover, a thick layer of drug free biodegradable materials were added on the top of the contact lenses, which already had five layers formed via LbL technique, as mentioned before. The formation of each layer was monitored by measuring absorbance, light transmittance and water contact angle of the lenses. Hydrogels were characterized by using FT-IR (Fourier Transform Infrared Spectroscopy), DSC (Differential Scanning Calorimetry) and SEM (Scanning Electron Microscope). Drug release concentration was measured by using UV-vis spectroscopy. It was determined that, p-HEMA hydrogel released around 60 % of the drug in 8 h. Hydrogels prepared by LbL methods (5 layers) have sustained drug release in which only 30 % of drug was released in 50 hours. It was demonstrated that the formed biodegradable material layers act as a barrier for drug release. Furthermore, addition of the drug-free layers on top of this surface modified hydrogels at different thickness gave rise to further reduction at the drug release. As a unique result, for the first time in literature, it is seen that addition of biodegradable material layer on the top of the surface modified hydrogels had allowed us to observe a "delay time". The observed delay time varied with the changing thickness of the layer applied. The delay time increased from 15 minutes to 45 minutes with increasing thickness. Thus, controlled drug release was accomplished. Moreover, addition of biodegradable materials layer on top of the contact lenses allowed us to see a delay time before the release of drug as well as it prevented the burst release. Also, side effects of the drug were eliminated which was released at the beginning.

This study shows clearly that combining the distinct properties of different materials by using LbL method has been an effective approach for the controlled release of ophthalmic drugs. The dose of drug, which each patient must receive according to his/her drug regimen, can be adjusted by varying the thickness or the changing the number of the layers applied.

REFERENCES

[1] C. Le Bourlais, L. Acar, H. Zia, P.A. Sado, T. Needham, R. Leverge, 1998, 'Ophthalmic Drug Delivery Systems - Recent Advances', Progress in Retinal Eye Research, 17(1),33-58.

[2] J.C. Lang, 1995, 'Ocular Drug-Delivery Conventional Ocular Formulations', Adv Drug Deliver Rev, 16(1), 39-43.

[3] Y. Kapoor, J.C. Thomas, G. Tan, V.T. John, A. Chauhan, 2009, 'Surfactant-laden soft contact lenses for extended delivery of ophthalmic drugs', Biomaterials, 30, 867–878.

[4] D. Gulsen, A. Chauhan, 2004,'Ophthalmic Drug Delivery through Contact Lenses' Investigative Ophthalmology & Visual Science, 45(7), 2342-2347.

[5] X. H. Hu, Y. S. Zhang, 2014, Synthesis and surface modification of chitosan containing hydrogel for ophthalmic drug delivery, 'Materials Technology: Advanced Performance Materials', 29(3).

468570 Chitosan-Coated Selenium Nanoparticles for Anti-Acne Applications

Abstract:

Introduction: The skin is the bodyâ€™s strongest form of protection against infectious and pathogenic microbes, and when this barrier is perforated by wounds, lesions or other chronic skin conditions, the body is vulnerable to more serious diseases. Among these skin diseases, acne vulgaris is one of the most prevalent, affecting up to 95% of all adolescents and persisting in up to 10% of adults. *Proprionibacterium acnes* has been suggested as the main contributor to acne vulgaris. Our lab has synthesized selenium nanoparticles to combat bacterial infection, and have found them to be effective against several bacterial strains, and relatively non-toxic to healthy fibroblasts and other mammalian cells. Here, to make such particles suitable for skin applications, we employed a new synthesis of chitosan-coated selenium nanoparticles as an effective topical treatment for *P. acnes*, as well as common skin denizens, such as *S. aureus*, while preserving the healthy skin tissue surrounding the acne lesions.

Materials and Methods: Chitosan-coated selenium nanoparticles (Ch-SeNP) were synthesized in a precipitation reaction, in which chitosan (dissolved in 1% acetic acid) was added dropwise to sodium selenite solution, followed by an addition of ascorbic acid. The resulting Ch-SeNP are separated via centrifugation and resuspended in DI water. Characterization was performed via dynamic light scattering and transmission electron microscopy. *P. acnes* (ATCC 11827) were cultured in modified reinforced clostridial broth medium or on tripticase soy agar plates with defibrinated sheep blood, while *S. aureus* was cultured in 3% tryptic soy broth or on agar plates. For *S. aureus*, kinetic growth curves were tracked by measuring the optical densities of treated cultures via spectrophotometry over 24 hrs. *P. acnes*, an anaerobic bacterium, required anaerobic conditions to grow, and were thus tested via a colony-forming unit assay in anaerobic gas chambers over 4 days. All experiments were repeated three times, and statistical significance were analyzed via a two-talked Studentâ€™s T-test.

Results and Discussion: Ch-SeNP were found to be as effective, if not more, as antibacterial agents as the previously studied BSA-coated SeNP on *S. aureus*, with lag times lengthening as Ch-SeNP concentrations increased. At a mid-range concentration of Ch-SeNP (0.023 mg/ml), *P. acnes* growth was completely absent compared to untreated colonies, suggesting that Ch-SeNP were strongly effective at preventing *P. acnes* proliferation.

Conclusions: Chitosan-coated selenium nanoparticles were found here to be exceedingly more stable in solution than their BSA-coated predecessors without sacrificing their antibacterial efficacy. Additionally, we found the Ch-SeNP to be effective in preventing the proliferation of the anaerobe *P. acnes*, suggesting a novel SeNP application in topical skin infections.

Acknowledgements: The authors thank Northeastern University and the College of Engineering for funding.

471959 Regulating the Timing and Sequence of Pro- and Anti-Inflammatory Cytokine Deliveries from Magnetically Responsive Biomaterials for Use in Wound Healing Applications

Abstract:

In the United States, approximately 6.5 million patients are diagnosed with chronic wounds. The treatment of chronic wounds costs more than $25 billion annually and this is expected to grow due to the increasing cost of healthcare, an aging population, and a rise in the occurrence of diabetes and obesity worldwide [1]. The process of wound healing includes four major steps: hemostasis (blood clothing), inflammation, proliferation, and tissue maturation [2]. Even a slight perturbation in this process can disrupt proper healing, leading to chronic wounds. Chronic wounds are often a result of arrest in the inflammation phase of healing [2]. Although inflammation critically initiates repair and helps clear infections, a prolonged inflammatory reaction can cause considerable harm to the injury site. After an appropriate duration of inflammation, this inflammatory response can be shifted to a more pro-healing response through the delivery of cytokines like interleukin 4 (IL-4) and interleukin 10 (IL-10). These anti-inflammatory cytokines alter the phenotype of macrophages from pro-inflammatory (M1) to anti-inflammatory (M2), suggesting a potentially powerful drug delivery strategy if these cytokines can be delivered in a delayed manner. We hypothesize that the transition of macrophage phenotype from pro-inflammatory (M1) to anti-inflammatory (M2) can be controlled through sequenced delivery of interferon gamma (IFN-γ), followed by IL-4 and/or IL-10. The goal of this research was to develop a wound-healing hydrogel system that initially delivers IFN-γ, followed by magnetically triggered delivery of IL-4 and/or IL-10.

Our biomaterial system was composed of two-compartments: (1) a porous gelatin outer compartment designed to recruit macrophages and establish an initial pro-inflammatory (M1) phenotype, and (2) a magnetically responsive alginate inner compartment which was designed to deliver IL-4 and/or IL-10 when magnetically triggered to shift the response to anti-inflammatory by promoting (M2) phenotype. This

magnetically responsive inner compartment was made by casting 7 wt% Fe_3O_4in 1 wt% alginate with 2.5 mM adipic acid dihydrazide (AAD) crosslinking in the presence of a magnet to generate biphasic gels. These biphasic gels were then swollen in water, frozen at -20°C, and lyophilized to generate porous, magnetically compressible biphasic gel structures. Gelatin outer compartments were loaded with 1000 ng of IFN-γ and designed to diffusively release IFN-γ early on. Magnetically deformable inner compartment gels were loaded within 100-1000 ng of IL-4 or IL-10 and were designed to deliver these anti-inflammatory proteins to macrophages that were recruited to the outer compartment at a time dictated by the application of the magnetic stimuli. A pulsatile 5 kGauss graded magnetic field was used in these experiments and applied at 1.5 Hz. Validation of these protein delivery profiles over time were quantified for IL-4, IL-10, and IFN-γ using ELISA.

We have demonstrated this system's ability to initially release a pro-inflammatory cytokine (IFN-γ). We have found that more than 95% of the loaded IFN-γ was released from the outer compartment in the first 6 hours. We have also shown the ability to retain anti-inflammatory cytokines (IL-4 and IL-10) for several days and magnetically trigger their release from the inner compartment. Furthermore, we were able to magnetically control the timing and rate of these deliveries. For example, we showed that when we load 1000 ng of IL-4 in the inner compartment, we were able to retain IL-4 in the gel for 4 days with low release rates prior to magnetic stimulation (29.6 ± 3.3 ng/day for 4 days) and magnetically trigger enhanced release rates on day 4 (345.8 ± 103.29 ng/day over 4 hours). Using gels loaded with 500 ng of IL-10, we again demonstrated excellent retention for 3 days prior to magnetic stimulation (0.17 ± 0.04 ng/day for 3 days) and magnetically enhanced release rates on day 3 (6.99 ± 4.24 ng/day over 3 hours). The timing of IL-10 delivery could be stimulated on day 5 instead of day 3. In this case, IL-10 was retained at a low release rate prior to magnetic stimulation (0.11 ± 0.04 ng/day for 5 days) and magnetically released at enhanced rates on day 5 (3.67 ± 0.776 ng/day for 4 hours). Finally, the rate of IL-4 release could be controlled by using different magnetic stimulation regiments. For example, the release rate of IL-4 when magnetically stimulated at 1.5 Hz for 1 hour every 12 hours for over days was only 20.23 ±7.33 ng/day whereas continuous magnetic stimulation at 1.5 Hz for 4 hours resulted in a much higher release rate (345.85 ±103.29 ng/day). We believe that this biomaterial system will be highly useful for understanding how the timing and sequence of cytokine deliveries impact wound healing outcome.

Keywords: Wound healing, Macrophages, Cytokine delivery, Biomaterial, Magnetic stimulation.

REFERENCES

1. Sen, Chandan K. et al. "Human Skin Wounds: A Major and Snowballing Threat to Public Health and the Economy." *Wound Repair and Regeneration* 17.6 (2009): 763–771.

1. Frykberg, Robert G., and Jaminelli Banks. "Challenges in the Treatment of Chronic Wounds." *Advances in Wound Care* 4.9 (2015): 560–582.

471748 Evaluation of Poly(curcumin) Microparticle Degradation and Activity in the Presence of Free Radical Generating Systems

Abstract:

Under a variety of pathophysiological conditions (e.g., radiation injury, inflammation, acute lung injury), cells are no longer able to maintain a balance between antioxidants and exogenous or endogenous radical generation, resulting in free radicals overproduction and oxidative stress.

Curcumin, a potent antioxidant, has three potential scavenging sites, but due to its hydrophobicity, poor solubility and bioavailability, it has possessed a low therapeutic effect *in vivo*. Free antioxidant drug delivery systems can also pose strain to the environment that can lead to pro-oxidant effects rather than the latter. To overcome these limitations, our group has previously synthesized a covalently crosslinked poly(curcumin beta amino), poly(curcumin), film, where the hydrolytically degradable backbone of the polymer network increases curcuminâ€™s bioavailability, utilizing a controlled release mechanism, whichÂ gives great potential to show positive efficacy on oxidative stress treatment and cellular protection. Poly(curcumin) films can be cryomilled into microparticles to employ versatility in size and for localized delivery of interest. An advantage of the poly(curcumin) network is its comonomer system composed of a hydrophilic (polyethylene glycol (400) diacrylate) and hydrophobic (curcumin multiacrylate (CMA)) component. The ratios of these components can be varied to modify the rate of drug release. The design of microparticle blends can provide unique regulation to future dosage and treatments to specific oxidative environments (e.g. different stages of oral mucositis) to maintain a constant therapeutic release of curcumin. The use of kinetic and predictive modeling can be used to model poly(curcumin) microparticles as a two compartment system to verify the contributing factors to microparticle degradation as well as the consumption or scavenging rates of the released curcumin. We hypothesize that curcumin release from microparticles of different compositions that are introduced to reactive oxygen species to the microparticle system will respond and consume differently based on different degradation rates. and accelerate the hydrolytic cleavage of the weak ester bonds, further accelerating the release and consumption of curcuminâ€™s

antioxidant activity due to its scavenging capabilities. Predictive models can allow for us to evaluate and understand our complex polymer network mechanistically and better understand the contribution of volume, hydrolysis, and drug affinity in our system.

470891 Simulating Intestinal Molecular Transport Using Tissue Engineered in Vitro Models

Abstract:

Introduction: The intestinal mucosal barrier lining the gastrointestinal tract â€" composed of an epithelial cell monolayer and the mucins they secrete â€" plays a critical role in maintaining metabolic and immunological homeostasis. Recently developed microfluidic gut-on-a-chip devices have significantly advanced the capability of in vitro models in recapitulating in vivo intestinal tissue physiology (1). However, remaining key challenges for such devices include (i) supporting both anaerobic and aerobic intestinal milieu, (ii) simultaneous human cell co-culture, and (iii) real-time in situ observation of their dynamic crosstalk.Â Here, we introduce a new bioengineered platform technology with high analytical power that can potentially allow for anaerobic and aerobic culture by creating a transverse compartmentalized device that retains previously identified anatomical and mechanical tissue complexities. Peristaltic motion that is critical in promoting gut epithelial cell growth and function was further simulated via pneumatic actuation.

Materials and Methods: Type I collagen infused polyacrylamide hydrogel precursor solution was casted in a 500μm thick PDMS sheet having pre-cut villous mimicking micropatterns on a silanized glass slide. (2) An actuating layer made with Ecoflex that has been used in soft-robotics (3) was attached outside of a hydrogel layer and sealed with another glass slide. We then increased the cellular and extracellular complexity by sequentially introducing (i) HT-29 intestinal epithelial cells, (ii) mucin extract from pig intestines, (iii) bacteria, and (iv) human peripheral blood-derived mononuclear cells (PBMC). By using a solenoid valve in tandem with a microcontroller (μC), the chamber was expanded and contracted at controlled pressures and frequencies using in house vacuum and air.

Results and Discussion: Covalent attachment of polyacrylamide hydrogels to silanized glass slides was achieved in a sealed, compartmentalized microfluidic device. Molecular diffusion tests using rhodamine dye showed that the hydrogels are permeable to gas and allow for small molecular diffusion. Different spatially patterned hydrogels can be microfabricated with villous-like structures that recapitulate the physiology of the gastrointestinal tract. The luminal compartment of the microfluidic was seeded with HT-29 colorectal cancer cells that preferentially adhered to the hydrogel, forming an epithelial barrier while the tissue compartment was filled with PBMCs derived from blood donors. Additionally, mucin and bacteria were introduced in the luminal compartment and continuously perfused in order to reduce bacterial density while cell-to-cell interactions were imaged in real-time. Actuation was controlled by an Arduino-based μC that can have very diverse control of operations.

Conclusions: We have developed a new type of microfluidic device that integrates soft biomaterials. We envision that such a modular in vitro microphysiological intestinal tissue model can serve as a translational platform to discover the biophysical etiology of disruption of the mucosal barrier.

468198 Silk-Extracellular Matrix Hydrogels for Cardiac Tissue Engineering

Abstract:

Cardiovascular disease is the leading cause of death in the U.S.[1] and rapidly becoming the leading cause of death worldwide.[2] The development and use of injectable biomaterials for cardiac repair can reduce the invasiveness of the biomaterial application.[3] Natural injectable materials, like solubilized cardiac extracellular matrix (cECM), contain appropriate binding sites for cells, yet the mechanics and porosity are difficult to control. In contrast, synthetic materials allow for finer control over mechanics and porosity but do not contain the appropriate diversity of cellular binding sites. Recently, we developed a tunable, elastic, silk hydrogel that is formed via di-tyrosine crosslinking using a horseradish peroxidase (HRP)/hydrogel peroxide (H_2O_2)-based enzymatic crosslinking approach.[4] Silk fibroin protein does not contain integrin binding sites for cells, but we have previously demonstrated that the addition of cECM to silk sponges enhanced cardiac cell phenotype and function over native silk fibroin alone.[5] In addition, we have shown that addition of ECM to silk sponges enhances native cell infiltration in a subcutaneous implant study.[5] Here, we aim to augment these elastic, silk hydrogels with cECM and assess the tunability of this material for both *in vivo* applications, as well as its utility in our bioreactor systems to modulate cardiac cell function.

To achieve these goals, we formed silk-based hydrogels from silk solution as previously reported.[4,6] Adult porcine hearts were obtained from the local butcher shop and decellularized using sodium dodecyl sulfate as previously described.[5] Left ventricular cardiac extracellular matrix (cECM) was solubilized in pepsin[7] and incorporated into silk solution prior to the initiation of the HRP/H_2O_2 reaction. Mechanics of acellular gels and cellularized gels were measured via unconfined compression or tension using dynamic mechanical analysis. To evaluate cellular response to silk-cECM hydrogel, cardiac fibroblasts (CFs, adult Sprague Dawley rats) and neonatal cardiomyocytes (CMs, 1-day old pups, Sprague Dawley) were isolated using procedures approved by the Tufts University IACUC. CFs were activated to a myofibroblast phenotype (typically present in infarct tissue) via treatment with 2.5 ng/ml TGF-beta for 48 hours. For both cell types, viability was assessed by live-dead stain and a DNA assay. Important fibroblast markers, key players in integrin signaling, and proteins important for cardiac maturity and function were assessed by Western blot. Hydrogels were cultured both as unconfined gels in tissue culture plates and within our custom built bioreactor, which is capable of delivering physiologically relevant electromechanical stimulation.[8-10]

Addition of cECM did not negatively impact the gelation of the silk-HRP hydrogels and the gelation kinetics can be tuned by altering the concentration of the starting silk solution or the molecular weight of the silk fibroin protein. In unconfined culture, we found that acellular gels demonstrated a time-dependent stiffening in culture, which is useful for the generation of infarct models to study how CF activation and CM viability and function change based on changes to tissue mechanics.

For the culture of activated cardiac fibroblasts, the addition of cECM to elastic silk-HRP gels enhanced and altered gene expression that was modulated by time-dependent stiffening of the hydrogels. Importantly, the addition of cECM augmented protein expression particularly in the proteins important in integrin mediated signaling. In addition, subcutaneous implant studies demonstrate the injectable nature of these formulations and relate the *in vitro* results to *in vivo* potential. On-going work is focused on the evaluation of neonatal CMs within the bioreactor system to determine the role of cECM composition, hydrogel mechanical properties, and electromechanical stimulation on CM maturation and function. Future work seek to tune the stiffening process to aid in functional muscle tissue restoration upon injection *in vivo*.

References:

1. D. Shepard, A. VanderZanden, A. Moran, M. Naghavi, C. Murray and G. Roth, *Circulation: Cardiovascular Quality and Outcomes*, 2015, **8**, 455-456.

2. D. Mozaffarian, E. J. Benjamin, A. S. Go, D. K. Arnett, M. J. Blaha, M. Cushman, S. de Ferranti, J. P. Despres, H. J. Fullerton, V. J. Howard, M. D. Huffman, S. E. Judd, B. M. Kissela, D. T. Lackland, J. H. Lichtman, L. D. Lisabeth, S. Liu, R. H. Mackey, D. B. Matchar, D. K. McGuire, E. R. Mohler, 3rd, C. S. Moy, P. Muntner, M. E. Mussolino, K. Nasir, R. W. Neumar, G. Nichol, L. Palaniappan, D. K. Pandey, M. J. Reeves, C. J. Rodriguez, P. D. Sorlie, J. Stein, A. Towfighi, T. N. Turan, S. S. Virani, J. Z. Willey, D. Woo, R. W. Yeh, M. B. Turner, C. American Heart Association Statistics and S. Stroke Statistics, *Circulation*, 2015, **131**, e29-322.

3. T. D. Johnson, R. L. Braden and K. L. Christman, *Methods Mol Biol*, 2014, **1181**, 109-120.

4. B. P. Partlow, C. W. Hanna, J. Rnjak-Kovacina, J. E. Moreau, M. B. Applegate, K. A. Burke, B. Marelli, A. N. Mitropoulos, F. G. Omenetto and D. L. Kaplan, *Advanced Functional Materials*, 2014, **24**, 4615-4624.

5. W. L. Stoppel, D. J. Hu, I. J. Domian, D. L. Kaplan and L. D. Black, *Biomedical Materials*, 2015, **10**.

6. D. N. Rockwood, R. C. Preda, T. Yucel, X. Q. Wang, M. L. Lovett and D. L. Kaplan, *Nature Protocols*, 2011, **6**, 1612-1631.

7. C. Williams, K. P. Quinn, I. Georgakoudi and L. D. Black, 3rd, *Acta Biomater*, 2014, **10**, 194-204.

8. K. Y. Morgan and L. D. Black, 3rd, *J Tissue Eng Regen Med*, 2014.

9. K. Y. Morgan and L. D. Black, 3rd, *Tissue Eng Part A*, 2014, **20**, 1654-1667.

10. K. Y. Morgan and L. D. Black, III, *Organogenesis*, 2014, **10**, 317-322.

470209 Polymer Nanoparticle Synthesis and Characterization for the Delivery of Osteoinductive Molecules

Abstract:

Localized delivery of osteoinductive factors (e.g., growth factors, peptides, and small molecules) remains a major challenge in bone tissue engineering to promote regeneration and repair of damaged tissues. Delivery of select bone morphogenetic proteins (BMPs) has shown great therapeutic potential (specifically, BMP-2 and BMP-7). However, current delivery methods require supraphysiological concentrations to obtain desired effects and experience inadequate retention at the site of delivery. In contrast, the use of small molecules to induce bone regeneration has not been fully explored and displays several advantages over the use of growth factors (including small size, high stability and non-immunogenicity). Local administration of osteoinductive small molecules can be achieved through incorporation within osteoconductive materials, such as, bone tissue engineered scaffolds. However, optimal carriers for small molecule delivery within these scaffolds remain partially unknown. For this purpose, polymer nanoparticles were synthesized and characterized for the delivery of osteoinductive small molecules. Nanogels were synthesized via solution polymerization. N-isopropylacrylamide (NIPAM) was co-polymerized with methacrylic acid (MAA) and either benzyl methacrylate (BZMA) or 2-hydroxyethyl methacrylate (HEMA) at varying concentrations using N,Nâ €™-methylenebisacrylamide as the crosslinker. Synthesis of P(NIPAM-co-MAA-co-BZMA) and P(NIPAM-co-MAA-co-HEMA) nanogels was confirmed by Fourier transform infrared spectroscopy. Nanoparticle size was determined using dynamic light scattering. Results show successful synthesis of the desired polymer formulations at varying monomer concentrations. Thus, nanogels can be further used to investigate the load and release of small osteoinductive molecules. Successful delivery of these molecules within tissue engineered constructs could have major impact in bioengineering and in specific therapeutic modalities in the clinical milieu.

469467 Transitional Liver Models for the Investigation of Chemical and Mechanical Cues on the Progression of Fibrosis

Abstract:

Liver fibrosis often results in hepatocellular carcinomas, multiple organ failure, coma, and is the tenth leading cause of death in the United States. Alcohol abuse, obesity, diabetes and hepatitis C viral infections are some initiating events that induce fibrosis. Fibrosis occurs when uncontrolled wound healing results in the excessive accumulation of extracellular matrix (ECM) components, resulting in scar tissue. The principal liver cell type, the hepatocytes, performs a majority of the organâ€™s metabolic and biotransformation functions. However, the nonparenchymal cells (NPCs) are primarily responsible for the initiation and progression of fibrosis. These include the liver sinusoidal endothelial cells (LSECs), Kupffer cells (KCs) and hepatic stellate cells (HSCs).

The Space of Disse is a protein rich interface that separates the NPCs from the hepatocytes. This membrane is primarily composed of ECM components and increases 6-10 fold in stiffness as a result of fibrosis. There is little information on how changes in the material microenvironment result in the progression of the disease. Since liver fibrosis occurs over time, we seek to investigate how the physical and chemical properties of the microenvironment gradually transition a healthy tissue into a diseased organ. Our focus is to design transitional liver models wherein the biomaterial properties are presented to cells as a gradient, thereby sending cues to gradually change the cellular profiles from healthy to fibrotic.

We have designed a biopolymeric membrane comprised of ECM components to act as a Space of Disse mimic. This membrane is assembled from alternating layers of collagen (COL; type 1) and hyaluronic acid (HA). The hydrated thickness of this membrane is approximately 700 nm, which is similar to the height of the Space of Disse *in vivo*. The membranes were cross-linked in order to vary the rigidity of the membrane to mimic healthy and fibrotic liver tissues. Increasing the cross-linking time resulted in COL/HA membranes whose elastic moduli ranged from approximately 21

kPa to 43 kPa.

Changes to the ECM composition also occur during hepatic fibrosis. Laminin, an ECM protein, increases in concentration in fibrotic livers. Therefore, we modified the COL/HA membranes with rhodamine-conjugated laminin. NPCs were seeded on unmodified and laminin-conjugated COL/HA membranes and monitored for changes in proliferation, cytokine secretion, migration, and morphology. The effect of membrane stiffness on hepatocytes was investigated by overlaying a monolayer of cells with COL/HA films of varying stiffness.

LSECs and KCs monocultures were seeded on laminin-conjugated or unmodified COL/HA membranes. LSECs exhibited an approximately 36% increase in proliferation on laminin-conjugated PEMs, whereas KC proliferation was approximately 66%. No significant proliferation of either cell type occurred on the unmodified membranes. These results clearly demonstrate that laminin induced proliferation of NPCs; especially the inflammatory KCs. Hepatocytes overlaid with stiffer PEMs exhibited cell death and rounded morphology. The relative viability of hepatocytes decreased by 29%. Combined, these results demonstrate the importance of the chemical and mechanical properties of the microenvironment in the development of liver fibrosis. Ongoing efforts are focused on designing membranes that exhibit chemical and mechanical gradients as well as conjugating additional proteins such as high mobility group box proteins (HMGB1). We are investigating how NPCs, specifically, HSCs are affected by HMGB1, a protein found in fibrotic hepatic tissues.

469115 Additive Manufacturing of Multicompenent Biomaterials

Abstract:

Additive manufacturing, including 3D printing and 3D bio-printing, are enabling the prototyping and production of geometrically-defined and patient specific biomaterials.[1] Despite the many recent advances in additive manufacturing technologies,[2,3] there remains limitations in the speed of production, access to a broad range of materials, and easy integration of multiple materials and cells within a single device. Here, we present a facile manner to produce rapidly multicomponent biomaterials with defined geometry incorporating both structural and mechanical support from polymeric constituents (*e.g.*, (meth)acrylates, polycaprolactone, poly(lactic-*co*-glycolic acid)) as well as cellular and tissue integration from hydrogels (*e.g.*, poly(ethylene glycol) based, collagen/elastin based, gelatin methacryloyl based). This process integrates the benefits of additive manufacturing with the benefits of hydrogel biomaterials by enabling cell-laden biomaterials with defined structure and mechanical integrity. In addition, the process can access a variety of applications on account of its scalability and versatility. We demonstrate the efficacy of this approach to fabricate cell-laden biomaterials for *ex vivo* models of functional tissue as well as for on scale tissues for regenerative medicine applications. Overall, this presentation will present a facile process of multicomponent biomaterial fabrication and illustrate how this contributes to the emerging field of additive manufacturing in biotechnology and medicine.

References

1. Murphy SV *et al. Nat Biotech* **32** (2014) 773-785
2. Appel EA *et al. Chem Soc Rev* **107** (2010) 18392-18397
3. Kang HW *et al. Nat Biotech* **34** (2016) 312-319

480142 Enhancing Cell Migration Through Dual Biomolecule Delivery From Electrospun Fibrous Scaffolds

Abstract:

Over one million meniscal surgeries are performed in the United States every year, where the most common treatment option following a meniscal tear is a partial meniscectomy. Meniscal tissue removal, however, results in a proportional increase in contact stresses on the tibial plateau and predisposes the patient to osteoarthritis. As with other dense fibrous connective tissues, the intrinsic healing of the meniscus is severely limited due to poor vascularity and hypocellularity. Previous research indicates that the high extracellular matrix density in the mature meniscus serves as a physical hurdle to cell migration and proliferation, ultimately limiting endogenous repair. To address this limitation, we have developed electrospun fibrous scaffolds to locally deliver collagenase, a matrix-degrading enzyme, in combination with platelet-derived growth factor-AB (PDGF-AB), a known chemokine, to enhance cell mobility and recruitment, respectively, to the wound interface. In vivo results indicate the synergistic delivery of collagenase and PDGF-AB increases cell concentration in the tissue close to the wound interface and within the fibrous scaffold. Furthermore, immunohistochemistry reveals the beginning of new tissue formation for treatment groups with combined biomolecule delivery.

472291 Design of Self-Assembled Vaccines to Program Activation of Multiple Adjuvant Pathways

Abstract:

Biomaterials have the potential to better control immune function through co-delivery of signals, targeting, and controlled release. However, addition of polymers increases the complexity of vaccine composition. Further, many polymeric carriers trigger stimulatory pathways even without other immune cues. While potentially a trait to be harnessed, these effects can also hinder rational design since carriers can alter the types of immune response that develop. Development of materials that juxtapose multiple antigens and adjuvants at high density while eliminating carrier effects could simplify vaccine composition and support tunable control of immune responses to infectious disease or cancer. Along these lines, recent studies have explored vaccines using combinations of molecular adjuvants based on toll-like receptor (TLR) ligands. These signaling pathways have evolved to detect patterns common in pathogens but uncommon in humans. Thus new vaccines are seeking synergistic benefits using combinations of TLR agonists (TLRas) that simultaneously activate several TLR pathways.

Polyelectrolyte multilayers (PEMs) offer fundamental features to provide a unique level of control over the delivery of multiple TLRAs. These nanoscale structures are prepared by electrostatic, layer-by-layer assembly of oppositely charged polymers. Recently, PEMs has been studied as vaccines using biocompatible polymers to encapsulate or adsorb antigens or adjuvants. We sought to design PEMs composed entirely of peptide antigens and combinations of TLRas (i.e., TLRa3, TLRa7, and TLRa9). These assemblies are termed immune-PEMs (iPEMs). iPEMs were assembled from a model antigen (SIINFEKL) and combinations of TLR3a polyinosinic-polycytidylic acid (polyIC), TLR7a polyuridylic acid (polyU), TLR9a Class B CpG oligonucleotide (CpG). Each of these TLRas is polyanionic, thus SIINFEKL was modified with arginine residues (SIIN*) to support electrostatic assembly. We initially assembled iPEMs containing a single TLRa, observing linear growth of SIIN*/polyIC, SIIN*/polyU, and SIIN*/CpG iPEMs on silicon substrates. We observed that iPEMs grow proportional to the number of layers deposited on the silicon substrates. Similarly, iPEMs containing two types of TLRas, or all three types of TLRas could be deposited with tunable control over the loading of these components. We next tested if SIIN*/polyIC IPEMs activate primary dendritic cells (DCs) and expand antigen-specific T cells in culture and in mice. We observed that SIIN and polyIC in iPEMs maintained the specificity and functionality of each component by promoting upregulation of DC maturation markers (i.e., CD40, CD80, and CD86) and inflammatory cytokine secretions (i.e, IFN-γ, IL-6 and IL-1β). These structures also drove expansion of SIINFEKL-specific $CD8^+$ T cells. Direct measurements of TLR3 signaling also revealed iPEMs containing polyIC activated TLR3, while iPEMs containing other TLRas or irrelevant nucleic-acid sequences did not.

To test whether SIIN*/polyIC iPEMs generate functional, antigen-specific T cell responses incorporated in iPEMs, mice were primed at day 0 and boosted on day 15. iPEMs significantly expanded antigen-specific $CD8^+$ T cells at day 7 ($p \leq 0.05$) and 22 ($p \leq 0.001$) compared to both naïve mice and admixed vaccines (i.e., free) containing the same dose of antigens and adjuvants. In tumor study, mice were primed and boosted (day 15 and day 28) with SIIN*/polyIC iPEMs, then challenged with 1×10^6 B16-OVA cells on Day 36 (N=6). In these studies, SIIN*/polyIC iPEMs conferred a 56% increase in median survival compared to admixed vaccines. Ongoing studies are assessing these same ideas using iPEMs assembled from multiple TLRas. This platform could thus offer modularity to selectively program the activation of TLR pathways in development of new vaccines and tumor therapies.

461245 Gold Nanorods for ANTI-Tumor Applications

Abstract:

Gold nanoparticles have attracted much attention due to their many advantageous properties. They have a size range of 1-100 nm and can be easily incorporated into biological systems. Also researchers have found that gold nanoparticles exhibit stability, low toxicity, facile fabrication and functional flexibility. They have also been investigated for their brilliant colors. Their unique optical properties result from the localized surface plasmon resonance (LSPR) effect. When a light wave front passes the gold nanoparticles, the oscillating electromagnetic field of light will cause a collective movement of the free electrons with respect to the positively charged metallic lattice, resulting in a standing oscillation of the electron density in resonance with the light's frequency. At this frequency, both the scattering and absorption of gold nanoparticles are strongly enhanced. This enhanced scattering can be used in areas like imaging and disease diagnosis, while the enhanced absorption can be applied in areas like photothermal therapy.

The LSPR band of gold nanoparticles can be tuned by changing their size, shape or structure. Researchers have been trying to tune the frequency to the near-infrared region (650-900nm, NIR region), within which the light has a maximum depth of penetration in tissues and blood. This makes the gold nanoparticles promising candidates for *in vivo* applications. For example, gold nanoparticles with LSPR frequency in the NIR region, if combined with antibodies or folic acid, can be used for combined imaging and photothermal therapy of cancer. Anti-tumor drugs can also be conjugated onto gold nanoparticles for photothermal-chemotherapy. These novel treatments can overcome the shortages of conventional ones in that functionalized gold nanoparticles can selectively target and kill the tumor cells while having little or no influence on the surrounding healthy cells.

Among different gold nanoparticles, gold nanorods have been of high interest due to their tunability and large red shift towards the NIR region. Using the seed-mediated growth method, the LSPR band of gold nanorods can be tuned to cover the entire NIR region by simply changing the amount of one reactant. Furthermore, gold nanorods can be readily conjugated with different molecules, like polyethylene glycol (PEG), folic acid, antibody, DNA, peptides and anti-tumor drugs, for various applications. In our study, we synthesized gold nanorods of different aspect ratios, tested their cytotoxicity and modified them with different molecules like PEG for future research. This *in vitro* cell study showed that bare gold nanorods can kill more than half of breast cancer cells (MDA-MB 231) at a concentration as low as 18.75 μg/ml while having a limited influence on human dermal fibroblasts up to 75 μg/ml. Therefore, there is a suitable concentration range of 18.75 – 75 μg/ml in which these novel gold nanorods can selectively kill the tumor cells, suggesting they should be further studied for a wide range of medical applications.

467549 Posh Inhibitor Peptide Amphiphile Micelles As a Novel Leukemia Therapeutic Modality

Abstract:

Acute lymphoblastic leukemia (ALL) is associated with high-risk and poor prognosis for which few effective treatments exist. Published data has shown that the scaffold peptide 'Plenty of SH3 Domains' regulates lymphocyte signaling pathways and its disruption by a novel peptide (TAT-POSH) leads to significant cell death in greater than 90% of T cell and B cell leukemias evaluated to date. While TAT has been shown to facilitate moderately enhanced cellular internalization, it carries out its function by macropinocytosis which may not be the best method for cytosolic payload delivery. Peptide amphiphile micelles (PAMs) are biomaterials that have been shown to be internalized by lipid bilayer intercalation which allows for significant cytosolic delivery. By combining POSH peptide with the micellar platform, a novel ALL treatment system capable of broad spectrum anti-neoplastic effects may be achievable.

POSH peptide amphiphiles were synthesized by covalently conjugating two hydrophobic aliphatic palmitic acid tails to POSH peptides which were found to self-assembly into small and mostly spherical micelles in water at very low concentrations. Since ALL is a blood-borne cancer, therapeutics need to be capable of being delivered intravenously so small, spherical micelles are an ideal device. Interestingly, spherical micellization often requires the use of an additional hydrophilic linker (*e.g.* PEG) to achieve the desirable micelle shape, but the highly hydrophilic and charged nature of the POSH peptide facilitated spherical micelle formation without additional components. Confocal microscopy revealed that fluorescently-labeled POSH PAMs are readily uptaken by even non-phagocytic cells indicating their intracellular payload potential. To determine the enhanced cancer cytotoxicity of POSH PAMs they were incubated with three different leukemic cell lines (murine LM-138, human Mec-1, and human Mec-2) and their effects were compared to the known cytotoxic peptide TAT-POSH. POSH PAMs was found to greatly outperform TAT-POSH inducing enhanced cytotoxicity at every dose and similar cytotoxicity at a sixteenth to an eighth of the dose. The capacity for POSH PAMs to improve cancer cytotoxicity over a gold-standard cell penetrating peptide like

TAT is a quite exciting discovery. These results provide strong evidence that POSH PAMs hold tremendous promise as an ALL treatment modality.

465529 Zwitterionic-Containing Triblock Peptide Amphiphiles Self-Assemble into Unique Higher-Order Micellar Structures

Abstract:

Peptide Amphilphiles (PAs) are a class of diblock biomaterials made of hydrophilic peptide head(s) conjugated with hydrophobic lipid tail(s) which self-assemble into Peptide Amphiphile Micelles (PAMs) in water. PAMs play a significant role in biomedical research having shown promise for cancer therapies, prophylactic vaccines, and regenerative medicine. After two decades of research, the fundamental thermodynamic principles that govern micelle formation are well understood. In specific, the critical packing parameter is an excellent tool for researchers to utilize to control basic PAM structure. Therefore, the design of first-order micellar structures such as spherical micelles, cylindrical micelles, bi-layers, and inverse micelles is straightforward. In recent years, some second-order micellar structures like twisted ribbons and helices have been discovered which has opened up the opportunity for the development of more complex design rules. In specific, unevenly distributed hydrogen bonds within a micelle and hydrogen bonding between different micelles are believed to be responsible for the formation of these second-order micellar structures.

To further investigate the effect of complexation in higher-order micelle formation, we have designed a class of triblock PAs which possess a bioactive peptide, zwitterionic block, and hydrophobic tail(s). These zwitterionic-containing triblock PAs interestingly form not only second-order micellar structures but some can even form third-order micellar structures where helical micelles wrap themselves together to form unique â€œbraided micellesâ€. In addition to their exotic morphology, braided micelles have shown significant preliminary promise for a variety of biomedical applications since they have been found to be more stable in the presence of proteins and undergo enhanced cellular uptake when compared to other micellar structures studied. Braided micelles and other previously unseen higher-order structures can be created by changing the chemistry of the three components, rearranging the block sequences, and/or changing environmental conditions. These results have helped us to begin to understand the physical phenomena that govern higher-order micelle structures and start to generate micellar design rules which will serve as a tool box that can be leveraged for a wide variety of future biomedical applications.

462547 Investigating Biomembrane-Nanoparticle Interactions with Giant Vesicles Fabricated from Inverted-Headgroup Lipids

Abstract:

The surface chemistry of the cell membrane plays an important role in how cells interact with particulate species. These interactions are dictated in large part by lipid headgroup charge. To investigate the nature of electrostatic interactions between lipid bilayers and nanoparticles in solution, we investigated nanoparticles interacting with lipids in which the zwitterionic head group was inverted from its natural configuration. This inverted lipid species, referred to as DOCP, is based on 1,2-dioleoyl-glycero-3-phosphocholine (DOPC) but has the positions of the phosphate and choline groups reversed.

These interactions were investigated by fabricating giant unilamellar vesicles (GUVs) with DOPC lipids and DOCP lipids respectively, and introducing nanoparticles to suspensions of both. In the presence of charged polystyrene nanoparticles, GUVs displayed various deformational modes depending on the charge sign and size of the nanoparticles. The differences in the responses of the two lipid species were quantified according to the size changes of GUVs. These differences illuminate how the relative positions of the negative and positive charge groups on the lipid interact with charged particles. They also reveal a particle size dependence of the nanoparticle-membrane electrostatic interaction. This work provides fundamental insights regarding the nature of this interaction, which has implications for the development of new drug delivery vehicles as well as understanding the environmental toxicity of nanomaterials.

469416 Enhancing Epoxy Network Toughness and Recoverability with Mussel-Inspired Catechol-Iron Crosslinks

Abstract:

Iron-catechol metal-ligand coordination bonds have previously been shown to impart both hardness and strength to the marine mussel byssal thread while simultaneously maintaining high extensibility, thus creating an extremely tough material with excellent energy-dissipation properties. These properties show a clear deviation from the commonly accepted trade-off between strength and extensibility observed in typical industrial polymers, making them desirable properties for polymer coatings and adhesives. Though these properties have been successfully realized in low-modulus hydrogel networks, this work aims to demonstrate that high-elastic modulus (~100 MPa), high-strain (~100-200%) materials can also be synthesized using a similar synthetic approach in epoxy networks. Low-cross-link density poly(ethylene glycol)/epoxy networks were synthesized with pendant catechol groups and subsequently swollen and treated in iron(III)-containing solutions at both low and high pH. The treated networks were then dried and chemically verified using various spectroscopies, and the mechanical properties were tested under tension. Larger than an order-of-magnitude increase in elastic modulus was observed with less than 25% decrease in extensibility as compared to the untreated samples. Due to the reversible nature of the iron-catechol coordination bonds, the networks also demonstrate the ability to quickly recover their initial mechanical properties after plastically deforming during cyclic tensile tests. Therefore, it is clear that the incorporation of iron-catechol coordination bonds in dry polymer materials may be a synthetic strategy for greatly increasing material strength without sacrificing extensibility.

bility.

463617 Maintenance of Neural Progenitor Cell Stemness in 3D Hydrogels Requires Matrix Remodeling

Abstract:

While neural progenitor cells (NPCs) hold significant therapeutic promise, relatively little is known about the material properties required to maintain NPC stemness and regenerative potential. Previous studies using 2D substrates have suggested that matrix stiffness can bias NPC differentiation, but studies in 3D materials have conflicting results about the relative importance of stiffness. To determine the material properties required to maintain NPC stemness, we designed an elastin-like protein (ELP) hydrogel system in which matrix stiffness and degradability are co-varied independently of hydrogel swelling, microstructure, and nutrient diffusivity. NPCs cultured in gels with low stiffness and high degradability significantly improved their maintenance of stemness compared to cells in gels with high stiffness and low degradability. Quantifiable indicators of stemness included increased proliferation rate, enhanced metabolic activity, expression of the NPC markers Nestin and Sox2, and efficient differentiation into neurons and astrocytes. To assess if these changes in stemness were due to matrix stiffness, encapsulated NPCs were treated with small molecule inhibitors of cytoskeletal tension and classical mechanotransduction pathways. Interestingly, inhibition of mechanosensing resulted in no changes in NPC behavior relative to uninhibited controls. Furthermore, preventing integrin-mediated matrix engagement by using non-cell-adhesive ELPs also had no effect on NPC phenotype. These results indicate that the observed changes in NPC stemness were independent of matrix stiffness. To investigate the role of NPC-mediated matrix remodeling, antibody arrays and protease inhibitor screens were used to identify a disintegrin and metalloprotease 9 (ADAM9) as the protease primarily responsible for ELP degradation by NPCs. Reducing ADAM9 activity with small molecule inhibitors or with shRNA-mediated knockdown resulted in decreased hydrogel degradation and a significant reduction in NPC stemness. These results suggested that matrix remodeling, and not matrix stiffness, was the required material parameter for maintaining NPC stemness. To test the generalizability of this result, a second engineered biomaterial was designed using poly(ethylene glycol) crosslinked with ADAM9-cleavable peptides to create a family of hydrogels with independent tuning of stiffness and ADAM9-mediated degradability. When ADAM9-degradability was held constant, no differences in Nestin or Sox2 expression were observed regardless of the gel stiffness. In stark contrast, Nestin and Sox2 expression were significantly degreased in gels with 0% compared to 100% ADAM9-cleavable crosslinks, even in the softest gels. Taken together, these results have identified matrix degradation as a previously unknown requirement for NPC stemness maintenance in 3D hydrogels. Furthermore, these data demonstrate that this effect is independent of classical mechanosensing through cytoskeletal contractility, which was previously shown to play a critical role in NPC-matrix interactions in 2D. In the future, biomaterials that exploit ADAM9-mediated remodeling may increase the regenerative potential of NPC therapies.

472880 Formation of Biomimetic Hybrid Materials Using Macromolecular Templates

Abstract:

Biominerals found in living organisms have diverse biological functions and unique properties arising from the particular structure, orientation, and morphology of the constituent minerals. A systematic investigation of the morphological evolution of biominerals grown in organic matrices, and the interaction of such matrices with the inorganic crystals are essential for understanding biomineralization processes and for developing bioinspired materials. We employ here hydrogels as a model organic matrices and a â€˜double-diffusionâ€™ method to mimic the formation of calcium oxalate, the major constituent of kidney stones. Gel-mediated crystallization provides a great way to vary the local supersaturation through controlling the diffusion rate of the reacting ions. We study the nucleation, growth, type of polymorph and orientation of the calcium oxalate crystals as affected by the gel density, reservoir concentrations, molar stock solution ratio, and additives. The obtained crystals are characterized for their morphology and crystal structure using Scanning Electron Microscopy (SEM) and X-ray Diffraction (XRD), respectively. We also investigate the possible occlusion of gelatin in calcium oxalate through Thermogravimetric Analysis (TGA). These findings help our understanding of the mechanism of gelatin incorporation in calcium oxalate crystals, and provide insights into the formation of organic-inorganic composite materials

467348 Probing Phase Transitions in Dynamic Biopolymer Complexation

Abstract:

In nature, biopolymers partition into dynamic compartments to facilitate and regulate their interactions. These dynamic compartments are referred to as membraneless organelles (*e.g.,* nucleolus, P bodies, stress and germ granules) and consist of biopolymer-rich interiors that rapidly assemble and disassemble to form liquid droplets, hydrogels or fibril structures. However, the physical interactions that affect the formation, dissolution, and regulation of these assemblages are poorly understood, yet vital in determining their function in normal and disease states. Therefore, there is a strong need to investigate the underlying mechanisms that drive dynamic biopolymer complexation and develop an understanding of their biochemical function. Interestingly, polyelectrolyte complexes produced *in vitro* using simple homopolymers are strikingly similar to the membraneless organelles found *in vivo*. Polyelectrolyte complexation is an entropically driven process, where electrostatic attraction between oppositely charged polymers results in a release of bound counter-ions and rearrangement of water molecules. Under defined conditions, oppositely charged polyelectrolytes can form complexes consisting of a dense polymer-rich phase (liquid coacervate or solid precipitate) in a polymer-depleted aqueous phase. Whether complexation results in a liquid or solid precipitate depends on the strength of electrostatic interactions, which are mediated by salt concentration, acidity/basicity of the monomers and their distribution along the polymer backbone. In this work, we investigate the forces that govern membraneless organelle formation by engineering model polypeptide analogs and systematically studying their phase transition behavior. From a broad perspective, our work holds the potential to develop a basic understanding of protein/polyelectrolyte complexation, impacting design of novel functional biomimetic materials.

459819 Molecularly Imprinted Polymer-Peptide Hybrid Materials for Engineered Protein Recognition

Abstract:

The recognition of protein biomarkers is of utmost importance in medical applications including imaging, drug delivery, and regenerative medicine. Molecularly Imprinted Polymers (MIPs) are a class of biomimetic materials formed through the polymerization of functional monomers around a molecular template. Following template extraction, nanocavities possessing a complementary geometric and chemical profile to the template remain. Herein, we present the synthesis of MIPs on the surface of nanomaterials for drug delivery applications. We strengthen specific protein-MIP interactions through the synthetic incorporation of engineered oligopeptides, which have been selected through molecular docking simulation for high-affinity protein binding.

Polycaprolactone (PCL) nanoparticles were formed by solvent displacement methods using a custom polymaleic anhydride-alt-1-octadecene-g-polyethylene glycol methacrylate (PMAO-g-PEGMA) surfactant according to the method developed by Culver *et al*[1]. PCL nanoparticles were selected in particular because of their biodegradable nature and ability to encapsulate a hydrophobic therapeutic agent. Imprinted and control non-imprinted polymers (NIPs) were synthesized through the polymerization of anionic, cationic, and hydrophilic functional monomers with 5% crosslinker and PMAO-g-PEGMA functionalized PCL in 0.1x PBS, in the presence (MIPs) or absence (NIPs) of 0.5 mg/mL trypsin, a model high isoelectric point protein template.

Oligopeptides which dock in high affinity with trypsin were selected through a systematic molecular docking study. One thousand random, water-soluble 6-mer peptides were selected using the soly-pep random sequence generator available online from Paris Diderot University[2]. These peptides were docked to an isolated volume of trypsin using GOLD[3], defined as a spherical volume centered at the active site. 28

promising sequences, which docked in high affinity with trypsin, lacked arginine and lysine residues, and contained a cysteine for thiol-coupling were isolated. Oligopeptide selectivity for trypsin was quantified through a systematic surface plasmon resonance (SPR) study. Oligopeptides were conjugated to dextran coated SPR sensors through thiol-coupling in order to mimic their orientation while coupled to MIP networks. The ligand-coated sensor was then sequentially exposed to trypsin, lysozyme, cytochrome c, and hemoglobin at a range of concentrations while transient protein adhesion was monitored.

MIPs in the absence of oligopeptides bound 59±22% more trypsin than corresponding NIPs across a range of solution concentrations up to 0.5 mg/mL. These MIPs also bound 48±17% more lysozyme and 47±22% more cytochrome c, which are high isoelectric point proteins with a smaller hydrodynamic diameter than trypsin. Hemoglobin, which is a low isoelectric point protein with a larger hydrodynamic diameter than trypsin, was excluded significantly from MIPs as compared to NIPs. Oligopeptides, which possessed micromolar affinity for trypsin as validated by SPR, were incorporated at 88% efficiency into MIPs and NIPs through a thiol coupling reaction. These peptide-MIP hybrid materials, which exhibit recognitive properties for trypsin, can serve as a model system for the development of future intelligent theranostic hydrogels.

This work was supported in part by the UT-Portugal Collaborative Research Program (CoLab). JRC is supported by an NSF Graduate Research Fellowship.

[1] H. R. Culver, Stephanie D. Steichen, M. Herrera-Alonso, and Nicholas A. Peppas, "A versatile route to colloidal stability and surface functionalization of hydrophobic nanomaterials," *Under Review.*

[2] S. Teletchea, J. Rey, V. Stresing, S. Hervouet, P. Tuffery, and D. Heymann, "SolyPep: a fast generator of soluble peptides," *In Preparation.*

[3] M. L. Verdonk, J. C. Cole, M. J. Hartshorn, C. W. Murray, and R. D. Taylor, "Improved protein–ligand docking using GOLD," *Proteins Struct. Funct. Bioinforma.*, vol. 52, no. 4, pp. 609–623, 2003.

472658 Novel Tethering Approaches for Supported Biomembranes with Composition and Orientation Control for Biomaterial Ligand Displays and Membrane Protein Assays

Abstract:

Integral membrane proteins are structures present within the lipid membranes of a cell that have a wide variety of functionality, ranging from cleavage to transport to cell-to-cell signaling. The proteins often contain large extramembrane domains leading to difficulties in reconstitution in isolated systems. These proteins require a confluent bilayer, proper protein and substrate orientation, and appropriate chemical environment, along with adequate spacing to relieve outside stresses on the membrane. Despite extensive studies across a variety of integral membrane proteins, a high throughput systematic assay has yet to be developed for such proteins. In these studies, a supported lipid bilayer containing tethering and cushioning moieties is developed to isolate the biologically relevant transmembrane orientations of the membrane bound enzymatic system. This work shows two biomembrane platforms used to build new multifunctional systems, planar biomembranes and also spherical proteolipobeads that facilitate high throughput analysis. We have characterized the structures, phase localization and compositions of enzymes in proteolipobead and planar systems by employing correlative optical and surface microscopy from the micro- to the nanoscale with superresolution microscopy. This work is directed at the at the intramembrane protease Î^{3}-secretase, an enigmatic enzyme with major public health relevance. It is proposed that a platform containing biologically active proteins in appropriate configurations and lipid compositions with proper support structures can be developed for the functional analysis of Î^{3}-secretase and other integral membrane proteins.

467964 Development of Injectable Microgels for Galns Enzyme Replacement Therapy

Abstract:

Introduction: In this study, we developed injectable microgels comprised of multiarm polyethylene glycol acrylate (PEG-Ac) with tunable degradation and controlled release features. We employed the microgels for the encapsulation and delivery of N-acetylgalactosamine 6 sulfate sulfatase (GALNS) enzyme with the broader goal of developing an effective enzyme replacement therapy. Deficiency of GALNS causes accumulation of glycosoaminoglycans and ultimately Morquio A disease. Although effective in alleviating some disease symptoms, current enzyme replacement therapy (ERT) has many limitations, such as a need for recurrent infusions and occurrence of immune responses.[1] Delivery of GALNS encapsulated in PEG microgels could offer an improvement over existing therapies, as it preserves enzyme activity, leads to longer circulation times and attenuates immune responses, while requiring a single injection.

The PEG microgels in this study were prepared by electrospraying by Michael's addition reaction. We were able to obtain microgels of size range 70-300 μm by varying the electrospraying set-up parameters. The microgels could be injected through a needle, while still preserving their integrity upon injection. Further, the degradation of the microgels could be tuned to control release of encapsulated proteins. Release kinetics for several model proteins varying in size and hydrodynamic radii was found to be dependent on the protein size, as well as mesh size and degradation rate of the hydrogel. Furthermore, we observed preservation of GALNS enzyme activity in the upon encapsulation in the microgels. The nanoporous microgel mesh size precluded protein release until sufficient microgel degradation occurred.

Conclusions: The developed drug delivery approach has potential as a modified-ERT system to improve treatment outcomes and quality of life of Morquio A patients.

References: 1. Montaño, A. M.; Tomatsu, S.; Gottesman, G. S.; Smith, M.; Orii, T. International Morquio A

Registry: clinical manifestation and natural course of Morquio A disease. *Journal of inherited metabolic disease* **2007,** 30, (2), 165-74.

471705 Computationally-Driven Design of a Heterobivalent and Gate to Improve Targeting Agent Selectivity

Abstract:

Targeted molecular imaging agents have greatly advanced the sensitivity and specificity of detection methods and increased the ability of clinicians to detect and characterize cancers. However, these imaging agents often target biomarkers that are not exclusively expressed on cancerous cells, resulting in false positives and high signal background. Many cancer types do not have a single biomarker that can consistently distinguish them from the normal tissue. Yet the overexpression of two biomarkers is a more unique phenomenon that could enable selective targeting for a diverse array of patients. The goal of this study is to develop a modular molecular targeting system, labeled a biological AND gate, that can discriminate between cells that have a moderate overexpression of two biomarkers over cells that have moderate or even high expression of only one.

The AND gate molecule is a heterobivalent ligand comprising two binding domains linked via a flexible peptide. The two binding domains target two disease-associated biomarkers with a weak affinity that precludes monovalent binding to singly overexpressing cells while enabling robust bivalent binding to dually overexpressing cells. Optimal AND gate design – including individual affinities of the binding domains, linker length, and linker structure – depends on several properties of the both the tumor and the biomarker, such as average cell surface expression, internalization rate, homogeneity across cell membrane, and potential agonistic/antagonistic effects. To guide quantitative design, a geometric, cell-surface binding model was developed and integrated with a literature biodistribution model (Thurber and Wittrup *J. Theor. Biol.* 2012). The hybrid model indicates that, given a surface expression of 100,000 biomarkers per cell (each), a 100-fold increase in specificity can be achieved over cells expressing 100,000 of only one biomarker with individual ligand affinities of 125-575 nM. Stronger affinities hinder selectivity while weaker affinities preclude even bivalent binding. The optimal flexible linker length, to maximize local ligand concentration after binding of the first domain, is 22 amino acids for a receptor radius of 20 Å. Differential results are observed for flexible and rigid linkers.

These optimal ligand properties were then engineered into an experimental heterobivalent ligand. In our model system, the binding domains are two non-antibody protein scaffolds (10th type III domain of fibronectin (Fn3) domains) that have been previously engineered to bind either epidermal growth factor receptor (EGFR, clone A') or carcinoembryonic antigen (CEA, clone C7) with initial affinities of 0.42 ± 0.06nM and 15 ± 15 nM, respectively. To obtain ligands with affinities within the 125-575 nM range, yeast surface display libraries, created by mutagenesis of each parental gene, were sorted by flow cytometry for decreased binding to the proper target. Mutants from the sorted sub-populations were produced and titrated on cancer cells, which revealed almost no correlation between yeast-displayed affinity to recombinant target and soluble affinity to the natively-expressed target. Nevertheless, several mutants for each target were identified; C7 mutants ranged from 2.5 to 5,600 nM and A' mutants from 15 to 3,000 nM. To determine scaffold modularity, *i.e.*, whether their affinities were affected by the fusion of a second Fn3 domain, the parental ligands A' and C7 were fused on either the N- or C- termini with a non-binding Fn3 clone. The fusion of a non-binding Fn3 to the C-termini (distal to the binding site) decreased the affinity of A' 50-fold to 21 ± 12 nM and C7 affinity held at 16 ± 10 nM. When the non-binding Fn3 was fused to the N-termini (near the binding site), A' affinity decrease was similar (18 ± 6 nM), while C7 showed a decrease in affinity to 89 ± 97 nM. Two mutants of A' were also tested in the fusion context; A' mutants A'.1 and A'.2, which have affinities of 28 ± 3 nM and 17 ± 4 nM, respectively, as single domains, displayed affinities of 28 ± 12 nM and 125 ± 93 nM, respectively, when fused to the non-binding Fn3 via their C-termini. An additional high-affinity Fn3 binder to EGFR (clone D) demonstrated no effect when fused via its C-terminus (0.4 ± 0.1 nM) compared to the monovalent affinity of 0.25 nM. These variations in fusion modularity demonstrate the importance of quantifying the *in vitro* affinity of fusion components – in the fusion context – and highlight dependence on unknown binding paratope, receptor binding epitope, binding orientation, and structural folding variations. Ongoing efforts include identifying the structural effect of the fusion via circular dichroism and SEC analysis.

For initial AND gate testing, three heterobivalent ligands were synthesized: A'-C7, A'.1-C7, and A'.2-C7, where the N-terminus of C7 was fused to the C-terminus of the three EGFR binding Fn3 via a glycine-rich linker to test their selectivity on native and modified cell lines of varying EGFR and CEA expression. Preliminary data of the A'-C7, A'.1-C7, and A'.2-C7 fusions on A431 cells (<2,000 CEA, 4,000,000 EGFR per cell) show that the fusion of C7 via the C-terminus further decreases the EGFR affinity of the fusion compared to the non-binding Fn3 fusion to 102 ± 31 nM, 443 ± 238 nM, and 1200 ± 1000 nM, respectively. However, titration of A'.1-C7 on lentiviral-transfected HEK-293T cells (250,000 CEA, <2,000 EGFR receptors per cell) results in an affinity of 69 ± 10 nM, showing that the C7 affinity is indistinguishable when fused to either A'.1 or the non-binding Fn3. This data shows that the effect of the fusion can be dependent on the clone you are fusing, which is crucial when accurately measuring the properties of the system, and the importance of cell lines with accurately characterized expression to quantify individual domain affinities within the heterobivalent ligand. Ongoing studies, evaluating binding selectivity and sensitivity on cell lines with varying levels of CEA and EGFR, will be discussed. Additional results to be shared include evaluation of ligands of varying affinities, additional scaffolds, and alternative fusion methods to create a more modular system.

461410 Supramolecular Pegylation of Biopharmaceuticals

Abstract:

Introduction: The practice of medicine has been transformed by the use of protein drugs, but these also introduce an array of attendant complications arising from poor chemical and/or structural stability, thus requiring extensive effort in the development of stabilizing excipients and formulation methods. Direct chemical modification with poly(ethylene glycol) (PEG) has demonstrated improvements in protein stability and circulation half-life, a practice that has been used clinically for a number of protein therapies. Herein, we demonstrate for the first time a platform approach to PEGylation of biopharmaceuticals using only supramolecular interactions by leveraging high-affinity host-guest interactions in order to endow PEG-like functionality on proteins non-covalently.

Materials and Methods: A supramolecular macrocycle with high binding affinity was synthesized and functionalized with PEG chains of controlled molecular weight *via* "click" chemistry. Leveraging host-guest binding of this macrocycle with native residues on proteins, this PEGylated compound was added in formulation with insulin, glucagon, and a therapeutic antibody. Functional performance of all three biopharmaceuticals was assayed both *in vitro* and *in vivo.*

Results and Discussion: The supramolecular PEGylation (SM-PEG) of insulin extends *in vitro* stability and preserves potency when aged under agitating conditions, extending protein life from approximately 14 hours alone to over 100 days when in formulation with SM-PEG. Moreover, in a diabetic mouse model the duration of insulin activity can be tuned by selection of PEG molecular weight. Extending this approach to glucagon and a therapeutic anti-CD20 antibody, the SM-PEG lends significant stability and preservation of activity to these proteins as well. This approach to supramolecular modification of therapeutic proteins provides an excipient-only strategy to modulate protein activity and stability without the need for covalent modification of the protein drug molecule, and has broad possible use as a platform for biopharmaceutical formulation.

Conclusions: In biopharmaceutical practice, assuring stability and modulating pharmacokinetics are two primary areas of interest, and covalent PEGylation has been among the more successful strategies evaluated in this regard. Using high binding affinities afforded by specific supramolecular host-guest interactions, proteins can be modified non-covalently with PEG chains of various molecular weights. This supramolecular PEGylation strategy dramatically extended the stability of three therapeutic proteins, and also prolonged the activity of insulin in a modular fashion dependent on molecular weight. This approach may be broadly applicable as a platform of "smart" excipients for formulation and shelf-life extension of biopharmaceuticals.

462185 *Non-Invasive Screening of Diseases Using an Orally Available Near-Infrared Imaging Agent*

Abstract:

Early detection of diseases is crucial to improving the chances of successful treatment, such as curative surgery of cancer or early intervention with disease-modifying treatments to prevent irreversible damage in rheumatoid arthritis. Current methods of detection involve either minimally invasive blood tests or medical imaging modalities. Blood tests can provide large amounts of molecular information but lack spatial information about the site of the disease, and imaging modalities can provide spatial information but only offer anatomical data. Molecular imaging modalities can bridge the gap between the aforementioned techniques, but the drawbacks include safety concerns with ionizing radiation and the associated high cost.

We aimed at developing a near-infrared orally available imaging agent to detect diseases non-invasively. This technique employs near-infrared fluorescence as a method of detection that makes the process safe (non-ionizing radiation) but limits the detection of diseases within 1 to 2 cm from the surface of the skin. This method of detection is easily portable, cheap and the use of an orally available imaging agent makes it completely non-invasive. Importantly, oral delivery avoids associated risks with parenteral delivery and could improve compliance in a screening setting.

As proof of concept, we chose a single-digit nanomolar integrin binding peptidomimetic as a model system. The small molecule ligand has the highest affinity for Î±$_{v}$Î²$_{3}$ integrin, which is highly expressed on certain forms on cancer and on macrophages present at sites of inflammation such as arthritic joints. The binder was conjugated to several near-infrared dyes in order to make a series of agents with varying physicochemical and optical properties to ascertain what properties are important for orally available imaging agents. *In vitro* and preliminary *in vivo* experiments were conducted to find the agent that had the best mix of oral bioavailability, *in vivo* optical properties and specificity. Based on these results the IRDye 800CW conjugate was used in an orthotopic breast cancer model to show that this

method of detection is effective following oral delivery. We were able to produce target-to-background ratios (TBR) of 4.7 and tumor-to-muscle ratios over 20 using our method of delivery and detection. A mechanistic model based on partial differential equations was also made to correlate the properties of the agents to TBR, in order to highlight the most important properties from a quantitative pharmacokinetic and transport perspective.

467547 Investigation of the Cellular Uptake and Cytotoxicity of Cell-Penetrating Peptides in *Candida* Fungal Pathogens

Abstract:

Cell-penetrating peptides (CPPs) have been widely studied as tools for delivering a variety of molecular cargo into cells, including DNA, RNA, proteins, and nanoparticles. Previous work with CPPs has mainly focused on their uptake in mammalian cells, but CPPs also have potential as drug delivery and research tools in other organisms, including fungi. We have investigated 12 different peptides for their ability to function as CPPs for the important human fungal pathogens *Candida albicans* and *C. glabrata*. The peptides were commercially synthesized with an N-terminal 5-carboxyfluorescein label serving as small-molecule cargo and as a fluorescent probe to report cellular uptake. We have used flow cytometry to screen the CPPs for their potential to efficiently enter the *Candida* cells. Among the 12 CPPs, we have identified several promising CPPs, including pVEC and SynB, that can enter both *C. albicans* and *C. glabrata* cells, and the overall uptake process depends on both the salt concentration in the buffer and on the concentration of the peptides. We have also identified CPPs with some specificity; for example, Hst-5 can only translocate into *C. albicans*, whereas TP-10 has higher uptake in *C. glabrata*. We found that generally the CPPs with higher uptake also have higher toxicity toward both *Candida* species, but SynB showed almost no antimicrobial activity at a concentration where nearly 100% of cells have high levels of peptide uptake. Furthermore, we explored the subcellular localization of CPPs at different concentrations with fluorescent microscopy and flow cytometry. Most of the promising CPPs showed a 3-stage behavior, where CPPs evenly distributed in the cytosol at high concentrations, localized inside vacuoles at moderate concentrations, and gathered around the membranes at lower concentration. Cecropin B, in contrast, shows â€œall-or-nothingâ€ behavior, with the peptide either present in the cytosol or exhibiting no apparent association or uptake. In addition, we have also investigated the uptake at different temperatures to aid understanding of the mechanism of uptake for the peptides. Some peptides (e.g., MPG and $(KFF)_3K$) are significantly affected by temperature, indicating a potential ATP-dependent uptake mechanism, whereas temperature has minimal impact on other peptides (e.g., pVEC and Cecropin B). Based on these results, we are now exploring the structure-function relationship for translocation of CPPs into fungal cells.

470913 Catalase-Laden Microdevices for Cell-Mediated Enzyme Delivery

Abstract:

Catalase-Laden Microdevices for Cell-Mediated Enzyme Delivery

Junfei Xia[1], Zhibin Wang[1], Yuanwei Yan[1], Zhijian Cheng[2], Li Sun[2], Yan Li[1], Yi Ren[2], Jingjiao Guan[1,*]

[1] Department of Chemical and Biomedical Engineering, FAMU-FSU College of Engineering, Florida State University

[2] Department of Biomedical Sciences, College of Medicine, Florida State University

Abstract

Enzymes have been used as drugs to treat a wide variety of human diseases and traumas. However, therapeutic utility of free enzymes is generally impeded by short circulation time, lack of targeting ability, immunogenicity, and inability to cross certain biological barriers. Various strategies have been developed to overcome these limitations. Among them is cell-mediated drug delivery, which is featured by integrating drugs with live cells and taking advantage of the unique capabilities of the cells to achieve controlled drug delivery. We developed simple and inexpensive methods based on soft lithography and layer-by-layer (LbL) assembly for fabricating disk-shaped microparticles termed microdevices and generating the cell-microdevice complexes. [1, 2] We herein applied these methods to the fabrication of enzyme-laden microdevices and corresponding cell-microdevices complexes for the first time. We used catalase as a model enzyme in this study. It is a water soluble protein with an isoelectric point of 5.8. It is thus negatively charged at pH 7. Poly(diallyldimethyl ammonium chloride) (PDAC) was chosen here as a polycation to form a bilayer with catalase. Catalase in the microdevices was catalytically active and active catalase was slowly released from the microdevices. The catalase-laden microdevices were attached to the external surface of live suspension and adhesion cells respectively to form cell-microdevice complexes. This technique is applicable to other therapeutic enzymes and therapeutic cells, and thus promises to find clinical applications for treating various human diseases and traumas.

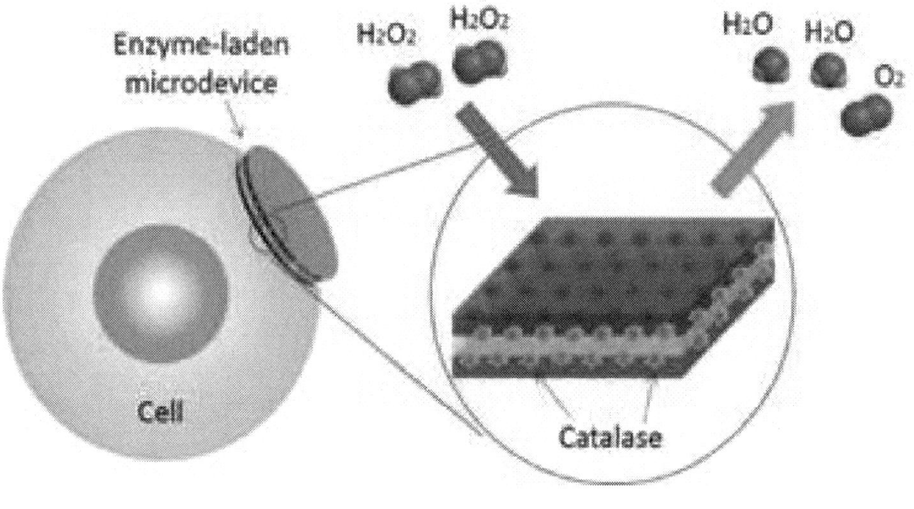

References

1. Junfei Xia, Zhibin Wang, Danting Huang, Yuanwei Yan, Yan Li, Jingjiao Guan*,*Asymmetric Biodegradable Microdevices for Cell-Borne Drug Delivery*, **ACS Applied Materials & Interfaces** (2015) 7: 6293-6299.

2. Zhibin Wang, Junfei Xia, Yuanwei Yan, Ang-Chen Tsai, Yan Li, Teng Ma, Jingjiao Guan*,*Facile functionalization and assembly of live cells with microcontact-printed polymeric biomaterials*, **Acta Biomaterialia** (2015) 11: 80-87.

472660 Ragweed Pollen-Based Oral Vaccine Delivery

Abstract:

Oral vaccine delivery although highly desirable, continues to be a difficult target. This is because the harsh acidic and enzymatic environment together with physical and biological barriers in the gastrointestinal (GI) tract pose unique challenges to deliver vaccines orally. We have developed pollen grains (PGs) as a novel method to deliver antigens across the GI mucosa. In this study ragweed pollens were used to deliver ovalbumin (OVA) orally.

Ragweed pollens were chemically processed to remove native biomolecules and to obtain clean pollen particles. Mice were then immunized with a formulation of OVA and ragweed pollens. Immune responses were characterized by determining the presence of serum anti-OVA IgG, IgG1, IgG2a, IgE, fecal IgA, salivary IgA, and vaginal IgA antibodies using enzyme-linked immunosorbent assay (ELISA). Our study shows that vaccine formulations with ragweed pollens stimulated significantly higher antibodies compared to control mice that received only OVA without ragweed pollens. Using ragweed pollens, high OVA-specific systemic immune response was induced compared to control mice. Furthermore, anti-OVA IgA antibodies were stimulated not only in the intestinal mucosa, but at the distal vaginal mucosal site. Overall, our study demonstrates the potential of ragweed pollen for a simple method of oral vaccination that can induce systemic and mucosal immunity. Further study is required to optimize the antigen dose and frequency of the delivery system.

472952 the Roles of Competing Dissolution, Diffusion and Transient Marangoni Convection Fluxes in Surfactant-Enhanced Spreading of Aerosols for Pulmonary Delivery

Abstract:

We are formulating aerosols for pulmonary delivery with surfactants to overcome aerodynamic limitations when treating obstructive lung diseases such as cystic fibrosis. The added surfactant is designed to establish transient Maragoni flows which can convect drug laterally after aerosol droplets are deposited onto airway surface liquid (ASL) and disease-associated mucus plaques. We have used a combination of aerosol droplet deposition experiments with mock systems (entangled (bio)polymer solution subphases mimicking the ASL, fluorophores mimicking drugs, multiple surfactant species) and numerical transport simulations (via COMSOL) to examine the competing roles of surfactant dissolution, diffusion and transient Marangoni convection fluxes in determining the extent of drug spreading. Large dissolution and diffusion fluxes lead to significant net drug transport, but poor spreading while small dissolution and diffusion fluxes lead to poor drug transport and poor spreading; extensive drug spreading occurs when dissolution and diffusion fluxes are similar in magnitude to the transient Marangoni convection flux. Several combinations of mock subphases, surfactants, and mock drugs will be discussed along with implications for delivery via both liquid and solid aerosols.

461551 Li-Rich Anti-Perovskite Superionic Conductor Films for All-Solid-State Li-Ion Batteries

Abstract:

Development of high-performance inorganic ionic conductors for all-solid-state lithium batteries has been considered as an ultimate solution to address the safety issues for transportation applications. The solid-state superionic conductors also enable the fabrication of thin and flexible batteries, which are urgently needed in burgeoning applications such as wearable electronics and implantable medical devices. However, the ionic conductivities of solid electrolytes are still much lower than those of the liquid electrolytes, thus limiting the applications of lithium batteries in many important areas. Here, we report the deposition of anti-perovskite Li_3OCl solid-state electrolyte films with significantly enhanced superionic conductivity via pulsed laser deposition using a composite target. A room-temperature ionic conductivity of Li3OCl films as high as 2.0 Ã— 10â^'4 S cmâ^'1 is achieved using such a processing scheme, and this value is more than two orders of magnitude higher than that of its bulk counterpart. The superior compatibility between the electrolyte film and metallic Li is revealed by cycling a Li/Li_3OCl/Li symmetric cell. The applicability of Li_3OCl film as an electrolyte for solid-state thin-film Li-ion batteries has been demonstrated for the first time by examining the full Li-ion batteries. Our findings illustrate that the anti-perovskite films hold great promise as alternative solid electrolytes for electric-energy storage devices.

466943 Carbon Foams from Polyhipe/Reduced Graphene Oxide Composites and Their Performance As Electrodes in Supercapacitor Devices

Abstract:

Carbon foams were produced from composites of macroporous poly(divinylbenzene) (poly(DVB) and reduced graphene oxide (rGO). Poly(DVB) was synthesized by polymerizing the continuous phase of water-in-oil high internal phase emulsions (HIPEs) stabilized at the interface by amphiphilic rGO, employed as a particulate emulsifier. The use of rGO as an emulsifier allows for the synthesis of stable macroporous composites of poly(DVB)HIPEs with a percolating network of rGO at the surface of the material. Both permeable and non-permeable carbon foams, or 'carboHIPEs', could be prepared by carbonization of these macroporous composites at 800 °C. The resulting carboHIPEs gave yields as high as 26 wt.% of the original material. Furthermore, carboHIPEs retain the pore structure of their macroporous precursor, while also producing a newly-formed microporous structure, leading to hierarchical porosity and huge increases in surface area upon carbonization. Surface areas of up to 1800 m^2/g and excellent electrical conductivities of up to 270 S/m are achievable, among the highest reported within the fields of both polyHIPEs and carboHIPEs. Using an rGO emulsifier when creating these structures allows for the production of true 'all-carbon' foams upon carbonization. These polyHIPE composites do not require modification, such as additional crosslinking, prior to carbonization, due to the inherently crosslinked structure of poly(DVB). It is demonstrated that the rGO derived carboHIPEs are good candidates as electrodes in supercapacitor applications such as electrical double-layer capacitor (EDLC) devices, where carboHIPEs derived from more conventional silica-stabilized HIPEs perform poorly. The use of a pourable, aqueous emulsion-template enables simple moulding, minimizes waste and avoids the strong acid treatments used to remove many conventional solid-templates. The capability of producing monolithic, porous carbon foams allows for the production of binderless devices, potentially simplifying the production process of supercapacitors. Devices demonstrated maximum specific electrode capacitance to the tune of 26 F g^{-1} at 10 mV s^{-1}, 5.2 Wh kg^{-1} of energy density, 280 W kg^{-1} of power density and coulombic efficiency of up to 99.6 %. In short, the carbonization of rGO-poly(DVB)HIPE composites enables the production of hierarchically porous carboHIPEs, suitable for a wide range of applications as sorbents and electrodes.

473013 Designing Electrolytes for Beyond Li-Ion Batteries Using Coupled High Throughput *Ab Intio* Calculations and MD Simulations

Abstract:

Transformative outcomes for transportation and electrical grid require high performance, low cost energy storage systems beyond Li-ion technology and innovations in electrodes and electrolytes, alike. Multivalent batteries, chemical transformation and redox flow batteries are promising candidates for next generation electricity storage technologies. An automatic High-throughput computational infrastructure has been constructed for the electrolyte genome project supported by the US Joint Center for Energy Storage Research (JCESR) coupled with experimental analysis with the aim to design electrolytes for beyond Li-ion battery technology.[1] A non-aqueous multivalent (e.g., Mg^{2+}, Ca^{2+} and Zn^{2+}) metal cell is one of the potential candidates for a post-lithium-ion battery. The theoretical volumetric capacity of a metal anode coupled with the lack of dendrite formation at a multivalent metal anode provide an attractive opportunity in energy storage.[1] However, development and commercialization of multivalent batteries are fraught with complications. Thus a fundamental understanding of molecular level properties of these electrolytes is required to improve the electrochemical stability and the charge transfer properties. In this work, we present a multi-scale modelling approach for multivalent salts in various solvents. We uncover a novel effect between concentration dependent ion pair formation and anion stability at reducing potential, e.g., at the metal anode.[2] We find that both Mg and Zn electrolytes are highly prone to ion pair formation, even at modest concentrations, for a wide range of solvents with different dielectric constants, which have implications for dynamics as well as charge transfer.[3] Specifically, we observe that, at Mg metal potentials, the ion pair undergoes partial reduction at the Mg cation center ($Mg^{2+} \rightarrow Mg^{+}$), which competes with the charge transfer mechanism and can activate the anion to render it susceptible to decomposition. It was observed that Zn electrolytes are also prone to similar $TFSI^-$ anion decomposition but with much higher

transition state barrier, which results in better reversible Zn deposition on Zn metal anode in non-aqueous Zn electrolytes as compared to Mg electrolytes. Another category of battery that has gained much attention recently includes chemical transformation systems such as Li-S batteries, which stores energy through chemical bonds instead of intercalation. Here, we study the effect of salt anion and the solvent on the solvation structure and dynamics of Li-polysulfide in the solution. We observe that counter anion (such as TFSI, TfO, FSI & TDI) and solvent interaction strength with Li^+ is critical in controlling polysulfide solubility.[4] Also, high mobility of counter anions can cause faster capacity loss due to enhanced polysulfide solubility and SEI layer formation. The third energy storage concept is of redox flow batteries, which have shown outstanding promise for grid-scale energy storage with improved grid stability. Here we study ionic liquid tethered ferrocene catholyte used as redox center utilizing the Fc/Fc^+ reaction for improving the performance of non-aqueous redox flow battery materials. We observed that solubility of Fc/Fc^+ redox center can be increased up to eight-fold in a mixture of carbonate electrolytes through the modification Fc with a polar tetraalkylammonium function group containing $TFSI^-$anion. It was also observed that at solubility limit, the precipitation of solute is initiated through agglomeration of contact-ion pairs due to overlapping solvation shells. This works shows that the combination of multi-scale modeling with experimental techniques provides unprecedented insight into the origin of the electrochemical, structural, and transport properties of electrolytes, which is crucial in designing electrolytes for beyond Li-ion batteries.

References

[1] X. Qu, A. Jain, N. N. Rajput,Y.Zhang, S.P.Png, M. Brafman, L.Cheng, E. Maginn, L.A. Curtiss, K.A. Persson, "Electrolyte Genome Project: A Big Data Approach in Battery Materials Discovery", Computational Materials Science, 2015.

[2] N. N. Rajput, X. Qu, N. Sa, A. K. Burrell, K.A. Persson, "The Coupling between Stability and Ion Pair Formation in Magnesium Electrolytes from First-Principles Quantum Mechanics and Classical Molecular Dynamics", Journal of the American Chemical Society, 2015.

[3] S. Han, N.N. Rajput, X. Qu, B. Pan, M. He, M. Ferrandon, C.Liao, K.Persson, A. Burrell, ACS Applied Materials & Interfaces, 2016, 8(5), pp 3021-3031

[4] M. Vijaykumar, N. Govind, E. Walter, S. D. Burton, A. Shukla, A. Devraj, J. Xiao, J. Liu, C. Wang, A. Karim, S. Thevuthasan,"Molecular structure and stability of dissolved lithium polysulfide species" Physical Chemistry Chemical Physics, 2014

[5] K. S. Han, N. N. Rajput, X. Wei, W. Wang, J. Z. Hu, K. A. Persson, K. T. Mueller, "Diffusional motion of redox centers in carbonate electrolytes", The Journal of Chemical Physics, 2014

464841 The Influence of Tetrabutylammonium (TBA) Salts on the Discharge and Charge Behavior of Li-O_2 Batteries

Abstract:

The rise of carbon dioxide (CO_2) emissions from fossil fuel use due to rapid industrial growth has exacerbated changes in the climate. Alternative energy sources such as solar and wind have been developed, but extensive commercial utilization has been hindered by their intermittency. To complement these energy conversion methods, energy storage media such as batteries and fuel cells are being developed. Of the battery chemistries, Li-air (O_2) appears promising because it can allow for an order of magnitude greater gravimetric energy density than current commercial Li-ion batteries. However, numerous challenges face Li–O_2 batteries that stem from electrolyte, electrode, and salt degradation, and the limited oxygen reduction and evolution kinetics. These challenges lead to limited current rates, poor capacity retention, and lower cycle life.

Additives such as water, fluorinated solvents, and salts like $LiNO_3$ have been explored to modify the intermediate (superoxide) and O_2 solubility to increase discharge capacities. During charging, catalysts or redox mediators have also been incorporated to aid in the decomposition of the discharge product – Li_2O_2; a bulk insulator that requires charging potentials above 4 V_{Li}.

Ammonium-based salts such as tetrabutylammonium (TBA) have been shown to support oxygen reduction and evolution reactions using cyclic voltammetry, and TBA-superoxide complexes are known to form during reduction and the complexes are easily removed during oxidation. However, these salts have not been studied in actual Li–O_2 batteries.

In this work, we studied the influence of TBA on the discharge and charge behavior of Li–O_2 batteries. We fabricate Li–O_2 cells using TBA as the only salt present in the electrolyte. Despite the lack of lithium salt, a typical discharge plateau (~2.6 V_{Li}) is observed – similar to the voltage obtained with lithium salt – and lithium peroxide (Li_2O_2) is formed as the primary discharge product. The mechanism for discharge was studied, and the influence of solvent on the discharge process was also explored. The presence of TBA also affects the charging behavior, leading to lower charging overpotentials (compared to lithium salts) and Li_2O_2 oxidation. We investigated the charging mechanism and show that oxidation of TBA during charge is responsible for the subsequent Li_2O_2 oxidation. Knowledge gained from this work can spur greater understanding of the discharge and charge mechanisms of Li–O_2 batteries.

467844 Freestanding V_2O_5-PEDOT Thin Film Electrode for Rechargeable Aqueous K-Ion Energy Storage

Abstract:

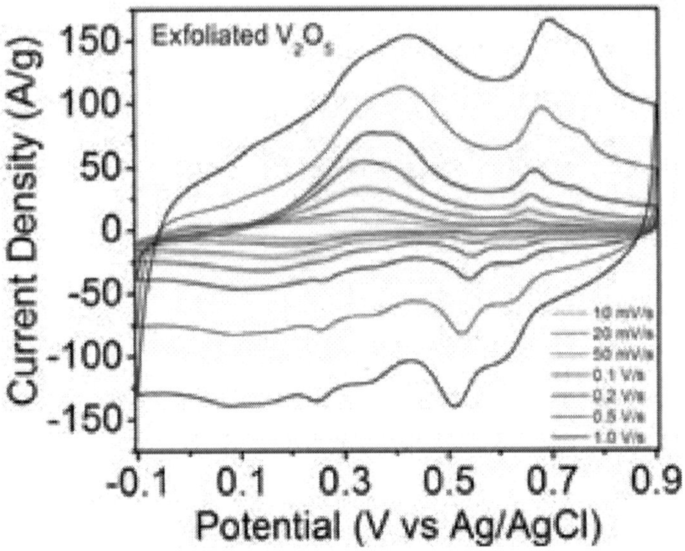

Rechargeable aqueous electrochemical energy storage (EES) devices are a promising technology for a wide range of energy storage applications due to their low cost and superior safety compared to Li-ion batteries. However, their low energy density has inhibited implementation. Materials with layered structures possess favorable characteristics for improving the energy density of aqueous EES devices. In particular, for aqueous EES using cations with larger ionic radii (e.g. Na-ions and K-ions), which are more abundant and therefore, more sustainable as well as less expensive than Li-ions. Herein we report the exfoliation of bulk vanadium pentoxide using the conductive polymer poly(3,4-ethylenedioxythiophene)-poly(styrenesulfonate) (PEDOT:PSS) via a wet chemical method. The resulting exfoliated V_2O_5-PEDOT:PSS metal oxide-polymer composite can be fabricated into freestanding thin film electrode with simple processing techniques. Structural characterization of the exfoliated V_2O_5-PEDOT:PSS composite was investigated with various techniques including synchrotron X-ray diffraction and neutron total scattering analyzed in real space with the Atomic Pair Distribution Function. A structural rearrangement from bulk vanadium pentoxide to a structure with large interlayer spacing and a coherence length on the nanoscale is observed upon exfoliation. Cyclic voltammetry measurements conducted in a 1M KCl electrolyte and 1V potential window show multiple redox features, which remain visible at all scan rates tested between 10 and 1000 mV/s. The exfoliated V_2O_5-PEDOT:PSS nanocomposite not only has the ability for high capacity K-ion storage (605 F/g at 10 mV/s, or 168 mAh/g 100s discharge time) but also the ability to operate at fast charge-discharge rates (203 F/g at 1 V/s, or 56 mAh/g at a 1s discharge time). This work demonstrates a scalable top down approach for the development of a freestanding, high capacity and high power nanocomposite electrode for aqueous K-ion energy storage.

458651 Theoretical Consideration of Nanostructured Magnetite As an Electrode Material for Li-Ion Energy Storage

Abstract:

Theoretical Consideration of Nanostructured Magnetite as an Electrode Material for Li-ion Energy Storage

C. N. Lininger[a], M. S. Hybertsen[b], A. C. West[a] [a]Department of Chemical Engineering, Columbia University, New York, NY 10027, USA

[b]Center for Functional Nanomaterials, Brookhaven National Laboratory, Upton, NY 11973, USA

The electrode material in a Li-ion battery must allow for reversible intercalation of lithium, and there are a range of materials capable of this behavior. Magnetite, Fe_3O_4, is a low cost, environmentally inert, iron oxide that allows for intercalation into its crystal lattice. Upon full reduction, magnetite has a high theoretical capacity against Li/Li^+ at 926 mAh/g as opposed to the current commercial cathode materials operating in the 100-300 mAh/g range.[1–3] Due to its low cost, low environmental impact, high theoretical capacity, and well defined nanoscale synthesis method, magnetite is a model material to push the state of the art in lithium-ion batteries.[5]

Magnetite is an inverse spinel transition metal oxide. The iron in magnetite takes on two valence states, Fe^{2+} and Fe^{3+}, spread over two coordination environments, tetrahedral 8a and octahedral 16d. The tetrahedral iron is all Fe^{3+}, while the octahedral iron centers are comprised of half Fe^{2+} and half Fe^{3+}.[7–9] The oxygen in magnetite forms a robust close packed cubic array at 32e sites, which was recently found to act as an invariant scaffold around which structural rearrangements occur during lithiation.[10] Due to the multivalent irons in magnetite, it is capable of accepting multiple electrons per iron during discharge. Multiple-electron transfer (MET) materials are of great interest to the battery community as the ability of a single cation center to accept multiple electrons greatly increases energy density.[10] However, as a result of facilitating multiple electrons, the metal center goes through oxidation state changes and structural rearrangements occur within the material. Phase transformations can be irreversible and are known to affect cyclability and limit the current rates that conversion-materials can be operated under.[10]Due to the complexity of MET materials, there exists a need for fundamental research into the electronic and ionic transport processes that are occurring in these MET materials during discharge.

Magnetite can be synthesized as a nanomaterial, with spherical nanoparticles ranging from 6-12 nm in a low-temperature synthetic method that does not involve extraneous synthetic processes.[4,5] Upon manufacture into an electrode, the magnetite nanoparticles are observed to agglomerate into micron-sized agglomerates.[6] We have developed continuum-level models for the electrochemical performance of this material including both nanoscale and agglomerate-scale physics. Neither exclusively nanoscale or agglomerate-scale simulations were able to accurately capture the experimentally observed electrochemical performance of magnetite.[6,11] In the agglomerate, ionic transport occurs through nanoporous pathways in-between the aggregated nanoparticles. On this length scale, the size of the diffusing species and the pores through which the diffusion occurs are both in the nano regime. On the nanoscale, solid-state diffusion occurs as lithium inserts into the crystalline magnetite. Depending on concentration, this causes structural rearrangements. The crystalline diffusivity can vary by orders of magnitude as a function of the state of discharge, due to these structural rearrangements that occur during lithiation.

We have used density functional theory to explore structural rearrangements that occur in magnetite during lithiation between zero and one equivalents. We have established that structural rearrangements include the movement of tetrahedral iron to an octahedral coordination environment at concentrations as low as half an equivalent of lithium. We have also established that at this concentration, the lithium has lower energy in octahedral coordinated sites as compared to tetrahedral coordination. While evidence pointed to such changes in coordination, in all previous literature, the lithium concentration at which it occurred had not been established.

Additionally, we are using DFT to explore i) the diffusivity of lithium in crystalline solid magnetite and ii) the lithium open circuit potential in the dilute concentration limit. The use of DFT to understand the kinetic resistances to the structural rearrangements that we have shown occur at low levels of lithiation will give insights into relaxation times observed for this material. Both the DFT determined solid-state diffusivity and the theoretically predicted open-circuit potential will aid in informing the limiting ionic and electronic processes occurring in magnetite for its uses as an electrode material in lithium-ion batteries.

This work was supported as part of the Center for Mesoscale Transport Properties, an Energy Frontier Research Center supported by the U. S. Department of Energy, Office of Science, Basic Energy Sciences, under award #DE-SC0012673. C. N. L. acknowledges the support of the National Science Foundation Graduate Research Fellowship under Grant No. DGE-11-44155. Any opinions, findings, and conclusions or recommendations expressed in this material are those of the authors and do not necessarily reflect the views of the National Science Foundation. This research used resources of the Center for Functional Nanomaterials, which is a U.S. DOE Office of Science Facility, at Brookhaven National Laboratory under Contract No. DE-SC0012704. The Yeti Shared HPC Cluster at Columbia University was also used, which includes support from Empire State Development's Division of Science, Technology, and Innovation under contract number C090171.

(1) Doeff, M. M. In *Encyclopedia of Sustainability Science and Technology*; Meyers, R. A., Ed.; 2013; pp. 5–49.

(2) Menard, M. C.; Takeuchi, K. J.; Marschilok, A. C.; Takeuchi, E. S. *Phys. Chem. Chem. Phys.* **2013**, *15*, 18539–18548.

(3) Ito, S.; Nakaoka, K.; Kawamura, M.; Ui, K.; Fujimoto, K.; Koura, N. *J. Power Sources* **2005**, *146*, 319–322.

(4) Zhu, S.; Marschilok, A. C.; Takeuchi, E. S.; Yee, G. T.; Wang, G.; Takeuchi, K. J. *J. Electrochem. Soc.* **2010**, *157*, 1158–1163.

(5) Zhu, S.; Marschilok, A. C.; Takeuchi, E. S.; Takeuchi, K. J. *Electrochem. Solid-State Lett.* **2009**, *12*, 91–94.

(6) Knehr, K. W.; Brady, N. W.; Cama, C. A.; Bock, D. C.; Lin, Z.; Lininger, C. N.; Marschilok, A. C.; Takeuchi, K. J.; Takeuchi, E. S.; West, A. C. *J. Electrochem. Soc.* **2015**.

(7) Thackeray, M. M.; David, W. I. F.; Goodenough, J. B. *Mater. Res. Bull.* **1982**, *17*, 785–793.

(8) Sickafus, K. E.; Wills, J. M. *J. Am. Ceram. Soc.* **1999**, *82*, 3279–3292.

(9) Menard, M. C.; Takeuchi, K. J.; Marschilok, C.; Takeuchi, E. S. **2013**, 18539–18548.

(10) Zhang, W.; Bock, D. C.; Pelliccione, C. J.; Li, Y.; Wu, L.; Zhu, Y.; Marschilok, A. C.; Takeuchi, E. S.; Takeuchi, K. J.; Wang, F. *Adv. Energy Mater.* **2016**.

(11) Knehr, K. W.; Brady, N. W.; Lininger, C. N.; Cama, C. A.; Bock, D. C.; Lin, Z.; Marschilok, A. C.; Takeuchi, K. J.; Takeuchi, E. S.; West, A. C. *ECS Trans.* **2015**, *69*, 7–19.

465161 Highly Flexible Self-Assembled V_2O_5 Cathodes Enabled By Conducting Diblock Copolymers

Abstract:

Structural energy storage materials combining load-bearing mechanical properties and high energy storage performance are desired for applications in wearable devices or flexible displays. Vanadium pentoxide (V_2O_5) is a promising cathode material for possible use in flexible battery electrodes, but it remains limited by low Li^+ diffusion coefficient and electronic conductivity, severe volumetric changes upon cycling, and limited mechanical flexibility. Here, we demonstrate a route to address these challenges by blending a diblock copolymer bearing electron- and ion-conducting blocks, poly(3-hexylthiophene)-*block*-poly(ethyleneoxide) (P3HT-*b*-PEO), with V_2O_5 to form a mechanically flexible, electro-mechanically stable hybrid electrode. V_2O_5 layers were arranged parallel in brick-and-mortar-like fashion held together by the P3HT-*b*-PEO binder. . This unique structure significantly enhances mechanical flexibility, toughness and cyclability without sacrificing capacity. Properties are probed as a function of composition using dynamic mechanical analysis, impedance spectroscopy, cyclic voltammetry, and galvanostatic cycling. It is found that optimum composition exists, where too little polymer results in brittle inflexible electrodes and too much polymer reduces the electrochemical activity. Only 5 wt % polymer is required to triple the flexibility (normal strain) of V_2O_5, and electrodes comprised of 10 wt % polymer have unusually high toughness (293 kJ/m^3) and specific energy (530 Wh/kg), both higher than reduced graphene oxide paper electrodes. Electrodes with commercial binder poly(vinylidene difluoride) shows no remarkable improvement. These results demonstrate a new approach for the development of structural energy storage materials.

472403 Combined Experimental and Theoretical Study of Oxygen Mobility in Hydrated Alpha-MnO2

Abstract:

Alpha-MnO_2 is of interest as a cathode material for conventional lithium cells and as an electrode/electrocatalyst for hybrid Li-ion/Li-O_2 cells. It has a framework structure with large 2x2 tunnels that can contain water; the H_2O component can be ion-exchanged with lithium to create a Li_2O-stablized electrode. When used in Li-O_2 cells, it is possible that the electrocatalytic process is associated with the removal from and reintroduction of lithium and oxygen into the host structure during charge and discharge, respectively. In this contribution, we present a study of the oxygen mobility via water intercalation in alpha-MnO_2. We present evidence of the dehydration of alpha-MnO_2 from temperature-resolved in-situ synchrotron powder x-ray diffraction, thermogravimetric analysis, and in-situ UV-resonance Raman spectroscopy; and structural, thermodynamic, and kinetic information from density functional theory calculations. We also present our analysis of the charge state of the hydration species from density functional theory, redox titration, and x-ray absorption near-edge spectroscopy. Finally, we discuss implications for improved sustainability in the practical usage of alpha-MnO_2 for reversible Li or Li-O_2 storage.

466876 Quantifying Gel Network Properties Via Poroelasticity (Invited Talk)

Abstract:

Due to their unique chemical and structural properties, polymer gels are employed in various membrane-based applications where selective transport is critical to device performance. Characterizing the chemical and structural properties of a gel is critical to understanding its transport behavior. However, quantifying these properties is nontrivial as it often requires multiple measurement techniques. In this talk, we demonstrate poroelastic relaxation indentation (PRI) as a single measurement tool to characterize the swelling, mechanical and transport properties of model poly(ethylene glycol)-based hydrogel systems. By applying the appropriate thermodynamic polymer network models and the linear theory of poroelasticity, we are able to use the results from PRI to extract the thermodynamic parameters, elastic modulus, water permeability and mesh size of these gels. We validate these results with small angle neutron scattering to illustrate the simplicity of the PRI measurement technique.

473397 Elastobuoyant Heavy Spheres in a Soft Gel : Understanding Large Deformations in Soft Elastic Polymeric Gels

Abstract:

Elastobuoyant Heavy Spheres in a Soft Gel : Understanding large deformations in soft elastic polymeric gels

Aditi Chakrabarti and Manoj Chaudhury*

Department of Chemical and Biomolecular Engineering, Lehigh University, Bethlehem, PA 18015, USA

Serge Mora

Laboratoire de Mécanique et de Génie Civil. UMR 5508,

Universit´e Montpellier and CNRS. 163 Rue Auguste Broussonnet. F-34090 Montpellier, France.

Yves Pomeau

University of Arizona, Department of Mathematics, Tucson, USA.

Soft solids undergoing large deformations exhibit various fascinating and puzzling mechanical behaviors, yet systematic studies and methods to understand the mechanics of such huge strains are lacking. Here we investigate this complex problem systematically with a simple experiment: by introducing a heavy bead of radius 'a' in an incompressible ultra-soft elastic medium, by varying the sphere radius and medium elasticity. For the softest gels, the sphere becomes elastobuoyant at a particular depth (δ) inside the gel, where its weight balances the medium's elastic strain energy. The scaling law that we find for the penetration depth of the bead at this regime is very different from what one would expect in the small deformation limit. We develop an original asymptotic analytic model to explain such large deformation behavior that is consistent with our experimental observations. Ultimately, we show that such huge deformations in ultra-soft materials can be utilized in many different ways, for example, driving three-dimensional self-assembly of particles inside such a substrate!

454327 Quantifying the Effects of Aging on the Viscoelastic Behavior of Flexible Polyurethane Foams

Abstract:

The thermal-mechanical properties of flexible polyurethane foams have been shown to evolve with age creating questions as to the appropriateness of reusing these materials in critical applications. For example, the room temperature shear modulus of a polyurethane foam sample approximately 30 years old was recently shown to have increased approximately 60% from its newly manufactured state. The increase in stiffness was accompanied by a broadening of the glass transition region which increases the viscoelasticity expected under normal service environments and temperatures. In many applications this change in properties could result transportation or other environments that exceed those originally specified.

Here we present thermal-mechanical characterizations of several densities of flexible polyurethane foam as well as a comparison of new and aged foam properties, where available. We also present efforts to quantify the effects of age (via chemical oxidation) on the thermal-mechanical behavior of these foams in an effort to link the degree of chemical oxidation to changes in mechanical properties. Linking chemical changes to mechanical property changes will allow predictions of acceptable service life in oxidative environments. Further, a discussion on current modeling efforts aimed at capturing flexible foam viscoelastic behavior using the simplified potential energy clock (SPEC) model is presented.

Sandia National Laboratories is a multi-program laboratory managed and operated by Sandia Corporation, a wholly owned subsidiary of Lockheed Martin Corporation, for the U.S. Department of Energy's National Nuclear Security Administration under contract DE-AC04-94AL85000.

469169 Effect of Layer Density on Irreversible Thermal Expansion and Interlayer Strength in Additively ManufacturedÂ Acrylonitrile Butadiene Styrene

Abstract:

Additive manufacturing (AM) has drawn interest from fields ranging from aerospace to tissue engineering and regenerative medicine (TERM) to metamaterials. With AM, specimens with complex internal geometries and tailorable structures can be manufactured. Despite the advantages and interest, broader use of AM is limited by poor mechanical properties, lack of reliable part printing, and lack of expertise in AM. In many cases, AM parts have inferior properties to identical parts made with traditional manufacturing methods and exhibit poor interlayer bonding and residual stress. Additionally, understanding the processing-structure-properties relationship is critically important to ensure reproducible, reliable parts. More fully understanding these relationships may lead to parts with comparable, and ultimately superior, properties as well as additional functionalities with less manufacturing waste. In this work, the impact of layer density on physical and mechanical properties of additively manufactured parts was explored. Specifically, acrylonitrile butadiene styrene (ABS) samples were printed using fused deposition modeling (FDM) at several layer heights and their irreversible thermal expansion, tensile strength, and flexural strength were measured. Irreversible thermal strain increases with increasing layer density, up to 21% strain. Tensile and flexural strength exhibited the same behavior, with maxima at a layer density of 5 mm^{-1}. While residual thermal stress increases with increasing layer density, macroscale print quality increases with increasing layer density. These competing phenomena resulted in the observed maxima in mechanical properties related to interlayer bonding. The results of this work have significant implications for the application of additively manufactured polymers, both in terms of quality control and for tailorable mulitfunctionality.

455899 Polyolefin Toughened Polypropylene: Phase Behavior and Mechanical Properties

Abstract:

A series of poly(cyclohexylethylene-*ran*-ethylene) (CE) copolymers were synthesized with statistical segment lengths designed to match that of isotactic poly(propylene) (iPP). Melt blending the CE compounds containing 50% to 70% by weight cyclohexylethylene repeat units with *i*PP at loadings between 5 to 20 wt.% resulted in a dispersion of CE droplets with number average diameters between 150 and 400 nanometers as determined by scanning electron microscopy. Linear dynamic mechanical elastic (G¢) and loss (G¢¢) moduli data were fit to the Palierne model providing estimates for the interfacial tension between *i*PP and CE, which was interpreted based on enthalpic (χ_H) and entropic (χ_S) contributions to the overall segment-segment interaction parameter, $\chi = \chi_H + \chi_S$, attributed to density and statistical segment length differences, respectively. The small particle sizes are explained by a reduced interfacial tension, driven by minimization of χ_S (» 0), counterbalanced by $\chi_H > 0$, establishing an optimal balance between phase separation and dispersion during blending at practical molecular weights. The optically clear blends exhibit superior tensile mechanical properties, with the strain at break increasing from 20% for *i*PP to 400% at a loading of 5 wt%, attributable to shear yielding of the matrix triggered by cavitation of the uniformly dispersed CE copolymer particles.

467948 Micromolding of UV Curable Coatings

Abstract:

In micromolding, a microstructured coating is created by pressing a stamp into a coating, UV curing with the stamp in place, and then removing the stamp. The process can be implemented in roll-to-roll systems and nanometer to micrometer features can be created; however, one limitation of UV-micromolding is low throughput mainly due to slow curing speed. Additionally, defects may be introduced into the microstructured coating because of strong adhesion between the mold and the coating.

In this research, we explored new formulation approaches based on thiol-ene-acrylate ternary systems, which are insensitive to oxygen during the reaction, cure fast and form highly homogeneous polymer networks. The curing rates and the limiting extents of cure of various thiol-ene-acrylate coating formulations, characterized by real-time Fourier transform infrared spectroscopy (FTIR), increase with the amount of acrylate oligomers in coating formulations. By changing the acrylate concentration in the coating formulation, cured resins with a wide range of mechanical properties have been achieved, from stiff highly cross-linked polymers to soft elastomers. At the same time, the surface energy of a cured coating is independent of the acrylate concentration.

Microchannel arrays composed of 50 μm wide and 25 μm deep channels were successfully replicated with our thiol-ene-acrylate coating systems using polydimethylsiloxane stamps fabricated from SU-8 photoresist masters. The pattern quality strongly depends on the duration of UV exposure and the mechanical properties of cured coatings. For a specific coating formulation, there exists a lower limit of the curing dosage for successful pattern replication. Well-defined surface microstructures are completely transferred to coatings with various elastic moduli. However, the mechanical properties of coatings still need to be optimized to avoid damage of surface features during the demolding step.

To understand the mechanism of debonding between the stamp and the structured coating surface, the effects of the demolding direction and the coating modulus on the demolding force were investigated by T-peel testing. Results reveal different demolding characteristics between the coatings demolded parallel and perpendicular to surface micro-channels. Furthermore, a smaller demolding force is observed during the demolding of low-modulus coatings.

468969 CO_2 Induced Plasticization in Glassy Polymeric Membranes for Gas Separation

Abstract:

The plasticization phenomena of glassy polymers induced by gases or vapors in glassy polymeric materials are of crucial importance in the field of gas separation membranes, and are typically recognized as an increase in gas permeability as the upstream pressure increases. The mechanism is often associated to the sorption of large amounts of CO_2, or other low molecular weight species (present e.g. as impurities in the feed streams), having the ability to produce significant swelling of the polymer matrix, and to reduce its mechanical rigidity, with a consequent depression of the glass transition. The transport properties of the penetrating gases can be thus altered, eventually leading to a loss of the separation performances of the membrane material.

This work aims to address the CO_2 induced plasticization effect on glassy polymers from different points of view, looking for a clear understand of this phenomenon and of the mechanism related. A commercial polyimide (Matrimid) has been thus selected as an interesting test case material, suitable for CO_2/CH_4 separation, which shows an apparent non-monotonous behavior in the CO_2 permeability going through a minimum point (the so-called plasticization pressure).

The influence of CO_2 sorption on polymer material properties has been investigated by direct dynamic mechanical analysis (Rheometrics DMTA 3E) in isothermal frequency sweep operation at 35°C on the CO_2-containing polymer sample at gas pressures in the range 0-30 atm. The permanent effects produced by the CO_2 at high pressure have been also analyzed by temperature sweep tests on Matrimid samples before and after gas conditioning. Sorption tests, carried out in a pressure decay apparatus, provided the required penetrant loading in the polymer sample at the various pressures, as well as the evaluation of its diffusion coefficient. Finally, the CO_2 transport properties of the membrane have been characterized by permeation experiments in the same pressure range.

The DMA analysis showed that CO_2 is able to lower appreciably but not dramatically the elastic modulus (approximately 10% decrease), while the relaxation dynamics revealed a significant tan d increase, which almost doubles its value. However, the very high glassy transition point of such polyimide (T_g = 313 °C) prevented any second order transition even at the maximum pressure investigated. These results are then compared with the previous data obtained on conventional polymer glasses such as polycarbonate (PC) and polymethylmethacrylate (PMMA), which presented opposite CO_2 permeation behavior, as they showed decreasing and increasing trend, respectively.

Nonetheless, the minimum in CO_2 permeability is clearly identified at 15 bar, and further increasing the upstream pressure significantly enhanced the gas transport, while the gas diffusion coefficient, calculated both from transient sorption and from permeation tests, increased monotonously.

Therefore, the obtained results suggested that increasing permeability behaviors, as well as non-monotonuos trends, are not directly related to softening effect produced by the large CO_2 uptake on the polymeric matrix, while are most likely correlated with the behaviors of solubility and diffusivity coefficient of the penetrant itself.

474022 Predicting Mechanical Properties of Organic Semiconductors from Molecular Dynamics Simulations

Abstract:

The mechanical properties of organic semiconductors are of critical importance for roll-to-roll production and thermomechanical reliability of organic electronic devices. Here, we describe the use of coarse-grained molecular dynamics simulations to predict the density, tensile modulus, Poisson ratio, and glass transition temperature for poly(3-hexylthiophene) (P3HT) and its blend with C_{60}. We show that the resolution of the coarse-grained model has a strong effect on the predicted properties. In particular, we find that a three-site model treating each 3-hexylthiophene unit by three coarse-grained beads predicts values of density and tensile modulus that are much closer to experimental measurements than a one-site model. The three-site model also correctly predicts the strain-induced alignment of chains as well as the vitrification of P3HT by C_{60} and the corresponding increase in the tensile modulus. We also observe a decrease in the radius of gyration and the density of entanglements of the P3HT chains with the addition C_{60} which may contribute to the experimentally noted brittleness of the composite material. Although extension of the model to poly(3-alkylthiophenes) (P3ATs) containing side chains longer than hexyl groups correctly predicts the experimental trend of decreasing modulus with increasing length of the side chain, obtaining absolute agreement could not be accomplished by a straightforward extension of the three-site model, indicating limited transferability of such models. Nevertheless, the accurate values obtained for P3HT and P3HT:C60 blends suggest that coarse graining is a valuable approach for predicting the thermomechanical properties of organic semiconductors of similar or more complex architectures.

461730 Adaptive, Point-to-Point Assembly of DNA Nanotubes Between Molecular Landmarks

Abstract:

In vivo, self-assembly processes often adapt what is assembled to the shape and arrangement of existing cellular nanostructures. This capacity for adaptation makes it possible to build integrated nanosystems consisting of multiple nanostructures elements that work together to perform complex functions. We demonstrate a synthetic self-assembly process in which a DNA nanotube self-assembles between molecular landmarks whose separation distance and orientation are unknown. Connections between DAE-E nanotubes occur when nanotubes grow from molecules at the start and destination locations that act as nanotube nuclei for growth from the two opposing nanotube ends. When the nanotubes grow long enough that their ends can contact one another, these ends can join to produce a stable connection. Time-lapse multicolor microscopy confirms that nanotubes connections generally form by the proposed process and follow the predictions of simple physical models. This process can occur under a variety of physical conditions and may make it possible to build wires between dynamic components through metalization of the nanotubes, or to build complex connected DNA architectures at the micron scale with novel mechanical properties.

472433 Photo-Triggered Self-Assembly and Actuation of DNA Nanostructures and Machines Using Photocaged Nucleotides

Abstract:

DNA nanotechnology has emerged as one of the most powerful methods for constructing complex nanoscale devices with precise addressability. The ability to actuate these devices on-demand would provide a powerful method for creating dynamic nanomachines that can reconfigure nanostructures, apply precisely defined forces, or spatiotemporally control self-assembly. Typically, DNA nanostructures have been dynamically modulated by either incorporating photoswitchable azobenzene nucleotides that, while effective, often suffer from slow kinetics and undesired reversibility. Another approach involves adding displacement strands that can bind to toehold regions and out-compete other interactions to change the state of a nanostructure. However, this approach is limited by the need to add an exogenous strand, which is not feasible for many applications, especially in biology. To circumvent these limitations, we have developed a series of photo-cleavable oligonucleotides that block hybridization until illuminated with UV light. By incorporating caged displacement strands within an existing nanostructure, we can prevent their action until illuminated. These strands serve as effective "internal toeholds" that can be activated on-demand. We have constructed a number of one-, two-, and three-dimensional assemblies that can form only upon irradiation, as well as a nanomechanical tweezer that switches between the closed and open state with UV light. In effect, this device can apply a nanomechanical force within a few seconds of illumination, paving the way for dynamic nanomachines that can exert controlled motion. We will also describe the design principles for more complex nanostructures that can find applications in targeted drug delivery or reconfiguration inside cells to influence biological processes.

461942 Biological Self-Recognition and Self-Assembly for the Next Generation of Hybrid Wires

Abstract:

Biological self-recognition and self-assembly for the next generation of hybrid wires

Xiao Hu[1], Chenbo Dong[1], Rigu Su[2], Quan Xu[2*], and Cerasela Zoica Dinu[1]*

[1]Department of Chemical Engineering, West Virginia University, WV, USA

[2]State Key Laboratory of Heavy Oil Processing, China University of Petroleum (Beijing), Beijing, China

Abstract

The next generation of nanowires that could help advance the integration of functional nanosystems into synthetic applications from photocatalysis to optical devices needs to demonstrate increased ability to promote electron transfer at their interface while ensuring quantum confinement. Current techniques used to synthesize the nanowires such as the interface lithography technology and chemical vapor deposition only allow for a limited control of the nanowireâ€™s structureâ€"property relationships thus limiting their implementation in applications and design of nano-functional devices. Herein we proposed to use the biological recognition and self-assembly properties of tubulin, normally involved in building the microtubule cellular structure, to create stable, free standing and conductive sulfur-doped carbon nanodots-based hybrid bio-nanowires. We evaluated the physico-chemical properties (e.g., composition, morphology, diameter etc.) of such user-synthesized hybrids using classical atomic and spectroscopic techniques, while the electron transfer rate was estimated using peak currents formed during voltammetry scannings. Such synthesized hybrid bio-nanowires have diameters ranging from 6 to 30 nm and were proved to be more stable than their biological skeletons. Our results demonstrate the ability to create individually addressable hybrid bio-nanowires capable to reduce energy losses at interfaces to be used for the advancement and implementation of nanometer-scale functional devices.

455693 pH Sensing through Silicon Nanoribbon/Carbon Nanotube Porin Hybrid Sensor

Abstract:

pH of local environments is an important parameter for detection of pathological conditions and understanding of cellular mechanisms. Silicon nanoribbon field effect transistor (FET) has been proved to be a high-throughput and sensitive platform for pH sensing. However, the spatial resolution of the silicon nanoribbon is limited by the lithography techniques. Recent studies have demonstrated sub-2nm diameter carbon nanotube (CNT) porins are fast and efficient proton conductors. Furthermore, the CNT porins also can be successfully incorporated into lipid bilayers. In this research, we focus on a bioelectronic approach toward designing pH-sensing devices. We will describe an organic/inorganic hybrid device that combines nanoribbon FET with self-assembled CNT/lipid bilayer for pH sensing. And we will also discuss its potential application in nanometer spatial resolution intracellular biosensing.

463312 Design of Membrane-Embedded Amphiphilic Nanoparticles from Atomistic Molecular Dynamics Simulations

Abstract:

Monolayer-protected nanoparticles (NPs) are a versatile materials platform useful for biological applications because features of the protecting monolayer can be engineered to tailor interactions with the biological milieu. Recently, amphiphilic NPs with surface properties that mimic typical globular proteins were shown to enter cells via a non-endocytic, non-disruptive process that could be of broad interest for applications in drug or gene delivery. Here, we use detailed atomistic molecular dynamics simulations to gain molecular insight into these experimental observations. We find that the amphiphilic NPs insert into the bilayer to obtain a configuration resembling a membrane-embedded protein due to favorable interactions between hydrophobic molecules on the NP surface and the hydrophobic bilayer core. We identify a kinetic pathway for insertion that mimics the early onset of vesicle-vesicle fusion. Our calculations also suggest that charged ligand end groups cross the bilayer on experimentally relevant timescales that are significantly more rapid than expected. Finally, we leverage this mechanistic understanding to design monolayer compositions optimized for bilayer insertion and cellular uptake. This work emphasizes that carefully tuning NP surface coatings to resemble existing biological materials permits favorable interactions with the cell membrane, enabling the creation of new nano-bio hybrid materials with properties similar to those of membrane proteins.

470872 Photo-Electrochemical Characterizations of Photosystem I (PS I) Assembly Under Bio-Mimetic Membrane Confinement

Abstract:

The robust structural and photoactive electrochemical properties of Photosystem I (PSI) make it an ideal candidate for incorporation into bio-hybrid photovoltaic devices. However, the first step towards the rational design of such devices require systematic and oriented assembly of PS I at bio-abiotic interfaces via suitable scaffolding. To this end, our goal is to incorporate and assemble individually oriented PS I complexes into synthetic membrane-bound structures that mimic the natural thylakoid membrane housing of PS I. We investigate the photocurrent generation from such organic-inorganic interfaces in membrane bound PS I systems. Detergent-mediated reconstitution of proteoliposomes is analyzed via absorption and fluorescence spectroscopy measurements. Successful immobilization of membrane bound PS I on gold substrates are mediated by the hydrophillic interfaces of self-assembled monolayers (SAMs). To this end, force-distance spectroscopy during atomic force microscopy (AFM) imaging of lipid bilayer membranes on thiolated Au substrates confirm monolayer formation. Furthermore, topography measurements using liquid-AFM techniques confirm the assembly of membrane-bound PS I. Finally, the current work presents photo-electrochemical (PEC) characterizations demonstrating enhanced photocurrent generation from PS I complexes assembled under membrane confinement as compared to their isolated immobilizations on electrodes. In turn, the aforementioned studies provide valuable insight into future incorporation of bio-mimetic membrane bound protein complexes into novel bio-hybrid devices.

471619 Modeling Protein Folding/Aggregation on Nanoparticle Based Biosensors in Complex Solvent Environments By a Coarse Grained Simulation System

Abstract:

Understand protein behavior in a complex environment is the key in rationalizing the design of many novel techniques such as protein biosensors based on novel nanoparticle (NP) materials. Protein folding and aggregation properties can be largely affected by the environments such as solid abiotic interfaces and complex co-solvents, which would directly affect the expected functionality or performance of the biosensors. Difficulties remain in applying current experimental methods to obtain a detailed picture of protein structures on a solid surface, so our ability is limited by these factors in the design of biosensor surface chemistry a priori for desired protein behavior. To address this point, in this work we present a coarse-grained simulation system that is able to model protein folding, misfolding, and aggregation on various NP surfaces and in the presence of co-solvents based on the Karanicolas and Brooks (KB) Go-like protein model. A flat-surface potential was first included and carefully parameterized based on a large experimental data set. This surface model has been shown in many studies to be able to reproduce and predict protein behavior at different types of surfaces as verified by experimental tests. Further effort in developing the surface potential was applied in describing surface curvature effects on protein behavior as a means of understanding nano-sized biosensor materials. Tanford's transfer free energy model was recently added to this simulation system as a descriptor of co-solvent effects. The energy of each part of the underlying folding/surface/solvation potential has been well balanced so that each element works collaboratively as a single complex environment for the protein. Together, this coarse-grained modeling system is expected to accurately predict the protein adsorption, folding/unfolding, aggregation, and protein-protein interactions in a complex environment as it would experience in differing cellular or solvent environments as well as in the context of a NP-based biosensor.

471963 Aggregation-Induced Emission of Hydrophobically-Modified Metal Clusters in Lipid Nanodiscs

Abstract:

Biocompatible lipid-based nanodiscs (also known as â€œbicellesâ€) showed superior cellular uptake compared with liposomes of a similar length scale [1]. In this presentation we will demonstrate that bicelles can entrap nanometer-sized monolayer protected gold clusters (Au-MPCs) or hydrophobically-modified quantum dot (ZnSe/ZnS QD) through self-assembly, retaining reasonably uniform sizes (< 50 nm) as the bare ones. Aggregation-induced emission is observed in such nanoparticle-in-nanoparticle complexes (NP^2) with nearly two orders of magnitude enhancement in luminescenceÂ in both cases of Au-MPC-NP^2 and QD-NP^2. The outcome of this research will provide the insight into the design of biosensing nanoparticles in a controllable and scalable manner.

[1] W. Aresh,Â Y. Liu, J. Sine, D. Thayer, A. Puri, Y. Huang, Y. Wang, M.-P. Nieh "The Morphology of Self-Assembled Lipid-Based Nanoparticles Affects Their Uptake by Cancer Cells" *J. Biomed. Nanotech.* (in press)

465791 Delta- and Proteorhodopsin-Based Bionanoelectronic Devices for Light-Controlled Conversion of Protonic to Electronic Currents

Abstract:

The burgeoning field of bionanoelectronics aims at interfacing biology with electronics through the use of bio-based or biocompatible nanomaterials. A particular area of interest is the controlled conversion of ubiquitous biological ionic gradients into electrical signals that can be read or processed by conventional electronics. Here, we describe the efficient integration of proton-transporting rhodopsins with palladium (Pd) contacts to achieve light-activated conversion of protonic currents into electronic signals. The process involves engineering archaeal or proteorhodopsins with a Pd-binding peptide in order to bring the proton exit site in close proximity to the electrode; using protein and folding engineering strategies to express high levels of functional rhodopsins that are activated at various wavelengths; optimizing the integration and orientation of these proteins in liposomes; and forming supported bilayers on Pd/PdHx microfluidic devices fabricated by photo- and soft lithography. We demonstrate that addition of a Pd-binding extension leads to a nearly one order of magnitude increase in electronic signal with light-controlled "on" and "off" responses that are complete in seconds, discuss the development of steady state currents and describe devices in which proton transport and the corresponding electronic responses are selectively activated by different wavelengths of light.

462828 Self-Assembly of Protein Nano-Shapes Using Synthetic Coiled Coils

Abstract:

Protein self-assembly is an emerging and powerful approach to create structures at the nanometer scale. Driven by protein-protein interactions, protein subunits can be designed to self-assemble into supramolecular nanostructures with desired dimension and geometry. Recently, computational methods have been used to design complex protein interfaces,[1] while rational protein design approaches using simple protein motifs such as coiled coils have been successfully applied to design artificial structures different from naturally occurring protein folds.[2] Coiled coils are simple, rod-shaped, superhelical protein motifs composed of two or more α-helical peptides. They self-assemble into oligomers stabilized by specific interactions between the α-helical peptide subunits. Because their sequence-to-structure relation is well characterized, coiled coils are potentially useful self-assembling modules for design of artificial nanostructures. For example, an approach of covalent linkage and folding of different coiled-coil segments has been used to create a protein tetrahedron which does not exist in nature.[3] Furthermore, such a rational approach can be extended to design other artificial protein shapes using new sets of coiled coils. In this study, we demonstrate self-assembly of triangular protein nanostructures using synthetic heterodimeric coiled coils. We recently reported interaction profiles of computational designed hetero-associating coiled-coil peptides. Among them, sets of coiled-coiled pairs that make specific binding interactions over weak undesired crosstalk with others are available, and these orthogonal pairs can potentially serve as specific synthetic modules for protein self-assembly.[3] In our design, orthogonal coiled coil pairs are covalently linked and self-assemble into a triangular shape. We investigated mixing and thermal annealing of these protein building blocks, and maximized the yield of self-assembled triangular protein structures by manipulating the annealing temperature pathway. The fraction of self-assembled proteins was purified from the mixture of annealed proteins, using size-exclusion chromatography. Atomic force microscopy revealed the triangular shape of self-assembled proteins, and their hydrodynamic diameter and the supercoiled folding were

characterized using dynamic light scattering and circular dichroism, respectively. Our approach is simple and broadly applicable to design of other artificial protein nanostructures, which provide great potential in the area of protein materials development.

1. N. P. King, J. B. Bale, W. Sheffler, D. E. McNamara, S. Gonen, T. Gonen, T. O. Yeates and D. Baker, *Nature*, 2014, **510**, 103-108.
2. H. Gradišar, S. Božič, T. Doles, D. Vengust, I. Hafner-Bratkovič, A. Mertelj, B. Webb, A. Šali, S. Klavžar and R. Jerala, *Nat. Chem. Biol.*, 2013, **9**, 362-366.
3. K. E. Thompson, C. J. Bashor, W. A. Lim and A. E. Keating, *ACS Synth. Biol.*, 2012, **1**, 118-129.

471188 Mechanical Properties of Ionically and Covalently Crosslinked Alginate Hydrogels (Invited Talk)

Abstract:

Polysaccharide networks are widely used in biomedical and food applications. Polysaccharides are semiflexible polymers and the networks consisting of these molecules can display interesting mechanical properties. In this presentation, results on alginate hydrogels, a model polysaccharide network, will be presented. Effect of two different types of crosslinking, ionic and covalent, on the mechanical properties will be considered. Calcium salts have been used to obtain ionically crosslinked gels and these gels consist of junction zones with finite dimensions. Diamines have been used for synthesizing covalently crosslinked gels. Shear and cavitation rheology techniques were utilized to investigate the mechanical properties of these gels. The large amplitude oscillatory shear (LAOS) rheology captures the strain-stiffening behavior for these gels. In addition, negative normal stress was observed, which has not been reported earlier for any polysaccharide networks. The magnitude of negative normal stress increases with the applied strain amplitude and can exceed that of the shear stress at large-strain. Deformation mechanism of alginate gels involves stretching and bending of the alginate chains. In addition, for ionically crosslinked gels, bending of the junction zones play an important role. Cavitation rheology results indicate that the pressurization of defect inside these gels resulted in fracture. The results presented here provide new understanding of the deformation behavior of alginate hydrogels.

454955 Structural Rearrangements during Stress Relaxation of a Tough, Physically Crosslinked Hydrogel

Abstract:

Synthetic hydrogels have typically been limited by their poor mechanical behavior, which limits their applications as compared to natural tissues. The seminar work of Gong and co-workers identified the use of double networks as a route to signficantly toughen hydrogels [1]. The enhanced toughness is attributed to a sacrificial network that dissipates energy on deformation. However, the energy dissipation is not reversible, but similar concepts of energy dissipation with healable crosslinks have been identified [2]. One simple route is through assembly of hydrophobic components of a amphiphilic copolymer, where these hydrophobic aggregates act as crosslinks to prevent the copolymer from dissolving in water [3]. However in this case, the energy required to break up the aggregates is less than that to break covalent bonds; under sufficient deformation, the aggregates will break up, dissipating energy, but the driving force to re-aggregate after the force is removed leads to re-formation of the network to generate extendable and tough hydrogels. Perfluorinated hydrophobes present to advantages: first, the energy of the association is larger, which leads to increased mechanical strength of the hydrogel over simple hydrocarbons. Second, the fluorine provide contrast in both x-rays and neutrons in order to examine the nanoscale structure with scattering [4]. Here we exploit the contrast difference between the aggregates and the water-rich hydrophilic segments in the copolymer to examine how the crosslink structure recovers from a large deformation with small angle neutron scattering (SANS). Use of different ratios of D2O and H2O allow the recovery of the hydrophilic chains and the aggregates to be probed independently. The extension of the hydrogel leads to anisotropic scattering that slowly recovers over several hours. We will describe methods to quantify the temporal evolution in the scattering profiles in order to relate the nanoscale structure to macroscopic stress relaxation.

[1] J.P. Gong, Y. Katsuyama, T. Kurokawa, Y. Osada. Adv. Mater. 2003, 15, 1155.

[2] J.-Y. Sun, X. Zhao, W. R. K. Illeperuma, O. Chaudhuri, K. H. Oh, D. J. Mooney, J. J. Vlassak, Z. Suo, Nature 2012, 489, 133.

[3] J. Hao, R. A. Weiss, Macromolecules 2011, 44, 9390.

[4] J. Tian, T. A. P. Seery, D. L. Ho, R. A. Weiss, Macromolecules 2004, 37, 10001.

472768 Strong Polymer Gels with Tunable Crystalline Domains

Abstract:

The ability to create synthetic materials that mimic the structural and mechanical properties of soft biological tissues remains a significant challenge. In this presentation, we focus on creating strong hydrogels with a high elastic modulus and compressive modulus by engineering crystalline domains into associative hydrogels of poly(lactic acid)-poly(ethylene oxide)-poly(lactic acid) (PLA-PEO-PLA) triblock copolymers. In aqueous media, these materials form associative gels with PLA domains serving as network junctions. We extend previous studies from our group and others by varying the stereochemistry of the PLA block to create polymers with PLA blocks with ratios of L/D lactide units varying from 100/0 to 50/50. We had previously found that the 100/0 systems (triblocks with poly(L-lactide) blocks) formed gels with nanoscale crystalline domains, and these gels displayed a high value of the elastic modulus which was strongly dependent on PLA block length. Interestingly, we find that the modulus of these gels does not seem to vary monotonically with L/D ratio, with systems at intermediate L/D values displaying higher moduli than the 100/0 systems. Collectively, our work shows that stereochemistry can be used to access novel structures and properties in relatively simple synthetic polymers, perhaps giving insight into new routes for creating complex soft materials.

465349 Quantifying Topology and Elasticity of Polymer Networks

Abstract:

Despite the ubiquity of applications, much of our fundamental knowledge about polymer networks is based on an assumption of ideal end-linked structure. However, polymer networks invariably possess topological defects: loops of different orders which have profound effects on network properties. Here, we develop a kinetic graph theory which demonstrates that all different orders of cyclic topologies are a universal function of a single dimensionless parameter characterizing the conditions for network formation. The theory is in excellent agreement with both experimental measurements of hydrogel loop fractions and Monte Carlo simulations without any fitting parameters. We demonstrate the superposition of the dilution effect and chain-length effect on loop formation. The one-to-one correspondence between the network topology and primary loop fraction demonstrates that the entire network topology is characterized by measurement of just primary loops, a single chain topological feature. Different cyclic defects cannot vary independently, in contrast to the intuition that the densities of all topological species are freely adjustable.

Quantifying different cyclic defects facilitates studying the correlations between the network topology and gel elasticity. Classical theories of gel elasticity assume acyclic tree-like network topology; however, all polymer gels inevitably possess cyclic defects: loops that have profound, yet previously unpredictable, effect on gel properties. Here, we develop a real elastic network theory (RENT), a modified phantom network theory that accounts for the impacts of cyclic defects. We demonstrate that small loops (primary and secondary loops) have vital effect on the modulus; whereas this negative impact decreases rapidly as the loop order increases, especially for networks with higher junction functionalities. Loop effect is non-local, which can propagate to its neighborhood strands. RENT provides predictions that are highly consistent with experimental observations of polymer network elasticity, providing a quantitative theory of elasticity that is based on molecular details of polymer networks.

470788 Ultra-Soft, Dry Polydimethylsiloxane Elastomers from Architecture-Driven Entanglement Free Design

Abstract:

Polydimethylsiloxane (PDMS) elastomers are the most widely used silicon-based organic polymeric materials. Conventional PDMS elastomers formed by crosslinking entangled linear polymers are intrinsically stiffer than an intrinsic limit, ~200 kPa, the value set by entanglements. Here we develop ultrasoft, solvent-free PDMS elastomers by crosslinking bottlebrush polymers rather than linear polymers. We design the chemistry to allow commercially available linear PDMS precursors to deterministically form bottlebrush polymers, which are simultaneously crosslinked, thus enabling a one-step, room temperature synthesis. The bottlebrush architecture prevents the formation of entanglements, resulting in elastomers with precisely controllable elastic moduli from ~1 to ~100 kPa, below the intrinsic lower limit of traditional elastomers. Moreover, the solvent-free nature of the soft PDMS elastomers enables a negligible contact adhesion compared to commercially available silicone products of similar stiffness. The exceptional combination of softness and negligible adhesiveness may greatly broaden the applications of PDMS elastomers in both industry and research.

453534 Nanovoid Formation and Mechanics: A Simulation Study of Poly(dicyclopentadiene) and Epoxy Cross-Linked Networks

Abstract:

Nanovoid formation and mechanics: A simulation study of poly(dicyclopentadiene) and epoxy cross-linked networks

Robert M. Elder, Daniel B. Knorr, Jr., Joseph L. Lenhart, Jan W. Andzelm, and Timothy W. Sirk*

US Army Research Laboratory, Aberdeen Proving Ground, MD 21005

ABSTRACT

Cross-linked polymer networks are widely used in protective and structural applications under extremes of temperature, pressure, and strain rate. Typical materials for these applications, like epoxy resins, have high stiffness but inferior toughness. Recent experiments showed that cross-linked poly(dicyclopentadiene) (pDCPD), a hydrocarbon, can overcome this trade-off between toughness and stiffness, resulting in excellent protective performance. Based on the physicochemical properties of pDCPD and epoxy, it was hypothesized that the superb toughness of pDCPD is related to the formation and growth of nanovoids at extreme strain rates. Voids can dissipate energy that might otherwise lead to failure. Here, we use atomistic molecular dynamics to study pDCPD and epoxy networks undergoing high strain rate deformation. We quantify the networks' mechanical properties (tensile modulus and yield strength), and the prevalence and characteristics of nanoscale voids. We identify molecular-level properties – such as the polar chemistry of epoxy – that influence the behavior of the networks. The strong non-covalent interactions (e.g., hydrogen bonds) that accompany polar moieties increase the modulus and yield strength of epoxy networks. Conversely, the weak non-covalent interactions accompanying the aliphatic (non-polar) chemistry of pDCPD can account for its lower modulus and strength. We examine the local chemical environment of nanovoids, finding that nanovoids are more prevalent in regions of non-polar chemistry. Consequently, pDCPD generally has more, larger nanovoids that grow more readily during deformation. Thus nanovoids may be related to the higher fracture toughness and more ductile behavior observed in pDCPD.

*Corresponding author: *timothy.w.sirk.civ@mail.mil*

464850 Simultaneous Scission and Gelation Reactions in Photocrosslinkable Polymers

Abstract:

Polymers used in medical devices, lithographic patterning, and other applications are often exposed to ultraviolet, e-beam, or gamma irradiation to obtain a sterile and/or crosslinked polymer. Since these stimuli can produce free radical intermediates on the polymer backbone, simultaneous gelation and scission reactions will occur. These reaction rates underlie the resulting mechanical properties, % oxidation, and % leachables, but in practice these reactions are challenging to directly quantify. To establish a simple approach for extracting stoichiometric likelihoods of scission or gelation reactions resulting from irradiation, we have adapted the Miller-Macosko statistical model to account for these effects. The simple, closed form expressions that result provide a means to quantitatively analyze sol fraction vs. dose data and gain mechanistic insight into these reactions. Here we will describe our approach in detail and apply it a model photopolymer, poly(styrene sulfonyl azide-alt-maleic anhydride).

469609 Multi-Component Flory-Rehner Theory: Swelling of Copolymer Networks and Swelling of Networks in Solvent Mixtures

Abstract:

While the swelling of homopolymer networks in pure solvents is well-understood, refinements to existing thermodynamic models are required to describe equilibrium swelling of three-component or higher systems quantitatively. Examples of multi-component swelling include the swelling of homopolymer networks in solvent mixtures and the swelling of random copolymer networks in pure or mixed solvents. The thermodynamic model established by Flory and Rehner in 1943[1] was initially aimed at describing swelling of homopolymer networks in pure solvents, but its underlying ideas can be extended to describe swelling in multi-component systems. A generalized extension of classical Flory-Rehner (FR) theory has been derived to describe swelling of homopolymer or copolymer networks in any number of solvents.

A specific case of swelling of random copolymer networks in a mixture of two solvents is examined to test the applicability of the multi-component FR theory. For this four-component (two monomer units, two solvents) scenario, it is shown that the six independent χ_{ij} (interaction parameters) can be reduced to two independent interaction parameters (θ_1 and θ_2) that can be obtained experimentally from fits to equilibrium swelling data in pure solvents. Key inputs to the model include the partial molar volumes of water and ethanol (V_1 and V_2), which are obtained from molecular simulations, the ethanol-water interaction parameter (χ_{12}), the concentration of elastically effective chains (v), and two copolymer-solvent thermodynamic interaction parameters (θ_1 and θ_2).

A model system consisting of 121 random copolymer networks of n-butyl acrylate (B) and 2-hydroxyethylacrylate (H) is designed in a combinatorial fashion, with systematic variations in crosslinker concentration and B:H mole ratio, in order to provide a robust set of data. Equilibrium swelling measurements are first made in pure water, pure ethanol, and ethanol-water mixtures. The volume fractions of ethanol and water in the swollen networks at equilibrium are measured by a combination of gravimetry and high-performance liquid chromatography (HPLC). Polymer-solvent thermodynamic interaction parameters (θ_1 and θ_2) are obtained by fitting the multi-component FR model to equilibrium swelling data for all copolymer networks swollen in pure water and pure ethanol. Using fitted values of θ_1 and θ_2, predicted volume fractions of ethanol and water at equilibrium in mixed solvents are computed, and they show excellent agreement with the experimentally measured values. Not surprisingly, the interaction parameters are found to be slightly concentration-dependent, much like traditional FR c parameters. The multi-component FR model extension is useful to model the behavior of networks in mixed-solvent systems, with liquid phase membrane separations being a key example.

1. P. J. Flory; J. Rehner, Statistical Mechanics of Cross-Linked Polymer Networks II. Swelling, *The Journal of Chemical Physics*, **1943**, *11* (11), 521-526.

469347 Unraveling Chemical Transformation of Reactively-Modified Interfacial Thin Films Using Neutron Reflectometry (Invited Talk)

Abstract:

Creating an appropriate display of chemical functionality is useful across a wide variety of applications for polymer thin films, ranging from adhesives and barrier coatings to membranes and sensing devices. In this presentation we describe our efforts to tease apart the complex connections between structure and structural rearrangement of interfacial polymer thin films and the penetration and distribution of small-molecule functionalizing agents installed by *in situ* modification. Model thin films based on the reactive polymer, poly(2-vinyl-4,4-dimethylazlactone) (PVDMA) are used for this purpose, and neutron reflectometry is brought to-bear as a key tool for metrology. Highly constrained analyses of neutron reflectivity prove that reactive modification is not uniform throughout the PVDMA thin film nor complete; rather, it depends on key structural parameters of the interfacial layer and the size of the small molecule functionalizing agent. This quantitative picture diverges substantially from insights generally gleaned from ellipsometric measurements of modified thin films, highlighting both key limitations in the latter method and the richness of results that may be obtained from the former technique. Overall, the findings provide broadly support efforts to produce polymer films in which various surface properties are tailored by *in situ* chemical modification.

468431 Thin Film Crystallization and Morphology of Poly(ε - caprolactone)

Abstract:

Poly(Îµ-caprolactone) (PCL) is a semi-crystalline, hydrophobic, biodegradable polymer with applications in biomedical engineering including anti-adhesion biomaterial films, drug delivery media, and others. Degree of crystallinity, crystal morphology, and crystallite size are known to affect the biodegradation profile of PCL fibers and films by enzymes, so morphological control is important to designing PCL coatings and films.Â By combining a flow coating technique to generate gradient thickness films and automated atomic force microscopy imaging, we are able to rapidly characterize morphology transitions as a function of film thickness. These transitions are driven by the reorientation of crystalline lamellae under thin film confinement.Â Initial studies examined these morphological transitions for several molecular weights of linear PCL in an effort to connect transition film thicknesses to molecular length scales. Subsequently, we blended PCL with a block copolymer and identified previously unreported novel crystal morphologies that result from the block copolymer directing and confining the crystallization of PCL during film casting. Our results show that thin film and molecular level confinement can be used to control crystallization and morphology in semi-crystalline polymer films making these materials tailorable for specific applications.

469105 Development of New Chemistries for Molecular Layer Deposition

Abstract:

High quality nanoscale polymer films fabricated with molecular precision are seeing increasing importance in applications such as microelectronics, sensors, biomedical devices, and energy technologies. Most deposition methods, such as dip-coating, spin-coating, and surface initiated-polymerizations, are solution based and lack nanoscale control. Molecular layer deposition (MLD) is an emerging vapor-phase technique that offers excellent control at the nanoscale. This powerful method creates conformal polymer films with molecular precision by sequentially depositing organic bifunctional precursors in a self-limiting layer-by-layer fashion. However, MLD proceeds in absence of solvent, pH cannot be varied, and catalysts cannot easily be employed, and this has so far limited the number of linking chemistries applicable to MLD, and hence also its applications.

Here our search for new linking chemistries compatible with the solid-gas phase nature of MLD will be presented. Our effort can be divided into three very different reaction types. Firstly, we have examined nucleophilic substitutions of alkyl iodides with different nucleophilic precursors, to e.g. enable controlled growth of nanoscale polyethylenimine. This reaction type is so far unexplored for MLD, but it is similar in nature to the nucleophilic acyl substation which is the most common MLD linking chemistry. The second reaction type is the pericyclic Diels Alder reaction. This reaction creates new carbon-carbon bonds without the need for a catalyst, and it is therefore a powerful addendum to the existing toolbox of MLD chemistries. The third reaction class is photo-activated MLD, in which the direct input of energy from UV-light is utilized to overcome otherwise infeasible reaction barriers. Photo-activated MLD may also result in new carbon-carbon bond forming reactions. We will present our successful and unsuccessful results pertaining to these new classes of MLD chemistries, including XPS, ellipsometry and IR analysis of the nanoscale polymer thin films. Additionally, the challenges we have encountered with each of these new chemistries will be discussed.

466629 Designing Non-Charging Surfaces from Non-Conductive Materials

Abstract:

Static charge generated from contact of solid surfaces is ubiquitous, and can give rise to a wide variety of undesirable consequences. These effects may cause little annoyances in daily lives (e.g., an electric shock when one touches a door knob), a reduction in efficiency in many industrial processes (e.g., charged particles that adhere to surfaces of vessels can hinder heat transfer), damage to equipment (e.g., electronics), and potentially hazardous situations (e.g., explosions). For many decades, the general strategy to address problems arising from contact electrification is to make surfaces conductive (e.g., by adding antistatic agents) in order to dissipate charge. However, this method may not be effective when the contacting material is insulating; in addition, modification of the materials to make them conductive may change the properties of the materials in undesirable ways. Here, we introduce a general strategy of fabricating non-charging surfaces â€" by copolymerizing a molecule that has the tendency to charge positively with another molecule that has the tendency to charge negatively. At an appropriate proportion of the two molecules, the copolymer can be fabricated to resist charging against a reference material. Importantly, this polymer naturally resists charging against the reference material and is insulating; that is, this strategy does not require the material to be conductive. We showed that the copolymer can prevent the adhesion of micron-sized particles due to attractive electrostatic force. In addition, the strategy is general: different copolymers can be fabricated to resist charging against many different contacting materials. Thus, it is possible to fabricate a class of non-charging materials with properties that are compatible to different applications.

472482 A Novel Crosslinker for Polyolefinic Thin Surface Coatings By Surface Initiated Ring Opening Metathesis Polymerization

Abstract:

Ring opening metathesis polymerization (ROMP) is a very emerging method of polymerization and is being researched intensively due to its benefits over current commercial polymerization methods. Surface initiated ring opening metathesis polymerization provides precise control over the properties of the surface coatings like thickness, rate of polymerization and surface density. But degradation of these films has been observed upon brief exposure to organic solvents in air. The root cause of degradation is the alternate carbon-carbon double bond in the polymer chains on the surface which are susceptible to rapid auto oxidation at these potential sites. Polymer chains break ultimately decreasing the thickness or density of surface coating. Here we have developed a crosslinker which can be introduced during Si-ROMP in very low concentration so as not affecting the intended properties of the coatings but also strengthening it. We have optimized the structure, size and concentration of crosslinker based on the general properties of Si-ROMP coatings of Polynorbornene. These coatings were characterized with and without crosslinker by contact angle goniometry, grazing angle reflection absorption infrared spectroscopy, Raman spectroscopy, nuclear magnetic resonance spectroscopy. Stability was also checked against organic solvent in air before and after introduction of crosslinker.

472600 Crossover in Topologically Driven Surface Segregation of Cyclic/Linear Polymer Blends

Abstract:

In comparison to the surface segregation that occurs in polymer blends due to chemical differences between the two types of repeat units or in blends with different chain sizes, where shorter chains are preferred at the surface, topologically driven surface segregation is more difficult to describe theoretically. We report experimental observation and theoretical consideration of a crossover in the surface segregation of blends of linear and cyclic polymers from control by a universal topological driving force for long chains to control by surface packing for short chains. For a 37k cyclic/linear polystyrene blend the surface is enriched in cyclic chains in quantitative agreement with self-consistent field theory. For a 2k cyclic/linear polystyrene blend the surface is enriched in linear chains, consistent with Wall Polymer Reference Interaction Site Model theory. While the surface segregation seen in blends of linear and branched chains can be described using a conventional surface potential, the topological driving force important here cannot be.

470822 Morphological Control By Localized Blending in Phase Separated Block Copolymer Thin Films

Abstract:

Block copolymer self-assembly is a promising bottom-up route to fabricating dense periodic arrays of monodisperse nanoscale features. In recent years, block copolymer lithography in thin films has attracted great interest as a tool for nanoscale device fabrication. However, one major limitation to the use of block copolymer lithography is the difficulty in tuning equilibrium morphologies and domain size in a continuous film. Here we will present a simple layer-by-layer approach to form block copolymer blends, with varying composition across the thin film, that independently controls both the morphology and domain spacing of nanostructures. This process results in equilibrium nanostructures of diverse shapes and sizes that will ultimately reduce the number of processing steps required to fabricate nanoscale devices from self-assembled block copolymer thin films.

455411 Effect of Backbone Flexibility on the Structure and Orientation of Polyurea Chains Grown By Molecular Layer Deposition

Abstract:

Effect of Backbone Flexibility on the Structure and Orientation of Polyurea Chains Grown by Molecular Layer Deposition

David S. Bergsman[1,*], Richard G. Closser[2], Christopher J. Tassone[3], Bruce M. Clemens[4], Dennis Nordlund[3], and Stacey F. Bent[1,2,4]

[1]Department of Chemical Engineering, [2]Department of Chemistry, [4]Department of Materials Science and Engineering, Stanford University, Stanford, California, USA 94305. [3]SLAC National Accelerator Laboratory, Menlo Park, California 94025

* bergsdav@stanford.edu

In many nanotechnology applications, such as microprocessors, sensors, and solar cells, there is an increasing need for the incorporation of nanoscale polymeric films. However, many commonly-used polymer deposition processes cannot meet the increasingly strict requirements regarding film thickness, conformality, composition, and crystallographic structure needed for these technologies. One method of meeting these demands is molecular layer deposition (MLD), in which polymer chains are grown from the vapor phase by exposing a surface to an alternating sequence of two or more monomers that only react with the previous monomer. Due to the self-limiting nature of these reactions, films deposited using this technique have been shown to be conformal on even high aspect ratio substrates, and the library of materials that can be deposited is continuing to expand. But despite the continuing advancements made in the development of this technique, many questions still remain about the mechanism behind the film growth and their resulting structures. In order to develop a better model for its growth behavior, MLD was used to deposit polyurea films onto silicon with bifunctional monomers whose backbone chemistries were varied in conformational flexibility. Backbones included phenyl, ethyl, and butyl groups for both the first and second monomers used in the A-B process. Trends in the thickness, bonding, crystallinity, and chain orientation were then measured using variable-angle spectroscopic ellipsometry, Fourier transform infrared spectroscopy, grazing incidence x-ray diffraction, and angle-dependent near edge X-ray absorption fine structure. These trends were then used to deduce structural information about the chains in these films. Growth rates of these chemistries ranged from about 4 angstroms/cycle for the phenyl-phenyl backbone to 1 angstrom/cycle for the butyl-butyl backbone, suggesting that the films primarily adopt non-fully extended configurations. Fourier transform infrared spectroscopy confirms the polymerization of the monomers, as well as demonstrating increased ordering of the urea groups for the phenyl-phenyl, butyl-butyl, and ethyl-butyl backbones. Grazing incidence x-ray diffraction further supports this ordering, and shows an out-of-plane paracrystalline peak with lattice spacing that decreases with increasing chain flexibility. Combined with molecular orientation data collected with x-ray absorption, it is believed that these chains form mixed domains, with some chains lying in horizontally stacks of paracrystalline segments and some chains adopting tilted configuration, growing away from the substrate. The implications of these results on the likelihood of double reaction terminations occurring will also be discussed.

474103 Hydrophobic Coating Based on Fluorinated Acrylate Monomer Using UV-LED Polymerization

Abstract:

Water repellent or hydrophobic coating offers various promising advantages such as anti-fouling and anti-corrosion surface properties. It is well known that fluorine based moiety can alter the surface wettability due to its low surface energy characteristic. In this study, UV-curable hydrophobic coating mixture was prepared based on aliphatic urethane acrylate oligomer, 2-ethylhexyl acrylate (2-EHA), methyl methacrylate (MMA), trimethylopropane triacrylate (TMPTA), heptadecafluorodecyl methacrylate (HDFMA) and hydroxyphenyl ketone photoinitiator (Chivacure 300). The reaction mixture was coated onto a glass substrate and cured via UV-LED irradiation. From the ATR-FTIR spectra, the C=C absorption peak at 1635 cm^{-1} was decreased after UV-LED irradiation, indicating the proceeding of polymerization. The effect of HDFMA content on the properties of UV-curable coating was investigated. In order to quantify the degree of hydrophobicity, water contact angle of the coating was determined. The addition of fluorinated acrylate monomer improved the water repellence behaviour and increased the water contact angle from 86.01° to 123.57°. As the HDFMA content increased, the surface tension of the coating also became lower, due to the presence of the fluorinated group. Moreover, the light transmittance of the coating layer was not significantly affected with increasing HDFMA content. It can be concluded that the fluorinated containing monomer gave the most contribution towards the hydrophobicity of the coating assisted by UV-LED curable system. Analogously, these results may attain various potential applications such as self-cleaning, optoelectronic and liquid-repellent coating. Importantly, the utilization of UV-LED in this research was a great initiative to develop green aspect for photopolymerization.

466658 Protein-Based Catechol Conjugates As Aqueous Adhesives and Networks

Abstract:

Polymers that adhere to wet surfaces that are biocompatible, degradable, and devoid of animal-sourced materials are desired for applications as tissue adhesives. Silk from the domesticated silkworm *Bombyx mori* is a naturally occurring biopolymer with charged hydrophilic terminal regions that endcap a blocky core dominated by hydrophobic domains. Silk was selected for modification for use as an aqueous adhesive because of its high molecular weight, tunable mechanical and degradation properties, aqueous processability, and for its hydrophobicity that was envisioned to afford control over the mechanical properties and aqueous swelling of the modified biopolymers. Taking inspiration from mussels that produce proteins rich in L-3,4-dihydroxyphenylalanine to adhere to a variety of organic and inorganic surfaces by manipulation of oxidation states, silk was functionalized with catechol groups by carbodiimide coupling of dopamine. The silk polymers were characterized to determine composition, adhesive strength using lap shear measurements, cell compatibility and support of proliferation, and their ability to form beta sheets, which physically crosslink silk.Â As both aqueous and organic phase syntheses are possible with silk, several solvent systems were investigated to determine the conditions that led to successful coupling of dopamine without protein aggregation. Adhesive bonding and cell compatibility of the resulting materials were measured. Human mesenchymal stem cells were found to attach and proliferate on the silks *in vitro*. The formation of beta sheets was not affected by the chemistry, thus could be exploited to control the adhesive networkâ€™s properties.Â This work developed new silk-catechol conjugates that can be triggered to crosslink by oxidation. Tuning the substitutions proved to be critical to the strength of the adhesive bond and aqueous swelling. All materials supported cell attachment and proliferation *in vitro*. Inducing beta sheet formation after bonding is currently under investigation as an additional route for controlling the mechanical properties of the adhesive networks. These materials thus represent a protein-based approach to water soluble, catechol-based adhesives, which we envision to have utility as biodegradable adhesives and sealants.

473007 Novel Zwitterionic Macro-Crosslinker Enhancing Durability of Non-Fouling Coating

Abstract:

For many applications involving surface exposure to biomolecules and microorganisms, biofouling is a serious problem that causes implantable device failure and unwanted microorganism attachment. Recently, surface modification with non-fouling polymers has been utilized to overcome the biofouling problem. Encouraging results have been obtained in various applications in different traditional substrates such as anti-thrombosis polyurethane catheters and coatings on metal ship hull to resist marine fouling by using non-fouling zwitterionic polymers. Zwitterionic polymer coatings are super-hydrophilic and attract a dense layer of water molecules to resist non-specific protein binding and bacteria adhesion. Many methods have been involved in fabricating zwitterionic coatings. Nevertheless, the resulting zwitterionic polymer coatings are typically unstable with only short-term function, regardless of how they are immobilized on the surface. A novel zwitterionic macro-crosslinker was developed and applied to fabricate durable non-fouling coatings on a polyurethane substrate. The zwitterionic macro-crosslinker coating exhibited superior durability over the traditional brush polymer coating and was able to retain its non-fouling property even after weeks of shearing in flowing liquid.

469253 Single Molecule Study of the Role of Aptamer Secondary Structure in the Dynamics of Aptamer-Ligand Binding Interactions

Abstract:

Nucleic acid aptamers are a rapidly expanding technology due to a high affinity for their target molecules, stability, low cost of production, and ease of chemical modification. Aptamers can be engineering to bind to a wide range of molecules, which results in a sequence that has high affinity for the target ligand after several rounds of selection. These aptamers adopt a number of secondary structures when the ligand is bound, including stem/loops bulges, g-quadruplexes, hairpins, and psuedoknots; yet, little is understood about the mechanism by which the aptamer-ligand binding occurs. For a number of aptamers, an adaptive recognition model of binding has been proposed, where the aptamer adopts an unfolded conformation in solution and folds upon addition of the target molecule. However, some aptamers appear to have a prefolded structure, as they undergo little to no conformational change upon target addition. We employed single molecule Forster resonance energy transfer (FRET) to measure hybridization dynamics between an aptamer and a complementary â€œcaptureâ€ strand immobilized on a fused silica surface. We found that the ligand disrupts the DNA duplex when a toehold is present, similar to DNA strand displacement reactions. This suggests that the mechanics of DNA strand displacement can be extended to improve the understanding and development of functional aptamer-ligand systems such as biosensors, molecular motors, and biocomputational circuits.

465250 Mixed Poly(ethylene glycol) and Poly(sulfobetaine) Brushes to Control Protein Adsorption and Denaturation on Biomaterial Surfaces

Abstract:

Polymer brushes have received considerable attention as coatings to improve the biocompatibility of materials in physiological environments or in contact with proteins for applications including biomaterials for tissue engineering and drug delivery, medical devices, and biosensors. Despite a widespread interest in applications of polymer brushes, a mechanistic understanding of the connection between brush properties and non-specific protein adsorption and denaturation remains elusive. We investigated the effect of grafting density of poly(ethylene glycol) (PEG) brushes on the interactions of the brush with fibronectin (FN) using high-throughput single-molecule tracking methods, which directly measure protein adsorption, diffusion, and unfolding within the brush. We observed that as grafting density increased, the rate of FN adsorption decreased; however, surface-adsorbed FN unfolded more readily and unfolded molecules were retained on the surface for longer residence times relative to folded molecules. While PEG brushes overall exhibited non-fouling properties, they did not prevent protein adsorption completely at the molecular level and, more interestingly, did not preserve the native conformation of protein molecules. Based on these results, we are currently investigating mixed polymer brushes containing PEG and poly(sulfobetaine), a zwitterionic polymer. We hypothesize that the inclusion of poly(sulfobetaine) will mediate the retention of unfolded molecules on PEG brushes by altering the local hydration within the brush. To test this hypothesis, mixed PEG/ poly(sulfobetaine) brushes have been prepared with varying ratios of PEG-to-poly(sulfobetaine), which have been characterized by water contact angle and FTIR-ATR. The adsorption of FN and retention of unfolded FN as a function of the PEG-to-poly(sulfobetaine) has been characterized in analogous single-molecule experiments as in the case of pure PEG. Our studies will provide new insights into the rational design of polymer brushes to improve their performance in biological environments as well as control cellular responses to non-specifically adsorbed protein in the physiological milieu.

461369 Development of Antithrombotic Nanoconjugate Targeting Collagen

Abstract:

Arterial thrombus formation, initiated from the platelet adhesion on exposed collagen, is considered to cause many arterial diseases of high mortalities, such as stroke and acute myocardial infarction. Blocking the collagen exposed on diseased blood vessel would prevent platelet adhesion while not affect the normal function of platelets. Our previous work has proposed a biomimetic design strategy of antithrombotic peptides, and obtained an effective inhibitor of integrin Î±2Î²1-collagen interaction and subsequent thrombus formation, LWWNSYY. However, its practical application suffered from the formation of clusters in physiological environment caused by its high hydrophobicity. Therefore, an antithrombotic nanoconjugate was designed in which LWWNSYY was immobilized to the surface of poly(glycidyl methacrylate) nanoparticles (PGMA NPs). The obtained LWWNSYY-PGMA nanoparticles (L-PGMA NPs) conjugate have proven capable of inhibiting platelet adhesion *in vitro*. High inhibition efficiency of L-PGMA NPs in thrombus formation was further confirmed *in vivo*. Therefore, L-PGMA NPs were developed as a high-efficiency antithrombotic nanomedicine targeted for collagen exposed on diseased blood vessel wall, which would be helpful for the prevention and cure of thrombotic diseases.

462679 Modelling and Controlling Degradation and Protein Release Profile from Degradable Polymer-Polymer Hydrogel Depot

Abstract:

Hydrogels are suitable candidates as protein encapsulation depot and subsequent release process due to their hydrophilic environment. Degradable hydrogels are attractive because they eliminate the need for removal by surgery at the end of application as well as accelerating the protein release. To optimize the hydrogel design, it is crucial to have a clear understanding of network properties changes during degradation which has a considerable effect on drug release. In this research we propose a comprehensive model to account for combined effect of bulk-degradation, swelling and mass loss on the protein release for chemically crosslinked polymer-polymer hydrogels. Instead of extending eroding probability of one chain to the whole network by multiplying it to precursors initial weight fractions like previous models, more stochastic microscopic approach -Monte Carlo simulation scheme- was applied to obtain the detaching rate of polymer chains from network based on pseudo first order kinetic for crosslinkage cleavage as well as changes in cross-linking density. Bray-Merrill equation is utilized to obtain degradation kinetic rate constant through fitting with experimental volumetric swelling profile. Another advantage of this approach is that the effect of swelling on hydrogel mass loss fraction, its concentration and then drug release is included by solving diffusion-convection equation for both free polymer chains and drug in deformed geometry in which diffusion coefficient of drug and free polymer chains are function of hydrogel concentration. Initial network parameters such as molecular weight, degree of modification with reacting groups, concentration of polymeric precursors and types of degradable linker are expected to affect the erosion behavior of the network and drug release. Simulating the effect of these parameters on the degradation behavior would facilitate the design of degradable hydrogel systems for controlled release applications.

463889 Viral Gene Vector-Releasing Biodegradable Polyelectrolyte Fiber-Sutures

Abstract:

Heparin, a sulfated polysaccharide biopolymer, has numerous important biological activities due to its interaction with diverse proteins. Especially, heparin has a strong interaction with adeno-associated virus (AAV) which have great potential as a powerful tool for gene therapy. However, there are limitations to the AAV delivery system in that it lacks localized delivery with high efficiency. Efficient AAV delivery carriers are required to expect a considerable therapeutic outcome. In this study, biocompatible, biodegradable, and high tensile sutures composed of polyelectrolyte fibers were developed as an efficient vehicle for AAV delivery. Interfacial polyelectrolyte complex (IPC) was formed by bringing chitosan and heparin solution into contact, and the polyelectrolyte fibers were made by physically extending the formed IPC away from the reaction solution. This fiber formation process is continuous because the IPC can be generated from the newly contacting interface. To fabricate the suture, 64 strands of polyelectrolyte fibers were spun and merged to one fiber. Then, the merged fibers were washed in PBS solution and air-dried. As a result, a suture with high tensile strength composed of 64 strands of chitosan and heparin polyelectrolyte fiber was prepared. The fabricated sutures were biocompatible, biodegradable, and had high mechanical strength. Furthermore, the sutures had enhanced interaction with a variety of biomaterials such as protein, cells, and AAV due to the unique properties of incorporated heparin groups. To investigate the potential of the developed suture as a gene delivery carrier, AAV were immobilized onto the suture and implanted in subcutaneous tissues of mouses. Interestingly, AAV r3.45 immobilized chitosan/heparin suture demonstrated improved localized cellular transduction compared with the conventional AAV encapsulated scaffolds. In conclusion, we have developed an efficient vehicle for localized AAV delivery that has considerable potential as a new biomaterial for tissue engineering and regenerative medicine. In addition, recent studies which further improves the polyelectrolyte system through catechol functionalization will be presented.

472245 Alginate/Chitosan Microparticles for Gastric Passage and Intestinal Release of Therapeutic Protein Nanoparticles

Abstract:

Proteins have significant potential to treat diseases due to their specific mechanisms of action compared to small molecule drugs. Protein nanoparticles offer a means to turn soluble protein, which has very low cellular uptake, into a colloidal protein carrier capable of enhanced uptake and subsequent intracellular activity. Previous work produced a stable therapeutic nanoparticle formulation encapsulating a bacterial effector protein, AvrA, which is capable of modulating inflammatory signals in an in vivo inflammatory bowel disease (IBD) mouse model. A major drawback is that these nanoparticles are instilled transrectally, limiting delivery to lower regions of the colon. To implement our protein nanoparticles as a clinically viable treatment for IBD with access to the entire intestinal tract and improved patient compliance, an oral formulation must be designed.

Localized intestinal delivery of protein therapeutics via the oral route is a highly attractive approach, however, the harsh acidic and enzymatic conditions in the gastrointestinal tract challenge retention of protein function. The approach is to deliver protein nanoparticles within gastro-protective microparticles made from alginate and chitosan that would release the nanoparticles in the intestine. A flow focusing microfluidic device was used to control alginate droplet diameters by tuning the continuous and dispersed phase flow rates. Alginate droplets encapsulating protein nanoparticles were then simultaneously crosslinked with calcium and layered with chitosan to form uniform, coated microparticles with high loading efficiency. Protein nanoparticles encapsulated within crosslinked alginate/chitosan microparticles retained approximately 50% of their activity after 1 hr incubation in simulated gastric fluid (SGF). Subsequent incubation in simulated intestinal fluid (SIF) released all remaining active protein within 3 hrs. Alginate vs. chitosan droplets, chitosan concentration, alginate concentration, number of layers, and crosslinker type were tuned to optimize protein protection in SGF and controlled release in SIF. Altogether, these hydrogel microparticles could serve as a potential oral delivery vehicle for protein therapeutics.

458095 Photodynamic Enhancement of Drug Delivery to Tumors

Abstract:

Pancreatic cancer is associated with one of the poorest prognosis of all malignancies, with an unchanged five- year survival rate of 5% or less. The complex desmoplastic microenvironment impeding effective drug delivery to pancreatic tumors and the compensatory ability of cancer cells to adapt alternative survival pathways suggest that treatments will need to attack several targets. Photodynamic therapy (PDT) may have a specific role in enhancing drug delivery to pancreatic tumors and ameliorating drug resistance mechanisms by the targeted destruction of specific pathways. We have developed a bi-directionally cooperative combination regimen of PDT and irinotecan for pancreatic cancer using two clinically approved, nanotechnology-enabled therapies with a clear molecular basis. In an orthotopic pancreatic cancer mouse model, a combination of photodynamic therapy (PDT) and FDA-approved liposomal irinotecan (nal-IRI, MM-398) synergistically reduced the primary and metastatic tumor burden, and prolonged animal survival. This presentation will discuss how PDT increases tumor permeability and irinotecan transport by several mechanisms including disruption of efflux pumps, vascular and stromal integrity. Our findings provide a framework for the translation of mechanistic insights into rational new therapies for several cancers. This study is supported by the following NIH grants: 1K99CA194269 (HH); 5K99CA175292/ 4R00CA175292 (IR); 4R01CA15841505, 5R01CA16099805 5P01CA08420312 (TH). JL was supported by the Herchel Smith fellowship and Harvard College Research Program.

471199 Highly Efficient Encapsulation of Small-Molecule N-Acetylcysteine within PLGA Nanoparticles to Restore Redox Balance in Oxidant-Stressed Environments

Abstract:

Introduction: Overproduction of reactive oxygen species (ROS) is a major factor in the pathogenesis of multiple sclerosis[1]. Localized antioxidant delivery is therapeutically relevant to counteract and neutralize ROS in such oxidant-stressed systems. Poly(lactic-co-glycolic acid) (PLGA) nanoparticles can provide a localized, controlled release of encapsulant, though reported encapsulation efficiencies of small-molecule therapeutics have been low: 12%[2] and 16%[3] for dopamine and N-acetylcysteine (NAC), respectively. Here we investigate how modifying the traditional double-emulsion, solvent evaporation method by doping external water phases with the small-molecule drug affects encapsulation efficiency, temporal release profiles, released drug activity, and released drug protective effect on oligodendrocyte progenitor cell (OPC) viability after H_2O_2 stress. We aim to deliver NAC, a potent, cheap, FDA-approved antioxidant, with high efficiency and temporal control.

Materials and Methods: NAC-loaded PLGA nanoparticles with varying lactide:glycolide (L:G) ratios and inherent viscosities were prepared by the double-emulsion, solvent evaporation method. The solvent for all water phases was deionized water. Briefly, the internal water phase (10% w/v NAC) was dispersed via sonication within the oil phase (15% PLGA in dichloromethane). The primary emulsion was dispersed in the outer water phase (1% PVA with or without 2.5% NAC) via sonication. The secondary emulsion was stirred in 0.3% PVA with or without 5% NAC for 4 hours. Particles were washed 3x in either DI water or 2.5% NAC and lyophilized. Release was determined by incubating particles at 37°C in phosphate-buffered saline and measuring supernatant NAC via Ellman's Reagent. LAP photoinitiator radical concentrations were measured via luminol to determine the dose response of NAC radical scavenging. OPC viability in response to oxidative conditions with and without the NAC antioxidant was measured via the CellGlo reagent.

Results and Discussion: Encapsulation efficiencies (relative to theoretical load without doped external phases) for NAC-loaded PLGA nanoparticles were greatest with external phase doping – reading greater than 300% of the case when undoped. With the doped particles, NAC was released for 14 days. NAC ranging from 0.5-1000 µM (the particle supernatant concentration range for each sample time) scavenged photoinitiated LAP radicals in a dose-dependent manner, with 0.5 µM and 1000 µM NAC scavenging 6.82% and 95.5% of radicals, respectively. Ongoing experiments show NAC-loaded nanoparticle antioxidant activity and protective effect on OPC viability.

Conclusions: Drug-doping of external water phases during the double emulsion, solvent evaporation method gave NAC-loaded PLGA nanoparticles with significantly higher encapsulation efficiencies than without doping, and up to 20-fold higher than previously reported values. NAC scavenged LAP radicals in a dose-dependent manner and was released from PLGA nanoparticles for at least two weeks. This indicates that these particles will provide a larger and more continuous small molecule dose longer than previously reported methods. Efficiently loaded antioxidant nanoparticles capable of controlled and localized delivery could have therapeutic potential in oxidant-stressed systems and bias progenitor cells such as OPCs toward a self-renewing fate.

References: [1]Lehnardt S. ***J. Neurosci,*** 2002. 22:2478-86. [2]Pahuja R. ***ACS Nano,*** 2015. 9:4850-71. [3]Lancheros R. ***Univ. Sci.,*** 2014. 19:161-168

472474 Fabrication and Characterization of PLGA Nanoparticle-Bacteria Conjugate Nanobeads for Increased Tumor Penetration

Abstract:

Cancer is the second cause of overall mortality in the world and systemic chemotherapy is a major therapeutic approach for nearly all types and stages of cancer. Grand challenges of limited efficacy and systemic toxicity of existing chemotherapeutic treatments can be addressed through better drug carrier targeting and enhanced penetration of drug carriers. Attenuated strains *Salmonella* Typhimurium have previously been shown to have the ability to preferentially colonize tumor tissues. However, clinical success in immunocompetent hosts has been rare due to insufficient tumor colonization by bacteria[1–3]. We hypothesize that a combinatorial therapy approach based on integrating bacteria with chemotherapeutics-loaded nanoparticles will amplify both forms of therapy. Thus, we have developed a Nanoscale Bacteria-Enabled Drug Delivery System (NanoBEADS)[4] in which the functional capabilities of bacteria are interfaced with and augmented by chemotherapeutic-loaded nanoparticles. Each NanoBEADS agent is constructed by interfacing a *Salmonella* Typhimurium VNP20009 (i.e. a live engineered tumor-targeting bacterium) with an ensemble of poly(lactic-co-glycolic acid) nanoparticles loaded with chemotherapeutics for the purpose of using the bacterium to transport the particle deep into the tumor. To create the NanoBEADS, *S.* Typhimurium VNP20009 bacteria were cultured in Luria-Bertani broth (1% tryptone, 0.5% yeast extract and 0.5% sodium chloride) and coated with biotin-conjugated rabbit polyclonal anti-*S.* Typhimurium antibody. The nanoparticles were fabricated using a modified nanoprecipitation method[5] and fluorescently tagged with 6,13-Bis(triisopropylsylylethynyl)pentacene (TIPS pentacene). With a size of approximately 110 nm in diameter and a zeta potential of about -30 mV, the carboxyl end groups allow for EDC coupling to streptavidin. The biotin-coated bacteria were then mixed with streptavidin coated nanoparticles for 30 minutes at room temperature to promote self-assembly of the particles to bacteria. Scanning electron microscopy (SEM) images of NanoBEADS show the average number of nanoparticles attached to each bacterium to be 35. In-vitro assays were carried out to compare the penetration of passively diffusing PLGA nanoparticles and NanoNBEADS in multicellular tumor spheroids.

1. Heimann, D. M. & Rosenberg, S. A. Continuous intravenous administration of live genetically modified salmonella typhimurium in patients with metastatic melanoma. *J. Immunother.* **26,** 179–80 (2003).

2. Toso Gill, V., Hwu, P., Marincola, F., Restifo, N, Schwartzentruber, D, Sherry, R, Topalian, S, Yang, J., Stock, F., Freezer, L., Morton, K., Seipp, C., Haworth, L., Mavroukakis, S., White, D., MacDonald, S., Mao, J., Sznol, M., and Rosenberg, S., J. & Toso, J. Gill, V., Hwu, P., Marincola, F., Restifo, N, Schwartzentruber, D, Sherry, R, Topalian, S, Yang, J., Stock, F., Freezer, L., Morton, K., Seipp, C., Haworth, L., Mavroukakis, S., White, D., MacDonald, S., Mao, J., Sznol, M., and Rosenberg, S. Phase I study of the intravenous administration of attenuated Salmonella typhimurium to patients with metastatic melanoma. *J Clin Oncol.* **20,** 142–152 (2002).

3. Pawelek, J. M., Low, K. B., Bermudes, D. & Pawelek, J. M., Low, K. B., Bermudes, D. Bacteria as tumour-targeting vectors. *Lancet Oncol.* **4,** 548–556 (2003).

4. Traore, M. A., Damico, C. M. & Behkam, B. Biomanufacturing and self-propulsion dynamics of nanoscale bacteria-enabled autonomous delivery systems. *Appl. Phys. Lett.* **105,** 173702 (2014).

5. Niu, X., Zou, W., Liu, C., Zhang, N. & Fu, C. Modified nanoprecipitation method to fabricate DNA-loaded PLGA nanoparticles. *Drug Dev. Ind. Pharm.* **35,** 1375–83 (2009).

460403 Interstitial Release of Cisplatin from Triggerable Liposomes Enhances Efficacy Against Triple Negative Breast Cancer Solid Tumor Analogues

Abstract:

In solid tumors, including Triple Negative Breast Cancer (TNBC), limited penetration of therapeutics leads to heterogeneous intratumoral distributions reducing efficacy. Liposomal cisplatin decreases toxicities associated with free cisplatin. However, the heterogeneous intratumoral distributions of liposomes combined with limited release of cisplatin from liposomes may also reduce efficacy. This study uses pH-responsive liposomes designed to release cisplatin within the acidic tumor interstitium (7.0 < pH â‰¤ 6.0) with the aim (1) to improve the penetration of the free drug within tumors on the assumption of greater diffusivities based on the free drug's much smaller size and (2) to increase the availability of the free agent near cancer cells.

Introduction: In advanced solid tumors, including Triple Negative Breast Cancer (TNBC) solid tumors, the limited penetration of therapeutics results in limited therapeutic effects. TNBC is a subgroup of breast cancer associated with a high chance of cancer reoccurrence outside the breast and has been found to exhibit sensitivity to DNA-damaging platinum-based compounds, which is the basis for increased clinical use of platinum-derived agents. To enable selective and effective treatment of TNBC solid tumors, we study a drug delivery carrier of CDDP that is expected to result in potentially deep penetration and homogeneous distribution of the drug within the tumor interstitium. The carrier is a tunable (pH-releasing) liposome encapsulating CDDP. Tumor selectivity by these liposomes is mediated by the EPR effect. Efficacy is mediated by the pH-triggered release of contents within the tumor interstitium (7.0 < pH < 6.5) by extravasated liposomes. The small size of CDDP relative to its nanometer-sized carrier should exhibit greater diffusivities and deeper tumor penetration.

Methods: To study the efficacy of these liposomes on TNBC cells two different TNBC cell lines were studied: MDA-MB-231 and MDA-MB-468. The efficacy of pH-releasing liposomes (LD_{50}) loaded with CDDP was compared to the efficacy of other constructs including conventional (non-p-releasing) liposomes, empty liposomes, and free cisplatin. Following incubation of 6 and 24 hours, in addition to cell viability, cisplatin per cell, cisplatin intercalated to DNA and (if applicable) retention of cisplatin within the liposomes, were measured. In multicellular spheroids exhibiting intracellular pH gradients measured in situ by confocal microscopy using a membrane impermeant pH-indicator, we evaluated the spatiotemporal distributions of liposomes and contents. In dose-response studies, upon completion of incubation with all constructs, spheroids were washed and their volume change (volume at time t/volume at initiation of treatment x100%) was monitored in time followed by an outgrowth assay.

Results: On MDA-MB-231 and MDA-MB-468 cell monolayers, extracellular acidification resulted in lower LD_{50} values only for pH-releasing liposomal cisplatin (101.9 ± 24.1 and 14.5 ± 3.6 μg/mL, respectively) which were significantly less than for non-pH-releasing liposomal cisplatin (374.2 ± 20.0 and 33.7 ± 5.0 μg/mL, respectively). In contrast, at neutral pH, the LD_{50}values of both liposome types were comparable. Killing efficacies correlated strongly with the nuclear DNA-associated cisplatin independent of the delivery modality or the extracellular pH. pH-releasing liposomal cisplatin decreased the volumes of multicellular spheroids by 46.4-70.0 % more than non-pH-releasing liposomal cisplatin. pH-releasing liposomes resulted in improved penetration and more homogeneous profiles within spheroids of a cisplatin-surrogate cell-permeant fluorophore compared to non-pH-releasing liposomes.

Conclusions: These findings demonstrate the potential of intratumorally CDDP-releasing nanometer-sized carriers to selectively and effectively control TNBC solid tumors.

Funding Sources:

American Cancer Society (ACS): RSG-12-044-01

National Science Foundation (NSF): DMR-1207022

AAUP-AFT Professional Development Fund

457602 Bioresponsive Polymer Coating on Targeted Drug Nanorods

Abstract:

Non-specific delivery and ineffective drug release at the target site are among the top challenges to effective treatment. The first challenge reflects the fact that interactions with carriers (*i.e.*, nanoparticles) can denature the active ingredients of drugs by shear stress. The second challenge is low affinity for specific receptor proteins overexpressed on the target cell surface. The third challenge of ineffective drug release can be overcome by exploiting the disease microenvironment. Here, we report a one-step process to prepare bioresponsive polymer coated drug nanorods (NRs) from liquid precursors using the solvent diffusion method. A thin layer (10.3 ± 1.4 nm) of poly(e-caprolactone) (PCL) polymer coating was deposited on the surface of camptothecin (CPT) anti-cancer drug NRs. The mean size of PCL-coated CPT NRs was 500.9 ± 91.3 nm length x 122.7 ± 10.1 nm width. The PCL polymer coating was biodegradable at acidic pH 6 as determined by Fourier Transform Infrared (FT-IR) spectroscopy. CPT drugs were released up to 51.5% when PCL coating dissolved into non-toxic carboxyl and hydroxyl groups. Trastuzumab (TTZ), a humanized IgG monoclonal antibody, was conjugated to the NR surface for breast cancer cell targeting. Combination treatments using CPT and TTZ decreased the HER-2 positive BT-474 breast cancer cells by 66.9 ± 5.3 % *in vitro*. These results suggest highly effective combination treatments of breast cancer cells using bioresponsive polymer coated specific drug delivery.

466655 Internalization of Nanoparticles Functionalized with Low Molecular Weight Protamine into Erythrocytes

Abstract:

Erythrocytes have been exploited extensively for their potential applications as delivery vehicles due to their remarkable biocompatibility, biodegradability and prolonged life-span in circulation. The mature human erythrocyte is a terminally differentiated cell that does not possess an endocytic uptake mechanism. As a result, the efficient migration of bioactive molecular cargos such as large macromolecular drugs, genetic material, or proteins through the plasma membrane of erythrocytes remains as a major challenge for intracellular delivery applications. A variety of techniques have been successfully developed to entrap proteins, drugs, and enzymes within erythrocytes. The majority of these techniques, however, require disruption of the cell membrane to create large pores or perturbations through which biomolecules can diffuse. These techniques include chemical methods such as drug-induced endocytosis, mechanical methods such as hypotonic osmosis/dialysis, and electrical methods such as electroporation. The main disadvantages associated with these methods are the loss of structural integrity and cellular components of the cells due to weakened cell membranes. As a consequence, these modified erythrocytes are easily recognized by the host immune system. To achieve the benefits associated with erythrocytes as a drug delivery vehicle it is vital to maintain complete structural and functional integrity of erythrocytes.

Toward this goal, we introduced a novel method for transporting nanoparticles (NPs) into intact erythrocytes by combining the advantages of both cell-penetrating and targeting peptides. The method employs an ERY1 (WMVLPWLPGTLDGGSGCR) peptide that has a high affinity for the erythrocyte membrane protein glycophorin A and a low molecular weight protamine (LMWP) peptide that has been shown to transport proteins and small molecules into erythrocytes. These peptides are conjugated to a NP formed between fluorescently-labeled bovine serum albumin and a grafted copolymer of poly(ethylene) glycol and poly-L-lysine.

The formation of nanoparticles was observed using SEM, TEM, and dynamic light scattering. The nanoparticles had a core/shell structure and their size ranged from approximately 20-100 nm. Confocal microscopy and flow cytometry data confirmed that uptake of LMWP functionalized NPs was significantly higher (approximately > 3-5 fold) than that of other NPs at 6 h. Under these conditions, synthesized NPs were not toxic to the erythrocytes and no significant hemolysis was observed. Catalytic activity of the encapsulated protein was not negatively affected upon encapsulation or crosslinking. In conclusion, NPs were efficiently internalized into RBCs without adversely affecting cellular function and this method has the advantage over existing methods in that it does not appear to alter the structural integrity of the cell. The above approach of utilizing a surface modified NP architecture to improve internalization into erythrocytes may serve as a universal platform to deliver therapeutic proteins/drugs into RBCs in future. Furthermore, this nanoparticle system capable of entering red blood cells and extending circulation time has enormously broad potential, including improved treatment of blood-specific diseases such as malaria as well as treatment of acute lymphoblastic leukemia.

466720 Enzyme Replacement Therapy Extended to the Brain through Nano-Polymersomes

Abstract:

Approximately 1 in 5,000 to 8,000 children are born annually with a lysosomal storage disease (LSD). Current treatment of LSDs, called enzyme replacement therapy (ERT), involves weekly to monthly intravenous(IV) infusions of missing endogenous enzymes. While ERT has proven effective in LSDs without central nervous system (CNS) involvement, the brain has remained untreatable due to the presence of the blood-brain barrier (BBB). The BBB is made up of endothelial cells with tight junctions, possessing a width between connections of 0.2 nm in size, which prevents passage of 98% of small molecule drugs, including enzymes, from the blood into the brain. Patients with neuropathic LSDs, which account for 50-70% of diseases in this category, present with severe CNS degeneration, ataxia, and premature death, with no treatment available on the market. The disease of focus, GM1 gangliosidosis, causes premature death between the ages of two and four with no hope available for parents of patients. Encapsulating enzyme into polymeric vesicles, called polymersomes, facilitates the transport of this missing enzyme into the brain through IV injections, effectively extending ERT therapy to the brain for the first time.

Polyethylene glycol-b-poly(lactic acid) (PEGPLA) has been proven to self-assemble into vesicles via both passive mixing and the solvent injection, using dimethyl sulfoxide (DMSO) and water. The mixing enzyme in GM1 gangliosidosis, β-galactosidase (βgal), is fluorescently tagged with reactive Alexa Fluor 488 at an efficiency of 13 dye molecules per mole of protein, with a finally protein concentration of 0.35 mg/mL attainable. Fluorescently tagged βgal is dissolved in stirring water prior to injection of DMSO and PEGPLA. Apolipoprotein E (ApoE) and CF350 attachment is facilitated through the introduction of homobifunctional PEG during injection. Polymersome diameter is characterized by dynamic light scattering (DLS). Self-assembly was stopped by freezing prior to lyophilization, with mannitol as a lyoprotectant. βgal loaded polymersomes were released *in vitro* in buffers of pH 7.4 and 4.8. Attachment was confirmed with DLS and Lowry assays. An immortalized feline fibroblast line, GM1/SV3, was used in *in vitro* studies to confirm particle uptake and treatment efficacy.

Previous studies indicated that incorporating 2 wt%/v mannitol during formation led to PSDs 54 ± 0.8% smaller after lyophilization. Therefore, 2 wt%/v mannitol was used to aid in the formation of brain-deliverable polymersomes encapsulated with βgal. PEGPLA polymersomes loaded βgal at 0.08 ± 0.03 mg βgal/mg polymersome with activity maintained at a value of 452 ± 82 a.u. DLS data indicates that polymersomes with βgal had diameters statistically similar to control. Release of βgal was doubled over 6 hours when loaded polymersomes are placed into citrate buffer (pH=4.1) when compared to tris buffer (pH=7.4), indicating increased enzyme release expected in the lysosome. First order kinetics of release were followed for three hours when βgal was released from polymersomes in acidic buffer prior to a plateau in mass released, indicating the potential for sustained long term release from polymeresomes. Activity of βgal was maintained after release. CF350 Amine was attached to polymersome surface, with an increase in diameter from 147.2 ± 24 nm to 210.5 ± 8.6 nm (Fig 1A). Fluorescence microscopy confirmed attachment (Fig 1C), with 79 ± 21% of polymersomes labeled with blue fluorescence. Furthermore, our carriers were capable of both encapsulating Alexa Fluor 488 and attaching CF350 at a high efficiency, with 87 ±12% of polymersomes presenting with both colors in flow cytometry. Attachment of ApoE was confirmed with an increase in polymersome diameter to 194.4 ± 3.5 nm and an increase in protein content from 0.06 ± 0.11 mg/mL to 1.03 ± 0.71 mg/mL. GM1SV3 cells, an immortalized cell line from felines with GM1 gangliosidosis, present low density lipoprotein receptors (LDLR) on the surface, as shown in confocal microscopy, indicating that they are a good model for ApoE mediated delivery through the BBB. ApoE increases the uptake of polymersomes into GM1/SV3 fibroblasts when compared to unlabeled control polymersomes.

This work highlights our ability to create polymersomes capable of protecting βgal in the blood stream, shown by limited release in a pH 7.4 environment, targeting the nanocarrier to the BBB, shown by the facilitated attachment of ApoE, and after transcytosis, releasing βgal in the lysosome of neural cells, shown by extended release in a pH 4.1 enviornment and cell studies in GM1SV3 cells. This combination of ERTand nanotechnology, demonstrating the capability of our carrier to transport enzymes to the brain while maintaining their activity, will create a paradigm shift in the treatment of CNS disease, providing treatment for currently untreatable and fatal diseases like GM1 gangliosidosis.

465081 Computational Tools for Studying Peptide Based Templating of Novel Materials (Invited Talk)

Abstract:

Peptide-based templating of bio/nanomaterials offers a low-energy and potentially low cost synthesis route to materials with exquisite morphology and properties (e.g., mimicry of the biosilica of diatoms). This talk will discuss recent applications of new simulation methods based on the metadynamics family of enhanced sampling methods to the study of interactions of peptides with hard interfaces as well as the guided self-assembly of peptides in solution. Specific applications include the controlled synthesis of various amorphous biosilica morphologies with amphiphilic peptides, investigation of pH-induced changes in the adsorption properties of small peptides on silica, and new computational methods to calculate study peptide unbinding kinetics and adsorption thermodynamics.

473377 Simulation of Biological and Nanostructured Interfaces to Discover New Materials

Abstract:

The mechanism of specific adsorption of polymers and biomacromolecules onto metallic and oxidic nanostructures will be explained in atomic resolution resulting from simulations with novel force fields and surface models in comparison to measurements. Variations in peptide adsorption on Pd and Pt nanoparticles depending on shape, size, and location of peptides on specific bounding facets are determined by soft epitaxial processes and induced charges. Predictions of specific nanocrystal growth and shape development are possible using computation and experiment and will be illustrated by examples. Also, computational estimates of reaction rates in C-C coupling reactions and in olefin hydrogenation will be shown using particle models derived from HE-XRD and PDF data, which illustrate the utility of computational methods for the rational design of new catalysts. On oxidic nanoparticles such as silica and apatites, it is shown how changes in pH lead to similarity scores of attracted peptides lower than 20%, supported by model surfaces of appropriate surface chemistry and data from adsorption isotherms. The results demonstrate how new computational methods can support the design of nanostructures, hydrogels and drug delivery vehicles, as well as the understanding of calcification mechanisms in the human body. The main features of the INTERFACE force field for accurate simulations of inorganic/organic and inorganic/biological interfaces will be discussed and explained by means of examples.

466234 Dissipative Particle Dynamics Studies of pH-Sensitive Tri-Block Copolymer Containing Zwitterionic Sulfobetaine As a Novel Anti-Cancer Drug Carrier

Abstract:

Dissipative particle dynamics (DPD) simulations were performed to study the microstructures and doxorubicin (DOX) loading/releasing in pH-sensitive amphiphilic triblock copolymer, composed of biodegradable hydrophobic poly(ε-caprolactone) (PCL), pH-sensitive poly (diethylaminoethyl methacrylate) (PDEA) and hydrophilic poly (sulfobetaine methacrylate) (PSBMA) / poly (ethylene glycol methacrylate) (PEGMA). Simulation results show that both PCL-PDEA-PSBMA and PCL-PDEA- PEGMA system would self-assemble into core-shell-corona micelles; with increasing copolymer concentration from 10% to 50% under the mass ratio of copolymer to drug as 10:3, the morphologies formed by PCL-PDEA-PSBMA and DOX remains spherical micelles even though adhesion at a certain direction is observed at high concentration, but PCL-PDEA-PEGMA undergoes the transition from spherical to cylindrical and finally to lamellar micelles. Though both two copolymer systems can self-assemble into core-shell-corona micelles under a proper condition, the inner structures are quite different. The shell layers formed by PEGMA micelles are inhomogeneous in size due to the amphiphilicity of PEG; while shell layers in PCL-PDEA-PSBMA micelles are homogenous because of strong hydrophilicity of the zwitterionic PCB. The phenomenon indicates that PSBMA is more stable than PEGMA. In particular, the micelles exhibit pH dependency as a result of the protonation of the PDEA block resulting in the instantaneous releasing of DOX. What's more, the releasing process follows the swelling-demicellization-releasing mechanism. In order to moderate the micelle's drug loading efficiency, in vivo stability and drug release effect, PCL-PDEA-PSBMA/PEGMA copolymer with the block ratio at 20:20:20, while the concentration remains 10% and DOX remains 3% is considered to be the best. The integration of simulation might be a valuable method for the optimization and design of biomaterials for drug delivery with desired properties.

461468 Investigation of DNA Conjugated Hydrogel Networks Using Discontinuous Molecular Dynamics Simulations

Abstract:

Hydrogels are three-dimensional, cross-linked networks of water-soluble polymer chains. They are used in a variety of applications including diapers, cosmetics, and wound dressings. The features that make hydrogels useful for these types of applications are their high water content, biocompatibility, and ability to store molecules within their networks. In order to better understand how hydrogels store molecules, we have developed a coarse grained model of a hydrogel for use with discontinuous molecular dynamics simulations (DMD). DMD is an efficient technique that allows molecular-level simulations of larger systems at longer timescales than traditional molecular dynamics. We are using DMD simulations to optimize design parameters for the preparation of hydrogels that can be used in drug delivery applications. We model two types of 4-armed poly(ethylene glycols) (PEG-A and PEG-B) precursors covalently functionalized on all ends with two different oligonucleotides and a third oligonucleotide chain (aptamer cross-linker) which is complementary to the sequence of the oligonucleotides in PEG-A and PEG-B. We also model adenosine molecules as single coarse grained molecules with strong attractive interaction toward guanines; adenosine is used as a break-down agent in the hydrogels. Preliminary simulation results starting from an initially uncrosslinked system show that the PEG-A, PEG-B, and cross-linker oligonucleotide bind to each other and form a network due to their Watson-Crick pairing. Once the hydrogel has been formed, it can be dissociated by adding adenosine molecules. Adenosines compete with cytosines to bind guanines, and the decrease of actual guanine-cytosine bindings due to this competition weakens the interactions between the oligonucleotides in PEGs and cross-linker, allowing the dissociation of the hydrogel network. This series of simulations shows that the DNA conjugated hydrogels potentially have drug loading and releasing abilities suitable for use in medical applications.

464471 Understanding Titanium Dioxide Surface Chemistry for Biomolecule Adsorption: Experiments and Molecular Dynamics Simulations

Abstract:

The behavior of surface water, especially the adsorption and dissociation characteristics, is a key to understanding and promoting photocatalytic and biomedical applications of titanium dioxide materials. A knowledge of these effects and a strategic optimization are important in many applications: examples include biochemical sensing and separation, cell imaging, *in vitro* and *in vivo* targeted drug delivery, cancer therapy, toxic species removal from air, remediation of soils and ground water. Using molecular dynamics simulations with the ReaxFF force field and experimental techniques, we have studied the following systems to understand how surface chemistry affects adsorption of biomolecules at TiO_2 surfaces: (a) water/TiO_2; (b) water/amino acid/TiO_2; (c) bovine serum albumin/water/TiO_2. The results show that TiO_2 surfaces demonstrate different reactivity for water dissociation. The resulted TiO_2 surface chemistry change, the surface hydroxyl groups due to water dissociation, significantly affects the interaction of amino acid and bovine serum albumin with TiO_2 materials.

1. Liangliang Huang, Keith E. Gubbins, Licheng Li and Xiaohua Lu, “Water on Titanium Dioxide Surface: A Revisiting by Reactive Molecular Dynamics Simulations”, *Langmuir*, 2014, **30**, 14832.
2. Chang Liu, Yanhua Guo, Qiliang Hong, Chao Rao, Haijuan Zhang, Yihui Dong, Liangliang Huang, Xiaohua Lu, and Ningzhong Bao, “Bovine Serum Albumin Adsorption in Mesoporous Titanium Dioxide: Pore Size and Pore Chemistry Effect”, *Langmuir*, 2016, **32**, 3995.

469536 Combining Simulation and Spectroscopy to Determine the Structure and Orientation of a Carbohydrate Binding Module (CBM) Inspired Model Peptide on Cellulose

Abstract:

Interfacial processes, such as the enzymatic conversion of an insoluble polysaccharide at the solid-liquid interface, are critically important to the design of new systems to convert renewable resources into green fuels and chemicals. Despite this fact, we often lack a molecularly detailed understanding of these interfaces that hinders the rational design of new and/or improved biocatalytic systems. Atomistic simulations of biomolecular adsorption using state-of-the-art multiscale molecular modeling tools have strong potential to combine with advanced, surface-specific spectroscopy to elucidate structural and mechanistic details at the interface. To this end, enhanced sampling methods based on the metadynamics family of methods have been combined with experimental data from sum frequency generation spectroscopy (SFG) to probe the equilibrium structure(s) and preferred orientation(s) of a 14-residue peptide fragment inspired by the carbohydrate binding module (CBM) of a cellulase. Experimental SFG data indicates the CBM fragment adsorbs to the cellulose surface and aligns along the fibers in a distinct orientation/conformation. We use a variety of molecular simulations to provide molecular level insight to this observation and add additional detail beyond the amide I backbone spectra, thus providing new mechanistic insights regarding biomass conversion. Additionally, simulation results are used to determine the energetic penalties for binding in alternate orientations, and to identify the specific reasons for the observed behaviors. In a broader sense, this integrated approach of combining computational methods with experiments aims to determine the role of the surface in modulating biomolecular structure, thereby permitting the rational design of new biocatalytic approaches.

469397 Computational Screening of Photoactive Cyclic Peptides for Self Assembly and Disassembly

Abstract:

Cyclic peptides have been widely studied for applications such as targeted therapeutics [1] and biologically compatible nanostructures [2]. The ability to control the assembly of peptide nanostructures in vivo is of interest for applications such as targeted drug delivery. We introduce the potential for photoactive assembly and disassembly to cyclic peptides by adding azobenzene linkers. Azobenzene undergoes a reversible photoinduced transition between an extended trans ground state configuration and a compact cis exited state configuration [3]. We seek cyclic azobenzene-peptide sequences that assemble when the azobenzene linker is in the trans configuration and disassemble when azobenzene linker is in the cis configuration.

The large potential parameter space for amino acid sequences makes experimental screening cumbersome. We computationally screened cyclic azobenzene-peptide sequences by computing dimer binding free energies for azobenzene-peptide sequences using well-tempered metadynamics [4]. The peptide portion is modeled using the CHARMM 36 force field. Modeling the azobenzene linker required the development of a new force field. We use the CHARMM general force field [5] to fit parameters for the ground state trans azobenzene linker and the exited state cis azobenzene linker. Separate simulations for the dimer binding free energy with trans and cis azobenzene are performed. Peptide sequences of varying length, side chain size, and side chain hydrophobicity are screened to find dimers that are stable when the azobenzene linker is in the trans configuration, and unstable with the azobenzene linker in the cis configuration.

References

[1] A. Tapeinou, M-T Matsoukas, C. Simal, T. Tselios, Biopolymers, **2015**, 104, 453â€“461.

[2] D. Mandal, A. N. Shirazi, K. Parang, Org. Biomol. Chem., **2014**, 12, 3544-3561.

[3] C. Renner, L. Moroder, ChemBioChem, **2006**, 7, 868-878.

[4] A. Barducci, G. Bussi, M. Parrinello, Phys. Rev. Lett., **2008**, 100, 020603.

[5] K. Vanommeslaeghe et al., J. Comp. Chem., **2010**, 31, 671-690.

454417 A Conformational Analysis of an Engineered Laminin-Mimetic, Elastin-like Fusion Protein Using Molecular Dynamics Simulations

Abstract:

Introduction: In the central nervous system, laminin is a vital component of the extracellular matrix that is responsible for a number of cellular functions such as adhesion, migration, survival, and differentiation, and is involved in early embryonic basement membrane assembly. The fifth globular domain located at the C-terminus of the laminin α2 chain (LG5) module plays a major role as a heparin and α-dystroglycan (α-DG) binding site[1]. Heparin is a highly anionic, sulfated glycosaminoglycan that helps bind exogenous growth factors that regulate and maintain neural stem cell differentiation[2]. In neural cells, the α-DG glycoprotein complex plays a fundamental role in facilitating new laminin polymerization at the cell surface and supporting cellular adhesion[3]. LG5 also has a biologically active site associated with the binding to integrin β1, which is the most abundant and multifunctional integrin subunit of laminin receptors. The LG5 domain provides the functional core of our simulated protein that regulates cell-protein-matrix interactions, whereas the self-assembling structural module of the protein is composed of a designer elastin-like polypeptide (ELP) sequence. ELPs undergo a thermally-triggered inverse phase transition where they phase separate at high temperatures[4]. This reversible lower critical solution temperature behavior can be imparted to soluble proteins fused to an ELP[5]. Due to their thermal responsive behavior, recombinant ELP fusion proteins can be easily purified using inverse transition cycling[6], vastly reducing the need for expensive chromatographic resins and improving scalability for large-scale production. In order to expand on new opportunities of exploiting the self-assembling properties of these ELPs as a route to developing novel biomaterials, a more meticulous investigation into the assembly process, as well as the interfacial properties of these proteins are needed.

Methods: We adopted the crystal structure 1DYK[7] for the LG5 domain. The ELP$[K_2I_2L_2K_2]_1$ sequence was built with the program Avogadro[8] using the peptide builder tool. We assumed an alpha-helix starting structure with Φ = -60° and Ψ= -40° angles. Atomistic MD simulations were performed using NAMD 2.10[9] and the CHARMM36 force field for proteins. Temperatures were maintained using a Langevin thermostat. The production times for the simulations were 100 ns.

Results and Discussion: The first discrete event in protein folding within the elastin-like polypeptide (ELP) domain is the loss of the α-helix secondary structure and a transition to β type secondary structures. At temperatures below 305 K, we observed a collapse of the ELP domain with respect to its initial starting conformation. At 310 K, there is a noticeable transient unfolding of the ELP domain starting at ~75 ns, highlighted by the loss of intra-strand hydrogen bonds. The sudden appearance of intra- β-strands at 310 K also suggests an inverse phase transition at that temperature. This distinct folding transition to higher ordered structures has been shown to predict self-assembly in *vitro* observed by other groups[10]. Biomaterials incorporating this self-assembling property allow them to be highly dynamic, and can self-heal following external shear forces. An analysis into the hydration properties of the ELP domain reveals that as the temperature increases, the number of surrounding water molecules decreases and the number of peptide-peptide hydrogen bonds increases. There is a slight decrease in number of surrounding water molecules over time in all simulated temperatures, however, there is an abrupt reduction of water molecules at 64 ns and 82 ns at 310 K. This change in the hydration structure is correlated with the formation of β-sheets within the trajectory.

Conclusions: We introduce a realistic design strategy that could transform current approaches to biomaterials development and address the particular challenges of characterizing the dynamic processes that occur in biological matrices. Our simulations provide both an immediate, high-resolution view of the conformational dynamics of our system, as well as a robust and extensible platform for future work directed toward our overarching goal of creating self-assembling, protein-based biomaterials as candidates for cell-delivery therapeutics.

References: [1] Urushibata, S. et al., *Arch Biochem Biophys* 2010, 497, (1-2), 43-54. [2] Nugent, M. A. et al., *Int J Biochem Cell B* 2000, 32, (2), 115-120. [3] Muntoni, F. et al., *Neurotherapeutics* 2008, 5, (4), 627-632. [4] Long, M. M. et al. *Biophys. J.* 1975, 15, (2), A71-A71. [5] Christensen, T. et al., *Biomacromolecules* 2013, 14, (5), 1514-1519. [6] Meyer, D. E. et al., *Nat Biotechnol* 1999, 17, (11), 1112-1115. [7] Hohenester, E. et al., *Molecular Cell* 1999, 4, (5), 783-792. [8] Hanwell, M. D. et al., *J. Cheminfo.* 2012, 4. [9] Phillips, J. C. et al., J Comput Chem 2005, 26, 1781. [10] Dong, H. et al., *JACS* 2007, 129, (41), 12468-12472.

472109 Tuning Physical Properties of Block Copolymers Containing Fatty Acid-Derived Long-Chain Polyacrylates (Invited Talk)

Abstract:

Vegetable oils and their fatty acids are promising replacements for petroleum sources due to their abundance, low cost, lack of toxicity, biodegradability and ease of functionalization that provides convenient routes to polymerization. In this study, the properties of long-chain polyacrylates were explored, derived from fatty acids found in vegetable oils. The thermal and mechanical properties of triblock copolymers containing the long-chain polyacrylates were readily tuned by variation of the acrylate composition and resulting distribution of side-chain lengths. Surprisingly, the alkyl side-chain length of the polyacrylate did not impact the thermodynamic interactions between the components of the triblock copolymers. This provides a route to manipulating the physical properties of the polymers through variation of the side-chain length without impacting the phase behavior and morphology. Shear alignment of the triblock copolymers produced highly aligned close-packed spherical morphologies. The development of structure-property relationships in these polymers will enable the widespread implementation of fatty-acid derived materials.

455754 Enhanced Yield Strength in Polyethylene-Glassy Block Copolymers

Abstract:

Polyethylene is the world's most widely produced synthetic polymer in terms of annual production volume, and has tremendous utility for many applications as a low-cost, easily processable tough plastic. However, polyethylene is poorly suited to certain uses due to the relative ease of permanent deformation under an applied load. For example, linear, high-density polyethylene typically exhibits a yield stress of approximately 30 MPa.[1] In a semicrystalline material, this yield point corresponds to the critical stress at which the crystal stems making up the crystal lamellae dislocate and "slip" in the direction of the applied force. Crystallite dislocation and eventual fragmentation are irreversible, and for most molded plastics this mode of permanent deformation effectively ruins the article.

The liquid-like amorphous fraction (T_g ~ -120 °C) filling the space between polyethylene crystal lamellae is incapable of providing mechanical reinforcement to the crystal fold surface, so the homopolymer's yield stress is dictated largely by the crystal thickness. Slow cooling from the melt or annealing slightly below the melting temperature can be used to increase crystal thickness, allowing for some control over the yield stress.[1] However, these improvements are modest: the yield stress of a 70 kg/mol narrow-distribution linear polyethylene differs by only ~30% between a sample slowly cooled from the melt over the span of three hours versus a sample rapidly quenched from the melt over the span of a few seconds.

Our alternative strategy for controlling the mechanical properties of a semicrystalline polymer is the attachment of a short amorphous block with a high glass transition temperature. Upon crystallization from the melt, the glassy block will be excluded to the amorphous fraction, stiffening the domains between adjacent crystal surfaces. Block copolymers containing a high-T_g block and a perfectly linear polyethylene block can be readily prepared via sequential ring-opening metathesis polymerization of substituted norbornene monomer and cyclopentene, followed by catalytic hydrogenation.[2] The present work makes use of hydrogenated poly(norbornyl norbornene) as the glassy block (T_g = 115 °C), although ongoing work will consider other candidate glassy blocks such as hydrogenated poly(cyclohexyl norbornene) (T_g = 86 °C, though substantially more miscible with polyethylene). We investigate a range of minority glassy block contents, spanning ~10% – 20% by weight, with the molecular weights of these polymers selected to ensure single-phase melts for ease of processing.[3]

By simultaneously varying the glassy block content and the crystal thickness/crystallinity via thermal history, the mechanical properties of the block copolymers can be altered dramatically from those of a polyethylene homopolymer. Polymers with intermediate glassy block content (15%), with crystallinity maximized by slow cooling from the melt in order to further concentrate the glassy block within the amorphous layer, exhibit remarkably large increases in strength and stiffness relative to polyethylene of comparable molecular weight and crystal thickness. These enhancements can be as large as a doubling of the yield stress and a tripling of the Young's modulus, vastly more pronounced than can be achieved by thermal treatment of a polyethylene article alone. These changes can be understood in terms of enhancing the rigidity of the amorphous layer with an increasing concentration of glassy block, and our current work aims to understand the extent of block mixing within the amorphous layer (i.e., phase-separated glassy domains versus a homogeneous glassy block/amorphous polyethylene mixture) in order to explain the mechanical role of the high-T_g material.

The observed increases in strength and stiffness of the glassy-polyethylene block copolymers are typically associated with a decrease in the ultimate strain. Prepared by slow cooling from the melt, linear polyethylene homopolymer exhibits an extension at break of ~1000%. However, under the same thermal history, the block copolymers with 15% glassy block fracture at 45% extension, reduced to 2% extension for the materials with 20% glassy block. The loss of mechanical toughness for these polymers can be explained by a lack of tie molecules spanning the inter-lamellar distance. The most brittle materials were those with the lowest molecular weights (40 kg/mol for the 20% glassy block case), owing to the requirement to avoid microphase separation in the melt. At these low chain lengths, the predicted mean end-to-end distance for a polyethylene chain[4] is significantly lower than the periodic domain spacing measured by small-angle X-ray scattering. Our ongoing work seeks to improve the material toughness by preparing the glassy block from a polymer that is more miscible with polyethylene than hydrogenated poly(norbornyl norbornene). This way, the block copolymer molecular weight can be increased to improve the tie molecule density while retaining a homogeneous melt.

In summary, the yield stress of polyethylene can be substantially increased by incorporation of a short, high-T_g block. Upon crystallization, the glassy block is excluded to the amorphous layer between crystal lamellae and helps to partially immobilize the crystal surface. By slowly cooling from the melt, the crystallinity of these block copolymers can be raised in order to further enrich the fraction of glassy block in the amorphous layer. The molecular weights of these polymers are tuned to avoid microphase separation in the melt, however the lowest molecular weight polymers become brittle under a slow cooling thermal history due to the lack of tie molecules between neighboring crystal lamellae.

This work was generously supported by the National Science Foundation, Polymers Program (DMR-1402180).

References

1. Kennedy, M. A.; Peacock, A. J.; Mandelkern L. *Macromolecules* **1994**, *27*, 5297.
2. Trzaska, S. T.; Lee, L.-B. W.; Register, R. A. *Macromolecules* **2000**, *33*, 9215.
3. Bishop, J. P.; Register, R. A. *Macromolecules* **2010**, *43*, 4954–4960.
4. Fetters, L. J.; Lohse, D. J.; Richter, D.; Witten, T. A.; Zirkel, A. *Macromolecules* **1994**, *27*, 4639.

472118 Structure Direction Via Sequence Control in Block Copolymers

Abstract:

Engineering of polymer-inorganic hybrid materials with structure control from the nanoscale to the mesoscale has remained elusive, especially in comparison to biological composites such as nacre and bone. However, the ability to synthetically design highly crystalline, hierarchically structured, and well-ordered inorganic and composite materials is desired in areas across biomaterials and energy. To this end, block copolymer self-assembly offers a rich diversity of phases with long range periodic order of structures from 0-D to 3-D. While block copolymers have previously been utilized to structure direct a variety of inorganic materials including metals and oxides, such templating typically has relied primarily on spatial confinement. In contrast, proteins and their smaller peptide components provide multiple self-assembly pathways from the folding of the amino acid-based polymer to biomineralization, in which a protein matrix directs crystal phase and orientation of inorganic materials. By combining the long-range periodic structure direction of block copolymers with the increased information content provided by sequence control in proteins, new highly controlled composite materials might be created. Synthesis of block copolymers with sequence controlled domains will be discussed. Through varying charge density and location within the amino acid sequence, we investigate the effect of sequence choice on block copolymer morphology and on the assembly of inorganic components.

466996 A Facile Approach for Simulating Complex Phases in Block Polymers Via Self-Consistent Field Theory

Abstract:

Self-consistent field theory (SCFT) is the primary theoretical framework to study the phase behavior of block polymers. Solving the highly nonlinear self-consistent field equations requires providing guess values of the potential fields that describes the interaction of a single test chain with the mean field created by the other chains in the melt. The equations are then solved iteratively to locate a local free energy minimum. Convergence (or lack thereof) depends strongly on the guess values of the potential fields; the fields should represent the structure well and should be close enough to the desired microphase. We developed a robust, physically informed approach to generate the guess values of the potential fields using information available from experiments: dimensions and type of the unit cell, shape, size and positions of the particles forming the microdomain, and space group symmetry of the microphase. To demonstrate the utility of our approach, we have generated converged SCFT results for the complex Frank-Kasper Ïƒ phase observed in poly(styrene-isoprene-styrene-ethylene oxide) (SISO) tetrablock terpolymer. Furthermore, we compare these SCFT predictions to experimental results for SISO tetrablocks, which report an array of sphere-forming phases: body-centered cubic, face-centered cubic, hexagonally closed packed, and Frank-Kasper Ïƒ and A15 phases.Â

456333 Photoswitchable Nanomaterials Via Ordered Self-Assembly of Monodisperse Oligodimethylsiloxanes

Abstract:

Phase separation in segmented macromolecules such as block copolymers has emerged as a powerful tool for the fabrication of well-ordered nanomaterials. Here, we present a class of monodisperse azobenzene-functionalized oligodimethylsiloxanes (AzoODMS) which self-assemble to form multi-domain nanomaterials with sub-7nm periodicities. By balancing the effects of phase separation and directional endgroup aggregation, we can tailor the nature of self-assembly and obtain materials with drastically different morphologies. We show the importance of these morphologies in conferring bulk material properties. Using the trans-cis photoisomerization of azobenzene motifs, we further demonstrate the reversible photoactivated solid-liquid transition of AzoODMS arranged in bilayers or the rapid motion of AzoODMS arranged in hexagonally-packed cylinders.

469088 Molecular Alignment of Phase-Separated, Segmented Polyurethanes in Thin Films Using Intrinsic Fluorescence

Abstract:

Interfaces in thin films introduce molecular alignment and nanostructure into polymers that are not observed in the bulk state. Segmented polyurethanes, based on a hard- and soft-phase morphological structure, are widely used and have been extensively studied. A number of techniques have been used to investigate the phase separation of these polymers such as DSC, DMA and SAXS. Here, we apply fluorescence for the first time to investigate the phase separation of segmented polyurethanes. Fluorescence has been used previously to study chemical and physical properties of polyurethanes such as curing kinetics, water uptake and strain behavior. Our novel use of intrinsic fluorescence takes advantage of the aromatic structure in the most common hard segments of polyurethanes in order to gain information about molecular alignment in thin film geometries. Molecular self-assembly is directed by both aromatic (Π–Π) interactions and hydrogen bonding. It is important to study how these two non-covalent interactions contribute to the molecular alignment of hard segments in thin film geometries and the long range order of surface and substrate effects.

Intrinsic fluorescence of aromatic rings typically results in two types of emission: monomer fluorescence and excimer fluorescence. Monomer emission occurs from the isolated single excited state of a phenyl ring whereas excimer emission occurs from two phenyl rings in a parallel, sandwich-like conformation with a separation distance of several Angstroms. Using a novel approach involving intrinsic fluorescence of a segmented polyurethane composed of a poly(propylene oxide) soft segment and 2,4-toluene diisocyanate hard segment, both monomer and excimer emission are observed. There are two pathways for excimer formation: dynamic or static. Using excitation spectra, we observe ground-state interactions of the hard segments in polyurethane thin films. In sufficiently thin ($\leq$ 200 nm) polyurethane films, on both interacting (quartz; with a potential for surface hydroxyl groups to hydrogen bond with carbonyl groups in polyurethane hard segments) and non-interacting (methylated glass; with no surface hydroxyl groups) substrates, ground-state dimers are formed and produce static excimer fluorescence.

Using the ratio of excimer emission to monomer emission (Ie/Im) as a function of film thickness, the effect of confinement on molecular orientation was examined. Upon changing the interactions with the substrate, the bulk film ratios were unaffected. On the quartz substrate, there was little or no temperature dependence of excimer fluorescence. In contrast, with the non-interacting methylated substrate, excimer fluorescence exhibited a significant temperature dependence. As the film is confined to thicknesses below 200 nm, the films on both substrates begin to align effectively to produce more excimer fluorescence from ground-state dimers than monomer fluorescence. This effect is seen to a larger degree on the methylated substrates than on the quartz substrates. The excitation spectra of polyurethane thin films on the methylated substrate are indicative of both dynamic and static excimer formation whereas the films on a quartz substrate show only static excimer formation. This difference in excimer fluorescence is hypothesized to be the result of the suppression of dynamic excimer formation by hydrogen bonding interactions between the quartz substrate and the polymer. Due to these interactions, a more rigid environment is introduced from the substrate. In addition, the effects of this rigid substrate environment are observed to propagate up to 200 nm from the substrate into the film. This effect is seen both above and below the glass transition temperature of the hard segment.

Interfaces in thin films lead to alignment and formation of ground-state dimers not seen in bulk polyurethanes. With these techniques, we were able to discern the non-covalent interactions that impact the molecular alignment of thin film polyurethanes. In the absence of hydrogen-bonding at the substrate, the molecular alignment of the hard segments due to aromatic interactions is significant in thin films. In the presence of hydrogen bonding at the substrate, the molecular alignment of aromatic rings into ground-state dimers is still present but the dynamic formation of dimers is substantially reduced.

472471 Complex Coacervate Core Micelles for the Dispersion and Stabilization of Organophosphate Hydrolase in Organic Solvents

Abstract:

Complex coacervation is a liquid-liquid phase separation that occurs in solutions of oppositely charged polyelectrolytes. This phenomenon can be exploited to from complex coacervate core micelles (C3Ms) that can be used in a variety of applications in aqueous media, including nanoparticle growth, gene delivery, and MRI contrast agent formulation; however, little work has been done to study the behavior of these C3M systems in hydrophobic environments. Here, we investigate the formation of C3Ms between a positively charged-neutral block copolymer, poly(oligo(ethylene glycol) methacrylate))-*b*-poly(4-vinyl N-methylpyridyl iodide) (POEGMA-*b*-qP4VP), and an oppositely charged homopolymer, poly(acrylic acid) (PAA) in two organic solventsâ€"ethanol, a water miscible solvent, and dimethyl methylphosphonate (DMMP), a solvent that is marginally miscible with water. We also investigate the ability of these C3Ms to both disperse and protect the organophosphate hydrolase (OPH) enzyme for use in decontamination of bulk chemical warfare agents.

The formation of C3Ms was first studied in aqueous solution, where dynamic light scattering (DLS) was used to determine the ideal buffer concentration and polymer mixing ratio for micelle formation. Small-angle neutron scattering (SANS) and transmission electron microscopy (TEM) confirmed the formation of micellar structures approximately 40 nm in diameter. The same techniques showed formation of similar structures after addition of OPH. C3Ms formed by POEGMA-*b*-qP4VP and PAA were then transferred into two organic solvents, ethanol and DMMP (which has similar density, water miscibility, molecular weight, chemical structure, and atomic content to chemical warfare agents of interest). The structures formed by the C3M system were compared to those formed by the POEGMA-*b*-qP4VP block copolymer alone were compared in both solvents using DLS. In both solvents, the structures formed by the block copolymer alone were different from those formed by C3Ms. Interestingly, while the structures formed by the block copolymer alone in DMMP were smaller than those formed by the block copolymer alone in ethanol, the C3Ms formed in DMMP were several nm larger than those formed in ethanol. Fits to SANS data taken in ethanol confirmed the formation of 14.4 ± 0.4 nm diameter structures by the block copolymer alone, and 40.0 ± 0.4 nm diameter structures by the C3M system. TEM was used to confirm the formation of C3Ms in DMMP. Similarly sized structures were observed after the addition of the OPH enzyme. To evaluate the ability of C3Ms to chemostabilize the OPH enzyme, activity against methyl paraoxon was assayed after incubation in both organic solvents. In ethanol, OPH in C3Ms retained $26 \pm 1\%$ of its activity, and in DMMP, OPH in C3Ms retained $35 \pm 3\%$ of its activity. This is a dramatic improvement over the activity retained by OPH alone after incubation in both of these solvents, which was less than 5%.

471626 Biosensor Coatings from Protein-Polymer Complex Coacervates

Abstract:

Engineering proteins into functional catalysts enables their application as biosensors and green catalysts. The biocatalyst must maintain mechanical integrity, protein stability and longevity, and access to the protein active site in order to be a viable technology. In addition, high protein density is needed to achieve high catalytic activity. The self-assembly of proteins in solid-state materials has the ability to achieve each of these requirements. Templating protein assembly with block copolymers offers a simple bottom-up method for the three-dimensional nanoscale patterning of proteins. Using ionic interactions between the protein and block copolymer, known as complex coacervation, to drive the self-assembly enables all aqueous processing of the two components. The enzyme alkaline phosphatase was immobilized via ionic interactions with a neutral-cationic block copolymer. The complex coacervate was flow coated on substrates, photo-crosslinked to immobilize the film, and assayed for enzyme activity. The immobilized phosphatase maintained activity over several cycles and was used to detect heavy metal contaminants. The nanostructure of the films before and after crosslinking was evaluated by AFM. Previous work has demonstrated that even though many proteins do not readily form coacervates in their native state, increasing the surface charge on proteins results in coacervation. While alkaline phosphatase did form a coacervate with polycations, the effect of protein surface charge on activity, loading in the films, and stability was evaluated.

464693 Microscopic Structures in Ion Exchange Fuel Cell Membranes

Abstract:

Ion exchange membrane (IEM) fuel cells transform chemical energy into electrical energy. Comparing to traditional petroleum based energy source, they generate water rather than more harmful carbon-based emissions on operation. In general, ion transport through the polymer membrane is dictated by the organization of the phase-segregated ion-transport domains at length scales ranging from nanometers to microns. And as a result, the rational design of any IEM requires a clear connection between the chemical arrangement of monomers within the copolymers and the resulting molecular-level organization of the ion-transport domains.

We therefore aim to provide detailed microscopic structural information for the design of fuel cell membranes, using polyimide-PEG (polyethylene glycol) random copolymer as the model system. Combining aromatic polyimide chains with PEG chains, the polyimide-PEG copolymers have great potential in making good IEMs. These copolymers are conveniently synthesized from the random co-polymerization of aromatic dianhydrides, aromatic diamines, and PEG oligomers (M.W. around 1 kDa) that are functionalized on both ends with amine groups. The flexibility in the choice of aromatic dianhydrides and aromatic diamines thus offer the resulting copolymer potentially tunable properties that are suitable in versatile operating conditions. As an IEM, the phase-segregated PEG domains provide ion transport channels, while the aromatic polyimide networks form mechanically and thermally stable matrix.

In this presented work, the polyimide-PEG random copolymers in the form of self-assembled membranes were synthesized and characterized using a variety of methods. The thermal stability and mechanical strength of the membranes were characterized using Thermogravimetric Analysis (TGA), Differential Scanning Calorimetry (DSC), and tensile tests. The microscopic structures were investigated using a combination of Transmission Electron Microscopy (TEM), Atomic Force Microscopy (AFM), as well as Small Angle X-ray Scattering (SAXS). In order to further interpret the SAXS results, theoretical analysis was collaboratively employed. By predicting the self-assembled microscopic structure of worm like random copolymer chains, this theoretical study helps us to quantitatively understand the spatial distribution of PEG domains. Membranes conductivity was measured using Electrochemical Impedance Spectroscopy (EIS) in conjunction with Cyclic Voltammetry (CV). In addition, electrolytes such as ethylammonium nitrate (EAN) were incorporated into the membranes and the resulted doped membranes were compared with those bare membranes in terms of both microscopic structures and performances. The observed structure-property relations in various polyimide-PEG membranes are discussed.

459781 Adsorption, Structural and Dynamic Properties of Ethanol-Water Mixtures in Graphene and Hexagonal Boron Nitride Slit Pores

Abstract:

The industries discharge a variety of pollutants, such as heavy metals, organic toxins, and oils, in water resources. Exposure of these contaminants in water causes adverse health effects on various forms of life [1-3]. Novel materials are needed for the effective removal of pollutants from industrial wastewater. Graphene and hexagonal boron nitride (hBN) sheets are promising materials for removal of organic pollutants [4]. In this work, the suitability of the sheets for the separation of ethanol-water mixture is investigated by studying the adsorption and structural behavior of ethanol-water mixtures in slit pore with variable width (7 to 13 Å) using molecular dynamics. The simulation results indicate that the selectivity of ethanol varies with pore width and with the nature of the surface. The selectivity of ethanol is highest for 9 Å pore and lowest for 7 Å pore, irrespective of the nature of the substrate. However, selectivity of ethanol is relatively higher for hBN compare to the graphene surface, for all slit-pore widths. At lower pore width, molecular sieving plays an important role for separation of ethanol; whereas at higher pore width separation is governed by the adsorption behavior. The diffusion coefficients of water and ethanol molecules substantially decrease with decrease in pore width for both graphene and hBN surfaces. The resident time of water and ethanol molecules decreases with increase in the slit-width. Furthermore, water and ethanol molecules confined in hBN pores show higher residence time and lower diffusion coefficient values compare to graphene slit pores.

References

[1] M. A. Shannon, P. W. Bohn, M. Elimelech, J. G. Georgiadis, B. J. Mariñas, A. M. Mayes, "Science and technology for water purification in the coming decades" Nature, 2008, 452: 301-310.

[2] M. A. Montgomery and M. Elimelech, "Water and sanitation in developing countries: including health in the equation" Environmental Science and Technology, 2007, 41: 17-24.

[3] F. Zhang, WB. Zhang, Z. Shi, D. Wang, J. Jin, L. Jiang, "Nanowire-haired inorganic membranes with superhydrophilicity and underwater ultralow adhesive superoleophobicity for high-efficiency oil/water separation" Advanced Materials, 2013.

[4] W. Lei, D. Portehault, D. Liu, S. Qin, Y. Chen, "Porous boron nitride nanosheets for effective water cleaning" nature communications, 2013.

466478 Quantitative Monitoring of Microphase Separation Behaviors in Cationic Liposomes Using Multiple Fluorescent Probes

Abstract:

Microphase separation behaviors of cationic liposomes have been investigated using a pH-sensitive fluorescent probe with 4-heptadecyl-7-hydroxycoumarin (HHC), 1,6-diphenyl-1,3,5-hexatriene, and 6-lauroyl-2-dimethylaminonaphthalene, and to estimate localized electrostatic potentials. Shifts of the apparent pKa values of HHC were observed in cationic liposomes in proportion to the amount of cationic lipids. Two pKa values were obtained with 1,2-dioleoyl-*sn*-glycero-3-phosphocholine (DOPC)/3ß-[*N*(*N'*,*N'*-dimethylaminoethane)-carbamoyl] cholesterol hydrochloride (DC-Ch) liposomes, while only one pKa value was generated with either DOPC/1,2-dioleoyl-3-trimethylammonium-propane (DOTAP) or DOPC/dimethyldioctadecylammonium-bromide (DODAB) liposomes. The physicochemical membrane property analyses, focusing on membrane fluidity and membrane polarity, revealed heterogeneity among DOPC/DC-Ch liposomes. By analyzing the pH titration curves using sigmoidal fitting, the localized electrostatic potentials were estimated. For DOPC/DOTAP=(7/3), the membrane was in the liquid-disordered phase and the density of cationic molecules was 0.41 cation/nm^2. For DOPC/DC-Ch=(7/3), the membrane was heterogeneous and the densities of cationic molecules in liquid-disordered and liquid-ordered phases were 0.25 and 1.24 cation/nm^2, respectively. We thereby conclude that the DC-Ch molecules can form nanodomains when these molecules are concentrated to 59%.

458670 A Multi-Fluid Model of Phase-Inversion Membrane Formation

Abstract:

Industrial processes for producing polymer-based materials operate away from equilibrium, making the final microstructure – and thus the properties of the material – dependent on process history. Describing such processes by traditional simulation methods has proven difficult because of the confluence of complex thermodynamics and non-trivial fluid mechanics. We have recently developed a "multi-fluid" simulation framework for examining the dynamics of multiphase polymer solutions that combines a field-theoretic approach to thermodynamics with continuum fluid mechanics. In this framework, we use a ternary polymer solution model to investigate the phase inversion process for making polymer membranes. After characterizing the model and methods, we investigate a number of important phenomena that occur across several length and time scales including (i) mass-transfer between the polymer film and the non-solvent bath, (ii) phase separation between polymer lean and polymer rich regions and (iii) long-time coarsening behavior. We will further discuss the implications of these results on the formation mechanisms of various microstructures found in polymer membranes.

470502 Charged Nanochannels By Random Copolymer Micelle Assembly

Abstract:

Membranes that can separate solutes by factors other than size (e.g. charge, hydrophobicity, chemical functionality) would enable the use of this energy-efficient, green technology in new applications. However, achieving membrane selectivity based on parameters other than size has been a challenge, especially when the molecules to be separated are small. Biological pores such as ion channels achieve exceptional selectivity by confining flow into very small pores lined with functional groups. Therefore, incorporating nanostructures that confine permeation into membrane selective layers can significantly enhance selectivity. In this study, we created membrane selective layers with nanopores lined with carboxyl groups using the deposition of micelles formed in methanol onto porous membranes. To prepare these membranes, we prepared an amphiphilic random copolymer that combines highly hydrophobic, fluorinated repeat units of 2,2,2-trifluroethyl methacrylate (TFEMA) with repeat units of methacrylic acid (MAA), PTFEMA-*r*-PMAA, by free radical polymerization, because it a simple, robust and scalable method. The copolymer contained 55 wt% TFEMA, and was insoluble in water. We found that this copolymer forms micelles in methanol with a multimodal size distribution, as shown by dynamic light scattering (DLS). When these micelles are coated onto the surface of a porous support membrane whose pores are smaller than the micelles, a selective layer of micelles packed together is formed in between which permeation occur. If the membrane pore size is too large, micelles go inside the membrane, clogging inner pores and resulting in a membrane with very low flux. Also, if the solvent evaporation time is too long, the micelle merge and create a dense selective layer with a very low permeability. By proper selection of conditions (i.e. copolymer concentration, solvent evaporation time, base membrane MWCO, etc.), we achieved membranes with pure water permeabilities up to ~6 L/m^2.h.bar and high selectivity (separation coefficient up to 24) between small molecules with similar size (~1nm) but opposite charges. For example, these membranes can retain a negatively charged dye by > 85% whereas a positively charged dye with a similar size was retained by < 20%. Membrane rejection can be modeled by Donnan exclusion theory. This is the first demonstration of a new approach to forming membranes from easily synthesized copolymers by micelle assembly. We believe this approach can be adapted to address various pore sizes

and pore surface chemistries, leading to novel membranes with promising selectivity.

469664 Development of Polybenzimidazole Nanofiltration Membranes with Biomimetic Surfaces

Abstract:

Aquaporins are membrane protein water channels present in cells, and they restrict the passage of contaminants without preventing the passage of water. Small molecules such as urea and boric acid, the removal of which is inefficient by conventional membranes are also rejected. Biomimetic membranes, an innovation in water filtration technology attempts to replicate this natural process of highly selective and efficient transport of different molecules across a semi-permeable membrane occurring at the cell level. Therefore, aquaporins have received worldwide attention because of their potential to form biomimetic membranes with high flux and selectivity for water reuse and desalination. However, challenges involved in the incorporation of aquaporin proteins in membranes limit their applicability. One of them is to attach aquaporins to the membranes without chemically altering or damaging the aquaporins during the binding to the membrane. The second challenge is to design and prepare an assembly that allows biomimetic membranes with aquaporins to sustain hydraulic water pressure gradients without losing their integrity and performance. The overarching objective of this project is to form a biomimetic membrane made of unaltered aquaporins dispersed in a polymeric membrane selective layer and capable of operation under high hydraulic pressure. Polybenzimidazole (PBI) membranes were surface modified with treated aquaporins in order to achieve higher water flux and selectivity. Membranes modified with aquaporins displayed lower flux declines and higher flux recovery values after backwash as compared to unmodified PBI membranes. Also, modified membranes showed improved rejection values for both protein and salt solutions of different concentrations. However, there was leakage of ions between the channels. Therefore, in order to improve the rejection of protons, ions and other impurities, aquaporin channels were aligned with the direction of water flow. Functional groups were installed on AqpZ for covalent attachment to the polymer matrix so that the proteins could be immobilized to the membranes and aligned in the direction of the flow.

465129 Bijel Derived, Fouling Resistant and Catalytic Ultra-/Micro-Filtration Membranes for Advanced Water Treatment Applications

Abstract:

Bicontinuous interfacially jammed emulsions (bijels) are a new class of soft materials [1] with promising applications in various areas ranging from healthcare, food, energy and reaction engineering due to their unique structural, mechanical and transport properties. We recently presented a new method to prepare bijels via solvent transfer-induced phase separation (STRIPS), which enables continuous fabrication of hierarchically structured composite fibers and membranes with a manifold of organic/inorganic building blocks [2]. These asymmetric membranes are composed of a porous polymer scaffold with inorganic nanoparticles on their surface and introduce a variety of important features: (i) Hollow fibers can be used as ultra- and micro- filtration membranes for the separation of model contaminants in water and organic solvents due to their controllable pore size and chemical stability. (ii) The membranes are resistant to biofouling due to the simplicity of chemical surface modifications. (iii) Stimuli-responsive membranes that can open/close their pores are fabricated, facilitated by the fact that broad ranges of polymeric building blocks can be integrated. And (iv) photocatalytic titanium dioxide membranes for organic contaminant oxidation are realized; other functionalities can be readily imparted due to the wide range of material choices for the nanoparticles on the membrane surface. Our work demonstrates a novel and versatile fabrication route for separation membranes with potential applications in water treatment, food processing or as separation membranes in the chemical industry.

[1] Herzig, E. M., et al. (2007). Nature materials, 6(12), 966-971.

[2] Haase, M. F., Stebe, K. J., & Lee, D. (2015). Advanced Materials, 27(44), 7065-7071.

461733 Boosting the Performance of Random Zwitterionic Copolymers Using Ionic Liquids during Membrane Formation

Abstract:

Polymer self-assembly is a promising tool for scalable manufacture of membranes while maintaining high permeability and controlled pore size. Tuning copolymer composition (monomers, additives) and processing methods can change copolymer behavior which can dramatically affect the self-assembly and hence the membrane performance (permeability, selectivity, fouling resistance). Studies based on block copolymers demonstrate the influence of additives (solvent, homopolymer etc.) on the copolymer self-assembly and membrane performance. But studies on additives in casting solutions of random copolymers and how they affect membrane performance have not been reported to our knowledge. Recently, we have introduced a new class of membranes with ~ 1 nm effective pore size whose selective layers are made of self-assembling zwitterionic amphiphilic random copolymers. These membranes derive not only their excellent fouling resistance but also their permeability and selectivity from this self-assembled nanostructure. These membranes have numerous applications in the biochemical and pharmaceutical industries, as well as wastewater treatment processes. These membranes are prepared simply by coating random copolymers of hydrophobic and zwitterionic monomers onto a porous support membrane. In this study, we have used ionic liquids and other additives in the coating solutions to boost and alter the performance of these membranes. Membranes prepared by coating copolymer solutions with sufficient amounts of selected additives on commercial ultrafiltration membrane supports exhibit permeances as high as 50 L/m^2.hr.bar, up to 10 times higher than membranes formed without additives. These membranes also exhibit a narrow pore size distribution, retaining the same size-based selectivity with a ~1 nm size cut-off demonstrated by filtering negatively charged dyes. Performance of these membranes depends on the amount of additive as well as the membrane manufacturing method (non-solvent, drying time etc.). This is attributed to phase separation and solvent diffusion kinetics that are in play during membrane formation by SNIPS (self-assembly and non-solvent induced phase separation) mechanism. These structure-property relationships serve as a guideline to develop new materials with high fluxes and sharp selectivity for various applications such as textile wastewater treatment, pharmaceutical purification and bioseparation applications.

472502 Ultra-Stretchable Iono-Elastomers with Mechano-Electrical Response (Invited Talk)

Abstract:

Structural materials with high conductivity and high stretchability are becoming an important area of research due to emerging technologies involving stretchable electronics. Strategies to achieve these materials involve sophisticated integration of elastomeric substrates with micro- or nano-structured organic or inorganic electronic materials via multi-step and costly processes. We report a novel strategy for facile synthesis of transparent iono-elastomers via crosslinking of a micellar ion gel with face-centered cubic (FCC) crystalline structure. Our system consists of an amphiphilic triblock copolymer, Pluronic-diacrylate (F127DA), dissolved in a protic ionic liquid, ethylammonium nitrate (EAN). Analogous to the non-acrylated Pluronic (F127) [1], F127DA/dEAN solutions form spherical micelles, which remain unchanged upon crosslinking in presence of a UV-curing initiator. Crosslinking of concentrated F127DA/dEAN solutions produces soft elastomers (ionoelastomers) with remarkably large elongation at break and high ionic conductivity. The tensile properties and in-situ microstructure evolution during uniaxial extension of the ionoelastomers are measured using a Linkam tensile stage mounted in a synchrotron SAXS beam line (12-ID-B) at the Advanced Photon Source. Uniaxial deformation induces a reversible transition from isotropic FCC lattice grains to a hexagonal close packed crystal structure in the F121DA/dEAN. These microstructural transitions result in reversible decrease in the electrical resistivity of the ionoelastomers when they are subjected to loading-unloading cycles. THis mechano-electrical response can be use to design motion sensors and actuators.

[1] C. R. López-Barrón, D. Li, N. J. Wagner, J. L. Caplan. Macromolecules, 2014, 4 (47), 7484-7495.

462893 High Performance Electrospun Polyethylene Fibers By Gel-Electrospinning

Abstract:

Electrospinning has been widely used over the last decade to fabricate polymeric fibers with submicron to micron diameters for various applications, such as filters, battery materials, and biomaterials. Despite their many unique properties, electrospun nanofibers tend to be weaker than conventional polymer fibers, especially when compared to engineered high performance materials such as SpectraÂ®. Inspired by a commercial gel-spinning process, we developed a novel procedure that we have dubbed â€œgel-electrospinningâ€, wherein a polymer gel is drawn by electrical forces to obtain high performance fibers with submicron diameters. Using this process, we fabricated ultrahigh molecular weight polyethylene (UHMWPE) fibers with diameter less than [value] and moduli greater than 100 GPa from solutions of *p*-xylene and tert-butyl ammonium bromide at elevated temperature. Here, we report the tensile modulus, ultimate tensile strength, and toughness of these fibers, along with their crystallinity and molecular orientation, to demonstrate the structure-property relationship among these parameters. The key to the production of these fibers is careful control of the temperature and rheology in several process zones.

470766 Diffusion NMR (DOSY) for Fast Absolute Molecular Weight Analysis of Polyethylene Furanoate (PEF)

Abstract:

Polyethylene furanoate (PEF) has the potential to replace polyethylene terephthalate (PET) with a bio-based and less permeable plastic for commercial application in bottles, packaging, textiles, medical grafts, etc. [1]. However, as promising the advantages and properties of this new type of polymer are, so difficult can be its characterization. As aromatic polyesters in general, PEF is insoluble in most common solvents. Established MW analysis of PEF and PET using size exclusion chromatography (SEC) requires hexafluoro-2-propanol (HFIP) as solvent, special HFIP-resistant equipment, solvent volumes up to liters, and analysis times up to hours. Besides, absolute molecular weights are essential for process and material development, as measurements relative to standards such as polystyrene can grossly overestimate the actual molecular weight. For absolute molecular weights, expensive multi-angle light scattering (MALS) equipment is required. In this work, we present diffusion NMR (DOSY) as an alternative and quick method to measure absolute molecular weight of PEF [2].

NMR DOSY applies pulsed field gradients to spatially-labelled molecules and measure signal intensity upon a magnetization de- and refocusing. The further a polymer molecule moves, as dictated by its size, during a given diffusion time (Δ) between de- and refocusing, the stronger the signal intensity decreases. Fitting of the decay over a sequence of pulse gradients of different strength with the Stejskal-Tanner equation [3] derives a diffusion coefficient D. The relationship between diffusion coefficient and molecular weight is readily calibrated using PET standards of known absolute molecular weight. Conduct of the measurement can be done even without greater knowledge with a pre-installed double stimulated echo pulse sequence (dstegpd2), which enables very reproducible analysis due to convection compensation.

DOSY performance is simply dependent on signal quality for the fitting, which can be tuned by sample concentration, as well as the number of repeated scans (*ns*) and pulse gradient increments. To study the effect of concentration, DOSY experiments were successfully performed in a range of 0.4 – 8 mg/mL of polymer in 0.75 mL deuterated trifluoroacetic acid (TFA-d), where Brownian motion regime without molecular interaction was always confirmed. However, sample concentration affects viscosity η and thus diffusion D, which can be avoided by staying within a 1 mg/mL concentration range, or compensated by a correction factor from a fit to the measured D-η dependence. Signal-to-noise increases with applied *ns* and gradient increments, but so does measurement time. This enables 1) the measurement of low concentration samples (e.g. 0.3 mg/mL) without losing data for fitting, and 2) very fast measurement of samples available at higher concentration (e.g. 5 mg/mL). In the latter case, DOSY analysis time per sample can be as short as 1 min without any significant change (<5%) in molecular weight estimation. Alongside with the little solvent use, this makes DOSY a unique and reliable method for semi-online measurement of kinetic samples or quality control during polymerization reactions. Finally, the results for PET and PEF derived from DOSY match those from laser scattering (MALS), and can therefore be considered absolute molecular weights (as shown in Figure 1) [4].

Besides molecular weight analysis, NMR DOSY can be used to assess further properties of polymer samples. In the case of ring-opening polymerization-based PEF, as synthesized in our labs, the cyclic monomers yield a different chemical shift than the PEF product (A1 vs A2 in Figure 2), which enables simultaneous molecular weight and conversion analysis via peak area comparison. Furthermore, DOSY can be used to analyze T1 and T2 relaxation to study flexible and rigid parts of the molecular backbone [5]. We are currently investigating this to compare with the predictions of molecular dynamics and quantum chemistry simulations.

References

[1] L Sipos, E De Jong, MA Dam, J M Gruter, *ACS Symposium Series* 1105, P Smith (Ed.), 2012, 1-11
[2] W Li, H Chung, C Daeffler, JA Johnson, RH Grubbs, *Macromolecules* 2012, 45, 9595-9603
[3] EO Stejskal, JE Tanner, *J. Chem. Phys.* 1965, 42, 1-5
[4] JG Rosenboom, G Storti, M Morbidelli 2016, in preparation
[5] H Berglund, H Baumann, S Knapp, R Ladenstein, T Härd, *JACS.* 1995, 117, 12883-12884

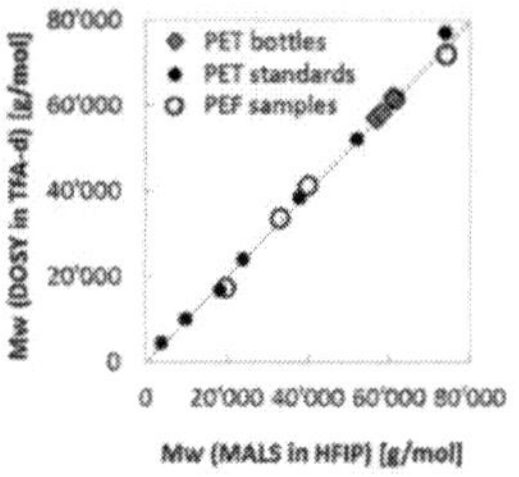

Sample	Mw (MALS) [g/mol]	Mw (DOSY) [g/mol]
PET standard 1	9,900	10,240
PET standard 2	37,770	38,540
PET standard 3	73,230	77,240
PET bottle	60,900	61,750
PEF product 1	19,420	17,670
PEF product 2	39,910	41,480
PEF product 3	72,420	72,270

Figure 1: Absolute molecular weights determined with MALS and DOSY for PEF and PET samples.

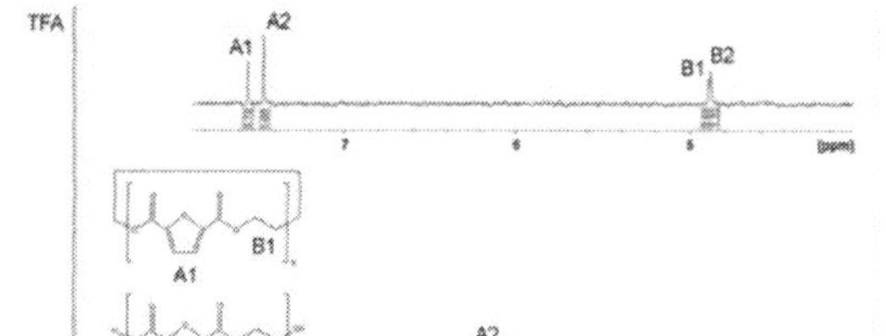

Figure 2: Monomer (A1) and PEF polymer (A2) chemical shifts, enabling simultaneous conversion and molecular weight analysis.

472280 Precise Control of Mesoscale Morphology in Photoreactive Polymer Blends Using Visible Light Beams

Abstract:

Controlling the structure of polymeric blends with microscale precision and high scalability is critical for their deployment in applications. We present a new process to organize multiphase morphology in photoreactive polymer blends by coupling light-induced polymer waveguide formation to polymerization-induced phase separation. By exposing a blend of photocrosslinker and linear chain polymer to an array of microscale beams, generated from a visible-light LED or white-light source, photopolymer fibers grow from the point of exposure across the blend medium. Concurrently, the underlying polymerization of the growing fibers induces spatially-local, mesoscale spinodal decomposition of the blend along the fibers' path lengths. Thereby, large-scale, multiphase morphology is produced that corresponds to the beams' size and arrangement. We show that polymer blend morphology can be controlled with microscale precision over unprecedentedly large depths (> 1 mm) and areas (> 10 cm^2) – well beyond the limits of uniform polymerization and holography – by varying the size and arrangement of the beams. Systematic variation of the relative weight fractions in the blend also tunes the final structure. While a single array of beams produces a 2D structure, 3D structures can be produced by propagating multiple arrays in different directions through a common surface of the medium. The generalizable nature of phase separation enables a wide range of polymer blend and hybrid blend systems to be self-organized. Hence, we anticipate that this method will become a versatile mechanism that serves the dual role of fabricating 2D and 3D mesostructured composite materials, while concurrently organizing the morphology of its components.

469742 Influence of Carbonate Molecular Structures on the Morphology and Properties of Non-Isocyanate, Segmented Polyhydroxyurethane Copolymers

Abstract:

Polyurethane (PU) is an important class of commodity polymer with a total global production estimated to reach 18 million tons in 2016. It has wide ranging applications such as in adhesives, coatings, elastomers, foams, implants etc. It is produced from step growth reaction between isocyanates and alcohols with relatively fast reaction kinetics.[1] In recent years, regulatory bodies across the world have tightened the scrutiny on the use, transport and handling of isocyanates in consumer products. These moves have fueled significant investigations into alternative chemistries to produce non-isocyanate PU (NIPU) or PU-like materials.

Polyhydroxyurethane (PHU) derived from cyclic carbonate aminolysis has emerged as a very promising candidate as non-isocyanate PU-like materials. This polymer contains urethane linkages and additional primary or secondary hydroxyl groups adjacent to the urethane bonds. In the last few decades, significant investigations have been devoted toward PHU synthesis with major focus on the production of cyclic carbonate monomers, reaction catalysis, synthesis of single-phase PHUs and crosslinked PHU networks from renewable resources.[1,2] Only two studies have successfully synthesized and investigated segmented, nanophase-separated PHU.[3,4] Segmented PUs are composed of alternating sequences of hard and soft blocks; the soft segment is typically a long, flexible molecule with a glass transition temperature (T_g) below room temperature while the hard segment consists of diisocyanate condensed with small molecule diol with T_g above room temperature.

Recently, Torkelson and coworkers[3] reported the synthesis and characterization of nanophase-separated, thermoplastic PHU elastomer from several polyether-based soft segments. Their study indicated that the hydroxyl groups in the PHU hard segments play a critical role in the development of nanophase separation due to their ability to form hydrogen bonding with soft segment ether oxygen. Employing the appropriate soft segment molecules that reduce or minimize the intersegmental hydrogen bonding is critical to obtain materials with robust elastomeric responses. This study only investigated one carbonate molecule as the hard segment. To date, the influence of isocyanate structure on the final properties of segmented PUs has been studied in numerous reports.[5] In contrast, no study has yet investigated the influence of carbonate molecular architecture in the hard segment on the final properties of segmented PHU copolymers. Such an investigation is needed to advance the understanding of PHU material properties in light of the rising needs of competitive NIPU materials. Structure-property relationship studies enable the design and development of PHU materials with improved properties which might compete with commercially available PU products in the market.

Here, we investigated the influence of hard-segment carbonate structure on the morphology and properties of segmented PHU copolymers. Three carbonates prepared from commercially available epoxy sources are investigated namely divinylbenzene dicyclocarbonate (DVBDCC), bisphenol A dicarbonate (BPADC) and cyclohexane dimethanol dicarbonate (CHDMDC). These carbonates were formulated with two different soft-segment compositions: a polytetramethylene oxide (PTMO)-based soft segment as system where intersegmental hydrogen bonding is permitted and polybutadiene-co-acrylonitrile (PBN)-based soft segment as system where hydrogen bonding is restricted to occur only within the hard segment. Characterizations via small-angle X-ray scattering, dynamic mechanical analysis, infrared spectroscopy, and tensile testing reveal interesting interplays between carbonate structure and soft-segment chemical make-up on the resulting morphologies and properties of segmented PHUs.

In PTMO-based formulation, SAXS shows that only DVBDCC-based PHU possess nanophase-separated morphology with interdomain spacing of 13.8 nm while CHMDC- and BPADC-based PHUs are single phase materials. Tensile testing shows that DVBDCC-based PHU have Youngâ€™s modulus of 60 MPa, tensile strength of 5.1 MPa and elongation at break 640 % while other PHUs are not mechanically robust to be tested. DMA results also show that only DVBDCC-PHU is nanophase-separated with soft segment glass transition temperature (T_g) of -65 Â°C and flow temperature of ~75 Â°C while other PHUs are single phase with single T_gs. The single phase structure of BPADC-PHU and CHDMD-PHU is likely due to a combination of the molecular flexibility of BPADC and CHDMDC relative to DVBDCC and hydrogen bonding of hydroxyl groups to oxygen atoms in the soft segment.

In contrast with PTMO-based formulation, all PHUs formulated with PBN-based soft segment are nanophase-separated with 10-16 nm interdomain spacings measured by SAXS regardless of carbonate structures employed.Â Tensile testing shows that DVBDCC-PHU possesses the best tensile properties with tensile strength of 3.4 MPa while BPADC-PHU has tensile strength of 1.1 MPa and CHMDC-PHU is too soft too test as the material softens near room temperature. DMA results of PHUs in this series show that they are nanophase separated with highest flow temperature obtained by DVBDCC-PHU at ~80 °C followed by BPADC-PHU and CHMDC-PHU at 48 °C and 28 °C. These results show that the switch to soft segment with no potential of intersegmental hydrogen bonding is necessary in formulating PHUs using more flexible carbonate molecules. This study highlights the important role of hard-segment carbonate structure and its interaction with soft segment on the resulting morphology and properties of segmented PHU copolymers.

References:

1) Guan, J.; Song, Y.; Lin, Y.; Yin, X.; Zuo, M.; Zhao, Y.; Tao, X.; Zheng, Q. *Ind. Eng. Chem. Res.* **2011**, 50, 6517-6527.

2) Blattmann, H.; Fleischer, M.; Bähr, M.; Mülhaupt, R. *Macromol. Rapid Commun*. 2014, 35, 1238-1254.

3) Leitsch, E. K.; Beniah, G.; Liu, K.; Lan, T.; Heath, W. H.; Scheidt, K. A.; Torkelson, J. M. *ACS Macro Letters* **2016**, 5, 424-429.

4) Nanclares, J.; Petrovic, Z. S.; Javni, I.; Ionescu, M.; Jaramillo, F. *J. Appl. Polym. Sci.* **2015**, 132, 42492.

5) Yilgor, I.; Yilgor, E.; Wilkes, G. L. *Polymer* **2015**, 58, A1-A36.

466591 Aging of Thermal Insulation Properties of Polyurethane Foams

Abstract:

Polyurethane foams are extensively used in many branches of industry. Our contribution is focused on the thermal insulation properties of rigid polyurethane foams, which have closed-cell structure. The practical use of rigid PU foams is based on their low thermal conductivity, low density, low water absorption and relatively good mechanical strength. Therefore, they are widely used as heat insulators in housebuilding, cooling systems, furniture, etc.

The good thermal insulation properties of rigid PU foams result from the combination of a fine, closed-cell foam structure and cell gases, which are poor heat conductors. Rigid PU foam is formed by the reaction of isocyanates with polyols in the presence of catalysts and foamed with chemical and physical blowing agents. As the chemical blowing agent, CO2 generated during the reaction is used; as the physical blowing agent, pentane or cyclopentane is often used. However, these blowing agents spontaneously diffuse out of the foam and are replaced by air. As a consequence, the heat conductivity of the foam increases. This phenomenon is called foam aging. A general understanding of this process would help to produce foams with better insulation properties over the whole lifespan of these products.

The mathematical modelling of aging is a challenging multiscale problem, which requires the combination of many models. We developed a new model, which is able to simulate the diffusion of gases through a series of successive polymer walls and gas cells. This model estimates the time evolution of the concentration profiles of individual gases inside the foam. Once the composition of gases in the foam is known, the gas conductivity and then the foam conductivity are calculated through connected models.

The mathematical modelling was complemented by experimental work. The solubility and the diffusivity of gases in PU were measured on gravimetric and pressure decay apparatuses. To properly simulate the morphology, PU foam samples prepared using carbon dioxide and cyclopentane as blowing agents were scanned by X-ray micro-tomography

and the cell size distribution and porosity were determined. The evolution of foam conductivity during accelerated aging of these foam samples at elevated temperatures was used to validate the mathematical model. Then, this evolution was compared with the predictions and a good agreement was observed.

Acknowledgement: Financial support of the 7th FP EU project NMP4-SL-2013-604271 â€œMODENAâ€ is acknowledged.

458174 Biorenewable Tough Blends of Polylactide and Acrylated Epoxidized Soybean Oil Compatibilized By a Polylactide Star Polymer

Abstract:

Polylactide (PLA), a commercially available thermoplastic derived from plant sugars, finds applications in consumer products, disposable packaging, and textiles, among others. The widespread application of this material is limited by its brittleness, as evidenced by low tensile elongation at break, impact strength, and fracture toughness. A multifunctional vegetable oil, acrylated epoxidized soybean oil (AESO), was investigated as a biodegradable, renewable additive to improve the toughness of PLA. AESO was found to be a highly reactive oil, providing a dispersed phase with tunable properties in which the acrylate groups underwent crosslinking at the elevated temperatures required for processing the blends. Additionally, the presence of hydroxyl groups on AESO provided two routes for compatibilization of PLA/AESO blends: 1) reactive compatibilization through the transesterification of AESO and PLA and 2) synthesis of a PLA star polymer with an AESO core. The morphological, thermal and

mechanical behavior of PLA/oil blends were investigated, in which the dispersed oil phase consisted of AESO, soybean oil (SYBO), or a 50/50 mixture of AESO/SYBO. The oil additives were found to toughen the PLA matrix, with significant enhancements in the elongation at break and tensile toughness values, while maintaining the glass transition temperature of neat PLA. In particular, the blend containing PLA, AESO, SYBO, and the PLA star polymer was found to exhibit a uniform oil droplet size distribution with small average droplet size and interparticle distance, resulting in the greatest enhancements of PLA tensile properties with no observable plasticization.

469957 Sustainable Anticorrosive Self-Healing Smart Coatings for Metal Protection

Abstract:

Corrosion and failure of petroleum-production processing equipment, pipelines, storage facilities, and transportation equipment is one of the largest—and most costly—fundamental problems facing the petroleum industry. In the US alone, preventable costs upwards of $US 85 billion—a value >5% of the US gross national product (GNP)—are incurred annually to the petroleum industry. Typical metal coatings are susceptible to damage in the form of small cracks. Once cracks have formed within polymeric coating materials, the integrity of the structure is significantly compromised and the underlying metals are exposed to the environment-leading to corrosion. Smart self-healing coatings for corrosion protection of metallic substrates (steel, magnesium, and aluminium, and their alloys) have attracted tremendous interest due to their capability to prevent crack propagation in the protective coatings by releasing active agents from microcapsules. Here we report a polymeric coating containing microcapsules with the ability to automatically heal micro-cracks in real-time. The polyurethane (PU) microcapsules containing self-polymerizing linseed oil was successfully synthesized by interfacial polymerization. In controlled healing tests, the damaged coatings successfully healed and continued to protect the structure without any intervention. We envision this smart material having great potential in the field of metal coating corrosion protection.

469548 Nanostructured Shape Memory Polymers Triggered By Unconventional Stimuli

Abstract:

Smart shape memory polymers (SMPs) can memorize and recover their permanent shape in response to an external stimulus, such as heat, light, and solvent. They have been extensively exploited for a wide spectrum of applications ranging from biomedical devices to aerospace morphing structures. However, most of the existing SMPs are thermoresponsive and their performance is hindered by slow response speed, heat-demanding programming and recovery steps. Here, by integrating scientific principles drawn from two disparate fields that do not typically intersect - the fast-growing photonic crystal and SMP technologies, we demonstrate a new type of SMP that enables unusual "cold" programming and instantaneous shape recovery triggered by a large variety of unconventional stimuli, such as static pressure, numerous vapors and liquids, lateral stress, and ultrasonic wave. The active components of the new SMPs are thin macroporous photonic crystal layers (only a few mm thick) which are fabricated by using self-assembled, 3-D highly ordered colloidal crystals as structural templates. This nanoscopic thin-film configuration renders orders of magnitude faster response speed than bulky SMP samples in traditional applications. Additionally, these nanostructured multi-stimuli-responsive materials differ greatly from existing bulk SMPs as they enable striking chromogenic effects and all-room-temperature operations for the entire shape memory cycle, promising for many applications ranging from reusable chromogenic vapor sensors to reconfigurable nanooptical devices. Moreover, this interdisciplinary integration provides a simple and sensitive optical technique for investigating the intriguing shape memory effects at nanoscale.

467635 Stimuli-Induced Color Variations of Polymer Brush/Gold Nanoparticle Composites

Abstract:

Surface coatings of stimuli-responsive polymer brushes that are chemically grafted to the substrate have received much attention due to numerous applications in material and life science. The surface-grafted brushes can be employed as templates for the immobilization of gold nanoparticles, thus creating composite materials with interesting optical properties that can be used for colorimetric sensing. Color changes are induced by employing stimuli-responsive polymers that undergo reversible swelling-shrinking transitions upon variations of outer stimuli (1,2). As the stimuli-induced color change of the composites depends on the particle loading and interparticle distance, it is crucial to control the assembly of the immobilized nanoparticles. Ways to control the assembly of nanoparticles in polymer brushes include variations of brush grafting density (3), brush thickness (4), polymer-particle interaction strength, particle size, and particle surface modification (1). Here we employed UV/vis spectroscopy, neutron reflectometry, scanning force microscopy and ellipsometry to study the assembly of gold nanoparticles in stimuli-responsive poly(N-isoacrylamide) brushes. We found that the particle assembly and temperature-induced color change depended on the particle surface functionalization. Whereas citrate-coated nanoparticles penetrated into the PNIPAM brushes and formed a 3D assembly, nanoparticles coated with 12-mercaptododecanoic acid molecules remained near the brush surface, thus forming a 2D assembly (1). Furthermore, we investigated the color change of the citrate-coated particles/PNIPAM composites in presence of different salts and found that the salt-induced color change depended on the type of salt.

References:

(1) Christau, S.; Möller, T.; Brose, F.; Genzer, J.; Soltwedel, O.; von Klitzing, R. Effect of gold nanoparticle hydrophobicity on thermally induced color change of PNIPAM brush/gold nanoparticle hybrids, *Polymer,* 2016, doi:

10.1016/j.polymer.2016.03.088

(2) Kesal, D.; Christau, S.; Krause, P.; Moeller, T.; von Klitzing, R. Uptake of pH-Sensitive Gold Nanoparticles in Strong Polyelectrolyte Brushes, *Polymer,* 2016, 8, 134; doi:10.3390/polym8040134

(3) Christau, S.; Möller, T., Yenice, Z.; Genzer, J.; von Klitzing, R. Brush/gold nanoparticle hybrids: Effect of grafting density on particle uptake and distribution within weak polyelectrolyte brushes. *Langmuir,* 2014, *30,* 13033-13041.

(4) Christau, S.; Thurandt, S.; Yenice, Z.; von Klitzing, R. Stimuli-Responsive Polyelectrolyte Brushes as a Matrix for the Attachment of Gold Nanoparticles: The Effect of Brush Thickness on Particle Distribution. *Polymers*, 2014, *6,* 1877.

457887 Unprecedented Improvement in Mechanical Property Achieved through Layer-By-Layer Assembly of Polymer and Oxidation-Free Graphene

Abstract:

Negatively charged graphene oxide (GO) can be combined with positively charged polymers to produce layer-by-layer assemblies with greatly improved mechanical and electrical properties, which are imparted by the almost parallel orientation of exfoliated GO platelets. However, desirable properties of these assemblies are not maximized unless GO is chemically reduced. It is possible to eliminate this extra reduction step by replacing GO with pristine (i.e., oxidation-free) graphene. But the lack of oxygen functional groups on pristine graphene makes its orientation and exfoliation within a polymer matrix extremely challenging. We have discovered a new way to prepare polymer/pristine-graphene multilayer thin films, which overcomes this problem. This method involves depositing polyvinylpyrrolidone-stabilized graphene platelets alternately with poly(acrylic acid) using hydrogen-bonding assisted layer-by-layer assembly. With this method, we can incorporate pristine graphene sheets into a polymer matrix with excellent orientation and exfoliation, resulting in an unprecedented improvement in mechanical property. A multilayer thin film containing just 3.9 vol% of highly exfoliated and structurally intact graphene increases the elastic modulus (E) of a polymeric multilayer thin film by more than 320% (from 1.41 to 4.81 GPa), while maintaining a visible light transmittance of â ‰ˆ90%. This magnitude of increase in elastic modulus exceeds all previously reported values for any graphene-filled polymer nanocomposite with a glassy (E > 1 GPa) matrix. This method provides a powerful new avenue for improving properties (mechanical, gas transport, etc.) of polymer nanocomposites by enabling the incorporation, exfoliation, and parallel orientation of 2D nanoparticles, which feature excellent intrinsic properties but are initially unsuitable for layer-by-layer assembly due to the lack of essential functional groups.

464855 Orthogonally Spin-Coated Bilayer Films for Photochemical Immobilization and Patterning of Polymer Monolayers

Abstract:

Versatile and spatiotemporally controlled methods for decorating surfaces with monolayers of attached polymers are broadly impactful to many technological applications. However, current materials are usually designed for very specific polymer/surface chemistries and/or do not rapidly respond to high-resolution stimuli such as light. We describe here the use of a polymeric adhesion layer, poly(styrene sulfonyl azide-alt-maleic anhydride) (PSSMA), that is capable of immobilizing a 1-7 nm thick monolayer of preformed, inert polymers via photochemical grafting reactions. Solubility of PSSMA in very polar solvents enables processing alongside hydrophobic polymers or solutions and by extension orthogonal spin-coating deposition strategies. Sequential brush grafting steps using photomasks were used to pattern different regions of surface energy on the same substrate. These patterns spatially controlled the self-assembled domain orientation of a block copolymer, combining top-down and bottom-up nanopatterning approaches.

460431 Assembly of Bioreducible Layer-By-Layer Films for Sequential and Dual Stage DNA Delivery

Abstract:

Polyelectrolyte layer-by-layer (LbL) thin films made by cationic and anionic polymers or bioactive molecules are promising biomaterials in biomedical applications. The layer structure enables control of the amount and location of DNA in the film to be released at different times. The LbL method of assembling polyelectrolyte multilayers (PEMs) has become one of the most promising coating methods capable of mimicking cellular microenvironments and releasing therapeutic agents from the surface of biomedical devices. Our research focuses on the study of the relationship between film assembly and disassembly for potential gene delivery applications. Such biomedical applications generally require the LbL films to disassemble in physiological conditions in order to release the bioactive molecules embedded in the film. We fabricated the LbL film with bioreducible poly(amido amine)s (PAAs) containing the disulfide bond and non-bioreducible polycations in addition to DNA. We show that by combining the two types of polycations, we can maintain the well-defined layer structure in the film and achieve sustained DNA release. The DNA release is based on the bioreducible property of the disulfide bond in PAA. The periodic insertion of non-bioreducible polymer such as poly(ethylenimine) (PEI) as a barrier layer can slow down the release rate to obtain a sequential DNA release pattern. We studied DNA release sequence from the LbL film by atomic force microscopy, fluorescence spectroscopy, and dynamic light scattering. We changed the LbL film disassembly behavior from bulk to sustained release by inserting the PEI barrier layer. The LbL film gene delivery property was demonstrated by transfection experiments in vitro using human embryonic kidney 293 cells and osteoblast precursor cells. By using two different DNA plasmids, we observed sequential DNA release and successful cell transfection. We will present results that directly link the LbL film structure with DNA release dynamics and cell transfection efficiency. Our work demonstrates a simple method for the design of LbL films for sequential, sustained, and dual stage DNA release for biomedical applications.

Surface Characterization of Polymethacrylate Coated Biochips with Water-in-Air Contact Angle Analysis

Samira Hosseini[1], Ivan Djordjevic[2], Leo. H. Koole3, Victor H Perez-Gonzalez[1], Roberto C Gallo-Villanueva[1] and Sergio O. Martinez-Chapa[1,*]

[1] School of Engineering and Sciences, Tecnológico de Monterrey, Monterrey, Mexico
[2] School of Materials Science and Engineering, College of Engineering, Nanyang Technological University. Singapore
[3] Faculty of Health, Medicine and Life Sciences, Maastricht University, the Netherlands

Abstract: - In this paper, the synthesis and development of the polymeric biochips have been described in detail. Different compositions of the copolymer poly methyl methacrylate-co-methacrylic acid poly(MMA-co-MAA) have been synthesized through free-radical polymerization reaction. Biochips aimed for virus detection have been developed by coating silicon wafers by using spin coating technique. A thin layer of coatings on the surface of biochips has been produced from polymer solution of different copolymer compositions. A detailed surface analysis of the biochips was conducted by using of water-in-air contact angle measurement in order to assess the wettability of the coated biochips. Hydrophilicity of the surfaces was further studied by conducting measurements in different pH of the solutions, which has revealed direct correlation between wettability of the analyzed platforms and the presence of carboxyl (-COOH) functional groups on the surface.

Key-Words: Water-in-air contact angle; surface functional groups; hydrophobicity; hydrophilicity; surface property.

1 Introduction

Enzyme-linked immunosorbent assay (ELISA) has found variety of applications in determination of food allergens, recognition of certain types of drugs, detection of cancer biomarkers and hormones and the most commonly detection of viruses in human blood [1]. ELISA methodology lacks from many aspects such as having a lengthy and laborious protocol, lack of reproducibility, inconsistency of the results, and high sample volume required for relatively more accurate detection [2]. The fundamental aspect of ELISA is based on heterogeneous processes occurring on interfaces between analyte and solid phase. Despite the standardization and commercialization of ELISA, the poor performance of the substrate itself can be listed as one of the major drawbacks of this conventional assay. Common materials of choice such as poly methyl methacrylate (PMMA) or polystyrene (PS) for mass production of ELISA analytical kits, are inert in their nature and do not carry desirable surface functional groups. Presence of favorable functional groups on the supporting substrate plays a vital role in efficient immobilization and subsequent virus detection, in particular, when the covalent immobilization is aimed.

In that perspectieve, there are interesting opportunities to use engineered polymer substrates with the tailored surface properties to improve the performance of the assay. Well-designed pltaforms would promote solid-liquid interaction and higher binding affinity between the biomolecules and supporting substrates [3, 4].

In regards to the mentioned limitations of the existing analytical kits and in order to overcome the problems, we have developed a polymeric surface with inherited surface carboxyl (–COOH) groups which can be further employed for detection of viruses via physical and covalent immobilization. By using methyl methacrylate (MMA) and methacrylic acid (MAA), copolymer compositions of poly methyl methacrylate-co-methacrylic acid poly(MMA-co-MAA) in different molar ratios of the monomers have been prepared. Polymerization of different compositions of the mentioned copolymer has been performed to tune the concentration of surface –COOH groups by controlling the reaction condition. If one aspect matters more than presence of desirable functional group on the surface, that would be the optimum concentration of them on the surface [5]. It is known that insufficient concentration of functional groups would lead to the proteins deactivation in close proximity to the surface. On the other hand, overly functionalized surface would also yield in protein deactivation as a result of steric repulsion [6]. In that perspective, polymeric biochips were developed with controlled concentration of surface –COOH groups by using spin coating technique. The surfaces of the biochips coated with different polymer compositions have been thoroughly analyzed by field emission scanning electron microscopy (FESEM) and water-in-air contact angle measurement (WCA).

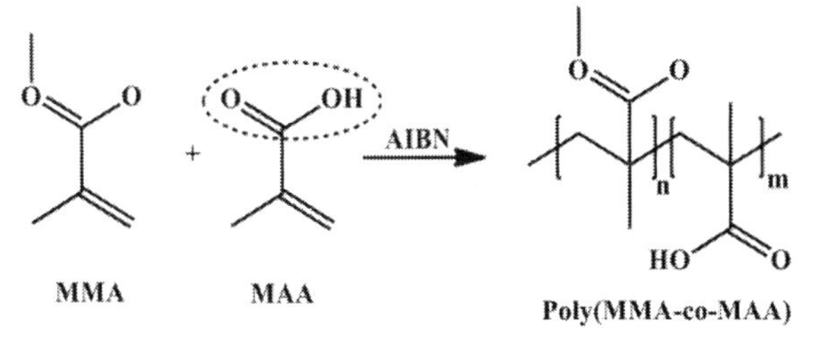

Scheme1. Chemical structure of poly(MMA-co-MAA) and pendant –COOH groups derived from MAA segments.

2 Materials and methods

Different copolymers from the monomers (MMA and MAA) have been synthesized through free-radical polymerization. Compositions of poly(MMA-co-MAA) were prepared by using different molar ratios of MMA/MAA as follows: poly(MMA-co-MAA-9:1), poly(MMA-co-MAA-7:3), poly(MMA-co-MAA-5:5) and pure PMMA as control. For the ease of discussion, further in the text, we refer to the composition as follows: pure PMMA, comp.(9:1), comp.(7:3) and comp.(5:5). Apart from pure PMMA, the numbers in brackets indicate initial MMA/MAA molar ratios. Surface –COOH groups in copolymer compositions are derived from MAA segment of the copolymer as it can be seen in Scheme 1. The polymerization reaction has been performed following the previously reported method by using THF as solvent and AIBN as an initiator [7]. In brief, a two-neck round-bottom flask was fitted with a condenser and has been sealed inlet for its use in reactants feeding. Pre-calculated amount of MMA in 50 ml of THF was added to the set up under stirring condition for 5 minutes. A mixture of MAA monomer and initiator (AIBN, 0.164 g) was slowly added into the reaction mixture. Reaction has been carried out for 8 hours at 90 ºC and stopped by adding reaction mixture into 1L of distilled water. Observed white color precipitation confirmed the formation of copolymer compositions. Mentioned precipitation has been filtered and thoroughly washed with distilled water and dried in the oven of 40°C [7].

Coated biochips have been prepared by using 5% polymer solutions in tetrahydrofuran (THF). The coating procedure was performed by spin coater (model WS-650MZ-23NPP, spinning time of 55s at 3000 rpm). Coated samples were dried in oven (60°C) overnight. The polymerization technique and coating procedure were described in our previous publication in a great detail [7]. Coated surfaces were analyzed by field emission scanning electron microscopy (FESEM, JEOL, JSM7600F. Coated surfaces of different composition have also been analyzed by water-in-air contact angle (Dataphysics Contact Angle System OCA instrument). The volume of the water droplets was 0.1 µl and the measurements have been done after 1 second of the interface [7]. The presented values are the average of 5 different measurements on the center and 4 corners of the sample (3 samples). Furthermore, contact angles for all of the coated biochips were measured by depositing droplets of the acidic (pH~4) and alkaline (pH~10) to study the behavior of the surface towards such different environments. It is important to note that standard deviations were negligible as a very minor spreading was observed for almost all of the measurements (± 2°).

3 Results and discussion

The frontal view and the cross-section images of the coated surfaces have been analyzed by SEM and results are presented in Fig 1. Since no significant difference was observed from one coated biochip to another, images recorded for coated biochips with comp.(9:1) are depicted as representatives. As it can be observed from Fig.1, biochips contain a very smooth and uniform coated layer with the thickness of 10-15 μm.

Fig 1. SEM images of: (a) top view of coated surface and (b) cross-section of coated surfaces with comp.(9:1) as representative.

Figure 2 presents the contact angle measurements of the coated samples in the natural environment (pH=7). It can be observed that the contact angle has gradually decreased by increasing the molar ratio of the MAA monomer in polymerization reaction (Scheme 1). The average contact angle of 78.03 ± 0.7° was recorded for pure PMMA which is in a close agreement with previously reported results [8, 9].

Relatively small changes in water contact angles from pure PMMA to different molar ratios of the poly(MMA-co-MAA) clearly indicate that the surfaces of different compositions have been chemically altered.

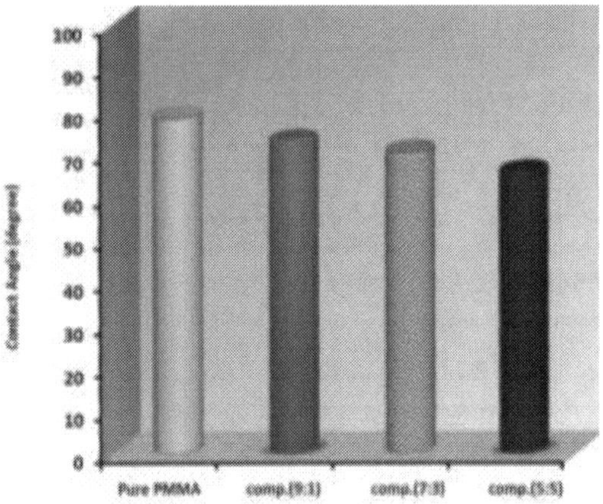

Fig 2. Water-in-air contact angles measured for the coated surfaces of biochips in neutral environment (pH~7).

This gradual decrease in contact angle is due to the increase in concentration of MAA monomers which, in turn, generates free –COOH groups on the coated surfaces. Presence of the –COOH has evidently caused an increase in surface hydrophilicity.

Figure 3 shows the digital images of representative contact angles measured for coated biochips of PMMA and poly(MMA-co-MAA). As it can be observed from figure 3 the lowest value for contact angle (64.9 ± 0.2°) was measured at the surface of comp.(5:5) where the highest concentration of –COOH groups can be expected.

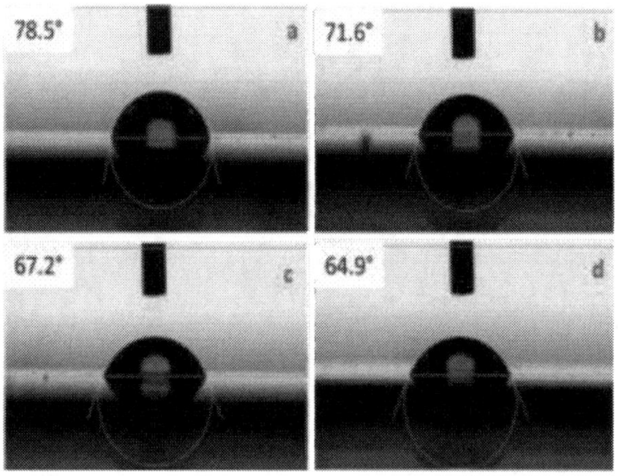

Fig 3. The images of the droplets on the surfaces of coated biochips with: (a) PMMA, (b) comp. (9:1), (c) comp.(7:3) and (d) comp. (5:5).

Further confirmation for the presence of surface –COOH groups have been provided by conducting the contact angle measurements in different environments than neutral such as acidic and alkaline mediums to study the response of the surface to the change in pH of the droplet.

Figure 4 presents the results of the measured contact angle when the deposited droplet was selected from acidic environment (pH~4). Assuming that surface –COOH groups have shown a dissociation rate similar to acetic acid (pKa = 4.76), it can be calculated that approximately 85 % of the surface functional groups may not be dissociated while almost 15 % of –COOH groups can be dissociated to form negatively charged, $-COO^-$ groups. Hence, it was predicted that the poly(MMA-co-MAA) surfaces exhibit relative hydrophobic properties at pH = 4 [10].

As it can be seen, Fig.4 also shows no sign of dramatic changes in the measured contact angles of the developed biochips. However, behavior of the surface has considerably been changed when the droplet of alkaline environment was deposited on the examined surface.

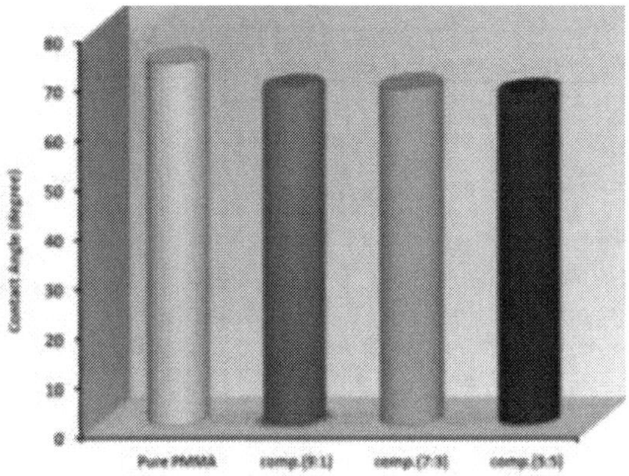

Figure 4. Water-in-air contact angle measured for the coated surfaces of biochips in acidic environment (pH~4).

Figure 5 has shown the measured WCA on the surfaces of the biochips of different compositions by using droplets of alkaline environment.

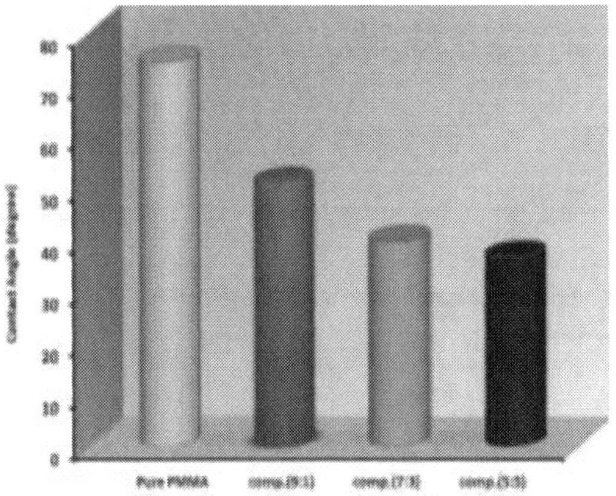

Fig 5. Water-in-air contact angle measured for the coated surfaces of biochips in alkaline environment (pH~10).

Examination of the biochips' surfaces in basic environment has resulted in a general decrease in WCA for all of the poly(MMA-co-MAA) coated surfaces. It is expected that the dissociation percentage of acid groups exceeds to 99.999 in contact with basic environment. Following this phenomenon the relative hydrophobic behavior of the coated surfaces turns to hydrophilic behavior as presented WCA values in Fig.5 suggest. This is quite understandable as multiple negatively charged $-COO^-$ groups at alkaline contact impose higher hydrophilicity to the surfaces of the tested biochips. As it can be seen and compared from the results presented in Fig.2, Fig.4 and Fig.5, such a dramatic change in WCA with pH has not happened when PMMA coated chips have been examined with droplets of acidic or alkaline environment. It is expected as there exist no –COOH groups inside the structure and at the outmost layer of the surface of this particular polymer chips (Scheme 1). Presented data in Fig 5, approvingly confirm the hypotheses that indicates: by introducing MAA monomer to the polymerization reaction, natural –COOH groups will be formed in the structure of the resultant copolymer materials; by varying the molar ratios of the monomers in synthesis procedure, the variation of the surface –COOH groups can be controlled as well. The gradual decrease in WCA upon going from pH = 4 and/or pH = 7 to pH = 10 reached to the lowest recorded angles for comp.(5:5) which is concentrated with the abundance of –COOH functionalities [10, 11]. The detailed WCA analysis performed in this study confirms the predetermined ratios of the monomers involved in polymerization reaction, to some extent.

Such a simple but essential surface characterization provides unequivocal evidence for existence of –COOH groups at the surface of the biochips of different compositions. This fact, to a large extent, overcomes the major concern about generation of effective surface functional groups that can play a vital role in protein immobilization and subsequent detection of viruses. Most importantly, generated functional groups in this study are expected to be resistant towards aging effect or reorientation phenomenon that happens over the time, as they are part of the chemical fingerprint of the molecule and would not be affected by any outsider forces. Lastly, our investigation recalls the important fact that instead of applying relatively uncertain methods for generation of desirable functionalities, a tunable functionalized platform with controlled surface characteristics can be chemically synthesized.

4 Conclusion

Polyacrylate coated biochips with different molar ratios of the monomers have been developed for the further application in virus detection. Free-radical polymerization reaction was performed for preparation of different compositions of the copolymers by which a thin layer of coating on the silicon wafer have been developed. Designed coated surfaces inherited the presence of surface carboxyl groups from the MAA segments of the copolymers. Variety of surface –COOH concentration has been achieved by variation of the molar ratios of the monomers. Surfaces of the developed biochips have been thoroughly characterized by WCA measurement in order to investigate the wettability of the developed surfaces. Results have shown gradual increase in the values of WCA as the concentrations of the surface –COOH groups were increased proving that the WCA is a direct function surface concentration of –COOH groups. Moreover, WCA has been measured for all of the developed platforms in contact with acidic and basic droplets. Recorded WCA values were significantly lower in alkaline contact which is due to the dissociation of the surface functional groups. Surface analysis with mediums of different pH has given a further evidence for the presence of –COOH groups on the surface. Surface functional groups are part of the chemical structure of the copolymers hence would not be affected by time or known phenomena such as aging effect. Developed biochips in our study have shown a great potential for its application in diagnostic systems. Presence of desirable surface functionality, in particular, is of a major importance as such functional groups can be effectively used in protein immobilization via either ways of physical or covalent attachment.

Acknowledgment

Authors would like to acknowledge the financial support of Tecnologico de Monterrey, Mexico, for the special grant (grant number: 002EICII01) awarded to the Sensors and Devices Focus Group, School of Engineering and Sciences, Tecnologico de Monterrey, Monterrey, Mexico.

References

[1] Kirsch J, Siltanen C, Zhou Q, Revzin A, Simonian A. Biosensor technology: recent advances in threat agent detection and medicine. Chemical Society Reviews 2013;42:8733-68.

[2] Bai Y, Koh CG, Boreman M, Juang Y-J, Tang IC, Lee LJ, et al. Surface Modification for Enhancing Antibody Binding on Polymer-Based Microfluidic Device for Enzyme-Linked Immunosorbent Assay. Langmuir 2006;22:9458-67.

[3] Hosseini S, Azari P, Farahmand E, Gan SN, Rothan HA, Yusof R, et al. Polymethacrylate coated electrospun PHB Fibers: An exquisite outlook for fabrication of paper-based Biosensors. Biosensors and Bioelectronics.

[4] Hosseini S, Ibrahim F, Djordjevic I, Rothan HA, Yusof R, Marel Cvd, et al. Synthesis and characterization of methacrylic microspheres for biomolecular recognition: Ultrasensitive biosensor for dengue virus detection. European Polymer Journal 2014;60:14-21.

[5] Hosseini S, Ibrahim F, Djordjevic I, Koole LH. Recent advances in surface functionalization techniques on polymethacrylate materials for optical biosensor applications. Analyst 2014;139:2933-43.

[6] Hosseini S, Ibrahim F, Djordjevic I, Rothan HA, Yusof R, van der Mareld C, et al. Synthesis and Processing of ELISA Polymer Substitute: The Influence of Surface Chemistry and Morphology on Detection Sensitivity. Applied Surface Science 2014;317:630–8.

[7] Hosseini S, Ibrahim F, Djordjevic I, Koole LH. Polymethyl methacrylate-co-methacrylic acid coatings with controllable concentration of surface carboxyl groups: A novel approach in fabrication of polymeric platforms for potential bio-diagnostic devices. Applied Surface Science 2014;300:43-50.

[8] Park IJ, Lee S-B, Choi CK. Surface Properties of the Fluorine-Containing Graft Copolymer of Poly((perfluoroalkyl)ethyl methacrylate)-g-poly(methyl methacrylate). Macromolecules 1998;31:7555-8.

[9] Brown L, Koerner T, Horton JH, Oleschuk RD. Fabrication and characterization of poly(methylmethacrylate) microfluidic devices bonded using surface modifications and solvents. Lab on a Chip 2006;6:66-73.

[10] Hosseini S, Ibrahim F, Rothan HA, Yusof R, Marel Cvd, Djordjevic I, et al. Aging effect and antibody immobilization on –COOH exposed surfaces designed for dengue virus detection. Biochemical Engineering Journal 2015;99:183–92.

[11] Schmidt SW, Christ T, Glockner C, Beyer MK, Clausen-Schaumann H. Simple Coupling Chemistry Linking Carboxyl-Containing Organic Molecules to Silicon Oxide Surfaces under Acidic Conditions. Langmuir 2010;26:15333-8.

465567 Designing Durable Icephobic Surfaces

Abstract:

Ice accretion has a negative impact on critical infrastructure, as well as a range of commercial and residential activities. Icephobic surfaces are defined by an ice adhesion strength τ_{ice} < 100 kPa. However, the passive removal of ice requires much lower values of τ_{ice}, such as on airplane wings or power lines (τ_{ice} < 20 kPa). Such low τ_{ice} values are scarcely reported, and robust coatings that maintain these low values have not been reported previously. Here we show that, irrespective of material chemistry, by tailoring the crosslink density of different elastomeric coatings, and by enabling interfacial slippage, it is possible to systematically design coatings with extremely low ice-adhesion (τ_{ice} < 0.2 kPa). These newfound mechanisms allow for the rational design of icephobic coatings with virtually any desired ice adhesion strength. We also propose a new model for interfacial slippage, based on Flory-Huggins theory coupled with a Hemi-wicking analysis. Our model not only allows for icephobic elastomers, polymers and monolayers to be systematially designed, but also makes a connection to previous icephobic systems based on lubrication. Utilizing our newfound mechanisms and model, we fabricate extremely durable coatings that maintain τ_{ice}< 10 kPa after severe mechanical abrasion, acid/base exposure, 100 icing/de-icing cycles, thermal cycling, accelerated corrosion, and exposure to Michigan wintery conditions over several months.

This work was published in *Science Advances* in March of 2016. DOI: 10.1126/sciadv.1501496.

470189 Surface Morphology, Chemistry, and Modulus of Novel Polyorganosiloxane-Based Coatings and Their Effect on Ice Adhesion

Abstract:

One of the ongoing effects of global climate change is the opening of previously inaccessible waterways in the Arctic. The availability of these areas to shipping and resource exploitation necessitates an increased coast guard and naval presence in these regions. Operating in conditions of extreme cold has leads to a number of issues, not the least of which is the build-up of ice on ship superstructures. As a result, the development of coatings that minimize ice adhesion is an on-going priority. We have studied the impact of coating properties on ice-adhesion for a series of novel polyorganosiloxane-based coatings. These coatings, based on the Hybridsil platform developed by Nanosonic, Inc (Pembroke, VA,) have exhibited exceptionally low ice adhesion and durability properties. The current work has focused on understanding the relationship between the surface morphology, chemistry, and modulus of the coatings on the ice adhesion behavior. Attenuated total reflectance Fourier-transform infrared spectroscopy (ATR-FTIR) was carried out to examine the chemical nature of the coating surface. In general, only small differences were observed in the IR spectra between samples. Atomic Force Microscopy (AFM) was used to image the coatings and to measure the modulus at different locations. Significant differences in film roughness and homogeneity are observed in the AFM imaging, both in the height images and the phase contrast images, depending on the specific formulation and method of coating application. These lead to significant differences in ice adhesion. In addition, force curves were obtained for a subset of samples and were analyzed to determine the film moduli. We observed differences in the film moduli both between different samples and at different locations for a single sample (i.e. between low and high points on the film.) These spatial variations also impact the strength of ice adhesion on the coatings. The coating chemistry is important in order to create a durable film with a low affinity for interactions with water ice; however, it is vitally important to understand and control the coating surface properties in order to achieve the desired low values for ice adhesion.

463448 Adaptable Elastin-like Protein – Hyaluronic Acid (ELP – HA) Hybrid Hydrogels with Tunable Stress Relaxation Rates for Cell-Matrix Interaction Studies

Abstract:

Synthetic hydrogels have been widely used as *in vitro* models to study cell-matrix interactions and elucidate signal transduction mechanisms in biomaterials science. However, these hydrogels are typically elastic with minimal viscous dissipation, whereas native extracellular matrices (ECMs) are viscoelastic with stress relaxation characteristics. Thus, to better mimic the native cellular microenvironment and to understand the role of stress relaxation in guiding cell behaviors, we designed a viscoelastic hybrid hydrogel system consisting of protein-engineered, elastin-like protein (ELP) and chemically modified hyaluronic acid (HA). This family of adaptable ELP-HA hydrogels was formed *via* dynamic covalent hydrazone linkages through the reactions between hydrazine groups on ELP and aldehyde or benzaldehyde groups on HA. By combining the precise sequence control of protein-engineered materials and the kinetic features of dynamic covalent chemistry, we were able to independently tune the stress relaxation rate without causing changes to several other important matrix parameters, including polymer concentration, initial stiffness, and cell-adhesive ligand density. Human intestinal organoids were encapsulated within this family of ELP-HA hydrogels to evaluate their potential as *in vitro* biomimetic scaffolds for cell-matrix interaction studies. While all cultures remained highly viable, matrix stress relaxation rate appeared to influence cell proliferation rate and the ability of the cells to properly polarize and organize into an organoid-like tissue with an epithelial-coated internal lumen.

Targeting Cancer Cells via Tumor-homing Peptide CREKA Functional PEG Nanoparticles

Aysu Ceren Okur[1+], *Pelin Erkoc*[1+], *Seda Kizilel*[1,2]*

[1]*Koç University,* Biomedical Sciences and Engineering, [2]*Chemical and Biological Engineering, Istanbul, 34450, Turkey*

e-mail: skizilel@ku.edu.tr
Fax: 90-212-338-1538

+: equal authorship

*: corresponding author

Keywords: water-in-water emulsion; hydrogel nanoparticles; CREKA; IKVAV

Abstract

Targeting cell microenvironment via nano-particle based therapies holds great promise for treatment of various diseases. One of the main challenges in targeted delivery of nanoparticles for cancer therapy includes reduced localization of delivery vehicles at tumor site. The therapeutic efficacy of drugs can be improved by recruiting delivery vehicles towards specific region of tumorigenesis in the body. Here, we demonstrate an effective approach in creating PEG particles via water-in-water emulsion technique where tumor-homing peptide CREKA was used for functionalization. Simultaneous conjugation of laminin peptide IKVAV into hydrogel network and influence of altered combinations of ligands on intracellular uptake of anticancer drugs by HeLa cells were investigated. CREKA conjugated hydrogel nanoparticles were more effective to improve apoptotic effects of the model drug Doxorubicin (DOX) compared to that of particles conjugated with other peptides. Fluorescence intensity analysis on confocal micrographs suggested significantly higher cellular uptake of CREKA conjugated PEG particles than internalization of nanoparticles in other groups. We observed that fibrin binding ability of PEG particles could be increased up to 94% through CREKA conjugation. Our results suggest the possibility of cancer cell targeting via CREKA-functional PEG nanoparticles.

1. Introduction

Targeting cancer cells via nanoparticle drug delivery vehicles contributed to significant advances in biomedical applications.[1-3] Functionalization of nanoparticles through peptides with specific affinities towards transmembrane proteins and extracellular matrix elements can be an effective strategy for tumor directed drug delivery. To improve internalization of nanoparticles into cancer cells via receptor-mediated endocytosis several peptides have been used that provided advantages such as low immunogenicity, stability in blood fluid and ease of use.[4-9] IKVAV (Isoleucine-Lysine-Valine-Alanine-Valine) is a cell adhesive peptide, derived from α_1 chain of laminin basement membrane protein and was first used to mimic neural microenvironment and trigger neuronal differentiation.[10, 11] However, studies about the role of IKVAV in tumor targeting and imaging are limited. IKVAV is known to interact with different cancer cell types through binding integrins $\alpha_3\beta_1$ and $\alpha_6\beta_1$.[12-15] Recently, $\alpha_6\beta_1$-integrin has been identified as a functionally related receptor in glioma stem-like cells (GSC). GSCs are chemotherapy resistant cell populations in glioblastoma multiforme, and interfere with the binding $\alpha_6\beta_1$ to laminin leading to decreased tumor formation. An IKVAV epitope carrying self-assembling peptide amphiphile was found to increase immobilized β_1-integrin at the GSC membrane, inhibited focal adhesion kinase (FAK) and consequently induced apoptosis in GSCs.[16] In another study, radiolabeled IKVAV was injected intravenously into mice with lung tumors, where peptide contributed to significant localization in lungs compared to other tissues.[10, 12] IKVAV also binds to nucleolin expressed on tumor cell surface, which is a new molecular target for cancer therapy.[17, 18]

Besides targeting of cell membrane proteins through the use of integrin binding peptides, a tumor-homing peptide CREKA (Cysteine–Arginine–Glutamic acid–Lysine–Alanine) has been considered for targeting of tumor ECM.[11] CREKA was first selected by *in* vivo phage display peptide libraries that home selectively to tumor blood vessels. Type IV collagen in basement vascular membrane is the receptor for CREKA binding.[6, 19] Due to leaky vasculature of cancer microenvironment, fibrin deposition occurs in tumor ECM, but not in healthy tissues. CREKA has specific affinity towards these deposited fibrin clots.[4-6] Previously, CREKA was conjugated to iron oxide nanoparticles, poly(ethylene glycol)–polylactic acid polimeric nanoparticles, poly(amidoamine) dendrimer nanoparticles and amphiphile micelles.[4, 6, 11, 20, 21] *In vivo* studies suggested significantly improved retention and anti-cancer effects of CREKA conjugated vehicles compared to controls.[6, 11, 21] Even though several studies exist related to tumor-homing potential of CREKA, the knowledge about molecular target of CREKA is limited to *in vivo* phage library screening. Comparison of alternative integrin-binding peptides with CREKA may provide insight into its binding site and may improve its targeting capacity. In addition, tumor-homing function of CREKA peptide in blood may be diminished due to short half-lives as a result of degradation of peptide by endogenous proteases.[22] Hence, finding efficient delivery vehicles for CREKA is important for cancer diagnostics.

Water-in-oil emulsion polymerization has been commonly employed for nano-scale synthesis of hydrogel particles. In these studies, generally hydrogel microparticles were used to encapsulate co-cultures of varying cell types.[23] However, the use of organic solvents, and surfactants used in emulsion processes is not desirable in biological systems.[24] Hydrogel particles can also be prepared in aqueous medium using water-in-water emulsion. For example, PEG hydrogel microparticles can be synthesized in dextran, a natural polysaccharide, utilizing the immiscibility of two aqueous phases of polymers above a certain concentration.[25-27] Application of water-in-water emulsion system as a drug delivery vehicle was limited to only few reports where VEGF and calmodulin proteins were encapsulated in PEG microparticles, and dispersed prepolymer phase were crosslinked in dextran and raffinose continuous phases by UV light photopolymerization. These hydrogel microparticles were then characterized with respect to their chemical compositions, morphologies and size distributions via light microscopy.[24, 25]

Here, our objective was to use water-in-water emulsion for the synthesis of hydrogel particles that have targeting potential towards cancer cells. PEG was chosen as a polymeric network due to its low immunogenicity and low surface charge that extends blood circulation period when used as a delivery vehicle.[28-31] Dextran was selected as the continuous phase in this system, a natural analogue of PEG due to its similar properties such as solubility in water, biocompatibility and anti-fouling. To achieve higher internalization into cancer cells, we reduced the sizes of particles produced by water-in-water emulsion technique down to nanoscale and used visible light photopolymerization. Next, cross-linked PEG particles were recovered from dextran continuous phase after rinsing (**Scheme 1**).

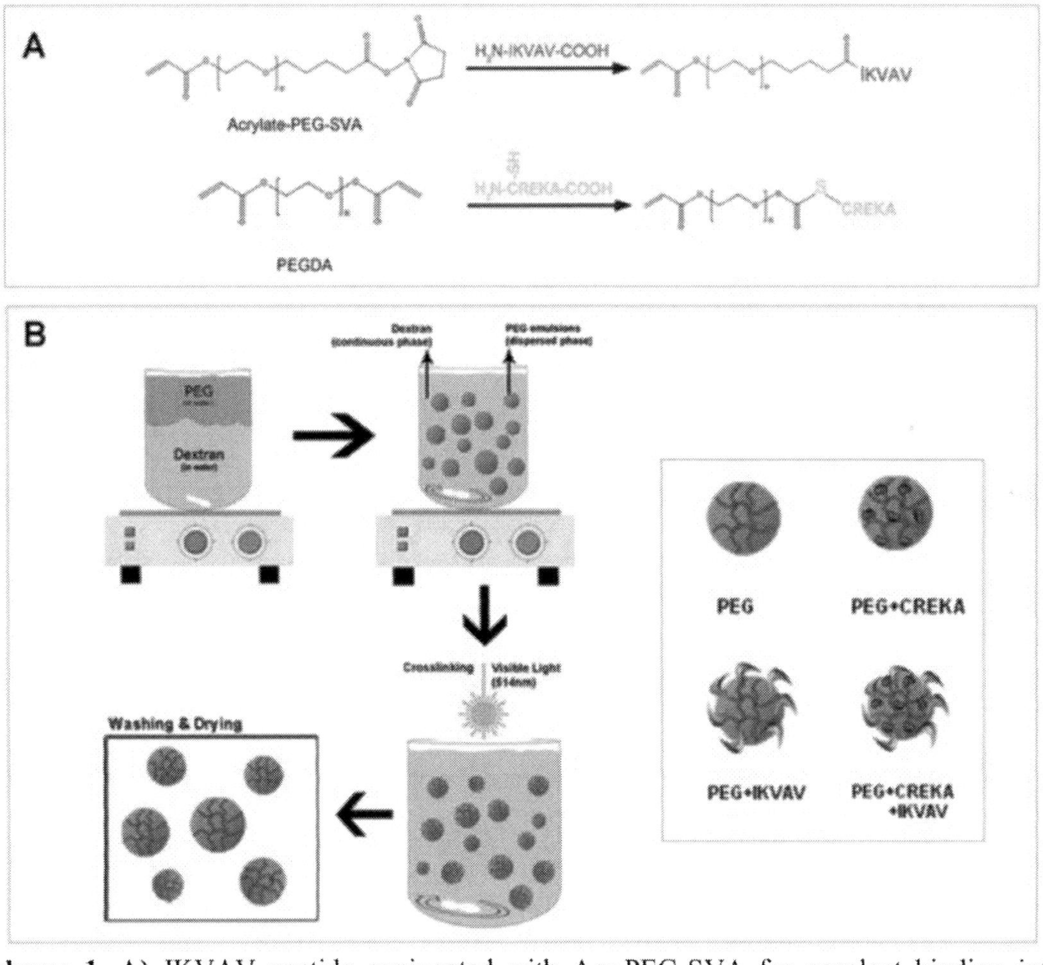

Scheme 1. A) IKVAV peptide conjugated with Acr-PEG-SVA for covalent binding into hydrogel structure, CREKA was bonded to PEGDA polymer chains via thiol groups at the terminal cysteine. **B)** Illustration of water-in-water emulsion approach and functionalization of hydrogel particles with various peptides and peptide groups.

Targeting potentials of PEG particles were investigated through conjugation with IKVAV and CREKA peptides into PEG hydrogel nanocarriers and their targeting efficiencies were further studied in cervical cancer cells (HeLa). Results suggested these delivery vehicles as promising cancer therapeutics to improve cellular uptake and efficacy of drugs. This study is original for simultaneous investigation of combinatory targeting potentials of ECM peptides and utilization of water-in-water emulsion for the synthesis of targeted nanocarriers.

2. Materials and methods

2.1. Materials

Potassium Chloride (KCl), Sodium Bicarbonate (NaHCO3), Fluorescein o-acrylate, Eosin-Y, 1-vinyl- 2-pyrrolidone (VP), Dextran (40 kDa), Poly(ethylene glycol) diacrylate (PEGDA, MW: 575 Da), Triton X-100 and dialysis tubing cellulose membrane were purchased from Sigma–Aldrich (St. Louis, MO). Phosphate Buffer Saline (PBS), Trypsin–EDTA and Penicilin/streptomycin were purchased from Gibco (Life Technologies, Grand Island, NY). Amicon Ultra Centrifugal Filter (NMWL: 60 kDa), cellulose and nylon filters (0.45 μm), in vitro fibrin binding assay, Triethanol-amine (TEA) were obtained from Merck. Doxorubicin Hydrochloride (DOX) was purchased from MP Biomedicals (Solon, OH). Dulbecco's Modified Eagle Medium (DMEM), L-Glutamine and 4',6-Diamidino-2-Phenylindole, dihydrochloride (DAPI) were purchased from Invitrogen (Carlsbad, CA). IKVAV (MW:528 Da), CREKA (MW:605 Da) were purchased from Elim Biopharmaceuticals (Hayward, CA). Acrylate PEG-Succinimidyl Valerate (Acr-PEG-SVA, Mw:3.4 kDa) was purchased from Laysan Bio (Arab, AL). Dialysis cassette (NMWL:3.5 kDa) and Fluoraldehyde™ o-Phthaldialdehyde Reagent Solution (OPA) were obtained from ThermoFisher Scientific (Waltham, Mass). Fetal Bovine Serum (FBS) was obtained by Bioindustries (Cromwell, CT). Cell-Titer-GLO (CTG) viability assay was obtained from Promega (Madison, WI). Cell culture plates were purchased from Greiner Bio-One (Solingen, Germany).

2.2. Preparation of emulsions and synthesis of PEG hydrogel particles

In order to obtain crosslinked and functional PEG hydrogel particles, we used water-in-water emulsion polymerization, where PEG prepolymer was dispersed in dextran phase. First the peptides and peptide-PEG-acrylate conjugates are prepared, then functional hydrogel particles were synthesized with CREKA and IKVAV peptides (**Scheme 1**). For CREKA functionalization, peptide (0.5 mM) was directly added into prepolymer solution, where acrylate groups in PEGDA bound via thiol ether linkage group with sulfur on cysteine. IKVAV (2.5 mM) was conjugated to acr-PEG-SVA from amine terminus in basic pH conditions. The conjugation reaction was carried out for 2 h at room temperature and the solution was dialyzed with dH_2O and lyophilized for storage.

Free amine percentages for both CREKA and IKVAV conjugation reactions were measured to calculate covalent binding efficiency of the peptides using fluorometric fluoraldehyde o-phthalaldehyde (OPA) assay. Briefly, standard solutions of each peptide with altered concentrations were prepared in PBS. Three different concentrations of acr-PEG-IKVAV conjugate was also prepared in PBS. PEGDA prepolymer soultions were prepared with and without CREKA and incubated at room temperature for 1h. All samples and standard solutions were reacted with fluoraldehyde reagent (1:10 v/v) and fluorescence was monitored at 340 nm excitation and 455 nm emission wavelengths using a plate reader (Biotek, Synergy H1).[32, 33] The standard solutions were used to prepare calibration curves.

Prepolymer solution was prepared in dH_2O with 518 mM PEGDA, 225 mM TEA (co-initiator) and 37 mM VP (accelerator) and 0.025 mM Eosin Y (initiator) at pH 8. PEG prepolymer was added to 20% (w/w) dextran and stirred to obtain prepolymer emulsion droplets for 45 minutes. Following exposure to green light (λ: 514 nm, P: 2.5 mW/cm^2, Argon ion laser-Coherent, Innova 70C), gels were rinsed with centrifugal filters and were lyophilized. These gels were then used for DOX loading and subsequent experiments. The conversion of PEG monomer in this system is about 80 %, and crosslinking density is 6×10^{-4}.[31, 34]

2.3. Structural, chemical and physical characterization of hydrogel particles

Particles were characterized in terms of chemical and physical structure, size and morphology. Field Emission Scanning Electron Microscope (FESEM) (Zeiss Ultra Plus, Bruker, MA) and Atomic Force Microscope (AFM) (The Dimension Icon, Bruker, MA) were used to characterize structure and morphology of the gels. For FESEM characterization, lyophilized particles were placed on conductive carbon layer and coated with 10 nm thick gold. In AFM measurements, Peak Force QNM mode in air, RFESP (MPP 21100-10) cantilever (Veeco probes, USA, k=3 N/m, fo: 86-94 kHz) was used and NanoScope Analysis 1.5 software was used to export images. Dynamic light scattering (DLS) (Zetasizer, Malvern ZS) was used to determine hydrodynamic radii of particles in water. To confirm peptide incorporation, Fourier Transform Infrared Spectrometer (FTIR) (Thermo Scientific iS10 FT-IR), with a single reflection diamond ATR was used. CREKA presence was also confirmed by X-ray Fluorescence Spectrometer (Bruker, Tiger S8).

2.4. Fibrin Binding Assay

Fibrin gels were synthesized in 96-well plates. Each well was treated with fibrinogen and thrombin solutions with 60 % and 40 % (v/v), respectively. Solutions were reacted at 37°C with continuous shaking for 4 hours. Next, 0.5 mM and 1 mM free CREKA peptides, particles without peptides (1 mg/ml), and particles covalently bonded with 0.5 mM and 1 mM CREKA were added on fibrin gels and incubated for 1h. Initial absorbance and absorbance after incubation period was measured at 220 nm with NanoDrop2000 Spectrophotometer (Thermo Scientific, Mass, US). Binding efficiency (%) of molecules to fibrin layer was calculated using Equation (1). First, absorbance of the wells was measured in the beginning, where the wells included CREKA, PEG or CREKA-PEG solutions. This absorbance was defined as total absorbance. Next, these solutions were transferred to the wells that included fibrin gels, and incubated for one hour. Next, fibrin gels were rinsed, and the rinsing buffer was combined with the liquids in the wells before rinsing step. The absorbance of this combined fluid was measured, and this is defined as unbound absorbance.

$$Fibrin\ \ Binding\ \ (\%) = \frac{Total\ \ Absorbance - Unbound\ \ Absorbance}{Total\ \ Absorbance} \quad (1)$$

2.5. Drug loading and release from hydrogel particles

Drug-loading was carried out with 10 mg of particles in 0.1 mg/ml DOX solution in an orbital-shaker with vigorous mixing at room temperature. Unloaded DOX was removed with ultracentrifugal filters. Particles were loaded into a cellulose membrane, and incubated in 10 ml PBS supplemented with 10 % FBS at 37°C. Samples were taken from the outside medium of cellulose membrane at predetermined time points for 6 days. DOX absorbance of collected samples was measured with NanoDrop at 480 nm. Loading percentage was calculated from the weight fraction of the drug in doxorubicin-loaded particles using Equation (2), and drug release was calculated using Equation (3) and Equation (4), where C_t is the concentration of DOX in the buffer at time t, V is the total volume of the buffer, V_s is the volume of the sample taken. Using the ratio of M_t values to total weight of DOX released, % mass release was calculated.

$$Loading\ \ efficiency = \frac{(M_0 - M_f)}{M_f} * \ 100 \quad (2)$$

$$M_t = C_t * V \ + \ \sum C_{t-1} * \ V_s \quad (3)$$

$$Mass\ \ Release\ \ (\%) = (M_t / M_\infty) * \ 100 \quad (4)$$

2.6. Cytotoxicity Assay on HeLa and Human Dermal BJ Cells

HeLa cells and human BJ skin fibroblasts were cultured in DMEM supplemented with 10% FBS, 2 mM L-glutamine, and 1% penicillin/streptomycin. Cells were maintained in a humidified incubator with 5% CO_2 at 37°C. To examine the cytotoxicity of gels, cells were seeded to 96-well plates at 5000 cells/well and pre-incubated overnight. Media in wells were removed and particles in different groups were suspended in DMEM at 0.1 mg/ml concentration were added. Effect of hydrogels on cytotoxicity was determined via CTG assay after 24 and 72hours incubation. Luminescence was normalized to control cells which were not treated with gels. Cytotoxicity of DOX-loaded hydrogel particles was also evaluated in HeLa cells, similarly.

2.7. Internalization of DOX- loaded particles

HeLa cells (5000 cells/well) were seeded on previously autoclaved glass slides placed in 24-well plates. Meanwhile, gel groups were incubated with DOX in DMEM with a ratio of 10:1 (w/w) for 24 hours. Next, DOX loading was determined, measuring the absorbance at 480 nm by spectrophotometer and calculating the ratio of absorbance after centrifuge to absorbance before centrifuge. Media of seeded cells were replaced with DOX-loaded particles suspended in DMEM. After 1hour incubation, cells were fixed as described previously.[35] Cell nuclei were stained with DAPI. Cellular uptake of DOX was visualized by laser scanning confocal microscopy (Eclipse 90-i, Nikon). As a sign of cellular internalization efficiency, DOX/DAPI intensity ratio of cell images captured on constant gain and intensities were quantified using Image J. Same method was also applied to human dermal BJ fibroblast cells to quantify DOX internalization in these cells.

2.8. SEM Imaging of Particle Treated HeLa Cells

HeLa cells (5000 cell/well) were seeded on glass slides and incubated with 0.1 mg/ml of hydrogel particles for 1 hour. Cells were fixed with 4% glutaraldehyde and 0.1 % osmium tetroxide at room temperature, following dehydration with increasing concentration gradient of ethanol and slides were dried with a critical point dryer in 40°C and 112 bars for 1 hour. These samples were fixed on carbon tape attached stubs and coated with 10 nm gold layer using a sputter coater (Cressington). Size and morphology of gel treated cells were analyzed via Zeiss Evo LS15 SEM (Bruker) at an acceleration potential of 5 kV.

2.9. Statistical analysis

Data were analyzed using GraphPad Prism5 software package. Oneway ANOVA with Tukey's Multiple Comparison post-test was performed as was indicated in the results. Differences were considered as statistically significant when the p-value was less than 0.05.

3. Results and discussion

3.1. Preparation and characterization of peptide precursors

PEG was covalently bonded to IKVAV peptide using Acr-PEG-SVA through the reaction between SVA and amine groups, where CREKA was bonded to PEGDA polymer chains via Michael addition between thiol groups at the terminal cysteine and acrylate groups in PEGDA (Scheme 1A).[30, 36] Conjugation reaction of IKVAV and functional groups were confirmed by FTIR (Fig. S1a) and amount of free amines present after conjugation was measured by OPA assay. IKVAV conjugation efficiency was calculated as 98% and CREKA binding efficiency to PEGDA was found as 99% (Fig. S1b). The presence of CREKA in the hydrogel network was confirmed by X-ray fluorescence spectroscopy (XRF), where the peak at 2.307 keV demonstrates Kα peak from sulphur atom coming from the cysteine amino acid of CREKA peptide (Fig. S2).

3.2. Synthesis and characterization of particles

Prepolymer solution was emulsified in dextran and was exposed to visible light for crosslinking. This method resulted in the formation of covalently cross-linked hydrogel particles in sub-micron size scale. Briefly, the dispersed droplets of prepolymer solution containing PEGDA, TEA, VP and eosin Y were obtained in dextran continuous phase. Upon visible light exposure dispersed PEG prepolymer droplets were covalently crosslinked to form submicron-scale hydrogel particles. Acr-PEG-IKVAV and CREKA were also added into the prepolymer solution to obtain functional hydrogel particles.[30]

Previous studies about the synthesis of hydrogels via water-in-water emulsion resulted in the formation of micron size particles, where characterization was based on optical microscope imaging.[24, 25, 27] Here, nano-sized hydrogel particles were synthesized through optimization of emulsions, where detailed characterization was carried out using FESEM, AFM and DLS. FESEM and AFM micrographs in **Fig. 1** revealed the size distribution within the range of 140-400 nm. FESEM and AFM images for the IKVAV incorporated particles were also found in submicron size (**Fig. S3-S4**).

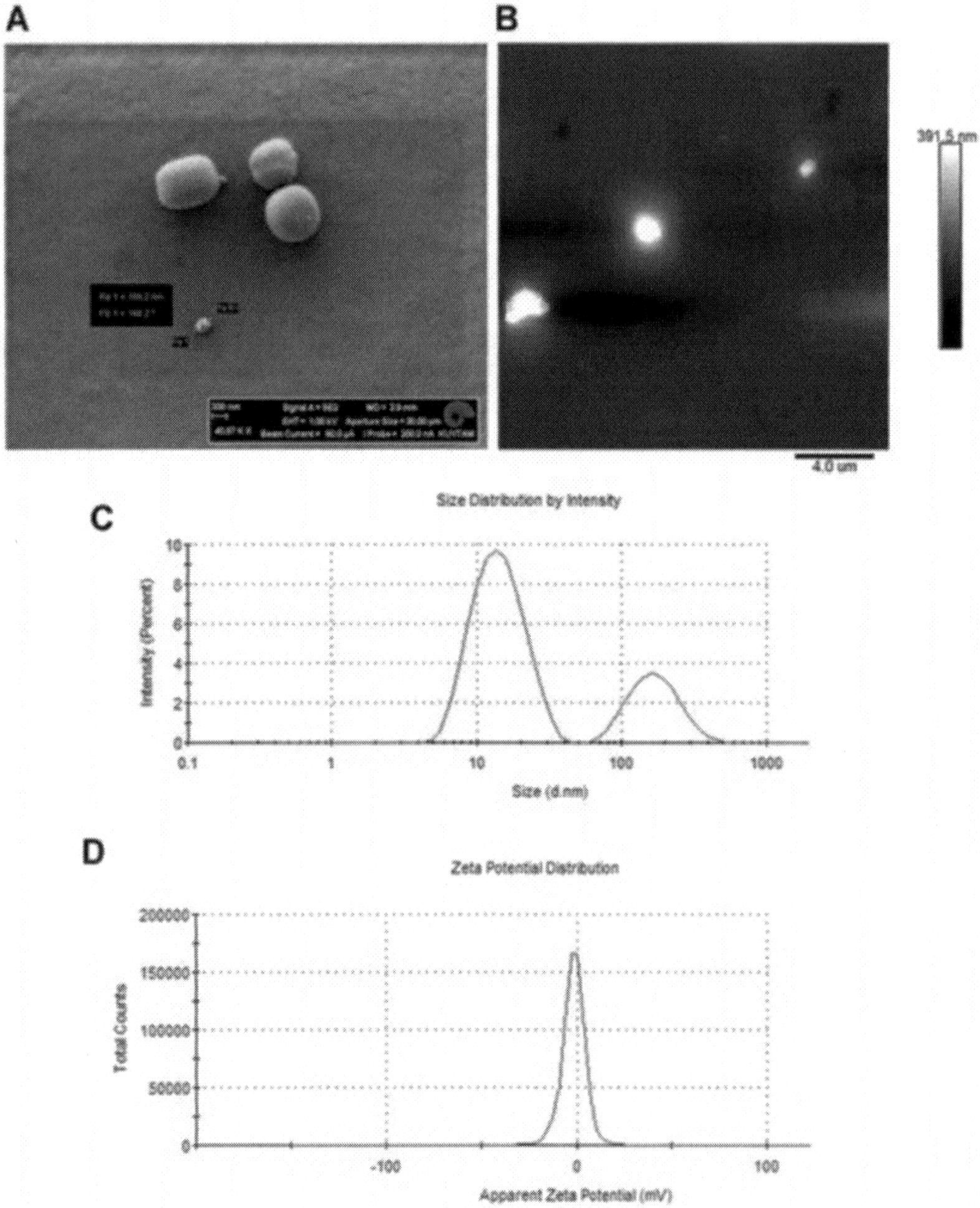

Fig. 1. A) FESEM image of lyophilized particles. The scale bar corresponds to 200 nm. B) AFM image of particles obtained by vacuum drying of aqueous suspension of particles on

glass slides. **C)** DLS characterization of particles depending on intensity percentage. **D)** Zeta potentials of particles.

DLS measurements revealed that hydrodynamic radii of the hydrogel particles were approximately 15 nm for 71 % and 178 nm for 25 % of particles (**Fig. 1C**). Peptide conjugated particles exhibited similar size distribution in DLS measurements (**Fig. S5-S6).** Nanoparticles smaller than 40 nm could not be imaged by FESEM, as this was beyond the limits of imaging capacity of FESEM for polymeric nanoparticles (**Fig. 1A**). Zeta potential of particles in aqueous environment was measured as -3.38 ± 5.79 (**Fig. 1D**). This slightly negative zeta potential value resulted from the lone pair electrons of oxygen atoms in PEG chains, and is consistent with the previous observations in the literature.[35] Previously, it was demonstrated that neutral particles could escape from Kupffer cell clearance and allocate intensely and homogenously in cancer bearing region.[21, 37, 38]

3.3. DOX loading and release from hydrogel particles

DOX molecules were loaded into swollen polymer matrixes, due to the high water uptake capacity of PEG hydrogel particles and electrostatic interaction between slightly negatively charged PEG particles and positively charged DOX, which represents weak base at physiological conditions.[39] The amount of DOX-loaded into particles was calculated as 62 %, based on the measurement of absorbance at 480 nm.

DOX release experiment was carried out to characterize the release profile of the drug from particles. Cumulative release profile of DOX and experimental data was fitted to cumulative Gaussian percentages of nonlinear regression method as was shown in **Fig. 2.**

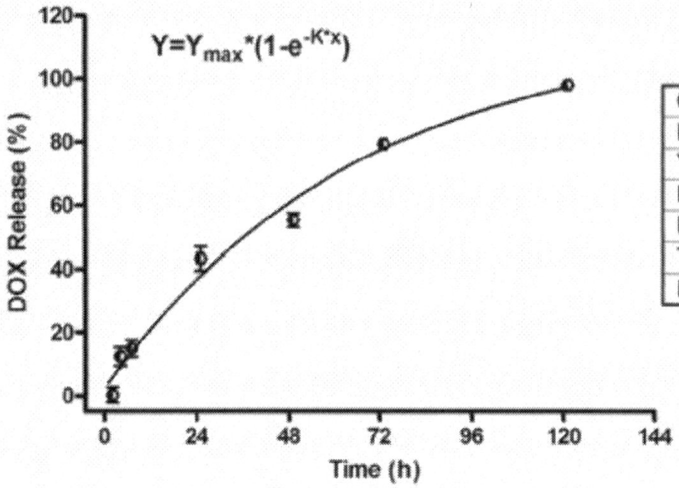

One-phase association	
Best-fit values	
Y0	4.125
Plateau	115.8
K	0.01503
Tau	66.52
Half-time	46.11

Fig. 2. DOX release versus time from DOX-loaded targeted hydrogel particles. Total drug release was calculated from absorbance values of DOX measured at 480 nm at predetermined time points. One-phase exponential association analysis was used for curve fitting to calculate release half-time of the drug. (n=3, R^2=0.9860).

DOX release reached to saturation in 144 hours. Higher release of DOX obtained during initial hours of the experiment may be due to the presence of DOX on hydrogel particle surface, which may result from the static interactions between PEG network and small DOX molecules that might have influenced the release profile.[40] One-phase association equation of nonlinear regression method was applied for the curve fitting of the *in vitro* drug release kinetics as follows:

$$Y = Y_{max} * \left(1 - e^{-K*X}\right) \tag{5}$$

$$Tau = 1/K \tag{6}$$

$$Half - time = \ln(2)/K \tag{7}$$

In equation 5, K represents the rate constant, expressed in inverse minutes; Tau is the time constant. Equation 5 assumes a pseudo-first order release kinetics, where certain fraction of the drug is released during each time interval where fewer number of drug molecules are released as time advances and the curve levels off.[41] Half-time of DOX release from gels was calculated as 46 hours. R^2 value was obtained as 0.9860 suggesting the good level of fitting obtained for experimental data.

3.4. Binding affinity of CREKA and CREKA-hydrogel particles to fibrin clots

In vitro fibrin binding assay was performed to investigate the targeting potential of CREKA functionalized particles (**Fig. 3**).

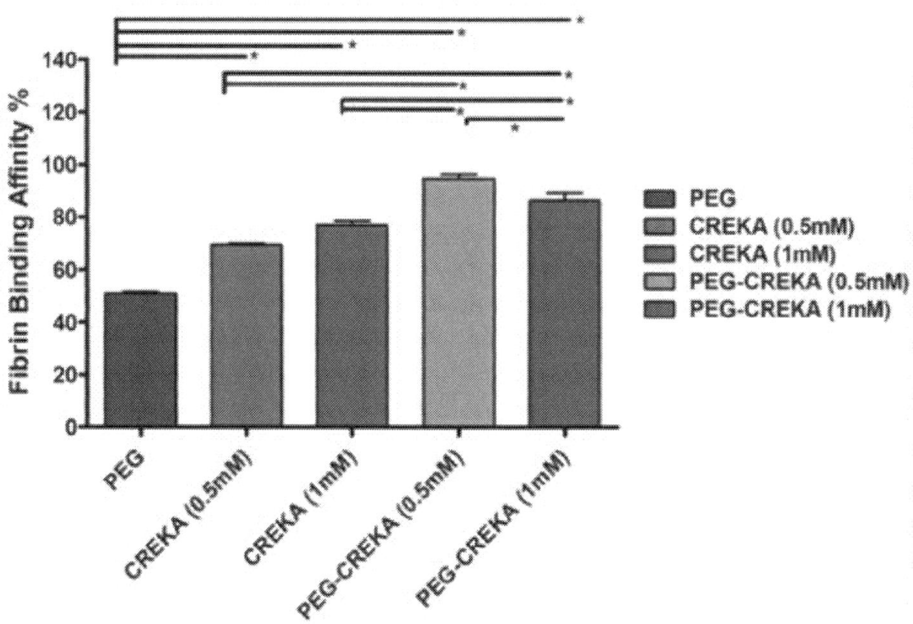

Fig. 3. Fibrin binding affinity of free and PEG bound CREKA peptides. Affinities of particles to fibrin clots were assayed in the presence and absence of two different concentrations (0.5 mM and 1 mM) of CREKA. One-way ANOVA followed by Tukey's multiple comparison test, Results indicated with * show statistical significance between the control group and other groups, $P < 0.05$, n=3.

In a previous study, fibrin-binding ratio related to bound-to-free form of CREKA were measured for CREKA only, CREKA-conjugated iron oxide nanoparticles (CREKA-IONP), and FITC-IONP from FITC absorbance as 0.69, 0.62 and 0.42, respectively.[20] We also obtained similar ratios, where fibrin-binding percentages were found approximately 70 % for free CREKA peptides and 45 % for particles without CREKA. Low level of non-specific affinity measured for PEG particles to fibrin can be explained simply by the transport of particles to underlying fibrin gel due to gravity. CREKA conjugated particles demonstrated high binding affinities towards fibrin for both concentrations of CREKA (0.5 mM and 1 mM) compared to that of CREKA peptide alone. The affinity for binding to fibrin at 0.5 mM was higher than 1 mM CREKA. The slight differences in affinity might suggest that particles reached to saturation at 0.5 mM CREKA during synthesis process. Since particles with 0.5 mM CREKA conjugation demonstrated the highest affinity of 94 % towards fibrin binding, we continued the rest of the experiments with 0.5 mM CREKA.

3.5. Cytocompatibility of particles

HeLa cells and non-cancerous healthy cell line human BJ dermal fibroblasts were incubated with 0.1 mg/ml hydrogel particles with or without peptide groups for characterization of cytocompatibility. Cell-Titer-GLO luminescent viability assay was used to investigate cytotoxicity of particles with altered combinations of IKVAV and CREKA. The luminescence values measured were normalized with respect to untreated control group. The viabilities measured for both HeLa cells and human BJ dermal fibroblasts were above 80 % and that particles were not toxic for both HeLa and BJ cells **(Fig. 4A, Fig. 4B)**.

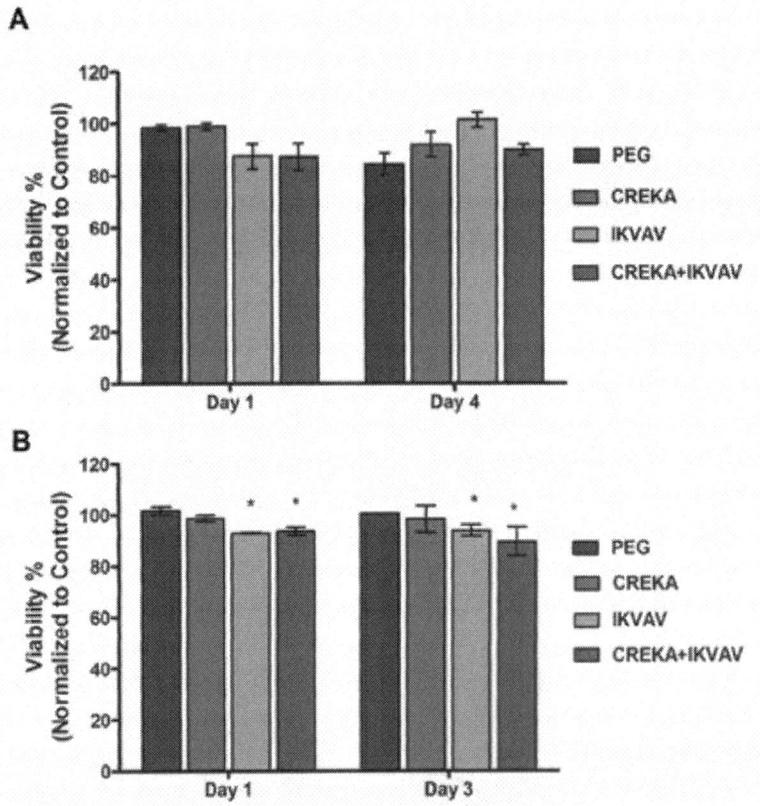

Fig. 4. A) Cell viability of HeLa and **B)** BJ cells. Viabilities were measured on day 1 and day 3 with experimental groups; untreated, PEG particles and peptide functionalized particles treated cells. One-way ANOVA followed by Tukey's multiple comparison test (*indicates statistical significance compared to the control group, $P < 0.05$, n=3).

3.6. Cellular uptake and in vitro cytotoxicity of DOX-loaded particles

Cellular internalization of particles by HeLa cells was also investigated via laser scanning confocal microscopy. Particles were loaded with DOX, a chemotherapeutic drug that has specific excitation at 480 nm and emission maximum at 560-590 nm.[35, 42] DOX-loaded particle treated cells were fixed with formaldehyde and their nuclei were stained with DAPI. Images in DOX and DAPI/DOX merged panels indicated that internalization of CREKA incorporated gels by HeLa cells was significantly higher than internalization of particles in other groups (**Fig. 5A** and **5B**).

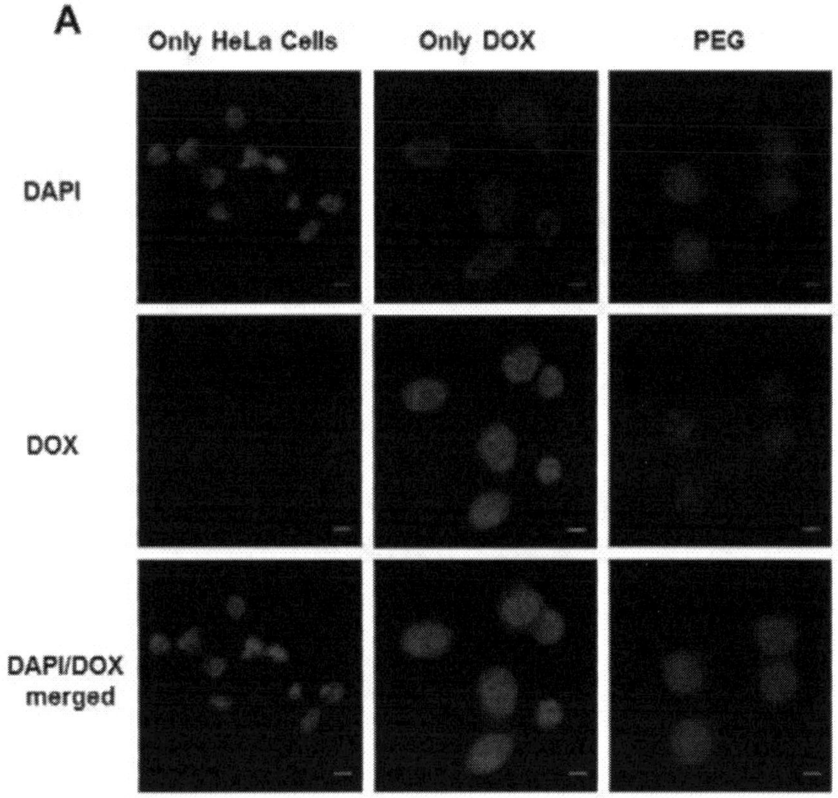

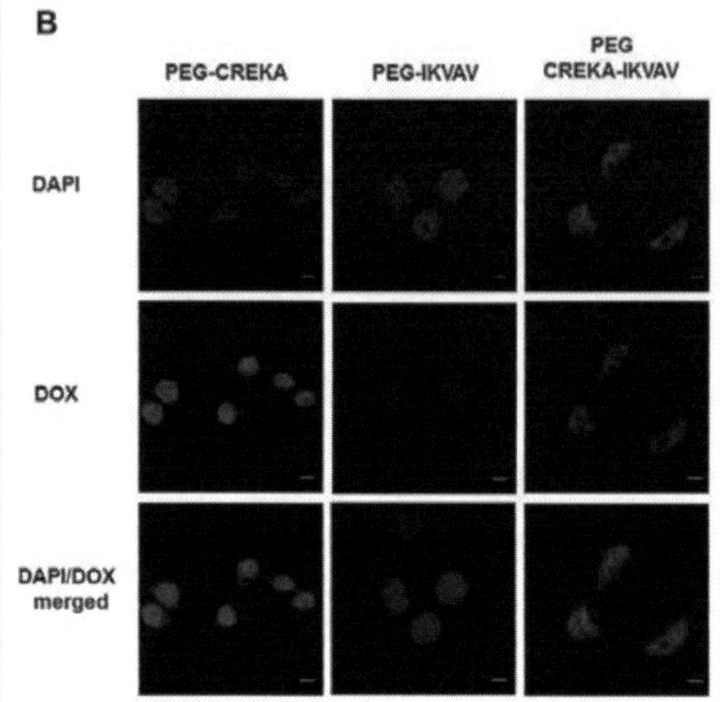

Fig. 5. Cellular internalization of free DOX and DOX-loaded particles visualized by laser scanning confocal microscope. First row illustrates cell nuclei staining with DAPI (blue); middle raw DOX fluorescence and third row, overlay images of first two rows (scale bar =10 μm).

It was also observed that simultaneous conjugation of CREKA with IKVAV did not improve internalization of particles by HeLa cells further (**Fig. 5B**). The reason for higher red fluorescence intensity in free DOX sample than the hydrogel treated samples, except CREKA bound ones, can be explained by the need for a longer uptake period due to larger size of DOX-loaded in hydrogels than its free form. Uptake efficacy was further quantified as the ratio of DOX (red) to DAPI (blue) fluorescence intensity (**Fig. 6**).

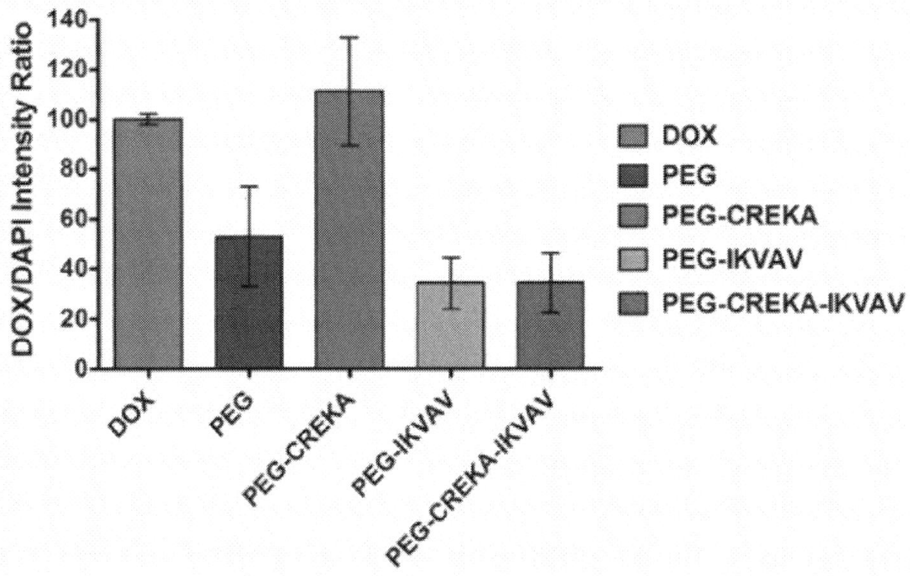

Fig. 6. Quantification of fluorescence intensity based on confocal microscopy images using Image J. One-way ANOVA was followed by Tukey's multiple sample comparison (*indicates statistical significance between PEG-CREKA and other groups, $P < 0.05$, n=3).

DOX-loaded HeLa cells were considered as control and results in the rest were normalized with respect to DOX fluorescence in only DOX incubated HeLa cells. DOX uptake was approximately 50% in the presence of hydrogel particles without any peptide functionality. This high amount of internalization may be related to dextran residues that remained on hydrogel particle surface. The contribution of dextran to internalization of particles was observed previously, in which higher internalization of dextran micelles was explained possibly by the probe role of dextran in fluid phase pinocytosis.[43, 44]

Internalization of particles by HeLa was significantly improved with CREKA conjugation into particles, compared to the conditions of DOX in the cell culture medium alone or IKVAV functional particles or particles with dual peptide functionality. Particle internalization was 111 % for PEG-CREKA, and 34 % for both PEG-IKVAV and PEG-CREKA-IKVAV functionalized particles (Fig. 5 and Fig. 6).

IKVAV target different domains of integrin, where CREKA targets collagen IV in fibrin structure. The differences in targeting domains of these peptides might be the reason for altered efficiencies observed for these peptides. An earlier study showed that HeLa cells expressed collagen IV as an extracellular matrix protein.[45] Hence, increased targeting potential of CREKA can be explained by increased secretion of collagen IV in HeLa cell microenvironment. We observed relatively lower levels of uptake in combinations of CREKA with IKVAV compared to the uptake observed with only CREKA functionalized particles. Reduced internalization of particles despite the presence of CREKA with hydrogel particles of dual peptide functionality could be attributed to the accessibility of CREKA peptide to fibrin. PEGDA chains were conjugated to thiol groups of CREKA peptide via Michael-type addition, where IKVAV was conjugated into the network through the acrylate group of acrylate-PEG-peptide molecules. The presence of long PEG chains in IKVAV-PEG-acrylate and the absence of long PEG chain in CREKA functionalized conjugates might have blocked interactions of CREKA with fibrin, and hence resulted in reduced uptake compared to internalization of CREKA only functionalized particles (**Scheme 1A**).[46] Overexpression of nucleolin in HeLa cells serves as a ligand for IKVAV.[47, 48] Expression of this receptor might have also contributed to the uptake of IKVAV containing hydrogel particles in our study. To evaluate target selectivity of the nanoparticles quantitatively, we carried out control experiments using healthy (non-tumor) cells and the gel nanoparticles with peptides. We investigated cellular uptake of nanoparticles by healthy cells, where we treated healthy human fibroblast cells with PEG nanoparticles that included altered combinations of peptides. Fluorescence images of HeLa and fibroblast cells incubated with DOX and DOX-loaded CREKA-PEG nanoparticles were shown in **Fig. S7**. Pseudo-color maps of the corresponding scattering intensity of cells were generated to highlight intensity differences between samples (**Fig. S7**). When the fluorescence intensities of cell nuclei for DOX-loaded nanoparticle treated HeLa cells and fibroblasts were compared, the cellular uptake was significantly higher in HeLa cells than fibroblasts and that alterations in peptide sequences did not change uptake significantly in fibroblast cells as they did in HeLa cells.

The cellular uptake was also evaluated in HeLa cells via viability assay in the presence of DOX-loaded nanoparticles with altered peptide sequences (**Fig. S8**). The cellular viability was reduced to approximately 20 % in the presence of nanoparticles after 1day treatment, however peptide incorporation to the hydrogels did not significantly affect the viability.

Particle uptake was finally characterized with FESEM for higher spatial resolution (**Fig. 7**). At high magnification it was observed that cells preserved their morphology and integrity after treatment with particles. Focal adhesions were visible around the cell ECM and apoptotic bodies or vesicles known as blebs that started to form near the surface of cells (**Fig. 7B, Fig. S9**).[49]

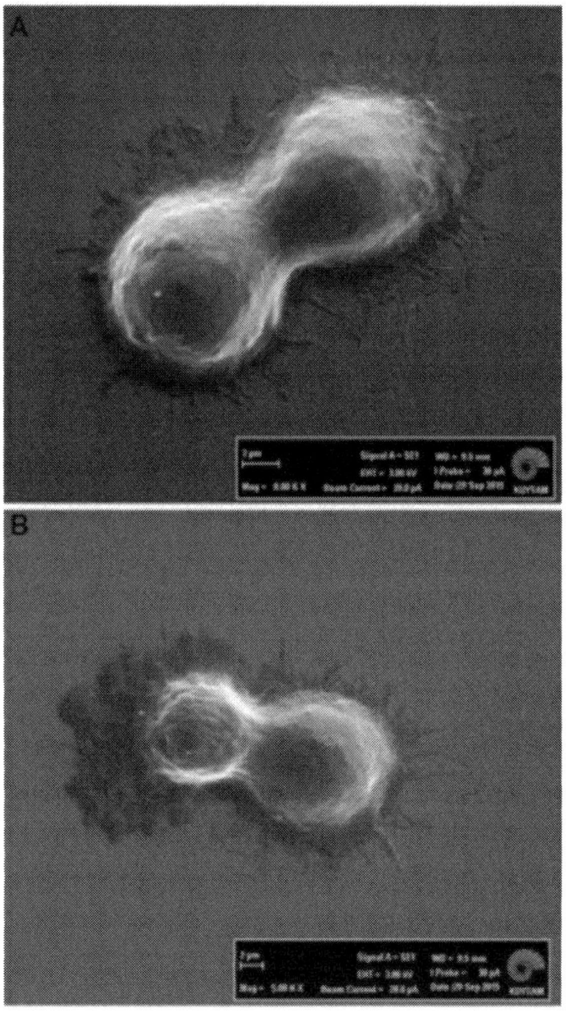

Fig. 7. SEM images of **A**) HeLa cells, **B**) HeLa cells treated with CREKA incorporated particles. Scale bars are equal to 2 μm. Adobe Photoshop CS4 software was used to obtain false colored SEM images of HeLa cells.

The occurrence of blebs might have resulted from the higher cell permeabilization due to the presence of hydrogel particles. However, blebs can also be observed in healthy cells during cell locomotion and division.[50] According to these images, cells preserved their morphology, and the hydrogel particles were observed on the surface of cells.

4. Conclusions

In this study, we developed emulsion-based synthesis of sub-micron biocompatible hydrogel particles without the use of organic solvents and investigated selective targeting potential of ECM derived peptides for cancer therapy. The potential of targeting efficacy of these carriers were studied with cancer cells using an anticancer drug, DOX. The targeting potentials of the two ligands were evaluated, where *in vitro* experiments suggested higher targeting potential of CREKA compared to that of IKVAV. In addition, combination of peptides was not superior with respect to internalization of particles. Incorporation of CREKA into hydrogel network improved internalization of the delivery vehicle, which would be significant to reduce systemic side effects of cancer drugs. This study suggests improved targeting that could be obtained with delivery vehicles towards cancer cells with CREKA functionalization compared to targeting that could be achieved with laminin peptide IKVAV. Targeted PEG nanocarriers developed here is promising to promote the efficacy of drugs, to decrease undesirable side effects of the drug, and to eliminate the resistance of cancer tissue towards drug.

Acknowledgements

The authors would like to thank Cansu Yıldırım and Barıs Yagcı for providing assistance with FESEM, XRF measurements at Koc University Surface Science and Technology Center (KUYTAM), and Tugba Bal for providing help with the preparation of peptides. We thank Dr. Tugba Bagcı Onder at School of Medicine at Koc University for providing human fibroblast cells.

References

[1] C.E. Ashley, E.C. Carnes, G.K. Phillips, D. Padilla, P.N. Durfee, P.A. Brown, T.N. Hanna, J. Liu, B. Phillips, M.B. Carter, N.J. Carroll, X. Jiang, D.R. Dunphy, C.L. Willman, D.N. Petsev, D.G. Evans, A.N. Parikh, B. Chackerian, W. Wharton, D.S. Peabody, C.J. Brinker, The targeted delivery of multicomponent cargos to cancer cells by nanoporous particle-supported lipid bilayers, Nat Mater, 10 (2011) 389-397.
[2] D.J. Irvine, Drug delivery: One nanoparticle, one kill, Nat Mater, 10 (2011) 342-343.
[3] S.E. Jin, H.E. Jin, S.S. Hong, Targeted delivery system of nanobiomaterials in anticancer therapy: from cells to clinics, Biomed Res Int, 2014 (2014) 23.
[4] E.J. Chung, Y. Cheng, R. Morshed, K. Nord, Y. Han, M.L. Wegscheid, B. Auffinger, D.A. Wainwright, M.S. Lesniak, M.V. Tirrell, Fibrin-binding, peptide amphiphile micelles for targeting glioblastoma, Biomaterials, 35 (2014) 1249-1256.
[5] P.P. Karmali, V.R. Kotamraju, M. Kastantin, M. Black, D. Missirlis, M. Tirrell, E. Ruoslahti, Targeting of albumin-embedded paclitaxel nanoparticles to tumors, Nanomedicine : nanotechnology, biology, and medicine, 5 (2009) 73-82.
[6] D. Simberg, T. Duza, J.H. Park, M. Essler, J. Pilch, L. Zhang, A.M. Derfus, M. Yang, R.M. Hoffman, S. Bhatia, M.J. Sailor, E. Ruoslahti, Biomimetic amplification of nanoparticle homing to tumors, Proc Natl Acad Sci U S A, 104 (2007) 932-936.
[7] J.S. Wadia, S.F. Dowdy, Transmembrane delivery of protein and peptide drugs by TAT-mediated transduction in the treatment of cancer, Adv Drug Deliv Rev, 57 (2005) 579-596.
[8] W.C. Chen, A.X. Zhang, S.-D. Li, Limitations and niches of the active targeting approach for nanoparticle drug delivery, European Journal of Nanomedicine, 4 (2012) 89-93.
[9] A. Accardo, D. Tesauro, G. Morelli, Peptide-based targeting strategies for simultaneous imaging and therapy with nanovectors, Polymer Journal, 45 (2013) 481-493.
[10] P.O. Zamora, D. Eshima, D. Graham, L. Shattuck, B.A. Rhodes, Biological distribution of 99mTc-labeled YIGSR and IKVAV laminin peptides in rodents: 99mTc-IKVAV peptide localizes to the lung, Biochim Biophys Acta, 1182 (1993) 197-204.
[11] J. Wu, J. Zhao, B. Zhang, Y. Qian, H. Gao, Y. Yu, Y. Wei, Z. Yang, X. Jiang, Z. Pang, Polyethylene glycol-polylactic acid nanoparticles modified with cysteine-arginine-glutamic acid-lysine-alanine fibrin-homing peptide for glioblastoma therapy by enhanced retention effect, Int J Nanomedicine, 9 (2014) 5261-5271.
[12] Y. Kikkawa, K. Hozumi, F. Katagiri, M. Nomizu, H.K. Kleinman, J.E. Koblinski, Laminin-111-derived peptides and cancer, Cell adhesion & migration, 7 (2013) 150-256.
[13] R.S. Bresalier, B. Schwartz, Y.S. Kim, Q.Y. Duh, H.K. Kleinman, P.M. Sullam, The Laminin Alpha-1 Chain Ile-Lys-Val-Ala-Val (IKVAV)-Containing Peptide Promotes Liver Colonization by human colon-cancer cells, Cancer Res., 55 (1995) 2476-2480.
[14] T.E. King, S.C. Pawar, L. Majuta, I.C. Sroka, D. Wynn, M.C. Demetriou, R.B. Nagle, F. Porreca, A.E. Cress, The role of alpha 6 integrin in prostate cancer migration and bone pain in a novel xenograft model, PLoS One, 3 (2008) e3535.
[15] J.M. Whitelock, M.L. Paine, J.R. Gibbins, R.F. Kefford, R.L. Ogrady, Multiple levels of post-transcriptional regulation of collagenase (matrix metalloproteinase 1) in an epithelial cell line, Immunol Cell Biol, 71 (1993) 39-47.
[16] M. Srikanth, S. Das, E.J. Berns, J. Kim, S.I. Stupp, J.A. Kessler, Nanofiber-mediated inhibition of focal adhesion kinase sensitizes glioma stemlike cells to epidermal growth factor receptor inhibition, Neuro Oncol, 15 (2013) 319-329.
[17] M.C. Kibbey, B. Johnson, R. Petryshyn, M. Jucker, H.K. Kleinman, A 110-kD nuclear shuttling protein, nucleolin, binds to the neurite-promoting IKVAV site of laminin-1, Journal of neuroscience research, 42 (1995) 314-322.

[18] T. Watanabe, K. Hirano, A. Takahashi, K. Yamaguchi, M. Beppu, H. Fujiki, M. Suganuma, Nucleolin on the Cell Surface as a New Molecular Target for Gastric Cancer Treatment, Biol Pharm Bull, 33 (2010) 796-803.
[19] A. Grayson, S.E. Zale, D. DeWitt, Therapeutic particles suitable for parenteral administration and methods of making and using same, Google Patents, 2013.
[20] A.M. Kruse, S.A. Meenach, K.W. Anderson, J.Z. Hilt, Synthesis and characterization of CREKA-conjugated iron oxide nanoparticles for hyperthermia applications, Acta Biomater, 10 (2014) 2622-2629.
[21] J. Zhao, B. Zhang, S. Shen, J. Chen, Q. Zhang, X. Jiang, Z. Pang, CREKA peptide-conjugated dendrimer nanoparticles for glioblastoma multiforme delivery, Journal of colloid and interface science, 450 (2015) 396-403.
[22] D. Zanuy, A. Flores-Ortega, A.I. Jimenez, M.I. Calaza, C. Cativiela, R. Nussinov, E. Ruoslahti, C. Aleman, In silico molecular engineering for a targeted replacement in a tumor-homing peptide, The Journal of Physical Chemistry B, 113 (2009) 7879-7889.
[23] C.L. Franco, J. Price, J.L. West, Development and optimization of a dual-photoinitiator, emulsion-based technique for rapid generation of cell-laden hydrogel microspheres, Acta Biomaterialia, 7 (2011) 3267-3276.
[24] W.J. King, M.W. Toepke, W.L. Murphy, Facile formation of dynamic hydrogel microspheres for triggered growth factor delivery, Acta Biomaterialia, 7 (2011) 975-985.
[25] N.A. Impellitteri, M.W. Toepke, S.K.L. Levengood, W.L. Murphy, Specific VEGF sequestering and release using peptide-functionalized hydrogel microspheres, Biomaterials, 33 (2012) 3475-3484.
[26] O. Franssen, W.E. Hennink, A novel preparation method for polymeric microparticles without the use of organic solvents, International Journal of Pharmaceutics, 168 (1998) 1-7.
[27] M. Parlato, A. Johnson, G.A. Hudalla, W.L. Murphy, Adaptable poly(ethylene glycol) microspheres capable of mixed-mode degradation, Acta Biomaterialia, 9 (2013) 9270-9280.
[28] S. Kizilel, E. Sawardecker, F. Teymour, V.H. Perez-Luna, Sequential formation of covalently bonded hydrogel multilayers through surface initiated photopolymerization, Biomaterials, 27 (2006) 1209-1215.
[29] K. Xiao, Y. Li, J. Luo, J.S. Lee, W. Xiao, A.M. Gonik, R.G. Agarwal, K.S. Lam, The effect of surface charge on in vivo biodistribution of PEG-oligocholic acid based micellar nanoparticles, Biomaterials, 32 (2011) 3435-3446.
[30] T. Bal, B. Kepsutlu, S. Kizilel, Characterization of protein release from poly(ethylene glycol) hydrogels with crosslink density gradients, Journal of biomedical materials research. Part A, 102 (2014) 487-495.
[31] S. Kizilel, Mathematical Model for Microencapsulation of Pancreatic Islets within a Biofunctional PEG Hydrogel, Macromolecular Theory and Simulations, 19 (2010) 514-531.
[32] G.Y. Wu, C.J. Meininger, Analysis of citrulline, arginine, and methylarginines using high-performance liquid chromatography, Method Enzymol, 440 (2008) 177-189.
[33] D.L. Hern, J.A. Hubbell, Incorporation of adhesion peptides into nonadhesive hydrogels useful for tissue resurfacing, Journal of Biomedical Materials Research, 39 (1998) 266-276.
[34] S. Kizilel, V.H. Pérez-Luna, F. Teymour, Modeling of PEG Hydrogel Membranes for Biomedical Applications, Macromolecular Reaction Engineering, 3 (2009) 271-287.
[35] C. Nazli, G.S. Demirer, Y. Yar, H.Y. Acar, S. Kizilel, Targeted delivery of doxorubicin into tumor cells via MMP-sensitive PEG hydrogel-coated magnetic iron oxide nanoparticles (MIONPs), Colloids and Surfaces B: Biointerfaces, 122 (2014) 674-683.
[36] T. Bal, C. Nazli, A. Okcu, G. Duruksu, E. Karaoz, S. Kizilel, Mesenchymal stem cells and ligand incorporation in biomimetic poly(ethylene glycol) hydrogels significantly improve insulin secretion from pancreatic islets, J Tissue Eng Regen Med, (2014).
[37] S. Stolnik, L. Illum, S.S. Davis, Long circulating microparticulate drug carriers, Advanced Drug Delivery Reviews, 16 (1995) 195-214.

[38] T. Neuberger, B. Schöpf, H. Hofmann, M. Hofmann, B. von Rechenberg, Superparamagnetic nanoparticles for biomedical applications: Possibilities and limitations of a new drug delivery system, Journal of Magnetism and Magnetic Materials, 293 (2005) 483-496.
[39] J.O. Kim, A.V. Kabanov, T.K. Bronich, Polymer micelles with cross-linked polyanion core for delivery of a cationic drug doxorubicin, Journal of Controlled Release, 138 (2009) 197-204.
[40] M. Goncalves, P. Figueira, D. Maciel, J. Rodrigues, X. Qu, C. Liu, H. Tomas, Y. Li, pH-sensitive Laponite((R))/doxorubicin/alginate nanohybrids with improved anticancer efficacy, Acta Biomaterialia, 10 (2014) 300-307.
[41] Q. Yang, K.R. Moulder, M.S. Cohen, S. Cai, L.M. Forrest, Cabozantinib Loaded DSPE-PEG Micelles as Delivery System: Formulation, Characterization and Cytotoxicity Evaluation, BAOJ pharmaceutical sciences, 1 (2015).
[42] P.E. Saw, J. Park, E. Lee, S. Ahn, J. Lee, H. Kim, J. Kim, M. Choi, O.C. Farokhzad, S. Jon, Effect of PEG pairing on the efficiency of cancer-targeting liposomes, Theranostics, 5 (2015) 746-754.
[43] R. Raveendran, G.S. Bhuvaneshwar, C.P. Sharma, Hemocompatible curcumin-dextran micelles as pH sensitive pro-drugs for enhanced therapeutic efficacy in cancer cells, Carbohydrate Polymers, 137 (2016) 497–507.
[44] C. Commisso, R.J. Flinn, D. Bar-Sagi, Determining the macropinocytic index of cells through a quantitative image-based assay, Nat. Protocols, 9 (2014) 182-192.
[45] J.J. Furth, T.H. Wroth, S. Ackerman, Genes for collagen types I, IV, and V are transcribed in HeLa cells but a postinitiation block prevents the accumulation of type I mRNA, Exp Cell Res, 192 (1991) 118-121.
[46] J.J. Furth, T.H. Wroth, S. Ackerman, Genes for collagen types I, IV, and V are transcribed in Hela cells but a postinitiation block prevents the accumulation of type I mRNA, Experimental cell research, 192 (1991) 118-121.
[47] C. Nazli, T.I. Ergenc, Y. Yar, H.Y. Acar, S. Kizilel, RGDS-functionalized polyethylene glycol hydrogel-coated magnetic iron oxide nanoparticles enhance specific intracellular uptake by HeLa cells, Int J Nanomedicine, 7 (2012) 1903-1920.
[48] X. Chen, D.M. Kube, M.J. Cooper, P.B. Davis, Cell surface nucleolin serves as receptor for DNA nanoparticles composed of pegylated polylysine and DNA, Molecular therapy : the journal of the American Society of Gene Therapy, 16 (2008) 333-342.
[49] L. Liu, J.K. Zhang, M.L. Li, X.H. Zhang, J.L. Zhang, Z.J. Li, L.K. Wang, J.H. Wu, C. Luo, Inhibition of HepG2 cell proliferation by ursolic acid and polysaccharides via the downregulation of cyclooxygenase-2, Mol Med Rep, 9 (2014) 2505-2511.
[50] L.F. Barros, T. Kanaseki, R. Sabirov, S. Morishima, J. Castro, C.X. Bittner, E. Maeno, Y. Ando-Akatsuka, Y. Okada, Apoptotic and necrotic blebs in epithelial cells display similar neck diameters but different kinase dependency, Cell Death Differ, 10 (2003) 687-697.

469855 Directing Vascular Regeneration *in-Situ*

Abstract:

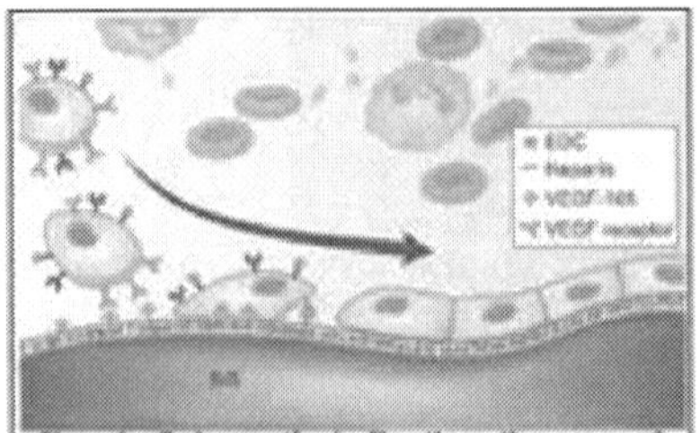

Fig. 1: Schematic indicating the use of VEGF to capture blood borne, VEGFR-expressing cells and repopulate the vascular graft lumen.

Recently our group demonstrated that immobilized VEGF can capture circulating endothelial cells from the blood *in-vitro*. Furthermore, we have demonstrated proof of concept by implanting a-cellular tissue engineered vessels (A-TEVs) comprised of SIS immobilized with heparin and vascular endothelial growth factor (VEGF) into the arterial system of sheep which remained patent (92%, n=12) for 3mo (Fig.1). Upon analysis, the lumen of these grafts was comprised of a fully functional endothelium as early as 1mo post implantation. This study sought to identify the type of cells that are captured by VEGF on the lumen of A-TEVs *in-vivo* and understand how these cells turn into an endothelial (EC) monolayer that is capable of maintaining patency *in-vivo*. A-TEV implantations were performed as previously published. *In-vivo Analysis:* Fixed explants of 1wk, 1mo, 3mo, and 6mo VEGF functionalized A-TEVs are assessed via IHC for MC and EC markers. *In-vitro Capture under Flow:* Capture of VEGFR expressing cells from blood under flow will be assessed in a microfluidic device with a flat surface comprised of chitosan, heparin, and VEGF (CHV) as previously published. *In-vitro differentiation:* Blood borne mononuclear cells that are captured on surface immobilized VEGF are coaxed to differentiate into EC with a combination of soluble and biophysical signals. A-TEVs were implanted as interpositional grafts into the arterial circulation of an ovine animal model. As early as 1mo post-implantation, the graft lumen was fully endothelialized as shown by IHC for EC markers, CD144 and eNOS. At the same time, luminal cells co-expressed leukocyte markers CD14 and CD163 (Fig. 3). To understand these results, we performed cell capture experiments under flow using microfluidic devices. Interestingly, blood mononuclear cells expressing high levels of VEGF receptors were captured on CHV surfaces with high specificity under a range of shear stresses. Initially, these cells expressed high levels of CD14 and CD16. Under the right conditions they were coaxed to differentiate into an EC phenotype as shown by expression of CD144 and eNOS (Fig.2), additional IC analysis, WB, qRT-PCR, and flow cytometry. We will also discuss the role of soluble signals and biophysical forces in transdifferentiation of blood cells into EC that maintain graft patency. We demonstrate the ability of VEGF functionalized surfaces to capture cells directly from the blood *in-vitro* and *in vivo*. In the presence of the right biochemical and biophysical signals these cells differentiate into EC like cells that maintain graft patency and vascular function. Our results shed light into the process of vascular tissue regeneration *in situ* using the body's regenerative capacity.

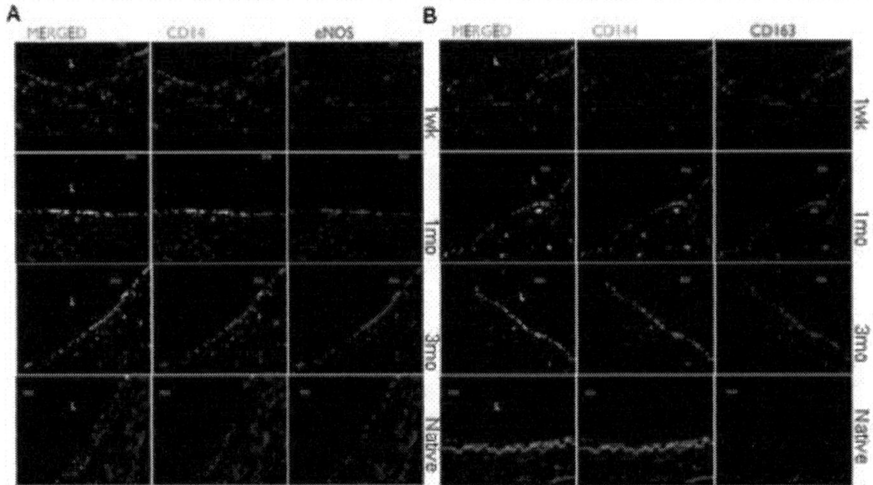

Fig. 2: VEGF functionalized A-TEVs develop a confluent endothelium. Explanted grafts at the indicated times (n=4 animals at each time point) were stained for EC markers eNOS (A) and CD144 (B) and MC/MΦ markers CD14 (A) and CD163 (B). Blue stain=DAPI. Note that the basal lamina of native arteries is highly auto-fluorescent (wavy structure under the EC monolayer).

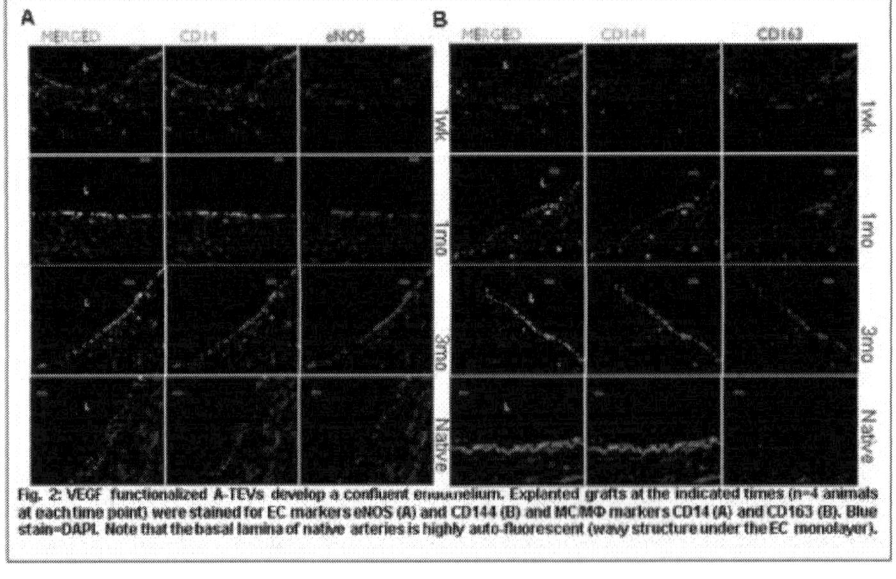

Fig. 2: VEGF functionalized A-TEVs develop a confluent endothelium. Explanted grafts at the indicated times (n=4 animals at each time point) were stained for EC markers eNOS (A) and CD144 (B) and MC/MΦ markers CD14 (A) and CD163 (B). Blue stain=DAPI. Note that the basal lamina of native arteries is highly auto-fluorescent (wavy structure under the EC monolayer).

468119 Heterogeneity in Valvular Interstitial Cell Phenotype Is a Predictor of Cell Activation and Acquisition of Degenerative Properties

Abstract:

Valvular interstitial cells (VICs), the resident cells in heart valves, play major roles in valvular pathophysiology. VICs are quiescent (qVIC) in healthy heart valves and maintain tissue homeostasis. They, however, become contractile and activated (aVIC) under chemically- or mechanically-induced degenerative conditions. It is well-known that qVIC transform to aVIC on stiffer substrates, at higher passage numbers and in the presence of paracrine signals. Regardless of quiescence or activation states, investigations of culture heterogeneity can provide insights into culture likelihood for degenerative transformation. In this study, we quantify six different VIC morphologies – rhomboid, spindled, tailed, round, triangle and multi-extension. We then correlate the distributions of these six morphologies with cell activation state. The main goal of this study is to demonstrate that the distribution of different VIC morphologies changes significantly with culture conditions and those changes, in turn, correlate with the amounts of degenerative cells (aVIC).

Porcine aortic VICs from passage P2 to P4 were cultured on soft polyacrylamide gels and on plastic culture dishes. In a separate experiment, the same cells were grown in the presence or absence of high numbers of progenitor VICs (pVICs). pVICs have the ability to differentiate in culture, and potentially become either qVIC or aVIC. Cell monolayers were imaged under phase and fluorescence microscopy of actin fibers. Three cell morphology parameters were analyzed: maximum cell span, span ratio and cell area. All six distinct morphologies of VICs were represented and their relative proportions in culture were calculated. Immunoblotting of protein lysates and immunofluorescence staining of monolayers were also employed in the determination of qVIC and aVIC content in all experiments.

An increase in α-smooth muscle actin (α-SMA) expression, the main marker for aVICs, was detectable with increasing passage number, increasing substrate stiffness and upon contact with pVICs. Round, rhomboid and triangle VICs were prominent in conditions favoring aVIC transformation (later passages, stiffer substrates and pVICs signals) whereas spindled and tailed morphologies were prominent in conditions favoring VIC quiescence. Span ratio, maximum span and cell area were found to be significantly different ($p<0.5$) between each pair of conditions tested.

The substantial increase of α-SMA expression, or aVIC content, indicates that VICs can sense their environment and become activated in response to increasing matrix stiffness. The substantial increase in α-SMA due to passage number indicates that VICs become activated over time as they multiply *in vitro*. The increase in α-SMA due to pVIC contact indicates that pVIC will differentiate to aVIC phenotype due to paracrine communication. Since earlier passages and soft substrates contained more spindled and tailed morphologies, these can be associated with physiological tissue conditions. Later passages and stiffer substrates expressed more aVICs, characterized by high numbers of round, rhomboid and triangle morphologies.

Our conclusion is that characterization of these six morphologies is arguably more important for engineering of valvular tissues than the traditional qVIC and aVIC classification. Follow-up experiments in progress include 3D characterization studies of heterogeneous populations as well as isolations of these six morphological cases. These studies will lead the way to proper maintenance of spindled and tailed morphologies, guide future differentiation studies of pVICs and potentially provide unique ways to target and inhibit degenerative proteins in aVIC.

470442 *A Novel Integrated Intestine-Liver-Brain Model to Investigate Ethanol Metabolism*

Abstract:

To date, *in vitro* models of the gastrointestinal (GI) tract and liver have been developed separately with very limited efforts to integrate them into a single system. The communications between the GI and liver modulate multiple metabolic functions, inflammatory responses and the biotransformation of chemicals. Ethanol-induced toxicity can have severe consequences on all three of these organs. In the liver, excessive ethanol consumption can lead to fibrosis, while significant cell death has been reported in intestines exposed to ethanol. Excessive alcohol consumption can be toxic to neurons in the brain. There is a need to understand how these organs function cohesively to metabolize ethanol. More importantly, the effects of the metabolites secreted by these organs on the brain are not fully understood.

We describe a novel integrated GI-liver model that is designed by integrating a primary rat ileum explant with a 3D organotypic multi-cellular liver model. Primary ileum tissues were surgically excised from rats. These explants were cultured in oxygenated culture medium that contained antibiotics. The ileum cultures remained stable with virtually no shedding of the intestinal epithelium. Ethanol (EtOH) was administered over 100-1000mM concentration range. These concentrations ranged from sub-lethal to highly toxic. Upon exposing the integrated ileum-liver cultures to 100mM EtOH for 24 hours, significant changes in hepatic function were observed. Albumin levels were approximately 2-fold lower in the integrated ileum-liver models when compared to liver models. A similar trend was also observed in urea production. The integrated ileum-liver models exhibited significantly lower urea levels compared to liver models alone.

EtOH administration is known to result in cell death and in the production of reactive oxygen species (ROS). Hepatocytes in the integrated cultures and liver models were imaged for ROS production upon EtOH administration. Even at a sub-lethal dose of 100mM EtOH, ROS production was significantly higher in the integrated cultures when compared to cells in the liver models alone. Interestingly, ROS production by the integrated ileum-liver culture was higher than liver models that had been treated with a combination of LPS (lipopolysaccharide) and 100mM EtOH.

Astrocyte cultures were exposed to EtOH in culture medium or spent medium obtained from the integrated ileum-liver models. The viabilities of astrocytes exposed to condition medium obtained from EtOH treated gut-liver models were approximately 2-fold lower compared to cultures that were exposed only to EtOH. These trends clearly demonstrate that the integrated gut-liver models exhibit increased sensitivity to EtOH.

Ongoing efforts are focused upon measuring the levels of inflammatory cytokines and production of reactive oxygen species when this multi-organ model (gut-liver-brain) is treated with EtOH. Integrating the GI, liver and brain will serve as a powerful platform to provide holistic information on toxicity, inflammatory response and organ crosstalk.

474170 Alginate Capsule Composition Influences the Pancreatic Differentiation of Human Embryonic Stem Cells

Abstract:

Introduction: Type 1 diabetes (T1D) is an autoimmune disease affecting millions of people worldwide wherein the beta cells of the pancreas are destroyed, resulting in insulin dependence. Shortage of donor islets combined with immune rejection limits islet transplantation from becoming a viable therapy. We have shown previously that calcium alginate encapsulation for differentiation of human embryonic stem cells (hESCs) results in efficient differentiation to insulin-producing cells. The unlimited supply of hESCs and the immunoisolation capability of alginate capsules, suggest this approach as an alternative treatment for T1D. Islet encapsulation studies demonstrate that capsule composition has a significant effect on immunoisolation. However, capsule composition can modify the differentiation of encapsulated cells. Hence in this study we are evaluating the effect of material properties on pancreatic differentiation of encapsulated hESCs.

Materials and Methods: A single cell suspension of ROCK inhibitor treated hESCs were suspended in 1.1% alginate and 0.2% gelatin followed by encapsulation by drop wise addition into a divalent cation bath. hESCs were encapsulated in barium alginate, crosslinked with 10, 15, 20, 50 or 100 mM $BaCl_2$. The hESCs were differentiated entirely under encapsulation and characterized after completion of pancreatic differentiation. Encapsulated hESCs were induced to the pancreatic islet like state following a stagewise differentiation protocol. Definitive endoderm (DE) was induced with ActivinA and Wnt3Afor 4 days followed by pancreatic progenitor induction with KAAD-cyclopamine and KAAD-cyclopamine with retinoic acid for 2 days each, respectively. DE stage differentiation was characterized by gene and protein expression of SOX17, FOXA2, EOMES, and CER. Pancreatic commitment was analyzed by PDX1 gene and protein expression.

Results: Increasing the concentration of $BaCl_2$ from 10 to 100 mM $BaCl_2$ resulted in an increase in substrate stiffness from 4 to 75 kPA, as determined by AFM analysis. Viability and proliferation of encapsulated hESCs were very sensitive to capsules stiffness, with only 4-7 kPa capsules supporting cell growth and colony formation. Diffusional differences were confirmed to be minimal under the tested conditions. DE stage differentiation was found to be enhanced as capsule stiffness was increased, as shown by both gene and protein analysis of SOX17, FOXA2, EOMES, and CER. Pancreatic commitment was analyzed by PDX1 gene and protein expression. *PDX1* gene expression was highest in the 4 kPa stiffness gel (10 mM $BaCl_2$), and was significantly reduced as stiffness increased. The same trend was observed with PDX1 protein expression, which showed the highest number of cells positive for PDX1 in the 10 mM $BaCl_2$ condition.

Discussion: Our study demonstrates that capsule materials properties can significantly modulate the efficiency of chemically induced pancreatic differentiation. Specific encapsulation conditions can significantly enhance the efficiency of chemical induction. In general, higher crosslinking was found to be detrimental to the viability and differentiation of hESCs. Research is currently underway for encapsulating hESCs in a high throughput alginate array platform. This will allow for a more thorough investigation of the effect of substrates stiffness on differentiation, as well as delineate the signaling pathways which govern pancreatic differentiation. Results of these studies will be presented and discussed.

464495 Smart Biomaterials to Study, Mimic, and Exploit Cellular Mechanobiology

Abstract:

Living cells are capable of processing a variety of mechanical signals encoded within their microenvironment, which can in turn act through the cellular structural machinery to regulate many fundamental behaviors. In this sense, cells may be regarded as "smart biomaterials" that dynamically and locally modulate their physical properties in response to environmental stimuli. I will discuss our recent efforts to understand and engineer these living microsystems, and to create new, bio-inspired materials that mimic sequence/structure/function relationships of cytoskeletal networks. Key areas of emphasis will include: (1) Applying material and genetic strategies to probe the timing and molecular mechanisms of stem cell mechanobiology; (2) Using single-cell micropatterning and femtosecond laser ablation to precisely quantify the microscale mechanics of actomyosin stress fibers, and (3) Engineering stimulus-sensitive intrinsically disordered protein brushes based on neuronal cytoskeletal networks.

462699 The Role of the Immune System in Modulating Tumor Cell Migration Following Radiation-Induced Changes in the Tissue Microenvironment

Abstract:

Despite aggressive surgical, radiological, and chemotherapeutic intervention, triple-negative breast cancer patients experience high rates of recurrence. The role of the tumor microenvironment in recurrence, however, is not well understood. We hypothesize that the irradiated tumor microenvironment influences tumor and immune cell behavior. In this study, we characterize the effects of normal tissue irradiation on tumor and immune cell migration to evaluate how tumor-stromal interactions modulate recurrence after therapy. This work represents the first step toward elucidating how stromal radiation response contributes to tumor and immune cell recruitment.

The effect of radiation on tumor cell migration to normal tissues was studied using an orthotopic breast cancer model. Nude mice were inoculated with luciferase-labeled 4T1 murine or MDA-MB-231 human triple-negative breast cancer cells in the mammary fat pad (MFP), and BALB/c mice with depleted CD4+ and/or CD8+ T cells were inoculated with 4T1 MFP tumors. The contralateral normal MFP was irradiated to a dose of 20 Gy with a 250 kVp cabinet x-ray machine when tumors were palpable. Cell migration was monitored with bioluminescence imaging (BLI) 10 d after irradiation. Irradiated and control tissues were evaluated using immunohistochemistry (IHC). Tissue sections were stained with F4/80 to determine the extent of macrophage infiltration. Flow cytometry was also performed on dissociated irradiated and control tissues to characterize immune cell populations. Of particular interest were CD11b+F4/80+ macrophages and CD11b+GR1+ myeloid-derived suppressor cells (MDSCs).

Radiation enhanced tumor cell migration to normal tissues *in vivo*. *Ex vivo* BLI analysis in nude mice revealed that normal tissue irradiation attracted tumor cells to the MFP and surrounding tissues, including the peritoneum and muscle ($p<0.001$). Tumor cells migrated to irradiated normal tissues in BALB/c mice with depleted CD4+ and CD8+ T cells ($p<0.001$). Depletion of CD8+ T cells alone significantly enhanced tumor cell migration in comparison to depleting CD4+ T cells alone ($p<0.01$). This indicates that CD8+ T cells may negatively regulate local recurrence following radiotherapy. An increase in macrophage infiltration was also found in irradiated tissue sections in the absence of T cells. Flow cytometry confirmed increases in macrophage and MDSC infiltration in irradiated areas.

This study establishes that normal tissue radiation response modulates tumor and immune cell migration after radiation. The increase in macrophage and MDSC infiltration after irradiation in the absence of T cells indicates a crucial role of the immune system in tumor cell migration. Our work suggests that the tumor stroma may facilitate tumor cell invasion and tumor regrowth following radiotherapy. Future studies will utilize these results to engineer improved *in vitro* tumor microenvironment models to probe the complex physical, chemical, and biological cues that influence cancer recurrence and metastasis.

470850 Hemica-Hydrogel Encapsylated Micro-Channel Array in Cancer Metastasis

Abstract:

Introduction: Cell migration is a complex process that involves cell interactions with tissues of varying stiffness and porosity. The current state of art for studying cell migration in vitro consists of 2D and 3D matrixes as well as polydimethylsiloxane (PDMS)-based microfluidic devices. These methods cannot be used simultaneously, alter pore-size independent of matrix stiffness or recreate microenvironments of physiologically relevant stiffness. We herein present the first device that bypasses these hurdles by using a polyacrylamide (PA)-based Hydrogel Encapsulated MIcro-Channel Array (HEMICA) and focuses on the study of cancer cell metastasis.

Methods: Silicone wafers with the desired micro-channel design are created via photolithography and used as a mold for PA. Access holes to the PA imprinted design are punched out. Glass cover-slips are activated and PA is cured on them creating a thin (60-100um) layer. Both gel layers are allowed to reach their respective maximum swelling. The upper layer with the design on its lower surface is then sandwiched to the bottom gel/glass via a NHS-ester ligand. The encapsulated micro-channels are coated with collagen type I using Sulfo-SANPAH as a ligand. The device is seeded with cells and subsequently submerged into media for live microscopy. To quantify cell speed, persistence and cell morphology we used FIJI and a custom MATLAB code. Statistical significance was assessed with two-way ANOVA test.

Results and Discussion: MDA-MB-231 breast cancer cells were seeded in HEMICA and studied in a wide range of microenvironments spanning full confinement (*WxH=3x10Î¼m*^{2}) to 2D spaces (*WxH=50x10Î¼m*^{2} *or 200x50Î¼m*^{2}). Cell speed shows a biphasic behavior with maximum speed in 10um width (similar to the dimension of the nucleus) and intermediate stiffness 13kPa (similar to the stiffness of human breast tissue). Cell persistence is lower in the softest gels with higher significance on the larger width channels, suggesting a correlation between stiffness and directed migration. Although large widths allow cells to adhere preferentially to one channel wall (contact guidance), we show that low stiffness (7kPa) reduces contact guidance based on the frequency cells cross the channel center-line and the probability of following a path within the central regions of the channels. In addition, intermediate stiffnesses increase cell polarization, especially on small widths, while low stiffnesses increase cell circularity indicating reduced cell-matrix adhesions. On the other hand, cell projected area increases with increasing channel width independent of stiffness. Moreover, the cell variance of area and circularity is higher in large widths of low stiffness, indicating major morphodynamic changes in the process of migration. In contrast, the variance of solidity (how protrusive cells are) is higher in confinement with intermediate stiffness, which implies that cells in full confinement change the number of their protrusions frequently in order to achieve an optimum mode of migration. Through our ongoing experiments we are identifying the mechanobiology of cancer cell migration by establishing the link between stiffness, myosin II, microtubules and traction forces in confinement.

Conclusion: We have created a novel and versatile device which mimics in vivo microenvironments. HEMICA can be used to study collective migration, cell-cell communication, cell-matrix interactions and cell population dynamics, offers organ-on-a-chip applications and allows post-migration studies. Our results demonstrate that the geometry of the cell microenvironment and tissue compliance promote different modes of migration ultimately affecting all stages of cancer metastasis.

464918 Tissue Origami: Directed Folding of Tissues By Programmed Cell Contractility Networks

Abstract:

A major barrier to the engineering of complex tissues is the difficulty in controlling tissue structure across a wide range of length scales. Further, proper function of synthetic biological tissues is hampered by mass transport limitations, and the lack of extracellular matrix (ECM) physicochemical anisotropies that guide appropriate cell behaviors in space and time. In biology, tissues are typically built over a range of length scales via self-organized morphogenesis processes in which extracellular matrix remodeling occurs at the same time as cell division, migration, and fate specification. We take a biomimetic approach to building tissues by engineering the spatial organization of sheets of ECM gels via contractile networks of cells. Such contractile networks are co-patterned with ÒpassengerÓ cell types that can be guided into prescribed 3D topologies that instruct cell self-organization processes. Thus, the folding trajectory and form is combined with control over the type and placement of different cell types on the sheet, enabling control over 3D tissue architecture in concert with cellular interfaces.

Introduction: Living tissues are characterized by the specific arrangement of biological cells in layers of ECM. Interfaces between cells and their matrix microenvironment are often folded into arrays of functional units such as the epithelial acini of the breast and the crypts and villi of the intestine. In many cases, the curvature and spatial organization of the ECM is relevant to tissue development and function. For example, the specification of endoderm cell fate at crypts in the gut is governed by cell-cell signaling networks that are influenced by the folding state of tissue layers. We seek to build folds of specific geometry in ECM-mimetic gels in order to understand the systems-level interactions between cells in folded tissues. Moreover, we rely on the intrinsic contractility of cells to sculpt the physicochemical anisotropy of the ECM to mimic in vivo tissues.

Materials and Methods: We build arrays of contractile tissues at the upper and lower surface of 250 micron-thick matrigel-collagen ECM-mimetic gels using DNA-patterned assembly of cells (Figure 1a). The type and location of cells on each surface of the gel prescribe the final folded geometry of the ECM. Collagen fibers are pre-labeled with fluorescent dye in order to track curvature of the sheet over time via confocal microscopy.

Results and Discussion: Regular arrays of contractile mammary epithelial cells placed on one side of circular ECM gels yield cap-shaped objects with maximum curvatures that increase over time. The curvature at a given timepoint increases monotonically as the spacing between tissues on the sheet is decreased (and more tissues are added, Figure 1b), scales linearly with cell contractility, and is partially blocked by blebbistatin. We found that anisotropic grids of fibroblast tissues produce consistent anisotropic folds along controlled axes, allowing us to fabricate a range of 3D topographies and predict them with a biophysical FEA model (Figure 1c). For example, we constructed the Miura-ori fold (Figure 1d), which resembles folds of the gut epithelium in development.

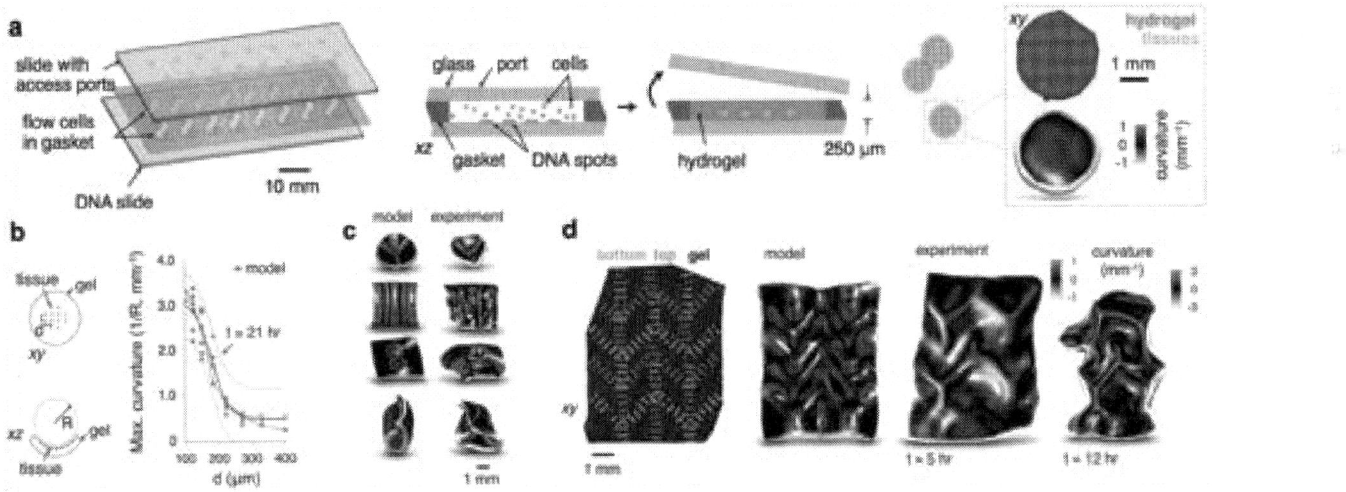

Figure 1. Engineering tissue curvature via cellular contractility. (a) microfluidic apparatus for building thin ECM sheets with contractile tissue actuators at prescribed positions in *xy*. (b) ECM curvature increases with tissue density. (c) Various folds can be prescribed in ECM sheets and predicted via FEA modeling. (d) The Miura-ori fold.

Conclusions: We report engineering control over the 3D shape of biomimetic ECM scaffolds by leveraging tissues as mechanical actuators. Our ability to design specific fold geometries in concert with cell compositions and spatial arrangement in ECM sheets could improve drug toxicity screens and enable studies of the necessary conditions for topological transitions in tissue development.

470860 Engineering the Gut-Liver-Brain Axis to Investigate Chemical Toxicity

Abstract:

The inter-organ crosstalk between the gut, liver and brain has been shown to affect many physiological functions. The accumulation of drugs, chemicals, toxins and their metabolites in the liver can reach the brain and result in impaired cognitive function. Brain diseases such as hepatic encephalopathy result from complex interactions along the gut-liver-brain axis. We have designed and assembled a multi-organ system to obtain fundamental insights into the intricate cross-talk between these three organs.

Primary intestinal tissues were surgically obtained from rat jejunum and cultured in medium containing antibiotics. Gram staining was performed to detect the presence of gut bacteria over a 3-day period. Minimal numbers of Gram-positive and Gram-negative bacteria were released over the 72 h period. Imaging was conducted to determine the viability of intestinal cells that were shed. Enterocyte and goblet cell phenotypes were confirmed via alkaline phosphatase activity and Alcian blue staining, respectively. The intestinal tissues expressed these markers even after 96 h.

Hepatocyte function in the integrated jejunum-liver models was measured through urea secretion and albumin production and was comparable to liver models alone. Acetaminophen, a well-known hepatotoxicant, was administered to integrated gut-liver cultures. Even at a 10mM concentration of acetaminophen, the decrease in urea production and albumin expression increased two-fold in the integrated cultures when compared to liver models only. These trends clearly demonstrate the enhanced sensitivity to acetaminophen exhibited by the integrated cultures, underscoring the need to combine organs.

The gut-liver-brain crosstalk was accomplished by exposing astrocyte cultures to spent medium from the integrated systems and liver models. Specifically, the spent culture medium was obtained from systems exposed to 10mM acetaminophen. The production of reactive oxygen species (ROS) in astrocytes was visualized using the DCFDA assay. ROS production was compared between astrocyte cultures treated with conditioned medium from the integrated gut-liver systems and liver models only. There was approximately 10-fold higher ROS in cultures treated with conditioned medium obtained from only liver models. This trend could be due to increased acetaminophen metabolism by the gut in the integrated cultures through the glucuronidation and sulfation pathways, thus decreasing the production of toxic metabolites from the cytochrome P450 2E1 (CYP2E1) mediated metabolism of this hepatotoxicant.

The design and assembly of multi-organ engineered tissues provides a powerful platform to investigate the interactions and crosstalk that occur during many physiological processes. We are currently focusing on elucidating the signaling pathways between these three organs to better understand how they function.

473906 Modulation of Lipid Composition As a Therapy in Alcohol Injured Liver Cells

Abstract:

Liver cells exposed to ethanol are altered in NADH/NAD^+ ratio, generation of reactive oxygen species, excessive fat accumulation and cell apoptosis. These changes in a healthy liver can result in alcoholic liver diseases (ALDs), which are major causes of liver related morbidity in the world. However, despite the well documented metabolism of ethanol in parenchymal liver cells, hepatocytes, current treatment options are limited to cessation of alcohol consumption. Studies in other cell types and organisms have demonstrated that, in addition to changes in metabolism, physical changes to cell membranes occur upon exposure to ethanol and these changes are dependent on cell membrane lipid composition. Here we present an alternative perspective on mitigating ethanol injury on hepatocytes based on modulating the membrane lipid composition. Considering the likely relationship of lipid composition with ethanol tolerance from a yeast system, we use cationic liposomes that contain DOPC (1,2-dioleoyl- sn-glycero-3-phosphocholine) to recover the functionality of ethanol injured primary rat hepatocytes. As a cationic lipid, DOTAP (1,2-dioleoyl-3-trimethylammonium-propane) is used to facilitate liposome cell interactions. Incorporation of the liposomes into the hepatocyte membranes were monitored by measuring DOTAP as a surrogate marker using liquid chromatography / mass spectrometry with a triple quadrupole (QqQ) mass analyzer and an electrospray ionization source (ESI). Cell function was assessed using secreted albumin as a marker. The results of these *in vitro* studies illustrate the potential of using cationic liposomes to mitigate ethanol injury *in vivo* using existing or novel means of delivering lipids to hepatocytes in the liver.

470712 Signal Transduction of the Physical Environment in the Neural Differentiation of Stem Cells

Abstract:

Neural differentiation is largely dependent on extracellular signals within the cell microenvironment. These extracellular signals are mainly in the form of soluble factors that activate intracellular signaling cascades that drive changes in the cell nucleus. However, it is becoming increasingly apparent that the physical microenvironment provides signals that can also influence lineage commitment and very low modulus surfaces has been repeatedly demonstrated to promote neurogenesis. The molecular mechanisms governing mechano-induced neural differentiation are still largely uncharacterized; however, a growing body of evidence indicates that physical stimuli can regulate known signaling cascades and transcription factors involved in neural differentiation. Understanding how the physical environment affects neural differentiation at the molecular level will enable research and design of materials that will eventually enhance neural stem cell differentiation, homogeneity, and specificity. Previous findings have shown that the neural phenotype in stem cells is strongly repressed by the Neuron Restrictive Silencer Factor (NRSF). NRSF coordinately represses up to 300 genes involved in neural differentiation. Recently, we found that the neural induction in mesenchymal stem cells (MSCs) by forskolin and IBMX is dependent on the downregulation of NRSF and derepression of its downstream targets. Therefore we set out to determine whether soft surfaces could induce neural phenotype in MSCs also through downregulation of NRSF. Indeed we found that several neural characteristics induced by soft surfaces including increase in neural marker expression (Tuj), biosynthetic enzymes (TH), and neuronal function are repressed by NRSF.

470794 NANOG Restores the Impaired Contractile Function of Senescent Mesenchymal Stem Cells

Abstract:

Contractility is an essential function of tissue engineered constructs containing smooth muscle cells such as blood vessels and bladders. Although, mesenchymal stem cells (MSC) have been extensively used as a source of smooth muscle cells (SMC), recent studies in our laboratory showed deficits in the myogenic differentiation capacity of senescent MSCs due to decrease in the formation of actin filaments. Unfortunately, current strategies for improving the contractile function have been unsuccessful in restoring this function after senescence, since senescent cells failed to respond to extracellular signaling factors. Interestingly recent work in our laboratory showed that ectopic expression of a single pluripotent transcription factor, NANOG, could reverse aging and restore the impaired contractile properties of aged MSC. In this study, we sought to identify the effects of NANOG on the actin filament organization and force generation capacity of MSC in a 3D microenvironment.

To this end, Human Hair Follicle Mesenchymal Stem Cells (hHF-MSCs) were isolated from a 73 year old donor and senesced by serial passaging in vitro. The cells were transduced with a tetracycline regulatable vector that tightly controls the expression of NANOG by the addition of the tetracycline analogue, doxycycline (Dox). The protein and mRNA levels of Î±-smooth muscle actin (ACTA2) were quantified by western blotting and real time PCR, respectively. Organization of actin filaments were visualized using immunocytochemistry techniques. To measure the biomechanical force and map the biochemical properties of cells in 3D tissues, we utilized micro-fabricated tissue gauges (ÂµTUGs) that contain a pair of flexible micropillars. The statistical significance between two mean values was analyzed using two-tailed unpaired Studentâ€™s t-test values of $p<0.05$ were considered statistically significant.

We found that, cellular senescence impaired the expression of myogenic genes, actin filament organization and the force generation capacity of MSC derived SMC. Notably, expression of NANOG restored all these functions,

ultimately reversing cellular senescence. Interestingly, while senescent cells exhibited impaired TGF-Î²1 pathway, NANOG completely restored this pathway and reversed the loss of contractile force generated by senescent 3D microtissues (Fig. 1). We will show that the mechanism involves activation of SMADs, which were impaired in senescent cells. We also investigated whether NANOG must be continuously present in order to elicit its effects. Despite the short half-life of NANOG protein (2.5Â±0.5 hr), the effects of NANOG on myogenic gene and protein expression and contractile function were sustained up to 10 days after termination of expression. This result may suggest interesting epigenetic regulation that may provide long-term recovery of the senescent phenotype of 3D tissues. We are currently investigating such epigenetic events.

In conclusion, ectopic expression of the embryonic transcription factor, NANOG, restored actin polymerization and the contractile function of senescent MSC-derived SMC, suggesting that NANOG may be used to enhance the contractile properties of tissue engineered constructs containing SMC, such as bladder and arteries. Given the effects of NANOG on SMC differentiation and contractile fucntion, it is possible that NANOG might also affect the contractile function of other types of muscle such as cardiac or skeletal muscle.

472164 Supported Biomembrane Microenvironments of Controlled Composition for Gamma-Secretase Substrate Cleavage Assays

Abstract:

The intramembrane protease gamma-secretase is a current target of therapeutic intervention, with pivotal pathological functions within Alzheimerâ€™s disease and cancer. Our primary objective is to expand methods for *in vitro* studies of the intramembrane proteases (IMPs) with the development of a microsphere-supported (proteolipobead) and planar biomembrane platforms. IMPs are found throughout all branches of life and their functions are extremely broad. Despite extensive studies, understanding of IMP regulation and catalysis has been hampered by membrane-associated enzymology. Rhomboids and Secretases are polytopic membrane proteases that are widely conserved in all organisms. The precise reaction mechanisms of the intermembrane proteases remain to be elucidated and furthermore, rigorous analysis of the kinetics of interfacial catalysis in these systems has not yet been undertaken.

High throughput of screening of drug candidates has been carried out in bulk assay systems using cell membrane fragments, solubilized enzymes and is underway in proteoliposomes. A critical barrier to further progress in the study and HTS of gamma-secretase is that such bulk systems do not allow for the direct in situ quantification of enzyme, substrates, or inhibitors or their relative distributions within the structures under assay. We have expanded in vitro models of gamma-secretase to include supported biomembranes, enabling: 1) characterization and verification of biomembrane loading of substrates, inhibitors and protein effectors, 2) studies of co-localization, phase-partitioning and lateral mobility of membrane-bound assay constituents, and 3) the development of flow cytometry-based assays of cleavage. In this work we have characterized the structures, phase localization and compositions of gamma-secretase proteolipobead and planar systems by employing correlative optical and surface microscopy from the micro- to the nanoscale. We have probed the formation of Lo domains in situ, monitored using 3D FRET phase detection. We have probed the phase partitioning of enzyme, substrates and cleavage products in PLB systems with superresolution microscopy. A combination of biomembrane mobility methods are being used to probe the diffusivities and mobile

fractions of lipid, substrates and gamma-secretase. The results of phase partioning and diffusivity measurements are being integrated into bottom-up spatial diffusion-reaction models to correlate microenvironment with enzyme function. As new drug targets in cancer and infectious disease involved in regulated intramembrane proteolysis (RIP) based cell signaling are uncovered, it is expected that other IMP enzymes can be subsequently studied and probed at high throughput with this platform.

AUTHOR INDEX

AUTHOR INDEX

AUTHOR INDEX

AUTHOR INDEX

AUTHOR INDEX

AUTHOR INDEX

AUTHOR INDEX

AUTHOR INDEX

AUTHOR INDEX

AUTHOR INDEX

AUTHOR INDEX

AUTHOR INDEX

AUTHOR INDEX

AUTHOR INDEX

AUTHOR INDEX

AUTHOR INDEX

AUTHOR INDEX

AUTHOR INDEX

AUTHOR INDEX

AUTHOR INDEX

AUTHOR INDEX

AIChE
120 Wall Street, FL 23
New York, NY 10005-4020

ISBN 978-1-5108-3440-8
9 781510 834408

CURRAN ASSOCIATES INC.
proceedings.com